W9-BIN-568

REFRIGERATION AND AIR-CONDITIONING

REFRIGERATION AND AIR-CONDITIONING

AIR-CONDITIONING AND REFRIGERATION INSTITUTE

1815 North Fort Meyer Drive
Arlington, Va. 22209

PRENTICE-HALL, INC., *Englewood Cliffs, New Jersey 07632*

Library of Congress Cataloging in Publication Data

Air-Conditioning and Refrigeration Institute.
Refrigeration and air-conditioning.

Includes index.
1. Refrigeration and refrigerating machinery.
2. Air conditioning. I. Title.
TP492.A27 1979 621.5′6 77-25335
ISBN 0-13-770164-0

Editorial production supervision
and interior design by: JAMES M. CHEGE

Page layout by: MARTIN J. BEHAN

Cover design by: RICHARD LO MONACO

Manufacturing buyer: CATHIE LENARD

© 1979 by Prentice-Hall, Inc., Englewood Cliffs, N.J. 07632

All rights reserved. No part of this book
may be reproduced in any form or
by any means without permission in writing
from the publisher.

10 9 8 7 6 5

Printed in the United States of America

PRENTICE-HALL INTERNATIONAL, INC., *London*
PRENTICE-HALL OF AUSTRALIA PTY. LIMITED, *Sydney*
PRENTICE-HALL OF CANADA, LTD., *Toronto*
PRENTICE-HALL OF INDIA PRIVATE LIMITED, *New Delhi*
PRENTICE-HALL OF JAPAN, INC., *Tokyo*
PRENTICE-HALL OF SOUTHEAST ASIA PTE. LTD., *Singapore*
WHITEHALL BOOKS LIMITED, *Wellington, New Zealand*

CONTENTS

PREFACE

Refrigeration, the first part of this text, was prepared to provide a basic textbook for use by students and instructors in secondary and post-secondary courses conducted to train future refrigeration service technicians in the fundamentals of refrigeration, and the installation, servicing, and repair of refrigeration equipment. It was written and edited under the direction of the Manpower Development Committee of the Air-Conditioning and Refrigeration Institute (ARI), the national industry trade association whose members manufacture more than 90% of all air-conditioning and refrigeration equipment produced in the United States. *Air-Conditioning*, the second part of this text, is published under the guidance of the same committee for use by classes in air-conditioning, which differs from refrigeration only in application, since both are governed by the same physical principles and require similar basic knowledge on the part of those who work with them.

This textbook is based on course outlines developed earlier in cooperation with the U. S. Office of Education's Vocational Education branch, and approved by that agency. Development of the course guide, and subsequently this textbook, were preceded by a number of surveys which showed that in the early 1960's, the industry was growing so rapidly that it would face a shortage of manpower for installation and servicing within a short time.

Public acceptance of, and demand for, refrigeration and air-conditioning in the years following World War II, and especially in the '60's, threatened to outstrip the pool of manpower available to install, service, and maintain the equipment already installed and being added daily to the nation's cooling requirements. As a result, the ARI developed a training program, promoting the establishment of courses in high schools and vocational schools. These classes turned out thousands of students ready to start in as beginning service technicians and helpers. The first publication under this program was the course guide for instructors already referred to, but the need for texts for the students, in addition to the guide for instructors, was felt, and ARI undertook the preparation of *Refrigeration and Air-Conditioning* to fill this need.

The reader should be aware that the period of

time in which the material for this text was in the process of being formulated and made ready for printing was one of dynamic change in the affairs of domestic and world economy, and the impact this had on the air-conditioning and refrigeration industry will be significant enough to perhaps create rapid changes in product development and application techniques in future years. Energy conservation and environmental considerations are pressing issues that must be examined and solved and, as solutions are presented, some of the data in this text may be subject to rapid obsolescence. Therefore, we recommend that the use of specific product information, application materials, etc., should always be checked with the particular manufacturer.

Arlington, Va.

AIR-CONDITIONING AND REFRIGERATION INSTITUTE

REFRIGERATION

1

INTRODUCTION TO REFRIGERATION

REFRIGERATION THROUGH THE AGES

Although man's earliest forebears probably knew about, and observed, the effects of cold, ice, and snow on their bodies and on things around them—such as the meat they brought home from the hunt—it is not until we reach the early history of China that we find any reference to the use of these natural refrigeration phenomena to improve the lives of people, and then only for the cooling of beverages. But other uses were developed; the Chinese are the first known civilization to harvest winter's ice and store it, packed in straw or dried weeds, for use in the summer months.

Natural ice and snow provided the only means of refrigeration for many centuries. The early Egyptians discovered that evaporation could cause cooling, so they learned to put their wine and other liquids into porous earthenware containers, placing them on the rooftops at night so that cool breezes caused evaporation and cooled the contents (Fig. 1-1).

Some of our own early colonists developed methods of preserving perishable food and drink with snow and ice. They built storage buildings (ice houses) in which they could keep ice, harvested during the cold New England winters, for use in the hot months (Fig. 1-2).

During colonial days and well into the nineteenth century ice was an important commodity in trade with foreign countries that produced no natural ice.

By the early 1900s industrial refrigeration using the mechanical cycle had been developed, and meat packers, butcher shops, breweries, and other industries were beginning to make full use of mechanical refrigeration.

With the growth of the electric industry and wiring of homes, household refrigerators became popular and replaced window and standing iceboxes, which required a block of ice daily.

This increased interest in household refrigerators was aided by the design of fractional horsepower electric motors to operate the compressors in mechanical "iceboxes." Since the early 1920s these appliances have been produced in large numbers

FIGURE 1-1 Even the early Egyptians knew that evaporation of a liquid cooled the surrounding air or substance around it. These porous jars are typical of those used by the Egyptians for cooling wine, beer and other liquids by placing them on the rooftops at night. They are among the earliest "refrigerators."

and have become a necessity for all rather than a luxury for a few.

Not only is food preserved in our homes today, but commercial preservation of food is one of the most important current applications of refrigeration. Commercial preservation and transport of food is so common that it would be difficult to imagine an unrefrigerated America.

More than three-fourths of the food that appears on American tables every day is produced, packaged, shipped, stored, and preserved by refrigeration. Millions of tons of food are stored in refrigerated warehouses; millions more are in frozen food warehouses, in private lockers, and in packing and processing plants.

Without the many types of modern refrigeration in stores, warehouses, aircraft, railway cars (Fig. 1-3), trucks, and ships, the storage and transportation of all types of perishables would be impossible.

We are not limited to the enjoyment of fruits, vegetables, and other produce grown locally at any given time of the year; we can have food products from other parts of the country and even from distant lands the year round.

Refrigeration has improved the economy of many areas by providing a means of preserving their products en route to remote consumers. It has aided in the development of agricultural regions through greater demand for the products, and it has helped the dairy and livestock-producing areas similarly.

Prior to 1941, most of the tires for automobiles, trucks, airplanes, and the like were made of natural rubber produced on plantations in southeast Asia. When the shipment of latex from the foreign rubber plantations to the United States was curtailed following the outbreak of World War II, American industry and the federal government established a cooperative synthetic rubber program based upon previous research. Scientists learned that some manufacturing processes could make the artificial rubber more durable and wear resistant through use of low temperatures; thus refrigeration became vital to still another industry.

There has been a rapid increase of new products since World War II. The petrochemical industry (plastics), textile plants, and data processing industry are heavy users of process refrigeration. Without refrigeration (and air conditioning) many of these new products could not be manufactured and/or used.

One of the more recent developments associated with the energy situation is the importation of liquefied natural gas from foreign sources. Temperatures of −270°F are required to change the gas into a liquid, which is then loaded onto a refrigerated tanker for shipment to a U. S. receiving port. All the while the liquid must be maintained at −270°F until it is ready for vaporization back into a gas.

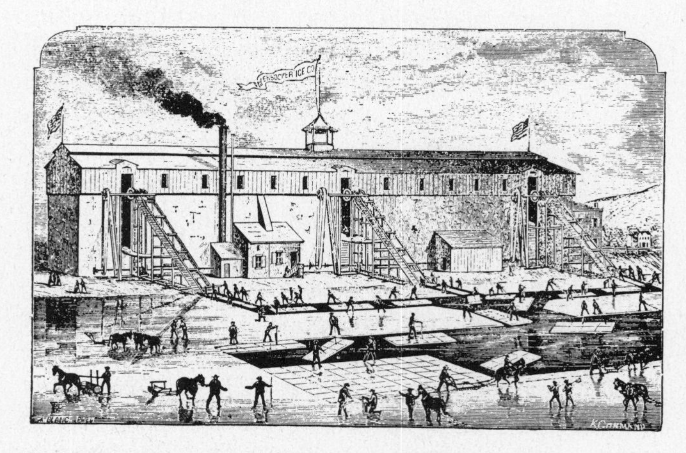

FIGURE 1-2 Throughout the ages, and almost up until the beginning of the Twentieth Century, natural ice, cut from lakes, ponds, and rivers was the sole source of refrigeration . . . . . and scenes like this were common each winter in all parts of the country.

Solar heating and solar refrigeration, along with air-conditioning, are expected to provide many opportunities for new products and for qualified application, installation, and service personnel.

In 1975 U. S. homes, plants, and commercial buildings added new refrigeration and air-conditioning equipment valued at more than $8 billion. The dollar total was actually much larger, since the cost of many accessory products, such as ductwork, grilles, insulation, and controls that go into a complete system was not included. Thus, refrigeration and air-conditioning constitute one of the major domestic industries.

World markets in this field are also experiencing rapid growth. Canada, Japan, West Germany, United Kingdom, France, Mexico, Iran, and Venezuela are among the major users of refrigeration and air-conditioning. And this market will continue to expand as foreign countries improve their standard of living and engage in more industrial development.

MANPOWER REQUIREMENTS

As a result of the expanding markets for refrigeration and air-conditioning products and changing technology, there will be a dramatic increase in the need for qualified personnel both at home and abroad.

A. R. I. recently published a *Manpower Survey Report* that estimated a great number of new jobs to be filled. Of these, most were for technician/mechanically skilled personnel. It was estimated that for every $1 million in installed value of equipment, the following employees are needed:

1 graduate engineer

2 technicians/master mechanics

11 refrigeration, air-conditioning, and heating mechanics

2 refrigeration, air-conditioning, and heating helpers

7 sheet-metal mechanics

1 sheet-metal helper

2 salesmen

In actual practice it's not uncommon for a technician/mechanic to be qualified or required to work on refrigeration, air-conditioning, or heating—or all three. Thus you can readily see the types of jobs available in this broad category.

These job opportunities not only result from a continuously expanding market for refrigerating machinery, but because of the fact that many installers and service mechanics who have been active for years will be retiring or moving up to supervisory positions.

OCCUPATIONAL OPPORTUNITIES

Where are the skills noted above used in the refrigeration industry (and this includes air-conditioning, since both achieve their purpose through application of the "refrigeration cycle")?

FIGURE 1-3 Trains made up of refrigerator cars carry refrigerated food from one part of the country to another.

Fields in which these employment opportunities exist include high-temperature cooling (usually comfort or process air conditioning), medium temperature (product storage above freezing), low temperature (quick freezing processes as well as storage below freezing), and extreme low temperature (utilized in medical and chemical applications and some industrial and scientific applications).

Equipment manufacturers hire a number of trained technicians for various jobs. Laboratory technicians may be involved in actually building prototypes of new products, or they may conduct tests on products for performance ratings, life-cycle testing, or compliance to ARI certification or UL safety standards or sound evaluation.

Factory service personnel prepare installation instructions, training material, renewal parts information, etc. Some equipment manufacturers employ large numbers of qualified service technicians for field assignment, which involves installation and start-up situations and rendering actual customer service.

The largest employer classifications are the wholesale/contracting trades. The refrigeration wholesaler must have technical personnel capable of servicing a dealer/contractor organization on the proper application, installation, and maintenance of the product lines they carry.

The dealer/contractor is the organization that installs and services the product for the end user. Depending on the extent of the contractor's business, a technician may be involved in installation only or in start-up or perhaps customer service. In smaller organizations he may do all three.

In the larger metropolitan areas there are refrigeration service companies that are involved in nothing but service. They specialize in "maintenance contracts" with the user, to provide scheduled inspections as well as emergency repairs.

It's common for large manufacturing or industrial plants or institutions that utilize considerable refrigeration or air-conditioning products to have in-house maintenance departments that do most of the routine service work.

Large food chain companies such as Kroger, A & P, Safeway, etc., employ skilled technicians to assist in or direct the installation and service of the refrigeration equipment. The modern supermarket cannot afford down time on either its food preservation equipment or its comfort air–conditioning. Other areas of employment include the electric utilities, schools, and certain industries.

Knowledge, training, and experience are the key ingredients to being a successful technician. This text is designed to help prepare you to take advantage of these broad opportunitites.

DEFINITIONS

Before getting started on the fundamentals of the course, a few basic definitions should be considered, since they are the groundwork for what is to come.

Heat is a form of energy transferred by virtue of a difference in temperature. Heat exists everywhere in greater or lesser degree. As a form of energy it can be neither created nor destroyed, although other forms of energy may be converted into heat, and *vice versa*. Heat energy travels in only one direction—from a warmer to a cooler object or area.

Cold is a relative term referring to the lack of heat in an object or space. Some definitions describe it as the absence of heat, but there is nothing known in the world today from which heat is totally absent (no process yet devised has been capable of achieving "absolute zero," the state in which all heat has been removed from a space or object). Theoretically this zero point would be 459.69 degrees below zero on the Fahrenheit thermometer scale, or 273.16 degrees below zero on the Celsius thermometer scale.

Refrigeration, or cooling, is the removal of unwanted heat from a selected space or object and its transfer to another space or object. Removal of heat lowers the temperature and may be accomplished by the use of ice, snow, chilled water, or mechanical refrigeration.

Mechanical Refrigeration is the utilization of mechanical components arranged in a "refrigeration system" for the purpose of transferring heat.

Refrigerants are chemical compounds that are alternately compressed and condensed into a liquid and then permitted to expand into a vapor or gas as they are pumped through the mechanical refrigeration system or cycle.

The refrigeration cycle is based on the long-known physical principle that a liquid expanding into a gas extracts heat from the surrounding area.

(You can test this principle yourself, simply wet a finger and hold it up. It immediately begins to feel cooler than the others, particularly if exposed to some air movement. That's because the liquid in which you dipped it is evaporating, and as it does, it extracts heat from the skin of the finger and the air around it.)

Refrigerants evaporate or "boil" at much lower temperatures than water, which permits them to extract heat at a more rapid rate than the water on your finger.

SYSTEM COMPONENTS—SOME FUNDAMENTALS

As pointed out above, the job of the refrigeration cycle is to remove unwanted heat from one place and discharge it into another. To accomplish this, the refrigerant is pumped through a completely closed system. If the system were not closed, it would be using up refrigerant by dissipating it into the air; because it is closed, the same refrigerant is used over again, each time it passes through the cycle removing some heat and discharging it (Fig. 1-4).

Imagine as a comparable situation a boat that is filled with water. We want to remove the water, so we use a pail. We scoop out the unwanted water and transfer it to the outside of the boat, but we hold onto the pail, and use it over and over again to continue bailing.

The closed cycle serves other purposes as well: it keeps the refrigerant from becoming contaminated and controls its flow, for it is a liquid in some parts of the cycle and a gas or vapor in other phases.

Let's take a look at what happens in a simple refrigeration cycle, and the major components of which it is made (this fundamental approach will be enlarged upon in subsequent chapters; see Fig. 1-5).

Two different pressures exist in the cycle—the evaporating or low pressure in the "low side," and the condensing, or high pressure, in the "high side." These pressure areas are separated by two dividing points; one is the metering device where the refrigerant flow is controlled, and the other is at the compressor, where the vapor is compressed.

The *metering device* is a good place to start the trip through the cycle. This may be an expansion valve, a capillary tube, or other device to control the flow of the refrigerant into the *evaporator*, or cooling coil, as a low-pressure, low-temperature refrigerant. The expanding refrigerant evaporates (changes state) as it goes through the cooling coil, where it removes the heat from the space in which the evaporator is located.

Heat will travel from the warmer air to the coils cooled by the evaporation of the refrigerant within the system, causing the refrigerant to "boil" and evaporate, changing it to a vapor. This is similar to the change that occurs when a pan of water is boiled on the stove and the water changes to steam—except that the refrigerant boils at a much lower temperature.

Now this low-pressure, low-temperature vapor is drawn to the *compressor* where it is compressed into a high-temperature, high-pressure vapor. The compressor discharges it to the *condenser*, so that it can give up the heat that it picked up in the cooling coil or evaporator. The refrigerant vapor is at a higher temperature than is the air passing across the condenser (air-cooled type); therefore heat is trans-

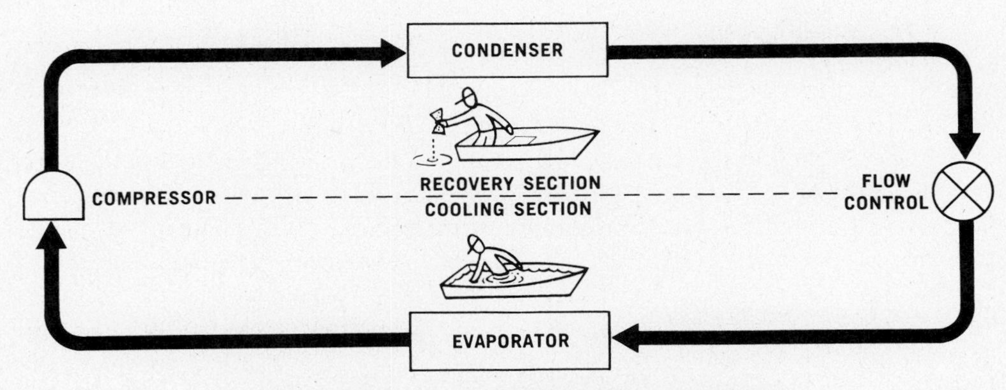

FIGURE 1-4 (Courtesy, Allied Chemical Corporation).

ferred from the warmer refrigerant vapor to the cooler air.

In this process, as heat is removed from the vapor, a change of state takes place and the vapor is condensed back into a liquid, at a high pressure and a high temperature.

The liquid refrigerant now travels to the metering device where it passes through a small opening or orifice where a drop in pressure and temperature occurs, and then it enters into the evaporator or cooling coil. As the refrigerant makes its way into the larger opening of the tubing or coil, it vaporizes, ready to start another cycle through the system.

The refrigeration system requires some means of connecting the major components—evaporator, compressor, condenser, and metering device—just as roads connect communities. Tubing or "lines" make the system complete so that the refrigerant will not leak out into the atmosphere. The *suction* line connects the evaporator or cooling coil to the compressor, the *hot gas* or *discharge* line connects the compressor to the condenser, and the *liquid* line is the connecting tubing between the condenser and the metering device. Some systems will have a *receiver* or storage tank immediately after the condenser and before the metering device, where the liquid refrigerant remains until it is needed for heat removal in the evaporator.

There are many different kinds and variations of the refrigerant cycle components. For example, there are at least a half dozen different types of compressors, from the reciprocating piston through a centrifugal impeller design, but the function is the same in all cases—that of compressing the heat-laden vapor into a high-temperature vapor.

The same can be said of the condenser and evaporator surfaces. They can be bare pipe, or they can be finned condensers and evaporators with electrically driven fans to pass the air through them.

There are a number of different types of metering devices to meter the liquid refrigerant into the evaporator, depending on size of equipment, refrigerant used, and its application.

The mechanical refrigeration system described above is essentially the same whether the system be a domestic refrigerator, a low-temperature freezer, or a comfort air-conditioning system. Refrigerants will be different and size of equipment will vary greatly, but the principle of operation and the refrigeration cycle remain the same. Thus, once a student understands the simple actions that are taking place within the mechanical cycle, he has gone a long way toward understanding how a refrigeration system works.

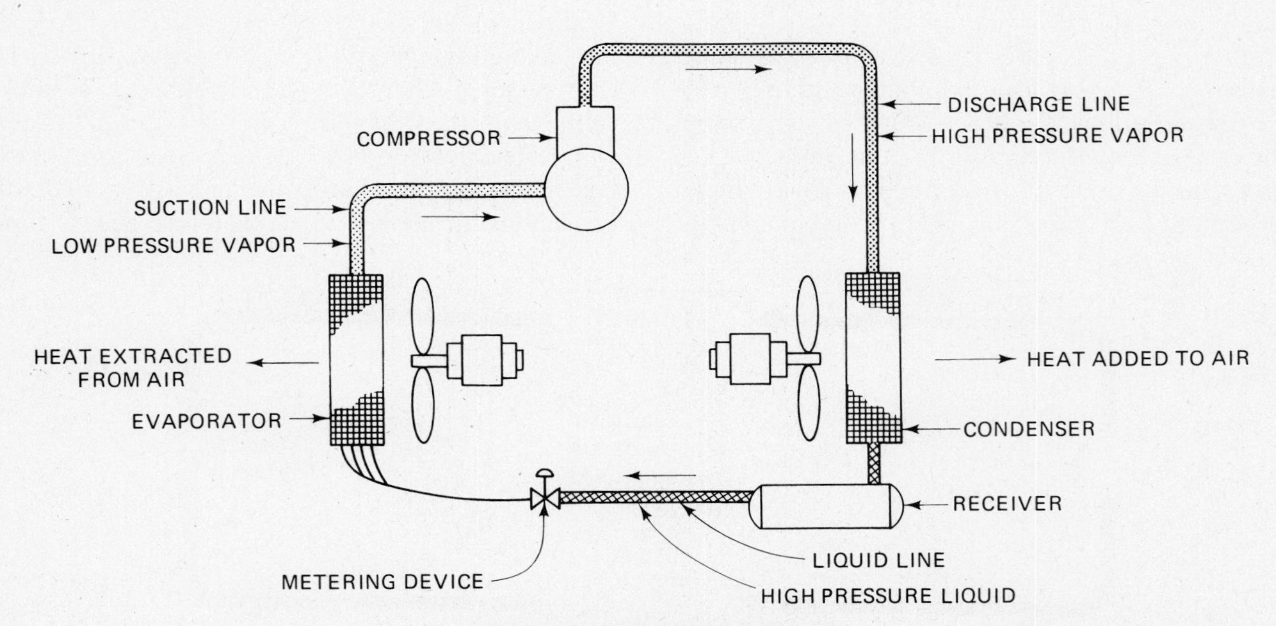

FIGURE 1-5 Simple refrigeration system.

PROBLEMS

1. The country of ______________ is the earliest known to use ice for refrigeration.
2. The mechanical refrigeration cycle was first used in the early ______________s.
3. Refrigeration and air-conditioning systems represent one of the major domestic and world industries. True or false?
4. For every $1 million in installed value of equipment, are there two, six, or 13 refrigeration and air-conditioning technicians needed?
5. Heat is the absence of cold. True or false?
6. "Absolute zero" is 0°F on the Fahrenheit thermometer. True or false?
7. Name the four major components of a mechanical refrigeration cycle.
8. Refrigerant in the system evaporates and in the process ______________ heat.
9. The hot gas line connects the ______________ and ______________.
10. The suction line connects the ______________ and ______________.

2

MATTER AND HEAT BEHAVIOR

STATES OF MATTER

All known matter exists in one of three physical forms or states: solid, liquid, or gaseous. There are distinct dissimilarities among these physical states, namely:

(1) Matter in a *solid* state will retain its quantity, shape, and physical dimensions. A cubic foot of wood will retain its weight, size, and shape even if moved from place to place.

(2) Matter in a *liquid* state will retain its quantity and size but not its shape. The liquid will always conform to the occupying container. If a cubic foot of water in a container measuring 1 foot on each side (Fig. 2-1) is transferred to a container of different dimension, such as a cylindrical tank or one of different rectangular dimensions, the quantity and volume of the water will be the same although the dimensions will change.

(3) Matter in a *gaseous* state does not have a tendency to retain either its size or its shape. If a 1-cubic-foot cylinder containing steam or some other gas is connected to a 2-cubic-foot cylinder on which a vacuum has been drawn, the vapor will expand to occupy the volume of the larger cylinder (Fig. 2-2)

Although these specific differences exist in the three states of matter, quite frequently, under changing conditions of pressure and temperature, the same substance may exist in any one of the three states, such as a solid, a liquid, or a vapor (ice, water, and steam, for example).

Solids always have some definite shape, whereas liquids and gases have no definite shape of their own, but will conform to the shape of their containers.

MOLECULAR MOVEMENT

All matter is composed of small particles known as molecules, and the molecular structure of matter (as studied in chemistry) can be further broken down into atoms.

Later on, in the section on basic electricity, you will learn how the atom is further divided into electrons, protons, and neutrons. In that area you will

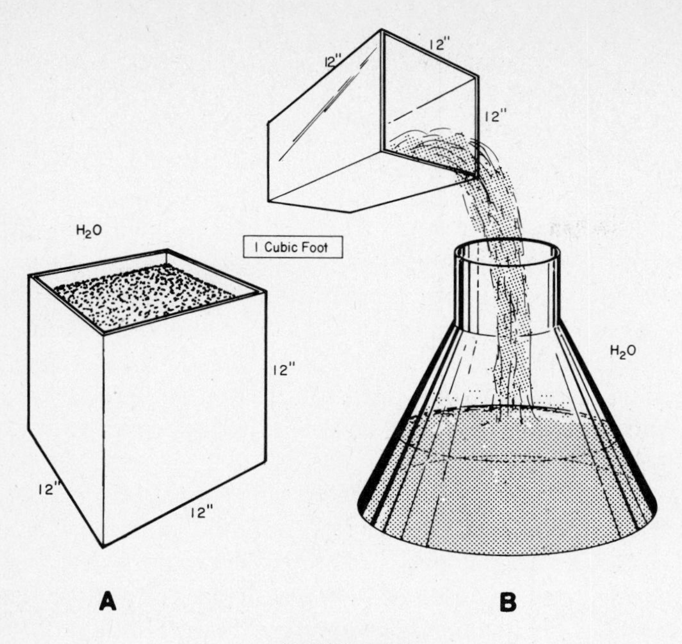

FIGURE 2-1.

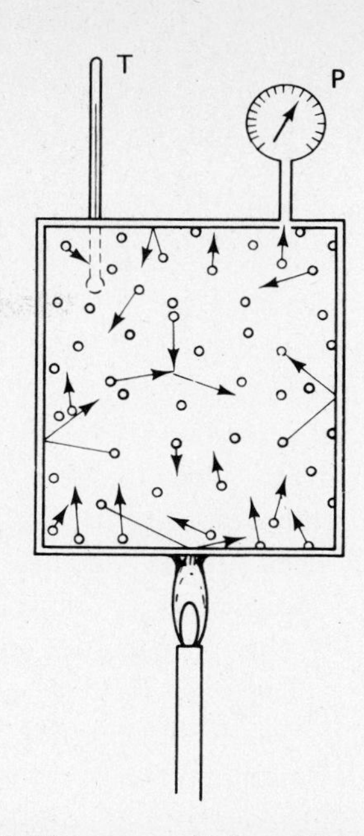

FIGURE 2-3 Molecular agitation with heat application.

study the electron theory, which concerns electric current flow—the movement of electrons through a conductor.

For the present we will concern ourselves only with the molecule, the smallest particle into which any matter or substance can be broken down and still retain its identity. For example, a molecule of water (H_2O) is made up of two atoms of hydrogen and one atom of oxygen. If this molecule of water were to be broken down more, or divided further, into subatomic particles, it would no longer be water.

Molecules vary in shape, size, and weight. In physics we learn that molecules have a tendency to cling together. The character of the substance of matter itself is dependent on the shape, size, and weight of the individual molecules of which it is made and also the space or the distance between them, for they are, to a large degree, capable of moving about.

When heat energy is applied to a substance (Fig. 2-3), it increases the internal energy of the molecules, which increase their motion or velocity of movement. With this increase in the movement of the molecules, there is also a rise or increase in the temperature of the substance.

When heat is removed from a substance, it follows that the velocity of the molecular movement will decrease and also that there will be a decrease or lowering of the internal temperature of the substance.

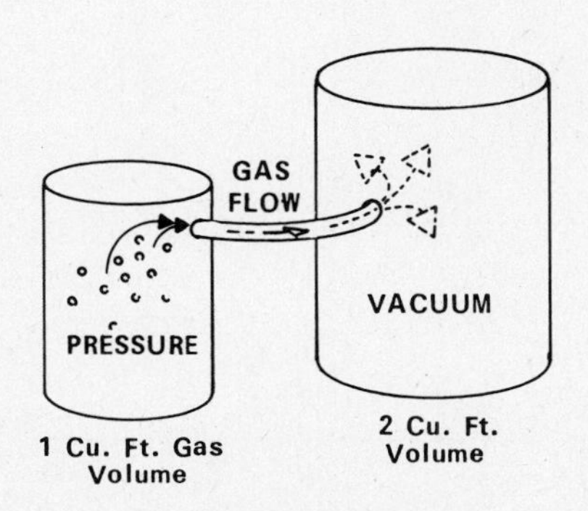

FIGURE 2-2.

CHANGE OF STATE

When a solid substance is heated, the molecular motion is chiefly in the form of rapid motion back and forth, the molecules never moving far from their normal or original position. But at some given temperature for that particular substance, further addition of heat will not necessarily increase the molecular motion within the substance; instead, the additional heat will cause some solids to liquefy (change into a liquid). Thus the additional heat causes a change of state in the material.

The temperature at which this change of state in a substance takes place is called its *melting point*. Let us assume that a container of water at 70°F, in which a thermometer has been placed, is left in a freezer for hours (Fig. 2-4). When it is taken from the freezer, it has become a block of ice—*solidification* has taken place.

Let us further assume that the thermometer in the block indicates a temperature of 20°F. If it is allowed to stand at room temperature, heat from the room air will be absorbed by the ice until the thermometer indicates a temperature of 32°F, when some of the ice will begin to change into water.

With heat continuing to transfer from the room air to the ice, more ice will change back into water; but the thermometer will continue to indicate a temperature of 32°F until all the ice has melted. *Liquefaction* has now taken place. (This change of state without a change in the temperature will be discussed later.)

As mentioned, when all the ice is melted the thermometer will indicate a temperature of 32°F, but the temperature of the water will continue to rise until it reaches or equals room temperature.

If sufficient heat is added to the container of water through outside means (Fig. 2-5), such as a burner or a torch, the temperature of the water will increase until it reaches 212°F. At this temperature, and under "standard" atmospheric pressure, another change of state will take place—*vaporization*. Some of the water will turn into steam and, with the addition of more heat, all of the water will vaporize into steam; yet the temperature of the water will not increase above 212°F. This change of state—without a change in temperature—will also will be discussed later.

If the steam vapor could be contained within a closed vessel, and if the source of heat were removed, the steam would give up heat to the surrounding air, and it would condense back into a liquid form—water. What has now taken place is *condensation*—the reverse process from vaporization.

Oxygen is a gas above −297°F, a liquid between that temperature and −324°F, and a solid below that point. Iron is a solid until it is heated to +2800°F, and it vaporizes at a temperature approximating 4,950°F.

Thus far we have learned, with some examples, how a solid can change into a liquid, and how a liquid can change into a vapor. But it is possible for a substance to undergo a *physical* change through which a solid will change directly into a gaseous state without first melting into a liquid. This is known as *sublimation*.

All of us probably have seen this physical change take place without fully recognizing the process. Damp or wet clothes, hanging outside in the freezing temperature, will speedily become dry through sublimation; just as dry ice (solid carbon dioxide, or CO_2) sublimes directly into a vapor under normal temperature and pressure.

Let us review these changes of state,

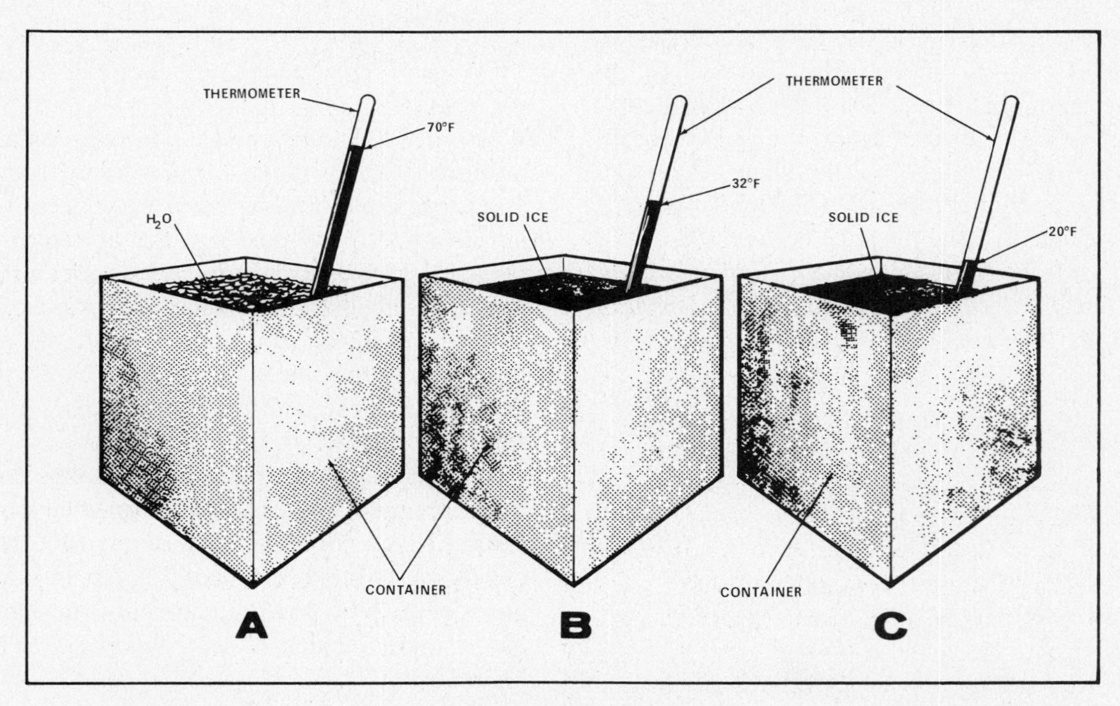

FIGURE 2-4.

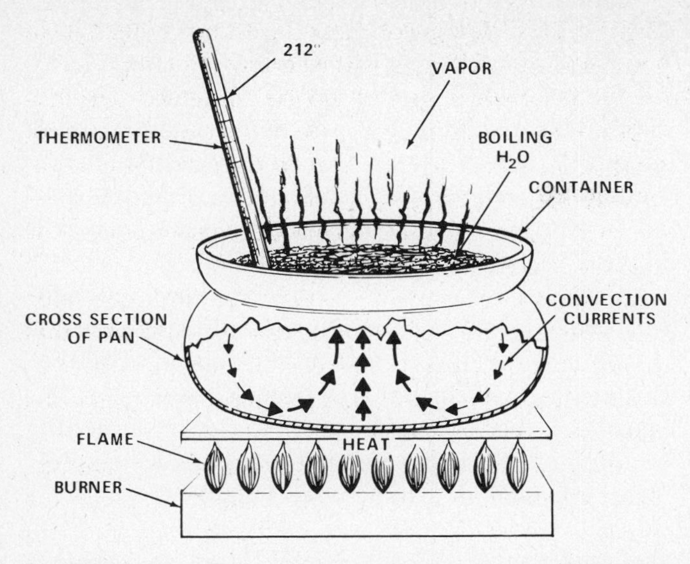

FIGURE 2-5 Convection currents caused by temperature differential.

SOLIDIFICATION—a change from a liquid to a solid.

LIQUEFACTION —a change from a solid to a liquid.

VAPORIZATION —a change from a liquid to a vapor.

CONDENSATION—a change from a vapor to a liquid.

SUBLIMATION —a change from a solid to a vapor without passing through the liquid state.

MEASUREMENTS

Most of us are acquainted with common measurements, such as those pertaining to length, weight, volume, etc.; but now we move into other types of measurements, such as those of heat intensity, heat quantity, and energy conversion units.

HEAT INTENSITY

You will recall that heat is a form of energy which is not measurable in itself; but the heat intensity, or temperature of a substance, can be measured. A unit of the intensity of heat is called the *degree*, measured on a temperature scale.

In the discussion of states of matter, temperature was discussed, as was the addition or removal of heat. Relatively, water is colder than steam; yet it is, at the same time, warmer than ice. Temperature scales were formulated through use of glass tubes with similar interior diameters and a reservoir for a liquid—such as mercury—that will expand and rise up in the tube when heated.

The *Fahrenheit thermometer* or scale is based on the relative positions of the mercury in the thermometer when water is at the freezing point and when water is boiling. The distance between these two points was divided into 180 equal portions or parts called *degrees*. The point where water either will freeze, or ice will melt, under normal atmospheric conditions, was labeled as 32°; whereas the location, or point, on the thermometer where water will boil was labeled 212°. The Fahrenheit thermometer has been the one most commonly used in most types of engineering work and in weather reports. A *Celsius thermometer*, formerly called a *Centigrade thermometer*, is used in chemistry and physics, especially in continental Europe, and is coming into more common usage with the move toward metrication.

The *Celsius thermometer* or scale was based on a decision to divide the distance between the freezing and boiling points into 100 equal portions or parts, with the freezing point set at 0° and the boiling point at 100°. A comparison of the Fahrenheit and Celsius scales is shown in Fig. 2-6.

Frequently students ask why the boiling point of water and the melting point of ice were used as the standards for both thermometers. These points or temperatures were chosen because water has a very constant boiling and freezing temperature, and of course water is a very common substance.

TEMPERATURE CONVERSION

Most frequently a conversion from one temperature scale to the other is made by the use of a conversion table, but if one is not available, the conversion can be done easily by a formula using these equations:

$$°F = 1.8°C + 32 \qquad \textbf{(2-1)}$$

$$°F = \tfrac{9}{5}°C + 32$$

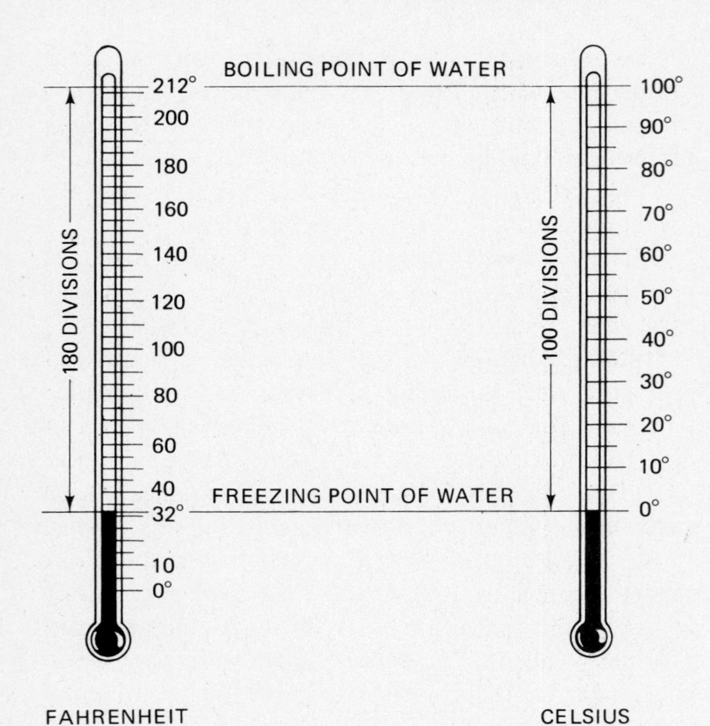

FIGURE 2-6 Comparison of the Fahrenheit and Celsius temperature scales.

$$°C = \frac{°F - 32}{1.8} \qquad \textbf{(2-2)}$$

$$°C = \tfrac{5}{9}(°F - 32)$$

EXAMPLE:

To convert a room temperature of 77°F to its equivalent on the Celsius scale:

(a) $°C = \frac{77 - 32}{1.8} = \frac{45}{1.8} = \underline{25°C}$

(b) $°C = \tfrac{5}{9}(77 - 32) = \tfrac{5}{9}(45) = \tfrac{225}{9} = \underline{25°C}.$

EXAMPLE:

To convert to the Fahrenheit scale a temperature of 30°C:

(a) $°F = 1.8 \times (30) + 32 = 54 + 32 = \underline{86°F}.$

(b) $°F = \tfrac{9}{5} \times (30) + 32 = 54 + 32 = \underline{86°F}.$

Thus far, in the measurement of heat intensity, we have located two definite reference points—the freezing point and the boiling point of water on both the Fahrenheit and Celsius scales. We now must locate still a third definite reference point—absolute zero. This is the point where, it is believed, all molecular action ceases. As already noted on the Fahrenheit temperature scale, this is about 460° below zero, —460°F, while on the Celsius scale it is about 273° below zero, or —273°C.

Certain basic laws, to be discussed in later chapters, are based on the use of absolute temperatures. If a Fahrenheit reading is given, the addition of 460° to this reading will convert it to degrees Rankine or °R; whereas if the reading is from the Celsius scale, the addition of 273° will convert it to degrees Kelvin, °K. These conversions are shown in Figure 2-7.

EXAMPLE:

What is the freezing point of water in degrees Rankine (°R)? Since the freezing point of water is 32°F, adding 460° makes the freezing point of water *492° Rankine*. (32° + 460° = 492°)

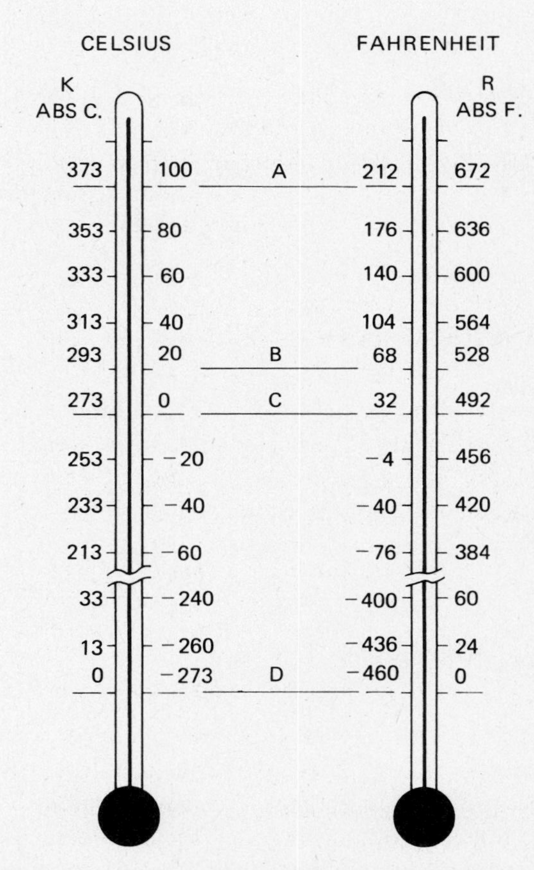

FIGURE 2-7 Fahrenheit, Celsius, Rankine and Kelvin thermometer scales: A. Boiling temperature of water; B. Standard conditions temperature; C. Freezing temperature of water; D. Absolute zero.

Heat quantity is different from heat intensity, because it takes into consideration not only the temperature of the fluid or substance being measured but also its weight. The unit of heat quantity is the *British thermal unit* (Btu). Water is used as a standard for this unit of heat quantity; a Btu is the amount of heat required to raise the temperature of 1 pound of water 1 degree Fahrenheit at sea level.

Two Btu will cause a change in temperature of 2 degrees Fahrenheit of 1 pound of water; or it will cause a change in temperature of 1 degree Fahrenheit of 2 pounds of water. Therefore, when considering a change in temperature of water, the following equation may be utilized:

$$\text{Btu} = W \times TD \qquad \textbf{(2-3)}$$

where

Change in *H*eat (in Btu) = *W*eight (in pounds) × *T*emperature *D*ifference

Example:

Calculate the amount of heat necessary to increase 10 pounds of water from 50°F to 100°F.

Since the Temperature Difference (TD) = 50°, then

$$\text{Heat} = \text{Btu} = W \times TD = 10 \times 50 = \underline{500 \text{ Btu}}$$

In the above example heat is added to the quantity of water, but the same equation is also used if heat is to be removed.

Example:

Calculate the amount of heat removed if 20 pounds of water are cooled from 80°F to 40°F.

$$\text{Btu} = W \times TD = 20 \times 40 = \underline{800 \text{ Btu}}$$

FIRST LAW OF THERMODYNAMICS

The first law of thermodynamics (that branch of science dealing with mechanical action of heat) states that energy can neither be created nor destroyed. It can only be converted from one form to another. Energy itself is defined as the ability to do work, and heat is one form of energy.

There are other forms of energy: mechanical, electrical, and chemical, which may be converted easily from one form to another. The steam-driven turbine generator of a power plant is a device that converts heat energy into electrical energy. Chemical energy may be converted into electrical energy through the use of a battery. Electrical energy is converted into mechanical energy through the application of an electric motor to drive or operate controlled devices. Electrical energy may be used or converted into heat energy by means of an arrangement of electric coils of wire, whether they be in an electric grille, toaster, heater, or the like.

Heat travels downhill, from the warmer to the cooler of two substances. This is called the *second law of thermodynamics*.

In summary, heat is a form of energy; heat can be transferred; the intensity of heat can be measured; and heat is present in all substances, above a temperature of absolute zero.

Heat and heat transfer are also expressed in different forms that are important to the refrigeration and air conditioning industry.

SPECIFIC HEAT

The specific heat of a substance is the quantity of heat in Btus required to change the temperature of 1 pound of the substance 1 degree Fahrenheit. Earlier we were presented with the information that a Btu was the amount of heat necessary to increase the temperature of 1 pound of water 1 degree Fahrenheit, or to lower the temperature of the same weight of water by the same unit of measurement on a thermometer.

Therefore, the specific heat of water is 1.0; and water is the basis for the specific heat table in Fig. 2-8. You will see that different substances vary in their capacity to absorb or to give up heat. The specific heat values of most substances will vary with a change in temperature; some vary only a slight amount, while others can change considerably.

Suppose that two containers are placed on a heating element or burner side by side, one containing water and the other an equal amount, by weight, of

Water	1.00
Ice	0.50
Air (dry)	0.24
Steam	0.48
Aluminum	0.22
Brass	0.09
Lead	0.03
Iron	0.10
Mercury	0.03
Copper	0.09
Alcohol	0.60
Kerosene	0.50
Olive oil	0.47
Glass	0.20
Pine	0.67
Marble	0.21

FIGURE 2-8 Specific Heats of Common Substances Btu/lb/°F.

olive oil. You would soon find that the temperature of the olive oil increases at a more rapid rate than that of the water, demonstrating that olive oil absorbs heat more rapidly than water.

If the rate of temperature increase of the olive oil was approximately twice that of the water, it could be said that olive oil required only half as much heat as water to increase its temperature 1 degree Fahrenheit. Based on the value of 1.0 for the specific heat of water, it would show that the specific heat of olive oil must be approximately 0.5, or half that of water. (The table of specific heats of substances shows that olive oil has a value of 0.47).

Equation (2-3) from the previous discussion can now be stated as:

$$\text{Btu} = W \times c \times TD \qquad \textbf{(2-4)}$$

where

c = specific heat of a substance

Example:

Calculate the amount of heat required to raise the temperature of one pound of olive oil from 70°F to 385°F.

Since the TD = 315° and c of olive oil = 0.47, then

$$\text{Heat} = \text{Btu} = W \times c \times TD$$
$$= 1 \times 0.47 \times 315 = \underline{148 \text{ Btu}}$$

The specific heat of a substance also will change, with a change in the state of the substance. Water is a very good example of this variation in specific heat. We have learned that, as a liquid, its specific heat is 1.0; but as a solid—ice—its specific heat approximates 0.5, and this same value is applied to steam—the gaseous state.

Within the refrigeration circuit we will be interested, primarily, with substances in a liquid or a gaseous form, and their ability to absorb or give up heat. Also, in the distribution of air for the purpose of cooling or heating a given area, we will be interested in the possible changes in the values for specific heat—more about this later.

Air, when heated and free or allowed to expand at a constant pressure, will have a specific heat of 0.24. Refrigerant 12 (R-12) vapor at approximately 70°F and at constant pressure, has a specific heat value of 0.148, whereas the specific heat of R-12 liquid is 0.24 at 86°F.

When dealing with metals and the temperature of a mixture, a combination of weights, specific heats, and temperature differences must be considered in the overall calculations of heat transfer. In order to calculate the total heat transfer of a combination of substances it is necessary to add the individual rates as shown in Eq 2-5. below.

$$\text{Btu}_t = (W_1 \times c_1 \times TD_1) + (W_2 \times c_2 \times TD_2) + (W_3 \times c_3 \times TD_3) \ldots \text{etc} \qquad \textbf{(2-5)}$$

Example:

How much heat must be added to a 10-lb copper vessel, holding 30 lbs of water at 70°F to reach 185°F if the specific heat of copper is 0.095 F.

Equation (2-5) may be used as:

$$\text{Btu} = (W_1 \times c_1 \times TD_1) + (W_2 \times c_2 \times TD_2)$$

where W_1, c_1, and TD_1 pertain to the copper vessel, and W_2, c_2, and TD_2 pertain to the water.

Therefore,

$$\text{Btu} = (10 \times 0.095 \times 115) + (30 \times 1.0 \times 115)$$
$$= 109.2 + 3450$$
$$= \underline{3{,}559.2 \text{ Btu}}$$

SENSIBLE HEAT

Heat that can be felt or measured is called *sensible heat*. It is the heat that causes a change in temperature of a substance, but not a change in state. Substances, whether in a solid, liquid, or gaseous state, contain sensible heat to some degree, as long as their temperatures are above absolute zero. Equations used for solutions of heat quantity, and those used in conjunction with specific heats, might be classified as being *sensible* heat equations, since none of them involve any change in state.

As mentioned earlier, a substance may exist as a solid, liquid, or as a gas or vapor. The substance as a solid will contain some sensible heat, as it will in the other states of matter. The total amount of heat needed to bring it from a solid state to a vapor state is dependent upon: (a) its initial temperature as a solid, (b) the temperature at which it changes from a solid to a liquid, (c) the temperature at which it changes from a liquid to a vapor, and (d) its final temperature as a vapor. Also included is the heat that is required to effect the two changes in state.

LATENT HEAT

Under a *change of state*, most substances will have a melting point at which they will change from a solid to a liquid without any increase in temperature. At this point, if the substance is in a liquid state and heat is removed from it, the substance will solidify without a change in its temperature. The heat involved in either of these processes (changing from a solid to a liquid, of from a liquid to a solid), without a change in temperature, is known as the *latent heat of fusion.*

Figure 2-9 shows the relationship between temperature in Fahrenheit degrees and both sensible and latent heat in Btus.

As pointed out earlier, the specific heat of water is 1.0 and that of ice is 0.5, which is the reason for the difference in the slopes of the lines denoting the solid (ice) and the liquid (water). To increase the temperature of the ice from 0°F to 32°F requires only 16 Btu

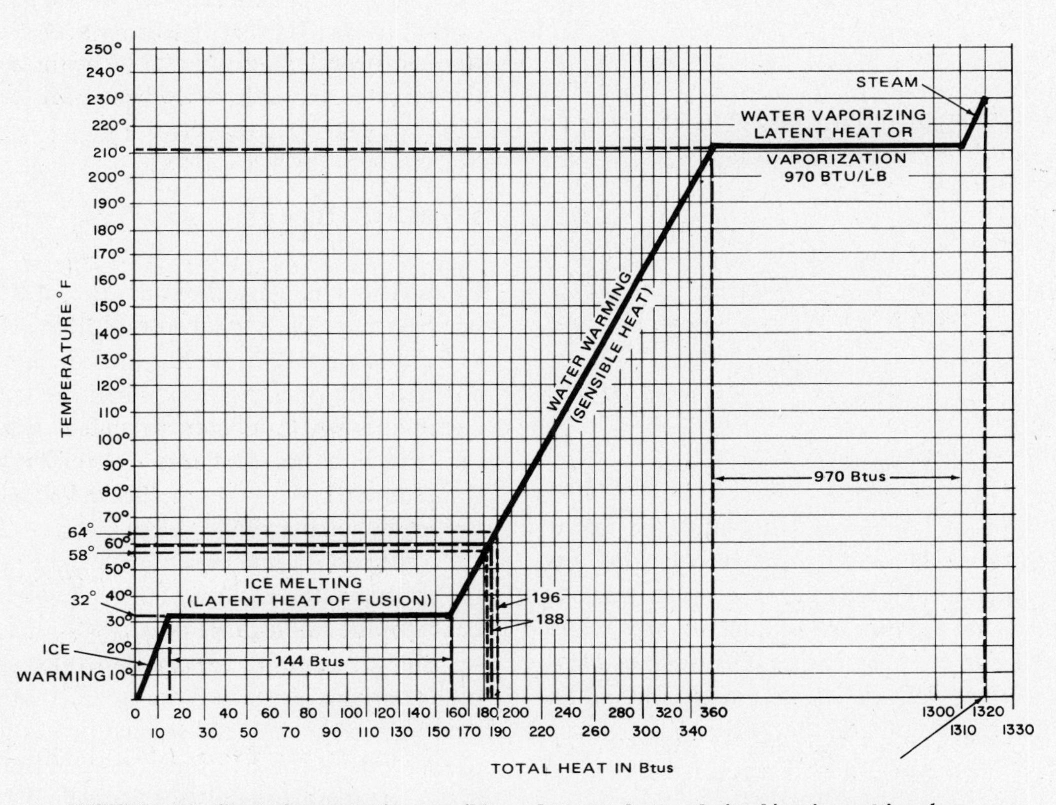

FIGURE 2-9 Chart demonstrating sensible and latent heat relationships in melting ice, changing ice to water and water to steam.

of heat; the other line shows that it takes only 8 Btu to increase the temperature of water 8°F (60° to 68°), from 188 to 196 BTU/lb.

Figure 2-9 also shows that a total of 52 Btu of sensible heat is involved in the 196 Btu necessary for converting 0°F ice to 58°F water. This leaves a difference of 144 Btu, which is the latent heat of fusion of water or ice, depending on whether heat is being removed or added.

The derivation of the word *latent* is from the Latin word for *hidden*. This is hidden heat, which does not register on a thermometer, nor can it be felt. Needless to say, there is no increase or decrease in the molecular motion within the substance, for it would show up in a change in temperature on a thermometer.

$$\text{Btu} = (W_1 \times c_1 \times TD_1) + (W_1 \times \text{latent heat}) + (W_2 \times c_2 \times TD_2) \quad \textbf{(2-6)}$$

EXAMPLE:

Calculate the amount of heat needed to change 10 lb of ice at 20°F to water at 50°F.
Utilizing Eq. (2-6):

$$\text{Btu} = (10 \times 0.5 \times \underset{(32^\circ - 20^\circ)}{12}) + (10 \times 144) + (10 \times 1.0 \times \underset{(50^\circ - 32^\circ)}{18})$$

$$\text{Btu} = (60) + (1{,}440) + (180)$$

$$\text{Btu} = \underline{1{,}680}$$

Another type of latent heat that must be taken into consideration when total heat calculations are necessary is called the *latent heat of vaporization*. This is the heat that 1 pound of a liquid absorbs while being changed into the vapor state. Or it can be classified as the *latent heat of condensation*; for, when sensible heat is removed from the vapor to the extent that it reaches the condensing point, the vapor condenses back into the liquid form.

The latent heat of vaporization of water as 1 pound is boiled or evaporated into steam at sea level is 970 Btu. That amount also is the heat that 1 pound of steam must release or give up when it condenses into water. Figure 2-9 also shows the relationship between temperature and both sensible heat and latent heat of vaporization.

The absorption of the amount of heat necessary for the change of state from a liquid to a vapor by evaporation, and the release of that amount of heat necessary for the change of state from a vapor back to a liquid by condensation are the main principles of the refrigeration process, or cycle. Refrigeration is the transfer of heat by the change in state of the refrigerant.

Figure 2-9 shows the total heat in Btu necessary to convert 1 pound of ice at 0° to superheated steam at 230°F under atmospheric pressure. This total amounts to 1,319 Btu. Only 205 Btu are sensible heat; the remainder is made up of 144 Btu of latent heat of fusion and also 970 Btu of latent heat of vaporization.

SECOND LAW OF THERMODYNAMICS

The *second* law of thermodynamics, as discussed earlier, states that heat transfers in only one direction—downhill; and this takes place through one of the three basic methods of heat transfer.

CONDUCTION

Conduction is described as the transfer of heat between the closely-packed molecules of a substance, or between substances that are touching or in good contact with one another. When the transfer of heat occurs in a single substance, such as a metal rod with one end in a fire or flame, movement of heat continues until there is a temperature balance throughout the length of the rod.

If the rod is immersed in water, the rapidly moving molecules on the surface of the rod will transmit some heat to the molecules of water, and still another transfer of heat by conduction takes place. As the outer surface of the rod cools off, there is still some heat within the rod, and this will continue to transfer to the outer surfaces of the rod and then to the water, until a temperature balance is reached.

The speed with which heat will transfer by means of conduction will vary with different substances or materials if the substances or materials are of the same dimensions. The rate of heat transfer will vary according to the ability of the materials or substances to conduct heat. Solids, on the whole, are much better conductors than liquids; and, in turn, liquids conduct heat better than gases or vapors.

Most metals, such as silver, copper, steel, and iron, conduct heat fairly rapidly, whereas other solids such as glass, wood, or other building materials transfer heat at a much slower rate and therefore are used as insulators.

Copper is an excellent conductor of heat, as is aluminum. These substances are ordinarily used in the evaporators, condensers, and refrigerant pipes connecting the various components of a refrigerant system, although iron is occasionally used with some refrigerants.

The rate at which heat may be conducted through various materials is dependent on such factors as: (a) the thickness of the material, (b) its cross-sectional area, (c) the temperature difference between the two sides of the material, (d) the heat conductivity (k factor) of the material, and (e) the time duration of the heat flow. Figure 2-10 is a table of heat conductivities (k factors) of some common materials.

MATERIAL	CONDUCTIVITY (k)
Plywood	0.80
Glass Fiber-organic bonded	0.25
Expanded polystyrene insulation	0.25
Expanded polyurethane insulation	0.16
Cement Mortar	5.0
Stucco	5.0
Brick (common)	5.0
Hard woods (maple, oak)	1.10
Soft woods (Fir, pine)	0.80
Gypsum Plaster (Sand aggregate)	5.6

FIGURE 2-10 Conductivities for common building and insulating materials. k values expressed in BTU (hour) (square foot) (°F difference) per inch thickness of material.

NOTE:

The k factors are given in Btu/hr/ft²/°F/in. of thickness of the material.

These factors may be utilized correctly through the use of the following equation:

$$\text{Btu} = \frac{A \times k \times TD}{X} \qquad (2\text{-}7)$$

where

A = Cross-sectional area in ft²

k = Heat conductivity in Btu/hr

TD = Temperature difference between the two sides

X = Thickness of material in inches.

Metals with a high conductivity are used within the refrigeration system itself because it is desirable that rapid transfer of heat occur in both evaporator and condenser. The evaporator is where heat is removed from the conditioned space or substance or from air that has been in direct contact with the substance; the condenser dissipates this heat to another medium or space.

In the case of the evaporator, the product or air is at a higher temperature than the refrigerant within the tubing and there is a transfer of heat downhill; whereas in the condenser the refrigerant vapor is at a higher temperature than the cooling medium traveling through or around the condenser, and here again there is a downhill transfer of heat.

Plain tubing, whether copper, aluminum, or another metal, will transfer heat according to its conductivity or k factor, but this heat transfer can be increased through the addition of fins on the tubing. They will increase the area of heat transfer surface, thereby increasing the overall efficiency of the system. If the addition of fins doubles the surface area, it can be shown by the use of Eq. (2-7) that the overall heat transfer should itself be doubled, when compared to that of plain tubing.

CONVECTION

Another means of heat transfer is by motion of the heated material itself and is limited to liquid or gas. When a material is heated, convection currents are set up within it, and the warmer portions of it rise, since heat brings about the decrease of a fluid's density and an increase in its specific volume.

Air within a refrigerator and water being heated in a pan are prime examples of the results of convection currents (see Figure 2-11). The air in contact

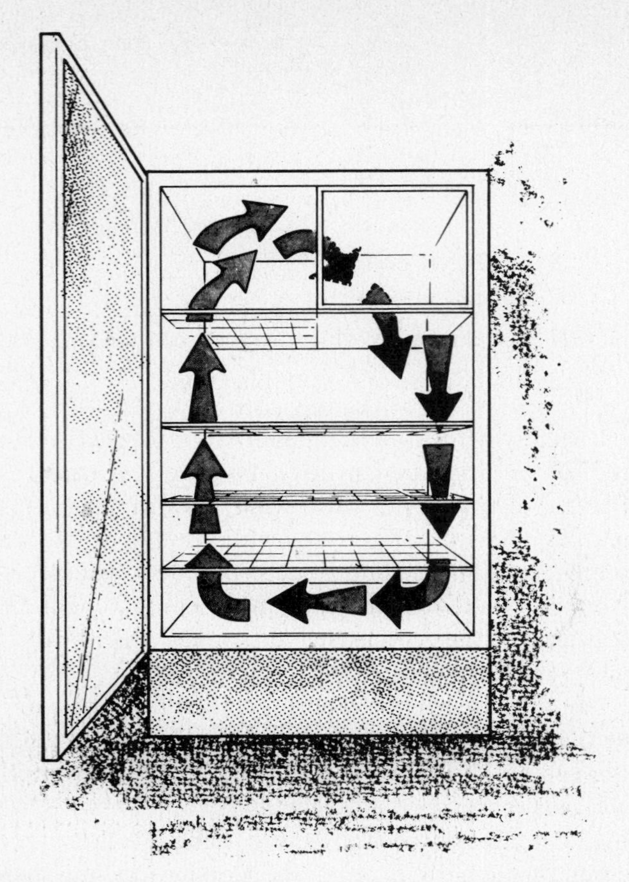

FIGURE 2-11 Convection currents caused by temperature differential.

with the cooling coil of a refrigerator becomes cool and therefore more dense, and begins to fall to the bottom of the refrigerator. In doing so, it absorbs heat from the food and the walls of the refrigerator, which, through conduction, has picked up heat from the room.

After heat has been absorbed by the air it expands, becoming lighter, and rises until it again reaches the cooling coil where heat is removed from it. The convection cycle repeats as long as there is a temperature difference between the air and the coil. In commercial-type units, baffles may be constructed within the box in order that the convection currents will be directed to take the desired patterns of air flow around the coil.

Water heated in a pan will be affected by the convection currents set up within it through the application of heat. The water nearest the heat source, in absorbing heat, becomes warmer and expands. As it becomes lighter, it rises and is replaced by the other water, which is cooler and more dense. This process will continue until all of the water is at the same temperature.

Convection currents as explained and shown here are natural, and, as in the case of the refrigerator, a natural flow is a slow flow. In some cases, convection must be increased through the use of fans or blowers and, in the case of liquids, pumps are used for forced circulation to transfer heat from one place to another.

RADIATION

A third means of heat transfer is through radiation by waves similar to light or sound waves. The sun's rays heat the earth by means of radiant heat waves, which travel in a straight path without heating the intervening matter or air. The heat from a light bulb or from a hot stove is radiant in nature and is felt by those near them, although the air between the source and the object, which the rays pass through, is not heated.

If you have been relaxing in the shade of a building or a tree on a hot sunny day and move into direct sunlight, the direct impact of the heat waves will hit like a sledge hammer even though the air temperature in the shade is approximately the same as in the sunlight.

At low temperatures there is only a small amount of radiation, and only minor temperature differences are noticed; therefore radiation has very little effect in the actual process of refrigeration itself. But results of radiation from direct solar rays can cause an increased refrigeration load in a building in the path of these rays.

Radiant heat is readily absorbed by dark or dull materials or substances, whereas light-colored surfaces or materials will reflect radiant heat waves, just as they do light rays. Wearing-apparel designers and manufacturers make use of this proven fact by supplying light-colored materials for summer clothes.

This principle is also carried over into the summer air-conditioning field where, with light-colored roofs and walls, less of the solar heat will penetrate into the conditioned space, thus reducing the size of the overall cooling equipment required. Radiant heat also readily penetrates clear glass in windows, but will be absorbed by translucent or opaque glass.

When radiant heat or energy (since all heat is energy) is absorbed by a material or substance it is converted into sensible heat—that which can be felt or measured. Every body or substance absorbs radiant energy to some extent, depending upon the temperature difference between the specific body or substance and other substances. Every substance will radiate energy as long as its temperature is above absolute zero and another substance within its proximity is at a lower temperature.

If an automobile has been left out in the hot sun with the windows closed for a long period of time, the temperature inside the car will be much greater than the ambient air temperature surrounding it. This demonstrates that radiant energy absorbed by the materials of which the car is constructed is converted to measurable sensible heat.

INSULATION

In the section on heat transfer by means of conduction it was pointed out that certain substances are excellent conductors of heat, while others are poor conductors, which may be classified as insulators. Therefore, any material that deters or helps to prevent the transfer of heat by any means is called and may be used as insulation. Of course, no material will stop completely the flow of heat. If there were such a substance, it would be very easy to cool a given space down to a desired temperature and keep it there.

Such substances as cork, glass fibers, mineral wool, and polyurethane foams are good examples of insulating materials; but numerous other substances are used in insulating refrigerated spaces or buildings. The compressible materials, such as fibrous substances, offer better insulation if installed loosely-packed or in blanket or batt form than if they are compressed or tightly packed.

The thermal conductivity of materials, the temperature to be maintained in the refrigerated space, the ambient temperature surrounding the enclosed space, permissible wall thicknesses of insulating materials, and the cost of the various types of insulation, are all points to consider in selecting the proper material for a given project. Most service personnel are not involved in the selection or the installation of insulating material in a refrigeration application, but they may come in contact with different types of insulation, and under various conditions.

Insulation should be fire and moisture resistant, and also vermin proof. Large refrigeration boxes or walk-in types of coolers are usually insulated with a rigid-type of insulation such as corkboard, fiber glass, foam blocks, and the like; whereas smaller boxes or receptacles might be filled or insulated with a foam type that flows like a liquid and expands to fill

up the available cavity with foam. Low-temperature boxes require an insulation that is also vapor-resistant, such as unicellular foam, if the walls of the refrigerated enclosure are not made of metal on the outside, so that water vapor will not readily penetrate through into the insulation and condense there, reducing the insulating efficiency.

REFRIGERATION EFFECT—"TON"

A common term that has been used in refrigeration work to define and measure capacity or refrigeration effect is called a *ton* or *ton of refrigeration.* It is the amount of heat absorbed in melting a ton of ice (2,000 lb) over a 24-hour period.

The *ton of refrigeration* is equal to 288,000 Btu. This may be calculated by multiplying the weight of ice (2,000 lb) by the latent heat of fusion (melting) of ice (144 Btu/lb). Thus

$$2{,}000 \text{ lb} \times 144 \text{ Btu/lb} = 288{,}000 \text{ Btu}$$

in 24 hours or 12,000 Btu per hour (288,000 ÷ 24). Therefore, *one ton of refrigeration* = 12,000 Btu/hr.

EXAMPLE:

A 10-ton refrigerating system will have a capacity of 10 × 12,000 Btu/hr = 120,000 Btu/hr.

SUMMARY

The change of state of matter can be effected by adding or taking away heat. Heat effect or intensity can be measured by the use of thermometers. Heat always travels from a warmer condition to a colder condition. Substances have different capacities to absorb heat. Heat exists in two forms: *sensible* and *latent.* The unit of measure to express heat quantity is the Btu. Heat can be transferred by several methods: conduction, radiation, and convection. An insulator is a substance that will retard the flow of heat.

In the refrigeration cycle we work with an enclosed system where the effects of heat and pressure are highly related, and in Chapter 3 we will examine the behavior of fluids and pressures.

PROBLEMS

1. Matter exists in one of three forms; name them.
2. What is the smallest particle of matter?
3. Is temperature a measure of heat quantity or intensity?
4. What is meant by the term *absolute zero*?
5. Define a British thermal unit (Btu).
6. Convert a room temperature of 68°F to the Celsius scale.
7. How much heat is required to raise 100 lb of water at 70°F to 120°F?
8. If 750 Btu are supplied to 15 lb of water at 72°F, what will be the resulting temperature?
9. What is the specific heat value for water?
10. Latent heat can be measured on a thermometer: true or false?
11. Heat travels by one of three means: name them.
12. Is insulation a good or poor conductor of heat?

3

FLUIDS AND PRESSURE

Webster's dictionary describes a fluid as "any substance that can flow, liquid or gas." Therefore a refrigerant may be classified as a fluid, since, within the refrigeration cycle, it exists both as a liquid and as a vapor or gas. Although, as previously mentioned, ice—a solid—is also used in heat removal, its usage in refrigeration has been overshadowed somewhat by the discovery of the versatility of the chemicals and chemical combinations used as refrigerants today.

FLUID PRESSURE

The weight of a block of wood or any other solid material acts as a force downward on whatever is supporting it. The force of this solid object is the overall weight of the object, and the total weight is distributed over the area upon which it lies.

The weight of a given volume of water, however, acts not only as a force downward on the bottom of the container holding it, but also as a force laterally on the sides of the container. If a hole is made in the side of the container below the water level (Fig. 3-1), the water above the hole will be forced out because of its force acting downward and sideways.

Fluid pressure is the force per unit area that is exerted by a gas or a liquid. It is usually expressed in terms of psi (pounds per square inch). It varies directly with the density and the depth of the liquid, and, at the same depth below the surface, the pressure is equal in all directions. Notice the difference between the terms used: force and pressure. *Force* means the total weight of the substance; *pressure* means the unit force or pressure per square inch.

If the tank in Fig. 3-1 measures 1 foot in all dimensions, and it is filled with water, we have a cubic foot of water, the weight of which approximates 62.4 lb. Therefore we would have a total force of 62.4 lb being exerted on the bottom of the tank, the area of which equals 144 square inches (12 in. $\times$ 12 in. = 144 in.^{2}). Using the equation:

$$\text{Pressure} = \frac{\text{Force}}{\text{Area}} \qquad \textbf{(3-1)}$$

or

$$P = \frac{F}{A}$$

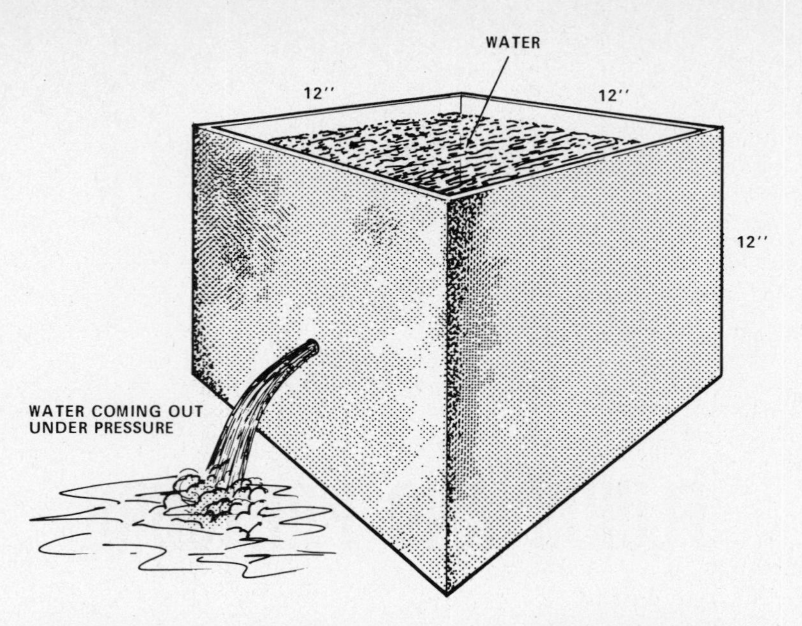

FIGURE 3-1 Water pressure in a container exerts pressure in all directions.

the unit pressure of the water will be:

$$\text{Unit pressure} = \frac{\text{Force}}{\text{Area}} = \frac{62.4}{144} = \mathbf{0.433\ psi}$$

This pressure of 0.433 psi is exerted downward and also sideways. If a tank is constructed as in Figure 3-2, the pressure of the water would cause the tube to be filled with water to the same level as it is in the tank. Fluid pressure is the same on each square inch of the walls of the tank at the same depth, and it will act at right angles to the surface of the tank.

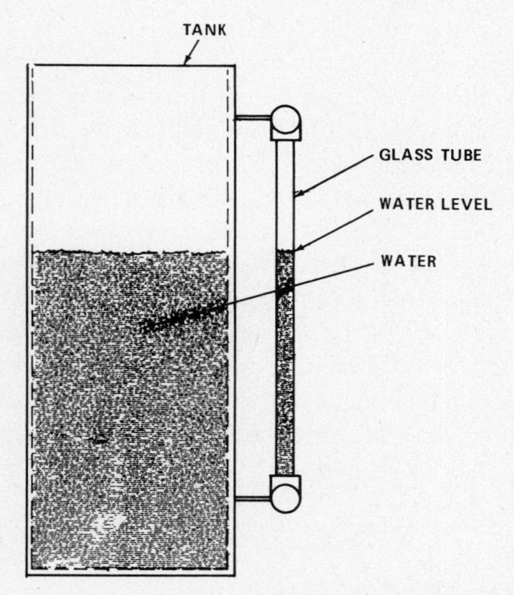

FIGURE 3-2 Fluid pressure same in tube as in tank.

EXAMPLE:

A tank is 4 feet square and is filled with water to its depth of 3 feet. Find:
(a) the volume of water, (b) the weight of the water, (c) the force on the bottom of the tank, and (d) the pressure on the bottom of the tank.

(a) Volume = 4 ft × 4 ft × 3 ft = **48 cu ft (ft^3)**

(b) Weight = 48 ft^3 × 62.4 lb per ft^3 = **2,995 lb**

(c) Force = weight = **2,995 lb**

(d) Pressure = force ÷ area = 2,995 ÷ 2,304 = **1.3 psi**
(Area = 4 ft × 4 ft × 144 $in.^2/ft^2$ = 2,304 $in.^2$)

HEAD

Pressure and depth have a close relationship when a fluid is involved. In hydraulics, a branch of physics that has to do with properties of liquids, the depth of a body of water is called the *head* of water. *Water pressure varies directly with its depth.* As an example, if the tank in Figure 3-1 was 2 ft high and filled with water, it would contain a volume of 2 ft^3 of water and would weigh 2 × 62.4, or 124.8 lb. Now the force of the water on the bottom of the tank would be 124.8 lb, the area of the bottom would still be 144 in.^{2}, and the unit pressure would be 0.866 psi (124.8 ÷ 144). This is twice the amount of pressure that was exerted when the head of water was only 1 ft. Therefore, in an open-top container, the pressure of the water will equal 0.433 psi for each foot of head.

If there is a decrease or increase in the head of a body of water, there will be a corresponding decrease or increase in the pressure involved, as well as in the weight of the water, providing the other dimensions stay the same. If the head of water was only 6 in. ($\frac{1}{2}$ ft), the pressure would equal $0.433 \times \frac{1}{2}$, or 0.217 psi; and if there were 10 feet of water in an open tank, the pressure would be 0.433 × 10 or 4.33 psi.

This relationship can be expressed in the equation:

$$p = 0.433 \times h \tag{3-2}$$

where

p = pressure in psi
h = head in feet of water

The tank in Fig. 3-1 has an area of 1 square foot with 1 foot of head; therefore the pressure on the bottom of the tank is 0.433 psi. If there is a fish pond covering an area of 50 square feet, and the depth of water in it is 1 foot, pressure on the bottom of the pond will still be just 0.433 psi, even though there is a larger volume of water. This points up the relationship between pressure and depth and demonstrates that there is not necessarily a relationship between pressure and volume.

With this relationship between pressure and depth established, we can transpose the equation so that the depth of water in a tank can be found if we know the pressure reading at the bottom of the tank.

If $p = 0.433 \times h$, then $h = p/0.433$ is true, by transposition.

EXAMPLE:

If a pressure gauge located at the bottom of a 50-ft high water tower showed a reading of 13 psi, (a) what is the depth of the water in the tank, and (b) what would the gauge indicate if the tower were filled with water?

(a) $h = \frac{p}{0.433} = \frac{13}{0.433} = \mathbf{30\ feet\ of\ water}$

(b) $p = 0.433h = 0.433 \times 50 = \mathbf{21.65\ psi}$

PASCAL'S PRINCIPLE

In the middle of the seventeenth century, a French mathematician and scientist named Blaise Pascal was experimenting with water and air pressure. His scientific experiments led to the formulation of what is known as Pascal's principle, namely, that pressures applied to a confined liquid are transmitted equally throughout the liquid, irrespective of the area over which the pressure is applied. The application of this principle enabled Pascal to invent the hydraulic press, which is capable of a large multiplication of force. In Fig. 3-3, the unit pressure is equal in all

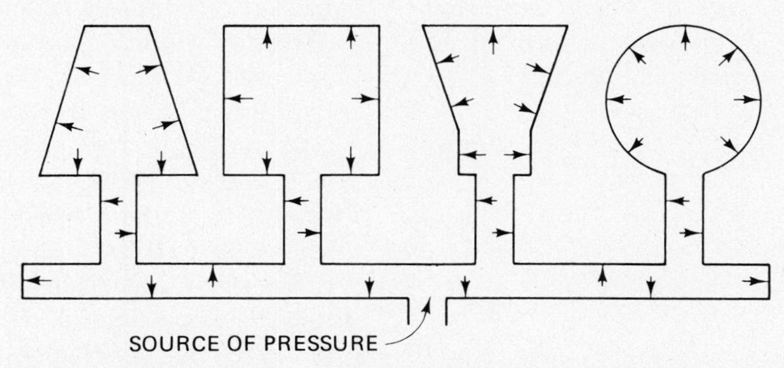

FIGURE 3-3 Pascal's principle of pressure equalization in various shaped containers.

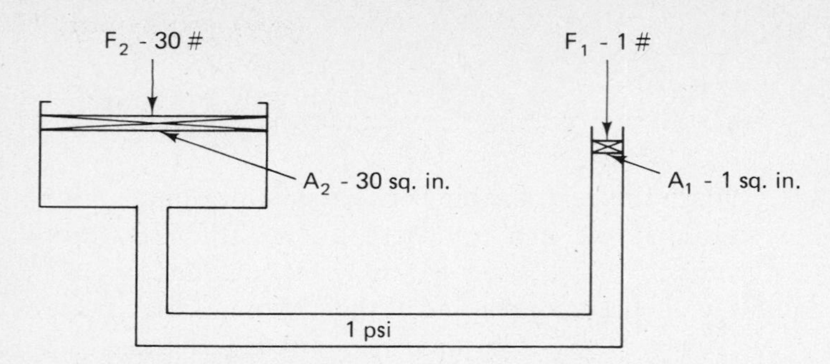

FIGURE 3-4 Pascal's principle of pressure transmission by hydraulic action.

vessels regardless of their shape, for pressure is independent of the shape of the container.

Fig. 3-4, illustrating this principle, shows a vessel containing a fluid such as oil; the vessel has a small and a large cylinder connected by a pipe or tubing, with tight-fitting pistons in each cylinder. If the cross-sectional area of the small piston is 1 in.^{2} and the area of the large one is 30 in.^{2} a force of 1 lb when applied to the smaller piston will support a weight of 30 lb on the larger piston, because a pressure of 1 psi throughout the fluid will be exerted.

Since we found earlier that pressure equals force divided by area, a force of 1 lb applied to an area of 1 in.^{2} will create a pressue of 1 psi. By transposition, force equals pressure multiplied by the area; therefore, a weight of 30 lb will be supported when a pressure of 1 psi is applied over an area of 30 in.^{2}

DENSITY

From a scientific or physics viewpoint, *density* is the weight per unit volume of a substance, and it may be expressed in any convenient combination of units of weight and volume used, such as *pounds per cubic inch* or *pounds per cubic foot*. An equation can be formulated which expresses this relationship:

$$D = \frac{W}{V} \tag{3-3}$$

where

D = density

W = weight

V = volume

As previously mentioned, the weight, or density of water is approximately 62.4 lb per ft^3, and it can be expressed as 0.0361 lb per cu in. (1 ft^3 contains 1,728 cu in., and $62.4 \div 1{,}728 = 0.0361$). The densities of some other common substances are listed in Fig. 3-5.

Substance	*Density, lb/ft^3*	*Specific Gravity*
Water (pure)	62.4	1
Aluminum	168	2.7
Ammonia (liquid, 60°F)	38.5	0.62
Brass	530	8.5
Brick (common)	112	1.8
Copper	560	8.98
Cork (average board)	15	0.24
Gasoline	41.2	0.66
Glass (average)	175	2.8
Iron (cast)	448	7.2
Lead	705	11.3
Mercury	848	13.6
Oil (fuel) (average)	48.6	0.78
Steel (average)	486	7.8
Woods:		
Oak	50	0.8
Pine	34.2	0.55

FIGURE 3-5 Density and specific gravity of some common substances.

The *specific gravity* of any substance is the ratio of the weight of a given volume of the substance to the weight of the same volume of a given substance (where solids or liquids are concerned water is used as a basis for specific gravity calculations that may have to be made, and air or hydrogen is used as a standard for gases).

$$\text{density (solid or liquid)} = \text{specific gravity} \times \text{density of water (in lb/ft}^3)$$

The specific gravity of water is considered as 1.0 and values for other substances are listed in Fig. 3-5.

Pressure within a fluid is directly proportional to the density of the fluid. Consider the tank in Fig. 3-1, which, if filled with water weighing 62.4 lb, has a force on the bottom of the tank of 62.4 lb and a unit pressure of 0.433 psi. If this tank were filled instead with gasoline, which has a specific gravity of 0.66, the force on the bottom would be only 66 % as great as when water is in the tank, and the pressure of the gasoline would be only 66 % as great. Therefore the relationship can be expressed as:

$$\text{pressure} = \text{head} \times \text{density}$$
$$p = h \times D \qquad (3\text{-}4)$$

where

p = pressure in lb per ft^2

h = head or depth below the surface in feet

D = density in pounds per cubic ft

or where

p = pressure in pounds per square inch

h = head or depth below the surface in inches

D = density in pounds per cubic inch

Care must be taken whenever this equation is used to make sure that the proper units of weight and measurement are placed in the equation.

SPECIFIC VOLUME

The specific volume of a substance is usually expressed as the number of cubic feet occupied by 1 lb of the

substance. In the case of liquids, it will vary with temperature and pressure. The volume of a liquid will be affected by a change in its temperature; but, since it is practically impossible to compress liquids, the volume is not affected by a change in pressure.

The volume of a gas or vapor is definitely affected by any change in either its temperature or the pressure to which it is subjected. In refrigeration, the volume of the vapor under the varying conditions involved is most important in the selection of the proper refrigerant lines.

The appropriate specific volumes and pressures for refrigerants will be covered later in Chapter 7, but as an example of the effect temperature has on a refrigerant vapor, refer to Fig. 3-6 for properties of Refrigerant 12 (R-12).

Note that at +5°F the specific *volume of vapor* is 1.46 ft^3/lb, whereas at 86°F it is only 0.38 ft^3/lb. Correspondingly there is an increase in pressure

	PRESSURE		VOLUME VAPOR	DENSITY LIQUID	HEAT CONTENT BTU/LB.	
Temp. F.	*Psia*	*Psig*	*Cu. Ft./Lb.*	*Lb./Cu. Ft.*	*Liquid*	*Vapor*
−150	0.154	29.61*	178.65	104.36	−22.70	60.8
−125	0.516	28.67*	57.28	102.29	−17.59	83.5
−100	1.428	27.01*	22.16	100.15	−12.47	66.2
− 75	3.388	23.02*	9.92	97.93	− 7.31	69.0
− 50	7.117	15.43*	4.97	95.62	− 2.10	71.8
− 25	13.556	2.32*	2.73	93.20	3.17	74.56
− 15	17.141	2.45	2.19	92.20	5.30	75.65
− 10	19.189	4.49	1.97	91.70	6.37	76.2
− 5	21.422	6.73	1.78	91.18	7.44	76.73
0	23.849	9.15	1.61	90.66	8.52	77.27
5	26.483	11.79	1.46	90.14	9.60	77.80
10	29.335	14.64	1.32	89.61	10.68	78.335
25	39.310	24.61	1.00	87.98	13.96	79.9
50	61.394	46.70	0.66	85.14	19.50	82.43
75	91.682	76.99	0.44	82.09	25.20	84.82
86	108.04	93.34	0.38	80.67	27.77	85.82
100	131.86	117.16	0.31	78.79	31.10	87.63
125	183.76	169.06	0.22	75.15	37.28	88.97
150	249.31	234.61	0.16	71.04	43.85	90.53
175	330.64	315.94	0.11	66.20	51.03	91.48
200	430.09	415.39	0.08	60.03	59.20	91.28

* Inches of mercury below one atmosphere.

FIGURE 3-6 Table of properties of liquid and saturated vapor of refrigerant R–12. Note pressures corresponding to standard evaporating temperature of 5°F. and condensing temperature of 86°F. (Courtesy, Freon Products, Div. of E. I. du Pont de Nemours and Company).

(psia) from 26.48 to 108.04 psia. Note also the change in density from 90.14 to 80.67 lb per cu ft.

Under the pressure columns there are two sets of data shown: psia for absolute pressure and psig for gauge pressure. The following discussion explains the difference between them.

ATMOSPHERIC PRESSURE

The earth is surrounded by a blanket of air called the atmosphere, which extends 50 or more miles upward from the surface of the earth. Air has weight and also exerts a pressure known as *atmospheric pressure*. It has been computed that a column of air, with a cross-sectional area of one square inch and extending from the earth's surface at sea level to the limits of the atmosphere, would weigh approximately 14.7 lb. As was pointed out earlier in the chapter, force means also the weight of a substance, and pressure means unit force per square inch; therefore standard atmospheric pressure is considered to be 14.7 psi at sea level.

This pressure is not constant; it will vary with altitude or elevation above sea level, and there will be variations due to changes in temperature as well as water vapor content of the air. This atmospheric pressure can be demonstrated by construction of a simple barometer, as shown in Fig. 3-7, using a glass tube about 36 in. long and closed at one end, an open dish or bowl, and a supply of mercury. Fill the tube with mercury and invert it in the bowl of mercury, holding a finger at the open end of the tube so that the mercury will not spill out while the tube is being inverted. Upon removal of the finger, the level of the mercury in the tube will drop somewhat, leaving a vacuum at the closed end of the tube. The atmospheric pressure bearing down on the open dish or bowl of mercury will force the mercury in the tube to stand up to a height determined by the pressure being exerted on the open surface of the mercury, which will be approximately 30 in. at sea level.

As previously shown in Fig. 3-5, the specific gravity of mercury is 13.6—that is, mercury weighs and exerts pressure 13.6 times that of an equal volume of

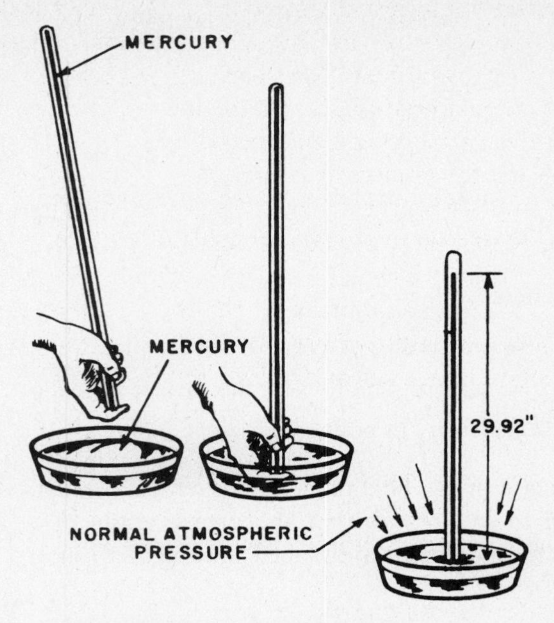

FIGURE 3-7 Simple barometer.

water. Since a 1-in. high column of water exerts a pressure of 0.0361 psi, a similar column of mercury will exert pressure that is $0.0361 \times 13.6 = 0.491$ psi. A 30-in. column of mercury will exert pressure of $0.491 \times 30 = 14.7$ psi, or an amount equal to atmospheric pressure. Conversely, atmospheric pressure at sea level (14.7) bearing down on the open dish of mercury divided by the unit pressure that is exerted by a 1-in. column of mercury (0.491) should cause the column of mercury in the tube to stand approximately at a height of 30 in. (actually 29.92 in.).

$$\frac{14.7}{0.491} = \text{30-in. column of mercury (Hg)}$$

Of course, water could possibly be used in this barometer instead of mercury; but the tube would have to be approximately 34 feet high, and this is not too practical.

$$\frac{14.7}{0.0361} = \text{407 in. of water (33.9 ft)}$$

MEASUREMENT OF PRESSURE

A *manometer* is one type of device utilized in the refrigeration and air-conditioning field for the measurement of pressure. This type of pressure gauge utilizes a liquid, usually mercury, water, or gauge oil, as an indicator of the amount of pressure involved. The water manometer or water gauge is customarily used when measuring air pressures, because of the lightness of the fluid being measured.

A simple open-arm manometer is shown in Fig. 3-8a and 3-8b. The U-shaped glass tube is partially filled with water, as shown Fig. 3-8a and is open at both ends. The water is at the same level in both arms of the manometer; because both arms are open to the atmosphere, and there is no external pressure being exerted on them.

Figure 3-8b shows the manometer in use with one arm connected to a source of positive air pressure that is being measured. The water is at different levels in the arms, and the difference denotes the amount of pressure being applied.

A space that is void, or lacking any pressure, is described as having a *perfect vacuum*. If the space has pressure less than atmospheric pressure, it is defined as being in a *partial vacuum*. It is customary to express this partial vacuum in *inches of mercury*, and not as negative pressure. In some instances it is also referred to as a given amount of absolute pressure, expressed in psia, and this will be covered a little later in this chapter.

If a partial vacuum has been drawn on the left arm of the manometer by means of a vacuum pump,

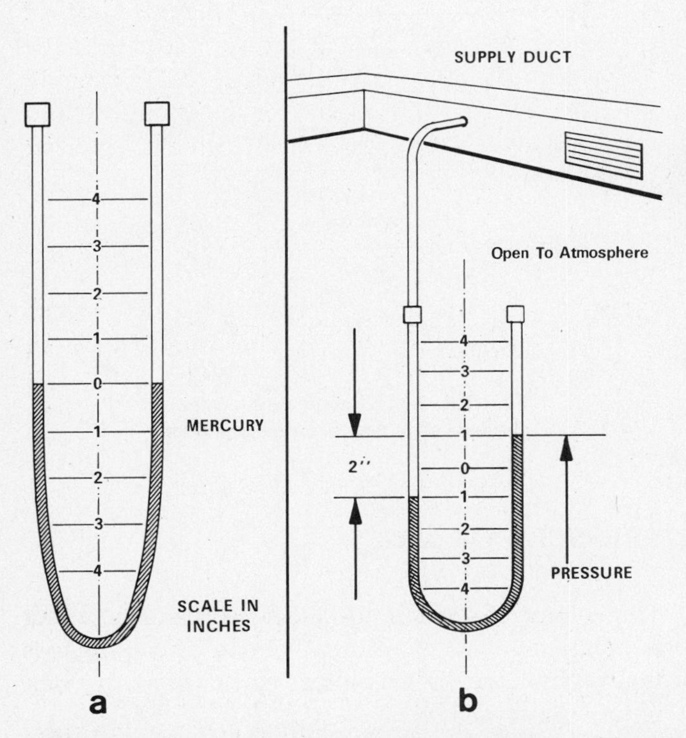

FIGURE 3-8 Water fill manometer used to measure air pressure.

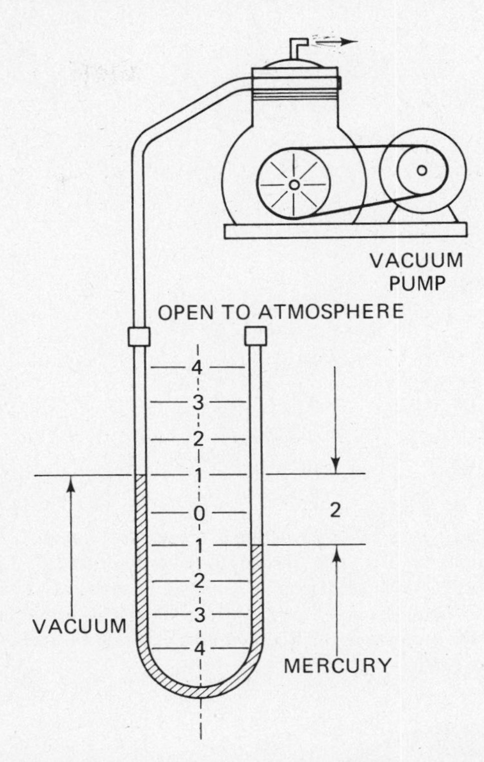

FIGURE 3-9 Mercury fill manometer measuring "vacuum" pressure.

as shown in Fig. 3-9, the mercury in the right arm will be lower, and the difference in levels will designate the partial vacuum in inches of mercury.

Pressure gauges most commonly used in the field by service technicians, to determine what is going on within the refrigeration system, are of the Bourdon tube-type. As is shown in Fig. 3-10, an internal view, the essential element of this type of gauge is the Bourdon tube. This oval metal tube is curved along its length and forms an almost complete circle. One end of the tube is closed, and the other end is connected to the equipment or component being tested.

As shown in Fig. 3-11a and 3-11b, the gauges are preset at 0 lb, which represents the atmospheric pressure of 14.7 psi. Therefore any additional pressure applied when the gauge is connected to a piece of equipment will tend to straighten out the Bourdon tube, thereby moving the needle or pointer and its mechanical linkage, thus indicating the amount of pressure being applied. Figure 3-11a is an example of a pressure gauge to indicate the amount of pressures above that of the atmosphere. Figure 3-11b is a compound pressure gauge, which has a dual function:

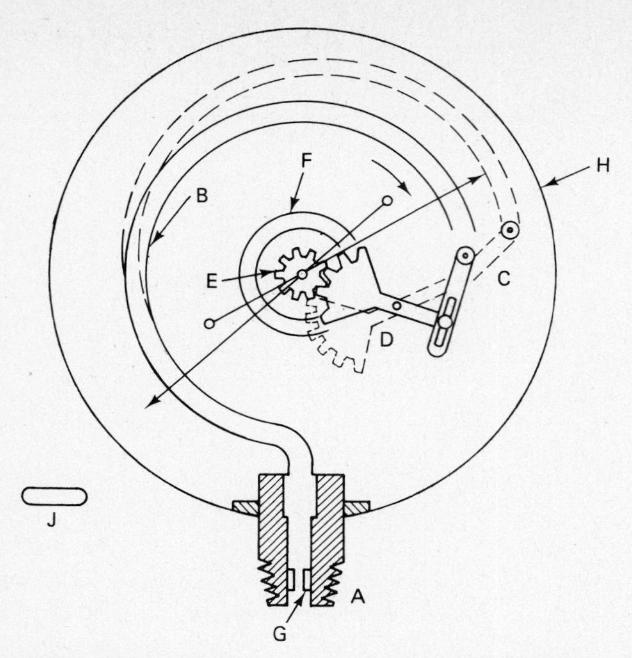

FIGURE 3-10 Internal construction of a pressure gauge: A. Adapter fitting, usually an 1/8″ pipe thread; B. Bourdon tube; C. Link; D. Gear sector; E. Pinter shaft gear; F. Calibrating spring; G. Restricter; H. Case; J. Cross-section of the Bourdon tube. The dotted line indicates how the pressure in the Bourdon tube causes it to straighten and operate the gauge.

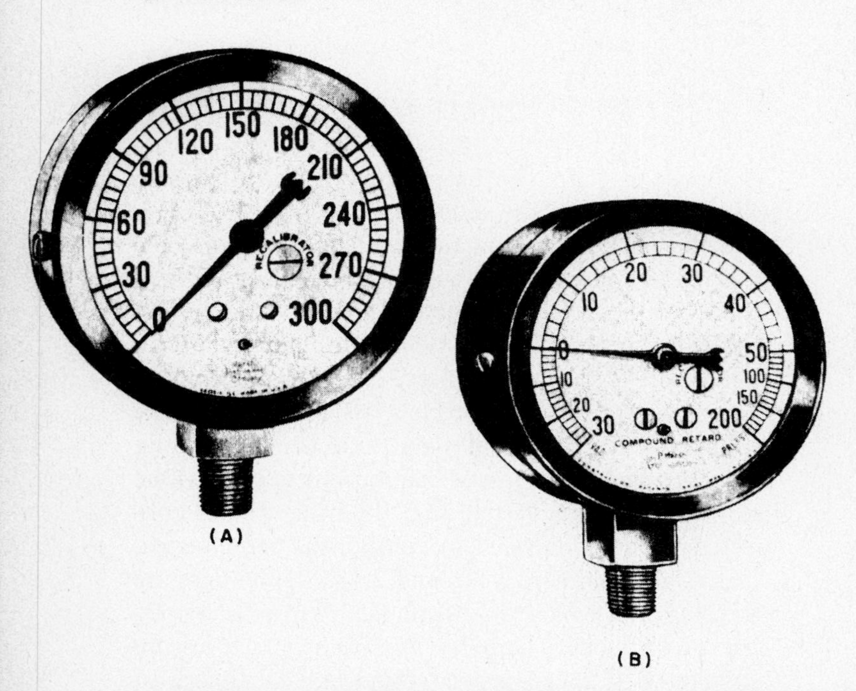

FIGURE 3-11 Typical pressure gauges used in refrigeration work (Courtesy, Marsh Instrument Company).

that of registering a pressure above atmospheric pressure, and registering one that may be below that of the atmosphere.

Pressures below atmospheric are customarily expressed in inches of mercury. There is an indication of the range between 0 gauge and 30 in. of mercury (Hg) on the compound gauge.

ABSOLUTE PRESSURE

Figure 3-12 shows a definite relationship among absolute, atmospheric, and gauge pressures. For many problems, atmospheric pressure does not need to be considered, so the customary pressure gauge is calibrated and graduated to read zero under normal atmospheric conditions. Yet when gases are contained within an enclosure away from the atmosphere, such as in a refrigeration unit, it is necessary to take atmospheric pressure into consideration, and mathematical calculations must be in terms of the absolute pressures involved.

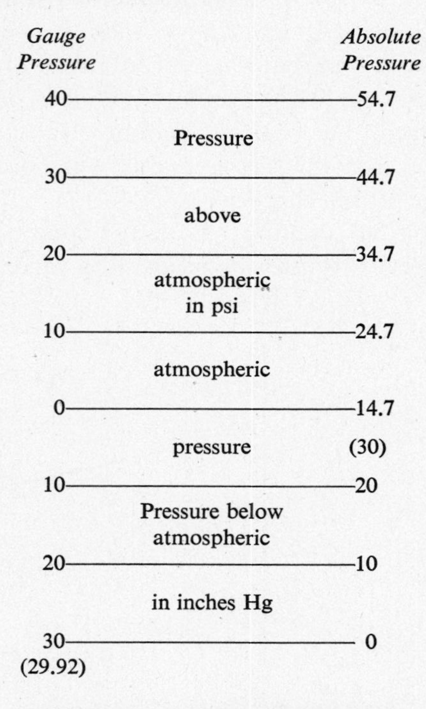

FIGURE 3-12 Relationship between absolute, atmospheric and gauge pressure

PRESSURE OF A GAS

The volume of a gas is affected by a change in either the pressure or temperature, or both. There are laws that govern the mathematical calculations in computing these variables.

Boyle's Law states that the volume of a gas varies inversely as its pressure if the temperature of the gas remains constant. This means that the product of the pressure times the volume remains constant, or that if the pressure of a gas doubles the new volume will be one-half of the original volume. Or it may be considered that, if the volume is doubled, the absolute pressure will be reduced to one-half of what it was originally.

This concept may be expressed as:

$$p_1V_1 = p_2V_2 \qquad \textbf{(3-5)}$$

where

p_1 = original pressure

V_1 = original volume

p_2 = new pressure

V_2 = new volume

It must be remembered that p_1 and p_2 have to be expressed in the *absolute pressure* terms for the above equation to be used correctly.

EXAMPLE:

If the gauge pressure on 2 cu ft of gas is increased from 20 psig to 50 psig while the temperature of the vapor remains constant, what will be the new volume?

$p_1 = 20 + 14.7$ Since

$= 34.7$ psia $\qquad p_1V_1 = p_2V_2$

$p_2 = 50 + 14.7$ Then

$= 64.7$ psia $\qquad V_2 = \frac{p_1V_1}{p_2}$

$V_1 = 2$ cu ft Therefore

$V_2 = ?$ $\qquad V_2 = \frac{34.7 \times 2}{64.7} = \mathbf{1.072\ cu\ ft}$

The same basic equation may be used to determine the given new pressure.

EXAMPLE:

If additional pressure is applied to a volume of 2 cu ft of gas at 20 psig so that the volume is lessened to 1.072 cu ft, and the temperature of the gas remains constant, what is the new pressure in psig?

$p_1 = 34.7$ psia (20 + 14.7) $\qquad p_2 = \frac{p_1V_1}{V_2}$

$V_1 = 2$ cu ft $\qquad p_2 = \frac{34.7 \times 2}{1.072} = 64.7$ psia

$V_2 = 1.072$ cu ft $\qquad p_2 = 64.7 - 14.7 = \mathbf{50\ psig}$

EXPANSION OF GAS

Most gases will expand in volume at practically the same rate with an increase in temperature, providing that the pressure does not change. And, if the gas is confined so that its volume will remain the same, the pressure in the container will increase at about the same rate as an increase in temperature.

Theoretically, if the pressure remains constant, a gas vapor will expand or contract at the rate of $\frac{1}{492}$ for each degree of temperature change. The result of this theory would be a zero volume at a temperature of −460°F, or at 0° Absolute.

Charles' Law states that the volume of a gas is in direct proportion to its absolute temperature, providing the pressure is kept constant; and the absolute pressure of a gas is in direct proportion to its absolute temperature, providing the volume is kept constant. That is:

$$\frac{V_1}{V_2} = \frac{T_1}{T_2} \qquad \textbf{(3-6)}$$

and

$$\frac{P_1}{P_2} = \frac{T_1}{T_2} \qquad \begin{matrix} T = \text{absolute temp} \\ P = \text{absolute pressure} \end{matrix} \qquad \textbf{(3-7)}$$

In order to clear the fractions, these may be expressed also as:

$$V_1T_2 = V_2T_1 \qquad \text{and} \qquad p_1T_2 = p_2T_1$$

EXAMPLE:

If the temperature of 2 cu ft of gas was increased from 40°F to 120°F what would be the new volume, if there was no change in pressure?

$$V_2 = \frac{V_1T_2}{T_1} = \frac{2 \times (120 + 460)}{(40 + 460)}$$

$$= \frac{1{,}160}{500} = \mathbf{2.32\ cu\ ft}$$

EXAMPLE:

If a container holds 2 cu ft of gas at 20 psig what will be the new pressure in psig if the temperature is increased from 40°F to 120°F?

$$p_2 = p_1T_2 = \frac{(20 + 14.7) \times (120 + 460)}{(40 + 460)}$$

$$= 40.25 \text{ psig}$$

$$40.25 - 14.7 = \mathbf{25.55\ psig}$$

In numerous cases dealing with refrigerant vapor, none of the three possible variables remains constant, and a combination of these laws must be utilized, namely the *general law of perfect gas*:

$$\frac{p_1V_1}{T_2} = \frac{p_2V_2}{T_2} \quad \text{or} \quad p_1V_1T_2 = p_2V_2T_1 \qquad \textbf{(3-8)}$$

in which the units of p and T are always used in the absolute.

EXAMPLE:

If a volume of 4 cu ft of gas at a temparature of 70°F and at atmospheric pressure is compressed to one-half its original volume and increased in temperature to 120°F, what will be its new pressure? Transposing Eq. (3-8),

$$p_2 = \frac{p_1V_1T_2}{V_2T_1}$$

Therefore,

$$p_2 = \frac{14.7 \times 4(120 + 460)}{2 \times (70 + 460)} = \frac{34{,}104}{1{,}060} = 32.17 \text{ psia}$$

$$32.17 - 14.7 = \mathbf{17.47\ psig}$$

PROBLEMS

1. Find the total force and also the unit pressure exerted on the bottom of a tank filled with water, if the tank measures 3 ft × 3 ft × 1 ft high.

2. If a tank with a flat base is filled with water to a level of 8 ft, what pressure is exerted on the bottom of that tank? And what pressure is exerted halfway between the bottom of the tank and the surface of the water?

3. What unit pressure is exerted on a roof of a building on which is located a cooling tower weighing 1,580 lb when in operation and filled with water, and the base of which measures 3 ft × 4 ft?

4. In a hydraulic press as shown in Fig. 3-4, what force must be exerted on the small piston having an area of 2 sq in. if a weight of 600 lb must be supported on the larger piston with an area of 16 sq in.? Neglecting friction, what pressure is within the system?

5. If the compound gauge on a compressor shows a reading of 10 in. Hg on the low side, and the pressure gauge reads 120 psi on the high side, what is the increase in pressure through the compressor?

6. If a tank with a base measuring 15 in. × 24 in. is filled with water, the total weight of the water is 312 lb. Find: (a) height of the tank; (b) unit pressure on the bottom; and (c) the pressure halfway up the tank.

7. If the volume of 10 cu ft of gas at a pressure of 25 psi is to be compressed to 2 cu ft with no change in temperature, what will be the new pressure in psi?

8. What would the final pressure be in psi if the volume in Prob. 7 was reduced to 1 cu ft?

9. With the pressure remaining constant, find the new volume of 4 cu ft of gas when its temperature is increased from 60°F to 250°F.

10. What would be the new pressure exerted by a given volume of gas if its temperature was increased from 50°F to 150°F, and the original pressure was 80 psig?

4

REFRIGERATION PIPING MATERIALS AND FABRICATION

Chapter 1 introduced the major components of a mechanical refrigeration cycle: the compressor, evaporator, condenser, and metering device. It also defined the tubing or piping necessary to connect these elements and form a sealed system so that the refrigerant did not escape. Chapter 4 will review the materials, tools, and methods most commonly used by technicians to form and assemble the refrigeration piping system.

REFRIGERANT PIPING MATERIAL

Most tubing used in refrigeration and air-conditioning piping is made of copper. However, aluminum is now widely used for fabrication of the evaporator and condenser internal coil circuits, although it has not become popular for field fabrication of the connecting refrigerant lines, principally because it cannot be worked as easily as copper and is more difficult to solder.

Steel piping is used in assembly of the very large refrigeration systems where pipe sizes of 6-in. diameter and above are needed. Threaded steel pipe connections are not used in modern refrigeration work, since they cannot be made leak proof. These systems are welded, and couplings are bolted to the equipment and/or where service joints are needed.

The term *tubing* generally applies to thin-wall materials, which are joined together by means other than threads cut into the tube wall. *Piping*, on the other hand, is the term applied to thick pipe-wall material (for example iron and steel), into which threads can be cut into the wall and which are joined by fittings that screw onto the pipe. Piping can also be welded. Another distinction between tubing and piping is the method of size measurement (see Fig. 4-1). Tubing sizes are expressed in terms of the outside diameter (O.D.), and pipe sizes are expressed as nominal inside diameters (I.D.). Thus in Fig. 4-1 the $\frac{1}{2}$-in. O.D. (Type *L*) copper tube will have an inside diameter of 0.43 in. The $\frac{1}{2}$-in. I.D. nominal steel pipe will have an inside diameter of 0.50 in. and an outside diameter of 0.75 in.

Because of the special treatment of aluminum tubing and welded steel piping the technique of fabrication will not be covered in this discussion.

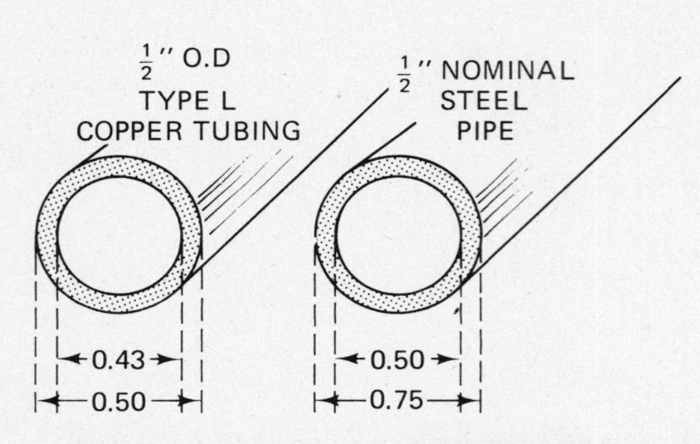

FIGURE 4-1 Method of sizing tubing and pipe.

COPPER TUBING

The tubing used in all domestic refrigeration systems is specially annealed copper. Copper tubing when formed has a tendency to harden, and this hardening action could cause cracks in the tubing ends when they are flared or formed. The copper may be softened by heating to a blue surface color and allowing it to cool. This process is called *annealing* and is done at the factory.

Copper tubing used in refrigeration and air-conditioning work is known as *ACR tubing*, meaning it is intended for use in air-conditioning and refrigeration work and has been specially manufactured and processed for this purpose. ACR tubing is pressurized with nitrogen gas to keep out air, moisture, and dirt, and also to provide maximum protection against the harmful oxides that are normally formed during brazing. The ends are plugged, and these plugs should be replaced after cutting a length of tubing.

COPPER TUBING CLASSIFICATION

Copper tubing has three classifications: K, L, and M, based on the wall thickness:

K–heavy wall–ACR approved

L–medium wall–ACR approved

M–thin wall–not used on refrigeration systems

Type M thin-wall tubing is not used on pressurized refrigerant lines, for it does not have the wall thickness to meet the safety codes. It is, however, used on water lines, condensate drains, and other associated system requirements.

Type K heavy-wall tubing is meant for special use where abnormal conditions of corrosion might be expected. Type L is most frequently used for normal refrigeration applications. Figure 4-2 provides a table of specifications for both type K and L tubing. Both

	DIAMETERS			
Type	*Outside Inches*	*Inside Inches*	*Wall Thickness Inches*	*Weight Per Foot Lbs.*
	1/2	0.402	0.049	0.2691
	5/8	0.527	0.049	0.3437
	3/4	0.652	0.049	0.4183
	7/8	0.745	0.065	0.6411
	1 1/8	0.995	0.065	0.8390
K	1 3/8	1.245	0.065	1.037
	1 5/8	1.481	0.072	1.362
	2 1/8	1.959	0.083	2.064
	2 5/8	2.435	0.095	2.927
	3 1/8	2.907	0.109	4.003
	3 5/8	3.385	0.120	5.122
	1/2	0.430	0.035	0.1982
	5/8	0.545	0.040	0.2849
	3/4	0.666	0.042	0.3621
	7/8	0.785	0.045	0.4518
	1 1/8	1.025	0.050	0.6545
L	1 3/8	1.265	0.055	0.8840
	1 5/8	1.505	0.060	1.143
	2 1/8	1.985	0.070	1.752
	2 5/8	2.465	0.080	2.479
	3 1/8	2.945	0.090	3.326
	3 5/8	3.425	0.100	4.292

FIGURE 4-2 Specifications of common copper tubing sizes.

K and L copper tubing are available in soft-or hard-drawn types.

SOFT COPPER TUBING

Soft copper tubing, as the name implies, is annealed to make the tubing more flexible and easy to bend and form. It is commercially available in sizes from $\frac{1}{8}$-in. O.D. to $1\frac{5}{8}$-in. O.D. and is usually sold in coils of 25, 50, and 100-foot lengths. The coils are dehydrated and sealed at the factory. Soft copper tubing may be soldered or used with flared or other mechanical-type fittings. Since it is easily bent or shaped it must be held by clamps or other hardware to support its own weight. The more frequent application is for line sizes from $\frac{1}{4}$ in. to $\frac{3}{4}$ in. O.D. Above $\frac{3}{4}$ in. O.D. forming becomes rather difficult.

HARD-DRAWN COPPER TUBING

Hard-drawn tubing is also used extensively in commercial refrigeration and air-conditioning systems. Unlike soft-drawn, it is hard and rigid and comes in straight lengths. It is intended for use with formed fittings to make the necessary bends or changes in direction. Because of its rigid construction it is more self-supporting and needs fewer supports. Sizes range from $\frac{3}{8}$ in. O.D. to over 6 in. O.D. Hard-drawn tubing comes in standard 20-foot lengths that are dehydrated, charged with nitrogen, and plugged at each end to maintain a clean moisture-free internal condition. The use of hard-drawn tubing is most frequently associated with larger line sizes of $\frac{7}{8}$ in. O.D. and above.

CUTTING COPPER TUBING

There are two methods of cutting copper tubing. The first uses the hand-held tube cutters shown in Fig. 4-3; they are suitable for cutting soft- or hard-drawn tubing. Hand-held cutters may be obtained in different models to cut from $\frac{1}{8}$ in. O.D. to as much as $4\frac{1}{8}$ in. O.D.

The hand-held cutter is positioned on the tubing at the proper cut point. Tightening the knob forces the cutting wheel against the tube. Then by rotating the cutter around the tube and continual tightening of knob, the cut is made. A built-in reamer blade is used to remove burrs from inside the tube after cutting.

A second (but less desirable) method of cutting larger size hard-drawn tubing is the use of a hack saw and a sawing fixture to help square the end and make more accurate cuts (See Fig. 4-4).

The saw blade should have at least 32 teeth per inch to ensure a smooth cut. Try to keep saw filings from entering the tubing to be used. Some mechanics

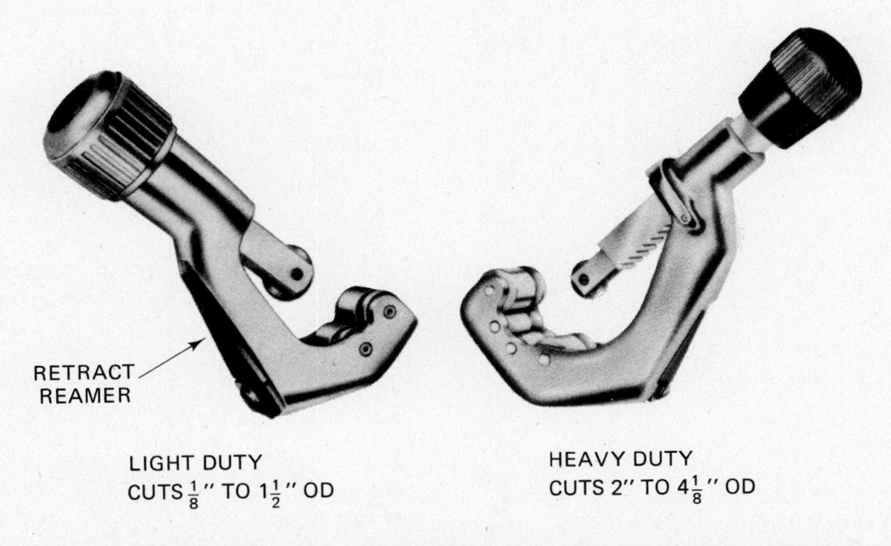

Figure 4-3 Typical tubing cutters (Courtesy, Gould, Inc., Valves and Fittings Division).

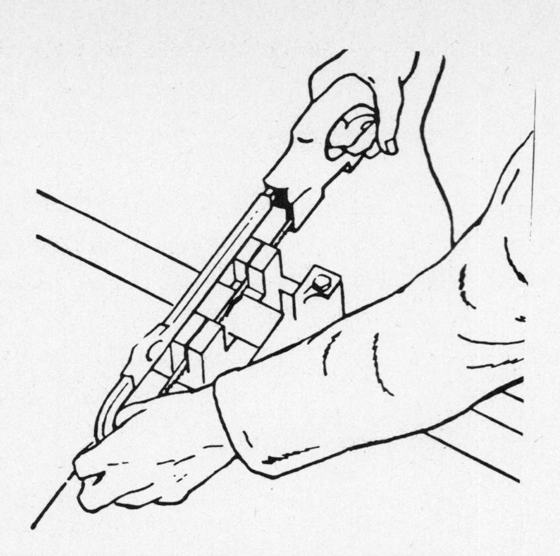

FIGURE 4-4 Cutting tube with hack saw and sawing fixture (Courtesy, Gould, Inc., Valves and Fittings Division).

also file the end of the tube to provide a smooth surface. Wipe the inside of the tube with a clean cloth.

For contractors who do considerable refrigeration piping there are portable power-operated machines that use abrasive wheels to cut the tubing and buff both the inside and outside of the tube; however, such machines are not required for the average installation.

TUBE BENDING

Where smaller sizes of soft-drawn tubing are used it is generally more convenient and economical to simply bend the tubing to fit the application requirements without using formed fittings. This can be done by hand without special tools—but it takes

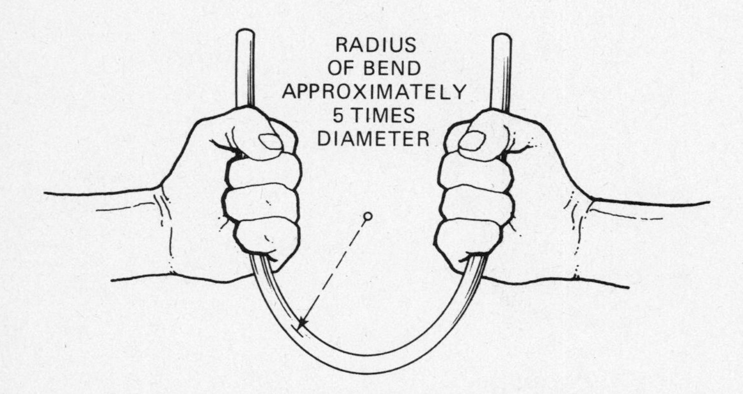

FIGURE 4-5 Recommended technique in bending tubing by hand. (Courtesy, Gould, Inc., Valves and Fittings Division).

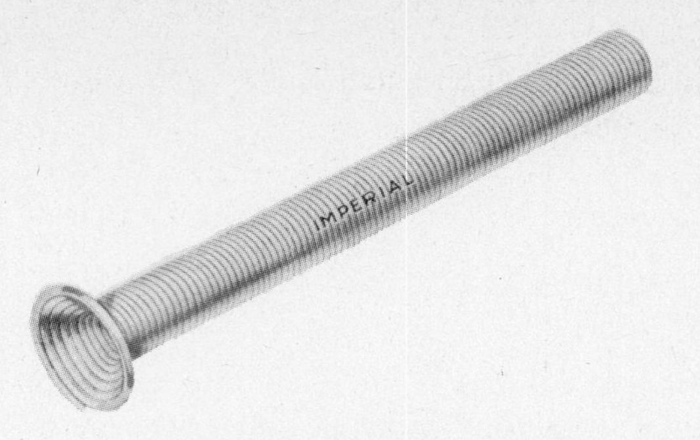

FIGURE 4-6 Spring type tube bender (Courtesy, Gould, Inc., Valves and Fittings Division).

practice so as not to make too sharp or too tight bends and so flatten the tube.

As a rule of thumb, the minimum bending radius in which a smaller tube may be curved is about five times the tube diameter, as illustrated in Fig. 4-5. Larger tubing may require a radius of up to 10 times the diameter.

To make a hand bend, start with a larger radius and gradually work the tubing into the proper shape as you decrease the radius. *Never* try to make the first bend as small as the final radius desired.

Tube-bending springs as illustrated in Fig. 4-6 are available to insert inside or on the outside of the tube

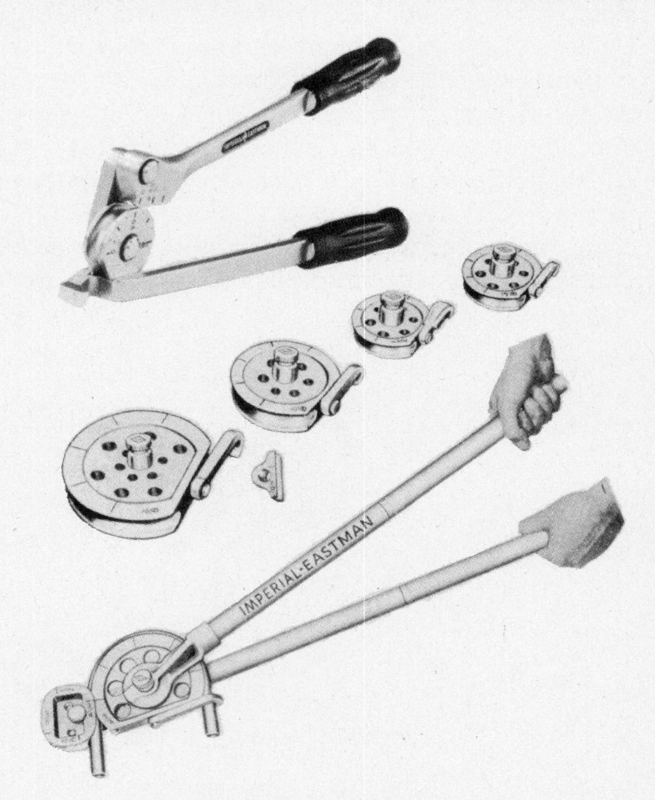

FIGURE 4-7 Lever type tube bender (Courtesy, Gould, Inc., Valves and Fittings Division).

so as to prevent the tube from collapsing. These springs are relatively inexpensive and come in sizes to fit most common tubing requirements.

The most accurate and reliable method of tube bending is a lever-type tube-bending tool kit as illustrated in Fig. 4-7. It bends both hard and soft tubing. Various sizes of forming wheels and blocks are furnished up to about $\frac{7}{8}$-in. O.D. Bends can be made at any angle up to 180°.

METHODS OF JOINING TUBING

As mentioned earlier, the walls of copper tubing are too thin for threading, so other means must be used to connect the tubing. These may be divided into two broad catagories:

(a) Mechanical Couplings—flared and compression fittings that are semi-permanent in that they can be mechanically taken apart.

(b) Heat Bonding—soldering and brazing, which means they are permanent.

MECHANICAL COUPLINGS

Flared Connections

Since about 1890 the flare connection has been one of the most widely used techniques to join soft-drawn copper tubing. A properly made flare is most important if leakproof joints are to be achieved—and this requires the right tools and practice.

Flares are made with special tools, which expand the end of the copper tubing into a cone shape as shown in Fig. 4-8. This cone is formed at a 45° angle that mates against the face of a flare fitting. The flare nut when tightened will press the soft copper against the machined fitting seat, thus forming a tight seal. The flare illustrated in Fig. 4-8 is called a *single thickness flare*. Others are called *double thickness flares*. Both forming techniques will be shown.

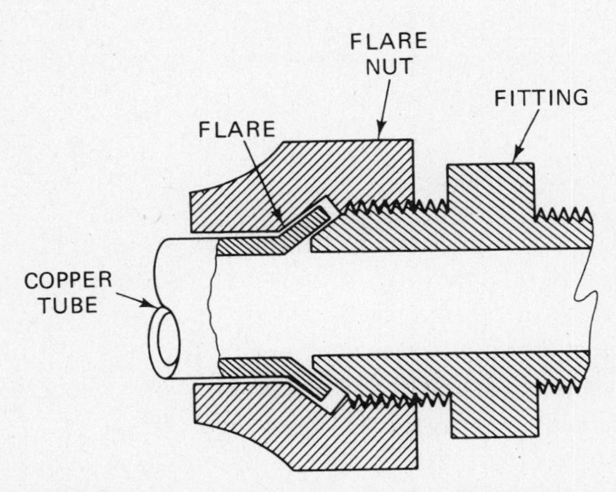

FIGURE 4-8 Cross-section of 45° angle flared fitting.

Single Thickness Flares

Figure 4-9 represents a typical flaring tool consisting of a flaring block or base and a slip-on yoke that holds the screw-driven flaring cone.

Note that the base-block dies slide to allow the tubing to be inserted into the particular size hole. The end wing nut is then tightened to hold the material in place during flaring.

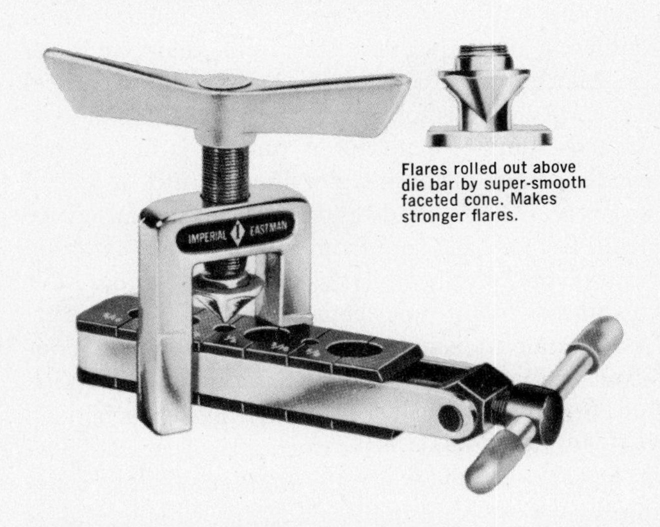

FIGURE 4-9 Typical flaring tool operation (Courtesy, Gould, Inc., Valves and Fittings Division).

Figure 4-10 represents a typical flaring operation. The chamber in the block is at a 45° angle. The tubing should extend slightly above the flare block approximately one-third the height of the flare. This is necessary to provide enough material to fill the opening after the flaring cone is driven into the tubing and also to ensure the proper surface area against the fitting face. If it is too small, the flare nut may not hold the tubing; if too large, the flare nut may not fit over it. This operation takes practice.

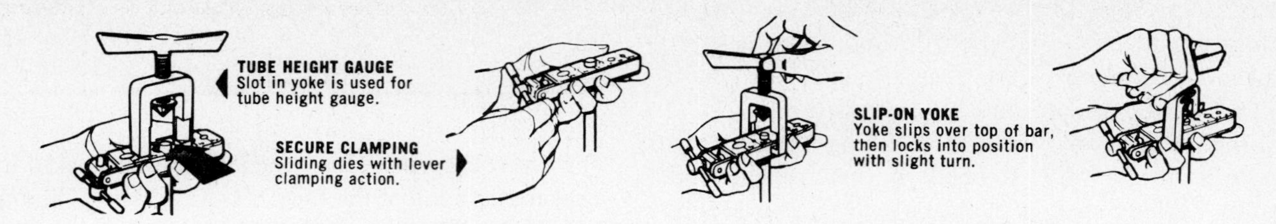

FIGURE 4-10 Typical flaring operation (Courtesy, Gould, Inc., Valves and Fittings Division).

Before making the flare it is important to prepare the tubing. Use a good grade of tubing (not all types of tubing are recommended for flaring). Follow proper cutting and deburring practices. Cut the tube squarely (use file if necessary) and remove internal and external burrs, using a deburring tool on the cutter.

To form the flare, first put the flare nut on the tubing. Insert the tubing into the proper size hole (adjusting the height above the block) and clamp with wing nuts. Put a drop of oil on the forming cone. Tighten the forming cone into the tubing using an initial one-half turn. Then back off one-quarter turn. Retighten three-quarters turn and then back off one-quarter turn. Continue this back and forth procedure until a flare is formed; with practice this motion becomes routine. Continuous turning is not recommended, since it doesn't give the operator a feel of the progress being made and may tend to harden the metal. Remove the flare from the block and carefully examine the completed flare to see that the sides have no splits or other imperfections.

Maintain tools in a clean and well-lubricated condition at all times for ease of operation and extended tool life. Never over-torque the feed mechanism when making a flare—flare washout will result.

NOTE—45° flares are the standard of the refrigeration and air-conditioning industry. In other industries, such as automotive, steel or brass tubing is used. These metals don't form as easily as copper, so 37° angle flares are used. Therefore flare fittings and tools are not interchangeable.

Double Flaring

A double flare consists of a twin-wall sealing surface. Double-flare fabrication is accomplished through use of a conventional flaring tool provided with double-flare adapters.

As illustrated in Fig. 4-11, the tubing is clamped in the flaring bar and allowed to protrude a measured distance above the top of the bar. The adapter is positioned in the tube, the yoke is engaged with the bar, and the feed screw is advanced until a positive resistance is encountered. This completes the preform operation, which folds the tube inward. The adapter is removed and the forming cone is advanced until moderate resistance is encountered; then final forming is done as described for a single flare. This operation completes the two steps required in forming a double flare.

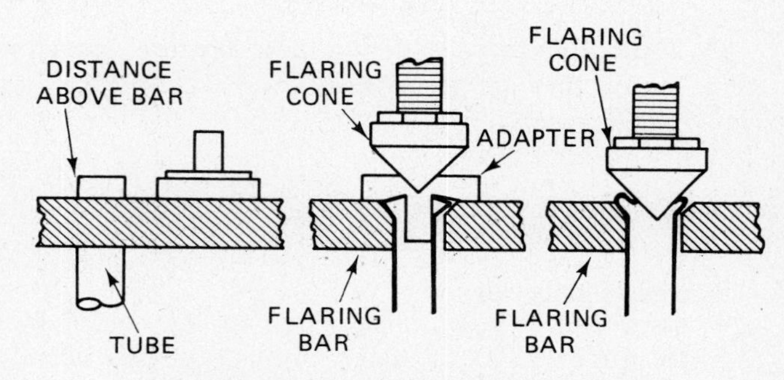

FIGURE 4-11 Method of making double flare.

Why double flares? They are mainly used on larger size tubing where a single flare may have a tendency to be weak due to excessive expansion. They offer greater resistance to fracture on installations that are subjected to excessive vibration. And they can be assembled and disassembled more frequently without flare washout.

Flared Fittings

To accommodate the many types of flared fittings needed in refrigeration and air-conditioning systems, a variety of elbows, tees, unions, etc., are available to select from, as illustrated in Fig. 4-12. The fittings are usually drop-forged brass and are accurately machined to form the 45° flare face.

Threads for attaching the flare nuts are S.A.E. National Fine Thread. Some fittings such as the flared half-union are made to join to national pipe threads.

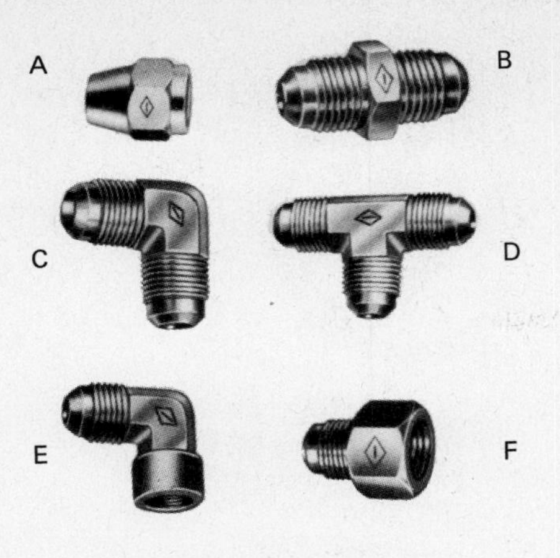

FIGURE 4-12 Common flared fitting (Courtesy, Gould, Inc., Valves and Fittings Division).

All fittings are based on the size of tubing to be used. The flare nuts are hexagon-shaped for easy wrench tightening. The fitting body also usually has a flat surface to take an open-end wrench attachment.

Compression Fittings

In recent years there has been a trend toward using compression-type fittings for joining refrigerant tubing. This method has gained popularity in the residential air-conditioning field because it reduces the field labor requirement in making flared connections or soldered piping. Figure 4-13 represents a typical mechanical coupling concept as manufactured by the Primore Co.

A gas-tight seal is obtained by simply connecting the tubing to the coupler with a coupling nut, crimp collar, and "0" ring to assure a leak-proof assembly. Once assembled, the joint may be disconnected and reassembled without lessening its sealing effectiveness.

As long as the tubing is straight and round in order to be inserted into the tube entrance, a proper connection can be made with minimum mechanical skills. This particular coupling comes in sizes from $\frac{1}{4}$ in. to $1\frac{1}{8}$ in. O.D. Adapters are available to join these couplings to other fittings such as soldered pipe or threaded fittings.

Heat Bonding

Heat-bonding of refrigerant tubing is called *soldering* or *brazing* and consists of joining two pieces of metal together with a third metal (or solder), which melts at a lower temperature than the pieces to be joined. When melted, the solder flows between the two pieces. The molten metal adheres to the surfaces of the two metals and forms a good bond

FIGURE 4-13 Mechanical tube connection (Courtesy, Primore Sales, Inc.).

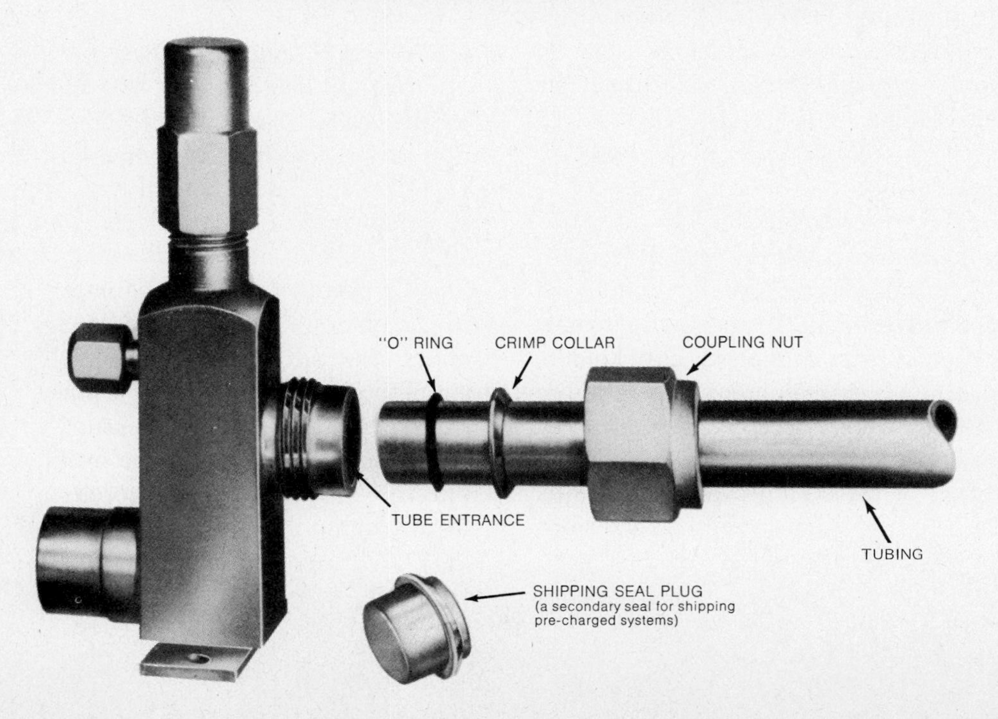

between them. The solder usually has less strength than the metals it joins, so for greatest strength of the solder joint the layer of solder must be very thin. The essential difference between soldering and brazing is the temperature at which the molten solder flows.

Welding differs from soldering or brazing in that it does not use another metal to act as the bonding material of the joint. The two pieces of material to be welded must be "puddled," that is, their edges or surfaces to be joined must be melted and allowed to mix together, so that when they cool they are a part of one another.

Soft Soldering

The most common solder is a mixture of tin and lead. If it is one-half tin and one-half lead, it is called *fifty-fifty* (*50–50*) solder. Fifty-fifty solder starts to melt at 360°F but does not become fluid until further heated to 415°F. The 50–50 tin-lead solders are called *soft solders* and are primarily used on plumbing and heating systems with working temperatures up to 250°F. *They are not recommended for refrigeration work.*

Another form of soft solder, but one that is somewhat harder than tin-lead solders, is composed of 95% tin and 5% antimony and is commonly referred to as *ninety-five-five* (95–5). It starts to melt at 450°F and is fully liquid at 465°F. It is easily worked with a small hand-held propane or acetylene torch-kit as illustrated in Fig. 4-14. Ninety-five-five is the recommended soft solder to use on refrigeration work, particularily on small-diameter soft-drawn copper tubing.

Hard Soldering

When larger hard-drawn copper tubing is used, or where local building codes require the use of hard solder, the hard solder joint makes a much stronger bond. Over the past twenty years or so, hard solders (or *silver solders* or *silver brazing alloys*, as they are now called) have been widely accepted by the refrigeration and air-conditioning industry for joining metal. The selection of these alloys over others was prompted by the need for high-strength, corrosion-

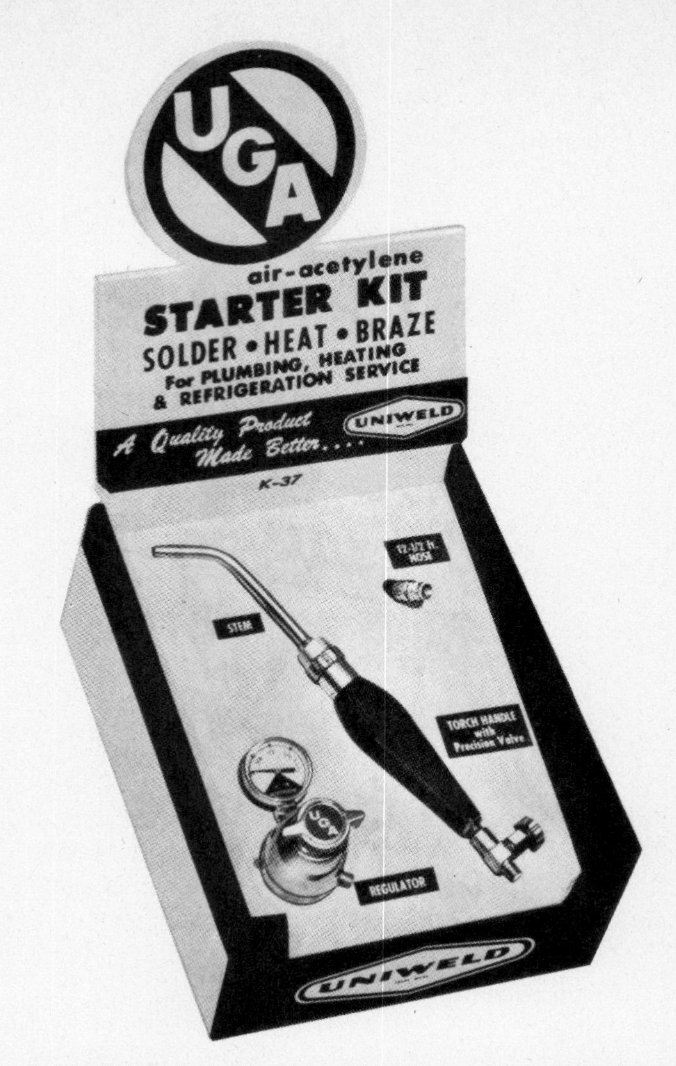

FIGURE 4-14 Typical hand torch kit (Courtesy, Uniweld Products, Inc.).

resistant, vibration-proof, and leak-tight joints. Silver brazing satisfied all these requirements, and also offered the additional advantage of joining similar or dissimilar metals with greater ease of application and at low brazing temperatures.

Silver solders in general flow at a temperature of approximately 1,100 to 1,200°F. Copper melts at 1,981°F, so this alloy flows at some 800°F below the melting point of copper, making it a very safe alloy to use on copper tubing and fittings. When properly made, the silver-solder joint will have a tensile strength above the material it joins.

Some of the lower temperature silver-solders may be applied with air-acetylene or air-propane torches, but the oxygen-acetylene torch illustrated later is most commonly used. Techniques of proper soldering and brazing, and pointers on brazing equipment are also discussed later.

Solder Fittings

For use with both soft- and hard-soldered systems, a number of *sweat fittings* are available, as illustrated in Fig. 4-15. Some sweat fittings are for connecting tubing to tubing, and others may have a connection on one end for national pipe threads.

These fittings are brass rod, brass forging, or wrought copper, and are accurately manufactured to permit the tubing to be inserted into the fitting opening with a snug fit, leaving only a very thin *solder clearance* for the flow of solder. If solder clearance is too large, the joint will be weakened.

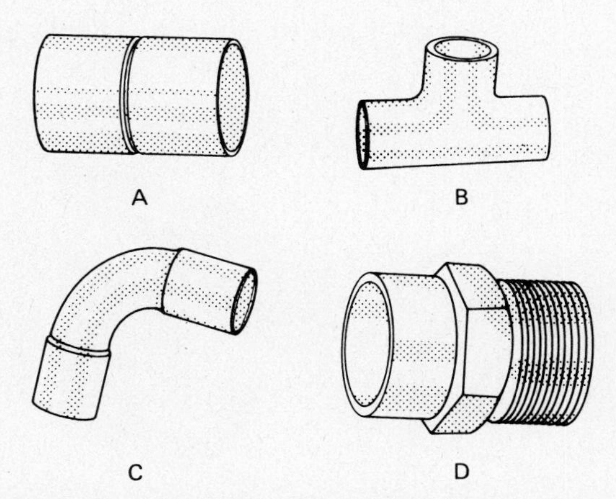

FIGURE 4-15 **Some common soldered or brazed copper tube fittings: A. Coupling with rolled stop, sweat × sweat; B. Tee, sweat × sweat; C. 90° elbow, sweat × sweat; D. Adapter, sweat × male pipe thread (m.p.t) (Courtesy, Mueller Brass Company).**

Swaging Copper Tubing

Sometimes in the assembly of copper tubing with the same diameter, some fabricators feel it's more reliable to join the two pieces of tubing by making a *swaged connection* with only one solder joint, as illustrated in Fig. 4-16a. This is done with a swaging tool (Fig. 4-16b), which is somewhat similar to the flaring block. The block holds the tube, and the correct sized punch is forced into the end of the tubing until the bell shape is produced. Some tools use the screw mechanism for force. Others require a blow by a hammer to force the punch into the tube.

Swaging, when properly done and soldered, reduces the number of soldered joints and thereby reduces leak hazards; however, it does take more time

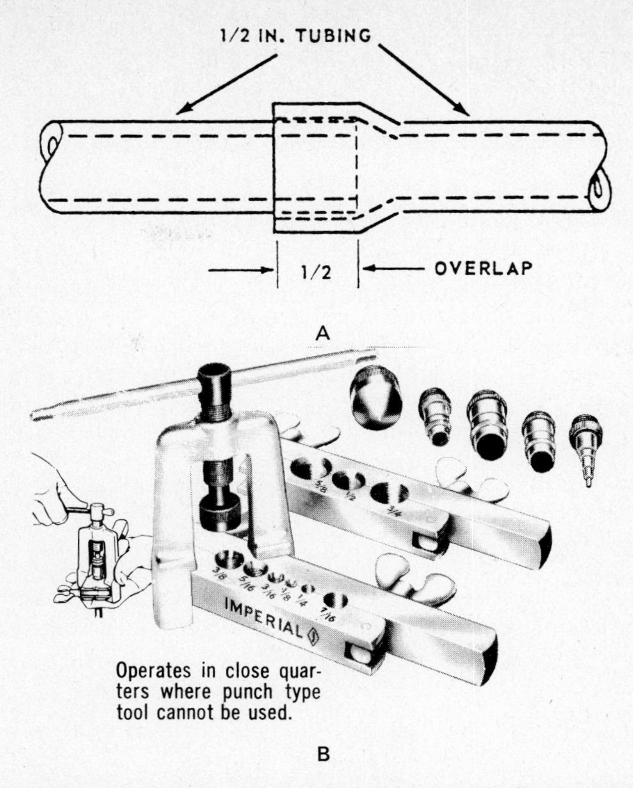

FIGURE 4-16 (a) Swaged connection. Two pieces of soft copper tubing are shown swaged and ready for soldering or brazing to make a joint connecting the two pieces of tubing. Note that both pieces of the tubing are of the same diameter; (b) swaging tool (Courtesy, Gould, Inc., Valves and Fittings Division).

than using preformed fittings, and thus the option becomes one of individual choice.

SOLDERING AND BRAZING TECHNIQUES

In the refrigeration industry, certain fundamental procedures of silver-alloy brazing have been arrived at through long experience. In many respects, these fundamentals are similar to those recommended for soft soldering, and a good soft-soldering operator generally makes a good silver-brazing operator.

There are six simple steps to follow in producing leak-tight, strong joints that are basic to both soft soldering and silver brazing:

(a) good fit and proper clearance
(b) clean metal
(c) proper fluxing
(d) assembling and supporting
(e) heating and flowing the alloy
(f) final cleaning

FOUR STEPS OF PREPARATION

(a) Cutting and fitting the pipe or tubing (Fig. 4-17).

- Cut the tube to proper length. Make sure the ends are cut square; a tube cutter is by far the best tool. If a hacksaw is used, tilt the tube downward so the cuttings fall out.
- Remove burrs with a reamer or half-round file, as shows in Fig. 4-18.
- Try end of tube in the fitting to be sure it has the proper close fit. Clearance should be uniform all around. Where necessary, on soft-drawn copper, a sizing tool may be used to round out the tubing.

(b) Cleaning the tubing and fittings (Figs. 4-19 and 4-20). This means that the surfaces to be joined must be *free of oil, grease, rust, or oxides.*

- In those cases where the tubing or fittings have a coating of oil or grease, a liberal application of cleaning solvent with a brush or clean rag will effectively remove this contamination.
- Clean socket of fitting and end of tubing with a clean wire brush and fine sand cloth. *Do not use steel wool* (shreds may adhere to the fitting and cause a void). *Do not use emery cloth,* since it often contains oils and undesirable abrasives that can cut too deeply.
- Do not handle the surfaces after cleaning.

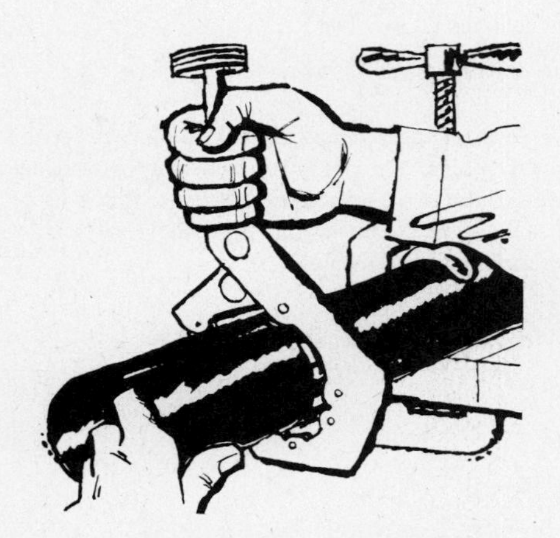

FIGURE 4-17 Cutting tube (Courtesy, Mueller Brass Company).

FIGURE 4-19 Wire brushing (Courtesy, Mueller Brass Company).

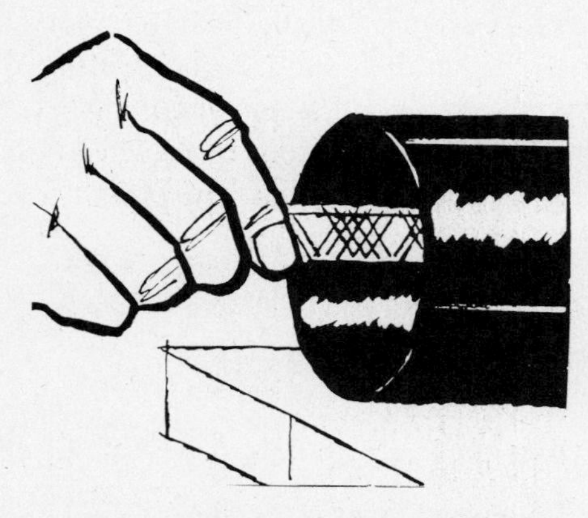

FIGURE 4-18 Deburring (Courtesy, Mueller Brass Company).

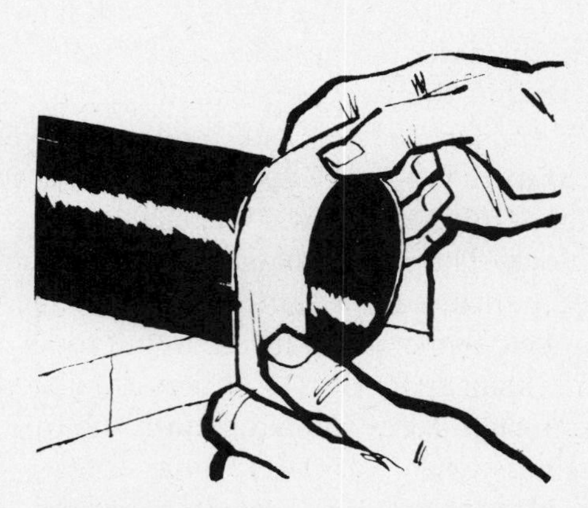

FIGURE 4-20 Sanding (Courtesy, Mueller Brass Company).

- Cleaning should be done just before soldering so that oxidation is reduced to a minimum.

(c) Proper Fluxing
Flux does not clean the metal. It keeps the metal clean once it has been mechanically cleaned as mentioned above. Flux is available in both paste and liquid form, but in general the paste form is preferred by most refrigeration installers.

- The first rule is to select the proper flux depending on whether the job is soft soldering or silver brazing. For brazing, use a good quality low-temperature silver brazing flux. *This is very important*, and requires special attention in refrigeration work; consult a local welding supply jobber.
- Always stir the flux before using (see Fig. 4-21), since when flux stands the chemicals tend to settle to the bottom, especially in hot weather. Use a brush—never apply soldering paste with your fingers; perspiration and oils may prevent solder from sticking.
- In most ordinary work, both the end of the tube and the inside of the fitting are fluxed (Fig. 4-22), but in refrigeration work the end of the tube is inserted partway into the fitting and the paste flux is brushed all around the outside of the joint.
- The tube is then inserted to full depth in the socket (Fig. 4-23).
- Where possible, revolve the fitting or tubing to spread the flux uniformly.
- *Caution*—too much flux can be harmful to the internal components and operation of the refrigeration system.
- *Note*—with certain silver brazing alloys, copper-to-copper brazing can be done without fluxing. On copper tube-to-brass fittings, flux is always required.

(d) Supporting the assembly.

- Before soldering or brazing, the assembly should be carefully aligned and adequately supported.
- Arrange supports so that expansion and contraction will not be restricted.
- See that no strain is placed on the joint during brazing and cooling.
- Plumber's pipe strap makes an excellent temporary support until permanent arrangements are made.

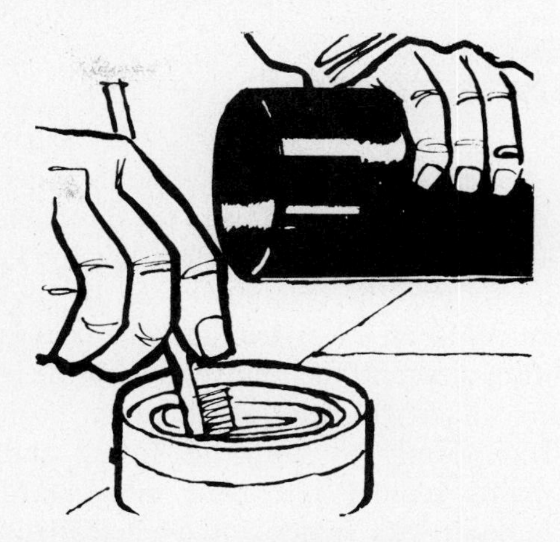

FIGURE 4-21 Stirring flux (Courtesy, Mueller Brass Company).

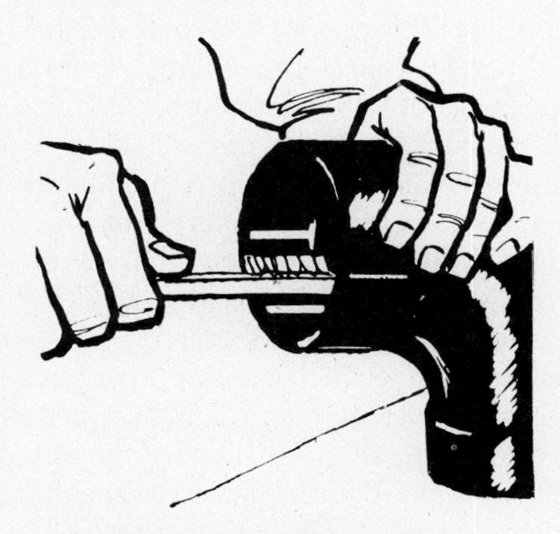

FIGURE 4-22 Applying flux (Courtesy, Mueller Brass Company).

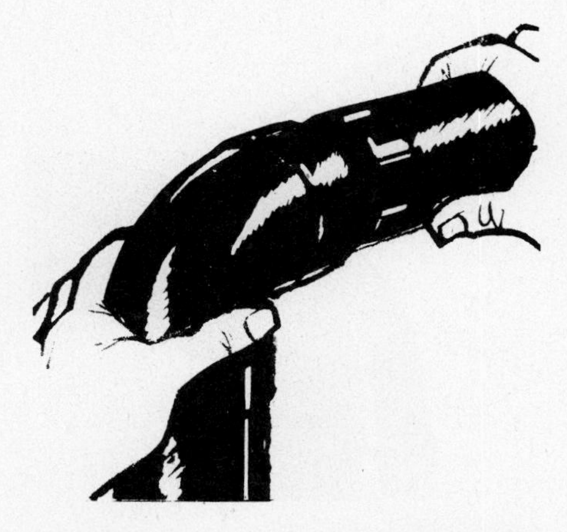

FIGURE 4-23 Joining (Courtesy, Mueller Brass Company).

HEATING AND FLOWING SOFT SOLDER

To soft solder, apply the flame to the shoulder of the fitting so that the heat will flow in the direction of the tube (Fig. 4-24). Be careful not to allow the flame to enter directly into the opening where the solder will be drawn. With an ell or tee, heat the heaviest part of the fitting first and then move toward the opening where the solder will enter. In this way, heat is distributed uniformly to both the fitting and tube. Occasionally, remove the flame momentarily and touch the joint with solder to see whether the metal is hot enough to melt solder.

When applying the soft solder (Fig. 4-25), never push the solder into the joint. When the joint is hot enough, merely touch the solder, and capillary attraction will draw it into the clearance space between the surfaces to be joined. When a ring of solder appears all around the circumference, you have made a joint that's absolutely leak-proof and tremendously strong. Wipe the joint clean with a clean cloth while the solder is still molten (Fig. 4-26). This is called *smoothing* the solder.

When working with large size fittings (2 in. or over) it is helpful to use two torches or, better yet, a Y-shaped torch tip (Fig. 4-27), which gives a double or quadruple flame. This insures a sufficient and even distribution of heat.

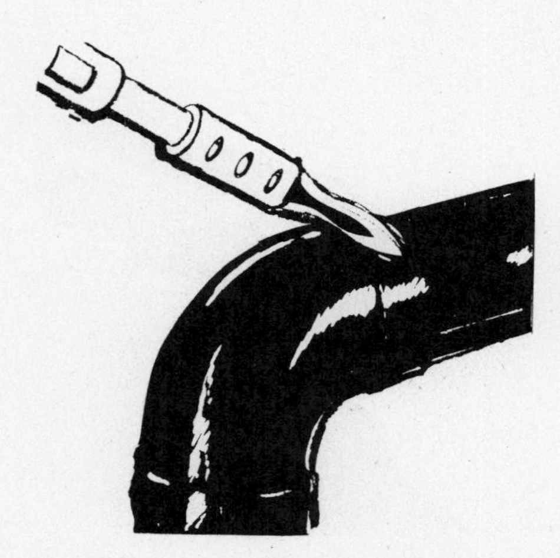

FIGURE 4-24 Applying flame (Courtesy, Mueller Brass Company).

FIGURE 4-25 Applying solder (Courtesy, Mueller Brass Company).

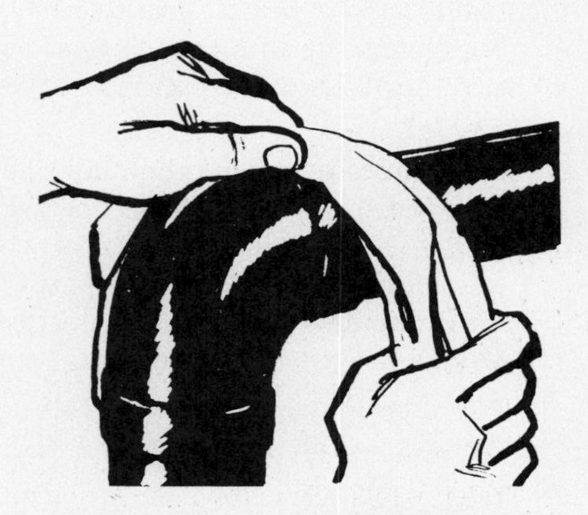

FIGURE 4-26 Smoothing solder (Courtesy, Mueller Brass Company).

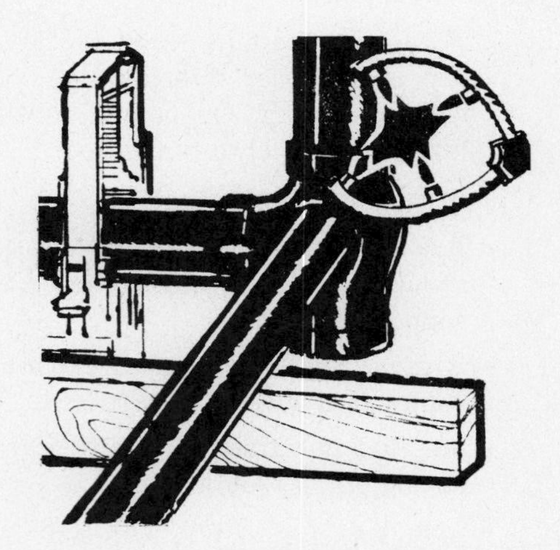

FIGURE 4-27 Large size work (Courtesy, Mueller Brass Company).

FIGURE 4-28 Tapping (Courtesy, Mueller Brass Company).

Also, when working with larger fittings 2 in. and over, each fitting should be rapped (with a small mallet) at two or three points around its circumference while the solder is being fed. (Fig. 4-28). This settles the joint and also releases any trapped gases that might hinder the flow of the solder.

HEATING AND FLOWING SILVER BRAZING ALLOY

- For rapid, efficient silver brazing, a soft bulbous oxyacetylene flame provides the best type of heat. Air-acetylene or air-propane torches have been used successfully for silver brazing fittings and tubing up to about 1 in. in size.
- Adjust the oxyacetylene for a slightly reducing (less oxygen) flame (*note*—brazing equipment and flames will be covered later).
- Start heating the tube about one-half to one inch away from the end of the fitting (Fig. 4-29). Heat evenly all around to get uniform expansion of the tube and to carry the heat uniformly to the end inside the fitting.
- When the flux on the tube adjacent to the joint has melted to a clear liquid, transfer heat to the fitting.
- Sweep the flame steadily back and forth from fitting to tube, keeping it pointed toward the tube. Avoid letting the flame impinge on the face of the fitting, as this can easily cause overheating.
- When the flux is a clear liquid on both fittings and tube, pull the flame back a little and apply alloy firmly against the tube and the fitting. With proper heating the alloy will flow freely into the joints.

The technique in making vertical joints (Fig. 4-30) is essentially the same: start with the preliminary heating of the tube, and then move to the fitting. (NOTE—this is slightly different from the soft-solder technique.) When the tube and fitting reach a black heat the flux will become viscous and milky in appearance. Continued heating will bring the material up to brazing temperature, and the flux will become clear; at that point apply the silver brazing alloy and sweat it.

Brazing larger diameter pipe requires using the above technique but selecting only a 2-in. segment at a time and overlapping the braze from segment to segment. This operation requires a degree of practice once the basic points are learned.

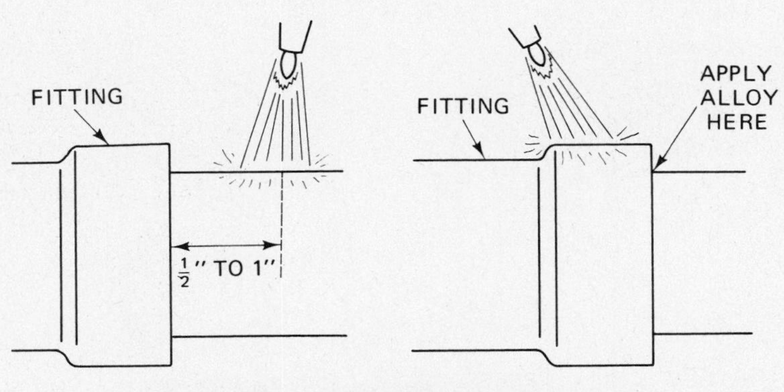

FIGURE 4-29 Brazing horizontal joints.

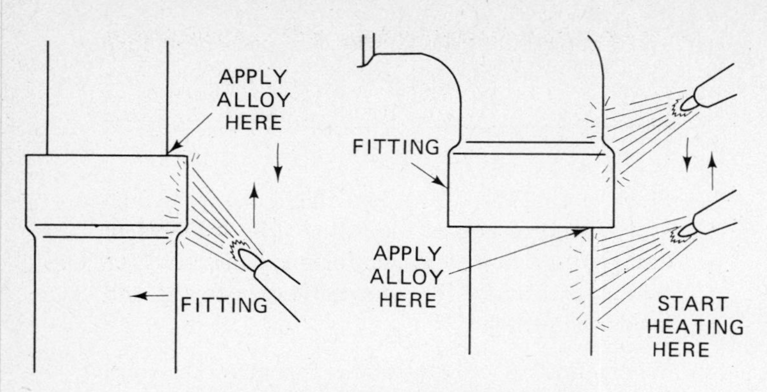

FIGURE 4-30 Brazing vertical joints.

Cleaning After Brazing. Do not quench. For cast fittings especially, allow to air cool until the brazing alloy has set; then apply a wet brush or swab to the joint to crack and wash off the flux. All flux must be removed before inspection and pressure testing. Use a wire brush if necessary.

SAFETY HINTS FOR SOLDERING AND BRAZING

- Never attempt to solder or braze while the system is under pressure or a vacuum.
- Many fluxes contain chloride or fluoride and should be handled carefully to prevent excessive skin contact or inhalation of fumes.
- Many silver solders contain cadmium in varying degrees. Cadmium fumes are very poisonous, so make sure the work space is well ventilated.
- Use safety glasses.

BRAZING EQUIPMENT

As mentioned earlier, it is possible to achieve melting temperatures for some low-temperature silver solders by using air-acetylene or air-propane torches, but where extensive installation or repairing of refrigeration equipment is involved, the use of oxyacetylene brazing equipment has proven to be the most satisfactory. The introduction of pure oxygen along with the acetylene produces a very hot flame. Figure 4-31 illustrates the components involved in a typical oxyacetylene rig.

The efficient use of oxyacetylene equipment depends on a constant metered flow of oxygen and acetylene in correct proportions. Therefore, the operator must become thoroughly familiar with his particular equipment.

Both the oxygen tank and acetylene will have pressure regulators and two sets of gauges: one to register tank pressure and one to indicate pressure

FIGURE 4-31 Oxyacetylene brazing equipment (Reprinted from *Basic Air-Conditioning*, by G. Schweitzer and A. Ebling, copyright Hayden Book Company, Inc.; used by permission of the publisher).

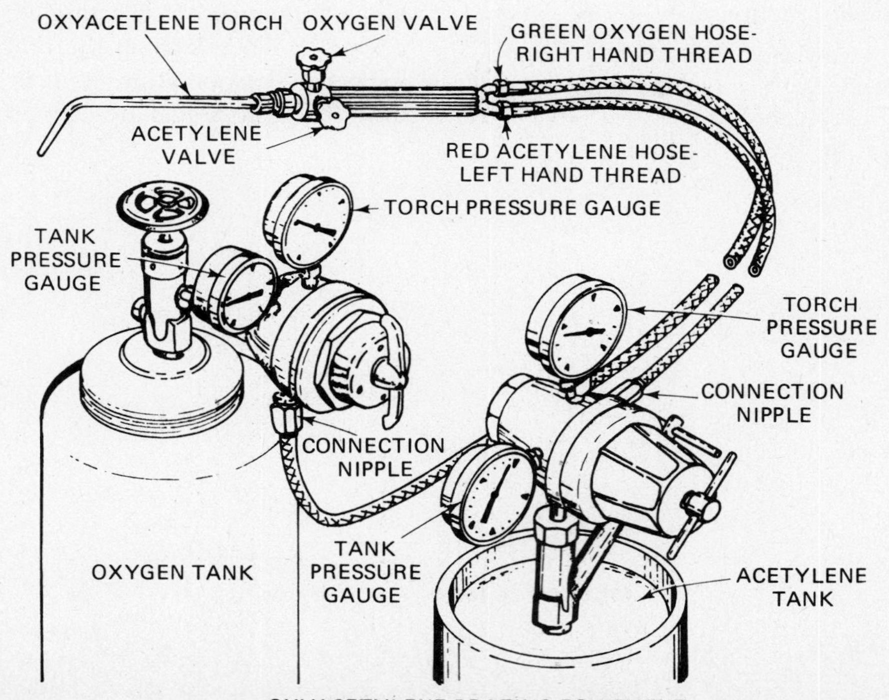

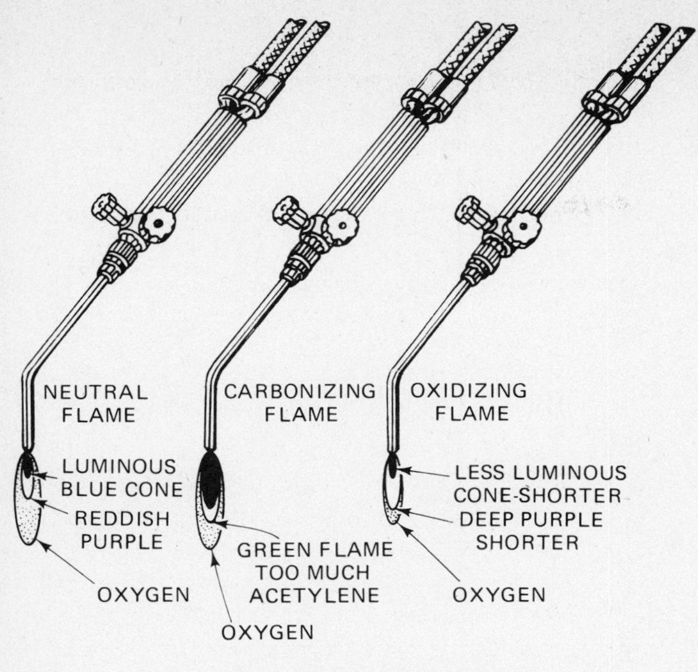

FIGURE 4-32 Oxyacetylene flames (Reprinted from *Basic Air-Conditioning*, by G. Schweitzer and A. Ebling, copyright Hayden Book Company, Inc.; used by permission of the publisher).

being furnished to the torch. The required torch pressure will vary with the particular torch and torch tip used.

While operating the oxyacetylene equipment never point the unlit torch toward any open flame or source of sparks. The acetylene is highly flammable—and the oxygen supports combustion very actively.

Lighting the torch calls for a torch lighter or sparker. *Do not use matches.* Open the torch acetylene valve approximately one-quarter turn. The torch oxygen valve is then "cracked" open as the spark or torch lighter is used to ignite the flame. Once ignition is achieved, the acetylene valve is adjusted to obtain the desired flame size. The oxygen valve is turned *slowly* to establish the type of flame, as illustrated in Fig. 4-32.

The correct flame is called a *neutral flame*; that is, it has a luminous blue cone with a touch of reddish purple at the tip. A carbonizing flame will be evidenced by a greenish flame caused by too much acetylene. It produces soot or carbon that will restrict the flow of solder. An oxidizing flame results from too much oxygen; it will cause pitting of the metal and tends to harden the joint, making it more susceptible to breakage from vibration.

SOLDERING IS AN ART

The beginner should test a few joints that he has soldered by reheating them and taking a joint apart to examine it. He can then determine whether some places in the joint were not soldered, and if this is true, what caused the failure and how to correct it.

There is probably no work a service technician is called upon to do that requires more skill than soldering. He may know all the principles of soldering, but he will find that considerable experience will be required before he becomes adept at making consistently tight, strong, and neat soldered joints.

PROBLEMS

1. What is the most common material used in refrigeration tubing?
2. Tubing differs from pipe in that it has a thick wall: true or false?
3. Copper tubing used in refrigeration work is classified as type ____________ of ____________.
4. Copper tubing is available in two forms: true or false?
5. Name two methods of joining tubing.
6. Flares used in refrigeration work are formed at what angle?

7. Name two thicknesses of flares.

8. Soldering techniques fall into what two broad methods?

9. The expression "95-5" soft solder refers to what ratio?

10. Silver solders generally start to flow at approximately what temperature?

11. Flux is needed to clean the metal surface: true or false?

12. Silver brazing is usually done with an ____________ and ______________ gas mixture.

5

REFRIGERATION TECHNICIANS' HAND TOOLS AND ACCESSORIES

The development of the technician's tool kit will of course depend on the scope of his needs. These will vary somewhat, depending on whether he is involved in installation or service repair, or both, and the kind of work; refrigeration, air-conditioning, heating, or all three. Regardless of job requirements, the careful selection, care, and knowledge of the use of his tools are important considerations for the technician. Poor workmanship or injury can frequently be traced to a lack, or improper use, of hand tools.

The term *tool kit* implies a tool box of some sort, and of course this would be true for small hand tools; in reality the technician will most likely be working out of a truck or van that provides storage for hand tools, power tools, soldering equipment, piping materials, refrigerant cylinders, etc.

The following discussion separates common hand tools from measuring or testing equipment such as thermometers, pressure gauges, electric meters, leak detectors, and the like, which will be covered later. It is also assumed that special equipment such as the tube cutters, benders, flaring tools, and brazing equipment previously presented will be available as needed.

A partial listing of common hand tools needed by a refrigeration technician is detailed below. Some do not need explanation; others require reference to their scope or use.

TYPICAL HAND TOOLS

- wrenches—of several sorts (as reviewed below)
- pliers—of several sorts (as reviewed below)
- spirit level
- tin snips
- screwdrivers—of several sorts (as reviewed below)
- hammers—ball peen and common claw
- mallets—with nonmetalic heads (plastic, wood, rubber)
- hacksaw—blades with 14, 18, and 32 teeth per inch
- brushes—of several sorts (as reviewed below)
- files—of several sorts (as reviewed below)
- tape measures and hand rule—(as reviewed below)
- micrometer and calipers—(as reviewed below)

- punches—for marking drill point
- chisel—$\frac{3}{4}$ in. flat cold chisel
- drills—of several sorts (as reviewed later)
- vises—machinist and pipe vise
- pocket knife
- flashlight
- 50-ft electrical extension cord
- stop watch

Of the listing above the wrenches are probably the most frequently used tool of all, so we will cover these in some detail. Several types of wrenches are listed, but not necessarily in general order of use or preference [see Fig. 5-1(a) through 5-1(j)].

- ratchet wrench
- socket wrenches
- box wrenches
- open-end wrenches
- adjustable wrenches
- allen wrenches
- nut driver

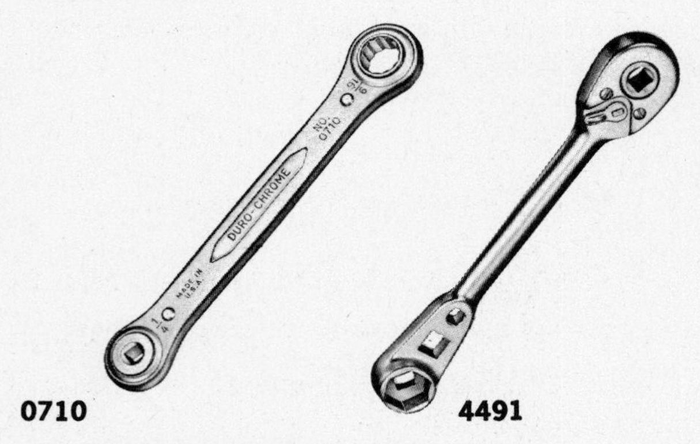

Rachet Wrench (Figure 5-1a)*

The steel ratchet wrench is especially adapted for use on small refrigerant cylinders and shut-off valves. The ratchet permits rapid changing of direction so the operator can switch movement to open or close a valve, etc. The wrench handle openings vary from $\frac{1}{4}$ in. to $\frac{1}{16}$ in. Some also have a $\frac{1}{2}$ in. hex socket cast into one end.

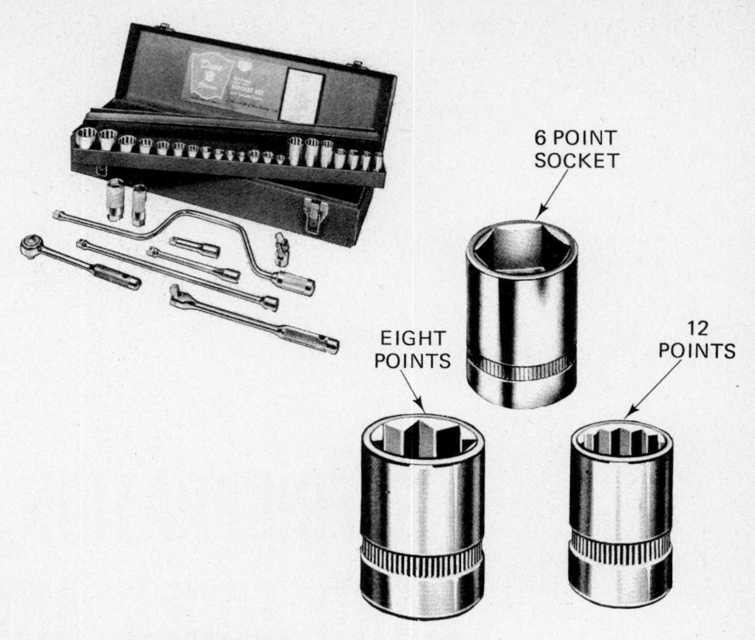

Socket Wrenches (Figure 5-1b)

Socket wrenches are used to slip over bolt heads. They are made of steel, and the array of sockets vary from square to six-point hexagonal shape or up to 12-point double hexagonal shape. Common sizes run from $\frac{5}{32}$ in. to 2 in. in English measurement. Metric sizes are also readily available. Swivel sockets are particularly useful to reach hard-to-get-to nuts and bolts. The more points a socket has the easier it is to use in a restricted area.

The handle of a socket wrench may vary from a straight fixed tee drive to a ratchet operation to a special torque wrench handle that includes a gauge to measure the force being applied. In certain operations the manufacturer may specify torque requirements or limitations. The force or torque is measured in foot-pounds or inch-pounds.

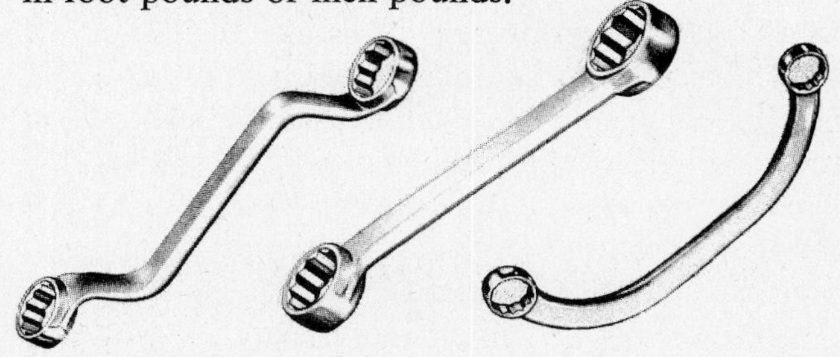

Box Wrenches (Figure 5-1c)

Box wrenches are useful in certain close-quarter situations. The ends are usually in the form of a 12-point double hexagonal shape as illustrated.

* Figures 5-1(a) to 5-1(f) and 5-1(i) and (j), courtesy, Duro Metal Products, Inc.

Ends may be of the same size or different sizes. The handle may be straight or offset. Common sizes range from $\frac{1}{4}$ in. to $1\frac{1}{2}$ in.

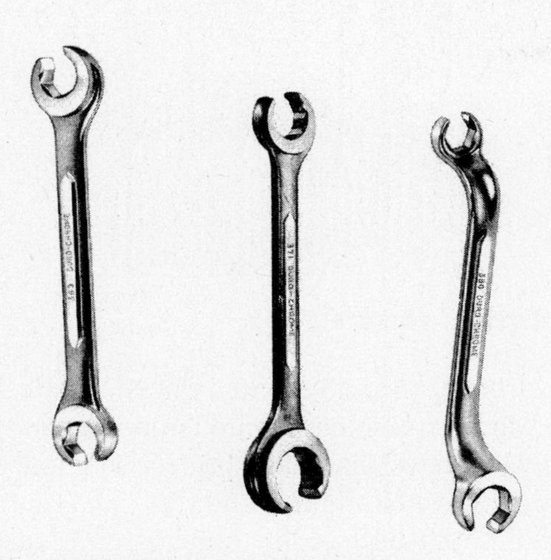

Flare-Nut Wrenches (Figure 5-1d)

The flare-nut wrench is a special variation of the box wrench in that the heads are slotted to allow the wrench to slip over the tubing and then onto a flare nut. After tightening, the wrench is removed in a reverse manner. The wrench heads are made to fit the standard S. A. E. flare nuts. Several sizes are needed.

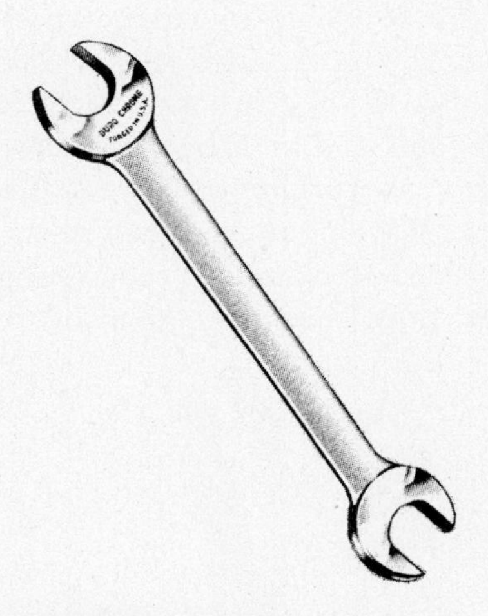

Open-End Wrenches (Figure 5-1e)

Open-end wrenches are needed where it is impossible to fit a socket or box wrench on a nut, bolt, or fitting from the top. The open-end wrench permits access to the object from the side. Of necessity the wrench has only two flats. The distance between these flats determines the size, which normally ranges from $\frac{1}{4}$ in. to $1\frac{5}{8}$ in.

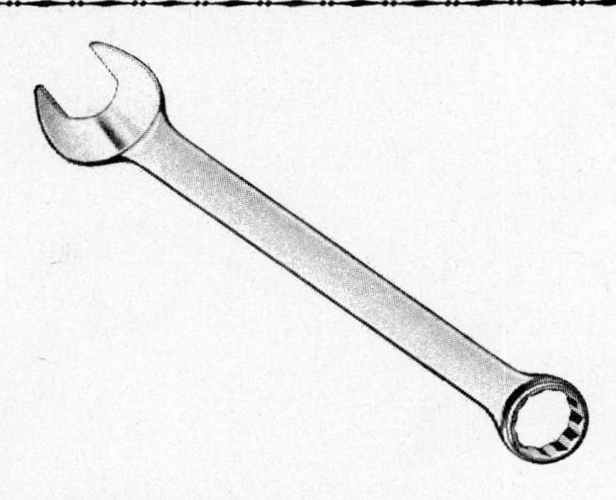

Combination Box and Open-End Wrenches (Figure 5-1f)

Some service mechanics prefer the combination box and open-end wrench to gain the advantage of each type. The ends are made to fit the same size nut or bolt so he can use the best selection.

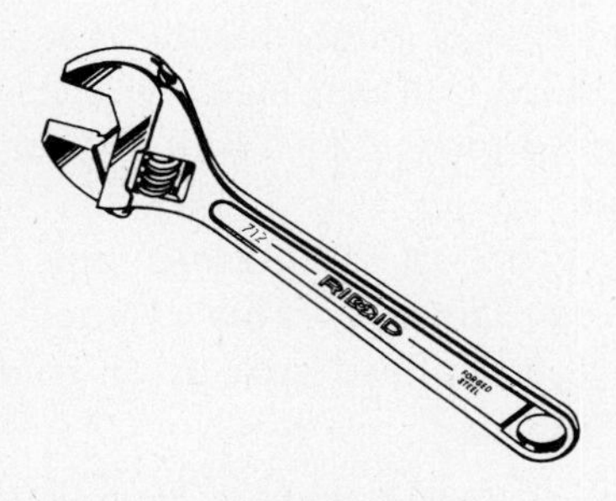

(Courtesy, Duro Metal Products Company).

Adjustable Wrenches (Figure 5-1g)

The familiar adjustable wrench as illustrated is useful where a regular open-end wrench could be used, but the adjustable screw permits fitting the flat to any size object within the maximum and minimum opening. The handle size indicates the general capacity. For example, a 4-in. size will take up to a $\frac{1}{2}$-in. nut. A 16-in. handle will take up to a $1\frac{7}{8}$-in. nut. Always use this type of wrench in a manner such that the force is in a down or clockwise direction when tightening a bolt; this keeps the force against the head. If the wrench were used in an opposite manner, it might suddenly loosen and injure the operator.

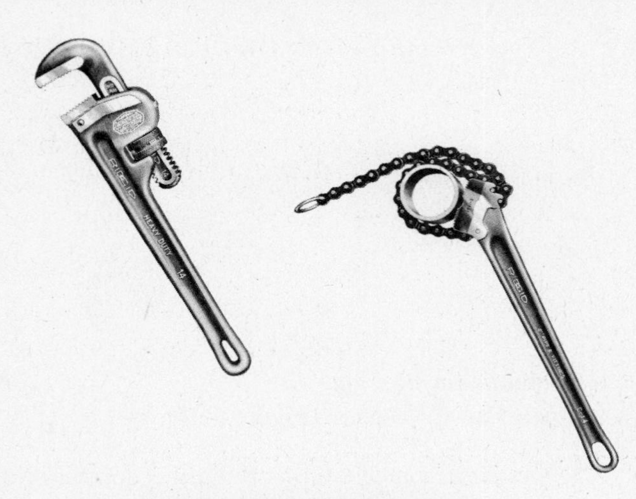

(Courtesy, Ridge Tool Company).

Pipe Wrenches (Figure 5-1h)

The pipe wrench is a common tool used in refrigeration installation and service work to assemble or disassemble threaded pipe. It is tough and can withstand a great deal of punishment. At least two sizes are recommended—an 8-in. size, which can handle up to 3-in. diameter pipe and a 14-in. size for up to 8-in. diameter pipe.

Another form of adjustable pipe wrench is called the *chain wrench*. This wrench can make work easier in a confined area or on round, square, or irregular shapes.

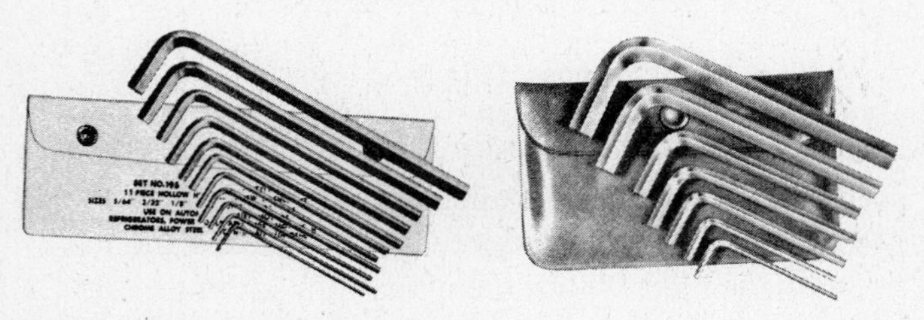

Allen Wrenches (Figure 5-1i)

Allen wrenches are necessary for removing or adjusting fan pulleys, fan blade hubs, and other components that are held in place or adjusted by allen set screws. The wrenches are tough alloy steel with six-point flat faces. The wrench goes inside the set screw and may be used at either end. Sizes range from about $\frac{1}{16}$ in. to about $\frac{1}{2}$ in.

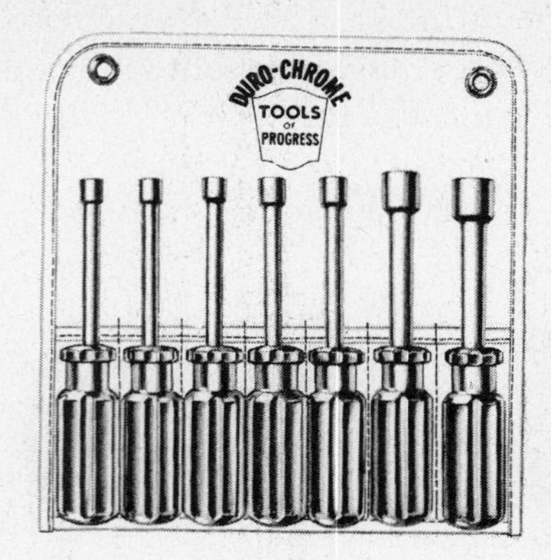

Nut Drivers (Figure 5-1j)

Nut drivers are actually a type of small socket wrench. Many equipment manufacturers have started using metal screws that have both a slot and a nut head; some have eliminated the slot. The nut driver consists of a plastic handle that has different drive sockets to fit over the screw or nut head. The nut driver is useful in tightening or removing sheet-metal screws that hold equipment panels in place or fasten control-box covers; it is also used to secure the control mechanism itself.

Pliers (Figure 5-2)

Pliers are also one of the most frequently used tools and are available in many types (Fig. 5-2). The familiar slip-joint pliers (Fig. 5-2a) are handy for general use. At least two sizes are recommended. The curved-joint and/or arc-joint pliers (Fig. 5-2b) are necessary for working on larger objects and for holding pipe. Locking or vise grip pliers (Fig. 5-2c) are often needed to clamp objects during soldering, for example, thus freeing the hand of the operator.

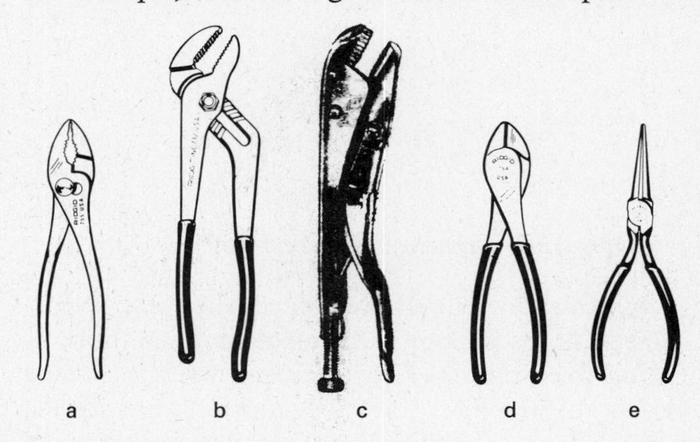

FIGURE 5-2 Pliers: (a) Slip-joint; (b) curved joint; (c) Vise grip; (d) Diagonal cutting; (e) Needle nose (Courtesy, Ridge Tool Company).

In electrical work several different styles of pliers are needed. The diagonal cutting plier (Fig. 5-2d) is used to cut wire (or rope). Needle-nose pliers (Fig. 5-2e) are used to form wire loops and grip tiny pieces firmly; some also have built-in side cutters. Pliers in general are not made to tighten or unscrew heavy bolts and nuts; but are useful tools to start or hold the item until another tool takes over.

Screwdrivers (Figure 5-3)

The well-equipped tool kit must have a variety of screwdrivers. The most common is the flat-blade design (Fig. 5-3a), and a complete set is recommended, from $\frac{1}{8}$-in. blades to the large $\frac{5}{16}$-in. size. Handle sizes vary with the blade dimension. Flat blades fit the common single slot screw and it's important to have a tight fit; otherwise it's possible to strip the screw slot.

Phillips-tip screwdrivers (Fig. 5-3b) are needed where Phillips-head screws are used. These are more common in the electrical phases of refrigeration work. Again a complete set is recommended for accurate fitting to screw heads.

Starting with these two standard items the technician will customize his needs with specialized items like screwdrivers with magnetized blades or blades with a screw grasping clip. Screwdrivers, in addition to their primary purpose, can be used (with discretion) for light-duty prying, wedging, or scraping, *but never pound a screwdriver with a hammer.*

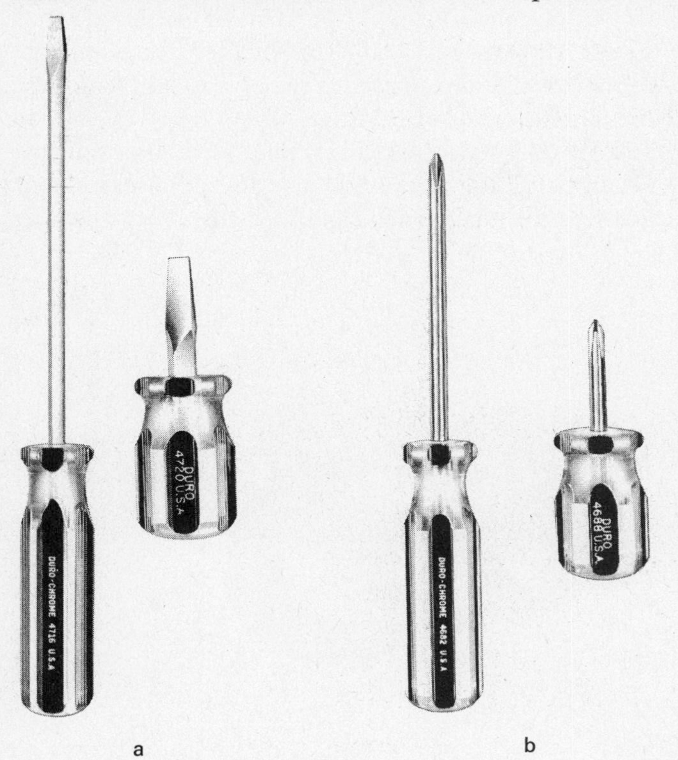

FIGURE 5-3 Screw drivers: (a) Square flat blade; (b) Phillips tip (Courtesy, Duro Metal Products, Company).

Brushes

Under the section on soldering the use of a wire brush (Fig. 5-4) to clean the inside of tubing and fitting was recommended. These brushes come in sizes from $\frac{1}{4}$ in. to $2\frac{1}{8}$ in. to fit the O. D. of the soldered fitting. The brush should have fine steel wire bristles, thickly set and firmly attached to the handle stem.

FIGURE 5-4 Brushes.

A solder flux brush is also recommended in applying paste. There is only one size and the brushes are very inexpensive. Frequently, they become too dirty to use and must be discarded. A paint brush is useful to brush dirt or dust out of a control box, for example. Or it may be used to apply cleaning solvent to an object.

Files

Files come in several shapes: flat or rectangular, round, half-round, triangular, square, etc. In refrigeration work the common flat file and the half-round are used in preparing tubing for soldering—squaring the end or removing burrs.

The ability of a file to cut metal or other surfaces depends on the tooth size, shape, and number of cuts or directions of cut. Figure 5-5 illustrates single-cut and double-cut direction files.

A single-cut file is used for finishing a surface, for example, in preparing copper pipe for soldering. A double-cut file is more coarse and would be used where deeper and faster metal removal is needed. A rasp is an extremely coarse double-cut file intended for very rough work. The selection of files will vary with the need.

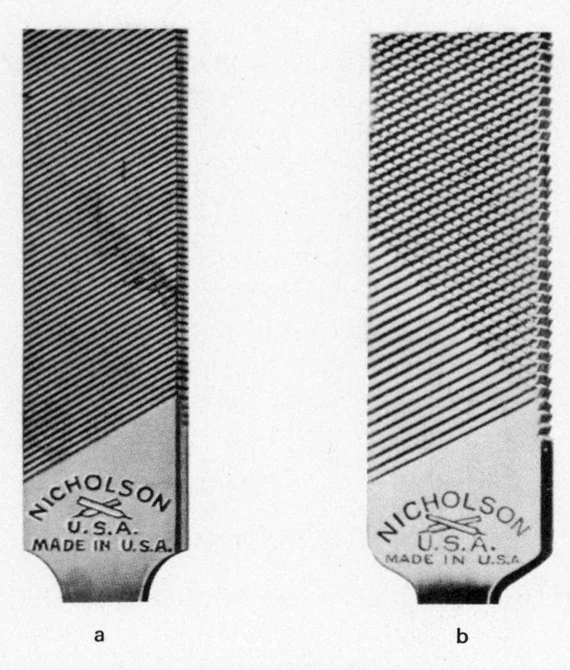

FIGURE 5-5 Files: (a) Single cut; (b) Double cut (Courtesy, Nicholson File Company).

Vises (Figure 5-6)

A machinist's vise (Fig. 5-6a) is quite useful in holding parts for drilling, filing, and other operations. A medium-size portable model is recommended for field service work; larger models may be needed in shop work.

The pipe vise (Fig. 5-6b) is a requirement both in the field and in the shop to hold tubing or pipe while cutting or threading operations are taking place. *Caution*—when holding copper or other soft metal be sure the jaws are soft so as not to mar the surface. Also do not over-apply clamping pressure so as not to squeeze tubing out of round.

Tapes and Hand Rules

Start with the familiar 6- or 8-foot flexible steel tape—it's invaluable for many measurements; tube diameters, short tube lengths, filter sizes, duct sizes, etc. Where long piping runs are to be assembled, a 50- or 100-foot steel or plastic tape is recommended for accurate measurements. A cloth tape is not recommended due to its stretch characteristics. Some technicians also like to have a 12-in. nonrusting steel ruler handy with graduations down to $\frac{1}{32}$ in. for more precise measurements.

Micrometers and Calipers

Sometimes a service technician must check dimensions of parts very accurately in thousandths of an inch. Such measurements require special tools, which are, however, commonly available and reasonably inexpensive.

For measuring the outside diameter or width of an object such as a fan shaft or metal thickness the use of a micrometer (Fig. 5-7a) is required. The object is placed inside the micrometer frame and the thimble is tightened to force the object lightly against the anvil. One turn of the thimble advances the spindle $\frac{1}{40}$ in. or 0.025 in. (twenty-five thousandths of an inch). The sleeve and spindle are so graduated as to permit the user to read measurements down to

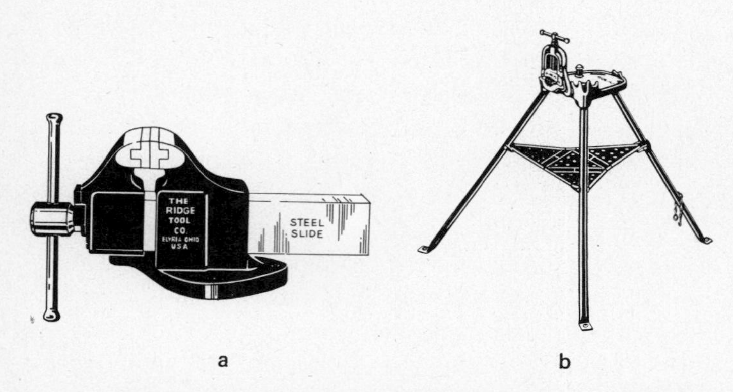

FIGURE 5-6 Vises: (a) Machinist vise; (b) Pipe (Courtesy, Ridge Tool Company).

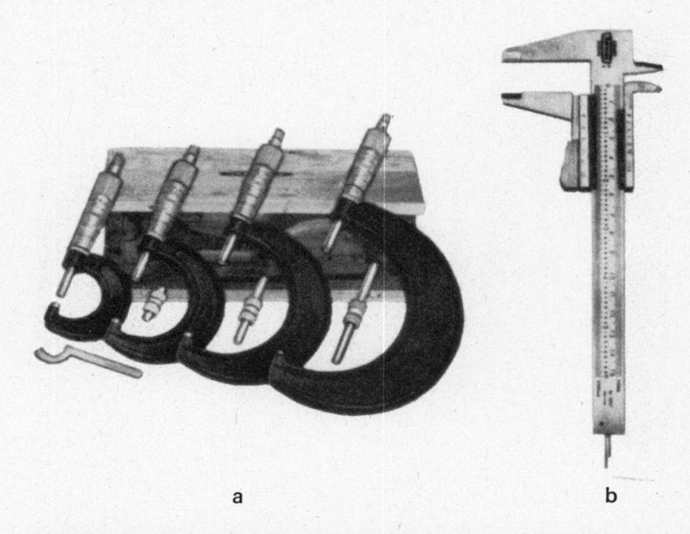

FIGURE 5-7 Typical micrometer and caliper measuring instruments: (a) Micrometer sets; (b) Caliper.

$\frac{1}{1000}$ (0.001) of an inch. Micrometers come in ranges, meaning the maximum permissible size that can be measured. A 1-in. micrometer measures from 0 to 1 in., a 2-in. micrometer measures from 1 to 2 in., a 3-in. measures from 2 to 3 in., and so on.

The caliper (Fig. 5-7b) is an instrument for measuring the inside diameter or dimension of an object such as a bearing opening or cylinder bore in a compressor casting. The jaw is placed inside the opening and the dimension is read from a vernier scale or dial, depending on the model. Again, these read to 0.001 inch accuracy. Some calipers can be used to measure both inside and outside dimensions.

Drills

Drills (Fig. 5-8) are a must and are frequently used by the refrigeration mechanic in the work of installation and repair. In the field it is assumed a hand-held portable electric drill will be used. A $\frac{1}{4}$-in. or $\frac{3}{8}$-in. chuck size is sufficient for small operations, such as drilling wood, plastic, thin metals, and light-duty masonry work.

Heavy-duty drilling in brick, cement, and thick steel is often required during installation work; for such operations a $\frac{1}{2}$-in. heavy-duty model drill is recommended. Such a drill can also double for rotary-hammer drills.

FIGURE 5-8 Electric drill, (Courtesy, The Black and Decker Manufacturing Company).

Either drill should have a variable speed control plus a reversing switch to back out stuck bits. Bit selection will of course depend on the nature of the material, but as a minimum, a set of high-speed alloy steel bits for metal is a must. Such bits are also fine for drilling wood and plastic. Masonry bits and wood-boring bits may be added as job requirements demand.

Accessory Items

In addition to hand tools, the refrigeration technician will need an array of accessory items. Most of these are expendable and will need periodic replacement. Following is a partial listing:

- abrasive sand cloth, which comes in rolls or sheets (several grades will be needed)
- steel wool and cleaning pads
- Cleaning solvent that is not dangerous to health or explosive. Among those *not recommended* are gasoline and carbon tetrachloride. Consult the refrigeration supply outlets for various acceptable solvents.
- roll of friction tape
- roll of rubber tape
- pipe "dope" for sealing threads

Although it is not a tool or accessory as such, the availability and use of safety protection is one of the most important considerations in establishing minimum requirements for personal protection and/or to meet and comply with OSHA (Occupational Safety and Health Act) or local codes. Minimum safety equipment should include:

- hard hat
- safety glasses
- safety shoes
- gloves
- fire extinguisher
- first aid kits
- check list of actions to take in case of an emergency
- common sense

PROBLEMS

1. A special type of box wrench used in refrigeration work is called a ______________.
2. Socket wrenches can be obtained with sockets for different points: true or false?
3. The more points a socket has the harder it is to use in a restricted area: true or false?
4. Set screws on a fan shaft are turned by ______________ wrenches.
5. Name two types of screwdriver blades.
6. Files are made into what two types?
7. A machinist vise is also good for holding pipe: true or false?
8. Common micrometers and calipers read to ______________ of an inch.
9. Gasoline and carbon tetrachloride are recommended cleaning solvents: true or false?
10. What does OSHA stand for?

6

THE COMPRESSION CYCLE COMPONENTS

The vapor compression refrigeration cycle as discussed in Chap. 1 is the most common method of heat energy transfer. There are four major components in the compression cycle: evaporator, compressor, condenser, and refrigerant flow control device.

EVAPORATORS

The evaporator or cooling coil is the part of the refrigeration system where heat is removed from the product: air, water, or whatever is to be cooled. As the refrigerant enters the passages of the evaporator it absorbs heat from the product being cooled, and, as it absorbs heat from the load, it begins to boil and vaporizes. In this process, the evaporator accomplishes the overall purpose of the system—refrigeration.

Manufacturers develop and produce evaporators in several different designs and shapes to fill the needs of prospective users. The blower coil or forced convection type evaporator (Fig. 6-1) is the most common design; it is used in both refrigeration and air-conditioning installations.

Specific applications may require the use of flat plate surfaces for contact freezing. Continuous tubing is formed or placed between the two metal plates, which are welded together at the edges, and a vacuum is drawn on the space between the plates. These plates also may be assembled in groups arranged as shelving, utilizing refrigerant in a series flow pattern (see Fig. 6-2).

Other shapes of plate-type evaporators are shown in Fig. 6-3. They are widely used in small refrigerators, freezers, and soda fountains, where mass production is economical, and the plates can easily be formed into a variety of shapes.

Plate-type evaporators also are assembled in groups or banks for installation in low-temperature storage rooms and are mounted near the ceiling as shown in Fig. 6-4. This type may be connected for either series or parallel refrigerant flow, depending upon usage requirements. Plate-type coils are also used in refrigerated trucks and railway cars for the transportation of refrigerated food and frozen food

FIGURE 6-1 Blower coil (Courtesy, Kramer Trenton Company).

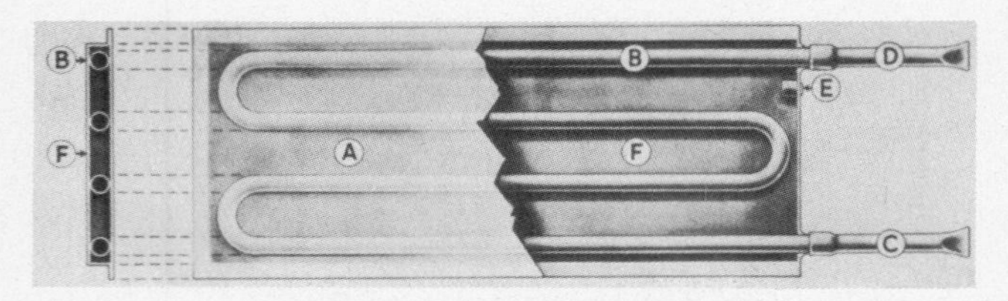

FIGURE 6-2 The Dole vacuum plate: A. Outside jacket of plate. Heavy, electrically welded steel. Smooth surface; B. Continuous steel tubing through which refrigerant passes; C. Inlet from compressor; D. Outlet to compressor. Copper connections for all refrigerants except ammonia where steel connections are used; E. Fitting where vacuum is drawn and then permanently sealed; F. Vacuum space in dry plate. Space in hold-over plate contains eutectic solution under vacuum. No maintenance required due to sturdy, simple construction. No moving parts; nothing to wear or get out of order; no service necessary (Courtesy, Dole Refrigerating Company).

products, and are of a design similar to that shown in Fig. 6-5. Frequently the space between the plates is filled with a solution that retains its refrigeration if the unit is not in operation for short periods of time.

The bare-tube type of coil may be used for the cooling of either air or a liquid, with the smaller evaporators being constructed of copper tubing. Steel pipe is used for evaporators in systems using ammonia as the refrigerant and in the larger evaporators containing other refrigerants.

An air film adheres to the outside surface of a coil, acting as an insulator and slowing down the heat-transfer process, which is dependent primarily on surface area and temperature differential. One of the methods used to overcome or compensate for the conduction loss due to the air film is to increase the surface area. This may be accomplished through the addition of fins to the evaporator pipe or tubing, as shown in Fig. 6-6. The addition of fins does not eliminate air film, for it furnishes more area to which air film will cling or adhere; but it does afford more surface area for heat transfer, without increasing the size of the coil to any great extent.

Another method of overcoming the heat-transfer loss caused by air film is through the addition of a fan or blower, which will cause rapid movement of air across the evaporator. Such a type of forced convection coil is shown in Fig. 6-7. Depending upon the design and usage of the coil, the fan may be located for the movement of air across the coil either by means of an induced or drawing action of the air or by a forced circulation or blowing action of the air across the evaporator coil.

The use of a fan improves the air flow and transfer of heat from the air to the refrigerant within the

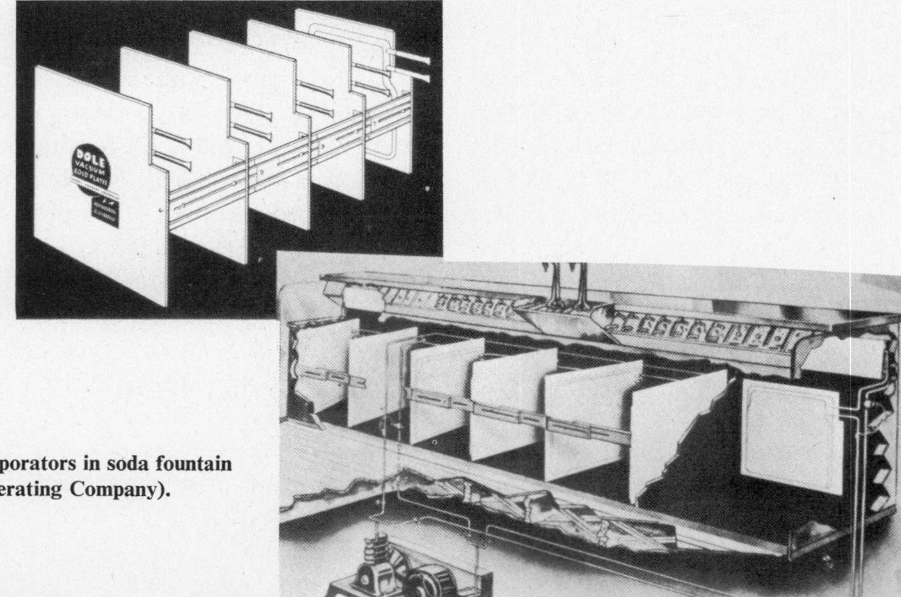

FIGURE 6-3 Plate evaporators in soda fountain (Courtesy, Dole Refrigerating Company).

FIGURE 6-4 Plate evaporators for storage rooms (Courtesy, Dole Refrigerating Company).

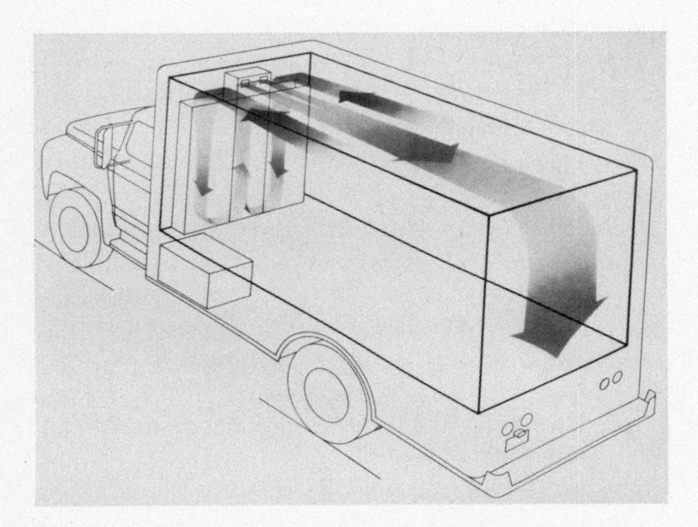

FIGURE 6-5 Refrigerated truck (Courtesy, Dole Refrigerating Company).

coil, since a greater proportion of the air will come in contact with the coil's surface area. Many coil manufacturers have designed their heat-transfer units with staggered rows of tubing, thus permitting, with the use of a blower, a large volume of air to come in contact with either the tubing surface or the fins connected to it, but a forced or induced air motion across the coil usually will result in a greater portion of the air giving up its heat to the refrigerant within the coil over a specific time period.

In the early days of mechanical refrigeration, cooling coils were constantly maintained at a temperature below freezing. Since these evaporators did

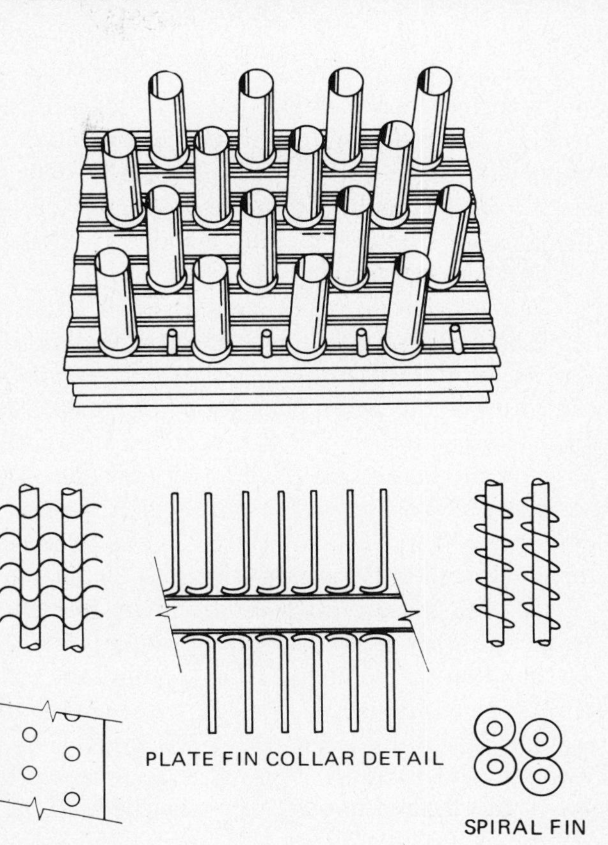

FIGURE 6-6 Finned tube evaporator (Reprinted by permission from ASHRAE handbook and product directory, 1975).

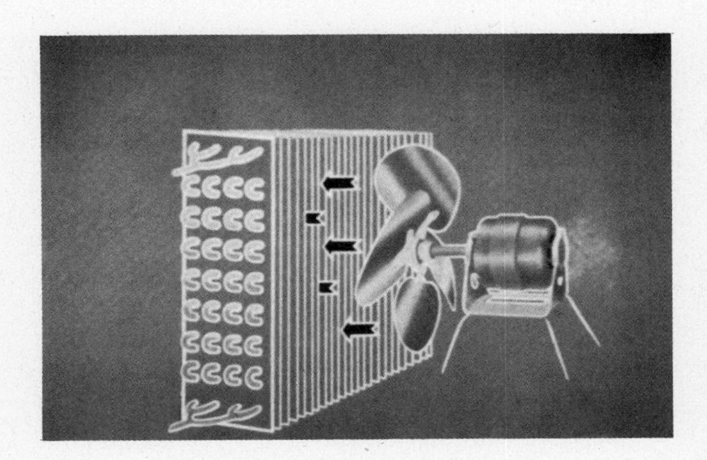

FIGURE 6-7 Forced air evaporator (finned) (Courtesy, Carrier Air-Conditioning Company).

not reach a temperature above 32°F, the frost accumulating on them did not have an opportunity to melt off while the equipment was in operation. The units had to be shut off and manually defrosted, since frost accumulating on the evaporator curtailed the amount of heat it could remove from the air passing across the coil.

In many of today's refrigeration applications, low temperatures must be maintained so that products may be kept in frozen storage condition. But defrosting of the pooling unit is performed by means other than manual, which will be explained in Chap. 16. Frost accumulation on the cooling unit comes from moisture in the air and the products in the refrigerated space. When this moisture is removed from the air, the humidity is lowered.

Conditions may be such that an extremely low temperature or a low moisture content of the air surrounding the cooling unit is not desirable. If the temperature in the refrigerated space needs to be maintained at approximately 35°F, a coil or evaporator in which the refrigerant is at a temperature below this desired temperature must be used. As the air comes in contact with the cooling coil at a temperature below 32°F, some frost will form on the surface of the evaporator. When the desired temperature is achieved, the control mechanism will stop operation of the refrigeration unit. With the surrounding air temperature at 35°F, this warm air will melt the frost on the cooling unit and thereby defrost the cooling unit. This will occur naturally, particularly if it is a forced air coil and the warmer temperature air is forced across the evaporator surface.

The off-cycle period of the refrigeration unit should be long enough to assure the complete defrosting of the cooling coil. If not, there is a possibility of only partial defrosting, which results in moisture collecting on the lower section of the unit. If this occurs, an icing condition on the coil may result. If this condition is allowed to continue, ice may cover the entire surface of the coil and may develop into a complete blockage of the coil.

Design conditions may be such that a high humidity must be maintained, to preserve freshness in the product being cooled, or to prevent a loss in weight, or deterioration. Examples of applications in which a high relative humidity is desired are a cold-storage room where fresh meat is kept, and a florist display cabinet with a moisture-laden atmosphere. Vegetable storage rooms or boxes also should be kept at a high humidity.

These conditions may be achieved through the use of nonfrosting evaporators, which are oversized coils used with thermostatic expansion valves as the metering devices. In order to maintain the refrigerated space temperature at approximately 35°F, the large-surface coil needs an internal refrigerant temperature of only 23°–25°F. This permits an external coil temperature of about 30°–32°F, allowing only a rare accumulation of frost, which will disappear rapidly when the space temperature requirement is satisfied and the compressor shuts off.

The condensate drain is shown in Fig. 6-8 beneath the nonfrosting coil, even though the drain is not put to any great use when the system is operating correctly. This type of coil is designed not to remove too much moisture from the air, so that a relative humidity of up to 85% may be maintained within the refrigerated space.

The evaporators described thus far have been of the *dry-expansion type*, as compared with the flooded type. The direct or dry-expansion type coil is designed for complete evaporation of the refrigerant in the coil itself, with only a vapor leaving the coil outlet. This vapor may even be superheated in the last part of the cooling coil. (*Superheating* means raising the

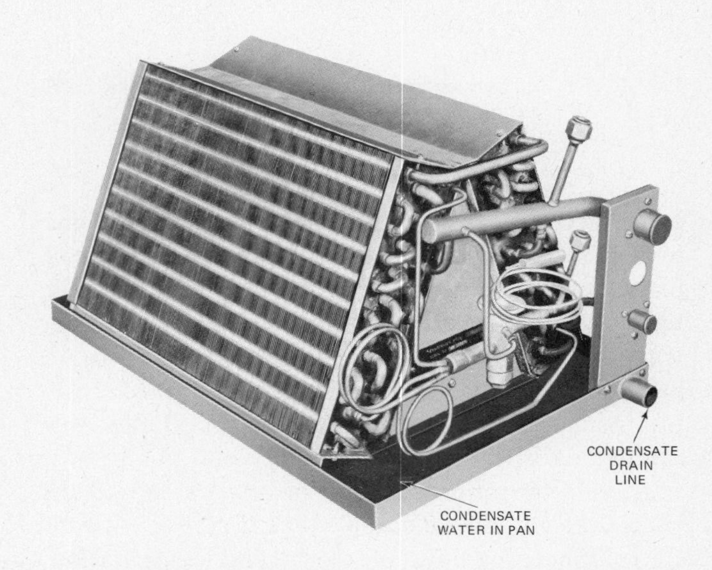

FIGURE 6-8 Condensate drain (Courtesy, York, Div. of Borg-Warner Corporation).

temperature of the refrigerant above that temperature required to change it from a liquid to a vapor.) It will reach the compressor in a superheated condition, picking up additional heat as it passes through the suction line. Figure 6-9 is a schematic showing a direct-expansion coil with a thermostatic expansion valve. The coil contains a mixture of liquid and gaseous refrigerant at all times when the unit is in operation. A constant superheat is maintained by the throttling of the valve, which is caused by the sensitivity of the thermal bulb to temperature changes at its location.

The characteristics of the dry or direct-expansion coil can be maintained by the automatic expansion valve, which maintains a constant pressure within the evaporator. This type of valve is usually used when a steady load is anticipated. Refrigerant controls will be more fully discussed later on in this chapter and again in Chap. 20.

The *flooded* type of evaporator is filled with liquid refrigerant. It is designed so that the refrigerant (liquid) level is maintained by a float arrangement located in an accumulator that is situated outside the evaporator coil itself. A typical design is shown in Fig. 6-10. Part of the liquid refrigerant evaporates in the coil, and this vapor goes to the accumulator. From there the vapor is drawn from the top into the suction line and then to the compressor, while any liquid left in the accumulator is available for recirculation in the evaporator coil. When the equipment is properly calibrated, the remaining liquid is minimal.

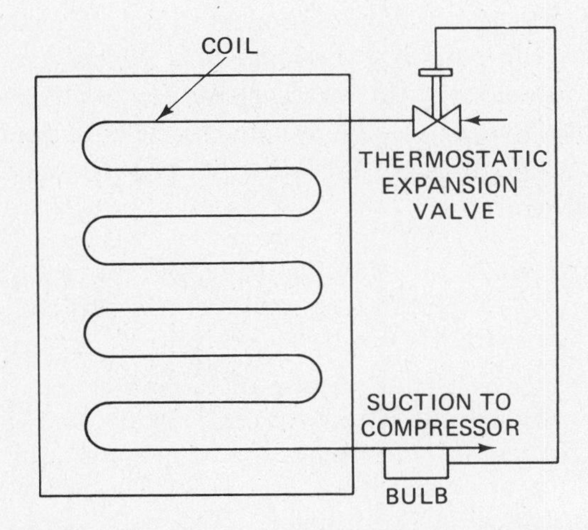

FIGURE 6-9 Dry expansion coil with thermostatic expansion valve (Reprinted by permission from ASHRAE handbook and product directory, 1975).

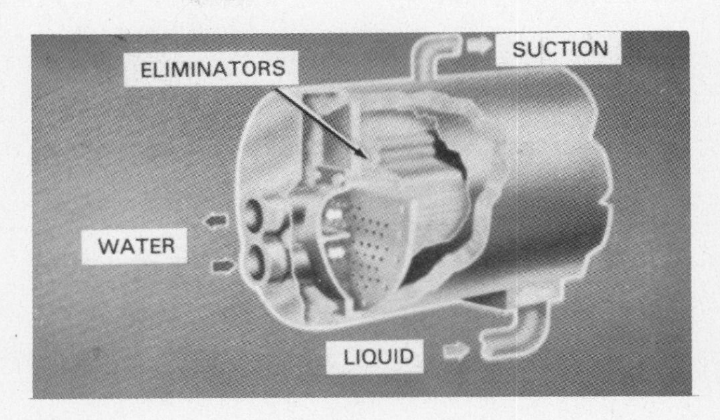

FIGURE 6-10 Flooded chiller (Courtesy, Carrier Air-Conditioning Company).

As the refrigerant in the flooded coil evaporates as a result of the heat it has absorbed, the liquid level lowers in the coil. As the float lowers with the liquid level, it permits more refrigerant to flow into the accumulator so that a fairly constant liquid level is maintained. A flooded coil has excellent heat transmission efficiency because its interior surfaces are liquid-wetted instead of being vapor-wetted.

Liquid-cooling coils vary in their design depending on their application and usage, just as do air-cooling coils. Since there is a greater heat transfer between liquids and metals than between air and metals, a submerged coil has the capability of removing several times as many Btu as an air-cooled coil under similar conditions. Submerged coils are used in a water-bath type of cooler, in which the "cold-holding" capacity is put to good use when cans filled with warm milk or other liquids are placed in the cooler.

Shell-and-tube and *shell-and-coil* are other types of arrangements for the cooling of one or more liquids, even in the cooling of brine solutions. A shell-and-coil water cooler is shown in Fig. 6-11. It is a direct-expansion type of system with the refrigerant circulated within the coil as the water is circulated within the shell at a temperature not much below 40°F to prevent freezing.

Tube-in-tube, sometimes classified as *double-pipe evaporator*, is a liquid-cooling coil that provides high heat-transfer rates between the refrigerant and the liquid being cooled. The path of refrigerant flow

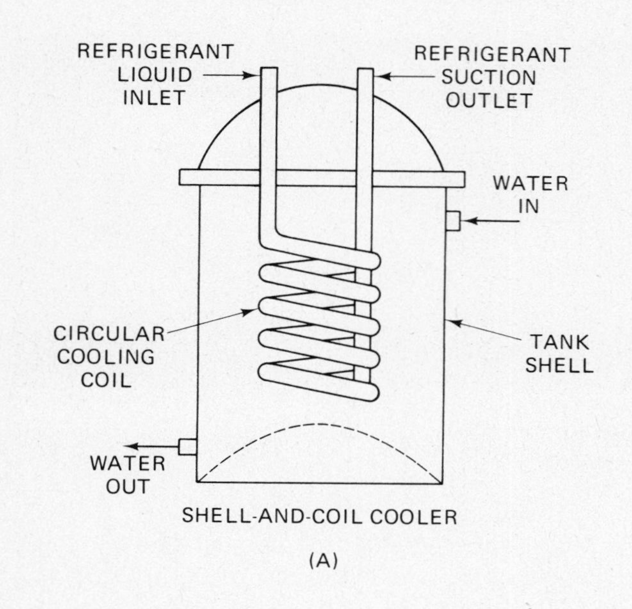

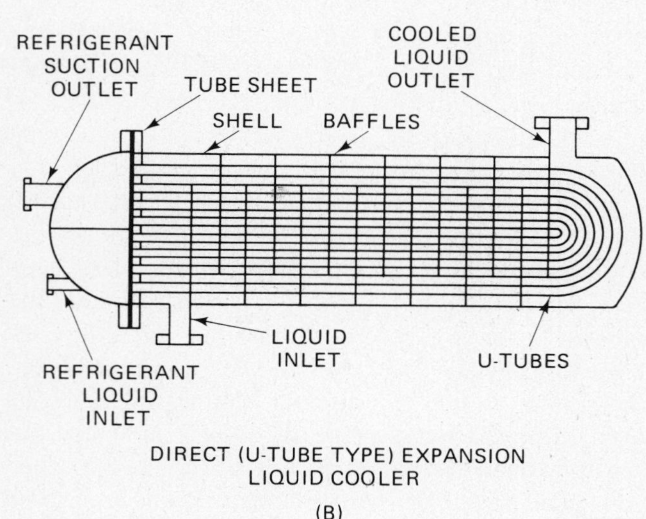

FIGURE 6-11 (a) Shell and coil water cooler; (b) Direct (U-tube type) expansion liquid cooler (Reprinted by permission from ASHRAE handbook and product directory, 1975).

may be through either of the tubes, although usually the brine or liquid to be cooled is circulated through the inner tubing, and the refrigerant removing the heat is between the two tubes. This type of heat-exchange coil is also used in condenser design, described later in this chapter.

A *Baudelot cooler*, shown in Fig. 6-12, has several applications. It may be used for cooling water or other liquids for various industrial uses, and it is frequently used as a milk cooler. The evaporator tubing is arranged vertically, and the liquid to be cooled is circulated over the cooling coils by gravity flow from the trough type of arrangement located above the coils. The liquid gathers in a collector tray at the bottom of the coil, from which it may be recirculated over the Baudelot cooler, or pumped to its destination in the industrial process.

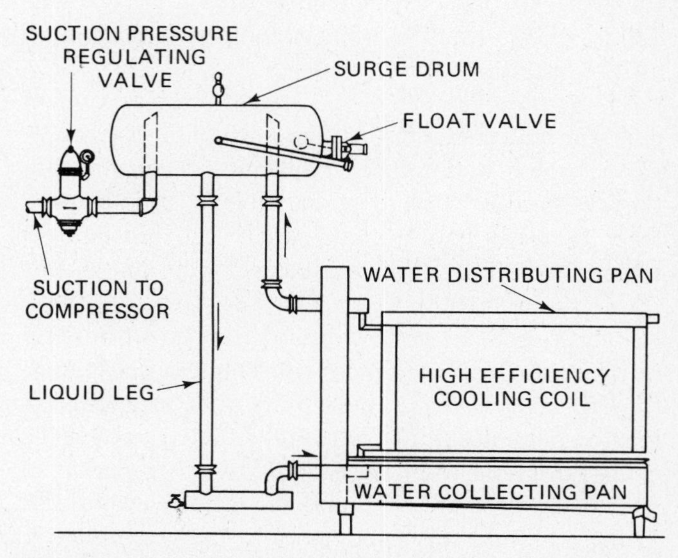

FIGURE 6-12 Boudelot cooler (Reprinted by permission from ASHRAE handbook and product directory, 1975).

COMPRESSORS

After it has absorbed heat and vaporized in the cooling coil, the refrigerant passes through the suction line to the next major component in the refrigeration circuit, the compressor. This unit, which has two main functions within the cycle, is frequently classified as the *heart* of the system, for it circulates the refrigerant through the system. The functions it performs are:

(a) receiving or removing the refrigerant vapor from the evaporator, so that desired pressure and temperature can be maintained.

(b) increasing the pressure of the refrigerant vapor through the process of compression, and simultaneously increasing the temperature of the vapor so that it will give up its heat to the condenser cooling medium.

Compressors are usually classified into three major types: reciprocating, rotary, and centrifugal.

The *reciprocating compressor* is used in the majority of domestic, smaller commercial and industrial condensing unit applications. This type of compressor can be further classified according to its construction, according to whether it is open and accessible for service in the field, or fully hermetic, and so not able to be serviced in the field.

Reciprocating compressors vary in size from that required for one cylinder and its operating piston to one large enough for 16 cylinders and pistons. The body of the compressor may be constructed of one or two pieces of cast iron, cast steel, or, in some cases, aluminum. The arrangement of the cylinders may be horizontal, radial, or vertical, and they may be in a straight line or arranged to form a **V** or a **W**.

Figures 6-13 and 6-14 show external views of common types of reciprocating compressors used in commercial applications. As the compressors differ in design and construction, so do the individual components within the compressors. But their main goal remains the same—the compression of the refrigerant vapor to a high temperature and high pressure, so that its heat content can be reduced and it will condense into a liquid to be used over again in the cycle.

Pistons within the compressors may have the suction valve located in the top of the piston; this is classified as a *valve-in-head* type, or the piston may have a solid head, with the suction and discharge valves located in a valve plate or cylinder head. A typical valve plate, showing the suction and discharge internal valves of a two-cylinder reciprocating compressor is shown in Fig. 6-15.

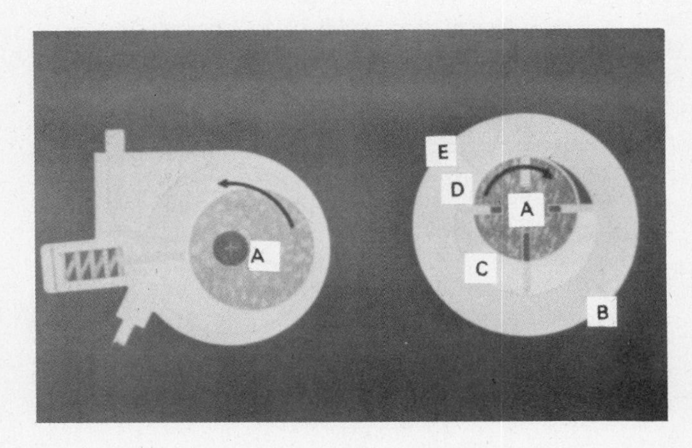

FIGURE 6-14 Typical reciprocating compressor (Courtesy, Carrier Air-Conditioning Company).

Figure 6-16 presents sketches of a compressor piston and the internal suction and discharge valves in different stages of the compression cycle.

Figure 6-17 shows an assembly consisting of the piston, wrist pin, connecting rod, and crankshaft. All components of the reciprocating piston arrangement are finely machined, balanced carefully to eliminate vibration, and fitted with close tolerances to assure that the compressor will have a high efficiency in pumping the refrigerant vapor. A different type of crankshaft, one of an eccentric design, is shown in Fig. 6-18. The connecting rod is assembled on an off-center eccentric fastened with balance weights. If

FIGURE 6-13 Typical reciprocating compressor (Courtesy, York, Div. of Borg-Warner Corporation).

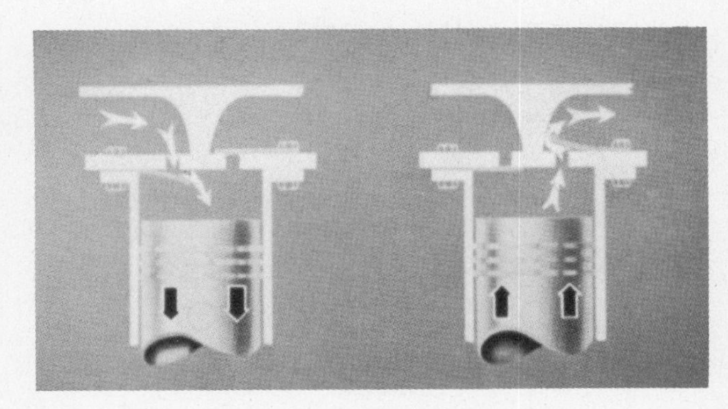

FIGURE 6-15 Gas flow reed valves (Courtesy, Carrier Air-Conditioning Company).

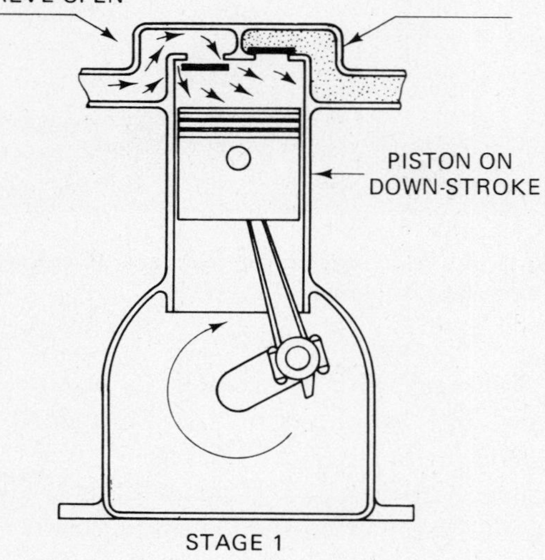

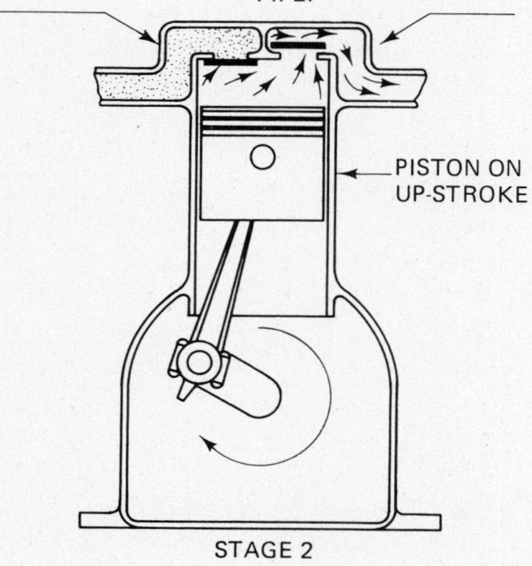

FIGURE 6-16 How differential pressures work the valves of the reciprocating compressor.

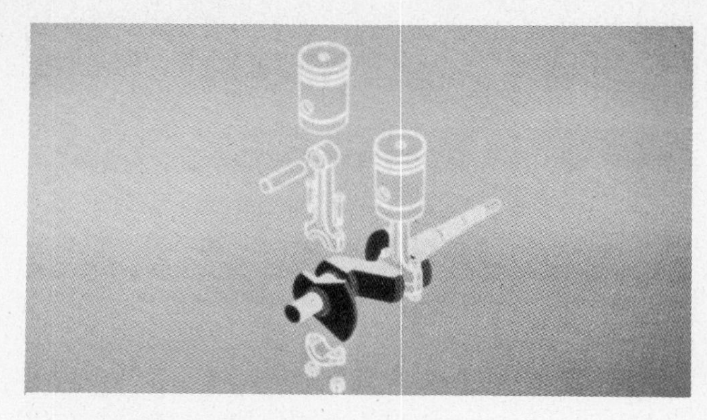

FIGURE 6-17 Crank type assembly (Courtesy, Carrier Air-Conditioning Company).

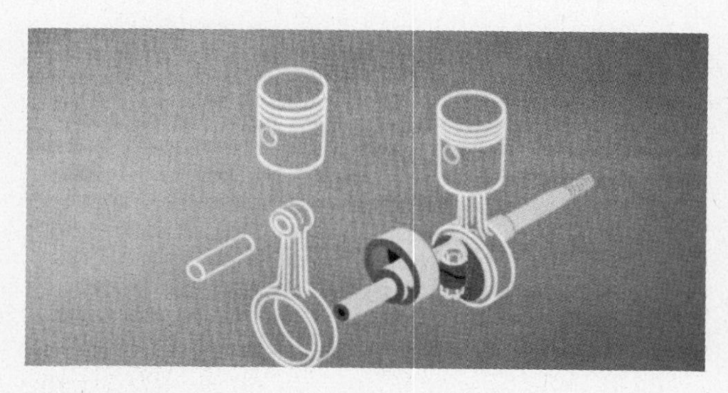

FIGURE 6-18 Eccentric type crankshaft (Courtesy, Carrier Air-Conditioning Company).

the crankshaft is not almost completely machined, it should be dynamically balanced.

The internal valves of a compressor receive quite a bit of wear and tear in normal operation, since they must open and close hundreds of times each minute the compressor is running. Small commercial units usually have a high-grade steel disc or reed type of valve, both of which are quieter operating, efficient, simpler in construction, and longer lasting than the nonflexing ring-plate type of valve. Figure 6-19 shows some of the various designs of internal compressor valves. The proper operation of the valves is very important to the overall efficiency of the compressor.

If the suction valves do not seat properly and allow refrigerant vapor to escape from the cylinder, the piston can't pump out all of the compressed vapor into the hot gas line. If the suction valve leaks, the compressed vapor, or part of it, will go into the suction line and heat up the low-pressure, low-temperature vapor there. If the discharge valve leaks, some of the high-pressure, high-temperature vapor

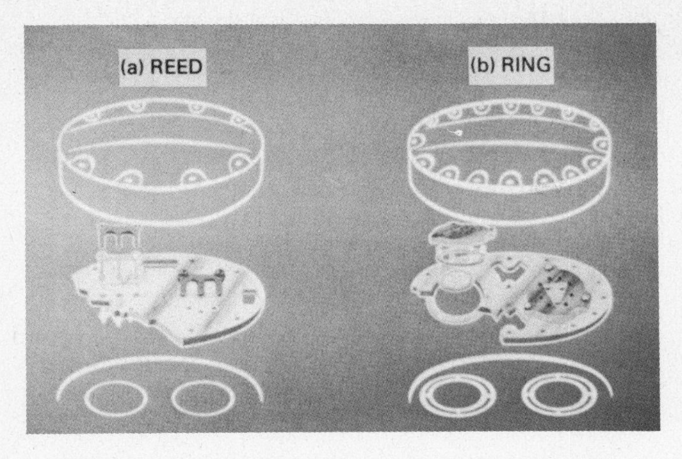

FIGURE 6-19 **Types of valves (Courtesy, Carrier Air-Conditioning Company).**

in the hot gas line will leak back into the cylinder on the down stroke of the piston, limiting the volume of suction vapor entering the cylinder.

In an open-type compressor, one end of the crankshaft extends through the crankcase housing for connection directly to an outside drive motor, or it may have a pulley attached for belt drive by an external motor. Some provision must be made to prevent the leakage of gas and oil around the crankshaft where it extends through the compressor shell; this is accomplished through the addition of a shaft seal.

One type of shaft seal is pictured in Fig. 6-20. The type of crankshaft shown has a seal shoulder built into it, against which a neoprene washer and self-lubricating seal ring are held stationary by the seal cover plate. A gas- and oil-tight seal is maintained between the seal ring and the crankshaft shoulder seal by the neoprene washer, which fits tightly on the

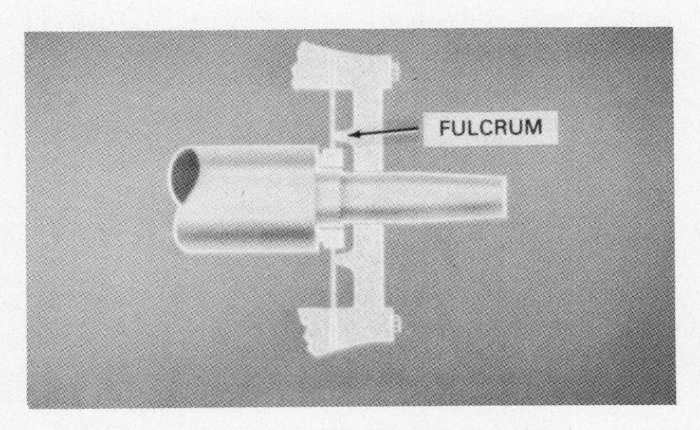

FIGURE 6-20 **Crankshaft seal (diaphragm type) (Courtesy, Carrier Air-Conditioning Company).**

shaft. Seals on reciprocating compressors are on the low-pressure or suction side. It is desirable that as nearly perfect a seal as possible be maintained, since if conditions require that the low-pressure side of the system operate in a vacuum, a leak at the seal or elsewhere in the low side would draw air and moisture into the system.

In most reciprocating compressors gaskets are utilized between mating parts to assure leak-proof conditions, because most surfaces are not so finely machined as to provide metal-to-metal leak-proof joints. Primarily, gaskets are used between the cylinder head and the valve plate, between the valve plate and the compressor housing, between the compressor body and the bottom plate (if there is one), and also between the exterior service valves and their mounting bases.

When the mating parts are tightly secured, they impress their form and outline on the gasket material, which is usually soft and resilient enough to take the impression and thus seal off any gas or oil from possibly leaking to the atmosphere and prevent taking in air. The gasket material must be such that there will be no chemical reaction when it comes in contact with the oil and refrigerant in the system. When gaskets need replacing after some component has been removed and possibly replaced, replacement gaskets should be of the same material originally used by the manufacturer and of the same thickness as those removed, whether they are of aluminum, cork, rubber, asbestos, or composition. A variation in thickness will affect the efficiency and operation of the compressor. Too thick a gasket between the compressor housing and the valve plate will increase the clearance space above the piston and cause a loss in the volumetric efficiency. Too thin a gasket may permit the piston to hit against the valve plate, damaging the compressor.

As already mentioned, reciprocating compressors of the open type need externally driven motors, which may be connected directly through the use of couplings. This causes the compressors to operate at the same speed as the driving motors. Or a compressor may have a flywheel on the end of the crankshaft, which is turned by means of one or more V-belts between the flywheel and a pulley mounted on the motor shaft. The speed at which the compressor will turn depends on the ratio of the diameters of the flywheel and the motor pulley. The speed of the motor may be calculated as shown below:

$$\text{Compressor RPM} = \frac{\text{motor RPM} \times \text{pulley diameter}}{\text{flywheel diameter}}$$

EXAMPLE:

At what speed will a compressor turn if it has a 10-in. flywheel and is driven by a 1,725-RPM motor with a pulley of 4-in. in diameter?

$$\text{Compressor RPM} = \frac{1{,}725 \times 4}{10} = \mathbf{690\ RPM}$$

The purpose of the hermetic compressor is the same as that of the open compressor, to pump and compress the vapor, but it differs in construction in that the motor is sealed in the same housing as the compressor. A typical fully hermetic compressor cutaway is shown in Fig. 6-21. Note the vertical crankshaft, with the connecting rod and piston in a horizontal position. The fully hermetic unit has an advantage in that there is no projecting crankshaft; therefore no seal is necessary, and there is no possibility of leakage of refrigerant from the compressor or of air being drawn in when the system is operating in a vacuum. A compressor of this design cannot be serviced in the field; internal repairs must be made at a region or area repair station or at the factory where it was manufactured.

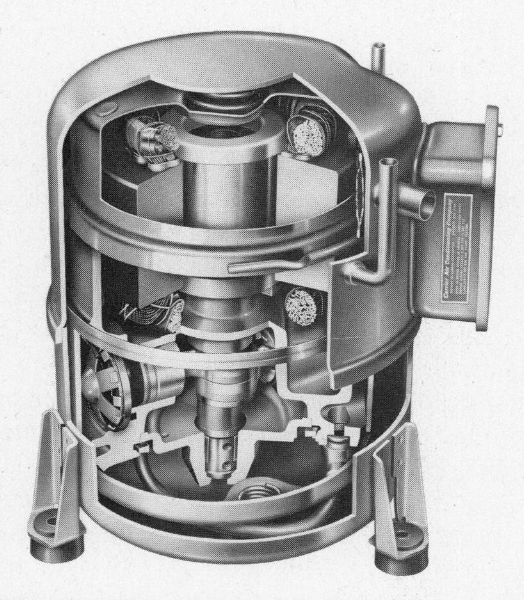

FIGURE 6-21 Typical reciprocating compressor (Courtesy, Carrier Air-Conditioning Company).

Some hermetic compressors are constructed with internally-mounted springs to absorb vibration caused by the pulsation of the refrigerant vapor being pumped by the pistons. Some hermetic compressors also have springs or hard-rubber vibration mounts located on the outside to absorb shock and vibration.

The bottom portion of the hermetic compressor acts as an oil sump, like the crankcase of an open-type compressor. As the oil circulates and lubricates the internal moving parts, it picks up some of the compressor heat caused by friction of the moving parts. The oil transfers some of this heat to the external shell of the compressor.

Most hermetic compressors are constructed so that the suction vapor is drawn across the motor windings before it is taken into the cylinder or cylinders. This, of course, helps to remove some of the heat from the motor windings and also helps to evaporate any liquid refrigerant that may have entered the compressor.

Suction and discharge mufflers are built into some of the smaller-sized hermetic compressors, to absorb or lessen the sound caused by the pulsing vapor as it is pumped through the compressor. A suction muffler can be seen in the cutaway view shown in Fig. 6-21.

Figure 6-22 shows a typical external-type discharge muffler arrangement used on some compres-

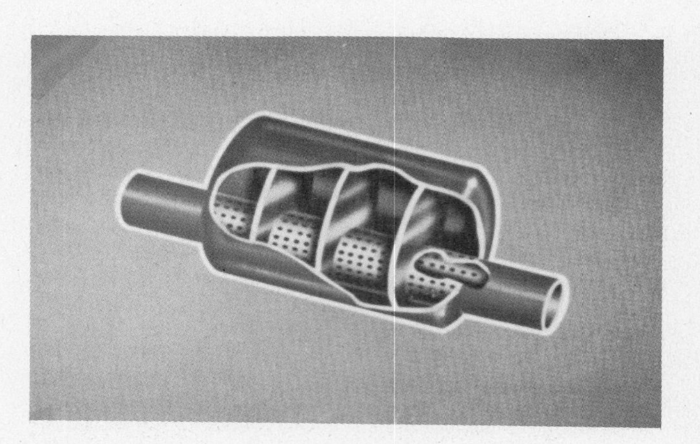

FIGURE 6-22 External discharge muffler (Courtesy, Carrier Air-Conditioning Company).

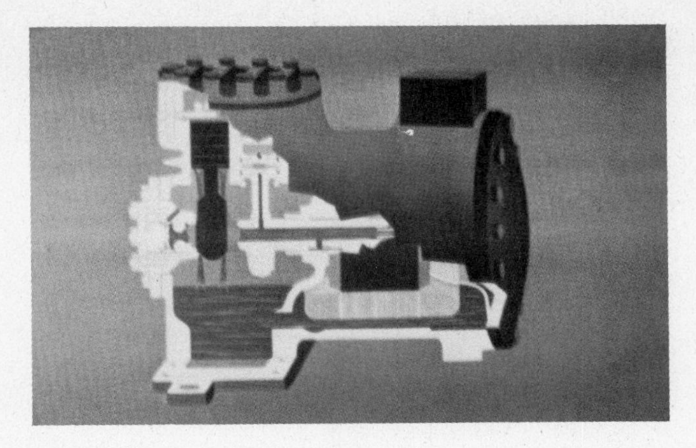

FIGURE 6-23 Serviceable hermetic compressor (Courtesy, Carrier Air-Conditioning Company).

sors. Since the crankshaft and motor shaft are the same unit, high-speed operation causes considerably more noise than the slower operating open compressors, thus requiring mufflers.

Another type of compressor is shown in Fig. 6-23. It combines the motor in the same shell as the compressor, but, unlike the fully hermetic unit, this type provides access to the compressor for repair in the field. This unit is called by several names such as *semi-hermetic*, *accessible*, and *serviceable-hermetic*.

Rotary compressors are so classified because they operate through application of a rotary, or circular, motion, instead of the reciprocating operation previously described. A rotary compressor is a positive displacement unit, and usually can be used to pump a deeper vacuum than a reciprocating compressor.

There are two primary types of rotary compressors used in the refrigeration field: the *rolling piston* type, with a stationary blade, and the *rotating blades* or *vanes* type. Both are similar in capacity, variety of applications, physical size, and stability, but they differ in manner of operation.

The rolling piston type, as shown in Fig. 6-24, has the roller mounted on an eccentric shaft. The blade is located in a keeper in the housing of the compressor. As the rolling piston rotates, vapor is drawn into the space ahead of the spring-loaded blade, as shown, and is compressed by the roller into a continually smaller space until it is forced out the discharge port, and the compression cycle begins again. The roller does not make a metal-to-metal contact with the cylinder, because a film of oil, in normal operation, provides a clearance between the two surfaces.

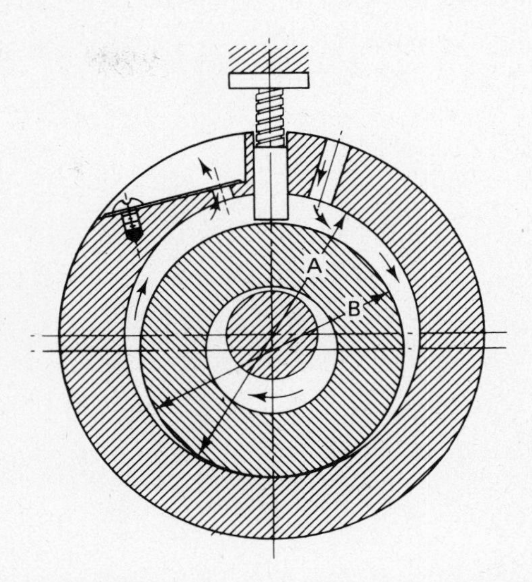

FIGURE 6-24 Rolling piston type rotary compressor (Courtesy, ASHRAE).

Figure 6-25 shows the other primary type of rotary compressor. This unit consists of a cylinder and an eccentric roller or rotor having several blades, held in place by either springs or centrifugal force. As the roller turns in the cylinder, suction vapor is trapped in the crescent-shaped space between two of the blades. As the roller continues to turn, the suction gas is compressed in volume, and its pressure and temperature are increased until it is discharged from the cylinder.

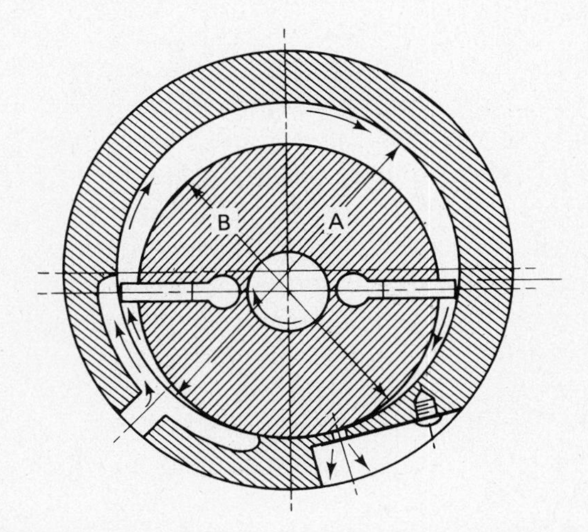

FIGURE 6-25 Rotating vane type rotary compressor (Courtesy, ASHRAE).

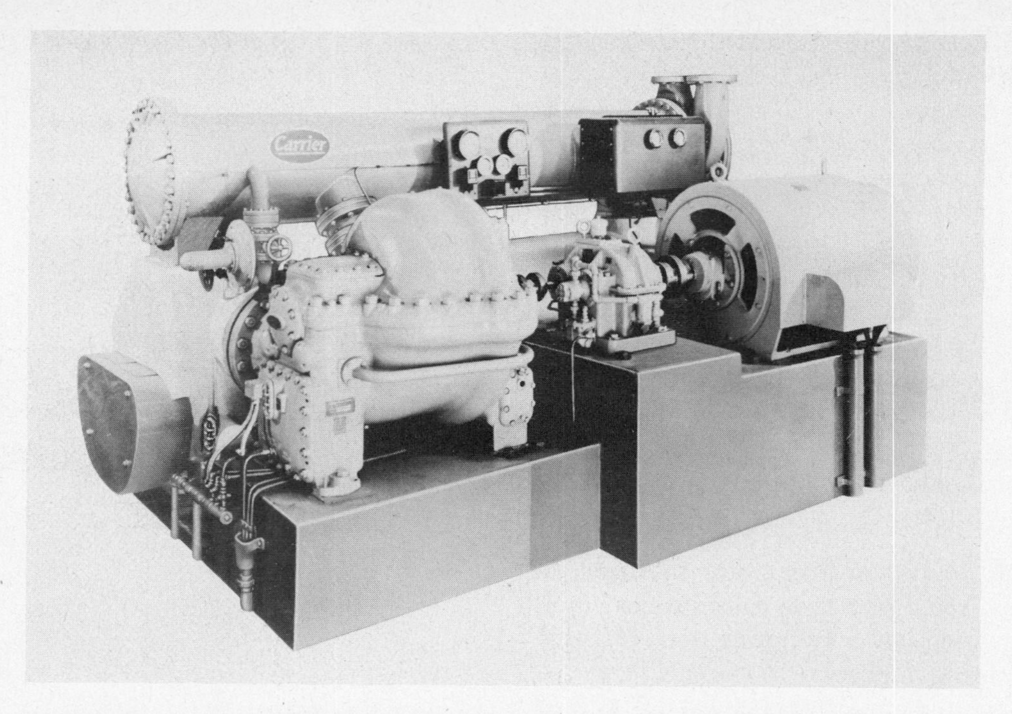

FIGURE 6-26 Complete centrifugal refrigeration unit (Courtesy, Carrier Air-Conditioning Company).

As mentioned earlier, a film of oil prevents leakage of vapor from the cylinder, or between the spaces separated by the blades, while the unit is in operation. To prevent hot gas from leaking back into the cylinder from the discharge port when it is uncovered by the roller, a check valve is usually placed in the discharge line. During the off-cycle or period of shutdown, warm vapor is prevented from leaking back into the evaporator by the check valve.

Rotary compressors are well-balanced units, and those enclosed in a hermetic shell are usually spring-supported or mounted. Usually they are very quiet in operation. Since hermetic rotary compressors are direct-driven, they operate at the motor speed—usually 3,450 rpm—and, although the sound level is in direct ratio to speed and horsepower, they operate comparatively quietly by any standards. Discharge mufflers are widely used to prevent the pulsations of gas being discharged from causing vibrations to be carried over to the discharge line and condenser. The principles of design and operation given here pertain to household and small commercial units, although, in general, the same principles also apply to larger rotary compressors, some of which have primary usage in the low-temperature field where they are used as low-stage or booster compressors.

A complete centrifugal refrigeration unit is shown in Fig. 6-26. Like the other types of compressors, it compresses the refrigeration vapor, as its name implies, through centrifugal action or force. This action is performed mainly by the impeller or rotor, as shown in Fig. 6-27. Vapor is drawn in at the intake near the shaft of the rotor and discharged from the exhaust openings at the outer edge of the rotor. With the rotation of the impeller, suction vapor is drawn rapidly into the impeller chambers, where it is forced to the outside of the housing sections through cen-

FIGURE 6-27 Impeller or rotor (Courtesy, Carrier Air-Conditioning Company).

trifugal action. To maintain the centrifugal force, the impeller is operated at a high rate of speed by an outside driving force, such as an electric motor, gasoline engine, or steam turbine. The pressure differential between inlet and outlet vapor is small. Therefore it is not a positive displacement unit like types previously described, and it is not capable of building up pressure against a closed valve in the system.

A centrifugal compressor may have one or more impellers. Compressors with several stages are constructed so that the discharge of one impeller or stage enters the suction inlet of the next. If the speed of the driving motor does not provide the desired operating speed of the compressor, speed-increasing gears or fluid couplings may be used to obtain optimum operating conditions. Since there are no pistons or internal suction and discharge valves, there is little wear and tear on the unit. The main bearings in the housing, supporting the drive shaft, are the components most subject to wear.

CONDENSERS

The next major component in the refrigeration system, following the compression stage, is the condenser. Basically, the condenser is another heat-exchange unit in which the heat picked up by the refrigerant in the evaporator—as well as that added to the vapor in the compression phase—is dissipated to some condensing medium. High-pressure, high-temperature vapor leaving the compressor is superheated, and this superheat customarily is removed in the hot-gas discharge line and in the first portion of the condenser. As the temperature of the refrigerant is lowered to its saturation point, the vapor condenses into a liquid for reuse in the cycle.

Condensers may be air-cooled, water-cooled, or cooled by evaporation. Domestic refrigerators usually have an air-cooled condenser, which depends upon the gravity flow of air circulated over it. Other air-cooled units use fans to blow or draw large volumes of air across the condenser coils.

Figure 6-28 depicts a typical small commercial condensing unit using an air-cooled condenser. It is dependent upon an ample supply of relatively "cool" air, for, in order to have a heat transfer from the refrigerant in the condenser to the coolant, the air must be at a lower temperature than the refrigerant.

FIGURE 6-28 **Small air cooled condensing unit (Courtesy, Copeland Corporation).**

Even when the surrounding temperature is above 100°F, the air is still cooler than the refrigerant in the condenser, which must give up some heat to return to its liquid state.

Air-cooled condensers are constructed somewhat like other types of heat exchangers, with coils of copper or aluminum tubing equipped with fins. Evaporators usually have filters in front to reduce clogging by dust, lint, and other matter, but condensers are not so equipped, and so must be cleaned frequently to prevent reduction of their capacities.

Remote air-cooled condensers usually have wider fin-spacing to prevent clogging as quickly as in those directly mounted on the condensing unit. Also, they can be located away from the compressor, which is a distinct advantage. Occasionally a complete condensing unit is placed somewhere inside the building in which it is to be used, where the heat dissipated from the condenser and motor can cause an increase in temperature within the storage or mechanical equipment room. As a result, the unit might have a higher operating discharge temperature and pressure, which would decrease its efficiency.

Figure 6-29 shows a remote air-cooled condenser, which may be located outdoors—beside a building, or on a flat roof. In such an open, outdoor location, an adequate supply of air as a coolant is readily available at the ambient outdoor temperature, thus

FIGURE 6-29 Remote air cooled condenser (Courtesy, Carrier Air-Conditioning Company).

avoiding undesirable temperatures in the building. The air movement across the coil is oreated either by a belt-driven centrifugal fan or a direct-drive propeller-type fan. The slow-speed wide-blade propeller fan moves the required volume of air without creating unreasonable noise.

This type of condenser may be assembled in any combination of units that may be required for the necessary heat removal. The air may either be drawn through or blown through the coils. In another design, a single condenser may have more than one circuit in its coil arrangement, so that it may be used with several separate evaporators and compressors.

In most installations of remote air-cooled condensers in this country, the difference between the ambient air temperature and the condensing temperature of the refrigerant is approximately 30°F. Therefore, if the outdoor temperature is 95°F, the refrigerant will condense at approximately 125°F.

Some difficulty may arise with remote air-cooled condensers when they are operated in low ambient temperatures, unless proper precautions are taken to maintain head pressures that are normal for the unit. Earlier in this chapter it was stated that a too-high condensing temperature and pressure lessens the overall operating efficiency of the unit. Conversely, a too-low condensing temperature and pressure will affect the efficient operation of the system by causing a reduction in pressure difference across the metering device, thus resulting in a loss of refrigerant flow into the cooling coil. Later in this chapter the effect of pressure drop through a metering device on the overall capacity and efficiency of a system will be discussed.

Some remote air-cooled condensers equipped with multiple fans have controls for the cycling of one or more of the fans during periods of low ambient temperatures. The flow of air across other types of condensers may be controlled by adjustable louvers. Still other manufacturers install controls to allow partial flooding of the condenser with liquid which, in turn, will lessen the condensing capacity. This is another means of keeping the head pressure within allowable limits.

Water-cooled condensers permit lower condensing temperatures and pressures, and also afford better control of the head pressure of the operating units. They may be classified as:

(1) shell-and-tube,

(2) shell-and-coil, and

(3) tube-in-a-tube types.

As pointed out in an earlier chapter, water usually is an efficient medium for transferring heat, since the specific heat of water is one Btu per pound per °F in temperature. If 25 lb of water increases 20°F, and if this heat is removed in one minute, 500 Btu are removed from the source of heat each minute. Also, if this rate of heat transfer continues for one hour (60 minutes), this means that the water is absorbing 30,000 Btu per hour.

If the above example involved a water-cooled condenser, it would mean that three gallons of water were being circulated per minute. Also, if the heat of compression amounted to 6,000 Btu per hour, it would mean that the evaporator load was 24,000 Btu per hour—a two-ton refrigeration load.

Water coming from a well or other undergound source will be quite a bit cooler than the ambient outdoor air. If cooling-tower water is used, its temperature can be lowered in the cooling tower, after it has picked up heat in the condenser, to within 5° to 8°F of the outdoor wet-bulb temperature. The use of a cooling tower and circulating pump permits the reuse of the water, except for a slight loss due to evaporation, and keeps the consumption and cost of water to a minimum.

The shell-and-tube type of water-cooled condenser consists of a cylindrical steel shell containing several copper tubes running parallel with the shell. Water is pumped through the tubes by means of the inlet and outlet connections on the end plates. The hot refrigerant vapor enters the shell at the top of the condenser, as shown in Fig. 6-30, and the liquid refrigerant flows as it is needed from the outlet at the bottom of this combination condenser-receiver.

The end plates are bolted to the shell of the condenser for easy removal to permit the rodding or cleaning of the water tubes of minerals that may be deposited on the inside of the tubes, causing a restricted water flow, a reduction in the rate of heat transfer, or both. A control of the water flow, that is, the number of times it travels the length of the condenser or the number of passes it makes, is built into the end plates of the condenser. If the water enters one end plate, passes through all the tubes once, and leaves the condenser at the other end plate, it is called a *one-pass condenser*. If the water inlet and outlet are both in the same end plate, it is a *two-pass* or some other even-numbered type of pass condenser.

If, instead of a number of tubes within the condenser shell, there are one or more continuous or assembled coils through which water flows to remove heat from the condensing vapor, it is classified as a *shell-and-coil* type of condenser. Figure 6-31 shows a condenser of this type. It is a compact unit and usually serves as a combination condenser-receiver within the circuit. Usually this type of condenser is used only on small-capacity units and when there is an assurance of reasonably clean water, for the only means of cleaning it is by flushing with a chemical cleaner.

The *tube-in-a-tube*, or *double-tube* as it is also known, may be classified as a combination air-and-water-cooled type of condenser. As pictured in Fig. 6-32, it has the refrigerant flowing through the outer tubing where it is exposed to the cooling effect of air flowing naturally over the outside of the outer tubes while water is being circulated through the inner tubes. Generally, water enters the bottom tubes of water-cooled condensers and leaves at the top. In this manner peak efficiency is obtained, for the coolest water is capable of removing some heat from

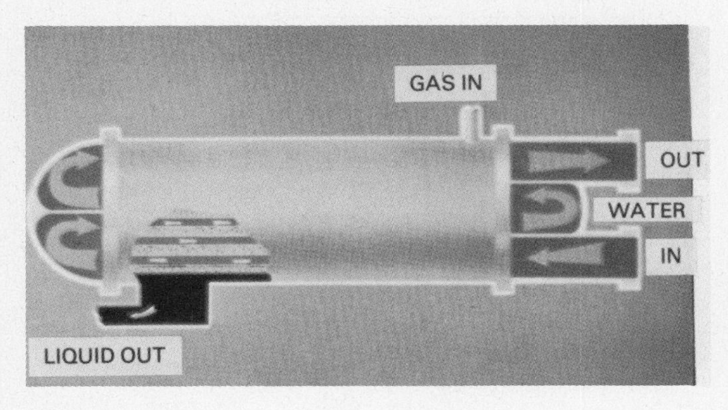

FIGURE 6-30 Shell and tube (Courtesy, Carrier Air-Conditioning Company).

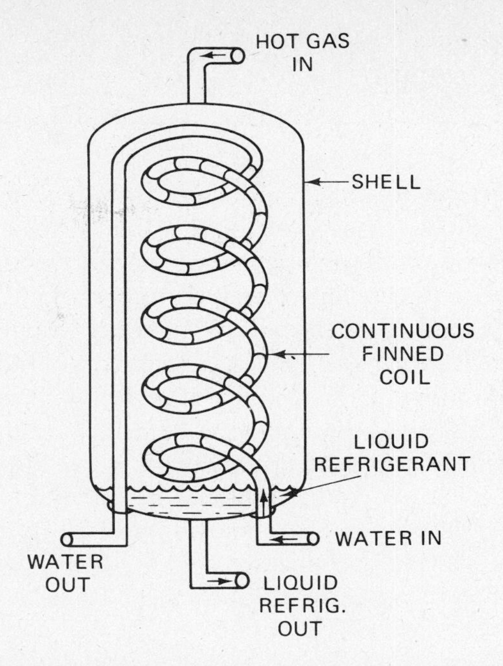

FIGURE 6-31 Shell and coil water condenser (Courtesy, Carrier Air-Conditioning Company).

FIGURE 6-32 Water cooled tube-in-tube condenser (Courtesy, Halstead & Mitchell, Division of Halstead Industries).

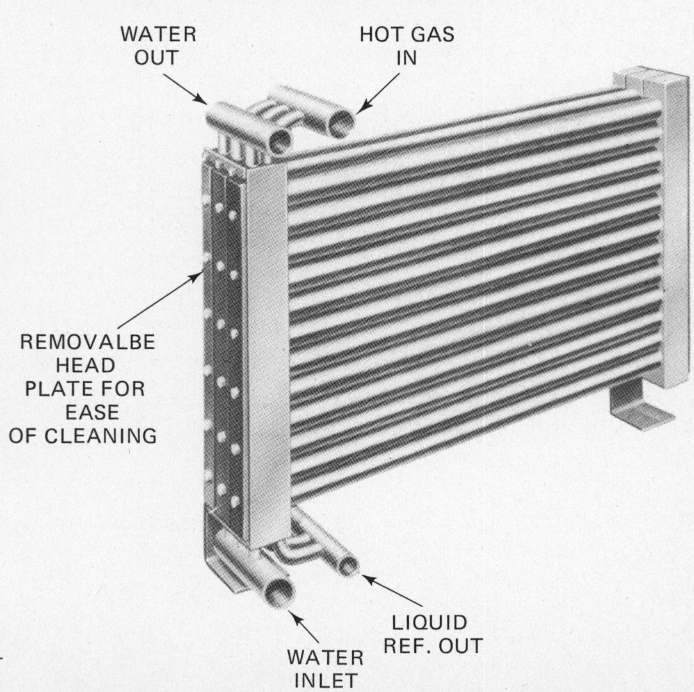

the refrigerant in a liquid state, thereby subcooling it. Then the warmer water still is able to absorb heat from the vapor, assisting in the condensation process.

When the ambient temperature is such that a satisfactory condensing temperature cannot be obtained with an air-cooled condenser, and when the water supply is inadequate for heavy usage, an *evaporative condenser* may be used to advantage. A diagram of this type of condenser is shown in Fig. 6-33, which shows the combined use of air and water for the purpose of heat removal from the refrigerant vapor within the condenser coil.

There is actually a double heat transfer in this unit: The heat of the vapor and the coil containing it is transferred to the water, wetting the outer surface of the coil, and then is transferred to the air as the water evaporates. The air can either be forced or drawn through the spray water.

When the air is blown through the unit, the fan and motor are in the dry entering air stream. When the system has a draw-through fan, it is essential that eliminators be installed before the air mover. Otherwise, there would quickly be an accumulation of scale on all of the air-mover components. Even with blow-through units, there is a possibility of some of the spray water blowing out of the evaporative condenser, and a set of eliminator plates should be installed to prevent this from occurring.

RECEIVERS

As mentioned earlier, some water-cooled shell-and-tube condensers also act as receivers, with liquid refrigerant occupying the space in the bottom of the condenser where there are no water tubes. If there is too much liquid in this type of condenser-receiver, some of the water tubes may be covered by the liquid level. This reduces the area of heat transfer surface in the condenser.

In systems other than those having condenser-receivers and those operating with a critical refrigerant charge, a receiver is needed, which is actually a storage container for the refrigerant not in circulation within the system. Receivers that are part of small, self-contained commercial units usually are

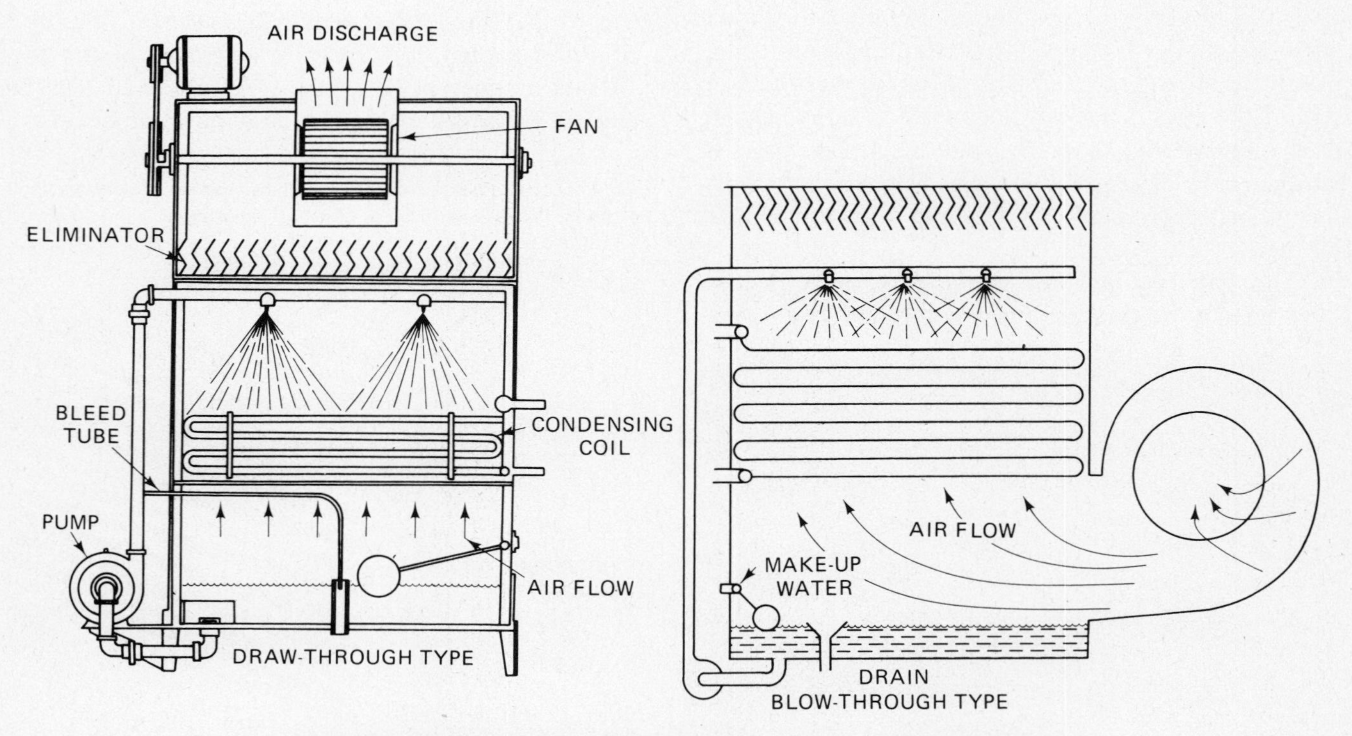

FIGURE 6-33 Functional view of evaporative condensers (Courtesy, ASHRAE).

large enough to hold the complete operating charge of refrigerant in the systems. This applies to a number of larger systems as well. Yet, in some cases, the receiver may not be large enough to hold the entire refrigerant charge if a pump-down becomes necessary for repair or replacement of a component. An auxiliary receiver would be necessary to provide pump-down capacity. If this is not provided, the surplus refrigerant would have to be pumped into an empty refrigerant drum or wasted to the atmosphere.

Precautionary measures are usually taken by manufacturers of receivers against the possibility of too much pressure or too high a temperature build-up in the receiver or in a combination condenser-receiver. These safety measures usually include the installation of pressure relief valves, customarily spring-loaded, that will open if excessive pressure should build up within the receiver. A fusable-plug type of relief valve may be installed, which is designed to melt at a preselected temperature and thus release the refrigerant if, for any reason whatsoever, that temperature is reached within the receiver.

REFRIGERANT FLOW CONTROLS

A fundamental and indispensable component of any refrigeration system is the *flow control*, or metering device. Its main purposes are:

(a) to permit the flow of refrigerant into the evaporator at the rate needed to remove the heat of the load

(b) to maintain the proper pressure differential between the high and low sides of a refrigeration system.

The metering device is one of the dividing points in the system.

The principal means of refrigerant flow control, in the early stages of refrigeration, was a basic hand valve. Knowing their work and their equipment, early operators of units in ice plants and similar operations having constant loads knew how far to open the hand valve for the work to be performed. However, in modern applications that have frequently varying loads, this is impractical because the hand-valve setting would have to be changed as the load changed.

The five main types of metering devices now used in various phases of refrigeration are:

(a) automatic expansion valve

(b) thermostatic expansion valve

(c) capillary tube

(d) low-side float

(e) high-side float

All are used to reduce the liquid refrigerant pressure.

The hand valve is obviously not suited for automatic operation, since any variation in requirements needs manual adjustment, so the automatic expansion valve came into being.

The *automatic expansion* or *constant-pressure valve*, shown in Fig. 6-34, maintains a constant pressure in the cooling coil while the compressor is in operation. In this diaphragm type of constant pressure expansion valve, the pressure in the evaporator effects the movement of the diaphragm, to which the needle assembly is attached.

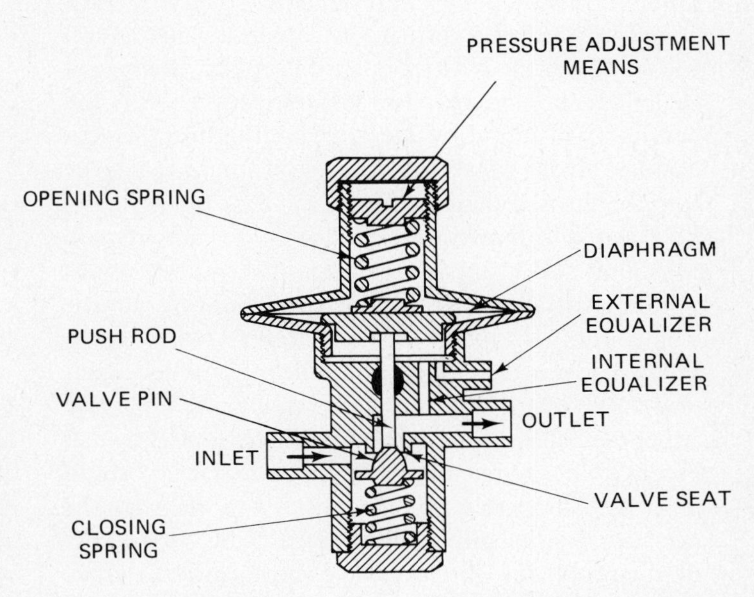

FIGURE 6-34 Constant pressure expansion valve (Courtesy, ASHRAE).

A condition of stability in refrigerant flow and evaporation is necessary for the correct operation of the constant pressure expansion valve. Like the hand valve, its use is limited to conditions of more or less constant loads on the evaporator, a situation that applies to the automatic expansion valve as well. In either valve there is a screw that applies pressure to the spring above the bellows or diaphragm. When the screw is adjusted clockwise, it causes more pressure on the bellows or diaphragm, forcing the valve to open more, admitting additional refrigerant to the evaporator, and resulting in higher operating pressure. If a lower operating pressure in the cooling coil is desired, the screw is turned counterclockwise, releasing pressure on the spring and therefore on the bellows or diaphragm. This allows the valve to close and curtails the flow of refrigerant. Following any adjustment, ample time should be allowed for a controlling device to settle down before any further change is made in its setting. For a given load on the evaporator coil being fed refrigerant, there is only one correct setting of the automatic expansion valve: when the coil is completely frosted. If the pressure is lowered, there will be a curtailment in the refrigerant flow, and the heat absorption capability of the coil will be lessened. If the pressure is raised, there will be an increase in the flow of refrigerant, with the possibility of liquid refrigerant flooding into the suction line, from which the refrigerant might reach the compressor and damage it.

Because all refrigeration loads do not remain constant, and someone cannot always be present at every installation to make compensating adjustments, another type of valve, the *thermostatic expansion valve*, was developed. Like the automatic expansion valve, the thermostatic expansion valve may be either the bellows type or the diaphragm type shown in Fig. 6-35. Both are equipped with a capillary tube and feeler bulb assembly, which transmits to the valve the pressure relationship of the temperature of the suction vapor at the outlet of the evaporator coil, where the feeler bulb is attached.

The basic purpose of the thermostatic expansion valve is to maintain an ample supply of refrigerant in the evaporator, without allowing liquid refrigerant to pass into the suction line and the compres-

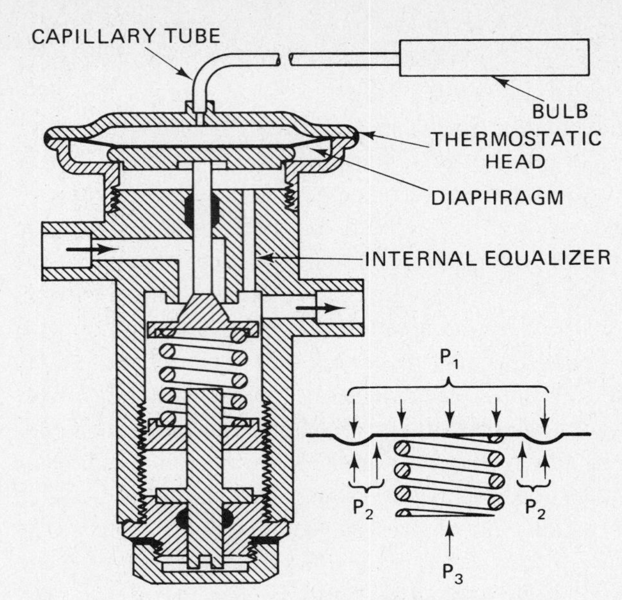

FIGURE 6-35 Diaphragm type thermostatic expansion valve (Courtesy, ASHRAE).

sor. When the metering device is a thermostatic expansion valve, its operation will depend upon superheated vapor leaving the evaporator, since a portion of the evaporator is used for the superheating of the vapor about 5°–10°F above the temperature corresponding to the evaporative pressure.

The capillary tube, which is based on the above principle, is the simplest form of refrigerant control or metering device and generally the least expensive. There are no moving parts to wear out or require replacing, since it is a small-diameter tube of the right length for the refrigeration load it is designed to handle. This metering device, like any other, is located between the condenser and evaporator, at the end of the liquid line or instead of a liquid line. One type of capillary refrigerant control is shown in Fig. 6-36. The advantages of this control have just been discussed; however, its disadvantages are that it is subject to clogging, requires an exact refrigerant charge, and is not as sensitive to load changes as other metering devices. Its internal cross-sectional area is so small that it takes only a minute dirt particle to plug the tube, or a small amount of moisture to freeze in it. A dryer and filter or strainer should be installed at the inlet to the capillary tube to prevent this clogging possibility.

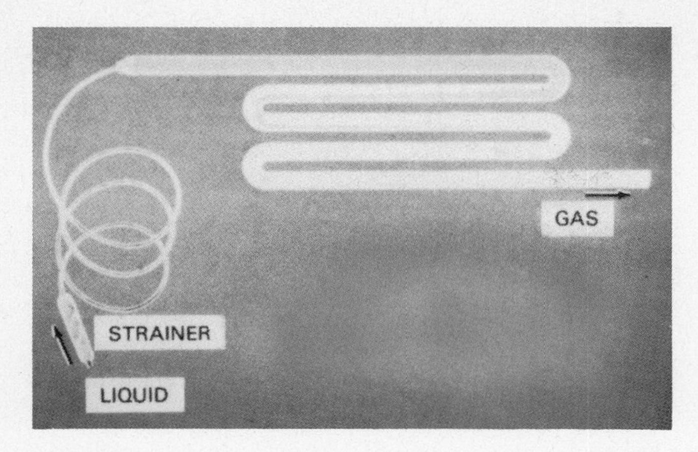

FIGURE 6-36 Capillary tubing (Courtesy, Carrier Air-Conditioning Company).

Another type of refrigerant control device is the *float arrangement*, which also meters the flow of refrigerant into the evaporator. The float itself is made of metal that will not cause a reaction with the refrigerant used in the system. It is constructed in the shape of a ball or an enclosed pan, which will rise or fall within the float chamber with the level of the refrigerant. It is connected through an arm and linkage to a needle valve, which opens and closes against a seat, allowing and curtailing the flow of refrigerant into the chamber.

A *high-side float*, as its name indicates, is located in the high-pressure side of the system. It may be of a vertical or horizontal design and construction, and it may be located near either the condenser or the evaporator. A typical high-side float is shown in Fig. 6-37. Its design is such that, as the float chamber fills with refrigerant, the buoyancy of the float lifts it and raises the valve pin away from the seat. This

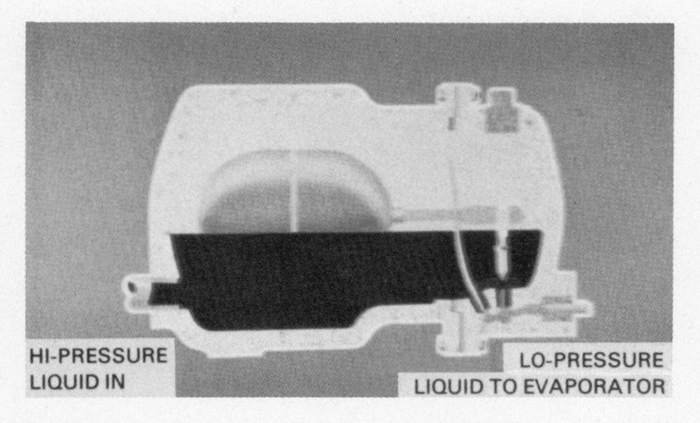

FIGURE 6-37 "Hi-side" float (Courtesy, Carrier Air-Conditioning Company).

permits the refrigerant to flow or be metered to the low-pressure side of the system and into the evaporator.

Since the float, through the pivot arrangement, is set to open at a given level, only a small amount of liquid stays in the high-side float chamber: most of the refrigerant in the system is in the evaporator. Therefore the system refrigerant charge is critical, to the extent that only enough refrigerant to maintain the proper level in the flooded evaporator is desirable, with no liquid flooding over into the suction line and compressor. If there is an overcharge, flooding will result, whereas a shortage of refrigerant will cause the evaporator to be starved and the system will be inefficient.

A *low-side float* metering device is one in which the float is located in the evaporator, or in a chamber adjacent to the cooling coil that is flooded, main-

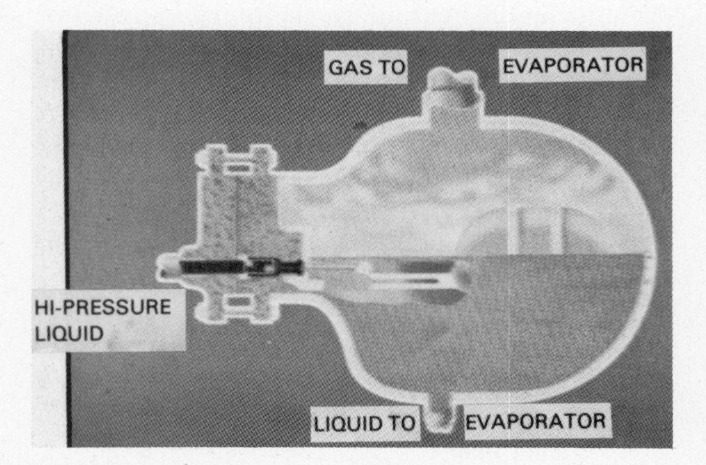

FIGURE 6-38 "Low side" float (Courtesy, Carrier Air-Conditioning Company).

taining a definite liquid level within the evaporator. It is constructed somewhat like the high-side float, with the exception that as the float rises it closes off the flow of the refrigerant. Its action is shown in Fig. 6-38: the high-pressure liquid is at the inlet to the float chamber, and the float assembly itself is in the low-pressure liquid, maintaining a definite level within the float chamber. As the load on the evaporator increases, liquid is evaporated and the liquid level in the evaporator and the float chamber drops.

As the float lowers, the needle is pulled away from the seat, allowing additional refrigerant to enter until the desired level is reached.

If the load on the evaporator decreases, less evaporation will take place, and the liquid level will be maintained, the float causing the needle to close against the seat. In this manner the low-side float assembly can maintain the correct flow of refrigerant as it is needed, by a fluctuating load condition. Usually, the same float arrangement cannot be used if a different refrigerant is desired in the system because of its operating characteristics, for the refrigerants will have different specific gravities as other characteristics change. A float with the correct buoyancy must be used.

A *check valve*, as shown in Fig. 6-39, is sometimes used to prevent the high pressure of one system or evaporator from backing up into a lower-pressure evaporator, when there are multiple coils in a system operating at different pressures. It may also be used to prevent an equalization of pressures when a system is not in operation. This type of valve opens easily if the flow is in the proper direction, but it will shut off the flow in the reverse direction, with the aid of a spring pressure or the weight of the internal valve itself.

A *solenoid*, or *magnetically operated*, *valve*, the operating principles of which will be explained in the chapter on electricity, is frequently used in refrigeration lines. Figure 6-40 shows an exterior view and a cross section of the internal construction of such a

FIGURE 6-39 Check valve (Courtesy, Carrier Air-Conditioning Company).

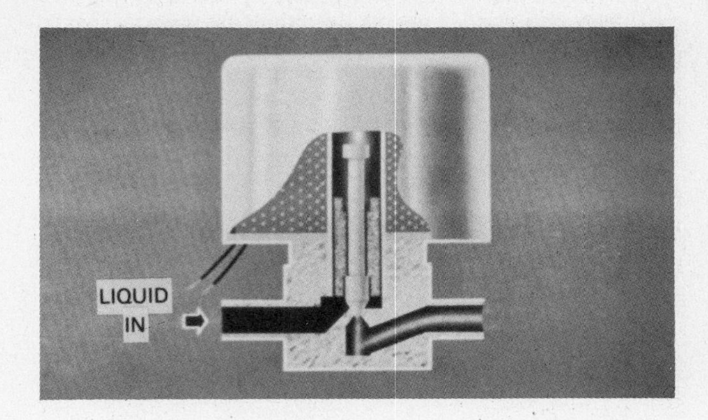

FIGURE 6-40 Solenoid valve (Courtesy, Carrier Air-Conditioning Company).

valve. When a positive shut-off of liquid flow is desired, for a pump-down cycle permitting the compressor to start up unloaded, a solenoid valve is installed in the liquid line ahead of the metering device. Its operation is controlled by a thermostat and an electrical circuit, or by some other means of actuating the valve. Most of these valves close by gravity and the weight of the plunger and valve, when the electrical circuit or other means of holding the valve open is broken.

REFRIGERATION LINES

Previous portions of this chapter have been devoted to the proper selection, balancing, and operation of the major components of a refrigeration system. But, regardless of how well these have been selected and balanced, system operation is dependent upon the means of moving the refrigerant, both as a liquid and a vapor, from one component to another within the refrigeration circuit.

Just as a highway must be constructed, maintained, and kept open between communities to provide adequate access for the vehicles that must use it, so must the piping in a refrigeration system be properly sized and installed so that there will be no restrictions to the flow of the refrigerant.

Refrigerant oil, necessary for the proper lubrication of the moving parts in the compressor and the metering device, must be readily miscible with the refrigerant in the liquid state. The oil will travel with the liquid as long as the liquid line is sized so that the refrigerant will travel through its entire length at a proper velocity.

If the system is self-contained, proper sizing and installation of the refrigerant lines is the manufacturer's respopsibility. But a system built up in the field, using products of several manufacturers, becomes the problem and responsibility of the individual designing the complete system, as well as those who install and connect the components. A liquid line that is sized too small or that has too many restrictions (fittings and bends) could easily cause too great a pressure drop in the line, which could result in a low in the capacity of the metering device when compared to the capacity required in the cooling coil.

Proper sizing of all refrigerant lines and the tables to be used in the selection of liquid, suction, and hot gas lines will be covered in Chap. 19.

Lines through which refrigerant vapor flows are the most critical, and these are the suction and hot gas or discharge lines. The velocity of the vapor should be at least 750 ft per minute in horizontal lines and above 1,500 ft per min in vertical lines, so that the refrigerant oil will be entrained with the vapor and returned to the compressor. If the lines are too large in size, the desired velocity cannot be maintained and the oil may not return, and the compressor may become short of its proper oil charge.

If the evaporator is located above the compressor, the oil usually will return to the compressor through gravity flow, provided no risers or possible traps are built into the line.

If the evaporator is situated below the compressor, or when the condenser is located a distance above the compressor, the vapor must have the proper velocity in order to carry the oil droplets with it. Oil traps may have to be built into the piping, or it may be necessary to use dual-pipes, if the capacity control of the compressor varies because of changing load conditions. These measures insure that, even in the case of minimum load conditions of 10–25% of full capacity, the refrigerant vapor will have adequate velocity to carry the oil along with it.

PROBLEMS

1. What are the most frequently used methods for compensating for conduction loss due to air film around the evaporator coil tubing?

2. Under what circumstances should nonfrosting evaporator coils be installed?

3. What are some types of cooling coils used for cooling liquids?

4. What are the three major classifications (by method of compression) of compressors?

5. What is the difference between a hermetic and an open compressor?

6. What is the basic purpose of the condenser?

7. What are the primary types of condensers?

8. What are the principal types of automatic metering devices to control the flow of refrigerant to the evaporator?

9. Which of the devices in Problem 6.8 is the simplest; why?

10. Why must refrigerant vapor move rapidly through the suction and hot-gas or discharge lines?

7

REFRIGERANTS AND LUBRICATION OF SYSTEMS

Refrigerants are the vital fluids in mechanical refrigeration systems. They absorb heat from a place where it is not wanted and reject it elsewhere. The evaporation of the liquid refrigerant removes heat, which is released by the condensation of the heated vapor.

Any substance that undergoes phase change from liquid to vapor and vice versa may function as the refrigerant in vapor-compression-type systems. However, only those substances that undergo these changes at commercially useful temperature and pressure levels are of practical value.

It can be said that there is no "universal" refrigerant. Since mechanical refrigeration is used over a wide range of the temperatures, some refrigerants are more suitable for high-temperature refrigeration, such as comfort cooling; others operate at lower temperature ranges, such as used for commodity storage, freezing process, and applications requiring even lower temperatures.

The choice of a refrigerant for a particular application frequently depends on properties not related to its ability to remove heat, for example, its toxicity, flammability, density, viscosity, and availability. Thus, the selection of a refrigerant for a particular purpose may be a compromise among conflicting properties.

Many different refrigerants have been used since the early days of refrigeration. Experimentation, research, and testing are still going on with various chemicals, or compounds and mixtures of chemicals. At one time or another air, butane, chloroform, ether, propane, water, and other organic and inorganic compounds have been used.

Ammonia is one of the old standby refrigerants. It was used in some of the earliest equipment, and its use continues in some of the larger commercial and industrial units. Ammonia requires the use of much heavier equipment than that needed by some of the other refrigerants in use today. Sulphur dioxide and methyl chloride were once widely used in household refrigerators and in small commercial condensing units. Because of toxicity and flammability of these substances, most systems using these refrigerants have been replaced, or are being replaced as they wear out.

Sulphur dioxide was, and still is, a fairly stable

refrigerant, and is nonexplosive and nonflammable. It has an extremely irritating and unpleasant odor, which, should a leak occur in the system, is easily discernible and readily located. Because of its toxicity however, even a slight leak can be hazardous to anyone in the vicinity of the equipment, and the fumes have been known to injure shrubbery and other plants.

Methyl chloride gives off a sweet-smelling vapor when it is exposed to the atmosphere, quite different from that of sulphur dioxide, but it is also toxic. The inhalation of too much of the vapor results in syptoms similar to those caused by an anesthetic. Although it is not considered harmful to flowers or other plants, it is a moderately flammable refrigerant. According to the Underwriters Laboratory Report MH-2375, its explosive limits in air are 8.1 to 17.2% by volume.

With the discovery of new chemicals and compounds, the practical advantages and disadvantages of each of the refrigerants in use were carefully appraised. Consideration is now given to the characteristics of the various refrigerants, both from a chemical and a physical standpoint. The availability and cost of each refrigerant is also of great importance. The characteristics and properties of several refrigerants will be found in the Appendix, under the *Refrigerant Table and Charts.* This data is presented by permission of the Freon Products Division, E. I. du Pont de Nemours & Co., Inc. In this chapter, pertinent and selected data of the refrigerants R-12, R-22, and R-502 only will be shown in charts and tables, since they are the most commonly used refrigerants.

The now universally used numbering system for refrigerants was developed by the Du Pont Company, the first to market many of the new breed of refrigerants. At the outset the letter *F* (for *Freon*, a registered Du Pont trademark) preceded the numbers. Later, as other producers of halocarbon refrigerants came into the picture, Du Pont made its numbering system available to the entire industry. As a result refrigerants now are known by *R* or *refrigerant* numbers, such as R-12, R-22, etc.

The numbering system is described in ASHRAE (American Society of Heating, Refrigerating, and Air-Conditioning Engineers) Standard 34-67, which has been adopted by the American National Standards Institute as ANSI Standard B79-67.

For the purposes of this chapter, only that portion of the system covering the most commonly-used refrigerants (designated in the text above) will be used. Students seeking further information on different types of refrigerants may refer to the standard cited above.

In the selection and use of a refrigerant in a specific and specialized project, the following characteristics must be taken into consideration:

(1) Chemical properties

- (a) flammability
- (b) explosiveness
- (c) toxicity
- (d) stability

(2) Physical properties

- (a) boiling point
- (b) freezing point
- (c) specific volume
- (d) density
- (e) critical pressure
- (f) critical temperature
- (g) latent heat content
- (h) oil miscibility
- (i) leak detection

FLAMMABILITY AND EXPLOSIVENESS

When a system is operating satisfactorily, there is no need to worry about the refrigerant within it. However, should a fire due to an outside source occur in the vicinity of any refrigeration component, there is the danger that the fire might spread, and the possibility of an explosion exists if a questionable refrigerant escapes from ruptured lines or tubes.

If a leak occurs, or if repairs must be made to some component, the flammability and possible explosiveness and toxicity of the refrigerant must be taken into consideration. Even though the system may be evacuated and all of the refrigerant contained in the receiver or condenser-receiver, the brazing, soldering or welding that may be necessary in the repair or replacement of a component may present hazards because enough vapor may remain in the component or connecting lines to cause danger from the heat or flame of the repair equipment.

The hydrocarbon refrigerants, among which are butane (R-600), ethane (R-170), and propane (R-290), are all highly flammable and explosive. Proper care must be taken with an open flame around a system using any of these refrigerants. The halocarbon refrigerants are considered to be nonflammable, but some become toxic when exposed to flame. Among these are trichlorofluoromethane (R-11), dichlorodifluoromethane (R-12), and monochlorodifluoromethane (R-22). R-502, which is an azeotrope refrigerant, is a mixture of 48.8% R-22 and 51.2% R-115 (chloropentafluoroethane), and is nonflammable.

There are some "tongue-twisters" in the chemical names of some refrigerants, and it was a relief when a numbering system was developed, which is more commonly used today instead of the chemical names. The number of the refrigerant is customarily preceded by the trade name of the manufacturer, but in general practice the number is preceded by the term *Refrigerant* or just *R*.

TOXICITY

The Underwriters Laboratories has made extensive tests of the toxicity of all refrigerants. Humans were not used in these tests; guinea pigs usually quickly show the effects of inhaling toxic gases and vapors.

Air is the exception to the rule that all gaseous substances are, to some degree, toxic. There are, of course, various degrees of toxicity. Carbon dioxide is exhaled whenever people breathe, and is therefore harmless to humans up to a certain level in the surrounding atmosphere. However, an individual exposed to an atmosphere containing about 8 to 10% carbon dioxide or more would soon become unconscious.

The tests made by Underwriters Laboratories pertain to the type of chemical or compound, the percent of the vapor being tested in the given atmosphere, and the duration of time that the subject is exposed to or breathes the mixture. Underwriters Laboratories has devised a number classification that ranges from 1 to 6, the lowest number designating the most toxic and dangerous refrigerant, and the highest number being the least toxic refrigerant. Table 7A lists the classification of refrigerants and the concentration and length of exposure that lead to serious injury.

TABLE 7A Hazard to life classification of gases and vapors by Underwriters Laboratories.

Group	*Limitations*	*Refrigerants*
1	Concentration of gases or vapors about 0.5–1% for about 5 min is capable of producing serious injury or death.	Sulphur Dioxide
2	Concentration of gases or vapors about 0.5–1% for about 30 min is capable of producing serious injury or death.	Ammonia
3	Concentration of gases or vapors about 2–2.5% for about 1 hr is capable of producing serious injury or death.	Chloroform
4	Concentration of gases or vapors about 2–2.5% for about 2 hr is capable of producing serious injury or death.	Methyl Chloride
5	Gases or vapors that are less toxic than Group 4 yet more toxic than those in Group 6.	Refrigerants 11, 22, 502, Butane, Propane
6	Concentration of gases or vapors about 20% for periods of about 2 hr has apparently no injury.	Refrigerants 12, 114

CRITICAL PRESSURE AND TEMPERATURE

Every refrigerant, whether a single element or a mixture of elements (a compound), has among its characteristics a pressure above which it will remain as a liquid, even though more heat is added. Each element or compound also has a temperature above which it cannot remain or exist in a liquid state, regardless of the pressure exerted on it. These points are the *vapor pressure* and the *critical temperature*, *respectively*, of the refrigerant, some of which are shown in Table 7B.

The pressures and temperatures shown in Table 7B are well above the pressures that can be expected

TABLE 7B Critical pressures and temperatures of some refrigerants

Refrigerant	*Critical Pressure psia*	*Critical Temperature* °F
R-12	596.9	233.6
R-22	721.9	204.8
R-502	591	179.9

to occur in the condenser, and also above the temperatures that accompany these operating pressures. These refrigerants are chemically stable in the ranges specified. They do not have a tendency, under normal conditions, to react with any material used in the manufacture of the components in the refrigeration system, nor do they ordinarily break down chemically. If the refrigerant in a system is unstable, it quickly becomes useless, since a chemical reaction will change the original characteristics or properties of the refrigerant. Decomposition of the refrigerant will add noncondensable gases, causing abnormal pressures and temperatures within the system, and the resulting sludge may cause mechanical troubles.

PHYSICAL PROPERTIES

It has been stated in an earlier chapter that a refrigerant absorbs much more heat when it changes state from a liquid to a vapor than when it absorbs heat as a liquid or as a vapor. Therefore the boiling point of a refrigerant is of great importance, for it must readily evaporate below the temperature of the product or space which is to be cooled. Refrigerants that do not have a relatively low boiling point require that the compressor be operated with a deep vacuum. This could cause a lowering of the system's capacity and efficiency.

The freezing point of a refrigerant is another important property, particularly in extreme low-temperature applications, for this point should be sufficiently lower than any anticipated evaporator temperature.

The latent heat content of the selected refrigerant usually is high, which is a desirable characteristic for large-capacity systems. With a high latent heat content, less refrigerant is circulated for each ton of refrigeration effect produced, and a compressor of lower horsepower could be used. In systems of low capacity, the use of a refrigerant with a low latent heat will require the circulation of more refrigerant than that of one with a higher latent heat. Such a refrigerant will also make the control of the system easier, since less sensitive control devices are needed in a system that circulates a large amount of refrigerant.

The density and specific volume of a refrigerant are two properties that have a reciprocal relationship. In comparing refrigerants for different applications, advantages and disadvantages must be taken into consideration. In order to utilize smaller lines, tubes, or refrigerant pipes (these terms are used interchangeably) a high-density, low specific volume refrigerant should be used. Such a system costs less to construct and install. In some commercial installations where there is a considerable vertical distance between the major components of the system, it may be desirable to use a refrigerant with a low density and a high specific volume. In this type of system, less pressure is required to circulate the refrigerant through the components and piping.

Substances that are compared must have a mutual relationship at some point. The three refrigerants in Table 7C are compared in the vapor stage at 5°F.

TABLE 7C Comparison of refrigerant vapors at 5°F

	SPECIFIC VOLUME	DENSITY
Refrigerant	(*cu ft/lb*)	(*lb/cu ft*)
R-12	1.46	0.685
R-22	1.24	0.80
R-502	0.825	1.21

Table 7D furnishes a comparison of the specific volume and density of the same three refrigerants in a liquid state at 86°F. The temperatures of 5°F for the vapor and 86°F for the liquid used in Tables 7C and 7D are classified as the standard cycle temperatures in the *ASHRAE Handbook of Fundamentals.*

Low oil solubility or miscibility is also an important characteristic of a refrigerant, but it can cause oil return problems if it is too low. Solubility, or miscibility, is the capability of the refrigerant in its liquid state to mix with the oil necessary for the

TABLE 7D Comparison of refrigerant liquids at 86°F

	SPECIFIC VOLUME	DENSITY
Refrigerant	*(cu ft/lb)*	*(lb/cu ft)*
R-12	0.0124	80.671
R-22	0.0136	73.278
R-502	0.0131	76.13

lubrication of the moving parts in the compressor. The oil will circulate with a highly soluble refrigerant in a liquid state from the condenser and/or receiver, through the liquid line and the metering device, and into the evaporator.

But, since the refrigerant vaporizes in the evaporator, a different situation exists, for oil and refrigerant vapor do not mix readily. The oil circulating through the system can continue on its way to the compressor only if the vapor is moving rapidly enough to entrain the oil and carry it along with it through the suction line to the compressor crankcase. If the vapor lines, either suction or hot gas, run in a horizontal path, it is customary to size them so that the vapor travels at a velocity rate not less than 750 fpm (feet per minute). When the vapor lines are run vertically upward, the velocity of the vapor should be not less than 1,500 fpm to assure that the oil will travel up the vertical line. An example of this would occur when the evaporator is located somewhere beneath the compressor or condenser unit.

Leak detection was and is fairly easy when ammonia or sulphur dioxide is being used, since their peculiar odors are discernible from all others. Yet, depending strictly upon odor for leak detection of these refrigerants may be deceiving as well as dangerous.

Ammonia leaks in a water-cooled condenser may be confirmed by the use of litmus or other test paper dipped in the outlet water from the condenser. Leaks in the rest of the system may be located by the use of a sulphur candle which, when lit, gives off a white cloud of smoke when exposed to ammonia fumes or vapor.

Conversely, leaks in a system using sulphur dioxide as the refrigerant are located by the use of a cloth or taper dipped in an ammonia and water solution of approximately 25 % strength. A white cloud of smoke will appear when the ammonia vapor from the solution nears the leak in the sulphur dioxide system. Frequently, in order to pinpoint a leak when there is quite a bit of vapor in the air, a solution of soap and water is swabbed or wiped over the suspected area. When a system is still under pressure, bubbles will form in the immediate area of a leak. This means of leak detection should not be used if a section of the system is under a vacuum, for the soap and water solution might be drawn into the system.

Figure 7-1 shows a halide leak detector, which has been used successfully on systems with halogenated refrigerants for a number of years. This leak detector consists of two major segments: the cylinder containing gas and the detector unit. The principle involved is that air is drawn through the search hose and across a copper reactor plate, which has been heated red hot from the gas flame. When the search hose is manipulated so that is comes in contact with the leaking refrigerant vapor, gas is drawn through the hose. When the gas or vapor comes in contact with the flame and the reactor plate, the flame changes to a bluish green or violet in color. For most leaks this type of detector works satisfactorily. How-

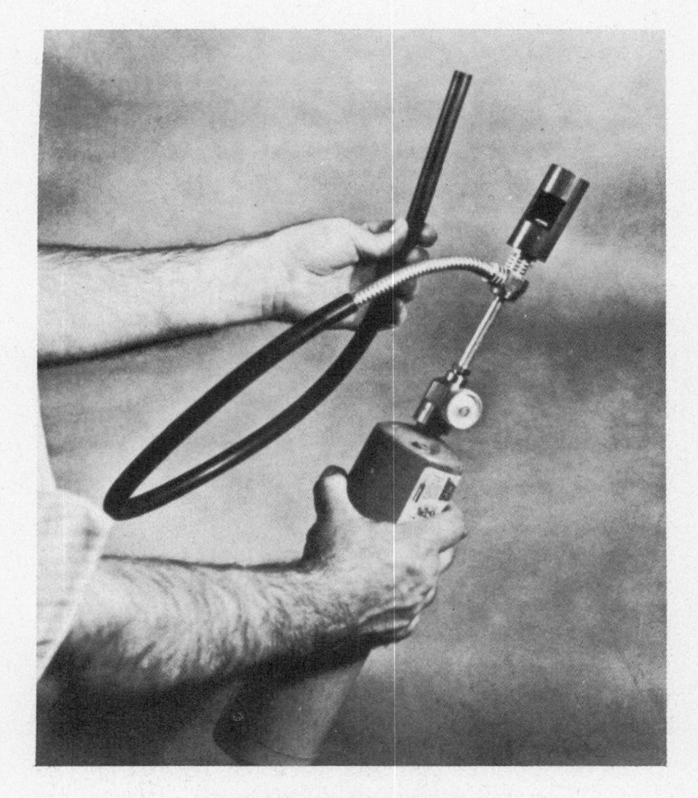

FIGURE 7-1 Halide leak detector.

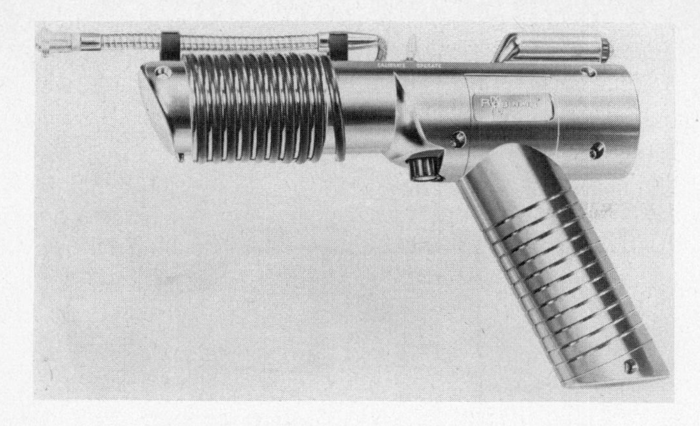

FIGURE 7-2 Electronic leak detector (Courtesy, Robinair Manufacturing Corporation).

ever, it is not as sensitive as the electronic leak detector shown in Fig. 7-2. This type of detector is capable of detecting leaks as small as 0.5 ounce per year. Its use is growing in the field because of its sensitivity; it is also available as a battery-operated unit. Some units have a dual type of control setting—a single HI-LOW selector—in addition to a balance-control dial.

This detector contains an internal pump that draws air into the probe and hose or tube. If there is any halogenated gas in the air drawn across the electrodes in the sensing element, a built-in signal light will flash frequently. The rate of the flashing will increase depending upon the amount of refrigerant vapor in the air. If there is a heavy concentration of refrigerant, the source of the leak can be fairly accurately pinpointed by correctly using the HI-LOW selector and also the balance control. This type of leak detector can be modified to give an audible signal of a leak.

CLASSIFICATION OF REFRIGERANTS

Table 7A listed some refrigerants grouped according to their toxic limitations by the Underwriters Laboratories. Refrigerants also are classified in the National Refrigeration Safety Code under ANSI Standard B-9, in which they are rated in three groups as to their toxicity and fire hazard characteristics. On the basis of low fire hazard or low toxicity, R-12, R-22, and R-502 are placed in Group I—the safest of the refrigerants. Group II refrigerants, which are toxic and somewhat flammable, includes such refrigerants as ammonia (R-717) and methyl chloride (R-40). Group III, which contains the flammable refrigerants, has among others butane (R-600) and propane (R-290).

For personal safety and to avoid possible damage to equipment, certain precautions should be taken. R-12, which is considered a safe refrigerant, may result in the formation of pungent acid vapors when it comes in contact with an open flame. Proper ventilation will prevent the possibility of breathing any minute quantities of phosgene vapor that might be produced. Therefore, when a halide leak detector is used on or around a system, the user should be careful that fumes from the torch are not inhaled.

If the refrigeration unit is located in a confined space, that area should be well ventilated before any time is spent there. Even a refrigerant of low toxicity can cause suffocation if the oxygen in the air is replaced by refrigerant vapor that has escaped from a unit. Liquid refrigerant spilled onto the skin can cause severe frostbite, because the liquid vaporizes rapidly at atmospheric pressure, removing heat from the skin. When refrigerant is handled, protective goggles should be worn because of the possibility of permanent damage should any liquid refrigerant get into the eyes.

Refrigerants should not be mixed in the field, either in a system or outside. To prevent the accidental mixture of refrigerants, the cylinders are color-coded for easy identification of contents, such as:

R-12—White *R-22—Green* *R-502—Orchid*

However, the words on the label should be the final means of identifying the contents.

The U. S. Department of Transportation has regulations pertaining to the manufacture, testing, filling, and shipment of all compressed gas containers subject to interstate transportation. The cylinders or containers must have identification markings stamped in the metal; they are also hydrostatically tested initially and, if returnable, are inspected at regular intervals. Most cylinders have built-in pressure relief valves, so that the pressure can be lowered, should the container be exposed to extremely high temperatures and in danger of exploding.

Many refrigerant cylinders used for storage or in the field are equipped with liquid-vapor valves, which permit withdrawal of either liquid or vapor. This type of valve makes it possible to charge a system or a service drum without inverting the storage cylinder, a necessary procedure before the combined valve came into use. One of the valves is depicted in Fig. 7-3, showing the two hand wheels. The vapor hand wheel is on the top of the valve in the customary position; the liquid hand wheel is located on the side and is attached to a dip tube, which extends to the bottom of the cylinder. The hand wheels are clearly marked to show whether they control the flow of refrigerant in a liquid or vapor state, and they also are color-coded: blue for vapor and red for liquid.

In addition to the larger rechargeable cylinders there are also smaller "disposable" cylinders as illustrated in Fig. 7-4, which are portable for field use. *Never* try to refill these containers.

FIGURE 7-4 Disposable freon cylinder (Courtesy, E.I. du Pont de Nemours and Company).

Refrigerant containers should not be heated to a temperature greater than 125°F, and a direct flame should never be applied to the cylinder. Containers should not be stored near hot pipes or other heating devices, nor should they be left in a closed service vehicle in the sun, since temperatures under such conditions can easily reach 160°F or more. If the container, filled with liquid, is allowed to become heated excessively, its walls will be stretched by the expanding liquid. No matter how strong the walls are, they can stretch just so much before they must burst under the excessive pressure. *No cylinder should ever be more than 80% full.*

FIGURE 7-3 Liquid-vapor valve (Courtesy, E.I. du Pont de Nemours and Company).

LUBRICATION

In refrigeration systems, moving parts in several components create friction, which can be destructive to metal surfaces. In addition, friction results in an increase in the temperature of the moving parts involved. Because proper lubrication reduces possible damage resulting from friction, it is an important aspect in the maintenance of the mechanical components.

The compressor requires proper lubrication for the bearings, pistons, and gears.

In the case of a reciprocating compressor, the space between the piston and the cylinder wall must be sealed off so that all refrigerant vapor will be forced out of the cylinder and into the hot gas discharge line. This sealing off is accomplished by the refrigerant oil as it is forced to travel along with the compressed refrigerant vapor. If the oil film does not seal off the space as the piston moves back and forth, some of the vapor leaks back into the compressor crankcase, resulting in a loss of efficiency.

As mentioned earlier, oil used in refrigerating systems mixes with and travels along with most refrigerants in the liquid state. It is imperative that oil from the crankcase be forced out of the compressor and into the condenser through the hot gas line. In order to maintain proper lubrication of moving parts and to keep the correct oil level in the compressor crankcase, the oil must complete the circuit along with the refrigerant, and then make its way back to the compressor.

Traveling with the liquid refrigerant, the oil reaches the evaporator, which is one of the components in which moving the oil presents a problem. If the oil does not travel from the evaporator to the suction line, the evaporator can become oil-logged, decreasing the heat transfer surface of the cooling coil.

Suction lines must be sized correctly to maintain the velocity of the vapor, in order to entrain the oil with it through the circuit back to the crankcase of the compressor. If the oil is not returned to the compressor, this component could soon be operating in a "dry" condition. When this happens, with no oil being pumped through the cylinder, the vapor seal will be gone and the compressor will lose efficiency. If this situation continues uncorrected for a lengthy period, there will be damage to the compressor.

Two principal methods are used for proper lubrication of compressors: (a) the splash system and (b) the force-feed or pressure system (Fig. 7-5). In the first method, lubrication is initiated by the revolving of the crankshaft in the oil within the crankcase. Fingers or throws on the crankshaft dip into the oil and throw it on the bearings or small grooves that lead to the bearings and seal. Oil also is thrown onto the pistons and cylinder walls, thus maintaining the vapor seal between these components. The importance of keeping the proper oil level in the crankcase cannot be stressed too strongly, along with the need for keeping the oil moving through the system along with the refrigerant.

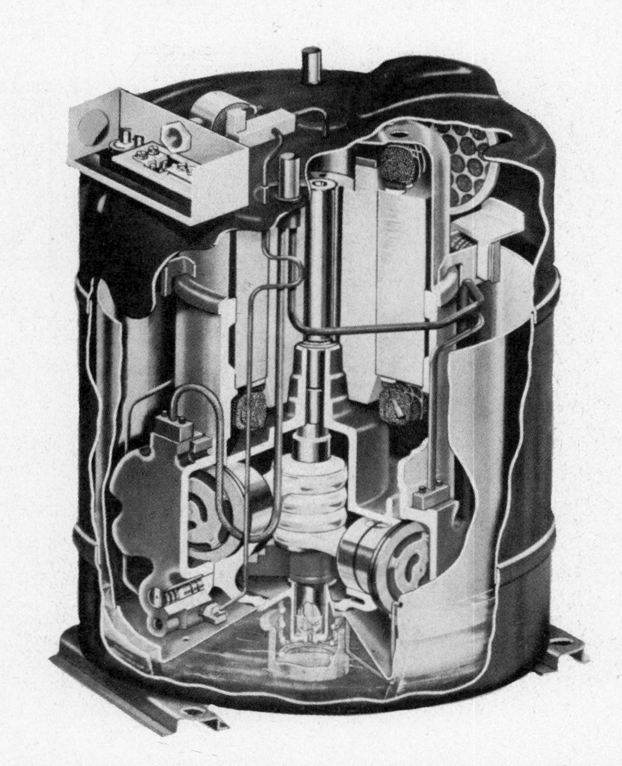

FIGURE 7-5 Lubricating pressure system for compressor (Courtesy, Westinghouse Electric Corporation).

A small pump is used in the pressure system to force the oil to the bearings, seal, piston pins, pistons, and the cylinder walls. A compressor with this type of lubricating system is, of course, more costly than one using the splash system, but the former supplies more protection and assurance of proper lubrication of the compressor, as long as there is an adequate supply of oil in the crankcase.

Some compressors are inherent "oil pumpers." That is, they pump oil out along with the refrigerant vapor at a rate faster than it can be returned through the system to the crankcase. Often, the manufacturer includes an oil separator on the condensing unit. If the compressor is to be used in a built-up system,

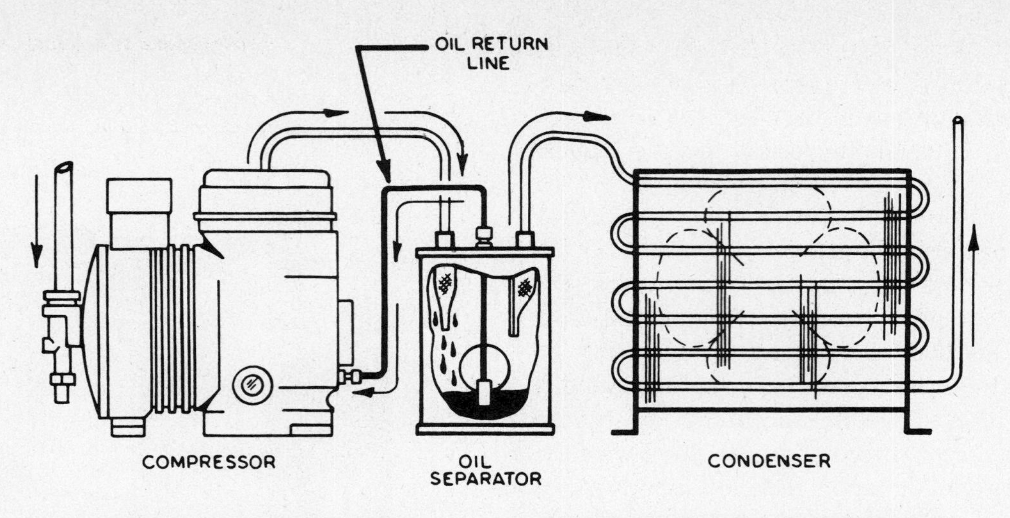

FIGURE 7-6 Cutaway of oil separator (Courtesy, E.I. du Pont de Nemours and Company).

the manufacturer recommends that such an oil separator be included in the installation.

Figure 7-6 shows a cutaway of an oil separator installed in a refrigeration circuit. It is important that the oil be returned to the compressor as soon as possible, so the oil separator is located between the compressor and the condenser. The high-temperature, high-pressure vapor, along with the oil forced out of the compressor, travels through the discharge line from the compressor until it reaches the oil separator. There its direction of flow is changed, and its rate of flow lessens, since the separator has a larger volume and cross sectional area than does the discharge line. Depending upon the design of the separator, it may contain impingement screens or other devices that will force the oil to drop into the reservoir of the separator while the refrigerant vapor continues on its path through the separator.

As shown in Fig. 7-6, most separators contain float and valve assemblies for the release of the oil back into the compressor. When a given amount of oil has accumulated in the reservoir chamber of the separator, the buoyancy of the oil will raise the float and the valve will open. The pressure of the refrigerant discharge vapor is greater than the pressure in the compressor crankcase, and this difference in pressure forces the oil to return to the crankcase.

FIGURE 7-7 Oil separators (Courtesy, AC & R Components, Inc.).

As the oil level in the separator lowers, so does the float, causing the needle valve to close and permitting the accumulation of more oil in the separator.

The separator usually is insulated so that it is kept warm; otherwise refrigerant vapor might condense in the separator when the unit is not in operation. If the unit is part of a system that tends to have long off-cycle periods, it might be advisable to install an electric heater either on or in the separator to keep the refrigerant in a vapor state. Figure 7-7 shows different models of oil separators used today in fully hermetic and semihermetic systems.

REFRIGERANT OIL REQUIREMENTS

As mentioned earlier, a refrigeration oil must have good lubricating qualities and the capability to seal off the low side from the high side in the compressor. While the oil lubricates the bearings in the compressor, it also acts as a cooling medium, removing from these bearings the heat caused by the friction of the moving components when the compressor is in operation.

The perfect oil for use with all refrigerants and under all conditions has not been developed as yet. Each of the refrigerant oils available has its good and not-so-good characteristics, and these must be balanced against the requirements of the installation and the use to which the particular system is to be put.

Here are some of the qualities of oil that are essential:

(1) It must remain fluid at low temperatures.

(2) It must remain stable at high temperatures.

(3) It must not react chemically with the refrigerant, metals, motor insulation (when used in hermetic compressors), air, or other contaminants.

(4) It must not decompose into carbon under expected operating conditions.

(5) It must not deposit wax when subjected to the low operating temperatures that must be met.

(6) It must be dry, and as free of moisture as possible.

For all practical purposes, oils available for refrigeration systems are of mineral origin. They can be separated into three main categories: (a) paraffin base, (b) naphthene base, or (c) a mixture of (a) and (b) called a mixed base. These different categories are derived from the crude oil found in different parts of the world. Proper refining processes remove the heavier paraffins and naphthenes from the crude oil.

Some of the characteristics of refrigerant oils are listed below, not necessarily in their order of importance:

(1) Viscosity

(2) Pour point

(3) Floc point

(4) Flash point

(5) Dielectric strength

(6) Fire point

(7) Corrosion tendency

(8) Oxidation resistance

(9) Color

The *viscosity* of a refrigerant oil, or any other liquid, is a measurement of its resistance to flow, or simply how thin or thick it is, under a given set of conditions. A measured sample of the liquid, at a specific temperature, flows through a calibrated orifice; the time it takes, in seconds, expresses its viscosity.

The *pour point* of an oil is the lowest temperature at which the oil will flow. Usually the temperature of the oil will be lowered to the point where it will no longer flow, and then 5°F is added to this temperature. A low pour point is an indication that the oil will not congeal at the lowest temperatures reached in the system, at its design operating conditions.

It has been found that all refrigerant oils contain wax in varying degrees. This wax will separate from other components in the oil when the temperature of the oil is lowered enough. When a refrigerant oil is dewaxed as much as possible (for the wax cannot be completely removed) tests are run to find the temperature at which the remainder of the wax will separate from the oil.

When the wax separates from the oil the mixture of oil and refrigerant becomes cloudy. As the temperature of the mixture is lowered further, fine suspended particles of wax will form into small balls or clusters. The temperature at which this formation is visible is called the *floc point* of the oil.

Since wax will collect at the colder areas within the refrigeration system (the expansion valve and the evaporator), there will be a loss of heat transfer efficiency in the evaporator, and the expansion valve or other type of metering device may very easily become restricted or clogged.

A particular oil may be used in high-temperature refrigeration or comfort conditioning, but it may not be satisfactory for use in low-temperature applications. Therefore the floc point is an important property to consider when choosing a refrigerant oil for a specific use.

Although refrigerant oils usually present no danger or fire hazard within various systems, it is important to know the *flash point* of the particular oil. This is the temperature at which oil vapor, when exposed to a flame, will flash into fire. This occurs at a specific temperature, and the oil becomes unstable and some of its components tend to separate. Therefore the flash point must be avoided.

Many compressors and motors are hermetically sealed together in housings or shells, and the refrigerant vapor from the evaporator passes across the insulated motor windings. In such cases the refrigerant oil must have a resistance to the flow of electric current, and the *dielectric strength* of a refrigeration lubricating oil is the measurement of this resistance.

The *fire point* of a refrigerant oil is associated with the flash point of the fluid, which was previously described. When the temperature is increased beyond the flash point of the oil vapor and the oil continues to burn during the test, the oil's fire point has been reached.

Sulphur compounds in a refrigerant lubricating oil are undesirable. Sulphurous acid forms when moisture mixes with a sulphur compound. This acid, which is not considered much of a factor in oils today, can be very corrosive to the metal components of a refrigeration system. A good lubricating oil should show a minimum of corrosive tendency when a strip of highly-polished copper is immersed in a sample of the oil and subjected to temperatures above 200°F. After a period of about three or four hours, the copper strip is removed from the oil sample. If it is pitted or more than slightly discolored, this is evidence that the oil contains too much sulphur.

Stability of the refrigerant lubricating oil was discussed in connection with the flash point of refrigerant oils. Still another indication of an oil's stability is its resistance to chemical reaction.

Oil to be used in most lubricating processes must be refined to remove unsaturated hydrocarbons. But the more an oil is refined, the lower its lubricating quality. In the early days of refrigeration, oil used in this process was continuously refined until it was almost colorless. The color of a good refrigeration oil ordinarily is light yellow, indicating that most of the hydrocarbons have been refined, without loss of its lubricating qualities.

REVIEW OF SAFETY

When a leak in a system is suspected, make certain that the room is thoroughly ventilated before starting to work on the unit. Always check for recommended operating pressures for each refrigerant. Check refrigerant R-number before charging. Make certain that the service cylinder is not overcharged when charging it from a storage cylinder. Make certain there are no lighted flames near the unit which is using a fluorocarbon refrigerant that is suspected of have a leak. Always check I.C.C. cylinder stamp for assurance of a safe cylinder. *Wear goggles at all times especially when charging or discharging to protect your eyes in case of a sudden leak.*

Always charge refrigerant vapor into a system. Liquid refrigerant entering a compressor may injure the compressor and may cause the unit to burst.

Liquid refrigerant on the skin may freeze the skin surface and cause a "frost bite." If this should happen, quickly wash away the refrigerant with water. Treat the damaged surface for the "frost bite."

If one accidentally gets refrigerant in the eyes and a doctor is not immediately available wash with mineral oil, (for refrigerants except ammonia), as the oil will absorb the refrigerant. Then wash with a boric acid solution. If the refrigerant is ammonia, wash immediately with water. Water may also be used for washing other refrigerants from the eyes.

PROBLEMS

1. In what two categories may refrigerant properties be classified?
2. If a leak occurs, what three characteristics must be considered?
3. What can be used for leak detection in any refrigeration system under pressure throughout?
4. The flame of a halide torch will burn what color in the presence of halogenated refrigerants?
5. Why is lubrication an important aspect in the maintenance of the mechanical components of a refrigeration system?
6. What is viscosity of lubricating oil?
7. What are the two principal methods of proper compressor lubrication?
8. What is the purpose of an oil separator?
9. What are the three main categories of refrigerant oils?
10. Why is the dielectric strength of an oil important?

8

PRESSURE-ENTHALPY DIAGRAMS

REFRIGERATION EFFECT

If a specific job is to be done in a refrigeration system or cycle, each pound of refrigerant circulating in the system must do its share of the work. It must absorb an amount of heat in the evaporator or cooling coil, and it must dissipate this heat—plus some that is added in the compressor—out through the condenser, whether air-cooled, water-cooled, or evaporatively-cooled. The work done by each pound of the refrigerant as it goes through the evaporator is reflected by the amount of heat it picks up from the refrigeration load, chiefly when the refrigerant undergoes a change of state from a liquid to a vapor.

As mentioned in Chap. 2, in order for a liquid to be able to change to a vapor, heat must be added to or absorbed by it. This is what happens—or should happen—in the cooling coil. The refrigerant enters the metering device as a liquid and passes through the device into the evaporator where it absorbs heat as it evaporates into a vapor. As a vapor, it makes its way through the suction tube or pipe to the compressor. Here it is compressed from a low-temperature, low-pressure vapor to a high-temperature, high-pressure vapor; then it passes through the high-pressure or discharge pipe to the condenser, where it undergoes another change of state—from a vapor to a liquid—in which state it flows out into the liquid pipe and again makes its way to the metering device for another trip through the evaporator. Shown in Fig. 8-1 is a schematic of a simple refrigeration cycle, describing this process.

When the refrigerant, as a liquid, leaves the condenser it may go to a receiver until it is needed in the evaporator; or it may go directly into the liquid line to the metering device, and then into the evaporator coil. The liquid entering the metering device just ahead of the evaporator coil will have a certain heat content (*enthalpy*), which is dependent on its temperature when it enters the coil, as shown in the Refrigerant Tables in the Appendix. The vapor leaving the evaporator will also have a given heat content (enthalpy) according to its temperature, as shown in the Refrigerant Tables.

The difference between these two amounts of heat content is the amount of work being done by each pound of refrigerant as it passes through the evaporator and picks up heat. The amount of heat absorbed by each pound of refrigerant is known as the *refrigerating effect* of the system, or of the refrigerant within the system.

This refrigerating effect is rated in Btu per pound of refrigerant (Btu/lb); if the total heat load is known (given in Btu/hr), we can find the total number of pounds of refrigerant that must be circulated each hour of operation of the system. This figure can further be broken down to the amount that must be circulated each minute, by dividing the amount circulated per hour by 60.

EXAMPLE:

If the total heat to be removed from the load is 60,000 Btu/hour, and the refrigerating effect in the evaporator amounts to 50 Btu per pound, then:

$$\frac{60{,}000 \text{ Btu/hr}}{50 \text{ Btu/lb*}} = \textbf{1,200 lb/hr or 20 lb/min}$$

Since 12,000 Btu/hour equals the rate of one ton of refrigeration, the 60,000 Btu/hour in the above example amounts to 5 tons of refrigeration, and the 20 lb of refrigerant that must be circulated each minute is the equivalent of 4 lb/min/ton of refrigeration. (One ton of refrigeration for 24 hours equals 288,000 Btu).

In this example, where 20 lb of refrigerant having a refrigerating effect of 50 Btu/lb is required to take care of the specified load of 60,000 Btu/h, the results can also be obtained in another manner. As previously mentioned, it takes 12,000 Btu to equal 1 ton of refrigeration, which is equal to 200 Btu/min/ton.

Therefore, 200 Btu/min when divided by the refrigerating effect of 50 Btu/lb, amounts to 4 lb/min. This computation can be shown by the equation:

$$W = \frac{200}{\text{N.R.E.}} \qquad \textbf{(8-1)}$$

where

W = Weight of refrigerant circulated per minute (lb/min)

200 = 200 Btu/min—the equivalent of one ton

N.R.E. = Net Refrigerating Effect in Btu/lb of refrigerant

Because of the small orifice in the metering device, a fact that will be more thoroughly discussed in a later chapter, when the compressed refrigerant passes from the smaller opening in the metering device to the larger tubing in the evaporator, a change in pressure occurs along with a change in temperature. This change in temperature occurs because of the vaporization of a small portion of the refrigerant (about 13%) and, in the process of this vaporization, the heat that is involved is taken from the remainder of the refrigerant.

From the Table of Saturated R-12 in the Appendix, it can be seen that the heat content of 100°F liquid is 31.10 Btu/lb and that of 40°F liquid is 17.27 Btu/lb; this indicates that 13.83 Btu/lb has to be removed from each lb of refrigerant entering the evaporator. The latent heat of vaporization of 40°F R-12 (from Appendix Tables) is 64.17 Btu/lb, and the difference between this amount and that which is given up by each lb of refrigerant when its liquid temperature is lowered from 100°F to 40°F (13.83 Btu/lb) is 50.34 Btu/lb. This is another method of calculating the refrigerating effect—or work being done—by each pound of refrigerant under the conditions given.

The capacity of the compressor must be such that it will remove from the evaporator that amount of refrigerant which has vaporized in the evaporator and in the metering device in order to get the necessary work done. The previous examples have shown that 3.07 cfm must be removed from the evaporator to perform 1 ton of refrigeration under the conditions specified; therefore, when the total load amounts to 5 tons, then 5 times 3.07 cfm, or 15.35 cfm, must be removed from the evaporator. The compressor must be able to remove and send on to the condenser the same weight of refrigerant vapor, so that it can be condensed back into a liquid and so continue in the refrigeration circuit or cycle to perform additional work.

If the compressor, because of design or speed, is unable to move this weight, some of the vapor will remain in the evaporator. This in, turn, will cause an increase in pressure inside the evaporator, accompanied by an increase in temperature and a decrease in the work being done by the refrigerant, and design conditions within the refrigerated space cannot be maintained.

* The 50 Btu/lb is for a particular evaporator and refrigerant and is not the same in all cases.

A compressor that is too large will withdraw the refrigerant from the evaporator too rapidly, causing a lowering of the temperature inside the evaporator, so that design conditions will not be maintained in this situation either.

In order for design conditions to be maintained within a refrigeration circuit, there must be a balance between the requirements of the evaporator coil and the capacity of the compressor. This capacity is dependent on its displacement and also on its volumetric efficiency. The measured displacement of a compressor depends the number of cylinders, their bore and stroke, and the speed at which the compressor is turning. Volumetric efficiency depends on the absolute suction and discharge pressures under which the compressor is operating. A thorough and elaborate presentation of these facts concerning displacement, as well as the variables pertaining to volumetric efficiency, will be offered in a later chapter, along with equations and other data.

CYCLE DIAGRAMS

Fig. 8-1 shows a schematic flow diagram of a basic cycle in refrigeration, denoting changes in phases or processes. First the refrigerant passes from the liquid stage into the vapor stage as it absorbs heat in the evaporator coil. The compression stage, where the refrigerant vapor is increased in temperature and pressure, comes next; then the refrigerant gives off its heat in the condenser to the ambient cooling medium, and the refrigerant vapor condenses back to its liquid state where it is ready for use again in the cycle.

Figure 8-2 is a reproduction of a Mollier diagram (commonly known as a Ph chart) of Refrigerant-12, which shows the pressure, heat, and temperature characteristics of this refrigerant. Pressure-enthalpy diagrams may be utilized for the plotting of the cycle shown in Fig. 8-1, but a basic or skeleton chart as shown in Fig. 8-3 might be used as a preliminary illustration of the various phases of the refrigerant circuit. There are three basic areas on the chart denoting changes in state between the saturated liquid line and the saturated vapor line in the center of the chart.

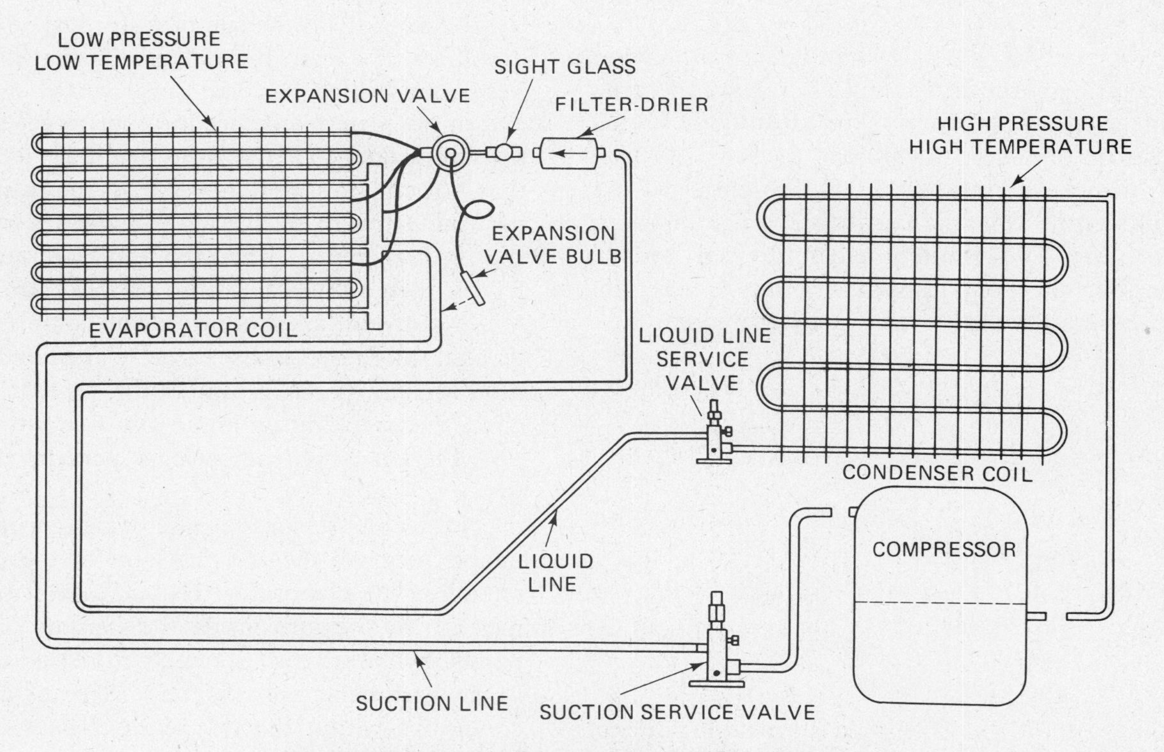

FIGURE 8.1 Schematic diagram of simple refrigeration cycle.

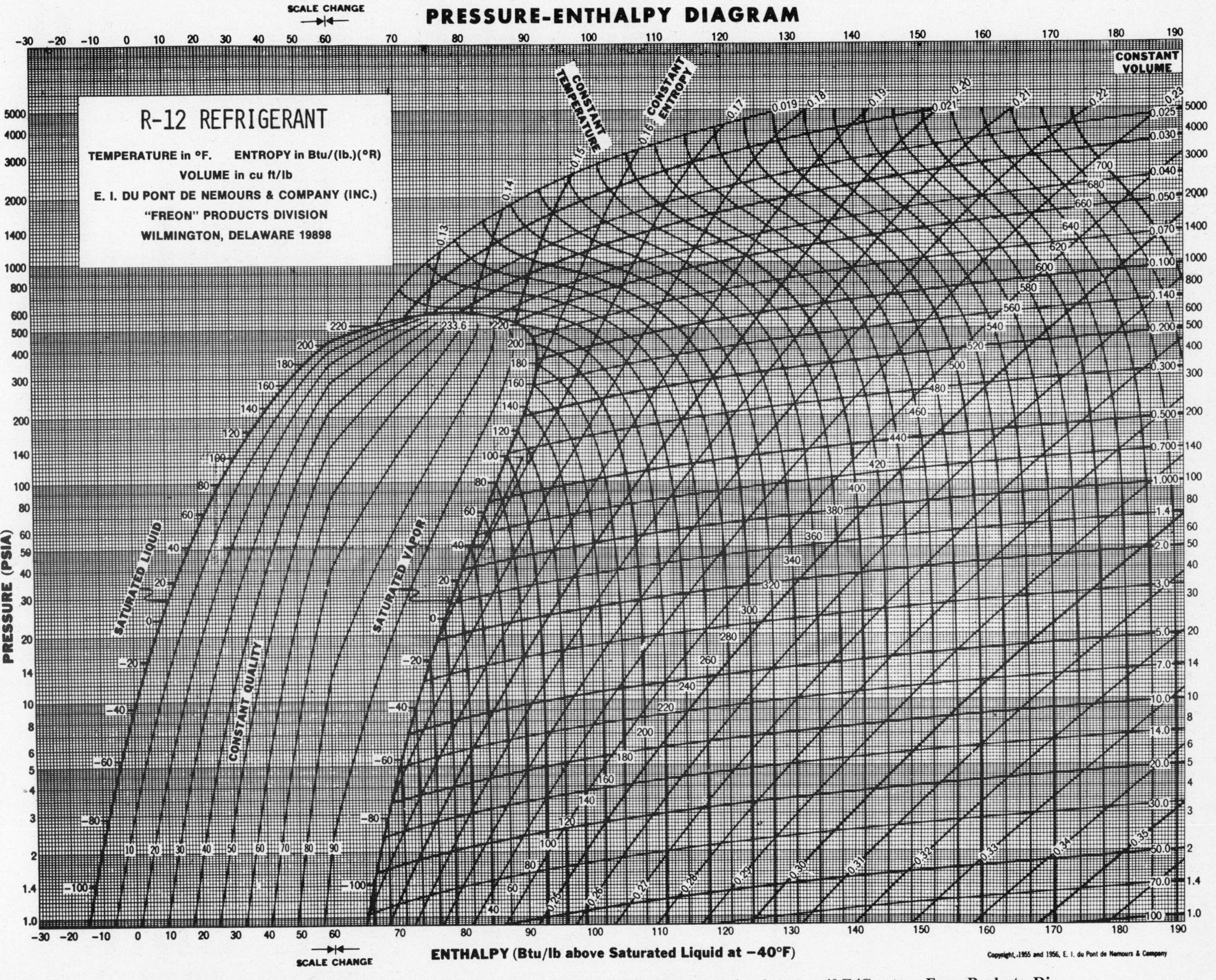

FIGURE 8-2 Pressure-enthalpy diagram for refrigerant R-12. Note that 0 of enthalpy scale is taken at −40 F (Courtesy, Freon Products, Div. of du Pont de Nemours and Company).

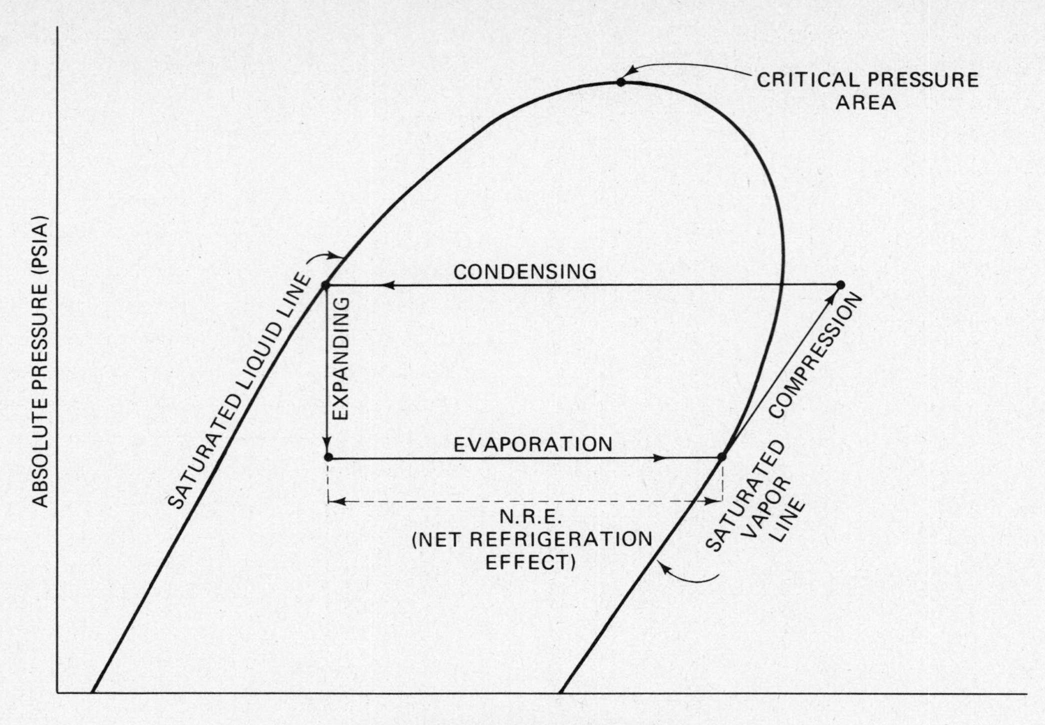

FIGURE 8-3 Enthalpy (Btu/lb).

The area to the left of the saturated liquid line is the subcooled area, where the refrigerant liquid has been cooled below the temperature corresponding to its pressure; whereas the area to the right of the saturated vapor line is the area of superheat, where the refrigerant vapor has been heated beyond the vaporization temperature corresponding to its pressure.

The construction of the diagram, or rather a knowledge and understanding of it, may bring about a clearer interpretation of what happens to the refrigerant at the various stages within the refrigeration cycle. If the state and any two properties of a refrigerant are known and this point can be located on the chart, the other properties can easily be determined from the chart.

If the point is situated anywhere between the saturated liquid and vapor lines, the refrigerant will be in the form of a mixture of liquid and vapor. If the location is closer to the saturated liquid line, the mixture will be more liquid than vapor, and a point located in the center of the area at a particular pressure would indicate a 50% liquid–50% vapor situation.

Referring to Fig. 8-3, the change in state from a vapor to a liquid—the condensing process—occurs as the path of the cycle develops from right to left; whereas the change in state from a liquid to a vapor—the evaporating process—travels from left to right. Absolute pressure is indicated on the vertical axis at the left, and the horizontal axis indicates heat content, or enthalpy, in Btu/lb.

The distance between the two saturated lines at a given pressure, as indicated on the heat content line, amounts to the latent heat of vaporization of the refrigerant at the given absolute pressure. The distance between the two lines of saturation is not the same at all pressures, for they do not follow parallel curves. Therefore there are variations in the latent heat of vaporization of the refrigerant, depending on the absolute pressure. There are also variations in pressure-enthalpy charts of different refrigerants and the variations depend upon the various properties of the individual refrigerants.

REFRIGERATION PROCESSES

Based on the examples presented earlier in the chapter, it will be assumed that there will be no changes in the temperature of the condensed refrigeration liquid after it leaves the condenser and travels through the liquid pipe on its way to the expansion or metering device, or in the temperature of the refrigerant vapor after it leaves the evaporator and passes through the suction pipe to the compressor.

Figure 8-4 shows the phases of the simple saturated cycle with appropriate labeling of pressures, temperatures, and heat content or enthalpy. A starting point must be chosen in the refrigerant cycle; let it be point *A* on the saturated liquid line where all of the refrigerant vapor at 100°F has condensed into liquid at 100°F and is at the inlet to the metering device. What occurs between points *A* and *B* is the expansion process as the refrigerant passes through the metering device; and the refrigerant temperature is lowered from the condensation temperature of 100°F to the evaporating temperature of 40°F.

When the vertical line *A–B* (the expansion process) is extended downward to the bottom axis, a reading of 31.10 Btu/lb is indicated, which is the heat content of 100°F liquid. To the left of point *B* at the saturated liquid line is point *Z*, which is also at the 40°F temperature line. Taking a vertical path downward from point *Z* to the heat content line, a reading of 17.27 Btu/lb is indicated, which is the heat content of 40°F liquid; this area between points *Z* and *B* will be covered later in this chapter.

The horizontal line between points *B* and *C* indicates the vaporization process in the evaporator, where the 40°F liquid absorbs enough heat to completely vaporize the refrigerant. Point *C* is at the saturated vapor line, indicating that the refrigerant has completely vaporized and is ready for the compression process. A line drawn vertically downward to where it joins the enthalpy line indicates that the heat content, shown at h_c is 81.44 Btu/lb, and the difference between h_a and h_c is 50.34 Btu/lb, which is the refrigerating effect, as shown in an earlier example.

The difference between points h_z and h_c on the enthalpy line amounts to 64.17 Btu/lb, which is the latent heat of vaporization of 1 lb of R-12 at 40°F. This amount would also exhibit the refrigerating effect, but some of the refrigerant at 100°F must evaporate or vaporize in order that the remaining portion of each pound of R-12 can be lowered in temperature from 100°F to 40°F.

The various properties of refrigerants, mentioned earlier, should probably be elaborated upon before we proceed with the discussion of the compression process. All refrigerants exhibit certain properties when in a gaseous state; some of them are: *volume*,

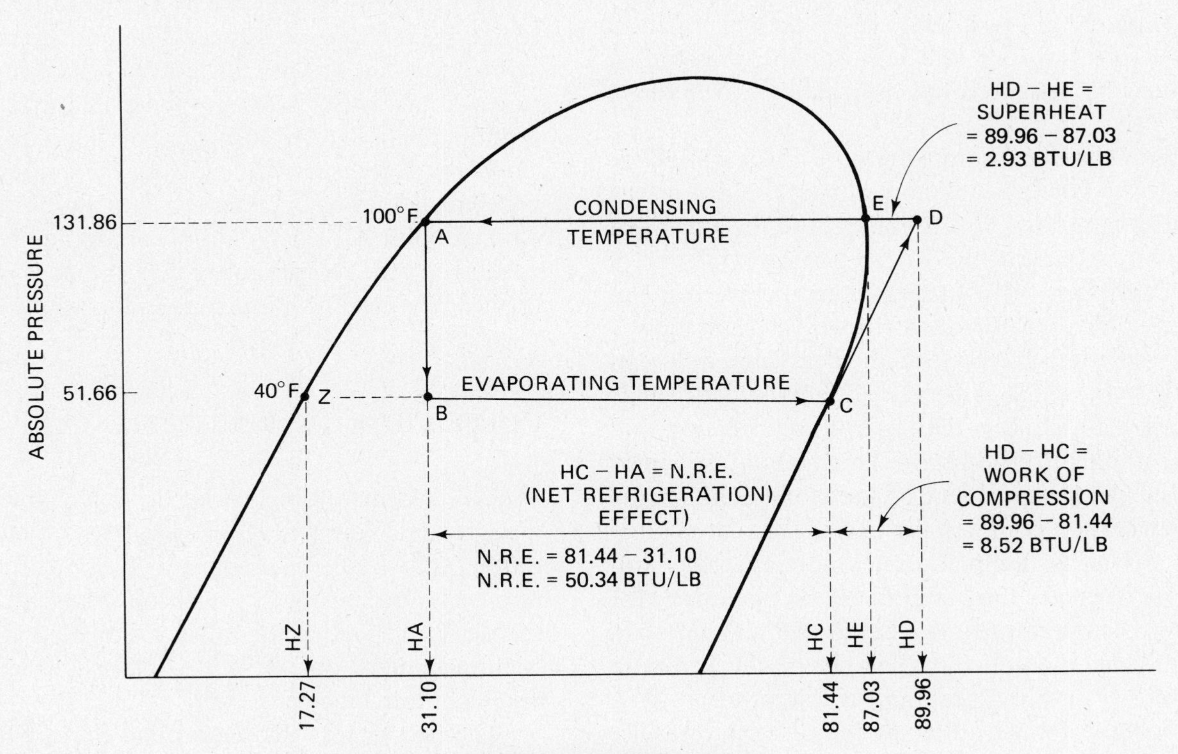

FIGURE 8-4 Enthalpy (Btu/lb).

temperature, *pressure*, *enthalpy* or *heat content*, and *entropy*. The last property—entropy—is really the most difficult to describe or define. It is the ratio of the heat content of the gas to its absolute temperature in degrees Rankine, and it relates to internal energy of the gas.

The Mollier chart plots the line of constant entropy, which stays the same providing the gas is compressed and no outside heat is added or taken away. When the entropy is constant, the compression process is called *adiabatic*; which means that the gas changes its condition without the absorption or rejection of heat either from or to an external body or source. It is common practice, in the study of cycles of refrigeration, to plot the compression line either along or parallel to a line of constant entropy.

In Fig. 8-3, line *C–D* denotes the compression process, in which the pressure and temperature of the vapor are increased from that in the evaporator to that in the condenser, with the assumption that there has been no pickup of heat in the suction line between the evaporator and the compressor. For a condensing temperature of 100°F, a pressure guage would read approximately 117 psig; but the Ph chart is rated in absolute pressure and the atmospheric pressure of 14.7 must be added to the psig, making it actually 131.86 psia.

Point *D* on the absolute pressure line is equivalent to the 100°F condensing temperature; it is not on the saturated vapor line, it is to the right in the superheat area, at a junction of the 131.86 psia line, the line of constant entropy of 40°F, and the temperature line of approximately 116°F. A line drawn vertically downward from point *D* intersects the heat content line at 89.96 Btu/lb, which is h_d; the difference between h_c and h_d is 8.52 Btu/lb—the heat of compression that has been added to the vapor. This amount of heat is the heat energy equivalent of the work done during the refrigeration compression cycle. This is the theoretical discharge temperature, assuming that saturated vapor enters the cycle; in actual operation, the discharge temperature may be 20°–35° higher than that predicted theoretically. This can be checked in an operating system by strapping a thermometer or a thermocouple to the outlet of the discharge service valve on the compressor.

During the compression process the heat that is absorbed by the vapor is a result of friction caused by the action of the pistons in the cylinders and by the vapor itself passing through the small openings of the internal suction and discharge valves. Of course, the vapor is also heated by the action of its molecules being pushed or compressed closer together, commonly called *heat of compression*. Some of this overall additional heat is lost through the walls of the compressor. A lot, therefore, depends upon the design of the compressor, the conditions under which it must operate and the balance between the heat gain and heat loss to keep the refrigerant at a constant entropy.

Line *D–E* denotes the amount of superheat that must be removed from the vapor before it can commence the condensation process. A line drawn vertically downward from point *E* to point h_e on the heat content line indicates the distance h_d–h_e, or heat amounting to 2.93 Btu/lb, since the heat content of 100°F vapor is 87.03 Btu/lb. This superheat is usually removed in the hot gas discharge line or in the upper portion of the condenser. During this process the temperature of the vapor is lowered to the condensing temperature.

Line *E–A* represents the condensation process that takes place in the condenser. At point *E* the refrigerant is a saturated vapor at the condensing temperature of 100°F and an absolute pressure of 131.86 psia; the same temperature and pressure prevail at point *A*, but the refrigerant is now in a liquid state. At any other point on line *E–A* the refrigerant is in the phase of a liquid-vapor combination, for, the closer the point is to *A*, the greater the amount of the refrigerant that has condensed into its liquid stage. And, at point *A*, each pound of refrigerant is ready to go through the refrigerant cycle again as it is needed for heat removal from the load in the evaporator.

COEFFICIENT OF PERFORMANCE

Two factors mentioned earlier in this chapter are of the greatest importance in deciding which of the refrigerants should be used for a given project of heat removal. Ordinarily, this decision is reached during the design aspect of the refrigeration and air-conditioning system, but we will explain it briefly now, and elaborate later.

The two factors that determine the coefficient of performance (CoP) of a refrigerant are refrigerating effect and heat of compression. The equation may be written as:

$$\text{CoP} = \frac{\text{refrigerating effect}}{\text{heat of compression}} \qquad \textbf{(8-2)}$$

Substituting values from the Ph diagram of the simple saturated cycle previously presented, the equation would be:

$$\text{CoP} = \frac{h_c - h_a}{h_d - h_c} = \frac{50.34}{8.52} = \textbf{5.91}$$

The CoP is therefore a rate or a measure of the efficiency of a refrigeration cycle in the utilization of expended energy during the compression process in ratio to the energy that is absorbed in the evaporation process. As can be seen from the above equation, the less energy expended in the compression process, the larger will be the CoP of the refrigeration system. Therefore, the refrigerant having the highest CoP would probably be selected—providing other qualities and factors are equal.

EFFECTS ON CAPACITY

The pressure enthalpy diagrams in Figs. 8-4 and 8-5 show a comparison of two simple saturated cycles having different evaporating temperatures, to bring out various differences in other aspects of the cycle. In order that an approximate mathematical calculation comparison may be made, the cycles shown in Figs. 8-3 and 8-4 will have the same condensing temperature, but the evaporating temperature will be lowered 20°F. Data can either be obtained or verified from the Table for R-12 in the Appendix ; but we will take the values of *A*, *B*, *C*, *D*, and *E* from Fig. 8-3 as the cycle to be compared to that in Fig. 8-4 (with a 20° evaporator). The refrigerating effect, heat of compression, and the heat dissipated at the condenser in each of the refrigeration cycles will be compared.

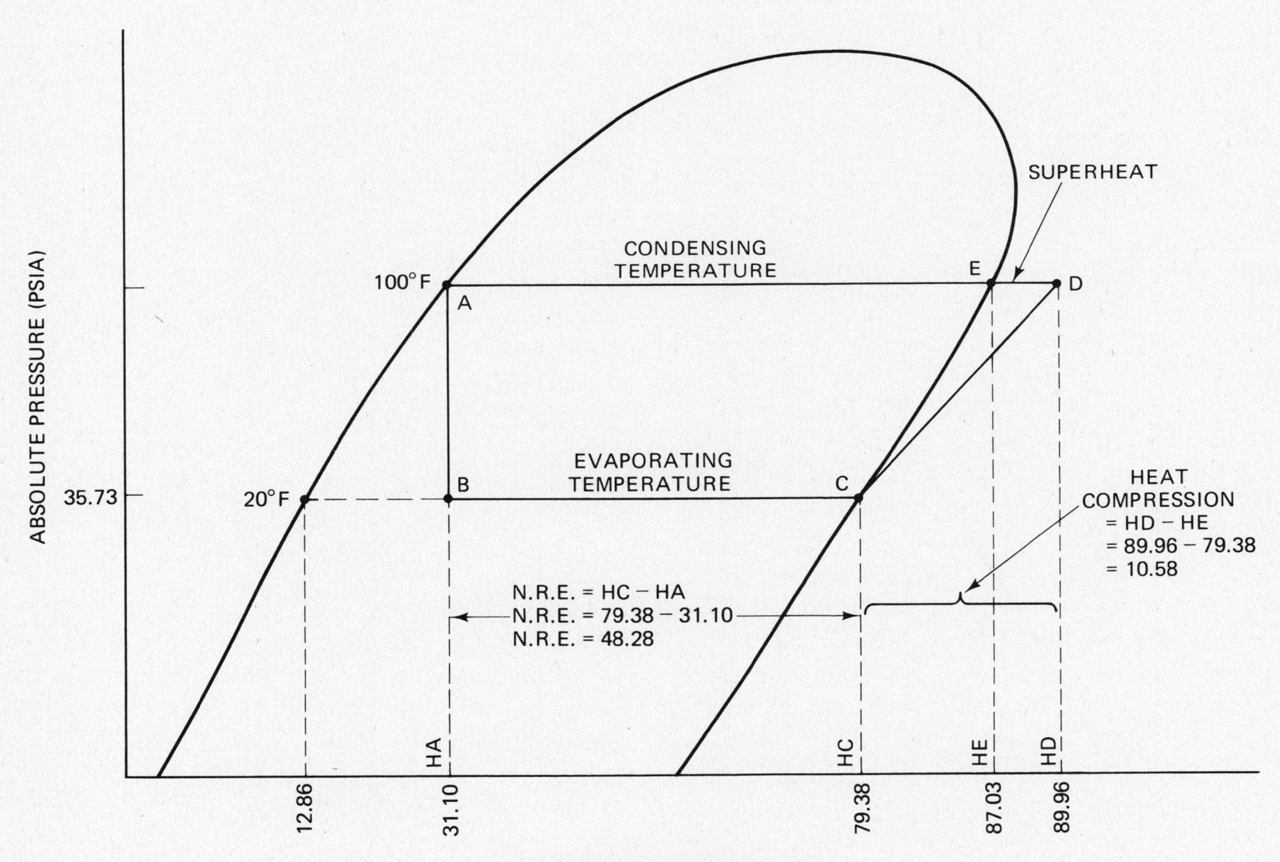

FIGURE 8-5 **Enthalpy (Btu/lb).**

The comparison will be based on data about the heat content or enthalpy line, rated in Btu/lb.

For the 20°F evaporating temperature cycle shown in Fig 8-5 the:

net refrigerating effect ($h_{c'} - h_a$) = 48.28 Btu/lb

heat of compression ($h_{d'} - h_{c'}$) = 10.58 Btu/lb

In comparing the above data with that of the cycle with the 40°F evaporating temperature (Fig. 8-4), we find that there is a decrease the in N.R.E. of 4% and an increase in the heat of compression of 28%. There will be some increase in superheat, which should be removed either in the discharge pipe or the upper portion of the condenser. This is the result of a lowering in the suction temperature, the condensing temperature remaining the same.

By utilizing Eq. 8-1, it will be found that the weight of refrigerant to be circulated per ton of cooling, in a cycle with a 20°F evaporating temperature and a 100°F condensing temperature, is 4.14 lb/min/ton:

$$W = \frac{200\ (\text{Btu/min})}{\text{N.R.E}\ (\text{Btu/lb})}$$

$$W = \frac{200\ \text{Btu/min}}{48.29\ \text{Btu/lb}}$$

$$W = 4.14\ \text{lb/min}$$

This of, course, would also necessitate either a larger compressor, or the same size of compressor operating at a greater rpm.

Figure 8-6 shows the original cycle with a 40°F evaporating temperature, but the condensing temperature has been increased to 120°F.

Again taking the specific data from the heat content or enthalpy line, we now find for the 120°F condensing temperature cycle that $h_a = 36.01$, $h_c = 81.43$, $h_d = 91.33$, and $h_e = 88.61$.

net refrigerating effect ($h_c - h_{a'}$) = 45.4 Btu/lb

heat of compression ($h_{d'} - h_c$) = 9.90 Btu/lb

condenser superheat ($h_{d'} - h_{e'}$) = 2.72 Btu/lb

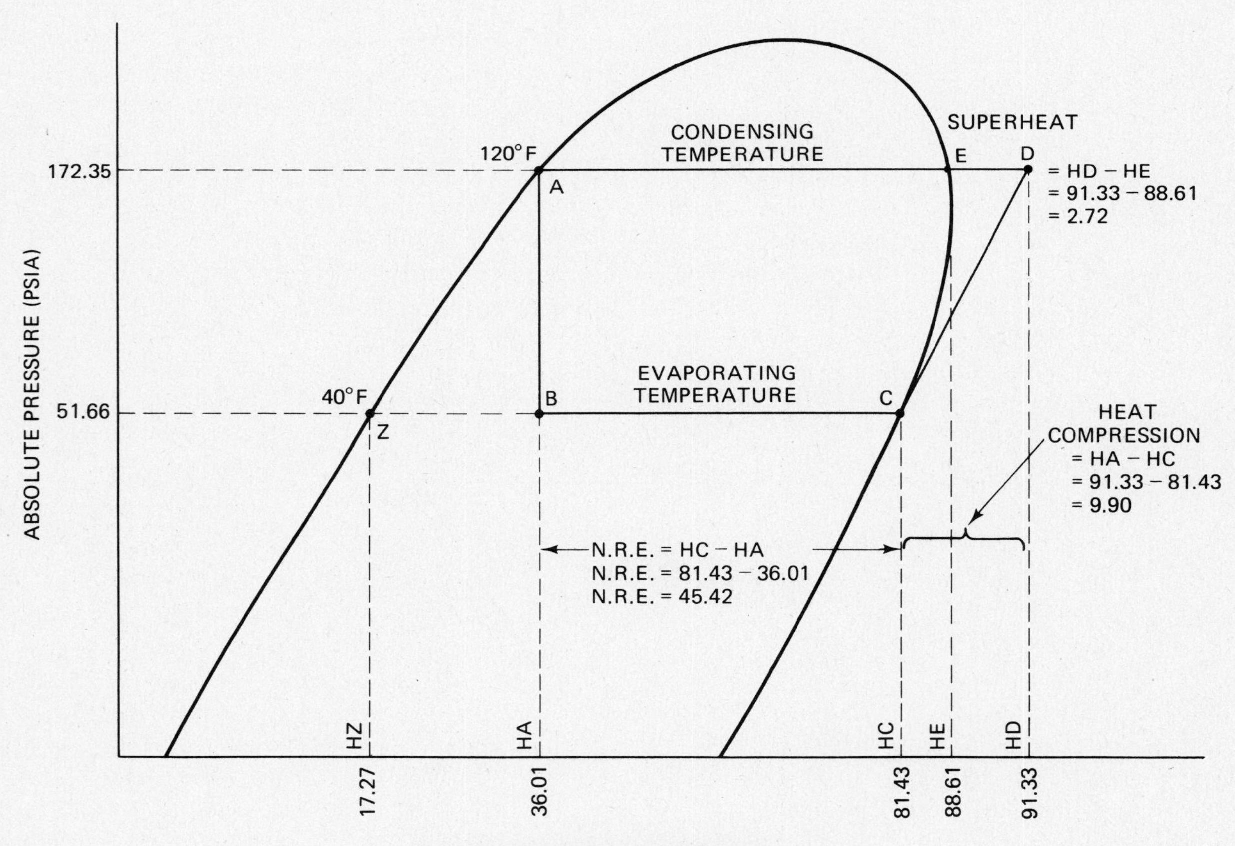

FIGURE 8-6 Enthalpy (Btu/lb).

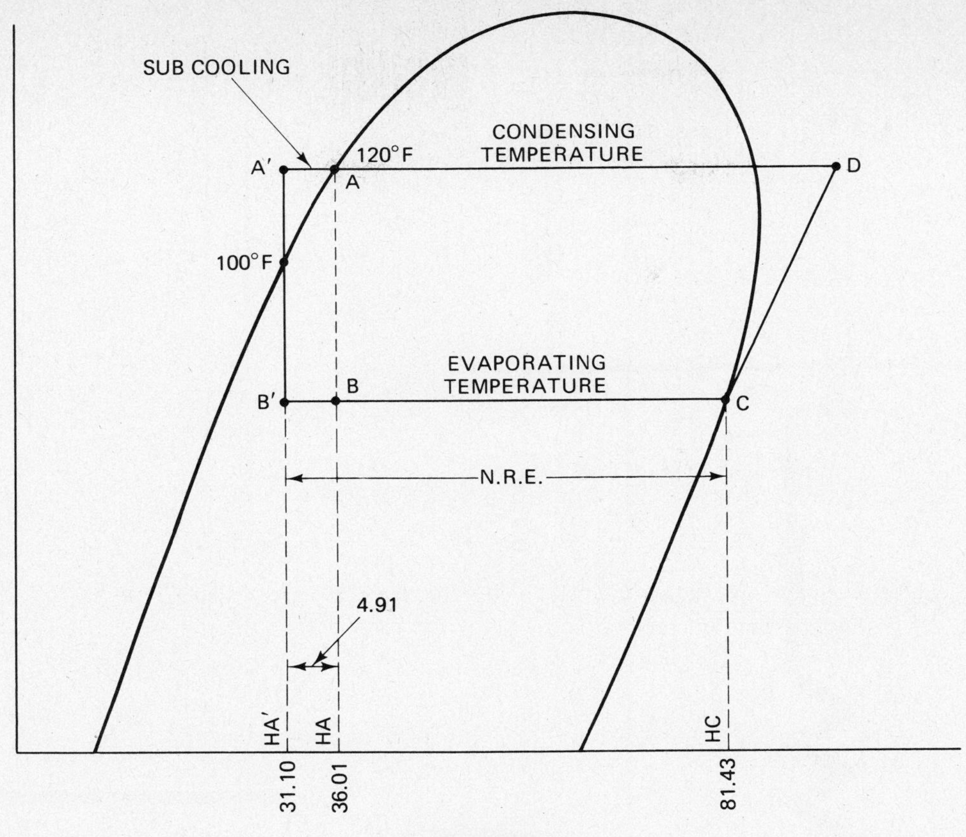

FIGURE 8-7.

In comparison with the cycle having the 100°F condensing temperature, it can be calculated that, by allowing the temperature of the condensing process to increase 20°F, there is a decrease in the N.R.E. of 9.8%, an increase of heat compression of 16.2%, and a decrease of superheat to be removed either in the discharge line or in the upper portion of the condenser of 7.1%.

Through the use of Eq. 8-1 it is found that with a 40°F evaporating temperature and a 120°F condensing temperature the weight of refrigerant to be circulated will be 4.4 lb/min/ton. This indicates that approximately 11% more refrigerant must be circulated to do the same amount of work as when the condensing temperature was 100°F.

Both of these examples show that, for the best efficiency of a system, the suction temperature should be as high as feasible, and the condensing temperature should be as low as feasible. Of course, there are limitations as to the extremes under which systems may operate satisfactorily, and other means of increasing efficiency must then be considered. Economics of equipment (cost + operating performance) ultimately determine the feasibility range.

Referring to Fig. 8-7, after the condensing process has been completed and all of the refrigerant vapor at 120°F is in the liquid state, if the liquid can be subcooled to point *A′* on the 100°F line (a difference of 20°F), the N.R.E. ($h_c - h_{a'}$) will be increased 4.91 Btu/lb. This increase in the amount of heat absorbed in the evaporator without an increase in the heat of compression will increase the CoP of the cycle, since there is no increase in the energy input to the compressor.

This subcooling may take place while the liquid is temporarily in storage in the condenser or receiver, or some of the liquid's heat may be dissipated to the ambient temperature as it passes through the liquid pipe on its way to the metering device. Subcooling also may take place in a commercial-type water-cooled system through the use of a liquid subcooler, which, in a low-temperature application, may well pay for itself through the resulting increase in capacity and efficiency of the overall refrigeration system.

Another method of subcooling the liquid is by means of a heat exchanger between the liquid and suction lines, whereby heat from the liquid may be transferred to the cooler suction vapor traveling from

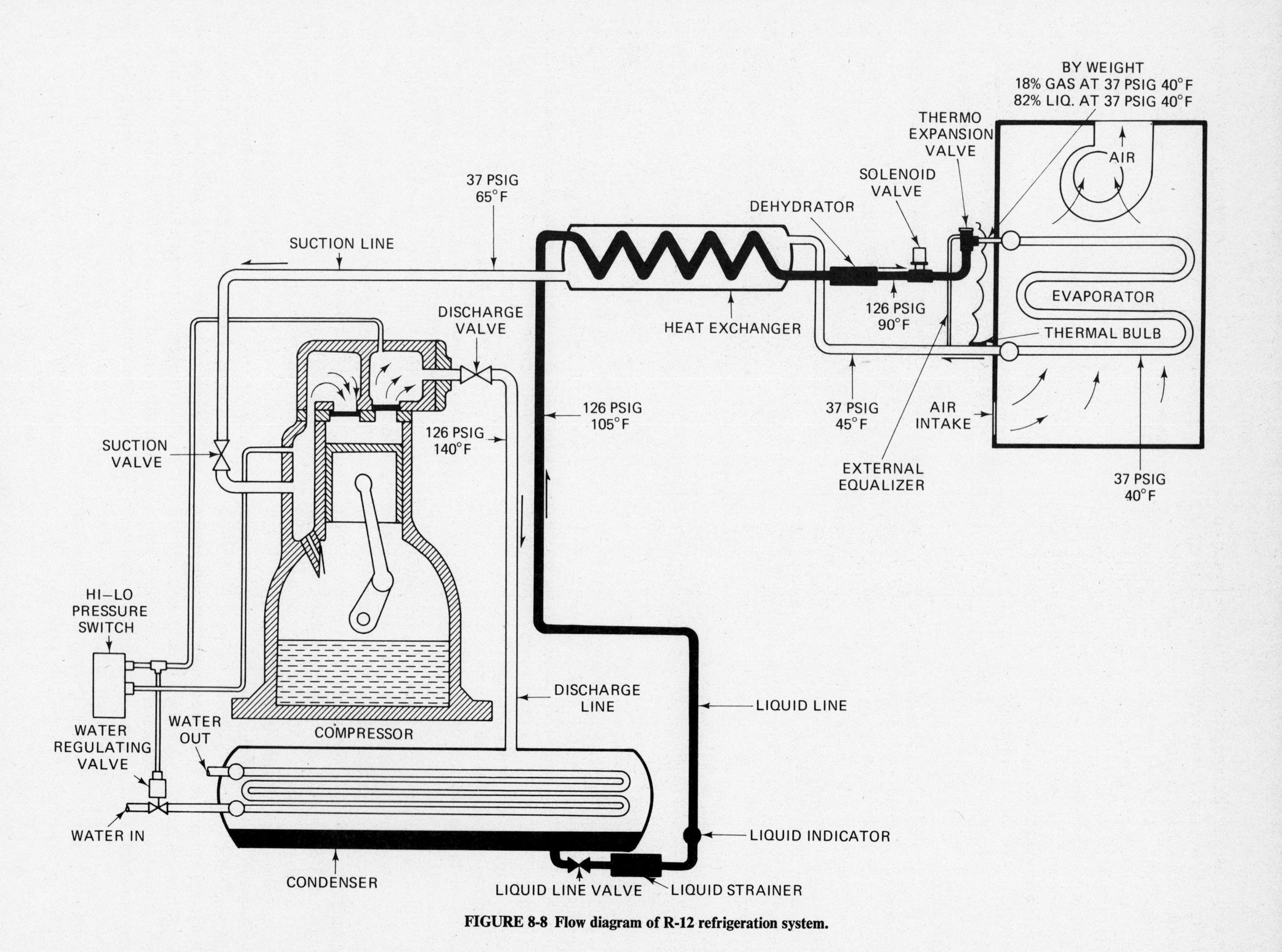

FIGURE 8-8 Flow diagram of R-12 refrigeration system.

the evaporator to the compressor. This type is shown in Fig. 8-8, a refrigeration cycle flow chart, using a liquid-suction heat exchanger. True, heat cannot be removed from the liquid and then added to the suction vapor without some detrimental effects to the overall refrigeration cycle; for example, the vapor would become superheated, which would in turn cause an increase in the specific volume of each pound of refrigerant vapor and consequently a decrease in its density. Thus, any advantage of subcooling in a saturated cycle would be negated; but, in an actual cycle, the conditions of a simple saturated cycle do not exist.

In any normally operating cycle, the suction vapor does not arrive at the compressor in a saturated condition. Superheat is added to the vapor after the evaporating process has been completed, in the evaporator and/or in the suction line, as well as in the compressor. If this superheat is added only in the evaporator, it is doing some useful cooling; for it too is removing heat from the load or product, in addition to the heat that was removed during the evaporating process. But if the vapor is superheated in the suction line located outside of the conditioned space, no useful cooling is accomplished; yet this is what takes place in the majority—if not all—of refrigeration systems.

Now, were some of this superheating in the suction pipe curtailed through the use of a liquid-suction heat exchanger, this heat added to the vapor would be beneficial, for it would be coming from the process of subcooling the liquid. As an example, suppose the suction temperature in the evaporator is at 40°F; the superheated vapor coming out of the evaporator may be about 50°F, and the temperature of the vapor reaching the compressor may be 75°F or above, depending on the ambient temperature around the suction. This means that the temperature of the vapor has been increased 25°F, without doing any useful cooling or work, because this heat has been absorbed from the ambient air outside of the space to be cooled.

If some or most of this 25°F increase in the vapor temperature were the result of heat absorbed from the refrigerant liquid, it would be performing useful cooling, since the subcooling of the liquid will result in an a refrigerating effect higher than it would be if the refrigerant reached the metering device without any subcooling. It is possible to reach an approximate balance between the amount of heat in Btu/lb removed by subcooling the liquid and the amount of heat added to the refrigerant vapor in the suction pipe without the heat exchanger.

PROBLEMS

1. What is meant by net refrigerating effect?

2. What two factors are involved in finding the N.R.E?

3. What is the equation used to find the weight of refrigerant to be circulated for a given load?

4. What is the difference between saturated and superheated vapor?

5. Upon what factors is the capacity of a compressor dependent?

6. What is a Ph chart of a system?

7. What are the divisions shown in a Ph chart or diagram?

8. What refrigerant properties are shown on the Ph charts?

9

REFRIGERATION MEASURING AND TESTING EQUIPMENT

In refrigeration (and air-conditioning) work you will use many different kinds of instruments and test equipment. Previous chapters have already detailed basic hand tools and certain test equipment such as refrigerant-leak testing devices. This chapter will review measuring and testing equipment needed to determine temperature and pressure conditions. Later chapters will review measuring and testing equipment associated with the electrical phases of refrigeration systems.

TEMPERATURE MEASUREMENTS

When analyzing a refrigeration system accurate temperature readings are important. The most common temperature measuring device is the pocket glass thermometer, illustrated in Fig. 9-1.

Note that it fits into a protective metal case. The thermometer head has a ring for attaching a string to suspend it if needed. The ranges of glass thermometers vary but −30° to +120°F is a common scale for refrigeration systems, and the thermometer is calibrated in 2° marks. Some have a mercury fill, but others use a red fill that is easier to read.

To check the calibration of a pocket glass thermometer insert it into a glass of ice water for several minutes. It should read 32°F plus or minus 1°. Should the fill separate, place the thermometer in a freezer, and the resulting contraction will probably rejoin the separated column of fluid. Another way to connect a separated fill is to carefully heat the stem, *not the bulb*. In most cases the liquid will coalesce as it expands.

Another form of pocket thermometer is the dial type, shown in Fig. 9-2. It too has a carrying case with pocket clip. The dial thermometer is more convenient or more practical in measuring air temperatures in a duct. The stem is inserted into the duct, but the dial remains visible. Again several ranges are available, depending on the accuracy needed and the nature of application. A common refrigeration range is −40°F to +160°F.

A different type of dial thermometer is the superheat thermometer, as illustrated in Fig. 9-3.

The highly accurate expansion-bulb thermometer is used to measure suction line temperature(s) in

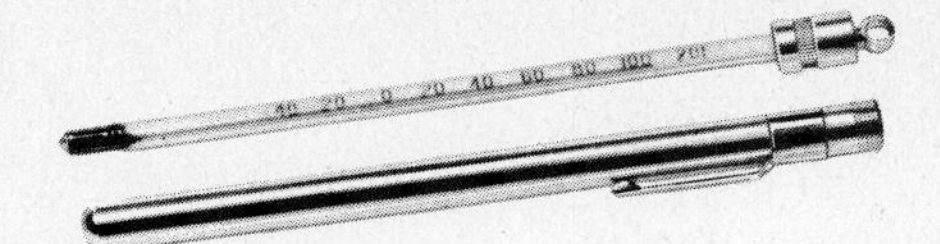

FIGURE 9-1 Glass thermometer (Courtesy, Robinair Manufacturing Corporation).

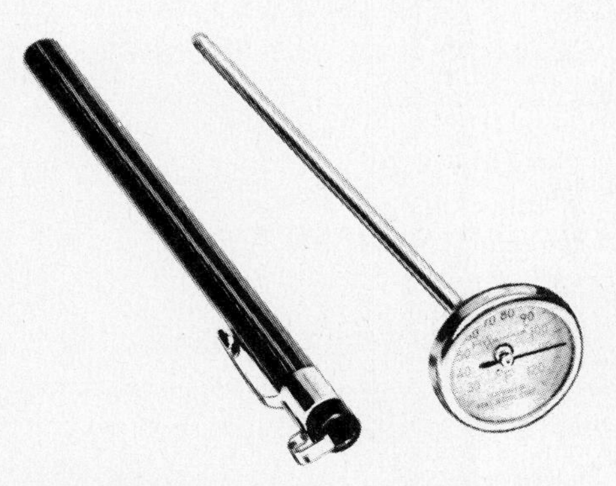

FIGURE 9-2 Pocket dial thermometer (Courtesy, Robinair Manufacturing Corporation).

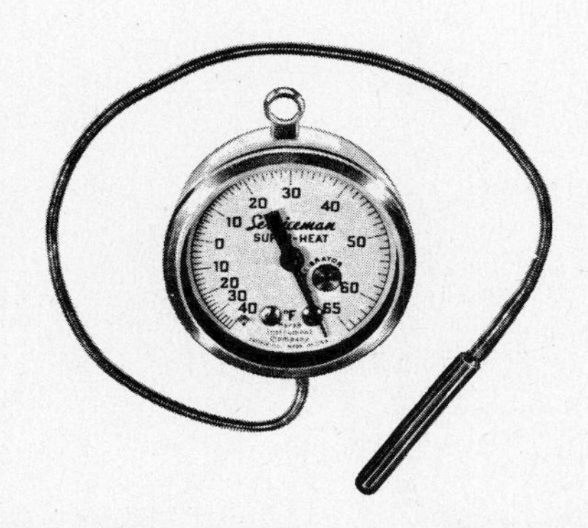

FIGURE 9-3 Superheat thermometer (Courtesy, Robinair Manufacturing Corporation).

order to calculate, check, and adjust superheat. A common range is −40° to +65°F. The sensing bulb is strapped or clamped to the refrigerant line and then covered with insulating material (such as a foam rubber sheet) to prevent air circulation over the bulb while you take a reading. Of course the superheat bulb thermometer can be used to measure air or water temperatures as well.

The thermometers just described have been the basic temperature-measuring tools for many years. However, they do have certain limitations; for example, the operator is required to actually be in the immediate area during the time of reading. An example would be trying to measure the inside temperature of your home refrigerator without opening the door.

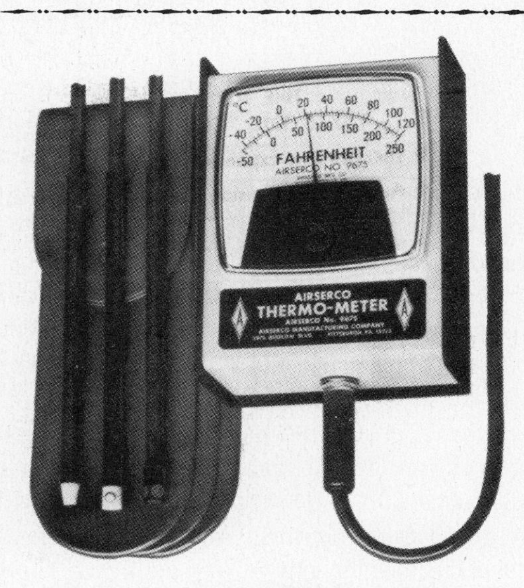

FIGURE 9-4 Thermo-meter (Courtesy, Airserco Manufacturing Company).

The thermometer illustrated in Fig. 9-4 is also a hand-held device, but it has a probe length of 30 in. to allow a degree of remote temperature measurement.

Different probes are available to measure surface or air temperatures. In refrigeration use, the surface probe determines superheat settings of expansion valves, motor temperature, condensing temperature, and water temperature. This is a useful instrument for many applications, but it too is limited, since only one reading in one place at a time can be taken.

As a result of the rapid development of low-cost electronic devices the availability and use of electronic thermometers is now very common. The electronic thermometer as illustrated in Fig. 9-5 consists

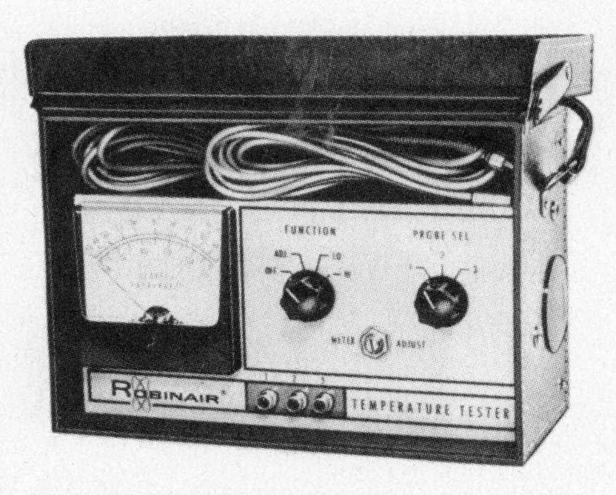

FIGURE 9-5 Electronic thermometer (Courtesy, Robinair Manufacturing Corporation).

of a tester with provision to attach one or several (three to six) sensing leads. The sensing lead tip is actually a thermistor element, which, when subject to heat or cold, will vary the electrical current in the test circuit because its resistance changes with temperature change.

The tester converts changes in electrical current to temperature readings. The sensing leads vary in length depending on the make of the unit, but extensions can be used to allow for remote testing.

Once the operator places the sensing probe(s) in the spots chosen to be tested he may switch from position to position and record temperatures without actually going into each test area, such as the refrigerator, walk-in cooler, freezer, or air duct. The sensing probe can also be used to check superheat.

Sometimes it may be necessary to record temperatures over long periods of time such as a day or even a week, in order to examine the changes in system conditions. Very expensive recording thermometers for highly sophiscated jobs are available, but for the average installation there are reasonably inexpensive, compact, and portable recording thermometers as shown in Fig. 9-6, which consist of a hand-wound chart-driving unit. A recording thermometer takes the guess work out of setting the system operating conditions or diagnosing and locating trouble areas—and it provides a permanent record of the results.

WIND IT — SET IT — FORGET IT!

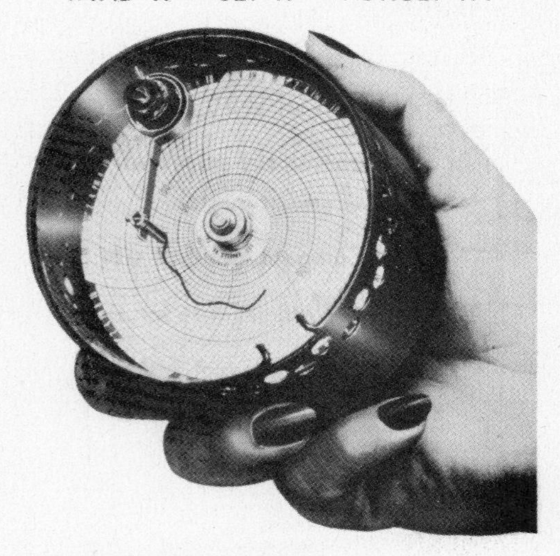

FIGURE 9-6 Portable recording thermometer (Courtesy, Airserco Manufacturing Company).

The final selection of temperature-measuring instruments will depend on the scope of work with which a technician is involved. If he is servicing refrigeration, plus air-conditioning and heating, he will need a wide variety of thermometers. It is important to remember that these are sensitive devices and require consistent care and calibration to provide accuracy and reliability.

PRESSURE MEASUREMENTS

Temperature measurements are usually taken outside the operating system. But it is also necessary for the service technician to know what is going on inside the system, and this he must basically learn from pressure measurements.

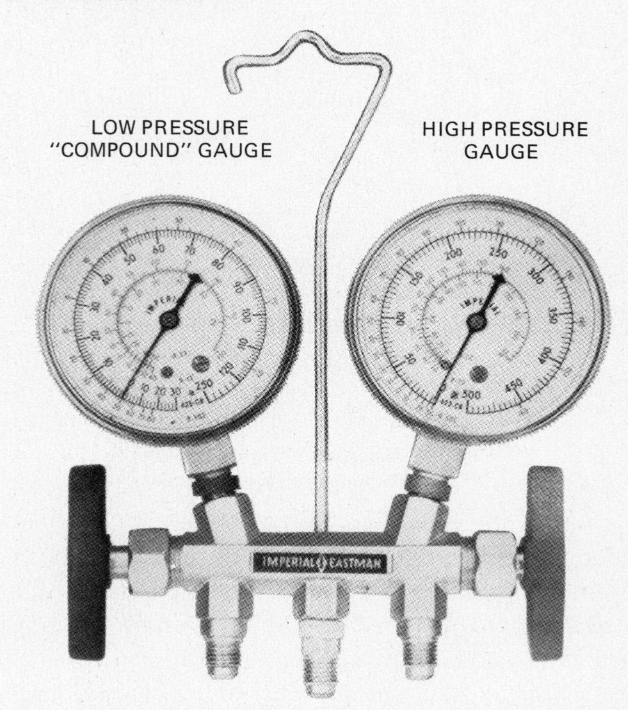

FIGURE 9-7 Pressure gauges Gould Inc., Valves and Fittings Division).

Chapter 3 discussed two pressure gauges as the measuring instruments necessary to obtain these readings (see Fig. 9-7). On the right in Fig. 9-7 is the *high-pressure gauge*, which measures high-side or condensing pressures. It is normally graduated from 0 to 500 psi in 5-lb graduations. The *compound gauge* (left) is used on the low side (suction pressures) and is normally graduated from a 30-in. vacuum to 120 psi; thus it can measure pressure above and below atmospheric pressure. This guage is calibrated in 1-lb graduations. Other pressure ranges are available for both gauges but these two are the most common.

Note that on the dial faces there is an intra scale, which gives the corresponding saturated refrigerant temperatures at a particular pressure. (Remember from Chap. 8 there is a definite relationship between pressure and saturated temperature for a given refrigerant). In Fig. 9-7 the dials are marked for R-12 and R-22. Gauges for other refrigerants and for metric measurements are also available.

A service device that includes both the high and compound gauges is called a *gauge manifold*. It enables the serviceman to check system operating pressures, add or remove refrigerant, add oil, purge noncondensibles, by-pass the compressor, analyze system conditions, and perform many other operations without replacing gauges or trying to operate service connections in inaccessible places.

The testing manifold as illustrated in Fig. 9-8 consists of a service manifold containing service valves. On the left the compound gauge (suction) is mounted and on the right the high-pressure gauge (discharge). On the bottom of the manifold are hoses which lead to the equipment suction service valve (left), refrigerant drum (middle), and the equipment discharge, or liquid-line valve (right).

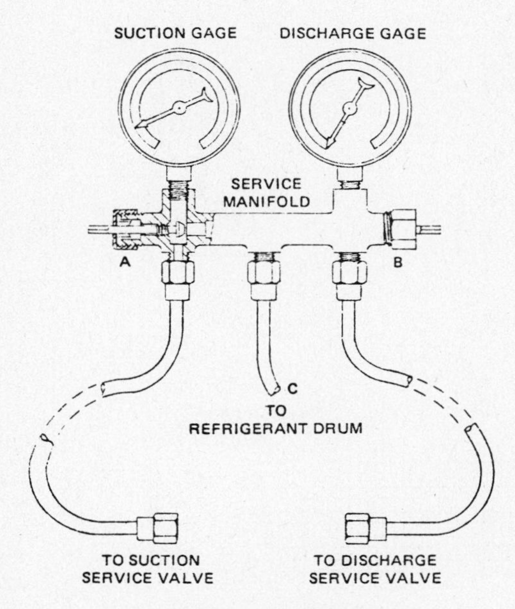

FIGURE 9-8 Testing manifold.

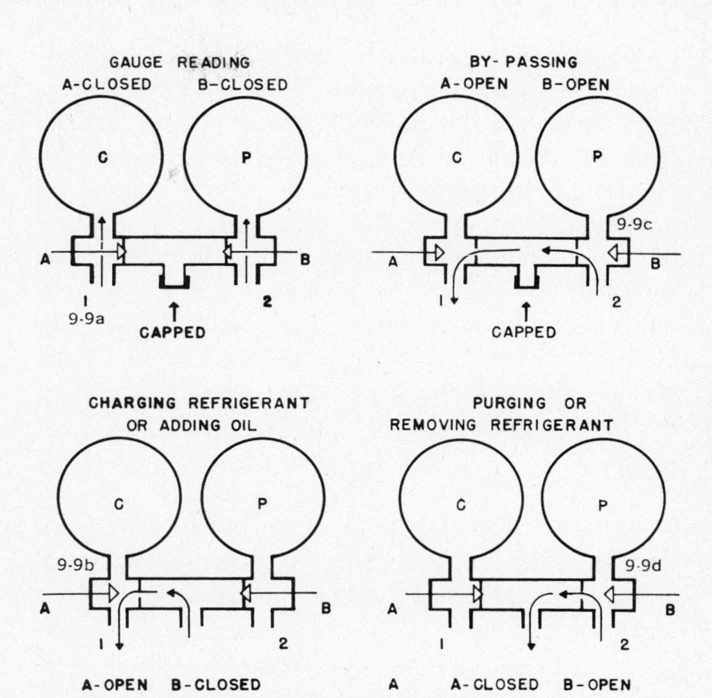

FIGURE 9-9 Manifold valve operation: C-Compound gauge; P-Pressure gauge; 1-Gauge line to suction service S.O.V.; 2-Gauge line to discharge service S.O.V. (Courtesy, BDP Company).

Many equipment manufacturers color code the low-side gauge casing and hose *blue* and the high-side gauge and hose *red*. The center or refrigerant hose is colored *white*. This system is very helpful to avoid crossing hoses and damaging gauges. A hook is provided to hang the assembly and free the operator from holding it.

By opening and closing the refrigerant valves on gauge manifold *A* and *B* (Fig. 9-9) we can obtain different refrigerant flow patterns. The valving is so arranged that when the valves are closed (front-seated) the center port on the manifold is closed to the gauges (Fig. 9-9A). When the valves are in the closed position, gauge ports 1 and 2 are still open to the gauges, permitting the gauges to register system pressures.

With the low-side valve (1) open and the high-side (2) closed (Fig. 9-9B), the refrigerant is allowed to pass through the low side of the manifold and the center port connection. This arrangement might be used when refrigerant or oil is added to the system.

Figure 9-9c illustrates the procedure for bypassing refrigerant from the high side to low side. Both valves are open and the center port is capped. Refrigerant will always flow from the high-pressure area to a lower-pressure area.

Figure 9-9d shows the valving arrangement for purging or removing refrigerant. The low-side valve is closed. The center port is open to the atmosphere or connected to an empty refrigerant drum. The high-side valve is opened, permitting a flow of high pressure out of the center port.

NOTE:

Purging large quantities of flurocarbon refrigerants to the open atmosphere is not recommended unless absolutely necessary.

The method of connecting the gauge manifold to a refrigeration system depends on the state of the system; that is, whether the system is operating or just being installed. For example, let's assume the system is operating and equipped with back seating in-line service valves (Fig. 9-10).

The first step is to purge the gauge manifold of contaminants before connecting it to the system:

(1) Remove the valve stem caps from the equipment service valves and check to be sure that both service valves are back-seated.

(2) Remove the gauge port caps from both service valves.

(3) Connect center hose from gauge manifold to a refrigerant cylinder, using the same type of refrigerant that is in the system, and open both valves on gauge manifold.

(4) Open the valve on the refrigerant cylinder for about two seconds, and then close it. This will purge any contaminants from the gauge manifold and hoses.

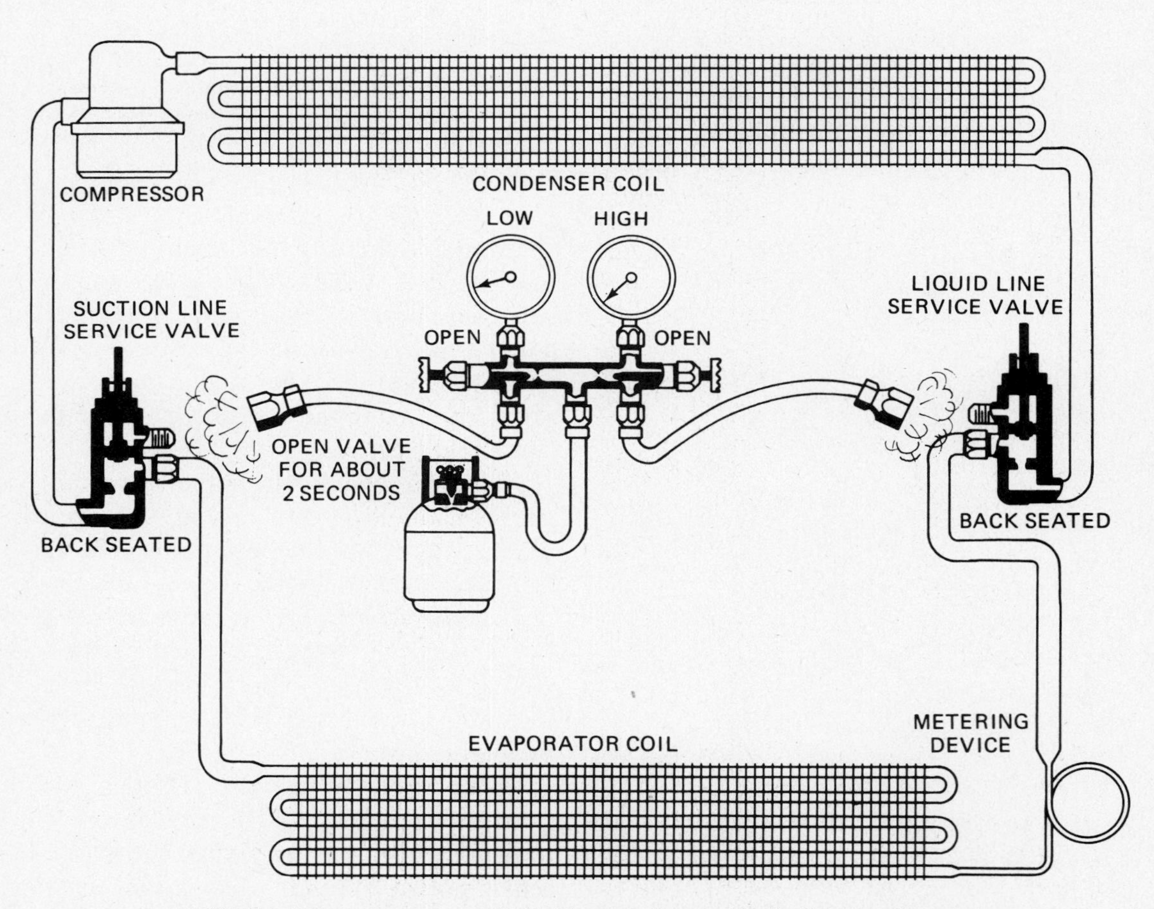

FIGURE 9-10 Purging gauge manifold (Courtesy, Bryant Air-Conditioning Company).

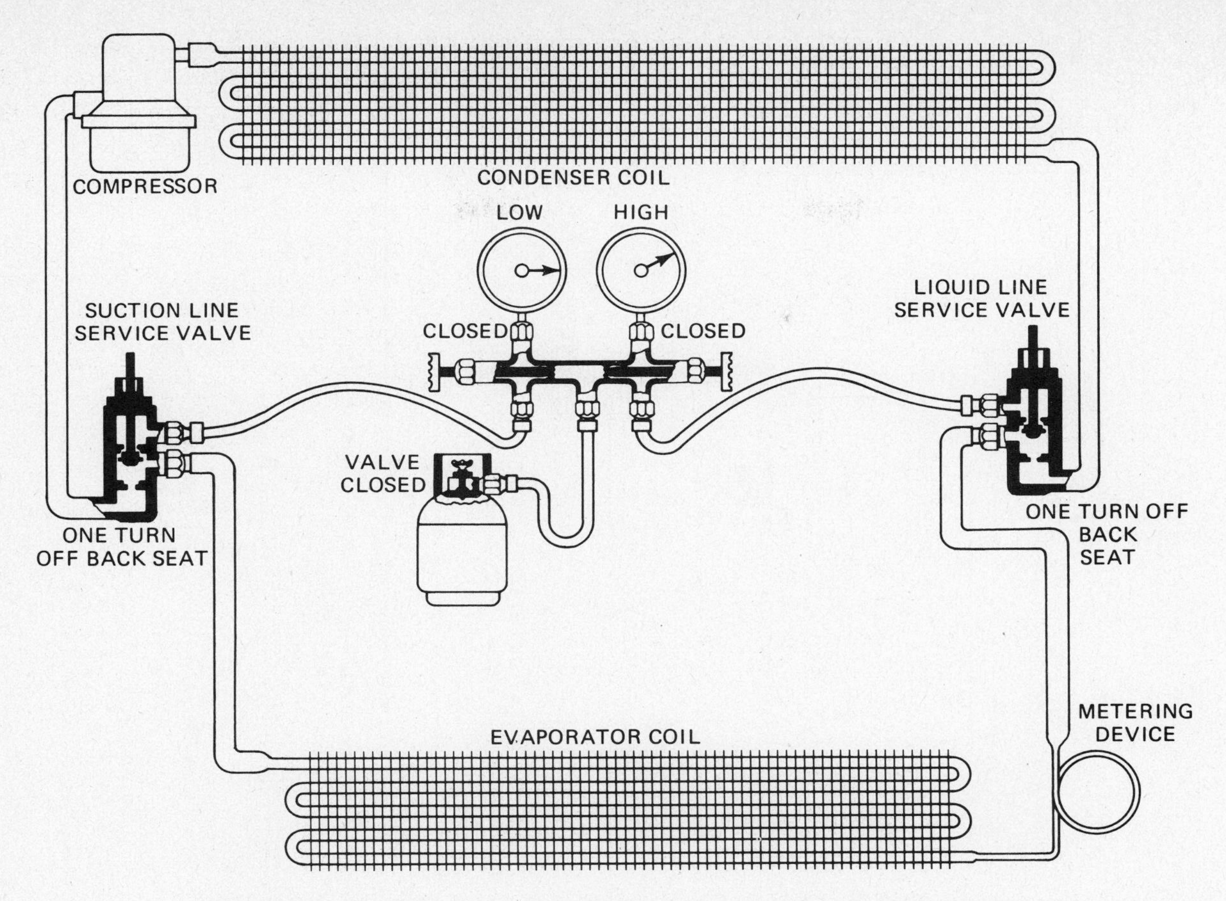

FIGURE 9-11 Connecting manifold (Courtesy, Bryant Air-Conditioning Company).

(5) Next, connect the gauge manifold hoses to the gauge ports—the low-pressure compound guage to the suction service valve and the high-pressure guage to the liquid line service valve, as illustrated in Fig. 9-11.

(6) Front-seat or close both valves on the gage manifold. Crack (turn clockwise) both service valves one turn off the back seat. The system is now allowed to register on each gauge. With the gauge manifold and hoses purged and connected to the system, we are free to perform whatever service function is necessary within the refrigeration cycle.

To remove the gauge manifold from the system, follow this procedure:

(1) Back-seat (counterclockwise) both the liquid and suction service valves on the in-line type.

(2) Remove hoses from gauge ports and seal ends of hoses with $\frac{1}{4}$-in. flare plugs to prevent hoses from being contaminated. (Some manifold assemblies have built-in hose seal fittings).

(3) Replace all gage port and valve stem caps. Make sure all caps *contain the gaskets provided with them* and are tight.

The manifold and gauges are necessary tools to perform many system operations. Once the system has been completed and cleaned of most of the air by purging, it must be tested for leaks. Or, whenever a component has been repaired or replaced, it is imperative that the entire system be checked for leaks.

LEAK TESTING

In most cases a low-pressure refrigerant may be used to build sufficient pressure in the system to check for leaks such as illustrated in Fig. 9-12.

Install the gauge manifold as described previously. Open both valves on the gauge manifold. Open the service valves on system. Open the valve on the refrigerant cylinder and pressurize the system to

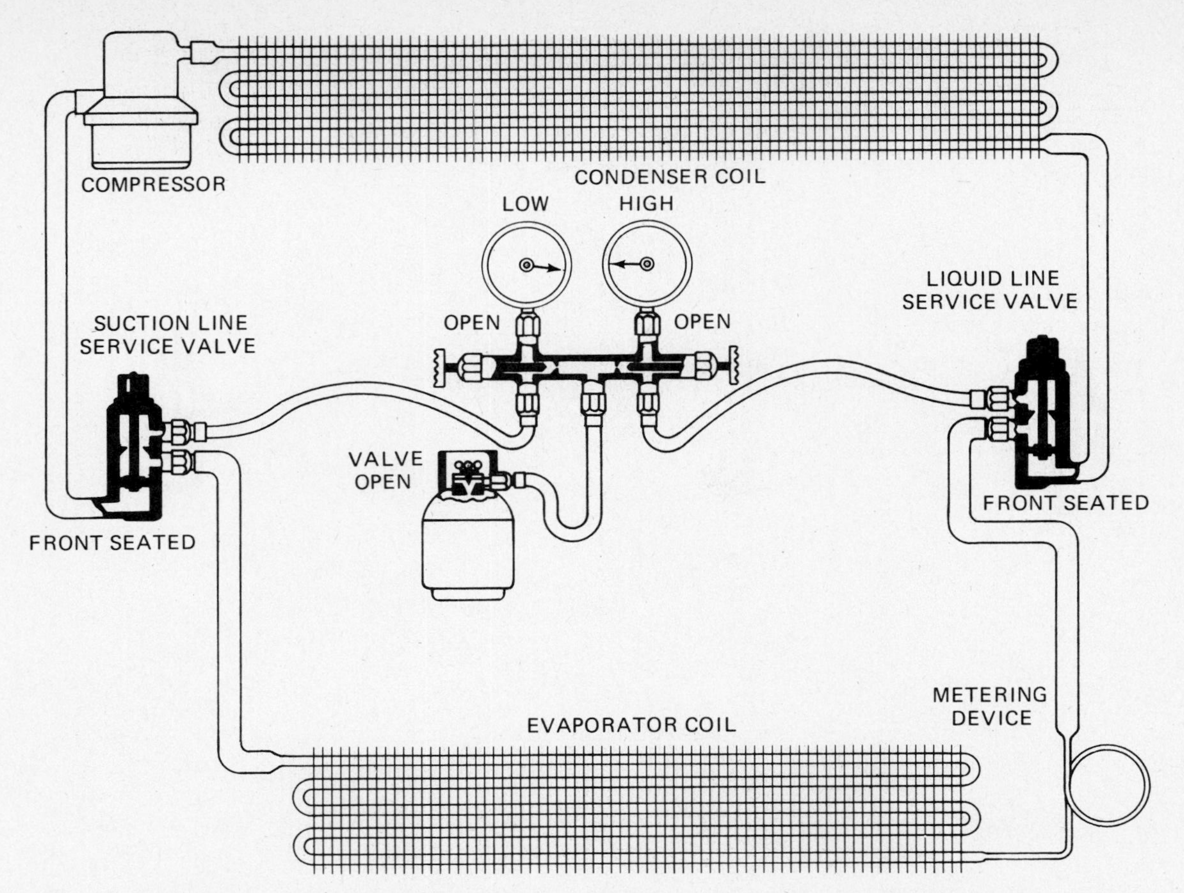

FIGURE 9-12 Using test manifold while leak testing (Courtesy, Bryant Air-Conditioning Company).

the cylinder pressure with refrigerant vapor (keep cylinder in upright position). Use any one of the three methods of leak testing as described in Chap. 7. Where the application or requirement of local codes requires a pressure test above the refrigerant vapor pressure, some other gas may be used for testing, for example, dry nitrogen. *Under no circumstances should oxygen be used.* Nitrogen would be introduced into the system through the center hose connection after disconnecting the refrigerant cylinder.

CAUTION:—the test gas cylinder must be equipped with a pressure gauge and regulator so that system test pressures do not exceed maximum allowable limits as prescribed by national or local codes or by the equipment manufacturer.

PURGING

Whenever a system is exposed to atmospheric conditions for a short period of time (less then five minutes for example), during component replacement, it is necessary to purge the system to remove any contaminants that may have entered the system. Similarly, during installation, if the refrigerant lines are left open for more than five minutes, the system should be purged.

The theory behind purging is to use a high-velocity charge of gaseous refrigerant to blow any contaminants from the system.

To purge a system which has just been installed, proceed as follows:

> Install the gauge manifold as illustrated in Fig. 9-13, with the low-side valve closed and not connected ot the suction service valve. Connect the center hose to the refrigerant drum. Connect the high-side hose to the liquid-line valve. Front-seat both service valves and open high-side manifold valve. Open valve (wide) on the refrigerant cylinder and allow a high-velocity charge of refrigerant vapor ($\frac{1}{2}$ to 1 lb or more depending on size of equipment) to enter the system. The refrigerant will push any contaminants through the system to the suction service valve where they will be purged through the gauge port.

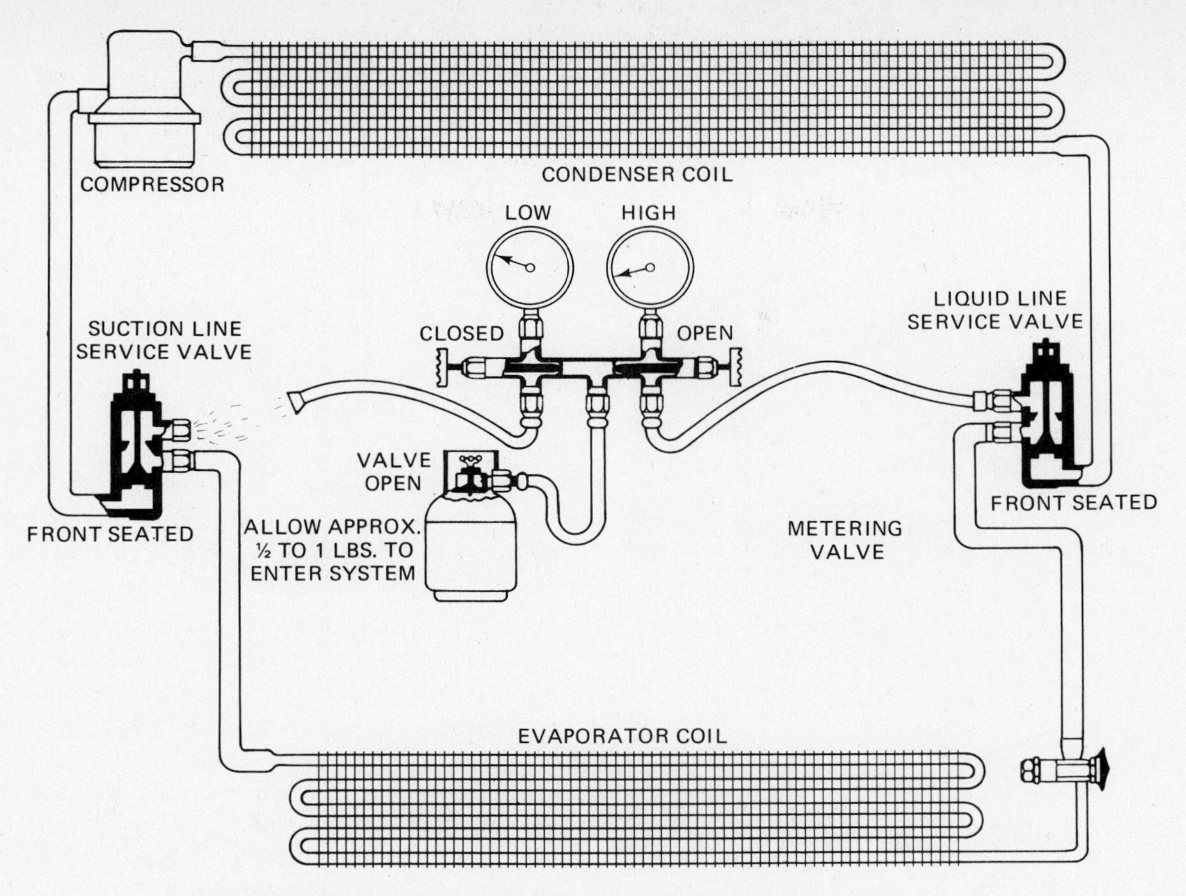

FIGURE 9-13 Purging (Courtesy, Bryant Air-Conditioning Company).

Whenever a defective component such as an expansion valve is to be removed, the system should be pumped down and that part of the system should be isolated by means of the service valves. Then when the new component is installed the lines should be purged from both sides. *Pumping down* means to store the refrigerant in the receiver or condenser.

EVACUATION

Proper evacuation of a unit will remove noncondensibles (mainly air, water, and inert gases) and assure a tight, dry system before charging. There are generally two methods used to evacuate a system: the *deep vacuum* method, and the *triple evacuation* method. Each has its advantages and disadvantages. The choice depends on several factors: type of vacuum pump available, the time that can be spent on the job, and whether there is liquid water in the system.

In refrigeration work, especially those systems that operate at very low suction pressures, the deep vacuum method is recommended, In higher temperature refrigeration systems and in air-conditioning work, triple evacuation is practiced. Both methods will be discussed.

The tools needed to evacuate a system properly depend on the method used. A good vacuum pump and vacuum indicator are needed for the deep vacuum method, and a good vacuum pump and compound gauge are needed for the triple evacuation method.

VACUUM PUMP

A vacuum pump, as illustrated in Fig. 9-14, is somewhat like an air compressor in reverse. Most are driven by a direct or belt-driven electric motor but gasoline engine-driven pumps are also available. The pump may be single or two stage depending on the design. Most pumps for normal field service are portable; they have carrying handles or are mounted on wheel dollies. Vacuum pump sizes are rated according to the free air displacement in cubic feet per minute or liters per minute metric. Specifications

FIGURE 9-14 Vacuum pump (Courtesy, Gould, Inc., Valves and Fittings Division).

may also include a statement as to the degree of vacuum the pump can achieve, expressed in terms of microns.

What is a micron? When the vacuum pressure approaches 29.5 in. to 30 in. on the compound gauge, the gauge is working within the last half inch of pressure, and the readout beyond 29.5 in. is not reliable for the single deep vacuum method. The industry has therefore adopted another measurement called the *micron*. The micron is a unit of linear measure equal to 1/25,400 th of an inch and is based on measurement above total absolute pressure, as opposed to gauge pressure, which can be affected by atmospheric pressure changes. Table 9A is a comparison of measurements starting at standard atmospheric conditions and extending to a deep vacuum.

Table 9A not only demonstrates the comparison in units of measure but dramatically shows the changes in the boiling point of water as the evacuation approaches the perfect vacuum. This is the main purpose of evacuation—to reduce the pressure or vacuum enough to boil or vaporize the water and then pump it out of the system. It will be noted the compound gauge could not possibly be read to such minute changes in inches of mercury.

TABLE 9A

BOILING POINT OF WATER		UNIT OF ABSOLUTE PRESSURE		UNITS OF VACUUM
F°	*C°*	*PSIA*	*Microns of Mercury*	*Inches of Mercury*
212	100	14.7	–	0
79	26	0.5	25,400	29.0
72	22	0.4	20,080	29.8
32	0	0.09	4,579	29.99
−25	−31	0.005	250	29.99
−40	−40	0.002	97	29.996
−60	−51	0.0005	25	29.999

HIGH-VACUUM INDICATORS

To measure these high vacuums the industry developed electronic instruments, such as is shown in Fig. 9-15. In general these are heat-sensing devices in that the sensing element, which is mechanically connected to the system being evacuated, generates heat. The rate at which heat is carried off changes as the surrounding gases and vapors are removed. Thus the output of the sensing element (either thermocouple or thermistor) changes as the heat dissipation rate changes, and this change in output is indicated on a meter that is calibrated in microns of mercury.

The degree of accuracy of these instruments is approximately 10 microns, thereby approaching a perfect vacuum as shown in Table 9A.

FIGURE 9-15 Electronic high vacuum gauge (Courtesy, Robinair Manufacturing Company).

DEEP VACUUM METHOD OF EVACUATION

The single deep vacuum method is the most positive method of assuring a system free of air and water. It takes slightly longer but the results are far more positive. Select a vacuum pump capable of pulling at least 500 microns and a reliable electronic vacuum indicator. The procedure is illustrated in Fig. 9-16, and is described below.

(1) Install gauge manifold as described earlier.

(2) Connect center hose to vacuum manifold assembly. This is simply a three-valve operation allowing you to attach the vacuum pump and vacuum indicator and a cylinder of refrigerant, each with a shut-off valve.

(3) Open the valves to the pump and indicator. Close the refrigerant valve. Follow pump manufacturer's instructions for pump suction line size, oil, indicator location, and calibration.

(4) Open (wide) both valves on gauge manifold and mid-seat both equipment service valves.

(5) Start vacuum pump and evacuate system until a vacuum of at least 500 microns is achieved.

(6) Close pump valve and isolate the system. Stop the pump for five minutes and observe the vacuum indicator to see if the system has actually reached 500 microns and is holding. If system fails to hold, check all connections for tight fit and repeat evacuation until system does hold.

(7) Close valve to indicator.

(8) Open valve to refrigerant cylinder and raise the pressure to at least 10 psig or charge system to proper level (covered later).

(9) Disconnect pump and indicator.

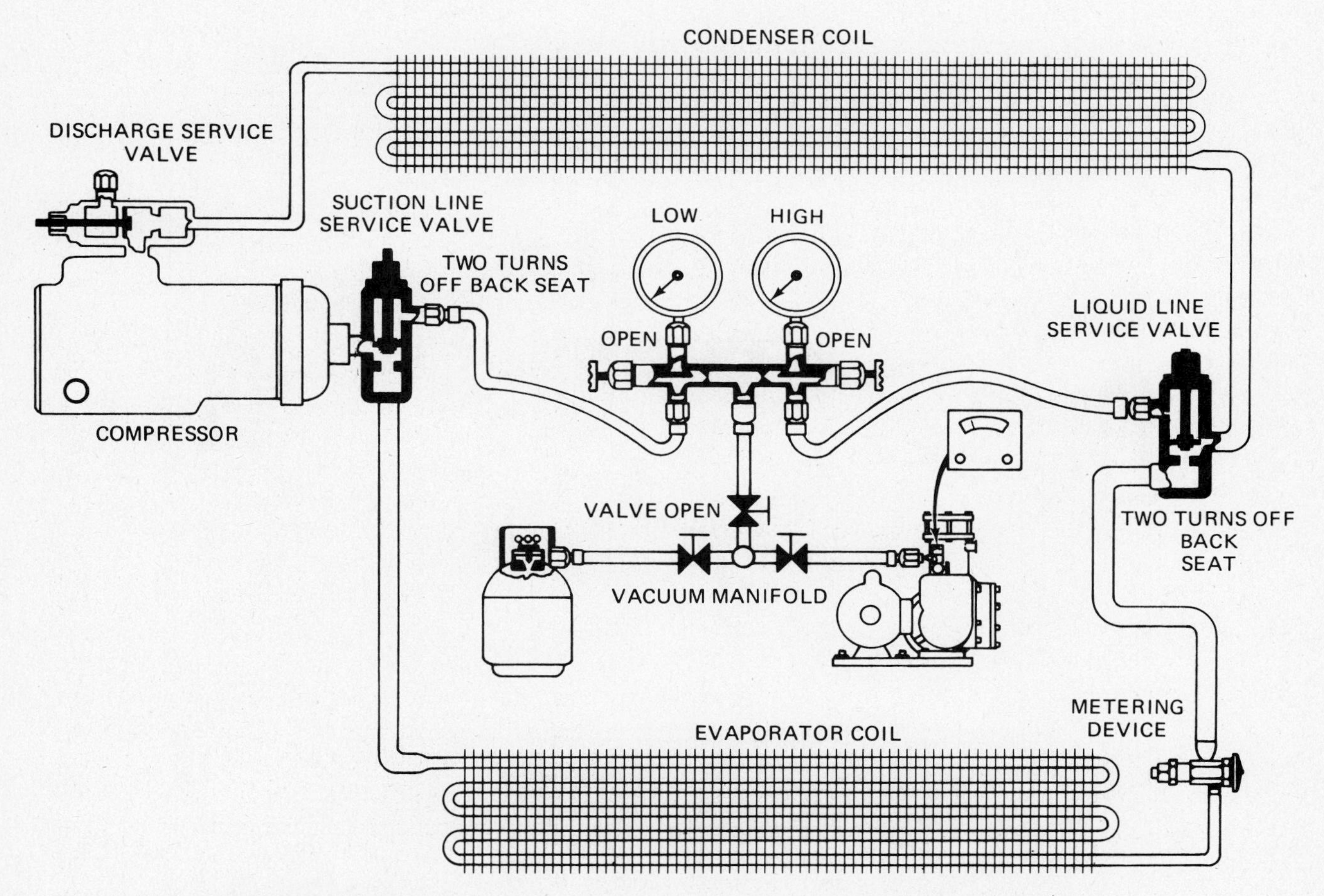

FIGURE 9-16 Deep vacuum evacuation assembly (Courtesy, Bryant Air-Conditioning Company).

TRIPLE EVACUATION

The triple evacuation method requires no specialized high-vacuum equipment. However, this method should not be used if liquid water is suspected in the system. An evacuation pump of sufficient capacity to pull 28 in. of mercury vacuum will be needed. Good quality refrigeration service gauges are important.

This method of evacuation is based on the principle of diluting the noncondensables and moisture with clean, dry refrigerant vapor. This vapor is then removed from the system, carrying with it a portion of the entrained contaminants. As the procedure is repeated, the remaining contaminants are proportionately reduced until the system is contaminant-free. Figure 9-17 illustrates the assembly procedure, described below.

(1) Install gauge manifold as described earlier.

(2) Connect center hose to vacuum manifold valves.

(3) Connect pump and refrigerant cylinder to manifold valves. Purge lines with refrigerant.

(4) Close refrigerant cylinder valve and open pump valve.

(5) Open (wide) both valves on gauge manifold and mid-seat both service valves.

(6) Start evacuation pump and evacuate system until a 28-in. mercury vacuum is reached on compound gauge. Let pump operate for 15 minutes at this level.

(7) Close pump valve and stop pump.

(8) Open refrigerant valve. Allow pressure to rise to 2 psig. Then close refrigerant valve. Allow refrigerant to diffuse through system and absorb moisture for five minutes before next evacuation.

(9) Close refrigerant valve. Open pump valve and repeat evacuation steps to again reach 28-in. mercury vacuum and hold for 15 minutes with pump running.

(10) Close pump valve and turn off pump. Open

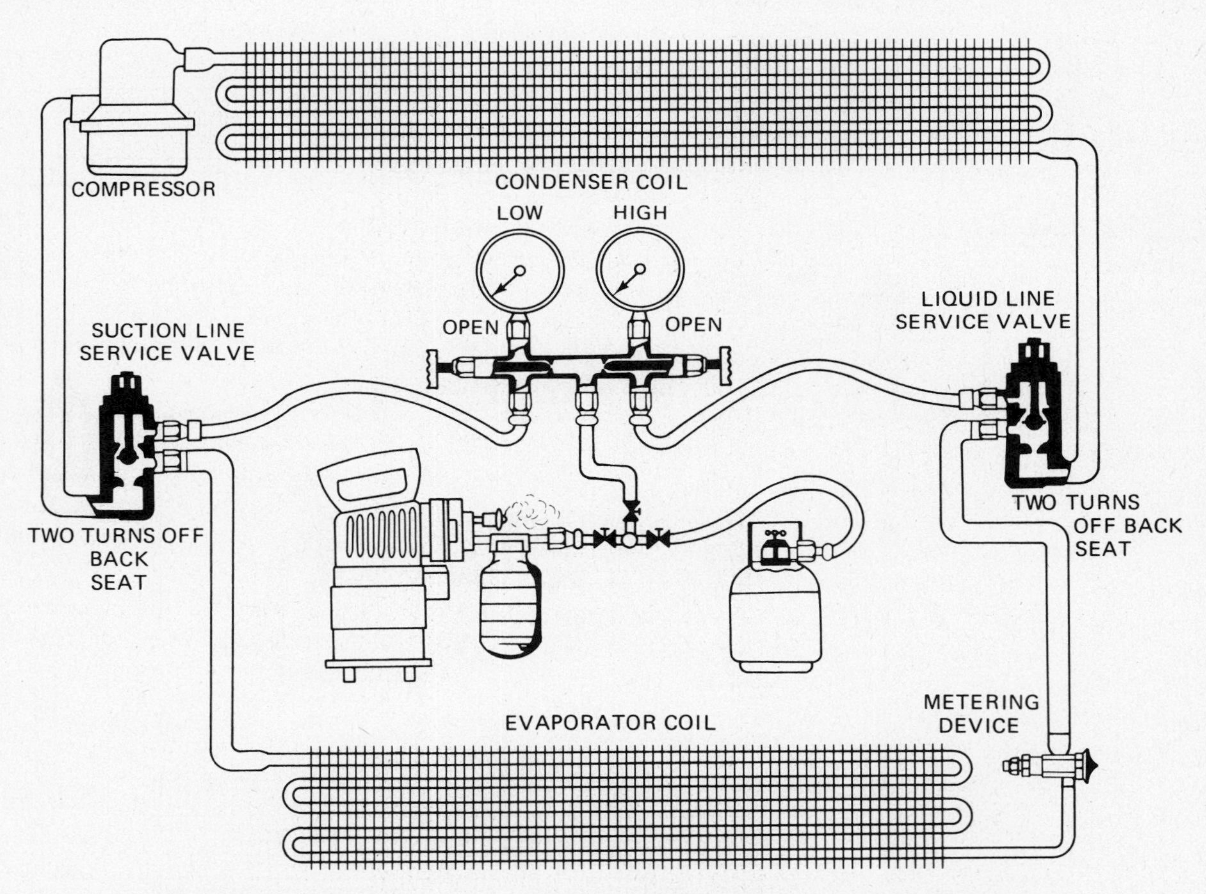

FIGURE 9-17 Triple evacuation assembly (Courtesy, Bryant Air-Conditioning Company).

refrigerant valve and charge to 2 psig, again holding for five minutes.

(11) Close refrigerant valve. Open pump valve. Start pump and evacuate again to 28-in. vacuum and hold 15 minutes.

(12) Stop pump and break vacuum, this time charging system to 10 psig or to proper level.

CHARGING THE SYSTEM

The quantity of refrigerant to be added to the system for initial charge or recharging depends on the size of the equipment and the amount of refrigerant to be circulated. In very large systems it is common practice to simply weigh the charge by placing the refrigerant cylinder or drum on a suitable scale and observing the reduction in weight in pounds. This method is fine for systems that have receivers or condenser volume ample enough to take a slight overcharge.

On smaller systems, and particularly those that are self-contained packaged units without receivers, the system refrigerant charge is *critical* to ounces, rather than whole pounds. In this case a "charging cylinder" is recommended as illustrated in Fig. 9-18. Refrigerant from the refrigerant drum is trans-

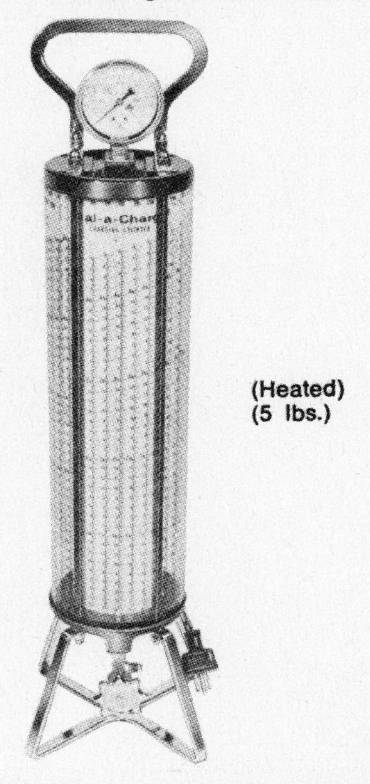

FIGURE 9-18 Charging cylinder (Courtesy, Robinair Manufacturing Corporation).

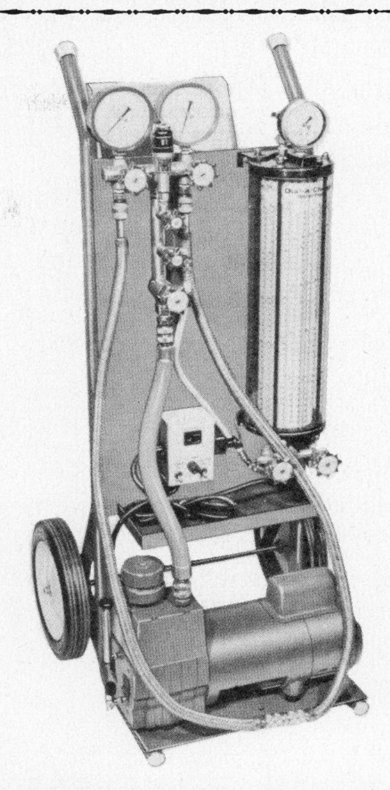

FIGURE 9-19 Mobile vacuum and charging station (Courtesy, Robinair Manufacturing Corporation).

ferred to the charging cylinder. The charging cylinder has a scale that is visible to the operator so that he may precisely measure the quantity of a specific refrigerant and compensate for temperature and pressure conditions. These cylinders are accurate to one-quarter of an ounce. Optional electric heaters are available to speed the charging operation.

Where considerable installation and service work is involved many contractors use a mobile evacuation and charging station, as illustrated in Fig. 9-19. It contains a vacuum pump, charging cylinder, and service manifold and gauges. More elaborate models may also include a vacuum indicator and space for refrigerant cylinder.

CHARGING TECHNIQUES

Refrigerant may be added in either the liquid or vapor form. Refrigerant is added in the vapor form, when the unit is operating, through the suction valve.

Refrigerant may be added in the liquid form, when the unit is off and in an evacuated condition, through the liquid-line service valve only.

Figure 9-20 illustrates the charging procedure for the vapor form (unit running). For simplicity we show only a refrigerant cylinder and assume the charge is weighed in during operation.

(1) Install gauge manifold.

(2) Attach refrigerant cylinder to center connection hose and open valve on low side of manifold.

(3) Place cylinder in upright position.

(4) Crack suction service valve two turns off back seat.

(5) Open valve on refrigerant cylinder and weigh in desired charge.

(6) When correct charge has been added, close low-side manifold valve and valve on refrigerant cylinder.

(7) Back-seat both suction and liquid-line service valves, remove hoses, and cap ports.

The charging procedure for the liquid form (unit not operating and evacuated) is detailed below and shown in Fig. 9-21.

(1) Install gauge manifold.

(2) Attach refrigerant cylinder, Invert cylinder unless it is equipped with a liquid-vapor valve, which permits liquid withdrawal in upright position.

(3) Open both suction and liquid service valves one turn off back-seat.

(4) Open valve on high side of gauge manifold.

(5) Open valve on refrigerant cylinder and add refrigerant.

(6) After correct charge is introduced, close valve

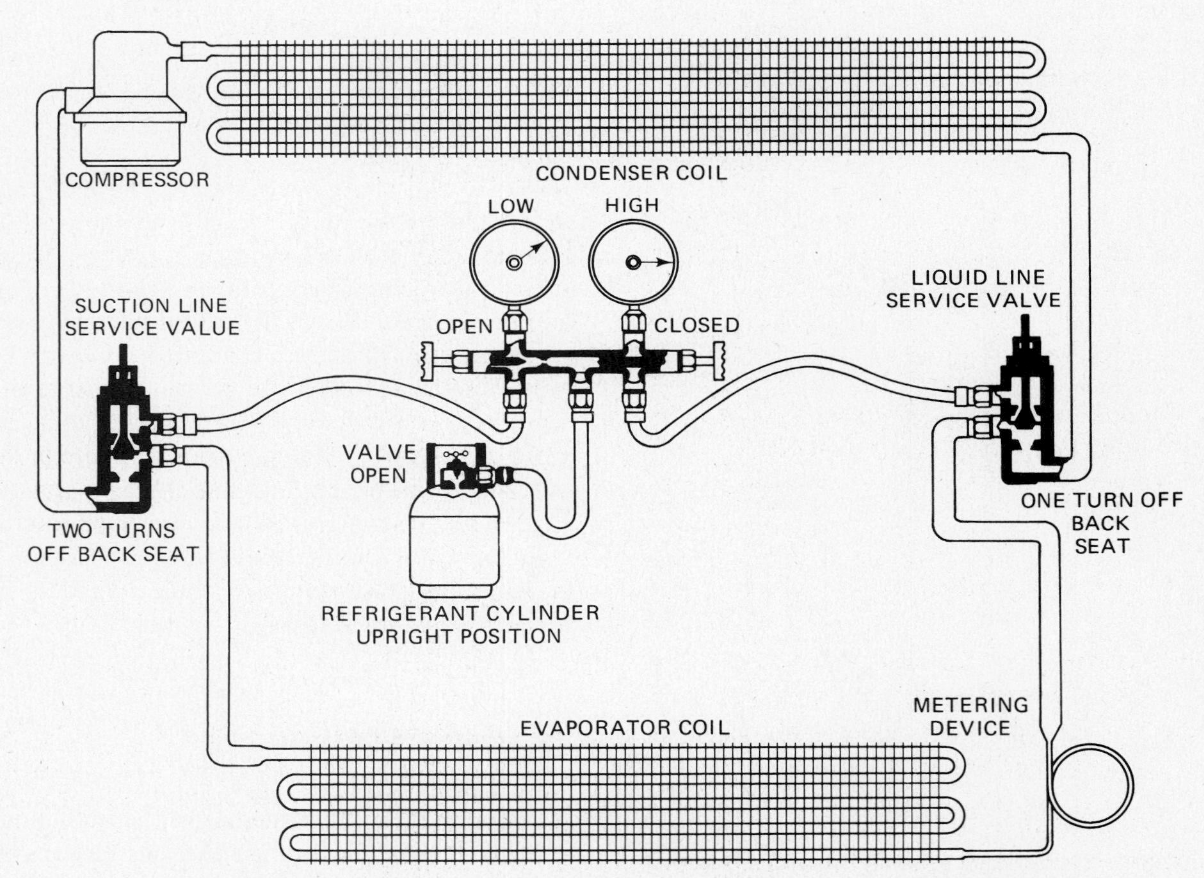

FIGURE 9-20 Vapor-form charging (Courtesy, Gould, Inc., Valves and Fittings Division).

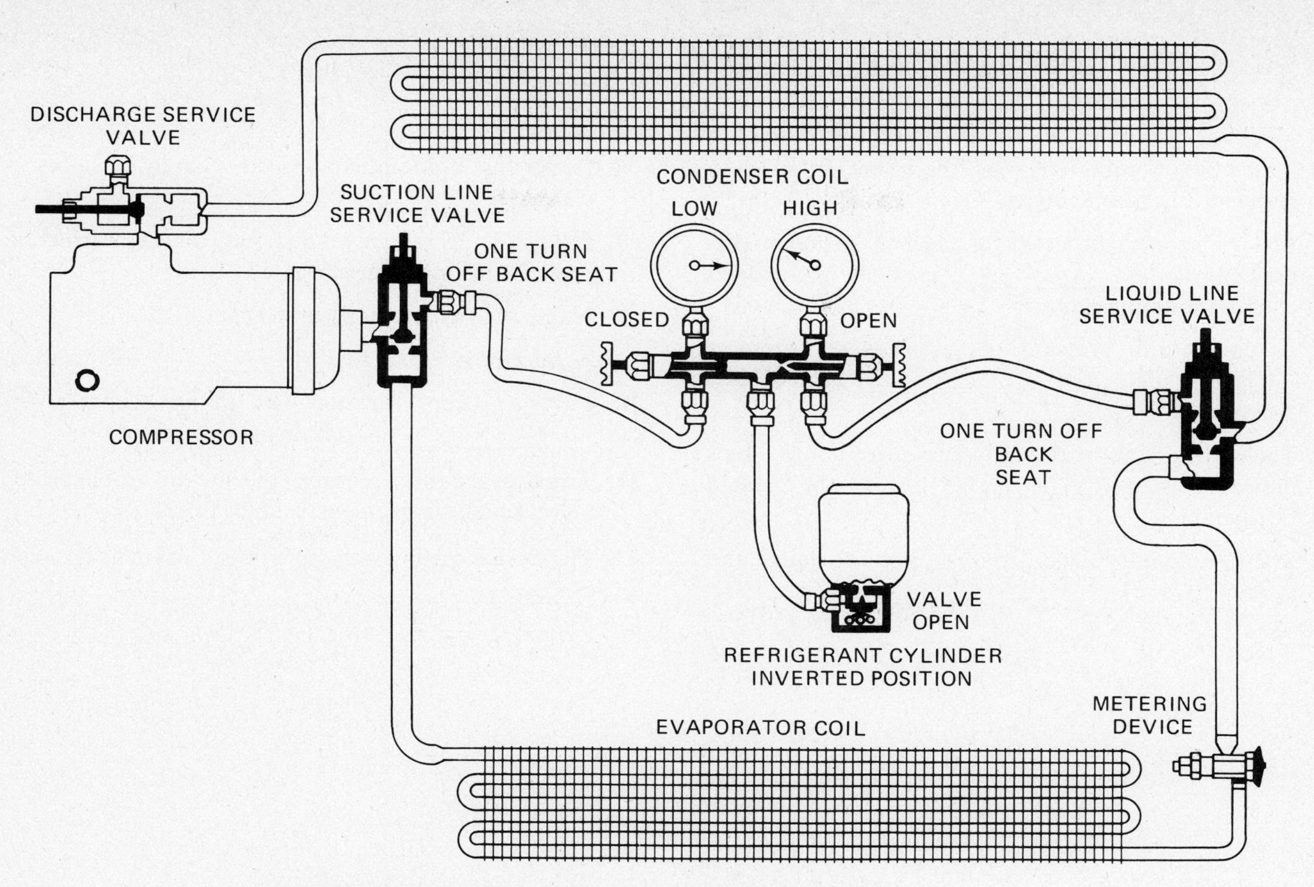

FIGURE 9-21 (Courtesy, Gould, Inc., Valves and Fittings Division).

on high side of manifold and close valve on refrigerant cylinder. Back-seat both suction and liquid service valves.

(7) Remove gauge manifold.

In both descriptions above the use of a charging cylinder would be recommended on smaller, critically charged systems where more accuracy is needed.

CHECKING THE CHARGE

Checking the charge of a new installation or of an existing unit is another function of the service manifold gauges. For example, the following procedure would be used for an air-cooled unit.

(1) Install gauge manifold.

(2) Allow system to operate until pressure gauge readings stabilize (approximately 15 minutes).

(3) While unit is operating, record the following information.

  (a) High-pressure gauge reading.

  (b) Dry bulb temperature of air entering the condenser coil.

  (c) Wet bulb temperature of air entering the evaporator coil. (This is done with a wet-wick thermometer.)

(4) A comparison of the above measurements with the head pressure charging table supplied with the unit will indicate if the system is adequately charged and operating properly.

The refrigeration measuring and testing equipment and their proper use as discussed above are the most basic tools for field service and installation. With experience or by need they may be supplemented with special devices to improve skills or save time.

PROBLEMS

1. Name two common pocket thermometers used by refrigeration technicians.
2. The electronic thermometer sensing probe makes use of the ______________ element to register changes in temperature.
3. A compound pressure gauge measures pressure ______________ and ______________ atmospheric pressure.
4. The refrigerant scales on pressure gauge dials indicate the ______________ for the stated refrigerant.
5. The device that includes both a pressure gauge and a service valve is called a ______________.
6. Removing contaminants from manifold hoses is called ______________.
7. Oxygen is a recommended gas for lead-detection pressure testing: true or false?
8. Name the two methods of evacuation.
9. A micron is equal to ______________ of an inch.
10. For a deep vacuum a pump capable of ______________ microns is recommended.
11. Triple evacuation requires the use of an electronic high vacuum instrument: true or false?
12. The device used to charge "critical" systems is called a ______________.

10

REFRIGERATION SYSTEM OPERATION

In previous chapters we have learned the major components of the refrigeration cycle, the theoretical operation of a cycle, the materials and tools needed to fabricate a system, and the instruments needed to test and check the operation. We will now put these elements to work by reviewing what might be expected, first during normal operation and then under abnormal conditions.

First, the field service technician does not usually carry with him a pressure-enthalpy diagram for each refrigerant. He will however carry with him a pocket-size pressure-temperature chart as illustrated in Fig. 10-1, which lists the saturated refrigerant pressure temperatures for various refrigerants. R-12, R-22, R-500, and R-502 are listed; they are the most common systems refrigerants. At any given saturated liquid temperature the technician can determine the equivalent pressure (or vacuum, if it is a low-temperature application). Now let's look at three systems and see what actually happens.

Figure 10-2 represents an air-cooled refrigeration cycle used in a freezer operation where the room storage temperature is to be held at 0°F and the outdoor (ambient) temperature is 95°F.

First identify the system components: compressor, condenser, evaporator, and expansion valve metering device. Next identify the glass (or dial) thermometers used to measure air temperature to evaporator and condenser. A six-station electronic thermometer is used to check temperatures at points *A–F*. Manifold gauges are attached to the respective service valves in the system. The system refrigerant for this example is R-22.

Let's first consider the evaporator conditions. Since the freezer room is to be held at 0°F it will be necessary to maintain a temperature differential between the air across the evaporator and the refrigerant temperature; This differential varies with the particular evaporator design. For this example we will assume a 10°F differential. The saturated refrigerant temperature at point *A* just as it enters the evaporator is −10°F. From the P/T chart in Fig. 10-1 note the corresponding pressure of R = 22 is 16.4 psig. As the evaporator picks up heat and the refrigerant boils, it will leave the evaporator at 0°F, meaning it has gained 10 degrees of superheat in

TEMPERATURE—PRESSURE CHART

	PSIG					PSIG			
TEMP.	*R-12*	*R-22*	*R-500*	*R-502*	*TEMP.*	*R-12*	*R-22*	*R-500*	*R-502*
−50	**15.4**	**6.1**	—	0.0	55	52.0	92.5	64.1	109.7
−45	**13.3**	**2.7**	—	2.0	60	57.7	101.6	71.0	115.6
−40	**11.0**	0.5	**7.9**	4.3	65	63.8	111.2	78.1	125.8
−35	**8.4**	2.5	**4.8**	6.7	70	70.2	121.4	85.8	136.6
−30	**5.5**	4.8	**1.4**	9.4	75	77.0	132.2	93.9	147.9
−25	**2.3**	7.3	1.1	12.3	80	84.2	143.6	102.5	159.9
−20	0.6	10.1	3.1	15.5	85	91.8	155.6	111.5	172.5
−15	2.4	13.1	5.4	19.0	90	99.8	168.4	121.2	185.8
−10	4.5	16.4	7.8	22.8	95	108.3	181.8	131.3	199.7
− 5	6.7	20.0	10.4	26.9	100	117.2	195.9	141.9	214.4
0	9.2	23.9	13.3	31.2	105	126.6	210.7	153.1	229.7
5	11.8	28.1	16.4	36.0	110	136.4	226.3	164.9	245.8
10	14.6	32.7	19.8	41.1	115	146.8	242.7	177.4	266.1
15	17.7	37.7	23.4	46.6	120	157.7	259.9	190.3	280.3
20	21.0	43.0	27.3	52.4	125	169.1	277.9	204.0	298.7
25	24.6	48.7	31.6	58.7	130	181.0	296.8	218.2	318.0
30	28.5	54.8	36.1	65.4	135	193.5	316.5	233.2	338.1
35	32.6	61.4	41.0	72.6	140	206.6	337.2	248.8	359.2
40	37.0	68.5	46.2	80.2	145	220.3	358.8	—	381.1
45	41.7	76.0	51.8	87.7	150	234.6	381.5	—	404.0
50	46.7	84.0	57.8	96.9					

BOLD FACE NUMERALS = Inches Hg. Below 1 ATM.

FIGURE 10-1 Pocket-size P/T table.

terms of temperature rise, but the vapor pressure will essentially remain the same if we assume there is no pressure drop in the coil. The feeler bulb from the thermostatic expansion valve reacts to changes in temperature at point *B* and will regulate refrigerant flow in an attempt to maintain a constant 10° superheat.

Depending on the size and length of the suction line, two things will occur. As the suction line runs outside the freezer room into a warmer location, it will gain some additional superheat (2°F is shown for this example). And, because of friction loss within the pipe, there will be some pressure reduction at point *C* as the gas enters the compressor suction valve.

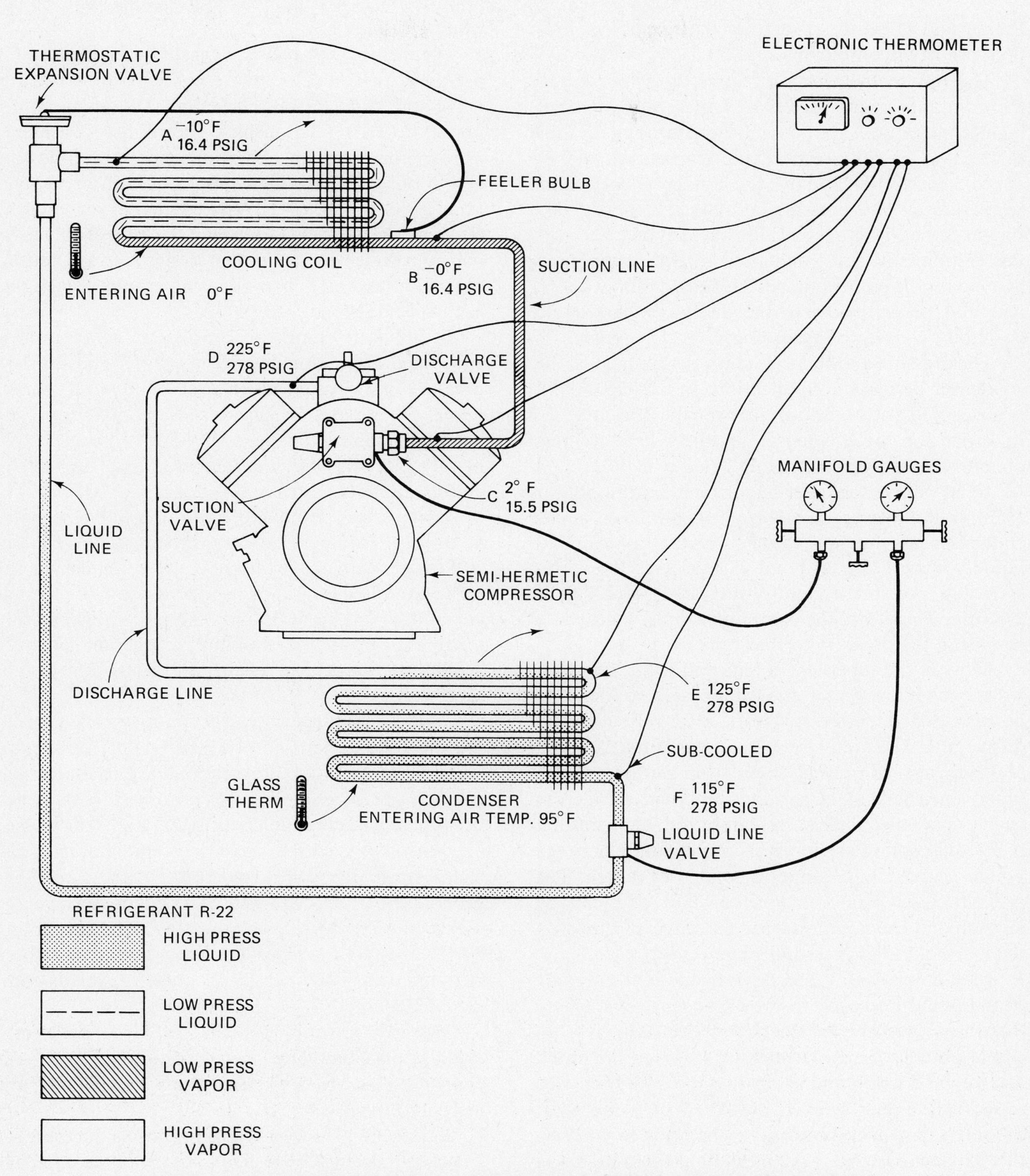

FIGURE 10-2 Freezer room application—refrigerant cycle with expansion valve and R-22 refrigerant.

Pressure at *C* can be read from the compound low-pressure gauge on our manifold.

The low temperature, low pressure vapor enters the compressor and is elevated to a high-pressure, high-temperature vapor as indicated at point *D*. Due to the heat of compression the temperature will be approximately 100°F above the condenser saturation temperature. The pressure at points *D* and *E* are shown to be the same, because the distance between the compressor and condenser in this example is assumed to be close coupled. This may not always be true and pressure loss in the discharge line would have to be considered accordingly.

The pressure at point *D* is really determined by the condenser design and application. In this example it is air cooled with outdoor air over the coil at 95°F. In order for the condenser to reject heat it must operate above 95°F. Generally, this will be around 30° to 40° above the entering air temperature. In our example the average condensing temperature in the condenser as the hot vapor becomes a saturated liquid will be 125°F at point *E*. For R-22 the corresponding pressure (refer to Fig. 10-1) is 277.9 or 278 psig. Therefore, the compressor must be capable of raising the pressure to that level.

Most of the superheat is removed from the vapor by the upper rows of the condenser, so that the bottom two or three rows are then full of liquid refrigerant. This fulfills two functions. First, a liquid seal is formed to prevent vapor from getting into the liquid line. Second, it permits the liquid refrigerant to be subcooled below its saturation temperature. A 10° subcooling as shown in our example between points *E* and *F* is a common condenser design. The pressure (278 psig) remains constant and can be measured at the liquid-line service valve and read on the manifold high-pressure gauge.

Liquid refrigerant then flows from point *F* by way of the liquid line back to the expansion valve where there is a pressure differential of over 250 psig. In actual practice there are pressure losses in the liquid line due to friction and also when the liquid must be raised to a higher level. If pressure loss is too great the refrigerant will flash into a gas prior to entering the expansion valve. Chapter 19 covers in detail procedures for determining refrigerant line sizes so as to minimize pressure loss, insure proper oil return to the compressor, and eliminate objectionable noise.

As a contrast to the low-temperature freezer application, let's now examine a high-temperature application as would be found in a comfort air-conditioning system.

Figure 10-3 shows a refrigerant cycle diagram for a typical air-cooled system employing an expansion valve and using R-500 refrigerant. The temperature-pressure (*T-P*) relationships shown are considered normal for a system operating under the following conditions: the entering evaporator coil air temperature is 80°F DB, 67°F WB; condenser air entering temperature is 95°F DB. Since the pressures shown will be affected by equipment design, always consult the manufacturer's pressure-temperature charts for each specific model.

Starting at point *A*, the liquid refrigerant is subcooled approximately 16° to 114°F and exerts a pressure of 218 psig. Assuming the liquid refrigerant line 25 ft long and properly sized, the pressure at the entering side of the expansion valve will be approximately 218 psig with the temperature of the refrigerant still about the same as at *A*, i.e., 114°F.

Downstream from the expansion valve at point *B*, the T-P readings will be approximately 44°F and 50.7 psig. With the expansion valve regulating the refrigerant flow for 12° of superheat, the *T-P* readings at point *C* will be 56°F and 50.7 psig. This pressure reading neglects the nominal refrigerant pressure drop of the vaporizing refrigerant through the evaporator coil.

At the suction gas intake at the compressor (point *D*), the *T-P* relationship of 61°F and 50.7 psig represents the temperature pickup of additional superheat caused by the suction line. No pressure loss is shown because we assume the cooling coil and compressor are reasonably close-coupled. At the hot gas discharge from the compressor (at *E*), these readings become 195°F and 218 psig. The condenser average *T-P* is 130°F and 218 psig at point *F*. The subcooling (16°F) then takes place at the last two and three bottom rows of the condenser and the cycle is completed at point *A*.

The two previous systems employed a thermostatic expansion valve metering device. Now let's compare these to a system using a capillary tube metering device.

Figure 10-4 shows a refrigerant cycle diagram for a typical air-cooled system employing a capillary tube and using R-22 refrigerant. The temperature-

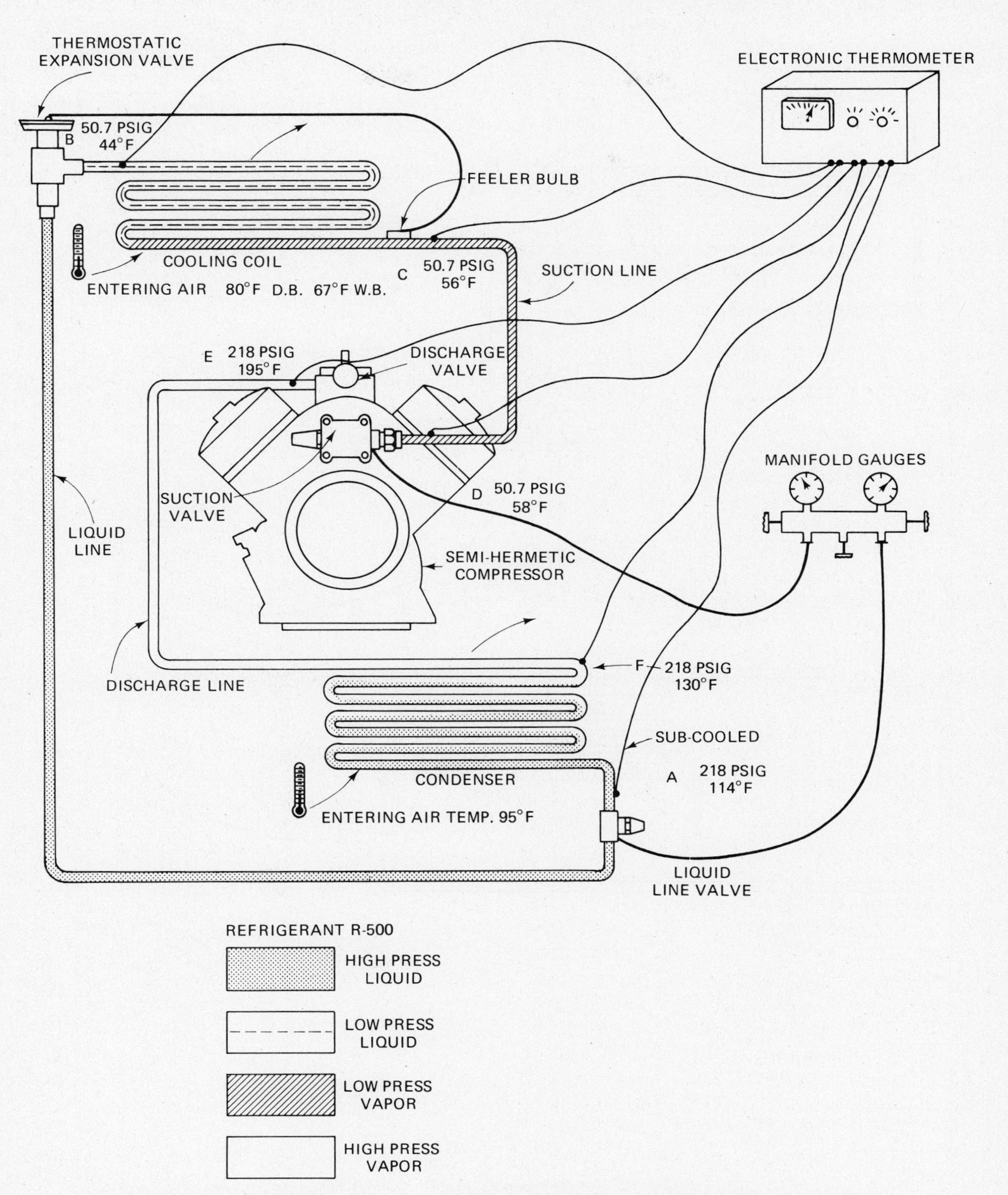

FIGURE 10-3 High temperature refrigeration application (air-conditioning) with thermostatic expansion valve.

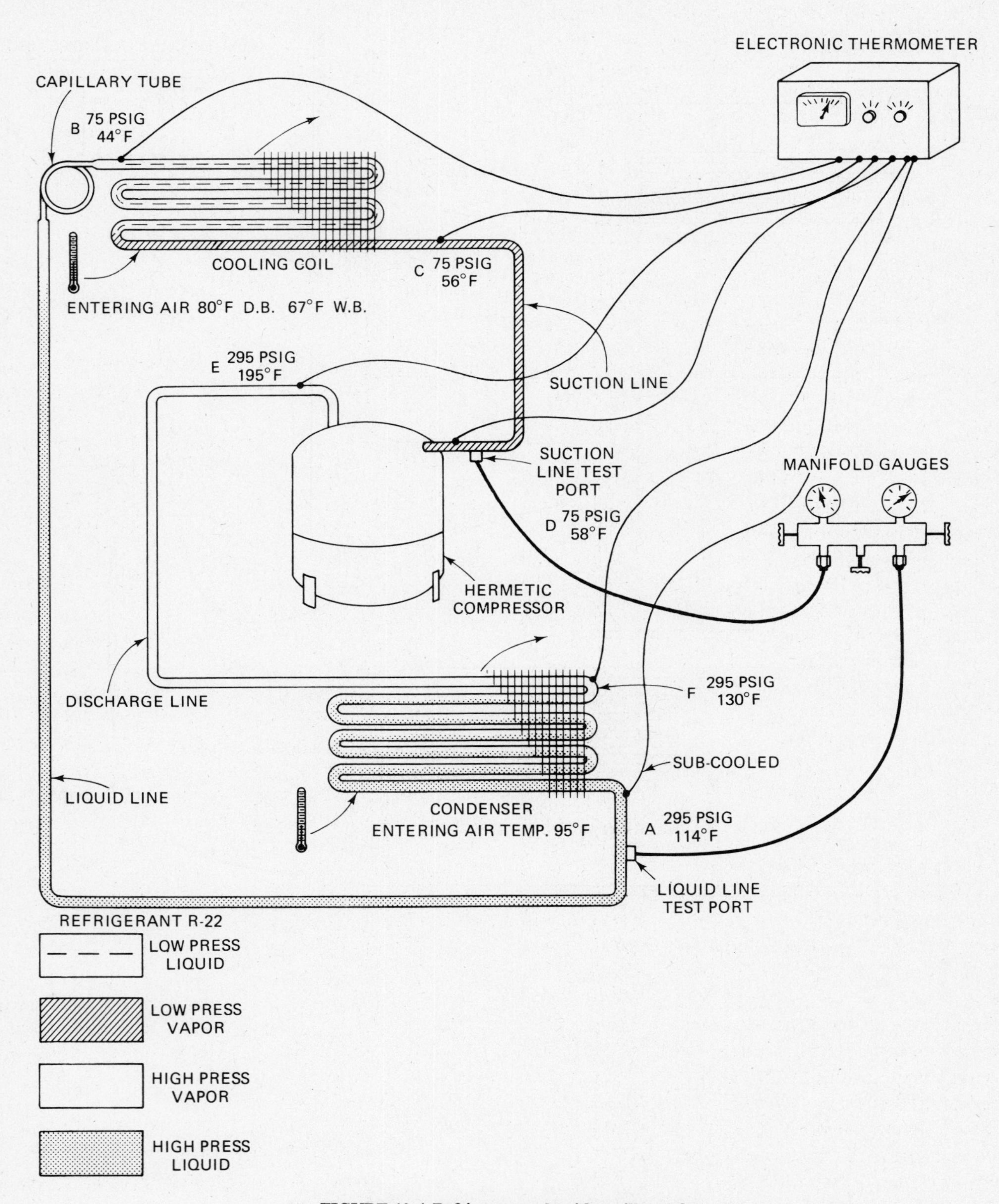

FIGURE 10-4 Refrigerant cycle with capillary tube.

pressure relationships shown are considered normal for a system operating under the following conditions: Entering evaporator coil air temperature is 80°F DB, 67°F WB; condenser air entering temperature is 95°F DB. Since these pressures will be affected by equipment design, always consult the manufacturer's pressure-temperature charts for each specific model. The system shown has a sealed hermetic compressor. Pressure-test connections (Schrader valves) are shown in the suction and liquid lines.

Starting at point *A*, the liquid refrigerant is subcooled approximately 16° to 114°F and exerts a pressure of 295 psig. Assuming the liquid refrigerant line is 25 ft long and properly sized, the pressure at the entering side of the capillary tube will be approximately 195 psig with the temperature of the refrigerant still about the same as at *A*, i.e., 114°F.

Downstream from the capillary tube at *B*, the *T-P* readings will be approximately 44°F and 75 psig. With the capillary tube restricting the refrigerant flow for 12° of superheat, the *T-P* readings at *C* will be 56°F and 75 psig. This pressure reading neglects the nominal refrigerant pressure drop of the vaporizing refrigerant through the evaporator coil.

Moving down to the suction gas intake at the compressor (point *D*), the *T-P* relationship of 61°F and 75 psig represents the temperature pickup of additional superheat caused by the suction line. At *E*, these readings become 195°F and 295 psig. The condenser average *T-P* is 130°F and 295 psig at *F*. The subcooling (16°F) then takes place at the last two and three bottom rows of the condenser and the cycle is completed at point *A*.

The above examples show what might be expected during normal system operation. Of course this varies with the refrigerant, the condensing medium (air or water or evaporative), and the type of application, and also with the make of equipment. Naturally the entire scope of differences cannot be presented here, but these variations will be discussed in later chapters and will become familiar through experience.

Regardless of the type of system there will be some common operating problems encountered, and the service technician must, like a doctor, be able to recognize the symptoms, diagnose the cause, and take corrective action. In most cases, the medical doctor is able to prescribe medicine or treatment immediately to relieve the patient. The refrigeration serviceman may have to arrive at a satisfactory diagnosis through the process of elimination of several possible causes, each of which may be the source of the complaint or the problem in the refrigeration system.

Craftsmen must be conscientious in their attempts to put the system back in proper operating condition and ethical in their dealings with the customer. There have been complaints from the field that some servicemen do not always measure up to the highest standards. For example, one may have added refrigerant to a system when there were indications of a shortage and, when this action did not correct the trouble, the serviceman was negligent and did not remove the excess refrigerant. (This, in itself, might be the cause of a future service complaint.) Some manufacturers of components such as expansion valves have had parts returned that were not defective. Rather, the strainers in some of the valves were merely dirty or clogged, but the serviceman would replace the valve, blaming the trouble on its operation.

In small refrigeration units, the major problems that occur are:

(a) the unit runs continuously with insufficient cooling, and

(b) the unit short cycles with insufficient cooling.

Of course, many other troubles may occur in the electrical circuitry, but these will be covered later.

Three main conditions in units that are operating but not cooling satisfactorily are:

(a) high head pressure,

(b) low suction pressure, and

(c) high suction pressure.

It is recommended that, if possible, the technician diagnose this problem without entering the sealed system. Some of the variables that could cause these problems and can be diagnosed by using gauges, are:

(A) High head pressure

(1) Dirty or partially blocked condenser

(2) Air or other noncondensable gases in system

(3) Overcharge of refrigerant

(4) Insufficient condensing medium (air, water, etc.)

(5) High temperature condensing medium

(6) Restricted discharge line

(B) Low suction pressure

(1) Insufficient air or heat load on evaporator coil

(2) Poor distribution of air over evaporator coil

(3) Restricted refrigerant flow

(4) Undercharge of refrigerant

(5) Faulty expansion valve or capillary tube

(6) Low head pressure

(C) High suction pressure

(1) Heavy load conditions

(2) Low superheat adjustment

(3) Improper expansion valve adjustment

(4) Poor installation of feeler bulb

(5) Inefficient compressor

(6) High head pressure on capillary tube systems

A-1. DIRTY OR PARTIALLY BLOCKED CONDENSER

An automobile engine will probably overheat if the radiator becomes clogged with leaves or insects. So will the operation of an air-cooled condensing unit be seriously affected if its condenser becomes partially blocked with paper, leaves, or other debris—particularly if the condenser is located outdoors. If the unit is located indoors, such as in the back room of a grocery store or restaurant, the condenser may not be subjected to leaves and other debris, but grease in the air collects on the fins, permitting dust and dirt to accumulate and prohibiting proper heat transfer.

This condition may be diagnosed during the visual check of the system by the service technician. External cleaning of the condenser fins and coil may be done with a stiff brush or, if a portable air tank is available, by the pressure of an air supply in the opposite direction to the normal air flow through the condenser. Accumulations of dirt and dust may have to be removed by the application of a soap and water solution, followed by flushing the condenser with water from a hose—again from the direction opposite to the normal air flow.

If grease has accumulated on a condensing unit in a restaurant or store, the condensing unit itself may have to be cleaned with a degreasing solvent applied with a brush or spray. This should be followed with a soap and water solution and external flushing with water.

Care must be taken that electrical connections are protected when the unit is being cleaned.

A-2. AIR OR NONCONDENSABLE GASES IN SYSTEM

If there is only relatively dry air in a refrigeration system, it is less harmful than moist air, but in either case oxygen may react with oil or metals to produce sludge, metal oxides, etc. The same applies if dry nitrogen or dry carbon dioxide has been used to pressure-test a system and has not been completely removed. However, moisture-laden air in the system indicates that it was opened for repair or component replacement and was not evacuated properly. Proper evacuation is absolutely necessary to eliminate both air and moisture.

Space in the condenser occupied by air or other noncondensables is not available for the proper function of that component, and can affect heat removal from the superheated vapor and condensation of the saturated vapor. A reduction of the heat transfer area in the condenser will make a greater temperature difference between the cooling medium and the condensing refrigerant necessary to permit removal of the required amount of heat from the refrigerant. At higher condensing temperatures, there will be a corresponding increase in head pressure.

The question now is how to determine if there is a noncondensable gas such as air in the condenser. To make a test, the temperature of the refrigerant in the condenser should be the same as the air surrounding it; so the compressor must be shut off (if it is in operation) and the refrigerant allowed to give up its heat to the surrounding air. This process can be speeded up if it is possible to bypass the controls and operate the condenser fan alone.

The difference in pressure within the condenser should not be more than 5 psig from the pressure corresponding to the temperature of the refrigerant being used. Assuming that R-12 is the refrigerant and that the ambient temperature (and that of the refrigerant in the condenser) is 95°F, the Refrigerant Table in the Fig. 10-1 shows a corresponding pressure of 108 psig. Therefore, the pressure within the condenser—as indicated on the gauge—should be 108 psig but not exceed 113 psig. If it does, the air or noncondensables must be purged from the unit.

Most small condensers do not have purge valves at the top, so the purging must be done through the gauge manifold. The purging should be done in small amounts, with a few moments of time elapsing between the brief periods of purging. This will permit the air or noncondensable gas to collect at the high point, which would be the gauge manifold, and allow it to be purged without losing too much refrigerant. It is impossible to purge without the loss of some refrigerant, since complete separation cannot be obtained. Purging should continue until the head pressure drops down to the proper point corresponding to the temperature of the refrigerant. Purging of capillary tubes or other critical charge systems is not recommended—only proper and complete evacuation procedures should be followed.

A-3. OVERCHARGE OF REFRIGERANT

If, as mentioned earlier in this chapter, additional, unnecessary refrigerant had been put into the system during a previous service call, or if the system was improperly charged during start-up (and then not removed) high head pressure may result, although, in a commercial system that has a receiver, the additional refrigerant will only raise the liquid level and should not affect the system unless it was greatly overcharged. An overcharge of refrigerant, like air or other noncondensable gases in the condenser, will occupy space in the condenser that is needed for proper heat transfer from the refrigerant vapor to the air used as a cooling medium, unless the system has a receiver. With a smaller area available, the increased temperature difference will cause an increase in the head pressure. When other possible causes of the high head pressure are eliminated, and a surplus refrigerant charge is suspected, some of the refrigerant must be removed from the system.

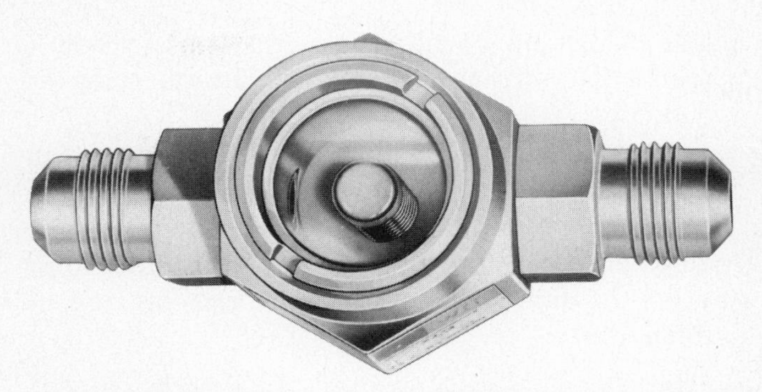

FIGURE 10-5 Sight glass (Reproduced by permission of Sporlan Valve Company).

If there is a sight glass, as illustrated in Fig. 10-5, installed in the liquid line just ahead of the liquid control, a full glass will indicate that there is either enough refrigerant in the system or a restriction ahead of the sight glass. If the possibility of a restriction has already been eliminated, the full sight glass may indicate a sufficient charge for proper operation, but it will not indicate an overcharge.

Refrigerant should be removed from the system until the sight glass indicates a shortage, that is, when bubbles of gas entering the liquid line are visible in the sight glass. The refrigerant removed from the system should be placed in an empty, dry drum or carefully purged to the atmosphere and new refrigerant added. Only enough refrigerant to clear up the gas bubbles should be charged back into the system.

A-4. INSUFFICIENT CONDENSING MEDIUM

As explained in A-1, a partially blocked condenser will result in inadequate heat transfer between the refrigerant and the cooling medium. Even if the condenser itself is not obstructed, there may be other reasons why insufficient air reaches or is available to the condenser. If the condensing unit is located too close to a wall, partition, or other obstacle, it is possible that not enough air can be drawn across the condenser.

Insufficient air moving across the condenser can also be the result of a loose or slipping belt between the motor and fan, a loose fan wheel on direct-drive equipment, or binding of the shaft of either the motor or fan because of bad shaft bearings or lack of lubrication.

A-5. HIGH-TEMPERATURE CONDENSING MEDIUM

If the temperature of ambient air surrounding the condensing unit increases, it follows that the operating head pressure will increase correspondingly. If the unit is located outdoors, there is of course no control of the outdoor dry-bulb temperature. The condensing unit may be protected from direct sun rays, although this factor is not too important. The unit should not be located indoors where it will be seriously affected by high ambient temperature.

If the condensing unit with a blow-through fan is located indoors and too close to a wall or other obstacle it is possible that the hot air exhaust from the condenser may be short-cycled back into the inlet of the fan. This would increase the temperature of the available air for removal of heat from the refrigerant. For optimum operation, a condensing unit located indoors must have some provision for the removal of condenser discharge air. A poorly-located outdoor condensing unit also can short-cycle air into the fan.

A-6. RESTRICTED DISCHARGE LINE

A kink that develops in the discharge line of a self-contained condensing unit, or in the hot gas line between a compressor and a remote air-cooled condenser would cause a restriction in the flow of the hot refrigerant vapor. An increase in pressure measured at the compressor, along with a corresponding increase in temperature, would result.

Similar results might occur in a recently-installed system where excess solder could cause a restriction to the flow of the discharge vapor. Usually a stoppage or major restriction of this type is diagnosed by the pulsation of pressure and the whistling sound of vapor trying to force its way past the restriction.

B-1. INSUFFICIENT AIR ON EVAPORATOR COIL

This is the most common cause of low suction pressure in a refrigeration or air-conditioning system. If the flow of air across the evaporator coil is reduced, the load on the coil is decreased. Since there is customarily a transfer of heat from the volume of warm air moving across the coil to the cool refrigerant within the coil, any decrease in the amount of air that passes through the coil will result in a loss of normal heat transfer. If the refrigerant is picking up less heat, its temperature will be lowered along with the suction pressure.

Insufficient air on the forced-air evaporator coil may be caused by dirty air filters, too-small return ducts, improper speed of the blower, a clogged cooling coil, a combination of these possible causes, or improperly adjusted duct dampers or registers. The service technician should check to see if there are filters in the air distribution system. If they are dirty, they should be cleaned or replaced. The cooling coil should be checked to make sure it is clean and free of dirt and lint, whether or not there is a filter in the system.

If the blower motor and/or blower shaft bearings have not been lubricated for some time and are not running freely, the flow of air through the cooling coil may be less than normal. An improperly adjusted blower belt could cause a slowdown in the blower speed and a reduction of the air flow across the coil. (It should be noted that a squirrel-cage blower running backward will deliver air in the proper direction, but at a greatly reduced volume).

B-2. POOR DISTRIBUTION OF AIR ON EVAPORATOR COIL

It is important that each circuit or portion of the coil receive fairly equal amounts of air so that the entire cooling load can be handled proportionately by the entire cooling coil. Without a proper distribution of air over the coil, the capacity and the efficiency of the cooling coil will be reduced. When there is ductwork connected to the inlet side of the coil, it is highly possible that an unbalanced condition of air distributed across the coil will occur if the air must make a turn in direction prior to entering the cooling coil. When air changes direction, such as in turning a corner prior to entering the cooling coil, most of the air will probably flow to the outside radius of the curve. In so doing, only a small portion of the air will flow across the coil section located at the inner part of the curve and so cause an unequal distribution of the load on the cooling coil.

It is possible that, because of the design and circuiting of the cooling coil, the liquid refrigerant in the circuit or sections of the coil nearest the inner radius of the curve will not be completely vaporized. In that case, the refrigerant will pass from the coil in a liquid form and will, in turn, cool the gas that is leaving the other sections of the coil to a temperature that is lower than normal. This lowering of temperature may cause the metering device to restrict the flow of liquid refrigerant to the other sections of the coil, thereby robbing the remainder of the coil of the refrigerant needed to handle the load properly. It may be necessary to install turning vanes in the sheet metal ductwork on the inlet side of the cooling coil.

B-3. RESTRICTED REFRIGERANT FLOW

For a cooling coil to vaporize enough refrigerant to satisfy the capacity of the compressor and to remove the proper amount of heat from the load, it must receive an adequate amount of liquid refrigerant. Any restriction to this flow of liquid refrigerant will mean a reduction in the capacity of the cooling coil for heat removal. There must be no restriction to this liquid flow between the outlet of the condensing unit and the inlet to the cooling coil. This includes the receiver, drier, sight glass, and refrigerant controls. If a receiver is used, the liquid-line valve on the receiver may be partially closed. There may also be a restriction in the liquid flow due to a crimp in the line, partially smashed tubing, a quick-connect coupling that is only partially open, a drier full of moisture, or an obstruction of some sort in the metering device. In any case, a restriction to the flow of liquid refrigeration could affect the operation of the system. It may cause enough pressure drop to reduce the boiling point of liquid available in the cooling coil. There will be a definite temperature drop across the point of restriction, which, depending on its location, may or may not be easily located.

B-4. UNDERCHARGE OF REFRIGERANT

A shortage of refrigerant in the system is usually indicated by a warm suction line along with a low suction pressure. If there is an undercharge in the system, the refrigerant vapor may not condense properly (in the condenser) before it is ready to reenter the cooling coil and remove additional heat from the load. A refrigerant that does not condense fully and enters the liquid line in a gaseous state may be indicated by a hissing noise at the metering device. In addition, as previously discussed, the refrigerant will not pick up as much heat when it is in a vapor state as it would if it entered the cooling coil in its proper state—as a liquid. The cooling coil and the suction line will probably be warm to the touch, because there is little or no liquid refrigerant being supplied to the cooling coil. A liquid indicator or sight glass installed in the liquid line will show a shortage of refrigerant by bubbles in the sight glass.

B-5. FAULTY METERING DEVICE

A metering device such as an expansion valve may have mechanical problems. This valve may stick in a nearly closed position, a fully closed position, or a fully open position. Sometimes dirt or frozen moisture will restrict the flow of refrigerant liquid through the valve or stop the flow of any liquid at all into the evaporator. In such a case, the compressor will short-cycle (that is, start and stop at frequent intervals) when the expansion valve is only partially closed and insufficient liquid is entering the coil.

With the expansion valve completely closed, the compressor will lower the pressure in the evaporator down to the cutout point of the low-pressure control, which will stop the compressor. If there is no low-pressure control switch in the system, the compressor will continue to run with no work being done until the motor windings heat up and are cut off by the electrical overload control operation.

B-6. LOW HEAD PRESSURE

Low head pressure on a system may be caused by an undercharge of refrigerant. It could also be the result of low ambient air temperature being circulated across the condenser coil. This low head pressure could result in a higher compression ratio and therefore a decrease in compressor efficiency, accompanied by a reduction in the expansion valve capacity. (Low refrigerant charge also can result in lower suction pressure, which may increase the compression ratio.) This reduction of capacity would be caused by a lowering of the pressure drop across the expansion valve with a decrease in the head pressure.

C-1. HEAVY LOAD CONDITIONS

A system might have a high suction pressure, and yet no part of the system is faulty. Possibly load conditions increased considerably, accompanied by an increase of high condenser entering air ambient conditions. In such a case, there would be a high head pressure along with the high suction pressure, with no fault in the mechanical operation of the system. These conditions may be remedied by the removal of the causes of the excessive load on the evaporator.

C-2. LOW SUPERHEAT ADJUSTMENT

A refrigeration system operating with an extremely low superheat setting can cause a high suction pressure. Liquid refrigerant will overflow into and evaporate in the suction line and might possibly reach the compressor. This could lead to serious complications, for not only would there be a loss of some cooling effect, but the compressor might be damaged as well. An improper setting of the expansion valve superheat should be corrected. The same situation might also be caused by other misadjustments of the expansion valve or from the location of the feeler bulb.

C-3. IMPROPER EXPANSION VALVE ADJUSTMENT

An expansion valve that is stuck in an open position, or is adjusted to allow too much liquid to flow to the coil, will lead to an excessive amount of sweating or frosting on the suction line. Occasionally the expansion valve may be only slightly misadjusted and no serious symptoms will be apparent. If the valve allowing only a little more liquid to pass through than it should, a small amount of sweating may appear on the suction line

C-4. POOR INSTALLATION OF FEELER BULB

Frequently, the expansion valve may open too much because the thermal bulb is not in good contact with the suction line or pipe. Such poor contact may be caused by a lack of insulation around the thermal bulb when the ambient temperature of the bulb and suction line is extremely high. The mounting of the thermal bulb and its actual location is very important. The bulb must be in good contact with the outlet of the cooling coil so that it can sense, thermally, exactly what is going on in the suction line and the evaporator.

The feeler or thermal bulb should be installed on the top of a horizontal section of the suction line, following the valve or equipment manufacturer's instructions. The liquid refrigerant in the feeler bulb and the suction line should be in close contact. If it is necessary to mount the thermal bulb on a vertical section of the suction line, the bulb should be located so that the capillary tube comes out of the top of the installed feeler bulb. In order to keep the bulb from being subjected to the influence of the air or other substances being cooled, it should be properly clamped to the suction line and insulated.

C-5. INEFFICIENT COMPRESSOR

If a high suction pressure exists in the system, along with a normal amount of superheat in the cooling coil, and all other possible causes of this condition have been eliminated, it is quite possible that the compressor's inefficiency is due to faulty valves.

The proper procedure for checking valves will depend on the particular equipment manufacturer's recommendations, but, generally, for a high-speed (3,600 rpm) compressor, the manufacturers recommend running the machine and timing the equalization of pressures when it is shut down (average time is about 40 seconds). A shorter time indicates valve leakage.

C-6. HIGH HEAD PRESSURE ON CAPILLARY TUBE SYSTEMS

Because the capillary is a fixed orifice metering device, the control of its refrigerant and charge is more critical than in those using expansion valves or float valves. The bore of a capillary tube is approximately 0.026 in. to 0.55 in., depending on its length. Its ability to pass liquid from the condenser to the evaporator can be greatly effected by the presence of foreign particles, changes in pressure, or undercharge of refrigerant.

Heat resulting from high pressure is the greatest factor in cap-tube failure. A plugged condenser or fan motor burnout will cause excessively high head pressure and the unit will cycle on high pressure control. The compressor then gets hot, and carbon forms on the discharge valves of the compressor from the breakdown of oil. When the carbon finds its way to the cap-tube, blocking the passage of refrigerant, trouble starts.

The resulting effect is self-compounding; liquid refrigerant backs up into the condenser, thus reducing its efficiency, and as a result head pressure rises higher and higher, as does the motor load. At the same time because of the reduction of suction gas to the hermetic compressor, the latter cannot properly cool the motor, and prolonged cycling and overheating may eventually result in total motor burnout.

PROBLEMS

1. In the suction line, is it normal to expect a drop in pressure or in temperature?
2. In order for an air-cooled condenser to reject heat to the atmosphere must the condensing temperature be 30° to 40° F above or below the outdoor ambient temperature?
3. Subcooling in the air-cooled condenser takes place in the top or the bottom rows?
4. In a capillary tube system the superheat is ______________.
5. A dirty condenser will produce a high or a low head pressure?
6. Reduced evaporator air flow causes high or low suction pressure?
7. If an expansion valve fails and is completely closed, does the compressor suction pressure rise or fall?
8. Location of the expansion valve feeler bulb has no major effect on the valve operation: true or false?

11

ABSORPTION REFRIGERATION

The mechanical refrigeration cycle as previously reviewed is based on the use of a compressor as a source of energy to transfer heat from one source to another. A different method of moving heat (or refrigerating) is called the *absorption refrigeration cycle*.

Absorption liquid chillers, as illustrated in Fig. 11-1, utilize heat as the energy source. Steam or hot water are the usual heat mediums, and they originate from several sources, such as:

(1) Existing heating boilers that are now used only during the winter.

(2) New boilers installed for heating and refrigeration air-conditioning.

(3) Low-pressure steam or hot water used in an industrial plant for process work.

(4) Waste heat recovered from the exhaust gases of gas engines or gas turbines.

(5) Low-pressure steam from steam turbine exhaust.

Utilizing waste heat to power the absorption chiller means a very inexpensive system to own and operate. Already in existence are systems that use solar energy to heat water. The results of these applications may cause even greater interest in the use of absorption equipment as part of energy conservation methods.

Other advantages are sound and vibration levels that are quite low compared to mechanical systems, making them ideally suited for installation in almost any part of a building or on the roof. Commercial units are available in capacities ranging from 25 to well over 1,000 tons. Residential air-conditioning absorption units are available in the 3- to 10-ton range.

OPERATION

The operation of the absorption system depends on two factors: a refrigerant (water) that boils or evaporates at a temperature below that of the liquid being cooled, and an absorbent (lithium bromide) having great affinity for the refrigerant.

The refrigerant (water) in an open pan will boil or evaporate at 212°F at sea level (14.7 psia) when heat is applied to it. By placing a tight lid on the pan we can cause the pressure and the evaporation temperature to increase. If, on the other hand, we create a vacuum in the closed pan of water, the evaporation will take place at a lower pressure and temperature (refer back to Chap. 9, Table 9A). Now we have only to replace the closed pan with an absorption chiller and the desired evaporator temperature will dictate the vacuum to be maintained.

For the absorbent, many types of salts could be used. Common table salt (sodium chloride) is an absorbent. You know what happens to the salt shaker during humid weather; it clogs up because the salt has absorbed moisture from the air.

Calcium chloride is another salt, frequently used on dirt roads or clay tennis courts to settle the dust. The calcium chloride absorbs moisture from the air, which keeps the surface of the road or court moist.

Lithium bromide is also a salt and it is in crystal form when it is dry. Tests were made on many types of salts before lithium bromide was selected as having the best overall characteristics for use in large capacity absorption chillers.

For use in the absorption chiller, lithium bromide crystals are dissolved in water; the mixture is known as *lithium bromide solution*. The amounts of lithium bromide and water in a solution are measured by weight and not by volume. The concentration of the

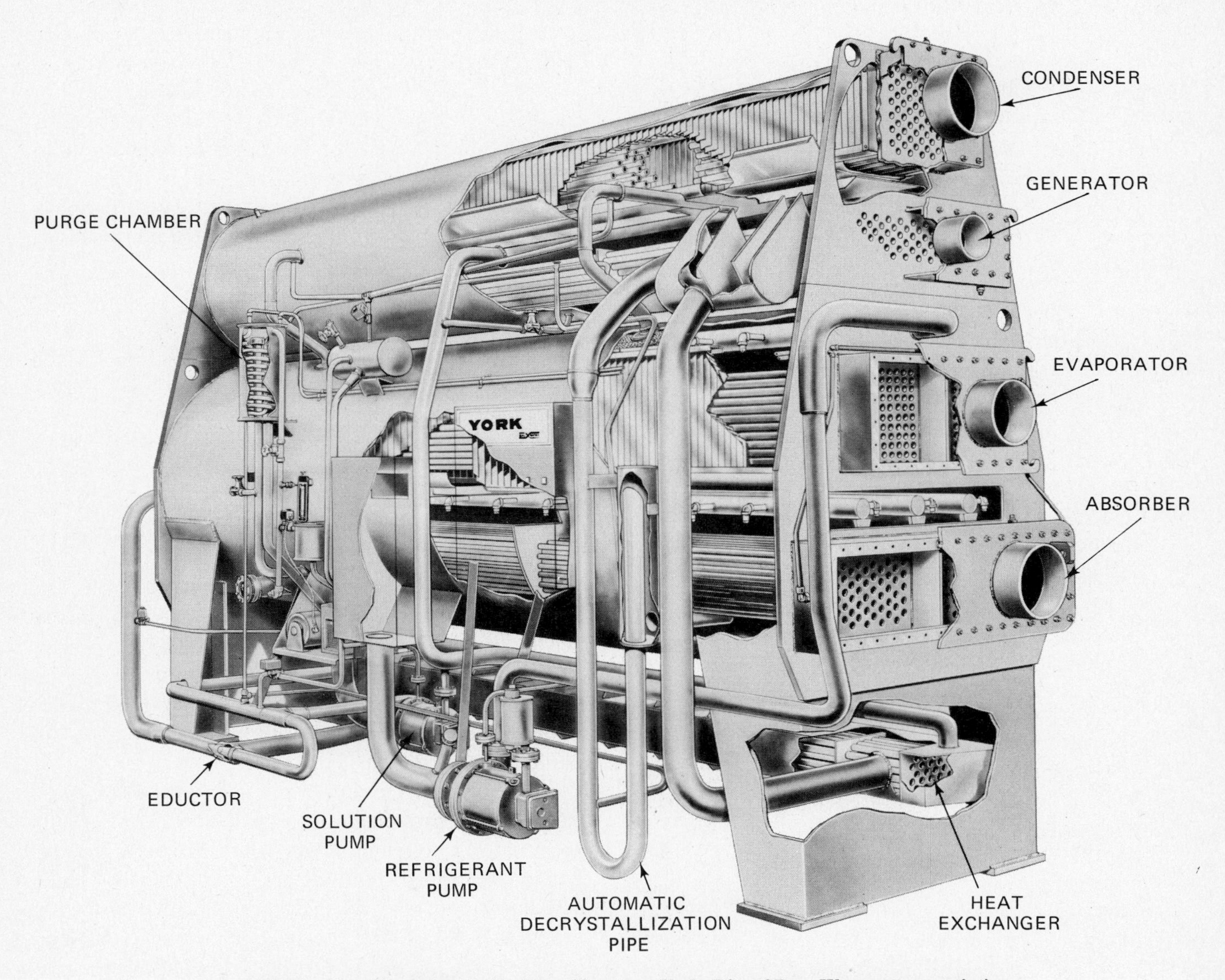

FIGURE 11-1 Absorption liquid chiller (Courtesy, York, Div. of Borg-Warner Corporation).

solution is stated in percent of lithium bromide in the total solution. A 65% concentration means that 65% of the total weight is lithium bromide.

EXAMPLE:

For each 100 lb of solution at 65% concentration there are 65 lb of lithium bromide and 35 lb of water.

$$100 \text{ lb (solution)} \times \frac{65}{100} = 65 \text{ lb lithium bromide}$$

$$100 \text{ lb (solution)} - 65 \text{ lb} = 35 \text{ lb water}$$

The absorbent (lithium bromide) serves as a vehicle to absorb and transport the refrigerant (water) from a part of the system where it is not needed to another part of the system where the refrigerant can be recovered to be used again.

ASSEMBLY

The absorption liquid chiller as shown in Fig. 11-1 consists of two-main shells. The upper shell contains the generator and condenser and is maintained at a vacuum of approximately one-tenth of an atmosphere (1.470 psia).

The lower shell contains the evaporator and absorber and is maintained at a vacuum of approximately one-hundredth of an atmosphere (0.147 psia).

COMPONENTS

All absorption chillers include four basic heat exchange components, which, when properly balanced, will cool a liquid to the desired temperature. The components are:

(1) evaporator
(2) absorber
(3) generator
(4) condenser

In addition there are auxiliary items which assist the four basic components to perform their functions. Typically they are:

(1) a heat exchanger
(2) two fluid pumps
(3) a purge unit
(4) a vacuum pump
(5) an automatic decrystallization device
(6) a solution control valve
(7) a steam or hot-water valve
(8) an eductor
(9) a control center

All this sounds fairly complicated until you take it one simple step at a time.

Throughout this discussion certain temperatures and pressures are mentioned to permit a better understanding of the work done in each part of the system. These example conditions are approximate and might vary for other leaving chilled water temperatures, cooling water temperatures, steam conditions, and so on.

EVAPORATOR

The purpose of the evaporator is to cool a liquid for use in process refrigeration work or in an air-conditioning system. Let's consider a typical application (see Fig. 11-2).

Chilled water enters the evaporator at 56°F and is to be cooled to 44°F. In order to accomplish this the lower shell is maintained at a pressure of 6 mmHg (0.117 psia). Under these conditions the refrigerant (water) will evaporate at 39°F, thereby providing a large enough temperature difference to cool the chilled water to 44°F.

Since the refrigerant (water) can evaporate more easily if it is broken up into small droplets, a pump recirculating system is used. The refrigerant (water) enters the top of the lower shell, part of it is evaporated as it comes in contact with the relatively warm tubes, and the liquid that is not evaporated is collected under the evaporator tubes. A pump recirculates this refrigerant through a spray header over the

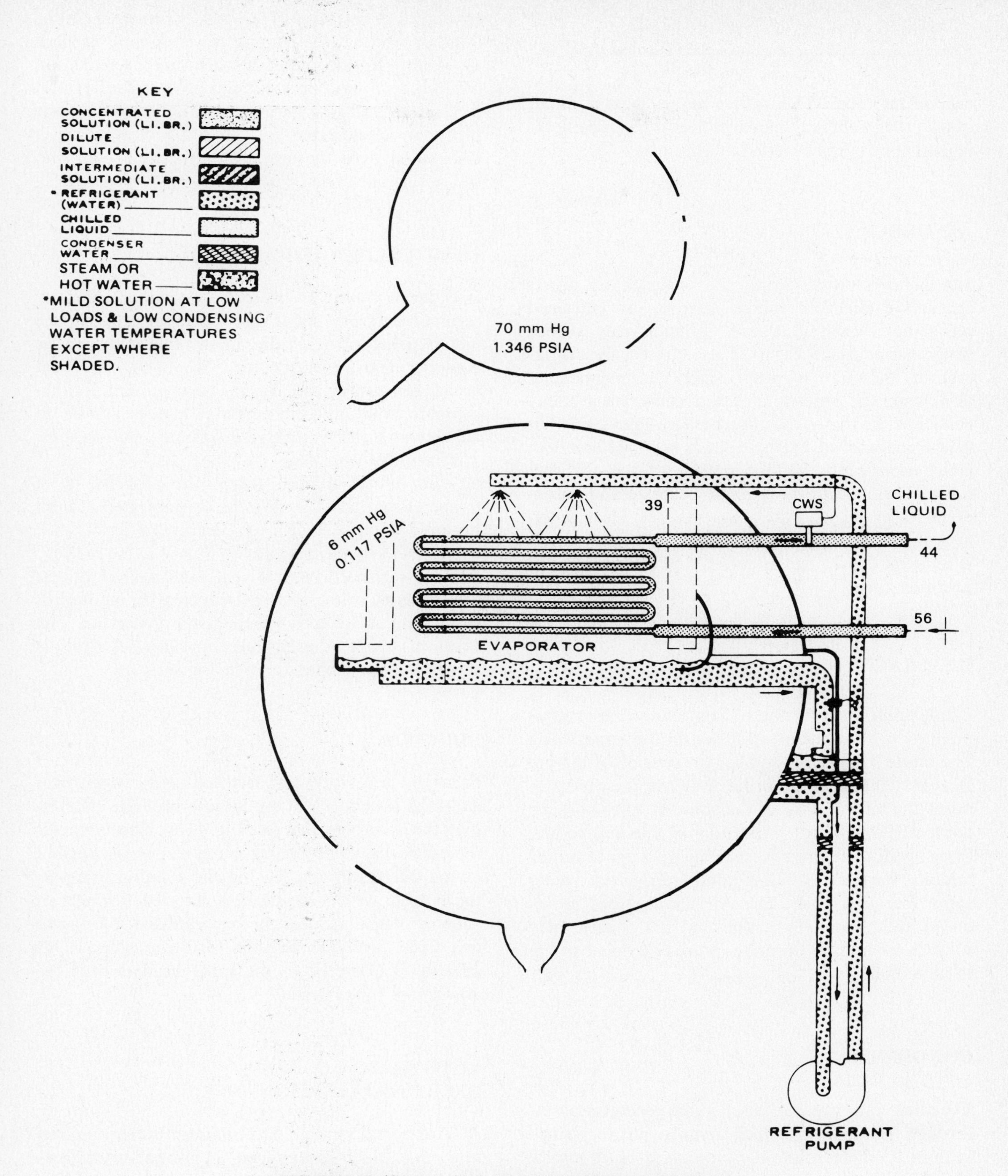

FIGURE 11-2 (Courtesy, York, Div. of Borg-Warner Corporation).

evaporator tubes. This system makes maximum use of the refrigerant and improves heat transfer by keeping the tube surface wetted at all times.

ABSORBER

Refrigerant vapor from the evaporator passes through eliminators, which remove any entrained refrigerant liquid. Lithium bromide can absorb water vapor more easily if its surface area is increased; therefore, a pump is used to circulate solution from the bottom of the absorber to a spray header at the top of the absorber (see Fig. 11-3). The vapor is absorbed by the lithium bromide solution, which is flowing over the outside of the absorber tubes. The mixture of lithium bromide and refrigerant water is called the *dilute solution*. Heat generated in this process is called *heat of absorption*. It is removed by condenser water, flowing through the absorber tubes.

GENERATOR

Dilute solution from the bottom of the absorber is pumped to the generator located in the upper shell. The dilute solution flows over the outside of the hot generator tubes. Steam or hot water in the generator raises the temperature of the solution to the boiling point and vaporizes a portion of the refrigerant. Once again there are two substances, a concentrated lithium bromide solution and refrigerant water vapor (see Fig. 11-4). The refrigerant vapor moves on to the condenser section and the concentrated solution of lithium bromide returns to the absorber to be reused.

CONDENSER

The refrigerant vapor released in the generator passes through eliminators, which remove any entrained lithium bromide solution. A pressure of 70 mmHg (1.346 psia) is maintained in the upper shell and causes the refrigerant to condense at 112°F on the condenser tubes (see Fig. 11-5). Condenser water is used, which after passing through the absorber tubes flows inside the condenser tubes. Condensed refrigerant flows by gravity and pressure differential through an orifice to the evaporator. This refrigerant plus that recirculated by the refrigerant pump is distributed over the evaporator tubes to complete the refrigerant cycle.

HEAT EXCHANGER

In order to make the absorption cycle more efficient, several accessory items are normally added; the first is a *heater exchanger* in the line between the absorber and generator see Fig. 11-6). The heat exchanger brings the warm, concentrated lithium bromide solution coming from the generator in contact with the relatively cool dilute solution coming from the absorber.

The dilute solution leaves the absorber at a temperature of 102°F, and the concentrated lithium bromide solution is at a temperature of 214°F as it comes from the generator.

The introduction of a heat exchanger improves the efficiency of the cycle by reducing the amount of steam or hot water required in the generator. The effect of the heat exchanger on both solutions is shown by the temperatures in Fig. 11-6.

EDUCTOR

A second accessory item is the *eductor*, which provides for circulation of the lithium bromide solution over the absorber tubes (see Fig. 11-6). This increases the efficiency of absorbing water vapor. A portion of the dilute solution leaving the solution pump at the bottom of the absorber is directed through an eductor, which induces the concentrated solution to mix with the dilute solution from the pump. This mixture is delivered to the spray headers over the absorption tube bundle.

AUTOMATIC DECRYSTALLIZATION PIPE

A third accessory item is the *automatic decrystallization pipe*, which is designed to prevent crystallization of the lithium bromide solution. Crystallization

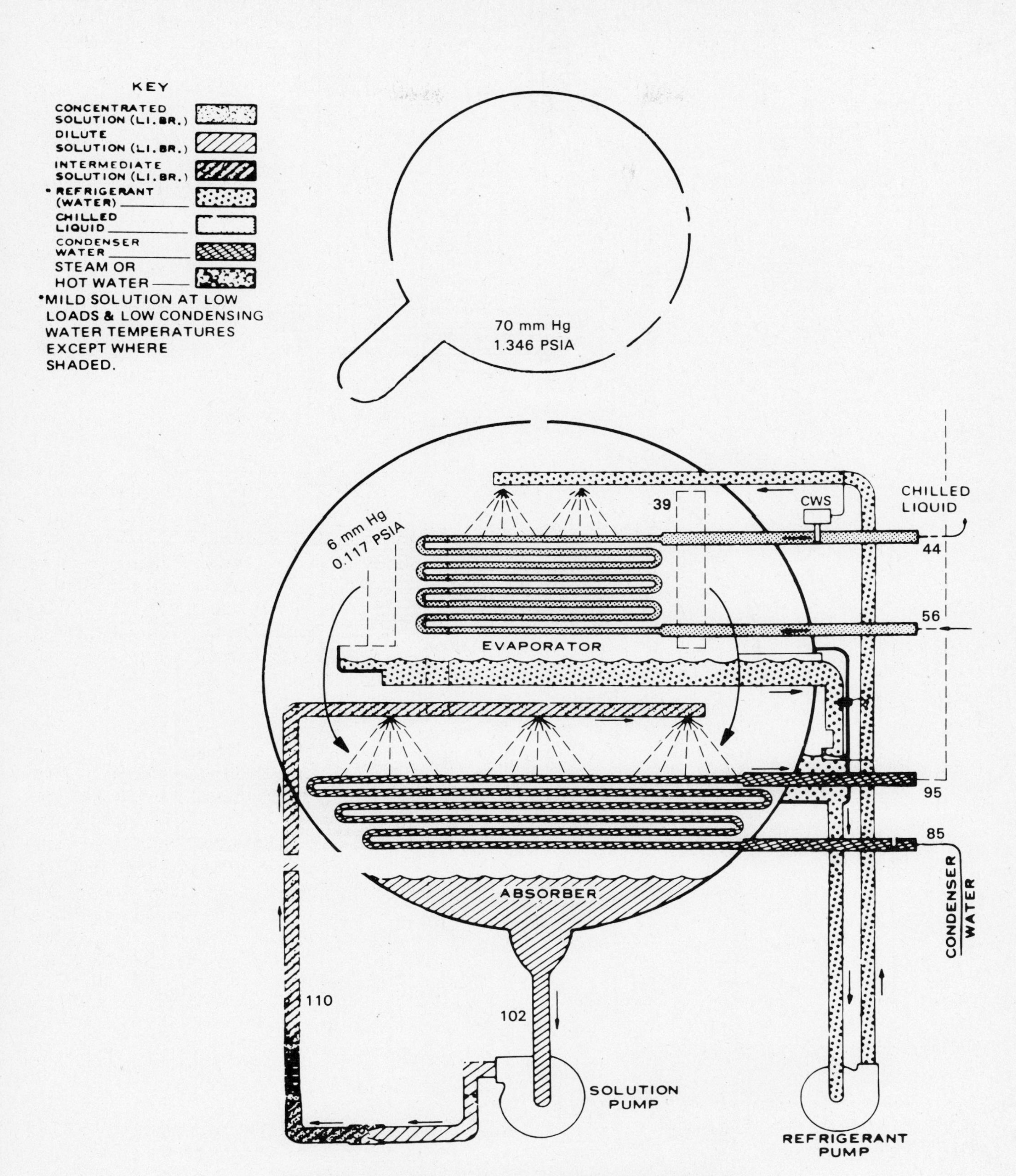

FIGURE 11-3 (Courtesy, York, Div. of Borg-Warner Corporation).

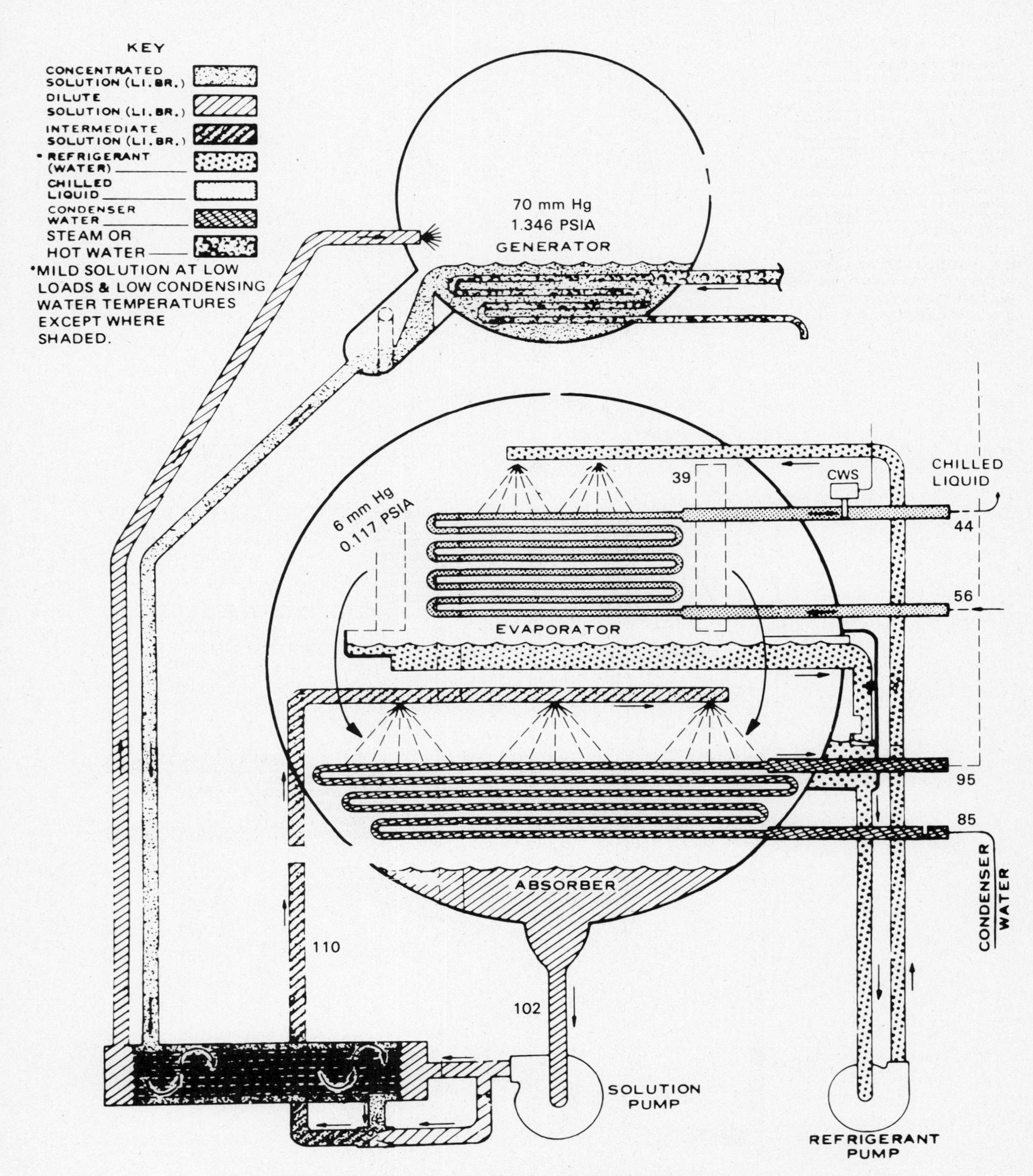

FIGURE 11-4 (Courtesy, York, Div. of Borg-Warner Corporation).

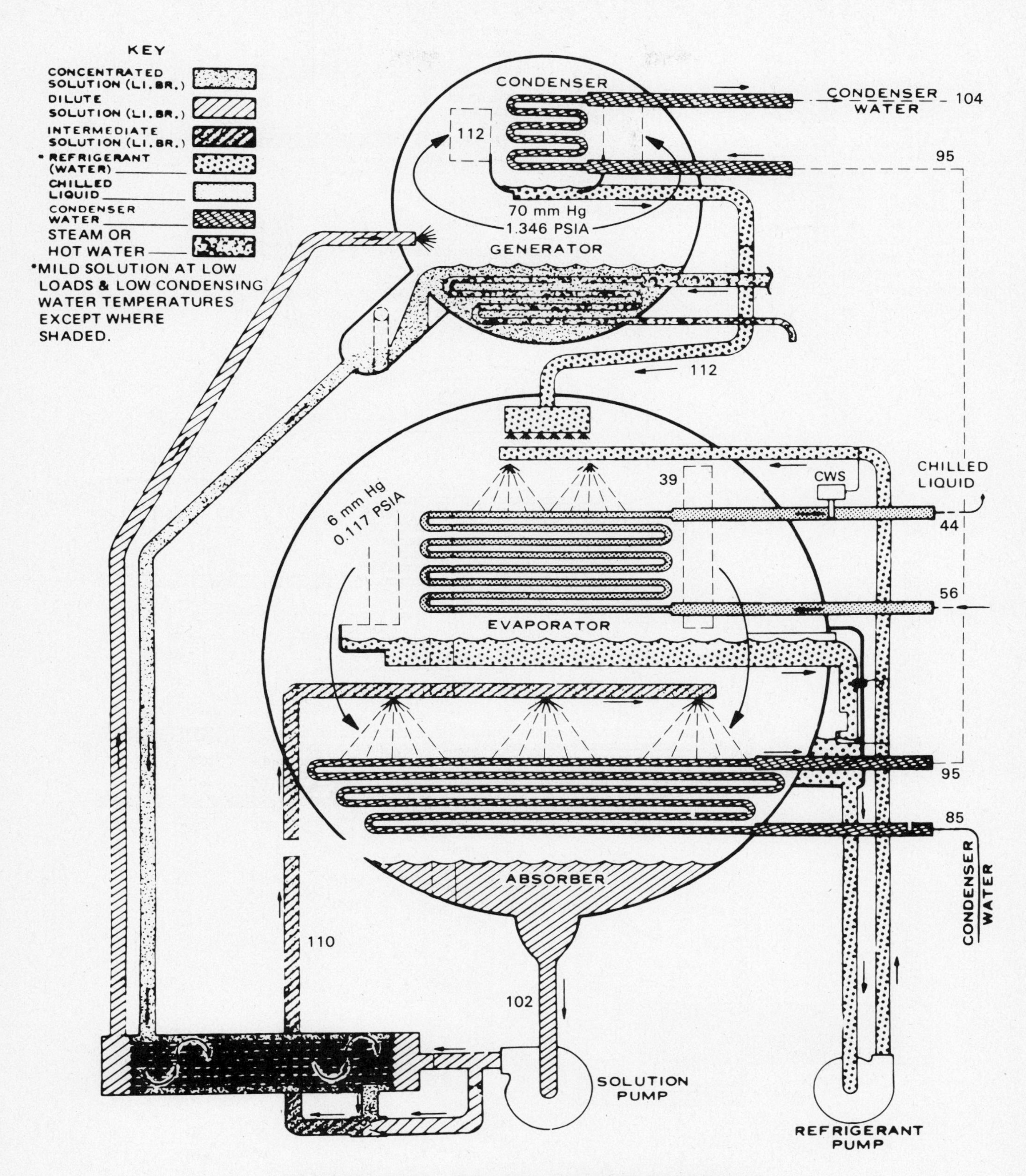

FIGURE 11-5 (Courtesy, York, Div. of Borg-Warner Corporation).

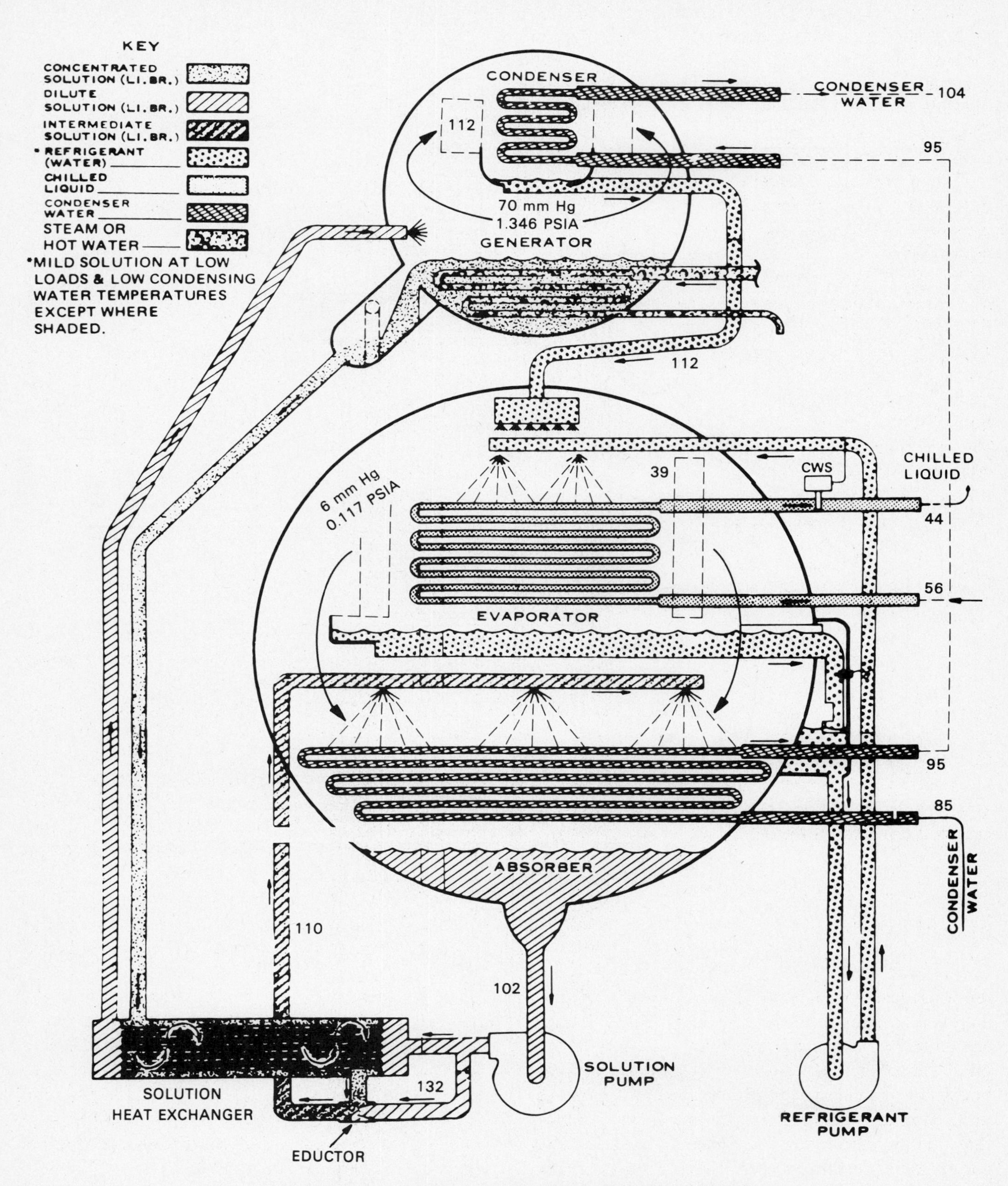

FIGURE 11-6 (Courtesy, York, Div. of Borg-Warner Corporation).

can occur when the lithium bromide solution becomes too concentrated (see Fig. 11-7).

Should the absorption chiller be shut down due to a prolonged power failure, the refrigerant and the lithium bromide solution temperatures would eventually reach the equipment room temperature, possibly causing the concentrated solution in the generator, the heat exchanger, and the connecting piping to crystallize.

When power is restored, heat from the steam or hot water will increase the temperature of the concentrated solution in the generator, causing it to liquefy. However, the concentrated solution in the piping and the heat exchanger will remain in crystalline form and the addition of heat is required to liquefy it.

The liquid solution from the generator will back up in the pipe on the generator side of the heat exchanger until it overflows into the automatic decrystallization pipe and then into the absorber. The solution temperature in the absorber will increase to about 214°F instead of 102°F. The solution pump will then pump 214°F concentrated solution through the heat exchanger. This additional heat will liquefy the solution in the heat exchanger and the piping.

When the flow from the generator to the absorber returns to normal, the flow through the automatic decrystallization pipe will cease and the system will return to normal operation.

In normal operation the automatic decrystallization pipe is designed so that there is a seal between the upper and lower shells to prevent an equalization of pressure.

To keep the decrystallization pipe ready for use at all times, a small amount of dilute solution is constantly flushed through it.

PURGE SYSTEM

A fourth accessory item is the *purge system*, which is designed to remove noncondensables. The noncondensable gases are collected in the water-cooled purge chamber, and they are removed by periodic operation of an electric motor compressor type purge unit. Manual operation of the purge unit assures that the operator will know the amount of noncondensables in the system. If the unit was purged automatically, the unit might have a large leak that would go undetected until extensive damage occurred.

CONTROL VALVE

The operation of the absorption chiller is controlled by a fifth accessory, a *steam* or *hot water control* valve (see Fig. 11-7). This valve modulates to control the flow to the generator tubes. It is activated by a sensing element in the chilled water line leaving the evaporator. In this way the energy supplied to the generator is only the amount required to produce a sufficient quantity of refrigerant to maintain the system chilled water temperature.

SOLUTION VALVE

The basic system includes the evaporator, absorber, generator, and condenser, as well as the heat exchanger, the eductor decrystallization pipe, the purge unit, and the control valve, all of which are standard equipment on most units. They are all needed for efficient operation of the unit. A valuable optional accessory item is the *solution control valve* (see Fig. 11-7). It is designed to provide the maximum in economical operation at part load.

Under normal conditions, at full load, twelve pounds of dilute lithium bromide solution is supplied to the generator for every pound of refrigerant that is boiled off.

At 25% load, without a solution valve, twelve pounds of solution is still circulated to the generator but only one-quarter pound of refrigerant is boiled off.

The addition of a solution valve restricts the flow of dilute solution to the generator in accordance with the reduced load requirements. The solution valve is actuated by a sensing element in the generator outlet box which maintains a constant temperature of the concentrated solution regardless of the load. The solution valve greatly improves the efficiency of the system at part load.

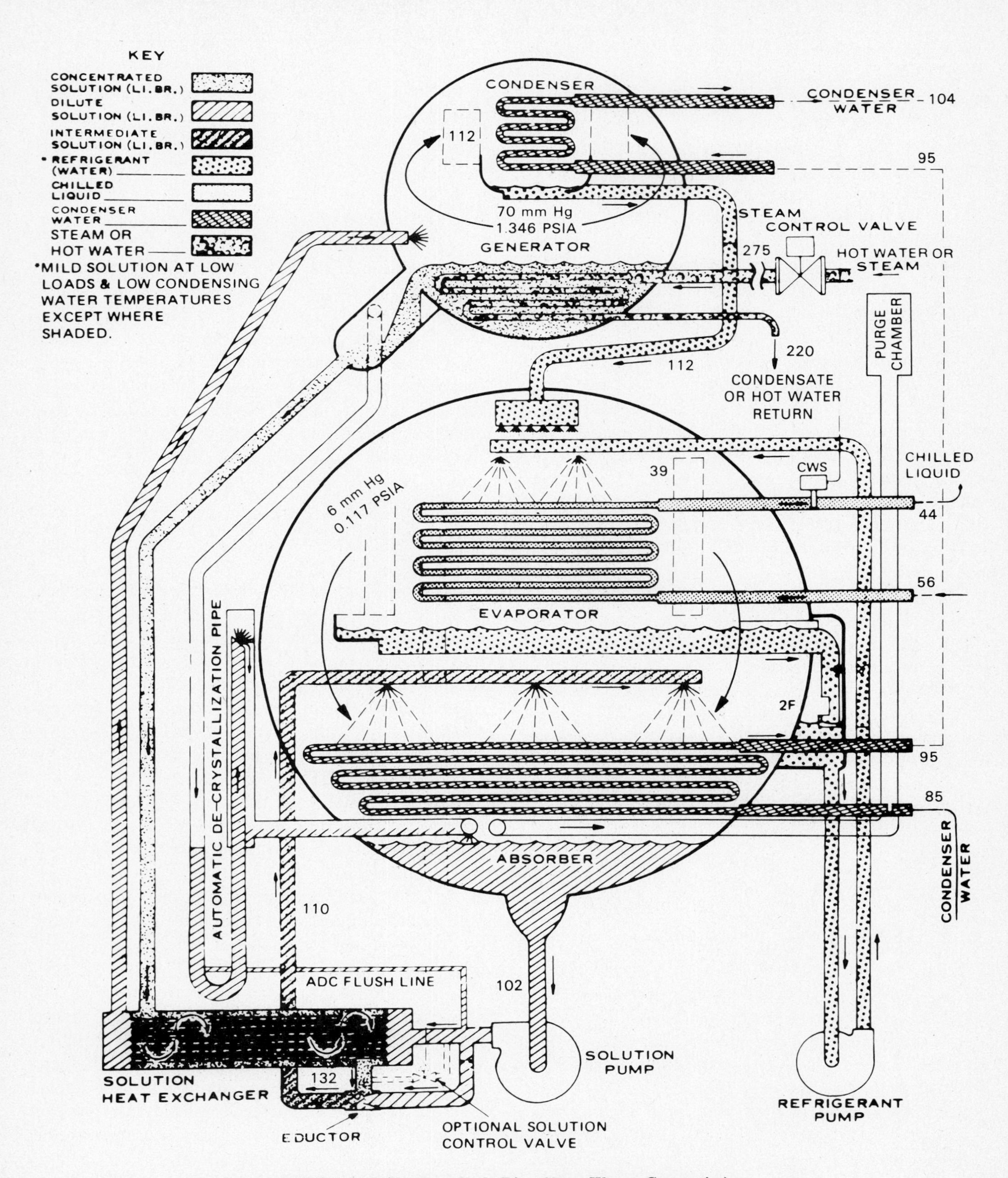

FIGURE 11-7 (Courtesy, York, Div. of Borg-Warner Corporation).

PROBLEMS

1. Absorption refrigeration units use ____________ as their energy source.
2. Are sound and vibration levels of absorption equipment higher or lower than conventional systems?
3. ____________ is the refrigerant in an absorption system.
4. The most common absorbent is ____________.
5. Name four main components of an absorption chiller.
6. The absorber contains (a) the evaporator or (b) the condenser?
7. The condenser is contained within the ____________.
8. The function of the purge unit is to remove ____________ gases.
9. Is the purge unit normally an automatic or a manual operation?
10. The control valve modulates the flow of steam or hot water to the generator: true or false?

12

BASIC ELECTRICITY

Electricity alone is a subject that would require several books to cover even the basics. However, in the next several chapters we will discuss electricity as it relates to the refrigeration, heating, and air-conditioning industry.

THE ELECTRON THEORY

All matter is composed of atoms—the smallest or basic components of molecules. Atoms, in turn, are composed of a heavy, dense nucleus containing *protons* and sometimes *neutrons*, surrounded by lighter particles called *electrons*. The proton carries a positive charge (+), the electron carries a negative charge (−), and the neutron (if present) is neutral. The attraction between the positively charged protons in the nucleus and the surrounding negatively charged electrons tend to hold them together in the unit we call the *atom* (see Fig. 12-1). The number of protons in the nucleus determines the type of element. The hydrogen atom, as illustrated in Fig. 12-1a, is the smallest, having one proton in the nucleus and one electron circling in orbit around it. Copper (Fig. 12-1b), on the other hand, has 29 protons and 34 neutrons.

Returning to the definition of positive or negative charges—a positive charge does not indicate an excess of protons; it means a deficiency of electrons. Thus, the terms *negative* and *positive*, simply mean more or less electrons are involved.

Materials that are charged with static electricity either attract or repel each other. Attraction takes place between unlike charges, because the excess electrons of a negative charge seek out a positive charge, which has a shortage of electrons (Fig. 12-2). Unlike charges (+ and −) attract. Like charges (− and −) or (+ and +) will repel each other.

Static electricity is the condition when the electrons are at rest but have a potential to move. *Dynamic electricity* is electrons in motion. The movement of electrons is called *current*.

Before the acceptance of the electron theory as

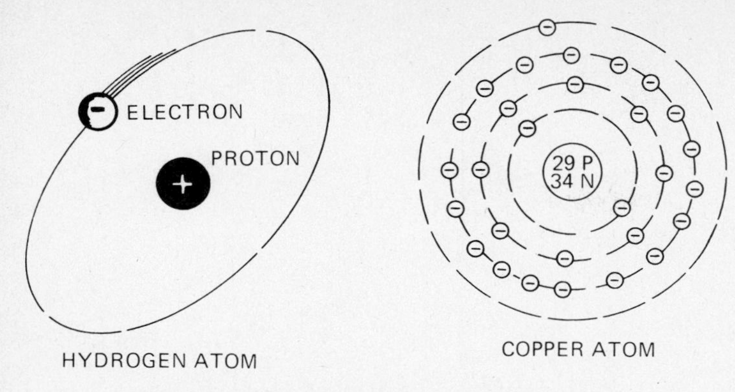

FIGURE 12-1 (a) Hydrogen atom; (b) Copper atom.

the basis of electrical behavior, it was thought that current flowed from positive to negative. Now it has been established that current (the flow of electrons) is actually from negative to positive.

SOURCES OF ELECTRICAL ENERGY

Energy is defined as the ability to do work, and since energy cannot be created or destroyed it must be converted from one form to another. Sources of electrical energy may be:

(a) chemical action
(b) friction
(c) heat
(d) light action
(e) pressure
(f) mechanical action
(g) nuclear action
(h) magnetism

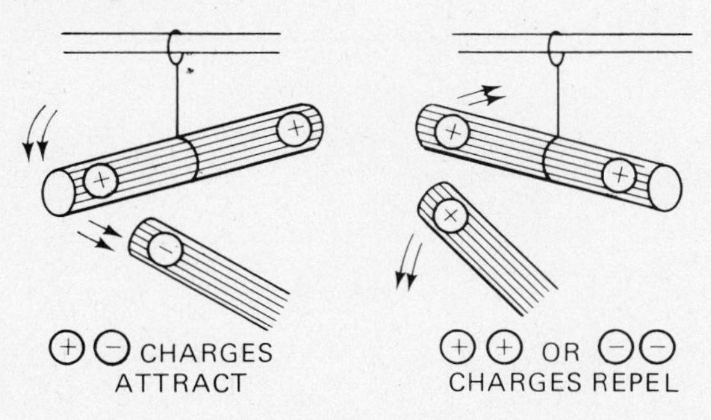

FIGURE 12-2 Attracting and repelling of charges.

Chemical Action

The prime example of converting chemical action to electricity is the common battery we used in our cars, flashlights, toys, etc. It may be a wet- or dry-cell action, and the resulting energy is dependent on the physical size of the battery.

Friction Action

Static electricity caused by friction is a condition most of us have experienced: for example, running a comb through our hair and picking up bits of paper, or walking on a rug and then getting shocked by touching another object. The principles of static electricity have not been applied to commercial generation of electricity but are widely used in other ways, such as dust collectors, electrostatic paint spraying, etc.

Heat Action

When two wires of unlike material are fused together at the ends and heated, a small amount of electrical current can be generated. Such a device is known as a *thermocouple* and is important in the refrigeration, heating, and air-conditioning field. The principle is used in temperature measurement devices or control devices, which react to the presence of heat.

Light Action

Light has energy, and when it strikes the surface of a substance it may dislodge some of the electrons from their normal orbits. If for example the material is a light-sensitive photocell, a small electrical current may be generated. A good example of this use is in the control system of an oil burner. On start-up, if the appearance of a flame is detected by the photocell, the oil pump is allowed to run. If no flame is present within a set period of time, the burner oil pump will be shut down.

Pressure

Some types of material, such as quartz, when compressed will produce an electric current in a specific direction. When pressure is removed, the flow of current changes direction. This effect can also be reversed, so that applying a current will cause the quartz crystal to bend or vibrate. Although the power produced by this pressure effect is very small, it can be amplified and has important practical applications in record-player pick-ups, microphones, etc.

Mechanical Action

Through electromagnetism (which will be reviewed later) large amounts of electricity are produced to serve our residential, commercial, and industrial needs, using rotating machines called *generators*, which are turned by various means: steam turbines fired by oil, gas, or coal, water turbines in hydroelectric power plants, and smaller systems that incorporate internal combustion engines to drive the generators. Thus, the mechanical energy of the generator is converted to electrical energy.

Nuclear Action

The atomic power plant is becoming an increasingly important source of energy for the production of electricity. Without getting into a technical discussion of atomic reactors, it need only be said that atomic reactions create a large amount of heat, which is then used to develop high-pressure steam to drive the mechanical turbine generators.

Magnetism

Earlier it was pointed out that objects of like charges repel each other, and unlike charges attract. You may recall from your study of physics that a natural magnet, such as a lodestone, is surrounded by a force field. The classic experiment is to place a bar magnet under a sheet of paper or glass on which rest iron filings (Fig. 12-3). The filings will align themselves along the lines of force that leave one end of the magnet and return to the other. One

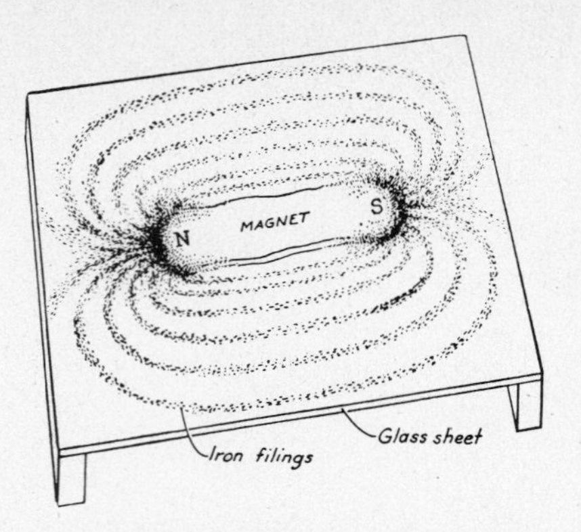

FIGURE 12-3 Magnet and iron filings.

end is called the *north* (*N*) *pole* and the other the *south* (*S*) *pole*.

Now if we take this bar magnet and pass it inside a coil of copper wire as illustrated in Fig. 12-4a, b, and c, we observe, on a sensitive galvanometer, the

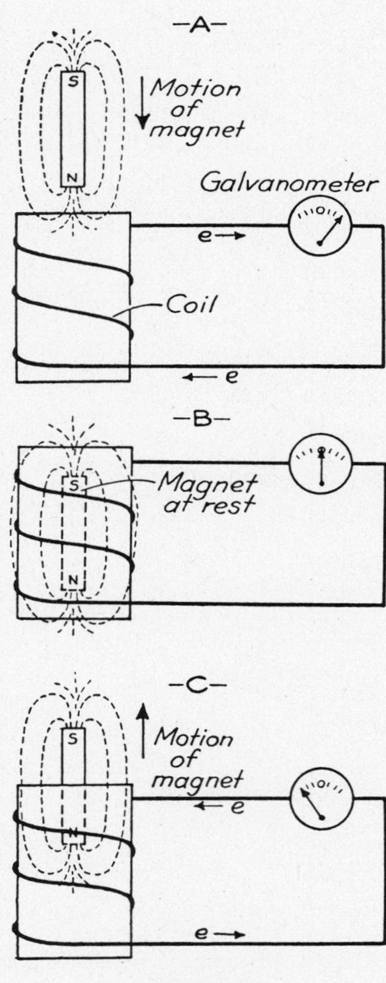

FIGURE 12-4.

development of a current as the copper wire (conductor) cuts the force field of the magnet.

In position *a*, the motion of the magnet is in a downward direction and the galvanometer needle will move to the right. When the magnet is at rest (position *b*), no current is established. With the magnet moving in an upward direction (position *c*) current again flows, but in the opposite direction to that shown in position *a*.

This phenomenon is fundamental to all practical power generation and electric motors. *When an electrical conductor cuts through a magnetic field, an electromotive force is set up between the ends of the conductor*. Later we will discuss the principles of electromagnetism as applied to generators and motors.

SIMPLE ELECTRICAL CIRCUIT

With a source of potential electrical energy established, let's examine a simple electric circuit.

The term *circuit* is to be taken literally, for unless electricity is able to follow a complete path from the source to a load and return to its source, it will not accomplish any work—such as lighting a lamp or running a motor. The two wires at the bottom of illustration Fig. 12-5, marked *L1* and *L2* in the diagram, go to the source of electricity for this circuit.

The electricity flows from and to the source, passing through the switch and through the bulb (load) on its way. We say that electricity flows. This popular conception of an electrical current as a flow of electrons reflects the common attempt to make electrical phenomena understandable by direct comparison to flowing water. (Note the term *current*, which again slips into our discussion). For our purpose, it is sufficient to say that when electrons move through a material, we have electricity.

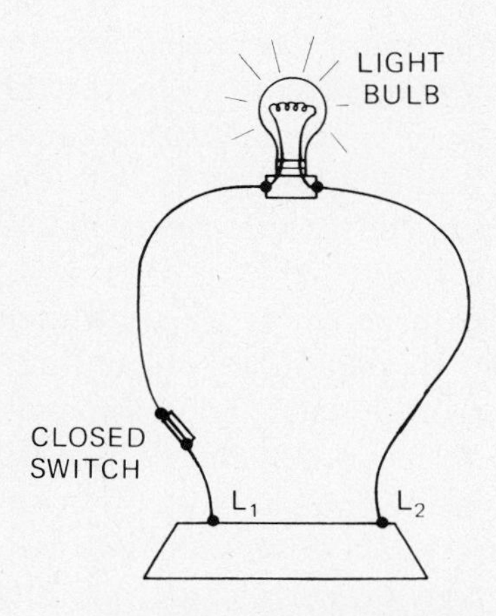

FIGURE 12-5 Simple electric circuit.

For electricity to flow, there must be, at the source, a buildup of electrons sufficient to create a pressure or force. This electron buildup may be accomplished chemically, as in a battery; magnetically, as in a generator; or by heat, as in a thermocouple. The proper term for this pressure is *electromotive force*, (*emf*). Since the unit of measurement for emf is the *volt*, the term *voltage* is synonymous with electromotive force in practical work.

As is true for all motion, the movement of electrons through a material is opposed by friction. Depending on the material, this opposition to electron flow, called *resistance*, may be great, moderate, or low.

Copper, for example, offers low resistance to the flow of electrons; therefore this metal is by far the most common electrical conductor used, whether in the form of small wires, large cables, or giant busbars. Aluminum is also a good conductor and is being used increasingly in electrical work. Tungsten, used in the filaments of incandescent lamps, offers a very high resistance to electron flow.

The unit of measurement for electrical resistance is the *ohm*. Devices, such as bulbs, with high resistance are called *resistors* and are denoted by a wavy line in wiring diagrams.

If the total resistance in a circuit is great enough to prevent any flow of electrons at a particular emf, no current can pass. When the voltage is high enough to overcome the resistance, electrons flow through the conductor and the resistor. For practical purposes, we are interested in knowing how many electrons flow past any point in a circuit during a given time. It is this *rate of flow* of electrons that is properly called *current*.

The unit of measurement for the rate of electron flow is the *ampere*, almost always called the *amp*.

Summarizing:

Electromotive force is expressed in volts.

Electrical resistance is expressed in ohms.

Current, the electrical rate of flow, is expressed in amps.

We apply this information to the simple electrical circuit in Fig. 12-6. A voltmeter shows the electromotive force is 120 volts; an ammeter indicates the current is 0.6 amps. Obviously, the emf is sufficient to cause current to flow through the total resistance offered by the wires and the bulb. This total resistance may be measured with an ohmmeter or it may be easily calculated; the calculations will be described later.

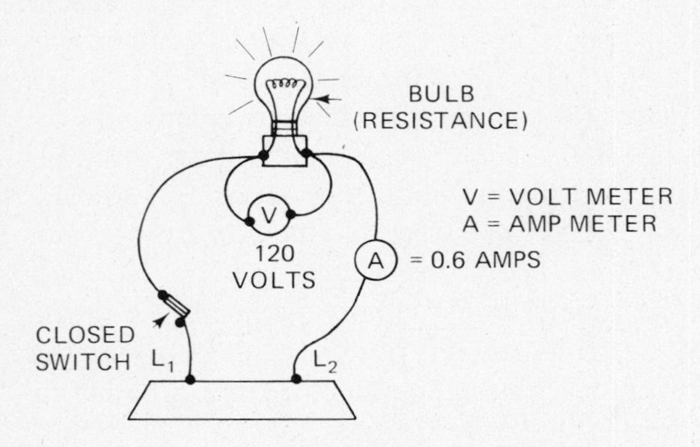

FIGURE 12-6.

With the same current flowing through the wire conductors, the switch, and the bulb, why do they not all get hot enough to emit light? Recall that different substances offer different resistances to the flow of electrons. Stated another way, a copper wire and a tungsten wire of the same diameter and length can both pass 0.6 amps under a pressure of 120 volts. But the electrons must work harder to get through the high-resistance tungsten than through the low-resistance copper. The harder work in this case results in heat and light.

An ohmmeter will show the resistance of the copper wire conductors to be practically zero; the resistance of a 25-watt bulb is about 570 ohms, when lighted. Obviously, the current can more readily flow through the copper than through the tungsten filament of the lamp.

If the resistor (light bulb) were not in the circuit, and only copper wire was connected to the emf, too much current—too many amps—would flow, and the copper wire would overheat and burn out (or a fuse would blow; the reasons for this will be explained later). We would have, in effect, a short circuit.

A short occurs when the current takes an accidental path that does not pass through the circuit's normal loads, causing the total resistance to drop near zero. If *L1* and *L2* were bare and accidentally touched, for example, or if a copper jumper wire were installed ahead of the resistor, as illustrated in Fig. 12-7, the current would take the path of least resistance and never touch the bulb. Since the resistance of the copper wire alone is low, so many amps would flow that the highly conductive copper would soon overheat.

An *open circuit* occurs when, for any reason, the path of the current flow is broken. Pulling the switch opens the circuit. A broken wire or broken bulb filament has the same effect.

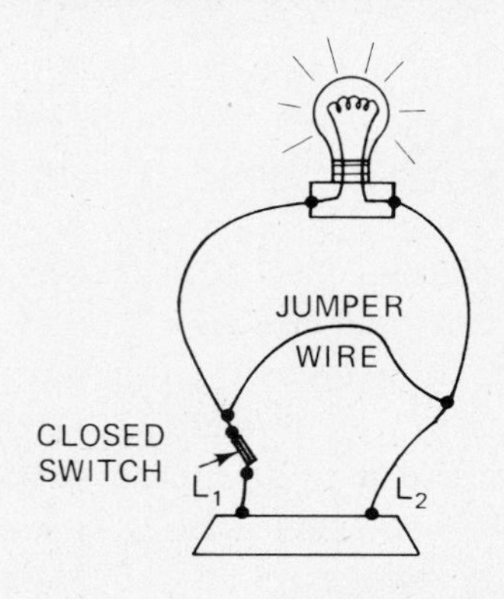

FIGURE 12-7.

Although some substances are nonconductors in a practical sense, any material may be made to pass some current if a sufficient emf (voltage) is applied. Thus, if the switch was open and a higher and higher voltage were supplied through *L1* and *L2*, at some point a spark would jump across the open switch contacts, indicating that current has passed through the air gap. Though air is normally a nonconductor, it will pass current if the voltage is high enough. (Artificial lightning can be made to travel hundreds of feet through the air.)

A circuit is said to be *closed* when all necessary contacts are joined to permit normal current flow through conductors and loads. The term must not be misunderstood to mean that the circuit is closed to current flow, as a valve might close off or shut off water flow. The circuit in Fig. 12-5 is closed, and current flows through properly made contacts and properly connected wires and load (resistor).

The relationship, indeed the interdependence, among electromotive force, resistance, and current is clearly indicated by the formal definition of a volt: the amount of electromotive force required to cause 1 ampere of current to flow through a resistance of 1 ohm.

The relationship is even more clearly defined by Ohm's law, named after the German scientist George Simon Ohm, who worked out these mathematical relationships early in the nineteenth century. Stated practically, Ohm's law shows that the greater the emf, the greater the current; the greater the resistance, the less the current.

Mathematically, the relationship is stated as: I equals E over R.

$$I = \frac{E}{R}$$

where I is the symbol for current in amps, E is the symbol for emf in volts, and R is the symbol for resistance in ohms.

Suppose, in the simple circuit in Fig. 12-8, the resistance symbol represents an electric iron, rather than a lamp bulb as it did in our previous discussion. With an ohmmeter, we determine the resistance to be 15 ohms. Either by taking the power company's word for it or by using a voltmeter, we determine the voltage is 120 V. How many amps will the iron draw?

$$I = \frac{E}{R} = \frac{120}{15} = 8 \text{ amps}$$

If the iron is replaced with a 100-watt bulb, we might find the bulb's resistance to be in the order of 145 ohms. With the same emf, how many amps will the light bulb draw?

$$I = \frac{E}{R} = \frac{120}{145} = 0.83 \text{ amp}$$

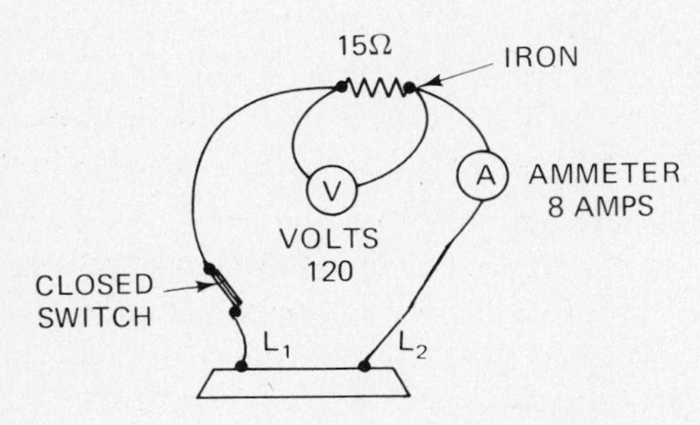

FIGURE 12-8.

It is obvious that the lower the resistance, at a given voltage, the higher the current that will be drawn. Theoretically, as the resistance approaches zero, the current approaches infinity.

If, in our simple circuit, we short the resistance out of the line, 120 volts will be pushing against almost zero resistance and hundreds of amps will flow—until a fuse or a burned wire opens the circuit. Assume the wire in our circuit has 0.2 ohm resistance; the current flow will be:

$$I = \frac{E}{R} = \frac{120}{0.2} = 600 \text{ amps}$$

The basic E-over-R formula may be altered, or transposed, to calculate any one factor, if the other two are known. For example, a volt-ammeter had been used to measure 1,000 amps at only 2 volts. What is the resistance?

$$R = \frac{E}{I} = \frac{2}{1{,}000} = 0.002 \text{ ohm}$$

Returning to the electric iron, assume the readings were 120 volts and 8 amps. What is R?

$$R = \frac{E}{I} = \frac{120}{8} = 15 \text{ ohms}$$

The formula may be transposed in one other way: $E = IR$.

In the 100-watt bulb example, suppose we had measured the resistance and the current but had no voltmeter. What is the emf in volts? (The current was 0.8 amps and the resistance 145 ohms.)

$$E = IR = 0.83 \times 145 = 120 \text{ V}$$

In electrical talk, the amount of voltage used up by each resistance in a circuit is called the *voltage drop* or the *IR drop* through that resistance.

There are several "helpful" tricks in remembering the formula for Ohm's law and its transpositions. We do not recommend such memory aids; If Ohm's law is used frequently enough in his work, the service technician has no need for a memory aid. And no good refrigeration technician trusts memory when making a calculation involving an almost forgotten formula; he looks it up.

Actually, most of the difficulty lies in remembering the transpositions, since the statement of the law as "*I* equals *E*-over-*R*" gives the first formula

$$I = \frac{E}{R}$$

To determine how to transpose, plug simple numbers into the basic formula. Substitute, for instance, 4 for *I*, 8 for *E*, and 2 for *R*. Thus:

$$4 = \frac{8}{2}$$

From this equation it is easy to detemine that 8 will equal 2 times 4 and 2 will equal 8 over 4.

Ohm's law, then, allows us to determine mathematically the third factor when any two are known by measurement or by reference to a specification sheet.

Although it is theoretically possible to get an increasing, even an unlimited number of amps from any voltage source—as resistance approaches zero—practical considerations, such as the heat needed to light the lamp or operate the iron, limit the current to a level that can be carried safely by the conductors and the devices on the line.

Here are some practical applications of the information so far:

(1) A 100-watt bulb manufactured for 120-volt operation will glow extra brightly on 240 volts for a few seconds, and then burn out. The resistance of the filament was taken into consideration in the design of its length and diameter, to make it compatible with 120 volts. At 240 volts this same resistance passes twice as much current —more than it is designed to carry. The filament overheats and burns out, opening the circuit.

(2) For a given material, resistance increases as the length of the conductor increases. This accounts for the voltage drop, the *IR* drop, on long extension lines. At a wall outlet, an electric iron may draw 9 amps at 120 volts. At the end of a 12-ft household extension wire, it draws 8 amps at 105 volts. In rural areas, it is not uncommon for the customer farthest from a pole transformer to have low voltage because of the voltage drop from the transformer through the long line to his home.

Ideally, *IR* drop should be kept under 3% of rated voltage. It is common to have a 5% drop; a voltage drop over 10% can cause malfunction and failure of motors, relays, and similar electrical devices.

(3) For a given material, resistance increases as the diameter of the conductor decreases. A lamp circuit in a home may call for No. 14 wire. Install a central air conditioner or electric range, and you must use No. 8 wire for the circuit. (The diameter of wires gets larger as the number gets smaller.) A larger diameter wire can carry more amps (45–60 amps for a range) than can a smaller wire (15 amps for a lighting circuit) without overheating. (Wire sizes will be more thoroughly discussed later.)

SERIES CIRCUITS

When more than one resistance device is placed in a circuit, the current can flow through more than one type of path on its way from and to the emf source. Depending on how the path is designed, the circuit will fall into one of three categories:

(1) series
(2) parallel
(3) series-parallel

Simplest to understand is the series circuit. It is familiar to most through the older type of Christmas-tree lights; when any one of them burns out, they all go out because of an open circuit. (We can see why all the lights go out by looking at Fig. 12-9.)

To travel from and to the emf source, the current must pass through each resistance in succession. Resistors in series are connected end-to-end across a voltage source.

Remove one lamp and the current stops because the circuit is no longer complete. Place a jumper around the defective bulb(s) and the rest will light.

Such a current stoppage can be useful. All switches, for example, must be in series with the devices they control. Protective devices, as fuses and overload protectors, are wired in series so equipment cannot operate when the safety device is electrically opened for any reason.

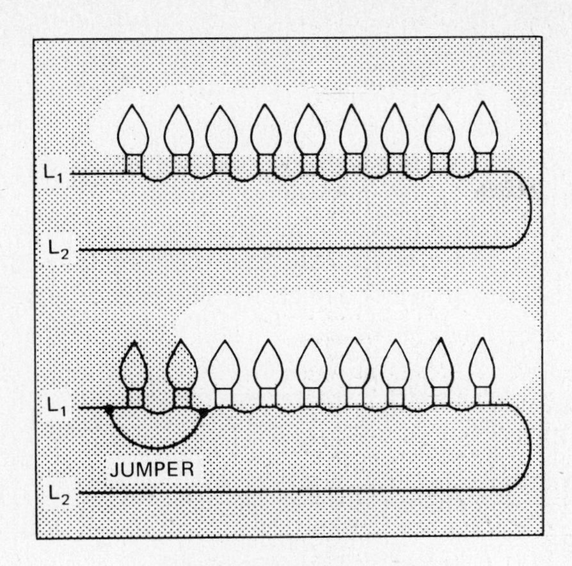

FIGURE 12-9 Series circuit.

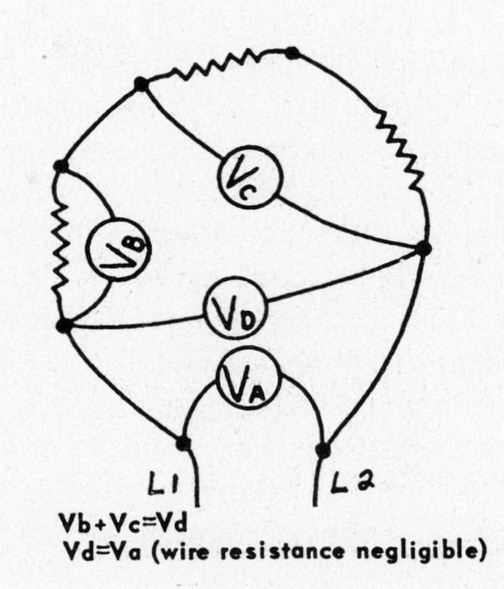

FIGURE 12-11 25 W bulb series diagram.

In the diagram in Fig. 12-10, taken from a full circuit diagram for a refrigeration unit, the fan motor overload protector (OL) is in series with the power supply. Should the protector open due to excessive current or overheating, no electricity can reach the fan motor windings. *B* and *BR* are color codes for the wires. What about volts, amps, and ohms in a series circuit? Remember that voltage may be read between any two points on a conductor or resistor and that the amount of voltage used up between any two points is termed the *voltage drop*.

If three equal resistances, such as three 25-watt bulbs, are wired in series Fig. 12-11, the voltage read by a meter at V_A will be the supply voltage, 120. A voltmeter across a single resistance, V_B, will show 40 volts. V_C, across two 25-watt bulbs, will show 80 volts, and V_D, across all three resistors will again show 120 volts. (The negligible resistance of the wires is ignored).

The practical effect of voltage drop in a series circuit can be seen in Fig. 12-12 that a simple circuit with a motor designed to operate off 120 volts. A 25-watt bulb placed in series with the motor causes enough voltage drop to prevent the motor from starting. Actual measurements on the illustrated circuit showed a 95-volt drop through the bulb; therefore only a 28-volt supply to the motor remains. Twenty-eight volts are not enough to operate a device designed for 120-volt application.

If one 25-watt and one 75-watt bulb are wired in series, the 25-watt bulb glows almost at full brilli-

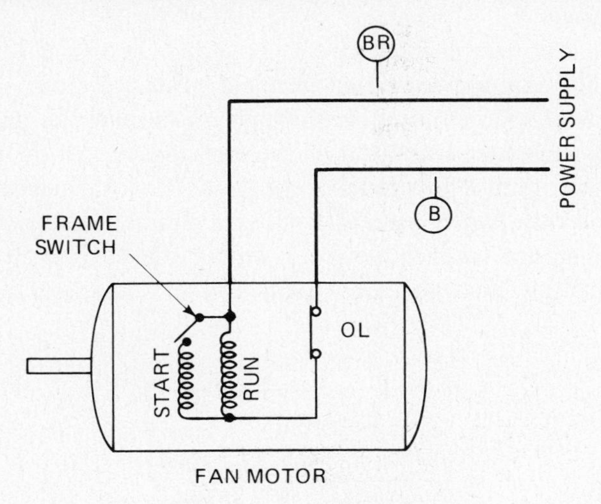

FIGURE 12-10 Fan motor.

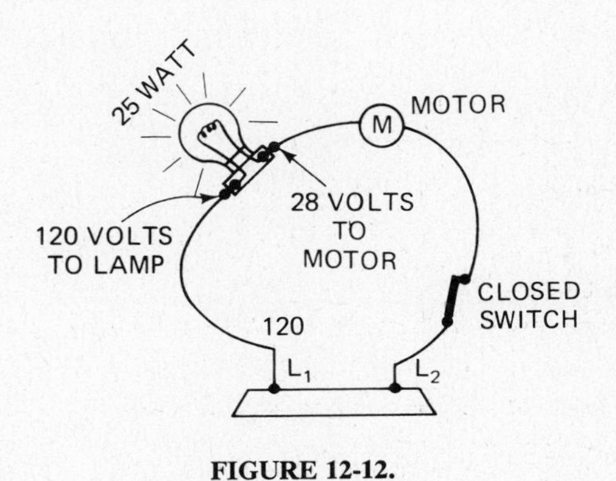

FIGURE 12-12.

ance, with a 110-volt drop, while the 75-watt bulb takes the remaining 10 volts without even glowing.

In a series circuit, then, the voltage drop through, or across, each resistance is only part of the total voltage and depends on the ohm-value of each resistor. The sum of these individual *IR* drops always equals the total applied voltage.

In a series circuit, the total resistance to current flow is the sum of all the individual resistances. The entire circuit may be treated as a simple circuit in Ohm's law calculations. For example, if a 75-watt bulb with 194 ohms resistance and a 25-watt bulb with 570 ohms resistance are in series, the total resistance of the circuit is $194 + 570 = 764\ \Omega$. The Greek capital letter omega (Ω), is the mathematical symbol for ohms.

For the entire circuit:

$$I = \frac{E}{R} = \frac{764}{120} = 0.16 \text{ amp}$$

Thus 0.16 amp will flow through each resistor. This amount of current is enough to heat the filament of the 25-watt bulb, but not enough to warm the filament of the 75-watt bulb or to start the motor in the example above. Using the same method and reasoning, it can be seen why a single 25-watt bulb will glow brilliantly at 120 volts, but three 25-watt bulbs in series will each glow only dimly.

The foregoing discussion has clearly indicated that the current, the amp draw, is the same in all parts of a series circuit. An ammeter placed anywhere in the circuit will read the same current (A) as at any other point. In Fig. 12-13, A_1 equals A_2 equals A_3 no matter how diverse the ohm values of the separate resistors.

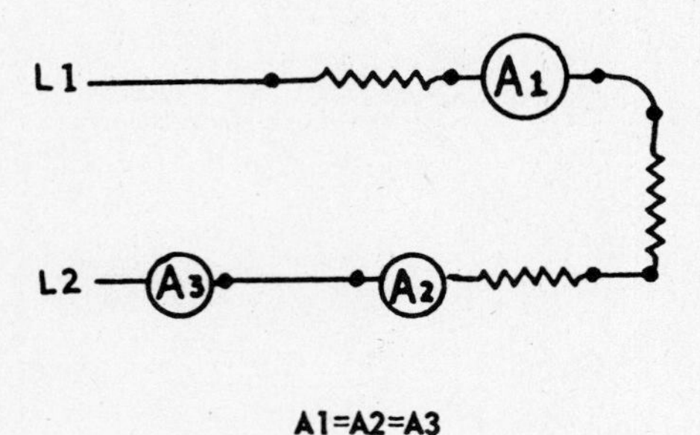

FIGURE 12-13.

PARALLEL CIRCUITS

Just as a series circuit is described as one in which the resistors are connected end-to-end, a parallel circuit is one in which the resistors are connected side-by-side across the voltage source.

The three bulbs in Fig. 12-14 are wired in parallel; this is a conventional line drawing for such a circuit, which clearly depicts the side-by-side arrangement of the resistances.

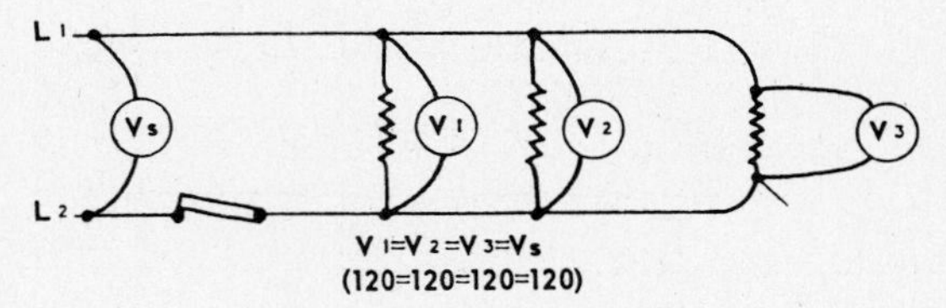

FIGURE 12-14.

Utility wiring in dwellings and most commercial-industrial structures is designed in parallel. The newer Christmas tree lights, those that do not all go out when one bulb fails, are in parallel.

Several characteristics of parallel circuits have immediate practical significance. You would notice first the full brilliance of the three 25-watt bulbs in parallel. The same three bulbs in series, across the same 120-volt source, would be considerably dimmer. A discussion of volts, ohms, and amps in a parallel circuit will indicate why this is so.

The voltage drop across each resistance in a parallel circuit is the same and each IR drop equals the emf of the source. A voltmeter across any of the resistors will read 120 volts, the same as directly across *L1* and *L2*. Mathematically, V_1 equals V_2 equals V_3 equals V_S.

Another parallel circuit is shown in Fig. 12-15. This diagram indicates that electrons (*e*) from the power source may flow through more than one path on their way from and to the emf source. All of the current must flow in *L1* and *L2*, but the total electron flow is divided among the three resistances. If the resistance of each device is equal, the current flow through each will be equal. If the resistances are

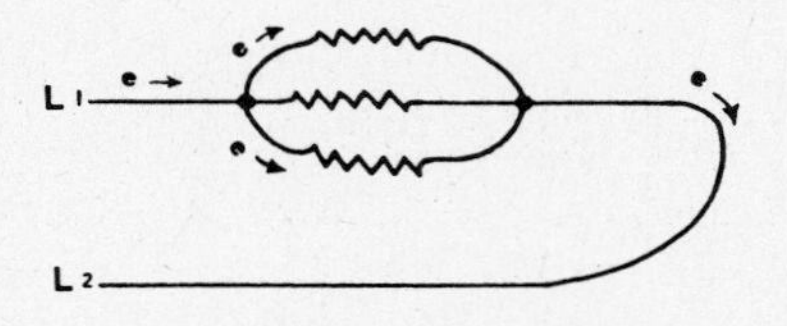

FIGURE 12-15.

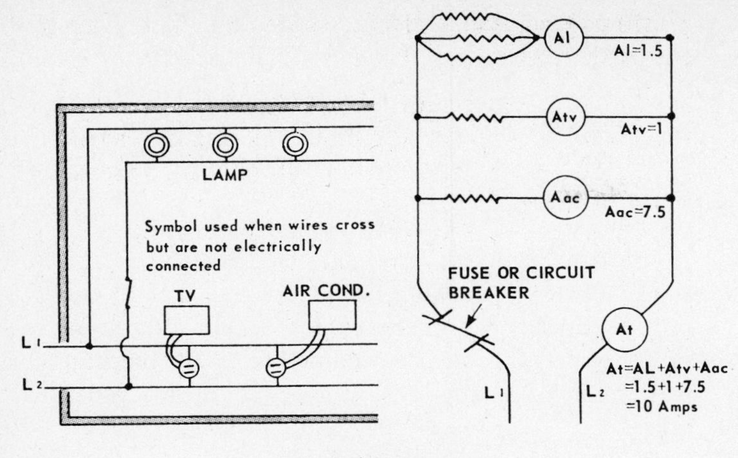

FIGURE 12-16.

unequal, the current flow through each will be unequal.

In both cases, the total current will equal the sum of the individual currents in each branch, or leg, of the parallel circuit.

Assume, for example, a household living-room circuit, fused for 15 amps (Fig. 12-16). All devices are in parallel when plugged into wall outlets, with three lamps, the TV, and an air conditioner operating simultaneously. Let us assume the lamps, all together, are drawing 1.5 amps, the TV is pulling 1 amp, and the air conditioner is drawing 7.5 amps.

Plug in a 10-amp iron so the housewife can iron while watching TV in the air-conditioned living room, and the 15-amp fuse will open under the 20-amp load.

Each leg of the circuit draws its portion of the total current, but all the electrons must come from and return to the emf source, through the fuse for the circuit. An ammeter placed at A_T will record this total amp draw of all the devices on the circuit. Remember, the amp draw through branches of a parallel circuit are equal only if the resistances in the several legs happen to have the same ohm-value.

Total resistance, in a parallel circuit, is *not* the sum of individual resistances. Instead, *total resistance in a parallel circuit is always less than that of the lowest individual resistance.*

Assume a two-device parallel circuit carrying an iron with a 10-Ω resistance and a toaster with a resistance of 23 Ω (Fig. 12-17). Using Ohm's law as in a series circuit, we would add the resistances and divide them into the voltage to get amps:

$$I = \frac{E}{R} = \frac{120}{33} = 3.6 \text{ amps}$$

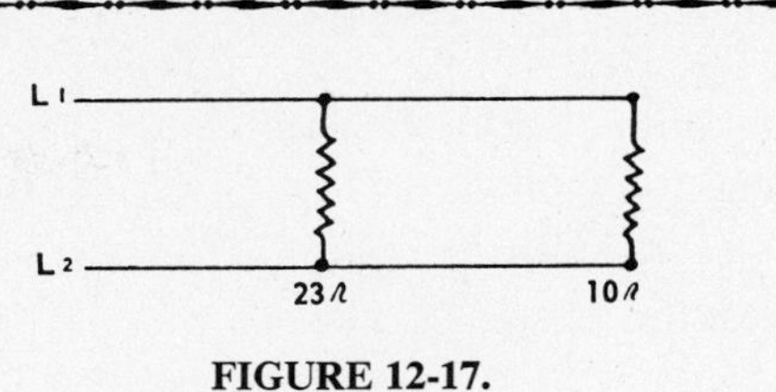

FIGURE 12-17.

Actual measurement, however, shows the toaster draws 5 amps and the iron draws 11 amps—a total of 16 amps. Using the Ohm's law transposition:

$$R = \frac{E}{I} = \frac{120}{16} = 7.5\,\Omega$$

7.5 ohms is the actual total resistance in this circuit, and this ohm value is less than the lowest individual resistance, 10 Ω.

The formula for finding the total resistance (R_T) in a parallel circuit is:

$$\frac{1}{R_T} = \frac{1}{R_1} + \frac{1}{R_2} + \frac{1}{R_3} + \cdots$$

In our example:

$$\frac{1}{R_3} = \frac{1}{23} + \frac{1}{10}$$

Usually, the easiest way to calculate is to convert the fractions to decimals: (Divide 23 into 1 and 10 into 1.)

$$\frac{1}{R_3} = 0.043 + 0.10$$

$$\frac{1}{R_3} = 0.143$$

$$0.143\, R_T = 1$$

$$R_T = \frac{0.143}{1} = 7 + \text{ohms}$$

Seven ohms approximates the 7.5 Ω obtained by calculation from field-measured amps and volts.

The calculated R_T can then be used in any calculations using Ohm's law.

Summarizing

In a parallel circuit:

Voltage across each resistor is the same and equals the applied voltage.

Total amp draw divides among the parallel branches in the circuit and may differ in each branch.

Total resistance is always less than that of the lowest individual resistance.

COMPARISON OF SERIES AND PARALLEL CIRCUITS

There is no "best" circuit. It is meaningless to dispute which type of circuit—series or parallel—is more useful, since each is indispensable to modern technology.

All but the most simple electrical circuits, indeed, make use of both types in what are logically called *series-parallel* circuits. On a refrigerated display case, the supply voltage is used to operate the compressor, the fans, the lighting circuit, and the defrost heaters. Depending on manufacturer's design, each component may be wired to operate independently or together with any one or more of the other devices. The choice is made, in effect, by the electrical circuitry—by proper use of series and parallel circuits either alone and in combination.

DIRECT AND ALTERNATING CURRENTS

All the information so far presented is valid for both direct current circuits and for alternating current circuits, the latter having resistance-type devices only on the lines. (The reason for this qualification about ac circuits will be explained later; it involves added resistance that is encountered in magnetic fields and is called *inductive reactance*.)

The rest of this discussion, dealing with such items as transformers, power distribution, solenoids, relays, and motors, will require an understanding of the distinction between ac and dc circuits and of differences in behavior between the two. Further, most of the remaining material will pertain to alternating current.

As defined earlier, electricity is said to flow whenever electrons move through a conductor. *If the electrons always move in the same direction through the conductor, the flow is called direct current (dc). If the electrons alternately move first in one direction, then in the opposite direction through the conductor, the flow is called alternating current (ac).*

Electrical current produced by chemical action, as in a battery, is always dc. This is also true of current produced by thermocouples. Either alternating or direct current may be obtained from the mechanical action of steam or hydroelectric turbine generators.

INDUCTANCE-CAPACITANCE-REACTANCE

Earlier it was stated that the discussion of Ohm's law and other such factors up to that point were valid for both direct current circuits and for **alternating current circuits with resistance-type devices only in the line**. The more perceptive reader may have noticed that the specific term *resistance* was applied to such things as incandescent lamps and heating elements but a coined term, like *ohm-value*, is used to refer to the opposition to current flow offered by a motor or other induction device.

An ohmmeter applied to the leads of a small shaded-pole motor may indicate that the pure resistance of the motor windings and leads is some 35 ohms. Using Ohm's law on an application across a 120-volt line:

$$I = \frac{E}{R} = \frac{120}{35} = 3.4 \text{ amps}$$

A small motor like this does not draw 3.4 amps. In fact, field measurement with a snap-around ammeter shows almost no deflection; clearly the current is under 1 amp.

The pure resistance of a solenoid coil was found to be 4 ohms. At 120 volts, this would operate at 30 amps theoretically. In actual operation, the coil draws 11.5 amps.

Obviously, something other than pure resistance must be opposing current flow in these instances. The increased opposition is due to *inductance*. In a later discussion of splitting the phase to enable a single-phase motor to start, it will be shown that merely creating more inductance in one coil than in the other drops current in the first far enough behind that in the second to get the shift of magnetic poles needed to obtain starting torque.

It is difficult at this point not to become too complex for the beginner and for those who need only an awareness, rather than a full understanding, of what follows. Even simplified explanations of alternating currrent circuits call for more mathematics than is advisable in a discussion at this level.

For now, it is accurate enough to say that in alternating current circuits, several factors combine to affect current flow in one way or another; in direct current circuits, only pure resistance is normally encountered.

Inductive reactance is that opposition to current flow offered by induction devices. Since inductance can only affect current flow while the current is changing (current changes generate an induced emf) it does not occur in dc circuits except at the moments of closing and opening the circuit.

If an ac circuit has only pure resistance, the current rises and falls at the same time as the voltage, and the two wave forms are said to be *in phase* with each other as illustrated in Fig. 12-18.

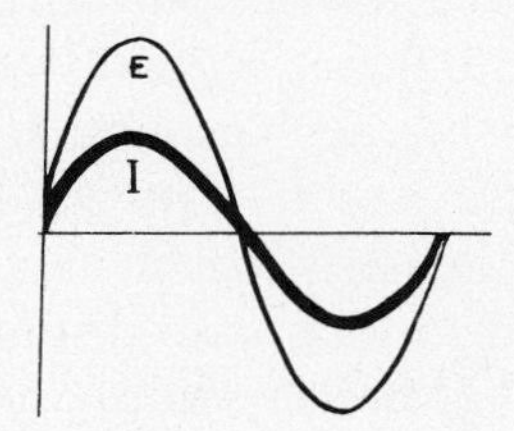

FIGURE 12-18.

Although it is only theoretically possible, if an ac circuit had only pure inductance, the opposed current *lags* behind the voltage a full quarter of a cycle, or 90° as shown in Fig. 12-19. In this instance, the voltage and current are said to be 90° out of phase.

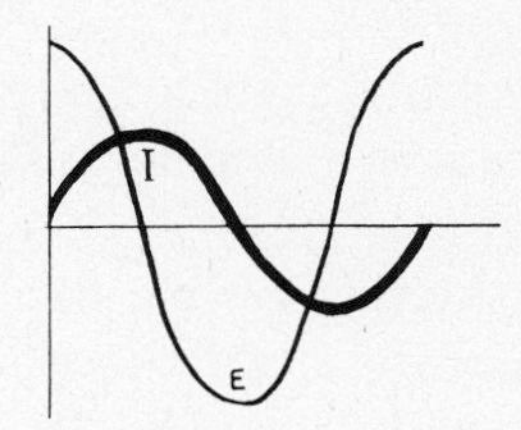

FIGURE 12-19.

The unit of measurement for inductance, symbol *L*, is the *henry*.

In a practical circuit containing both inductive reactance and resistance, the current wave will lag the voltage by an amount between 0° and 90°. Equal resistance and inductive reactance would produce a **45° phase angle**.

Just as inductive reactance opposes any change in current in an ac circuit, *capacitance* opposes any change in *voltage*. (Again, this phenomenon only affects dc circuits at the moments the current is turned on and off.) When the voltage increases, capacitance tries to hold it down; when the voltage decreases, capacitance tries to hold it up.

The electrical devices used to add capacitance to a line are, of course, the familiar *capacitors*, frequently called *condensers* by electronic and automotive people.

The actual action of capacitance in a circuit is to store a charge and to increase its charge if the voltage rises, discharge if the voltage falls.

Capacitance also offers an opposition to current flow—termed *capacitive reactance*—but the capacitive reactance (opposition to current flow) **decreases** as the capacitance (opposition to voltage change) **increases**.

As a result, in a theoretical circuit containing only pure capacitance, and no resistance, voltage can only exist **after** current flows. In such a theoretical circuit, the current wave **leads** the voltage wave by 90°, as illustrated in Fig. 12-20.

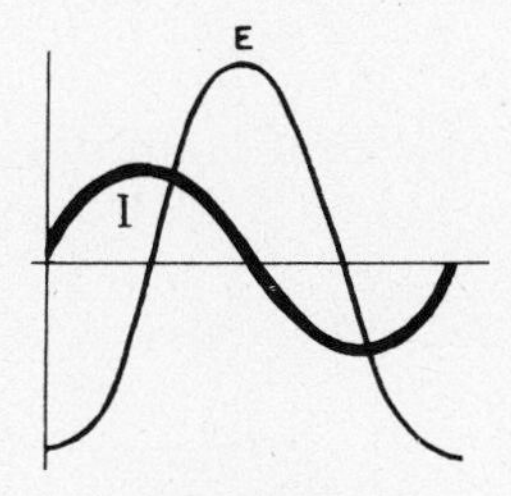

FIGURE 12-20.

The unit of measurement for capacitance, symbol *C*, is the *farad,* more practically the *microfarad*—one-millionth of a farad.

In a practical circuit containing both resistance and capacitance, the phase angle is between 0° and 90°, and the current always leads the voltage.

This is exactly opposite of what occurs in an inductive circuit where the current lags the voltage.

The fact that inductance and capacitance have opposing effects in a circuit is the explanation for the use of capacitors with a motor or on a power line to improve power factor, an item which will be discussed later.

In an alternating current circuit, then, the total opposition to current flow offered by resistance, inductive reactance, and capacitive reactance is termed *impedance*, symbol Z, and is expressed in ohms. The term *total opposition* here **is not** synonymous to an arithmetic sum of the several factors; we are ignoring the mathematics involved in converting henries and farads to ohms and the calculation of the total effect of varying inductances and capacitances.

Z, after it is determined, can be substituted for R in the normal Ohm's law statement:

$$I = \frac{E}{R} \quad \text{or} \quad I = \frac{E}{Z}$$

POWER FACTOR

In direct current circuits, and in alternating current circuits containing pure resistance only, power—in watts—is equal to the product of volts and amps.

$$P = EI$$

Thus, an incandescent lamp drawing 0.8 amp on a 120-volt line would be

$$P = EI = 120 \times 0.8 = 96 \text{ watts}$$

or, practically, a 100-watt bulb.

A 120-volt circuit containing several resistance devices with a total amp draw of 15:

$$P = EI = 120 \times 15 = 1{,}800 \text{ watts or } 1.8 \text{ kilowatt}$$

In this type of circuit, power factor is practically 100%; that is, the circuit actually expends very close to the calculated 1,800 watts. *Power factor* may be defined as the ratio of consumed power to supplied power, or as the percentage of time that the product of volts and amps equals actual power.

A 100% PF can only exist when the voltage and current are in phase as explained earlier. When either inductive reactance or capacitive reactance causes the current to lag or lead the voltage (get out of phase), the product of volts and amps gives only the *apparent power*, rather than the *actual power*.

Again ignoring the fundamental mathematics of the phenomena for the practical, the point is that for a typical commercial circuit, high in inductance, the utility must supply some power which does not actually perform work and which is not measured on the normal wattmeter, thus depriving the company of revenue.

If, for example, the calculation of measured volts and amps equals 2,000 watts, but the device on the line actually shows a 1,600-watt consumption, the power factor is:

$$PF = \frac{1{,}600}{2{,}000} \times 100 = 80\%$$

This means that only 80% of the supplied power is doing measurable work. The remaining 20% is *magnetizing current*, which makes possible the functioning of induction devices but which does no work itself and therefore is not normally recorded.

Induction devices, such as motors and fluorescent lights, never have a 100% power factor. In fact, fluorescent lighting uses so much magnetizing current that utilities tend to look with disfavor on this type of illumination. However, since induction devices are necessary to our technology, power companies in general have established a 90% PF as a practical minimum which must be maintained on their lines.

Depending on design and application, motors may have an inherent power factor as low as 60%. In terms of wave forms, this means that the inductive reactance is causing the current to lag considerably behind the voltage. Previous discussion showed that capacitive reactance tends to make the current lead the voltage. Logically, then, a low power factor due to inductance can be raised by placing capacitance in the line.

And in practice this is what happens. A running capacitor is placed on a motor; a bank of capacitors is installed in an industrial plant; pole capacitors are strategically located on distribution lines by utility companies. The resulting capacitance, acting in opposition to the inductance, establishes a more favorable power factor than would be possible with only inductance acting on the line.

PROBLEMS

1. The atom is composed of ____________ and ____________.
2. The neutron has a ____________ charge.
3. Current (the flow of electrons) is from a negative to a positive, true or false?
4. A battery produces electricity by ____________ action.
5. Emf stands for ____________.
6. The unit of measure for emf is ____________.
7. *Resistance* opposes the flow of electrons: true or false?
8. The unit of measure for resistance is the ____________.
9. The rate of flow of electrons (current) is called the ____________.
10. Name two basic types of electrical circuits.
11. Ohm's law is stated as $I = \frac{E}{R}$: true or false?
12. In series circuits the total resistance is the sum of all individual resistances: true or false?
13. Prob. 12 is also true for parallel circuits; true or false?
14. Resistance to current flow in an induction device (motor) is called ____________ ____________.
15. Capacitance opposes any changes in ____________.
16. A device that adds capacitance to a line is called a ____________.

13

ELECTRICAL GENERATION AND DISTRIBUTION

AC CURRENT GENERATION

Although batteries and thermocouples are called chemical and thermal generators of electricity, respectively, the term *generator*, used alone, always refers to a mechanical-magnetic machine, such as a hydroelectric or steam turbine, which is used to generate almost all the electrical power consumed for domestic and industrial purposes.

Essential to this last type of power generation is induced electromotive force and the resulting induced current; the voltage and current are caused by the relative motion of a magnet and an electrical conductor.

Assume the single coil or loop of wire shown in Fig. 13-1 is rotating in a magnetic field in the direction of the arrow. The following events occur as the loop is rotated (see Fig. 13-2).

(1) In position 1, two sides of the loop are parallel to the magnetic field, no lines of force are being cut, and no current is produced in loop. Emf (volts) is zero.

(2) In position 2, the coil has moved through 90°, the wires are now at right angles to the magnetic field, and maximum number of lines are being cut. This means that the emf is maximum in one direction.

(3) At position 3, the loop has completed half of complete turn, and again motion is parallel to the field and the emf again is zero.

(4) At position 4, the loop again is cutting maximum lines, and the emf again is maximum but in the opposite direction.

(5) The loop now is returned to position 1 and the cycle is completed.

These five steps describe the fundamental operation in the generating of alternating current. The smooth curve in Fig. 13-3, which traces the rise and fall of generated emf, illustrates one complete cycle. During the generating of typical household voltage, say, the curve traces the increase from zero to 120 volts in one direction, the drop from 120 to zero, the increase from zero to 120 in the opposite direction, and the drop again to zero. A cycle that takes place

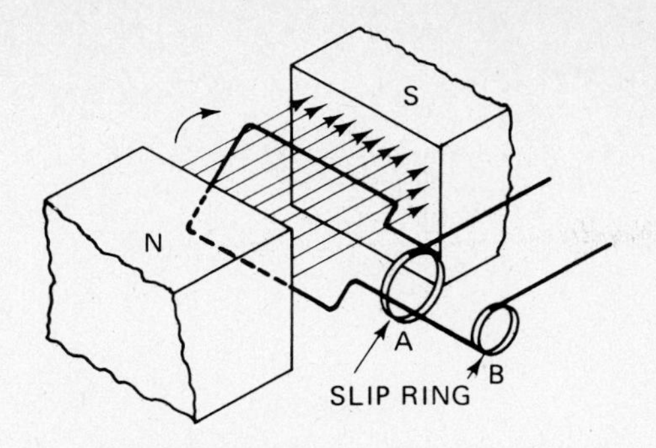

FIGURE 13-1 Rotating wire loop.

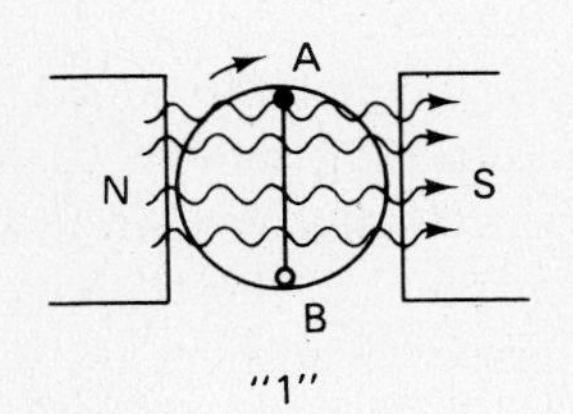

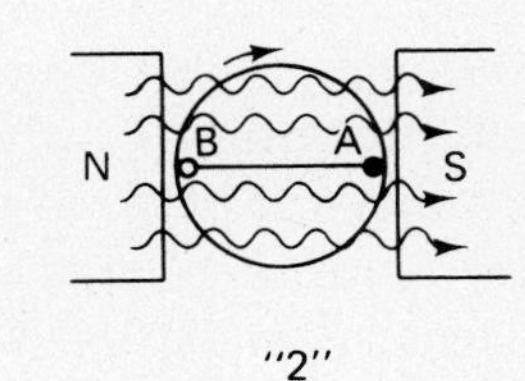

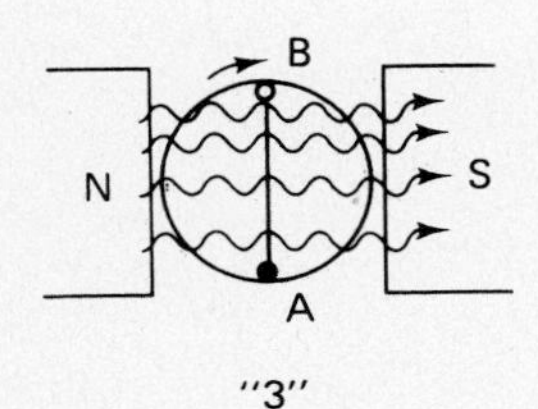

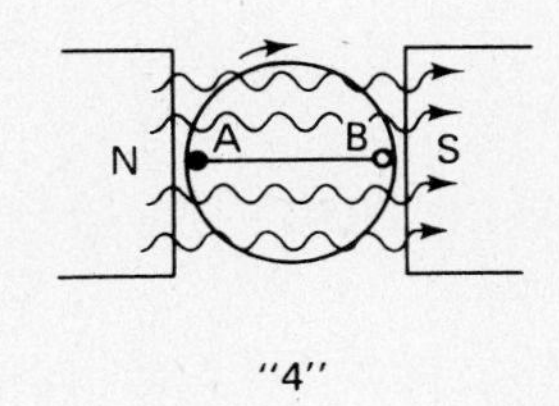

FIGURE 13-2.

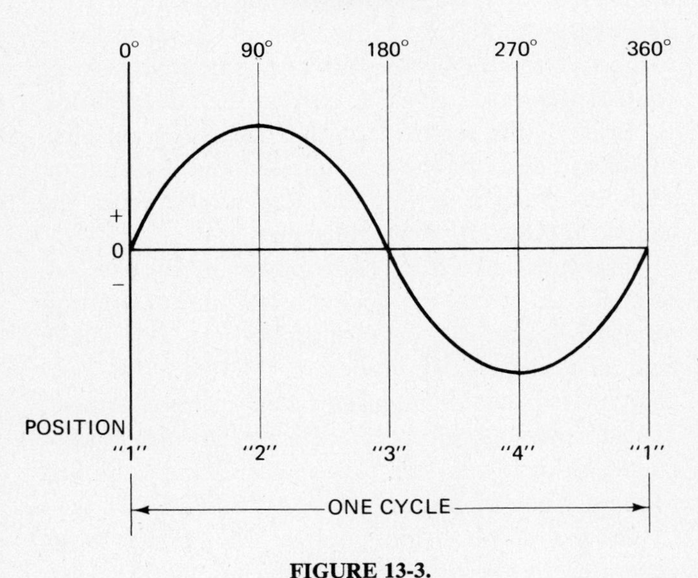

FIGURE 13-3.

in 1/60 of a second is called 60-hertz (cycle) current, meaning 60 cycles per second. The curve is also referred to as a *sine wave* or an *alternating current wave form*—these are mathematical terms that define the shapes of curves.

The current wave, as illustrated in Fig. 13-3, is said to be a *single-phase current*, meaning only one loop is cutting the magnetic field. Addition of another loop at right angles to first loop produces two separate voltages, one in each loop. Voltages produced are 90° apart; this is termed a *two-phase current*. Addition of a third separate loop results in a *three-phase current*, as illustrated in Fig. 13-4.

Each loop is at a 120° angle to the other. Single and three-phase systems of alternating current are the most popular.

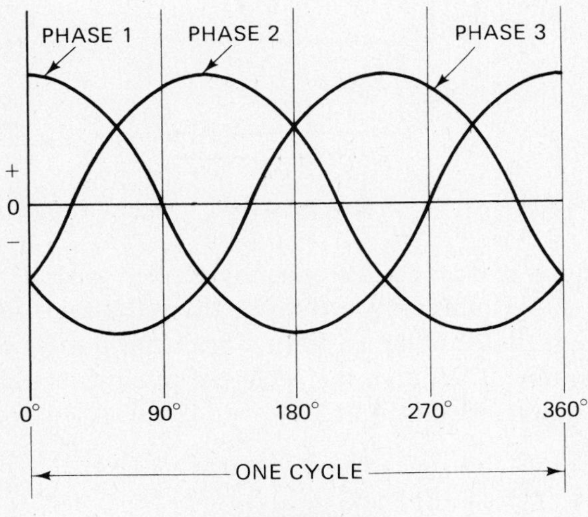

FIGURE 13-4.

Depending on the size of wires in the loop, the number of loops in a coil, and the speed of rotation, several to many thousand volts can be generated from this mechanical-magnetic induction apparatus. The speed of rotation also bears directly on the number of cycles (hertz) generated per unit of time. Almost universal in this country is 60-hertz ac. To generate 60-hertz current, modern generators rotate at 3,600 rpm (revolutions per minute). To produce the 25-hertz current, used until recently in most of Canada, the generator rotates at 1,500 rpm. In most parts of the world 50-hertz current is common, and in aircraft 400-hertz equipment is used to get the same work out of lighter and smaller machines than would be possible with 60-hertz apparatus.

DIRECT CURRENT GENERATION

If the voltage or current curve of a battery were plotted against time, as was the ac wave form, it would look like Fig. 13-5. The slanted line leading to the flat line denotes the moment of time required for direct current to build up to full value when the circuit is closed. Since chemical generation always causes electrons to build up on one of the materials involved and thus flow only in one direction at a constant rate (battery is full strength), the dc curve is really a straight line very unlike the ac wave form.

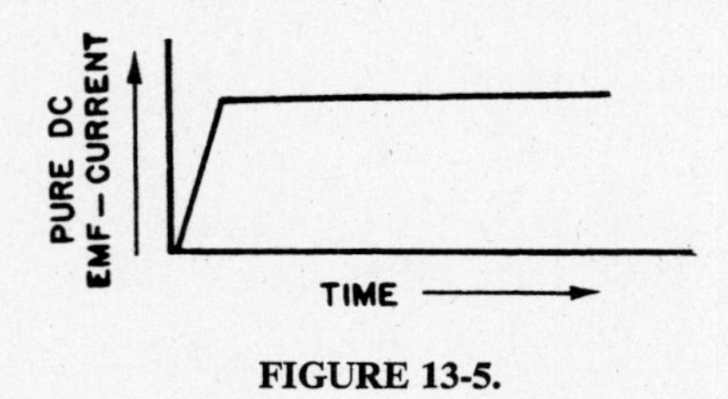

FIGURE 13-5.

Such steady dc is frequently termed *pure dc* or *true dc* to differentiate it from the pulsating direct current that is obtained from a mechanical-magnetic generator. The curve for such generated direct current looks like Fig. 13-6.

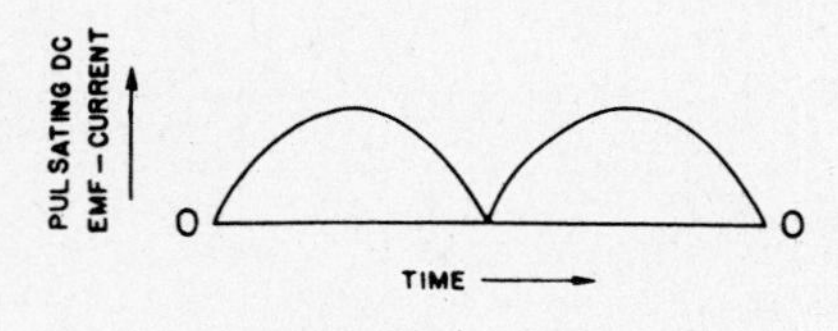

FIGURE 13-6.

Since a rotating coil in a magnetic field always generates alternating current, whence the direct current?

The transition from ac to pulsating dc is accomplished by altering the ac slip ring arrangement to a split-ring setup as illustrated in Fig. 13-7.

The effect of the split ring is to cut off the induced emf to the conductors just before it reverses direction on each cycle. As a result, the current can build up to two peaks per cycle, but both in the same direction—no curve below the zero line as in the case of alternating current. A single loop conductor causes wide peaks or pulsations, so in actual practice many loops are used to smooth out the current.

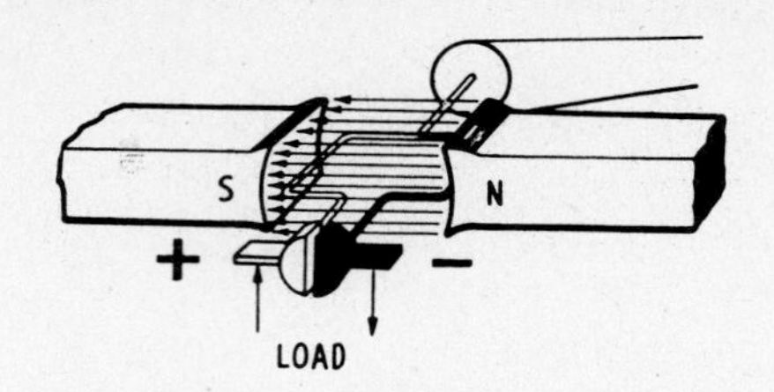

FIGURE 13-7.

Although a few cities still have some streets (usually downtown) on direct current service, the most widely-used type of electrical power used in the United States is alternating current, 60 hertz. The chief reason for the predominance of ac over dc is the greater ease and economy with which ac may be transmitted from relatively remote generating plants and distributed to consumers.

POWER DISTRIBUTION

In the beginning dc current was used to supply lighting and motor loads to the fledgling electrical industry. However, the major disadvantage of dc, its inability to be transmitted for any great distance without excessive power loss and expensive voltage control apparatus, made ac current the better choice.

Briefly, this segment of the discussion on electricity will describe the transmission and distribution steps involved in getting electrical power from the generating station to the consumer.

Turbines, driven by hydropower or by gas, oil, or coal-fired steam, produce the mechanical rotation needed to cause relative motion between a magnetic field and electrical conductors. The induced electricity is fed outside into giant step-up transformers. Transformers are electrical devices that either raise (step up), or lower (step down), the voltage supplied to them. Their operation will be discussed later.

In the step-up transformer, the 18,000 volts generated by the turbines is raised to 120,000 volts, or whatever voltage the system is designed for. (Specifics in the following discussion apply to a typical power system. Although transmission and distribution voltages may differ in certain areas, the basic principles of generating and handling power are universal.)

The next stage after the step-up transformer is the mat—a type of shipping department. In its maze of steel towers, cables, and relay equipment, the electricity begins its journey to the customers along the transmission lines. The term *transmission lines* refers to the conductors that carry the electricity from the generating station through utility-owned switching stations to substations.

In the substation the transmission voltage is lowered to 4,800 in a step-down transformer. From the substation, distribution lines carry the electricity either directly to industrial customers who purchase power at a primary rate and do their own stepping down or to strategically located step-down transformers, each of which serves a group of industrial or domestic customers.

An ac network can reduce the 4,800 volts from the substation to the proper utilization voltages and deliver those to individual customers in a downtown or commercial district.

In residential zones, the step-down is usually accomplished by pole transformers, and the power enters the home through the service entrance (meter, fuse box arrangement).

Figure 13-8 shows how different types of power are obtained. The primary wires, running along the tops of the utility poles, carry the current at distribution voltages, say 4,800 volts. Where three-phase power is to be taken off, all three wires are carried on the poles; when single phase will be used, either three or only two primaries may be used.

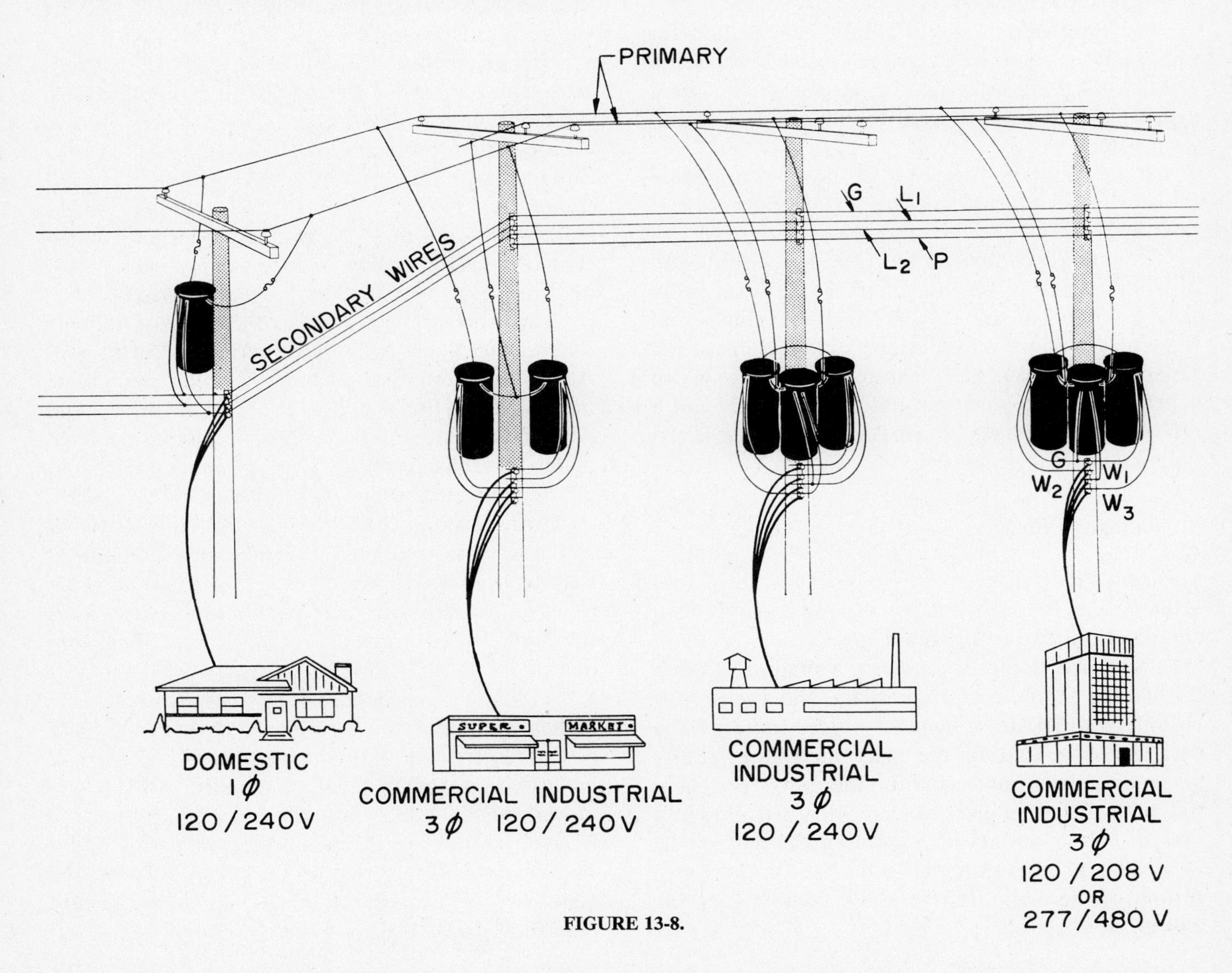

FIGURE 13-8.

Nominal Utilization Voltage	*60-hertz Service*
120 and 240–1 ϕ	120/240: 3-wire
120 and 240–1 ϕ 240–3 ϕ	120/240: 4-wire
120 and 208–1 ϕ 208/3 ϕ	120/208: 4-wire
120 and 240–1 ϕ	240: 2-wire
240–1 ϕ	240: 2-wire
120 and 240–1 ϕ 416–3 ϕ	240/416: 4-wire
120 and 240–1 ϕ 460–3 ϕ	265/460: *4-wire

* also termed 227/480

At the far left in Fig. 13-8, two wires enter the pole transformer where the secondary is tapped so as to give single-phase, 120/240-volt power over three wires, one neutral. The three secondaries from a single pole may travel from pole to pole and serve several houses—as many as the capacity of the transformer will allow.

Next, three wires are run from the primary into two transformers (which may be in a single shell). From the four wires issuing from the transformers, both single-phase and three-phase power are obtained at 120/240 volts. Again, secondaries can run from pole to pole so that one transformer serves several consumers.

The third sketch shows another connection for obtaining the same secondary service as in the setup just discussed. The power company chooses a method according to its overall need to balance its distribution system.

At far right in Fig. 13-8 is shown the manner in which the 120/208 or the 277/480 network is obtained.

From this brief review of electrical generation and distribution, it is obvious that the several utility voltages (120/240, 120/208, 265/460) and single- and three-phase power must all be obtained from the generating station and transmission lines which generally produce and transmit only one emf, not a multitude of voltages. Normally, three-phase alternating current is generated.

Voltages Utilized

From the three-phase (3 ϕ) and three-wire primary lines (wires) the following voltage and phase combinations can be obtained.

These are domestic 60-hertz voltages. Outside the United States both the voltage and cycles may be different. The term *nominal voltage* indicates the desired voltage and not the exact line voltage at any given time. It is impossible for the power company to hold exactly to a given voltage; they are allowed a plus or minus variation, in some areas as high as 10%. This point is extremely important in the application of hermetic refrigeration compressors, as will be seen later.

The trend is toward higher and higher voltages and for good reason. First, all normal distribution and utilization wiring is rated at 600 volts, so the insulation is the same for specific classes, whether the voltage is 120 or 480.

Second, there is little difference in the switches, relays, and other distribution and utilization details.

Third, motors that are wound for higher voltages, although the geometry of their windings may be different, will not cost very much more than regular motors.

Fourth—and most important—in larger installations the customer may realize savings up to 35% in wiring costs over a 120-volt base system.

The explanation of this last factor lies in the fact that when voltage is doubled, current is cut in half. At the lower amp draw, smaller diameter conductors may be used. For a house, the total savings would be insignificant, but for a large industrial or commercial establishment, the savings can come to hundreds, or even thousands, of dollars.

(Although the amp draw is less, the customer still consumes and pays for the same amount of electrical power. Power, in watts, is equal to volts-times-amps. The product of this multiplication is the same when volts are doubled and amps are halved.)

A major consideration in the move toward higher voltages is the very great increase in the use of fluorescent lighting, which operates off 277-volt ballasts. In a large installation supplied with 277/480 power, the 480-volt, 3 ϕ supply is for equipment, the 277-volt, 1 ϕ power is for fluorescent lighting, and a small transformer is used on the premises to get the 120-volt, 1 ϕ needed for small appliances, office machines, and control systems.

CONDUCTORS

A *conductor* is the path through which the electrical charges are transferred from one point to another. *Insulation*, such as the rubber or plastic coverings on the surface of conductors, confines the flow of electrical charges along a desired path.

Both conductors and insulators conduct current, but in vastly different amounts. The current flow in an insulator is so small that for all practical purposes it is zero.

Materials that readily conduct current, such as copper or aluminum wires, bus bars, and connectors, are the primary components of electrical equipment, circuits, and systems. The use of conductors and their insulation is governed primarily by the *National Electrical Code* (*NEC*). The *NEC* lists the minimum safety precautions needed to safeguard persons and buildings and their contents from hazards arising from the use of electricity for light, heat, refrigeration, air-conditioning, and other purposes. Compliance with the *NEC* and proper installation and maintenance result in a system that is more likely to be free from electrical hazards.

Wires and cables are the most common conductors, and the *NEC* specifies that the conductor size be selected to limit the voltage drop and also the maximum current that is safe for conductors of different sizes having different insulations and wired under different conditions.

CONDUCTOR SIZES

The system or measurement used to size a conductor is the *mil* or *circular mil*, as illustrated in Fig. 13-9.

A mil is one-thousandth (0.001) of an inch. A circular mil is the area of a circle one mil in diameter.

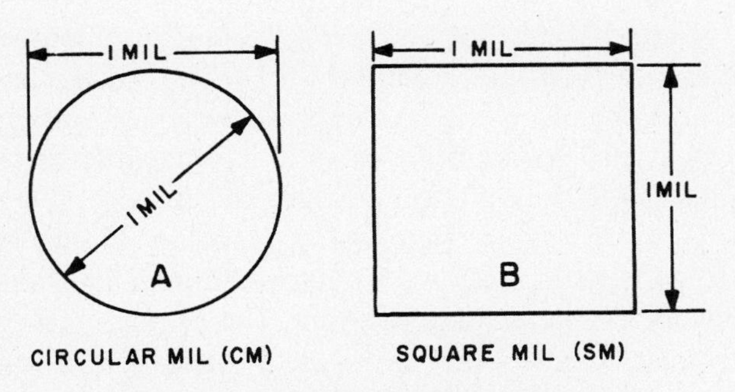

FIGURE 13-9 Numbering units for conductors.

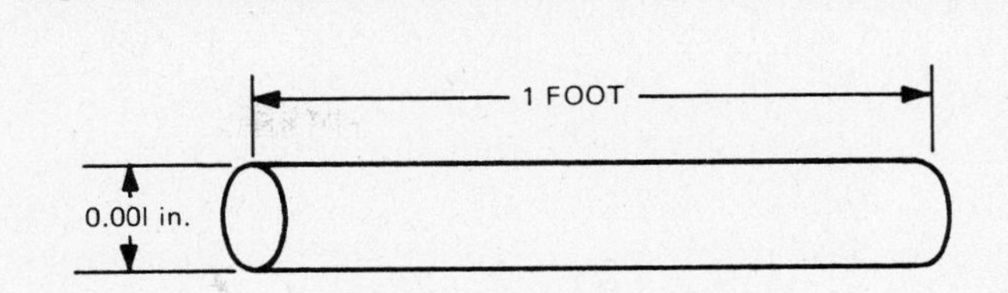

FIGURE 13-10 Circular-mil-foot.

Expressing wire sizes in mils is much simpler than in fractions of an inch. For example, the diameter of a wire is said to be 25 mils instead of 0.025 inches.

The mil-foot as illustrated in Fig. 13-10 is a convenient unit for comparing the resistance and size, called *resistivity*, of one conductor material with another. The resistivity valves are used to rate the current-carrying capacity of various conductor sizes as specified in the *NEC*.

WIRE SIZES

Wire sizes as published in the *NEC* are based on American Wire Gage (AWG) numbers. This is a convenient way of comparing the diameter of wires. A wire gage is shown in Fig. 13-11. It measures wires using a range calibrated from 36 to 0. To measure wire size, insert the *bare wire* into the

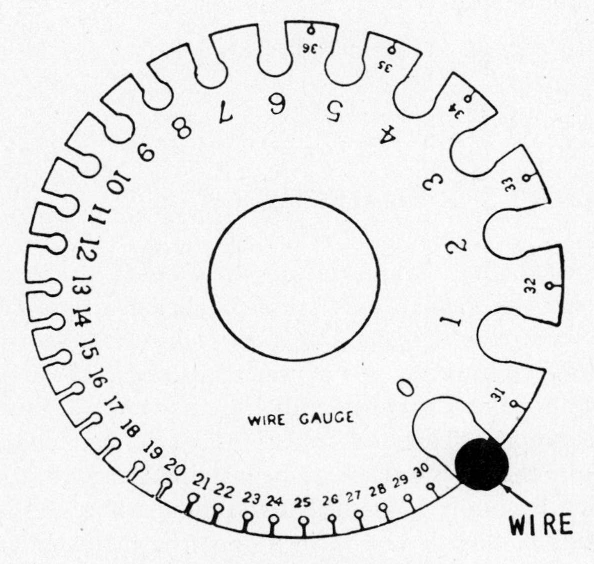

FIGURE 13-11 Wire gauge.

smallest slot that will just accommodate it. The gauge number corresponding to that slot indicates the wire size. Lamp cord is commonly a No. 18 AWG. No. 14 copper wire is the size-commonly used in household lighting circuits that carry 15 amps current. NOTE: Wire size numbers do *not* indicate the amp capacity. This value is obtained from the *NEC* book and depends on the type of conductor, the type of insulation, and surrounding conditions.

CONDUCTOR CLASSIFICATIONS

A basic classification of wires and cables can be made according to the degree of covering used:

- Bare Conductors have no covering. The most common use of bare conductors is in high-voltage electrical overhead transmission lines.
- Covered conductors are not insulated, but provide protection against weather and resistance to heat.
- Insulated conductors have an insulating covering over the metallic conductor that electrically isolates the conductor and allows grouping of conductors.
- Stranded conductors are composed of a group of wires in combination, usually twisted together.
- A cable can be either two or more bare conductors or insulated conductors.

In general the term *cable* is applied to the large conductors.

CONDUCTOR PROTECTION

In residential wiring (inside the house) the wire covering or insulation is usually sufficient to protect the wire from damage and from being grounded or shorted. However, in commercial and industrial buildings there is considerably greater risk of physical damage. Therefore most local electrical codes require the wires or cable to be installed in a *raceway*, which is simply a channel for holding wires, cables or bus bars, etc. The raceway may be metal or an insulating material.

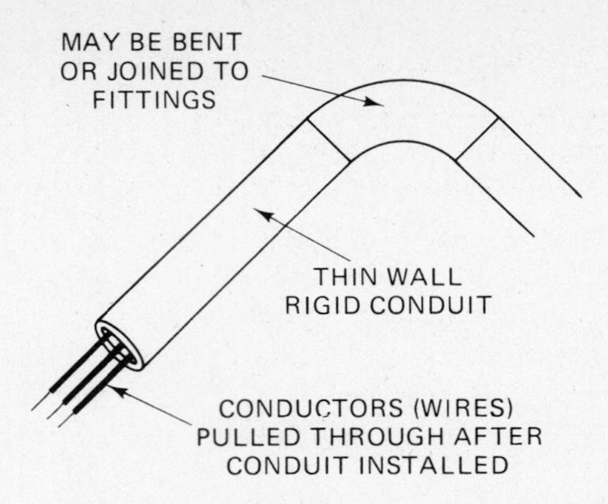

FIGURE 13-12.

Rigid metal conduit is one of the best known raceways (Fig. 13-12). Rigid conduit is made of steel or aluminum pipe that incloses all wires. A lubricant is used to help pull the wires through the conduit during installation. Such conduits are used for concealed wiring in buildings where the wiring is buried in concrete or masonry, since they provide good protection for wiring subject to mechanical damage or in hazardous locations. Rigid conduit bends can be made with a tube bender much like that used on copper refrigerant tubing.

Flexible conduit (Fig. 13-13), often called *Greenfield*, is used where rigid conduit is impractical. Flexible conduit is formed with a single strip of galvanized steel, spirally wound on itself and interlocked to provide strength and flexibility. It can be easily formed into numerous bends and turns. It absorbs vibration and can be used where adjustments in length (as when shifting a motor) are needed. Greenfield is not approved for wet or hazardous locations.

FIGURE 13-13 Flexible conduit.

Liquid-tight, flexible metal conduit, sometimes called *sealtight*, is suitable where oil, water, and certain chemicals and corrosive atmospheres might present problems. Essentially it is a flexible metal conduit with an outer liquid-tight jacket.

Armored cable (Fig. 13-14), generally known as *BX*, consists of insulated wire covered with an armor that looks like flexible conduit. The armor must be stripped back about 8 in. from the end before it is attached to a junction box.

Removing armor from BX cable.

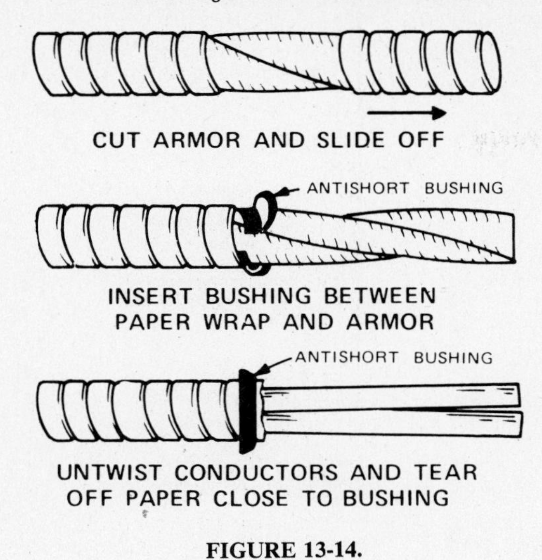

FIGURE 13-14.

FUSES AND PROTECTIVE DEVICES

Why fuses?

For protection, yes. But why the need for protection?

Simply, whenever electrons flow through a material, some heat is generated, and the temperature of the conductor rises. In a low-resistance conductor like copper, temperature rise is relatively low; in a high-resistance conductor like tungsten (lamp filaments) or nichrome (heating elements), temperature rise is more pronounced.

The point to remember is that, for each conductor of a certain material and a certain dimension, the temperature rise limits the amount of current that can flow safely. A quarter-inch, foot-long piece of copper will conduct more electricity without overheating than a piece of tungsten of the same dimensions. But the quarter-inch, foot-long copper can't carry as much current without overheating as a half-inch piece of copper of the same length.

Similarly, the 15-amp branch circuits in a home are wired with copper conductors capable of passing 15 amps without overheating. Should a short develop, resistance falls practically to zero and hundreds or thousands of amps flow through the wire. Without a fuse, the temperature rise caused by this excessive current will overheat the 15-amp conductors, at last burning them out, possibly causing a fire.

BASIC PRINCIPLE OF FUSE OPERATION

Essentially, a fuse is a strip of metal that will melt at a comparatively low temperature, placed in series with the circuit or devices it is designed to protect (see illustration, Fig. 13-15). When, for any reason, more current passes than the circuit is designed to carry, the metal strip melts apart, opening the circuit before the excessive current is able to overheat the conductors dangerously or burn out devices designed to operate at a much lower amp draw.

This is the basic principle of operation for all thermally operated fuses, whether of the plug or cartridge type. Where the entire electrical load on a circuit is resistance load—incandescent lamps, heaters, toasters—a simple strip fuse, plug, or cartridge is sufficient protection. On such circuits, the ampere rating of the fuse must not be greater than that of the conductors.

When motors are on the line, the situation is a bit different. When a refrigeration unit or air conditioner starts, for example, the amp draw on startup may be five to seven times what it is while in operation. For the part of a second that the motor requires to reach running speed, the conductors can carry the increased current safely. But a simple strip fuse, say a 15-amp fuse—designed to protect a motor with a 10-amp running current—may blow in less than half a second on a 60-amp starting current, causing an annoying current stoppage.

If a large fuse (60-amp) is used instead of a 15-amp fuse, the new fuse will still protect against a short in the circuit, but it will not protect against an overload. An overload, electrically speaking, is any condition that causes an electrical device to draw more current than normal, though it continues to operate. A bad bearing that results in a drag on a motor, can cause the motor to pull higher than normal amps in operation.

An oversize fuse in the circuit would make it possible to overheat the conductors dangerously, even though not enough current was drawn to blow the fuse chosen to protect against starting current.

Time-delay fuses are the answer to such problems. Supplied in both plug and cartridge types, the time-

Plug Fuses

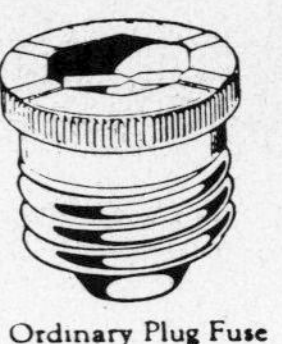

Ordinary Plug Fuse

FUSETRON dual-element plug fuse

Fustat

Cartridge Fuses

SUPER-LAG Renewable Fuse (Cut-away view)

Ordinary One-time Fuse (Cut-away view)

How a Fuse Works

If a current of more than rated load is continued sufficiently long, the fuse link becomes overheated, causing center portion to melt. The melted portion drops away but—

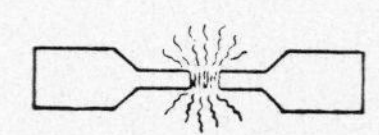

Due to the short gap, the circuit is not immediately broken and an arc continues, and—

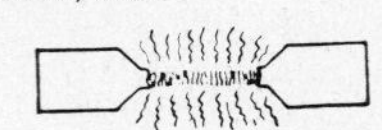

burns back the metal at each end, until—

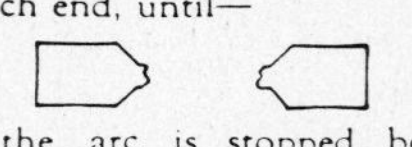

the arc is stopped because of the very high increase in resistance, and because the material surrounding the link tends to mechanically break the arc. The center portion melts first because it is farthest removed from the terminals which have the highest heat conductivity.

On a short-circuit, the entire center section is instantly heated to an extremely high temperature, causing it to volatilize (turn into a vapor). This vapor has a high current-carrying capacity, which would permit the arc to continue, but the arc-extinguishing filler cools and condenses the vapor and stops the arc. The degree or extent of the volatilization of metal is dependent upon the capacity and resistance of the circuit and upon the design of the fuse.

FIGURE 13-15 Fuse (Courtesy, Bussman Manufacturing, A McGraw-Edison Company Division).

delay fuse has a dual element. One element is the simple metal link that will open the circuit instantly in the event of a short. The second element makes use of a very small solder pot and a spring to make normal electrical contact. If current draw is excessive for a long enough period to melt the solder, the spring action breaks electrical contact, opening the circuit.

The pot and spring assembly may be designed for different protection ranges. One, for example, is so designed that it will allow overloads as high as 500% for 10 seconds without opening the circuit. Another will take a 200% overload for 110 seconds. This delay gives the motor ample time to start.

When a fuse blows, one of the following is the cause:

A short.

An overload.

Insufficient time-delay in the fuse.

Poor contacts on, in, or near the fuse.

Overheating of the fuse. (High ambients or too many fuses in a panel.)

Wrong fuse size.

Vibration.

A Cardinal Rule: *Never replace a blown fuse without determining and correcting the cause of the open circuit.*

Also operating on the principle of heat generation due to current flow, but applying the principle mechanically, is the *thermal overload protector*, used on hermetic compressors. Its basic component is a bimetallic strip. This is a thin strip of metal actually fabricated from two thinner strips of unlike metals pressed together. One metal has a higher rate of expansion than the other when heated (see Fig. 13-16).

Exposed to heat, the two metals expand at different rates, causing the strip to warp or buckle, since one or both ends are held fast. The bimetal strip can be designed and shaped so as to warp enough to open an electrical circuit—with which it is in series—at a predetermined temperature. (The temperature rise may be due to current flow or to high internal heat in the compressor.)

When the bimetal element cools off, it reverts

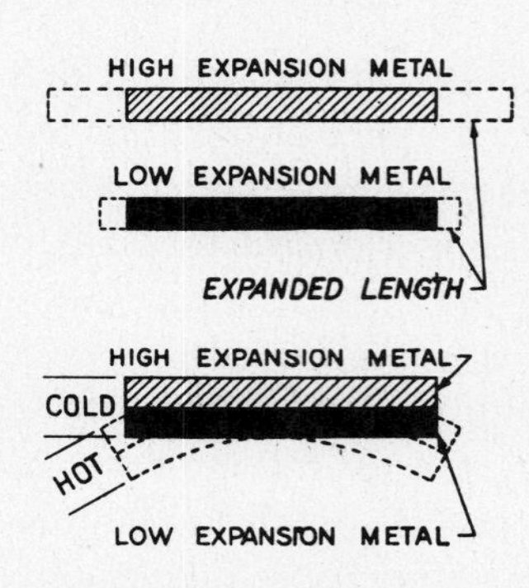

FIGURE 13-16 Bimetal strips.

to its original size and configuration, and closes the circuit again (automatic reset). If the cause of the first open was transient, the unit will operate normally. If an existing condition is causing chronic high temperature or current, the overload protector will repeatedly open the circuit until the condition is corrected.

Bimetallic elements are also applied in circuit breakers, motor starters, and relays, all of which will be discussed later.

PROBLEMS

1. Induced current results from rotating a wire coil in a magnetic field: true or false?
2. Does Alternating current produce a straight line or a sine curve?
3. 60-hertz current means porducing one cycle every 60 seconds: true or false?
4. Single and three-phase current are the most popular: true or false?
5. Common residential voltages are ______________.
6. Three-phase power is normally the voltage used by ______________.
7. A material that readily conducts current is called a ______________.
8. *NEC* stands for ______________.
9. Conductor sizes are measured in ______________.
10. A channel for holding wires, cables, etc., is commonly called a ______________.
11. Are fuses installed to protect the electrical appliances (lights, motors, etc.) or electrical conductors (wires)?
12. Time-delay fuses are used on circuits using motors: true or false?

14

ELECTRICAL COMPONENTS AND METERS

In Chap. 12 simple electric circuits were discussed. But, in the field of refrigeration, we must go beyond simple circuitry; for the serviceman and technician must be acquainted with such additional components as transformers, relays, contactors, starters, motors, motor protectors, and the meters needed to test out the circuits.

TRANSFORMERS

Information has been presented about the effect of current passing through a coil of wire, setting up a magnetic field around the coil. In an electromagnet, the magnetic field affects the core in the center of the coil, and an emf is induced in the coil and core center. An emf also can be induced in another coil, if it is placed adjacent to or in the vicinity of the first or primary coil. Figure 14-1 shows two adjacent coils, with an electric current passing through the first or primary coil and a magnetic field being set up around both coils. An emf is being induced in the secondary coil from the effects of the current passing through the primary coil, and a magnetic field is being set up around both coils.

The amount of voltage or emf induced in the coil depends on the number of turns within the secondary coil. The strength of the field between the two windings or coils also is dependent upon the number of turns in the primary coil, which is carrying the alternating current that sets up the magnetic field between both windings.

This process can be stated as a rule that: *A secondary voltage* (E_s) *is equal to the primary voltage* (E_p) *when it is multiplied by the ratio of the secondary turns* (N_s) *to the primary turns* (N_p). This is shown in Eq. 14-1.

$$E_s = E_p\left(\frac{N_s}{N_p}\right) \qquad \textbf{(14-1)}$$

EXAMPLE:

A given transformer has 250 turns in the primary winding and 125 turns in its secondary winding. If a source emf of 240 volts is applied to the primary coil, what will be the value of the voltage available in the secondary coil?

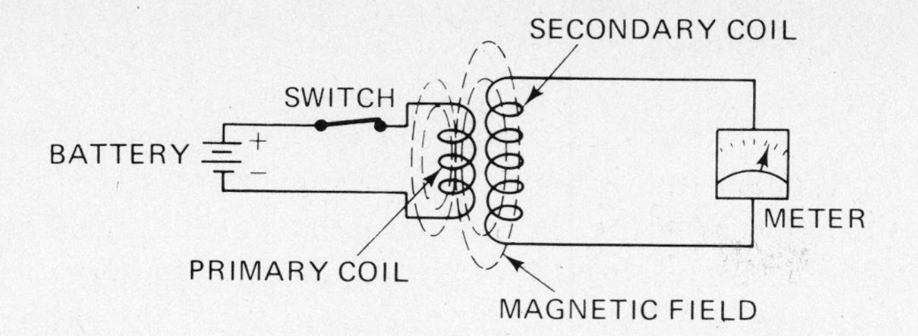

FIGURE 14-1 (Courtesy, NESCA).

SOLUTION:

Using Eq. 14-1, the solution would be:

$$E_s = 240(\tfrac{125}{250}) = 240(0.5) = \textbf{120 volts}$$

It will be noted that the number of turns in the secondary coil are only half the number of turns in the primary coil; therefore the secondary voltage is half the primary voltage. Another equation that might be set up is:

$$\frac{\text{primary source voltage}}{\text{secondary induced voltage}} = \frac{\text{total number of primary turns}}{\text{total number of secondary turns}}$$

or it can be simplified to:

$$\frac{E_p}{E_s} = \frac{N_p}{N_s} \qquad \textbf{(14-2)}$$

If, as shown in Fig. 14-2, the source voltage to the primary coil is 120 volts and there are 250 turns in the primary coil with only 50 turns in the secondary coil, the resulting secondary voltage would be 24 volts. The equation would be:

$$E_s = 120(\tfrac{50}{250}) = 120(0.20) = \textbf{24 volts}$$

A type of transformer in which the secondary coil has a smaller number of turns—and therefore

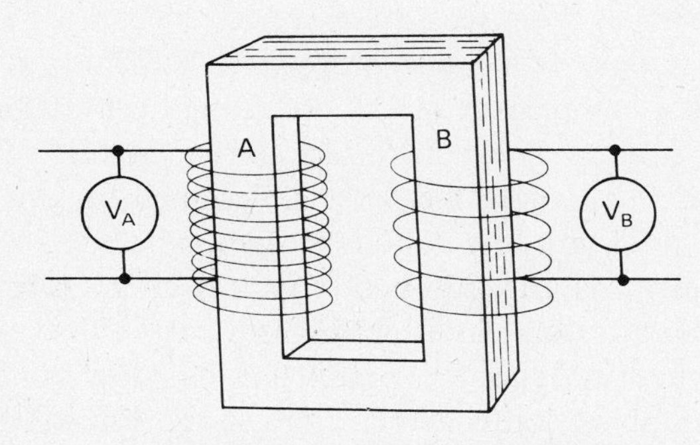

FIGURE 14-2 (Courtesy, NESCA).

less voltage—than the primary coil is classified as a *step-down* transformer. This is the type of transformer utilized in low-voltage control circuits used in refrigeration, air-conditioning and heating. A *step-up* type of transformer is one in which the secondary coil has a larger number of turns than the primary. The step-up transformer is utilized when a higher voltage is desired but is not readily obtainable from the primary source. If the primary coil has 100 turns of wire and the secondary coil has 1,000 turns, the secondary voltage will be 10 times the primary voltage.

Since power is the result of effective voltage multiplied by effective current, or $W = EI$, the relationship of power in the secondary coil is closely maintained to that in the primary coil. This relationship is expressed as:

$$E_p I_p = E_s I_s \qquad \textbf{(14-3)}$$

where

E_p equals the primary voltage

I_p equals the primary current

E_s equals the secondary voltage

I_s equals the secondary current

EXAMPLE:

If the primary voltage in a transformer is 120 volts and has a current of 1 ampere, and the secondary coil has a value of 24 volts, what current is available in the secondary?

Utilizing Eq. 14-3,

$$I_s = \frac{E_p I_p}{E_s} = \frac{120 \times 1}{24} = \frac{120}{24} = \textbf{5 amps}$$

In both the step-up and step-down transformers, another equation is used. The current in amperes flowing through a primary coil, multiplied by the number of turns in the coil, will equal the ampere-turns of the secondary coil. This is expressed in Eq. 12-4:

$$I_p N_p = I_s N_s$$

Therefore, in a step-down transformer, the secondary current will be greater than the current carried in the primary, since the number of turns are fewer in the secondary coil. If, in the above example, there are 250 turns in the primary coil and 50 turns in the secondary coil, Eq. 12-4 shows that:

$$1 \times 250 = 5 \times 50$$
$$= 250 \text{ ampere-turns in each coil.}$$

The primary winding of a transformer is always connected to the load being handled. The two principal types of transformer construction—the core-type and the shell-type—are shown in Fig. 14-3.

Figure 14-3a shows the core-transformer in which the copper windings surround the laminated iron core. Figure 14-3b pictures the shell-type transformer, in which the iron core surrounds the copper windings of the two coils.

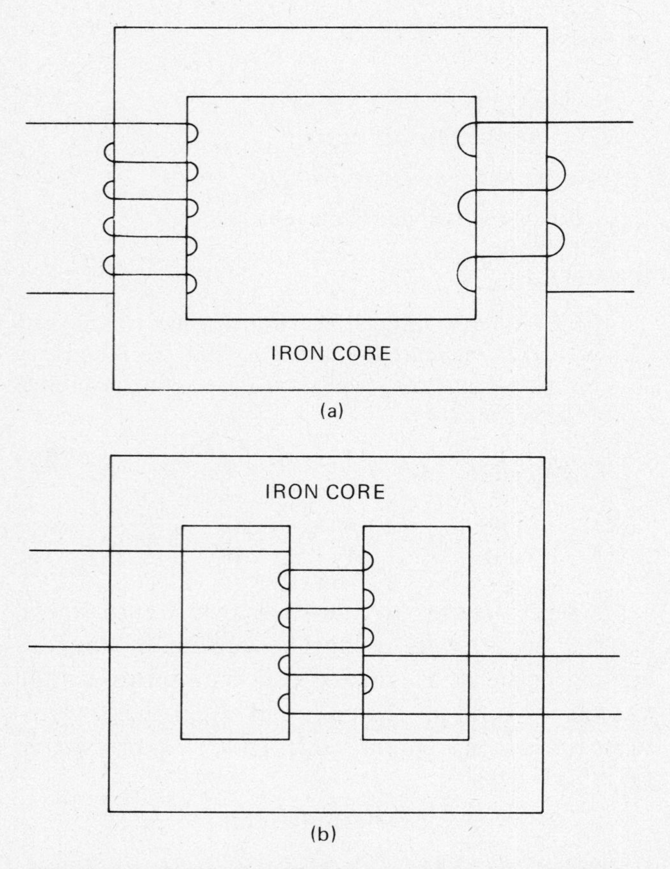

FIGURE 14-3 (a) Core type; (b) Shell type (Courtesy, Tappan Air-Conditioning Company).

In summary, the voltage relationship between the primary and secondary windings of a transformer is in direct ratio to the number of turns of wire in each winding. If one winding has twice as many turns as the other winding, it will naturally have twice the voltage. Conversely the current circulating in the two windings will be inversely proportional to the ratio of the two coils. The designation of the ratio in the transformer, such as that used in the example explaining Eq. 14-1, would be classified as a two-to-one step-down transformer, since the primary voltage is 240 volts and the secondary voltage is 120 volts.

RELAYS AND CONTACTORS

Relays are necessary in many control situations or circuits. They are required when:

(1) there is a difference in the voltage of the controlling device and the component being controlled.
(2) the controlling device is unable to handle the current required for the correct operation of the controlled component.
(3) the controlling device must control more than one electrical control circuit.

Examples of these requirements might be:

(1) an isolation of a low-voltage control circuit from the line voltage circuit used in a refrigeration system.
(2) a light-duty control circuit used to operate or control a heavy-duty load.
(3) where there is a seasonal changeover and the heating circuit of a system is controlled separately from the cooling cycle, and both are to be controlled by a single-contact type of device.

Figure 14-4 shows a relay, operated by a 24-volt power supply from a step-down transformer, controlling a compressor motor operating off a 220-volt power source. The controlling device in this schematic is the cooling thermostat, designed for operation in a 24-volt low-voltage system. The controlled component is the 220-volt compressor motor. In this diagram the relay is acting as a switch that connects the two circuits, permitting the low-voltage thermostat to control operation of the line-voltage motor.

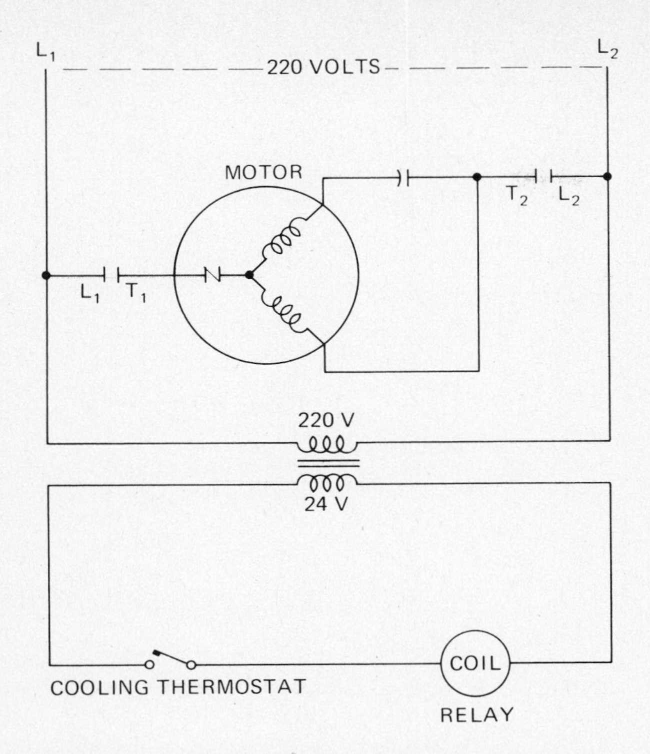

FIGURE 14-4 (Courtesy, Tappan Air-Conditioning Company).

Closing the contacts in the thermostat completes the 24-volt circuit and allows the relay coil to be energized. The core in the magnetic circuit pulls the armature down, and this closes contacts in the line-voltage circuits, thus starting the compressor. When the cooling thermostat is satisfied, the internal contacts open, breaking the low-voltage circuit to the coil in the relay. The armature is released from the core, and the contacts in the line-voltage circuit open, shutting off the compressor. This is classified as a normally-open (NO) relay, in which the bottom contact on the line-voltage circuit is stationary and the top contact movable.

By changing the line-voltage contacts and locating the armature in the relay, this would be a normally-closed (NC) type of relay. Figure 14-5 shows an NC relay, with the top contact stationary and the bottom contact movable. The bottom contact is held against the stationary contact by a spring. When the low-voltage circuit is energized, the armature is drawn down to the core of the relay coil. This moves the bottom contact away from the stationary contact and breaks the line-voltage circuit.

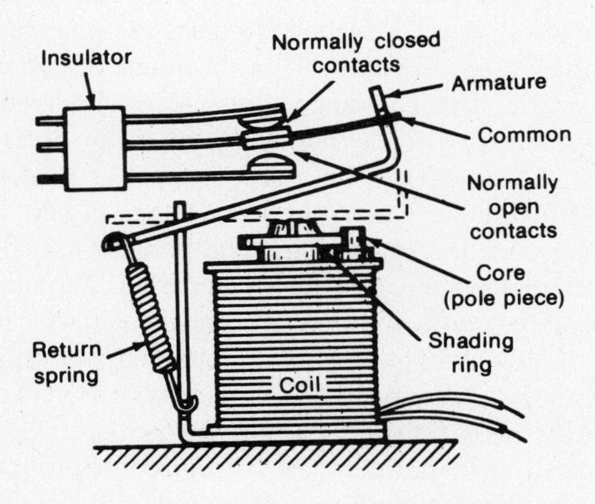

FIGURE 14-5 Potential starting relay.

Many arrangements and types of relays are utilized, but two main categories are *starting relays* and *thermal-overload relays*. The thermal overload type of relay will be covered in the section on circuit protection.

A starting relay is essential to motor-starting equipment when small, single-phase motors are used.

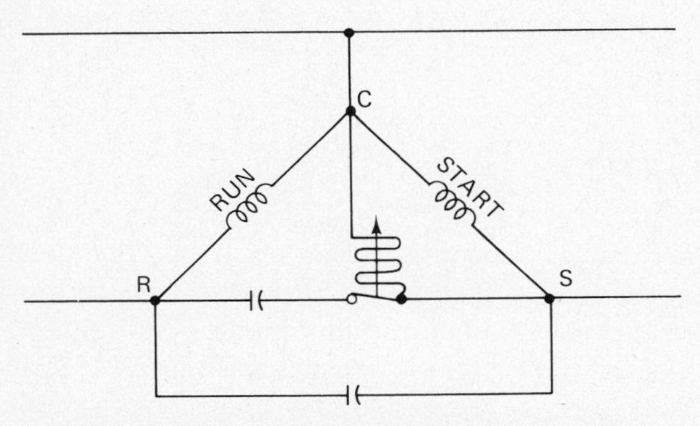

FIGURE 14-6 (Courtesy, Tappan Air-Conditioning Company).

In this case, the starting relay consists primarily of a set of normally-closed contacts, connected in series with the starting winding of the motor. When the control contacts close, both the starting winding and the running winding of the motor are energized. The coil of the electromagnet or relay is in series with an auxiliary winding of the motor, so that when the motor comes up to a predetermined speed, enough voltage is induced in the auxiliary winding to cause current to flow through the coils. The magnetic force or field attracts the spring-loaded armature, which opens the relay starting contacts. This is shown in Fig. 14-6. When the starting contacts open, the starting winding is automatically taken out of the electrical circuit, and the motor continues to run on the run winding. When the control device is satisfied and its contacts open, the flow of power to the motor is interrupted, and it will stop. This power interruption permits the closing of the starter contacts due to the spring-loaded armature, and therefore the motor is ready to start a new cycle when the controlling device calls for operation of the system again. Care must be taken in the replacement of this type of relay, which is precisely sized to motor requirements.

Contactors may be classified as heavy-duty relays, since they perform the same function as relays. That is, they control an electrical circuit by means of a magnetic coil in a low-voltage circuit, which, in turn, operates a line-voltage component. Because of its heavier construction, the contactor is capable of handling larger amounts of current in comparison to the relay. Usually the relay is limited to a maximum of 10 or 15 amperes, whereas the capacity of the contactor may easily be in excess of 500 amperes.

STARTERS

When automatic control of an electrical component is not needed, a manual motor starter may be used, as in the case of a fan or pump motor operating continuously, or one that is stopped and started at infrequent intervals. The design may range from a simple toggle or push-button switch to elaborate devices, which will provide protection for the motor. It may also limit the amount of current flowing to the component; or it may control, through additional components, the operation of the system. This can easily include time delay and sequential action of the components in the circuit.

Figure 14-7 is a schematic diagram of an electric circuit; it shows the control of an electric motor through use of a contactor, operated by means of a push-button start-and-stop type of switch. This diagram also shows overload protectors installed in line 1 and line 3 of the electrical circuit. Basically, starters are nothing more than contactors containing built-in overload protection.

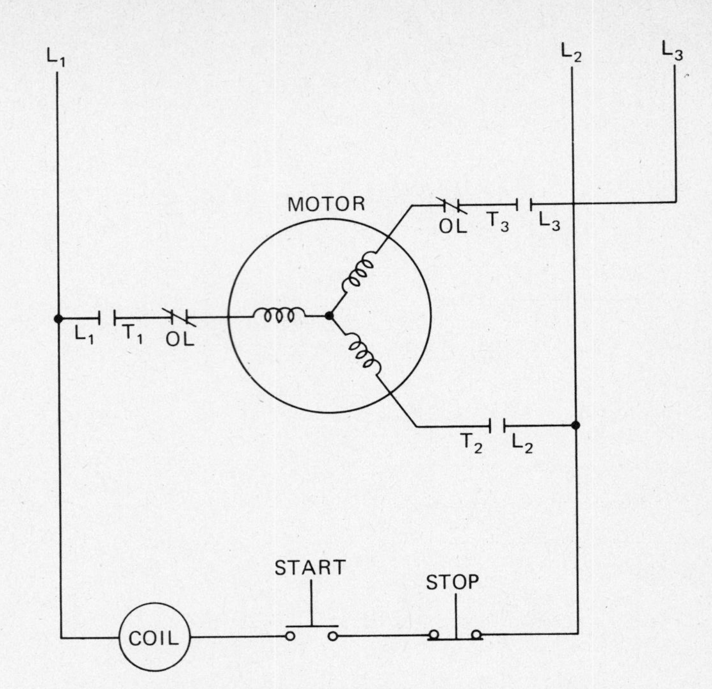

FIGURE 14-7 (Courtesy, Tappan Air-Conditioning Company).

The use of a manual starter, as shown, requires the presence of an operator near the equipment location. In Fig. 14-7, when the start button is pushed, the pair of internal contacts will close, thereby energizing the holding coil, which is situated between the terminals L_1 and L_2. The electromagnetism of the holding coil will keep the sets of contacts in a closed position, thus starting the motor, which will continue to run until the operator pushes the stop button, breaking the circuit to the holding coil. Of course the circuit can be broken by a power failure, or if one of the overload cutouts is exposed to excessive current.

If automatic control is required or desired in a given installation, a magnetic across-the-line starter should be used. Figure 14-8 is a schematic diagram of a magnetic starter arranged for a two-wire type of control, such as for a thermostat or a low-pressure switch. These devices have relatively light contacts, but, through the use of the low-voltage circuit and holding coil, the line-voltage components—such as a compressor motor—may be actuated. When the low-voltage control circuit is completed it will activate the low-voltage holding coil, which will pull in the contacts on the main line-voltage system. If excessive current is flowing, the heaters in the overload relay will open the contacts on the control circuit. If the voltage is low, the holding coil will open the

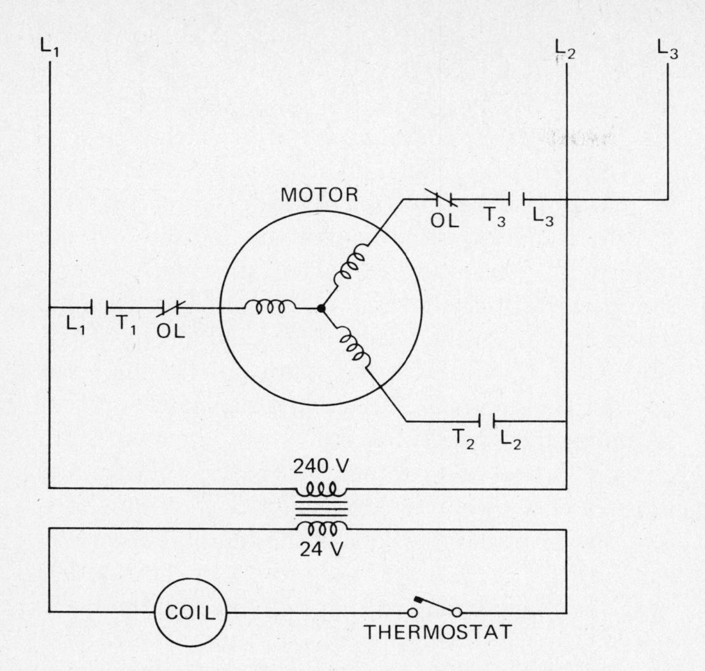

FIGURE 14-8 (Courtesy, Tappan Air-Conditioning Company).

main contacts. When the voltage returns to normal, the holding coil again will close the main contacts and restart the motor.

For compressors that start unloaded, and for some fans and pumps, magnetic starters classified as *reduced voltage starters* are available. These permit the starting of motors at a voltage lower than normal, through the use of added resistance or auto-transformers placed in the circuit temporarily through the operation of a relay.

MOTORS

Motors convert electrical energy into mechanical energy. Motors, like electrical circuits, are divided into two general categories, depending on the type of current involved—direct or alternating. Since most applications in refrigeration involve alternating current, this presentation will deal primarily with ac motors.

Alternating-current motors are classified in two main divisions (depending upon the type of power used) as:

(a) Single-phase
(b) Polyphase (three-phase)

Single-phase motors are less efficient than three-phase motors. They are used chiefly where the demand is for fractional horsepower units or where single-phase electrical service is the only type available. In some types of special applications, single-phase motors ranging up to several horsepower may be used, but generally the single-phase field stops at around 3 hp.

Naturally, all types of motors are not alike, since they are designed for different jobs. It is possible to turn over a small motor with a touch of a finger, whereas a motor used with a belt-driven compressor will need considerably more than the pressure of a finger to turn its shaft.

There seems to be more difficulty in understanding the operation of small, single-phase motors than that of three-phase motors. Learning why there is a need for different kinds of motors will help toward an understanding of the composition and operation of single-phase motors. Single-phase motors may be categorized as:

(a) split-phase
(b) capacitor-start
(c) permanent-split capacitor
(d) capacitor-start, capacitor-run
(e) shaded-pole

According to Webster's Dictionary, *torque is a force, or combination of forces that produces or tends to produce rotation.* In the case of an electric motor it is the ability to exert power, resulting in the turning of an object "load" such as a compressor fly wheel. The greater the capability a motor has to turn a heavy object, the greater the torque of the motor. Of course, it costs more to construct a motor with a high starting torque than one with a low starting torque, which accounts for the many types of motors produced. It would be uneconomical to produce a high-torque motor to turn a household fan, for example.

The two main parts of a motor are called the *rotor* and the *stator*, as shown in Fig. 14-9. The rotor is the rotating or revolving part, and is sometimes called an *armature*. The stationary part is known as the *frame* or the *stator*. Keep in mind the principles of magnet-

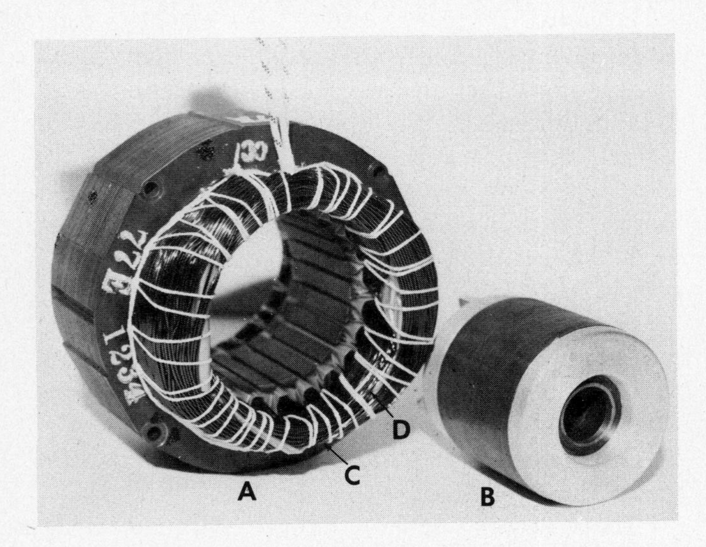

FIGURE 14-9 Stator and rotor from a $^1/_4$ hp split phase hermetic motor (Courtesy, General Electric Company, Hermetic Motor Department). A—Stator; B—Rotor; C—Running winding; D—Starting winding.

ism, i.e., "like" magnetic poles repel and "unlike" poles attract, as you study Figs. 14-10a and 14-10b, which show a simple electric motor. Fig. 14-10a shows two magnetized poles mounted on the outside of the motor shell or stator, while between these magnetic poles is a permanent magnet placed on a shaft. This permanent magnet in the center corresponds to the rotor. The north magnetic pole of the stator attracts the south pole of the permanent magnet or the rotor, and the south magnetic pole of the

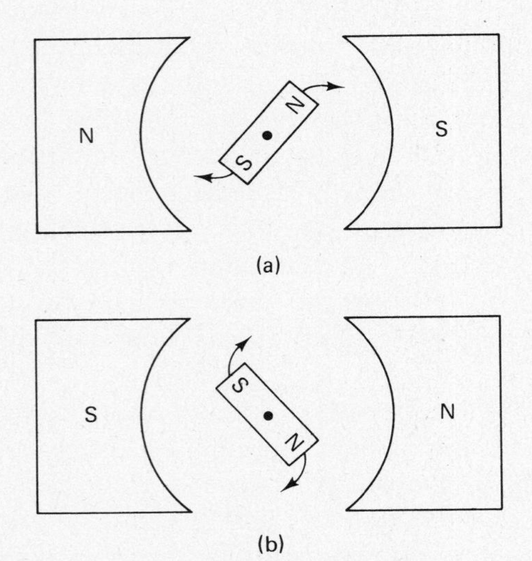

FIGURE 14-10 (Courtesy, Tappan Air-Conditioning Company).

stator attracts the rotor's magnetic north pole. This attraction occurs because, when current passes through the coils making up the magnetic poles of the stator, magnetic fields are set up in the coils.

Since this is a diagram of a simple alternating-current motor, the direction of the flow of current will reverse completely and reverse the polarity of the stator coils as the flow of current alternates. Therefore, the north coil in Fig. 14-10a will soon become the south pole of the stator as depicted in Fig. 1-10b. When this occurs, with the south pole of the rotor pivoting on the shaft, and with the alternating current reversing the polarity of the stator, the two south poles will repel one another. The rotor will continue to pivot on the shaft as long as the current flow continues to reverse in the coils of the stator. This briefly explains the operation of the motor, or the rotation of the permanent magnet that is pivoted on the shaft. The motor will operate in this way until the current is shut off, at which time the demonstration motor will reach the condition shown in Fig. 14-11. As you will see, the north pole of the stator attracts the south pole of the rotor and the permanent magnet or magnetized bar will remain as it is. Even when the current

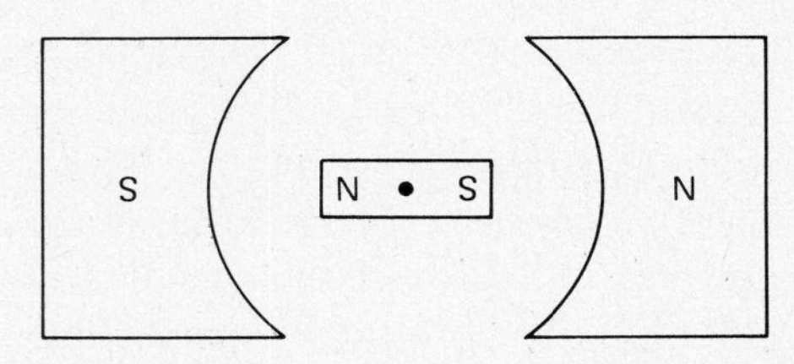

FIGURE 14-11 (Courtesy, Tappan Air-Conditioning Company).

flow is reversed due to alternating current, the rotor will stay where it is because it is not at an angle comparable to that in Fig. 14-10a or 14-10b. The attraction or repulsion of the poles cannot cause rotation when the bar is in the position shown in Fig. 14-11.

If we insert another set of poles in the stator or shell, as shown in Fig. 14-12a the new set of magnetic poles will offer enough attraction to start rotation of the permanent magnet again. But difficulty may be encountered if the rotor should stop in a position midway between the two poles, as shown in Fig. 14-12b. There would be an equal attraction between the south pole in the rotor and the two north poles in the stator as well as between the rotor north pole and the stator south poles. This problem must be overcome, since corresponding poles on the stator have the same magnetic strength at the same time. Each is using

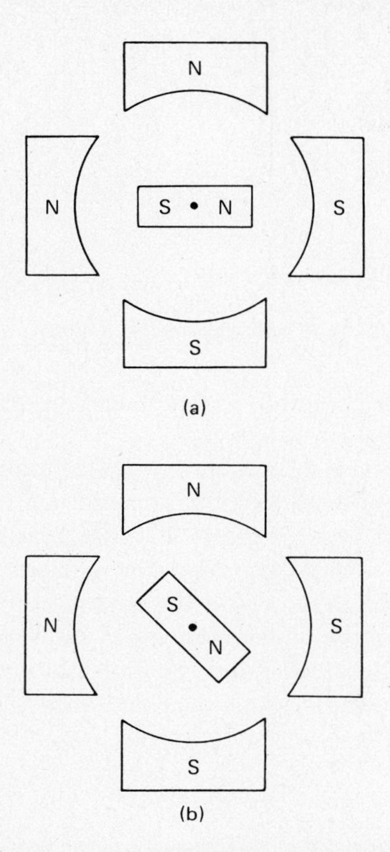

FIGURE 14-12 (a)—Courtesy of Tappan Air-Conditioning Company; (b)—Courtesy, NESCA.

the same source of electrical current, for we are dealing with a single-phase source.

If we put a resistance in one of the paths, temporarily obstructing the flow of electrons to that pole in the stator (as shown in Fig. 14-13), the electrons will not arrive at their destination in the poles at the same time. This will cause the electrons flowing to one pole to be out of step with those flowing to the other, because a second phase has developed. We now have a two-phase current that will help to start the motor operating again by aiding in the pulling or pushing of the rotor. The second phase is needed by all single-phase motors so that they can start rotating. The main difference among single-phase motors is the method used for producing and controlling this second phase of electrical current.

This second phase may be better understood by a study of Fig. 14-14. In Fig. 14-14a, two electrons are running together up and down the "hills" that range between the maximum north polarity and the maximum south polarity. When the electrons keep pace with one another, we have what is called *single-phase current*. If they travel up and down these hills 60 times a second, the current is a 60-hertz current. This is the condition mentioned earlier, in which the rotor

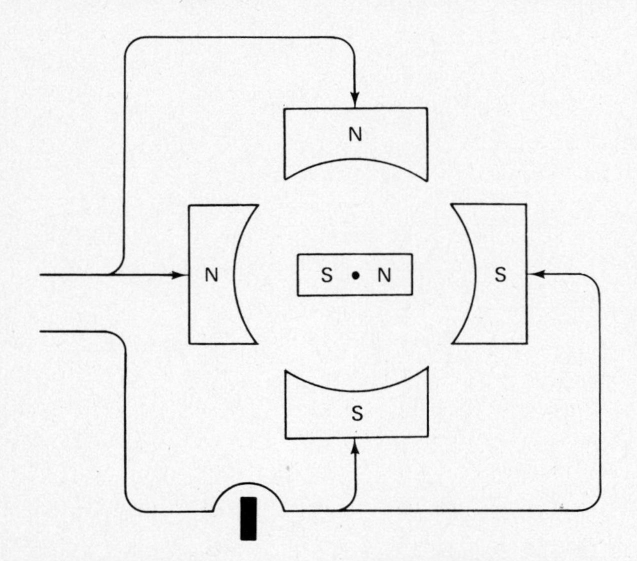

FIGURE 14-13 (Courtesy, Tappan Air-Conditioning Company).

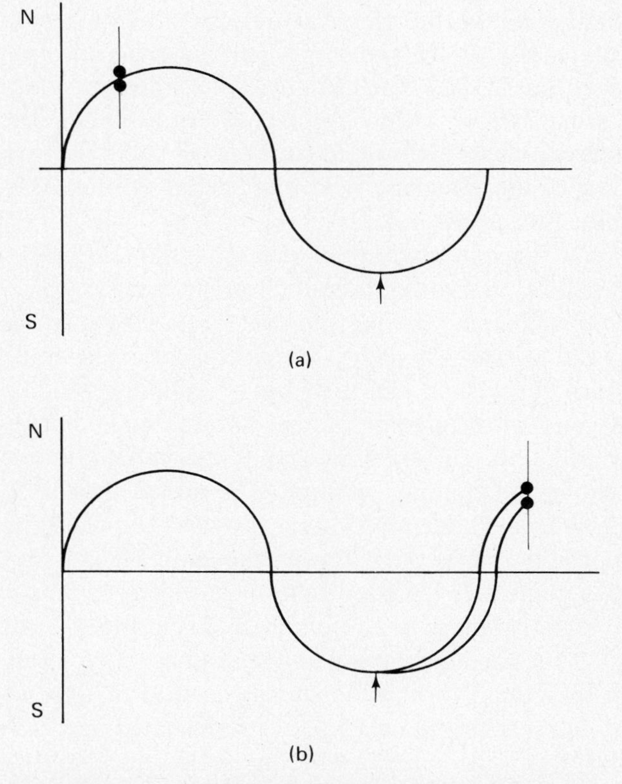

FIGURE 14-14 (Courtesy, Tappan Air-Conditioning Company).

stopped in a midway position between poles, as shown in Fig. 14-12b.

In Fig. 14-14b, with a resistance placed in one of the paths, the equally capable electrons have been placed in an unbalanced condition. The electrons are in corresponding positions on different hills, with one trailing behind the other. Because of the insertion of a resistance into one of the paths, the single-phase current has been changed to a current having two phases.

All of the five categories of single-phase motors mentioned earlier need this second phase of electrical current to start. The main difference among the categories is the way in which the second phase is produced. (Conceivably, since all of these motors have a second phase, they could be classified as two-phase motors instead of single-phase.)

SPLIT-PHASE MOTORS

If the single-phase motor had but one set of windings on the stator, the rotor would not always start to rotate, but would sometimes just sit there and hum when energized. If turned by hand when it fails to start, the rotor will pick up speed until it reaches its operating rpm. This hand-turning will begin the rotation of the magnetic field and set up the necessary induction in the motor.

In the split-phase type of motor, the self-starting capability is brought about through the addition of a second set of windings on the stator. This is known as the *starting winding*; it is made up of considerably more turns than the primary or running winding. Because of the greater number of turns in the starting winding, the current flowing in it lags behind the current in the running winding. The added number of turns in the starting winding provides the resistance placed in one of the paths, as shown in Fig. 14-13. It also duplicates the condition of the electrons flowing in an unbalanced condition, as shown in Fig. 14-14b.

This starting winding in a split-phase motor stays in the circuit during the starting period or until the motor reaches about 75% of its full-rated speed. At this point, the starting winding is disconnected from the circuit by a centrifugal switch or a relay (thermal, current, or potential) to disconnect the starting winding.

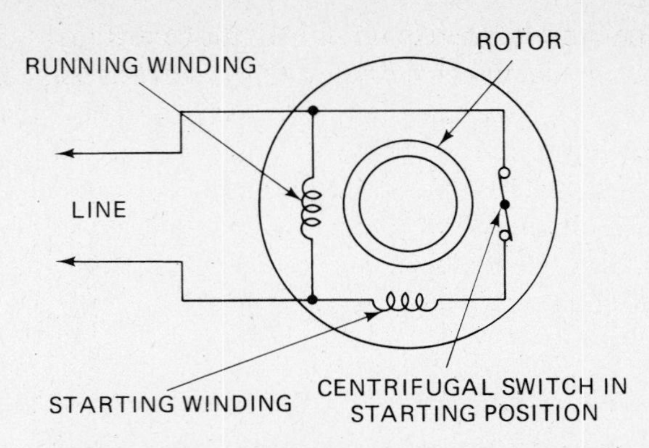

FIGURE 14-15 Split-phase motor (Courtesy, NESCA).

Split-phase motors have low starting torques and are used on small, fractional-horsepower refrigeration units. This type of motor is often used in a system that has a capillary tube for refrigerant control, because the pressures in the system equalize on the off-cycle of the unit. Figure 14-15 shows the arrangement of the starting and running windings in a split-phase motor, in conjunction with a centrifugal switch.

CAPACITOR-START MOTOR (CAPACITOR-START, INDUCTION-RUN MOTOR)

The capacitor-start motor, shown in Fig. 14-16, has a motor winding arrangement similar to that of a split-phase motor. An additional component, a

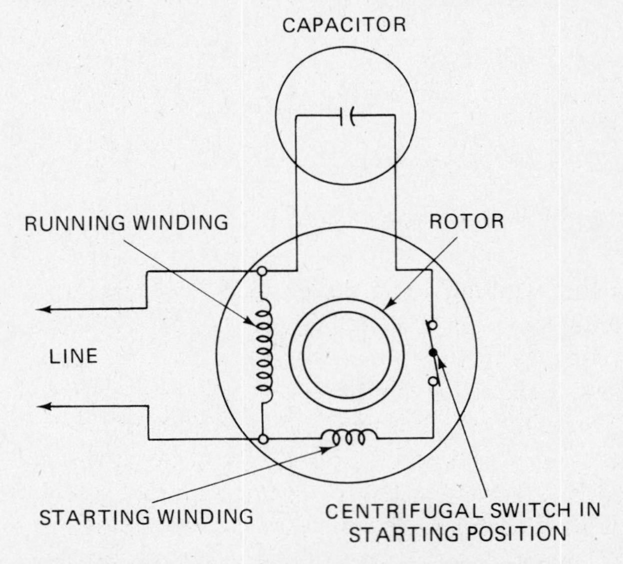

FIGURE 14-16 Capacitor-start motor (Courtesy, NESCA).

capacitor, is wired in series with the centrifugal switch and the starting winding. In this motor, the capacitor causes a "leading" current, and thus the starting winding is out of phase with the running winding. Again, in this type of motor, when the rotor reaches approximately 75% of its rated speed, the centrifugal switch will take the starting winding and the capacitor out of the electrical circuit.

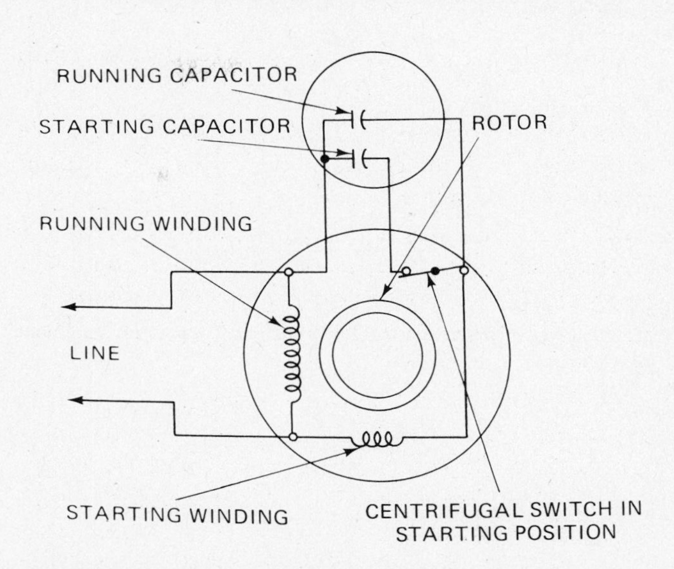

FIGURE 14-18 Capacitor-start, capacitor-run motor (Courtesy, NESCA).

PERMANENT-SPLIT MOTOR CAPACITOR

If a capacitor is used in the starting winding circuit, it is simpler if the capacitor can be left in the circuit at all times. Of course, in such a case, the correct capacitor must be selected for use with the starting winding. It must be able to pass enough current to the starting winding to provide adequate torque, but it must not allow enough current to overheat the starting winding while the motor is in operation. A diagram of this type of motor, which has a good starting torque, is shown in Fig. 14-17. The diagram shows a capacitor connected in series with the starting winding, which is energized at all times when the motor is in operation. Consequently, no centrifugal switch is needed with this type of motor: which is classified as a *permanent-split capacitor motor*, or a *PSC motor*.

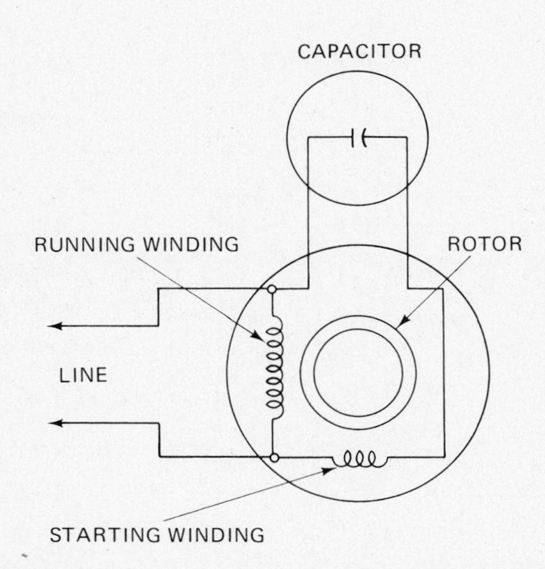

FIGURE 14-17 Permanent-split capacitor (Courtesy, NESCA).

CAPACITOR-START, CAPACITOR-RUN MOTOR (TWO-VALUE CAPACITOR MOTOR)

The capacitor-start, capacitor-run type of motor is quite similar in design and operation to the capacitor-start, induction-run motor. Figure 14-18 shows this type of motor using two capacitors. In this case, a centrifugal switch is wired in series with the large-capacity type of capacitor, and this series circuit is parallel to the running capacitor. In this motor, like others utilizing a centrifugal switch, when the motor reaches approximately 75% of its rated speed, the centrifugal switch will break the circuit, putting the larger capacitor out of use. This type of motor is adapted to conditions in which a strong starting torque is necessary, such as in a compressor that must start under full load conditions.

SHADED-POLE MOTOR

A shaded-pole motor, shown in Fig. 14-19, does not have a starting winding but is otherwise similar to a split-phase induction motor. In the motors previously described, some effort must be used to get the

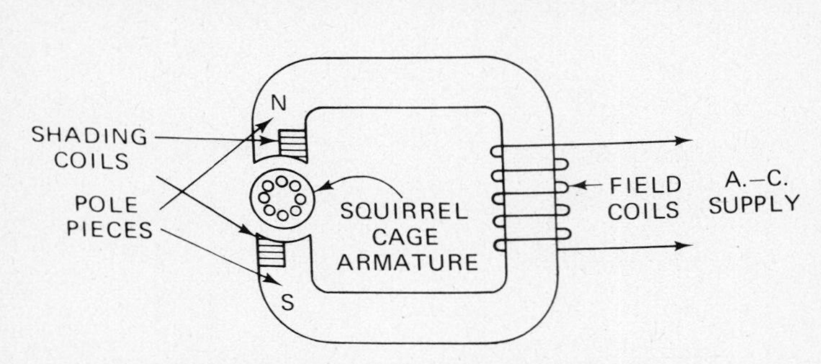

FIGURE 14-19 Shaded-pole motor (Courtesy, Tappan Air-Conditioning Company).

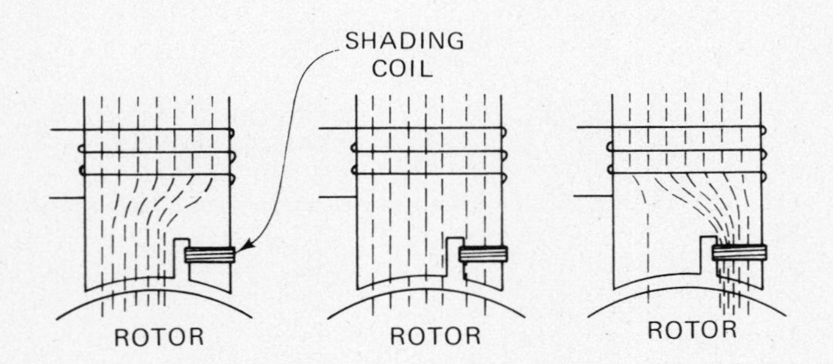

FIGURE 14-20 (Courtesy, Tappan Air-Conditioning Company).

motor turning, since it is only in the turning of the rotor that the lines of force are cut and set up their own magnetic field to maintain rotation. Figure 14-19 shows that the poles of the motor are split or grooved so that a smaller pole replaces part of the main pole. Around this smaller pole is a copper loop or strap called a *shading coil.* Instead of depending upon a separate starting winding to create a second phase, the shading coils produce the rotation of the rotor through a shifting of a portion of the magnetic field. Figure 14-20 suggests the principles involved in shifting this magnetic field.

MOTOR PROTECTORS

Built-in motor protectors are provided to permit the motor to perform beyond its normal duty up to a predetermined safe temperature limit without cutting off its power supply. Most regulations and codes are strict about overload protectors, since danger to persons or property could arise from a malfunctioning motor, or equipment connected with it.

There are two general types of built-in protectors. The first is activated primarily by temperature, and the second is activated by motor current or temperature, or both.

The temperature-responsive protectors (Fig. 14-21a) may be either of bimetallic or thermistor type, both of which operate at a fixed motor temperature, or of the rod-and-tube type, which has an anticipatory or "rate-of-rise" temperature sensitivity.

Bimetallic protectors use bimetallic strips or discs to control a normally-closed switch, and are designed to require minimum space so that in most cases they can be placed on the motor windings. Thermistors are semiconductors, the electrical resistance of which varies with temperature. They are extremely sensitive to relatively minute temperature changes, but the resistance variation must be amplified through an electronic circuit or mechanical relay to actuate a holding coil.

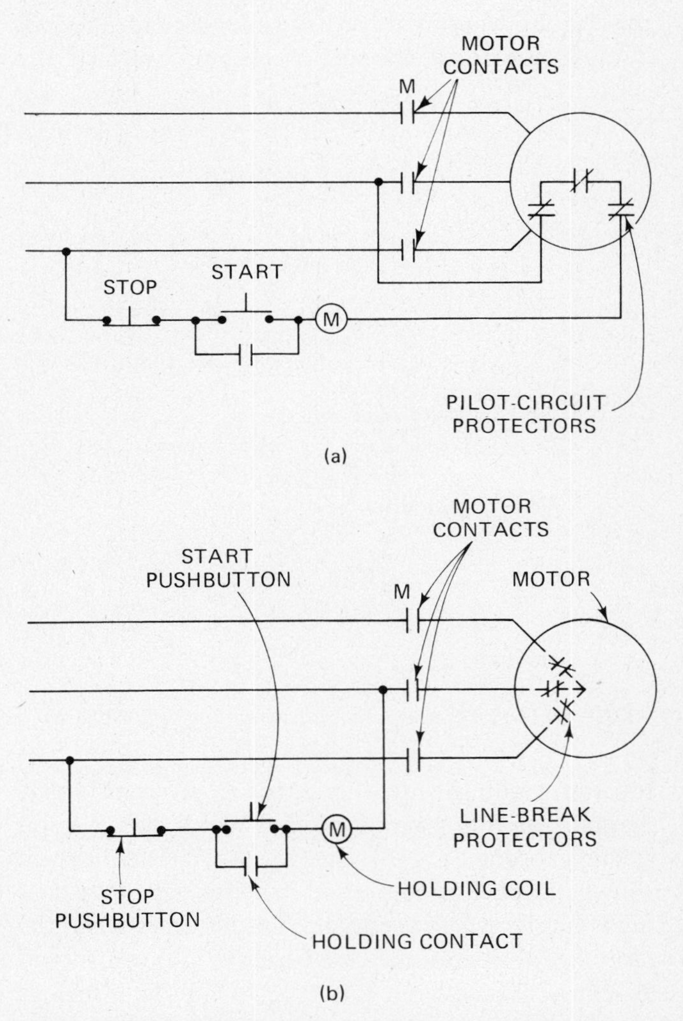

FIGURE 14-21 (a) Temperature-responsive (pilot-circuit); (b) Dual-responsive (line-break) thermal protector (Courtesy, NESCA).

Rate-of-rise protectors are almost always of a rod-and-tube type and operate through a pilot circuit. An external metal tube and an internal metal rod are connected so that they operate as a differential-expansion thermal element to actuate a self-contained snap-action switch. The metal of the tube has a higher coefficient of expansion than that of the rod. The element is installed in the motor winding where it will be affected by winding heating. On slow temperature changes, both tube and rod heat at the same rate, and the switch is set to open when the tube reaches a predetermined temperature. When the temperature change is rapid, the tube heats at a faster rate than the rod, so that the differential-expansion movement required to operate the switch is reached at a lower tube temperature than for a slow rate of change. This temperature anticipation makes it possible for the tube-and-rod protector to provide protection for both running and stalled conditions.

Dual-responsive protectors (the location of one type of which is shown in Fig. 14-21b) are used to protect against both slow and fast temperature increases, and are thus used for both running overloads and locked-rotor conditions. These protectors are essentially bimetallic thermostats with built-in heating elements in series with the motor winding. They may be meant for "on-winding" attachment or for attachment within the end bracket (Fig. 14-22).

The heating element may be the bimetallic strip or disc itself, a separate heater, or both. Heating of the bimetallic element is a function of both the temperature around the protector and the internal heat generated by the motor current through the element.

Resistances of the disc and heater are coordinated with the temperature setting of the disc to establish an operating point suited to the winding temperature limit. In the case of running overload, most of the heat reaching the thermal element is produced by the motor windings. During locked-rotor conditions, the rate-of-rise is too fast to cause the protector to operate quickly enough in response to motor heat alone. In this situation, the heat generated within the protector augments the winding heat, resulting in a response that is rapid enough to prevent damage.

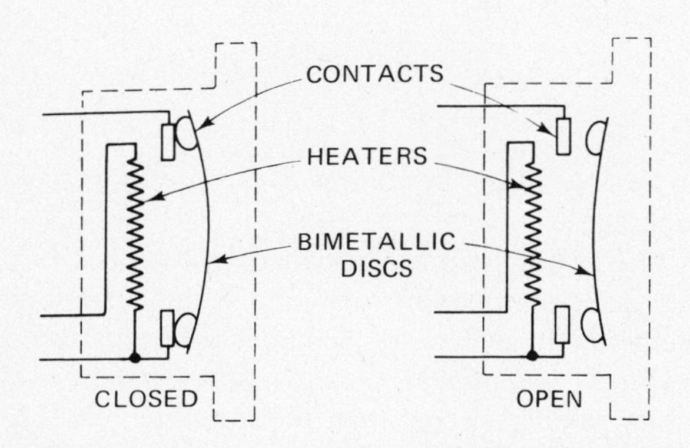

FIGURE 14-22 Thermal protector (dual-responsive) for end bracket mounting (Courtesy, Gould, Inc., Electric Motor Division).

For single-phase motors up to about 5 hp, thermal protectors are used. For three-phase motors, a single protector with three heating elements connected in series with the motor windings at the neutral point is used. The protector opens all three phases, as shown in Fig. 14-21b.

ELECTRICAL METERS

Most of the service problems in the field of refrigeration are found in the control circuitry. Knowledge of the electrical system and electrical meters enables the service technician to locate the trouble easily.

Many service technicians have an aversion to using test instruments and do not understand how they work or the need for their proper care. If the service technician makes a sincere effort to learn the proper use of test instruments, these valuable tools of the refrigeration trade can work for his benefit as well as the customer's. Test instruments should be just as familiar to the service technician as the conventional refrigeration tools in his tool box. These test instruments are delicate in nature and proper care should be taken in their handling and use.

If the voltage is to be checked, the instrument to use is a voltmeter. If the electrical current, or the flow of electrons, measured in amperes, is to be checked, an ammeter must be used. If the system or its components are to be checked for electrical resistance, for shorts in the circuit, or for continuity of the electrical circuit, the instrument to use is an ohmmeter. These test instruments are shown in Fig. 14-23a and b. Each is constructed for a specific purpose. They may have a selection of ranges in order to insure accurate readings. Some instruments, such as shown in Fig. 14-24, can be used for the testing of all three electrical properties: voltage, current, *and* resistance. Usually instruments of this type do not have as wide a range

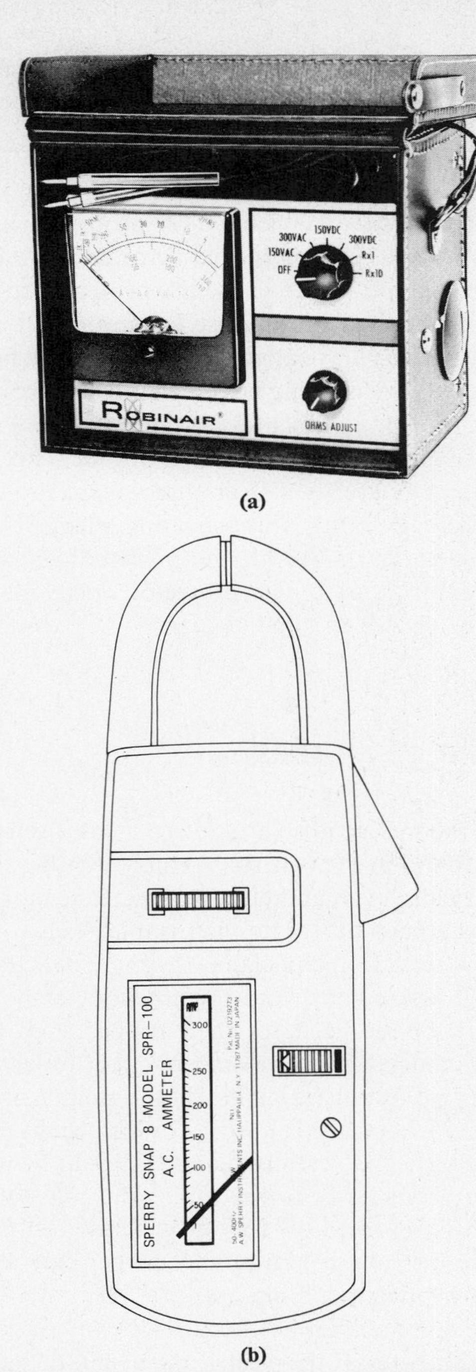

(a)

(b)

FIGURE 14-23 (a) Volt-ohm-meter (Courtesy, Robinair Manufacturing Corporation).; (b) Ammeter (Courtesy, A. W. Sperry Instruments Inc.).

FIGURE 14-24 Combination volt-ohm-ammeter (Courtesy, Airserco Manufacturing Company).

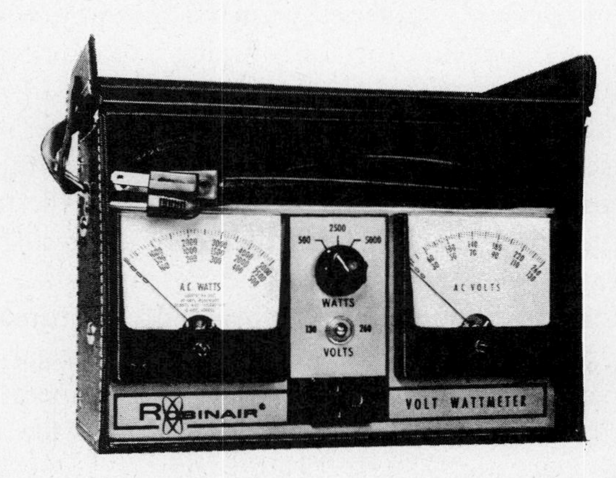

FIGURE 14-25 Volt wattmeter (Courtesy, Robinair Manufacturing Corporation).

as the single-purpose instruments shown in Fig. 14-23.

The *volt-watt meter* (Fig. 14-25) enables the technician to obtain simultaneous readings of the voltage and wattage when servicing and checking an electrical appliance. The refrigeration technician may have to check out appliances other than those directly related to refrigeration, and he may find appliances that have ratings in watts rather than amperage. In such a case, where a rating in amps is desired, the watt reading can be divided by the voltage of the circuit to give the amperage.

The technician may be called on to check out an electrical circuit having low voltage and current measurements. In such cases, he may use test instruments known as *milliammeters* and *millivoltmeters*, which record smaller values than the more generally used voltmeter and ammeter. These units are pictured in Fig. 14-26.

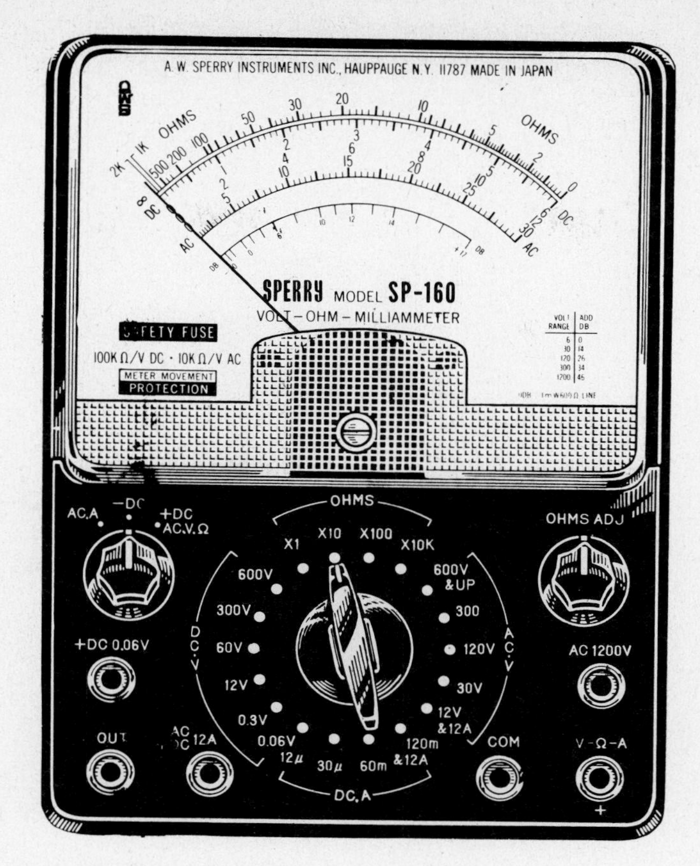

FIGURE 14-26 **(Courtesy, A. W. Sperry Instruments Inc.).**

PROBLEMS

1. What are the two types of transformers?
2. What is the difference between them?
3. How are amperage and voltage affected by the relationship of the number of turns on primary and secondary coils?
4. What are the two main purposes of relays?
5. What are the two major types of alternating current motors?
6. What is torque?
7. What are the starting windings in a split-phase motor?
8. What instrument does a service technician use to check (a) voltage in an electrical circuit; (b) rate of flow of current; (c) electrical resistance?
9. What is the purpose of millivoltmeters and milliammeters?
10. What is the purpose of a built-in motor protector?

15

REFRIGERATED STORAGE

Commercial refrigeration is the broadest in scope of all the segments of the refrigeration industry, encompassing the range between domestic and industrial applications. At the lower end of the size scale we find such units as the physically compact, relatively maintenance-free household refrigerators and freezers. At the upper end of the commercial range are systems that cool, chill, or freeze millions of cubic feet of product and, due to their complexity, require the presence of an operating engineer and other skilled personnel. Between these two extremes are a multitude of hardware system types, both mobile and stationary, which play a part in the production, adaptation, or preservation of most of the commodities used throughout the world today.

Some of the first commercial storage refrigerators appeared in the early part of the twentieth century and were installed in large meat markets, breweries, and other similar storage facilities. Although of a semi-automatic design, they required full-time attendance by an operator and were expensive to maintain. Although not satisfactory by today's standards, this equipment did serve a definite need at the time. A few years later, smaller units were designed for storage of ice cream. These units utilized some of the early refrigerants, which were described in Chap. 7. The newly designed ice cream storage used the old ice-and-salt mixture type of cooling.

Soon there was an influx of smaller commercial refrigeration units used as open or closed display cases, reach-in cabinets, and walk-in coolers, for fresh and frozen foods.

These units were used in grocery stores, meat markets, super markets, restaurants, drug stores, and the like. Different storage temperatures and conditions were required for various meats, poultry, fish, vegetables, fruits, candy, and ice cream, as well as medicines stored in drug stores.

Figure 15-1 shows a typical reach-in cabinet designed to store and refrigerate many different types of products. Figure 15-2 shows a typical walk-in cooler that may be used in many facilities such as grocery stores, markets, or restaurants.

Since numerous applications are covered by the broad field of commercial refrigeration, a wide variety of cabinets are constructed for this use. The

FIGURE 15-1 Typical reach-in refrigerator for commercial use (Courtesy, The Delfield Company, Division of ALCO).

reach-in cabinets used for keeping perishable produce and groceries at their required temperatures have one or more doors permitting access by the customers. Some cabinets are designed so the customer may have access to the product from the front of the cabinet while the store personnel have access from the rear of the cabinet to replenish the supply of fresh or frozen foods. Early boxes were constructed of wood and had

FIGURE 15-2 Typical walk-in cooler (Courtesy, Bally Case & Cooler, Inc.).

limited glass area. Modern cabinets usually have fiberglass or porcelain exteriors with thin layers of modern insulation material, and some have extensive glass display areas.

Some boxes depend on the gravity flow of air for the proper circulation in or around the product, while other units, particularly larger sizes, may use a circulating fan for air distribution within the cabinet. The refrigeration unit for this type of box may be inside the framework of the box itself or may be in another location.

When the cabinet is to be used for storage of meat or fresh produce, good control of temperature and humidity must be maintained, so that dehydration or deterioration of the product will not occur. In such an

FIGURE 15-3 Typical evaporator (Courtesy, Bohn Heat Transfer Division).

application, the temperature probably will be maintained between 34°F and 40°F, and the relative humidity may be kept at 75% or higher. If the product is such that there is no concern about dehydration or deterioration, colder temperatures may be maintained and control of relative humidity will not be critical. Figure 15-3 shows a typical blower type evaporator coil that may be used in a walk-in cabinet installation.

Figure 15-4 shows a display case often used in florist shops. In this type of application the air temperature may be kept higher than is customary in grocery or meat market boxes, since a temperature of about 55°F is best for storage of flowers.

Cabinets for frozen food storage and display are usually maintained at or below 0°F. Proper precautions must be taken to assure that the access doors are not frozen in a closed position so that they cannot be

FIGURE 15-4 Refrigerated floral display case (Courtesy, Bally Case & Cooler, Inc.).

opened by the customers. It is sometimes necessary to place an electric resistance heat coil around the door opening to prevent freezing.

Another operation requiring refrigeration is the cooling of bottled or canned beverages in automatic coin-operated dispensers. Some of these dispensers are designed to handle only one product, whereas others can dispense several different kinds of beverages. Figure 15-5 shows a unit that dispenses several kinds of beverages.

Table 15A (page 183) gives excerpts from the *Handbook & Product Directory 1974 Applications,* by courtesy of the American Society of Heating, Refrigerating and Air-Conditioning Engineers. This table lists many commodities and their recommended storage temperatures and other information pertinent to the products.

Since refrigeration plays such an important part in the preservation of food and other perishable products, one should recognize evidences of food spoilage and know recommended methods for its control. Some products may show effects of deterioration or food spoilage through a change in their appearance, odor, taste, loss of weight, or a change in the chemistry of the product itself.

If not properly cared for after being harvested, fruits eventually become overripe and soft because the chemical processes inside the fruits continue. Overripe fruit may still be fit for consumption by humans, but it cannot be commercially handled in a convenient manner without extensive damage to the fruit. Vegetables, like fruits, also must receive proper care after harvesting to prevent deterioration.

Eggs will lose weight through evaporation of the water content and emit an offensive odor when they spoil. Broken eggs must be salvaged immediately if they are to be processed into a frozen or powdered form that can be used by a bakery or other food industry.

Milk, if not refrigerated, will be "soured" in a very short time by growth of bacteria, producing a disagreeable odor and distinctively bitter taste. This obviously renders it unsuitable for direct consumption as a beverage. However, since milk that has been modified by bacterial growth may be used to produce other food products, the soured condition is not automatically categorized as spoilage. Other dairy products, such as butter or cheese, derived from milk or some of its components, will be decomposed by

FIGURE 15-5 Typical refrigerated coin-operated beverage cooler (Courtesy, Vendo Company).

(Table 15A)
Commodity Storage Requirements
Storage Requirements and Properties of Perishable Products

Commodity	Storage Temp, F	Relative Humidity, %	Approximate Storage Life*	Water Content, %	Highest Freezing Point, F	Specific Heat Above Freezing[a] Btu/lb/F	Specific Heat Below Freezing[a] Btu/lb/F	Latent Heat (Calculated)[b] Btu/lb
Alfalfa meal	0 or below	70–75	1 year, plus	—	—	—	—	—
Apples (Chapter 31)	30–40	90	3– 8 months	84.1	29.3	0.87	0.45	121
Apricots	31–32	90	1– 2 weeks	85.4	30.1	0.88	0.46	122
Artichokes (Globe)	31–32	95	2 weeks	83.7	29.9	0.87	0.45	120
Jerusalem	31–32	90–95	5 months	79.5	27.5[d]	0.83	0.44	114
Asparagus	32–36	95	2– 3 weeks	93.0	30.9	0.94	0.48	134
Avocados	45–55	85–90	2– 4 weeks	65.4	31.5	0.72	0.40	94
Bakery products, frozen (Chapter 37)								
Bananas (Chapter 32)[7]	—	85–95	—	74.8	30.6	0.80	0.42	108
Beans (Green or snap)	40–45	90–95	7–10 days	88.9	30.7	0.91	0.47	128
Lima	32–40	90	1 week	66.5	31.0	0.73	0.40	94
Beer, keg[c]	35–40	—	3– 8 weeks	90.2	28.0[d]	0.92	—	129
bottles, cans	35–40	65 or below	3– 6 months	90.2	—	—	—	—
Beets								
Bunch	32	95	10–14 days	—	31.3	—	—	—
Topped	32	95–100	4– 6 months	87.6	30.1	0.90	0.46	126
Blackberries	31–32	95	3 days	84.8	30.5	0.88	0.46	122
Blueberries	31–32	90–95	2 weeks	82.3	29.7	0.86	0.45	118
Bread (Chapter 37)	0	—	3 weeks to 3 months	32–37	—	0.70	0.34	46–53
Broccoli, sprouting	32	95	10–14 days	89.9	30.9	0.92	0.47	130
Brussels sprouts	32	95	3– 5 weeks	84.9	30.5	0.88	0.46	122
Cabbage, late	32	95–100	3– 4 months	92.4	30.4	0.94	0.47	132
Candy (Chapter 38)	0–34	40–65	—	—	—	—	—	—
Canned foods[c]	32–60	70 or lower	1 year	—	—	—	—	—
Carrots								
Topped, immature	32	98–100	4– 6 weeks	88.2	29.5	0.90	0.46	126
Topped, mature	32	98–100	5– 9 months	88.2	29.5	0.90	0.46	126
Cauliflower	32	95	2– 4 weeks	91.7	30.6	0.93	0.47	132
Celeriac	32	95–100	3– 4 months	88.3	30.3	0.91	0.46	126
Celery	32	95	1– 2 months	93.7	31.1	0.95	0.48	135
Cherries, sour	31–32	90–95	3– 7 days	83.7	29.0	0.87	—	120
Frozen	0 to −10	—	1 year	—	—	—	0.45	—
Sweet	30–31	90–95	2– 3 weeks	80.4	28.8	0.84	—	—
Cocoa	32–40	50–70	1 year, plus	—	—	—	—	—
Coconuts	32–35	80–85	1– 2 months	46.9	30.4	0.58	0.34	67
Coffee (green)	35–37	80–85	2– 4 months	10–15	—	0.30	0.24	14–21
Collards	32	95	10–14 days	86.9	30.6	0.90	—	—
Corn, sweet	32	95	4– 8 days	73.9	30.9	0.79	0.42	106
Cranberries[c]	36–40	90–95	2– 4 months	87.4	30.4	0.90	0.46	124
Cucumbers[c]	50–55	90–95	10–14 days	96.1	31.1	0.97	0.49	137
Currants	31–32	90–95	10–14 days	84.7	30.2	0.88	0.45	120
Dairy products								
Cheese, Cheddar	30–34	65–70	18 months	37.5	8.0	0.50	0.31	53
Cheddar	40	65–70	6 months	37.5	8.0	0.50	0.31	53
Processed	40	65–70	12 months	39.0	19.0	0.50	0.31	56
Grated	40	60–70	12 months	31.0	—	0.45	0.29	44
Butter	40	75–85	1 month	16.0	−4 to 31	0.50	—	23
Butter	−10	70–85	12 months	16.0	−4 to 31	—	0.25	23
Cream	−10 to −20	—	6–12 months	55–75	31.0	0.66–0.80	0.36–0.42	79–107
Ice cream	−20 to −15	—	3–12 months	58–63	21.0	0.66–0.70	0.37–0.39	86
Milk, fluid whole								
Pasteurized, Grade A	32–34	—	2– 4 months	87.0	31.0	0.93	—	125
Pasteurized, Grade A	−15	—	3– 4 months	87.0	31.0	—	0.46	125
Condensed, sweetened	40	—	15 months	28.0	5.0	0.42	0.28	40
Evaporated	70	—	12 months	74.0	29.5	0.79	0.42	106
Evaporated	40	—	24 months	74.0	29.5	0.79	0.42	106
Milk, dried								
Whole	70	low	6– 9 months	2–3	—	—	—	4
Non-fat	45–70	low	16 months	2–4.5	—	0.36	—	4
Whey, dried	70	low	12 months	3–4	—	0.36	—	4
Dates (Chapter 38)	0 or 32	75 or less	6–12 months	20.0	3.7	0.36	0.26	29
Dewberries	31–32	90–95	3 days	84.5	29.7	0.88	—	—
Dried foods	32–70	low	6 months to 1 year, plus	—	—	—	—	—
Dried fruits (Chapter 38)	32	50–60	9–12 months	14.0–26.0	—	0.31–0.41	0.26	20–37
Eggplant	45–50	90–95	7–10 days	92.7	30.6	0.94	0.48	132

Storage Requirements and Properties of Perishable Products (*Concluded*)

Commodity	Storage Temp, F	Relative Humidity, %	Approximate Storage Life*	Water Content, %	Highest Freezing Point, F	Specific Heat Above Freezing[a] Btu/lb/F	Specific Heat Below Freezing[a] Btu/lb/F	Latent Heat (Calculated)[b] Btu/lb
Sweet Potatoes	55–60	85–90	4– 7 months	68.5	29.7	0.75	0.40	97
Tangerines	32–38	85–90	2– 4 weeks	87.3	30.1	0.90	0.46	125
Tobacco, hogsheads	50–65	50–55	1 year	—	—	—	—	—
Bales	35–40	70–85	1– 2 years	—	—	—	—	—
Cigarettes	35–46	50–55	6 months	—	—	—	—	—
Cigars	35–50	60–65	2 months	—	—	—	—	—
Tomatoes								
Mature green	55–70 }[c]	85–90 }[c]	1– 3 weeks }	93.0	31.0	0.95	0.48	134
Firm ripe	45–50 }[c]	85–90 }[c]	4– 7 days[c] }	94.1	31.1	0.94	0.48	134
Turnips, roots	32	95	4– 5 months	91.5	30.1	0.93	0.47	130
Vegetable seed	32–50	50–65	—[c]	7.0–15.0	—	0.29	0.23	16
Yams	60	85–90	3– 6 months	73.5	—	0.79	—	105
Yeast, compressed baker's	31–32	—	—	70.9	—	0.77	0.41	102

* Not based on maintaining nutritional value.

[a] Calculated by Siebel's formula. For values above freezing point $S = 0.008a + 0.20$. For values below freezing point $S = 0.003a + 0.20$. Recent work by H. E. Staph, B. E. Short and others at the University of Texas has shown that Siebel's formula is not particularly accurate in the frozen region, because foods are not simple mixtures of solids and liquids and are not comple ely frozen even at −20 F.

[b] Values for latent heat (latent heat of fusion) in Btu per pound, calculated by multiplying the percentage of water content by the latent heat of fusion of water. 143.4 Btu.

[c] See text in this chapter or under appropriate commodity chapter.

[d] Average freezing point.

[e] Eggs with weak albumen freeze just below 30 F.

[f] Lemons stored in production areas for conditioning are held at 55 to 58 F; in terminal markets they are customarily stored at 50 to 55 F but sometimes 32 F is used.

Acknowledgment is due the following men for assistance with certain commodities: R. L. Hiner, L. Feinstein, and A. Kotula, meat and poultry; J. W. White and C. O. Willits, honey and maple sirup; E. B. Lambert, mushrooms; H. Landani, furs and fabrics; M. K. Veldhuis, orange juice; A. L. Ryall, plums and prunes; L. P. McColloch, tomatoes (all the former are United States Department of Agriculture staff members); J. W. Slavin, fish, United States Department of Commerce; and T. I. Hedrick, dairy products, Michigan State University.

bacterial activity—even products that utilize other types of bacteria in their formation. The chemical change in this decomposition process will produce a sour odor and unpalatable taste in milk by-products as well as in unprocessed milk.

Beef, pork, and other types of meat products show very little apparent change in physical appearance when they first begin to deteriorate. As the deterioration process continues, however, they give off a slight odor, which increases in intensity as the deterioration progresses. When certain stages of deterioration are reached, a slimy covering or coating appears on the surface of the meat, which is a definite indication of spoilage.

All of the above examples of food spoilage are caused by a series of chemical changes within the products. These changes may be brought about through natural forces or by external contamination. Either way the result can be a complete breakdown or decay of the products.

Bacteria are found in the air, in the soil, and in water, as well as in almost every known substance on earth. For our purposes the two important types of bacteria are:

(a) those which cause food deterioration, and

(b) those used in the preservation of foods.

This latter type was described in the paragraph concerning sour milk. It is also used in the processing of other foods such as wine, vinegar, and certain types of cheeses and butter. The first type of bacteria, however, is the most important to the refrigeration storage or processing plant operator.

Bacteria are subject to changes in temperature and moisture conditions and are even sensitive to normal lighting and changes in light conditions.

Since bacteria, under certain conditions, cause the decay or deterioration of food products, in any study of food preservation one must become conversant with the different types of bacteria that are found in food products. Each type has a minimum and maximum temperature range for ideal growth. Therefore the temperature to which bacteria are subjected must be closely controlled. Some become dormant or are dissipated if subjected to temperatures below 0°F, whereas other bacteria need the application of high-temperature steam under pressure before they are destroyed. Usually bacteria prefer darkness. As an example of the widespread presence of bacteria, more than 200 different types have been found in milk and milk products alone. Although the different types of bacteria are responsible for most of the spoilage of foodstuffs, another member of the plant kingdom—mold—also contributes.

Molds are the principal causes of deterioration and spoilage in citrus and other fruits. The different types of mold thrive particularly well in damp locations; darkness adds to their growth. As mentioned earlier, molds also are a form of plant life, and different types of molds cause the spoilage of citrus fruit and dairy products, soft rot of apples in storage, and deterioration of bread and grain.

It is obvious that control measures must be taken to suspend or prevent food from spoiling. Through experience and experimentation, different methods have been developed to preserve foodstuffs. Some of them are listed in Table 15A; one must follow the recommended temperatures and conditions to preserve different types of foodstuffs. Table 15A lists the *optimum* conditions and temperatures that will give the desired results. The variety or type of each product, the purpose for which it is being stored, and also the conditions under which it is being stored, dictate the conditions required for the most desirable storage environment.

If vegetables or fruits are refrigerated immediately upon harvesting, they will be preserved in storage for a longer period than if they are permitted to continue their ripening process after harvesting, before being refrigerated. This one difference in the preparation and storage of vegetables or fruits could easily result in variations of several months of continued satistactory quality of the product. Some of the products listed in Table 15A have a high moisture content; therefore conditions of high relative humidity are recommended for their proper storage.

PROCESSING

Other areas where commercial refrigeration is important are in the manufacture of candy and the processing of food products—particularly those to be sold in a frozen state. Processing plants for name-brand frozen foods are often located in the areas where those foods are produced. Thus, it is natural that processing centers for fruits would be located in a fruit-producing area of the country. By the same token processing centers for different types of vegetables are located in truck-farming areas. Processors of frozen fish have plants located along the coast, from which they ship their products throughout the country.

Not too many years ago, before the widespread acceptance of home freezers, it was a common practice for communities to have locker plants which would process, freeze, and store various products either for the producers of these products or for the consumers.

Figure 15-6 shows the layout of a typical frozen food locker plant, in which these steps or phases of operation are followed:

(1) ***The Receiving Area,*** which may or may not be refrigerated, where the incoming foods are weighed and listed.

(2) ***Chill and Age Areas,*** which must be refrigerated and maintained at approximately 35°F, where different types of meat, poultry, and even fish are prechilled before being cut into the desired size for packaging.

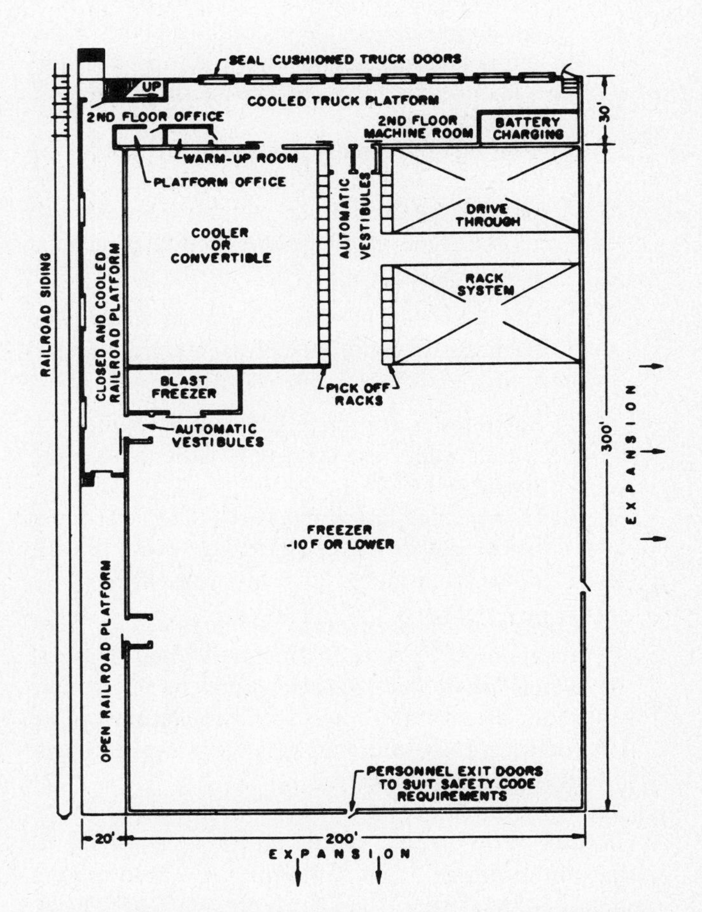

FIGURE 15-6 Typical frozen food locker plant (Courtesy, ASHRAE).

(3) ***Preparation Area,*** which may or may not be refrigerated, where the meat, fowl, and fish are processed for distribution to either the freezer or cure room. This includes the washing and blanching of fruits and vegetables when these products are handled or processed in the locker plant.

(4) ***Cure or Smoke Area,*** which is refrigerated and maintained at approximately 35°F. If some of the product is to be cured, smoked, or salted, such as bacon, hams, or sausage, the meat is taken into the cure room after it has been cut to size and/or processed.

(5) ***Packaging or Wrapping Area,*** which may or may not be refrigerated. This is the area where, before freezing, all of the pieces or sections of meat and other products are packaged and/or wrapped. It is here also that the product is identified by labeling each package as to its content, weight, date, and owner.

(6) ***Quick-freeze Area:*** all processing plants make an effort to freeze food as rapidly as possible. This is done by one of several methods such as:

(a) Placing the food packages on cold-plate shelves, which lower the temperature of the product by the contact method. (An example of this type of cold-plate shelf arrangement is shown in Fig. 15-7.)

(b) Quick freezing may be achieved by placing the packages of food on shelves in a blast-freezer unit.

(c) Some foods are immersed in subzero liquids, while some are sprayed with a cooling liquid.

In any case, the temperature of the product is brought down as quickly as possible to a range from minus 10°F to minus 40°F.

(7) ***Locker or Storage Area,*** which is usually kept at about 0°F. It contains individual lockers, which are rented. Also included in this locker room area is a storage space to accommodate food for customers who wish to store packages in their own home freezers.

Frozen food locker plants still are operating in some communities. The function of these plants ranges from the storage of commodities prepared elsewhere to the complete local processing and storage of many different kinds of foods.

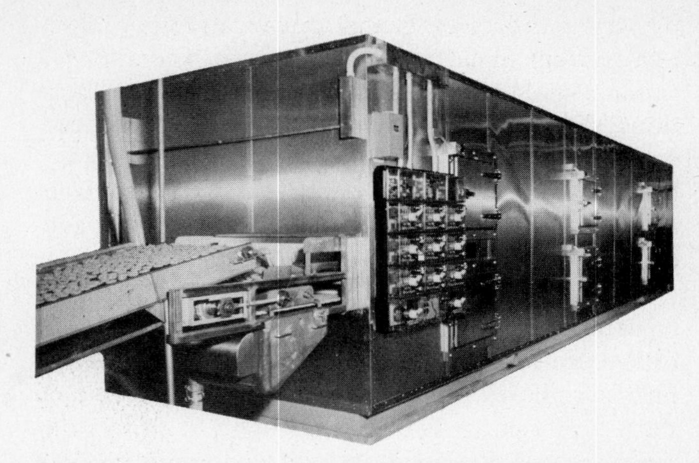

FIGURE 15-7 Cold plate shelf freezer (Courtesy, Refrigeration Engineering Corporation).

TRANSPORTATION AND DISTRIBUTION

Prior to the use of mechanical refrigeration, producers of perishable products had to depend the local distribution of their products, or rely on ice to keep them in saleable condition. Railway cars used for the shipment of perishable commodities were equipped with bunkers at each end in which ice was packed to refrigerate products in the car during shipment. There were hatches or access openings to the bunkers on the roofs of the cars so that the ice could be replenished when needed. If the weather was hot and the distance was long, ice would be replenished several times before the product arrived at its destination. Floor drains were provided to permit drainage of the water from the melting ice.

In many cases, a blower was used for the circulation of cool air across the product in the railway car. The motor for the blower was driven either by a small gasoline engine or else there was an electric motor driven by a battery. A grill was located in the bottom of the bunker, which permitted the warmer return air from the product to pass across the ice and thereby give up its heat. The cold air was then forced down across the product directly or through ductwork for equal distribution.

The recent trend has been toward the installation of mechanical refrigeration in railway cars used for the transportation of perishable products. These refrigeration units are driven directly from the turning axles of the cars when they are in motion, or by electric motors or gasoline engines when the cars are in a freight yard for any length of time.

Refrigerated trucks and trailers today are more evident than they used to be on the highways. The mechanical refrigeration systems in these vehicles are quite similar to those used in other commercial applications, except that the compressors may be driven by electrical motors, in engine-motor combinations, or even by individual gasoline or diesel engines. In the type using an electric motor, the motor not only will operate from the engine-driven electric generator, but its voltage, cycle, and phase are such that it is capable of being plugged into a wall outlet in a garage when the truck engine is not operating. The engine-motor combination is such that, when the truck is in operation, the transmission power takeoff of the vehicle operates the refrigeration unit. But, when the truck is parked overnight with the engine shut off, a standard motor is utilized for its operation. The gasoline or diesel engine units operate automatically, as the system load requires. Some truck and trailer units utilize a dual refrigeration system with a standard refrigeration unit for standby purposes, when the truck engine is not operated for a long time.

Some truck and trailer units use eutectic plates as the evaporator within the storage compartment. This type of system can be cooled down to a low temperature during the night and maintain a "cooling" storage capacity for use the following day. Examples of this are milk delivery trucks, which may be cooled during the evening when the trucks are being loaded with their products, so they will stay cold the following day while deliveries are made. Some eutectic plates are capable of being charged or cooled continuously while the truck is being used.

Many of the air-freight lines have planes equipped to handle refrigerated products in transit. Marine refrigeration units are basically the same as other commercial units. An improvement is that fishing vessels now are equipped with automatic ice makers, which can keep the bunkers aboard the ship filled with ice, if the catch requires storage. In the past, fishing boats would load up with ice from land sources, in the anticipation that the day's catch would be large enough to utilize the ice that was brought aboard.

ICE MANUFACTURE

As mentioned in Chap. 1, the use of ice to refrigerate food, beverages, and other products dates back to early civilizations. The shipment of ice by the clipper ships at times was a prosperous venture, but at times could be a disaster. These early ships used natural ice formed in rivers, lakes, and ponds in the cold areas of the country. With the advent of mechanical refrigeration, dependence on natural ice became a thing of the past. Since its inception, the manufacture of artificial ice has become an important part of the refrigeration industry. Large-scale production of ice is usually in block form, in sizes varying from 50 to 400 pounds. The variation in the sizes of the block depends upon the size of the cans in which the ice or water is frozen.

Before an ice plant is established, a survey of the market determines the sizes of blocks of ice to be produced. The smaller capacity freezing cans usually are approximately 2.5 feet high, with various inside dimensions. Cans accommodating larger blocks might have heights approximating 5 feet, with inside measurements about 1 by 2 feet. A typical current commercial operation coordinates the manufacture of ice with the use of refrigerated self-service ice-cube vending machines. The blocks are cut into convenient sizes of 10, 25, or 50 pounds to be placed in the vending machines. These smaller blocks may be wrapped or unwrapped.

This ice, of course, must be maintained at proper storage temperatures, as must the packages of ice cubes dispensed in the same manner.

The demands for cubed, crushed, sized, or flake ice are great. To cool a beverage without too-rapid dilution, a smaller surface of ice is needed in the beverage, and ice cubes would be best.

If a beverage or other liquid is to be cooled quickly, such as a glass of water in a restaurant, flake ice will serve the purpose satisfactorily. The greater the area of the ice surfaces, the more quickly heat is removed from the liquid in which it is immersed.

If an ice cube is one square inch on each side, it has six square inches of surface in contact with the beverage for heat transfer. Therefore, if a glass contains four ice cubes, the total heat transfer area is 24 square inches. If there is a hole in the center of each cube, the area is increased, but the volume of ice is less.

Figure 15-8 shows a typical ice-cube maker capable of manufacturing about 8,000 ice cubes in a 24-hour period, with a storage bin below it, which holds about 400 pounds of ice cubes.

The manufacture of this ice is automatic, and its operation is controlled by a thermostat located in the storage bin, which is actuated by the level of ice remaining in the bin.

Flake ice machines, such as shown in Fig. 15.9, range in capacity from as small as 200 pounds daily to those that produce approximately a ton of ice in a 24-hour period. The cylinder is sprayed with, or rotates in water, depending upon its design—which also governs whether it rotates in a horizontal or a vertical position. A film of ice forms on the cylinder or drum and, when it becomes approximately 1/16th of an inch in thickness, a cutting bar or roller breaks it up, allowing it to drop into the storage bin.

FIGURE 15-8 Typical ice cube maker (Courtesy, Liquid Carbonic Corp.).

FIGURE 15-9 Typical flake ice machine (Courtesy, Liquid Carbonic Corp.).

A glass filled with flake ice would probably hold at least 50 flakes measuring 1 in. by 1 in. by 1/16 in. each. Even though they are thin, each flake would have two square inches of heat transfer area (one square inch on each side). The glass containing the flake ice would have approximately four times as much ice area in the beverage as the glass containing the ice cubes; therefore the beverage with the flake ice would be cooled more quickly. Of course, the flake ice would melt much more quickly, too, thereby adding water to the liquid contents of the glass.

Ice for light commercial use is manufactured in many sizes and shapes, such as shell ice, tube ice, and even a thin film, which may be formed into small, briquette-type cubes. Some of the ice cube machines have attachments for crushing the cubes, if desired. Crushed or flake ice is used in restaurants and cafeterias for the chilling and display of salads and some desserts.

STORAGE LIFE

Table 15A lists the approximate storage life of a number of perishable foods, particularly those that have been freshly picked or harvested. It also lists the storage life and temperatures of several types of frozen foods. It should be noted that there is a definite difference in storage temperatures, depending upon the length of time the product is to be stored. The longer the storage time, the lower its storage temperature.

(Table 15B)

Commodity Storage Requirements

Approximate Rates of Evolution of Heat by Certain Fresh Fruits and Vegetables When Stored at the Temperatures Indicated[a]

Commodity	Btu per ton per 24 hrs		
	32 F	40 F	60 F
Apples	500– 900	1,100– 1,600	3,000– 6,800
Asparagus	5,900–13,200	11,700–23,100	22,000–51,500
Avocados	—	—	13,200–30,700
Bananas[b]	—	—	4,600– 5,500
Beans, green or snap	—	9,200–11,400	32,100–44,100
Beans, lima	2,300– 3,200	4,300– 6,100	22,000–27,400
Beets, topped	2,700	4,100	7,200
Blueberries[c]	1,300– 2,200	2,000– 2,700	7,500–13,000
Broccoli, sprouting	7,500	11,000–17,600	33,800–50,000
Brussels sprouts	3,300– 8,300	6,600–11,000	13,200–27,500
Cabbage	1,200	1,700	4,100
Carrots, topped	2,100	3,500	8,100
Cauliflower	3,600– 4,200	4,200– 4,800	9,400–10,800
Celery	1,600	2,400	8,200
Cherries	1,300– 1,800	2,800– 2,900	11,000–13,200
Corn, sweet	7,200–11,300	10,600–13,200	38,400
Cranberries	600–700	900– 1,000	—
Cucumbers	—	—	3,300– 7,300
Grapefruit	400– 1,000	700– 1,300	2,200– 4,000
Grapes, American	600	1,200	3,500
Grapes, European	300–400	700– 1,300	2,200– 2,600
Lemons	500–900	600– 1,900	2,300– 5,000
Lettuce, head	2,300	2,700	7,900
Lettuce, leaf	4,500	6,400	14,400
Melons, cantaloupes	1,300	2,000	8,500
Melons, honeydews	—	900– 1,100	2,400– 3,300
Mushrooms[d]	6,200– 9,600	15,600	—
Okra	—	12,100	31,600
Onions	700– 1,100	800	2,400
Onions, green	2,300– 4,900	3,800–15,000	14,500–21,400
Oranges	400– 1,000	1,300– 1,600	3,700– 5,200
Peaches	900– 1,400	1,400– 2,000	7,300– 9,300
Pears	700–900	—	8,800–13,200
Peas, green	8,200– 8,400	13,200–16,000	39,300–44,500
Peppers, sweet	2,700	4,700	8,500
Plums	400–700	900– 1,500	2,400– 2,800
Potatoes, immature	—	2,600	2,900– 6,800
Potatoes, mature	—	1,300– 1,800	1,500– 2,600
Raspberries	3,900– 5,500	6,800– 8,500	18,100–22,300
Spinach	4,200– 4,900	7,900–11,200	36,900–38,000
Strawberries	2,700– 3,800	3,600– 6,800	15,600–20,300
Sweet Potatoes	—	—	4,300– 6,300
Tomatoes, mature green	—	1,100	6,200
Tomatoes, ripe	1,000	1,300	5,600
Turnips	1,900	2,200	5,300

[a] Data largely from Table 1 of *USDA Handbook* No. 66, 1954. Acknowledgement is also due to the following for other commodities: Avocados, J. B. Biale; Brussels sprouts, J. M. Lyons and L. Rappaport; Cucumbers, I. L. Eaks and L. L. Morris; Honeydew melons, H. K. Pratt and L. L. Morris; Plums, L. L. Claypool and F. W. Allen; Potatoes, L. L. Morris; all from the Univ. of California. Asparagus, W. J. Lipton, USDA; Cauliflower, lettuce, okra, and onion, H. B. Johnson USDA; Sweet corn, S. Tewfik and L. E. Scott, Univ. of Maryland.
[b] Bananas at 68 F, 8,400–9,200.
[c] Blueberries at 50 F, 5,100–7,700; at 70 F, 11,400–15,000.
[d] Mushrooms at 50 F, 22,000; at 70 F, 58,000.

Of course the condition or quality of the product has a definite relationship to the length of time it can be stored. If the product is to be presented for sale to the consumer immediately, it must be picked at its peak of ripeness. This also applies if the quality product is to be quick-frozen.

If they are to be shipped any appreciable distance, many vegetables and fruits are precooled before shipping and chilled while being transported. If this is not done, the respiration process will speed up in the fruits or vegetables and they will soon begin to decay. Meats, as pointed out earlier, are affected by the growth of bacteria, but they do not undergo the respiration process. Fruits and vegetables continue to absorb oxygen and release carbon dioxide even after harvesting.

This process of absorbing oxygen and releasing carbon dioxide, through the chemical process of starch conversion into sugar within the living substance, generates heat. The increase of heat within the substance brings about an acceleration in the action of the enzymes in the product. This action should be reduced as much as possible so that the quality of the product will be maintained for presentation to the public.

Table 15B lists the amounts of heat, rated in Btu/ton/day, given off by various fruits and vegetables at different temperatures. This table will be used in Chap. 21, when calculations of heat loads, which occur under varying circumstances, are discussed. It is readily seen that this amount of heat is considerable, particularly at high temperatures. Therefore the product should be cooled to its lowest acceptable temperature as soon as possible.

The storage life of various products is important to fulfill the law of supply and demand. Usually, when there is a large supply of a product, even with an adequate demand, the price of such products may drop. This applies, for example, to fruits and vegetables "in season." When the product is not readily available or "out of season," the price demanded is usually somewhat higher. If people can have such things as strawberries, juicy apples, and even corn-on-the-cob during the winter months, they are often willing to pay a premium price for these delicacies.

Producers of many commodities may place some of their produce in storage to be available for sale during the "off season." An egg farmer may, when his hens are producing a surplus of eggs, place some of these products in storage so that they will be available for sale when production is lower.

MIXED STORAGE

To maintain quality merchandise under various storage conditions, separate storage areas or facilities for each of the products is desirable. This may not be feasible in many cases. It is possible to store several products in the same area if they have common recommended storage conditions. Usually, if mixed storage is used, it may call for compromise temperatures that may not be the optimum for some of the products. Some of the products may be stored at temperatures below point that is considered critical.

As already pointed out, higher storage temperatures may shorten the life of some of the products in mixed storage. Usually this does not create too much of a problem, for the mixed storage is ordinarily only for a short period.

The big problem in mixed storage is the transfer or absorption of odors from one product to another. All dairy products, along with eggs, fruit, and even some kinds of nuts, easily pick up odors and flavors from other products. Onions and potatoes transfer or give off their odors and flavors to other products that may be stored with them in the same facility.

Customarily, larger storage facilities, such as those provided by wholesalers or jobbers, will be adequate to permit long-term storage of individual products in their own facilities. Or, facilities will be provided for the storage of vegetables in one area, fruit in another, and dairy products in still another.

PROBLEMS

1. What refrigerant was used in the earliest commercial refrigerators?

2. How is dehydration of food products in a refrigerator cabinet prevented?

3. Approximately what temperature should be maintained in cabinets for frozen food storage or display?

4. Generally speaking, there are two primary types of

bacteria that affect foods in refrigerated storage. What are they?

5. Refrigerating fruits or vegetables immediately upon harvesting preserves them for longer periods than if there is a time lag between harvest and refrigeration. Why?

6. Before installation of mechanical refrigeration in railway cars for transport of fresh produce, what method was used?

7. What are eutectic plates?

8. Does a liquid, such as a beverage in a glass, cool more quickly when large cubes of ice are immersed in it or when a large number of flakes of ice are used?

9. Why is it desirable to store different types of produce under different conditions of refrigeration and relative humidity?

10. What term is applied to the storage of different food products in the same storage area?

16

EVAPORATORS

Earlier, in the chapter on the compression cycle, an evaporator was described as that part of a refrigeration system in which the refrigerant is converted from a liquid to a vapor through the process of evaporation. This takes place as the heat from the produce or load is absorbed by the refrigerant in the evaporator. As was previously discussed, evaporators fall into three types:

(a) bare tube,

(b) finned-tube and

(c) plate.

Most commercial display cases, walk-in coolers, reach-in boxes, and florists' refrigerators utilize the finned-tube design of coil. An evaporator of this type has a definite advantage over the bare-tube type of evaporator, shown in Fig. 16-1. The heat load handled by the evaporator reaches the cooling coil one or more of the three means of heat transfer: conduction, convection, or radiation. But this heat is transferred to the refrigerant through only one of these means—conduction. The extra area of the fins in addition to that of the bare tube permits a higher degree of heat transfer from the air surrounding the coil, as can be seen in Fig. 16-2. Therefore, the greater the surface area for the conduction of the heat from the product to the refrigerant in the evaporator, the greater the possible heat transfer. If the heat from the product reaches the evaporator but is not absorbed by the refrigerant within the coil, the box or area will reach a higher temperature than is desired. Increasing the surface area of the evaporator increases the capacity of the evaporator.

Many commercial evaporators are designed to use natural convection or flow of air through the coil. Of course, the capacity of this type of coil is based primarily on unrestricted flow of air. If the coil is improperly located, or if its design and/or installation is such that air circulation around and through it is restricted, the coil can't operate at peak efficiency.

As the air passes through the cooling coil it gives up its heat and is therefore cooled. As it is cooled, it contracts into a smaller volume, which weighs more than an equal volume of warmer air. In this way con-

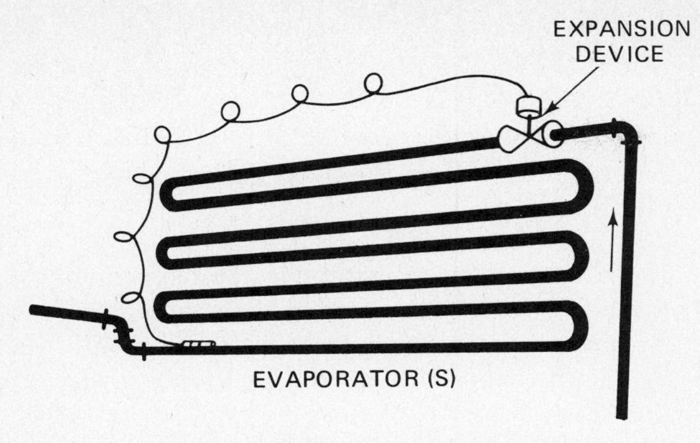

FIGURE 16-1 Bare tube type of evaporator.

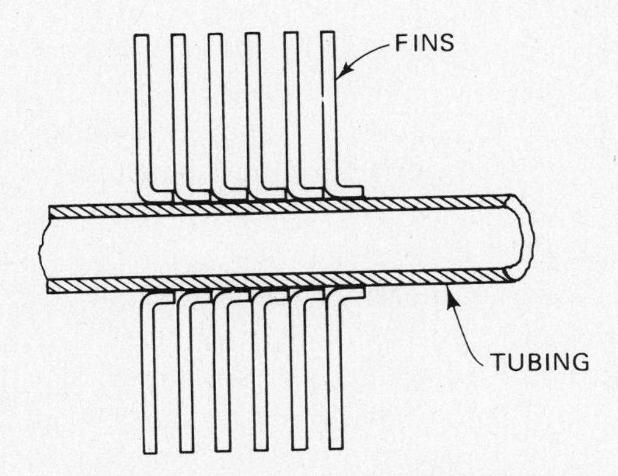

FIGURE 16-2 Flanged fins added to tubing.

vection currents of air are set up which carry away the heat from the product being cooled.

The circulation of this convection air can be, and frequently is, assisted by various means, one of which is the use of metal or composition baffles. These baffles are placed so that the air currents are forced to move in certain paths to give the optimum heat transfer.

If there is any doubt that the convection currents can produce the required flow of air across the evaporator coil to remove satisfactory amounts of heat from the air and product, forced circulation of air across the cooling coil is indicated. Evaporators utilizing forced circulation of air usually have finned tubes, supplemented by one or more axial (propeller-type) fans to accomplish the circulation of air. In such a case, baffles are not absolutely necessary to provide a standard pattern for required circulation, but are often used nonetheless. Figure 16-3 shows a typical commercial evaporator with an air-circulating fan.

With forced-air circulation, it is necessary that the fan be capable of:

FIGURE 16-3 Typical commercial evaporator with air circulating fan (Courtesy, Kramer Trenton Company).

(a) circulating enough air to remove the heat from the product

(b) distributing air at a satisfactory velocity around the room or conditioned space.

(c) assuring that there are no "dead spaces"

If the velocity of air across some products is greater than recommended there is a tendency toward too rapid dehydration, as related in an earlier chapter.

REFRIGERATING EFFECT

Transfer of heat from the product to the air and then to the refrigerant within the cooling coil has been described. The amount of heat each pound of refrigerant picks up from the product and the air as it travels through the evaporator is called the *refrigerating effect*. The liquid refrigerant entering the metering device and cooling coil has a certain heat content at its given temperature, as does the refrigerant vapor leaving the cooling coil at its lower temperature. The difference in heat content of these two stages is the amount of heat absorbed by the pound of refrigerant as it circulates through the cooling coil. Therefore, the refrigerating effect (or simply R.E.) is rated in the terms of Btu per pound of refrigerant circulated.

The heat absorbed by the refrigerant depends on two main conditions of the refrigerant and the temperatures at these conditions:

(a) the temperature of the liquid refrigerant entering the refrigerant control

(b) the evaporating temperature, or the temperature of the refrigerant vapor leaving the evaporator.

Under the following conditions, what would be the refrigerating effect?

> The refrigerant enters the metering device in a liquid state at the condensing temperature of 100°F, and it leaves the evaporator in a vapor state at 40°F.

Referring to the Refrigerant Table in the Appendix, which lists the properties of R-12 in a saturated vapor state, one finds that the enthalpy (heat content) of the liquid refrigerant at 100°F is 31.16 Btu/lb; whereas the enthalpy of 40° refrigerant vapor leaving the cooling coil 82.71 Btu/lb. Therefore the difference between these two figures amounts to 51.55 Btu/lb, or the amount of heat that each pound of refrigerant absorbs from the product and air under the given conditions.

Using the above example is can be seen that if the amount of heat to be removed from a project amounts to 30,930 Btu, 10lb of refrigerant must be circulated each minute. For, under the conditions given in the example, the heat-removal capacity of the refrigerant amounts to 51.55 Btu/lb circulated through the evaporator. When this capacity of one pound of refrigerant is divided into the total Btu to be removed, we find this equation:

$$\frac{30.930 \text{ Btu/hr}}{51.55 \text{ Btu/lb}} = 600 \text{ lb/hr}$$

Then

$$\frac{600 \text{ lb/hr}}{60 \text{ min/hr}} = 10 \text{ lb/min of refrigerant to be circulated}$$

As already pointed out, the two variables that will alter the refrigerating effect per pound of refrigerant circulated involve the *entering liquid temperature* and the *leaving vapor temperature*. Therefore, by lowering the entering liquid temperature, the refrigerating effect will be increased. This means that fewer pounds of refrigerant will have to be circulated to do the work required. Also, by raising the evaporating temperature while the condensing temperature remains the same, the amount of refrigerant needed to be circulated will be lowered. This can be proved by the following:

> If the condensing temperature—the temperature of the liquid refrigerant entering the evaporator—can be lowered from 100°F to 86°F, the enthalpy will be decreased to 27.72 Btu/lb. With the enthalpy of the leaving vapor remaining the same, there will be an increase of 3.44 Btu/lb in the refrigerating effect, producing a new R.E. of 54.99 Btu/lb.

The same required load of 30,930 Btu as in the previous example divided by the new R.E. of 54.99 Btu/lb results in a requirement of 562 lb/hr. When we divide by the 60 minutes in an hour, we find approximately 9.37 lb of refrigerant must be circulated each minute.

Conversely, if the evaporator temperature can be increased so that the vapor leaving the cooling coil is at 50°F instead of 40°F, there will be an increase of 1.07 Btu/lb in the enthalpy of the vapor. This bears out that, with the variables previously mentioned, the lower the temperature of the liquid entering the metering device and coil, the greater will be the refrigerating effect. It follows that the higher the evaporating temperature the greater the R.E.

We are studying the evaporator first in the commercial refrigeration cycle primarily because that is where the heat is removed from the product. The heat is absorbed by the refrigerant, which vaporizes in the coil and then proceeds to the compressor, which compresses the refrigerant vapor before it continues in the cycle to the condenser.

DIRECT-EXPANSION COIL CAPACITY

The capacity of any direct-expansion (DX) coil is dependent upon:

(1) the temperature of the refrigerant circulated

(2) the temperatures (dry and wet bulb) of the air being circulated through the coil

(3) the volume of air being circulated

As shown earlier, if the temperature of the liquid entering the cooling coil is varied, the refrigerating effect is also varied. This affects the capacity of the

			Air Velocity (FPM)											
			400			500			600			700		
Suct Temp	*Row*	*Fin No.*	*MBH*	*LDB*	*LWB*	*MBH*	*LDB*	*LWB*	*MBH*	*LDB*	*LWB*	*MBH*	*LDB*	*LWB*
35°F	4	8	18.0	49.0	48.0	20.4	51.0	49.8	22.3	52.7	51.3	23.9	54.0	52.4
	5		20.4	46.0	45.4	23.4	48.0	47.3	25.9	49.7	48.8	28.0	51.1	50.1
	6		22.3	43.8	43.4	25.8	45.6	45.2	28.9	47.2	46.7	31.6	48.6	48.0
	8		25.0	40.5	40.3	29.5	42.1	41.9	33.4	43.6	43.4	36.9	45.0	44.8
	4	10	19.6	47.0	46.4	22.2	49.0	48.3	24.3	50.8	49.9	26.1	52.2	51.2
	5		21.9	44.1	43.8	25.2	46.1	45.7	28.1	47.8	47.3	30.3	49.3	48.7
	6		23.7	41.9	41.8	27.6	43.8	43.6	30.9	45.5	45.3	33.8	47.0	46.7
	8		26.1	39.1	38.9	31.0	40.8	40.6	35.3	42.1	42.0	39.1	43.6	43.4
	4	12	20.7	45.5	45.1	23.6	47.6	47.1	26.0	49.3	48.8	27.9	50.7	50.1
	5		23.0	42.9	42.6	26.6	44.8	44.6	29.6	46.5	46.2	32.2	48.0	37.6
	6		24.7	40.9	40.7	28.9	42.7	42.5	32.5	44.4	44.2	35.6	45.9	45.7
	8		26.8	38.2	38.0	32.0	39.7	39.5	36.6	41.2	41.0	40.7	42.5	42.3
40°F	4	8	15.2	52.0	50.9	17.2	53.7	52.3	18.8	55.0	53.5	20.1	56.2	54.5
	5		17.3	49.6	48.8	19.7	51.1	50.3	21.8	52.5	51.5	23.5	53.7	52.6
	6		18.9	47.5	47.1	21.9	49.2	48.6	24.4	50.4	49.8	26.5	51.7	51.0
	8		21.3	44.8	44.6	25.0	46.1	45.9	28.3	47.5	47.1	31.2	48.5	48.2
	4	10	16.5	50.2	49.5	18.7	52.0	51.1	20.5	53.4	52.4	22.0	54.6	53.4
	5		18.6	47.8	47.4	21.3	49.5	49.0	23.6	50.9	50.4	25.5	52.1	51.5
	6		20.1	46.0	45.8	23.4	47.7	47.3	26.1	48.9	48.6	28.5	50.2	49.8
	8		22.2	43.6	43.4	26.3	45.0	44.8	30.0	46.3	46.1	33.1	47.3	47.1
	4	12	17.5	49.1	48.6	19.9	50.7	50.2	21.8	52.2	51.5	23.4	53.4	52.7
	5		19.5	46.7	46.4	22.5	48.5	48.1	24.9	49.8	49.5	27.0	51.0	50.6
	6		21.0	45.1	44.9	24.5	46.5	46.4	27.4	48.0	47.7	30.0	49.1	48.9
	8		22.9	42.9	42.7	27.3	44.1	43.9	31.1	45.3	45.1	34.5	46.5	46.3
45°F	4	8	12.3	55.0	53.7	13.8	56.3	54.8	15.1	57.4	55.7	16.1	58.4	56.5
	5		14.0	52.8	52.0	15.9	54.2	53.2	17.6	55.3	54.2	18.9	56.2	55.0
	6		15.4	51.2	50.7	17.7	52.5	51.9	19.6	53.7	52.9	21.3	54.5	53.8
	8		17.3	49.0	48.8	20.3	50.0	49.8	22.9	51.0	50.8	25.2	52.0	51.6
	4	10	13.3	53.5	52.7	15.0	54.8	53.9	16.4	56.0	54.9	17.6	56.9	55.7
	5		15.0	51.4	51.0	17.2	52.9	52.3	19.0	54.1	53.3	20.5	54.9	54.2
	6		16.4	49.9	49.7	18.9	51.2	50.9	21.1	52.4	52.0	22.9	53.5	52.9
	8		18.2	48.0	47.8	21.5	49.1	48.9	24.3	50.2	50.0	26.8	51.1	50.9
	4	12	14.1	52.4	51.9	16.0	53 9	53.2	17.5	55.0	54.2	18.8	55.9	55.1
	5		15.8	50.6	50.2	18.1	51.9	51.5	20.0	53.1	52.6	21.7	53.9	53.5
	6		17.0	49.3	49.1	19.8	50.4	50.2	22.1	51.5	51.3	24.2	52.5	52.2
	8		18.7	47.4	47.2	22.2	48.4	48.2	25.3	49.4	49.2	28.0	50.3	50.1

FIGURE 16-4 Coil capacity table. Type DX-coil ratings.

cooling coil, and, if the temperature of the air on the coil remains the same, any variation in the suction temperature will also change the temperature difference between the refrigerant and the air. If this temperature difference decreases, the rate at which the refrigerant evaporates also will decrease.

The same decrease in refrigerant evaporation will occur if the quantity of air across the cooling coil is decreased, since the lesser amount of air will cool to a lower temperature and reduce the temperature difference between the refrigerant and the air.

Figure 16-4 is a direct-expansion coil capacity table. It shows that, as the suction temperature is increased the coil capacity is lowered. This is caused by the decrease in the temperature difference between the refrigerant and the air.

As will be pointed out later, the identical condition affect the cooling coil and the compressor differently: as the capacity of the coil increases, the capacity of the compressor decreases. Therefore, components that are selected and installed in the field (not a prepackaged condensing unit) require careful balancing of the evaporator and the compressor components (with regard to their individual capacities) inorder to find out the point or points at which each will have the same capacity.

DEFROSTING

Whenever moisture is removed from the air or other product being cooled or frozen, frost will accumulate on the cooling element and must be removed periodically. Frost acts as an insulator, reducing the heat transfer between the air and the refrigerant in the cooling coil.

In the early days of refrigeration, when bare-tube or plate-coil evaporators were used in storage areas in locker plants, defrosting of these coils had to be done manually. It was not uncommon, particularly when the coils were located overhead at a height of 10 feet or more, for an employee to put on a raincoat and hat and use a long stick to knock down the accumulation of frost on the pipes or plate-coils. When he had finished, he would use a shovel or a broom to finish the job of defrosting by picking up the fallen frost or ice. This manner of defrosting was necessary because it was desired to continue the operation of the refrigeration equipment during the defrost period.

This type of manual defrosting is still being used in some small home or farm freezers, or even in some small commercial refrigeration units having plate-type evaporators. In these situations this method is satisfactory, because there are no fins or other obstructions on this type of cooling coil. These units cannot be shut off in order to permit the defrosting of the coils unless the contents or products are removed from the freezer to prevent melting.

In commercial installations where the room temperature is above 35°F, the finned-tube type of cooling coils does not present too great a problem. It is necessary to shut off the operation of the refrigeration unit temporarily and, with the continued circulation of room air at 35°F or above, the frost will be removed from the fins and cooling coil at a fairly rapid rate. This temporary shut-down of the refrigeration unit can be done manually or with some type of controlling device. This is called *off-cycle defrost*, for the refrigeration cycle has been temporarily stopped.

When the room air temperature is beolw 32°F, the accumulation of frost will not melt off the evaporator during the off-cycle, and some means of artificial defrosting becomes necessary. This may be accomplished through the application of warm air brought in through duct work, or by warm water, brine, or heat applied by some other means, such as an electrical heating element or a system of hot gas defrost. Unless there is ductwork already being utilized for the distribution and circulation of air within the space being cooled, it is not feasible to use the warm air method, since it necessitates the installation of complicated ductwork and accessories.

Of course, each of the methods of defrosting requires that the individual system be out of operation long enough to permit the defrosting of the coil. This usually takes a coil temperature between 36°F and 40°F. Therefore it is not practical to shut down the unit for any length of time in an application where the room temperature must be 32°F or below.

If two or more evaporators are connected to the same condensing unit but are at different locations, the condensing unit itself does not need to be out of operation in order to defrost the coils. The coils can be shut down individually and defrosted one at a time while the remaining coil or coils and the condensing unit continue to operate at the desired temperature. This may be accomplished by manually closing a valve in the liquid line of the coil being defrosted. When the defrosting of the individual coil is completed, the valve is opened, permitting the normal operation of that cooling coil. The other coil or coils are defrosted in the same manner until all coils in the system are defrosted.

This process can be done automatically by the use of a clock-timer to shut down part or all of a system for a given period of time at regular intervals. The length of the defrost period, along with the frequency of defrosting, depends on the individual installation.

It is recommended that the length of the defrost cycle be adjusted carefully so that the refrigeration system is put back into operation as soon as possible after defrosting. The more frequent the defrost cycles the smaller the accumulation of frost, thus shortening the defrosting period.

Water or brine may be used to defrost a cooling coil. Water is usually used when the evaporator temperature is above −30°F. The method is to spray the water over the coils of the evaporator unit. If the evaporator temperature is below −30°F, brine or an antifreeze solution should be sprayed over the coil instead of water. When brine is used, precautions must be taken to prevent corrosion of the coils. A water spray defrost system also requires precautions, so that the water does not spill over from the drain pan onto the floor and freeze. Care must be taken to drain the lines properly, so that there will be no residue of condensate water to freeze when the unit is again put into operation.

A number of manufacturers are producing cooling coils equipped with electric heating coils, either inserted into the copper tubing or placed in contact with each row of finned tubing in a forced convection system. Where an adequate supply of water is not readily available or the temperature of the available water is too low for satisfactory defrosting, an electrical defrosting system is frequently used.

The principle of hot gas defrosting is the use of the hot discharge refrigerant gas from the compressor as a medium for defrosting the cooling coil. A schematic of a basic hot gas defrost system is shown in Fig. 16-5. An auxiliary or bypass line is installed between the hot gas line and the refrigerant line entering the cooling coil after the metering device. The valve installed in this line can be operated either manually or by a solenoid. If hand operated, the defrosting takes place only when an operator is present. If it is operated by a solenoid, a timer is installed in the electrical circuit, and the system is automatically defrosted periodically at the selected times.

This system has some weak points, which make its use impractical if there is only one cooling coil in the system. Theoretically, its method of operation is that the compressor continues to run when the defrosting is accomplished. When the valve in the bypass line is opened, hot gas should flow from the discharge line out of the compressor to the evaporator. This hot gas going to the coil will cause the frost to melt off its exterior surface. But, when the refrigerant gas gives up its heat, it condenses into a liquid, and there

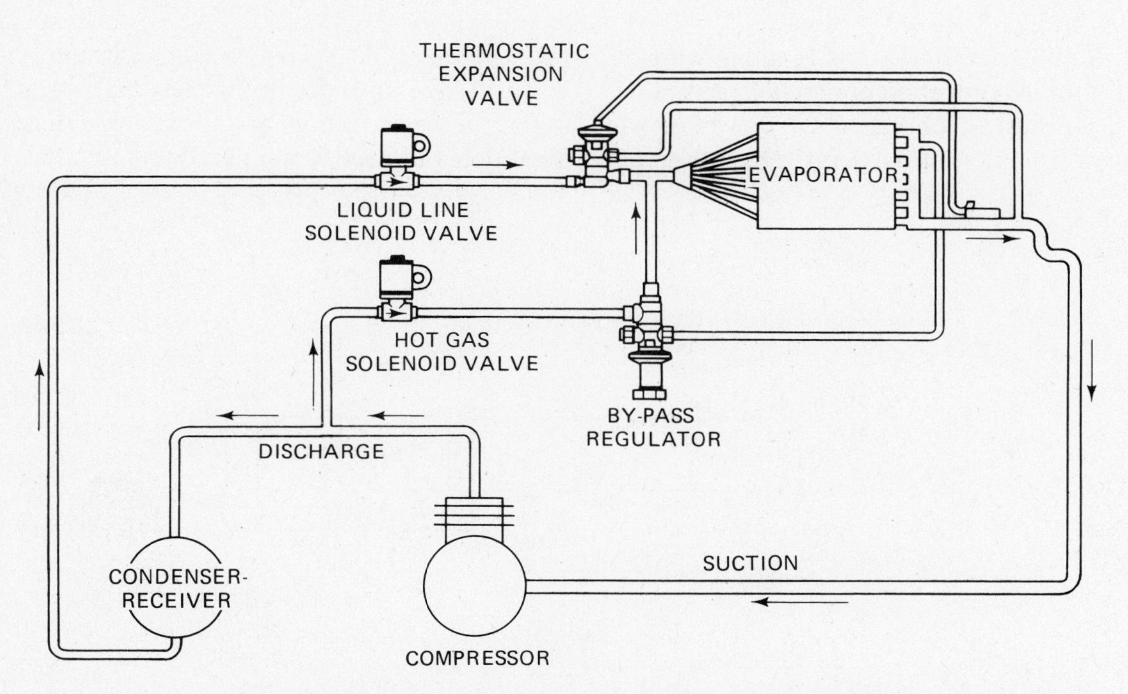

FIGURE 16-5 Typical schematic diagram of "hot gas" bypass from compressor discharge to evaporator inlet (Courtesy, ASHRAE).

is a possibility that a large slug of liquid refrigerant will flow down the suction line and damage the compressor.

Also, since no liquid enters the metering device and is vaporized in the evaporator while the system is on the defrost cycle, less refrigerant vapor is compressed by the compressor, and therefore less hot refrigerant vapor will be available for defrosting purposes.

Figure 16-6 shows a refrigerating system utilizing multiple evaporators in which two are illustrated. In this case the hot gas defrost method is feasible since, while one coil is being defrosted, the other continues to operate at its normal temperature. As mentioned previously, when the hot gas passes through the evaporator on the defrost cycle, it gives up its heat to melt the frost. In so doing, it condenses into a liquid state and—since only one evaporator coil is being used—it is quite possible that liquid slugs will damage the compressor.

OIL CIRCULATION

Oil travels in the refrigeration cycle, along with the refrigerant, for the lubrication of moving parts. Since the compressor must be lubricated, the flow of oil will be considered from that initial point. The refrigerant vapor comes in direct contact with the oil clinging to the cylinder walls as the pistons are lubricated. Some of this oil is carried along with the refrigerant vapor as it passes into the hot gas discharge line between the compressor and the condenser.

If only a small amount of oil travels along with the refrigerant, it will pass into the condenser and receiver and through the liquid line into the evaporator, returning to the compressor crankcase before that segment of the system is pumped short of oil.

But some compressors normally pump relatively large volumes of oil, and unless provisions are made for its speedy return to the crankcase, serious damage can occur to the compressor. A precautionary measure is the correct installation of refrigerant lines having the proper pitch to correctly-sized lines, with oil loops provided in the piping system where needed. These factors will be considered and elaborated on in Chap. 19 on refrigerant lines. Also included in Chap. 19 will be oil traps and oil separators, under the section dealing with accessories to the refrigerant system.

Small amounts of refrigeration oil will not be harmful to the cooling coil, but large amounts collecting in the circuits or passages of the evaporator will cause an increase in the cooling coil temperature Such an increase means less work will be done in the evaporator. There will be less total cooling if the suction pressure remains constant, so the entire system will be less efficient.

If the oil is allowed to remain in the evaporator, it will take up space in the coils that should be used for the vaporization of the refrigerant, and refrigeration will decrease in efficiency.

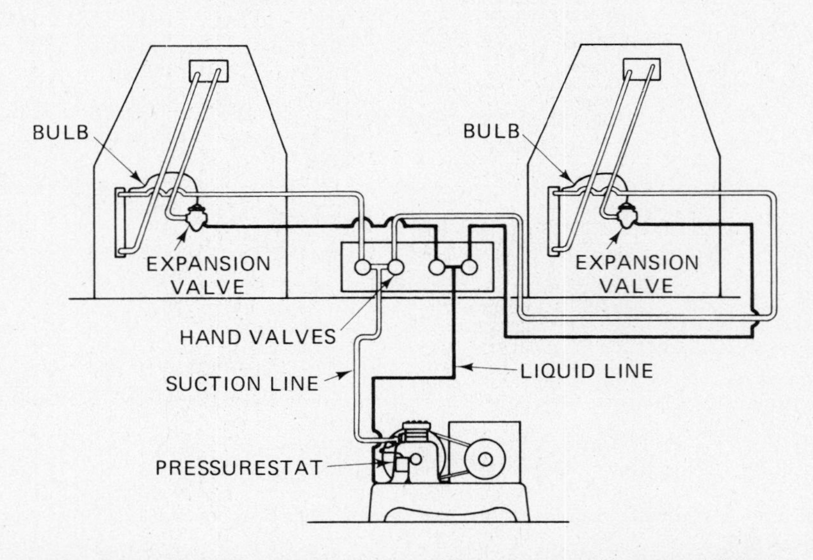

FIGURE 16-6 Refrigerating system with multiple evaporators (Courtesy, ASHRAE).

PROBLEMS

1. What are the major types of evaporators?
2. What is the advantage of a finned-tube evaporator over a bare-tube type?
3. Which of the three means of heat transfer—radiation, convection, or conduction—is the one by which heat is transferred to the refrigerant in an evaporator?
4. What is the purpose of installing baffles in a refrigerated space?
5. When is it desirable to use a fan or blower to provide forced circulation of air over or through an evaporator?
6. What is "refrigerating effect"?
7. Why must frost be removed when it collects on cooling coils?
8. What is the disadvantage of using a hot-gas defrost system in an application having but one cooling coil?
9. Why is it essential that refrigeration oil traveling with the refrigerant in a system be returned to the compressor crankcase?
10. What is the result of too much oil remaining in the evaporator?

17

COMPRESSORS

HISTORY

Early models of refrigeration compressors were largely the vertical, single-acting compressor typical of the ammonia machine shown in Fig. 17-1. Since ammonia was the most popular refrigerant in the old days, these compressors were very heavy in order to withstand very high pressures, and, in comparison to modern compressor designs, the early ones ran at relatively low speeds. Advances in valve design, compressor shaft seals, bearings, and lubrication systems resulted in a gradual increase in the design speed. This allowed the compressors to become smaller for a given horsepower (hp), since increased displacement was obtained from higher speed operation.

The introduction of new refrigerants also had considerable bearing on the designs and development of compressors. For example, when using ammonia, all those parts of the system exposed to refrigerant had to be constructed of steel. The introduction of sulphur dioxide and methyl chloride as refrigerants did make the use of nonferrous metals possible in some cases. The advent of the halogenated hydrocarbon refrigerants, however, had perhaps the greatest effect on compressor design. It became possible to use nonferrous metals such as aluminum. Simultaneously with the introduction of Refrigerant 12, the hermetically sealed type of compressor became popular.

The development of compressors for commercial refrigeration and air-conditioning applications has been influenced considerably by the use of compressors in household refrigerators. Hermetically sealed compressors and capillary tube refrigerant feed devices were first introduced and field-proven in household refrigerator applications. In the early thirties hermetically sealed compressors began to become the standard of the household refrigerator manufacturers. Within a few years, belt-driven compressors practically disappeared from the household refrigerator field. The manufacturers of ice cream cabinets, beverage coolers, water coolers, etc., were next to adopt hermetically sealed compressors.

In 1935 the first hermetic compressor for air-conditioning service was introduced and by the early 1940s most air-conditioning manufacturers had

FIGURE 17-1 Single acting ammonia compressor (Courtesy, York, Division of Borg-Warner Corporation).

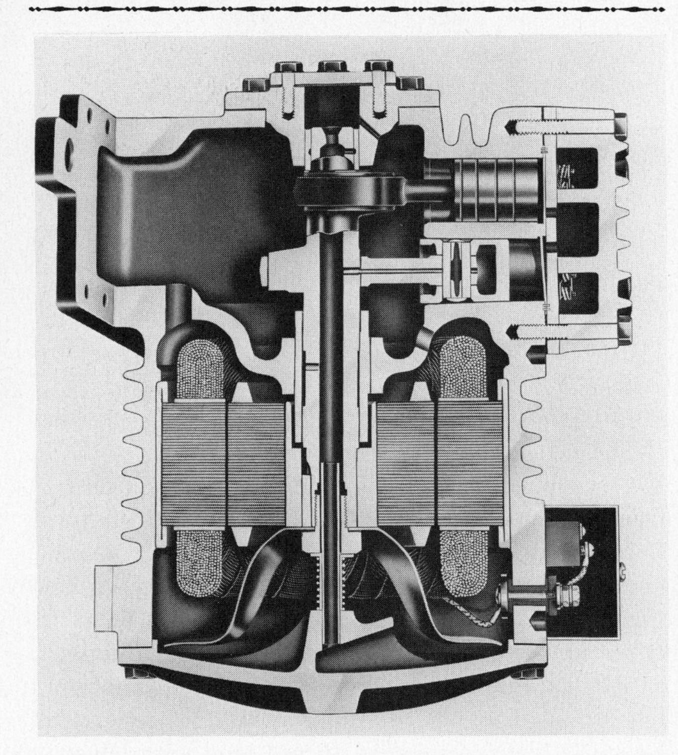

FIGURE 17-2 Reciprocating compressor (Courtesy, Copeland Corporation).

switched to the hermetic compressor for their products. The trend towards the use of the hermetic compressor for both commercial refrigeration and air conditioning has continued.

DESIGNS

The compressor is often called the heart of any refrigeration system. Chapter 6 noted the four designs of compressors in general terms; let us restate those as:

- Positive displacement compressors
  - — reciprocating
  - — rotary
  - — helical (screw)
- Kinetic compressors
  - — centrifugal

Positive displacement compressors such as illustrated in Fig. 17-2 are so classified because the maximum capacity is a function of the speed and volume of cylinder displacement. Since the speed is normally fixed (i.e., 1,750 or 3,500 rpm for typical hermetic reciprocating compressors), the volume or weight of gas (refrigerant) pumped becomes a mechanical ratio of strokes per minute times the cylinder(s) volume.

The kinetic compressor (centrifugal), shown in Fig. 17-3 and sometimes called the *turbo compressor*,

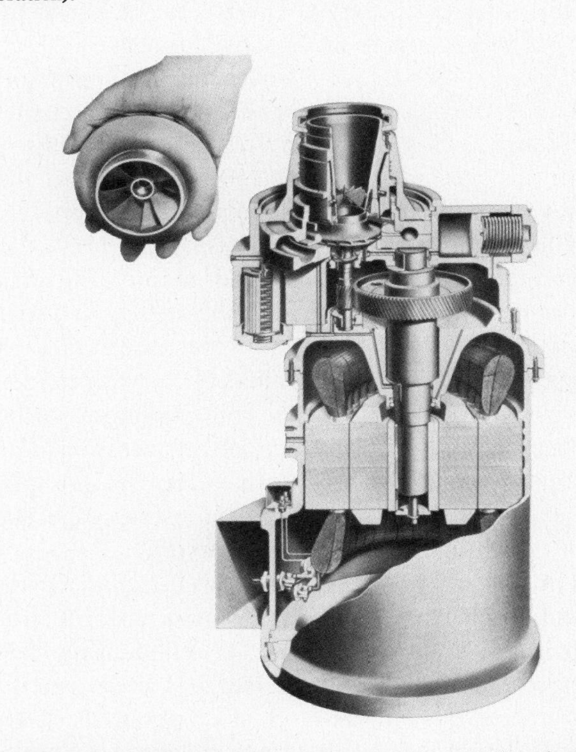

FIGURE 17-3 Centrifugal compressor (Courtesy, Westinghouse Electric Corporation).

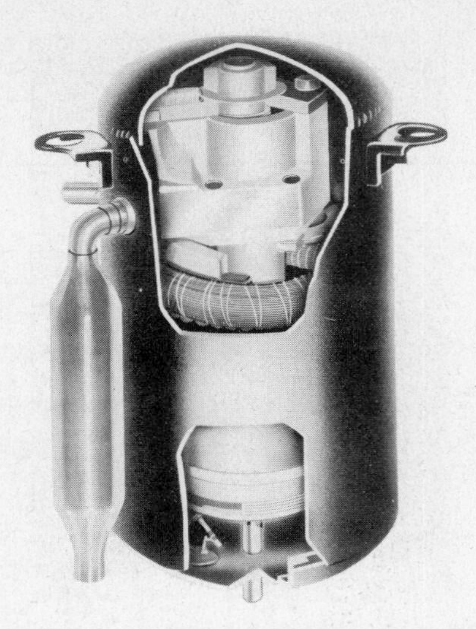

FIGURE 17-4 Rotary compressor (Courtesy, Fedders Corporation).

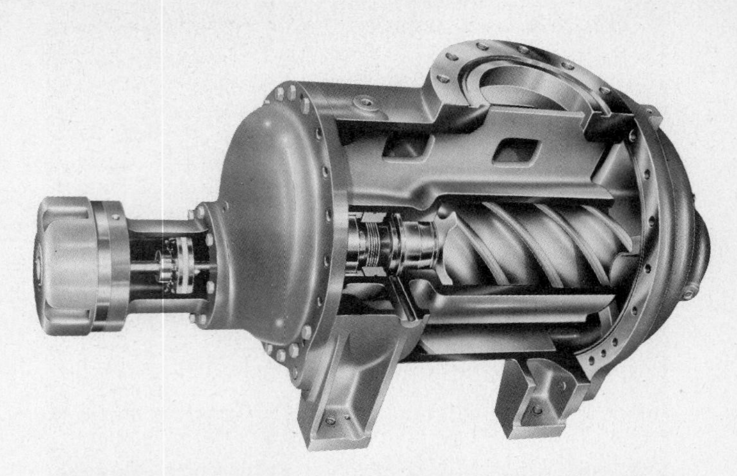

FIGURE 17-5 Screw compressor (Courtesy, York, Division of Borg-Warner Corporation).

is a member of a family of turbo machines including fans, propellers, and turbines, where the pumping force is subject to the speed of the impeller and the angular moment between the rotating impeller and the flowing fluid (refrigerant). Because their flows are continuous, turbo machines have greater volumetric capacities, size for size, than do positive displacement machines. However, at the present the design and cost of such compressors do not lend themselves to smaller applications (50 tons and less). Centrifugal machines currently start at the 80- to 100-ton range and extend upward to 8,000 tons and over.

Among the positive displacement compressors, the reciprocating compressors have gained the widest acceptance and application from fractional horse-power size up to the 100- to 150-ton range. At that point the crossover to centrifugals becomes apparent.

The rotary compressor (Fig. 17-4) as previously described in Chap. 6 was used mainly in small, fractional horse-power sizes associated with refrigerators, etc. More recently, however, the rotary has gained popularity in the nominal $1\frac{1}{2}$ to 5-ton residential air-conditioning field. The rotary compressor has not been adapted to commercial refrigeration duty, perhaps because it is usually inefficient in pumping against very high discharge pressures, especially when operating at low suction pressure.

The helical (screw) compressor (Fig. 17-5) is also a positive placement design and will perform satisfactorily over a wide range of condensing temperatures. The screw compressor has been used for refrigeration duty in the U. S. since about 1950. The original design was invented and patented in Sweden in the early 1930s.

Figure 17-6 demonstrates the compression cycle of a screw compressor:

(a) Gas is drawn in to fill the interlobe space between adjacent lobes.

(b) As the rotors rotate the interlobe space moves past the inlet port, which seals the interlobe space. Continued rotation progressively reduces the space occupied by the gas, causing compression.

(c) When the interlobe space becomes exposed to the outlet port, the gas is discharged.

Capacity control is achieved by internal gas recirculation, thus providing smooth capacity down to as low as 10% of design capacity. Current sizes of screw machines in operation range from 100 tons up to 700 tons, based on nominal A. R. I. conditions for chilled-water systems. Like the centrifugal equipment, screw machines are not currently being used in small tonnage refrigeration or air conditioning.

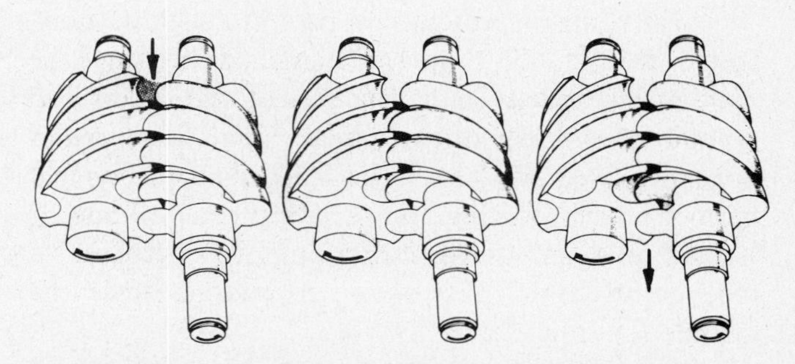

FIGURE 17-6 Screw compressor compression cycle (Courtesy, York, Division of Borg-Warner Corporation).

Thus, the reciprocating compressor is used in over 90% of those units ranging from fractional horsepower through 100 tons and therefore should be the starting point for any novice learning about compressors. The balance of this chapter will be devoted to this particular design.

TYPES OF RECIPROCATING COMPRESSORS

As previously mentioned, the distinction in reciprocating machines is between the open type of compressor and the hermetic. By *open type* we mean a compressor driven by an external motor (Fig. 17-7), either belt-driven or direct-connected. This type requires a shaft extending through the crankcase of the compressor and, thus, a shaft seal. As opposed to this type compressor is the hermetically sealed

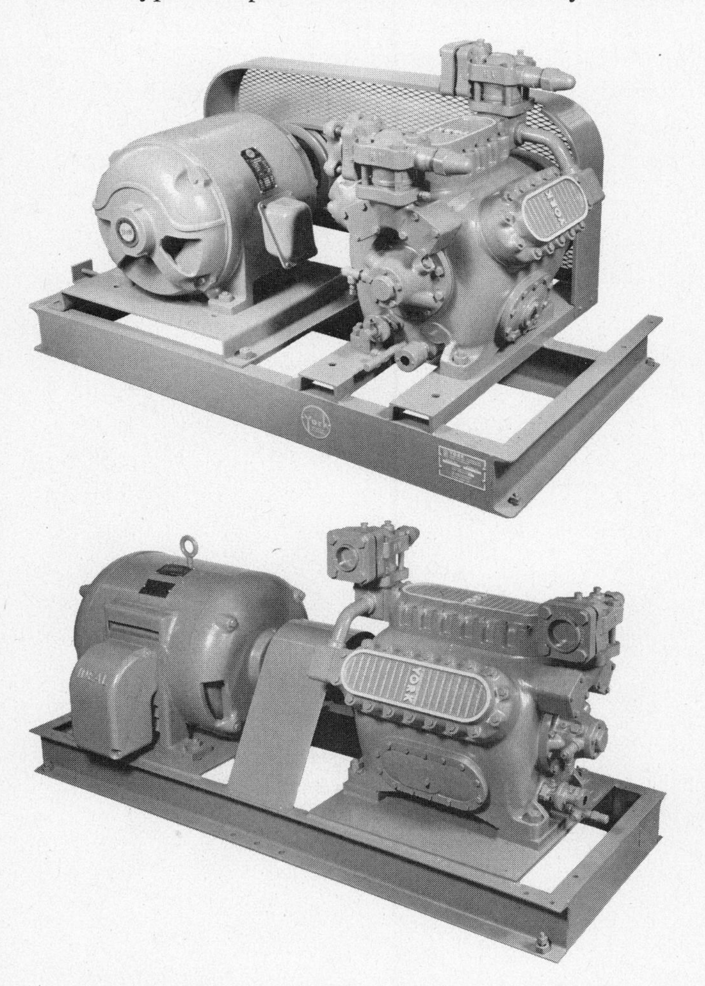

FIGURE 17-7 Open drive compressors (Courtesy, York, Division of Borg-Warner Corporation).

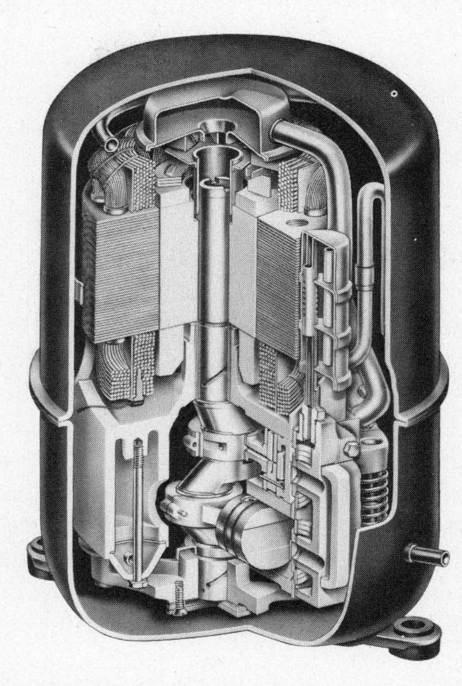

FIGURE 17-8 Hermetically sealed compressor (Courtesy, Tecumseh Products Company).

compressor (Fig. 17-8), in which the motor is incorporated in the same housing as the compressor. Thus the hermetically sealed compressor has no shaft extending through the crankcase and therefore requires no seal. Each type has some advantages over the other.

A belt-driven open-type compressor is quite flexible. Its speed may be varied so that a single compressor can often be used for two or three different hp size units. By merely changing the size of the motor pulley, in most instances this same compressor may be used not only with different sizes of motors but also for high, medium, or low-temperature applications. This is the most outstanding advantage of an open-type compressor as compared to a hermetic type. Other advantages are that this type of compressor may be used with motors available for odd voltages and frequencies, for which hermetic-type compressors are not produced. This has been an important factor in the overseas markets where voltages and phases differ from those in the U. S., for example, where 50-cycle (hertz) or direct current is the only available power. Hermetic compressors are not available in dc, but open motors are.

Open compressors are always field-serviceable, which is not true of all hermetics. In the event of a motor burnout, it is probably easier to replace the motor on an open-type system than on a hermetic system, where the motor is exposed to the refrigerant. In the event that a motor burns out in a hermetically sealed system, the entire compressor must be replaced and returned to a repair shop or factory to be dismantled and reconditioned. The refrigerant must be discharged and the system cleaned in order to eliminate any possibility that acid, which might have resulted from the burning of insulation, can circulate through the system.

There are instances however, where hermetically sealed compressors have distinct advantages compared to the open type. Perhaps the greatest of these is the elimination of the shaft seal. Shaft seals are vulnerable to dirt, temporary failure of lubrication, anything abrasive that might accumulate in the system (such as scale), and physical damage due to rough handling, etc. Although seals produced today are much improved over those of 15 to 20 years ago, they are still a potential source of trouble, especially on a low-temperature system where the low-side pressure may be at considerable vacuum. In such a case, a leak at the seal allows air and moisture vapor to enter the refrigerant system, which is more serious than the loss of refrigerant from the system.

Other advantages of the hermetic type of compressor are that it is smaller, more compact, more free of vibration, and has its motor continuously cooled and positively lubricated. There are no belts requiring frequent adjusting and eventual replacement.

To achieve its continuous cooling, the open compressor motor is cooled by the air surrounding the motor housing; if the motor is located in an area of high ambient temperature or poor ventilation, proper dissipation of heat may be a problem. In the hermetic design motor, heat is dissipated by passing cool suction gas through or around the stator winding to pick up the motor heat, and the refrigerant thus carries this heat to the condenser, where it is dissipated. Another advantage in this process is that suction gas is further superheated to help prevent any liquid refrigerant from entering the suction openings, and this is important in low-temperature work to dry the incoming gas.

This discussion of the relative merits of open and hermetic compressors may lead the reader to assume that the open compressors are still widely used in the industry. They actually represent a small part of the total number sold, and they are concentrated in the area of commercial refrigeration or process cooling, where refrigerant temperatures and conditions are different and more varied than found in the comfort air-conditioning field. There are a great number of open machines still operating, and in the course of service and maintenance it's quite possible for a technician to encounter one in need of repair or replacement, so it is important to understand the background of the open-type compressor.

HERMETIC FIELD-SERVICEABLE VERSUS TOTALLY SEALED

Most early hermetic compressors were of the field-serviceable design as illustrated in Fig. 17-9. Frequently called the *bolted hermetic*, it can be almost totally torn down on the job and refitted with new parts, a characteristic that is a distinct advantage in

FIGURE 17-9 Bolted hermetic compressor (Courtesy, Copeland Corporation).

FIGURE 17-10 Welded hermetic compressor (Courtesy, Tecumseh Products Company).

larger tonnage machines where sheer weight involved in removing a complete unit makes such actions physically and economically undesirable.

On the other hand the completely sealed hermetic, or *welded shell hermetic* (Fig. 17-10) is not field serviceable and, whatever the internal problem, be it motor failure, valve breakage or anything else, the unit must be returned to a repair station or factory. An exchange compressor is then installed.

The range of welded hermetics starts in fractional horsepower sizes and generally goes through the nominal $7\frac{1}{2}$-ton sizes; however, there are compressors on the market of up to 20 tons in a single shell.

Does it make sense to replace a complete compressor? The answer is "yes," and it's not necessarily a technical decision. When the growth explosion in residential air-conditioning came about in the late 1950s it became apparent there could never be enough skilled technicians in the industry to handle the field repair of millions of compressor installations. Nor did it make sense for manufacturers to stock locally all the parts needed to make field repairs; the financial impact would have been staggering.

Mass production and standardization of compressors brought improved quality to the industry. At the same time there were also technical changes in system design from field-fabricated refrigerant lines to the precharged variety. System reliability and compressor life expectancy improved rapidly to the point where failure rates represented a very small percentage, and economically it made sense to replace complete compressors in sizes up to the nominal $7\frac{1}{2}$-ton capacity. Between $1\frac{1}{2}$ and 5 tons, which is the bulk of the residential air-conditioning market and many commercial refrigeration applications, it is apparent the *welded hermetic* in reciprocating and rotary will dominate the market. Therefore, from a service technician's point of view it is not totally necessary to be able to tear down and reassemble a compressor. It is more important that he should concentrate on techniques of proper application, installation, and trouble-shooting, so as to minimize failures and maintenance requirements.

The mechanic who becomes involved in heavier commercial work and larger air-conditioning units that do require field repair can advance his knowledge as the situation requires. At this stage, the ability to tear down and reassemble a compressor teardown becomes more of an art than a science, and interpreting the manufacturers' procedures and design is important.

Chapter 6 detailed the internal working parts of a typical reciprocating compressor as well as types of motors and associated starting equipment used with this compressor. However, several factors affecting the operation and life expectancy of compressors weren't covered.

Nearly all motors used for refrigeration applications are *induction motors*, so named because the current in the moving part of the motor is induced, since the moving component has no connection to the source of current. The stationary part of an induction motor is called the *stator*, and the moving part, the *rotor*. The stator windings are connected to the power source, while the rotor is mounted on the motor shaft, the rotation of the rotor providing the motor with its driving power source.

MOTOR TEMPERATURE

The first law of thermodynamics states that energy can be neither created or destroyed, but may be converted from one form into another. The motor receives electrical energy from the power source, but because of friction not all of the energy can be turned into mechanical output energy. The balance of the input energy is converted to heat energy, and unless this heat is dissipated, the temperature in the motor windings will rise until the insulation is destroyed. If a motor is kept free from contamination and physical damage, heat is practically the only enemy that can damage the windings.

The amount of heat produced in the motor depends both on the load and on motor efficiency. As the load is increased, the electrical input to the motor increases. The percent of the power input converted to heat in the motor depends on motor efficiency, the heat decreasing with an increase in efficiency, and increasing as motor efficiency decreases.

The temperature level a motor can tolerate depends largely on the type of motor insulation and the basic motor design, but the actual motor life is determined by the conditions to which it is subjected during use. If operated in a proper environment, at loads within its design capabilities, a well-designed motor should have an indefinite life. Continuous overloading of a motor resulting in consistently high operating temperatures will materially shorten its life.

Since heat is the worst enemy of motors, the hermetic compressor enjoys a great advantage by utilizing suction gas to dissipate it effectively. By designing a hermetic motor for a specific application and controlling the motor temperature closely, a motor may be matched to a given load, and the motor output can be at its maximum capacity while maintaining a generous safety factor considerably above that available with standard open-type motors.

NAMEPLATE AMPERAGE

On open-type standard motors, NEMA (National Electrical Manufacturers Association) horsepower ratings are used to identify a motor's power output capability. Because of industry practice, this nominal horsepower classification has carried over into hermetic motor identification, but it may be misleading when applied to this type of motor. With controlled cooling and motor protection sized for the exact load, a hermetic motor may be operated much closer to its maximum ability, so a particular motor may be capable of much greater power output as a hermetic motor than an equivalent open motor. The amperage draw and watts of power required are the best indicators of hermetic motor operation.

Most industry hermetic compressors carry on their nameplates ratings for both locked-rotor and full-load amperes. The designation *full load amperage* persists because of long industry precedent, but in reality a much better term is *nameplate amperage*. On most industry compressors with inherent protection or internal thermostats, nameplate amperage has been arbitarily established as 80% of the current drawn when the motor protector trips. The 80% figure is derived from standard industry practice in sizing motor protection devices at 125% of the current drawn at rated load conditions.

This does not mean that every hermetic compressor may be operated continuously at a load greatly in excess of its nameplate rating without fear of failure. Motor amperage is only one factor in determining a compressor's operating limitations. Discharge pressure and temperatures, motor cooling, and torque requirements are equally critical. Safe operating limits have been established by manufacturers for each compressor and are published on compressor or unit specifications.

VOLTAGE AND FREQUENCY

Although electricity distributed in the United States is 60 cycles (hertz), distribution voltages are not standardized. Single phase voltages may be 115, 208, 220, 230, or 240, and most utilities reserve the right to vary the supply voltage plus or minus 10% from the nominal rating. Three-phase voltage may be 208, 220, 240, 440, 460, and 480, again plus or minus 10%. Unless motors are specifically designed for the voltage range in which they are operated, overheating may result.

Most open motors, whether used on refrigeration compressor, fans, etc., may be operated at the voltage on the motor nameplate, plus or minus 10%, without danger of overheating; however, this is not the case on single-phase hermetic compressor motors, where the windings have been more critically sized to the exact load and operating conditions. A case in point is the PSC (permanent split capacitor) motor, which does not have a starting capacitor and does have a relatively low starting torque. The PSC motor is used where it is assumed refrigerant pressures will equalize between the off and on cycles. Low-voltage starting of PSC motors can be very detrimental. And it's common practice to design hermetic compressors with only a 5% undervoltage and 10% overvoltage rating. So, when the nameplate states 208–230 dual voltage, the minimum operating voltage permitted is 5% off 208 or 197 volts. On the high side the maximum is 230 plus 10% or 253 voltage. Much too frequently electrical contractors who do not understand refrigeration compressors assume the 10% plus and minus rating. Should the actual voltage be a 208 volt supply and the power company drops the voltage 10% to 187 volts, the hermetic compressor will be in trouble. The point here is to make sure what the power company supply voltage is before installing the compressor. Three-phase compressors are not as critical in this respect as single-phase, because each winding is 120 degrees out of phase with the other windings, which results in a motor with a very high starting torque.

In many parts of the world, the electrical power supply is 50 cycle (hertz) rather than 60 cycle (hertz). If both the voltage and frequency supplied to a motor vary at the same rate, the operation (within narrow limits) of a given motor at the lower frequency condition is all right, in some cases. For example, a 440-volt, three-phase, 60-Hz motor will operate satisfactorily on 380-volt, three-phase, 50-Hz power. In the export of domestic compressors to foreign countries it is common practice for the manufacturer to approve standard 60-Hz machines for 50-Hz application, but with reduced capacity. In some cases, specially wound 50-Hz motors are required to meet local codes or specifications.

COMPRESSOR WIRE SIZES

Low or under voltage conditions are not always because the power company permits variations from standard rating. Too frequently the power source wiring to the compressor, condensing unit, etc., may be improperly sized, so that line losses exceed normal limits and the terminal voltage to the compressor is too low. This situation is further complicated by the use of copper and aluminum conductors, which have different rating characteristics. Most manufacturers list recommended wire sizes to the unit and the maximum length between the unit and main power source so as not to exceed a 3% voltage drop. Should excessive line loss be coupled with minimum power supply voltage, the resulting impact on the compressor can be significant. The point here is to ensure that proper electrical data is furnished to the electric installer—if wiring is not a part of the refrigeration installation contract.

REDUCED VOLTAGE STARTING

Previous discussions introduced common starters such as starting relays, capacitors, contactors, and across-the-line starters. These devices are meant to be used with full voltage, which is the least expensive method of starting a compressor designed for full voltage starts. However, because of some power company limitations on starting current, a means of reducing the inrush starting current on larger horsepower motors is occasionally necessary (particularly in other countries and also in some sections of the United States and Canada), in order to prevent light flicker, television interference and undesirable side effects on the equipment because of the momentary voltage dip. The reduced voltage start allows the power company voltage regulator to pick up the line voltage after part of the load is imposed, and thus avoids the sharper voltage dip that would occur if the whole load were thrown across the line. Some electrical utilities may limit the inrush current drawn from their lines to a given amount for a specified period of time. Others may limit the current drawn on startup to a given percent of locked rotor current.

Unloading the compressor can be helpful in reducing the starting and pullup torque requirement and will enable the motor to accelerate quickly. (Unloading is covered later in this chapter). Whether the compressor is loaded or unloaded, however, the motor will still draw full starting amperage for a small fraction of a second. Since the principal objection is usually to the momentary inrush current drawn under locked rotor conditions when starting, unloading the compressor will not always solve the problem. In such cases some type of starting arrangement is necessary that will reduce the starting current requirement of the motor.

Starters that accomplish this are commonly known as *reduced voltage starters*, although in two of the most common methods the line voltage to the motor is not actually reduced. Since manual starting is not feasible for refrigeration compressors, only magnetic starters will be considered.

There are five types of magnetic reduced voltage starters, each of which is suited for specific applications:

- part winding
- Star-Delta
- autotransformer
- primary resistor
- reduced voltage step-starting accessory

As the starting current is decreased, the starting torque also drops, and the selection of the proper starter may be limited by the compressor torque requirement. The maximum torque available with reduced voltage starting is 64% of full voltage torque, which can be obtained with an autotransformer starter, 45% for part winding, and 33% for Star-Delta—which means that for Star-Delta an unloaded start is essential if the compressor is to start under reduced voltage.

COMPRESSOR PERFORMANCE

The performance of a machine is an evaluation of the ability of the machine to accomplish its assigned task. Compressor performance is the result of design compromises involving certain physical limitations of the refrigerant, compressor, and motor, while attempting to provide:

(1) the gretaest trouble-free life expectancy

(2) the most refrigeration effect for the least power input

(3) the lowest cost

(4) a wide range of operating conditions

(5) a suitable vibration and sound level

Two useful measures of compressor performance are capacity, which is related to compressor displacement, and performance factor.

System capacity is the refrigeration effect that can be achieved by a compressor. It is equal to the difference in total enthalpy between the refrigerant liquid at a temperature corresponding to the pressure of the vapor leaving the compressor and the refrigerant vapor entering the compressor. It is measured in Btu/lb.

The *performance factor* for a hermetic compressor indicates combined operating efficiency of motor and compressor:

$$\text{Performance factor (hermetic)} = \frac{\text{Capacity, in Btu/h}}{\text{Power input, in watts}} = \text{Btu/watt}$$

Recently the performance factor has become important to the industry because of the spotlight on energy conservation. It is now termed *EER* (*Energy Efficiency Ratio*), and the actual performance of refrigeration and air-conditioning units is being certified and listed in ARI directories so that users, specifiers, installers, and power companies can evaluate the relative efficiency of various machines.

There are three other definitions and measurements of performance for a compressor that are used primarily by compressor design engineers and are generally not of practical use to the refrigeration technician; however, it is well to be familiar with them.

- ***Compressor Efficiency*** considers only what occurs within the cylinder. It is a measure of the deviation of the actual compression from the perfect compression cycle and is defined as work done within the cylinder.
- ***Volumetric Efficiency*** is defined as the volume of fresh vapor entering the cylinder per stroke divided by the piston displacement.
- ***Actual Capacity*** is a function of the ideal capacity and the overall volumetric efficiency.
- ***Brake Horsepower*** is a function of the power input to the ideal compressor and to the compression, mechanical, and volumetric efficiencies of the compressor.

This text and the functions of a refrigeration technician are more concerned with real or actual capacity and power input of the compressor or condensing unit over a defined range of operating conditions.

Manufacturers of compressors perform exhaustive tests (Fig. 17-11) on their compressors for ratings, which must either be in accordance with ASHRAE and/or ARI conditions. There are two types of tests for compressors. The first determines capacity, efficiency, sound level, motor temperature, etc. The second, equally necessary, determines the probable life expectancy of the machine. Life-expectancy testing must be conducted under conditions simulating those under which the compressor must operate for years. Safety and adherence to codes are major factors in all this work.

From this information the manufacturer can present or publish the performance and application data needed to use the product properly.

Capacity ratings are published in either tabular form or curves which include:

(1) compressor identification—the number of cylinders, bore and stroke, etc.

(2) Degrees of subcooling, or a statement that data has been corrected to zero degrees subcooling

(3) compressor speed

(4) type of refrigerant

(5) suction gas superheat

(6) compressor ambient

(7) external cooling requirements (if necessary)

(8) Maximum power or maximum operating conditions

(9) Minimum operating conditions under full load and unloaded operation.

Figure 17-12 represents a typical capacity and power input curve for a hermetic reciprocating compressor. First note the stated facts: Refrigerant 22, 10° liquid subcooling, 20° superheated vapor, and 1,750 rpm compressor speed. Capacity is shown on the left vertical axis and stated in Btu/h. Power input in kw is shown on the right vertical axis. On the bottom is a range of evaporating temperatures. Condensing temperatures are plotted on the diagonal curves. (Note this is the refrigerant condensing temperature, not to be confused with air- or water-cooled condenser nomenclature. The compressor doesn't know what kind of condenser is being used—it only knows what condensing temperature and pressures it must produce.

To determine the refrigerating capacity, let us assume a 25°F evaporator temperature and 115°F condensing temperature. Read up from 25° on the bottom to the 115° capacity curve to point *A*, then across to the left; the final reading is approximately 105,000 Btu/h.

To determine the power input, read up from 25°F to the power curves at the intersection of 115°F and point *B*, then across to the right. The reading is approximately 11.5 kw power input.

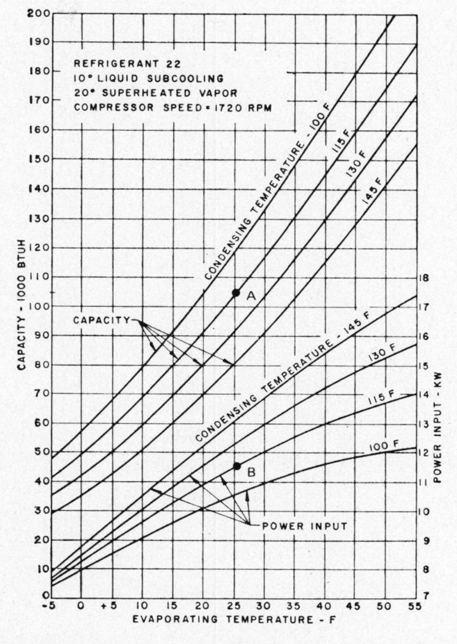

FIGURE 17-12 Typical capacity and power curves for a hermetic reciprocating compressor (Courtesy, ASHRAE).

FIGURE 17-11 Compressor test room (Courtesy, Carrier Air-Conditioning Company).

UNIT MODEL		H32SM-22E			H61SN-22E H61SP-22E			H62SP-22E H62SQ-22E			H92SQ-22E H92SR-22E H92SS-22E		
SAT. DISCH. TEMP. AND PRESSURE	SATURATED SUCTION TEMP. F	TONS	KW	HEAT REJ. MBH	TONS	KW	HEAT REJ. MBH	TONS	KW	HEAT REJ. MBH	TONS	KW	HEAT REJ. MBH
110 F (226.4 PSIG)	15	8.0	11.8	136	13.3	19.3	244	16.0	23.0	269	24.0	34.8	405
	20	9.1	12.4	151	15.2	20.3	251	18.3	24.2	301	27.4	36.8	452
	25	10.4	13.0	168	17.3	21.3	279	20.7	25.4	334	31.1	38.5	503
	30	11.7	13.5	186	19.5	22.3	309	23.4	26.5	370	35.1	40.0	556
	35	13.1	13.9	204	21.9	23.0	340	26.2	27.4	406	39.4	41.3	612
	40	14.6	14.2	223	24.4	23.5	372	29.3	27.9	445	43.9	42.0	668
	45	16.3	14.4	244	27.2	23.8	406	32.6	28.4	487	48.9	42.7	730
	50	18.1	14.6	266	30.2	24.1	443	36.2	28.7	531	54.3	43.1 R	796
115 F (242.7 PSIG)	20	8.7	12.7	147	14.5	20.8	244	17.4	24.8	292	26.2	37.6	440
	25	9.9	13.3	163	16.5	21.9	272	19.8	26.1	325	29.8	39.4	490
	30	11.2	13.9	181	18.7	23.0	320	22.4	27.3	361	33.6	41.2	542
	35	12.6	14.4	200	21.0	23.8	332	25.2	28.4	398	37.8	42.7	597
	40	14.1	14.8	219	23.5	24.5	364	28.2	29.2	437	42.3	43.9	655
	45	15.7	15.1	239	26.2	25.0	398	31.5	29.9	478	47.2	44.8	717
	50	17.5	15.3	261	29.1	25.4	434	35.0	29.8	520	52.5	45.4	783
120 F (259.9 PSIG)	20	8.3	12.9	143	13.8	21.2	237	16.6	25.2	284	24.9	37.9	426
	25	9.5	13.6	160	15.8	22.5	265	18.9	26.7	317	28.4	40.0	475
	30	10.7	14.3	176	17.9	23.6	294	21.5	28.1	352	32.2	41.9	527
	35	12.1	14.9	195	20.2	24.6	325	24.2	29.3	389	36.3	43.7	582
	40	13.5	15.3	213	22.5	25.4	355	27.1	29.9	426	40.6	45.1	639
	45	15.1	15.8	234	25.2	25.6	388	30.3	30.7	467	45.4	46.4	701
	50	16.7	16.1	254	28.1	26.1	425	33.7	31.2	510	50.6	47.2	766
130 F (296.8 PSIG)	30	9.7	14.9	166	16.2	24.2	276	19.5	Q 2[illegible].0	331	29.2	43.8	498
	35	11.0	15.6	184	18.4	P 2[illegible].4	306	22.1	30.4	367	33.1	S 4[illegible].9	551
	40	12.4	16.3	204	20.6	26.5	336	24.8	31.7	404	37.1	47.8	606
	45	13.9	16.9	224	23.2	27.6	371	27.8	32.9	444	41.7	49.7	667
	50	—	—	—	25.9	28.5	407	31.1	34.0	487	16.6	51.3	732
135 F (316.6 PSIG)	30	9.3	15.1	162	15.4	24.5	267	18.5	29.4	321	27.8	44.4	483
	35	10.5	16.0	180	17.5	26.0	297	21.1	31.7	360	31.6	47.0	537
	40	11.9	16.7	199	19.8	27.2	329	23.7	32.5	394	35.6	50.0	595
	45	—	—	—	22.2	28.5	362	26.6	33.9	443	40.0	51.1	652
	50	—	—	—	24.8	29.6	397	29.8	35.3	476	44.7	53.1	715

IMPORTANT: LARGER-SIZE MOTORS ARE REQUIRED FOR CERTAIN OPERATING TEMPERATURES AS INDICATED BY THE SHADED MOTOR IDENTIFICATION LETTERS P, Q, R OR S. IF FUTURE OPERATION AT THESE TEMPERATURES IS ANTICIPATED, ORDER THE COMPRESSOR WITH THE LARGE MOTOR. ALL RATINGS ARE APPLICABLE TO UNITS WITH THE LARGER MOTORS.

FIGURE 17-13 Capacity ratings for compressor units—Refrigerant-22; 2 5/8″ bore hermetic single stage compressor and condensing units.

Observe at a constant condensing temperature the rapid fall-off of capacity with lowering evaporator temperatures, which of course is due to the lower density of the gas being pumped by a constant displacement machine. Note, however, the power input curves do not fall as rapidly, demonstrating the high levels of work required to raise lower pressure vapors to proper condensing pressures. Thus the relative requirements for commercial refrigeration and air-conditioning are quite different.

Obviously it is not practical, or perhaps even technically possible, to use this particular compressor over a wide range of conditions. Thus the industry offers units with variations in speed (belt-driven models), or bore sizes, or length of stroke, and/or larger motors to meet specific ranges of application.

The manufacturers will offer capacity rating tables similar to Fig. 17-13. This example is a hermetic single-stage compressor with $2\frac{5}{8}$-in. bore used on R-22 refrigerant. Assume the required load is 26.5 tons at 115° discharge (condensing temperature) and a 40° saturated suction temperature (evaporator temperature less refrigerant piping losses). Note that model H61 can only produce 23.5 tons. The H61 (from the manufacturer's data) is a six-cylinder compressor with a short stroke. The H62 model, which is also a six-cylinder model but with a longer stroke, at the same conditions can produce 28.2 tons, *but* the higher power input requires a motor change (see the footnote). Note shaded arrow marked *Q*.

The heat rejection column specifies the amount of Btu/h the condenser must handle (437,000), be it water-cooled, air-cooled, or evaporative. At this point the compressor doesn't know what condenser will be utilized; it only knows it is pumping against a 242.7 psig condensing pressure corresponding to 115°F temperature for R-22.

Remember that these curves and tables reflect only compressor performance. Future discussions deal with the condenser options and will illustrate curves and/or tables for water, air, and evaporative condensing mediums.

COMPRESSOR BALANCING

Commercially available components will seldom exactly match the design requirements of a given system, and, since system design is normally based on estimated peak loads, the system may often have to operate at other than design conditions. More than one combination of components may meet the performance requirements. The efficiency of the system usually depends on the point at which the system reaches stabilized conditions or balances under operating conditions. The capacities of the three major system components, the compressor, the evaporator, and the condenser, are each variable but interrelated. Because of the many variables involved, the calculation of system balance points is extremely complicated. A simple, accurate, and convenient method of forecasting system performance from readily available manufacturers' catalog data can be graphically displayed in a component balance chart similar to that in Fig. 17-14, which is for an air-cooled condenser system. Evaporator capacity curves from the coil manufacturer are based on entering air temperatures with variable evaporating temperatures. Similar heavy lines can be drawn for balance conditions for the condenser-compressor balance points at various ambient temperatures. Assume the evaporator loading is 55,000 Btu/h at 40° entering air; this requires a 26° evaporating temperature or a 14° temperature difference between entering air and the refrigerant evaporating temperature (point *A*). Also assume the compressor/condenser is operating at 100°F outdoor ambient temperature; at the same 26° evaporating temperature its capacity would be approximately 59,000 Btu/h (point *B*). So there is a difference of 4,000 Btu/h. If the outdoor ambient temperature remains constant at 100°F and the evaporator load also remains constant, the result will be a depression in evaporating temperature and thus an increase in evaporator TD and capacity, while the compressor capacity falls correspondingly until the system reaches stabilized conditions at about 57,000 Btu/h and 24° evaporator temperature (point *C*).

This example ignores refrigerant line losses and other influences but does provide a good illustration nonetheless. The point is that *if* the original equipment selected requires a major shift in balance points, the selection is not a good one and must be reexamined. A severe change in evaporating coil temperatures can measurably affect the coil's latent and sensible capacity and frosting and defrosting relationship to a point where the food or products may be affected by over or undercooling, or too much humidity or too much dehydration.

The example also assumes a constant evaporator load, which seldom happens, and as the food, produce, or whatever is cooled the load drops and with it the balance point will also drop. So, where wide variations in load are anticipated it is necessary to provide some kind of capacity modulation of the system.

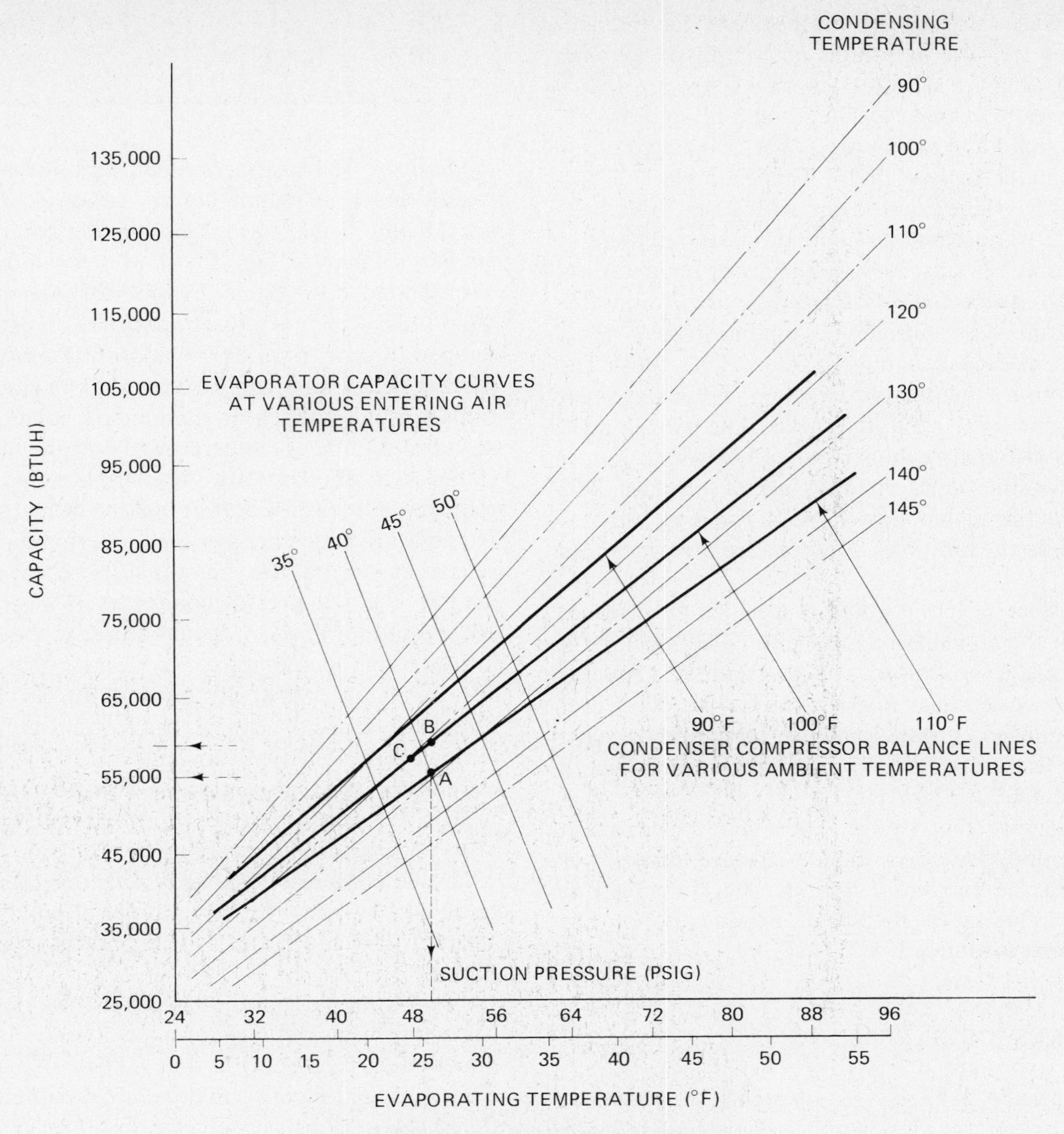

FIGURE 17-14 System balancing chart.

CAPACITY CONTROL

In order to provide a means for changing compressor capacity under fluctuating load conditions, larger compressors are frequently equipped with compressor unloaders. Unloaders on reciprocating compressors are of two general types. In the first (Fig. 17-15), suction valves on one or more cylinders are held open by some mechanical means in response to a pressure-control device. With the refrigerant suction valves open, refrigerant is forced back into the suction chamber during the compressor stroke, and the cylinder performs no pumping action.

A second means of unloading is to bypass internally a portion of the discharge gas into the compressor suction chamber. Care must be taken to avoid excessive discharge temperature when this is done.

A hot-gas bypass may also be achieved outside the compressor (Fig. 17-16). The solenoid in the bypass line can be controlled by temperature or pressure, depending on the nature of the application. As the controller calls for capacity reduction, the

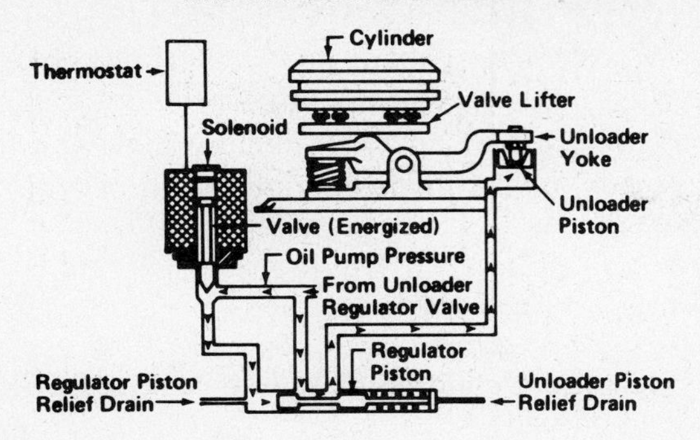

FIGURE 17-15 Electric unloader mechanism (Courtesy, Airtemp Corporation).

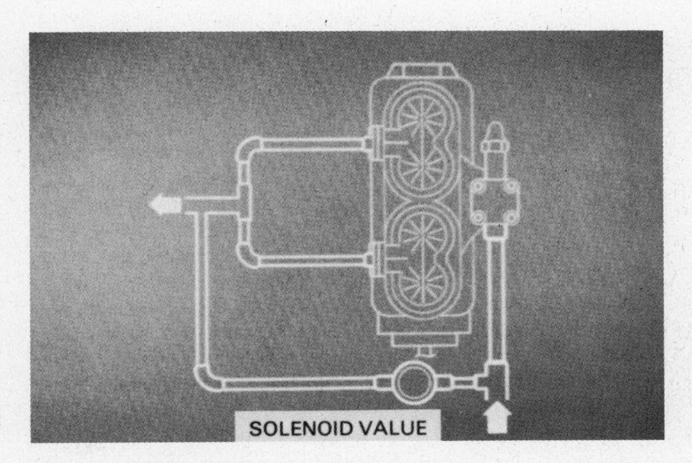

FIGURE 17-16 Hot gas bypass (Courtesy, Carrier Air-Conditioning Company).

solenoid opens, allowing some hot gas to go directly to the suction line.

Specific techniques recommended by compressor manufacturers employ different hardware, but the end result is essentially the same. The stages of unloading naturally depend on the size of machine, the number of cylinders, and the application requirements. *One point that must be considered is the decreased amount of suction vapor and entrained oil returning from the system. In hermetic compressors those quantities must be sufficient for proper lubrication and prevention of motor overheating.*

TWO-STAGE COMPRESSORS

The discussion on reciprocating compressors so far has concentrated on single-stage units, but because of the high compression ratio encountered in ultralow temperature applications, two-stage compressors (Fig. 17-17) have been developed for increased efficiency when evaporating temperatures are in the $-30°F$ to $-80°F$ range. Two-stage compressors are divided internally in low (or first) and high (or second) stages. Suction gas enters the low-stage cyclinders directly from the suction line and is discharged and metered into the interstage manifold so that it can provide adequate motor cooling and prevent excessive temperatures. The superheated refrigerant vapor at interstage pressure enters the suction ports of the high-stage cylinder and is then discharged to the condenser at the condensing pressure.

FIGURE 17-17 Two-stage compressor (Courtesy, Copeland Corporation).

COMPRESSOR MAINTENANCE

Long life is, of course, desirable in any product. Compressors built today are expected to provide many years of constant trouble-free and quiet operation. In many applications the compressors are called upon to run 24 hours a day, 365 days per year. Such continuous operation, however, is often not as hard on a compressor as is a cycling operation, where temperatures constantly change and the oil is not maintained at a constant viscosity.

The compressor must not only be designed to withstand normal operating conditions but also, at times, some periodic abnormal conditions such as liquid slugging, excessive discharge pressure, etc. The industry's compressor manufacturers have done an creditable job in designing and producing machines that take extra punishment, *but* no manufacturer has ever shipped a complete system, and this is where the professional refrigeration technician plays such an important roll, since most compressor failures stem from system faults and not from operating fatigue. The degree of technical skill and plain old common sense technicians use to install, operate, and maintain the equipment will ultimately determine the actual life expectancy of a system, particularly that of the compressor. Although the technician may not actually select the equipment, he can call attention to deficiencies before they become serious problems.

Subsequent chapters will deal with the installation, startup, and servicing of the complete system, but it will be helpful at this point to review some of the factors that shorten the life of a compressor, reduce its efficiency, or create undesirable conditions, factors that are related to the reciprocating design. The extent to which a unit can be corrected on the job depends on the particular compressor (whether it is serviceable or sealed).

Loss of Efficiency

This can result from a number of things:

- If liquid refrigerant enters the compressor, the efficiency and resulting capacity will be seriously affected. Besides physical damage, a shortage in capacity can also result from liquid slugging.
- Leaking discharge valves reduce pumping efficiency and cause the crankcase pressure to rise rapidly.
- Leaking suction valves seriously affect compressor efficiency (and capacity) especially of lower temperature applications.
- Loose pistons cause excessive blow-by and lack of compression.
- Worn bearings—especially loose connecting rods or wrist pins—prevent the pistons from coming up as far as they should on the compression stroke. This has the effect of increasing the clearance volume and results in excessive reexpansion.
- Belt slippage on belt-driven units.

Motor Overloading

When a compressor is not performing satisfactorily, the motor load sometimes provides a clue to the trouble. Either an exceptionally high or exceptionally low motor load is an indication of improper operation.

- Mechanical problems such as loose pistons, improper suction valve operation, or excessive clearance volume usually lead to a reduction in motor load.
- Another common problem is a restricted suction chamber or inlet screen (caused by system contaminants). The result is a much lower actual pressure in the cylinders at the end of the suction stroke than the pressure in the suction line as registered on the suction gauge. If so, an abnormally low motor load will also result.
- Improper discharge valve operation, partially restricted ports in the valve plate (which do not show up on the discharge pressure gauge), and tight pistons will usually be accompanied by high motor load.
- Abnormally high suction temperatures created by an excess load or other problem will cause a high motor load.
- Abnormally high condensing temperatures, created by problems associated with the condenser, will also lead to a high motor load.
- Low voltage at the compressor, whether the source is the power supply or excessive line loss, will contribute to high motor loading, particularly if any of the other problems are also present.

Noisy Operation

This condition usually indicates something is wrong. There might be some abnormal condition outside the compressor or something defective or badly worn in the compressor itself. Obviously, if it is due to some cause outside the compressor, nothing is gained by changing the compressor. Therefore, before changing a compressor one should first check for these possible causes:

- Liquid slugging—make sure only superheated vapor enters compressor.
- Oil slugging—possibly oil is being trapped in the evaporator or suction line and is intermittently coming back in slugs to the compressor.
- Loose flywheel (on belt-driven units).

- Improperly adjusted compressor mountings. In externally mounted hermetic-type compressors, the feet of the compressor may be bumping the studs, the hold-down nuts may not be backed off sufficiently, or springs may be too weak, thus allowing the compressor to bump against the base.

Compressor Noises

Noises coming from internal sources could be any of the following:

- ***Insufficient Lubrication.*** The oil level may be too low for adequate lubrication of all bearings. If an oil pump is incorporated, it may not be operating properly, or it may have failed entirely. Oil ports may be plugged by foreign matter or oil sludged from moisture and acid in system.
- ***Excessive Oil Level.*** The oil level may be high enough to cause excessive oil pumping or slugging.
- ***Tight Piston or Bearing.*** A tight piston or bearing can cause another bearing to knock—even though it has proper clearance. Sometimes in a new compressor such a condition will "wear in" after a few hours of running. In a compressor that has been in operation for some time, a tight piston or bearing may be due to copper plating, resulting from moisture in the system.
- ***Defective Internal Mounting.*** In an internally spring-mounted compressor the mountings may be bent, causing the compressor body to bump against the shell.
- ***Loose Bearings.*** A loose connecting rod, wrist pin, or main bearing will naturally create excessive noise. Misalignment of main bearings, shaft to crankpins or eccentrics, main bearings to cylinder walls, etc., can also cause noise and rapid wear.
- ***Broken Valves.*** A broken suction or discharge valve may lodge in the top of a piston and hit the valve plate at the end of each compressor stroke. Chips, scale, or any foreign material lying on a piston head can cause the same result.
- ***Loose Rotor or Eccentric.*** In hermetic compressors a loose rotor on the shaft can cause play between the key and key-way, resulting in noisy operation. If the shaft and eccentric are not integral, a loose locking device can be the cause of knocking.
- ***Vibrating Discharge Valves.*** Some compressors, under certain conditions, especially at low suction pressure, have inherent noise, which is due to vibration of the discharge reed or disc on the compression stroke. No damage will result, but if the noise is objectionable, some modification of the discharge valve may be available from the compressor manufacturer.
- ***Gas Pulsation.*** Under cretain conditions noise may be emitted from the evaporator, condenser, or suction line. It might appear that a knock and/or a whistling noise is being transmitted and amplified through the suction line or discharge tube. Actually, there may be no mechanical knock, but merely a pulsation caused by the intermittent suction and compression strokes, coupled with certain phenomena associated with the size and length of refrigerant lines, the number of bends, and other factors.

The recognition of compressor problems and corrective options comes with the experience, skill, and knowledge of the specific product's operating characteristics, which come with the development of a matured and professional refrigeration technician.

PROBLEMS

1. Name four designs of compressors.
2. A centrifugal compressor is a positive placement design: true or false?
3. Name two types of reciprocating compressors.
4. Sealed hermetic compressors require more ventilation than open compressors; true or false?
5. Name two types of hermetic compressors.
6. What is the greatest enemy of a motor winding?

7. Compressor wire must be sized to limit voltage drop to ____________ %.

8. E.E.R. stands for ____________.

9. Unloaders on reciprocating compressors control ____________.

10. Are two-stage compressors needed for the high or the low compression ratio for low temperature applications?

11. Compressor maintenance is an important function of its ____________.

12. Noisy operation is a signal that compressor troubles may soon develop; true or false?

18

CONDENSERS

GENERAL DISCUSSION

The condenser is another component in the system; it transfers heat from a place where it is not wanted to a place where it is unobjectionable: it transfers the heat from the refrigerant to a medium which can absorb it and move it to a final disposal point.

We have already discussed three other major components of a refrigeration system: the refrigerant, the evaporator, and the compressor. By adding the condenser we complete a typical system, as illustrated in Fig. 18-1. The receiver is associated with the condenser and will be included in this discussion.

In Chapter 6 it was pointed out that the mediums used in the condenser heat transfer process were (a) air, (b) water, or (c) a combination of both air and water. Let's reidentify and restate them in terms of types of condensers (Fig. 18-2).

- air-cooled
- water-cooled
- evaporative

There are four basic types of water-cooled condensers:

(1) double pipe
(2) open vertical shell and tube
(3) horizontal shell and tube
(4) shell and coil.

The first two types are only briefly discussed because they are rarely used. The horizontal shell and tube and the shell and coil condensers are discussed at length, because they are used extensively and represent by far the greatest percentage of today's installations.

The diagram on the right in Fig. 18-3 is one form of double-pipe condenser. Water flows through the inner pipe and the refrigerant flows in the opposite direction between the inner and outer tubes. This arrangement provides some air cooling in addition to water cooling. Its advantages are its high efficiency and its flexibility as to size and its adaptability in arrangement. Its construction, though efficient, leaves much to be desired because the great number

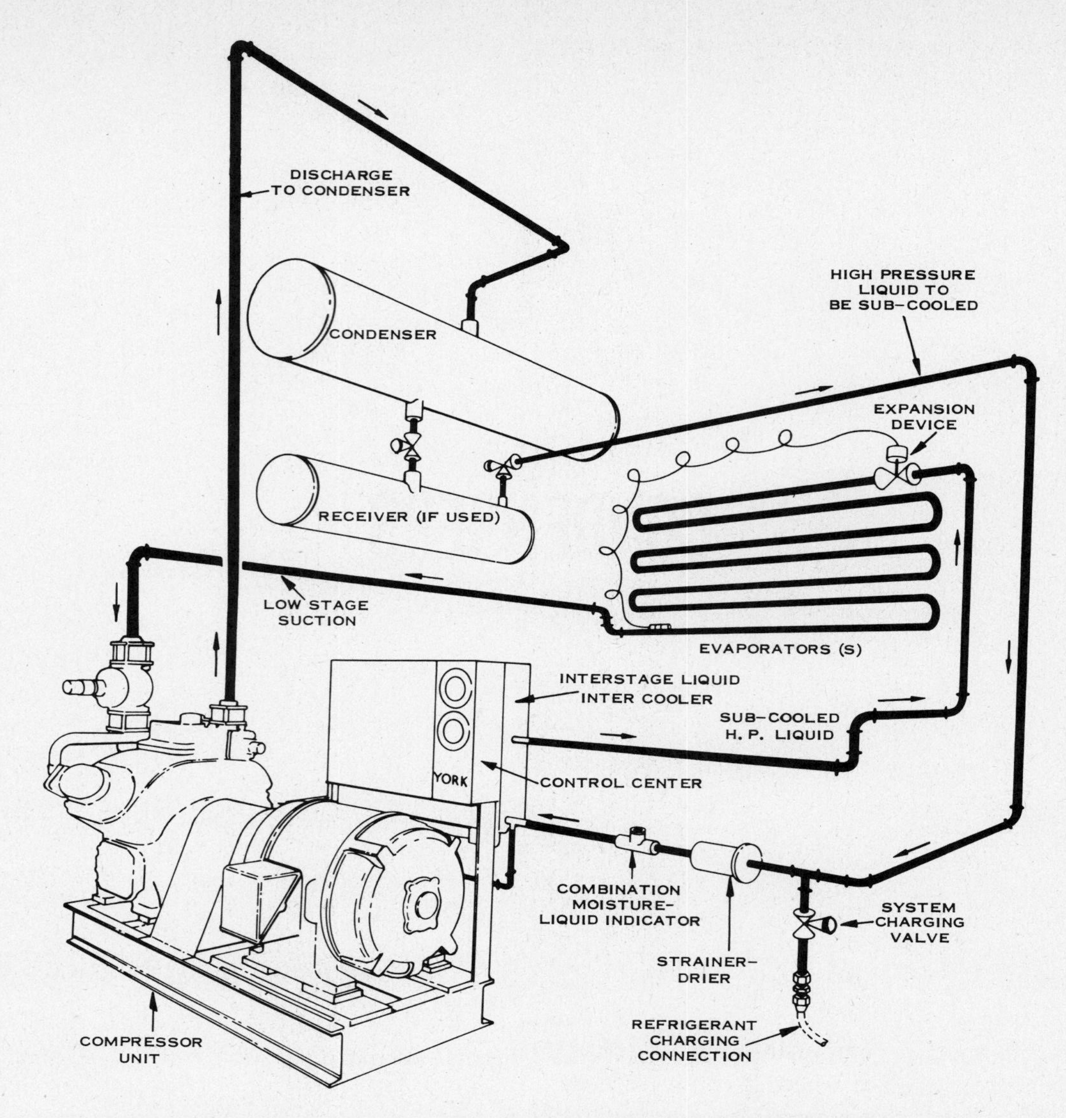

FIGURE 18-1 Refrigerant system (Courtesy, York, Division of Borg-Warner Corporation).

of flanges and gaskets makes leakage possible. Double-pipe condensers may be constructed according to designs other than the ones shown. In some, flanges and gaskets are eliminated.

The diagram at the left in Figure 18-3 illustrates an open-type vertical shell and tube condenser. Water is distributed over the head of the condenser, enters each tube, and flows down the inner surface. This condenser is normally found in medium or large ammonia plants. Its advantages are its low maintenance cost and the accessibility of its tubes for cleaning during operation. This condenser is the least efficient of the four types, requiring a larger unit for equivalent capacity.

Figure 18-4 shows the horizontal shell and tube condenser. This is similar in construction to the vertical shell and tube condenser except that water heads have been added, making a closed water circuit. This allows water to be redirected through the condenser more than once. This illustration shows the water making four passes through the condenser. By being able to redirect the water through the condenser, it is possible to obtain increased efficiency. A further increase in efficiency can be obtained by using finned condenser tubing in place of the bare tubes shown here. This type of condenser is extensively used in ammonia installations of all sizes and in medium and large installations using other refrigerants. The horizontal shell and tube condenser is very efficient and can be constructed at a reasonable cost. It is easily cleaned, and normal maintenance costs are relatively low.

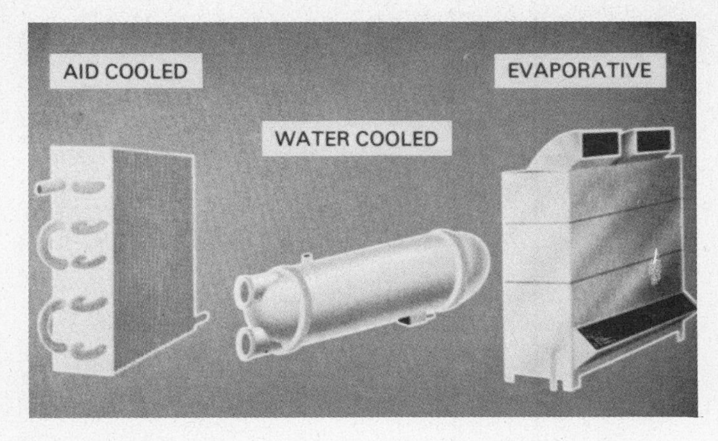

FIGURE 18-2 Types of condensers (Courtesy, Carrier Air-Conditioning Company).

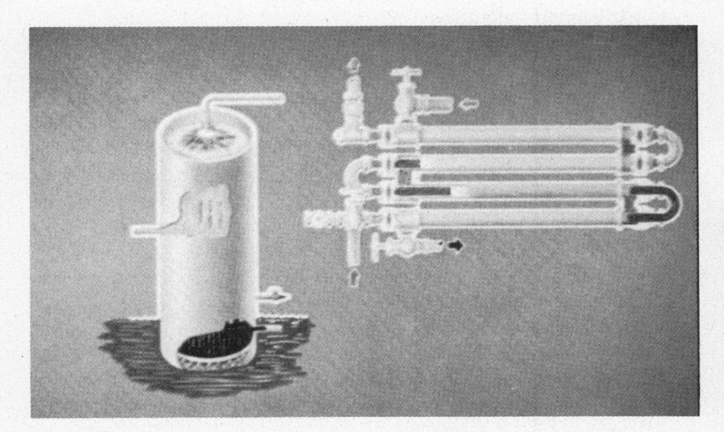

FIGURE 18-3 Double pipe, open shell and tube (Courtesy, Carrier Air-Conditioning Company).

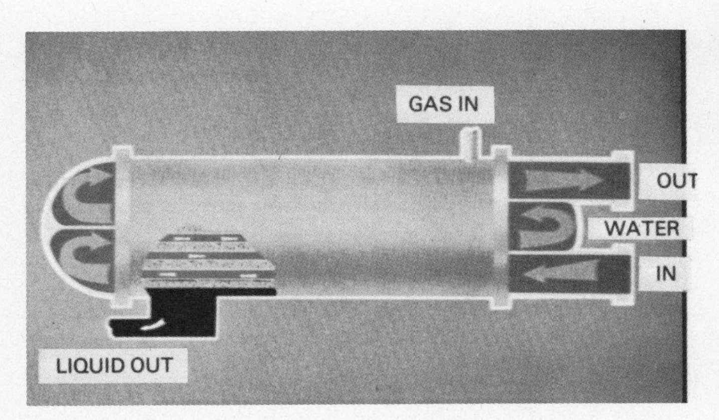

FIGURE 18-4 Shell and tube (Courtesy, Carrier Air-Conditioning Company).

Figure 18-5 shows a vertical shell and coil condenser. It consists of a welded shell containing a coil of finned water tubing. The coil within the shell is continuous and without joints. This condenser can also be made in a horizontal type. Either type may be of flanged, rather than welded, construction. This is one of the most efficient, compact, and cheapest of all condensers; consequently it is extensively used on small packaged equipment. However, it must be cleaned chemically, and when leaks develop, the cost of a single leak repair may exceed the cost of a new condenser.

Figure 18-6 shows what happens to the refrigerant in the condenser. It is also known as a *temperature Btu chart*, since the vertical scale is in degrees Fahrenheit and the horizontal scale is in Btu heat content of the refrigerant. On this chart, all the area to the right of line 1 represents the refrigerant in a gaseous state. The area to the left of line 2 represents the refrigerant in a liquid state. The area between lines 1 and 2 represents a gas-liquid mixture. For this

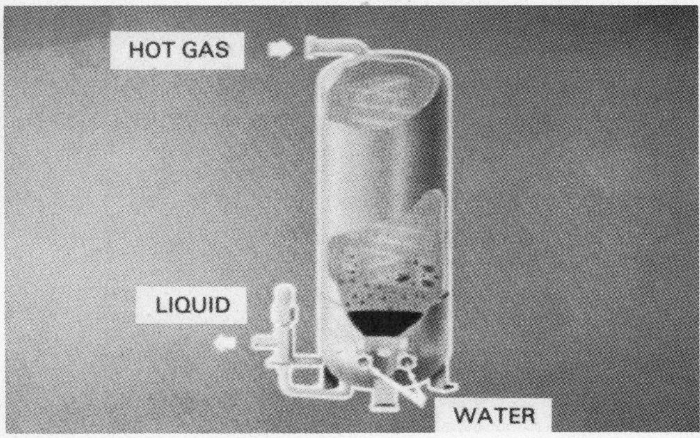

FIGURE 18-5 Vertical shell and coil (Courtesy, Carrier Air-Conditioning Company).

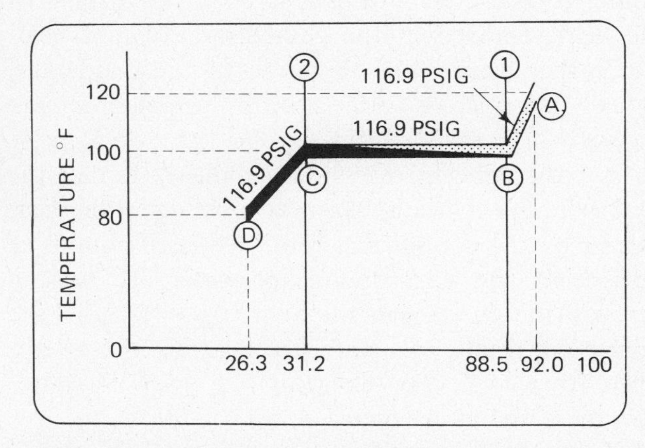

FIGURE 18-6 Heat content Btu/lb (Courtesy, Carrier Air-Conditioning Company).

illustration Refrigerant 12 at 116.9 pounds per square inch gauge is used. The corresponding condensing temperature is 100°F.

Hot refrigerant gas from the compressor enters the condenser at 120°F, represented by point *A*. The gas at this point has a 20° superheat. As it comes in contact with the tubes and/or fins, which are cooled by the cooling medium, the hot gas begins to give up its heat. Since the gas is superheated, this results in a reduction in temperature only. The temperature drops from 120°F to 100°F, which is represented by the line from *A* to *B*. The gas has reached the saturation temperature corresponding to its pressure and is ready to condense. Heat removal continues, and all the gas condenses to a liquid at point *C*. Note that the temperature has not changed between *B* and *C*. All heat removal in this area is latent heat; therefore, there is no temperature change. Since all the refrigerant is now a liquid, further cooling will subcool it below the condensing temperature of 100°F. In normally operating condensers, there would be some degree of subcooling. In this example, the liquid is subcooled 20°F to point *D* and leaves the condenser at 80°F.

An interesting point shown in this chart is the small amount of sensible heat removed compared to the latent heat. The sensible heat removal from *A* to *B* is only 3.5 Btu/lb of refrigerant. From *C* to *D* only 4.9 Btu of sensible heat has been removed. This is a total of 8.4 Btu of sensible heat removed compared to 57.3 Btu of latent heat removed between points *B* and *C*. It is thus easily seen that most of the heat removed in the condenser is latent heat. To further clarify the operation of the condenser, actual quantities of water used by a condenser are shown in Fig. 18-7.

A basic concept of condenser theory is that the heat given up by the refrigerant must equal the heat gained by the cooling medium. While 10 pounds of refrigerant and 43.8 pounds of water are passing through the condenser, the heat removed from the refrigerant results in a water temperature rise of 15°, from 70° to 85°. The chart in Fig. 18-8 illustrates how these figures are used.

Figure 18-8 shows the mathematics used in checking the heat lost by the refrigerant against the heat gained by the water. Each pound of refrigerant passing through the condenser loses 65.7 Btu. As there were 10 pounds of refrigerant passing through the condenser the total heat lost by the refrigerant should be 10 times 65.7, or 657 Btu. During the same period of time, 43.8 pounds of water pass through the condenser with a temperature rise of 15°F. The heat gained by the water is then 43.8 pounds times the specific heat, which is one (1), times the 15° temperature rise. This equals 657 Btu. In actual operation some heat is lost through radiation, but this quantity of heat is insignificant and is usually not considered in calculations. The concept of heat balance is true for all condensers, including those which are air-cooled and evaporative-cooled.

In natural draft air-cooled condensers (Fig. 18-9), air circulates over the condenser by convection. As the air comes in contact with the warm condenser it absorbs heat and rises. This allows the cooler air

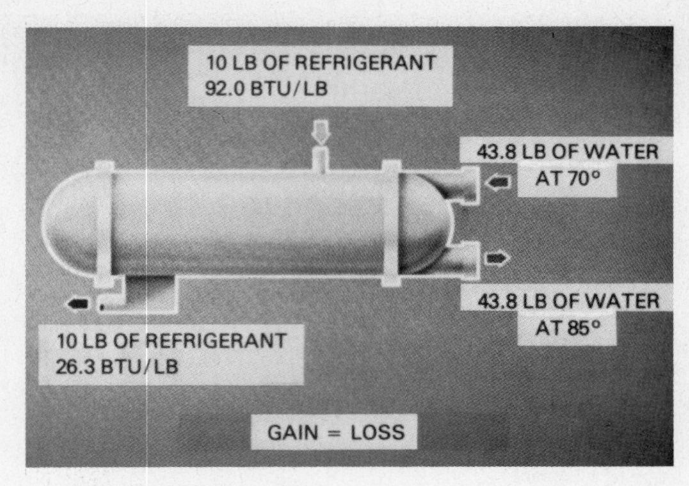

FIGURE 18-7 Gain-loss water condenser (Courtesy, Carrier Air-Conditioning Company).

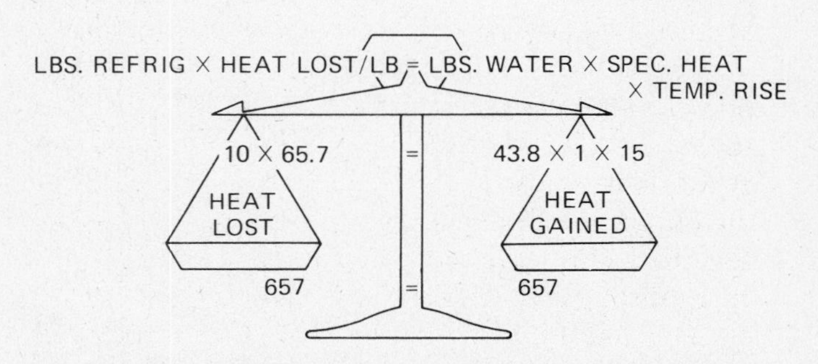

FIGURE 18-8 Heat balance of water-cooled condenser (Courtesy, Carrier Air-Conditioning Company).

FIGURE 18-9 Tube and fin plate condenser (Courtesy, Carrier Air-Conditioning Company).

underneath to rise to where it may, in turn, also absorb condenser heat.

The natural draft air-cooled condenser comes in one of two types. The tube and fin type is shown on the left in Fig. 18-9. On the right is a typical plate-type condenser, in which the plates are pressed into an outline of the condenser coil and seam-welded together. This leaves an interior space in the form of tubes through which the hot refrigerant gas passes.

The natural draft air-cooled condenser has a very limited use. Because the air moves very slowly it is not capable of removing heat rapidly from the condenser. Therefore, relatively large surfaces are required. One of its most common uses is in household refrigerators. It is cheap, easy to construct, and requires very little maintenance.

Condenser capacity can be increased by forcing air over the surfaces. (Fig. 18-10). This illustration shows a forced-air condenser. A fan has been added to increase the air flow.

Some of the earlier condensers of this type were of bare-tube construction. However, condensers are now usually of the tube and fin construction shown. Unlike the natural draft condenser, the forced-air type is more practical for larger cooling loads. The major limiting factors are economics and available space.

Either a propeller type fan (as shown) or a centrifugal type is used on air-cooled condensers. The selection of the fan depends on design factors such as air resistance, noise level, space requirements, etc.

Actual quantities and temperatures of air and refrigerant are indicated in the illustration for the purpose of showing heat balance. The refrigerant quantity and heat loss per pound are the same as for the water-cooled condenser. However, air is the condensing medium. Ten pounds of refrigerant enters with a heat content of 92 Btu per pound and leaves with a heat content of 26.3 Btu per pound. During the same period of time 3,910 cubic feet of air pass over the condenser with a temperature rise of 10°F, since the air enters at a temperature of 85°F and leaves at a temperature of 95°F.

The chart in Fig. 18-11 shows the actual heat balance. The amount heat given up by the refrigerant is determined by multiplying pounds of refrigerant times the heat loss per pound. The heat gained by the air is calculated by multiplying the pounds of air by the specific heat of air, times its temperature difference. Through substitution we find that the refrigerant gives up 657 Btu of heat. It is necessary

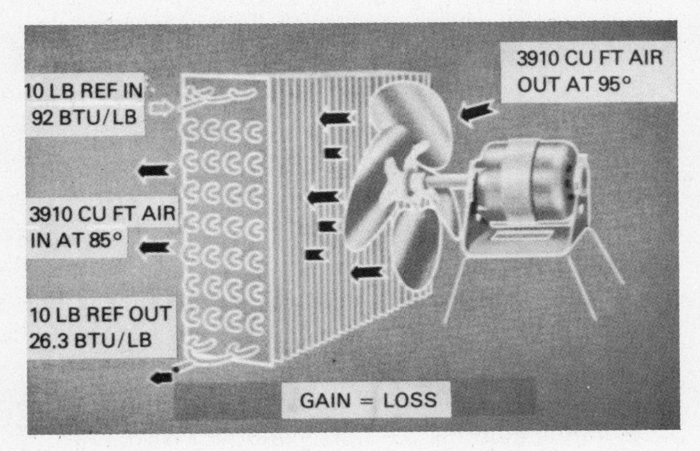

FIGURE 18-10 Condenser capacity, gain or loss (Courtesy, Carrier Air-Conditioning Company).

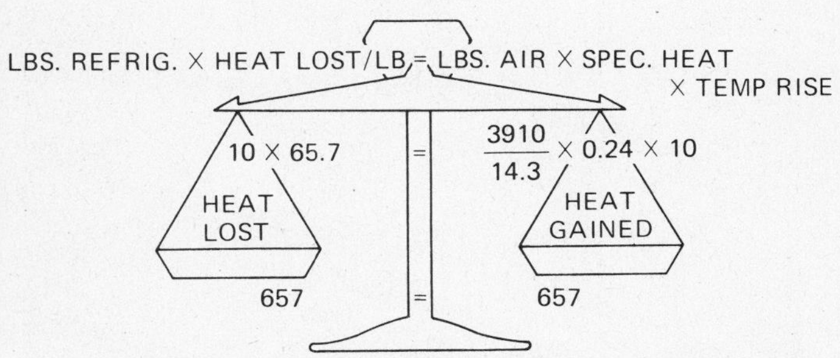

FIGURE 18-11 Heat balance of water-cooled condenser (Courtesy, Carrier Air-Conditioning Company).

to convert cubic feet of air to pounds by dividing by the specific volume of air; the result is 14.3 cubic feet per pound. The weight of air is multiplied by 0.24 (its specific heat), and then by 10 (the temperature rise in degrees). As the calculations show, this equals 657 Btu, again indicating that heat lost by the refrigerant must equal heat gained by the condenser medium.

Figure 18-12 illustrates the typical evaporative condenser. Note that it has some of the features of both air- and water-cooled condensers. In the evaporative condenser, heat is absorbed from the coil by the evaporation of water. In the case of both air- and water-cooled condensers, no evaporation takes place.

In operation, water is pumped from a pan in the base of the unit through a series of spray nozzles and then flows over the refrigerant coil. At the same time air enters through the inlet at the base, passes up through the coil and water spray, and then through eliminators to remove free water, next through the fans, and is then discharged from the unit. Water lost through evaporation is replaced through a water supply line, and the water level in the pan is controlled by a float valve. As impurities always remain behind when water is evaporated, a small continuous bleed-off is used to reduce this concentration.

As mentioned in previous chapters, large quantities of heat are required to change a liquid to a vapor. Water is no exception. When the spray water contacts the warmer refrigerant coil, evaporation takes place. The heat necessary to evaporate this water comes from the coil. At the same time an equal amount of heat is given up by the hot refrigerant gas inside the coil, and the refrigerant condenses. While this process is going on, the fan is removing the moisture-laden air from around the wetted coil and replacing it with air that has the ability to absorb more moisture.

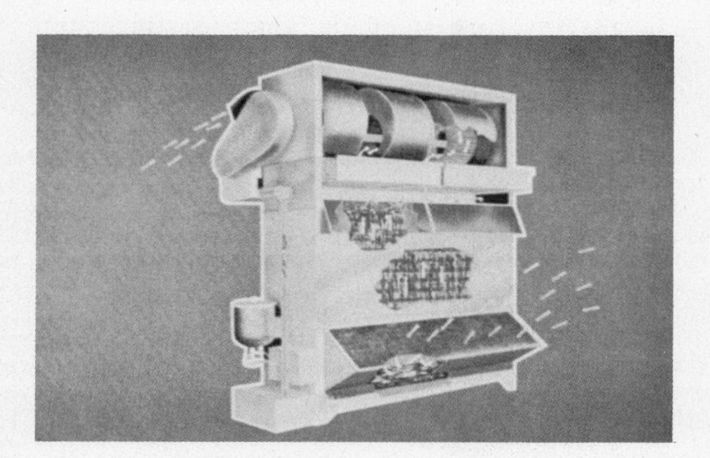

FIGURE 18-12 Typical evaporative condenser (Courtesy, Carrier Air-Conditioning Company).

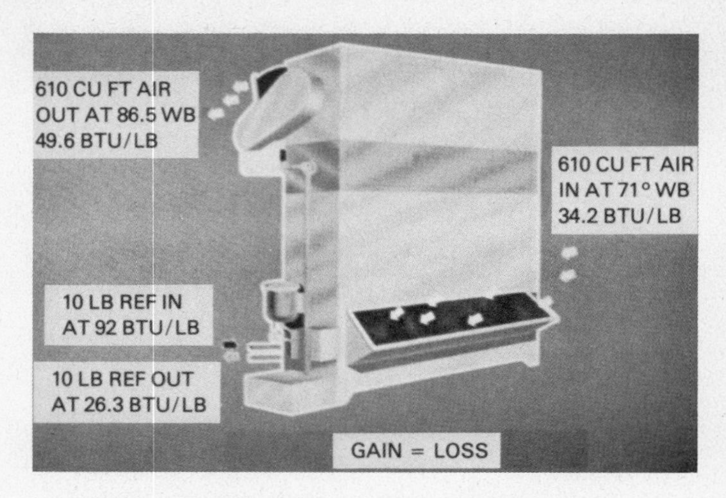

FIGURE 18-13 (Courtesy, Carrier Air-Conditioning Company).

The capacity of an evaporative condenser is determined by the amount of heat the entering air is capable of absorbing. Therefore, the more heat contained in the entering air the lower the capacity; the less heat in the entering air the higher the capacity. From this it is seen that the capacity of the evaporative condenser depends on the heat content of the entering air. The heat content of air is calculated from its wet-bulb temperature. This temperature is easily determined by placing a wetted wick over the bulb of an ordinary thermometer and holding the thermometer over the entering air stream. The wet-bulb temperature is the lowest reading noted. After the wick begins to dry the temperature will rise to the dry-bulb temperature.

Since the wet-bulb temperature can be used to calculate the heat content of the air, it follows that the difference between the wet-bulb temperature of the entering air and that of the air leaving, along with the air quantity, determines the capacity of any evaporative condenser.

In Fig. 18-13 actual air and refrigerant quantities and conditions have been added. Ten pounds of refrigerant enter with a heat content of 92 Btu per pound and leave with a heat content of 26.3 Btu per pound. During the same period of time 610 cubic feet of air pass over the condenser. From an entering air wet-bulb temperature of 70°F, it has been determined that the total heat content of this entering air is 34.2 Btu per pound. From the wet-bulb temperature of the leaving air, it has been determined that the total heat content is 49.6 Btu per pound.

Figure 18-14 then represents the actual heat balance through the evaporative condenser. The heat given up by the refrigerant is shown as pounds of refrigerant multiplied by the heat loss per pound. Therefore, the heat loss is 10 pounds of refrigerant times the loss of 65.7 Btu per pound, or 657 Btu loss through the condenser. The heat gained by the air is calculated by multiplying the pounds of air by the heat gained per pound. By dividing the cubic feet of air by its volume, 14.3 cubic feet per pound, the pounds of air can be determined. If this quantity is multiplied by 15.4 Btu, which is the difference between the heat in the air leaving and the air entering, we derive the total heat loss. The calculation shows that 657 Btu are picked up by the air, which is equal to the heat given up by the refrigerant.

In earlier applications of water-cooled condensers to refrigeration and air-conditioning it was common practice to pipe the condensers to top the city water supply and then waste the discharge water to a drain connection as illustrated in Fig. 18-15. An automatic water valve was placed in the line and the flow of incoming water was controlled by the condenser operating head pressure through a pressure tap. The temperature of the incoming water would naturally affect the condenser performance and flow rate of any heat load. Water temperatures in city water mains rarely rise above 50° to 60°F even in summer and frequently drop to much lower temperatures in winter. Condensers piped for city water

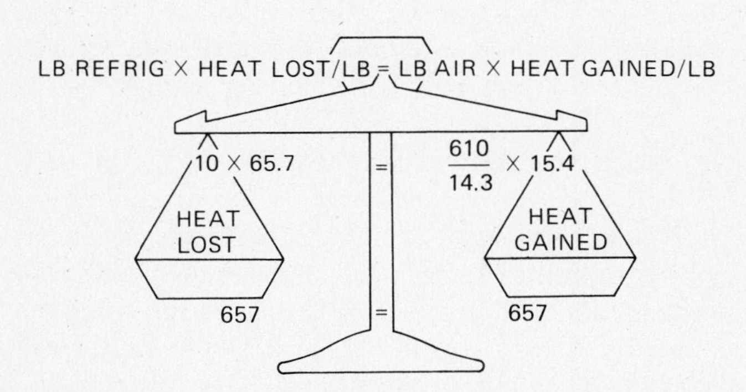

FIGURE 18-14 Heat balance scale (Courtesy, Carrier Air-Conditioning Company).

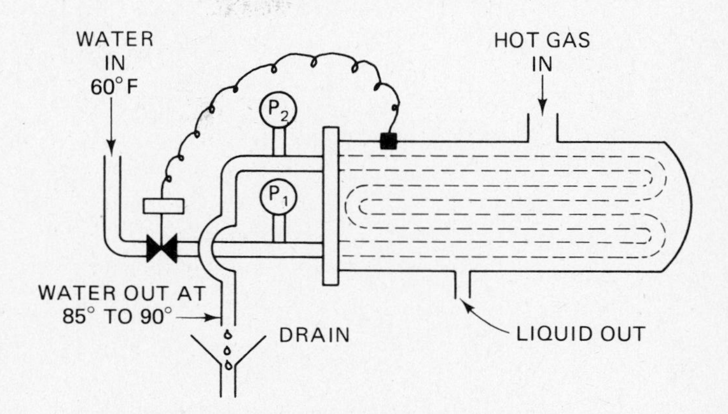

FIGURE 18-15 Series flow condenser on city water.

flow were always arranged for series flow, circuited for several water passes so as to achieve maximum heat rejection to the water, which is then wasted. Thus condensers drawing on city water could use only 1 to 2 gallons per minute per ton of refrigeration. The multipass circuit created high water-pressure drops (P1–P2) of 20 pounds or more; however most city pressures were able to supply the minimum pressure requirement (usually 25 psig). In time the cost and scarcity of using city water (unless drawn from a lake or wells and returned) became prohibitive and even outlawed by many local city codes. Ordinances restricted the use of water for refrigeration and air-conditioning to the point where such activities were forced to use all air-cooled equipment, water-saving devices such as the evaporative condenser described above, or the so-called water tower.

Evaporative condensers are of course efficient water conservation devices, but they do have a disadvantage in that extensive on-site refrigeration piping and a receiver are needed. And this tends to increase job costs and reliability risks subject to the skills of the installers. It also precludes the manufacturers assembling, testing, and shipping complete refrigeration units with minimum installation requirements. The result has been a trend toward more packaged air-cooled equipment for smaller systems and to the use of water towers where large quantities of heat must be dissipated.

Air-cooled operation will be covered later in this chapter. At this point we will concentrate on the

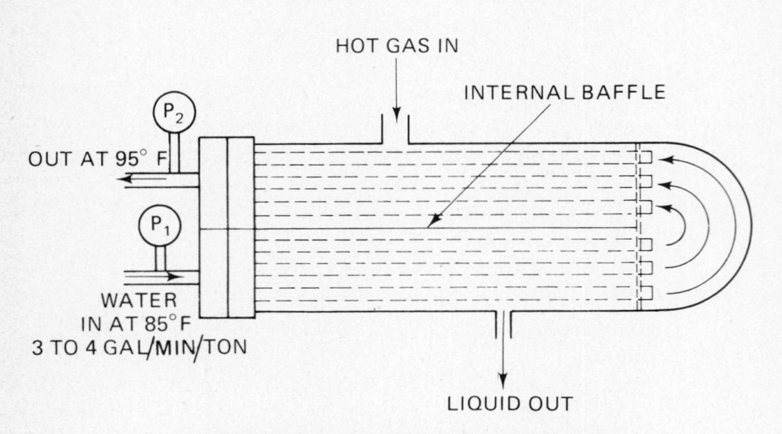

FIGURE 18-16 Condenser on tower operation.

water-cooled condenser and its application to water towers. First, the condenser (Fig. 18-16), when used with a recirculated flow such as in a water tower system, is usually designed for parallel tubes with fewer water passes to accommodate a greater water quantity (3 to 4 gal/min/ton) and lower pressure drop (P1–P2) at 8 to 10 psi (18 to 23 ft head of water) pressure drop. The nominal cooling tower application will involve a water temperature rise of 10°F through the condenser, with a condensing temperature approximately 10°F above the water outlet temperature.

Another consideration in using the water-cooled condenser for open recirculated flow is the fouling factor, which affects heat transfer and water pressure drop. Fouling is essentially the result of a scale buildup on the inside of the water tubes, some of which comes from chemical solids (calcium) but most from biological contaminants (algae, fungi, etc.) and entrained dirt and dust from the atmosphere. The progressive buildup of scale creates an insulating effect that retards heat flow from the refrigerant to the water. As the internal diameter of the pipe is reduced, so is the water flow, unless more pressure is applied. Reduced water flow naturally cannot absorb as much heat and thus condensing temperatures rise—as do operating costs.

In selecting condensers the application engineers will usually allow for the results of fouling so that the condenser will have sufficient excess tube surface to maintain satisfactory performance in normal operation, with a reasonable period of service between cleanings. For conditions of extreme fouling and poor maintenance, higher fouling factors are used. Proper maintenance depends on the type of condenser, meaning that mechanical or chemical cleaning or both may be needed or employed to remove scale deposits.

The function of the water tower is to pick up the heat rejected by the condenser and discharge it into the atmosphere, which it does by evaporation.

Why is the temperature of the water in a lake cooler than the surrounding air during a hot summer season? *Surface evaporation* is the reason, and surface evaporation is also what takes place when one stands in front of an electric fan and feels the cooling effects of moisture evaporating from the surface of the skin.

The evaporation of water from any surface removes heat in the water vapor that is produced. This heat is called the *latent heat of vaporization*. Air, when absorbing heat from water in this manner, is capable of cooling water below the atmospheric temperature. When one pound of water is evaporated it takes with it approximately 1,000 Btu in the form of latent heat. And this removal of latent heat by air is the cooling effect that makes it possible to cool the water in a cooling tower to a temperature below that of the surrounding air as read on an ordinary thermometer.

In air-conditioning and refrigeration work, the *dry-bulb* and *wet-bulb* temperatures are the basis for system design. Closely related to wet-bulb and dry-bulb temperature is relative humidity.

Relative humidity is the ratio of the quantity of water vapor actually present in a cubic foot of air to the greatest amount of vapor that air could hold if it were saturated. When the relative humidity is 100%, the air cannot hold any more water, and therefore water will not evaporate from an object in 100% humid air. But when the relative humidity of the air is less than 100%, water will evaporate from the surface of an object, be it a lake, a wet sponge, or the *drops of falling water in a cooling tower*.

When the relative humidity is 100%, the wet-bulb temperature is the same as the dry-bulb temperature, because when the wet-bulb thermometer, whose bulb is covered with a wet cloth, is whirled, water cannot evaporate from the cloth. But when the humidity is less than 100%, the wet-bulb will be less than the dry-bulb temperature, since water will evaporate from the cloth and will carry latent heat away from the bulb, thus cooling it below the dry-bulb temperature. The drier the air, the greater will be the difference between the dry-bulb and wet-bulb temperature and the easier it will be for water to evaporate.

It follows then, that cooling-tower operation does not depend upon the dry-bulb temperature. The ability of a cooling tower to cool water is a measure of how close the tower can bring the water temperature to the wet-bulb temperature of the surrounding air. The lower the wet-bulb temperature (which indicates either cool air, low humidity, or a combination of the two), the lower the tower can cool the water. It is important to remember that no cooling tower can ever cool water below the wet-bulb temperature of the incoming air. In actual practice the final water temperature will always be at least a few degrees above the wet-bulb temperature, depending upon the design conditions. The wet-bulb temperature selected in designing cooling towers for refrigeration and air-conditioning service is usually close to the average maximum wet-bulb for the summer months at the given location.

By way of summary, the reason cooling towers cool is because air, passing over exposed water surfaces, picks up small amounts of water vapor. The small amount of water that evaporates (about 1% for each 10°F of cooling) removes a large amount of heat from the water left behind.

How do cooling towers cool? There are two types of cooling towers: *mechanical draft* and *natural draft* (Fig. 18-17). A mechanical draft tower utilizes a motor-driven fan to move air through the tower, the fan being an integral part of the tower. A natural draft tower is one in which the air movement through the tower is dependent upon atmospheric conditions (wind).

When the water to be cooled arrives at the mechanical draft cooling tower it contains heat that has been picked up in the condenser of the refrigeration or air-conditioning unit. This heat usually amounts to about 250 Btu per minute for each ton of refrigeration. The water enters the tower at the top or upper distribution basin. It then flows through holes in the distribution basin and into the tower, filling staggered splash bars, which retard the fall of the water and break it into small droplets, as shown in Fig. 18-18. Meanwhile, the fan is pulling air through the filling. This air passes over and bathes each droplet, and the resulting evaporation transfers heat from the warm water into the air. Finally, the falling water is cooled and collects in the

(a)

(b)

FIGURE 18-17 (a) Mechanical draft cooling tower; (b) Natural draft cooling tower (Courtesy, Marley Cooling Tower Company).

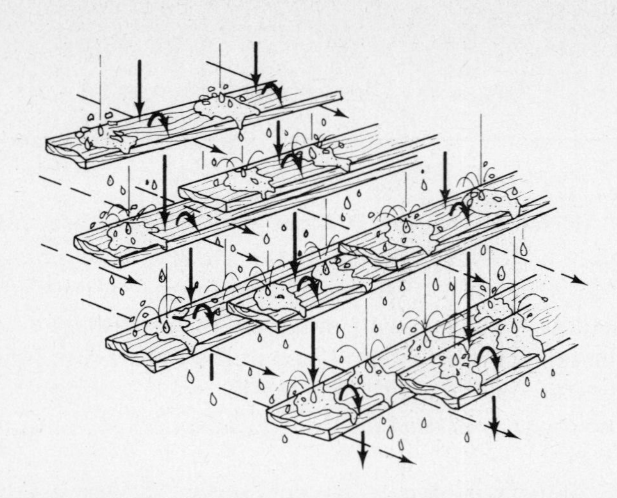

FIGURE 18-18 Typical filled section of a cooling tower; filling provides cooling surface and water breakup; air moves across water drops and filling surfaces (Courtesy, Marley Cooling Tower Company).

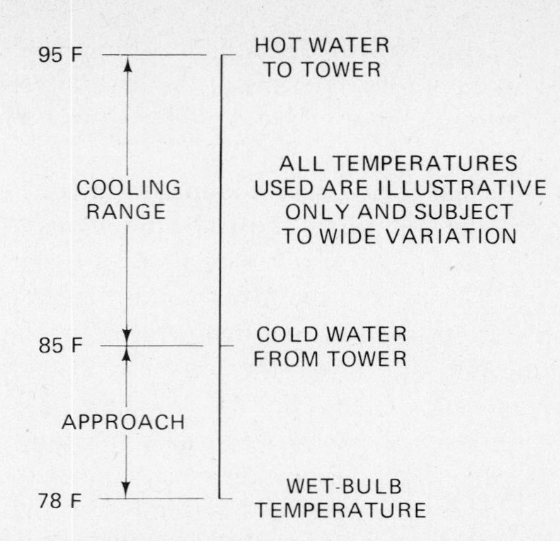

FIGURE 18-19 Diagram showing definition of "cooling range" and "approach" (Courtesy, Marley Cooling Tower Company).

lower (cold water) basin of the tower. It is then pumped back to the water-cooled condenser to pick up more heat.

When the cooling water arrives at the natural draft cooling tower, it is piped to the top where it enters spray nozzles. These nozzles break the water into fine droplets. Natural breezes moving through the louvered sides of the tower provide the fanning action as the droplets fall. Cooled by the time it reaches the bottom of the tower, the water is collected in a basin and pumped back to perform its cooling job again.

The natural draft tower depends upon the spray nozzles to break up the water. This tower has no filling or fan, and it will be seen that its size, weight, and location requirements (compared to mechanical draft towers) have reduced its use considerably. However, in normal refrigeration service work there may be occasions when the technician is called upon for service or maintenance on such units, and thus it is important to be familiar with the operation of natural draft towers.

What terms and definitions apply to cooling towers (see Fig. 18-19)?

- ***Cooling Range*** is the number of degrees Fahrenheit through which the water is cooled in the tower. It is the difference between the temperature of the hot water entering the tower and the temperature of the cold water leaving the tower.
- ***Approach*** is the difference in degrees Fahrenheit between the temperature of the cold water leaving the cooling tower and the wet-bulb temperature of the surrounding air.
- ***Heat Load*** is the amount of heat "thrown away" by the cooling tower in Btu per hour (or per minute). It is equal to the pounds of water circulated multiplied by the cooling range. For example, a tower circulating 18 gallons per minute with a 10° cooling range will have a capacity of:

$$18 \text{ gal/min} \times 60 \text{ min} \times 8.35 \text{ lb/gal} \times 10°\text{F} = 90.180$$

$$\text{Btu/h} = 7.5 \text{ tons}$$

Figure 18-20 represents a typical mechanical draft tower piping arrangement to the condenser of a packaged refrigeration or air-conditioning unit.

- ***Cooling Tower Pump Head*** is the pressure required to lift the returning hot water from a point level with the base of the tower to the top of the tower and force it through the distribution system. This data is found in the manufacturer's specifications and is usually expressed in feet of head (1 lb of pressure = 2.31 feet of head).
- ***Drift*** is the small amount of water lost in the form of fine droplets retained by the circulating air. It is independent of, and in addition to, evaporation loss.
- ***Bleed-Off*** is the continuous or intermittent wasting of a small fraction of circulating water to prevent the buildup and concentration of scale-forming chemicals in the water.
- ***Make-Up*** is the water required to replace the water that is lost by evaporation, drift, and bleed-off.

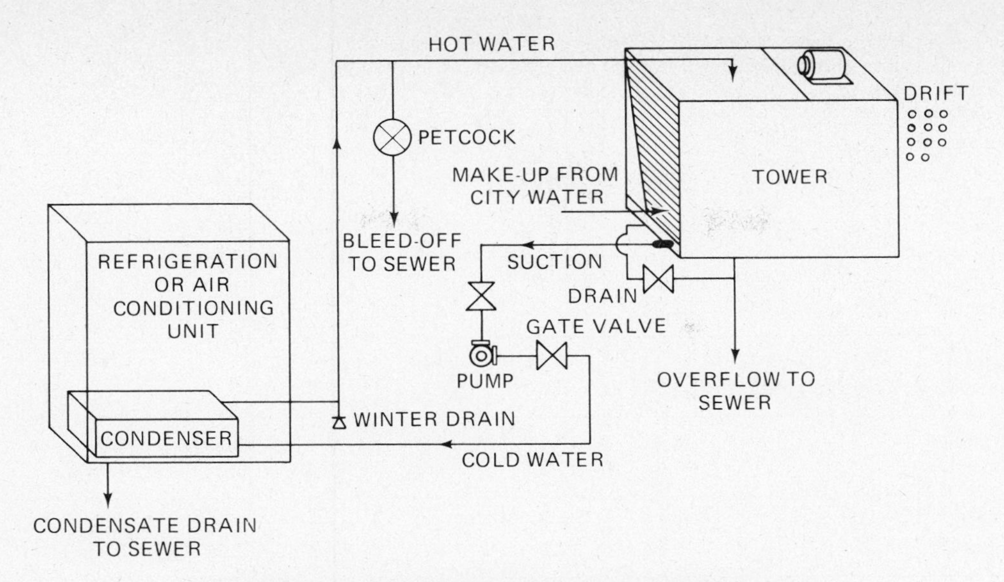

FIGURE 18-20 (Courtesy, Marley Cooling Tower Company).

We will now consider the piping system design. The first step is to determine the water flow to be circulated, based on the heat load given above. Normally towers run between 3.5 to 4.0 gal/min (gallons per minute) per ton. Water supply lines should be as short as conditions permit. Standard weight steel pipe (galvanized), type "L" copper tubing, and CPVC plastic pipe are among the satisfactory materials, subject to job conditions and local codes.

Piping should be sized so that water velocity does not exceed 5 feet per second. Table 18A lists approximate friction losses in standard steel pipe and type "L" copper tubing. Plastic pipe will have the same general friction loss as copper. The data is based on clear water, reasonable corrosion and scaling, and velocity flow at or below the 5 ft/sec range.

EXAMPLE:

100 ft of $1\frac{1}{4}$ in. standard steel pipe would have a pressure loss of 4.31 ft at 12 gal/min 50 ft would have a pressure loss of $4.31 \times 50/100 = 2.15$ ft, and 200 ft would have a pressure loss of $4.31 \times 200/100 = 8.62$ ft

In using the table, select the smallest pipe size that will provide proper flow and velocity, to keep installation costs to a minimum. Friction loss is expressed in feet of head per 100 feet of straight pipe length (see example above).

The entire piping circuit should be analyzed to establish any need for proper valves for operation and maintenance of the system. A means of adjusting water flow is desirable; shut-off valves should be placed so that each piece of equipment can be isolated for maintenance.

Valves and fittings (elbows, tees, etc.) create added friction loss and pumping head. Table 18B lists the approximate friction loss expressed in equivalent feet of pipe.

Following is an example of calculating pipe sizing:

EXAMPLE:

Determine the total pump head required for a 5-ton installation requiring 75 feet of steel pipe, 10 standard elbows, four gate valves and a net tower static lift of 60 inches. Water circulation will be 15 gal/min and the pressure drop across the condenser is 13 psi (data obtained from manufacturer).

SOLUTION

Assume $1\frac{1}{4}$ in. pipe, since the velocity of 15 gal/min is less than 5 ft/sec.

		Equivalent $1\frac{1}{4}$ in. *Pipe Length*
75 feet $1\frac{1}{4}$ in. standard steel pipe	=	75.0 feet
(10) $1\frac{1}{4}$ in. standard elbows × 3.5	=	35.0 feet*
(4) $1\frac{1}{4}$ in. open gate valve × 0.74	=	2.96 feet*
		112.96

From Table 18A, we find that for 100 feet of $1\frac{1}{4}$ in. pipe, the loss is 6.35 feet; for

$$112.96 \text{ ft, loss} = \frac{112.96}{100} \times 6.35 = 7.17 \text{ ft}$$

* See Table 18B.

TABLE 18A Approximate friction losses in standard steel pipe and type "L" copper tubing. (figures are shown as head loss in feet per 100 feet of pipe.)

		¾ in.		1 in.		1¼ in.		1½ in.		2 in.	
Water Flow GPM	*Type of Pipe or Tubing*	*Veloc-ity ft/sec*	*Head Loss ft/100 ft*	*Veloc-ity ft/sec*	*Head Loss ft/100 ft*	*Veloc-ity ft/sec*	*Head Loss ft/100 ft*	*Veloc-ity ft/sec*	*Head Loss ft/100 ft*	*Veloc-ity ft/sec*	*Head Loss ft/100 ft*
6	Std. steel	3.61	14.7	2.23	4.54						
	Copper-L	3.98	11.5	2.34	3.13						
9	Std. steel	5.42	31.1	3.34	9.72	1.93	2.75				
	Copper-L	5.96	24.2	3.50	6.63	2.30	2.38				
12	Std. steel			4.46	16.4	2.57	4.31	1.89	2.04		
	Copper-L			4.67	11.3	3.06	4.04	2.16	1.73		
15	Std. steel			5.57	24.9	3.22	6.35	2.36	3.22		
	Copper-L			5.84	17.1	3.83	6.12	2.70	2.62		
22	Std. steel					4.72	13.2	3.47	6.25	2.10	1.85
	Copper-L					5.21	12.5	3.96	5.57	2.28	1.40
30	Std. steel							4.73	11.1	2.87	3.29
	Copper-L							5.41	9.44	3.11	2.45
45	Std. steel									4.30	6.96
	Copper-L									4.66	5.20

Note 1—Data on friction losses based on information published in *Cameron Hydraulic Data* by Ingersoll Rand Company. Printed by permission.
Note 2—Data based on clear water and reasonable corrosion and scaling.

TABLE 18B Approximate friction losses in fittings and valves in equivalent feet of pipe.

Pipe Size Inches	*Gate Valve Full Open*	*45° Elbow*	*Long Sweep Elbow or Run of Std. Tee*	*Std. Elbow or run of tee reduced ½*	*Std. tee through side outlet*	*Close return bend*	*Swing Check Valve Full Open*	*Angle Valve Full Open*	*Globe Valve Full Open*
¾	.44	.97	1.4	2.1	4.2	5.1	5.3	11.5	23.1
1	.56	1.23	1.8	2.6	5.3	6.5	6.8	14.7	29.4
1¼	.74	1.6	2.3	3.5	7.0	8.5	8.9	19.3	38.6
1½	.86	1.9	2.7	4.1	8.1	9.9	10.4	22.6	45.2
2	1.10	2.4	3.5	5.2	10.4	12.8	13.4	29	58

Note: Data on fittings and valves based on information published by Crane Company. Printed by permission.

Pressure loss due to piping and fittings
= 7.17 feet
Pressure loss due to condenser = 13 × 2.31
= 30.00 feet†
Pressure loss due to static lift-cooling
tower pump head = 5.00 feet
Total head 42.17 feet

Pipe size is adequate, since velocity is less than 5 feet per second.

† To convert pounds per square inch pressure to feet of head, multiply by 2.31.

How is the pump selected? The selection is based on the gallons of water per minute (for example, 15 gal/min) and the total head (42.17), or 43 feet of head. The pump manufacturer's catalog will rate the pump capacity in gal/min versus feet of head and the horsepower size needed to do the job. There are many types of pumps on the market, but generally the slower speed unit (1,750 rpm) is recommended for quiet operation. Units operating at 3,450 rpm usually cost less, but are noisier.

The pump generally should be installed as follows:

- The pump should be located between the tower and the refrigeration or air-conditioning unit so that the water is "pulled" from the tower and "pushed" through the condenser. See Fig. 18-20 for a typical piping diagram. It is good practice to place a flow-control valve (a gate valve is satisfactory) in the pump discharge line.
- The pump should be installed so that the pump suction level is lower than the water level in the cold-water basin of the tower. This assures pump priming.
- If noise is not objectionable, the pump may be located indoors so as to eliminate outside wiring and to permit use of an *open* or *drip-proof* motor.
- If circumstances require use of open or drip-proof motor outdoors, units should be sheltered by adequate housing covers.
- The pump should be accessible for maintenance, and should be installed so as to permit complete drainage for winter shut-down.

What are some conventional methods of wiring a cooling tower? The most desirable wiring and control arrangement varies depending upon the characteristics of the available power supply and the size of the equipment being installed. In every case, the objective is to provide the specified results with optimum operating economy and protection to the equipment involved.

For small refrigeration and air-conditioning equipment, the ideal arrangement is based upon a sequence beginning with the cooling tower fan and pump. The starter controlling the fan and pump would then activate the compressor motor starter through an interlock. This method, illustrated in Fig. 18-21, assures sufficient condenser water flow

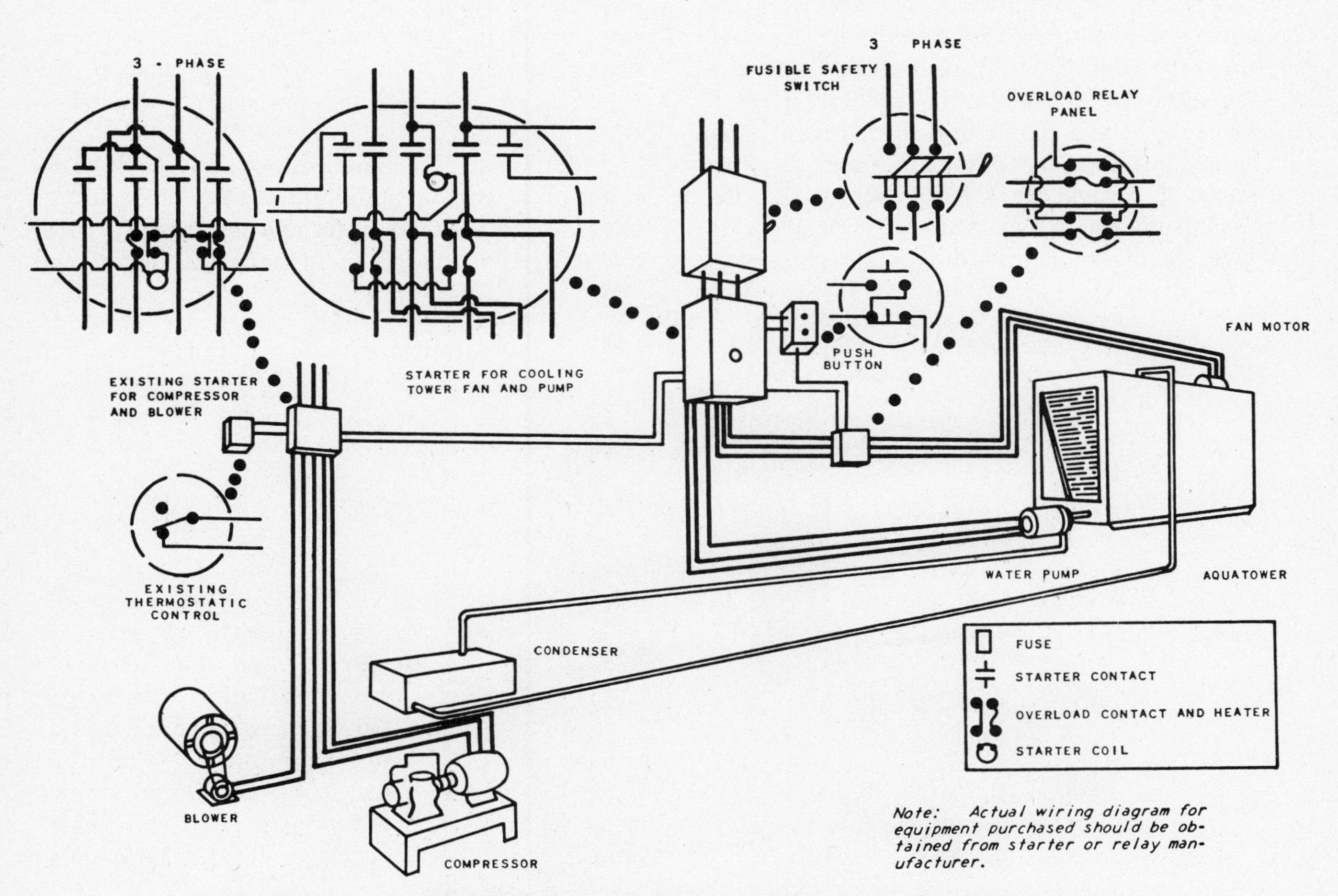

FIGURE 18-21 Typical wiring diagram showing separate starter for 220 volt three-phase electric motors driving cooling tower fan and pump. Compressor will not start until cooling tower fan and pump are operating (Courtesy, Marley Cooling Tower Company).

so that compressor short-cycling is eliminated in the event of pump or tower-fan motor failure. In other words, the compressor cannot run unless the tower fan and pump are operating.

There are other, more economical methods based on using the compressor starter to activate the pump and fan, but water-temperature or flow-sensing devices should be incorporated as protection for the compressor.

Where multiple refrigeration units are used on a common tower, the first unit that is turned on activates the cooling tower.

Winter operation or low ambient operation of a cooling tower is subject to special treatment for temperatures above freezing. Water that comes off the tower too cold may cause thermal shock to the condenser and result in a very low condensing temperature. One method of raising water-off temperature is to cycle "off" or reduce the speed of the tower fan (assuming it is a mechanical draft tower). This will reduce evaporation considerably and result in heat rejection, causing the water temperature to rise. If this isn't sufficient, a bypass valve can be placed in the tower piping to permit dumping warm water directly into the tower base, thus bypassing the spray or water distribution system. A combination of both fan and bypass may be needed to assure stable water-off temperatures during cold weather.

Prolonged below freezing conditions usually requires total shutdown and draining of the tower sump and all exposed piping.

The foregoing discussion on water-cooled condensers was presented as if most refrigeration and/or air-conditioning equipment had separate condensers and compressors connected as in Fig. 18-1. True, separate components are sold for built-up systems, but these are relatively few. Most condensers are assembled at the factory as part of a water-cooled condensing unit (Fig. 18-22) or part of a complete water-chilling unit. Therefore, the performance rating of the equipment becomes a system rating similar to the capacity rating table (Fig. 18-23), wherein the temperature of the water leaving the condenser and the saturated suction temperature to the compressor determines the unit tonnage, kw power consumption, resulting condensing temperature, and condenser flow in gal/min. As previously stated, 95°F water off the condenser is normal for tower operation.

Additionally, the manufacturer's data will list the water pressure drop across the condenser (in feet of head) for various flow rates and number of passes; the refrigerant charges in pounds (maximum and

FIGURE 18-22 Condensing unit (Courtesy, York, Division of Borg-Warner Corporation).

YORK **Ratings and Engineering Data**

Condenser Leaving Water Temp. (F)	Saturated Suction Temp. (F)	CONDENSING UNIT MODEL NUMBER											
		JS43L-12W413						JS53M-12W523					
		TONS CAP.	KW	Cond. Temp. (F)	Heat Rej. MBH	Cond. GPM	Cond. Δ P	TONS CAP.	KW	Cond. Temp. (F)	Heat Rej. MBH	Cond. GPM	Cond. Δ P
80	20	27.0	25.0	86.8	409	81.8	9.5	31.6	30.9	86.4	485	97.0	8.6
	30	33.5	26.6	88.3	493	98.6	13.3	39.5	32.8	87.8	586	117.2	12.2
	40	41.0	28.0	90.2	588	117.6	18.3	48.1	34.5	89.6	695	139.0	16.6
	50	—	—	—	—	—	—	58.3	36.2	91.8	824	164.8	22.7
85	20	26.0	26.0	91.6	401	80.2	9.1	30.5	32.1	91.2	476	95.2	8.3
	30	32.4	27.8	93.1	484	96.8	12.9	38.1	34.4	92.7	574	114.8	11.7
	40	39.8	29.4	94.9	578	115.6	17.8	46.7	36.3	94.4	684	136.8	16.1
	50	48.1	31.1	97.1	683	136.6	24.3	56.7	38.2	96.4	810	162.0	22.0
90	20	25.1	26.8	96.4	393	78.6	8.8	29.5	33.2	96.0	467	93.4	8.0
	30	31.5	29.0	97.9	477	95.4	12.6	37.0	35.9	97.5	567	113.4	11.4
	40	38.5	30.9	99.6	568	113.6	17.2	45.3	38.1	99.1	674	134.8	15.7
	50	46.8	32.8	101.7	673	134.6	23.6	55.0	40.2	101.1	797	159.4	21.4
95	20	24.2	27.6	101.1	384	76.8	8.5	28.5	34.2	100.9	459	91.8	7.8
	30	30.4	30.1	102.5	468	93.6	12.2	35.7	37.2	102.2	555	111.0	11.0
	40	37.4	32.3	104.3	559	111.8	16.7	44.0	39.9	103.8	664	132.8	15.3
	50	45.5	34.5	106.4	664	132.8	23.0	53.5	42.3	105.8	786	157.2	20.8
100	20	23.3	28.3	106.0	377	75.4	8.2	27.4	35.1	105.7	449	89.8	7.5
	30	29.4	31.2	107.3	460	92.0	11.8	34.5	37.6	107.0	542	108.4	10.5
	40	36.0	33.7	109.0	547	109.4	16.1	42.5	41.6	108.6	652	130.4	14.8
	50	44.0	36.1	111.0	651	130.2	22.1	51.9	44.5	110.4	775	155.0	20.3
105	20	22.3	28.9	110.8	367	73.4	7.8	26.3	35.8	110.5	438	87.6	7.1
	30	28.3	32.2	112.2	450	90.0	11.3	33.2	39.8	111.8	534	106.8	10.3
	40	35.0	35.0	113.7	540	108.0	15.7	41.2	43.3	113.3	642	128.4	14.4
	50	42.7	37.8	115.7	641	128.2	21.5	50.5	46.7	115.1	765	153.0	19.8

ΔP = Pressure Drop (Ft. H_2O)

FIGURE 18-23 Ratings and engineering data (Courtesy, York, Division of Borg-Warner Corporation).

minimum) when the condenser is used as a receiver or with an external receiver; and the pump-down capacity which is usually equal to about 80% of the net condenser volume but does not exceed a level above the top row of tubes.

Pump-down is used to store or contain all the system refrigerant in the condenser, so as to be able to perform service or maintenance operations on other components and not lose the refrigerant. Shut-off valves on the condenser permit this operation.

Water-cooled condensers are usually equipped with a purge connection to vent noncondensable gases and have provisions for a relief valve to meet certain national and or local code requirements. Setting of the relief valve varies somewhat with the refrigerant being used.

Peak performance of a water-cooled condenser and cooling tower system depends heavily on regular maintenance. Growths of slime or algae, which reduce heat transfer and clog the system, should be prevented. Sodium pentachlorophenolate, available under various trade names, is effective in killing these organisms. Copper sulphate is also satisfactory, but it is corrosive if too concentrated. Chlorine and potassium permanganate are also satisfactory. We recommend that you familarize yourself with the companies, chemicals, and cleaning techniques recommended. The success of any water treatment lies in beginning it early and using it regularly. Once scale deposits have formed it can be costly to remove them.

In addition to chemical treatment, regular draining and flushing of the tower basin is recommended. Also, the float valve should be adjusted to cause a small amount of overflow that trickles down the drain; this is called *bleed-off*, or sometimes *blow-*

down. It is the continuous or intermittent removal of a small amount of water (1% or less) from the system. Dissolved concentrates are continually diluted and flushed away.

The water problems mentioned above are also common in the evaporative condenser but are not quite as critical, since all the water is contained in the sump. There is no water-cooled condenser or extended water lines. Scale builds up on the exterior surface of the evaporative condenser refrigerant coil. Regular mechanical cleaning, chemical treatment, and bleed-off are also recommended.

Water treatment and maintenance can be an expensive and time-consuming process, and as water restrictions have become widespread the refrigeration and air-conditioning industry has turned to alternate systems.

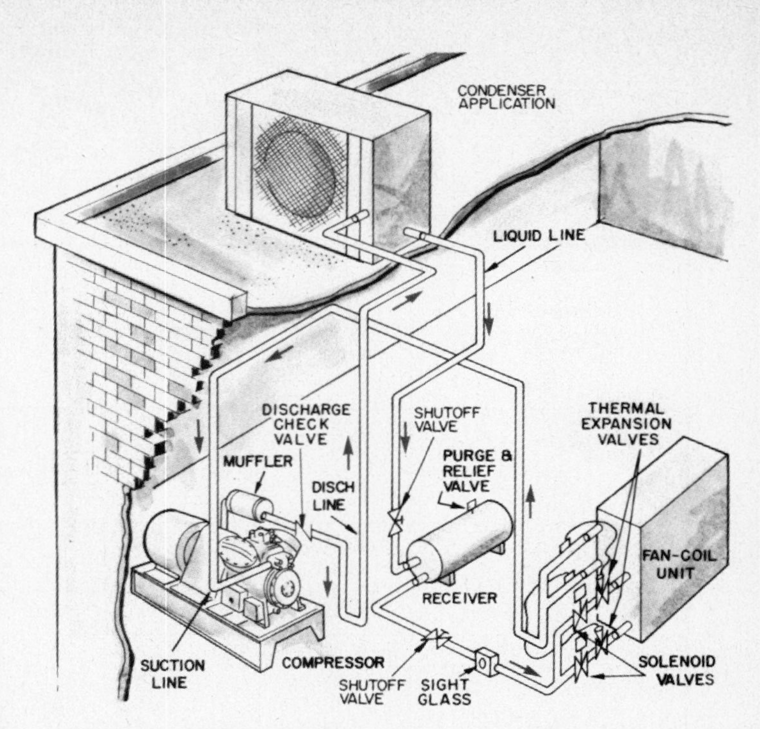

FIGURE 18-24 Remote air cooled condenser installation (Courtesy, Carrier Air-Conditioning Company).

AIR-COOLED CONDENSING

In addition to being free of water-maintenance problems, the use of air-cooled equipment has other advantage: low ambient or winter operation can be accomplished with relative ease; location of equipment is flexible; the elimination of extensive water lines reduce the need for plumbing tradesman and means the refrigeration technician can do all the mechanical work.

On the other hand, air-cooled condensing has some disadvantages too. First, an air-cooled condensing system is not as efficient as water-cooled in terms of Btu/watt (EER) ratios. Second, if the compressor and condenser are separated, the need for extended on-site refrigeration piping adds to the cost of, and technical skills required in, the installation, as well as the need for increased reliability where prevention of leaks, oil return problems, vibration, and noise must be considered.

It is apparent, however, that the advantages outweigh the disadvantages, because the bulk of commercial refrigeration and air conditioning, from fractional horsepower sizes to 100 tons and more, are using air-cooled condensing.

Note the expression *air-cooled condensing*; there is a great difference in hardware between an air-cooled condenser and an air-cooled condensing unit. Earlier, it was pointed out that in a remote air-cooled condenser installation such as shown in Fig. 18-24 the compressor is located inside the building, serving one or more evaporators or fan coil units, and the air-cooled condenser is on the roof. Figure 18-24 shows the use of a separate receiver; this subject will be covered later. The condenser itself as illustrated in Fig. 18-25 is nothing more than a fin-tubed coil within a cabinet or frame. The fan may be a propeller (as shown) or a centrifugal blower, which moves air across the coil face. This type of installation can and is still being done today, but several factors should be considered when choosing such a system:

- The need for on-site assembly of refrigeration lines with associated evacuation, leak testing, charging, etc.

FIGURE 18-25 Air cooled condenser (Courtesy, Carrier Air-Conditioning Company).

- The separation of the compressor motor and condenser fan motor adds to the cost of power and control wiring.
- The compressor unit takes up building space that may prehaps be needed for other purposes.
- The compressor unit noise is contained within the building.
- Separate components are more expensive than complete factory packages.

These factors are not critical disadvantages, but they have led to the development of more and more self-contained air-cooled condensing units. The remote air-cooled condenser is an excellent solution for converting existing water-cooled equipment to air-cooled operation.

In selecting a remote air-cooled condenser the following information is necessary:

(1) design suction temperature
(2) compressor capacity at design suction temperature
(3) refrigerant being used
(4) geographic location of installation
(5) type of condenser required

Items 1, 2, and 3 have already been discussed and are based on evaporator duty and compressor performance ratings. Item 4 is concerned with summer dry bulb design conditions relative to the suction temperature; this information will help determine the proper operating temperature difference in selecting a condenser. Item 5 deals with a number of factors, such as: type of fan (propeller or centrifugal); type of air flow (vertical or horizontal); type of coil (single or split circuit); and accessories like wind deflectors, winterstats, low ambient fan controls, etc.

Manufacturers' catalogs contain ratings based on specifications for the above information as well as selection methods.

The self-contained air-cooled condensing unit varies from small tonnage commercial refrigeration units (Fig. 18-26) used on meat cases, dairy cases, commercial refrigerators, etc., to units for large refrigeration loads (walk-in coolers and freezers, process freezing plants and air-conditioning duty).

The physical characteristics of larger self-contained air-cooled condensing units vary widely from one manufacturer to another. Figure 18-27 represents a typical unit used in commercial refrigeration work. It consists of a complete cabinet or enclosure to contain the condenser coil(s) and air movement fan(s); a compressor and control compartment and in many cases an external receiver are suspended underneath. Some of the advantages to this arrangement of the equipment are:

- All the heavy-duty mechanical equipment is outside (on the roof or at ground level) and doesn't take up valuable floor space. Noise and vibration are removed from the area inside.
- All major internal power and controls are preselected and prewired to a weatherproof panel.

FIGURE 18-26 Small self-contained condensing unit (Courtesy, Copeland Corporation).

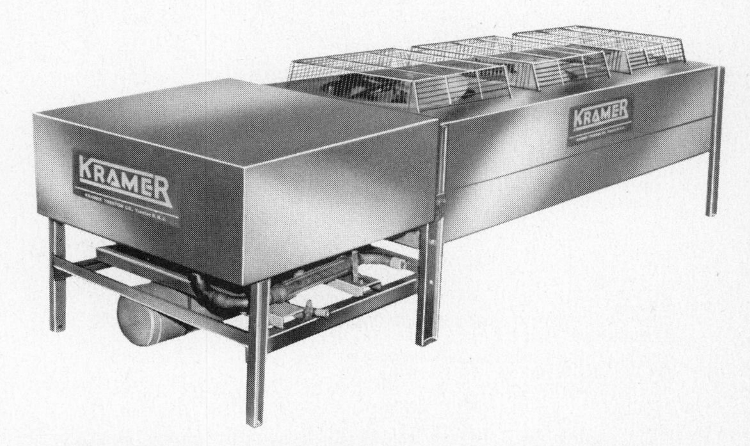

FIGURE 18-27 Large self-contained condensing unit (Courtesy, Kramer Trenton Company).

- Refrigeration piping is reduced—the discharge line is prepiped at the factory.
- The complete assembly can be factory tested for leaks and operation run-in to improve reliability.

Figure 18-28 is a schematic piping diagram of a multiple evaporator system on one air-cooled condensing unit. Instead of the evaporators commonly used in commercial refrigeration, it could be a fan coil unit for air-conditioning purposes. The essential difference would be the condensing unit selection procedure and the necessary controls to accomplish the application requirement.

Condensing units for commercial refrigeration of rooms purposes must have an approximate operating range from +35°F down to −20°F when matched with appropriate evaporators. Outdoor conditions vary from 115°F down to zero degrees and below. Refrigeration piping for low-temperature work also requires extra care.

Condensing units for comfort air-conditioning duty are nominally rated at 95°F outdoor ambient temperature to function with evaporators operating at 40°F with incoming air at 67° wb. However, there is an increasing need for comfort air-conditioning to operate when outside ambients fall below 75°F. High internal heat loads from people, lights, and mechanical equipment may demand cooling when outside conditions go down to 35°F and below.

So in the case of both commercial refrigeration and air-conditioning, low ambient operation is an important need. In earlier discussions we learned that in order to have proper refrigerant vaporization in a direct expansion (DX) evaporator coil it was necessary to maintain a reasonable pressure differential across the expansion device. In a normal air-cooled condenser operating between 80° and 115°F ambient, condensing pressures are sufficiently high, but in winter they can fall off 100 psi or more; thus the pressure across the expansion device will be insufficient to maintain control of liquid flow. Evapo-

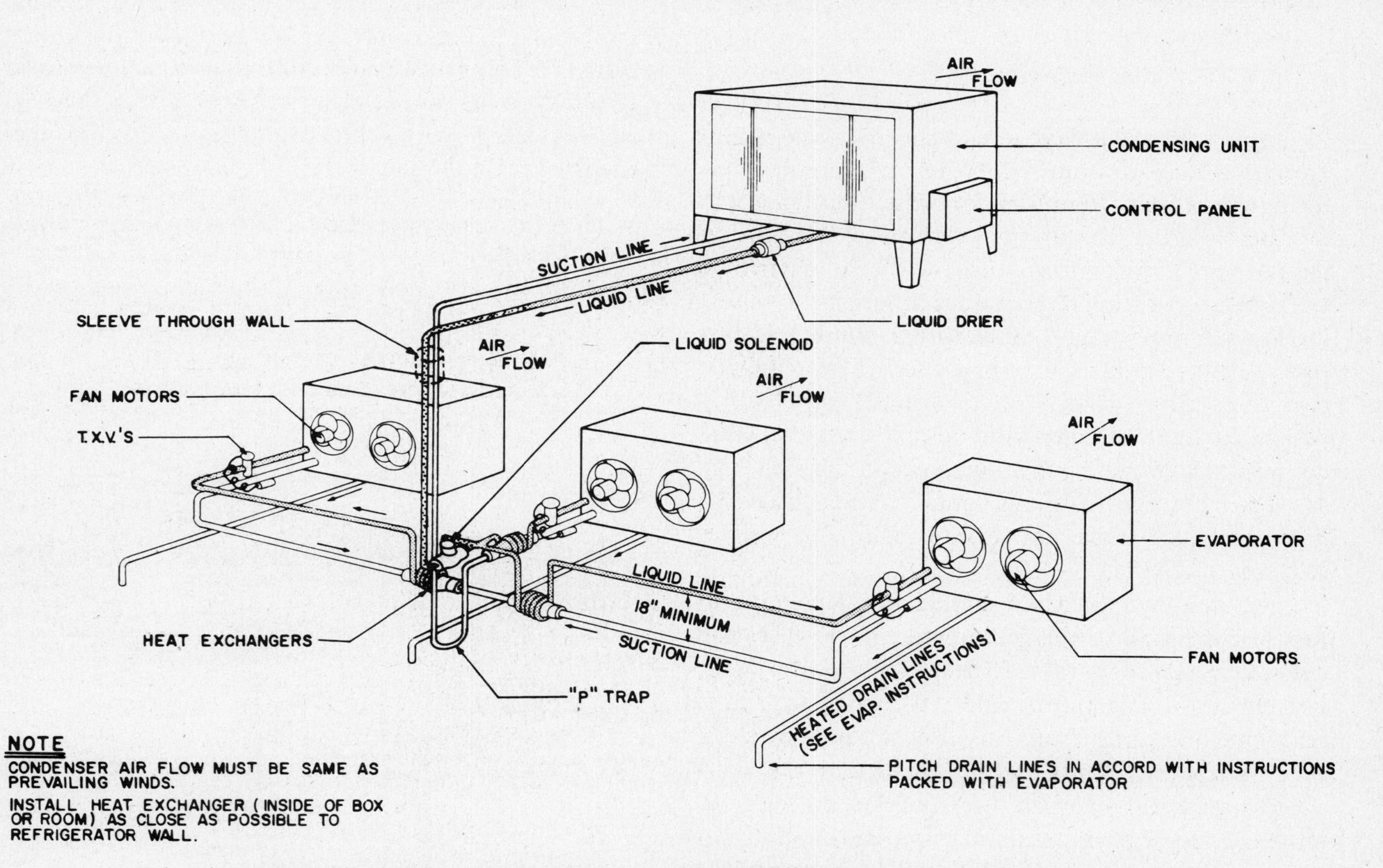

FIGURE 18-28 Schematic piping diagram for condensing units using two or more evaporators (Courtesy, Kramer Trenton Company).

rator operation becomes erratic. The thermal expansion valve will alternately open and close, first causing flooding back of liquid refrigerant and then starving the coil, and the valve closes. The capillary tube (if used) is worse, because it is a fixed metering device, and as pressure difference falls the flow of refrigerant is severely reduced. Below 65°F outdoor air a capillary tube system is in real trouble.

The solution to low ambient operation is to raise the head pressure artificially in order to maintain proper expansion across the evaporator. There are several different ways of doing this.

Where multiple condenser fans are employed on a single coil, the control system of the condensing unit can be equipped with devices to switch or cycle "off" the fans in stages. These controls are usually air stats that sense outdoor ambient temperature or pressure controls that sense actual head pressure. As fans are turned off the air flow across the coil is reduced, and thus its heat rejection is also reduced. Condensing temperatures therefore rise. All fans can be turned off, and then the coil acts as a static condenser with only atmospheric air movement. Its capacity at this point is usually ample to continue operating below freezing conditions.

Where only one condenser fan is used, air capacity can be reduced by employing a two-speed motor or solid-state speed controls, which have infinite speed control.

Another technique of restricting air flow across the condenser coil is the use of dampers on the fan discharge. These are used on nonoverloading centrifugal-type fans, not propeller fans. Dampers modulate from a head pressure controller to some minimum position, at which time the fan motor is shut off.

One common and important characteristic of the air flow restriction methods is that the full charge of refrigerant and entrained oil is in motion at all times, insuring positive motor cooling and oil lubrication to the compressor.

The other technique of artificially raising the head pressure is to back the liquid refrigerant up into the condenser tubes. The normal free-draining condenser has very little liquid in the coil; it is mostly all vapor. But if a control valve is placed in the liquid outlet of the condenser (Fig. 18-29) and is actuated by ambient temperature, some of the discharge gas is allowed to bypass the condenser and enter the liquid drain. This restricts drainage of the liquid refrigerant from the condenser, flooding it exactly enough to

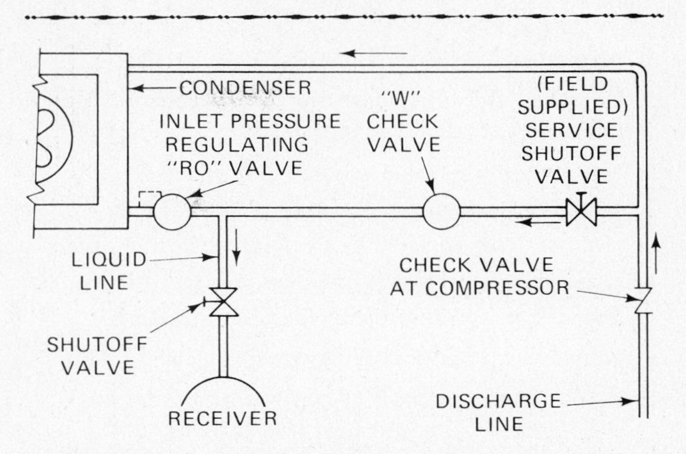

FIGURE 18-29 (Courtesy, Carrier Air-Conditioning Company).

maintain the head and receiver pressure. This method is not as common as restricting the air flow, because it is associated with systems that use an external receiver. Critically charged DX systems for air-conditioning normally do not employ receivers.

Receivers, as mentioned in Chap. 6 and shown in Fig. 18-1, are pressure vessels used to store refrigerant. They are most often used in commercial refrigeration systems, where the amount of refrigerant circulated varies widely and as a result can flood a water-cooled or air-cooled condenser to the point where its efficiency is impaired. So an external storage vessel is needed. Receivers are also needed on evaporative condensers, which likewise have limited storage capacity in their condenser surfaces.

Condensers and receivers should never be filled to more than 80% of their volume. The remaining 20% must be left empty to allow for expansion. The total system charge should always be checked against the storage capacity.

Maintenance of air-cooled condenser coils, as compared to water towers or evaporative condensers, is relatively easy. The coil face should be kept clean of leaves, sticks, blowing paper, etc. If the coil is in an industrial area it should be inspected for accumulations of grease and dust or chemical substances. Remove lint with compressed air or a vacuum cleaner. Grease or chemicals should be removed with a nonflammable, nonpoisionous solvent that will not attack aluminum or copper. Condenser fins can be corroded by salt air and over a period of time become less efficient. Fin damage from flying objects can usually be repaired with a fin comb.

PROBLEMS

1. Name three types of condensers.
2. In a shell and tube condenser, water generally flows in the tubes: true or false?
3. Name two types of air-cooled condensers.
4. In an evaporative condenser is most of the heat removal from sensible or from latent heat?
5. Is the capacity of a cooling tower basically a function of the wet-bulb or dry-bulb temperature?
6. Name two types of cooling towers.
7. Bleed-off is a method of controlling the buildup of scale forming concentrates: true or false?
8. The amount of make-up water to a tower depends on evaporation, bleed-off, and ______________.
9. Water velocity in tower piping generally should not exceed ______________ feet per second.
10. Using Table 18A, what size copper tube would be needed to flow 30 gpm and not exceed 5 ft/sec velocity?
11. Using Table 18B, what is the equivalent in feet of friction loss in a 1-in. close return bend?
12. The selection of water pumps is based on water flow and ______________.

19

REFRIGERATION PIPING AND ACCESSORIES

In previous discussions it was pointed out that the reliability of a field-assembled refrigeration and or air-conditioning system was greatly influenced by the proper design and installation of refrigeration piping and accessories. Proper refrigeration piping is as essential to the successful operation of the system as your veins and arteries are to your body. In both cases these pipes or tubes circulate a vital liquid or gas throughout the working parts of the system. Improper layout or sizing can change the efficiency of the various components, and thus affect the system's capacity.

The piping layout is usually made by an applications engineer, but the refrigeration technician who installs and services the system is also concerned in this layout because of the possiblity of difficulties and system faults. Also, the applications engineer's layout may be diagrammatic only, with no regard for the distances involved (horizontal or vertical) and the need for certain accessories as a result of specific on-site situations. Therefore, the technician is frequently in the position of having to interpret the engineer's intent and then applying sound technical modifications to complete the installation.

This chapter is designed to provide an understanding of the fundamentals of good piping layout in a simple refrigeration system. Piping for multiple systems is an advanced art that can be achieved with experience and guidance from other authoritative sources, such as equipment manufacturers, RSES. and ASHRAE.

METHODS AND MATERIALS FOR REFRIGERATION PIPING

In commercial refrigeration systems using the halogenated hydrocarbon-type refrigerants such as R-12, R-22, and R-502, copper tubing is normally used although welded steel pipe and fittings are often specified, especially in sizes above three inches. We will restrict this discussion to small systems using copper—either flared or soldered fittings as presented in Chap. 4. We will, however, note here the widespread use of mechanical-type fittings in air-conditioning work. These are called *quick-connect couplings*, *compression fittings*, and *flared connections*, and are associated with precharged equipment. The use of soldered joints is eliminated, as well as the need for extensive on-site leak testing, evacuating, and charging with refrigerant. This development paralleled and accelerated the rapid growth of residential air-conditioning, where system application is actually very similar. In home air-conditioning, evaporator temperatures are relatively constant and the system is always working at positive pressures, so that any small leaks will not permit water vapor to enter the unit. Also, refrigerant line sizes, application, and refrigerant charges are fairly predictable, so the manufacturer can build and ship precharged equipment with a high degree of reliability. Precharged lines are now made of copper and steel.

Commercial refrigeration applications don't enjoy the same simplicity of design and operation for two reasons. First, line sizes are larger and second, evaporator refrigerant temperatures vary widely, from +35°F down to −40°F and below. Structural location distances are more variable, and, particularly, commercial refrigeration systems on medium- and low-temperature work have refrigerant pressures at or near vacuum conditions. The presence of moisture from leaks or in the charging process can be disastrous. Oil return problems at low-temperature conditions are also more critical. Therefore, the use of precharged systems is almost impossible and is, indeed, nonexistent in commercial refrigeration work. The point being stressed is that learning the skills of refrigerant piping for commercial work qualifies the technician to handle most air-conditioning situations where built-up system piping is needed.

FUNCTIONS OF REFRIGERANT PIPING

The piping that connects the three major mechanical components of the system (Fig. 19-1) has two major functions: first, it provides a passageway for the circulation of refrigerant through the system; and second, it provides a passageway through which oil can be returned to the compressor. It should fulfill these two functions with a minimum of pressure drop and maximum protection for the compressor.

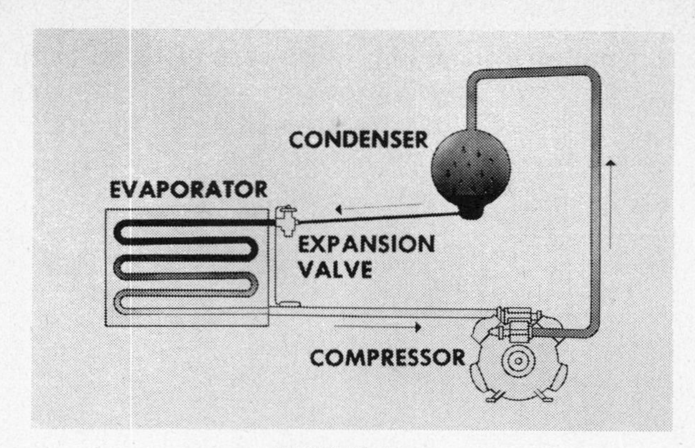

FIGURE 19-1.

The second function, the return of oil to the compressor, has been mentioned before, but let us review the topic here. Modern reciprocating compressors, like automotive engines, use forced-feed lubrication in the crankcase (Fig. 19-2). Some of the oil in the crankcase gets on the cylinder walls in the refrigeration compression and is blown out with the compressed gaseous refrigerant through the discharge valve ports. Some compressors pump much less oil than others, depending upon the design and manufacturing methods. However, there is no way to design a compressor so that none of the oil escapes into the refrigerant piping. This oil serves no useful purpose except to lubricate the compressor. However, the presence of oil in the piping influences the piping layout. A piping system that is not correctly selected or installed can cause;

(1) Burned-out compressor bearings due to lack of oil returning to the compressor for lubrication.

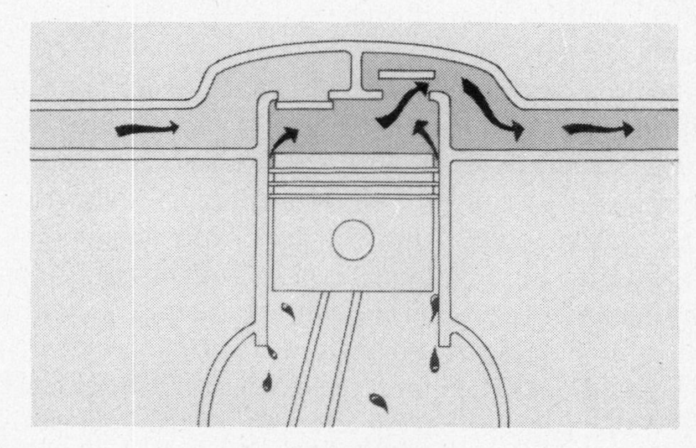

FIGURE 19-2.

(2) Broken compressor valves due to liquid refrigerant or large quantities of oil entering the compressor. The compressor is designed to pump vapor and will not pump liquid.

(3) Loss of capacity by restricting the flow of refrigerant through the system.

BASIC PIPING PRECAUTIONS

A refrigeration technician should learn this subject well enough to avoid costly mistakes. There are five basic rules to keep in mind when piping a system:

(1) ***Keep It Clean!*** Cleanliness is a key factor in the actual installation. Dirt, sludge and moisture will cause breakdown in the system and must be avoided. Neat work will save many service difficulties. A lintless cloth drawn through the tubing will remove most foreign particles.

(2) ***Use As Few Fittings As Possible.*** (See Fig. 19-3.) Fewer fittings mean less chance for leaks and needless pressure drops. Refrigerant is expensive and can leak through extremely small openings.

(3) ***Take Special Precautions in Making Every Solder Connection.*** Use the right solder for each application and follow the soldering technique as recommended by the equipment manufacturer.

(4) ***Pitch Horizontal Lines in the Direction of Refrigerant Flow.*** (See Fig. 19-4.) Because oil may cling to the inner walls of the tubing, horizontal lines should be pitched in the direction of refrigerant flow. This pitch, which allows oil to flow in the right direction, should be at least $\frac{1}{2}$ in. per 10 feet of run. Pitch also avoids backward flow during shutdown.

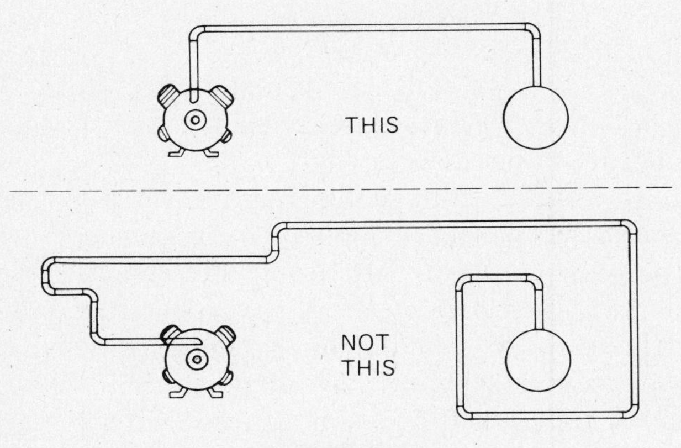

FIGURE 19-3.

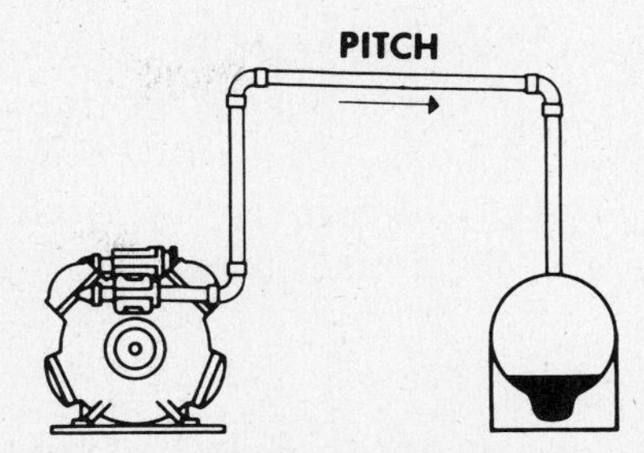

PITCH HORIZONTAL LINES ½ INCH PER TEN FEET

FIGURE 19-4 Pitch horizontal lines $\frac{1}{2}$″ per 10 ft (Courtesy, The Trane Company).

OIL IN THE PIPING

As noted some oil is always discharged into the piping by the compressor and it must be returned to the compressor to prevent damage.

The two refrigerants most commonly used in refrigeration and air-conditioning are R-12 and R-22. In their liquid forms these refrigerants will mix with oil and carry it along through the piping. However, in their gaseous states, they are poor carriers of oil. Oil under pressure in hot gas discharge lines turns into a mist. *The gaseous refrigerant and the oil will not mix.* If the piping is incorrectly designed, oil will collect on the walls of the tubing and drain down to low points in the system. However, if the gaseous refrigerant travels through the system fast enough, the oil will be entrained and carried along with it; therefore, *the velocity of refrigerant flow* is critical to oil flow, and also must be considered from the standpoint of noise and vibration.

HOT GAS LINES AND LIQUID LINES

Having covered the basic functions of these lines in previous discussions, you are now ready to study their individual problems. First, the installation of hot gas (discharge lines) is associated only with built-up systems that have separate components.

Condensing units, self-contained equipment, chillers, etc., will have all this work done at the factory. However, to install built-up jobs or to service existing systems you should be familiar with refrigeration piping techniques for hot gas lines.

Hot Gas Lines

The basic function of hot gas lines is to conduct compressed gas and entrained oil to the condenser, without creating excessive pressure drop. When the compressor and condenser are on approximately the same level, the hot gas line can go directly to the condenser inlet with the proper pitch in the horizontal line ($\frac{1}{2}$ in. per 10 ft of run).

Where the condenser is located above the compressor as in Fig. 19-5 and the vertical riser is not more than eight feet, slope the horizontal run at the recommended pitch away from the compressor in the direction of flow.

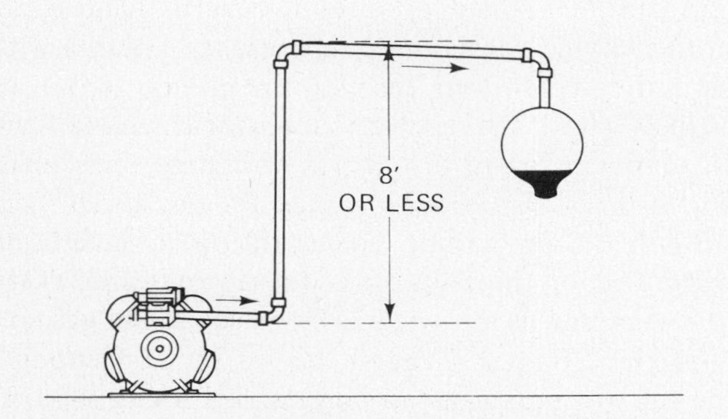

FIGURE 19-5 8 ft or less (Courtesy, The Trane Company).

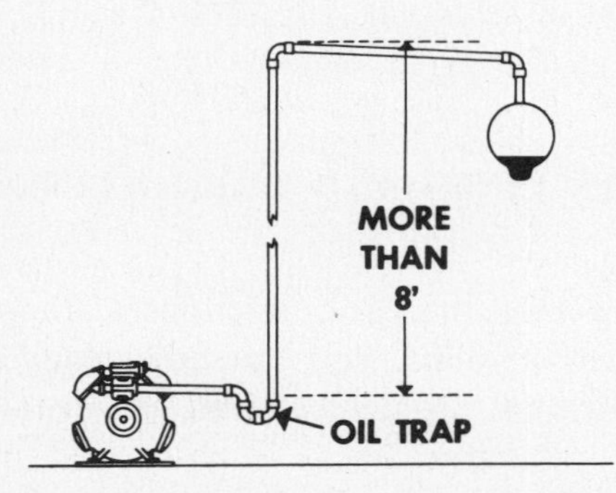

FIGURE 19-6 More than 8 ft (Courtesy, The Trane Company).

Where the vertical riser to the condenser is more than eight feet above the compressor (Fig. 19-6), the compressor must be protected from the possibility of oil draining down the tubing and spilling back onto the compressor discharge valve. You can see how this would happen when the compressor shuts off and there is no refrigerant flow. A simple oil trap (or loop) installed in the hot gas line near the compressor will prevent this oil from entering the compressor.

If the compressor is located where its temperature can be lower than that of the condenser or receiver, a check valve should be installed in the hot gas line near the condenser. This will prevent migration of refrigerant to the compressor during the "off" cycle.

Simple oil traps (Fig. 19-7) may be made with two standard elbows and a street ell. The trap serves two functions. It provides a place for the collection of oil, and it provides a means for the refrigerant vapor to entrain the oil and carry it up the riser. Oil will collect in the bottom of the trap. As it fills, the passage for the refrigerant becomes smaller. The velocity of the vapor will, therefore, increase and cause the oil to become entrained.

Hot Gas Velocity

In sizing and arranging hot gas lines, select tubing with a diameter small enough to provide sufficient velocity through the line to carry oil into the condenser. On the other hand, the diameter must be large enough to prevent excessive pressure drop. Here are two velocity figures to remember:

(1) ***In Horizontal Lines*** the gas velocity should be at least *750 feet per minute* to keep the oil moving in the direction of flow.

(2) ***In Vertical Risers*** a velocity of at least *1,500 feet per minute* is required.

Later on we will present tables and charts for determining the proper selection for both hot gas and suction lines.

In addition to these minimum velocities we also have a recommended limit on maximum velocity. Too high a velocity may cause objectional noise, vibration, and pressure drop. For most refrigeration and comfort air-conditioning jobs the velocity should not exceed 3,000 ft/min. By using velocities near or within the 3,000 ft/min limit, rather than the minimums, savings on pipe costs can be achieved.

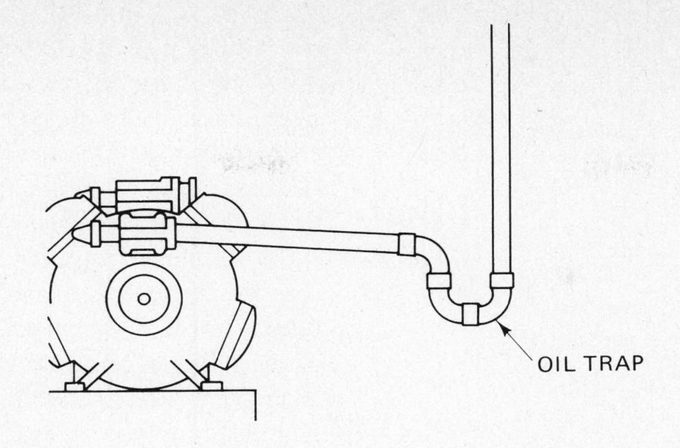

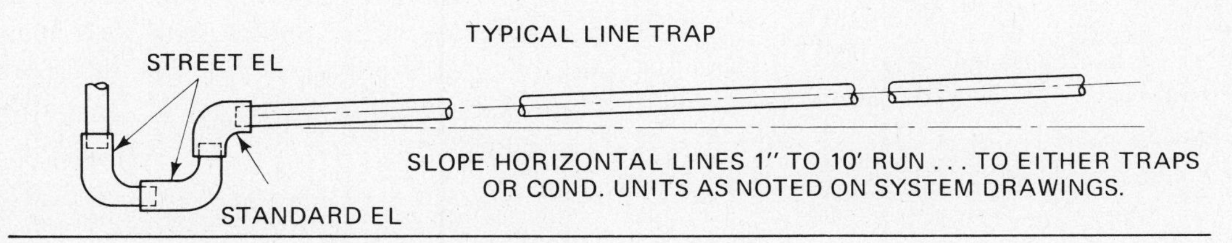

FIGURE 19-7 Typical line trap.

Pressure Drop

In hot gas lines the pressure drop is normally limited to three pounds per square inch (psi). Remember this figure. Greater pressure drop will cause unnecessary work for the compressor and increase operating costs.

You now have four figures to associate with hot gas lines as illustrated in Fig. 19-8. From Chap. 17 you will recall that some machines are equipped with capacity reduction devices to operate at partial capacity to balance the load. During unloaded or part load operation, a smaller amount of refrigerant is discharged from the compressor than at full load. *When a single vertical hot gas line is sized for full load, the velocity at part load will not be sufficient to carry the entrained oil.* On the other hand, if the vertical riser is sized for only part load, the pressure drop under full load will be excessive. Therefore, in a system having capacity control and a vertical rise of eight feet or more, a *double riser* should be installed in the hot gas line as illustrated in Fig. 19-9. The *smaller* riser is sized for not less than 1,500 ft/min velocity at *minimum* load. The larger riser is sized for not less than 1,500 ft/min velocity through both risers at full load. The oil trap at the bottom of the larger riser is a necessary part of this arrangement.

MINIMUM VELOCITIES:
HORIZONTAL LINES ...... 750 fpm
VERTICAL RISERS .... 1500 fpm
MAXIMUM VELOCITY ...... 3000 fpm
MAXIMUM PRESSURE DROP.... 3 psig

FIGURE 19-8.

When operating at part load, the hot gas velocity is not sufficient to carry oil up the larger riser. As a result the oil drops back down the riser and collects in the oil trap. When sufficient oil has collected to seal the trap as in Fig. 19-9(a), the hot gas flows up the smaller riser at a velocity that will assure carry-over of the oil mist. When the capacity of the system increases enough to break the oil seal in the trap, both lines again carry the hot gas and oil.

You may be wondering why the oil is not drained from the oil trap directly back into the compressor crankcase; in fact, why worry about carrying it all through the system?

By installing an oil separator as shown in Fig. 19-10 the amount of circulated oil can be minimized. This is an accessory that should be considered if there are unusually long runs involved. As the name implies, it causes most of the oil and refrigerant gas to separate. Oil accumulating at the bottom actuates a float valve, which allows the oil to escape into the compressor suction chamber (lower pressure) and then through a check valve into the crankcase. Remember oil separators add to the job costs and

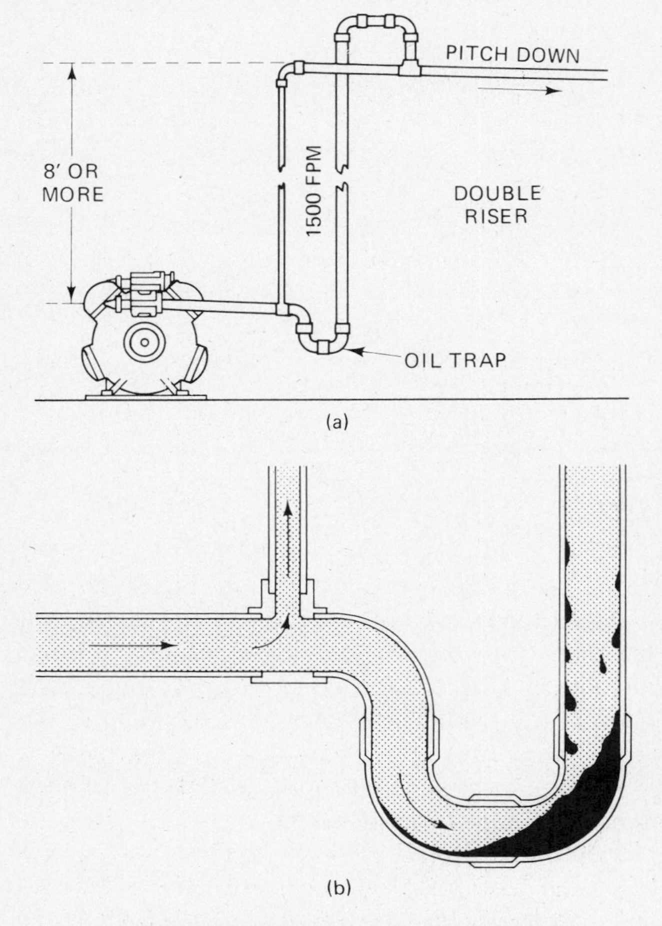

FIGURE 19-9 (Courtesy, The Trane Company).

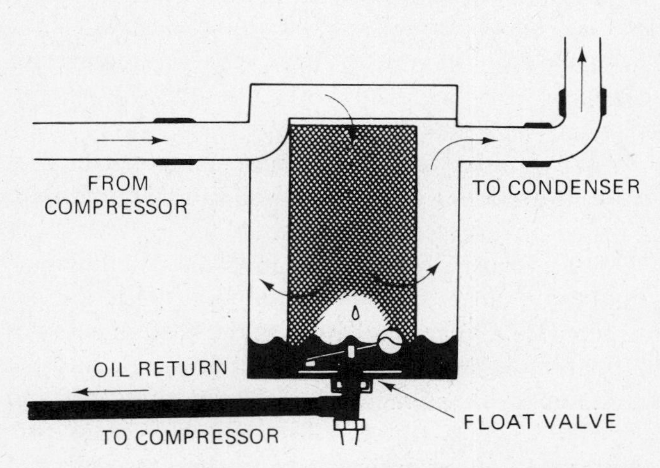

FIGURE 19-10 Oil separator (Courtesy, The Trane Company).

are not warranted for most normal installations. If an oil separator is used on a compressor with unloading or capacity control, double risers are not required. Oil separators are not a cure-all for oil troubles, because some oil will get past the separator and collect in low spots in the system.

A muffler is another accessory item that may be recommended by the compressor manufacturer to reduce noise or gas pulsations. The preferred location of a muffler is in the down-flow side of the hot gas loop (Fig. 19-11). In this position oil accumulation or refrigeration condensation is at a minimum. Also, the muffler is not in the way of normal service to the compressor. The muffler may also be placed in a horizontal section of hot gas line, providing the outlet connection comes off at the bottom to avoid trapping oil.

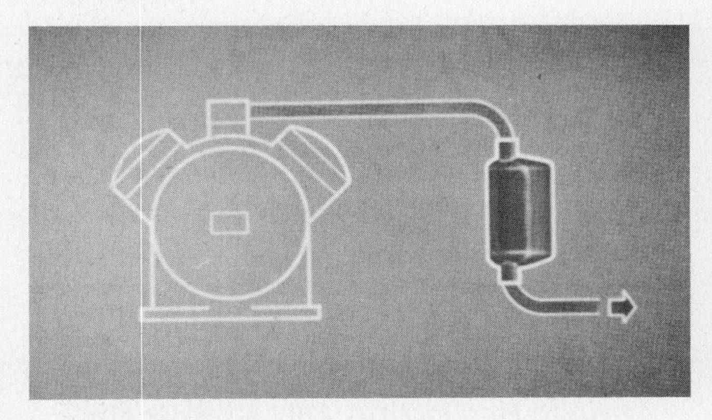

FIGURE 19-11 Muffler location (Courtesy, The Trane Company).

Liquid Line

Some oil will reach the condenser where it will mix with the liquid refrigerant at the bottom of the condenser. This mixture is then piped to the expansion device (TWX valve or capillary tube). Because liquid is more dense than gas, the liquid line can have a smaller diameter than the hot gas line.

To understand what happens in the liquid line, one must understand more about refrigerants. If the pressure on a liquid refrigerant decreases while its temperature remains the same or is increased, some of this refrigerant will eventually flash into a gas. If this occurs, the expansion valve operates inefficiently and the system therefore losses some of its capacity.

The pressure drop caused by friction in the liquid line should not exceed 3 psi. Horizontal lines can usually be sized well within this limit. Difficulty may be encountered in long vertical risers as illustrated in Fig. 19-12. The pressure on a particle at the bottom of the riser is greater than the pressure on a particle at the top of the riser. This is merely because of the weight of the liquid. This pressure is approximately 0.55 lb/in.^{2} for each vertical foot of pipe for R-12 and slightly less for R-22. Therefore, in a 10-foot riser of R-12, the pressure on the bottom will be $5\frac{1}{2}$ psi greater than the pressure at the top. This is an important point to remember, because the pressure on the liquid refrigerant determines the temperature at which it will boil or change to a gas. If refrigerant at the bottom of this riser happens to be just at the balance point between liquid and gas, it will boil into a gas instantly if the pressure on it is decreased. Therefore, before the refrigerant starts up

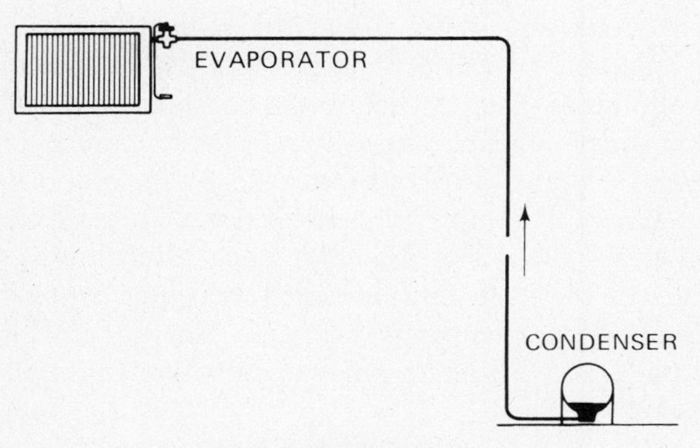

FIGURE 19-12.

the riser, it must be subcooled enough so that it will not change to a gas when its pressure decreases to that which exists at the top of the riser. Accessories such as valves, sight glasses, strainers, and driers are a source of additional pressure drop. In practice, subcooling of 10°F is generally sufficient for elevations up to 25 feet. Check the manufacturer's specifications on the equipment being installed for the liquid line size and the degree of subcooling provided to verify that there is enough lifting capacity to prevent refrigerant flashing.

So far the text has covered the hot gas line and the liquid line, along with problems caused by:

- the presence of oil in the piping
- variable capacity compressors
- vertical risers

Suction Line

This line carries cool refrigerant vapor or gas and compressor oil from the evaporator to the inlet of the compressor. Because the refrigerant in this line is a gas, which does not mix with oil, the considerations here are much the same as they were for the hot gas line. The piping must be such that oil will not return in large quantities or slugs, which may cause broken valves. The system should also be designed and installed so that the refrigerant does not return to the compressor in liquid slugs.

To learn how to prevent liquid slugging we first consider what happens in a simple system when the compressor stops operating when its cooling requirements are satisfied. The evaporator is still filled with refrigerant, part liquid and part gas. There will also be some oil present. The liquid refrigerant and oil may drain by gravity to points where, when the compressor starts running again, the liquid will be drawn into the compressor and cause liquid slugging. The piping design must prevent liquid refrigerant or oil from draining to the compressor during shutdown.

When the compressor is above the evaporator there is no problem. If the compressor is on the same level or below the evaporator, as in Fig. 19-13, a rise to at least the top of the evaporator must be placed in the suction line. This inverted loop is to prevent liquid draining from the evaporator into the compressor during shutdown. The sump at the bottom of the riser promotes free drainage of liquid refrigerant away from the thermostatic expansion valve bulb, thus permitting the bulb to sense suction gas superheat instead of evaporating liquid temperature.

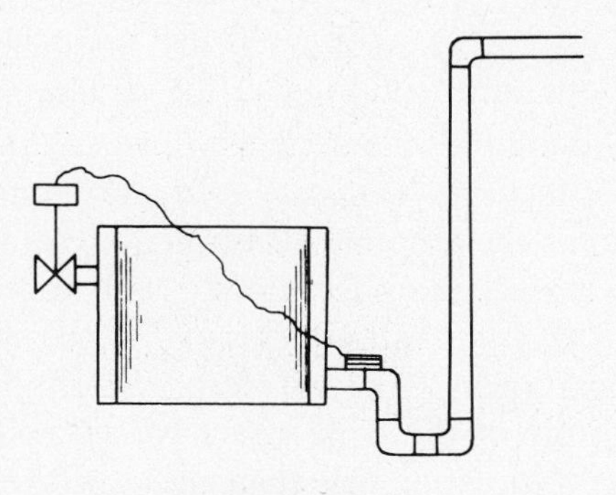

FIGURE 19-13.

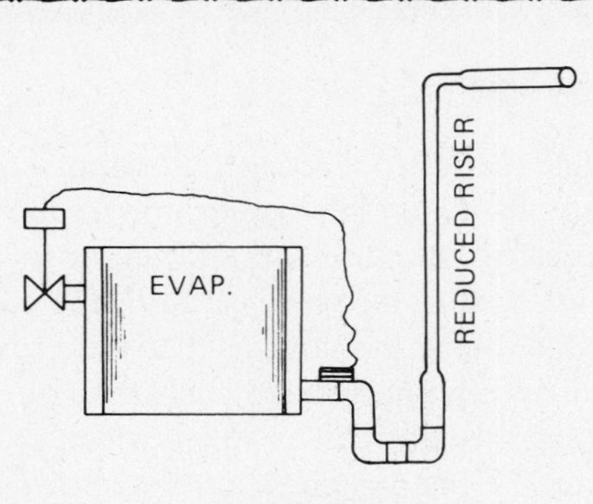

FIGURE 19-14.

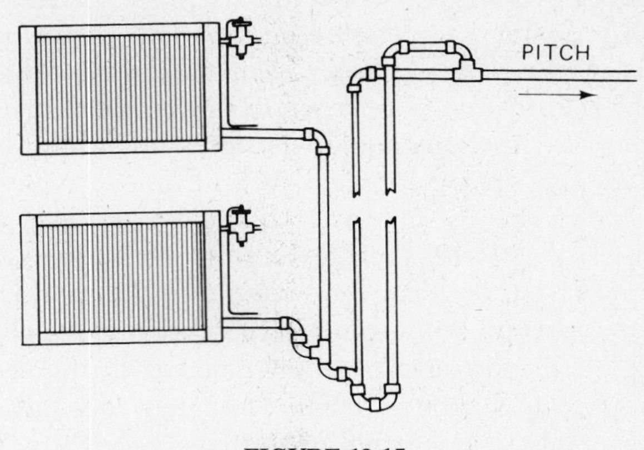

FIGURE 19-15.

Although it is important to prevent liquid refrigerant from draining from the evaporator to the compressor during shutdown, it is just as important to avoid unnecessary traps in the suction line near the compressor. Such traps would collect oil, which on startup might be carried to the compressor in the form of slugs, thereby causing serious damage.

Where the system capacity is variable because of capacity control or some other arrangement, a short riser will usually be sized smaller than the remainder of the suction line (Fig. 19-14) for a velocity of not less than 1,500 ft/min to insure oil return up the riser. Although this smaller pipe has a higher friction, its short length adds a relatively small amount to the overall suction line friction loss.

In general, the pressure drop for the total suction line should be a maximum of 2°F or 1 psi (pound per square inch) to avoid loss of system capacity. The compressor cannot pump or draw gas nearly as effectively as pushing or compressing it. (You'll recall the pressure drop could be approximately 3 psi for the hot gas line.)

If the system employs capacity control, it will be necessary to provide a double riser in the suction line, as illustrated in Fig. 19-15. Its operation is similar to the hot gas double riser. When maximum cooling is required, the system will run at full capacity, and both risers will carry refrigerant and oil. On part load, as the amount of refrigerant being evaporated decreases, the gas velocity will also decrease to a point where it will not carry oil upward through the vertical risers. The oil trap, which is located at the bottom of the large riser, will fill with oil. All the refrigerant vapor will then pass up through the smaller riser, carrying oil with it. As the system load increases and more refrigerant is passed through the evaporator, this increased pressure will break the oil seal in the trap and carry oil upward through both risers. *Velocity through both risers should be sized for not less than 1,500 ft/min at minimum load.*

Figure 19-15 also illustrates the use of multiple evaporators installed below the compressor. Notice that the piping is arranged so refrigerant cannot flow from the upper evaporator into the lower evaporator.

Where a number of vertical risers are necessary on either suction or discharge lines, as illustrated in Fig. 19-16, it is recommended that line traps be installed approximately every 20 feet so that the storage and lifting of oil can be done in smaller stages.

How are refrigerant lines sized to give sufficient velocity and still avoid excessive pressure drop? To begin with, the choice and internal dimensions of refrigerant tubing must be determined. Copper tubing is available in three standard weights known as K, L, and M. K is heavy duty, L is average duty,

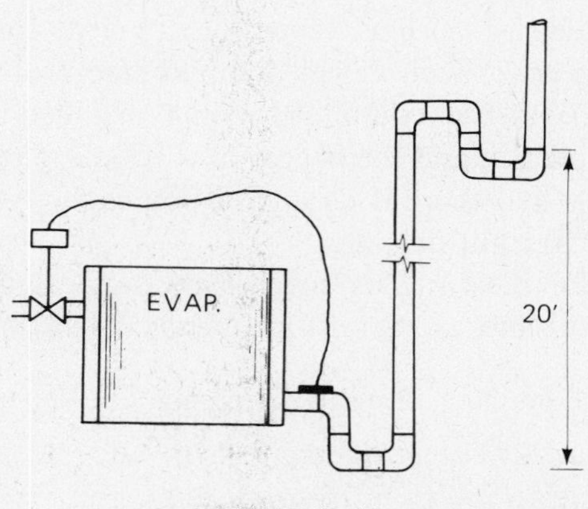

FIGURE 19-16.

DIMENSIONS OF COPPER TUBING

	DIAMETERS		TRANSVERSE AREA OF BORE	
TYPE	*Outside Inches*	*Inside Inches*	*Sq. In.*	*Sq. Ft.*
	$\frac{1}{2}$	0.430	0.1452	0.001008
	$\frac{5}{8}$	0.545	0.2333	0.001620
	$\frac{3}{4}$	0.666	0.3484	0.002419
	$\frac{7}{8}$	0.785	0.4840	0.003361
	$1\frac{1}{8}$	1.025	0.8252	0.005730
L	$1\frac{3}{8}$	1.265	1.257	0.008728
	$1\frac{5}{8}$	1.505	1.779	0.01235
	$2\frac{1}{8}$	1.985	3.095	0.02149
	$2\frac{5}{8}$	2.465	4.772	0.03314
	$3\frac{1}{8}$	2.945	6.812	0.04730
	$3\frac{5}{8}$	3.425	9.213	0.06398

FIGURE 19-17.

and M is light duty. K has the thickest wall, L is next, etc. L is recommended for ordinary refrigeration work and will be used in this discussion. Figure 19-17 is a table of nominal outside sizes and their inside diameters for type L tubing.

Figure 19-18 is a chart for R-12, published by the Trane Company, that reflects tubing sizes, hot gas velocity, and the load in tons of refrigeration (at 40°F suction and 105°F condensing temperature, standard conditions for compressors ratings). To save time we will assume the load is 5 tons and the discharge tubing connection is $1\frac{5}{8}$ in. O.D. (refer to Fig. 19-18). Follow the vertical line up from 5 tons to where it intersects the diagonal line representing $1\frac{5}{8}$ in. O.D. At that point read to the left and note the velocity is slightly less than 500 FPM; obviously the tubing is too large. What would the velocity be at 10 tons with the same $1\frac{5}{8}$ in. O.D? It would be nearly 1,000 ft/min, and at 15 tons it would be approximately 1,500 FPM.

In most cases the compressor discharge connection has been factory sized, based on average job conditions; however, it is important to recheck the velocity, since local conditions sometimes require an increase or decrease in size. Suppose you had a load of 15 tons at 40°F suction and 105°F condensing temperature, using R-12. What size tubing will give sufficient velocity in horizontal piping? Moving up the vertical line in the chart, at 15 tons you will find that to provide a minimum of 750 ft/min velocity, $2\frac{1}{8}$ in. O.D. tubing could be used. You could also

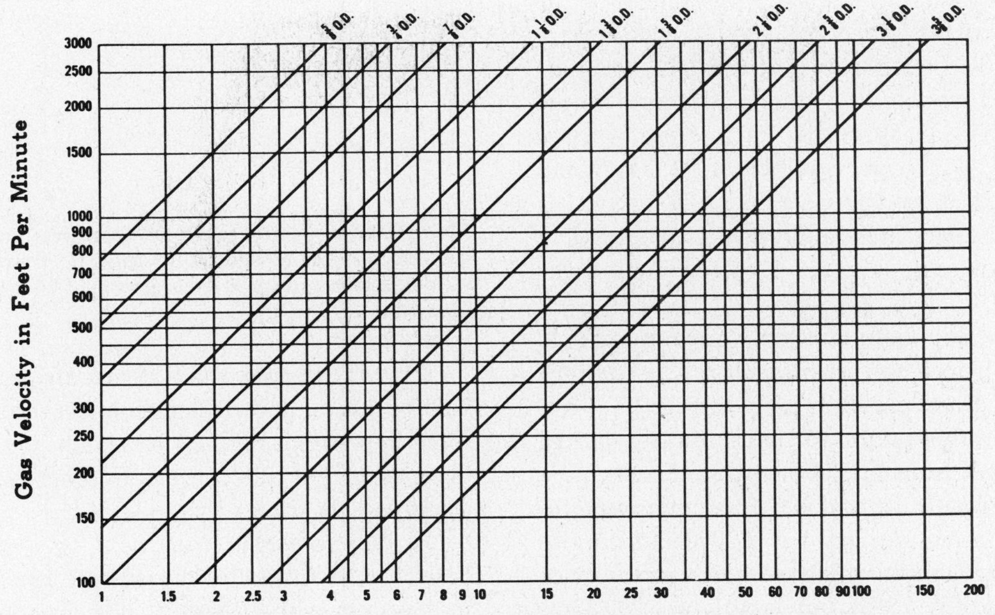

FIGURE 19-18 Hot gas velocities for refrigerant 12 only (Courtesy, Trane Company).

use $1\frac{5}{8}$ in. or $1\frac{3}{8}$ in. diameters, because they will provide a velocity of less than 3,000 ft/min.

For vertical risers in the hot gas line, you would have to select an outside diameter (O.D.) which would provide a minimum velocity of 1,500 ft/min. At 1,500 ft/min you'll find the vertical line for a 15-ton load, indicating that an O.D. of $1\frac{5}{8}$ in. could just barely be used. A smaller, $1\frac{3}{8}$ in. size would also be acceptable.

Whenever possible, the same size tubing is selected for both vertical and horizontal runs. In this case the $1\frac{5}{8}$ in. size would be acceptable, provided its pressure drop is within a 3 psi limit (to be covered later).

Figure 19-18 is based on the conditions of 40°F suction and 105°F condensing temperature. For other conditions a correction factor table (Fig. 19-19a) must be used. For a system operating at 30°F suction and 95°F condensing, the table shows the hot gas velocity correction is 1.12. The calculated tonnage of the system must be multiplied by this factor. You then use the corrected tonnage on the hot gas velocity chart to determine the size of the tubing. As an example, suppose the load is 10 tons; you would then read the R-12 chart at 11.2 tons.

The sizing of suction lines is very similar to the hot gas method, but since the suction gas is at a lower pressure and temperature it will have a different density, and thus its flow rate is different. Figure 19-19b is the suction chart for R-12. Here you see that for a load of 25 tons, a velocity of 1,500 ft/min can be maintained by using $3\frac{1}{8}$ in. O.D. tubing. If data on other operating conditions are needed, a correction factor table (Fig. 19-19c) for suction gas velocity is used.

For any suction temperature, you can quickly determine the proper correction factor. The tonnage capacity found in the previous table in Fig. 19-19b is then divided by this factor to determine the actual capacity of the suction line. For example, assume a 20°F suction temperature and 105°F condensing temperature. The correction factor is 1.29. Returning to the previous problem in which we had a capacity of 25 tons, we multiply 25 by our factor of 1.29. Thus we would enter the table at 32.3 tons. Velocity tables for other refrigerants would be used in the same way as the R-12 table.

CONDENSING TEMP. F	SUCTION TEMPERATURE F								
	−10°	0°	+10°	20°	30°	35°	40°	45°	50°
85°	1.37	1.29	1.26	1.23	1.21	1.20	1.18	1.17	1.16
90°	1.31	1.27	1.23	1.20	[illegible]	1.15	1.13	1.12	1.11
95°	1.25	1.22	1.18	1.14	1.12	1.10	1.08	1.07	1.06
100°	1.22	1.18	1.14	1.11	[illegible]	1.06	1.04	1.03	1.02
105°	1.17	1.13	1.09	1.06	1.02	1.01	1 00	.98	.97
110°	1.13	1.09	1.05	1.02	.99	.97	.96	.95	.94
115°	1.08	1.04	1.00	.97	.95	.93	.91	.90	.88

(a)

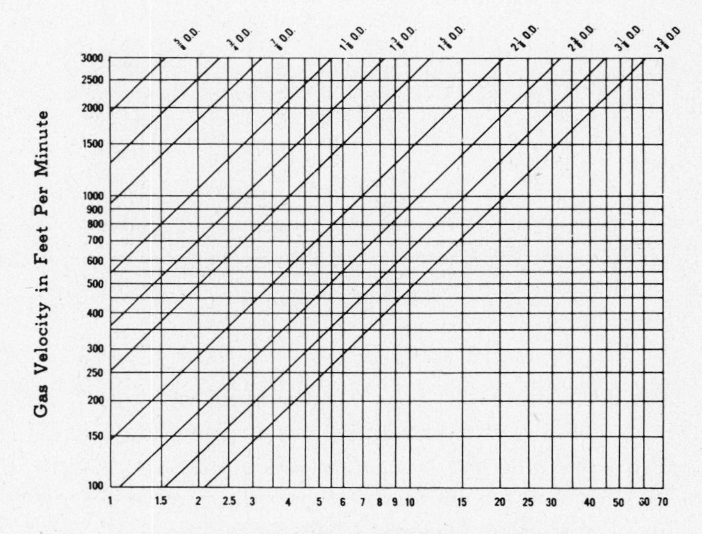

Load in Tons at 40 degrees Suction and 105 degrees Condensing Temperature

(b)

SUCTION TEMP. DEGREES	-20	-10	0	10	20	30	40	50
SUCTION PRESSURE LBS. GAGE	0.6	4.5	9.2	14.7	21.1	28.5	37.0	46.7
CORRECTION FACTOR	2.34	2.00	1.73	1.49	1.29	1.14	1.00	0.89

(c)

FIGURE 19-19 (a) Hot gas velocity correction factors; (b) Suction gas velocities for refrigerant 12 only; (c) Correction factors for various suction pressures—refrigerant 12 only (Courtesy, Trane Company).

Pressure drop in the hot gas refrigerant lines must also be given full consideration to avoid loss of capacity in the system. Figure 19-20 is a hot gas line pressure drop table that reflects various lengths and diameters of tubing. It shows the tons of capacity at pressure drops of 1 to 5 psi. *Note the 3 psi limit is printed in bold type*. It shows that if you use $1\frac{3}{8}$ in. O.D. tubing in a system with 10 feet of hot gas piping, at design pressure drop, this $1\frac{3}{8}$ in. tubing will be able to handle a load of 18.8 tons. If different sizes of tubing are selected for use in the same line (because of velocity),

		Outside Diameter of Pipe—Inches							
		$\frac{5}{8}$	$\frac{7}{8}$	$1\frac{1}{8}$	$1\frac{3}{8}$	$1\frac{5}{8}$	$2\frac{1}{8}$	$2\frac{5}{8}$	$3\frac{1}{8}$
Length of Pipe Feet	*Pressure Drop Psi*	*Inside Diameter of Pipe—Inches*							
		0.545	0.785	1.025	1.265	1.505	1.985	2.465	2.945
	1	0.97	2.9	7.0	10.3	16.0	34.0	61.1	123
	2	1.6	4.1	9.6	15.0	24.0	50.6	90.4	184
10	**3**	**2.0**	**5.2**	**11.8**	**18.8**	**30.2**	**63.5**	**113**	**228**
	4	2.4	6.1	12.1	23.3	35.2	74.4	142.8	268
	5	2.5	6.9	14.4	25.3	40.1	94.5	151.2	305
	1	0.86	2.3	4.6	8.1	13.0	27	48.6	87
	2	1.3	3.2	6.9	11.9	19.1	40.3	72	130
20	**3**	**1.6**	**4.2**	**8.6**	**15.0**	**24.0**	**50.4**	**90**	**162**
	4	1.8	4.6	10.1	17.6	28.1	59.2	105.3	190
	5	2.0	5.0	11.4	20.1	31.9	67.1	119.7	215
	1	0.7	2.0	4.1	7.3	11 7	24.5	44.0	71
	2	1.1	2.9	6.2	10.8	17.3	36.4	65	106
30	**3**	**1.4**	**3.4**	**7.3**	**13.2**	**20.2**	**45.5**	**78.0**	**132**
	4	1.6	4.5	9.1	16.0	25.4	53.5	95.5	155
	5	1.8	5.4	10.0	18.0	28.8	60.6	109	176
	1	0.6	1.6	3.2	5.7	9.1	19.0	34.1	55
	2	0.9	2.3	4.8	8.4	13.4	28.2	50.5	82
50	**3**	**1.1**	**2.9**	**6.0**	**10.5**	**16.8**	**35.4**	**63.2**	**102**
	4	1.3	3.5	7.1	12.4	19.7	41.6	74.5	120
	5	1.4	3.9	8.0	14.1	22.4	47.1	84.4	136
	1	0.4	1.1	2.2	3.8	6.1	12.9	23.1	37
	2	0.6	1.6	3.2	5.7	9.1	19.0	34.1	55
100	**3**	**0.8**	**2.0**	**4.1**	**7.1**	**11.4**	**24.0**	**42.8**	**69**
	4	0.9	2.3	4.8	8.4	13.4	28.2	50.5	82
	5	1.0	2.6	5.4	9.5	15.2	32.0	57.2	92
	1	0.3	0.9	1.7	3.1	4.9	10.3	18.4	30
	2	0.5	1.3	2.6	4.5	7.2	15.2	27.2	44
150	**3**	**0.6**	**1.6**	**3.2**	**5.7**	**9.1**	**19.0**	**34.1**	**55**
	4	0.7	1.9	3.8	6.7	10.7	22.4	40.1	65
	5	0.8	2.1	4.3	7.6	12.1	25.4	45.5	73

Above table includes allowance for pressure drop in an average number of fittings.
Recommended pressure drop at maximum load is shown in heavy type.

FIGURE 19-20 Tonnage capacities of discharge lines delivering hot refrigerant-12 vapor from compressor to condenser (Courtesy, Trane Company).

their combined pressure drop should not exceed the maximum recommended pressure drop.

To check suction lines, you can check the pressure drop by using a similar table (Figure 19-21). *Note that the recommended limit of 1 psi is printed in bold type.* Reading from the left, you'll find that for a 10-ft suction line with a 1-psi drop, a system capacity of 5.3 tons could be handled by using $1\frac{3}{8}$ in. O.D. tubing. A 20-foot length of $1\frac{3}{8}$ in. tubing can only handle 4.9 tons, and so on. Suppose you had tentatively selected a $2\frac{5}{8}$ in. O.D. tubing for a 20-ton load at 1,800 ft/min. To check this $2\frac{5}{8}$ in. size, let's assume the suction lines are 50 feet long. For the recommended 1 psi, a $2\frac{5}{8}$ in. O.D. line at 50 feet can handle 23.0 tons, which is adequate.

Correction factors for various suction pressures are shown at the bottom of Fig. 19-21. Multiply the correction factor by the assumed load and use this data on the table. As an example, at 30°F suction the correction factor is 1.14, and in the previous example the load of 20 tons would be corrected to 22.8. The line size of $2\frac{5}{8}$ in. would still be suitable.

TONNAGE CAPACITIES OF SUCTION LINES DELIVERING REFRIGERANT-12 VAPOR FROM COOLING COIL TO COMPRESSOR

To be used only for Refrigerant-12 vapor at 40 degrees suction temperature. For other suction temperatures, find correction factor. Then, to determine diameter of suction line, multiply factor by actual load. Or, to find capacity of a given suction line, divide capacity by the correction factor.

LENGTH OF PIPE, FEET	PRES-SURE DROP, PSI	OUTSIDE DIAMETER OF PIPE — INCHES								
		5/8	7/8	1 1/8	1 3/8	1 5/8	2 1/8	2 5/8	3 1/8	3 5/8
		INSIDE DIAMETER OF PIPE — INCHES								
		0.545	0.785	1.025	1.265	1.505	1.985	2.465	2.945	3.425
10	1	**0.6**	**1.5**	**3.1**	**5.3**	**8.5**	**17.8**	**32.0**	**51.5**	**76.5**
	2	0.8	2.2	4.4	7.9	12.6	26.4	47.5	76.5	114.0
	3	1.1	2.8	5.7	9.9	15.9	33.5	59.8	97.4	144.5
	4	1.2	3.3	6.7	11.7	18.7	39.3	70.5	114.0	170.0
	5	1.4	3.6	7.5	13.2	21.4	45.0	80.3	131.0	193.0
20	1	**0.5**	**1.4**	**2.9**	**4.9**	**7.9**	**16.6**	**29.9**	**48.0**	**71.5**
	2	0.8	2.1	4.1	7.4	11.8	24.7	44.3	71.5	107.0
	3	1.0	2.6	5.3	9.2	14.8	31.3	56.0	91.0	135.0
	4	1.1	3.1	6.2	10.9	17.6	36.8	65.8	107.0	158.5
	5	1.3	3.4	7.0	12.5	20.0	42.6	75.0	122.0	181.0
30	1	**.5**	**1.3**	**2.6**	**4.6**	**7.3**	**15.3**	**27.6**	**44.4**	**66.0**
	2	.7	1.9	3.8	6.8	10.9	22.8	41.0	66.0	98.5
	3	.9	2.4	4.9	8.5	13.7	29.0	51.6	84.0	125.0
	4	1.1	2.9	5.8	10.1	16.2	34.0	60.8	98.5	146.5
	5	1.2	3.1	6.5	11.5	18.5	38.8	69.3	113.0	167.0
50	1	**0.4**	**1.1**	**2.2**	**3.8**	**6.1**	**12.8**	**23.0**	**37**	**55**
	2	0.6	1.6	3.2	5.7	9.1	19.0	34.1	55	82
	3	0.8	2.0	4.1	7.1	11.4	24.1	43.0	70	104
	4	0.9	2.4	4.8	8.4	13.5	28.3	50.7	82	122
	5	1.0	2.6	5.4	9.6	15.4	32.3	57.7	94	139
100	1	**0.3**	**0.8**	**1.4**	**2.6**	**4.2**	**8.7**	**15.6**	**26**	**37**
	2	0.4	1.1	2.2	3.8	6.1	12.9	23.1	37	56
	3	0.5	1.4	2.7	4.8	7.7	16.3	29.2	47	71
	4	0.6	1.6	3.2	5.7	9.1	19.2	34.3	55	83
	5	0.7	1.8	3.7	6.5	10.4	21.8	39.1	63	94
150	1	**0.3**	**0.6**	**1.2**	**2.0**	**3.3**	**7.0**	**12.4**	**20**	**30**
	2	0.3	0.9	1.8	3.1	4.9	10.3	18.4	30	44
	3	0.4	1.1	2.2	3.9	6.2	13.0	23.2	37	56
	4	0.5	1.3	2.6	4.6	7.3	15.3	27.4	44	66
	5	0.5	1.4	3.0	5.2	8.2	17.4	31.1	50	76
200	1	**0.2**	**0.5**	**1.0**	**1.8**	**2.8**	**5.9**	**10.5**	**17**	**26**
	2	0.3	0.8	1.4	2.6	4.2	8.8	15.6	26	38
	3	0.3	0.9	1.9	3.3	5.3	11.1	19.7	31	48
	4	0.4	1.1	2.2	3.9	6.2	13.0	23.2	37	56
	5	0.4	1.2	2.6	4.4	7.1	14.8	26.5	43	64

ABOVE TABLE INCLUDES ALLOWANCE FOR PRESSURE DROP IN AN AVERAGE NUMBER OF FITTINGS.
RECOMMENDED PRESSURE DROP AT MAXIMUM LOAD IS SHOWN IN HEAVY TYPE.

CORRECTION FACTORS FOR VARIOUS SUCTION PRESSURES (REFRIG.-12)

SUCTION TEMP., DEGREES	−20	−10	0	5	10	15	20	25	30	35	40	45	50	55
SUCTION PRESSURE, LBS. GAGE	0.6	4.5	9.2	11.8	14.7	17.7	21.1	24.6	28.5	32.6	37.0	41.7	46.7	52.0
CORRECTION FACTOR	2.34	2.00	1.73	1.60	1.49	1.39	1.29	1.21	1.14	1.06	1.00	0.95	0.89	0.83

FIGURE 19-21 (Courtesy, The Trane Company).

If the condensing temperature is different than 105°F a correction factor from Fig. 19-22 is used to multiply the assumed load just as was done above.

Liquid lines are a bit simpler to size, since velocity is not a problem; pressure drop and vertical lift are the problems. Lift is basically a function of condenser subcooling, as previously discussed. The size and length of the line *do* affect pressure drop. The correct diameter for the liquid-line piping can be determined from a single table that is very easy to use. Figure 19-23 is the pressure-drop table for R-12 liquid lines. As was previously mentioned, a pressure drop of more than 3 psi in the liquid line may cause flashing, this limit is shown in bold type. An example: 10 feet of $\frac{7}{8}$ in. O.D. tubing at 3 psi can handle a 19.6 ton load. No correction factors are needed when this table is used with R-12. In its liquid stage the refrigerant capacity is not affected by different suction or condensing temperatures.

As in all previous tables, the liquid-line pressure-drop figures assume an average number of fittings. However, they *do not* include pressure drops created by the expansion device and liquid distributor. Their pressure drops are normally assumed and allowed for by the equipment manufacturer in his capacity rating.

The foregoing discussion is condensed from material published by the Trane Company and is intended only for the purpose of instruction, not for actual use in field piping. We recommend consulting the specific

SUCTION GAS VELOCITY
CORRECTION FACTORS

APPLICABLE TO FIGURE 19.21a

CONDENSING TEMP. °F	SUCTION TEMPERATURE °F								
	−10°	0°	+10°	20°	30°	35°	40°	45°	50°
85°	2.58	2.05	1.65	1.34	1.10	1.00	0.91	0.82	0.76
90°	2.64	2.10	1.69	1.37	1.13	1.02	0.93	0.84	0.78
95°	2.70	2.15	1.74	1.41	1.15	1.05	0.95	0.87	0.80
100°	2.77	2.21	1.78	1.44	1.18	1.07	0.98	0.89	0.81
105°	2.84	2.27	1.83	1.48	1.21	1.10	1.00	0.91	0.83
110°	2.94	2.33	1.88	1.52	1.24	1.13	1.02	0.93	0.85
115°	2.39	2.39	1.93	1.56	1.27	1.16	1.05	0.95	0.87

TO DETERMINE SUCTION GAS VELOCITY. MULTIPLY LOAD IN TONS BY FACTOR FROM FIGURE 19.22. USING THIS NEW VALUE AS LOAD. READ VELOCITY DIRECTLY FROM FIGURE 19.19b.

FIGURE 19-22 (Courtesy, The Trane Company).

manufacturer's recommendations, depending on the nature of the equipment and application, since there are some differences in techniques from one manufacturer to another.

Earlier, the term *accessory* was used to describe an oil separator, which is only one of many devices used in refrigeration work. A refrigeration accessory is an article or device that adds to the convenience or effectiveness of the system. The essential items of the basic refrigeration system are the compressor, condenser, metering device, and evaporator. An accessory gives the basic system certain conveniences or allows it to attain a degree of performance that is impractical or virtually impossible with commercially available basic system components. These accessories come in many shapes and sizes and serve many functions, but their primary purpose is to add to the convenience or effectiveness of the refrigeration system.

Following is a list of common accessories (there are others, but they are so specialized or so infrequently used that they will not be listed).

- Oil separator
- Muffler
- Heat exchanger
- Strainer-drier
- Suction accumulator
- Crankcase heaters
- Sight glass
- Moisture indicator
- Water valve
- Solenoid valve
- Check valve
- Evaporator pressure regulator
- Relief valve
- Fusible plugs

TONNAGE CAPACITIES OF LIQUID LINES DELIVERING LIQUID REFRIGERANT-12 FROM RECEIVER TO EXPANSION VALVE

LENGTH OF PIPE, FEET	PRESSURE DROP, PSI	OUTSIDE DIAMETER OF PIPE — INCHES						
		½	⅝	⅞	1⅛	1⅜	1⅝	2⅛
		INSIDE DIAMETER OF PIPE — INCHES						
		0.430	0.545	0.785	1.025	1.265	1.505	1.985
10	1	2.1	3.9	10.6	21.7	38.2	60.6	127
	2	3.1	5.9	15.5	32.0	56.3	89.8	188
	3	**3.9**	**7.4**	**19.6**	**40.1**	**70.6**	**112.5**	**237**
	4	4.6	8.7	23.1	47.2	83.4	132.0	278
	5	5.2	9.9	26.2	53.6	94.2	150.0	315
20	1	1.9	3.6	9.7	19.8	35.0	55.5	117
	2	2.8	5.4	14.2	29.3	51.4	82.0	173
	3	**3.6**	**6.8**	**17.9**	**36.7**	**64.6**	**103.0**	**217**
	4	4.2	8.0	21.2	43.1	76.0	121.0	255
	5	4.7	9.1	24.0	49.0	87.1	137.0	288
30	1	1.7	3.3	8.9	18.1	32.0	50.8	107
	2	2.6	4.9	13.0	26.8	47.0	75.0	158
	3	**3.3**	**6.2**	**16.4**	**33.6**	**59.1**	**94.2**	**198**
	4	3.9	7.2	19.3	39.4	69.4	110.0	233
	5	4.3	8.3	21.9	44.6	78.7	125.0	264
50	1	1.5	2.8	7.6	15.5	27.3	43.4	91.0
	2	2.2	4.2	11.1	22.9	40.2	64.1	135.2
	3	**2.8**	**5.3**	**14.0**	**28.7**	**50.5**	**80.5**	**169.2**
	4	3.3	6.2	16.5	33.7	59.4	94.4	198.9
	5	3.7	7.1	18.7	38.3	67.3	107.1	225.3
100	1	1.0	2.0	5.1	10.5	18.4	29.4	61.9
	2	1.5	2.8	7.6	15.5	27.2	43.4	91.0
	3	**1.9**	**3.6**	**9.5**	**19.5**	**34.3**	**54.6**	**114.8**
	4	2.2	4.2	11.1	22.9	40.2	64.1	135.2
	5	2.6	4.8	12.7	25.9	45.6	72.7	153.0
150	1	0.9	1.5	4.1	8.3	14.7	23.5	49.2
	2	1.2	2.3	6.0	12.3	21.7	34.6	72.7
	3	**1.5**	**2.8**	**7.6**	**15.5**	**27.2**	**43.4**	**91.0**
	4	1.8	3.3	8.9	18.2	32.0	51.1	107.1
	5	2.0	3.8	10.1	20.7	36.3	57.9	121.6

ABOVE TABLE INCLUDES ALLOWANCE FOR PRESSURE DROP IN AN AVERAGE NUMBER OF FITTINGS.
RECOMMENDED PRESSURE DROP AT MAXIMUM LOAD IS SHOWN IN HEAVY TYPE.

FIGURE 19-23 (Courtesy, The Trane Company).

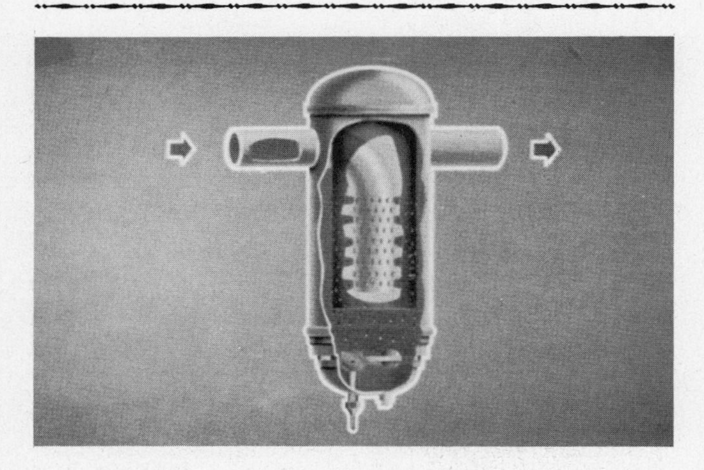

FIGURE 19-24 Oil separator (Courtesy, Carrier Air-Conditioning Company).

The purpose of the oil separator (Fig. 19-24) is to reduce the amount of oil in circulation through the system and thereby increase efficiency. All refrigeration systems will have some oil passing through them. In some cases the amount of oil in circulation can affect the evaporator heat transfer characteristics, create false float action, or even affect expansion valve operation. In these cases, an oil separator, by reducing the oil circulating within the system, can improve evaporator efficiency or reduce float and valve problems.

Systems with improperly sized or trapped lines quite frequently do not return oil to the compressor, thus creating compressor lubrication problems. The insertion of an oil separator into such a system will not correct these problems. An oil separator is not 100% efficient and some oil will pass through it. The installation of an oil separator on a system that is trapping oil will merely delay the problem; it will not in any way solve the problem.

In the separator illustrated, the hot gas-oil mixture from the compressor enters at the left and flows down and out through the perforated pipe. The mixture strikes against the screen where the oil usually separates from the gas. The oil drains down the screen into the small sump at the bottom of the separator. The gas passes through the screen and leaves the separator at the upper right. When the oil level rises in the sump, the float ball is raised and the oil is returned through an orifice to the crankcase.

The relative position of the oil separator in the refrigeration system was illustrated in Fig. 19-10. The separator is normally placed in the discharge line as close to the compressor as practical. The oil return line, as shown, leads directly to the crankcase.

The internal construction of the separator will vary greatly; however, its external appearance and its location in the system make the oil separator comparatively easy to identify.

When an oil separator is used, certain precautions must be taken. A cool separator will condense refrigerant gas into liquid, which, if allowed to return to the compressor crankcase, can cause damage to the compressor. Care should also be taken to keep the float orifice clean, as this orifice is subject to any oil sludge that might leave the compressor discharge. If the float were to stick open, hot gas might flow into the compressor crankcase. If the float were to stick closed, no oil would be returned to the compressor.

The purpose of the muffler (Fig. 19-25) is to dampen or remove the hot gas pulsations set up by a reciprocating compressor. Every reciprocating compressor will create some hot gas pulsations. Although a great effort is made to minimize pulsations in system and compressor design, gas pulsation can be severe enough to create two closely related problems. The first is noise, which, though irritating to users of the equipment, does not necessarily have a harmful effect on the system. The second problem is vibration, which can result in line breakage. Frequently, these problems occur simultaneously.

The illustration is a good example of muffler construction. It is designed to eliminate gas pulsations by allowing them to dissipate within the muffler

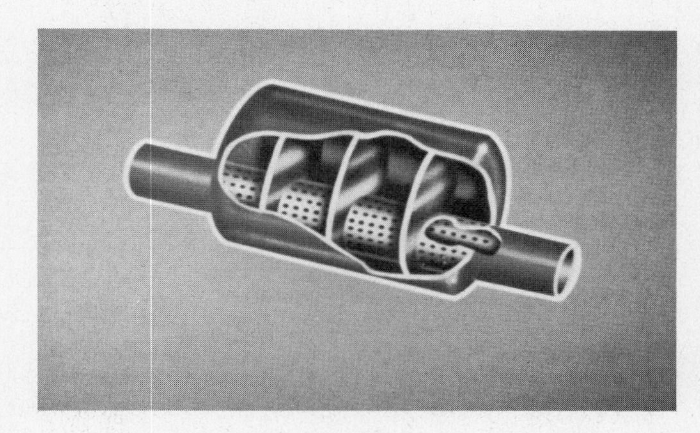

FIGURE 19-25 Muffler (Courtesy, Carrier Air-Conditioning Company).

itself. The pulsations force the hot gas through the holes in the pipe and into the chambers, where they are absorbed.

The muffler is inserted in the discharge line as close to the compressor as is practical. In welded hermetic compressors, the muffler is frequently within the compressor shell itself. Because it is usually constructed within the shell, the muffler is a natural trap. It will easily trap oil and may also trap liquid refrigerant. The muffler should be installed either in a downward flow or a horizontal line.

Selecting the proper size and location for a muffler can be a difficult engineering problem. Improper field selection of a muffler will sometimes lead to a condition of increased rather than decreased vibration.

The heat interchanger (Fig. 19-26) is a device to transfer heat from the liquid refrigerant to the suction gas. The exchanger has two principal uses. The first is to reduce the temperature or subcool the liquid refrigerant flowing from the condenser to the metering device. This reduction in temperature is necessary in systems that have high pressure drops to prevent flashing in the liquid line. These pressure drops may be caused by excessively long lines or by lengthy liquid risers.

The second use of the exchanger is to assure that the suction gas flowing into the compressor is dry. On systems with rapid load fluctuations, it is not unusual to find liquid "slopping" over from the evaporator. The heat interchanger will allow lower evaporator superheat settings, since some liquid carryover need not be considered dangerous.

The triple-pipe heat exchanger illustrates the principle of the heat interchanger. It is a counterflow device with warm liquid entering at the right and wet or cool gas entering at the left. As the liquid is at a higher temperature than the gas, the heat will flow out of the liquid and into the gas. This subcools the liquid to reduce flashing and superheats the gas to prevent flooding. Other forms of heat interchangers are the double pipe and the shell and coil.

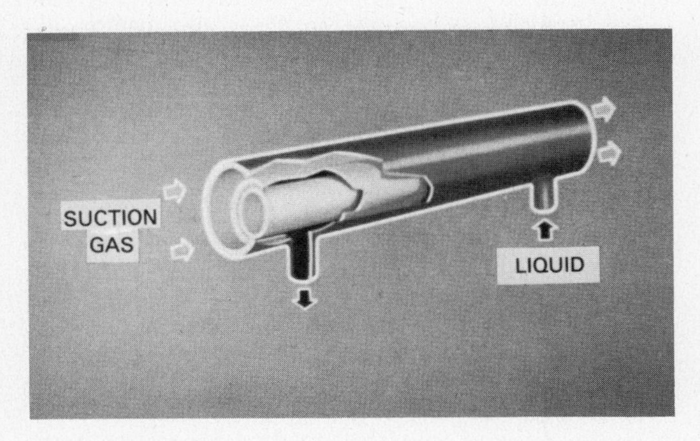

FIGURE 19-26 Heat interchanger (Courtesy, Carrier Air-Conditioning Company).

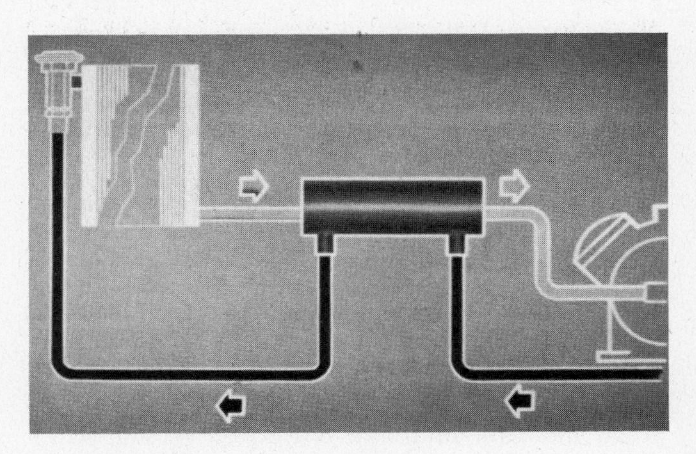

FIGURE 19-27 Heat interchanger location (Courtesy, Carrier Air-Conditioning Company).

The location of the heat exchanger (Fig. 19-27) will depend on its intended usage. If the heat exchanger is being used for the purpose of subcooling the liquid, it will be found as close to the condenser as practical piping practices will allow. A heat exchanger that is being used as protection against slopover may be mounted in the suction line quite close to the evaporator. Because both liquid and suction lines must be brought to the heat exchanger, equipment layout may have more bearing on the position of the heat exchanger than either of the other two factors.

Though heat exchangers serve many useful purposes, care must be exercised in their use, particularly with hermetic compressors. Hermetic compressor motors are cooled by suction gas and frequently have well-defined upper limits on suction gas temperatures. Because heat exchangers tend to increase suction gas temperatures, indiscriminate field use without proper engineering investigation can be dangerous.

The dangers of moisture within a refrigeration system have already been discussed. It will simply be repeated that no moisture should be present, particularly in those systems using halogenated hydrocarbon refrigerants. However, if moisture does enter the refrigeration system, it must be removed. One method of removing moisture is the strainer-drier shown in Fig. 19-28. This accessory consists of a shell through which the liquid refrigerant will pass. Inside the shell is a material known as a desiccant. As the moisture-laden refrigerant passes through the drier, the desiccant removes a portion of the moisture. Each passage through the drier removes additional moisture until the refrigerant is sufficiently dry or until the drier has reached its moisture-holding capacity. When this happens, the drier must be replaced.

The strainer-drier also performs a second service by filtering any solid particles from the flowing liquid refrigerant. These particles are filtered out by the desiccant core.

The strainer-drier is almost always found in the liquid line of the refrigeration system, as in Fig. 19-29. Because the volume of the liquid is much smaller than that of the gas, a smaller drier may be used. This, of course, results in lower cost. Also, the metering device is protected from solid particles when the strainer is in this location.

Two important facts should be noted about driers. The first is that they should be replaced when the desiccant is saturated. The second is that they create some pressure drop; therefore, they must be properly sized to prevent excessive pressure drop with resulting flash gas.

FIGURE 19-28 Strainer drier (Courtesy, Carrier Air-Conditioning Company).

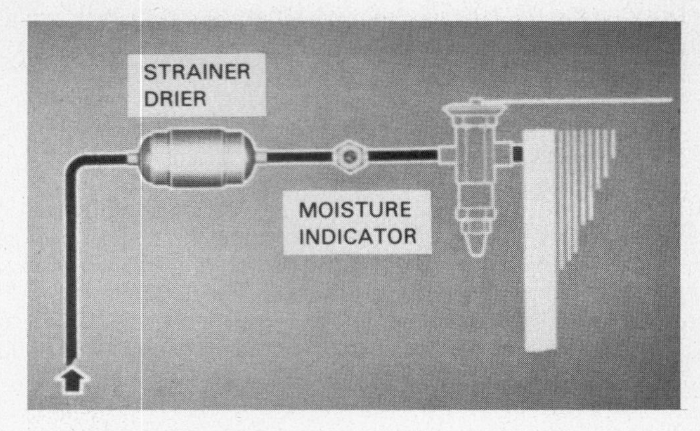

FIGURE 19-29 Liquid line accessories (Courtesy, Carrier Air-Conditioning Company).

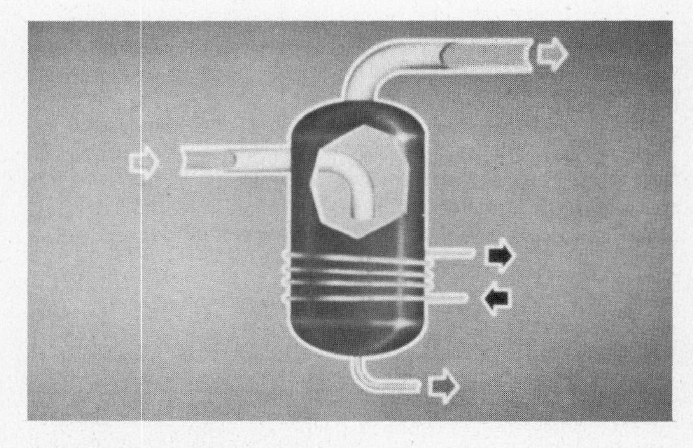

FIGURE 19-30 Suction accumulator (Courtesy, Carrier Air-Conditioning Company).

The suction accumulator (Fig. 19-30) is a simple device serving a very useful function. On some evaporators, the metering device action is not rapid enough to maintain pace with load changes. Also, capillary tubes are not designed to "shut off" under light evaporator loads. In both cases some liquid will occasionally leave the evaporator through the suction line. This could damage the compressor. The accumulator is nothing more than a trap to catch this liquid before it can reach the compressor. This surplus liquid is boiled or evaporated in the trap and returned to the compressor as a gas.

Sometimes space limitations require that the whole evaporator be used for boiling liquid. Here the suction accumulator can be used to advantage, as it allows full use of the evaporator without fear of spilling liquid into the compressor suction part. Its construction is not complicated. It consists of a container to collect and evaporate liquid refrigerant.

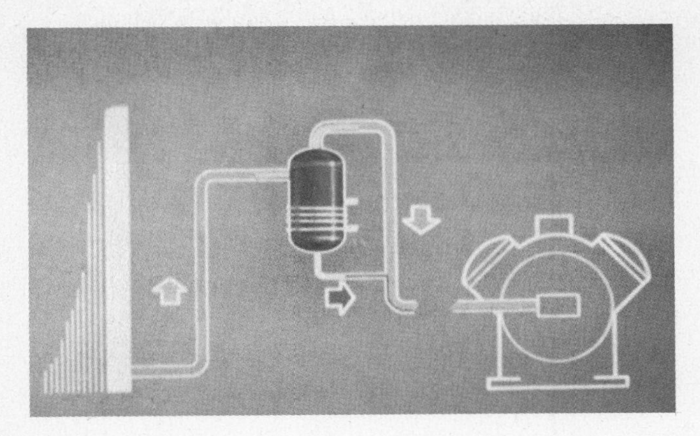

FIGURE 19-31 **Accumulator location (Courtesy, Carrier Air-Conditioning Company).**

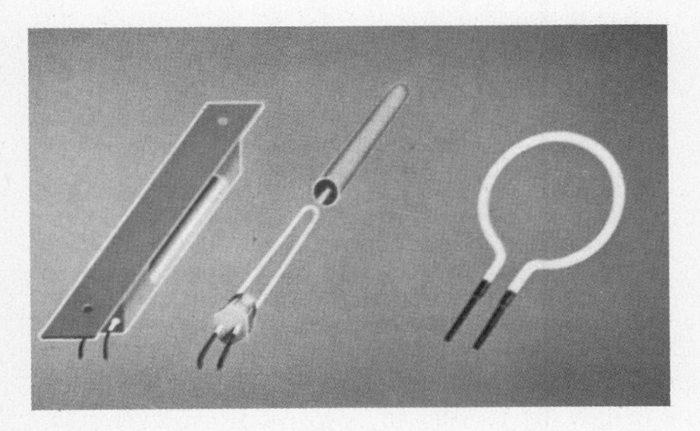

FIGURE 19-32 **Crankcase heaters (Courtesy, Carrier Air-Conditioning Company).**

The use of the liquid line to obtain heat to evaporate the liquid and the oil return line are optional features.

The suction accumulator is generally found quite close to the evaporator from which the liquid comes (Fig. 19-31). Occasionally it will be found in the main suction line of a multievaporator system, thereby protecting against slopover from all evaporators.

The suction accumulator must be properly sized or it may fill with liquid and cause compressor damage. There must also be provisions for oil return; this is essential, as it is a natural oil trap. The suction accumulator is used primarily in packaged equipment using a capillary feed.

The crankcase heater (Fig. 19-32) is used to prevent the accumulation of liquid refrigerant in the compressor crankcase during shutdown. Due to the affinity of oil for refrigerant, refrigerant may migrate to the crankcase during shutdown. Condensation of refrigerant in the compressor crankcase may also occur when the temperature of the compressor is lower than that of the rest of the system.

If refrigerant does gather in the crankcase, a reduction in suction pressure at startup will cause the liquid to evaporate or boil. This boiling causes the oil-refrigerant mixture to foam, and some of this foam leaves the crankcase and passes into the cylinder. There it may create liquid slugs, which will damage the valves. The refrigerant also reduces the lubricating value of the oil.

Accumulation of liquid in the crankcase during shutdown can be minimized by keeping the crankcase warmer than the rest of the system. The crankcase heater is designed for that purpose. Figure 19-32 shows three heaters designed for three different applications. The heater on the left is fastened to the bottom of the crankcase, the heater in the center may be inserted into the crankcase, and the one on the right is designed to be wrapped around the crankcase.

Compressors that perform well in low-temperature work, low ambient operation, or heat-pump application where cold oil conditions are encountered, will most likely be shipped from the factory with the crankcase heat installed. Or, the compressor will have a plugged opening (threaded) into which a heater may be inserted during installation.

The sight glass (Fig. 19-33) in a refrigeration system permits the installer or serviceman to observe the condition of the refrigerant at that particular point. The sight glass usually consists of a glass opening in the liquid line of the system. Frequently

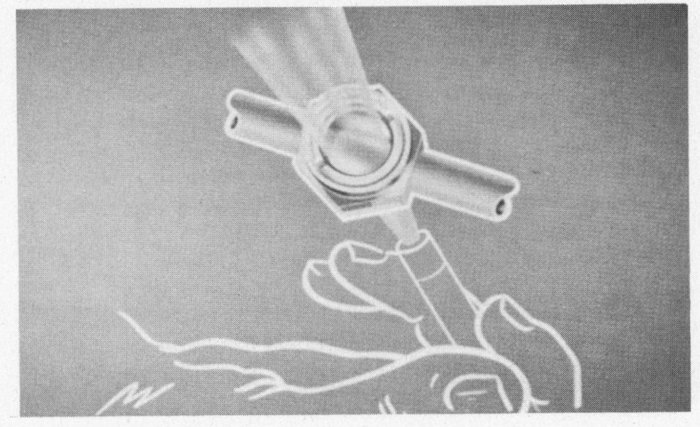

FIGURE 19-33 **Sight glass (Courtesy, Carrier Air-Conditioning Company).**

a glass on each side of the line is used to assure illumination of the line.

When the line is completely filled with liquid refrigerant, there is almost no obstruction when looking through the line. However, if any gas is in the liquid line, it will show immediately in the form of bubbles passing the sight glass. It should be pointed out that in a sight glass of this type, the glass will also show clear when only gas is present.

At first glance, the moisture indicator (Fig. 19-34) may appear to be a simple sight glass, a natural assumption since the indicator is normally a part of the sight glass.

Contained within the glass, but exposed to the liquid refrigerant, is a small colored dot as shown in Fig. 19-34. This dot is the moisture indicator. It is of a special chemical composition that will change color depending upon the amount of moisture in the refrigerant. When the amount of moisture is within the limits set by the manufacturer, the dot will be one color. However, if too much moisture is present, the dot will change color. When a maintenance man or serviceman sees that the color of the dot has changed, he knows that steps must be taken to remove the moisture from the system before harm is done. The moisture indicator will not be accurate unless it is completely filled with liquid refrigerant.

Depending on the purpose for which it is intended, the sight glass or moisture indicator may be found in more than one location. If the glass is to be used as an aid in determining whether the system is properly charged, it will be found at the condenser or receiver outlet. If it is to be used to assure that no flash gas is present prior to the liquid entering the metering device, it will be found immediately before the metering device, as shown.

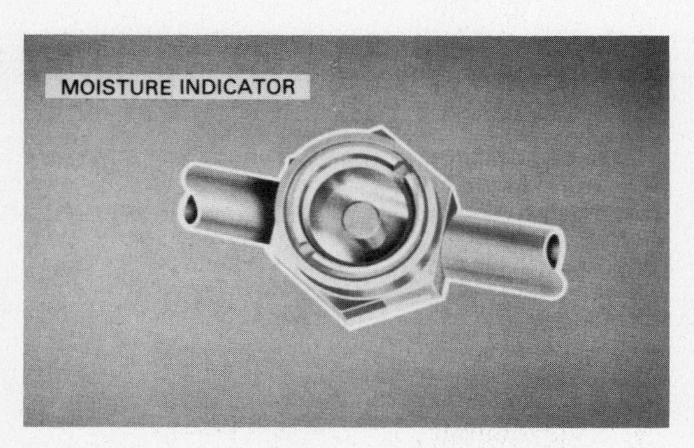

FIGURE 19-34 Moisture indicator (Courtesy, Carrier Air-Conditioning Company).

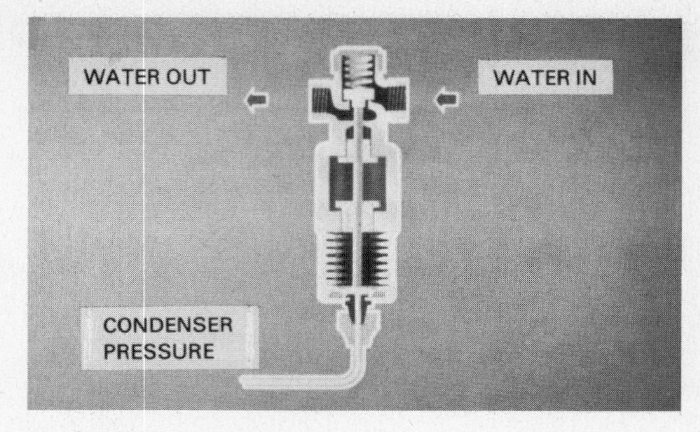

FIGURE 19-35 Water valve (Courtesy, Carrier Air-Conditioning Company).

The water valve (Fig. 19-35) is designed to control head pressure by governing the flow of water to a water-cooled condenser. The action of the water valve is controlled by the refrigerant pressure within the condenser. The gas in the condenser enters the valve at the bottom and exerts a pressure on the bellows. As the pressure increases it opens the valve to allow more water to flow. As pressure decreases the bellows contracts, closing the valve port. The proper water flow and condenser pressure are adjusted by setting the spring tension in the valve head.

Installation is simple but care should be taken that the direction of water flow through the valve is correct. The water valve is usually considered virtually trouble free; an occasional change of seat is all that should be required.

The solenoid valve (Fig. 19-36) may be used to control the flow of a gas or a liquid. It is frequently used in refrigerant liquid lines to control the flow of refrigerant to the evaporator. It is sometimes used in suction lines to isolate the evaporators of multi-evaporator, two-temperature systems. Another important use of the solenoid is as a pilot for a much larger valve.

The solenoid valve is an extremely dependable device when installed as recommended by the manufacturer. Poor installation practices such as failing to set the valve up right, poor wiring practices, and warping the valve body by excessive heat during

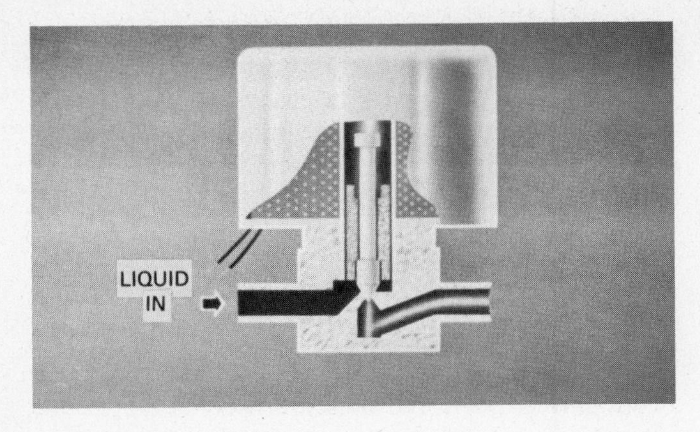

FIGURE 19-36 Solenoid valve (Courtesy, Carrier Air-Conditioning Company).

brazing can make a solenoid valve a source of constant trouble. Proper installation is thus most important for this device.

The check valve (Fig. 19-37) is designed to allow the flow of liquid or gas in one direction only. When the fluid flows in the direction of the arrow, the force of the fluid will lift the gate from its seat and allow the fluid through. When the fluid tries to flow in the opposite direction, the gate closes, thereby stopping the flow. This device is useful in preventing the return of liquid to a compressor through the discharge line during shutdown. It is also found in reverse-cycle systems and is frequently inserted in the suction lines on two-temperature systems to prevent equalizing of pressures during shutdown.

The check valve has many uses and is considered a reasonably trouble-free device. By its very nature, however, it creates a pressure drop within the line in which it is used. Therefore, it should not be used indiscriminately.

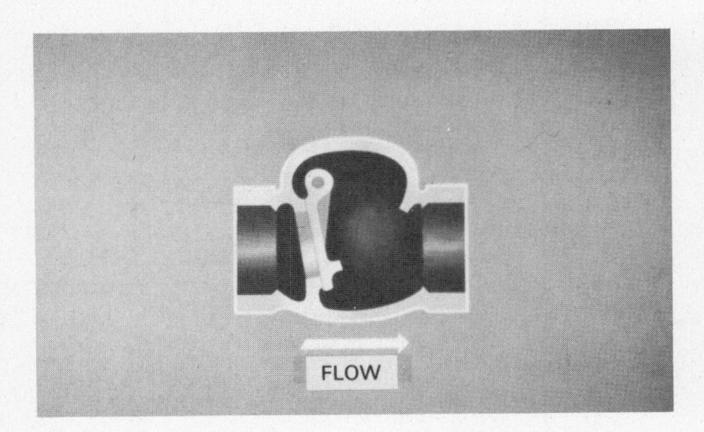

FIGURE 19-37 Check valve (Courtesy, Carrier Air-Conditioning Company).

The evaporator pressure regulator (Fig. 19-38) is frequently referred to as a *back pressure regulator*. It is designed to maintain a constant pressure or temperature in the evaporator regardless of how low the compressor suction pressure falls. It is always found in the suction line as close to the evaporator as is practical. A complete explanation of the kinds of regulators and their operation can be found in Chap. 20.

The relief valve is designed to protect the system against refrigerant pressures great enough to do physical damage. The valve shown in Fig. 19-39 is held closed by a spring. The valve will automatically close when the pressure has been reduced to within the required limits, and will remain closed

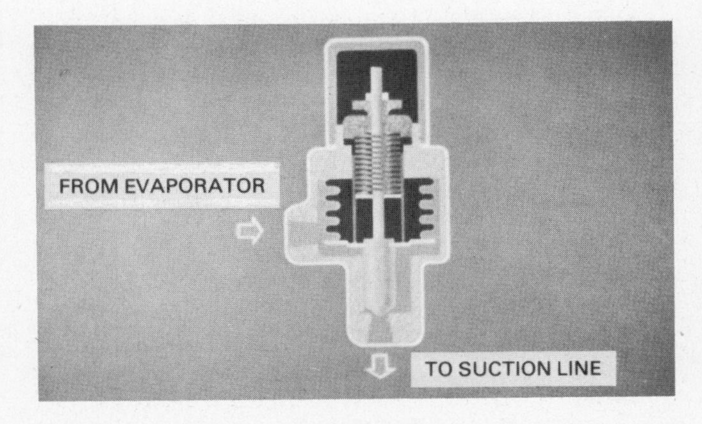

FIGURE 19-38 Evaporative pressure regulator (Courtesy, Carrier Air-Conditioning Company).

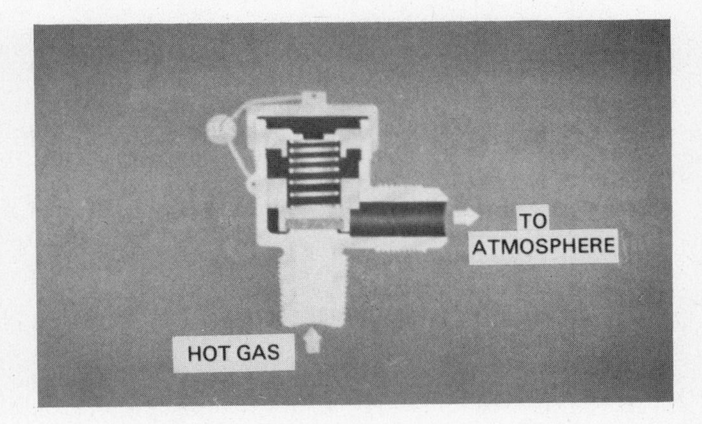

FIGURE 19-39 Pressure relief valve (Courtesy, Carrier Air-Conditioning Company).

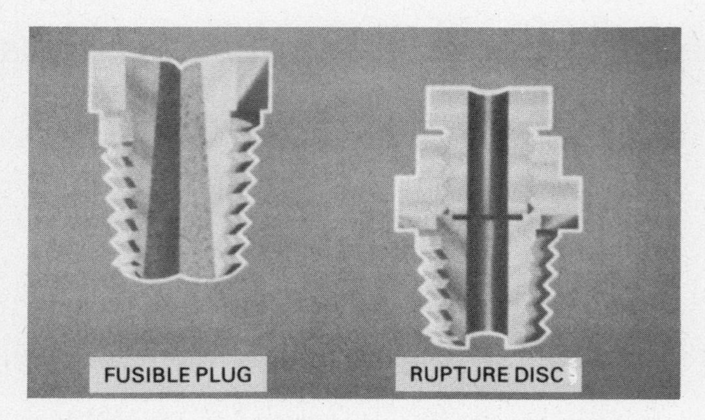

FIGURE 19-40 Safety fittings (Courtesy, Carrier Air-Conditioning Company).

until high pressure again requires relief. If the valve does open there will naturally be a loss of refrigerant, and the system subsequently must be checked. Since these valves are often required by law they often have some type of seal device so that they cannot be tampered with by unauthorized persons.

The fusible plug or rupture disc fittings, as illustrated in Fig. 19-40, are not literally controls, but are safety devices designed to protect the system against extreme pressures. The fusible plug fitting on the left contains a core of soft metal with a low melting point. In case of fire, the soft metal would melt and allow the gas to escape to the atmosphere before hazardous pressures could be built up.

The rupture disc fitting on the right in Fig. 19-40 is designed to serve the same purpose as the fusible plug, but in a different manner. The rupture disc consists of a thin piece of metal that is meant to break or rupture at a pressure below that which might create a dangerous condition.

Many shell vessels such as water-cooled condensers and receivers are factory equipped with fusible plugs or relief devices, since the systems are located indoors, where there are people. Air-cooled systems are not normally so equipped, and these devices would be installed at or near the outside condenser to "blow" into the open atmosphere.

PROBLEMS

1. Refrigerant in a vapor state is a good or poor carrier of oil.
2. The velocity of refrigerant flow has no effect on oil entrainment: true or false?
3. The recommended minimum refrigerant velocity in vertical hot gas risers is ______________.
4. Pressure drop in hot gas lines should not exceed ______________ psi.
5. Liquid-line pressure drop is not affected by vertical rise: true or false?
6. Pressure drop in liquid lines should not exceed ______________ psi.
7. Subcooling is a method of overcoming liquid-line pressure loss in vertical risers: true or false?
8. Pressure drop in suction lines should be no more than ______________ psi.
9. Should suction lines be pitched toward or away from the compressor?
10. An oil separator will cure all oil return problems: true or false?
11. A heat exchanger is a useful device to subcool liquid refrigerant: true or false?
12. A moisture indicator reveals the presence of water vapor in the system by changing ______________.

20

CONTROL DEVICES

In previous chapters we discussed four major components in the construction of a simple compression refrigeration cycle—the evaporator, the compressor, the condenser, and the refrigerant piping (with certain accessories). But the system will not operate unless there are a number of associated controls (Fig. 20-1) to direct its actions. Controls are much like the brain and nervous system of the body: they direct and regulate its actions.

Controls may be roughly classified according to their function in the system.

- ***Basic Operating*** These are the devices that get the system into operation. Examples are the metering devices: expansion valves, capillary tubes, or float controls.
- ***Regulators*** These are related to controls that add automation, convenience, and generally improve overall efficiency. Examples are the thermostat, solenoid valve and water valves.
- ***Safety Requirements*** Controls that function to protect the system in normal and abnormal operation and those that may be required by codes or regulatory agencies. Examples are high and low-pressure cutouts, oil-pressure cutouts, electrical overloads, pressure relief valves, etc.
- ***Application Improvement*** These are related to devices that are used to improve or alter the application. Examples are capacity reduction devices, defrost controls, back pressure regulators, etc.

Many controls serve dual functions and they cannot be classified until their main purpose is known.

The form these controls take may be mechanical, electrical, or a combination of the two. Under the term *electrical* would also be included *electronic devices*. *Mechanical* would also include *pneumatic* (air) operation, although its application to commercial refrigeration systems as compared to air conditioning is rather limited.

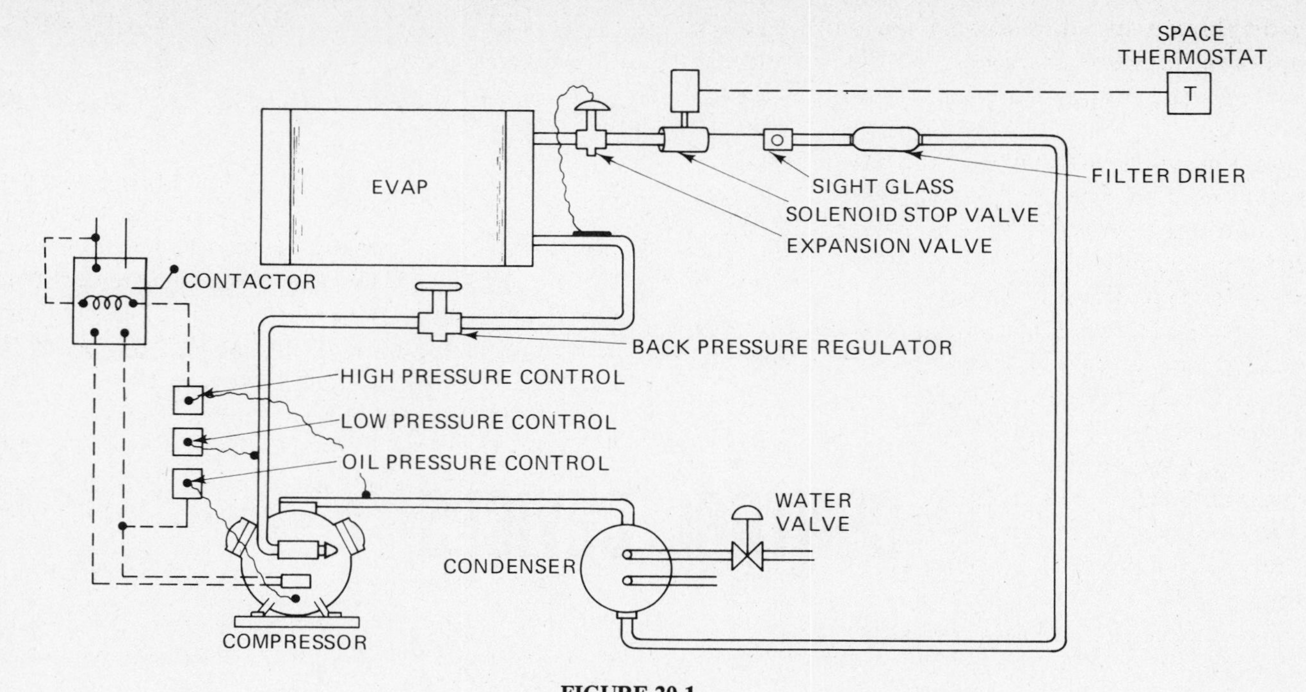

FIGURE 20-1.

METERING DEVICES

Going directly to the basic operating functions, we find the metering device is a proper starting point. Every compression cycle must have a means of regulating the flow of high-pressure liquid refrigerant from the liquid line into the evaporator. The five main types of metering devices used in various phases of refrigeration work are:

(1) automatic expansion valve
(2) thermostatic expansion valve
(3) capillary tube
(4) low-side float
(5) high-side float

As pointed out earlier, the use of hand valves in early refrigeration systems was a means of metering, but it is not suitable to application in modern automatic equipment and thus will not be covered.

The first important development after manual valve operation was the *automatic expansion valve*. This is not a good descriptive name, for some other types of expansion valves are also automatic. It could more accurately be called a *constant evaporator pressure expansion valve*, for it maintains constant outlet pressure regardless of changes in inlet liquid pressure, load, or other conditions.

Figure 20-2 is an illustration of an automatic expansion valve. This metering device is designed to maintain a constant pressure in the evaporator. Its primary motivating force is evaporator pressure, which exerts a force against the bottom of the diaphragm. An adjustable spring exerts a pressure on the top of the diaphragm. As the evaporator pressure increases, it overcomes the spring pressure and moves

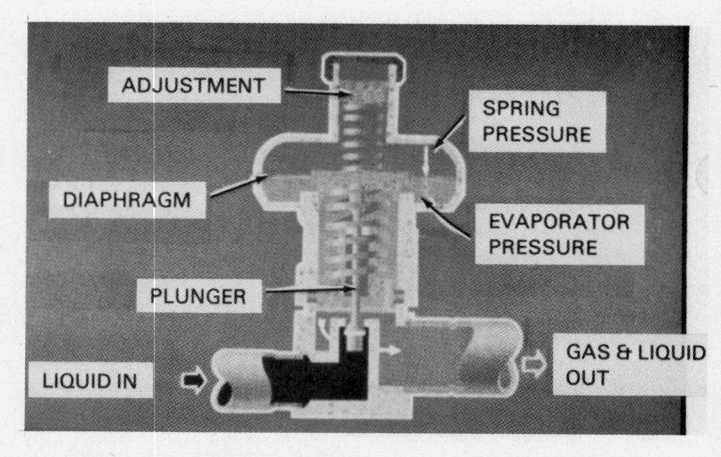

FIGURE 20-2 Automatic expansion valve (Courtesy, Carrier Air-Conditioning Company).

the diaphragm up, thus closing the valve. As the evaporator pressure decreases, the spring pressure overcomes the evaporator pressure and pushes the valve open.

As this valve maintains a constant evaporator pressure it also attempts to maintain a constant evaporator temperature. It is important to understand that this valve has a reverse action according to varying load conditions. As the load on an evaporator increases, the back pressure normally rises due to the increased rate of evaporation. To pick up the load this rise in back pressure should be accompanied by an increased rate of flow of liquid refrigerant to the evaporator. Where there is an automatic expansion valve, increased pressure closes the valve. By closing the valve when an increase in load is present, the supply of refrigerant is shut off rather than increased.

This means that this valve should only be used where the load is relatively constant. It is primarily found on such items as household refrigerators, small air conditioners, water chillers, and the like.

Before we explain the thermostatic expansion valve a brief review of *superheat* will be helpful.

Figure 20-3 shows a bare-tube evaporator using a hand valve. In this evaporator no pressure drop is shown; in other words, the pressure is the same throughout the evaporator. When the valve is opened slightly, a small amount of refrigerant enters the evaporator. Heat passes into the refrigerant through the tube, causing the refrigerant to boil. If only a small quantity of refrigerant is flowing it will all be boiled away at some point, such as *A*. (Throughout this example we will assume the evaporator pressure to be 21 psig; the saturation temperature at 21 psig is 20°F. From point *A*, the refrigerant is a gas, and any heat absorbed results in superheat. Superheat is the difference between the actual gas temperature and the saturation temperature, i.e., the boiling temperature corresponding to the gas pressure. At *B*, the gas temperature is 30°F and we have 10°F of superheat. At the last return bend the gas temperature is 40°F and the gas has 20°F of superheat. At point *C*, the difference between the actual gas temperature of 50°F and the saturation temperature of 20°F is 30°F superheat.

When the valve is opened wider (Fig. 20-4), the refrigerant flow increases and the point at which the last liquid boils will move to *B*. The reduced evaporator surface available to superheat the refrigerant results in a lower gas superheat. The gas is shown here leaving at 40°F, which represents 20°F of superheat.

The ideal condition would be that shown in Figure 20-5, where the valve is opened sufficiently to give 0°F, or *no* superheat, in the refrigerant leaving the coil. This chart shows the last of the liquid being boiled at point *C*.

If the hand valve were opened wide enough it would be possible for the refrigerant flow to be so great that liquid could flood back to the compressor. This is a dangerous condition, since the compressor may be damaged.

FIGURE 20-3 30°F superheat (Courtesy, Carrier Air-Conditioning Company).

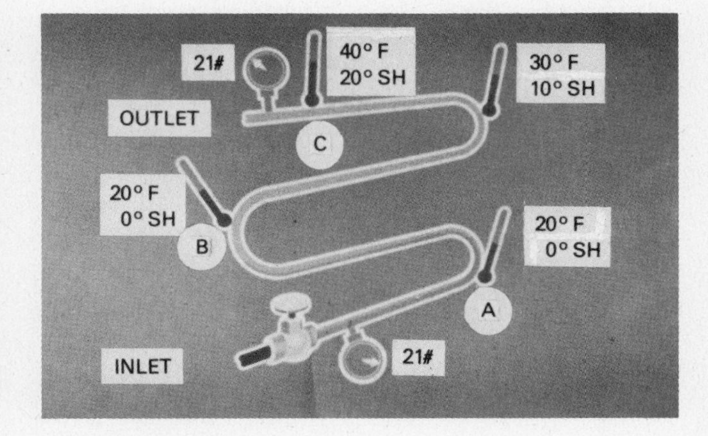

FIGURE 20-4 20°F superheat (Courtesy, Carrier Air-Conditioning Company).

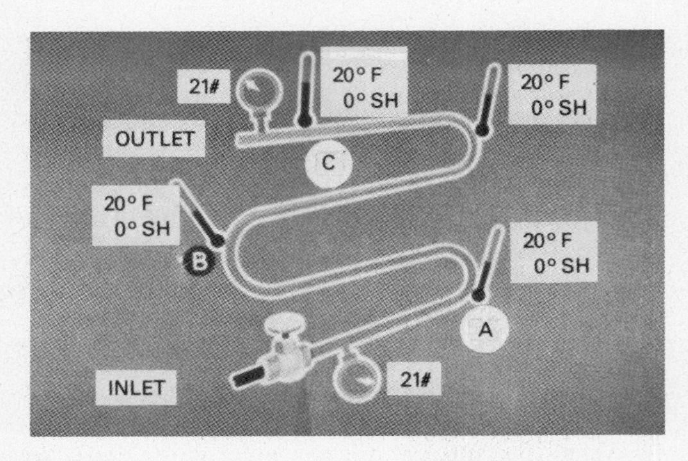

FIGURE 20-5 0°F superheat (Courtesy, Carrier Air-Conditioning Company).

The practical application of the thermostatic expansion valve approaches the condition shown in Fig. 20-6. In this example, the point of complete vaporization is at the last return bend of the evaporator, which allows the gas to pick up 10°F of superheat before it leaves the coil.

Because of variations in loads and a lag in the control of the metering devices, 10°F is the minimum practical superheat for air conditioning. In refrigeration and special applications lower superheat settings are often used.

By far the most widely used of all metering devices is the TEV (thermostatic expansion valve). This valve controls the flow of refrigerant by maintaining a relatively constant superheat at the end of the evaporator coil. Though widely used, this valve is the most difficult to understand of all metering devices.

FIGURE 20-6 10°F superheat (Courtesy, Carrier Air-Conditioning Company).

Figure 20-7 shows a cutaway of the thermostatic expansion valve, with the major components labeled. The forces operating on the plunger are emphasized in this diagram. They are spring pressure and evaporator pressure on the bottom of the diaphragm and bulb pressure on the top of the diaphragm. The evaporator pressure is admitted to the bottom of the diaphragm through the internal port in the valve. This is referred to an *internally equalized valve*.

Figure 20-8 is a schematic representation of the thermostatic expansion valve. Here again the three operating pressures are emphasized. The bulb pressure is on the top of the diaphragm and the spring pressure and evaporator pressure are on the bottom. When the bulb pressure is greater than the sum of the

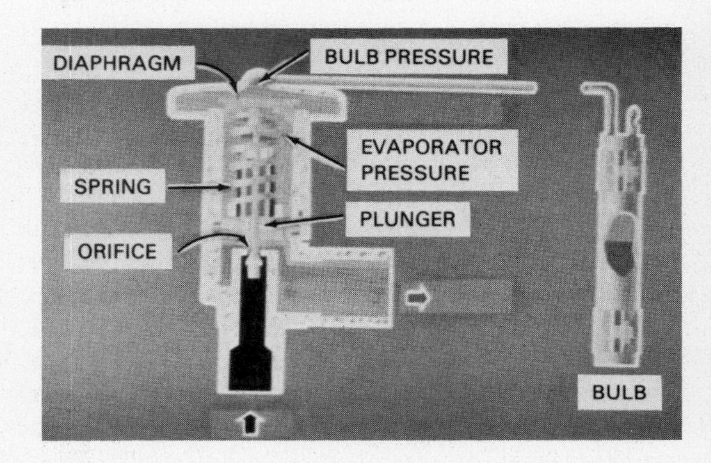

FIGURE 20-7 Thermostatic expansion valve (Courtesy, Carrier Air-Conditioning Company).

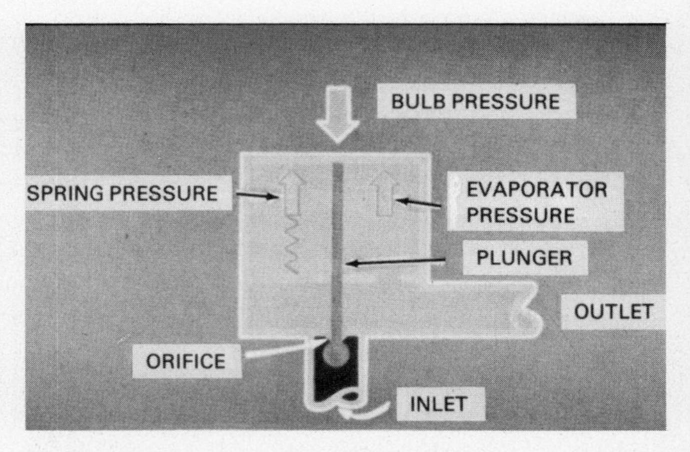

FIGURE 20-8 Schematic thermostatic expansion valve (Courtesy, Carrier Air-Conditioning Company).

spring pressure and evaporator pressure, the plunger will be pushed down, opening the orifice. When the bulb pressure is less than the sum of the spring pressure and evaporator pressure, the plunger will be pushed up, closing the orifice.

Figure 20-9 is a pictorial representation of a thermostatic expansion valve with pressure and temperatures added. Shown is a thermostatic expansion valve set for 10°F superheat with the pressure above the diaphragm equal to the sum of the two pressures below the diaphragm; that is, spring pressure plus evaporator pressure equals bulb pressure. Refrigerant-12 is used in this illustration. The evaporator pressure is 37 psig as shown by the guage. Assuming no pressure drop, this pressure is on the bottom of the diaphragm. The hand-set spring pressure is also exerted on the bottom of the diaphragm and is equal to 9.7 lb as shown in the inset. The total pressure on the bottom of the diaphragm is 9.7 + 37 or 46.7 lb.

The gas leaving the evaporator has a 10°F superheat. This indicates that the gas temperature at the end of the evaporator will be 10°F higher than the temperature corresponding to the evaporator pressure. The evaporator pressure is 37 psig, which corresponds to 40°F. Since the gas leaving the evaporator has a 10°F superheat, the temperature at the end of the evaporator will be 40°F plus 10°F, or 50°F, as shown by the thermometer. The temperature of the thermostatic expansion valve bulb will, therefore, be 50°F. If there is liquid R-12 in the bulb at 50°F, its pressure will be 46.7 psig. This pressure will be transmitted to the top of the diaphragm through the capillary. Because the pressure on the top and the bottom of the diaphragm are equal the valve is in equilibrium, and a continuous, steady rate of refrigerant flow to the evaporator is the result. The valve would remain in this position as long as there was no change in the rate of heat flow to the evaporator.

To compensate for pressure drop in an evaporator, the externally equalized valve is used (Fig. 20-10). In this valve the internal equalizing port is eliminated and the pressure under the diaphragm is being taken from the end of the coil near the thermal bulb as shown. All other conditions are the same, but the evaporator pressure under the diaphragm has dropped to 37 psig, giving a total pressure under the diaphragm of 37 + 9.7, or 46.7 psig. With 46.7 psig above the diaphragm, as shown in the inset, the valve will be in equilibrium at 10°F superheat.

In this valve operation the pressure drop in the coil has been ignored. The externally equalized valve obtains its operating pressures from the point on the coil at which the coil superheat is measured and is completely unaffected by the pressure at the head of the coil. Whenever a pressure drop of several pounds is encountered, an externally equalized thermostatic expansion valve should be used.

The thermostatic expansion valve is the most versatile of all metering devices. It can be used either as a primary metering device or as a pilot device for evaporator control on almost any application. Because of its complexity, however, it must be thoroughly understood by both application and

FIGURE 20-9 Schematic valve in equilibrium (Courtesy, Carrier Air-Conditioning Company).

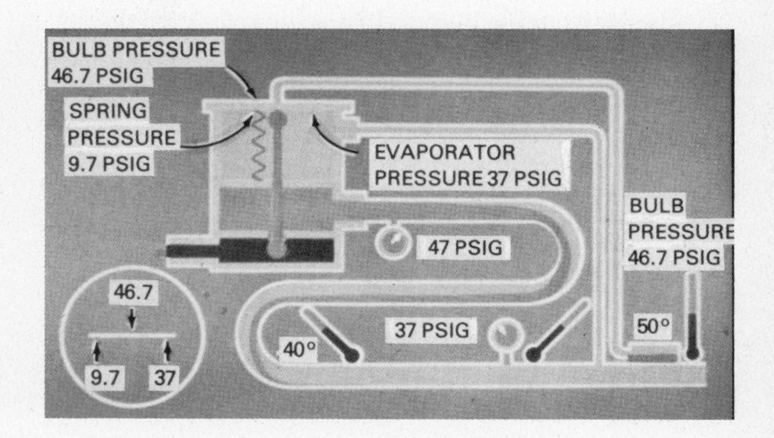

FIGURE 20-10 Externally equalized (Courtesy, Carrier Air-Conditioning Company).

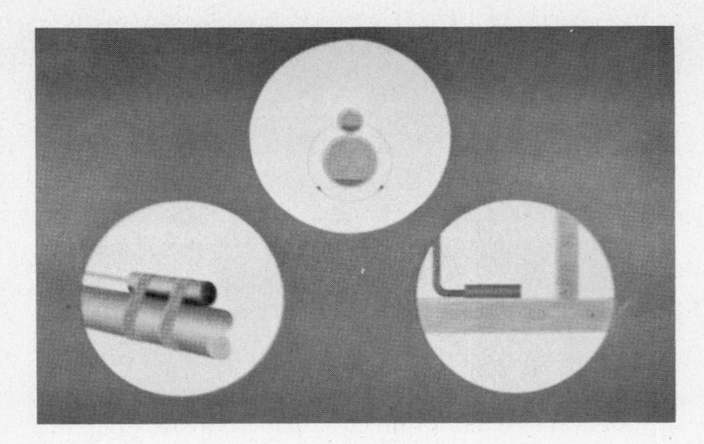

FIGURE 20-11 Points of importance (Courtesy, Carrier Air-Conditioning Company).

service engineers in order that satisfactory results be obtained.

Three important points to remember are shown on Fig. 20-11. The first is that the touching surface between the thermostatic expansion valve bulb and the suction line must be as clean and tight as possible to assure good heat transfer. The temperature of the thermal bulb must be as close as possible to the leaving gas temperature.

The center chart in Fig. 20-11 shows the importance of the bulb position on the suction line. Oil or liquid refrigerant may flow in the bottom of the line, giving a false temperature reading. The bulb must be placed so as to get a gas temperature reading, not an oil or liquid refrigerant temperature reading. Application or size of line will dictate the position of this valve on the line but *it should never be placed on the bottom.*

The chart on the right in Fig. 20-11 shows the position of the equalizing line. It should be downstream from the bulb so that a slight leakage through the packing will not affect the temperature at the bulb. These are only three of many important points relative to the application of the thermostatic expansion valve. The thermostatic expansion valve is an excellent metering device, but requires proper understanding and handling.

ADJUSTING THERMOSTATIC EXPANSION VALVES

The best way to adjust the TEV is first to open it a little too much (too low superheat), so as to frost out beyond the bulb. Then gradually close the TEV (increase the superheat) until the frost is barely back to the bulb. Change the adjustment in small steps and wait at least 15 minutes of running time before making another adjustment.

The main purpose in using a TEV is to get all the refrigeration possible from an evaporator by keeping it fully active. If the evaporator is not fully active, do not hesitate to adjust the TEV for lower superheat (open the valve), should that seem to be the cause of the partial activity; that is the purpose of the adjustment screw on the TEV. The TEV should be readjusted to a higher superheat (closed a little) if overfeeding of the evaporator, a frosted suction line, or liquid slugging by the compressor appears to be causing a low superheat.

As a rule, the TEV should rarely need adjustment after it has been properly installed. If at that time it is set at the correct superheat to make the coil fully active, and later the coil becomes only partially active, it is highly unlikely that readjusting the TEV is the proper procedure, since it is probably something other than the TEV superheat that has changed.

CAPILLARY TUBE

The simplest of all metering devices is the capillary tube, shown in Fig. 20-12. It is nothing more than a deliberate restriction in the liquid line. Because of its small tube size it creates a considerable pressure drop. A piece of capillary tube can be cut to a pretested length that will, under given conditions, create a desired pressure drop. This type of metering device is generally used only on small equipment with fairly

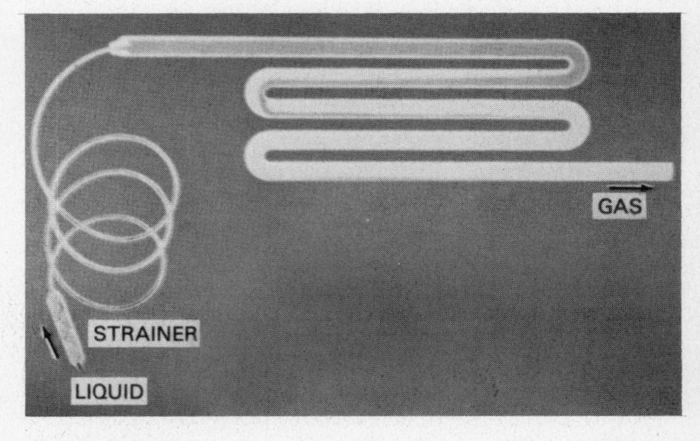

FIGURE 20-12 Capillary tubing (Courtesy, Carrier Air-Conditioning Company).

constant loads, such as domestic refrigerators, home food freezers, room air conditioners, and packaged residential air conditioners. It is generally not found in commercial refrigeration, where control of the evaporator is much more critical due to more widely fluctuating load conditions.

FLOAT CONTROLS

Figure 20-13 shows a *low-pressure float*. The name is derived from the fact that the float ball is located in the low-pressure side or *low-side* of the system. The primary means of control is the level of the liquid in the float chamber as shown.

This type of control is always used with a flooded evaporator. The float ball may be located directly in the evaporator or it may be in a float chamber adjacent to the evaporator. If the float chamber is used, both the top and bottom of the chamber must be connected to the evaporator so that the liquid level in both remains the same at all times.

As the load on the evaporator increases, more liquid is boiled away and the liquid level in the evaporator and float chamber drops. The float drops with it until the orifice opens and admits more liquid from the high-pressure side. As the load on the evaporator decreases less liquid is boiled away and the float will rise until the orifice is closed. The simplest type of float mechanism is shown here, but float valve construction can take many forms.

The low-pressure float is considered one of the best metering devices available for the flooded system. It gives excellent control and its simplicity makes

it almost trouble free. It can be found in any flooded application, large or small, and it can be used with any refrigerant. On the larger systems it is generally used as a pilot device.

Figure 20-14 is an illustration of a high-side float. This float is located on the high side of the system and is immersed in high-pressure liquid, which is the primary control. This type of metering device can only be used in a system that has a "critical" charge of refrigerant. As fast as the hot gas is condensed it flows to the metering device. As the liquid level in the float chamber rises, the float opens and allows the refrigerant to flow to the evaporator. This control allows liquid to flow to the evaporator at the same rate that it is condensed; therefore, there can be no provision in the system for the automatic storage of liquid refrigerant other than in the evaporator. Thus an overcharge of refrigerant would result in liquid flooding back to the compressor and an undercharge of refrigerant would result in a starved evaporator. Frequently suction accumulators are used to make the refrigerant charge less critical.

Having discussed metering devices to set the system in motion, we will now cover the most common cycle controls. The basic cycle control is a device that starts-stops, regulates and/or protects the refrigeration cycle and its components. Though it may take almost any form and be operated by different forces such as temperature or pressure, the function of the cycle control as defined is always the same.

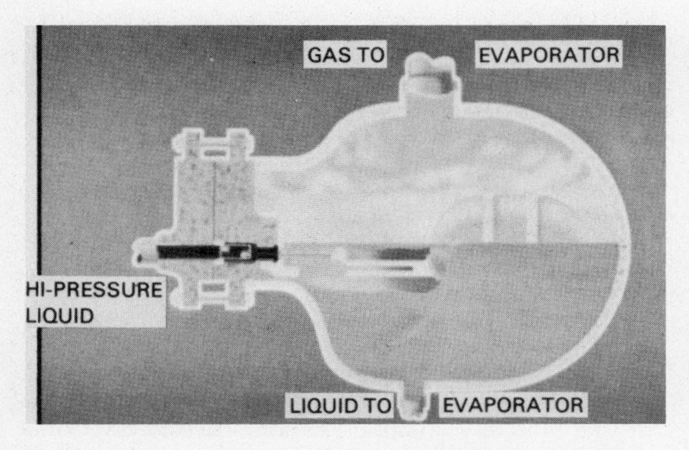

FIGURE 20-13 Lo-side float (Courtesy, Carrier Air-Conditioning Company).

FIGURE 20-14 Hi-side float (Courtesy, Carrier Air-Conditioning Company).

There are two categories of cycle controls, primary and secondary. A primary control actually starts and stops the cycle, either directly or indirectly, as dictated by temperature or humidity requirements. Secondary controls regulate and/or protect the cycle when required to do so by either a primary control or conditions within the cycle. Many controls can serve both functions in the same system, so the control cannot be defined as primary or secondary until its main purpose is known. The first group of controls discussed include those most frequently used as primary controls.

There are three types of primary controls. The first is operated by temperature and is called a *thermostat*. The second is operated by pressure and is called a *pressurestat*. The third is operated by moisture or humidity and is called a *humidistat*.

Each of these controls can be used to regulate cycle operation. For example, when a home refrigerator becomes too warm, that is, the temperature becomes too high for food storage, the thermostat senses this and starts the compressor.

Pressurestats are often used to control temperature conditions in a display case by controlling evaporator pressure. When the evaporator pressure and corresponding temperature becomes high, the pressurestat operates and starts the compressor.

In storage rooms where humidity is all-important, the humidistat is designed to start the refrigeration cycle when the humidity rises to a predetermined point.

Conversely, all the above controls will stop cycle operation when the conditions are satisfied.

Thermostats respond to temperature; they can do this because of the warping effect of a bimetal strip or of fluid pressure. The thermostat illustrated in Fig. 20-15 is a bimetal thermostat. The bimetal element is composed of two different metals bonded together. As the temperature surrounding the element changes, the metals will expand or contract. As they are dissimilar metals, having different coefficients of expansion, one will expand or contract faster than the other. In this drawing the open contacts are illustrated on the left. No current is flowing. If the temperature surrounding the bimetal element is raised, both metals *A* and *B* will start to

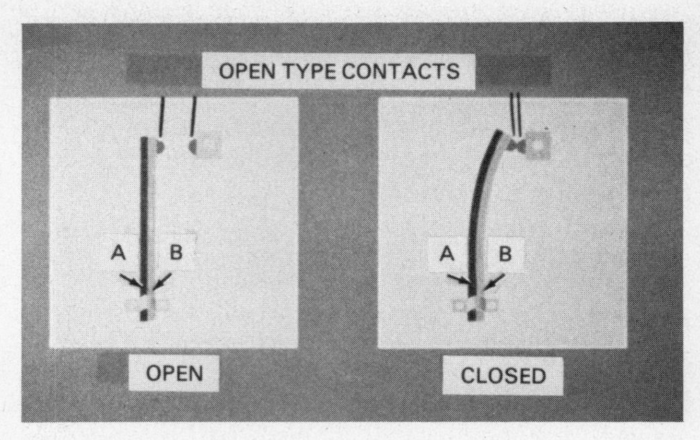

FIGURE 20-15 Bimetal thermostat (Courtesy, Carrier Air-Conditioning Company).

expand. However, metal *A* is chosen because it will expand faster than metal *B*. This will cause the bimetal to bend and close the contacts as shown on the right. As the temperature drops, *A* would contract faster than *B*, thereby straightening the element and opening the contacts.

This thermostat is widely used, especially in residential heating and cooling units, as well as in refrigeration systems that work above freezing conditions, where the stat is actually located in the controlled space. It is simple and cheap to construct, yet is dependable and easily serviced.

One of the common variations of the bimetal thermostat is a mercury bulb thermostat (Fig. 20-16). Though the bimetal principle is used, the contacts are enclosed in an airtight glass bulb containing a small amount of mercury as shown. The action of the bimetal element tilts the bulb. By tilting the bulb to the left, mercury in the bulb will roll to the left and

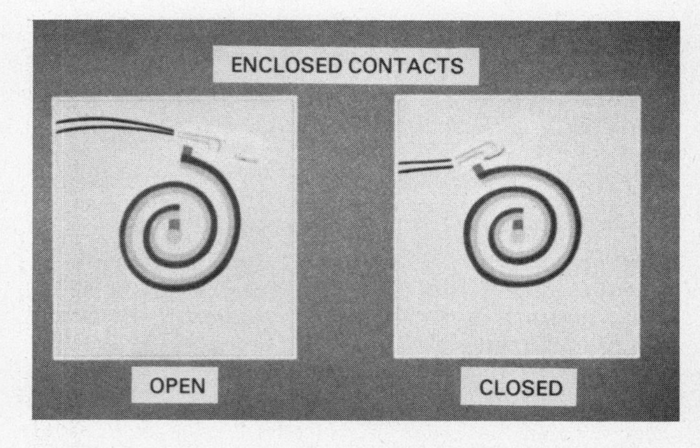

FIGURE 20-16 Bimetal mercury bulb thermostat (Courtesy, Carrier Air-Conditioning Company).

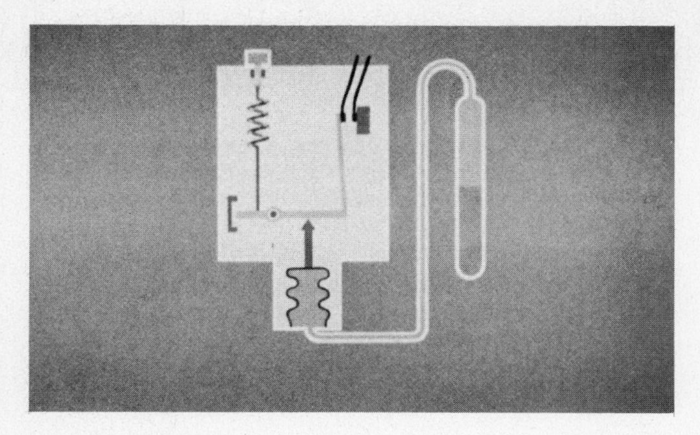

FIGURE 20-17 Bellows type (Courtesy, Carrier Air-Conditioning Company).

complete the electrical circuit. By tilting the bulb to the right, the mercury will flow to the other end of the bulb and break the electrical circuit.

This thermostat is used extensively in space heating and cooling. Though a little more expensive than the bimetal open contact thermostat, its operation is more dependable because dirt cannot collect on the contacts.

The second method of thermostat control is the fluid pressure type shown in Fig. 20-17. With a liquid and gas in the bulb, the pressure in the bellows will increase or decrease as the temperature at the bulb varies.

The thermostat illustrated is a heating thermostat. As the pressure in the bulb increases with the rise in bulb temperature, the bellows expands and, through a mechanical linkage, opens the electrical contacts. As the pressure in the bellows decreases with a lowering of bulb temperature, the bellows contracts and closes the electrical contacts. This thermostat is sometimes referred to as a *remote bulb thermostat.* The controlling bulb can be placed in a location other than that used for the operating switch mechanism. For example, with a cooling thermostat in a refrigerated room, the bulb can be placed inside the room and the capillary run through the wall to the operating switch mechanism located outside the room. Thermostat adjustments can be made without entering the refrigerated room, and the switch mechanism is not exposed to the extreme conditions inside the room.

Another version of the pressure-actuated thermostat is the diaphragm type (Fig. 20-18). In this control the diaphragm is completely filled with liquid. The liquid expands and contracts with a change in temperature. The movement of the diaphragm is very slight, but the pressures that can be exerted are tremendous. Because the coefficient of expansion of liquid is small, relatively large volume bulbs are used on this control. This results in sufficient diaphragm movement and also insures positive control from the bulb.

Pressure controls can also be divided into two categories. They are the *bellows type* and the *Bourdon-tube type.*

By far the most common is the *bellows type,* illustrated in Fig. 20-19. The bellows is connected directly into the refrigerant system through a capillary tube. As the pressure within the system changes,

FIGURE 20-18 Diaphragm type (Courtesy, Carrier Air-Conditioning Company).

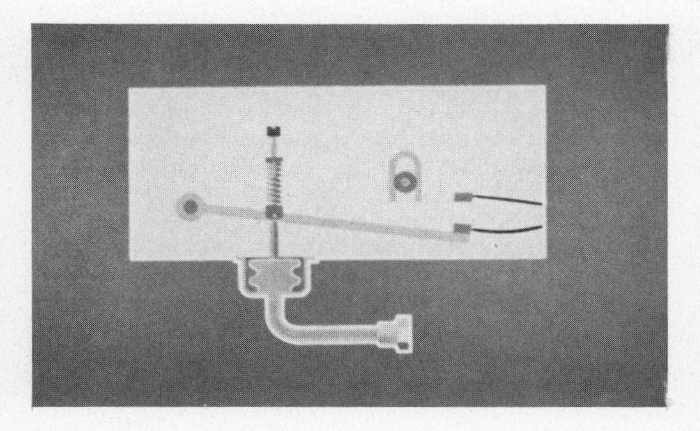

FIGURE 20-19 Bellows type (Courtesy, Carrier Air-Conditioning Company).

so does the pressure within the bellows, causing the bellows to move in and out in conjunction with pressure variation. As shown, the electrical connections are made as the pressure rises. This type of control is used normally as both a low and high-pressure control. That is, it can be connected into the high or low side of the system. Because of its simplicity, dependability, and adaptability, this control is found on almost every air-conditioning or refrigeration system.

Figure 20-20 is a *Bourdon-tube type* pressure control. Notice that a mercury bulb switch is shown here. The Bourdon-tube type control is ideally suited for mercury bulb operation and is frequently found in applications requiring enclosed contacts.

As the pressure inside the tubing increases, the tube will tend to straighten out. This in turn moves the linkage attached to the mercury bulb, causing it to move over center, thus moving the mercury from one end of the bulb to the other and making electrical contacts.

The third type of primary control is the *humidity control* or *humidistat* (Fig. 20-21). Hygroscopic elements are used on these controls, the most common being human hair. As the air becomes more moist and the humidity rises, the hair expands and allows the electrical contacts to close. As the hair dries it contracts, thus opening the electrical contacts.

This type of control is susceptible to dirt and dust in the air. Though accurate, it must be carefully maintained.

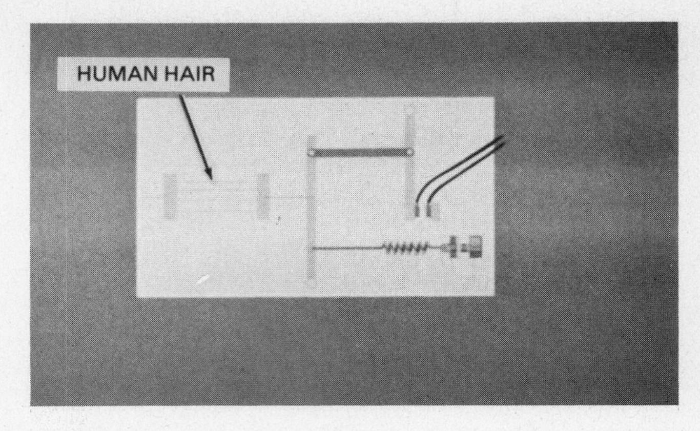

FIGURE 20-21 Humidity control (Courtesy, Carrier Air-Conditioning Company).

Another form of humidistat using a nylon element is shown in Fig. 20-22. The nylon is bonded to a light metal in the shape of a coil spring. The expanding and contracting of the nylon creates the same effect as is found in the spiral bimetal used in thermostats.

There are almost as many possibilities for correct settings on controls as there are applications. Therefore, all controls have some way of being adjusted to compensate for the different conditions under which they may be required to operate. These may be field adjustable or fixed at the factory. The first such adjustment is *range*. Range is the difference between the minimum and the maximum operating points within which the control will function accurately. For example, a high-pressure control may be used to stop the compressor when head pressure becomes too high. This control may have an adjustment that will allow a maximum cutout, or stopping point, of 300 psig and a minimum cutout of 100 psig. There-

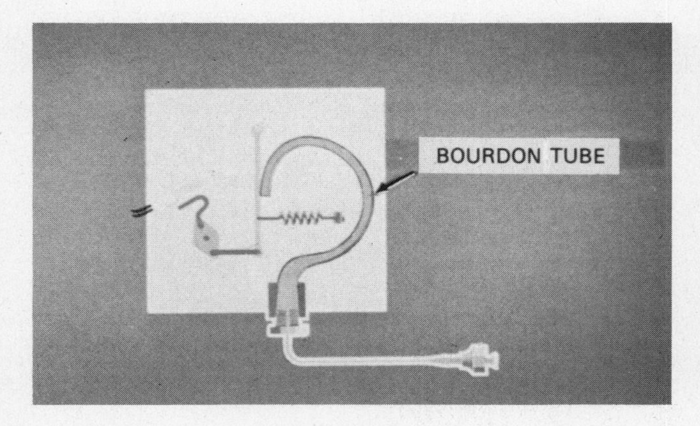

FIGURE 20-20 Bourdon tube pressure control (Courtesy, Carrier Air-Conditioning Company).

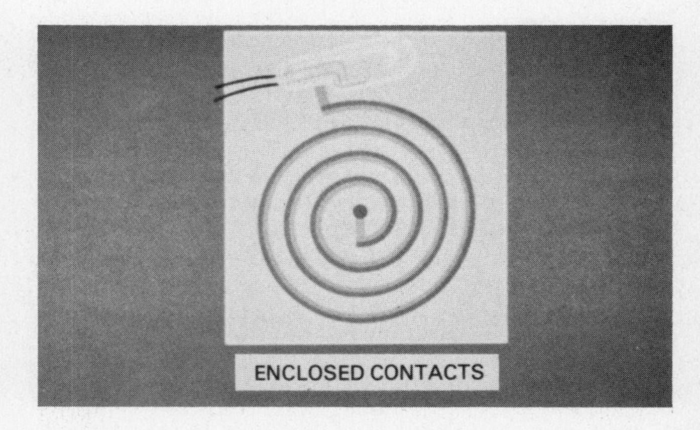

FIGURE 20-22 Nylon humidistat (Courtesy, Carrier Air-Conditioning Company).

fore, the control would have a range of 200 psig. It may be set at any cutout point within this range.

A control should never be set outside its range, as it will always be inaccurate and frequently will not even function.

Figure 20-23 shows one of the simpler methods of range control. As the pressure in the bellows increases, the lever is pushed to the left, thus opening the contacts. By varying the spring pressure, we can increases or decrease the bellows pressure necessary to open the contacts, thus raising or lowering the system pressure at which the control will operate. Many controls used this principle and usually have some external adjustment for field use.

If a control cuts out, or breaks the circuit, it is just as important that it cuts in, or remakes the circuit. The cutout and cutin point cannot be the same. The difference between these points is the *differential*. Differential can be defined as the difference between the cutout and cutin points of the control. For example, if the high-pressure control discussed previously cuts out or breaks the electrical circuit at 250 psig and cuts in or remakes it at 200 psig, the pressure-stat differential is 50 psig.

Referring again to Fig. 20-23, the control also has a differential adjustment. By opening or closing the effective distance between the prongs of the operating fork, the pressure at which the control will cut in can be varied. As the pressure on the bellows increases, the operating fork moves to the right, and the left prong of the fork will tilt the bulb and remake the electrical circuit. When the adjustable stop on the left prong of the fork is moved away from the prong on the right, the pressure in the bellows must decrease further before the bulb will be tilted and the electrical circuit remade. Thus, by varying the effective distance between the prong and stop, the difference between the cutout and cutin point can be varied.

The preceding paragraph explained one way to regulate both range and differential adjustment. There are, however, many ways in which each can be adjusted, depending upon the application, size, and manufacturer's preference.

There is another feature of electrical controls that should be mentioned briefly; this is *detent* or *snap action*. For technical reasons, all electrical contacts should be opened and closed quickly and cleanly. Detent, or snap, action is built into most controls to accomplish this purpose. Figure 20-24 shows four common examples. The magnet in the upper left should require little explanation. The pull of the magnet merely accelerates the closing and opening rate of the contacts. The closer the magnet is to the bimetal in the closed position, the more positive the snap action.

The bimetal disc, as shown in the upper right, is normally in a concave position. As the temperature rises, the two metals expand at different rates until the disc snaps to a convex position, thus sharply breaking the electrical contacts.

The mercury bulb in the lower left is another method of obtaining detent action. As the element moves the bulb over the top center position, the heavy mercury runs from one end of the bulb to the other. This shifting of weight causes the bulb to move quickly from one side of the center to the other, thereby rapidly making or breaking contacts.

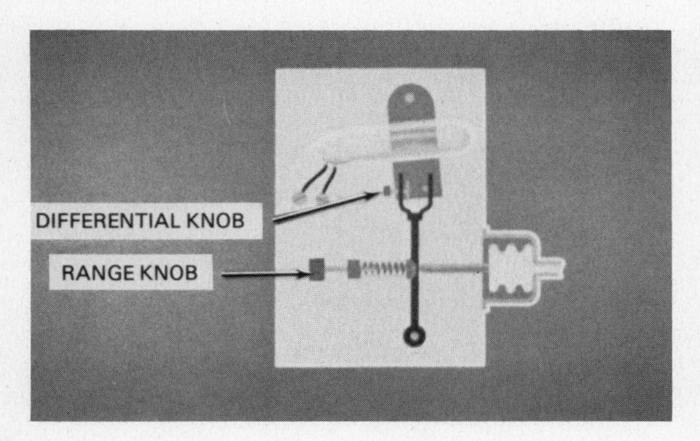

FIGURE 20-23 Range control (Courtesy, Carrier Air-Conditioning Company).

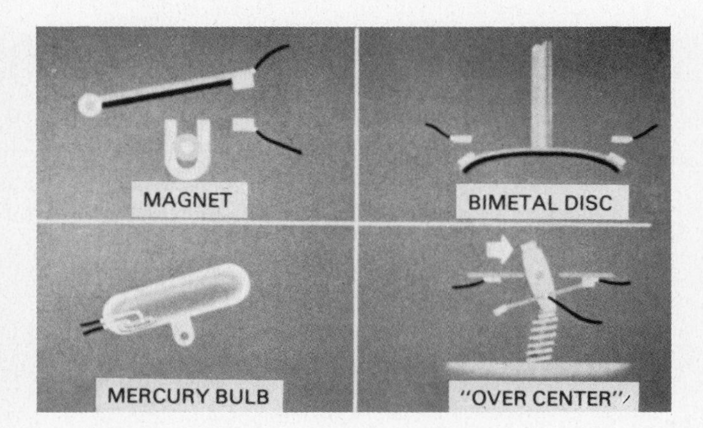

FIGURE 20-24 4 Control illustration (Courtesy, Carrier Air-Conditioning Company).

Shown in the lower right is a fourth method of obtaining detent action. Here the action is induced by using a compressed spring, such as is commonly found in a household toggle switch. Force is applied to the operating arm in the direction of the arrow. This rotates the toggle plate, which starts to compress the spring. As the switch approaches the center position the spring is at maximum compression. The moment the switch passes dead center position the compressed spring will force a rapid completion of the switching action. This explanation of compressed spring action concludes the discussion of primary controls.

Secondary controls can be divided into two categories: operating and safety controls. Operating controls can be expected to regulate the cycle during normal operation. The safety control is designed to protect components of the refrigeration system and should not operate unless some malfunction occurs.

Below is a summary of controls normally found in each category; some have been covered in previous discussions.

Secondary controls

Operating	*Safety*
Metering devices	Electrical overloads
Contactor	Safety thermostat
Water valves	High-pressure
Solenoid valve	Fusible plug
Capacity controls	Rupture disc
Back-pressure valve	Low-pressure
Check valve	Oil safety switch

Figure 20-25 is an illustration of a simple contactor. Basically, a contactor consists of a holding coil and one or more sets of contacts designed to carry heavy electrical loads. The normal thermostat, pressurestat, or other control, is not designed to carry high amperage current through the contacts. Therefore, when electrical loads are too heavy to be handled directly through the control contacts, a contactor must be used if the automatic feature of the primary control is to be retained.

The armature is free to move up or down in the holding coil. As can be seen in Fig. 20-25, when the coil is energized the armature is drawn up, thus closing the contacts and allowing current to flow. Though all contactors make or break electrical circuits, they come in many sizes, and their mechanical action and physical appearance vary with the manufacturer.

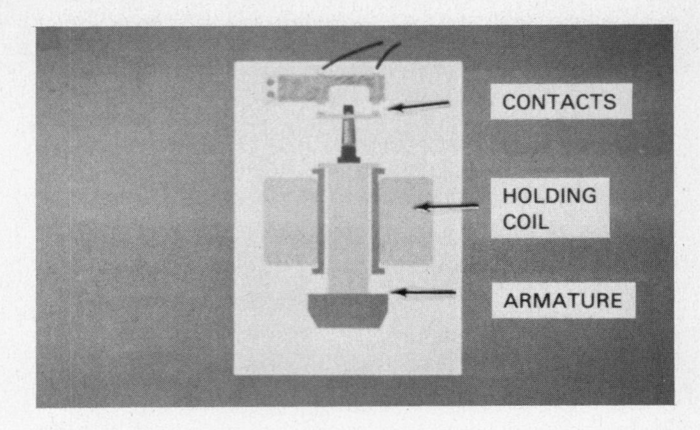

FIGURE 20-25 Contactor (Courtesy, Carrier Air–Conditioning Company).

The water valve (Fig. 20-26) is designed for systems using water as the condensing medium. Its purpose is simple; it controls the flow of water through the condenser to maintain a constant head or high-side pressure.

The action of the water valve is controlled by the refrigerant pressure within the condenser. The gas in the condenser enters the valve at the bottom and exerts a pressure on the bellows. As the pressure increases, the valve stem connected to the bellows is forced up, thus opening the valve and allowing more water to flow. As the pressure on the bellows decreases, the spring at the top of the valve forces the valve closed, thus cutting off the supply of water. An

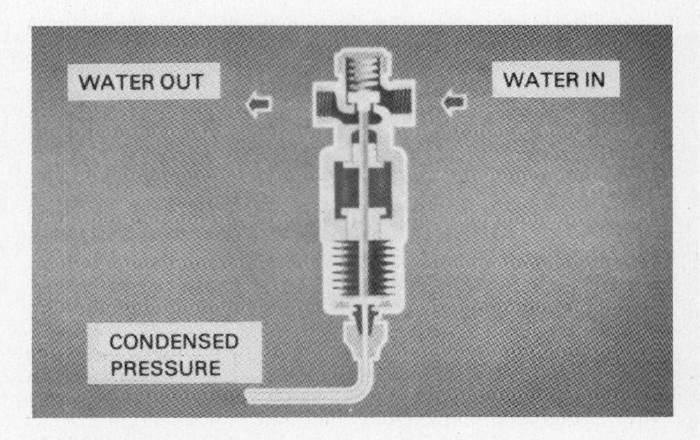

FIGURE 20-26 Water valve (Courtesy, Carrier Air–Conditioning Company).

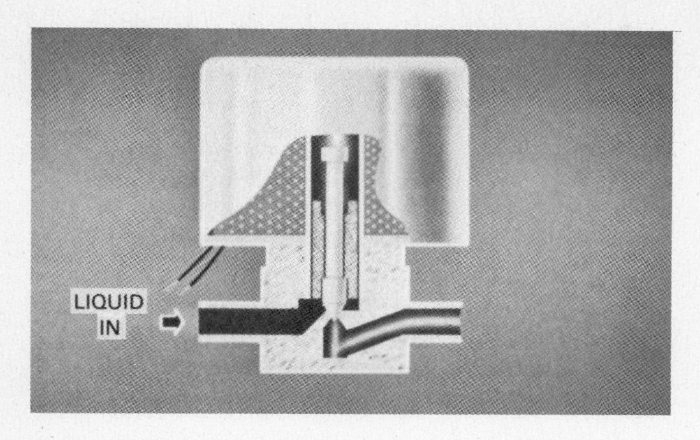

FIGURE 20-27 Solenoid valve (Courtesy, Carrier Air-Conditioning Company).

increase in pressure allows cooling water to flow through the condenser, and a decrease in pressure closes down the flow of water to the condenser. The pressure at which the condenser is maintained can be controlled by adjusting the pressure of the spring shown at the top of the valve.

Solenoid valves, like the one in Fig. 20-27 are used in many ways to control fluid flow, but one of the most common is a refrigerant liquid-line valve. Here a solenoid valve is used to stop the flow of liquid to the evaporator when refrigeration requirements are satisfied. When refrigeration is again required, the solenoid valve is opened and the liquid flows to the evaporator.

The operating principle of a solenoid valve is simple. A coil of wire is placed around a tube containing a movable plunger. When current flows through the wire a magnetic field is created. This pulls the plunger up into the tube. As the plunger rises, it opens the valve and allows liquid to flow.

The valve shown in Fig. 20-27 is a normally closed valve, because the valve is in the closed position unless current is flowing. Solenoid valves are also made in the normally open type. These valves will close only when the coil is energized.

Figure 20-28 is an illustration of a back-pressure valve or evaporator pressure regulator. This valve is designed to maintain a constant pressure or temperature in the evaporator regardless of the compressor suction pressure. Evaporator pressure enters the port at the left. As the evaporator pressure increases the pressure under the bellows increases. This pressure overcomes the spring tension and raises the valve plunger from the seat. This allows evaporator gases to flow to the suction line. As the gas leaves the evaporator the pressure in the evaporator is reduced. As the pressure under the bellows decreases, the spring pressure closes the valve. The operating pressure at the valve may be regulated by changing the spring pressure.

This type of valve is useful where control of temperature and humidity is important. It assures that the evaporator will not drop below a specified temperature. It is also used where two or more evaporators on the same system are operating at different temperatures.

Another type of back-pressure valve frequently found in larger systems is the pilot-operated valve shown in Fig. 20-29. Though it works on the same principle as the valve previously discussed, the control point is not at the valve itself. In this control,

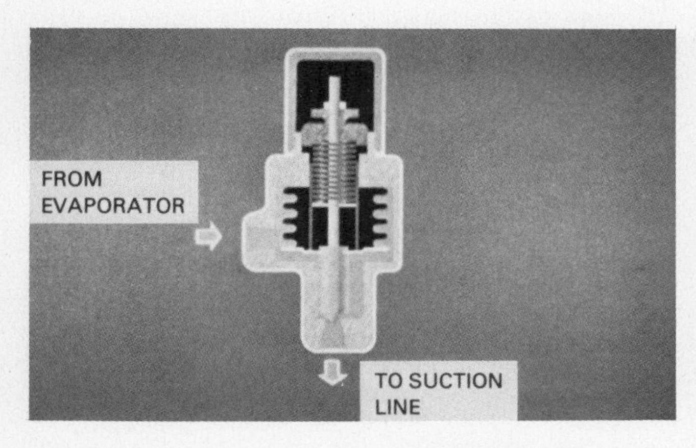

FIGURE 20-28 Evaporative pressure regulator (Courtesy, Carrier Air-Conditioning Company).

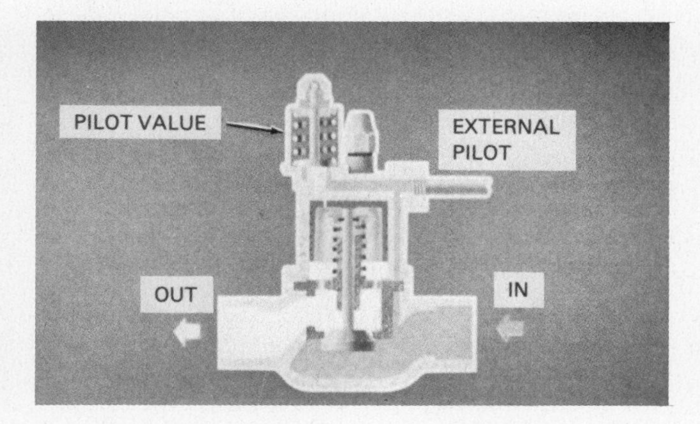

FIGURE 20-29 Evaporative pressure regulator—pilot operation (Courtesy, Carrier Air-Conditioning Company).

the control pressure enters through the external pilot connection. As the pressure increases, it forces the seat of the pressure pilot valve up away from the pilot port. This allows the gas to flow through the pilot port down to the top surface of the piston, creating sufficient pressure to lower the piston and open the valve. As the pressure at the external pilot decreases, the pressure pilot adjusting spring closes the valve and shuts off the pilot port. The gas pressure on top of the piston then bleeds through the piston orifice, and the main spring forces the piston up, closing the valve. The pressure to operate this valve should be taken from the evaporator because it gives better performance than the suction line.

The third type of evaporator pressure regulator is the *snap-action* or *two-temperature valve* (Fig. 20-30). This valve operates on the same principle as the two valves previously discussed. However, the mechanical action is designed so that the valve will be fully opened or fully closed.

The gas at evaporator pressure flows through the body orifice into the bellows. As the pressure increases, the bellows expands and moves the diamond-shaped finger up. As the finger moves up, the spring-loaded wheels ride up on the finger to the maximum finger width. If the finger continues to tilt, the wheels go over the wide point and the spring pressure tends to force them together. This force causes the wheels to ride down the diamond slope to push the valve plunger down and open the valve. The valve will stay open until the evaporator pressure is reduced enough to let the finger-wide point drop below the wheels, at which time the spring pressure, by forcing the wheels together, will lift the plunger and close the valve.

The valve is used in two-temperature applications where there are two or more evaporators operating on the same compressor. The two-temperature valve is always placed in the suction line of the highest temperature evaporator. It is so set that it will maintain a higher average evaporator pressure or temperature in this evaporator than will be maintained in the evaporator connected directly to the compressor suction.

The most important feature of the snap-action valve is that it does not maintain a constant pressure. Since the valve is wide open or closed, it cycles the evaporator in the same manner as it would were it directly connected to its individual condensing unit, controlled by a pressurestat or a thermostat. In cases where frost may be a problem, the high temperature in the evaporator can be adjusted so as to be above freezing, thus providing automatic defrost.

All the secondary controls just discussed are operating controls; that is, controls that regulate the cycle during normal operation.

The electrical heating overload control (Fig. 20-31) is the first of the secondary safety controls to be discussed. The safety control protects cycle components and is not expected to operate until a malfunction of the system occurs.

In all electrical circuits, some kind of protection against excessive current must be provided. The

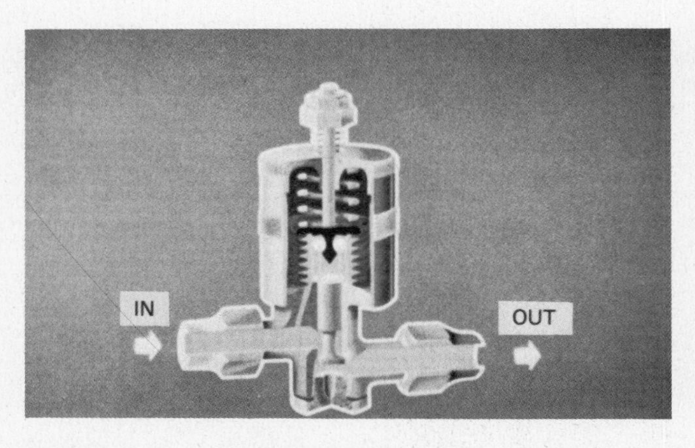

FIGURE 20-30 Two temperature valve (Courtesy, Carrier Air-Conditioning Company).

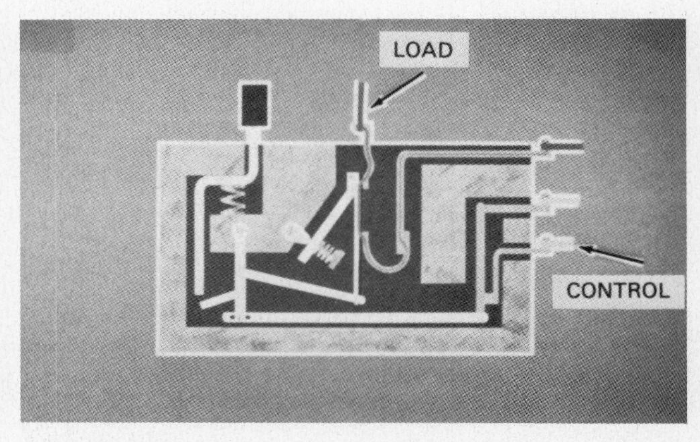

FIGURE 20-31 Electric heating overload (Courtesy, Carrier Air-Conditioning Company).

common household fuse is a good example of this type of protection. In motor circuits, however, a reset type of protector is frequently used. The overload shown is one method by which this is accomplished. It is used in conjunction with the contactor previously discussed. When a contactor is used, there is both a control circuit, in which the primary control is inserted, and a load circuit, which is opened and closed by the contactor. When excessive current is drawn in the load circuit, this device will break the control circuit.

The control circuit to the contactor coil passes through the controls at the lower right. The load circuit is passing through the bimetal shown in the center. If the current through the bimetal element becomes too high, the bimetal element bends to the left. This action forces the contact arm to the left and breaks the contacts. This breaks the control circuit and allows the contactor to open, thereby interrupting the power to the load. The bimetal will now cool and return to its original position, but because of the slot in the bottom arm the contacts will not be remade. The control circuit will remain broken until the reset button is pushed. This moves the arm to the right and closes the control circuit contacts. Because this control must be reset by hand, it is usually referred to as a *manual reset overload.*

A second type of electrical overload protector is the bimetal disc (Fig. 20-32). When the temperature is raised, the disc will warp and break the electrical circuit. On the left is shown the simplest device in which heat from an external source activates the disc. On the right, external heat is supplemented by resistance heat from an electrical load. The device on the right will react much faster on electrical overload than the device on the left because the supplemental heat is within the device itself.

The bimetal disc overload protector is widely used in the protection of electrical motors.

The current relay (Fig. 20-33) is another type of electrical overload that can be manually or automatically reset. The greatest advantage of this type of relay is that it is only slightly affected by ambient temperatures, thereby avoiding nuisance trip-outs.

The current relay is made up of a sealed tube completely filled with a fluid and holding a movable iron core. When an overload occurs, the movable core is drawn into the magnetic field, but the fluid slows its travel. This provides a necessary time delay to allow for the starting in-rush of momentary overloads. When the core approaches the pole piece, the magnetic force increases and the armature is actuated, thus breaking the control circuit.

On short circuits, or extreme overloads, the movable core is not a factor, because the strength of the magnetic field of the coil is sufficient to move the armature without waiting for the core to move. The time-delay characteristics are built into the relay and are a function of the core design and fluid selection.

The only difference between the thermostatic primary control previously discussed and the safety thermostat is the application in which they are used; the controls themselves may be identical.

FIGURE 20-32 Bi-metal disc overload (Courtesy, Carrier Air-Conditioning Company).

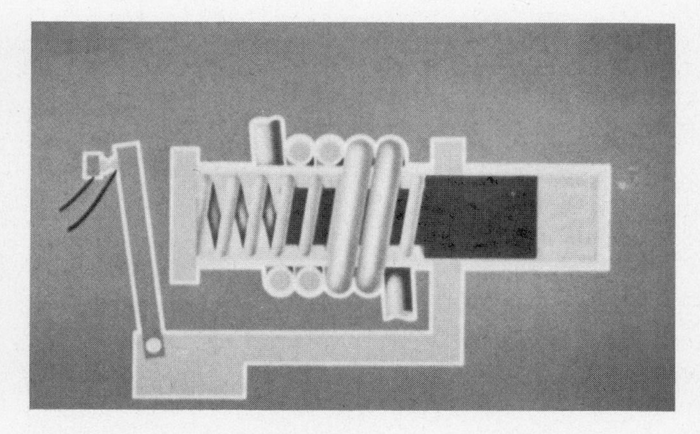

FIGURE 20-33 Current relay (Courtesy, Carrier Air-Conditioning Company).

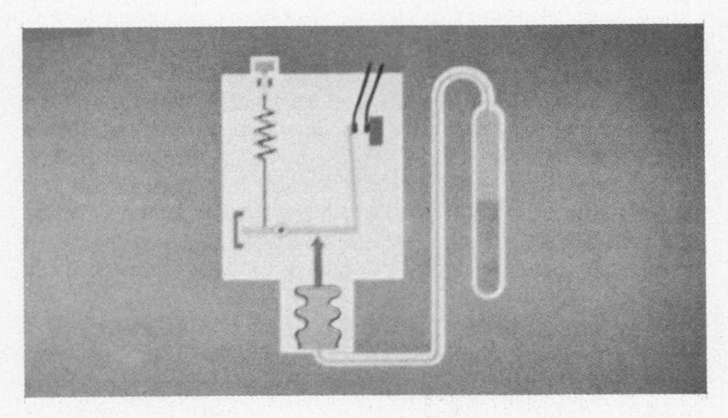

FIGURE 20-34 Bellows type (Courtesy, Carrier Air-Conditioning Company).

A good illustration of a thermostat used as a safety control can be found in most water-chilling refrigeration systems (Fig. 20-34). In water-chilling systems it is important that the water is not allowed to freeze, because it could then do physical damage to the equipment. In such systems a thermostat may be used with the temperature-sensing element immersed in the water at the coldest point. The thermostat is set so that it will break the control circuit at some temperature above freezing. This will stop the compressor and prevent a further lowering of water temperature and possible freezing. The pressure control used for safety purposes is physically the same as that used for operational purposes.

The high-pressure cutout is probably the best example of a pressure control used for safety purposes (Fig. 20-35). It can be set to stop the compressor before excessive pressures are reached. Such conditions might occur because of water supply failure on water-cooled condensers or because of fan motor stoppage on air-cooled condensers.

The operation of the low-pressure control is the same as discussed under the subject of primary controls. Mechanically there is no difference between the low-pressure control used as a primary control and one used as a safety control. The low-pressure control is used to stop the compressor at a predetermined minimum operating pressure. As a safety device, the low-pressure control can protect against loss of charge, high compression ratios, evaporator freeze-ups, and entrance of air into the system through low-side leaks

The oil safety switch is activated by pressure. However, it is operated by a pressure differential rather than a straight system pressure. It is designed to protect against the loss of oil pressure. Inasmuch as the lubrication system is contained within the crankcase of the compressor, a pressure reading at the oil pump discharge will be the sum of the actual oil pressure plus the suction pressure. The oil safety switch measures the pressure difference between the oil pump discharge pressure and suction pressure and shuts down the compressor if the oil pump does not maintain an oil pressure as prescribed by the compressor manufacturer. For example, if a manufacturer indicated that a 15-lb net oil pressure was required, a compressor with a 40-lb back pressure must have at least a 55-lb pump discharge pressure. At any lower discharge pressure from the pump this switch would stop the compressor. An examination of the operating force inset on the right (Fig. 20-36) will show that since suction pressure is present on both sides,

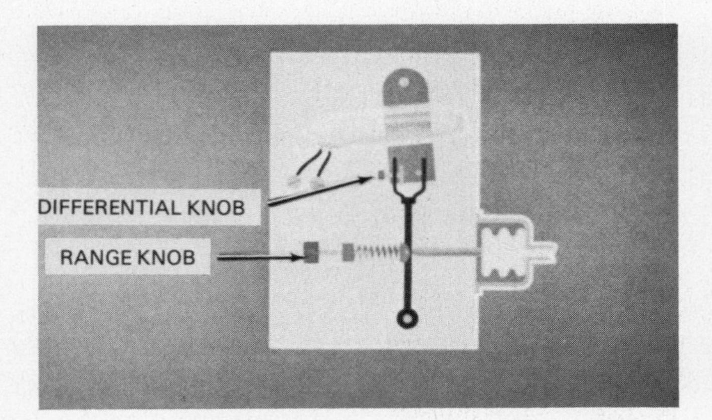

FIGURE 20-35 Control adjustments (Courtesy, Carrier Air-Conditioning Company).

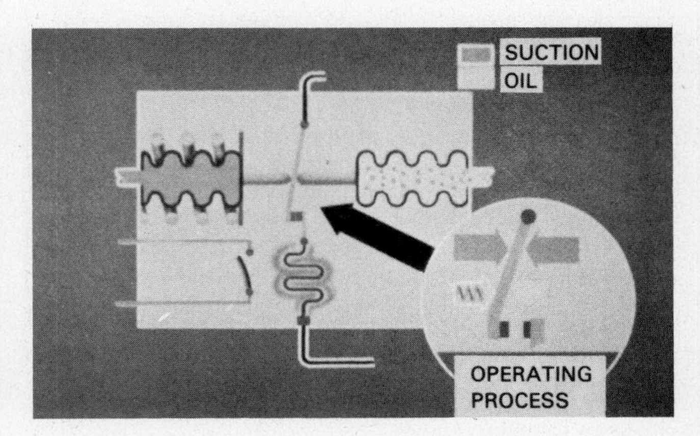

FIGURE 20-36 Oil safety switch (Courtesy, Carrier Air-Conditioning Company).

its effect is cancelled. The control function depends on the net oil pressure overcoming the predetermined spring pressure. The spring pressure must equal the minimum oil pressure allowed by the manufacturer.

When a compressor starts, there is no oil pressure. Full oil pressure is not obtained until the compressor is up to speed. Therefore, a time-delay device is built into this type of control to allow the compressor enough time to start. One of the methods used to create this time delay is a small resistance heater. When the compressor is started, because the pressure switch is normally closed, the resistance heater is energized. If the oil pressure does not build up to the cutput setting of the control, the resistance heater warps the bimetal, breaks the control circuit, and stops the compressor. The switch in Fig. 20-36 is automatically reset; However, these switches are normally reset manually.

If the oil pressure rises to the cutout setting of the oil safety switch within the required time after the compressor starts, the control switch is opened and deenergizes the resistance heater, and the compressor continues to operate normally.

If oil pressure should drop below the cutin setting during the running cycle, the resistance heater is energized and, unless oil pressure returns to cutout pressure within the time-delay period, the compressor will be shut down. The compressor can never be run longer than the predetermined time on subnormal oil pressure. The time-delay setting on most of these controls is adjustable from approximately 50 to 125 seconds.

There are as many control methods for operating refrigeration systems as there are engineers to design them, but there are two simple methods that are standard and should be discussed briefly. The first, illustrated in Fig. 20-37, is the simple thermostat start-stop method.

As the thermostat calls for cooling and closes the contacts, the control circuit is completed through the contactor coil. This control circuit includes a klixon, a pressure control, and an overload electrical heater as secondary controls. When the contactor coil is energized it closes the contacts and completes the power circuit to the motor. The thermostat is the primary control and the motor cycles as the thermostat dictates. Any of the secondary controls will break the control circuit and stop the compressor in the same manner as the thermostat.

This is one of the simplest of all control methods and therefore is widely used. Some small refrigeration systems and almost all small air-conditioning systems use this principle of control.

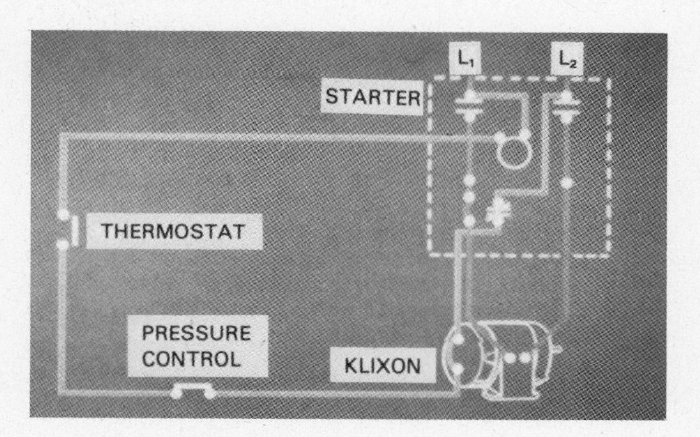

FIGURE 20-37 **Thermostat-start-stop (Courtesy, Carrier Air-Conditioning Company).**

A slightly more complex but still widely used control method is shown in Fig. 20-38. This is known as pump-down control. In this method the thermostat operates a solenoid valve. When the thermostat calls for cooling it completes a circuit through the solenoid valve, which is in the liquid line ahead of the metering device. By opening the valve, high-pressure liquid flows to the evaporator.

A pressure control is connected to the low side of the system. As the pressure rises in the evaporator coil, the pressure control closes and completes the circuit through the safety device to the contactor coil. The compressor starts and the system operates normally. When the thermostat is satisfied, the

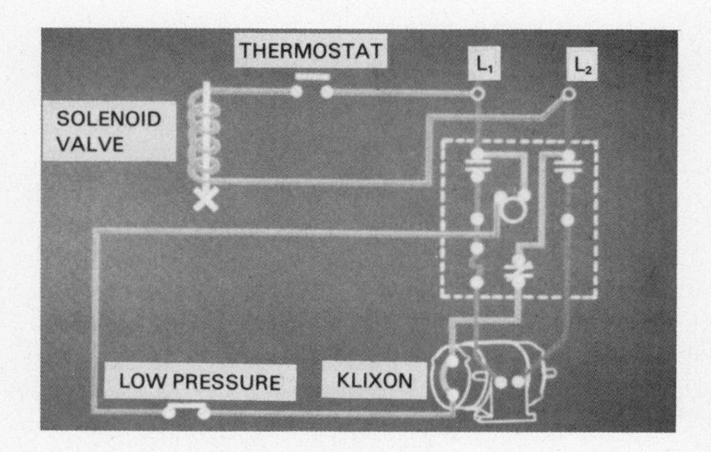

FIGURE 20-38 **Pump down control (Courtesy, Carrier Air-Conditioning Company).**

solenoid coil is deenergized and the flow of liquid is cut off to the evaporator. Because the compressor is still running, the evaporator pressure is reduced to the cutout point of the low-pressure control. The control contacts open and the compressor stops.

This system is designed to keep liquid refrigerant from filling the evaporator during shutdown. Regardless of the position of the thermostat, the low-pressure control will operate the compressor upon sensing a rise in pressure in the evaporator.

To review briefly: There are two basic categories of controls: primary and secondary. Primary controls actually start and stop the cycle, either directly or indirectly, as dictated by temperature or humidity requirements. There are three kinds of primary controls: thermostats, pressurestats, and humidistats.

Secondary controls regulate and/or protect the cycle when required to do so by a primary control or conditions within the cycle. There are two basic types of secondary controls: operating and safety. Operating controls include metering devices, contactors, water valves, solenoid valves, back-pressure valves, and check valves. Safety controls include electrical overloads, safety thermostats, high and low-pressure cut-outs, and all safety switches.

PROBLEMS

1. Controls are divided into four classifications; name them.

2. There are five main classifications of metering devices; name at least three.

3. The main function of the thermostatic expansion valve is to maintain a relatively constant ______________.

4. The bulb of the TEV should always be placed on the bottom of the suction line: true or false?

5. Most room thermostat sensing elements are made of ______________.

6. Many pressure-sensitive controls employ the ______________ tube principle.

7. Fusible plugs would be classified as ______________ controls.

8. An oil-pressure control stops the compressor immediately if it senses no oil pressure: true or false?

9. The back-pressure valve is designed to maintain a constant pressure at the ______________.

10. Some controls are equipped for either ______________ or ______________ reset.

21

THE REFRIGERATION LOAD

The total refrigeration load of the system as expressed in Btu/h comes from many heat sources. Figure 21-1 represents a cutaway view of a refrigerated storage room in a supermarket. Note the sources of heat caused by:

(1) Heat transmission

- The temperature difference of 60° between the 95°F outside air and the room temperature of 35°F, which can cause much heat conduction.
- The sun effect on the roof and walls results in radiation heat buildup.

(2) Air infiltration

- Air that enters a room as a result of opening and closing the doors during normal working procedures.
- Air that enters a room through cracks in the construction or around door seals.
- Air that may be purposely introduced for ventilation reasons.

(3) Product loads, which are the results of heat(s) contained within the product being stored. In some cases it is purely dry or sensible heat, such as cooling a can of juice from room temperature down to 35°F, or it may be a combination of dry heat (sensible) and moisture (latent heat) from produce; should the product be frozen, there are additional requirements connected with the latent heat of freezing. Some heat is also the result of chemical changes such as the ripening of fruit.

(4) Supplementary loads are caused by such things as electric lights, motors, tools, and also arises from human beings.

Although the refrigeration design engineers are basically responsible for estimating the loads and for the construction planning of the installation and application of equipment, the refrigeration technician should understand how these heat sources affect the operation of the system so that he can adjust the equipment to perform in a manner consistent with the design engineer's recommendations.

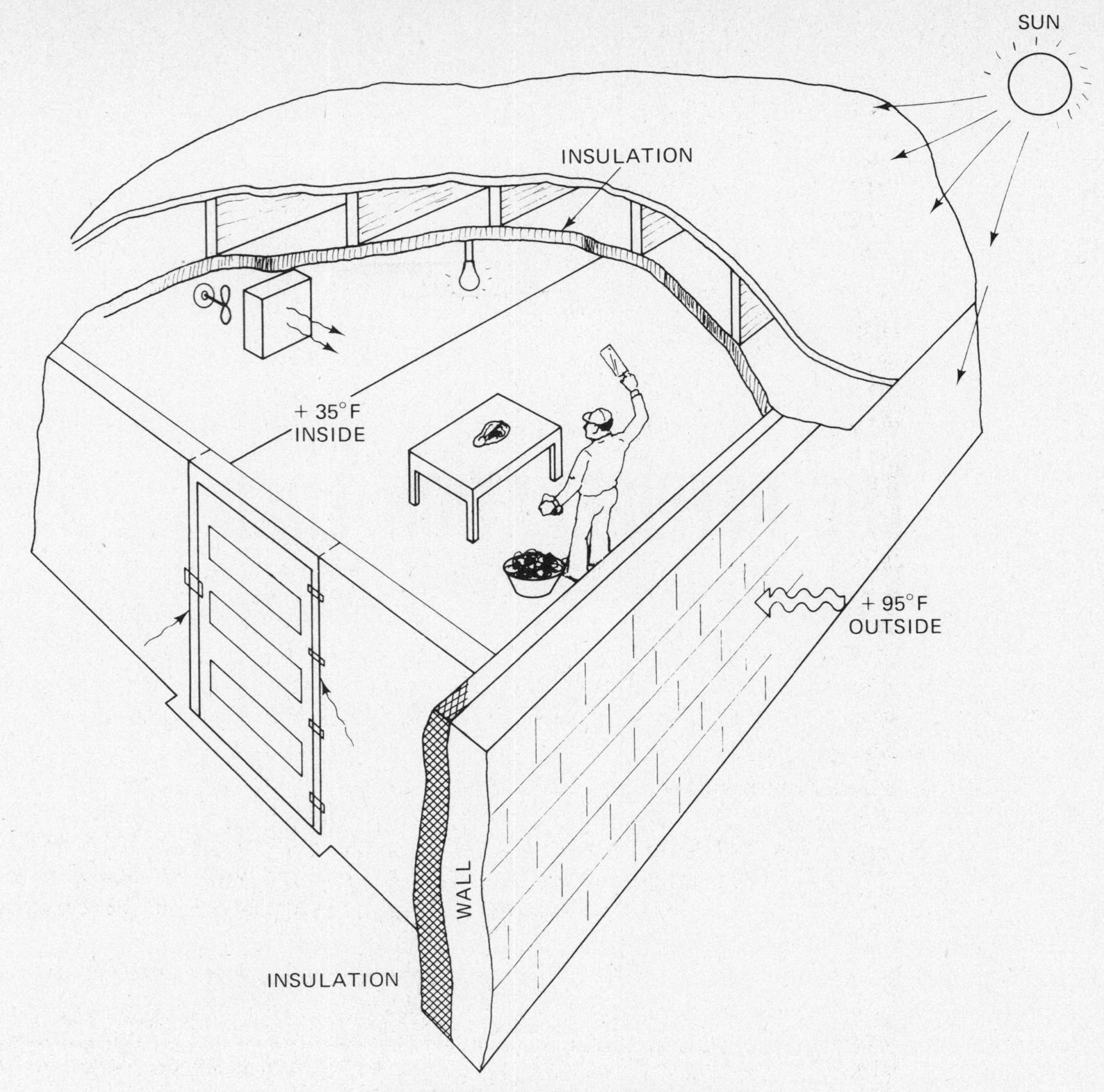

FIGURE 21-1 Refrigerated store room.

HEAT TRANSMISSION

The heat gain through walls, floors, and ceilings will vary with the type of construction, the area that is exposed to a different temperature, the type and thickness of insulation, and the temperature difference between the refrigerated space and the ambient air.

Thermal conductivity varies directly with time, area, and temperature difference and is expressed in Btu/h, per square foot of area, per degrees Fahrenheit of temperature difference, per inch of thickness.

It is readily apparent that in order to reduce heat transfer, the thermal conductivity factor (based on material composition) should be as small as possible and the material as thick as economically feasible.

Heat transfer through any material is also subject to or affected by the surface resistance to heat flow, which is determined by the type of surface (rough or smooth); its position (vertical or horizontal); its reflective properties; and the rate of air flow over the surface.

Extensive testing has been done by many laboratories to determine accurate values for heat transfer through all common building and structural material. Certain materials (like insulation) have a high resistance to flow of heat, others are not so good.

In order to simplify the task of calculating heat loss the industry has developed a measuring term called *Resistance* (*R*), which is the resistance to heat flow, either of one inch of material, or of a specified thickness, or of an air space, an air film, or an entire assembly. Its value is expressed as degrees Fahrenheit temperature difference per Btu per hour per square foot. A high R value indicates low heat-flow rates. The resistances of several components of a wall may be added together to obtain the total resistance:

$$RT = R_1 + R_2 + R_3 \ldots$$

Table 21A lists some R values for common building materials in order to illustrate the differences in heat-flow characteristics. For more extensive listings of R values refer to the ASHRAE *Handbook of Fundamentals*.

The actual quantity of heat transmission (Q) through a substance or material is then calculated by the formula

$$Q = U \times A \times TD$$

Q = heat transfer, Btu/h

U = overall heat transfer coefficient (Btu/h)(ft²)(°F TD)

$U = \frac{1}{R_t}$ value (If an assembly $R_t = R1 + R2 + R3$ for various components)

A = area in ft²

TD = Temperature difference between inside and outside design temperature and refrigerated space design temperature.

For example calculate the heat flow through an 8-in. concrete block (cinder) wall (Fig. 21-2), 100 square feet in area, having 60°F temperature difference between the inside and outside.

R value for an 8-in. concrete block wall (see Table 21A) is 1.72.

U therefore is $= \frac{1}{R} = \frac{1}{1.72} = 0.58$ Btu/h/°F/ft².

$Q = 0.58 \times 100\ \text{ft}^2 \times 60°\text{F} = 3480$ Btu/h heat flow into the space.

Now add 6 in. of fiberglass insulation to the wall (Fig. 21-3) and recalculate the transmission load.

U factor $= \frac{1}{R_t}$

$R_t = R_1$ (concrete block) $+ R_2$ (6-in. insulation)

$R_t = 1.72 + 18.72$ (R value for 1-in. insulation = 3.12) (R_2 then equals 6×3.12 or 18.72)

$R_t = 20.44$

TABLE 21A Typical heat transmission coefficients

MATERIAL	DENSITY lb/ft³	MEAN TEMP. °F	CONDUCTIVITY k	CONDUCTANCE C	RESISTANCE (R)	
					PER IN.	OVERALL
Insulating Materials						
Mineral wool blanket	0.5	75	0.32		3.12	
Fiberglass blanket	0.5	75	0.32		3.12	
Corkboard	6.5– 8.0	0	0.25		4.0	
Glass fiberboard	9.5–11.0	−16	0.21		4.76	
Expanded urethane, R11		0	0.17		5.88	
Expanded polystyrene	1.0	0	0.24		4.17	
Mineral wool board	15.0	0	0.25		4.0	
Insulating roof deck, 2 in.		75		0.18		5.56
Mineral wool loose fill	2.0– 5.0	0	0.23		4.35	
Perlite, expanded	5.0– 8.0	0	0.32		3.12	
Masonry Materials						
Concrete, sand and gravel	140		12.0		0.08	
Brick, common	120	75	5.0		0.20	
Brick, face	130	75	9.0		0.11	
Hollow tile, two-cell, 6 in.		75		0.66		1.52
Concrete block, sand and gravel, 8 in.		75		0.90		1.11
Concrete block, cinder, 8 in.		75		0.58		1.72
Gypsum plaster, sand	105	75	5.6		0.18	

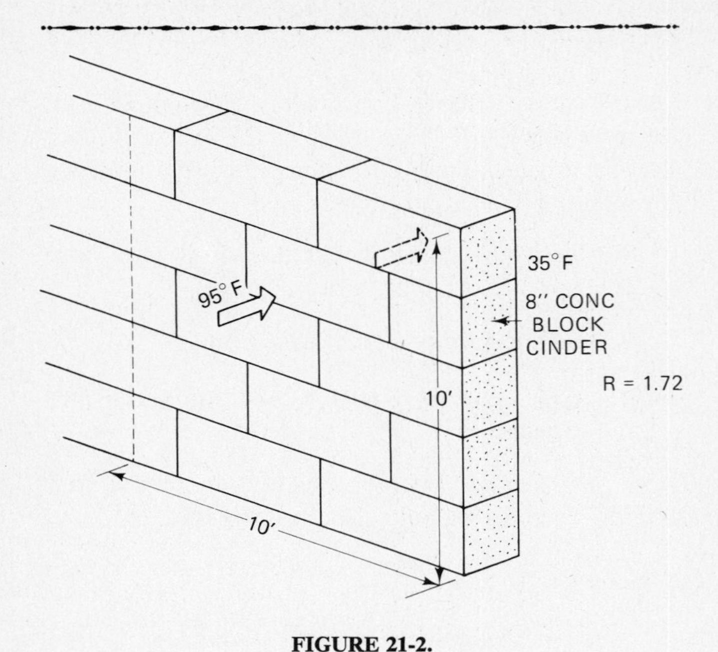

FIGURE 21-2.

35°F
95°F
8" CONC
BLOCK
CINDER
R = 1.72
10'
10'
(ADD 6" INSULATION R = 3.12 PER INCH)

FIGURE 21-3.

Therefore

$$U = \frac{1}{20.44} = 0.049$$

$$Q = 0.049 \times 100 \text{ ft}^2 \times 60°\text{F} = 294 \text{ Btu/h}$$

The above example demonstrates the cumulative effect of R values on determining the total wall resistance, but it also dramatically shows the heat reduction that can be achieved through proper insulation, in this case, from 3,480 Btu/h down to 294 Btu/h, which would drastically reduce the size of the refrigeration equipment needed and the resulting operating cost.

Insulation is the most effective method of reducing heat transmission. There are types made of material to meet the requirements of every application, although some are better than others. General classifications of the available forms of insulation are (1) loose fill, (2) flexible, (3) rigid or semirigid, (4) reflective, and (5) foamed-in-place.

Loose fill or "blown" insulation (Fig. 21-4) is used primarily in residential structures for reducing heat flow. Flexible insulations such as fiberglass, in batts or roll type (Fig. 21-5), are also common in new residential and commercial construction and come with a material such as kraft paper to act as a vapor barrier. Some are available with reflective surfaces to help reduce radiant heat effects.

Rigid and semi-rigid insulations (Fig. 21-6) are made of such materials as corkboard, polystyrene, foam glass, polyurethane, which are manufactured in various convenient dimensions and forms such as boards, sheets, or blocks. Some have a degree of structural strength, others do not. It is in this category that we find more widespread application to commercial refrigeration: walk-in coolers, freezers, display cases, etc. Because of their density and cell composition, they offer a built-in vapor barrier to moisture penetration.

FIGURE 21-4 Without insulation.

FIGURE 21-5 With 6″ iusulation.

Foamed-in-place insulation (Fig. 21-7) is widely used for filling cavities that are hard to insulate and also to cover drain pans, etc., where an effective temperature control and water seal is needed. Foamed insulation is also used in on-site, built-up refrigerated rooms, in connection with the rigid insulations mentioned above.

Regardless of what type of insulation is used, moisture control is most important. Figure 21-8 represents the gradual temperature changes within an insulating material from 90°F on the warm outside side to 40°F on the cold inside. At 90°F the warm-air side will have a dew-point temperature of 83° (*dewpoint temperature* is the temperature where condensation from the vapor to a liquid occurs). As illustrated, when there is no effective vapor seal (barrier) on the warm side, water will start to condense inside the insulation. Water is a good conductor of heat; about 15 times as fast as fiberglass. Thus if water gets into the insulation its insulating value is greatly reduced, not to mention the physical problems it causes in the construction.

Therefore, insulation must remain dry when first installed and be sealed perfectly so that it stays dry.

FIGURE 21-6 Installing rigid insulation board.

FIGURE 21-7.

Vapor seals can be formed from various materials; metal casings, metal foil, plastic film, asphalt coverings, etc. Some are more effective than others and the selection will depend on the application. The ability of a material to resist water vapor transmission is measured in *perms*, a term related to permeability. Rating tables for various materials are available from industry sources. In general vapor barriers of one perm or less have been found satisfactory for residential comfort heating and cooling work. But in low-temperature commercial refrigeration applications like freezers, perm ratings of 0.10 and below

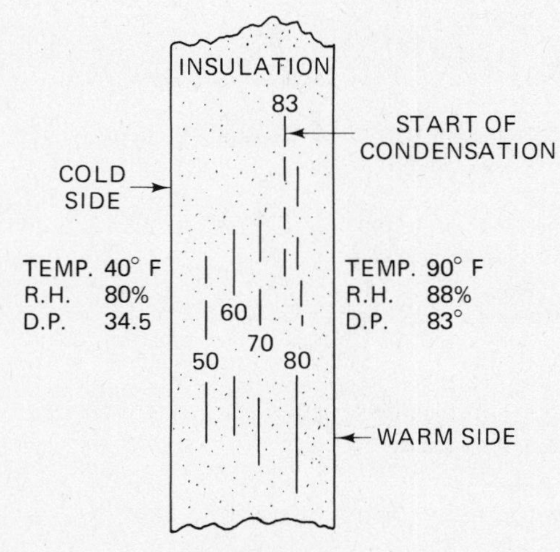

FIGURE 21-8.

may be needed. As with insulation and heat flow, the resistance to vapor flow is a function of the composite of all the materials, as constructed, not just the rating of the vapor barrier itself.

The effectiveness of both insulation and vapor barriers will be greatly reduced if any openings, however small, exist. Such openings may be caused by poor workmanship during construction and application, but they may also result from negligence in sealing around openings for refrigerant lines, drain lines, electrical wiring, etc., all of which are part of the refrigeration technician's responsibility.

SUN EFFECT

The primary radiation factor involved in the refrigeration load is the heat gain from the sun's rays. If the walls of the refrigerated space are exposed to the sun, additional heat will be added to the heat load. For ease in calculation or estimating the load, an allowance can be made by increasing the temperature differential by the factors shown in Table 21B, which are the degrees Fahrenheit to be added to the normal temperature difference between indoor and outdoor design conditions.

TABLE 21B Allowance for sun effect

Type of Structure	*East Wall*	*South Wall*	*West Wall*	*Flat Roof*
Dark-Colored Surfaces				
Slate, tar, black asphalt shingles, and black paint	8	5	8	20
Medium Colored Surfaces				
Brick, red tile, unpainted wood, dark cement	6	4	6	15
Light-Colored Surfaces				
White stone, light-colored cement, white asphalt shingles, white paint	4	2	4	9

DESIGN TEMPERATURES

Recommended outside design conditions are the results of extensive studies by the National Weather Bureau. For air-conditioning and refrigeration applications the maximum load occurs during the hottest

SUMMER OUTDOOR DESIGN DATA

(Design dry bulb and wet bulb temperature represents temperature equalled or exceeded during 1% of hours during the four summer months.)

(Extracted from 1967 ASHRAE Handbook of Fundamentals, Reprinted by Permission)

Location	Dry Bulb °F.	Wet Bulb °F.
ALABAMA		
Birmingham	97	79
Mobile	96	80
ALASKA		
Fairbanks	82	64
Juneau	75	66
ARIZONA		
Phoenix	108	77
Tucson	105	74
ARKANSAS		
Fort Smith	101	79
Little Rock	99	80
CALIFORNIA		
Bakersfield	103	72
Blythe	111	78
Los Angeles	94	72
San Francisco	80	64
Sacramento	100	72
COLORADO		
Denver	92	65
CONNECTICUT		
Hartford	90	77
DELAWARE		
Wilmington	93	79
D. C.		
Washington	94	78
FLORIDA		
Jacksonville	96	80
Miami	92	80
Tampa	92	81
GEORGIA		
Atlanta	95	78
Savannah	96	81
HAWAII		
Honolulu	87	75
IDAHO		
Boise	96	68
ILLINOIS		
Chicago	94	78
Springfield	95	79
INDIANA		
Fort Wayne	93	77
Indianapolis	93	78
IOWA		
Des Moines	95	79
Sioux City	96	79
KANSAS		
Dodge City	99	74
Wichita	102	77
KENTUCKY		
Lexington	94	78
Louisville	96	79
LOUISIANA		
New Orleans	93	81
Shreveport	99	81
MAINE		
Portland	88	75
MARYLAND		
Baltimore	94	79
MASSACHUSETTS		
Boston	91	76
Worcester	89	75
MICHIGAN		
Detroit	92	76
Grand Rapids	91	76
MINNESOTA		
Duluth	85	73
Minneapolis	92	77
MISSISSIPPI		
Biloxi	93	82
Jackson	98	79
MISSOURI		
Kansas City	100	79
St. Louis	96	79
MONTANA		
Billings	94	68
Helena	90	65
NEBRASKA		
Omaha	97	79

FIGURE 21-9 (Courtesy, ASHRAE).

weather. However, it is neither economical or practical to design equipment for the hottest temperature that might possibly occur, since that peak temperature might last for only a few hours over a span of several years. Therefore, the design temperature chosen is less than that. Figure 21-9 is a segment of the ASHRAE summer outdoor design data chart, which lists the recommended design dry-bulb and wet-bulb temperatures for states and major cities.

AIR INFILTRATION

Any outside air entering the refrigerated space must be reduced to the storage temperature, thus increasing the refrigeration load. In addition, if the moisture content of the entering air is above that of the refrigerated space, the excess moisture will condense out of the air, and the latent heat of condensation will add to the refrigeration load.

Because of the many variables involved, it is difficult to calculate the additional heat gain due to air infiltration. The traffic in and out of a refrigerator usually varies with its size and volume. Therefore, the number of times doors are opened will be related to the volume rather than the number of doors.

Some engineers use the air change method of estimating infiltration; this method is based on the average number of air changes in a 24-hour period compared to the refrigerator volume, as illustrated in Fig. 21-10. Note this is for rooms above 32°F average use. Heavier use would require increasing the values by 2. For storage at 0°F or below the usage will normally be less, and the values are reduced.

Another means of computing infiltration is by means of the velocity of air flow through an open door. Charts are available that list average infiltration velocity depending on door height and temperature difference. If the average time the door is opened each hour can be determined, the average hourly infiltration can be calculated. Once the rate of infil-

AVERAGE AIR CHANGES PER 24 HR. FOR STORAGE ROOMS DUE TO DOOR OPENINGS AND INFILTRATION

(above 32° F.)

Volume cu. ft.	Air Changes per 24 hr.	Volume cu. ft.	Air Changes per 24 hr.
200	44.0	6,000	6.5
300	34.5	8,000	5.5
400	29.5	10,000	4.9
500	26.0	15,000	3.9
600	23.0	20,000	3.5
800	20.0	25,000	3.0
1,000	17.5	30,000	2.7
1,500	14.0	40,000	2.3
2,000	12.0	50,000	2.0
3,000	9.5	75,000	1.6
4,000	8.2	100,000	1.4
5,000	7.2		

Note: For heavy usage multiply the above values by 2.
For long storage multiply the above values by 0.6.

FIGURE 21-10 (Courtesy, ASHRAE).

HEAT REMOVED IN COOLING AIR TO STORAGE ROOM CONDITIONS

(BTU per cu. ft.)

Storage room temp F	Temperature of Outside Air, F							
	85		90		95		100	
	Relative Humidity, Percent							
	50	60	50	60	50	60	50	60
65	0.65	0.85	0.93	1.17	1.24	1.54	1.58	1.95
60	0.85	1.03	1.13	1.37	1.44	1.74	1.78	2.15
55	1.12	1.34	1.41	1.66	1.72	2.01	2.06	2.44
50	1.32	1.54	1.62	1.87	1.93	2.22	2.28	2.65
45	1.50	1.73	1.80	2.06	2.12	2.42	2.47	2.85
40	1.69	1.92	2.00	2.26	2.31	2.62	2.67	3.06
35	1.86	2.09	2.17	2.43	2.49	2.79	2.85	3.24
30	2.00	2.24	2.26	2.53	2.64	2.94	2.95	3.35

Storage room temp F	Temperature of Outside Air, F							
	40		50		90		100	
	Relative Humidity, Percent							
	70	80	70	80	50	60	50	60
30	0.24	0.29	0.58	0.66	2.26	2.53	2.95	3.35
25	0.41	0.45	0.75	0.83	2.44	2.71	3.14	3.54
20	0.56	0.61	0.91	0.99	2.62	2.90	3.33	3.73
15	0.71	0.75	1.06	1.14	2.80	3.07	3.51	3.92
10	0.85	0.89	1.19	1.27	2.93	3.20	3.64	4.04
5	0.98	1.03	1.34	1.42	3.12	3.40	3.84	4.27
0	1.12	1.17	1.48	1.56	3.28	3.56	4.01	4.43
−5	1.23	1.28	1.59	1.67	3.41	3.69	4.15	4.57
−10	1.35	1.41	1.73	1.81	3.56	3.85	4.31	4.74
−15	1.50	1.53	1.85	1.92	3.67	3.96	4.42	4.86
−20	1.63	1.68	2.01	2.09	3.88	4.18	4.66	5.10
−25	1.77	1.80	2.12	2.21	4.00	4.30	4.78	5.21
−30	1.90	1.95	2.29	2.38	4.21	4.51	4.90	5.44

From 1967 ASHRAE Handbook of Fundamentals, Reprinted by Permission

FIGURE 21-11 (Courtesy, ASHRAE).

tration in cubic feet per hour has been determined by either method, the heat load can be calculated from the heat gain per cubic foot, given in the chart in Fig. 21-11.

Assume the volume of a refrigerated room is 1,000 ft^3 and the storage temperature is 40°F, with an outside temperature of 95°F and 60% relative humidity. From Fig. 21-10 note that a 1,000-ft^3 volume would average 17.5 air changes per 24 hours, or an infiltration of 17,500 ft^3 in 24 hours (1,000 × 17.5). From Fig. 21-11 note that for a room at 40°F with 95°F outside temperature and 60% relative humidity the Btu/ft^3 of heat is 2.62. Therefore, in 24 hours the load is 2.62 × 17,500, or 45,850 Btu. In one hour it is 45,850 divided by 24, or 1,910 Btu/h.

If in some systems positive ventilation is provided by means of supply or exhaust fans, the ventilation load will replace the infiltration load (if greater), and the heat gain may be calculated on the basis of the ventilating equipment air volume.

PRODUCT LOAD

The product load is any heat gain due to the product in the refrigerated space. The load may be the result of a product coming into the refrigerator at a higher temperature than that of the storage area, from a chilling or freezing process, or from the heat of respiration of perishable products. The total product load is the sum of the various types of product load of the particular application.

In order to calculate the refrigeration product load for food products, solids, and liquids, it is essential to know their freezing points, specific heats, percent water, etc. Figure 21-12 is a sample of food products data taken from ASHRAE information.

SENSIBLE HEAT ABOVE FREEZING

Most products are at a higher temperature than the storage temperature when placed in the refrigerator. Since many foods have a high water content, their reaction to a loss of heat is quite different above and below the freezing point. Above the freezing point the water exists in liquid form, while below the freezing point, the water has changed its state to ice.

As mentioned in earlier chapters, the specific heat of a product is defined as the Btu required to raise the temperature of 1 lb of the substance 1°F. The specific heats of various food products are listed in Fig. 21-12, for both above and below freezing temperatures.

The heat to be removed from a product to reduce its temperature above freezing may be calculated as follows:

$$Q = W \times C \times (T_1 - T_2)$$

Q = Btu to be removed

W = weight of the product in pounds

C = specific heat above freezing (Fig. 21-12)

T_1 = initial temperature, °F

T_2 = final temperature, °F (freezing or above)

For example, the heat to be removed in order to cool 1,000 pounds of veal (whose freezing point is 29°F) from 42°F to 29°F can be calculated as follows:

$$Q = W \times C \times (T_1 - T_2)$$

Q = 1,000 lb × 0.71 specific heat (veal) × 13 (42° − 29°)

Q = 9,230 Btu

LATENT HEAT OF FREEZING

The latent heat of freezing for water is 144 Btu/lb, as mentioned in earlier chapters. Most food products have a high percent of water content. In order to calculate the heat removal required to freeze the product, only the water need be considered. Water content is also shown in Fig. 21-12.

Since the latent heat of freezing for water is 144 Btu/lb, the latent heat of freezing for a given product can be calculated by multiplying 144 Btu/lb by the percentage of water content. To illustrate, veal is 63% water, and its latent heat is 91 Btu/lb (63% × 144 Btu/lb = 91 Btu/lb).

The heat to be removed from a product for the latent heat of freezing may be calculated as follows:

$$Q = W \times hof$$

Q = Btu to be removed

W = weight of product in lb

hof = latent heat of fusion, Btu/lb

The latent heat of freezing 1,000 pounds of veal at 29°F is:

$$Q = W \times hof$$
$$Q = 1{,}000 \text{ lbs} \times 91 \text{ Btu/lb}$$
$$Q = 91{,}000 \text{ Btu}$$

SENSIBLE HEAT BELOW FREEZING

Once the water content of a product has been frozen, sensible cooling again can occur in the same manner as above freezing, with the exception that the ice in the product causes the specific heat to change. Note in Fig. 21-12 that the specific heat of veal above freezing is 0.71, while the specific heat below freezing is 0.39.

The heat to be removed from a product to reduce its temperature below freezing may be calculated as

FOOD PRODUCTS DATA

Product	Average Freezing Point F	Percent Water	SP ht, Btu/(lb)(F deg)		Latent Heat of Fusion Btu/lb	Heat of Respiration Btu per (24 hr) (ton) at Temp. Indicated	
			Above Freezing	Below Freezing		°F	BTU
VEGETABLES							
Artichokes	29.1	83.7	0.87	0.45	120	40	10,140
Asparagus	29.8	93	0.94	0.48	134	40	11,700-23,100
Beans, string	29.7	88.9	0.91	0.47	128	40	9700-11400
Beans, Lima	30.1	66.5	0.73	0.40	94	40	4300-6100
Beans, dried		12.5	0.30	0.24	18		
Beets	31.1	87.6	0.90	0.46	126	32	2700
						40	4100
Broccoli	29.2	89.9	0.92	0.47	130	40	11,000-17,000
Brussels sprouts	31	84.9	0.88	0.46	122	40	6600-11,000
Cabbage	31.2	92.4	0.94	0.47	132	40	1700
Carrots	29.6	88.2	0.90	0.46	126	32	2100
						40	3500
Cauliflower	30.1	91.7	0.93	0.47	132	40	4500
Celery	29.7	93.7	0.95	0.48	135	32	1600
						40	2400
Corn (green)	28.9	75.5	0.79	0.42	106	32	7200-11,300
						40	10,600-13,200
Corn (dried)		10.5	0.28	0.23	15		
Cucumbers	30.5	96.1	0.97	0.49	137		
Eggplant	30.4	92.7	0.94	0.48	132		
Sausage (franks)	29	60	0.86	0.56	86		
Sausage (fresh)	26	65	0.89	0.56	93		
Sausage (smoked)	25	60	0.86	0.56	86		
Scallops	28	80.3	0.89	0.48	116		
Shrimp	28	70.8	0.83	0.45	119		
Veal	29	63	0.71	0.39	91		
MISCELLANEOUS							
Beer	28	92	1.0				
Bread		32-37	0.70	0.34	46-53		
Bread (dough)		58	0.75				
Butter	30-0	15	0.64	0.34	15		
Candy			0.93				
Caviar (tub)	20	55				40	3820
Cheese (American)	17	60	0.64	0.36	79	40	4680
Cheese (Camembert)	18	60	0.70	0.40	86	40	4920
Cheese (Limburger)	19	55	0.70	0.40	86	40	4920
Cheese (Roquefort)	3	55	0.65	0.32	79	45	4000
Cheese (Swiss)	15	55	0.64	0.36	79	40	4660
Chocolate (coating)	95-85	55	0.30	0.55	40		
Cream (40%)	28	73	0.85	0.40	90		
Eggs (crated)	27		0.76	0.40	100		
Eggs (frozen)	27			0.41	100		
Flour		13.5	0.38	0.28			
Flowers (cut)	32						480/sq. ft. Floor Area
Furs—Woolens				0.40			

← VEAL

FIGURE 21-12 (Courtesy, ASHRAE).

follows:

$$Q = W \times C_i \times (T_f - T_3)$$

Q = Btu to be removed

W = weight of product in pounds

C_i = specific heat below freezing (Fig. 21-12)

T_f = freezing temperature

T_3 = final temperature

For example, the heat to be removed in order to cool 1,000 lb of veal from 29°F to 0°F can be calculated as follows:

$Q = W \times C_i \times (T_f - T_3)$

$Q = 1{,}000 \text{ lb} \times 0.39 \text{ specific heat} \times (29 - 0)$

$Q = 1{,}000 \times 0.39 \times 29$

$Q = 11{,}310$ Btu

Total Product Load is sum of the individual calculations for the sensible heat above freezing, the latent heat of freezing, and the sensible heat below freezing. In the preceding example of 1,000 lb of veal cooled and frozen from 42°F to 0°F, the total product load would be:

sensible heat above freezing	9,230 Btu
latent heat of freezing	91,000 Btu
sensible heat below freezing	11,310 Btu
Total product load	111,540 Btu

If several different commodities, or crates or baskets, etc., are to be considered, than a separate calculation must be made for each item for an accurate estimate of the total product load. Note in all the above calculations the factor of time is not considered, either for initial pull down or storage life.

STORAGE DATA

For most commodities there are temperature and relative humidity conditions at which their quality is best preserved and their storage life is at a maximum. Recommended storage conditions for such things as perishable food products, flowers, etc., are published in the *ASHRAE Guide*. Figure 21-13 is a sample illustration. Apples have a long storage life (two to six months), while fresh figs can be stored only for several days. Note that no information is given for bananas, for which conditions vary widely due to the internal heat generated during the ripening process.

In the above calculations no allowance was made for the containers used to store the commodity. Some are in cartons, some are in wooden boxes, and some are stored loose, on pallets. Naturally these containers are also cooled along with the product, and so a storage container allowance factor is usually added to the total load.

SUPPLEMENTARY LOAD

In addition to the heat transmitted to the refrigerated space through the walls, air infiltration, and the product load, any heat gain from other sources must be included in the total cooling load estimate.

Any electric energy dissipated in the refrigerated space through lights and heaters (defrost), for instance, is converted to heat and must be included in the heat load. One watt equals 3.41 Btu, and this conversion ratio is accurate for any amount of electric power.

Electric motors are another source of heat load. For a motor that is actually in the refrigerated space, the following table gives the approximate Btu per horsepower per hour of heat generated.

Motor Horsepower	*Btu/hp/h*
$\frac{1}{8}$ to $\frac{1}{2}$	4,250
$\frac{1}{2}$ to 3	3,700
3 to 20	2,950

Motors outside the refrigerated space but coupled to a fan or pump that is inside will produce a lesser load, but should also be considered.

People give off heat and moisture, and the resulting refrigeration load will vary depending on the duration of occupancy in the refrigerated space, the temperature, the type of work, and other factors. The table below lists the average heat load due to occupancy, but for stays of short duration the heat gain will be somewhat higher.

STORAGE REQUIREMENTS AND PROPERTIES OF PERISHABLE PRODUCTS

Commodity	Storage Temp. F	Relative Humidity %	Approximate Storage Life
Apples	30-32	85-90	2- 6 months
Apricots	31-32	85-90	1- 2 weeks
Artichokes (Globe)	31-32	90-95	1- 2 weeks
Jerusalem	31-32	90-95	2- 5 months
Asparagus	32	90-95	2- 3 weeks
Avocados	45-55	85-90	4 weeks
Bananas	—	85-95	—
Beans (Green or snap)	45	85-90	8-10 days
Lima	32-40	85-90	10-15 days
Beer, barrelled	35-40		3-10 weeks
Beets			
Bunch	32	90-95	10-14 days
Topped	32	90-95	1- 3 months
Blackberries	31 32	85-90	7 days
Blueberries	31 32	85-90	3- 6 weeks
Bread	0		several weeks
Broccoli, sprouting	32	90-95	7-10 days
Brussels sprouts	32	90-95	3- 4 weeks
Candy	0-34	40-65	—
Carrots			
Prepackaged	32	80-90	3- 4 weeks
Topped	32	90-95	4- 5 months
Cauliflower	32	90-95	2- 3 weeks
Celeriac	32	90-95	3- 4 months
Celery	31-32	90-95	2- 4 months
Cherries	31-32	85-90	10-14 days
Coconuts	32-35	80-85	1- 2 months
Coffee (green)	35-37	80-85	2- 4 months
Corn, sweet	31-32	85-90	4- 8 days
Cranberries	36-40	85-90	1- 4 months
Cucumbers	45-50	90-95	10-14 days
Currants	32	80-85	10-14 days
Dairy products			
Cheese	30-45	65-70	
Butter	32-40	80-85	2 months
Butter	0 to —10	80-85	1 year
Cream (sweetened)	—15	—	several months
Ice cream	—15	—	several months
Milk, fluid whole			
Pasteurized Grade A	33	—	7 days
Condensed, sweetened	40	—	several months
Evaporated	Room Temp	—	1 year, plus
Milk, dried			
Whole milk	45-55	low	few months
Non-fat	45-55	low	several months
Dewberries	31-32	85-90	7-10 days
Dried fruits	32	50-60	9-12 months
Eggplant	45-50	85-90	10 days
Eggs			
Shell	29-31	80-85	6- 9 months
Shell, farm cooler	50-55	70-75	
Frozen, whole	0 or below	—	1 year, plus
Frozen, yolk	0 or below	—	1 year, plus
Frozen, white	0 or below	—	1 year, plus
Whole egg solids	35-40	low	6-12 months
Yolk solids	35-40	low	6-12 months
Flake albumen solids	Room Temp	low	1 year, plus
Dried spray albumen solids	Room Temp	low	1 year, plus

Commodity	Storage Temp. F	Relative Humidity %	Approximate Storage Life
Endive (escarole)	32	90-95	2- 3 weeks
Figs			
Dried	32-40	50-60	9-12 months
Fresh	28-32	85-90	5- 7 days
Fish			
Fresh	33-35	90-95	5-15 days
Frozen	—10-0	90-95	8-10 months
Smoked	40-50	50-60	6- 8 months
Brine salted		90-95	10-12 months
Mild cured	28-35	75-90	4- 8 months
Shellfish			
Fresh	33	90-95	3- 7 days
Frozen	0 to —20	90-95	3- 8 months
Frozen-pack fruits	—10-0	—	6-12 months
Frozen-pack vegetables	—10-0	—	6-12 months
Furs and Fabrics	34-40	45-55	several years
Garlic, dry	32	70-75	6- 8 months
Grapefruit	50	85-90	4- 8 weeks
Grapes			
American type	31-32	85-90	3- 8 weeks
European type	30-31	85-90	3- 6 months
Honey	—	—	1 year, plus
Hops	29-32	50-60	several months
Horseradish	32	90-95	10-12 weeks
Kale	32	90-95	2- 3 weeks
Kohlrabi	32	90-95	2- 4 weeks
Lard (without antioxidant)	45	90-95	4- 8 months
Lard (without antioxidant)	0	90-95	12-14 months
Leeks, green	32	90-95	1- 3 months
Lemons	32 or 50-58	85-90	1- 4 months
Lettuce	32	90-95	3- 4 weeks
Limes	48-50	85-90	6- 8 weeks
Logan blackberries	31-32	85-90	5- 7 days
Meat			
Bacon—Frozen	—10-0	90-95	4- 6 months
Cured (Farm style)	60-65	85	4- 6 months
Cured (Packer style	34-40	85	2- 6 weeks
Beef—Fresh	32-34	88-92	1- 6 weeks
Frozen	—10-0	90-95	9-12 months
Fat backs	34-36	85-90	0- 3 months
Hams and shoulders—Fresh	32-34	85-90	7-12 days
Frozen	—10-0	90-95	6- 8 months
Cured	60-65	50-60	0- 3 years
Lamb—Fresh	32-34	85-90	5-12 days
Frozen	—10-0	90-95	8-10 months
Livers—Frozen	—10-0	90-95	3- 4 months
Pork—Fresh	32-34	85-90	3- 7 days
Frozen	—10-0	90-95	4- 6 months
Smoked Sausage	40-45	85-90	6 months
Sausage Casings	40-45	85-90	
Veal	32-34	90-95	5-10 days
Mangoes	50	85-90	2- 3 weeks
Melons, Cantaloupe	32-40	85-90	5-15 days
Persian	45-50	85-90	1- 2 weeks
Honeydew and Honey Ball	45-50	85-90	2- 4 weeks
Casaba	45-50	85-90	4- 6 weeks
Watermelons	36-40	85-90	2- 3 weeks

FIGURE 21-13 (Courtesy, ASHRAE).

Cooler Temperature °F	Heat Equivalent/Person/Btu/h
50	720
40	840
30	950
20	1,050
10	1,200
0	1,300
−10	1,400

The total supplementary load is the sum of the individual factors contributing to it. For example, the total supplementary load in a refrigerated storeroom maintained at 0°F in which there are 300 watts of electric lights, a 3-hp motor driving a fan, and two people working continuously would be as follows.

300 watts × 3.41 Btu/h	1,023 Btu/h
3 hp motor × 2,950 Btu/h	8,850 Btu/h
2 people × 1,300 Btu/h	2,600 Btu/h
Total supplementary load =	12,473 Btu/h

TOTAL HOURLY LOAD

For refrigerated fixtures, display cases and refrigerators, prefabricated coolers, and cold-storage boxes that are produced in quantity, the load is normally determined by testing by the manufacturer; the refrigeration equipment is preselected and sometimes already installed in the fixture.

If it must be estimated, the expected load should be calculated by determining the heat gain due to each of the factors contributing to the total load. There are many short methods of estimating heat loads for small walk-in storage boxes, and they vary in their degree of accuracy. The most accurate use forms and data available from the manufacturer for such purposes, and each factor is considered separately as well.

Refrigeration equipment is designed to function continuously, and normally the compressor operating time is determined by the requirements of the defrost system. The load is calculated on a 24-hour basis, and the required hourly compressor capacity is determined by dividing the 24-hour load by the desired number of hours of compressor operation during the 24-hour period. A reasonable safety factor must be provided to enable the unit to recover rapidly after a temperature rise and to allow for any load that might be larger than originally estimated.

In cases where the refrigerant evaporating temperature does not drop below 30°F, frost will not accumulate on the evaporator, and no defrost period is necessary. The compressor for such applications is chosen on the basis of an 18 to 20-hour period of operation.

For applications with storage temperatures of 35°F or higher and refrigerant temperatures low enough to cause frosting, it is common practice to defrost by stopping the compressor and allowing the return air to melt the ice from the coil. Compressors for such applications should be selected for 16 to 18 hours of operation.

On low-temperature applications, some positive means of defrosting must be provided. With normal defrost periods, 18-hour compressor operation is usually acceptable, although some systems are designed for continuous operation except during the defrost period.

An additional 5% to 10% safety factor is often added to load calculations as a conservative measure to be sure the equipment will not be undersized. When data concerning the refrigeration load is uncertain, this practice may be desirable, but generally the fact that the compressor is sized on the basis of 16 or 18-hour operation in itself provides a sizeable safety factor. The load should be calculated on the basis of the peak demand at design conditions, and usually the design conditions are selected on the assumption that peak demand will occur no more than 1% of the hours during the summer months. If the load calculations are reasonably accurate and the equipment is properly sized, an additional safety factor may actually result in the equipment being oversized during light load conditions, a situation that could lead to operating difficulties.

PROBLEMS

1. The refrigeration load generally comes from four sources. What are they?
2. Product loads may be made up of either _____________ or _____________ heat or both.
3. Does insulation have a high or low resistance value?
4. How much heat will be transmitted through 10 ft^2 of a 4-in. thick common brick wall with no insulation if the temperature difference is 70°F?
5. Insulation is available in several different forms; name three.
6. Design temperatures represent the highest recorded temperatures of a particular location: true or false?
7. Opening and closing the refrigerator door creates a _____________ load.
8. The specific heat of a product determines the heat removal rate in refrigerating or freezing: true or false?
9. The specific heat of a product is the same above or below the freezing point; true or false?
10. The latent heat load of a product is related to the percentage of its _____________ content.

22

INSTALLATION AND STARTUP

The importance of this phase in the successful application of a refrigeration system cannot be overemphasized. No matter how well the equipment has been designed and manufactured or how good the application engineer's planning, poor installation can easily ruin the best system.

A good installation requires more than just technical or mechanical skills; it requires personal integrity to do the best job possible for the user and the development of an attitude called "pride in workmanship." In other words, it demands a professional approach.

The installation of the equipment depends of course on the type of product and system involved. The manufacturer's instructions prescribe the specific procedures to be followed, but there are some factors that are common to almost any situation.

CODES, ORDINANCES, AND STANDARDS

In earlier discussions we have occasionally referred to "compliance to national and local codes and/or regulations." This statement really has several meanings. Some codes or standards are related to the design and performance of the product, its application, or safety considerations. Other codes are directed at the installation phases. The most important national standards for refrigeration products are established by the following organizations (Fig. 22-1).

A. R. I.

The A. R. I. (Air-Conditioning and Refrigeration Institute) has its headquarters in Arlington, Virginia. As mentioned in the introduction of this text, the A. R. I. is an association of manufacturers of refrigeration and air-conditioning equipment and allied products. Although it is a public relations and information center for industry data, one of the institute's most important functions is to establish product or application standards by which the associate members can design, rate, and apply their hardware. In some cases, the products are submitted for test and are subject to certification and listing in nationally published directories. The intent is to provide the user with equipment that meets a recognized standard.

FIGURE 22-1.

ASHRAE

ASHRAE (American Society of Heating, Refrigeration, and Air-Conditioning Engineers) is an organization started back in 1904 as the American Society of Refrigeration Engineers with some 70 members. Today its membership is composed of thousands of professional engineers and technicians from all phases of the industry. ASHRAE also creates equipment standards for the industry, but its most important contribution probably has been the publication of a series of books that have become the reference bibles of the industry. These include the *Guide and Data Books for Equipment, Fundamentals, Applications, and Systems.*

ASME

ASME (American Society of Mechanical Engineers) is primarily concerned with codes and standards related to the safety aspects of pressure vessels.

UL

UL (Underwriters Laboratories) is a testing and code agency which is primarily concerned with the safety aspects of electrical products, although its scope sometimes also includes an overall review of the product. Most of us are familiar with the UL seal on household appliances (irons, toasters, etc.), but it also approves refrigeration and air-conditioning equipment. Its scope of activity has now even expanded into large centrifugal refrigeration machines.

UL approval for certain types of refrigeration and air-conditioning products is almost mandatory in order for it to be locally accepted by electrical inspectors. Compliance with UL is the responsibility of the manufacturer, and the approved products are listed in a directory sent to all local agencies. Installation in accordance with the approved standards is the responsibility of the installer, and violation of these standards can obviously cause a safety hazard as well as possibly voiding the user's insurance coverage should an accident or fire result. So rule number 1 of installation procedures—*conform to UL approval standards.*

NFPA

Closely associated with the work of UL in terms of electrical safety is the *National Electrical Code* (Fig. 22-2), sponsored by the NFPA (National Fire

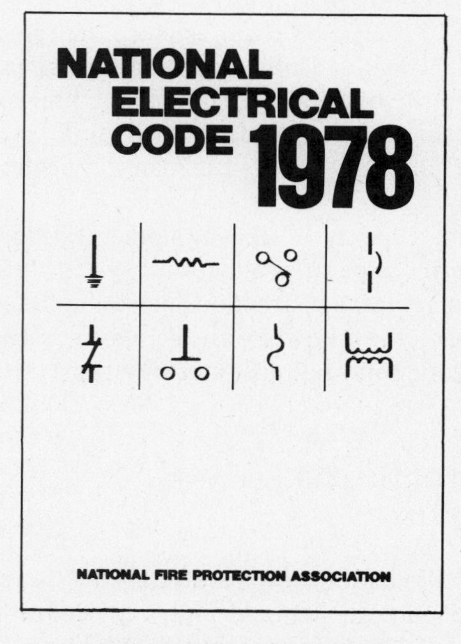

FIGURE 22-2 (Courtesy, NFPA).

Protection Association). The original code was developed in 1897 as a united effort of various insurance, electrical, architectural, and allied interests. Although it is called the National Electrical Code, its intent is to guide local parties in the proper application and installation of electrical devices. It is the backbone of most state or local electrical codes and ordinances, so rule number 2 of installation—*get a copy of the National Electrical Code Book and become thoroughly familar with its scope and where to find any information it contains.*

There are other national agencies involved in the industry and their names and activities will be learned through exposure and experience.

Compliance with *local codes* is the second phase of providing a good installation. Most states, cities, and counties have or are in the process of adopting local codes, which may be based on suggested national standards and also supplemented by local interpretation. These ordinances are usually divided into (1) electrical, (2) plumbing and refrigeration, and (3) other codes such as sound control. Become familiar with the requirements so the chances of violations and changes are minimized. Corrections are usually expensive and are the responsibility of the installer.

Local codes, ordinances, or regulations are not always administered by civil agencies, but can represent special situations such as power company restrictions on current draw. So rule number 3 of providing a good installation is *to become familiar with all applicable local codes and regulations.* If you are not sure, find out. Don't wait until the inspector puts a red tag on the installation.

Remember! *Standards* serve as guidelines to improve the performance and reliability of a system, but *codes and ordinances* are specific and mandatory rules to be complied with.

EQUIPMENT PLACEMENT

Despite what may seem to be a great many possibilities for positioning major cycle components during any installations three factors must be considered in the placement of equipment. First, when air-cooled condensing equipment is installed, ample space must be provided for air movement to and from the condenser. Second, all major components must be installed so that they may be serviced readily. When an assembly is not easily accessible for service, the cost of service becomes excessive. Third, vibration isolation must always be considered, not only in regard to the equipment itself, but also in relation to the interconnecting piping and sheet-metal ductwork. When selecting the placement of all the major components of a refrigeration system, these factors must be considered if a satisfactory installation and proper operation is to be assured. All manufacturers supply recommendations of the space required; these recommendations should be followed.

Figure 22-3 shows a properly placed air-cooled condensing unit. Although the unit has been installed in an inside corner, sufficient room has been left for the passage of incoming air around and over the unit.

Noise is also an important factor in the placement of air-cooled condensing equipment. Noise generated within the unit will be carried out by the discharge air. It is poor practice to "aim" the condenser discharge air in a direction where noise may be disturbing, such as neighboring office windows.

When positioning major system components, care must be taken to assure service accessibility. Figure 22-4 is an example of a common error. A shell and tube water-cooled condenser has been placed in such a position that the entire condensing unit must be moved to replace a single condenser tube. So, be sure to allow room for replacement of items such as compressors, fan motors, fans, and filters. High service costs are often attributable to the poor placement of system components.

When installing major components, it is important to control vibrations set up by rotating assem-

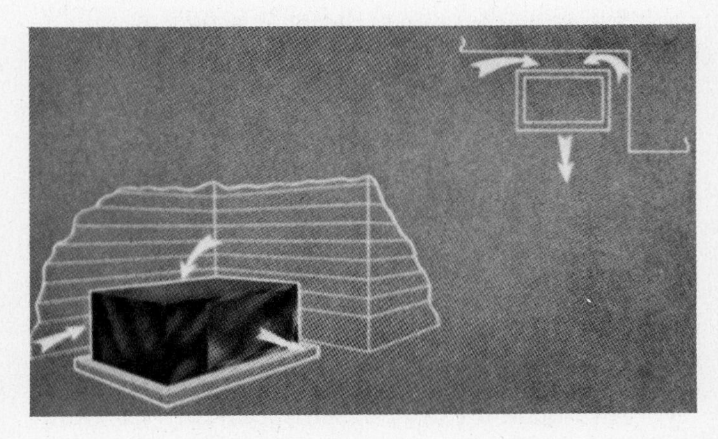

FIGURE 22-3 Air movement (Courtesy, Carrier Air-Conditioning Company).

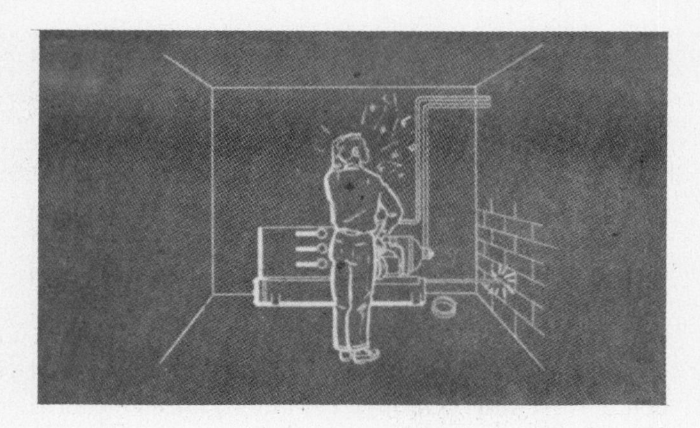

FIGURE 22-4 Service accessibility (Courtesy, Carrier Air-Conditioning Company).

blies such as compressors, fans, and fan motors. These vibrations can break refrigerant lines, cause structural building damage, and create noise.

Vibration isolation is required on all refrigeration and air-conditioning equipment where noise or vibration may be disturbing. Almost all manufacturers use some form of vibration isolation in the production of their equipment. This is usually enough for the average installation; however, unusual circumstances may require that more stringent measures be taken.

The compressor is considered to be the chief source of system vibration. Since it is good practice to isolate vibration at its source, compressor vibration isolation is essential during installation. One method of minimizing compressor vibration at the source is to bolt the compressor firmly to a solid foundation. An example of this is the use of poured concrete pads. Stud bolts are positioned in the concrete when it is poured and the compressor or condensing unit is bolted firmly to the pad, as shown in Fig. 22-5.

When the compressor or condensing unit is to be installed on the roof or in the upper stories of a multistory building, vibration isolators as shown at the upper right in Fig. 22-5 may be used. This type of isolator is usually available from the manufacturer of the unit and, in many cases, it is standard equipment.

An isolation pad designed specifically for vibration dampening is shown at the lower left in Fig. 22-5. This type of material is designed to dampen the vibration from a given amount of weight per square inch of area. As the right amount and type of this material can be properly selected only when both weight and vibration cycle are taken into consideration, a competent engineer should be consulted.

A final method of vibration isolation, used with small hermetic compressors, is shown at the lower right in Fig. 22-5. In this case the compressor is spring-mounted within the hermetic shell.

Figure 22-6 illustrates a vibration isolator that is inserted in the discharge line. This isolator is designed to absorb compressor discharge line pulsation before it creates noise or breaks refrigerant lines. The isolator consists of a flexible material banded and held in place by a metal mesh. The mesh will allow some linear movement of the flexible material, but no expansion. This type of isolator is normally placed in the compressor discharge line as close to the compressor as practical. It is particularly effec-

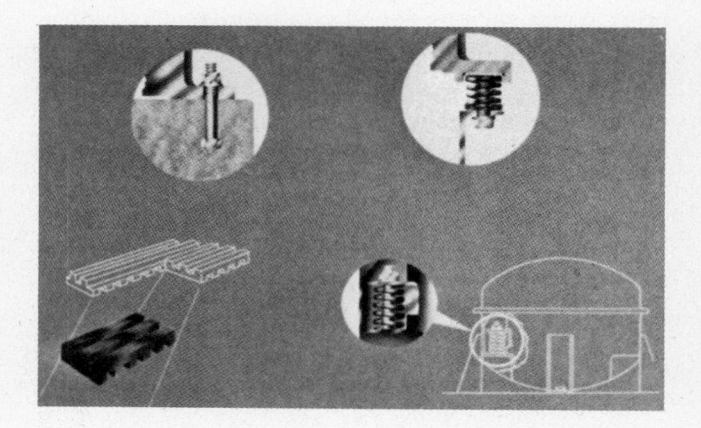

FIGURE 22-5 Vibration isolation compressor (Courtesy, Carrier Air-Conditioning Company).

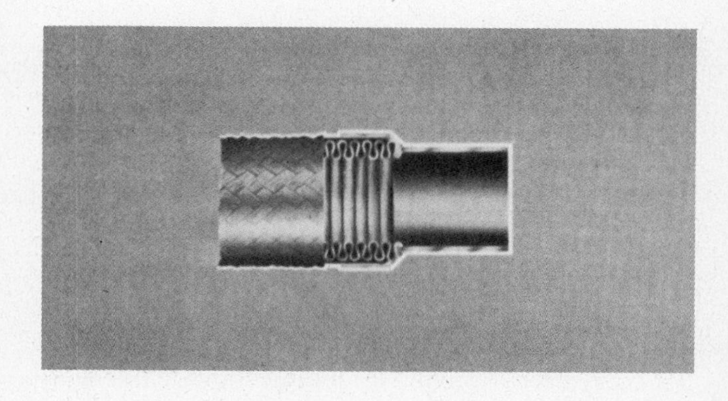

FIGURE 22-6 Vibration isolation discharge line (Courtesy, Carrier Air-Conditioning Company).

tive on installations where the compressor and condenser are on different bases, yet quite close together.

The eliminator should be placed in the line so that the movement it absorbs is in a plane at right angles to the device. Do not place this unit in a position that will put tension on it, as its useful life will be shortened.

Some system components, such as the evaporator, may be suspended from the ceiling. As the air-handling unit containing the evaporator will also normally contain a fan and fan motor, it too is a possible source of vibration. Most manufacturers isolate the fan and fan motor inside the unit with rubber mounts. If this supplies sufficient vibration isolation, the unit may be directly connected to the ceiling, as shown in Fig. 22-7. When more isolation is required, the same method used with compressors is effective. Coil springs support the air-handling unit for additional isolation.

In addition to the placement of major system components, most refrigeration system installations have three other major phases. They are

(1) erection of piping (both refrigerant and water)
(2) making electrical connections
(3) erection of ductwork

The erection of refrigerant piping is a primary responsibility of the refrigeration installer, along with the placement of major system components.

Though not always the responsbility of the refrigeration installer, electrical and duct work are an important part of most installations. The refrigeration installer should be familiar with good electrical and air duct installation techniques, since quite often, particularly with small equipment, the refrigeration installer is called upon to do the electrical wiring and make connections to ductwork.

The art of making flared and soldered connections in copper tubing has been discussed, as well as the procedures for sizing lines, installing traps, etc., but several important points should be remembered while actually erecting the system piping.

When hard copper tubing is selected for refrigerant piping, it is recommended that low-temperature silver alloy brazing materials be used. These alloys have flow points ranging from 1,100° to 1,300°F.

To attain these temperatures, oxyacetylene welding equipment is required. Figure 22-8 shows the equipment necessary for this type of brazing. Both oxygen and acetylene tanks with gauges and reducing valves are needed. Also shown is the torch used with the tanks. At the right is a bottle of dry nitrogen also equipped with gauges and reducing valves. The use of nitrogen is recommended, since it serves to keep the interior of the pipe clean during the brazing process. During low-temperature brazing the surface of the copper will reach a temperature at which the metal will react with oxygen in the air to form a copper oxide scale. If this scale forms on the interior surface of the pipe, it might be washed off by the refrigerant and circulated in the system. The scale can clog strainers or capillary tubes and will damage orifices.

This scaling can be prevented by replacing the air in the pipe by nitrogen. As nitrogen is an inert gas and will not combine with copper even under high

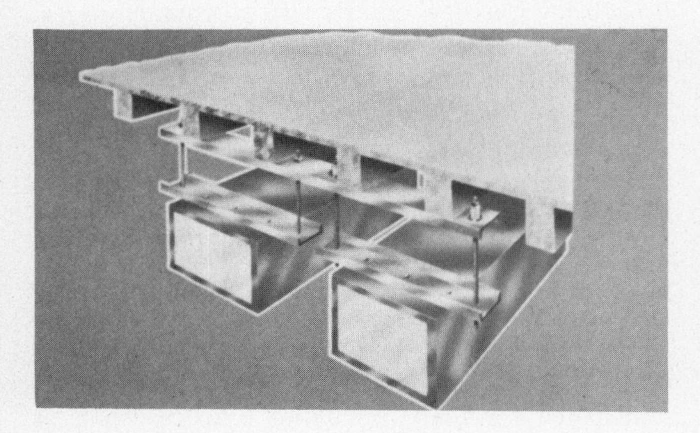

FIGURE 22-7 Vibration isolation evaporators (Courtesy, Carrier Air-Conditioning Company).

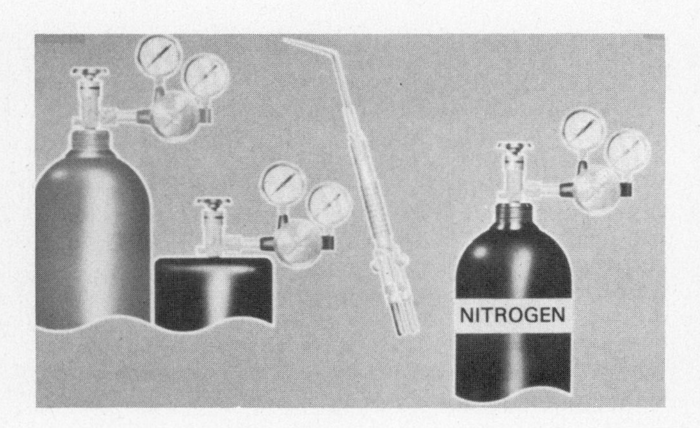

FIGURE 22-8 Brazing equipment (Courtesy, Carrier Air-Conditioning Company).

temperature conditions, the pipe interior will remain clean during brazing and no scale is formed, though discoloration may occur with overheating.

Figure 22-9 shows a method by which nitrogen may be introduced into the refrigerant tubing during the brazing operation. Instead of connecting the liquid line to the "king" valve or liquid-line stop valve, connect the liquid line directly to the nitrogen bottle as shown. By use of the reducing valve on the nitrogen bottle, a slight pressure is admitted to the tubing. This is just enough pressure to assure that air will be forced from the pipe. As shown in the insert, if the nitrogen flow can be felt on the palm of the hand it is sufficient. This nitrogen pressure is kept in the tubing throughout the entire brazing operation, thus assuring an oxygen-free pipe.

Clean pipe is essential in refrigeration installation; therefore, nitrogen is an extremely important part of the brazing operation, as it assures scale-free interior pipe walls.

The temperatures during brazing operations can warp metals and burn or distort plastic valve seats. It is important that brazing heat does not reach metal or plastic components that might be damaged. Figure 22-10 shows the results of brazing heat damage and also one method of assuring this. The valve at the upper left was not protected from brazing heat; the plastic seat has been damaged. Also, the seat holder has been warped, as shown in the insert. This valve obviously will not operate properly and would

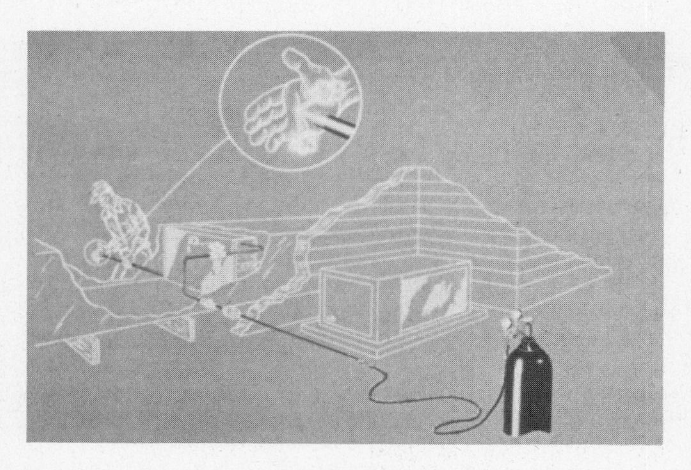

FIGURE 22-9 Brazing with nitrogen (Courtesy, Carrier Air-Conditioning Company).

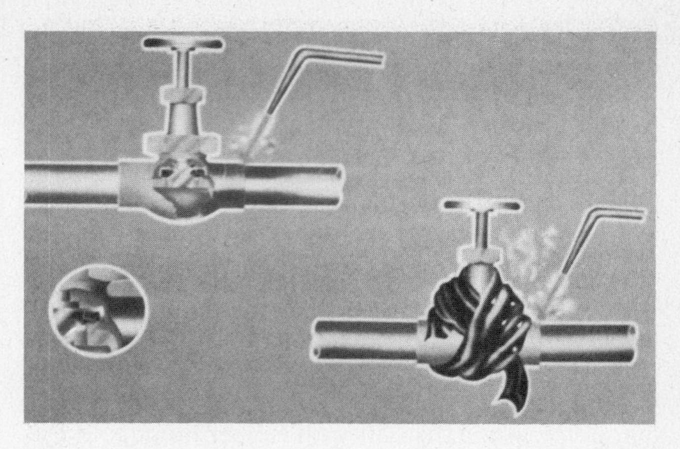

FIGURE 22-10 Overheat protection (Courtesy, Carrier Air-Conditioning Company).

require immediate replacement. However, the valve at the bottom right has a wet rag wrapped around the body. The rag absorbs the heat that flows to the valve body during the brazing operation. By keeping the rag wet, the valve and its component parts are protected from heat damage.

In many cases the suction line of the refrigerant cycle installation will run through a nonrefrigerated or a non-air-conditioned area. The outside surface temperature of this pipe is frequently below the dew point of the surrounding air. In this case the moisture in the air will condense on the exterior of the pipe. This will not only create problems where this continuous moisture drips, such as rotting wood, separating floors, etc., but it can be annoying. When this condensation is expected or does occur, the pipe should be insulated; thus the condensation is prevented.

The insulation must be of good quality, so that the temperature of its exterior surface will never drop below the dew point of the surrounding area. It must also be well-sealed, so that the air and moisture it contains cannot flow through the insulation to the pipe, thus causing condensation underneath the insulation.

There are many different types of insulation available; some are designed for specific uses. The insulation for a specific refrigeration installation should be chosen by a competent technician.

The primary purpose of the pipe hanger is to support the pipe properly but it may also serve as a vibration isolator. If vibration problems are not anticipated, ordinary plumbing practices such as shown in Fig. 22-11 may be used. The only purpose of this hanger is to support the pipe firmly.

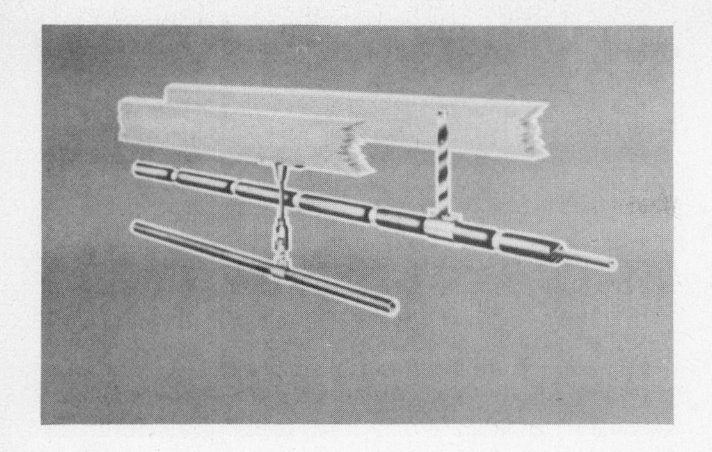

FIGURE 22-11 Pipe hangers (Courtesy, Carrier Air-Conditioning Company).

The hanger shown on the left also has a height adjustment feature. This enables the installer to level or pitch the pipe if required for oil flow. This type of hanger can be used with or without insulation.

At the right in Fig. 22-11 is an example of a typical hanger that might be found on a small commercial-refrigeration installation. This hanger will serve as a vibration isolator as well as a pipe supporter. This example is being used on an insulated line. A short length of light-gauge rust-proof metal has been wrapped around the insulated pipe. A metal strap has been attached to this length of metal; in some cases it is merely wrapped around the length of sheet metal. The free end of the strap is then fastened to the joist or ceiling.

The insulation in such a hanger will act as the vibration isolator. The purpose of the short length of metal is to prevent the thin strap from cutting the insulation.

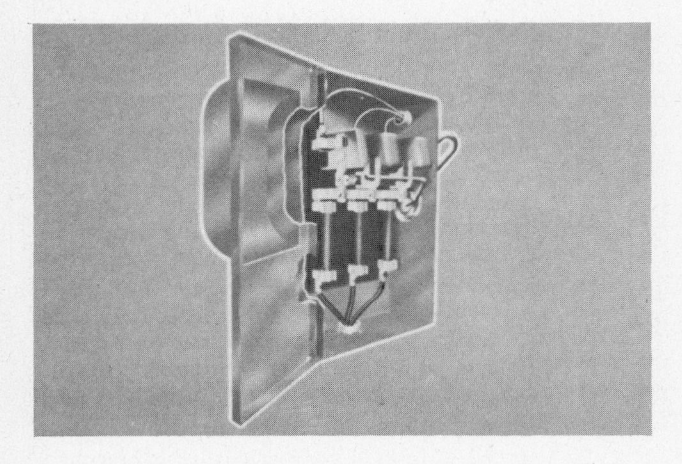

FIGURE 22-12 Fused disconnect switch (Courtesy, Carrier Air Conditioning Company).

The service technician who installs the refrigeration cycle is sometimes responsible for the final wiring connections between the installed unit and the fused disconnect switch shown in Fig. 22-12. All electrical power to the refrigeration unit must pass through this switch. When this switch is pulled, or opened, all electrical power to the unit must be disconnected. This same disconnect switch also contains fuses, which will interrupt the flow of current whenever a severe electrical overload occurs. This mechanism is a protection against fires and explosions and also against electrical shocks to people.

Electrical codes, both national and local, are made to protect property and life; they should always be followed.

All refrigeration cycle installations having electrical connections are governed by national or local electrical codes. For example, electrical codes require that the fused disconnect switch in Fig. 22-12 always be placed within sight of the unit that receives the power passing through the switch.

When electrical circuits are to be connected by the refrigeration installer, good wiring practices should be followed and every effort should be made to assure good electrical contacts. Lug-type connectors are recommended.

When braided wire is used, a single wire may separate and create a potential electrical hazard. Loose wires might contact other wires or "ground," causing electrical problems. Braided wire should be looped into the size required and soft solder should be allowed to flow over and coat the braided loop. This will assure good contact and eliminate the hazard of wire separation.

The refrigeration system installer is frequently required to make the final connection between the evaporator or air-handling unit and the ductwork (Fig. 22-13). This final connection is in the form of a canvas connection, which will eliminate any vibration carry-over from the air-handling unit to the ductwork. Good weatherproof canvas is required and must be installed correctly. If, as shown in the upper left in the illustration, it is too loose, it will drop into the airstream and obstruct normal air flow. It may also flap and create noise. If it is too tight it will stretch, harden, and disintegrate over a period of time, causing leaks. If the canvas is damp and installed too tightly, the ductwork might be pulled out of alignment.

The center of Fig. 22-13 shows the proper application: the canvas is loose enough to absorb vibration and not loose enough to interfere with air flow.

When the refrigeration system has been completely assembled and all electrical and duct connections completed, there are still a number of important steps that must be taken before the equipment is started.

The unit must be leak-tested and charged, and all belts must be checked for tension and alignment. There must be an electrical motor check of the direction of rotation, and the bearings must be oiled. The power source must be checked to be absolutely sure that the correct power will be applied to the unit. None of these steps is time-consuming or difficult to perform, but they are all extremely important and cannot be neglected.

Field leak-detection of the halogenated hydrocarbons is generally done using the halide torch or electronic leak detectors.

After all interconnecting tubing has been assembled, some refrigerant is introduced into the system as a gas. Though a leak test could be made at this time, some refrigerants do not exert sufficient pressure at room temperature to assure trustworthy results. By using nitrogen, the pressure in the system may be built up to approximately 250 psig, at which

FIGURE 22-13 Canvas duct connection (Courtesy, Carrier Air-Conditioning Company).

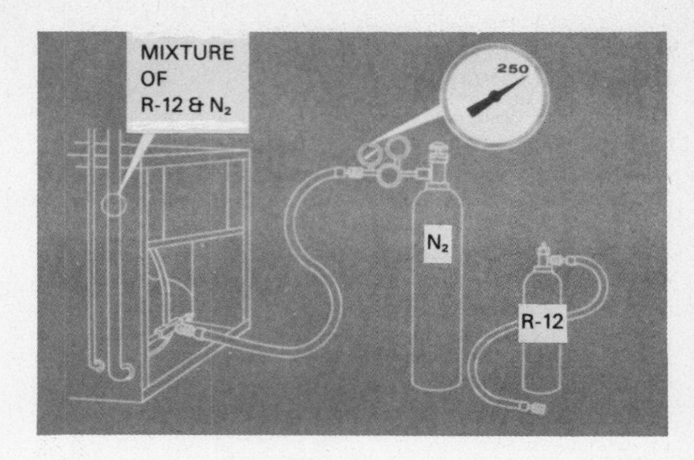

FIGURE 22-14 Leak test (Courtesy, Carrier Air-Conditioning Company).

pressure a true leak test can be made (Fig. 22-14). The mixture of refrigerant and nitrogen inside the unit will cause a reaction on the detector if a leak is present. Since nitrogen is an inert gas, the system should be purged with refrigerant gas after it has been determined to be leak-free.

Following the leak test the system is ready for charging or dehydration. In a small system operating at medium evaporator temperatures, elaborate dehydration procedures will not be required; a simple refrigerant gas purge of the system will be sufficient. If some moisture has inadvertently entered the system, it will show on the moisture indicator as previously described. A drier replacement program should then be started and continued until the system is dry.

If moisture is known to be in the system or if the system is designed for low-temperature application, dehydration by evacuation before charging is recommended. When in doubt, dehydrate. Dehydration by evacuation is accomplished by the use of a vacuum pump as detailed in Chap. 9.

CHARGING THE SYSTEM

The system may be charged with refrigerant in its normal state, or as a liquid or gas. Though a number of factors may affect the method of charging, the most important is the quantity of refrigerant involved.

Relatively speaking, most refrigerant containers, regardless of size, have a small outlet. In small units, sufficient gas vapor may be passed through this outlet to complete the full charging of the unit in a reasonable length of time. On large units, however, the time required for the proper amount of vapor to pass through this small outlet may take so long that gas charging is impractical. When this is the case, liquid charging is used. The small orifice will pass a much greater weight of liquid refrigerant in any given amount of time.

The point at which the refrigerant will actually enter the system is determined by one basic consideration: whether the unit is to be charged with refrigerant in a gas or liquid form.

Under normal charging conditions the refrigerant cylinder will be at ambient temperature and corresponding pressure. As most field refrigerant charging is done with the unit in operation, the refrigerant cylinder pressure will usually be below the head pressure of the condenser and above the back pressure of the evaporator.

The metering device orifice is not large enough to make gas charging on the high side of the device practical because of the time consumed; therefore, gas is normally charged into the system after the metering device. This charging can be done at any convenient point between the metering device and the compressor.

If the refrigerant is to be charged as a liquid, the charging should take place ahead of the metering device to protect the compressor from damage due to liquid flooding.

With open-type and serviceable hermetic systems that are small enough to make gas charging practical or where only a small amount of refrigerant is required, charging is usually done through the suction service valve on the compressor. Since welded hermetic systems are normally small enough so that gas charging is practical, pinch-off tubes are made available at the compressor for this purpose. These pinch-off tubes are connected into the suction side of the compressor and are designed to allow the compressor to pump gas directly from a refrigerant drum.

The original charge in units of this type is placed in the unit at the factory during production, but occasionally the charge must be replaced in the field. When this is necessary, the pinch-off tube may be cut and a connection placed on the tube. The unit is then charged through the tube, which is repinched and brazed. Occasionally, valves are placed permanently on this pinch-off line. This is practical where the pinch-off tube is either very short or inaccessible.

There are also devices that are designed to puncture a refrigerant line for charging purposes. The device then remains on the line and serves as a valve for either future charging or as a gauge connection.

Another device used for charging the hermetic circuit is the *tire* type of valve. This valve contains a plunger, which, when depressed, opens the circuit. A special adapter on the charging hose will depress the plunger when the hose is firmly attached to the valve. Refrigerant will then flow through the valve into the system. When the charging hose is removed, the plunger returns to its original position and reseals the refrigerant circuit. Always replace the valve cap after servicing. There are four ways to determine whether the proper amount of refrigerant is introduced into the system. These methods are: charging by sight glass; charging by pressures; weighing the charge in; and using a frost line.

All four methods are used extensively in commercial refrigeration work, and the choice of method depends upon the size and type of system involved. "Critical-charge" methods, common to small residential air-conditioning systems, *are not included*; they are critical to ounces of refrigerant charge.

The first method to be discussed will be charging by sight glass. In a properly charged system there should always be a solid flow of liquid refrigerant (no bubbles) to the metering device. A sight glass (as previously discussed) will indicate the presence of gas by bubbles in the glass.

To charge properly by the sight-glass method, refrigerant is added in the usual manner. Following the addition of a reasonably large percentage of the estimated charge of gas or liquid, the refrigerant cylinder is shut off and the system is allowed to "settle down" for a period of time. When the operating pressures have stabilized, the sight glass is again observed. If bubbles are still present, additional charge is added slowly until the bubbles disappear. If, after stabilizing the pressures, no bubbles are present when the unit is operating under maximum load conditions, the unit is properly charged.

A common method of determining whether the proper amount of charge is being introduced into the system is by "weighing" the charge. This is the most accurate method of adding a full charge when the required charge is known. The method is quite simple, as shown in Fig. 22-15.

With the system evacuated and ready to receive the refrigerant, the refrigerant bottle is weighed. On the left the scales read 190 lb, indicating that the bottle and refrigerant weigh a total of 190 lb. If the system requires a charge of 40 lb, as determined by the manufacturer or design engineer, gas or liquid is released from the bottle into the system. When 40 lb of refrigerant have been removed, the total weight of refrigerant and bottle should be 190 — 40, or 150 lb, as shown at the right in Fig. 22-15.

This method is used mostly on packaged equipment and then only when it requires a complete charge; the method is of little use when only a portion of the charge is required. Rarely is the exact charge known under such conditions.

A third way of charging, which may be used with factory-designed and balanced packages, is the head pressure method. From test data, the factory determines the proper head pressures under various evaporator loads or temperatures. This information is furnished to the installer either as a graph or in tabular form, which consists of a list of possible low-side pressures along with the proper head pressures supplied by the test data of the unit.

After charging with an estimated amount as directed by the factory, the unit is run for enough time to allow the pressures to settle and stabilize.

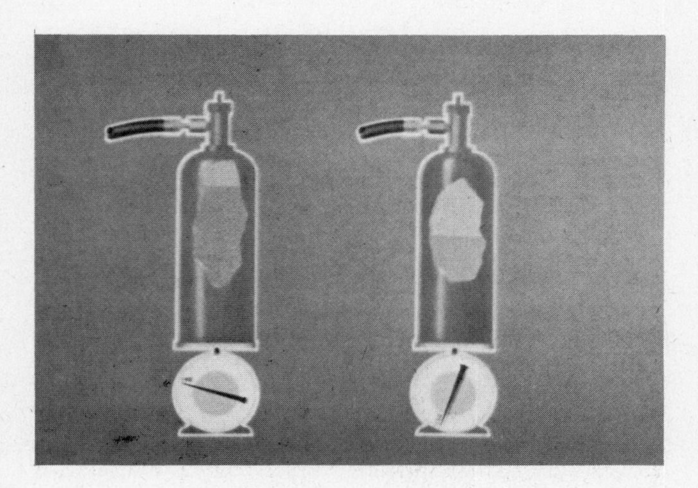

FIGURE 22-15 Weighing charge (Courtesy, Carrier Air-Conditioning Company).

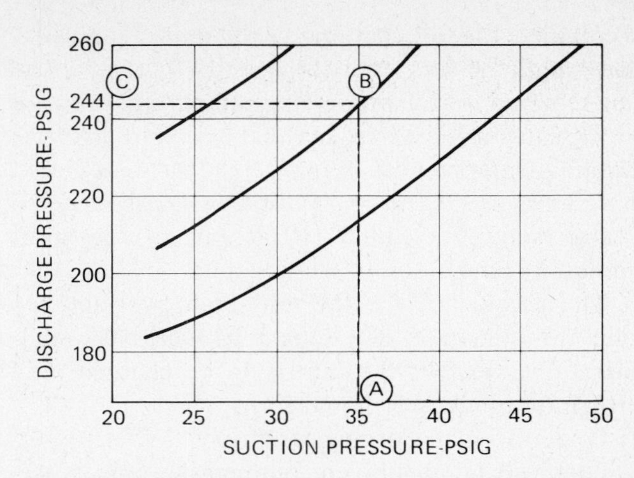

FIGURE 22-16 Charging chart (Courtesy, Carrier Air-Conditioning Company).

Pressure readings are then taken and compared to the manufacturer's chart.

If the actual head pressure is below the pressure that the chart indicates is correct, refrigerant is added. If the actual head pressure is above that indicated by the chart, refrigerant should be removed.

Figure 22-16 shows how the head pressure method is used. An original charge as suggested by the factory has been introduced into the unit. The unit is run long enough for pressure stabilization, and the back pressure is found to be 35 psig. This point is on the horizontal axis at point *A*. The chart is then read upward vertically until the intersection with the outdoor temperature line at *B*. From *B*, read horizontally to the left until point *C* is reached. This is the head pressure under which the unit should be operating when the back pressure is 35 psig. The chart shows this figure to be 244. If the discharge pressure gauge reading had been below this figure, refrigerant would have to be added until the gauge pressure is brought up to 244 psig. If the discharge pressure gauge reading had been above this figure, refrigerant would have to be removed until the pressure was brought down to 244 psig.

The fourth and final way of charging to be discussed is the frost-line method. This method can only be used in small hermetic systems that use capillary tubes; such systems are not too common in commercial refrigeration work. When a system of this type is operated without evaporator load, the back pressures will normally drop below the freezing temperature and frost will form on the coil.

In Figure 22-17 the evaporator load has been removed by placing a piece of cardboard over the

FIGURE 22-17 Frost line (Courtesy, Carrier Air-Conditioning Company).

coil face, thus shutting off the air flow. Since the load has been removed from the evaporator, the refrigerant will not evaporate as rapidly, and some will pass through the evaporator and evaporate in the suction line. Tests have shown that under these conditions a properly charged unit will normally frost to within a few inches of the compressor. By factory testing, this final frost point can be determined, and that information is given to the installer.

By recreating the frost line on the suction line, the installer can determine the proper charge. If the frost line does not reach the point designated by the factory tests, more refrigerant should be added. If the frost goes beyond the point designated by the factory, refrigerant should be removed.

In most systems some overcharging can be tolerated, but an undercharge is rarely acceptable. Overcharging will create high head pressure and high temperature, with all the resulant problems, such as motor overloading, sludge formation, and compressor valve failure. High head pressures can also result in poor load control, with liquid refrigerant flooding to the compressor.

Although the biggest problem with undercharging is that of capacity, it may also create frost conditions on the evaporator in higher temperature refrigeration equipment, and may also cause high evaporator superheats. As some hermetic compressor motors depend on suction gas for cooling, they can be damaged by overheating due to high suction gas temperatures.

Both overcharging and undercharging should be avoided, since either condition can do serious harm or destroy system components.

OIL CHARGING

There are several ways by which the proper amount of oil can be measured into the system. In a new system it can be measured or weighed in as is the refrigerant itself. Unit installation instructions include the compressor oil requirements, in either weight or liquid measurements. This method is also applicable following a compressor overhaul, when all the oil has been removed from the compressor; however, it should be used only when the system has no oil in it.

A second method of determining proper oil charge is the dip stick. This is used primarily with small vertical-shafted hermetic compressors, but some larger, open types of compressors may have openings designed for the use of a dip stick. The manufacturer's recommendations of the correct level should always be followed.

The third way to arrive at the correct oil charge is by using the compressor crankcase sight glass. When determining the correct oil charge by this method, the system should be allowed to operate for a period of time under normal conditions before the final determination of the proper oil level is made. This procedure will assure proper oil return to the crankcase; it will also allow the oil lines and reservoirs to fill and, where halogenated hydrocarbons are used, give the refrigerant an opportunity to absorb its normal operating oil content.

FIGURE 22-18 Oil charging (Courtesy, Carrier Air-Conditioning Company).

When a conpressor is replaced, the new unit should be charged with the same amount of oil as the old unit.

Oil is normally introduced into a refrigerant system by one of two methods (Fig. 22-18). It may be poured in as shown on the left, providing the compressor crankcase is at atmospheric pressure. This method is normally used prior to dehydration, since it will expose the compressor crankcase interior to air and the moisture the air contains.

On the right in Fig. 22-18 the method normally used with an operating unit is shown. In this case the crankcase is pumped down below atmospheric pressure and the oil is drawn in. When this method is used, the tube in the container subjected to air pressure should never be allowed to get close enough to the surface of the oil to draw air. As shown, the tube is well below the level of the oil in the container.

Oil charging of replacement welded hermetic compressors should be done according to the manufacturer's recommendations and depends on whether the replacement compressor has been shipped with or without an oil charge.

There are three precautions to take in charging or removing oil. The first is to use clean, dry oil. Hermetically-sealed oil containers are available and should be used. Second, pressure must be controlled when the crankcase is opened to the atmosphere. Too much pressure can force oil out through the opening rapidly and create quite a mess.

Third, system overcharging should be avoided. Not only will this create the possibility of oil slugs damaging the compressor, but it also can hinder the performance of the refrigerant in the evaporator. Oil overcharging will also cause liquid refrigerant to return to the compressor from the evaporator.

INITIAL STARTUP

Care must be taken to prevent damage to the refrigeration system during initial startup (Fig. 22-19). Valve positions must be checked to make certain that only proper and safe pressures will occur at the compressor. The compressor discharge valve must always

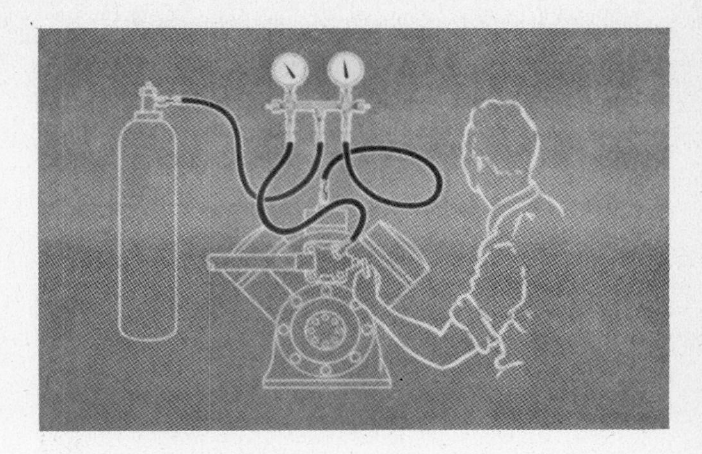

FIGURE 22-19 Refrigerant valve positioning (Courtesy, Carrier Air-Conditioning Company).

be opened prior to start. Low-side valves must be adjusted in such a way as to assure neither excessively high nor low pressures in the compressor suction. Pressures should be carefully observed and regulated until the unit is operating normally. When starting a new system, try to keep the operating pressures as close to normal as practical.

When the installed equipment is belt-driven, both the alignment and tension of the belt must be checked prior to startup. Improperly aligned belts show excessive wear and have an extremely short life. Loose belts wear rapidly, slap, and frequently slip. Belts that are too tight may cause excessive motor-bearing and fan-bearing wear. Belt alignment is simple to check; a straightedge laid along the side of the pulley and flywheel will show any misalignment immediately.

The belt tension should also be checked. The belt stretches very little, so the belt should be set at the right tension during initial startup. The correct tension allows about one inch of deflection on each side of the belt when squeezed by the fingers.

Many parts of newly-installed equipment will require simple maintenance prior to original startup. A good example of this type of maintenance is motor lubrication. Every motor that is not permanently lubricated should receive the proper amount of oil as recommended by the manufacturer. Instructions enclosed with all equipment indicate what maintenance of this type is required.

It is good installation practice to check electric motor nameplates against the available voltage prior to initial startup. Figure 22-20 shows the voltage at the disconnect switch being checked for comparison with the electric motor nameplate. It is also good

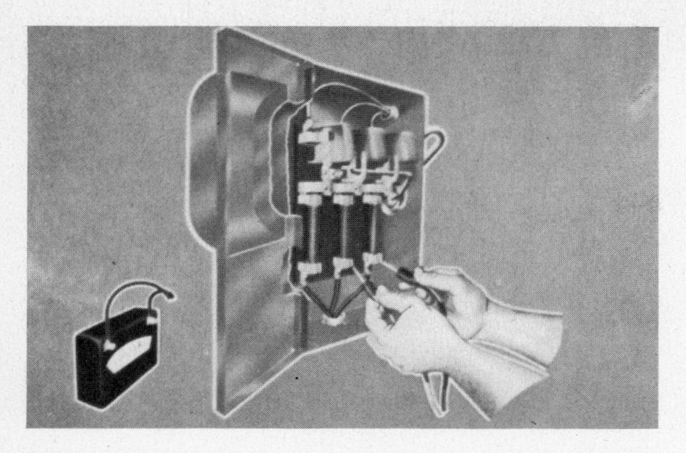

FIGURE 22-20 Electric power check (Courtesy, Carrier Air-Conditioning Company).

practice to determine if a phase unbalance exists in polyphase units prior to the final wiring connection.

When the system has been thoroughly checked for leaks, properly charged, valves positioned for startup, belts aligned, motors oiled, and wiring checked, then the unit can be started.

Following startup the unit must be checked to make certain that it is doing the job it was designed to do efficiently. This includes determining electrical load characteristics. No installation should ever be considered complete until it has been proven it can do the job it was designed to do.

Following the completion of tests to determine that the unit is performing satisfactorily, the final cleanup should be started. All excess tubing, wire, scrap metal, etc. should be removed and the unit left in a condition that will indicate the installer's pride of workmanship. The equipment should be an asset to the area in which it is located.

The final step in any installation is to instruct the customer in the operation of the equipment. There are probably as many service complaints because of a lack of customer understanding of the new equipment as from any other reason. The customer should be clearly and concisely shown how his equipment can be made to operate properly and efficiently. Maintenance requirements should be explained, including such things as filter changes and motor oiling. A customer who is thoroughly familiar with the way his equipment should operate and what he can do to keep it that way is going to be a satisfied customer.

PROBLEMS

1. The *National Electrical Code Book* is published by ______________.
2. UL approvals are not generally required on refrigeration or air-conditioning equipment: true or false?
3. Compressor vibration is always taken care of by the equipment manufacturer and requires no attention in the field: true or false?
4. Dry nitrogen is used in the brazing process to reduce ______________.
5. Suction lines running outside the refrigerated area will generally require insulation to eliminate ______________.
6. To prevent vibration transmission in metal ducts a ______________ connection is used.
7. A charging connection on welded hermetic compressors is called a ______________.
8. The frost-line method of checking the refrigerant charge applies to all systems: true or false?
9. The head-pressure method of checking the refrigerant charge can be used regardless of the ambient outdoor temperature: true or false?
10. Which is more critical, overcharge or undercharge of refrigerant?
11. Checking the oil level by the sight-glass method does not require the compressor be operated before checking: true or false?
12. Oil overcharging is not critical to the system: true or false?

23

TROUBLE-SHOOTING*

Even though the reliability of equipment and systems in modern refrigeration work is quite high, there will be an occasional breakdown or failure. Some breakdowns are caused by defective equipment, some by poor installations, and some because of poor maintenance. The last point is probably the main offender.

No matter what the cause, the skilled refrigeration technician must be able to diagnose that cause and correct the situation. Frequently time is also an urgent consideration in commercial refrigeration breakdowns because there may be many dollars worth of perishable foods involved. So, one of the most important skills to be learned is the art of trouble-shooting.

Remember, trouble-shooting is not so difficult. The equipment will do its best to indicate what is wrong and where. For example, just look at the compressor in Fig. 23-1. Each component has its own way of attracting attention: the belt will squeak, a valve will chatter, a leak will hiss or show an oil spot, or a motor will smoke. Each is saying the same thing—H E L P! It is the technician's job to recognize these symptoms and interpret them properly, just as the family doctor does when he examines his patient.

CONDENSERS

First, the air-cooled condenser. As portrayed in Fig. 23-2 the most common "disease" of air-cooled condensers is high head pressure. The most frequent causes of this condition are:

(1) ***Reduced Air Quantity.*** This may be due to:

- dirt on the coil
- restricted air inlet or outlet
- dirty fan blades
- incorrect rotation of fan

* This chapter is based on Chapter 16 of Carrier Air-Conditioning Company's General Training—Refrigeration Program, copyrighted 1964 by Carrier Corporation. Reproduced with permission of Carrier Corporation.

FIGURE 23-1 (Courtesy, Carrier Air-Conditioning Company).

FIGURE 23-2 (Courtesy, Carrier Air-Conditioning Company).

fan speed too low

fan motor going out on overload

prevailing winds

(2) ***Noncondensables in the Refrigeration System.*** This may be due to:

poor installation or service techniques

leak on the low-side, system in vacuum

Specific symptom: pressure in the system will not correspond to ambient temperature on shutdown. Only noncondensables will cause this. (Example—the pressure of an idle R-12 system in a room at 80°F should be 84.2 psig.)

(3) ***Short-Circuiting of the Condenser Air.*** This is usually caused by:

poor site selection

temporary obstruction

wind conditions causing a portion of the discharge air to enter the inlet opening.

(4) ***Refrigerant Overcharge.*** This is always caused by careless charging procedures.

Specific symptom: a definite temperature drop can be felt at the liquid level on the condenser return bends during operation. The bleed-off of excess refrigerant reduces head pressure but will not cause bubbles in the liquid-line glass.

The general symptoms for all four of the above situations will be cycling on the high pressure switch and the possible opening of the compressor motor overload.

Problems in a water-cooled condenser are not always as easy to diagnose as those in an air-cooled condenser. Generally, the flow of the condensing medium (water) is controlled by a water-regulating valve actuated by discharge pressure. Conditions that would cause a rise in head pressure on an air-cooled condenser merely increase the water flow rate on a water-cooled condenser.

Any one of the following problems on a water-cooled condenser with a water-regulating valve will cause an increase in head pressure—unless the maximum water flow has been reached. The condenser must be almost at the drowning point before any symptoms show up; the customer will probably complain of high water costs before this point is reached. Specifically, these problems are:

(1) Scaled condenser tubes. This reduces the efficiency of the condenser by decreasing the heat transfer rate from the hot gas to the water. In effect the scale is actually an insulator between the hot gas and the water.

Symptoms: High volume of water, low water temperature rise. Depending on design and selection, the saturated condensing temperature (discharge pressure converted to temperature) should not be more than 5°–10°F above the leaving water temperature.

(2) Refrigerant overcharge. Most water-cooled condensers are designed to provide some refrigerant storage space in the shell below the bottom row of tubes. If the system is overcharged so that the level in the condenser covers some of the tubes, the capacity of the condenser will be reduced.

Symptoms: The same as with scaling—high water quantity, low water temperature rise. If an overcharge does exist, the removal of some refrigerant will reduce water quantity and increase water temperature without causing gas bubbles in the liquid line.

(3) Noncondensables. Air in a water-cooled condenser dilutes the hot gas and seriously reduces the capacity of the condenser. Again, the symptoms are high water quantity and low water temperature rise. If air is present, the pressure in system will not correspond to the ambient temperature on shutdown. Purging will decrease the water quantity and increase the water temperature rise.

In cases where condenser water is supplied from a cooling tower (Fig. 23-3) or other sources at a fixed rate, the water-cooled condenser shows about the same symptoms as an air-cooled condenser.

Air in the system, excess refrigerant, or scale in the tubes will raise the head pressure and cause the compressor to stop or to cycle on the high-pressure cutout or the compressor motor protector.

In addition, failure of the tower fan, the circulating pump, or plugged water screens will result in high head pressure.

Symptoms: The compressor is stopped on the high-pressure switch or compressor motor overload.

Evaporative condensers use both air and water as condensing mediums, but their behavior is more like that of an air-cooled condenser. The major symptom of trouble in an evaporative condenser system is high head pressure. This usually results in system shutdown by the high pressure switch or the compressor overload. Causes can be:

reduced air quantity

pump failure

dirty water screens

water failure or scaled coil tubes

Remember, entering wet-bulb, not dry-bulb, temperature controls the capacity of an evaporative condenser.

To review, from a trouble-shooting standpoint, all condensers fall into the two groups shown:

liquid cooled

air cooled and liquid-air cooled

In the first group is the water-cooled condenser with an automatic water-regulating valve. This type will not show high head pressure as a symptom until maximum water flow is reached. Up to that point the basic symptom is increased water flow caused by the water valve trying to keep the head pressure down.

The second group includes air-cooled and evaporative condensers, as well as water-cooled condensers with cooling towers.

On condensers where the air and/or water volume remains constant the first symptom of efficiency loss is usually an increase in head pressure. This is because the cooling medium volume cannot increase to compensate for the loss in condensing efficiency. Sometimes the condensers in this group have condensing pressure-control devices such as air dampers, water bypass valves, or fan-cycling controls. However, their purpose is to keep head pressure up at light loads or low outside ambient temperatures.

Receivers, being simple storage tanks, seldom are a source of trouble. Occasionally, they are subject to vortexing, just as water that swirls down a bathtub drain (Fig. 23-4). Vortexing results in gas mixing with the liquid at the receiver outlet, thus reducing expansion valve and system capacity. This condition can be easily spotted at the liquid-line sight glass as a flash gas and is usually caused by a low charge. The possibility of vortexing is greater in a receiver designed with a liquid outlet extending downward from the bottom center of the vessel.

FIGURE 23-3 (Courtesy, Carrier Air-Conditioning Company).

FIGURE 23-4 (Courtesy, Carrier Air-Conditioning Company).

Dirt or trash may restrict the receiver outlet also, causing bubbles in the liquid-line sight glass. If this is the case, no amount of additional refrigerant will clear up the bubbles.

EVAPORATORS

Evaporator problem diagnosis depends a great deal on the medium being cooled, so for purposes of this discussion of trouble-shooting we will divide evaporators into the two categories:

Evaporators cooling air.
Evaporators cooling liquid.

In the first category are bare, fin, and plate coils with either forced or gravity circulation. Practically all evaporators in this category are of the direct expansion type. The refrigerant feed is usually through expansion valves or capillaries.

In the second category we find flooded and direct-expansion chillers and plate-type contact coolers. Here the refrigerant feed is usually an expansion valve or a low-side float.

There are two areas where trouble may be encountered in evaporators that cool air:

Air supply and distribution.
Refrigerant supply and distribution.

As illustrated in Fig. 23-5, an evaporator just cannot do a proper job of cooling without sufficient air. A shortage of air can be caused by the following:

Dirty filters.
Dirty coils.
Restricted ductwork.
Fan running backwards.
Slipping fan belt.
Improperly set fan pulley.
Improperly adjusted dampers.

Generally speaking, if the quantity of air has been reduced enough to seriously impair system performance, it will be obvious without air-measuring devices. Symptoms of low air quantity are:

Low suction pressure.
Frosted coil.
Iced coil.
Abnormally low air temperature.

If poor system performance is caused by reduced air quantity, returning air quantity to normal should clear up any or all of the above symptoms.

High air quantity problems are rare, but when they occur, noise is the usual symptom, along with occasional compressor overloading due to high suction pressure.

An evaporator cannot do a proper job with uneven air flow. Poor air distribution is generally caused by improper ductwork or coil placement. If this is the case, call for an engineering check. Air

FIGURE 23-5 (Courtesy, Carrier Air-Conditioning Company).

will take the path of least resistance, and this is usually the shortest distance. Remember, however, that turbulance causes resistance, so that even the shortest path may have great resistance.

Uneven air distribution over the coil can cause a lowering of capacity and will be evident by:

Lower than normal suction pressure due to uneven coil loading.

Possible refrigerant floodback due to lightly loaded refrigerant circuits.

Some baffling may be required to get even air velocity over the coil. In Fig. 23-6, the valve must have the proper supply of refrigerant to feed the coil. A shortage of refrigerant to the coil may be caused by any of the following:

System refrigerant shortage.

Plugged drier.

Restricted line or fitting.

Faulty expansion valve.

Symptoms of a refrigerant shortage to the coil will be:

Low suction pressure.

Possible frosting or icing of coil.

Bubbles in sight glass (depending on location).

Temperature drop in the liquid line (if due to restriction.)

FIGURE 23-6 (Courtesy, Carrier Air-Conditioning Company).

High expansion-valve superheat.

Flash gas due to long vertical liquid line.

If system shortage is the difficulty, the addition of refrigerant should raise suction, clear up any frost, and clear the sight glass. If liquid restriction exists, the temperature drop at restriction will persist until the restriction is cleared, in spite of the addition of more refrigerant.

When the refrigerant charge is ample and no restriction exists, but low suction and high superheat continues, look for a faulty expansion valve or an improper bulb application.

A "sick" distributor just can't perform correctly. Poor refrigerant distribution to the coil can be caused by:

Plugged distributor tube.

Restricted distributor tube.

Wrong size distributor orifice.

Symptoms of poor distribution are: some circuits sweating, some dry, or any combination of frosting, sweating, and dryness.

Make sure the air flow over all portions of the coil is fairly uniform before placing the blame on refrigerant distribution. Poor air distribution over a coil can give the same symptoms as poor refrigerant distribtion.

In liquid chilling devices, problems in cooling the liquid are almost identical regardless of whether the chiller is of the direct expansion or flooded type. A chiller must have a reasonable load to keep it operating comfortably. A shortage of chilled water will result in a lowering of system capacity. The most noticeable symptoms will be:

Excessively high rise in temperature between supply and return water.

Low suction temperature due to light load.

Possible trip-out of the protective thermostat due to low leaving water temperature.

Check the chilled water pump suction and discharge pressure and compare them to the pump rating for proper water quantity.

Poor water distribution within the chiller will show as a low temperature drop through the chiller and also as low suction pressure. To find the source of trouble check the baffling on DX chillers, the water box position, and the baffling on a flooded chiller.

A chiller will shrink in size (capacity) when starved for gas. A lack of system capacity will be caused by refrigerant shortage. On both DX and flooded chillers, this will result in a low temperature drop through the cooler, low suction pressure, and possible safety thermostat tripping.

A starved chiller may be caused by:

Actual shortage of refrigerant

Plugged drier.

Line restriction.

Faulty expansion valve.

COMPRESSORS

Although trouble-shooting of both hermetic and open compressors problems (Fig. 23-7) will be discussed here, no attempt will be made to cover the electrical portion other than to mention electrical components. These two types of compressors have much in common, except in their electrical layouts. Here are the problems that may occur in either open or hermetic compressors:

Mechanical seizure

Noise

Failure to pump properly

Overheating

Seizure is a major difficulty in compressors. Before rebuilding or replacing a seized compressor, the cause of the seizure should be determined in order to prevent a recurrence. A factory defect can show up at any time in the life of a compressor, but is most apt to appear shortly after startup. It is best to eliminate all possible job causes first, since they can cause a repeat failure. A compressor that is low on oil or out of oil indicates a lubrication failure, which could be caused by four conditions:

- Natural oil traps in the system.
- Refrigerant floodback to the compressor during operation due to a bad expansion valve or poor bulb location.
- Flooded start due to refrigerant accumulation in the compressor during shutdown.
- Refrigerant shortage, which causes oil trapping. The reason: lack of sufficient mass flow of vapor.

FIGURE 23-7 (Courtesy, Carrier Air-Conditioning Company).

If the compressor oil charge is normal on a seized hermetic compressor, it is safe to assume the compressor was mechanically defective. Seizure is fairly common with a burnout, so consider it the result of a burnout, not the cause.

On an open compressor, a tear-down of the unit and a knowledge of the quantity of oil in the crankcase will help determine the cause of failure.

The compressor in Fig. 23-8 is trying to tell us something. Excessive noise in a compressor generally indicates wear due to lack of lubrication. Noise can also be caused by the same conditions that lead to seizure (see the preceding paragraph).

Rattling of the compressor on startup, sometimes called *slugging*, is caused by an accumulation of liquid refrigerant in the compressor crankcase during the "off" cycle. The sudden reduction in

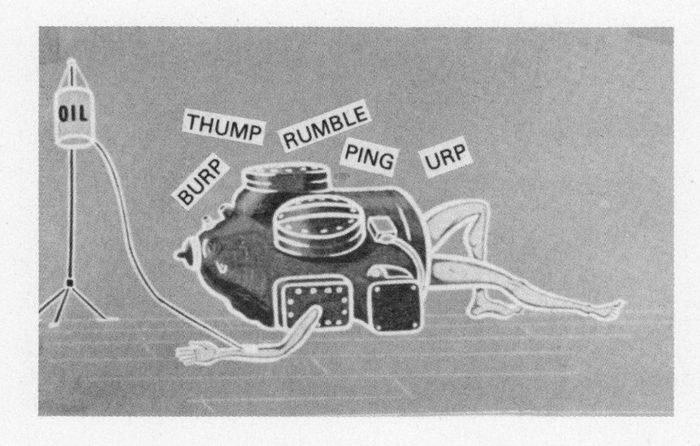

FIGURE 23-8 (Courtesy, Carrier Air-Conditioning Company).

crankcase pressure as the compressor starts causes the refrigerant oil mixture to foam. This foam passing through the valves causes the rattle, which continues until the refrigerant is distilled out and the oil level stabilizes. Crankcase heaters or pumpdown control will eliminate initial slugging.

A lowering in system capacity can be the result of a failure of the compressor to pump properly (Fig. 23-9). If we compared this compressor to a human being, we would say he was suffering from low blood pressure. A serious deficiency in compressor pumping capacity is obvious. The suction pressure will be high, the discharge pressure will be low, the discharge line temperature will be much lower than normal, and the compressor will be hotter. Less serious losses in pumping capacity are more difficult to pinpoint. Two good, simple checks are:

(1) With the compressor running, close the suction service valve. At normal head and with no refrigerant in the crankcase, the compressor should pull down to a range of 15 in. to 25 in. vacuum in less than a minute.

(2) Close the discharge valve quickly and tightly immediately after stopping the compressor. Pressure in the discharge side of the compressor should not drop more than 3 to 4 lb in a minute's time.

If the compressor is the open or semihermetic type and fails either test, remove the cylinder head and inspect the valves. Look for blown gaskets or broken or badly worn valves.

FIGURE 23-9 (Courtesy, Carrier Air-Conditioning Company).

Many welded hermetics have no service valves, so a judgment about pumping efficiency must be made on the basis of pressures and temperatures. When no gauge ports are provided, diagnosis is on the basis of temperature alone. Remember that hermetics are cooled by the gas flow through them. If the gas flow is down, due to poor pumping ability, the compressor will run hotter than normal.

A system will fail to cool when the compressor shuts down in response to its temperature device. Several things may cause a compressor to overheat:

- low voltage (hermetic)
- high voltage, low suction (hermetic)
- shortage of refrigerant (hermetic)
- high load (high suction pressure)
- high head pressure
- low oil charge
- very high compression ratio

CONTROLS

Controls also can contribute their share of headaches to the system. Although there may be a number of controls on a particular unit, this discussion will be confined to metering devices and pressure switches, since both actually tie into or form a part of the refrigeration cycle. Metering devices will be covered first.

As illustrated in Fig. 23-10, the expansion valve is often mistakenly held responsible for operating failures in the refrigeration cycle. The three problems most commonly blamed on expansion valves are:

- overfeeding (flooding)
- underfeeding (starving)
- erratic (hunting)

Flooding may be caused by the valve sticking in an open position; however, it may also be caused by the following:

- improper superheat on the valve
- improper valve bulb location on the suction line
- wrong type of valve for refrigerant in the system
- loose expansion valve bulb
- light load
- excess oil in the system

FIGURE 23-10 (Courtesy, Carrier Air-Conditioning Company).

Starving may be caused by the valve sticking closed, or by the partial loss of charge in the power element. The coil may also be starved for the following reasons:

wrong type of valve for refrigerant used
shortage of refrigerant
improper superheat setting
plugged drier
plugged refrigerant distributor
improper valve bulb location
plugged equalizer line

By their very nature, all expansion valves will "hunt" to some degree. Remember, a valve tries to control the superheat of the gas leaving the evaporator by varying the flow of liquid refrigerant entering the evaporator several feet upstream, so the response is not immediate. Excessive "hunting" can be caused by internal parts that stick or bind. Aside from faults in the valve itself, the following can cause excessive "hunting":

oversize expansion valve
very light load
long refrigerant circuit
rapid changes in condensing pressure or temperature
rapid load changes
intermittent flashing in the liquid line

Capillaries, though simple in themselves, behave in such a way as to make system problems hard to diagnose. In all other metering devices, the pressure drop takes place within a short distance, through an orifice. In a capillary the pressure drop from high to low side is spread out over the entire length of the tube. The only way a capillary can malfunction is to become partially or completely restricted. However, many other conditions in a system affect the performance of the capillary, yet give the same symptoms.

Low suction and a starved evaporator can be caused by:

shortage of refrigerant
too much oil in circulation
plugged drier
low head pressure

Flooding of the evaporator and increased suction pressure can be caused by:

high head pressure
excessive subcooling
overcharge of refrigerant

Remember that for any given pressure difference across it, a capillary can pass more weight of liquid than weight of gas, because the liquid has a much greater density. This fact makes a capillary very sensitive to excess oil, excess subcooling, or flashing.

Excess oil in circulation displaces some refrigerant in the capillary tube, and thus reduces the actual weight of refrigerant flowing through the tube and starves the coil.

Excessive subcooling reduces flashing in the capillary to such a point that too much refrigerant flows through the tube, and flooding results.

Excessive flashing due to shortage of gas, liquid-line restriction, or capillary in contact with a hot surface will reduce the amount of liquid the tube can pass and result in a starved coil.

The operation of other types of metering devices such as automatic expansion valves or low and high-side floats is much more simple and straightforward than that of either thermostatic expansion valves or capillaries. Sticking or leaking can occur in all three, but the symptoms of starving or flooding the coil are quite obvious.

Most low and high-pressure switches (Fig. 23-11) are of the automatic reset type. This is to prevent

FIGURE 23-11 (Courtesy, Carrier Air-Conditioning Company).

FIGURE 23-12 (Courtesy, Carrier Air-Conditioning Company).

service calls for an occasional nuisance tripping of a switch; however, these trips sometimes become a system "bug." Such a nuisance trip of a low pressure switch will usually occur on startup. The expansion valve may be slow in adjusting to the proper feed. During this period the suction pressure may pull down and the system may be stopped once by the low-pressure switch before balancing out and continuing to run.

Occasionally a high-pressure nuisance trip occurs because of heat buildup in an air-cooled unit exposed to the sun during the *off* cycle. When the system is started, the high-pressure switch may open once before the condenser fan can clear the accumulated heat from the unit. The fact that the compressor comes up to speed faster than the fan motor also tends to cause a high-pressure trip-out.

With automatic reset or closing type of pressure switches, it is sometimes difficult to tell if they have opened and are the cause of a problem. A simple way to check is to place a small fuse across the pressure switch terminals. The fuse must have a lower value than the current carried in the switch circuit so it will blow in case the switch opens. A check of the fuse on the next call will tell whether the switch has opened.

ACCESSORIES

Next we will discuss accessories. The main purpose of a strainer drier in a system is to pick up moisture (see Fig. 23-12). However, it must be remembered that this device can also release moisture back into the system. The strainer drier's water-retention capability will vary with the dessicant being used. However, the moisture capacity of all driers is reduced as the liquid refrigerant temperature rises. A drier installed in a 70°F liquid line during the winter may suddenly release moisture back into the system in the summer, when the liquid line temperature rises to 100°. This could cause a mysterious freeze-up of the expansion valve on a commercial job (but only if the drier was saturated at the 70° winter temperature).

No amount of moisture will restrict a strainer drier, but contamination in the system will. A partly-restricted strainer drier will act as a metering device and cause a pressure drop. Along with this pressure drop is a temperature drop, which can generally be felt with the hand. Sweating and even frosting may occur on lower temperature systems. Seldom is a strainer drier plugged so tightly that all flow is stopped.

Capillary tubes are very sensitive to premature flashing of the liquid. A very slight restriction in a strainer drier, too small to be perceptible to the touch, can cause sufficient flashing to greatly reduce the capacity of a capillary and thus cause a serious drop in system capacity.

Like the "lovely lady" Fig. 23-13, solenoid valves may hum because of low voltage, a loose connection, or a sticking plunger. They may fail to open because of sticking, an open coil, or excessive pressure difference across the valve. Check the maximum operating pressure difference on the valve nameplate before concluding that the valve is stuck.

Leakage frequently occurs with a liquid-line solenoid, and this can be determined by feeling the line on both sides of a closed valve when a pressure difference exists. If the valve leaks, a temperature drop can be felt across the valve.

The purpose of a water-regulating valve is to maintain a uniform head pressure while the system is in operation and to stop the flow of water when

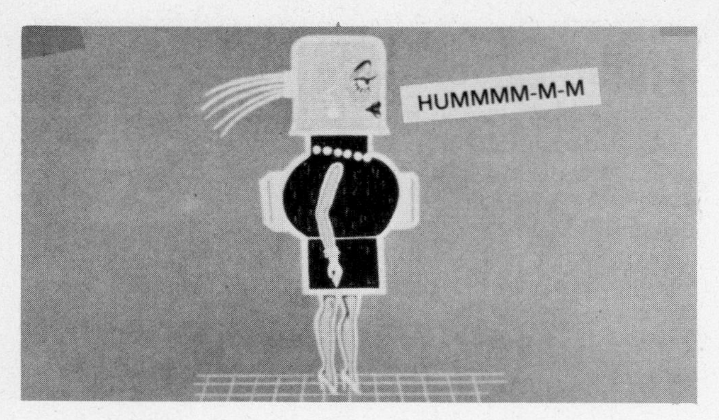

FIGURE 23-13 (Courtesy, Carrier Air-Conditioning Company).

the system is shut down. Water valves may stick and fail to shut off tightly, but don't blame the water valve for a large flow of cool water during operation. This may be caused by:

noncondensables in the system keeping the head pressure up

scaled or dirty condenser

valve set too low

Remember, a water valve requires a differential between operating and shutoff pressure. It cannot maintain an operating head pressure of 125 psig and be expected to shut off tightly at 120 psig. Most valves will require 10 to 20 psig difference between operating and shutoff pressures.

Fig. 23-14 shows the eye acting as a sight glass or window in a refrigeration system. It is important, when looking through a sight glass, that a completely full glass is not mistaken for a completely empty one; this is an easy mistake to make. Many, but not all, glasses have flow indicators to prevent such errors.

Bubbles in the glass indicate either a shortage of refrigerant in the system or a pressure drop due to a restriction, liquid lift, or undersized liquid line. Be sure you determine the cause of the bubbles before adding refrigerant; otherwise you may merely overcharge the system.

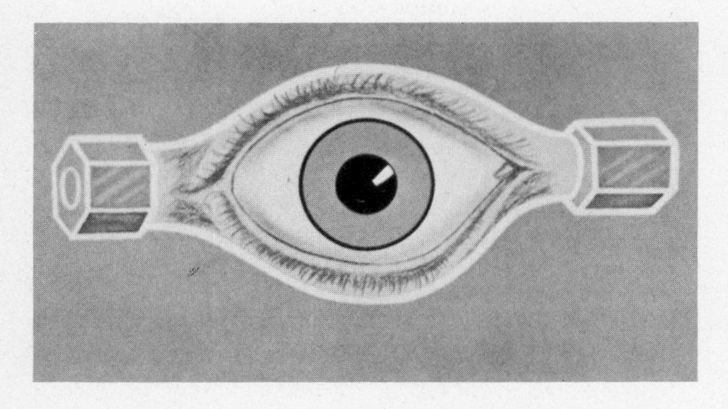

FIGURE 23-14 (Courtesy, Carrier Air-Conditioning Company).

Some sight glasses are equipped with moisture-indicating devices, which change color when safe moisture limits are exceeded. In effect, these devices change color at a certain relative humidity of the liquid refrigerant. This gives them two characteristics worth remembering. First, the indicator will not read properly unless it is completely submerged in liquid refrigerant. Second, the color change always occurs at the same relative humidity, but as the temperature of the liquid refrigerant goes up it takes more moisture in parts per million (ppm) to cause the same relative humidity. For example, 10% relative humidity in R-12 at 80°F represents 10 ppm, while 10% relative humidity in R-12 at 100°F represents 16 ppm.

A moisture indicator element that has been subjected to excess water will not read correctly, since the salts in the indicator may have been leached out. An element subjected to burn-out acids will also be damaged. Do not keep changing driers in an effort to get the proper color on a moisture indicator. Check the indicator on a fresh drum of refrigerant with a known low moisture content. (Most refrigerant manufacturers specify 10 ppm of moisture or less in their refrigerant.)

Now we will briefly review the major problems encountered in the basic system components shown:

(1) Condensers—receivers

high head pressure
noncondensables
overcharge
short-circuiting of air
scaling
water shortage
restricted flow

(2) Evaporators

air shortage
air distribution

FIGURE 23-15.

refrigerant shortage
refrigerant distribution
water shortage
water distribution

(3) Compressors

seizure
noise
poor pumping
overheating

(4) Basic cycle controls

metering devices overfeeding
underfeeding or erratic metering devices
nuisance tripping of pressure switches

(5) Accessories

plugged strainer driers
leaking solenoid valves
sticking water valves
care in observing sight glasses

No discussion on trouble-shooting would be complete without reference to a trouble diagnosis chart (see Fig. 23-15). Most manufacturers provide a chart or similar guide with their equipment to help the service technician diagnose specific problems that may be peculiar to their machinery.

PROBLEMS

1. The most common problem of air-cooled condensers is ______________.

2. Problems with water-cooled condensers are generally easier to diagnose than those of air-cooled condensers; true or false?

3. High head pressure in a water-cooled condenser with an automatic water valve will show up immediately; true or false?

4. Evaporative condenser performance is more affected by a high dry-bulb temperature than wet-bulb temperature; true or false?

5. A shortage of air across an evaporator will result in coil frosting or icing, together with ______________ suction pressure.

6. If a compressor seizure occurs after an extended period of operation, it is an indication that there are external system problems; true or false?

7. Low voltage to a hermetic compressor will probably cause it to shut down due to ______________.

8. Name three common problems of thermostatic expansion valves.

9. Excessive oil in the system has more or less effect on a capillary tube than on an expansion valve?

10. Driers can release moisture back into the system, depending on certain conditions; true or false?

11. A cold solenoid valve would indicate ___________.

12. A clear sight glass always indicates a properly charged system.

AIR-CONDITIONING

1

INTRODUCTION

HOW OLD IS AIR-CONDITIONING?

Air conditioning is as old as man himself. The primitive people shown in Fig. 1-1, who wore the skins of animals, were in a crude sense controlling the escape or containment of their own body heat and effecting a change in their personal comfort. Seeking shelter from the sun or finding refuge in caves from cold or heat were basically actions that changed their environment. The discovery and use of fire was perhaps the most important advance in that era.

Later history and artifacts show that the Egyptian ruling class used slaves (Fig. 1-2) equipped with palm branches to fan their masters. Thus the art of evaporative cooling provided some relief from the desert heat. History also recalls the Romans, who engineered ventilation and panel heating into their famous baths. The Romans also brought ice from the northern mountains to chill wine, and possibly also to chill water for bathing.

Moving into the middle ages, the remarkable Leonardo da Vinci built a water-driven fan to ventilate rooms of a house belonging to a patron friend. Other early innovations included rocking chairs with bellows action to produce spot ventilation for the occupant and clock mechanisms that activated fan devices above beds.

By today's standards, these examples of comfort conditioning seem rather crude, and perhaps, if fully explored, some would be humorous.

EARLY TECHNICAL DEVELOPMENTS

The art of ventilation and central heating progressed rapidly during the nineteenth century. Fans, boilers, and radiators had been invented and were in fairly common use. Early warm-air furnaces were coal-fired cast iron with gravity air distribution. Soon mechanical fans were used for forced circulation of air through ducts. Modern concepts of furnaces bear little resemblance to some of those "iron monsters." Size, weight, and ventilation of combustion products have been drastically changed, but most important

FIGURE 1-1 Ancient couple around fire.

were the developments that led to the gradual conversion from coal to oil and gas, and from manual to automatic firing.

Early texts on Refrigeration discussed the applications of using ice for preservation of food and the initial development of the concept of mechanical/chemical refrigeration in 1748 in Scotland by Dr. William Cullen.

FIGURE 1-2 Egyptian slave cooling master.

It was in 1844 that Dr. John Gorrie (1803–1855), director of the U. S. Marine Hospital at Apalachicola, Florida, described his new refrigeration machine. In 1851 he was granted U. S. Patent 8080. This was the first commercial machine in the world built and used for refrigeration and air conditioning. Gorrie's machine received world-wide recognition and acceptance. Many improvements to Dr. Gorrie's work followed and by 1880 the development of reciprocating compressors with commercial application to ice making, brewing, meat packing, and fish processing were fairly common. Refrigeration engineering became a recognized profession and in 1904 some 70 members formed the ASRE (American Society of Refrigeration Engineers).

The real "father of air-conditioning" (Fig. 1-3) was Willis H. Carrier (1876–1950) as noted by many industry professionals and historians. Throughout his brilliant career, Carrier contributed more to the advancement of the developing industry than any other individual. In 1911 he presented his epoch-making paper dealing with the properties of air.

FIGURE 1-3 Mr. Willis Carrier (Courtesy, Carrier Air-Conditioning Company).

These assumptions and formulas formed the basis for the first psychrometric chart and became the authority for all fundamental calculations in the air-conditioning industry.

Carrier continued his work and invented the first centrifugal refrigeration machine in 1922 and later pioneered the use of induction systems for multiroom office buildings, hotels, apartments, and hospitals. During World War II, he supervised the design, installation, and startup of a system for the National Advisory Committee for Aeronautics (NACA) in Cleveland for cooling 10,000,000 ft^3 of wind tunnel air down to −67°F (−19.45°C). Carrier died in 1950, having witnessed the real turning point in the industry's growth.

These were only a few of the steps along the way toward development of modern air-conditioning as we know it today.

COMFORT COOLING COMES OF AGE

Comfort air conditioning had its first major use in motion picture theaters during the early 1920s. Famous New York City movie houses like the Rivoli, the Paramount, the Roxy, and Loew's Theaters in Times Square (Fig. 1-4) were among the first. By the late twenties several hundred theaters throughout the country had been air conditioned. These were custom-designed, custom-manufactured, field-installed systems, which meant most of the assembly and erection of components was done on the job site.

Towards the end of the decade came the introduction of the first self-contained room air-conditioner. Not only was this an important technical achievement, but it became the industry's first attempt to "package" products that could be manufactured in so-called mass production and be factory tested and operated prior to shipment to the ultimate user.

The next milestone was the development of safe refrigerants. In 1930 Thomas Midgley of the Du Pont Company developed the now-famous fluorocarbon Freon refrigerant. In 1931 Freon-12 (Fig. 1-5) was introduced as a commercial refrigerant. Fluorocarbon refrigerants permit uses where more flammable or toxic material would be hazardous. Additionally, the operating characteristics of F-12 opened new vistas for compressor and system component designs. A

REMEMBER WHEN . . . summer meant straw hats and Saturday night talkies in a movie palace cooled with "frosted air". As air-conditioning caught on, this 1929 marquee was followed by more flamboyant signs featuring penguins and igloos. Temperatures in today's theatres are less of a novelty than in the past, a Loew's executive stated, because people are accustomed to air conditioning in their homes and places of work. In addition, industry product certification programs of the air-conditioning industry in effect for 13 years have helped air-conditioning engineers in assuring cool comfort in homes and public places.

The Bettmann Archive

FIGURE 1-4.

FIGURE 1-5 Freon 12 (Courtesy, du Pont de Nemours Company).

FIGURE 1-6 Old bolted hermetic compressor (Courtesy, Copeland Corporation).

whole family of Freon refrigerants soon followed where specific operating variations were needed. In 1955, other firms joined Du Pont in manufacturing the fluorocarbon refrigerants, and by 1956 a new numbering system was adopted to reflect the now-current R-12, R-22, etc., designations for these refrigerants.

About 1935 the industry introduced the first hermetic compressor for air-conditioning duty. Its size was considerably larger than today's equivalent capacities. Its motor speed was 1,750 rpm compared to the common 3,600 rpm of today. The outer shell was bolted rather than completely welded (Fig. 1-6).

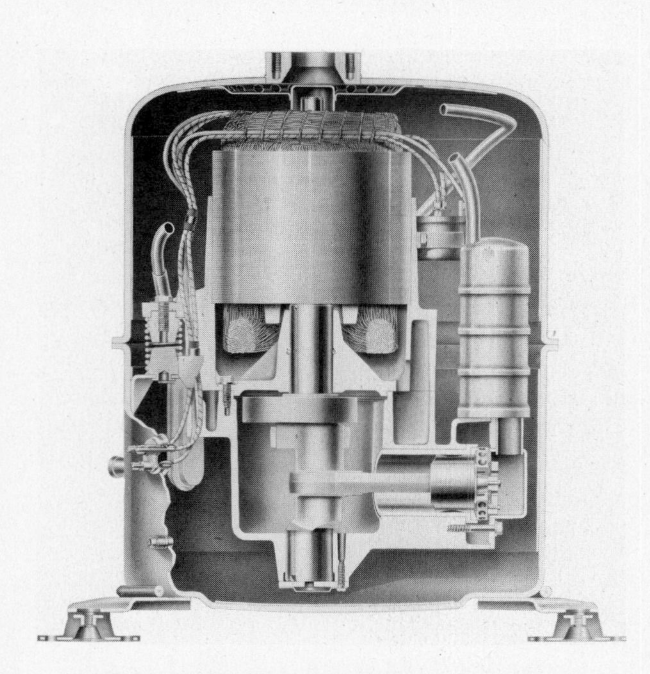

FIGURE 1-7 Three-ton compressor (Courtesy, Airtemp).

FIGURE 1-8 Applied machining systems (Courtesy, Airtemp Corporation).

But the concept of passing cool suction gas to and over the motor windings started a trend in sizes up to $7\frac{1}{2}$ tons that has now become almost universal among compressor manufacturers. Full-welded hermetics (Fig. 1-7) are currently offered in capacities up to 20 tons in a single shell. The advantages of the hermetic concept include reduced size and weight; less cost to the manufacturer; no seal failure problems; less noise; no belt maintenance; location not critical insofar as ventilation requirements for motor heat dissipation; and the suction gas is superheated by absorbing motor heat, so better oil separation is assured. Reliability reports from the industry support these technical advantages.

Post-World War II products consisted mainly of applied machinery systems (Fig. 1-8) for large buildings, store conditioners, and window air-conditioners.

FIGURE 1-9 Window air-conditioning unit (Courtesy, York, Div. of Borg-Warner Corporation).

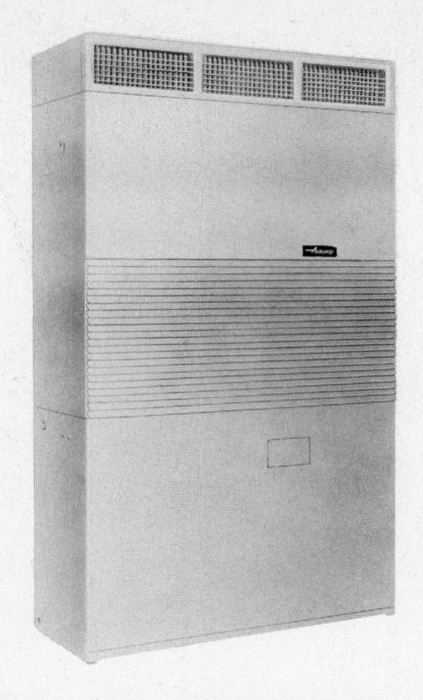

FIGURE 1-10 Typical water cooled self-contained unit (Courtesy, Airtemp).

Window units (Fig. 1-9) were used extensively to cool residences, smaller offices, small stores, and just about any conceivable application where access to window or through-the-wall mounting could be achieved. However, the "main street" commercial comfort conditioning market, the drug store, restaurants, dress shops, barber shops, grocery stores, etc., were handled by the *self-contained store conditioner* (Fig. 1-10). These units were primarily water cooled and were usually located in the conditioned space. Air distribution generally came from grilles on the discharge hood, although they could be ducted to some extent. Some disadvantages of these units included the need for a municipal water source and cost considerations, or cooling tower application problems, excessive noise, high air velocity, and drafts; they also occupied valuable selling space in retail establishments. But even with these deficiencies, these units are credited with creating the public appreciation and awareness for central comfort air conditioning. Some manufacturers offered variations that could be adapted to residential use, but the number of installations were relatively few.

The next major breakthrough that really started the industry sales sky-rocketing was the introduction in 1953 of air-cooled operation instead of water for condensing purposes. New technology in component and system design permitted pushing head pressures higher and higher so that machines could operate safely and reasonably efficiently up to 115°F outdoor ambient conditions. Early packaged units (Fig. 1-11) were mostly horizontal in configuration, for mounting in attics or on a slab at ground level. Installation consisted of mounting the unit, electrical hook-up, and a simple duct system to distribute air into the space. These units were equally adaptable to small commercial applications when mounted on the roof. Among the many advantages of these systems were; no refrigerant piping; little or no plumbing needed; a factory-charged, tested, and sealed refrigeration

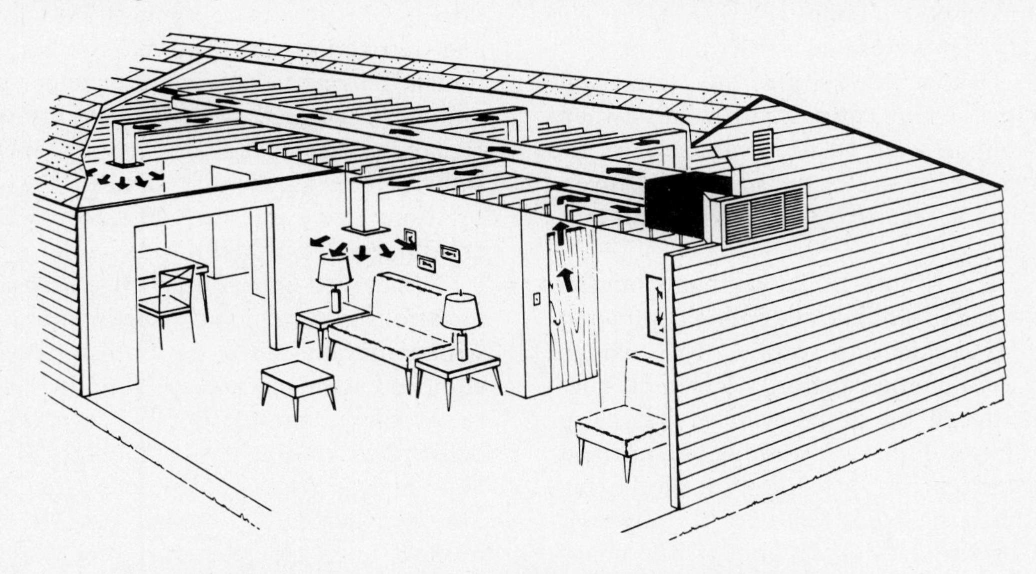

FIGURE 1-11 Horizontal packaged unit (Courtesy, Airtemp Corporation).

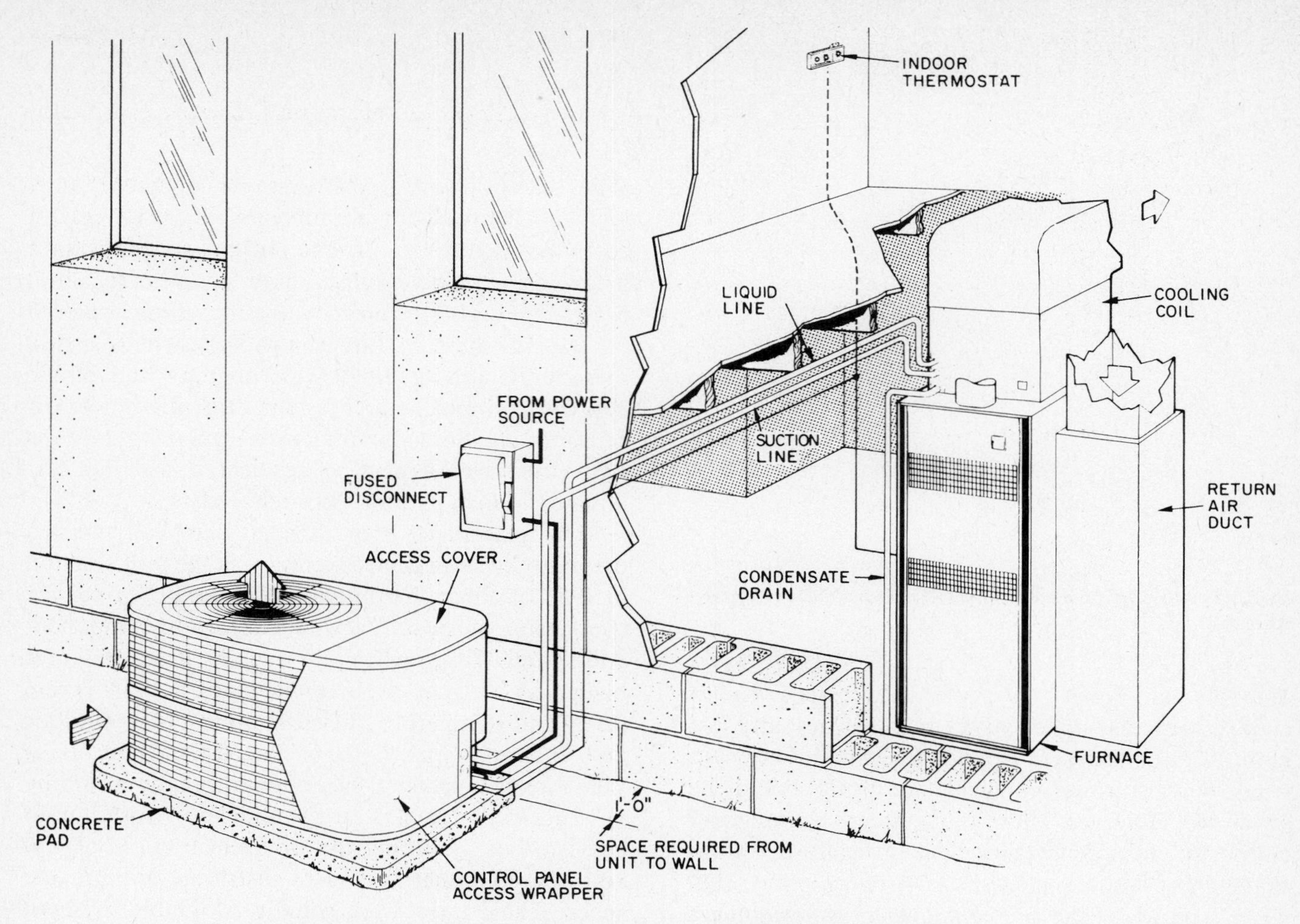

FIGURE 1-12 Split system (Courtesy, Carrier Air-Conditioning Company).

circuit; and minimum electrical wiring was required; all of which reduced on-site labor and material costs and greatly increased reliability of operation. The main disadvantage associated with this product design was its lack of flexibility in adapting to all kinds of residential applications, particularly where heating and cooling were to be combined. As the expression goes, "necessity is the mother of invention"; the industry reacted quickly with the split-system concept.

The split system (Fig. 1-12), as the name implies, consists of two parts—an indoor cooling component and an outdoor condensing section. The two are connected by liquid and suction refrigerant lines. The indoor unit can be varied widely, first as an evaporator coil with upflow, downflow, or a horizontal air flow arrangement, depending on the heating system, or it can be equipped with a blower to provide its own air-handling capability. The adaptability of these systems made them an immediate success and there was extremely fast growth in their use. Early versions, however, required on-site lubrication and brazing of refrigerant lines and evacuating and refrigerant charging before becoming operational. This added measurably to installation costs and also required more highly trained installers. Reliability depended on the installer's technical skills. Once again, the industry responded with solutions and came up with the precharged refrigerant line and quick-connect couplings.

Precharged lines (Fig. 1-13) are cut and fabricated in various lengths, and the suction line is preinsulated with continuous covering. Both ends of the lines are equipped with mechanical couplings having a thin metal diaphragm to form a seal. The lines are dehydrated, evacuated, and charged with precise amounts of refrigerant. When installed at the job, the lines are mated to connections on both indoor and outdoor units. One part of the coupling has a cutting edge that pierces the diaphragm, folds it back, and

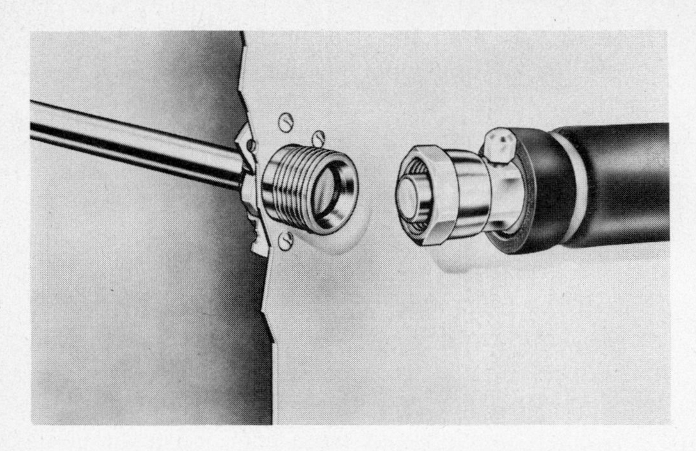

FIGURE 1-13 Precharged lines (Courtesy, Aeroquip).

opens the line. Thus a complete refrigeration system is put together with a minimum of labor and minimum risk of contamination. The reliability of these systems also increased measurably. Later chapters will define other variations of mechanical fittings.

During this same period of development of the package and split-system cooling equipment, the industry also converted many of these cooling-only systems to reverse-cycle operation and called them *heat pumps* (Fig. 1-14), meaning that heat could be pumped into or out of the space, depending on which mode was needed. Unfortunately, because of pressure on electric utilities to build winter heating loads, early heat pumps were not fully developed and lacked reliability. To compound the matter, heat pumps were sold in extreme climates where operational performance was marginal. Finally, the number of qualified service personnel available was minimal or even nonexistent. The results were far from satisfying to all concerned. However, experience and technical improvements have restored confidence in the heat pump, and with the world's attention turning to energy conservation, it will probably gain more popularity in the future.

The final major product innovation that occurred during the late fifties and early sixties and still accounts for the most dramatic growth rate of all packaged year-round conditioners was the *rooftop* combination gas-heating and electric-cooling unit (Fig. 1-15). It started with small two, three, four, and five-ton systems being installed on the rooftops of low-rise commercial structures of all descriptions, as well as residential applications. The trend rapidly spread. Equipment became more sophisticated and sizes changed rapidly; today's capacities range upwards of 100 tons of cooling. Both gas and electricity are available for heating. To coin a phrase, "Show me a shopping center and I'll show you a rooftop unit."

This review of the products that contributed to the coming-of-age for comfort air-conditioning by no means covers all the important technical advancements. In subsequent chapters, we will explore further systems and products currently offered by the industry. Also, before leaving the history of the

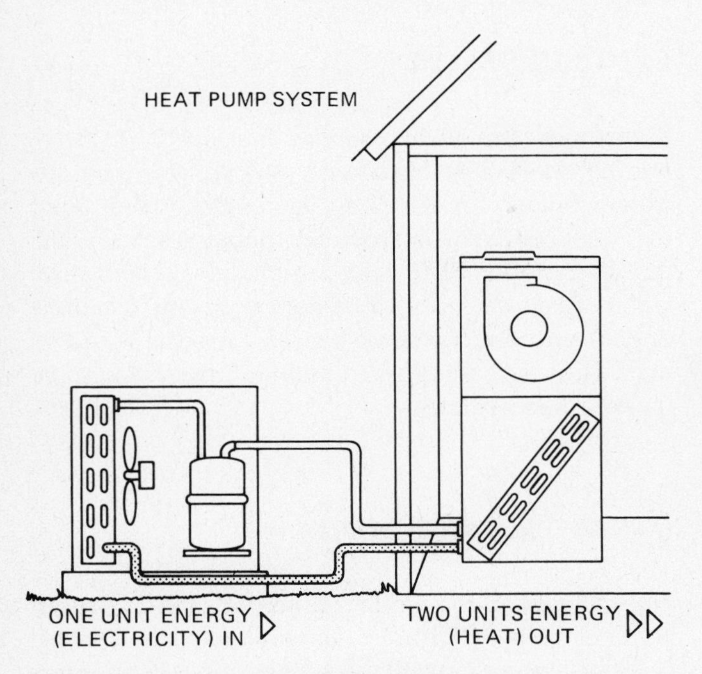

FIGURE 1-14 Heat pump system (Courtesy, Carrier Air-Conditioning Company).

FIGURE 1-15 Rooftop combination gas heating and cooling unit (Courtesy, York, Div. of Borg-Warner Corporation).

FIGURE 1-16 Engineering experiment station, University of Illinois campus (Courtesy, NESCA).

development of air-conditioning, we should note that *not all* the knowledge that led to such rapid growth came from manufacturers or suppliers of components. One of the most notable examples came from a trade association effort.

In the late 1950s the Field Investigation Committee and the Research Advisory Committee of the National Warm Air Heating and Air-Conditioning Association, in cooperation with the University of Illinois, established the Engineering Experiment Station, which was in reality a residence on campus (Fig. 1-16). Many important studies were conducted on heat-flow characteristics, the effects of insulation, design, and the efficiency of air distribution systems, etc. The results of this effort helped greatly to advance the art of sizing selection and application of heating and air-conditioning and became the basis for a number of design manuals. The name of the association has now been changed to NESCA (National Environmental Systems Contractors Association). In later chapters the use of their manuals will be reviewed in depth.

HISTORICAL MARKET GROWTH

Those who contemplate a career in air conditioning are naturally interested in the industry's past track record of growth and most frequently will ask, "Just how big is this industry?" That's a difficult question to answer because of its many facets; however, we will attempt an explanation.

ARI keeps track of industry shipments and breaks them down into several classifications. First we will discuss *unitary* shipments, which for this report covers all of the packaged-type of air-conditioners for cooling only, heating and cooling, heat pumps, etc. Since 1950 the record of sales, represented by Fig. 1-17, shows a rise through 1974. Several observations can be made. First note the dramatic growth since 1958. Most will agree this must be related to the introduction of air-cooled systems into the residential market. By 1973 it is estimated that 75% of the total were residential units. The rate of growth demonstrated by the slant of the curve is even more dramatic when compared to growth curves of other industries or even the gross national product we hear so much about from economists or commentators. It will be noted that there was a slight dip in 1974; industry shipments totalled 2,490,000 units, down 13% from 1973. This can be directly related to the economy of 1974. Air-conditioning sales can be tied closely to the construction industry, residential and commercial, both of which were down measurably for that period. Industry forecasters indicate the established historical upward trend will return with equal vigor as the economy improves. Nevertheless, the installed value of *unitary* equipment for 1974 was placed at over 3½ billion dollars. Approximately 45 million American homes have central air-conditioning, but this is still far less than all of the occupied residences in the country. The potential for future growth, despite 1974's setbacks, is unlimited.

CENTRAL STATION EQUIPMENT

Sometimes also called *applied machinery* or *field-erected equipment*, this category of product includes water chillers, air handling equipment, ducts, etc., that together make up the larger tonnage central plant installation. In 1973 ARI estimated the installed value of this category to be approximately 2 billion dollars, and since water chiller sales were up by 6.5% in 1974, it is reasonable to assume the total market increased accordingly.

OTHER PRODUCT MARKETS

Not included in the unitary or central station figures as reported by ARI are other products that can be classified in the broad, year-round air-conditioning field. Room air-conditioner sales in 1974 were estimated at 5 million units. Package terminal units

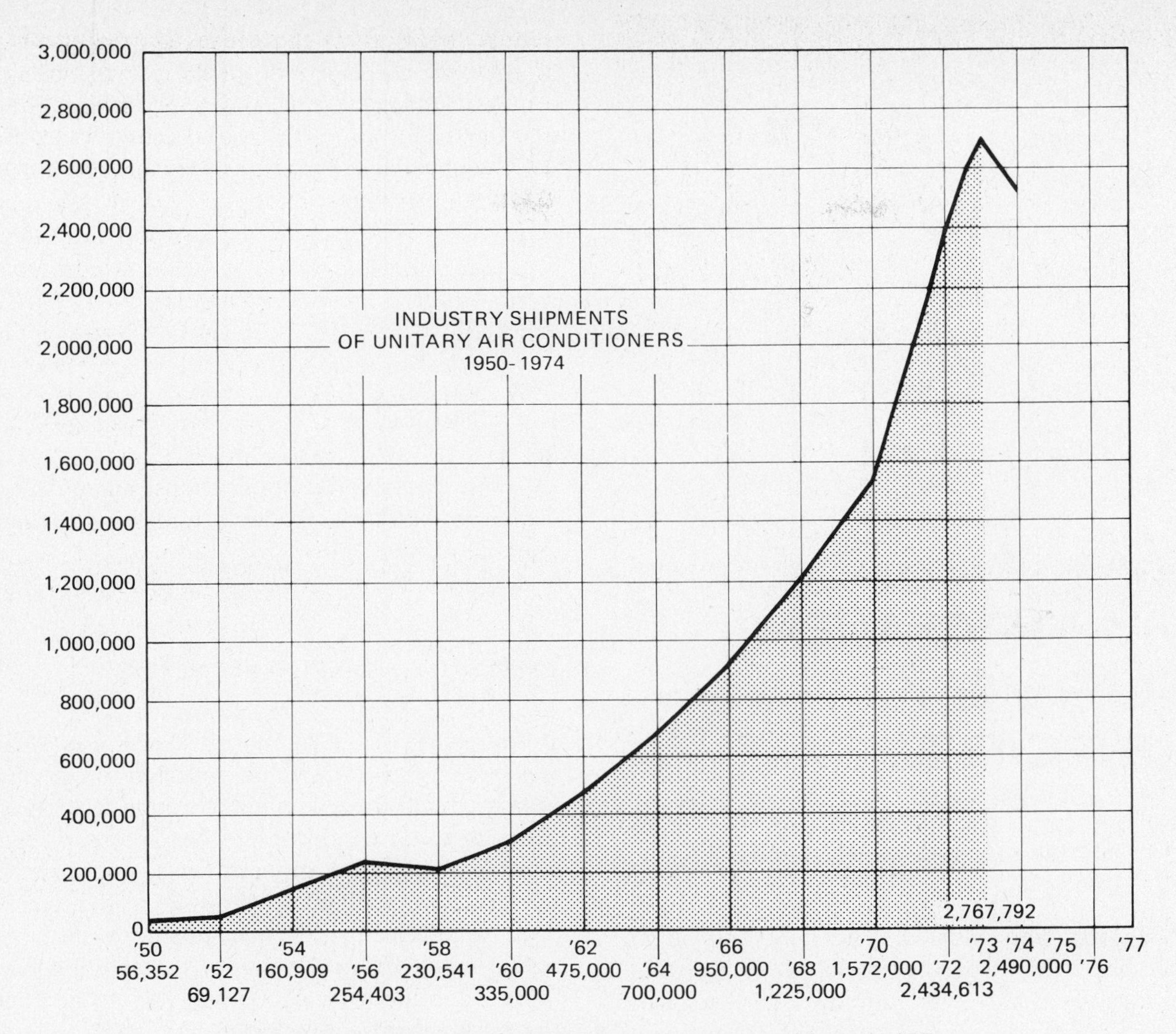

FIGURE 1-17 Industry shipments of unitary air-conditioners (Courtesy, ARI).

(through-the-wall units commonly used in motels) were estimated at 200,000 units. Sales of electronic air cleaners were slightly over 200,000 units.

Central heating equipment in 1974, as reported by GAMA (Gas Appliance Manufacturers Association), showed gas furnace sales at approximately 1.5 million units, oil furnaces at 275,000 units, electric furnaces at 400,000 units, and electrical duct heaters at 100,000 units.

Additionally, there is a group of products that are used in nonducted electric resistance heating, and estimates by the EEI (Edison Electric Institute) place this segment in excess of 400,000 units sold.

Finally, there are a great number of related accessories sold separately that are used to make up complete installations, such as grilles, registers, controls, filtering equipment, etc. There is no accurate method of determining or estimating their sales total, but it is probably in the millions of dollars.

In summary, the *total annual installed value* of this industry *domestically* probably approaches 8 billion dollars, and that figure may be understated. In addition, there is a significant and developing market overseas.

THE INTERNATIONAL AIR-CONDITIONING MARKET

Of the 1974 unitary air- conditioners shipments, 100,000, or 4%, went to foreign markets. U.S. manufacturers exported to some 153 countries. Major users were Canada, Japan, West Germany, Australia, the United Kingdom, Venezuela, France,

1970 - 1980

MANPOWER
SURVEY
REPORT

FIGURE 1-18 Manpower survey report.

Iran, Belgium, and Mexico. The U.S. and Japan produce about 90% of the world's *hvac* (heating, ventilating, and air-conditioning) output, with Europe claiming the rest.

One major air-conditioning manufacturer estimates the total 1974 foreign consumption to be approximately $800 million, which includes the 100,000 units mentioned above. Factoring these out leaves a foreign manufactured supply of around 550 million dollars.

By 1978 the same information source puts the total value at $1.6 billion and by 1983 at close to 3 billion dollars. Obviously, this is an optimistic forecast, but it is reasonable in terms of the manufacturing facilities being erected in foreign lands.

AIR-CONDITIONING USAGE BY TYPE OF MARKET

In 1973 U.S. manufacturers estimated 76% of unitary production went into residential consumption. The balance of 24% was spread across broad commercial uses. In 1974 these applications were estimated to be closer to a 70%–30% mix. Of the 70% residential use, approximately one-half went into new construction and the balance into existing home modernization.

Commercial applications were broken down into seven major groups:

Schools	10.0%
Hotels	8.5%
Stores, supermarkets, shopping centers	14.0%
Hospitals	14.0%
Industrial plants	14.5%
Multiple dwellings	17.0%
Office buildings	22.0%

Of these commercial applications, some 57% were served by unitary products and 43% by central station systems.

INDUSTRY MANPOWER NEEDS

In March of 1973, ARI published a *Manpower Survey Report* (Fig. 1-18), which reflected projections of industry manpower demands through 1980. A survey was sent to air-conditioning contractors, mechanical contractors, service organizations, and commercial refrigeration companies, who were asked to estimate their requirements for *new* positions, not normal replacement. From this survey it was estimated that there would be over 50,000 new jobs to be filled. Of these, two-thirds were for technicians/mechanics in the fields of heating, air-conditioning, refrigeration, and sheet metal work.

FIGURE 1-19 Laboratory technician.

To put it another way, for every $1 million in installed value of equipment, the following employees would be needed:

1 graduate engineer
2 technicians/master mechanics
11 air-conditioning, heating, and refrigeration mechanics
2 air-conditioning, heating, and refrigeration helpers
7 sheet-metal mechanics
1 sheet-metal helper
2 salesmen

OCCUPATIONAL OPPORTUNITIES

Manufacturers hire a number of trained technicians for various jobs. Laboratory technicians actually build prototypes of new products, or they conduct tests on products for performance ratings, life-cycle testing, or compliance to ARI certification or UL safety standards or sound evaluation (Fig. 1-19). Factory service personnel prepare installation instructions, training material, parts information, etc. Some manufacturers employ a large number of service technicians for field assignment. They are active in installation and startup situations or in rendering actual customer service.

The contractor trades have the largest demand for personnel. To explain this, we will divide the contractor industry into two groups. Group I are *h*eating, *v*entilating, and *a*ir-conditioning (h.v.a.c.) dealer/contractors who deal primarily with the residential and light commercial market. These businesses need personnel for installation and service (Fig. 1-20), as well as technically competent sales personnel who can estimate jobs, make duct layouts, and if so inclined, do the actual selling. Much of this type of contractor work is so-called self-designed, because there are no detailed plans and specifications; the contractor provides that service to his customer.

The other category is the mechanical contractor who works in the planning and specification market, which usually involves larger jobs that have been engineered by a consultant. He too must have qualified personnel to plan the work but, more important, he must have skilled installers who possess a broad knowledge of refrigeration, steam fitting, general plumbing, sheet metal, and control systems.

In the distribution function are h.v.a.c. and refrigeration wholesalers, distributors, supply houses, and factory branches where technicians are needed to help service the purchasing dealers. They act as countermen, service specialists, or salesmen, but must be well-grounded in the technical arts.

Finally, there is a wide range of other opportunities, including operating and maintenance personnel for large institutions like universities, hospital complexes, office buildings, government buildings, military facilities, and industrial plants, all of which do their own in-house service and repair. This list is extensive and an excellent source of job opportunities.

Training and experience are the key ingredients to being a successful technician, and it starts with knowledge. Many programs are offered by vocational

FIGURE 1-20 HVAC service personnel.

technical schools and community colleges for both the younger and the adult student. Manufacturers and trade associations are a continuing source of development training.

TRADE ASSOCIATIONS

With the rapid growth and variety of interests, it was only natural that trade associations would evolve to represent specific groups. The list includes manufacturers, wholesalers, contractors, sheet-metal dealers, and service organizations. Each is important and makes a valuable contribution to the field. Space does not permit a detailed examination of all their activities, but throughout this book many of these associations will be acknowledged as specific subjects are covered. There is one organization, however, that is responsible for this book and related course outlines, and it is proper at this time to discuss that particular group.

The ARI (Air-Conditioning and Refrigeration Institute), with offices in Arlington, Virginia (D. C. complex), was formed in 1953 and is the national trade association of manufacturers of residential, commercial, and industrial air-conditioning and refrigeration equipment, as well as of machinery, parts, accessories, and allied products for use with such equipment. There are currently some 180 member companies.

The objectives are briefly:

(a) public relations
(b) collection and dissemination of information
(c) establishment of equipment and product standards and safety codes.
(d) development of markets for industry products.
(e) relations with government agencies, other trade associations, and interested groups.

Item (b) has already been demonstrated in the section on the historical market growth. Most of the statistical information came from ARI data.

In item (c), the product sections break down into logical equipment categories, and member company representatives staff these sections, subsections, and committees (Fig. 1-21). Important decisions such as

FIGURE 1-21 Company representatives at product section meeting.

FIGURE 1-22 ARI certification seals.

FIGURE 1-23 International air-conditioning, heating, and refrigeration exposition.

equipment standards and certification programs (Fig. 1-22) not only vitally affect equipment design and field application but give assurance to customers, engineers, contractors, and specifiers.

Under item *c* is some of the most important work done by ARI to examine and report on pending legislation (national and local) that may not be beneficial to the industry or the consumer.

ARI, in cosponsorship with ASHRAE, holds an annual International Air-Conditioning-Heating-Refrigerating Exposition (Fig. 1-23), which may draw 15,000 to 20,000 people in the field, depending on key city location. Product exhibits and technical and business seminars highlight the event.

The Manpower Development Committee is the section that works closely with the U.S. Office of Education and the American Vocational Association in formulating training recommendations and materials directed at high schools, vocational schools, and junior colleges.

PROBLEMS

1. Name some ways early man tried to control his environment.
2. Who is considered to be the father of modern air conditioning?
3. The Du Pont Company developed the first "safe" refrigerant, called ______________.
4. Hermetic compressors for air-conditioning were first introduced in about the year ______________.
5. A broad distinction in air-conditioning products is ______________ and ______________.
6. ARI stands for ______________.
7. ARI is made up of ______________.
8. The total worldwide air-conditioning market in 1975 was approximately $______________.
9. Approximately what percent of the total market goes into residential installations?
10. For every $1 million in installed equipment, how many air-conditioning and heating mechanics are needed?

2

AIR-CONDITIONING BENEFITS

WHAT IS AIR-CONDITIONING?

That depends on what point of view is being considered. Ask the man on the street and he would most probably answer "keeping cool." Ask the owner of a printing plant and he would respond with a statement that might mean closely controlling temperature and humidity so that the behavior of the paper could be held within certain tolerances. One answer is from the standpoint of human comfort, the other is about a commercial consideration.

A dictionary definition might read "the process that heats, cools, cleans, and circulates air, and controls its moisture content on a continuous basis." So let's examine each definition as it affects the environment.

HUMAN COMFORT CONTROL

The human body (Fig. 2-1) is a heat-generating device. Its normal temperature is 98.6° Fahrenheit. It can regulate or control this condition by four methods; convection, radiation, conduction, and evaporation. When in an environment where room conditions are too warm (but less than 98.6°), it will transfer heat to the air passing over the skin by convection. Simultaneously, it gives up heat by conduction to clothing, bedding, or whatever is in contact with the skin. Additionally, it throws off heat by means of radiation to the cooler surrounding objects. If these three aren't sufficient, sweat glands

FIGURE 2-1.

FIGURE 2-2.

FIGURE 2-3.

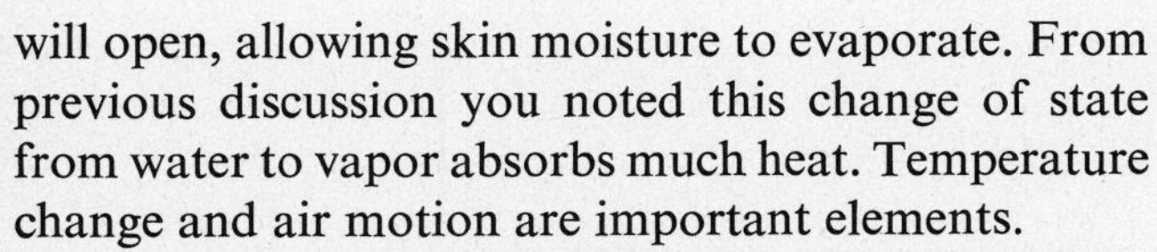

will open, allowing skin moisture to evaporate. From previous discussion you noted this change of state from water to vapor absorbs much heat. Temperature change and air motion are important elements.

In colder surroundings, radiation, conduction, and convection take place more rapidly, thus requiring clothing to insulate and hold body heat. Evaporation becomes minimal as the amount of skin perspiration decreases.

The body is also sensitive to impurities (Fig. 2-2). Dust, smoke, plant pollen, etc., cause irritation to the nose, lungs, and eyes, so this indicates another need for clean air.

Finally, the body requires "fresh air" (Fig. 2-3) to renew its oxygen supply or to dilute undesirable odors.

Stated simply, the body should have a comfortable and healthful atmosphere; and five properties of air must be treated:

(1) temperature (cooling or heating)

(2) moisture content (humidifying or dehumidifying)

(3) movement of the air (circulation)

(4) cleanliness of the air (filtering)

(5) ventilation (introduction of outside air)

The temperature of the air is indicated by the feeling of being hot or cold, and it can be measured by means of an ordinary household thermometer (Fig. 2-4), commonly referred to as a *dry-bulb thermometer*.

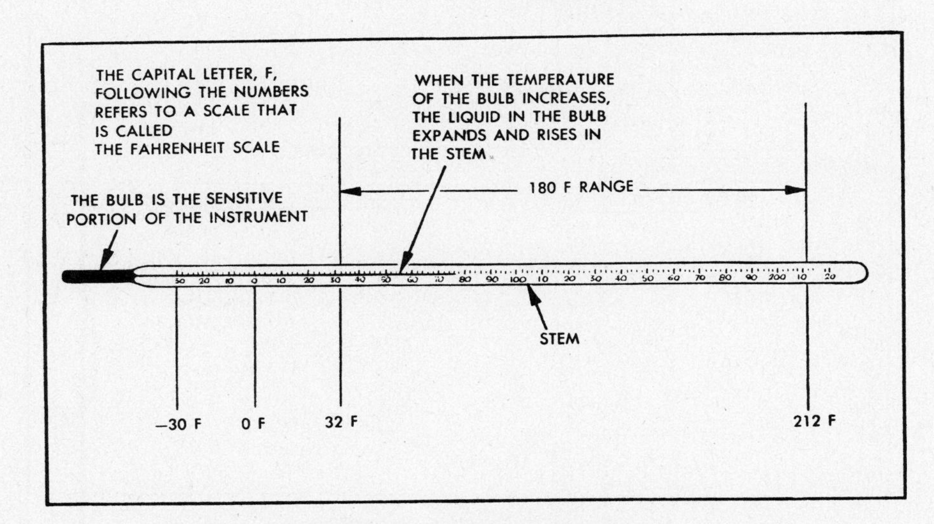

FIGURE 2-4 Ordinary thermometer.

The moisture content of the air is indicated by a feeling of dryness in winter or that muggy, sticky feeling in summer. Sometimes you can be uncomfortable from humidity in spite of the temperature. Humidity refers to the moisture that has evaporated into the air and exists as an invisible gas. To measure this feeling of humidity and express it in specific terms, a *wet-bulb thermometer* (Fig. 2-5) is used. Actually it's nothing more than an ordinary thermometer with a wick or sock placed over the bulb. By wetting the sock and passing air over it, moisture will be evaporated until it balances with the moisture content of the air and no more evaporation takes place. Heat absorbed during the evaporation process lowers the temperature of the bulb; the lower reading is the *wet-bulb temperature*.

For the convenience of the technician, a *sling psychrometer* (Fig. 2-6) is used to take dry-and wet-bulb temperatures. It's simply two matching thermometers (scales and calibration the same) mounted on a common frame attached to a handle that will permit the holder to rotate, whirl, or sling the instrument around. It is rotated about two to three times per second until repeated readings become constant—perhaps a total of a minute.

By observing both the dry-bulb and wet-bulb thermometer readings, we have an indication of the *relative humidity*, meaning the actual amount of moisture in the air compared to the maximum

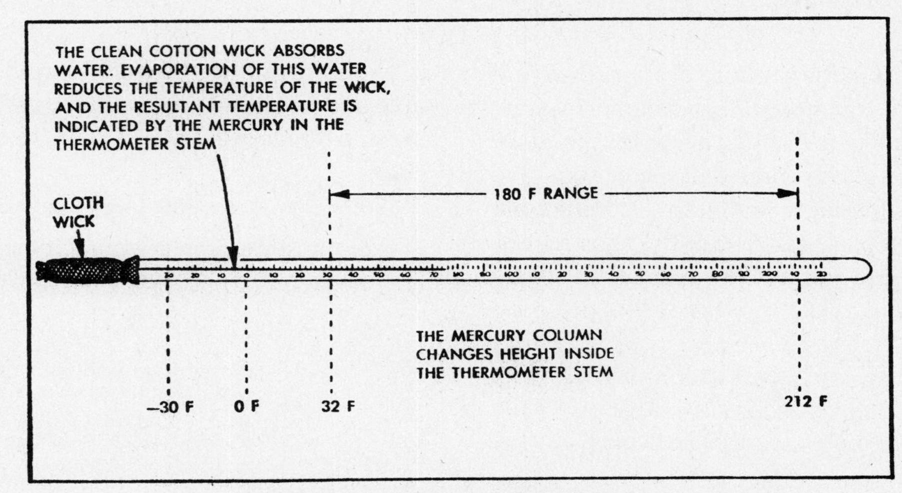

FIGURE 2-5 Wet-bulb thermometer.

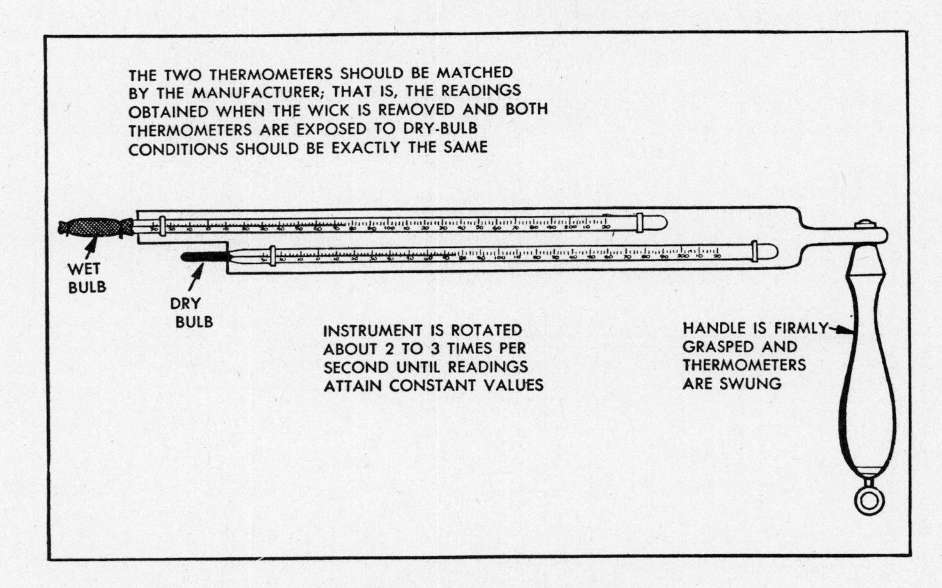

FIGURE 2-6 Sling psychrometer.

TABLE 2A Psychrometric table: percent relative humidity from dry-bulb temperature and wet-bulb depression

DB temp.	*WB Depression*																													
1	1	2	3	4	5	6	7	8	9	10	11	12	13	14	15	16	17	18	19	20	21	22	23	24	25	26	27	28	29	30
32	90	79	69	60	50	41	31	22	13	4																				
36	91	82	73	65	56	48	39	31	23	14	6																			
40	92	84	76	68	61	53	46	38	31	23	16	9	2																	
44	93	85	78	71	64	57	51	44	37	31	24	18	12	5																
48	93	87	80	73	67	60	54	48	42	36	34	25	19	14	8															
52	94	88	81	75	69	63	58	52	46	41	36	30	25	20	15	10	6	0												
56	94	88	82	77	71	66	61	55	50	45	40	35	34	26	24	17	12	8	4											
60	94	89	84	78	73	68	63	58	53	49	44	40	35	31	27	22	18	14	6	2										
64	95	90	85	79	75	70	66	61	56	52	48	43	39	35	34	27	23	20	16	12	9									
68	95	90	85	81	76	72	67	63	59	55	51	47	43	39	35	31	28	24	21	17	14									
72	95	91	86	82	78	73	69	65	61	57	53	49	46	42	39	35	32	28	25	22	19									
76	96	91	87	83	78	74	70	67	63	59	55	52	48	45	42	38	35	32	29	26	23									
80	96	91	87	83	79	76	72	68	64	61	57	54	54	47	44	41	38	35	32	29	27	24	21	18	16	13	11	8	6	1
84	96	92	88	84	80	77	73	70	66	63	59	56	53	50	47	44	41	38	35	32	30	27	25	22	20	17	15	12	10	8
88	96	92	88	85	81	78	74	71	57	64	61	58	55	52	49	46	43	41	38	35	33	30	28	25	23	21	18	16	14	12
92	96	92	89	85	82	78	75	72	69	65	62	59	57	54	51	48	45	43	40	38	35	33	30	28	26	24	22	19	17	15
96	96	93	89	86	82	79	76	73	70	67	74	61	58	55	53	50	47	45	42	40	37	35	33	31	29	26	24	22	20	18
100	96	93	90	86	83	80	77	74	71	68	65	62	59	57	54	52	49	47	44	42	40	37	35	33	31	29	27	25	23	21
104	97	93	90	87	84	80	77	74	72	69	66	63	61	58	56	53	51	48	46	44	41	39	37	35	33	31	29	27	25	24
108	97	93	90	87	84	81	78	75	72	70	67	64	62	59	57	54	52	50	47	45	43	41	39	37	35	33	31	29	28	26

amount that air could hold at that dry-bulb temperature. If the dry-bulb and wet-bulb readings were the same, the relative humidity would be 100%. The difference between dry-bulb and wet-bulb readings is called the *wet-bulb depression*.

Table 2A is a simple table that reflects the relative humidity in typical home conditions. For example, if the dry-bulb temperature were measured at 72°F and the wet-bulb at 61°F (a depression of 11°F), the relative humidity would be 53%.

Proper relative humidity is necessary so the air will be dry enough in summer to absorb body perspiration for comfort. In winter the air should not be so dry that skin, nose, and throat have that dryness sensation. Also, too much humidity can cause mold, mildew, and rust.

COMFORT ZONE

It's logical to ask what the desired temperature-humidity relationship is. The answer is that there is no one specific condition. People react differently to different conditions. ASHRAE (the American Society of Heating, Refrigerating, and Air-Conditioning Engineers) conducted a research study over many years, checking the reactions of large numbers of people to establish a range of combined temperatures, humidities, and air movement that provided the most comfort. This is known as the *comfort zone*. Each combination is known as an *effective temperature* (*ET*). It was found, for example, that with a given air velocity, a number of different combinations of dry-bulb temperatures and relative humidity readings would give the same feeling of comfort to over 90% of the people involved. Thus, a comfort zone (Fig. 2-7) could be constructed. From the shaded zone of effective temperatures, it can be determined what dry-bulb temperature and relative humidity will produce that result. Note one obvious fact: the higher the humidity the lower the dry-bulb temperature can be.

The comfort zone chart is a good selling point with average people, since it explains how temperature and humidity should be controlled and thus shows the need for year-round air-conditioning. The chart is representative of the conditions found in homes, theaters, offices, etc., where periods of long occupancy occur. However, it is not completely accurate for conditions in retail stores, banks, drug stores, and similar situations, where short duration of occupancy coupled with rapid temperature changes and air motion will indeed change the experienced effective temperature. Therefore, when designing systems you must consult the specific manufacturer, trade association, or local gas and electric utilities for recommendations. Until recently, it was

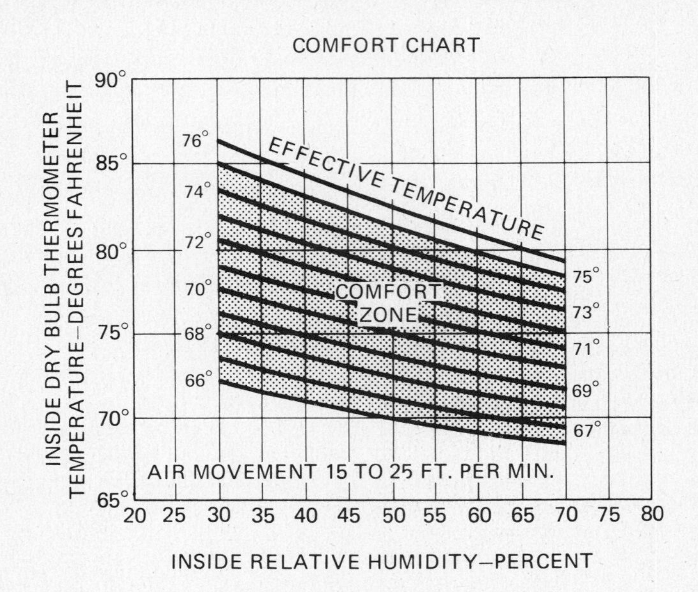

FIGURE 2-7 Comfort zone.

general practice to design for the following indoor conditions; for winter, 72° to 75°F dry-bulb temperature with relative humidity at 35 to 40%; for summer, 75° to 78°F dry-bulb temperature and 50 to 55% relative humidity.

With the nation's attention on energy conservation and the ever-increasing cost of fuels, the above design conditions are subject to change. In 1974 the federal government recommended lowering the indoor winter dry-bulb temperature to 68°F and increasing the summer design temperature to 80°F for government buildings, and it encouraged private industry and homeowners to adopt similar practices.

AIR MOTION

Air motion (Fig. 2-8) is another factor in comfort considerations. The comfort zone as presented above was based on an air movement of 15 to 25 ft/min (feet per minute) velocity. The effective temperature drops sharply as velocity is increased. This would seem to be desirable for summer air-conditioning, but conditioned air being introduced into a room is usually 15° to 20° below room conditions; and if velocity should approach 100 ft/min, cold drafts would be noticeable.

Forced warm-air heating systems are even more subject to drafts, particularly when the blower first comes on. The skin seems to react more quickly to warm air currents, and a good rule of thumb is not to exceed 50 ft/min velocity in the comfort zone. Recent industry discussion would suggest limiting *year-round* velocity to 70 ft/min. Ways of planning for good air distribution will be covered more completely in later chapters.

Too little air circulation should also be avoided, because people tend to feel "closed in." This may be a disadvantage of the nonducted type of heating systems that depend on gravity circulation and have no mechanical means of filtration. The situation becomes even more critical where the structure is well-insulated and infiltration of outdoor air very small.

Cleanliness and ventilation are the last two necessities for proper air treatment and are to some extent related and work against each other. We breathe 36 pounds of air each day, compared to eating 3.8 pounds of food and drinking 4.3 pounds of water. Clean air is important to both our health and comfort. Ordinary air (Fig. 2-9) is contaminated with impurities such as dust, pollen, smoke, fumes, and chemicals. All of these should be filtered out of both the indoor air and the outdoor air that enters the structure.

The efficiency of filtration depends on the type of system, and we will review these options later. Some have the capability of removing more than 95% of all impurities. However, even with the finest filtering equipment a degree of ventilation air (Fig. 2-10) is needed to eliminate that stale, dead-air feeling, and also to dilute odors and supply oxygen for breathing and make-up air for vented appliances (exhaust fans, dryers, furnaces, water heaters, etc.). The amount of outside air needed depends on the conditioned space. Table 2B lists some typical applications and the ASHRAE recommendations,

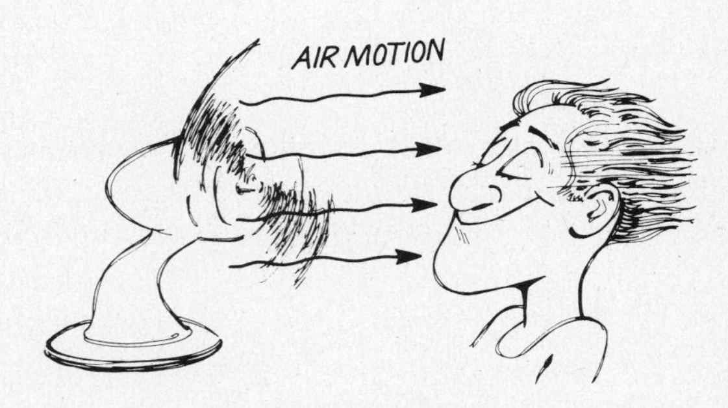

FIGURE 2-8 Air motion.

FIGURE 2-9 Cleanliness of air.

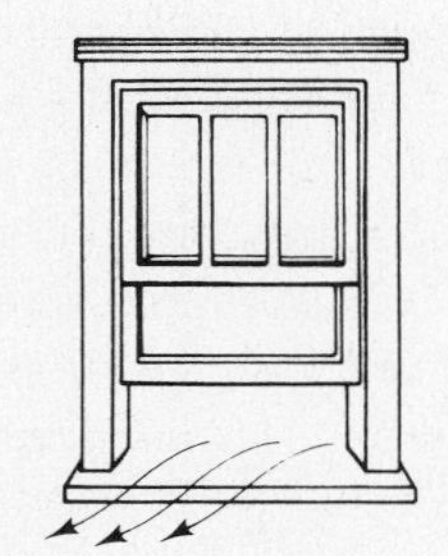

FIGURE 2-10 Ventilation.

based on cubic feet of air per minute (C.F.M) per person. Outdoor air should always be provided and properly filtered before entering the conditioned space. Many local codes and ordinances specify minimum standards for schools and public buildings.

TABLE 2B ASHRAE Ventilation table

		CFM PER PERSON	
Application	*Smoking*	*Optimum*	*Minimum*
Bank	Some	10	7½
Cocktail bar	Heavy	40	25
Offices	Some	15	10
General	None	25	15
Private	Considerable	30	25
Residence	Occasional	20	10
Restaurant	Considerable	15	12
Retail store	Very little	10	7½
Theater	Some	15	10

This review of the characteristics of air has not considered any economic impact. As with all human needs there must be compromise, depending on the individual's needs and his financial capability.

A window air conditioner basically performs the five functions, but cannot be compared to a true total comfort, year-round central air-conditioning system in function or form. Thus the industry offers a wide variety of products to suit many potential markets. In addition to the pure pleasure comforts, air-conditioning offers other important advantages.

MEDICAL CONSIDERATIONS

Air-conditioning may contribute to better health as a result of controlling air temperature, humidity, cleanliness, ventilation, and motion. For example, it may help eliminate heat rash, particularly among infants. Some physicians believe that air-conditioning can provide an environment that is easier for people suffering from ailments such as heart trouble to accept. Just moving around in extreme heat may put an undue strain on vital organs, the same as the strain from heavy manual labor, like shoveling snow in cold weather. As a result, doctors sometimes recommend air conditioning for their patients.

An effective air filtering system may relieve the sufferings of asthmatic and allergic patients. Such a system might include an electronic air cleaner as part of the home central system. Ordinary house dust, in addition to dust in the outdoor air, can contribute to allergy problems. House dust is a complex mixture that develops from the breakdown and wearing out of material in the house such as cotton, linens, wool, furniture stuffings, and carpet. Pollen allergy irritants, however, come from the outdoors. In the height of the hayfever season the pollen count in a home with central air-conditioning may be reduced considerably if an effective air cleaning system is used.

The U. S. Public Health Service has called air pollution a contributing cause of cancer and a serious irritant to lung and respiratory tissue. In 1952, a heavy chemical-laden smog in London was reported to be the immediate cause of death for four thousand persons. The problem is also staggering throughout the U.S., and thousands of specialists in industry and government are engaged in research looking for solutions. In the meantime, the homeowner and employer can protect himself and his personnel with air-cleaning apparatus.

Although it has not been proven conclusively, there is some belief that proper humidity may also help protect health. Adequate humidity in the air can help the membranes in the nose and respiratory tract to remain moist. This can alleviate the effect of bacteria and viruses. Lack of humidity also promotes dryness of those household dust sources mentioned above and thereby keeps the dust airborne.

HUMIDIFICATION PROTECTS PROPERTY

An atmosphere that is too dry can adversely affect furniture, clothing, shoes, books, documents, plaster, musical instruments—virtually everything in the home, including structural members. Let's look at a few examples: Lack of adequate humidity causes glue to dry out in tables and chairs and other furniture, and joints separate and crack. Door panels shrink and expose unpainted surfaces. Plaster or dry-wall joints dry out, which may lead to unsightly cracks. Musical instruments may lose tone. Hardwood floors may shrink and separate. Rug fibers become brittle and break.

Still another phenomenon experienced with winter dryness is the presence of, or generation of, static electricity. Most of us have touched a door knob and jumped from a jolt of static electrical shock. Although not harmful, it is disconcerting to say the least. A properly sized humidifier installed in a forced-air duct system can minimize or eliminate all these conditions.

The security benefits are more or less a recent consideration but, with the ever-increasing crime rate, many homeowners are installing mechanical and electrical security protection devices that require all doors and windows to be closed and locked. Without central air-conditioning, these devices would not be practical.

The social benefits of having central air conditioning are perhaps the least definable, but there are a number of homeowners who do a considerable amount of entertaining at home because of their social position and also for business reasons. Such concentrations of many people, along with smoking at cocktail and dinner parties, demand full treatment for all five properties of air.

OTHER BENEFITS

Schools

Schools have been a prime market for year-round air conditioning, not only for the personal comfort it affords, but studies have demonstrated the learning process is definitely improved for the student and teacher. Interestingly, at the time this text was being prepared, two articles appeared in the *Air-Conditioning, Heating and Refrigeration News* trade paper which said in part, "Clearwater, Florida—more than 1000 teachers belonging to the Pinellas (County) Classroom Teachers' Association here have ranked air conditioning as its top priority." From Fort Worth, Texas—". . . air conditioning for all schools here is the chief interest for a bond issue by the Certified Teachers' Association (CTA), which believes it is a necessity for a good learning environment." As population increases, the need for year-round school operation becomes apparent.

Commerce and Industry

Commerce and industry have used air-conditioning in several ways: first to increase personal productivity and second to provide space process-cooling for specific needs.

Worker productivity in air-conditioned areas has improved, in terms of less absenteeism, less labor turnover, less noise distraction, less trips to the water fountain, more efficient production, fewer mistakes, and less time lost due to heat fatigue and accidents. In general, better morale and better relationships between employer and employee result. The degree of benefit, of course, is subject to the type of surroundings—office worker as compared to garment seamstress, as compared to assembly line operator. No one would think of constructing a modern office building today without air-conditioning. Trade unions have also been instrumental in the growth of comfort air-conditioning for their members, most notably in the textile industry.

The uses of process cooling would make a list almost as long as this book; however, just to name a few: electronic computer rooms must be controlled closely as to temperature and humidity. Tapes and card decks need uniform conditions. Current computer designs require that large amounts of heat be removed. Printing plants must have a controlled atmosphere to maintain tolerances of paper shrinkage, accurate register in color printing, and efficient paper feed through the presses. Clean rooms, for all types of tools and gauges and highly critical precision manufacturing, must maintain close tolerances of conditions to protect dimensional accuracy, fits, allowances, gauging, and manufacturing limits. Temperature, humidity and cleanliness are very important. A telephone exchange building is something most of us accept as a part of telephone services. The myriad number of electrical contacts that are made and broken when phone numbers are dialed is fantastic. The slightest speck of dust or wide variations in temperature and humidity can cause an erratic or even a broken contact. The drug industry is one of the largest users of industrial air conditioning; the list goes on and on.

MARKETING TOOL FOR BUILDER/DEVELOPERS

In the early 1950s, residential builders were reluctant to install central air conditioning as standard equipment in new homes, and for good reason. The Federal Housing Administration and many local savings and loan companies were not favorably inclined toward increasing selling prices, which might result in mortgage payments that would possibly overextend the lower-income home buyers' ability to meet monthly payments. Air-conditioning was given a lower priority compared to certain other appliances in the new home. As a result, builders installed the heating system and offered cooling as a later option. Fortunately, there were builders like Levitt and Sons of Levittown, Pa., Ryan Homes of Pittsburgh, and Fox and Jacobs of Dallas who built large tract housing developments with central air-conditioning as standard equipment. The systems were pre-engineered by personnel from the manufacturer and builder and installed by trained technicians. Most important, mass purchasing brought installation costs to a minimum. Thus, a trend was started. Other builders and the broad financial community jumped on the bandwagon, and soon all were making statements like "any house that is built without central air-conditioning is already obsolete."

The explosion of apartment and condominimum townhouse development in today's housing market could not have reached the heights it has without central air-conditioning. Common walls and limited access to the outdoors makes indoor living a way of life and air-conditioning an absolute necessity, as well as a good sales tool.

This chapter has given an overview of the properties of air-conditioning from the physiological, sociological, and commercial standpoints. Now let's consider the mechanical aspects of air and its behavior: the art of psychrometrics.

PROBLEMS

1. What are the five control functions of air conditioning?
2. What is normal body temperature?
3. Moisture in the air is called ______________.
4. What is the comfort zone?
5. What is the recommended indoor heating temperature?
6. What are the recommended cooling temperature and humidity?
7. Air motion is measured in ______________.
8. A person breathes ______________ lb of air each day.
9. Common air pollution is made up of ______________.
10. Electronic air cleaners are said to be able to remove an excess of ______________% of all impurities.

3

PSYCHROMETRICS

Who uses psychrometrics? Consulting engineers, research and product development engineers, and application engineers, for example, need an in-depth understanding and working knowledge of the art, for their professional existence requires practical application of the theories. Does a technician really need to know all of the information? Generally, no, unless he is involved in one of the engineering or research activities mentioned above. What most technicians do need to know are only the basic principles, definitions of terms, the existence of psychrometric charts, and the relationship of the elements involved so as to be able to report information or make assumptions about equipment performance.

For example, unitary packaged products, particularly the residential kind, are predesigned by factory engineers with specific performance characteristics. There is little the technician can do to change their operation except possibly to vary the fan speed. Performance tables are available that rate equipment based on entering dry-bulb and wet-bulb conditions, outdoor ambient conditions, sensible-to-latent-heat ratios, total output, and so on. What the service person needs to know are the terms and definitions being used and how to understand the information being presented. The following discussion is designed to provide these basic skills.

BASIC TERMS

Atmosphere

The air (Fig. 3-1) around us is composed of a mixture of dry gases and water vapor. The gases contain approximately 77% nitrogen and 23% oxygen, with the other gases totaling less than 1%. Water vapor (Fig. 3-2) exists in very small quantity so it's either measured in *grains* or *pounds*. (It takes 7,000 grains to make one pound.)

Dry-Bulb Temperature (see Chap. 2)

The temperature as measured by an ordinary thermometer.

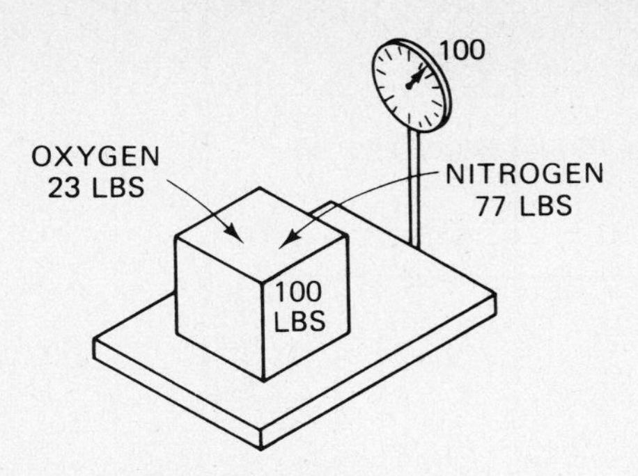

FIGURE 3-1 Atmosphere.

Wet-Bulb Temperature (see Chap. 2)

The temperature resulting from the evaporation of water from a wetted sock on a standard thermometer.

Dew-Point Temperature

The saturation temperature at which the condensation of water vapor to visible water takes place. An example is the sweating on a glass of ice water. The cold glass reduces the air temperature below its dew-point, and the moisture that condenses forms beads on the glass surface.

Specific Humidity

The actual weight of water vapor in the air expressed in grains or pounds of water per pound of dry air, depending on which data is used.

Relative Humidity

The ratio of actual water vapor in the air compared to the maximum amount that could be present at the same temperature, expressed as a percent (%).

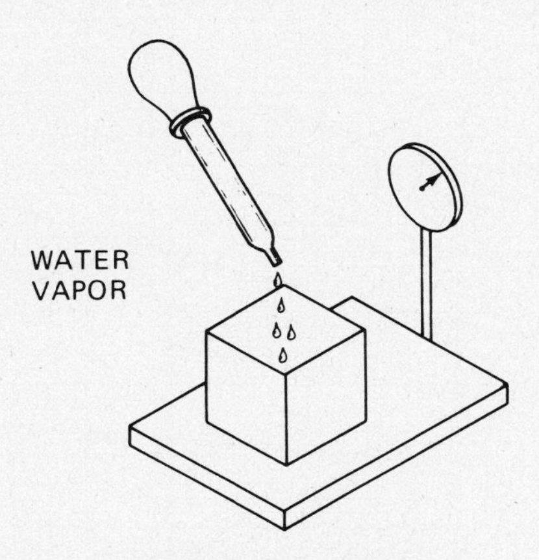

FIGURE 3-2 Water vapor.

Specific Volume

The number of cubic feet (ft^3) occupied by one pound of the mixture of air and water vapor.

Sensible Heat

The amount of dry heat expressed in Btu per pound of air; it is reflected by the dry-bulb temperature.

Latent Heat

The heat required to evaporate the moisture a specific amount of air contains. This evaporation occurs at the wet-bulb temperature. It, too, is expressed in Btu per pound of air.

Total Heat

The total heat content of the air and water vapor mixture is also known as *enthalpy*. It is the sum of both sensible and latent heat values, expressed in Btu per pound of air.

The psychrometric chart in Fig. 3-3 is probably the best way of showing what happens to air and water vapor as these properties are changed. The chart is published by ASHRAE and is the one most commonly used in the industry. Some manufacturers have developed their own charts which vary only in style and construction, but the relationship of the air properties are all the same.

To make this chart all we do is start with the ordinary temperature scale called the *dry-bulb temperature*. Just extend the thermometer scale as shown in Fig. 3-4. Note on the actual chart that these lines are not truly perpendicular. This is done so that other lines will come out straight instead of curved.

Next, the vertical scale is set up according to the amount of water vapor mixed with each pound of dry air. This scale (Fig. 3-5), called the *humidity ratio*, is expressed in pounds of moisture per pound of dry air. We know that air can hold different amounts of moisture depending on its temperature; if it is holding all the moisture it can (100%), it is termed *saturated*.

ASHRAE PSYCHROMETRIC CHART NO. 1

NORMAL TEMPERATURE

BAROMETRIC PRESSURE 29.921 INCHES OF MERCURY

COPYRIGHT 1963

AMERICAN SOCIETY OF HEATING, REFRIGERATING AND AIR-CONDITIONING ENGINEERS, INC.

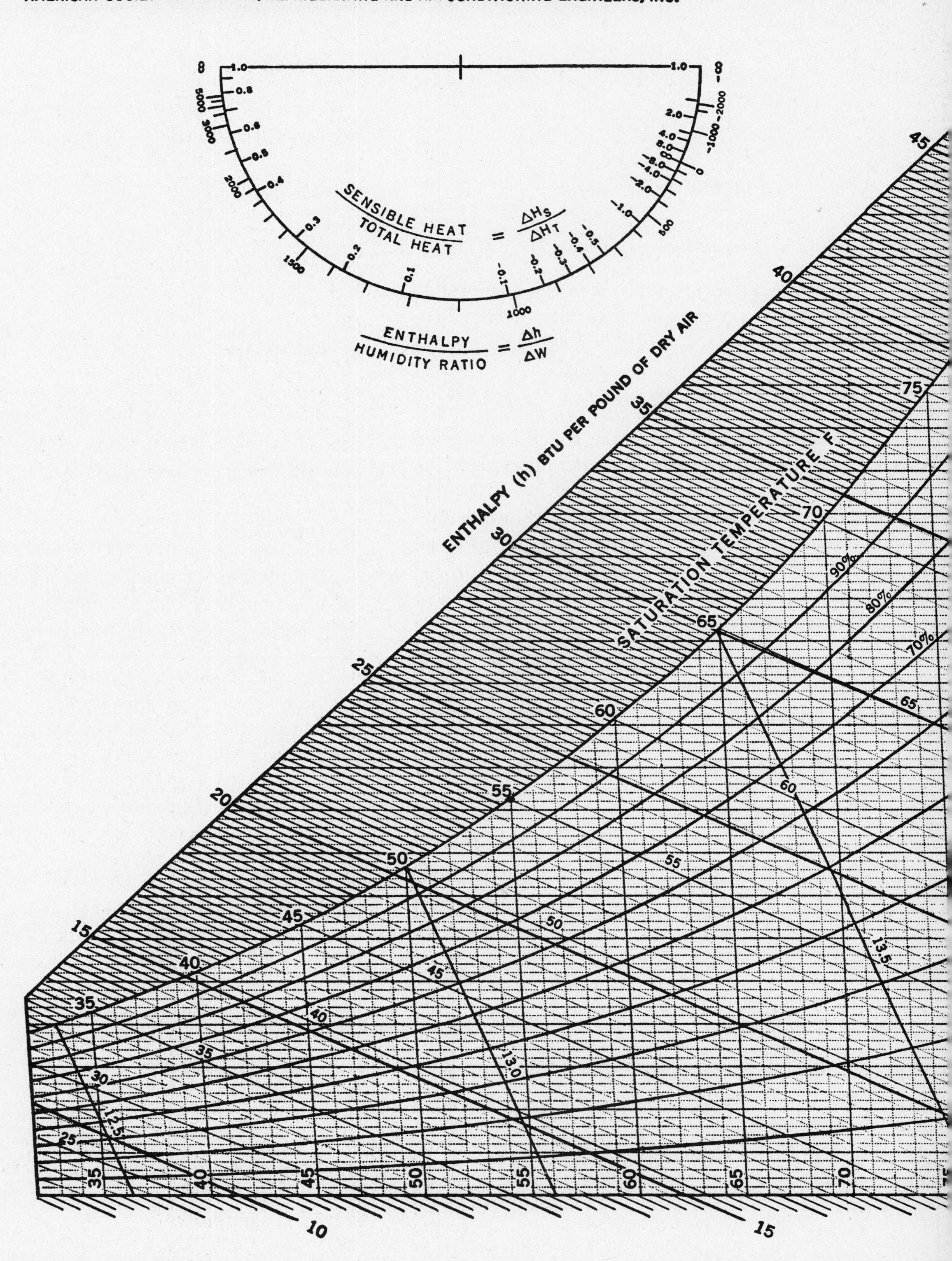

FIGURE 3-3(a) Psychrometric Chart No. 1.

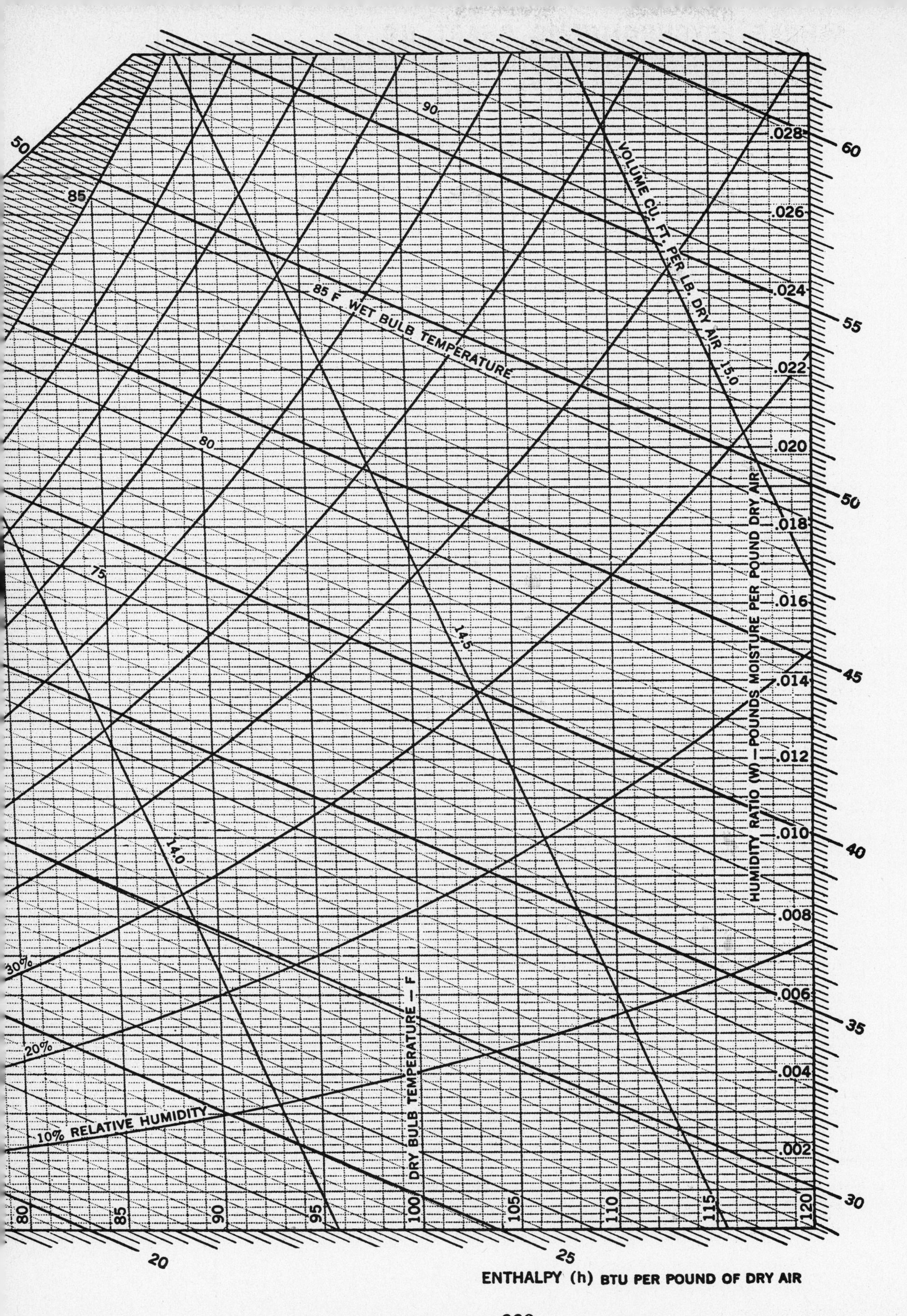

85 F WET BULB TEMPERATURE
VOLUME CU. FT. PER LB. DRY AIR
15.0
14.5
14.0
90
85
80
75
50
30%
20%
10% RELATIVE HUMIDITY
DRY BULB TEMPERATURE — F
80
85
90
95
100
105
110
115
120
HUMIDITY RATIO (W) — POUNDS MOISTURE PER POUND DRY AIR
.028
.026
.024
.022
.020
.018
.016
.014
.012
.010
.008
.006
.004
.002
60
55
50
45
40
35
30
20
25
ENTHALPY (h) BTU PER POUND OF DRY AIR

ENTHALPY* IN BTU PER POUND OF DRY AIR

Wet Bulb Temperature F	TENTHS OF A DEGREE .0	.1	.2	.3	.4	.5	.6	.7	.8	.9
35	13.01	13.05	13.10	13.14	13.18	13.23	13.27	13.31	13.35	13.40
36	13.44	13.48	13.53	13.57	13.61	13.66	13.70	13.75	13.79	13.83
37	13.87	13.91	13.96	14.00	14.05	14.09	14.14	14.18	14.23	14.27
38	14.32	14.37	14.41	14.46	14.50	14.55	14.59	14.64	14.68	14.73
39	14.77	14.82	14.86	14.91	14.95	15.00	15.05	15.09	15.14	15.18
40	15.23	15.28	15.32	15.37	15.42	15.46	15.51	15.56	15.60	15.65
41	15.70	15.75	15.80	15.84	15.89	15.94	15.99	16.03	16.08	16.13
42	16.17	16.22	16.27	16.32	16.36	16.41	16.46	16.51	16.56	16.61
43	16.66	16.71	16.76	16.81	16.86	16.91	16.96	17.00	17.05	17.10
44	17.15	17.20	17.25	17.30	17.35	17.40	17.45	17.50	17.55	17.60
45	17.65	17.70	17.75	17.80	17.85	17.91	17.96	18.01	18.06	18.11
46	18.16	18.21	18.26	18.32	18.37	18.42	18.47	18.52	18.58	18.63
47	18.68	18.73	18.79	18.84	18.89	18.95	19.00	19.05	19.10	19.16
48	19.21	19.26	19.32	19.37	19.43	19.48	19.53	19.59	19.64	19.70
49	19.75	19.81	19.86	19.92	19.97	20.03	20.08	20.14	20.19	20.25
50	20.30	20.36	20.41	20.47	20.52	20.58	20.64	20.69	20.75	20.80
51	20.86	20.92	20.97	21.03	21.09	21.15	21.20	21.26	21.32	21.38
52	21.44	21.50	21.56	21.62	21.67	21.73	21.79	21.85	21.91	21.97
53	22.02	22.08	22.14	22.20	22.26	22.32	22.38	22.44	22.50	22.56
54	22.62	22.68	22.74	22.80	22.86	22.92	22.98	23.04	23.10	23.16
55	23.22	23.28	23.34	23.41	23.47	23.53	23.59	23.65	23.72	23.78
56	23.84	23.90	23.97	24.03	24.10	24.16	24.22	24.29	24.35	24.42
57	24.48	24.54	24.61	24.67	24.74	24.80	24.86	24.93	24.99	25.06
58	25.12	25.19	25.25	25.32	25.38	25.45	25.52	25.58	25.65	25.71
59	25.78	25.85	25.92	25.98	26.05	26.12	26.19	26.26	26.32	26.39
60	26.46	26.53	26.60	26.67	26.74	26.81	26.87	26.94	27.01	27.08
61	27.15	27.22	27.29	27.36	27.43	27.50	27.57	27.64	27.71	27.78
62	27.85	27.92	27.99	28.07	28.14	28.21	28.28	28.35	28.43	28.50
63	28.57	28.64	28.72	28.79	28.87	28.94	29.01	29.09	29.16	29.24
64	29.31	29.39	29.46	29.54	29.61	29.69	29.76	29.84	29.91	29.99
65	30.06	30.14	30.21	30.29	30.37	30.45	30.52	30.60	30.68	30.75
66	30.83	30.91	30.99	31.07	31.15	31.23	31.30	31.38	31.46	31.54
67	31.62	31.70	31.78	31.86	31.94	32.02	32.10	32.18	32.26	32.34
68	32.42	32.50	32.59	32.67	32.75	32.84	32.92	33.00	33.08	33.17
69	33.25	33.33	33.42	33.50	33.59	33.67	33.75	33.84	33.92	34.01
70	34.09	34.18	34.26	34.35	34.43	34.52	34.61	34.69	34.78	34.86
71	34.95	35.04	35.13	35.21	35.30	35.39	35.48	35.57	35.65	35.74
72	35.83	35.92	36.01	36.10	36.19	36.29	36.38	36.47	36.56	36.65
73	36.74	36.83	36.92	37.02	37.11	37.20	37.29	37.38	37.48	37.57
74	37.66	37.76	37.85	37.95	38.04	38.14	38.23	38.33	38.42	38.52
75	38.61	38.71	38.80	38.90	38.99	39.09	39.19	39.28	39.38	39.47
76	39.57	39.67	39.77	39.87	39.97	40.07	40.17	40.27	40.37	40.47
77	40.57	40.67	40.77	40.87	40.97	41.08	41.18	41.28	41.38	41.48
78	41.58	41.68	41.79	41.89	42.00	42.10	42.20	42.31	42.41	42.52
79	42.62	42.73	42.83	42.94	43.05	43.16	43.26	43.37	43.48	43.58
80	43.69	43.80	43.91	44.02	44.13	44.24	44.34	44.45	44.56	44.67
81	44.78	44.89	45.00	45.12	45.23	45.34	45.45	45.56	45.68	45.79
82	45.90	46.01	46.13	46.24	46.36	46.47	46.58	46.70	46.81	46.93
83	47.04	47.16	47.28	47.39	47.51	47.63	47.75	47.87	47.98	48.10
84	48.22	48.34	48.46	48.58	48.70	48.83	48.95	49.07	49.19	49.31
85	49.43	49.55	49.68	49.80	49.92	50.05	50.17	50.29	50.41	50.54

* Interpolated to tenths of degrees from 1963 edition ASHRAE Guide and Data Book. Published by the American Society of Heating, Refrigerating and Air Conditioning Engineers, Inc.

FIGURE 3-3(b) Enthalpy in Btu per pound of dry air (Courtesy, ASHRAE).

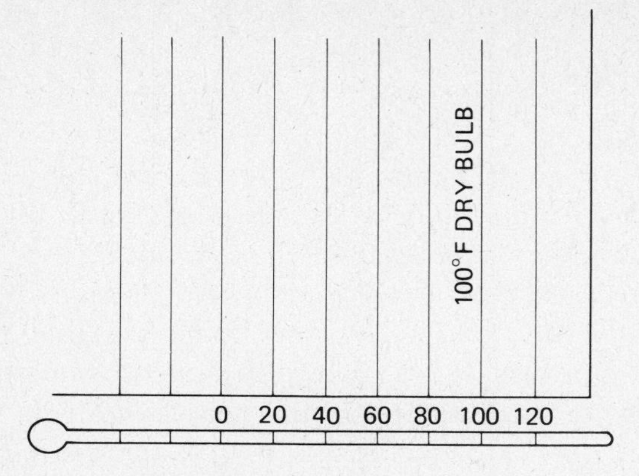

FIGURE 3-4 Thermometer scale.

From the *ASHRAE Guide and Data Book* we can find out exactly how much moisture air can hold at saturated conditions. Following is a simple table taken from this reference book:

Saturated Temperature °F DB	*Humidity Ratio* Lb/lb *of dry air*
70°	0.01582
72°	0.01697
75°	0.01882
78°	0.02086
80°	0.02233
82°	0.02389
85°	0.02642

Returning to the psychrometric chart construction, we can now plot saturation points (Fig. 3-6) for each condition of dry-bulb temperature, and when these are connected they form a curve or *saturation line.*

Assume an air sample (point *A*, Fig. 3-7) with a dry-bulb temperature of 80°F, holding 0.011 lb of moisture. If we were to heat the air without adding moisture, the point would move to the right on the horizontal line, showing an increasing dry-bulb temperature but an unchanging moisture content.

If we were to add moisture (humidify) without changing the dry-bulb temperature, the point would move vertically up. If the moisture were reduced (dehumidifying), it would move vertically down. If temperature and moisture were added, the point would move up and to the right, and if the air were cooled (without changing its moisture content), the point would move horizontally to the left.

Continuing the example, if the air sample is cooled, it eventually reaches the saturation line (point *B*, Fig. 3-8) where it cannot hold any more

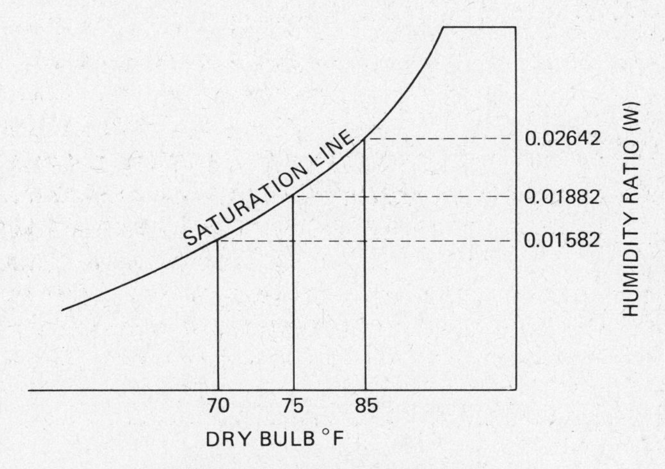

FIGURE 3-6 Saturation points.

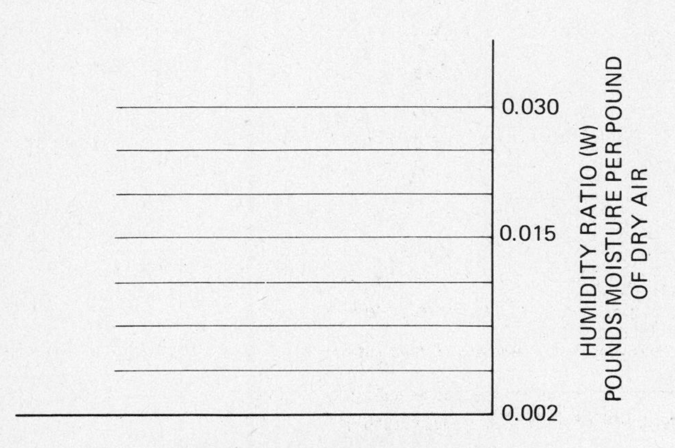

FIGURE 3-5 Humidity ratio scale.

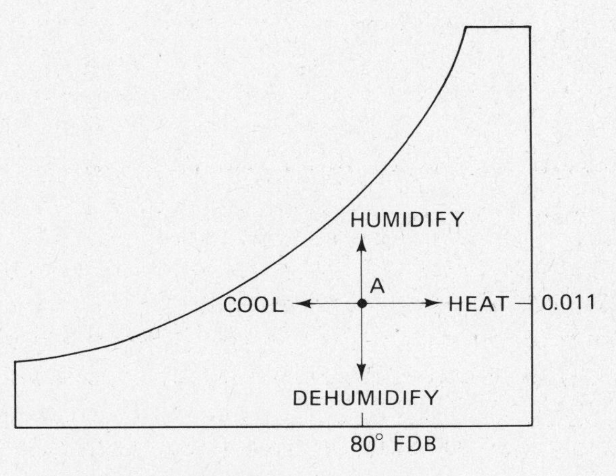

FIGURE 3-7 Air sample.

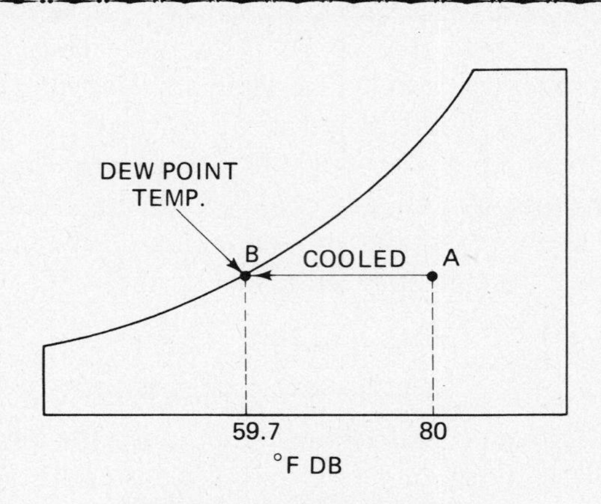

FIGURE 3-8 Saturation line (B).

water vapor, and on further cooling some water would start to condense. That temperature is just below 60°F, or about 59.7°F. This is known as the *dew-point temperature* of the sample. It can be read from the vertical dry-bulb index temperature. In summary, at point *B*, we have a 59.7°F dry-bulb temperature, a 59.7° dew-point temperature, and a moisture content of 0.011 lb of moisture per lb of dry air.

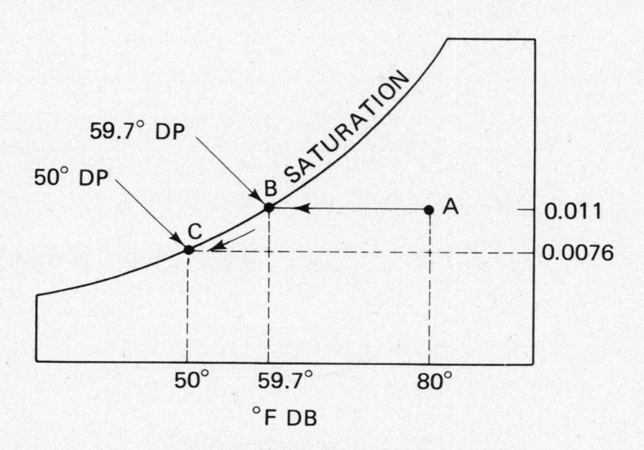

FIGURE 3-9 Saturation line (C).

Now if the sample is further cooled, for example to 50°F dry-bulb, moisture will condense out and follow along the saturation line to point *C* (Fig. 3-9), where it will have a dew point of 50°F and a humidity ratio of only 0.0076. Thus the sample has lost 0.0034 lb of moisture. It has been cooled and dehumidified.

A practical example of this process is a cold supply air duct (as shown in Fig. 3-10) running through a moist unconditioned area. Will the duct sweat and need to be insulated? Assume the air temperature inside the duct is 55°F and the unconditioned air surrounding the duct is at 95°F with 0.0142 lb of moisture content. This condition means that the outside air would have a saturated (dew-point) temperature of 67°F. Thus, as the 55°F duct temperature cools the air touching its surface to below the 67°F dew point, condensation will likely occur. Depending on conditions, it will be necessary to take some corrective action using appropriate insulation to prevent sweating.

The next element in our chart is the construction of relative humidity lines for partly saturated conditions (Fig. 3-11). We know the relative humidity is 100% at the saturation line. Lines for 80%, 60%, 40%, etc., can be plotted, since we know specific moisture contents in relation to temperatures. As an example, one pound of air at 75°F dry bulb will hold 0.01882 lb of moisture (point *A*) at saturation (100% relative humidity). Point *B* (50% relative humidity) can be located at approximately 0.0094 lb moisture ($\frac{1}{2}$ of 0.01882 lb). The same method can be used for each dry-bulb temperature, and eventually a

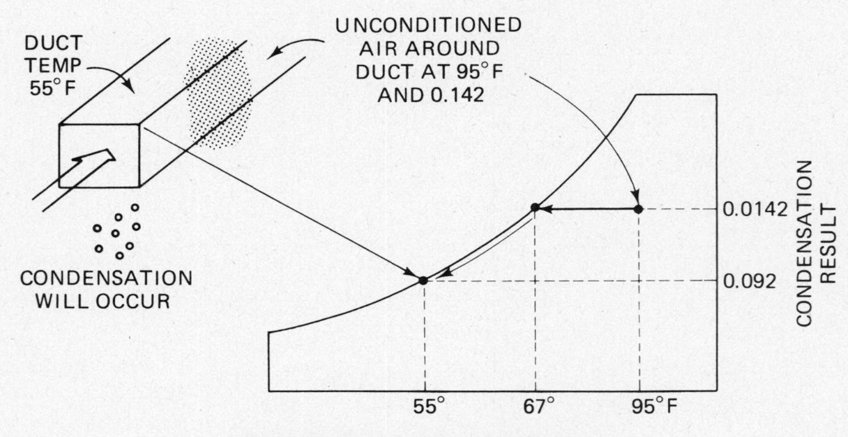

FIGURE 3-10 Cold supply air duct.

connecting line is drawn that represents a 50% relative humidity for any chosen condition of dry-bulb temperature. Similar lines can be drawn for different relative humidity conditions. We already know from Chap. 2 how useful it is to be able to express relative humidity, since it affects human comfort.

Unfortunately, it's not practical or convenient to measure the amount of moisture content or dew point of the air except under laboratory conditions, so we need to plot another element that will give us an easier method. It was noted in Chap. 2 that the wet-bulb temperature also reflects the amount of moisture in the air. The rate of evaporation on the sling psychrometer determined the wet-bulb depression below the dry bulb temperature. And from the table in Table 2A, we can determine the relative humidity.

For example, Table 2A showed that for an 80° dry-bulb temperature and an 11° wet-bulb depression (69°F actual measured WB), the relative humidity is 57%. Transferring this information to our psychrometric chart, we can plot point *A* (Fig. 3-12). If we were to cool the dry-bulb temperature to 76° and the wet-bulb temperature actually stayed at 69° on the sling psychrometer, we now have a WB (wet-bulb) depression of only 7°F and, from Table 2A, a relative humidity of 70%. Point *B* can now be located. By connecting points *A* and *B*, we create a constant wet-bulb line. This process could be repeated over and over until a complete grid of wet-bulb lines fill the chart. Wet-bulb temperature is read at the saturation temperature line, because at that point it can hold no more moisture and becomes the same as the dry-bulb and dew-point temperatures.

This completes the construction of the simplified psychrometric chart (Fig. 3-13). Although it is not 100% accurate, this description should help you understand the relationship of the lines on the real chart. Fortunately, precise and accurate information has gone into the construction of the ASHRAE chart, and it may be used with confidence. *Remember*, if any two of the five properties of air are known, the other three can be found on the psychrometric chart by locating the point of intersection of the lines representing the two known conditions.

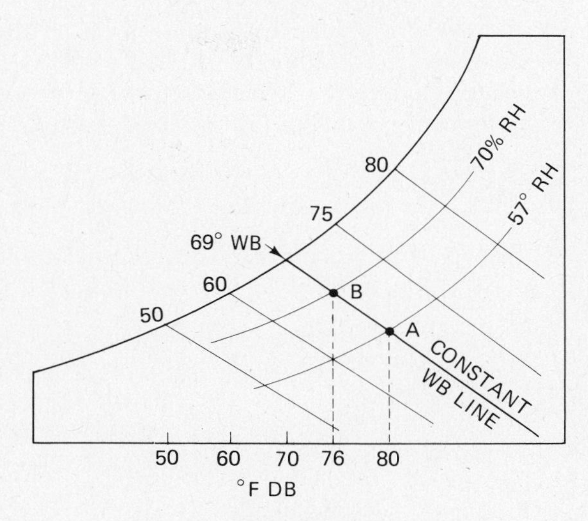

FIGURE 3-12.

Before going on, here are a few practice problems using the included ASHRAE chart.

(1) The owner of a building wants to maintain an indoor condition of 80°F DB and 50% relative humidity. Find the wet-bulb and dew-point temperatures. (Answer: 66.7°F WB, 59.8°F DP.)

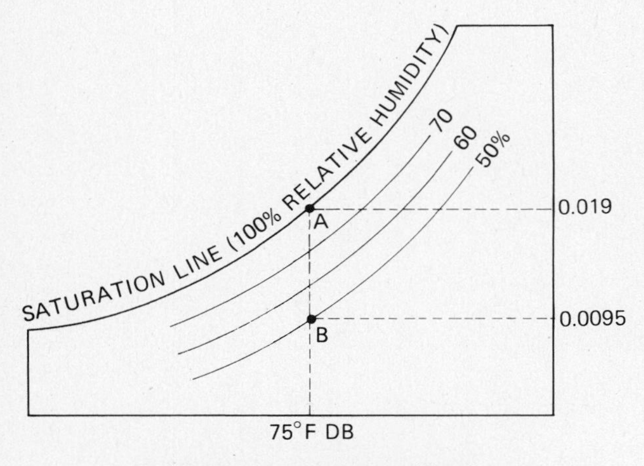

FIGURE 3-11 Relative humidity lines.

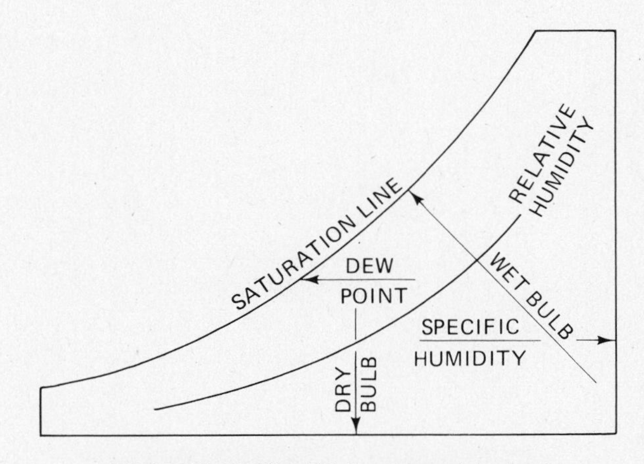

FIGURE 3-13. Simplified psychometric chart.

(2) Assume a wet-bulb temperature of 60°F and a dry-bulb temperature of 72°F. What is the relative humidity? The humidity ratio? (Answer: 50% relative humidity, 0.0084 lb of moisture.)

(3) Assume a DB of 68.6°F and a WB of 60.4°F. What is the relative humidity? (Answer: 63%.)

(4) What is the WB of air that has a relative humidity of 80% and a DB of 70°F. (Answer: 65.7°F.)

(5) The outdoor design conditions in Portland, Oregon, are 90° DB and 68° WB. Find the dew-point temperature and the relative humidity. (Answer: 56.0°F DB and 32% relative humidity.)

AIR MIXTURES

Before plotting an air-conditioning problem on the psychrometric chart, we must first know the initial temperature of the air to be cooled (or heated as the case may be). In most air-conditioning systems there will be ventilation air from outdoors mixed with room air returning to the unit. In Fig. 3-14 the system is handling 4,000 ft³/min (cubic feet per minute) total air. Also, 1,500 ft³/min of outside air at 95°F DB and 78°F WB is mixed with 2,500 ft³/min return (room) air at 80°F DB and 67°F WB. What is the air mixture temperature? First determine the percent of ventilation air in the mixture. We can do

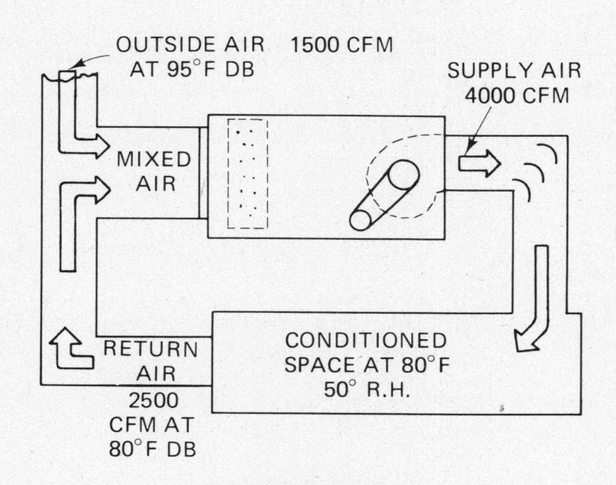

FIGURE 3-14.

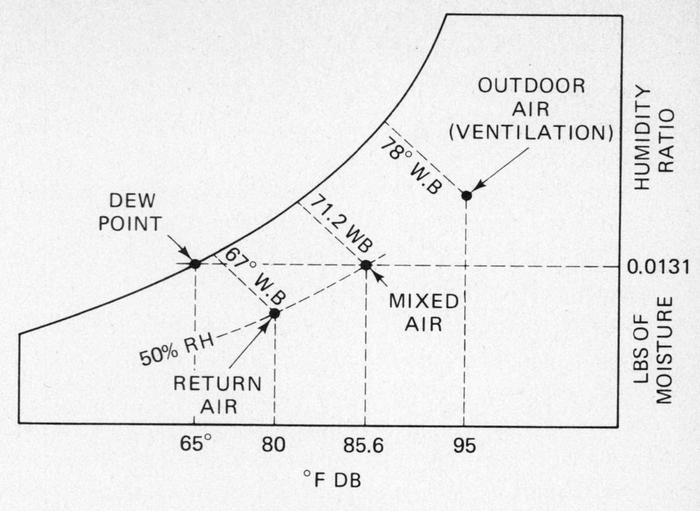

FIGURE 3-15.

this by dividing 1,500 by 4,000 = 0.375. Ventilation air is 37.5% and recirculated return air is 62.5%. The next step is very important! Multiply the dry-bulb temperature of each condition of the air by its percentage in the mixture. If the outdoor DB is 95°F and it is 37.5% of the mixture, it contributes 35.6 F DB degrees (95° × 0.375) to the mixture. The return air is 80° DB, so it contributes 80° × 0.625, or 50°F DB to the mixture. The total mixed air entering is thus 35.6° + 50°F or 85.6°F DB. Plotted on the psychrometric chart, it would be as represented by Fig. 3-15. The resulting air mixture wet-bulb temperature would be 71.2°F, the dew-point temperature would be 65°, the relative humidity 50%, and the humidity ratio 0.131 lb of moisture.

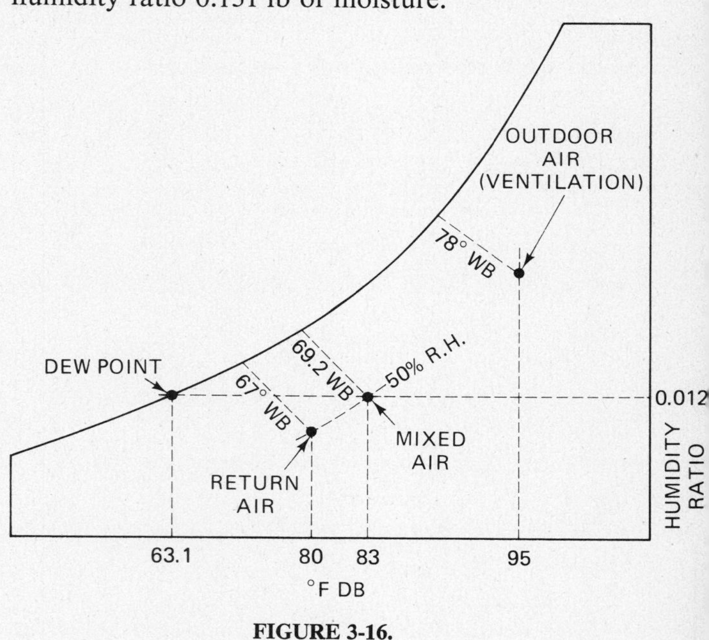

FIGURE 3-16.

Let's plot another practice problem:

Example:

- Outdoor design 95°F DB, 78°F WB
- Indoor design 80°F DB, 67°F WB
- Total air 20,000 ft³/Min
- Outside air 4,000 ft³/Min

Find the resulting conditions of the mixture.

Solution

$$\frac{4{,}000}{20{,}000} = 0.20 \quad \text{(20\% outside air, thus 80\% return air)}$$

$$95^\circ \text{ DB} \times 0.20 = 19.0^\circ\text{F}$$

$$80^\circ \text{ DB} \times 0.80 = 64.0^\circ\text{F}$$

$$\text{Mixed temp.} = \overline{83.0^\circ\text{F}} \text{ DB}$$

Refer to Fig. 3-16. Plot point *A* (outdoor air) and point *B* (return air). Draw a straight line between them. Locate 83.0°F DB on this line; note as point *C*. From point *C*, record a wet-bulb temperature of 69.2°F, a dew-point temperature of 63.1°F, a relative humidity of 50%, and the humidity ratio as 0.012 lb of moisture.

Now let's use the chart for other air-conditioning processes. First we will look for latent heat and sensible heat changes. A latent heat change occurs when water is evaporated or condensed and the dry-bulb temperature does not change. This shows as a vertical line on the chart, in Fig. 3-17. On the other hand, sensible heat shows as a change of temperature with no change in the amount of water vapor, as indicated by the horizontal line on the psychrometric chart in Fig. 3-18.

To illustrate a sensible heat process (Fig. 3-19), let's heat air by passing it over a heating coil. If the air starts out at point *A* (60°F dry-bulb and 56°F wet-bulb temperatures), its dew point is 53°F as obtained from the chart. After heating to point *B*, 75°F dry-bulb temperature (Fig. 3-20), the dew point remains the same, for no water vapor has been added. The

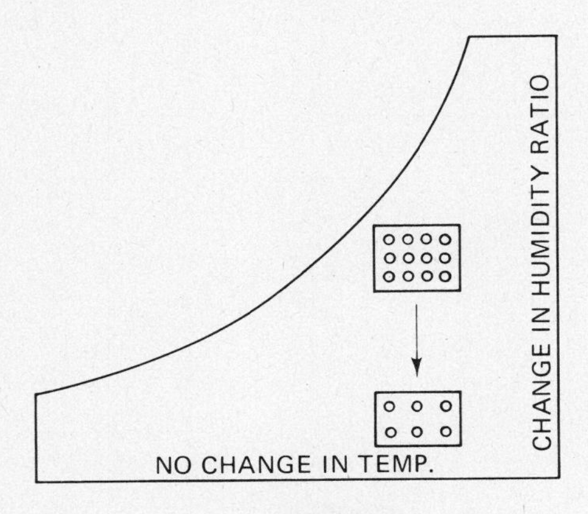

FIGURE 3-17.

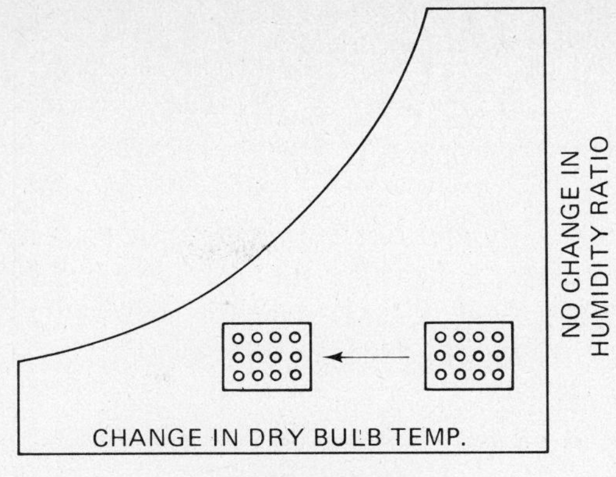

FIGURE 3-18.

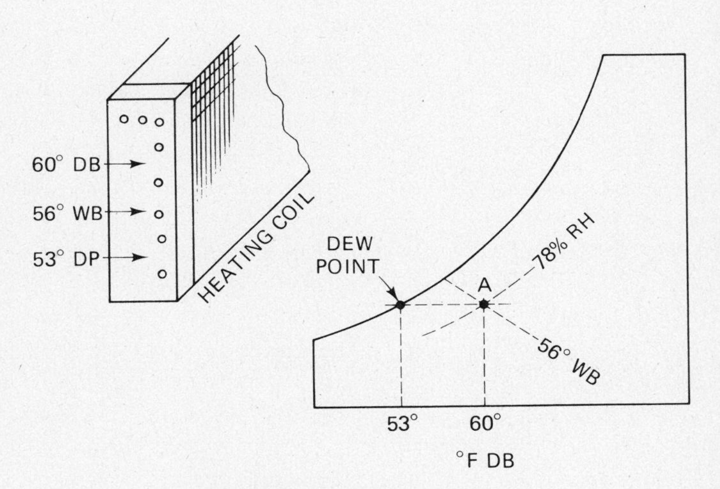

FIGURE 3-19 Heat process.

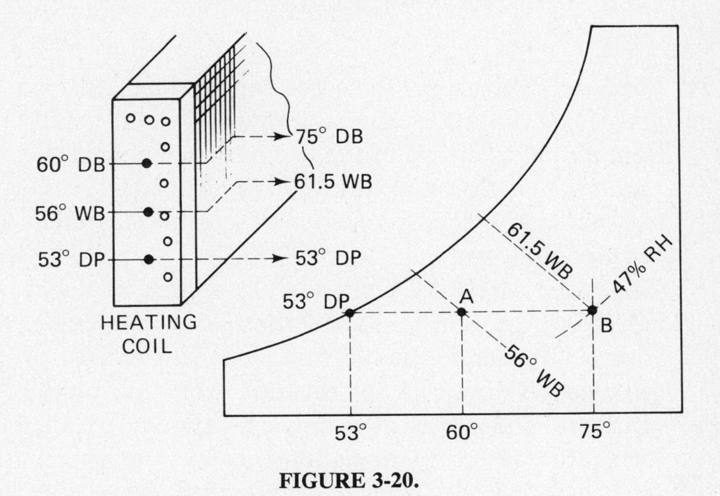

FIGURE 3-20.

wet-bulb temperature, however, has increased to 61.5°F. Also, note that the relative humidity has decreased. This explains why the relative humidity in the early morning is high, but decreases as the day gets warmer.

If the process is reversed and the 75°F dry-bulb temperature (a dew point of 53°F) is cooled back to 60°F DB, we have a sensible cooling process. The wet-bulb temperature drops, but the dew point remains the same (no moisture change).

However, if cooling is combined with dehumidification, the air tends to follow the sloping line to the left in Fig. 3-21. The amount of sensible heat and latent heat determines whether the line has a gentle

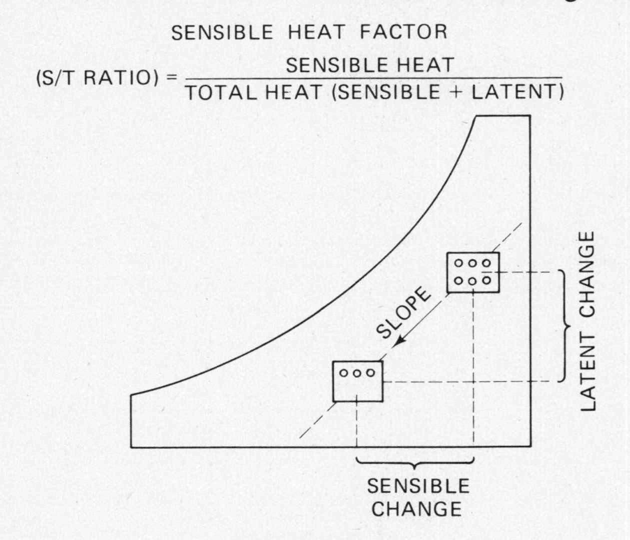

FIGURE 3-21.

slope or steep slope. This combination of sensible and latent cooling occurs so frequently in air-conditioning that the slope of the line has been named the *sensible heat factor* or the *S/T Ratio*, which is the sensible heat divided by the sum of sensible plus latent heat, or total heat.

On the ASHRAE chart there is a circle at upper left. Note the sensible heat ratios inside the circle, with 1.0 on the horizontal base line. This is used to estimate the slope of a line between two points on the chart and thus determine the S/T ratio by parallel lines rather than by calculation.

If no latent heat change occurs, then the sensible heat factor is 1.0 and the line drawn between the two

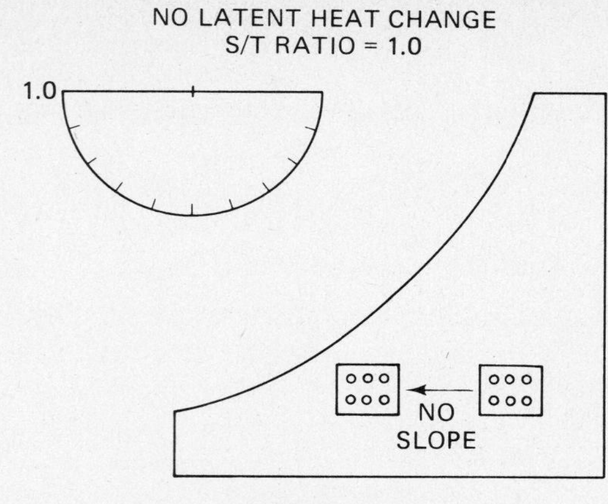

FIGURE 3-22.

points is horizontal (Fig. 3-22). If the sensible heat factor is 0.7, the line starts to slope measurably (Fig. 3-23). This means 70% of the total heat is sensible and 30% latent. Putting it another way, if you had a 10-ton cooling unit, 7 tons would be for sensible and 3 tons for latent heat removal. Packaged air-conditioners are predesigned with a fixed ratio usually of about 0.75 S/T. Large, built-up systems can be individually selected for a wider variety of conditions.

Another term that's used is the *total heat content*, or *enthalpy*, of the air and water vapor mixture. The enthalpy is very useful in determining the amount of heat that is added to or removed from air in a given process. It's found on the psychrometric chart in Fig. 3-24 by following along the wet-bulb temperature lines, past the saturation line, and out to the enthalpy scale. Point *A* represents air at 75°F DB and 0.0064 lb of moisture at about 35% relative humidity. Its enthalpy is 25.0 Btu per pound of dry air. Heat and humidify the air to point *B*, 95°F DB,

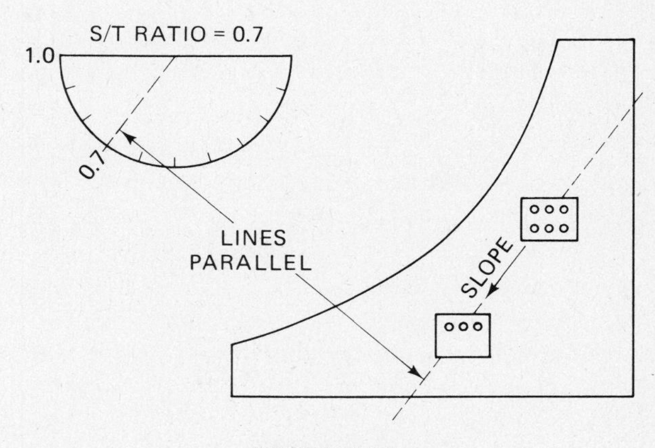

FIGURE 3-23.

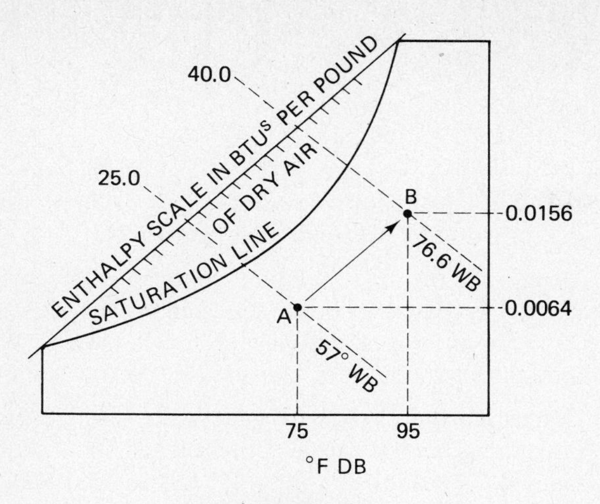

FIGURE 3-24.

0.0156 lb moisture and about 44% relative humidity, where the enthalpy is now recorded at 40 Btu per pound of dry air. The increase in total heat would thus be 40 − 25, or 15 Btu per pound of dry air.

If a triangle is drawn as shown in Fig. 3-25, the vertical distance represents the amount of moisture added; that is latent heat. The horizontal distance represents the sensible heating of air. The enthalpy at the intersection of the vertical and horizontal lines is 30 Btu per pound. Therefore, the amount of latent heat added is the difference between 40 and 30, or 10 Btu per pound. The sensible heat added is the difference between 25 and 30, which equals 5 Btu per pound.

The final property to be examined is *specific volume*, which is the number of cubic feet (ft³) occupied by one pound of dry air. For example, in Fig. 3-26 one pound of air at 75°F dry bulb has a volume of about 13½ ft³ at sea level. If the air is heated to 95°F, it takes up about 14 ft³, for the air is not as dense at the higher temperature. If cooled to 55°F, it would occupy only 13 ft³, for it is more dense at lower temperatures.

The lines for these specific volumes are shown on the psychrometric chart in Fig. 3-27 as almost vertical lines that slant to the left. Specific volume is used primarily for checking fan performance and for determining fan motor sizes for low and high-temperature applications.

As mentioned in the beginning of this chapter, the degree to which you will need and use psychrometrics depends on your particular activity. In most cases an air-conditioning technician will use only dry-bulb and wet-bulb information to determine system performance; however, should you need more information than was presented in this discussion, there are excellent materials available from industry sources for additional reading and understanding.

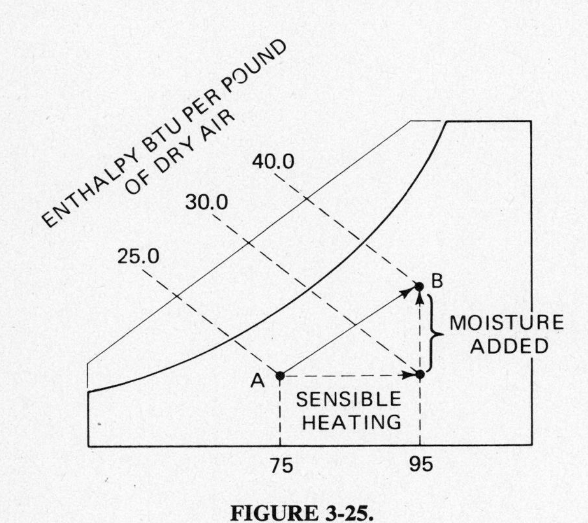

FIGURE 3-25.

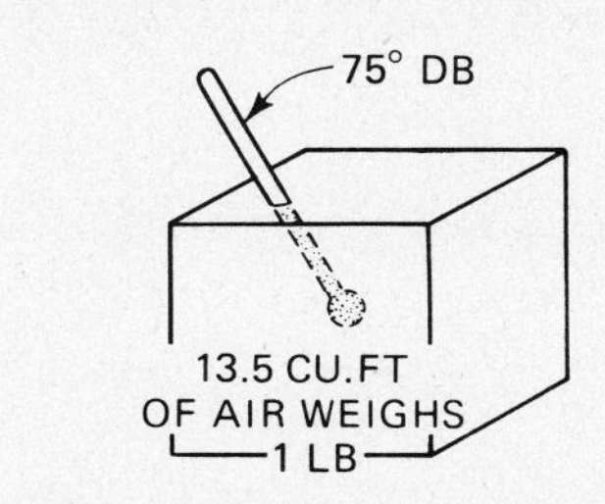

FIGURE 3-26.

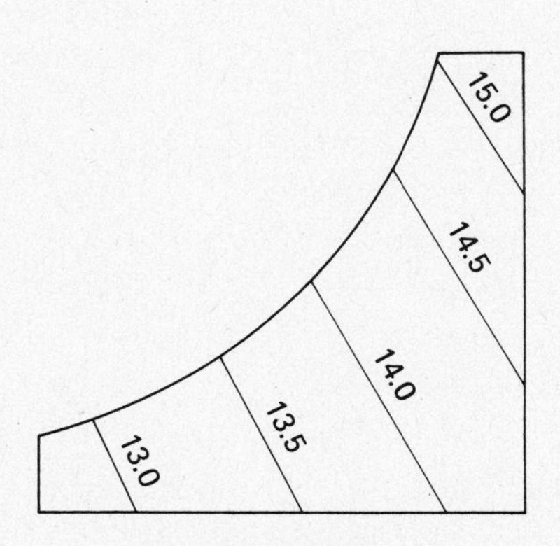

FIGURE 3-27.

PROBLEMS

1. What is *psychrometrics*?
2. The atmosphere is made up of ______________.
3. Wet-bulb temperature is measured by a ______________.
4. What is the dew-point temperature?
5. Specific humidity is a reference of ______________.
6. Relative humidity at saturation is ______________.
7. Total heat is made up of ______________ and ______________.
8. The form used to plot air conditions is called a ______________.
9. What is the humidity ratio at 80°F dry-bulb temperature and 67° wet-bulb temperature?
10. Determine the final mixture conditions of an air handler based on an outdoor design of 90°F DB and 76° WB, an indoor design of 78° DB and 66° WB, total air 16,000 ft^3/min, outside air 4,000 ft^3/min.
11. Determine the change in total heat, sensible heat, latent heat, and moisture content when air is cooled from 90° DB and 78° WB to 75° DB and 63° WB.

4

BUILDING CONSTRUCTION

The type of air-conditioning equipment selected, the air distribution system, and installation practices depend to a considerable extent on the structure itself. And so it is important that the technician have an understanding of building construction techniques in order to provide answers to the application variations. Obviously, building construction covers a wide range, all of which we cannot cover here. However, we will touch on the fundamental areas and review basics.

RESIDENTIAL

Most individual houses fall into one of three design categories. These are the one-floor plan or ranch type, the two-story, and the split-level house as indicated in Fig. 4-1 (a, b, c; page 350). These houses may be built over basements, crawl spaces, or on concrete slabs, depending on geographic considerations and marketing requirements. The foundation will affect the selection of the type of air distribution and determine installation considerations. For instance, it is comparatively easy to install a forced-air-conditioning system for a one-floor house over a full basement, since there is usually plenty of room for equipment and ductwork. This is true whether it's a new or old home. Not so with a two-story home, for getting ducts to and from the upper floor presents complications. Also, the comfort conditions between first and second floors can vary widely from winter to summer. The split level is probably the most difficult to work with properly.

Some parts of the house are often removed and disconnected from the main core; an example would be a bedroom over the garage. Such a room exposes extensive areas of walls, ceiling, and floor to unconditioned temperatures. An interesting reading assignment would be to go to a library and thumb through the many house-plan books to see the many variations in style and configurations being offered; essentially however, all are related to the three basic designs.

The foundation is the first consideration in erecting a structure, so let's begin our examination at

FIGURE 4-1 (a) One floor ranch type.

FIGURE 4-1 (b) Two story.

FIGURE 4-1 (c) Split level.

the ground level. Basements (Fig. 4-2) are most often found in home designs for the east, midwestern, and midsouthern parts of the country. They are not common in areas such as Florida, California, Texas, etc., because of the soil structure, water table, and drainage considerations. Most basement walls are constructed of 8 in. × 8 in. × 16 in. concrete blocks over poured concrete footings. Footings must be below the frost line, which means the depth below grade (ground level) where water doesn't freeze and cause the ground to heave.

Waterproofing is necessary. Walls are usually covered with a troweled coat of cement, followed by an asphalt coating which forms a tough durable skin that resists water penetration. Most important is a drain-pipe system to carry water away from the foundation. This can be drain tiles or plastic pipe with holes around the entire foundation. The run-off water must then go to a positive outfall or storm sewer. The height of the concrete blocks should be sufficient to meet or exceed FHA requirements of 7 ft 6 in. ceiling clearance. Some builders use poured concrete for the walls.

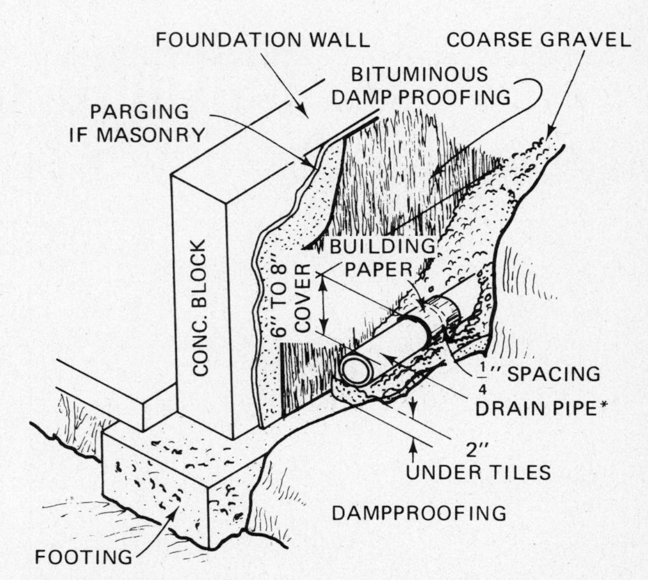

FIGURE 4-2 Basement foundation.

More recently there is a trend toward using treated plywood that has been government certified to be used below grade. Figure 4-3 represents this technique. Concrete footings are still needed, but the

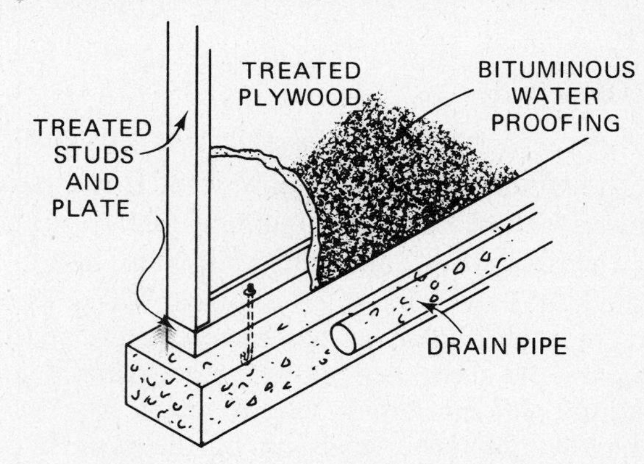

FIGURE 4-3 Treated plywood foundation.

plywood walls are supported by framing studs. In a way, it's like a two-story approach. Good waterproofing and drainage are also necessary.

Heat gain (for summer air conditioning) is not a factor for basement floors and walls that are underground, since the ground temperature, even in midsummer, is usually cooler than house conditions. In winter there is some heat loss to the outside earth depending on geographic location. The load calculations take these factors into account. Walls above grade are subject to the full extent of outdoor temperature and sun effect.

The crawl-space type of foundation (Fig. 4-4) is essentially what the name implies, a space between ground level and the floor just large enough for a person to crawl into. Some are constructed to such minimum standards that they will present difficulties to a large person. The footings (poured concrete) are similar to that for a basement below the frost line. Walls can be of concrete block, poured concrete, or treated plywood as previously explained. Drain tile is recommended as well as waterproofing of the wall. The earth floor is covered by a moisture barrier of roll roofing, sheet plastic, or equivalent. A protective covering of washed gravel is optional. Several specific points about crawl spaces should be noted. The distance between any wooden members (beams, joists, etc.) and the moisture barrier should not be less than 18 in. There should be a number of louvered vents to allow proper ventilation air during the summer; otherwise, wooden members may be damaged by high humidity and are subject to rot. These vents must have dampers that close in the winter when humidity is not a problem and cold air penetration could materially increase heating costs and cause discomfort from cold floors. Outside walls should be insulated with a vapor-proof insulating board such as polystyrene. (The subject of insulation is covered in Chap. 5.)

A third type of foundation (Fig. 4-5) is known as *slab-on-grade* and is meant for basementless houses. These are very popular in the more temperate regions. The final grade of the site determines whether or not a slab-on-ground can be used. The finished grade must fall sharply away from the house so that ground water cannot stand at a level high enough to flood the gravel fill under the floor. Soil composition is also a factor. After all the sod and organic material are removed, a fill of four inches of coarse washed gravel or crushed rock should be made. The entire area is then covered with a moisture barrier, 55-lb roll roofing with edges lapped, or six-mill polyethelene film. Steel reinforcement should be installed to prevent temperature and expansion cracks. Edge insulation is required around the entire perimeter of the slab; it should not be moisture absorbent. The most commonly used edge insulation is polystyrene foam boards. It may be installed either as an L-shaped piece extending back two feet under the floor slab or vertically inside the foundation wall, at least two feet below finished grade. Depending on the type of air-conditioning system, the inclusion of air ducts, water pipes, or electrical conduit will be formed into

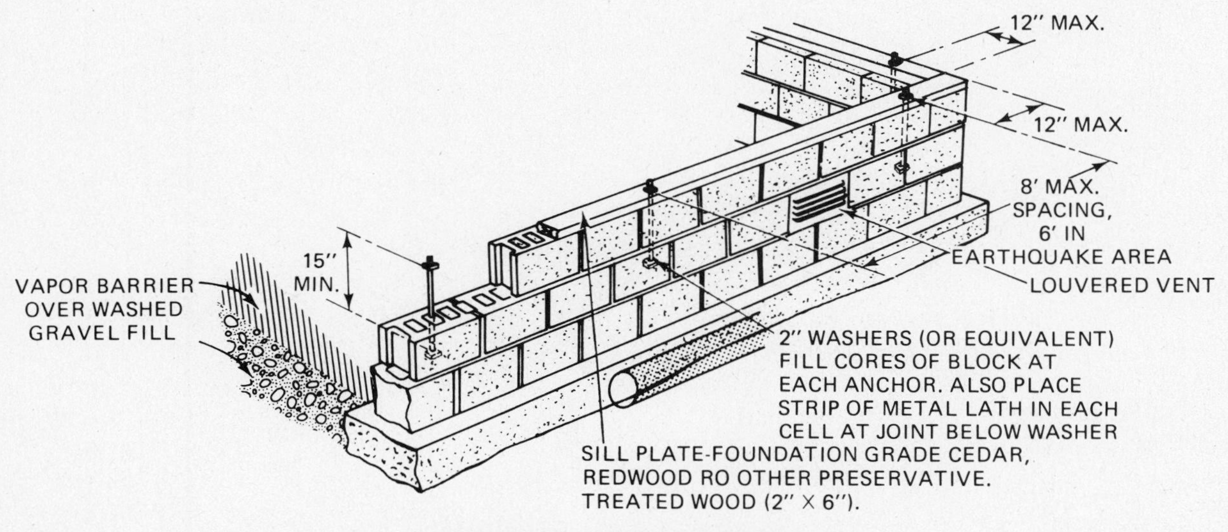

FIGURE 4-4 Crawl space foundation.

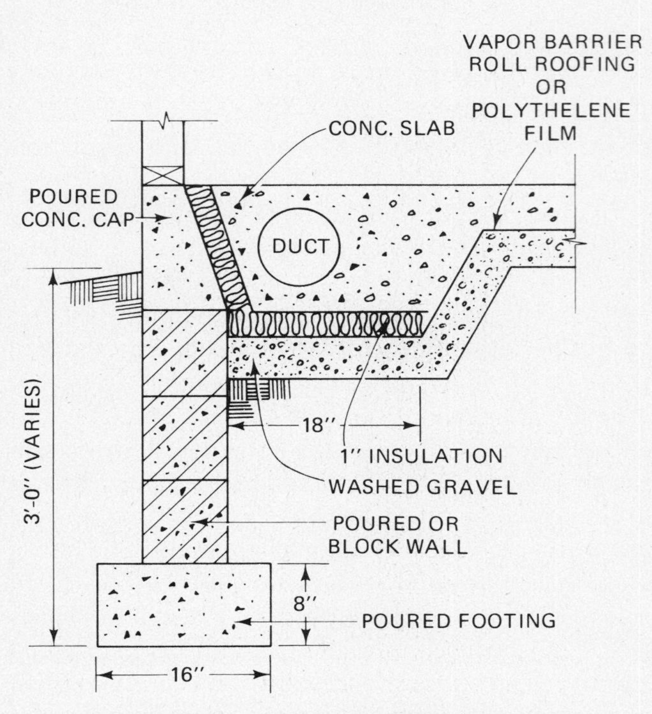

FIGURE 4-5 Slab-on-grade for basementless houses.

the concrete slab. (Specifics on heating or cooling systems will be covered later.) Heat loss through the slab edges can be severe, depending on geographic location, so be informed on the type and value of edge insulation being used.

Another variation of foundation and one that is most frequently found in so-called second homes, vacation homes, and recreational homes, is the post or pier support (Fig. 4-6), where the floor system is exposed to outside weather conditions. Treated posts are placed in concrete fill and support a network of girders and joists forming the floor system. Naturally, the floor must be heavily insulated if the house is to be lived in year round. Ductwork, if used, is normally run in the conditioned space to minimize operating expense. The treated wood foundation is the best choice in more remote areas where a limited amount of excavation equipment is available. Don't be misled into thinking this system is to be identified with low-cost housing; it has been applied in many expensive homes.

Now that we have discussed foundations, let's move on to the subject of floor framing. Figure 4-7 is a detailed drawing of a typical outside wall construction, showing first the sill or bed plate being anchored to the concrete block by means of bolts

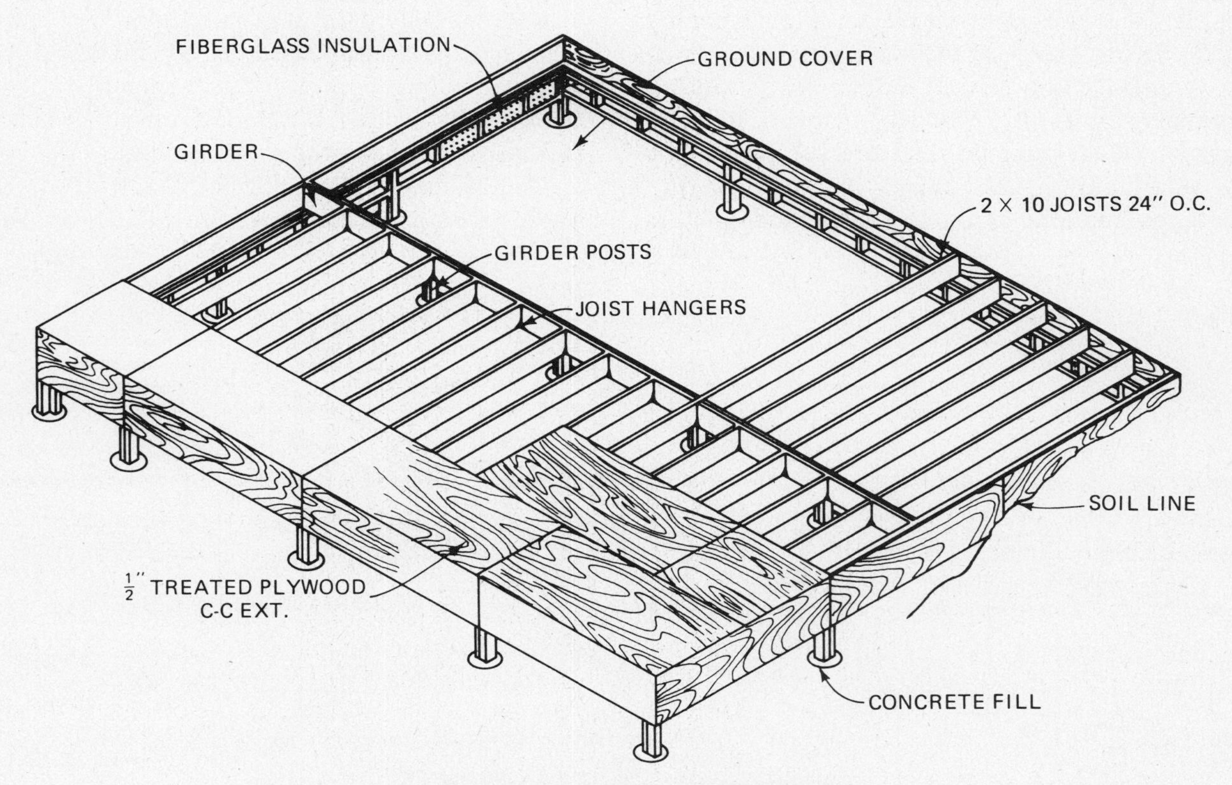

FIGURE 4-6 Post or pier support foundation.

which extend into the block and are fastened or mortared securely. Where termites are anticipated, a metal termite shield is applied between the block and wood sill plate.

In the center of the structure there is a main beam or girder or load-bearing wall (Fig. 4-8) to support the floor joists. These joists span the width from center to outside wall, where they rest on the sill plate. Several methods may be used to secure them at the center beam, such as shown in Fig. 4-8. Where joists rest on top of the beam they are lapped and nailed. Where they are under a load-bearing wall, the joists will be doubled for added strength. A cross-bridging of wood on steel is used to keep the joists (spanning more than 10 feet) from warping or twisting.

If a flush, finished ceiling below is desired, the joists are tied into the center beam by means of a ledger board and are notched to rest on it. Also, metal joist supports (Fig. 4-9) are available to be nailed into the beam.

Recently the use of metal beams (girder) and floor joists have become popular. They offer some obvious advantages; they are not affected by moisture, won't warp or twist from drying out, and are not subject to termite damage. Their disadvantages include the fact that they are not commercially available in all locations and cannot be cut and fit on the job site as easily as wood.

At this point, it would be a good idea to explain lumber sizes. *Nominal size* means the size of a piece of lumber when it is first cut—before seasoning or surfacing. For example, when a board is first cut, a 2-by-4 is 2 in. thick and 4 in. wide. But after it shrinks

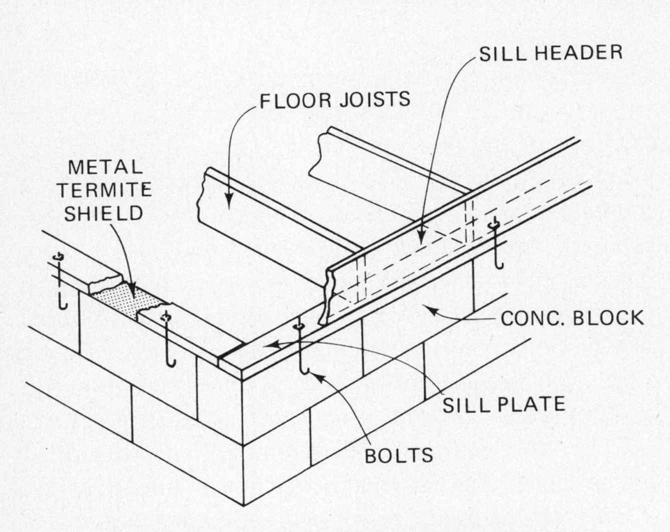

FIGURE 4-7 Outside wall construction detail.

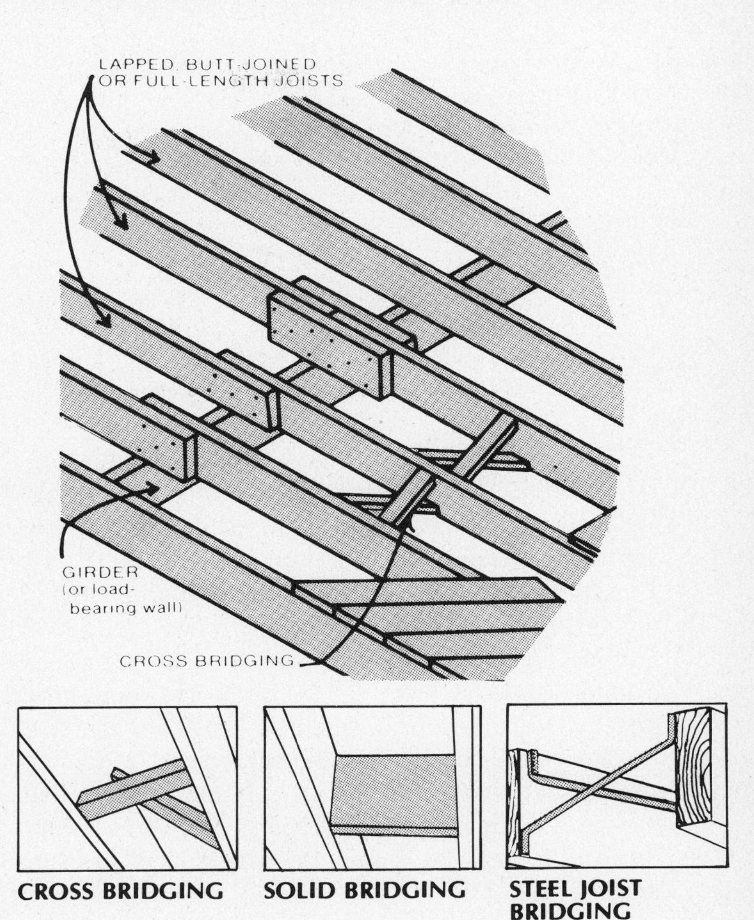

FIGURE 4-8.

from seasoning and is planed smooth on all four sides, its final dimensions are $1\frac{1}{2}$ in. by $3\frac{1}{2}$ in. Below is a table for *dimensional lumber sizes* (in inches).

Nominal Size Inches (Original Cut)	*Actual Size Inches (Minimum Dimension)*	*Nominal Size Inches (Original Cut)*	*Actual Size Inches (Minimum Dimension)*
1 × 2	$\frac{3}{4} \times 1\frac{1}{2}$	2 × 2	$1\frac{1}{2} \times 1\frac{1}{2}$
1 × 3	$\frac{3}{4} \times 2\frac{1}{2}$	2 × 3	$1\frac{1}{2} \times 2\frac{1}{2}$
1 × 4	$\frac{3}{4} \times 3\frac{1}{2}$	2 × 4	$1\frac{1}{2} \times 3\frac{1}{2}$
1 × 5	$\frac{3}{4} \times 4\frac{1}{2}$	2 × 6	$1\frac{1}{2} \times 5\frac{1}{2}$
1 × 6	$\frac{3}{4} \times 5\frac{1}{2}$	2 × 8	$1\frac{1}{2} \times 7\frac{1}{4}$
1 × 8	$\frac{3}{4} \times 7\frac{1}{4}$	2 × 10	$1\frac{1}{2} \times 9\frac{1}{4}$
1 × 10	$\frac{3}{4} \times 9\frac{1}{4}$	2 × 12	$1\frac{1}{2} \times 11\frac{1}{4}$
1 × 12	$\frac{3}{4} \times 11\frac{1}{4}$	4 × 4	$3\frac{1}{2} \times 3\frac{1}{2}$
		4 × 6	$3\frac{1}{2} \times 5\frac{1}{2}$
		4 × 10	$3\frac{1}{2} \times 9\frac{1}{4}$

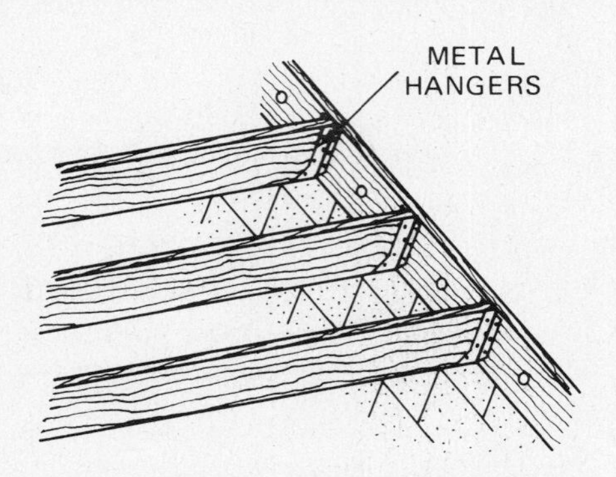

FIGURE 4-9 Flush finished ceiling detail.

Going back to the outside wall construction (Fig. 4-10), we see that the floor joists rest on top of the sill plate and are nailed to a sill header with a conventional spacing of 16 in. on centers. Joists are 2 in. × 6 in., 2 in. × 8 in. or 2 in. × 10 in., depending on load and span. Thus, for joists that are a nominal 2 in. wide ($1\frac{1}{2}$ in. actual size), the open distance between them is $14\frac{1}{2}$ in. (Note: some floor systems use a box-beam method on 24-in. centers.) In either case, both 16-in. on-center and 24 in. on-center spacing fit into a 4-ft building module that has been standard for many years. Plywood sheets, dry-wall board, etc., come 4 ft wide and in multiple lengths.

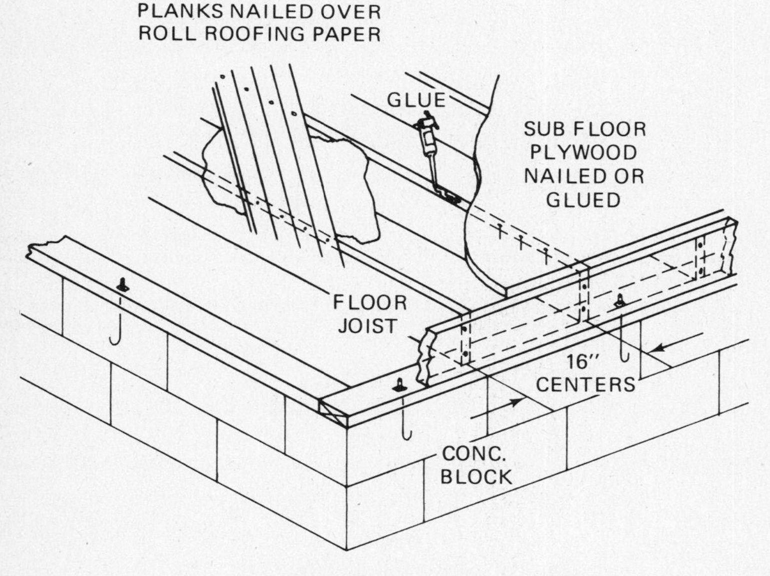

FIGURE 4-10 Floor system detail.

We are now ready to apply the subflooring. In the old days these were board planks spaced and nailed on a diagonal pattern. Modern construction uses plywood sheets, 4 ft × 8 ft of interior sheathing grade C or D. These panels are nailed or glued into position. Gluing has become increasingly popular because it takes the guess-work out of nailing, increases strength, and makes construction less noisy. Also, its cost is reported to be lower than other methods.

With the subfloor in place and providing a working platform for the carpenters, outside and interior

FIGURE 4-11 Subfloor in place, providing working platform; precut 2 × 4 studs on center are fabricated in section and then lifted into place.

walls may be erected. Precut 2-in. × 4 in. studs on either 16-in. or 24-in. centers are fabricated in large sections on the floor and then lifted into place (Fig. 4-11). Openings for doors, windows, etc., have been preplanned and framed in. Wall sections are trued, squared, braced, and nailed into place. Here, too, the use of light-weight metal for studs is gaining in popularity. Note in Fig. 4-12 the use of double 2 × 4s on top of the vertical studs to form a strong plate on which the roof members or second floor will be supported. Double studs and cross headers are used around window and door openings for added strength and nailing surfaces. These openings are slightly larger than the window or doors themselves

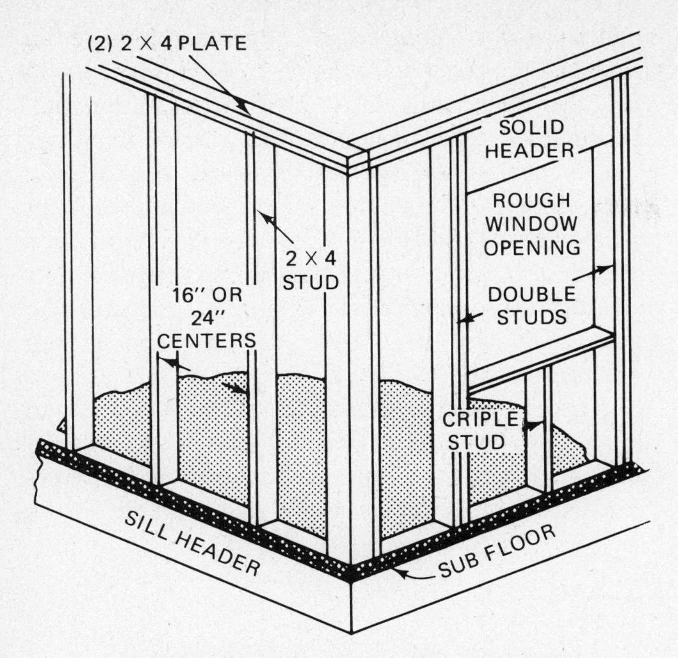

FIGURE 4-12 Wall and window framing detail.

and so they are called *roughing-in-dimensions* by the carpenter.

If a second floor is considered, the wall framing will continue with second-floor joists supported on the stud-wall plate and nailed to a header as in Fig. 4-13. A subfloor is applied and the 2 × 4 vertical wall is repeated. This method is used where no offset or overhang is desired.

Many home designs utilize an overhang (Fig. 4-14) to gain space and to offer shading for the lower story. Here the second-floor joists rest on the vertical wall plate but extend out some 18 inches to 2 feet. The ends are tied together with a joist header. The subfloor is applied and vertical walls erected just as explained for the first floor. This method of *platform framing* above is also called *boxed sill* or *western framing* and is the one most commonly used.

There is also another technique called *balloon framing* (Fig. 4-15). No header or sole plate is used. Joists are nailed to the studs. This would also be true for the second floor as well. Balloon framing allows the use of continuous wall studs. It also provides an open cavity between joists, which permits easier installation of duct work. Balloon framing is

2 × 4 STUD
SOLE
SUB FLOOR
CEILING JOIST
HEADER
DOUBLE 2 × 4 PLATE
2 × 4 STUD
SOLE PLATE
SUB FLOOR
FLOOR JOISTS
SILL PLATE
CONC. BLOCK

FIGURE 4-13 Second floor framing detail.

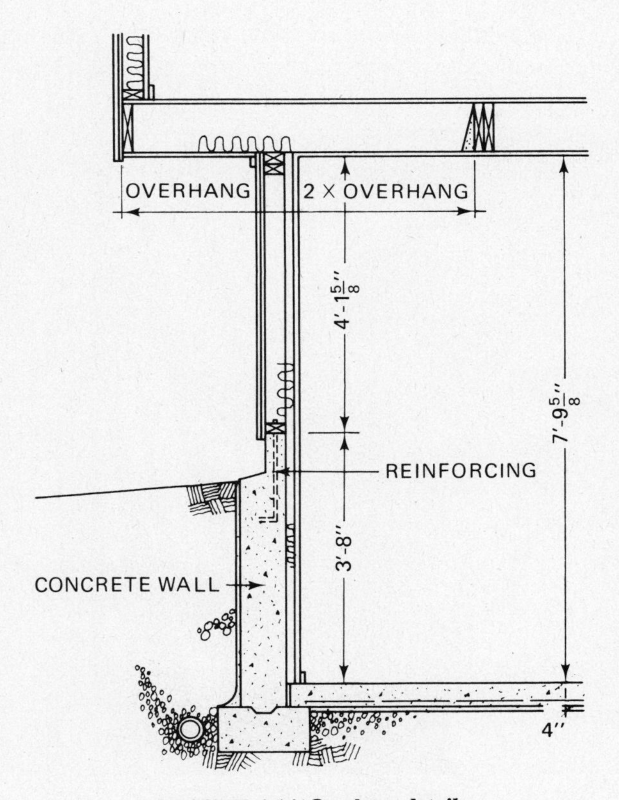

FIGURE 4-14 Overhang detail.

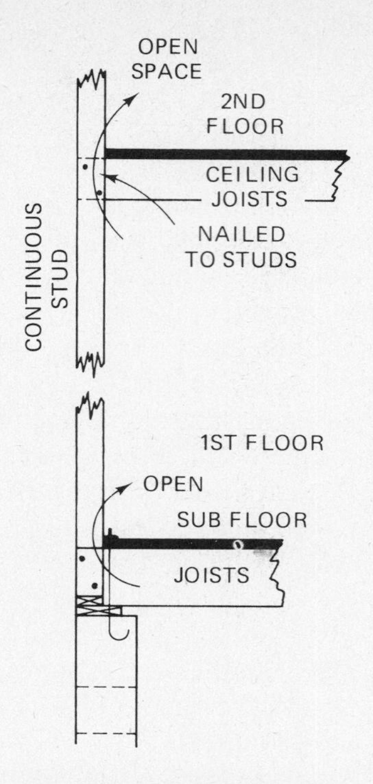

FIGURE 4-15 Balloon framing.

obviously not usable where offsets are needed. Open joist spaces can cause excessive drafts if not properly insulated.

The exterior walls are now ready for sheathing (Fig. 4-16). Sheathing is a very important part of the building process, and it consists of covering the structural members of the framework. Not only does it strengthen the frame, but it also provides a measure of insulation. The corners of the outside walls are usually covered with plywood panels to provide a strong bracing. The balance of the outside walls get a treatment of impregnated sheathing made of wood fibers that are asphalt impregnated to form a highly moisture-resistant product. It remains permeable to water vapor, however, thereby allowing the exterior walls to breathe.

FIGURE 4-16 Sheathing exterior walls (Courtesy, American Plywood Association).

In order to shelter the structure from wet weather, the roof system usually goes up next. In the single-story ranch house, almost all builders use a prefabricated truss system (Fig. 4-17). The example shown

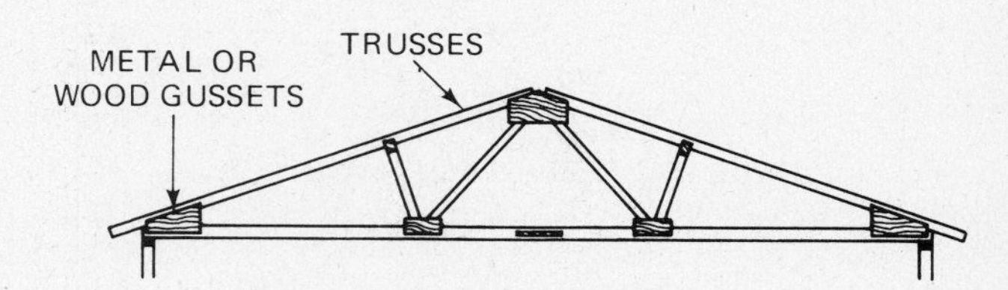

FIGURE 4-17 Prefabricated truss system.

is known as a Fink truss with reinforcing gussets. When a roof truss is used, erection time at the site is cut considerably, the average roof can be framed in just 60 minutes, thus saving considerable on-site labor and time. Because of the triangular design, lighter wood members can be used without sacrificing strength. The trusses are lifted by means of a lightweight, truck-mounted hoist. Trusses do not require the support of interior walls during initial framing, and thus installation is simplified.

The older method of roof framing (Fig. 4-18) begins with ceiling joists that span the perimeter walls; the size and length of the span determines whether inside load bearing-walls are needed to support the weight. Joists are usually spaced on 16-in. centers. A ridge beam, usually 2 in. × 8 in., is held up by a temporary support at each end. Next come the rafters on 16-in. centers; they are precut at the proper angle to be flush with the ridge beam. At the other end they are notched over the wall plate to form the eaves mount.

The *pitch* of a roof (Fig. 4-19) is the angle formed by the vertical rise compared to the span distance and is expressed in the number of feet it rises for the number of feet it runs (vertical rise compared to horizontal run). For example, a 3/12 roof will rise 3 feet for each 12 feet of horizontal run. Ranch-style

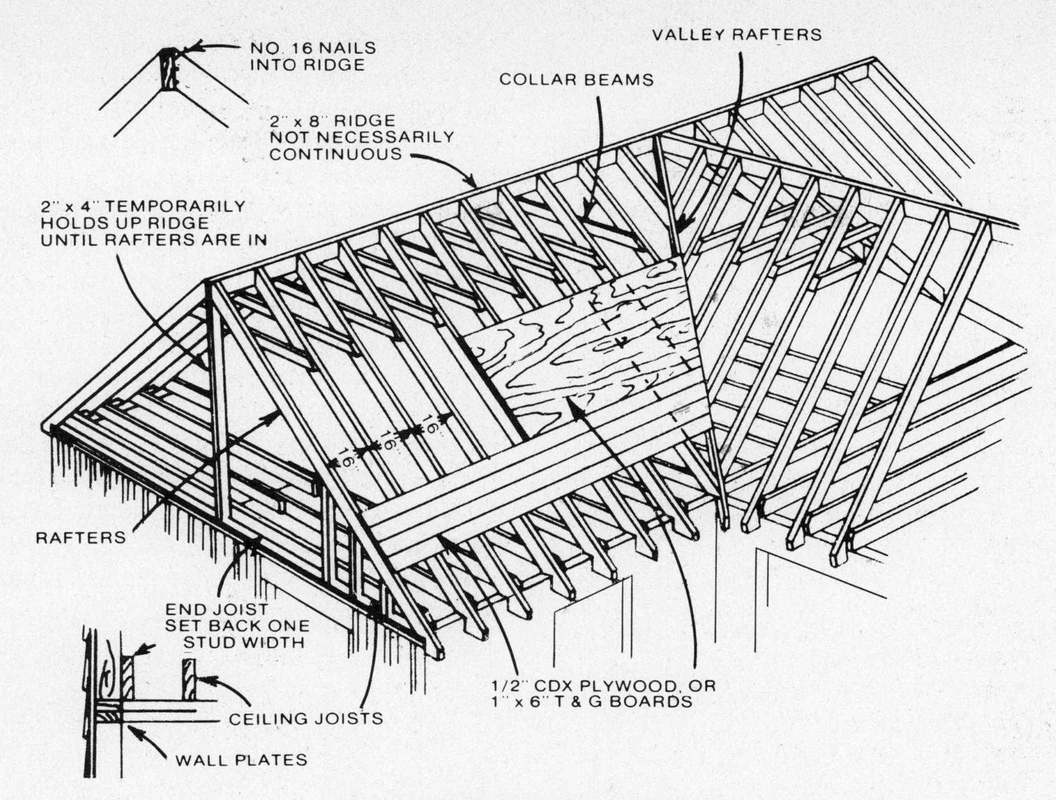

FIGURE 4-18 Roof framing.

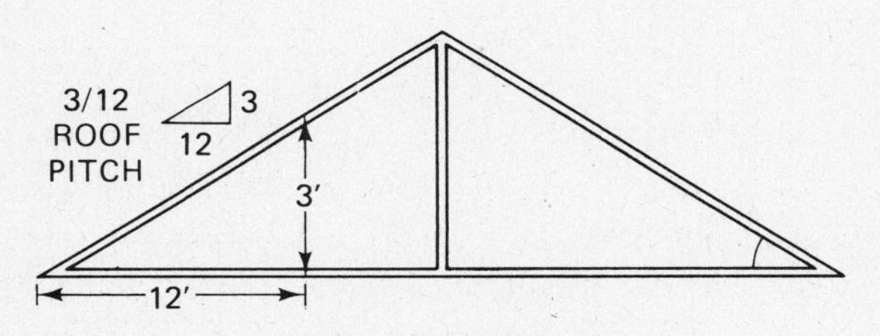

FIGURE 4-19 Roof pitch.

houses usually slope 4/12. Story-and-a-half houses with living space in the socalled second floor run much steeper to allow head room in the finished area. The roof pitch is also affected by the *snow load*, and in northern regions greater pitches are used to prevent too much center load of snow.

Once the truss or rafters are in place, roof decking can be applied. Plywood sheets $\frac{3}{8}$ in. thick or heavier are nailed to the rafters. They go on fast and form the surface for nailing of the finished roof. A covering of 55 lb roll roofing felt paper is applied and lapped to give temporary water protection. Some applications of roof decking still use 1 × 6-in. tongue and groove boards instead of the plywood sheets, but the installation costs of the former are prohibitive.

Next, it is customary to install exterior windows and doors for added weather protection. The variety of windows and doors is too great to cover all aspects, but there are some that should be considered (Fig. 4-20). Fixed windows, sometimes called *picture windows*, are for viewing only. These come with wood or metal frames and single and double-paned glass (more on this later). The most common ventilation window is the double-hung variety where both sashes move. There is also the single-hung type where only the lower sash moves. Sliding windows move horizontally; on some, both panes slide, in others, one pane is fixed. Awning windows are popular in the coastal and temperate areas, since they can remain open even during a rain. Casement windows are hinged on the side and open out with a crank or arm operation. You may be able to identify other variations, but the questions that are important to a technician are: How much solar or light-emitting area is involved? Do they have double glass or are they fitted with storm panes? The most effective method of preventing loss or gain of heat is the *double pane* or *Thermopane* (trade name) window (Fig. 4-21). It has two panes of glass with an air space between; the air space has been dehydrated, evacuated, and sealed. Still air is an excellent insulator, so heat transfer is greatly reduced compared to single-pane windows. Storm sashes are very helpful over single

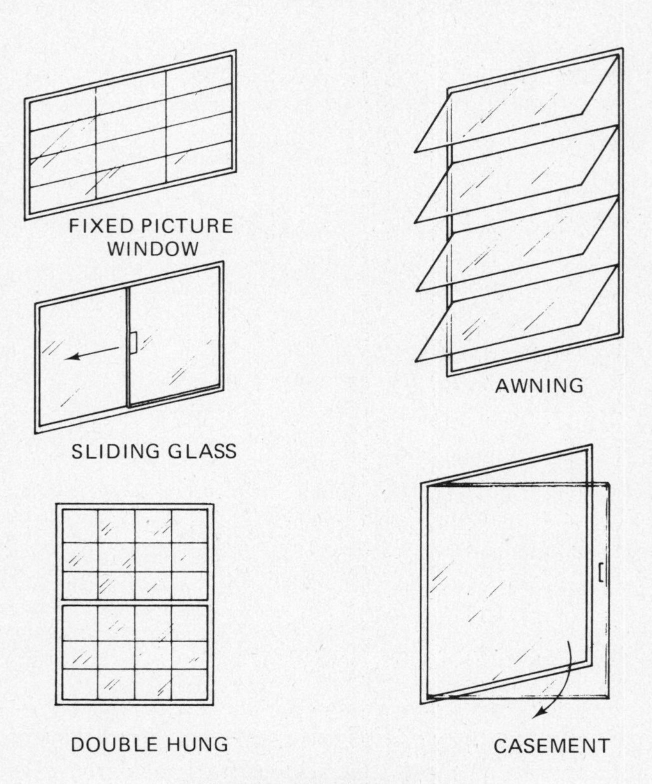

FIGURE 4-20 Window types.

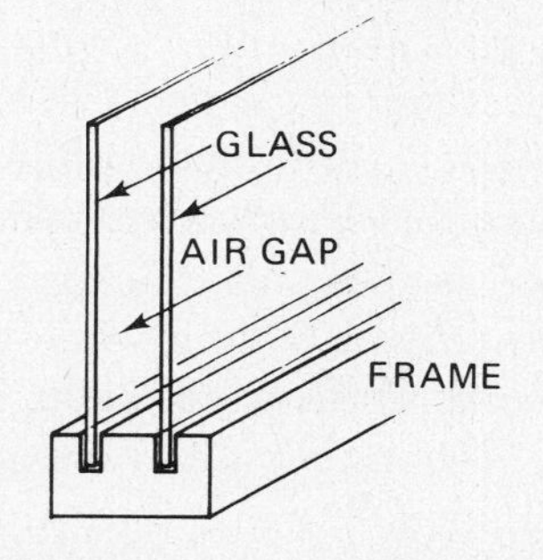

FIGURE 4-21 Double pane window.

pane windows but are not as effective as double glass unless tightly sealed.

Doors also come in many varieties (Fig. 4-22). Many are made of wood, solid or with glass insets for extra light. Recently introduced are doors of wood veneer over foam cores, metal doors with internal insulation, and doors of molded plastics. Proper weather-stripping and storm doors are important to minimize drafts and operating costs. Sliding glass doors, sometimes called *patio doors*, have large glass areas that are affected by solar infiltration and heat transfer. They can be either single or double pane and employ tempered glass for safety. To distinguish single from double glass, simply strike a match and hold it near the glass. If you see a double reflection, it's double glass.

The exterior wall covering is also important to the heat-flow characteristics of the structure. First are the masonry products (Fig. 4-23). Modern brick walls are actually brick veneer as contrasted to the solid brick of older houses. The brick veneer wall rests on the concrete foundation, and bricks are spaced about one inch from the sheathing, thus leaving an air space. Since the wall is only one tier thick, it must be tied somehow to the building frame. So, corrugated metal wall ties are nailed to the framework and bent into the mortar joints. Fieldstone walls are laid in a similar manner, but because of their random size and width require more labor to erect.

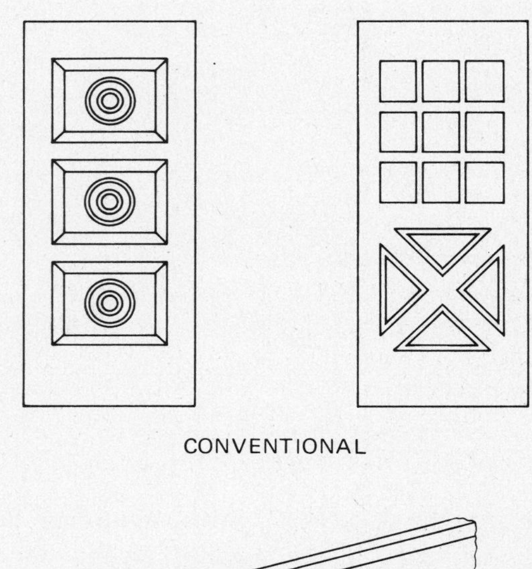

FIGURE 4-22 Door types.

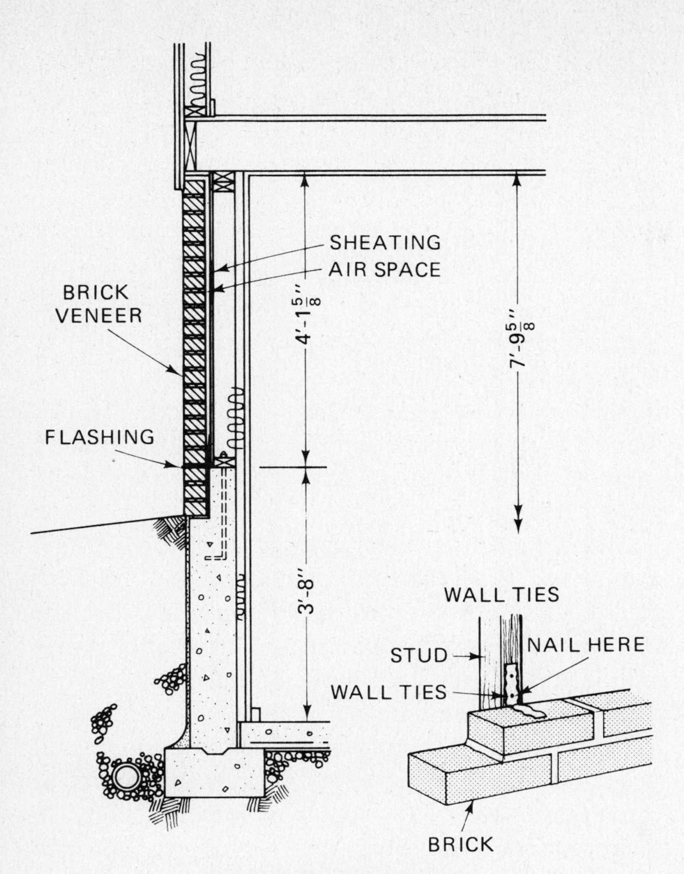

FIGURE 4-23 Exterior walls.

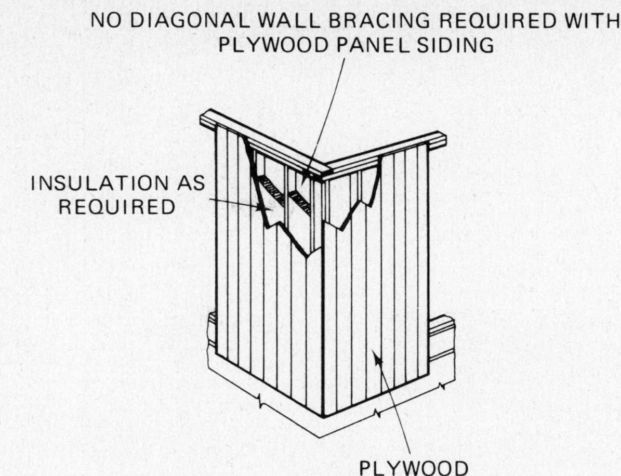

FIGURE 4-24 Plywood siding.

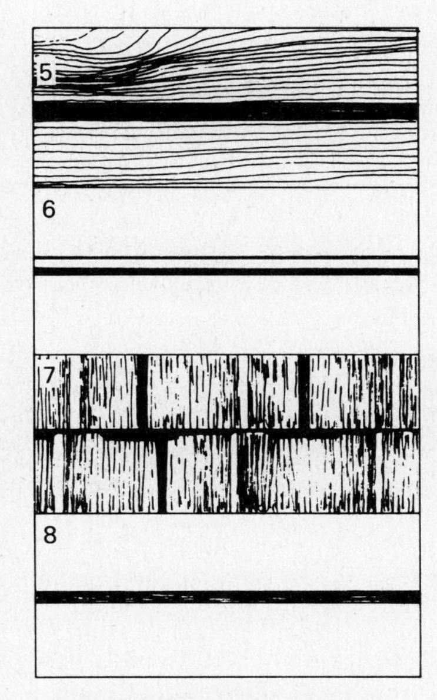

FIGURE 4-25 Lap siding.

Stucco finishes are quite popular in the more temperate zones. Stucco is a very thin coat of cement applied over concrete blocks or on a metal mesh lath nailed to the wood stud wall. Stucco finishes can be hand-troweled or sprayed on by high-pressure guns.

The family of wood products used for outside finishes includes many types: following are the most common. Plywood siding (Fig. 4-24) in 4-ft × 8-ft, 9-ft, or 10-ft sheets of plywood comes plain or with many variations of grooves and surface finishes. Some are applied over the sheathing and some may be applied without sheathing where edges are lapped and sealed. Textured surfaces can be stained to present most interesting grain effects, or they can be completely covered and opaque if desired.

Hardboard siding (Fig. 4-25), which is also a wood product, is not quite the same as plywood. It too is man-made, from wood chips that are converted into wood fibers which are then formed under heat and pressure into flat, dense panels. The resulting product is very strong and durable. It can be made into 4-ft × 8-ft, 9-ft, or 10-ft sheets, with the same grooved-panel effects and rough-sawn textures as plywood. More common perhaps is the family of lap sidings that are cut in 9-in. or 12-in. widths, 12 to 16 ft long, and 7/16 in. thick. These are applied in an overlapping horizontal manner.

Natural woods (southern pine, cypress, and redwood) are also milled into 6-in. and 8-in. widths of various lengths and applied in the same horizontal patterns as hardboard designs. Another interesting and decorative natural wood product is the cedar shake (Fig. 4-26). Cedar is naturally durable. It resists rot and decay in almost any climate, without any kind of preservative.

In the nonwood family of sidings we have several popular coverings. Aluminum siding (Fig. 4-27)

FIGURE 4-26 Cedar shake home (Courtesy, Red Cedar Shingle and Handsplit, Shake Bureau).

FIGURE 4-27 Aluminum siding on house (Courtesy, The Aluminum Association, Inc.).

offers strength, light weight, performance, and beauty. It's made of alloy sheets, cleaned, etched, and chemically coated on both sides. Then the exposed surface is finished with a baked-on decorative coating. Regular panels come 8 in. wide by 12 ft, 6 in. long and finished in white, which can be repainted. Colored panels are also available. Aluminum siding also comes with a $\frac{3}{8}$ fiberboard or polystyrene foam insulation bonded to the back side. Vent holes are provided on the bottom edge of the panels to permit moisture release.

Another siding that is primarily used in rural construction and remodeling is the 12 × 24-in. asbestos shake. It is similar to roofing material but comes in wood grain textures. Frequently, it is nailed over other surfaces that perhaps are no longer capable of withstanding weather exposure.

Now let's look at interior finishing materials. First, the floor. Earlier, we had only gotten to applying the subfloor; now, depending on the choice of the final covering, let's examine the technique to be used. Hardwood floors (Fig. 4-28) are formed from tongue-and-grooved oak or pine. There are three styles of wood flooring: strip, plank, and block. Oak comes in random strips from $1\frac{1}{2}$-in. to $2\frac{1}{2}$-in. wide strips. Pine, on the other hand, comes in planks that have a wider face. Very popular are the pine planks that have simulated wooden pegs in imitation of early American house construction. Block flooring is a method using by-products of short pieces of oak formed together in square or rectangular blocks. They may be applied quickly. The parquet pattern is most notable.

When carpeting, tiles, and/or sheet goods (vinyl or linoleum) are to be used, an underlayer (Fig. 4-29) over the subfloor is required. Plywood and particle board are the popular materials. Particle board has a smooth and very hard surface that resists heel or furniture pressure; it is also relatively inexpensive.

Techniques for inside wall finishes fall into several types. Dry wall (Fig. 4-30), or gypsum board, is formed into sheets 4 ft wide, from 8 to 16 ft long, in thicknesses of $\frac{1}{4}$ in., $\frac{3}{8}$ in., $\frac{1}{2}$ in., and $\frac{5}{8}$ in. Half-inch thickness is the most common for single-layer application. Quarter-inch and three-eighths-inch thicknesses are used over existing wall coverings as a finishing

FIGURE 4-28 Laying hardwood floor (Courtesy, National Oak Flooring Manufacturers Association).

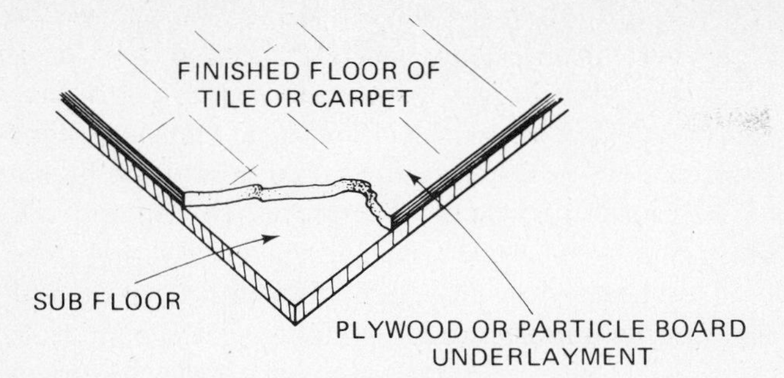

FIGURE 4-29 Underlayment detail.

surface. Five-eighths-inch thickness is used where sound deadening or fire resistance is needed. The dry wall is nailed or glued to the studs, and the joists are finished with tape and joint-finishing compound to provide a smooth, continuous surface. Dry wall can be painted or wallpapered.

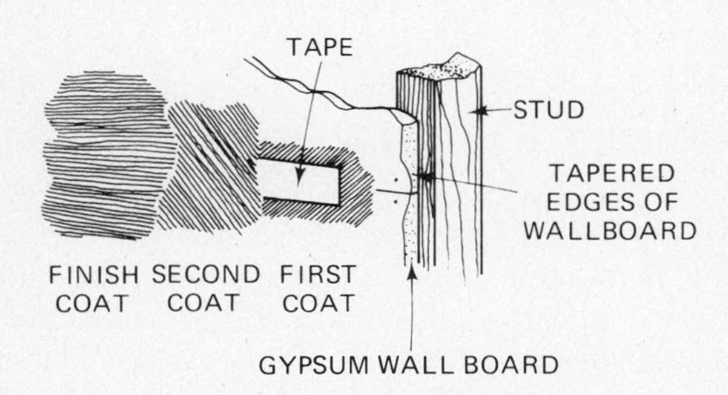

FIGURE 4-30 Inside wall finishing.

Plastering has become very rare, at least in residential work, and it is only mentioned because older homes (usually pre-World War II) were finished this way. The technique was first to nail a plaster lath to the wood studs and ceiling joists. Then a base coat (termed *brown coat*) about $\frac{3}{4}$ in. thick was troweled onto the lath. A thin finishing coat of pure white plaster mix was then applied. Much time, material, and labor were involved; however, a well-plastered house had excellent sound-deadening and insulating qualities.

Paneling (Fig. 4-31) has become an extremely popular interior wall finish, because it can go on quickly and comes in prefinished surfaces of all descriptions. The panel sheets are approximately $\frac{1}{4}$ in. thick with a very thin veneer of real wood on the surface. Most popular natural woods (oak, pine, cherry, walnut, maple, etc.) are available in prefinished surfaces and ready to hang.

Hardboard panels come with a factory printed wood-grain effect or textured appearance that is difficult to distinguish from the real wood appearance. Many of the rough-textured surfaces commonly associated with outside finishes are now being used indoors. Hardboard panels also come in lively decorative colors and patterns.

FIGURE 4-31 Paneling (Courtesy, American Plywood Association).

Paneling, either wood or hardboard, has minimum insulating and sound-deadening qualities, and other materials will be needed to improve thermal and sound control.

The last element of our structure is the roof covering. The finished roof covering (Fig. 4-32) most commonly used is self-sealing asphalt shingles, each measuring 12 in. × 36 in. They weigh 235 to 240 lb per square (a square equals 100 ft^2). Self-sealing adhesive is applied on the face; the sun's heat softens this adhesive, fusing each single tab to the underlying shingle. Premium grade asphalt shingles are available that weigh from 340 to 405 lb per square; these will last longer and, because of this thickness, can be made to simulate wood shingle patterns. Asphalt shingles come in a wide variety of colors from white to jet black. The lighter the color the better it reflects the sun's heat, which improves summer cooling but hinders warming during the winter. So color selection is geared to geographic and economic considerations. White roofs can present appearance problems where dirt and mildew cause splotched and streaked areas. Some shingles have a fungus-resistant metal oxide coating that destroys algae and fungus on contact.

Wood shakes are coming back into popularity. Cedar shakes and slate were the two roofing materials most available for early American homes. Slate is now practically nonexistent except to repair existing roofs. Cedar wood shakes have excellent insulating value and natural resistance to decay. They are most popular in the vacation or second-home market. Compared to asphalt shingles, wood shakes are more expensive in material and labor.

There are two other roof coverings that are used primarily in very temperate climates. One is a concrete tile that is cast into an arch shape. These tiles are nailed to the roof in an overlapping fashion to shed rain runoff. The arch effect creates air pockets that help insulate and ventilate the roof deck.

Another type is the built-up roof deck with tar and gravel sealer. This is usually associated with a system that uses the roof decking as a ceiling below. Decking is a layer of insulation material that is covered first by hot asphalt (tar) and then by washed gravel stones; the stones are embedded in the tar.

FIGURE 4-32 Applying asphalt shingle (Courtesy, Asphalt Roofing Manufacturers Association).

A final note on roof coverings. Some older homes had tin roof panels with locking edges for water-tight construction. Seldom are they used in modern building construction, except for barns, sheds, and the like.

This discussion has been directed at residential building techniques and highlights of the materials used. By no means did it cover all possible options, and a visit to a typical home building project in your community would enable you to see these techniques and materials in their various stages.

NOTE:

In this chapter we deliberately have not covered the subject of insulation in any depth. The topic will be reexamined in Chap. 5 as a factor of heat flow.

COMMERCIAL CONSTRUCTION

Now we will discuss commercial construction. In many small commercial buildings we find little difference in materials and methods from those used in residential work. Foundations are similar, as are frame walls with brick veneer outside, dry walls inside, and conventional truss roof with asphalt shingles. These buildings might house barber or beauty shops, used-car sales offices, small retail stores, drug stores, and the like.

Low-rise shopping centers are one of the largest consumers of air-conditioning products, and their construction characteristics are very similar to one another all across the country. A typical shopping center building (Fig. 4-33) consists of a slab-on-ground foundation with footings at each common building wall. Exterior walls are concrete block, as are the separating walls between stores, for fire protection. Usually the front and side walls get a decorative treatment of brick, stucco, or other material. The roof structure consists of steel bar joists spanning the block walls, or, if the span is too wide, it is center-supported by steel columns and I-beams as needed. On top of the bar joists are metal sheets (decking), which are welded to the bar joists and to each other, forming a complete deck overhead. On top of the deck is a layer of dense insulation material up to several inches thick. Finally, for water pro-

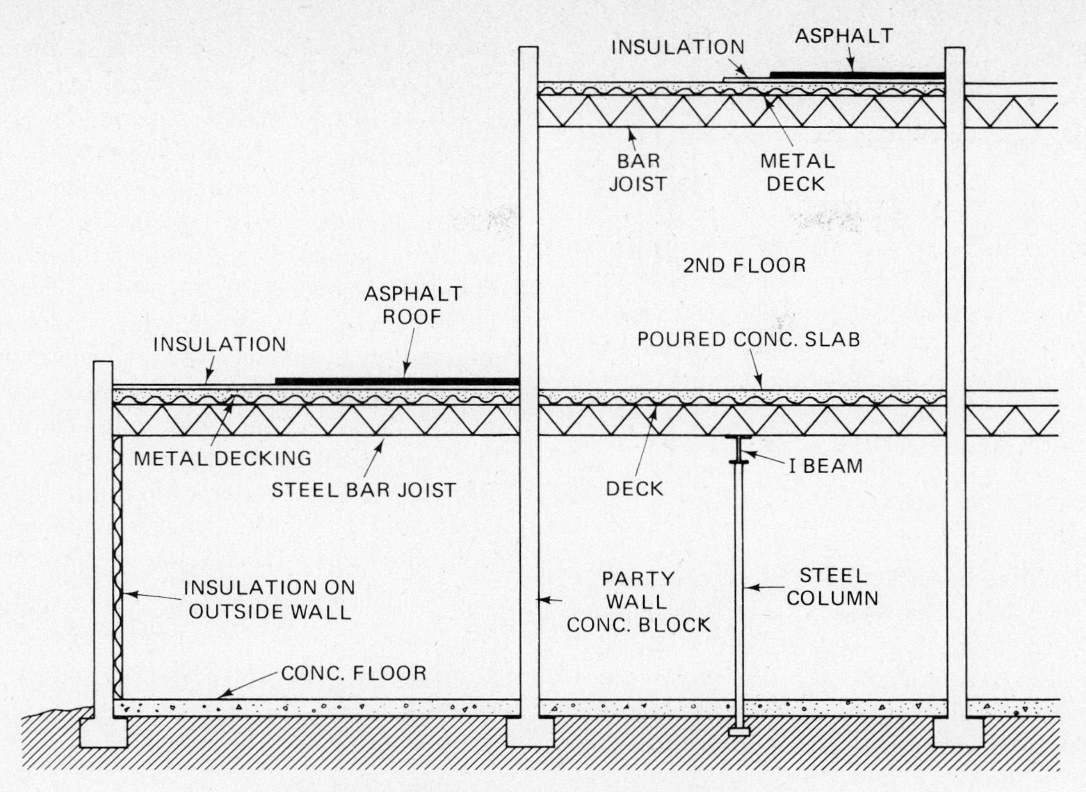

FIGURE 4-33 Shopping center building construction.

tection, a treatment of hot asphalt is applied. The roof slope along with drain pipes carry away water runoff.

Where a second floor is to be constructed, concrete can be poured on the metal decking to form a slab. Depending on the loading per square foot, the strength of the bar joists, support columns, and beams would then be increased accordingly to hold the added weight. A similar technique can also be used for the roof, where a slab of light-weight aggregate concrete is poured on steel decking. Insulation is then placed above the concrete and finally the asphalt sealer is applied.

The combination of insulation and asphalt sealer is called a *built-up-roof*, and roofing manufacturers offer a bonded guarantee against leaks and decomposition for many years, providing the roof is not damaged other than by natural causes. That statement is *very important* to the air-conditioning contractor where roof-mounted equipment is involved. Proper roof curbs and sealing precautions are spelled out by the air-conditioning equipment and roof manufacturers and should be closely followed.

This ceiling treatment (Fig. 4-34) is the one most commonly used. Suspended from the bar joists by means of hanger rods are T-bars to hold lay-in panels of light-weight accoustical ceiling materials. Fluorescent lighting strips and incandescent light fixtures are also suspended from this grid. The cavity between the bar joist and ceiling provides a groove trench passageway for air-conditioning ducts, plumbing, fire sprinklers, pneumatic tubes, and electrical wiring, so it's important that the construction work be planned and coordinated to minimize errors, changes, and delays.

Inside the building, party walls can be treated in many ways, from painted block to dry wall to panel-

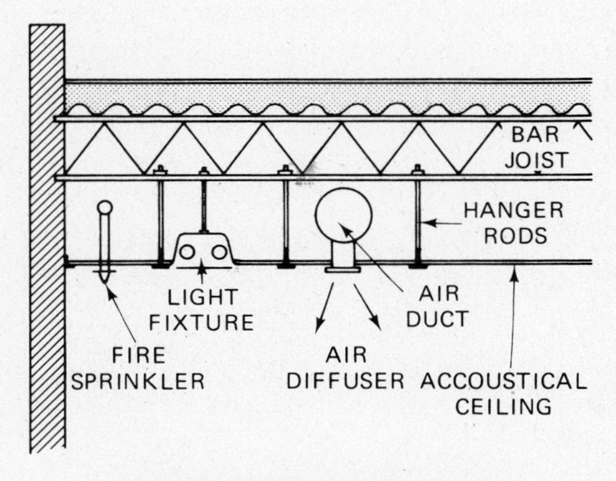

FIGURE 4-34 Ceiling construction.

ing, as previously described. Since these are retail stores, racks, shelving, etc., will, of course, be included. No insulation is applied between party walls because it is assumed that all the stores will be conditioned. Outside walls are usually given some insulation treatment to cut down heat transfer.

HIGH-RISE CONSTRUCTION

High-rise building construction must of necessity be different from that described above, and because of the many variations, and also because the entire building system is designed by architects and mechanical engineers complete with detailed drawings and specifications, we will not go into detail on the subject here. It is, however, important that the air-conditioning technician be able to read blueprints, know the symbols for plumbing, air-conditioning equipment, ductwork, grilles, registers, fire dampers, and electrical and pneumatic (air) control systems.

There are two distinctly different types of high rise construction (Fig. 4-35). First is the technique of massive steel column and beam construction. The skeleton of steel is welded and bolted together. Bar joists and decking and concrete forms are used to fill in the floors later. The other approach is to start at ground level and, with the aid of temporary forms, an entire network of concrete columns and reinforced concrete floor is poured. The forms are dismantled and reerected on the next floor; thus the building goes up one floor at a time in contrast to the steel structure. Both techniques can be and frequently are applied together to achieve certain architectural or structural objectives. One modern and interesting treatment of the exterior walls in high-rise buildings is the *curtain wall.* Unlike most masonry products, the curtain wall is made of relatively thin panels of steel, aluminum alloys, or reflective glass, which are hung or attached like a curtain to the frame structure. They do not necessarily tie into the floor slab at all. These panels are complete with windows or glass as required and have a core of insulation material. There is little structural strength in a curtain wall, and if packaged terminal units (through-the-wall type) are used, care must be exercised to ensure a positive wind and water seal around the unit. (Wind effect can be 75 mph or higher at the upper stories.) Don't depend on the curtain wall for support.

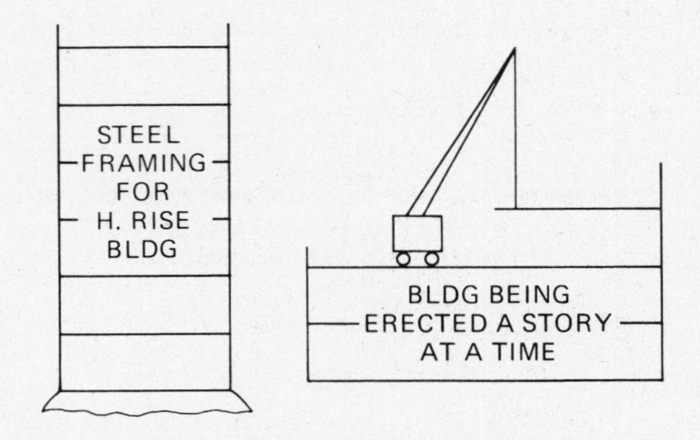

FIGURE 4-35 Types of building constructions.

FLOOR PLANS AND MEASURING TOOLS

As mentioned above, the technician should have a working knowledge of blueprints and, if necessary, be able to sketch and draw plans of existing structures. This is true for the salesman, estimator, installer, and even the service man; only the degree of day-to-day use makes the difference. Let's start with a list of tools and equipment:

(a) 6-ft tape rule
(b) 50-ft reel tape measure
(c) clip board
(d) architect's scale
(e) pencil and eraser
(f) grid sketch paper (approximately 8 to 10 squares per inch)
(g) pad of tracing paper (thin and transparent)
(h) roll of masking tape

The architect's scale (Fig. 4-36) is available at drug stores, school supply stores, etc. It has three faces on which are printed scales of $\frac{3}{32}$ in., $\frac{1}{8}$ in., $\frac{3}{16}$ in., $\frac{1}{4}$ in., $\frac{3}{8}$ in., $\frac{1}{2}$ in., $\frac{3}{4}$ in., 1 in., $1\frac{1}{2}$ in., and 3 in. to the foot. One-quarter in. per foot means $\frac{1}{4}$ in. on a blueprint is equal to one (1) foot of actual length. Thus, a 40-foot house would be 10 in. on a drawing, etc.

Most house plans are produced in $\frac{1}{4}$ in. scale for the floor plan and elevation drawings. The plan view is as if one were looking down on the house with the roof removed. Elevation views are those showing front, sides, and rear walls. Details of a wall section, etc., are frequently shown in $\frac{1}{2}$-in. or 1-in. scale, depending on the degree of accuracy or exposure needed. To use the architect scale, for example, to measure on a $\frac{1}{4}$-in. scale, place it alongside the line to be measured (Fig. 4-37). First read the whole

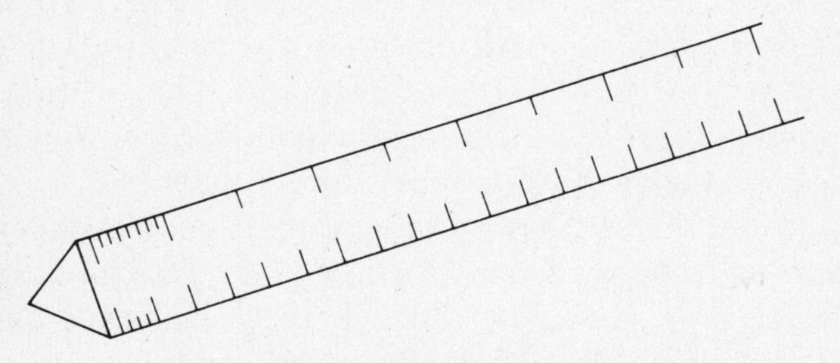

FIGURE 4-36 Architect's scale.

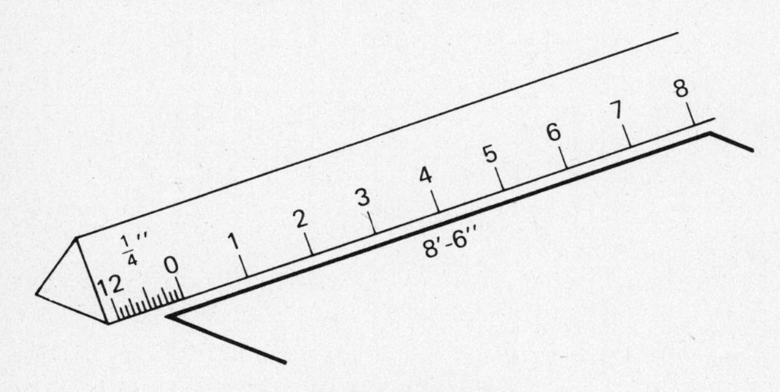

FIGURE 4-37 $\frac{1}{4}''$ scale.

length at 8 feet plus a part of a foot that is read from the inch scale as 6 in. Because two scales are printed on the face, make sure you read from the correct one.

Using the scale in that fashion, you can measure information from a house plan (Fig. 4-38) such as the length and width of the whole house and each room, window sizes, door openings, attic height, etc. This information would later be used in figuring the heat load and would be recorded on the survey form. Or, if a permanent record is needed and you can't keep the original plan, place a sheet of thin tracing paper over the drawing. Use masking tape to hold it in place; then make a single-line sketch of the information needed. It is not necessary to duplicate the exact drawing in every detail.

Drawings for larger commercial buildings are usually shown in $\frac{1}{8}$-in. scale for the plans and then in larger ratios. Plans and specifications are available; seldom would you have to trace blueprints.

If you are required to make a plan of an existing house or equipment room, etc., then a clip board and grid sketch paper are very useful (Fig. 4-39). Start at one corner using measuring tapes to find distances. Draw the total structure outside dimensions in single-line form. If interior walls, doors, and windows are needed, these are measured and penciled in. It's usually convenient to work to the center of the walls through door openings and then make a note of the wall thickness and construction. Interior measurements should add up to approximately the outside

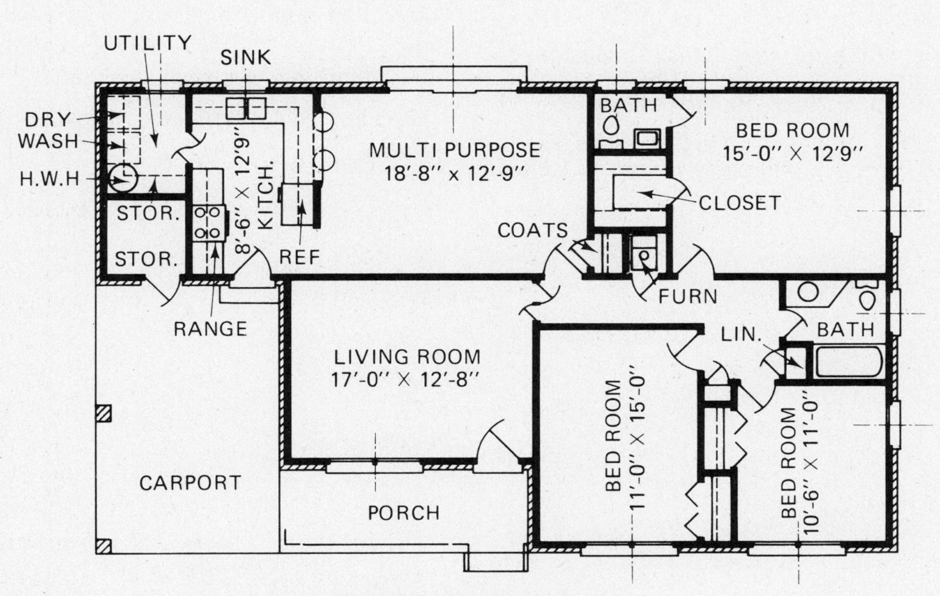

FIGURE 4-38 House plan.

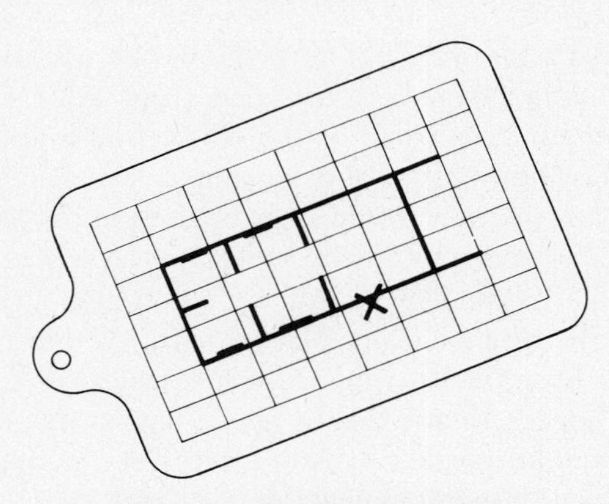

FIGURE 4-39 Clip board and grid paper.

measurements; otherwise, you have made an error. Sometimes, of course, you are not always able to measure the outside; then interior estimates are totaled and used. A good working assignment would be to make a complete sketch of your house or apartment. Then with a little experience behind you, go out to a model home or house under construction and repeat the process as if it were a "real" assignment. Come back and finish your drawing. At first you will forget important information; but with the aid of survey forms and more experience, you will be doing a better job in a relatively short time.

Now that we have a basic understanding of certain materials and construction terms, let's proceed to examine heat flow through these materials and other related data that affect size and selection of equipment.

PROBLEMS

1. Houses usually fall into three categories. Name them.
2. Name four types of foundations.
3. Timbers that span the foundation and support the floor are called ______________.
4. Vertical wall framing members are called ______________.
5. A 2 × 4-in. stud will have what actual dimensions?
6. Brick veneer walls are made of solid brick: true or false?
7. Prefabricated roof supports are called ______________.
8. Which is steeper, a 3/12 or 4/12 roof?
9. ______________ glass windows have low heat loss.
10. In commercial work a roof made of insulation with an asphalt sealer is called a ______________. Many are ______________.
11. Most commercial stores have what is known as a ______________ ceiling for concealing ducts, plumbing, electrical work, etc.
12. The rule used to measure blueprints is an ______________.

5

HEAT LOAD FACTORS

In previous chapters, we established the benefits of air-conditioning, human comfort standards, psychrometric properties of air, building construction techniques, and materials. Now let's put some of these fundamentals to practical use in phases of air-conditioning such as factors that affect heat loads in a structure. Then, in Chaps. 6 and 7 we will discuss the actual practice of estimating cooling and heating loads.

The year-round system must cope with source factors that affect both winter and summer conditions. To a great extent they are opposite in character. In winter we must add heat and humidification, and in summer we want to remove heat and dehumidify. But these opposites have two things in common: sensible heat and the other latent heat as previously defined under psychrometrics.

Sensible Heat Load Factors are derived from:

- heat transmission through walls, floor, ceiling, doors, windows, etc.
- heat produced (or lost) by bringing in outdoor ventilation air
- heat produced by people (occupants)
- heat from solar (sun) effect
- heat from lights, kitchen appliances, dryers, motors, etc.

Sensible heat is expressed in Btu.

Latent Heat Factors are derived from:

- moisture given off by humans
- moisture from cooking appliances, dryers, and bathing
- moisture introduced by ventilation air

Latent heat is also expressed in Btu.

In the descriptions above, it should be noted that some are influenced by external conditions and some are from internal sources; the analysis is demonstrated pictorially by Figure 5-1. With this composite representation in mind, let's now examine the individual sources.

EXTERNAL LOADS

Heat Transmission

The heat gain through walls, floors, and ceilings will vary with the type of construction, the area exposed to a different temperature, the type of insulation, the thickness of insulation, and the temperature difference between the refrigerated space and the ambient air.

Thermal conductivity varies directly with time, area, and temperature difference and is expressed in Btu per hour, per square foot of area, per degree Fahrenheit of temperature difference, per inch of thickness.

It is readily apparent that in order to reduce heat transfer, the thermal conductivity factor (based on the composition of the material) should be as small as possible and the material as thick as economically feasible.

Heat transfer through any material is also subject to or affected by the surface resistance to heat flow, and this is determined by the type of surface, rough or smooth; its position, vertical or horizontal; its reflective properties; and the rate of air flow over the surface.

Extensive testing has been done by many laboratories to determine accurate values for heat transfer through all common building and structural material. Certain materials (like insulation) have a high resistance to flow of heat; others are not so good.

In order to simplify the task of calculating heat loss, the industry has developed a measuring term called the *resistance* (R) value which is the resistance to heat flow of one inch of material, or of a specified thickness, or of an air space, an air film, or an entire assembly. Its value is expressed as degrees Fahrenheit temperature difference per Btu per hour per square foot. High R values indicate low heat flow rates. The resistances to several components of a wall may be added together to obtain the total resistance:

$$R_t = R_1 + R_2 + R_3 \ldots$$

Figure 5-2 lists some R values of common building materials in order to illustrate the differences in heat flow characteristics. For more exhaustive lists of R values, refer to the ASHRAE *Handbook of Fundamentals*. The actual quantity of heat transmission (Q)

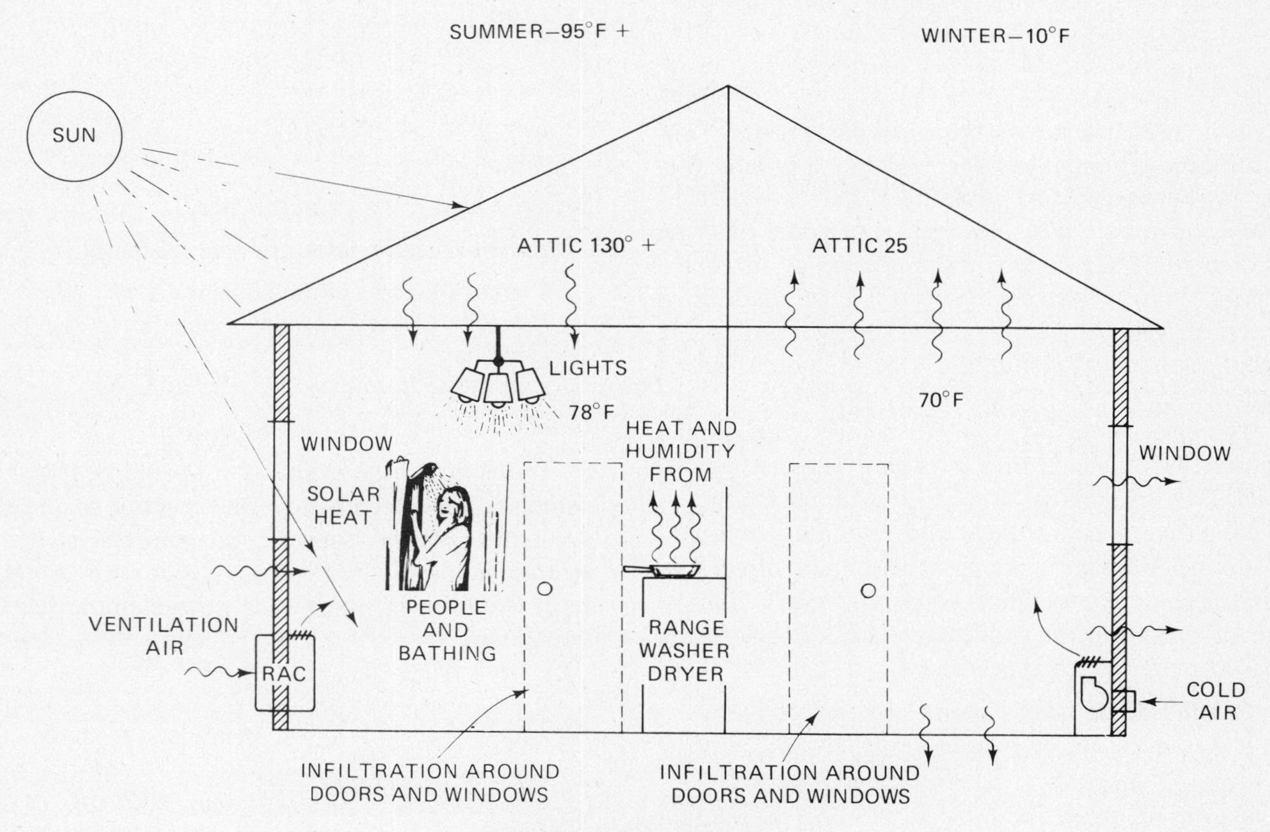

FIGURE 5-1 Heat load factors.

MATERIAL	DENSITY lb/ft³	MEAN TEMP °F	CONDUCTIVITY k	CONDUCTANCE C	RESISTANCE (R) PER IN.	RESISTANCE (R) OVERALL
Insulating Materials						
Mineral wool blanket	0.5	75	0.32		3.12	
Fiberglass blanket	0.5	75	0.32		3.12	
Corkboard	6.5– 8.0	0	0.25		4.0	
Glass fiberboard	9.5–11.0	−16	0.21		4.76	
Expanded urethane, R11		0	0.17		5.88	
Expanded polystyrene	1.0	0	0.24		4.17	
Mineral wool board	15.0	0	0.25		4.0	
Insulating roof deck, 2 in.		75		0.18		5.56
Mineral wool loose fill	2.0– 5.0	0	0.23		4.35	
Perlite, expanded	5.0– 8.0	0	0.32		3.12	
Masonry Materials						
Concrete, sand and gravel	140		12.0		0.08	
Brick, common	120	75	5.0		0.20	
Brick, face	130	75	9.0		0.11	
Hollow tile, 2 cell, 6 in.		75		0.66		1.52
Concrete block, sand and gravel, 8 in.		75		0.90		1.11
Concrete block, cinder, 8 in.		75		0.58		1.72
Gypsum plaster, sand	105	75	5.6		0.18	

FIGURE 5-2 Typical heat transmission coefficients.

through a substance or material is then calculated by the formula

$$Q = U \times A \times TD$$

Q = heat transfer, Btu/h

U = overall heat transfer coefficient, (Btu/h)/(ft²)/(°F TD)

$U = \frac{1}{R_t}$ value (For an assembly $R_t = R_1 + R_2 + R_3 \ldots$ for various components)

A = area in ft²

TD = Temperature difference between inside and outside design temperature and refrigerated space design temperature.

FOR EXAMPLE:

Calculate the heat flow through an 8-in. concrete block wall (Fig. 5-3) 100 ft² in area, having a 60°F temperature difference between inside and outside.

R value for an 8-in. concrete block wall is 1.72.

U therefore is $= \frac{1}{R} = \frac{1}{1.72} = 0.58$ Btu/h/°F/ft²

$Q = 0.58 \times 100$ ft² $\times$ 60°F = 3480 Btu/h heat flow into the space.

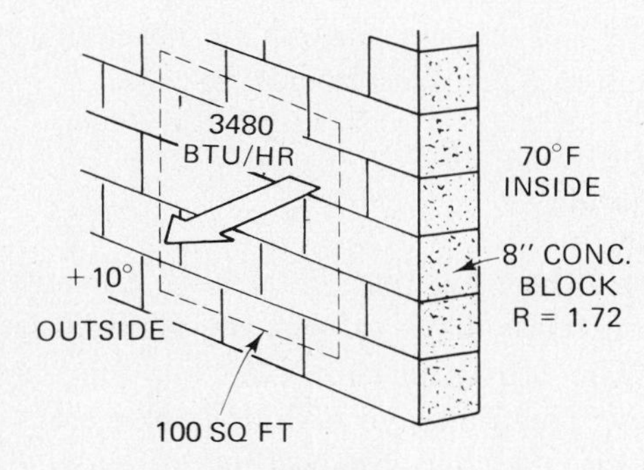

FIGURE 5-3.

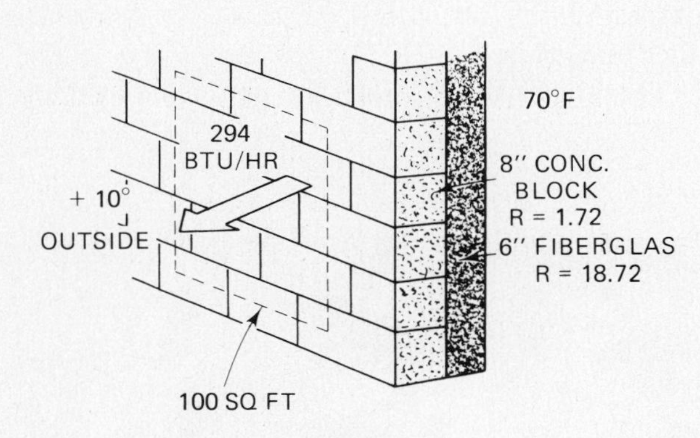

FIGURE 5-4.

Now add six inches of fiberglass insulation to the wall (Fig. 5-4) and recalculate the transmission load.

$$U \text{ factor} = \frac{1}{R_t}$$

$R_t = R_1$ (concrete block) $+ R_2$ (6 in. insulation)

$R_t = 1.72 + 18.72$ (R value for 1-in. insulation $= 3.12$; R_2 then equals 6×3.12, or 18.72)

$R_t = 20.44$

Therefore $U = \frac{1}{20.44} = 0.049$

$Q = 0.049 \times 100 \text{ ft}^2 \times 60°\text{F} = 294 \text{ Btu/h}$

This example demonstrates the cumulative effect of R values in determining the total wall resistance and dramatically shows the heat reduction that can be achieved through proper insulation; in this case, from 3,480 Btu/h down to 294 Btu/h, which would drastically reduce the size of the necessary heating equipment and resulting operating cost.

Insulation

It is in the transmission of heat movement that insulation really pays off—in the walls, floors, and ceiling. The actual function of insulation is to resist heat flow. The ability to do this is also expressed in its R value, which is the method developed by the National Mineral Wool Insulation Association for comparing the insulation value of its products to other materials.

The higher the R value, the better the insulation. For example: $3\frac{1}{2}$ in. thick fiberglass insulation has an R value of 11. To achieve the same degree of resistance with other materials it would take:

(1) a wood wall one foot thick
(2) a brick wall $4\frac{1}{2}$ feet thick
(3) a reinforced concrete wall 10 feet thick
(4) a stone wall 17 feet thick

This sounds incredible, but the facts do substantiate the R value. The R number is marked on the outside of each batt or blanket of bulk insulation and on each bag of loose fill. When several numbers appear, the value depends on where the insulation is used—wall, ceiling, or floor. Let's examine the types of insulation, their use, and their application.

Bulk insulation comes in three popular forms (Fig. 5-5): *flexible blankets* come in large rolls, *batts* are cut into lengths of 96 in. (8 ft) or less, and *loose fill* comes in bags containing loose fibrous material.

Both the roll blankets and batt insulation come with a strong, vapor-resistant facing of asphalt-impregnated kraft paper on one side. They also have a one-inch stapling flange at the edges for quick and easy installation. Another option is bright aluminum foil applied to one side of the fiberglass insulation; the foil reflects back the radiant heat that comes through the walls. The transmission R value for both types is based on thickness—for example, $2\frac{1}{4}$ in. thickness is rated R-7, $3\frac{1}{2}$ in. is rated R-11, $3\frac{5}{8}$ in. at R-13, and 6 in. is rated R-19. NOTE: No credit can be given to the reflective backing insofar as the rating system is concerned because there may be no constant solar load.

(a)

(b)

FIGURE 5-5 Types of insulation.

Fiberglass is also available in unfaced insulation, which is designed to hold itself in place between studs and joists until the interior finished surface is installed. It is useful around pipes and vents and also inside for a combination of thermal and sound control. Remember though, it has no vapor barrier if used in outside walls.

Pouring insulation is mainly for the do-it-yourselfer, and to insulate existing structures where the use of roll and batts may not be possible. The loose fill is also blown into stud spacing (Fig. 5-5a), joists, etc., by professional contractors. *R* factors depend on material and depth, so check the specifications; the value may differ for different manufacturers.

On the average, *R* values for poured or blown wool are as follows:

R Value	*Minimum Thickness*
R-22	10 in.
R-19	$8\frac{3}{4}$ in.
R-11	5 in.

Remember, poured or blown wool has no vapor barrier, and other methods of moisture control should be used. Polyethylene film or tar paper may be considered if they can be obtained. For an existing structure, there are vapor-resistant paints that can help, but the "perm rating" must be 1 perm or less. (Perm is a rating of how much vapor will pass through a substance.)

Reflective insulation (Fig. 5-6) is made of one or more layers of aluminum foil and functions both as an insulator and vapor barrier. Because it insulates by reflecting or bouncing back heat, it is used primarily in those areas where summer cooling (not heating) is the main consideration. To be effective, it requires an air space of at least $\frac{3}{4}$ in. facing the reflective surface. Again, check the *R* value of the particular brand being applied.

Rigid foam insulation panels (Fig. 5-7) are adaptable to a wide variety of needs. They are constructed of polystyrene or polyurethane material and come in panels. One particular size ($\frac{3}{4}$ in. × $13\frac{5}{8}$ in. × 48 in.) can be fitted between stud spacing or furring strips on 16 in. centers. Basement walls where paneling might be used are an excellent application, because the poly panels will not absorb moisture, rot, or corrode, and won't support fungus or mold. They may be glued in place using a water-base adhesive.

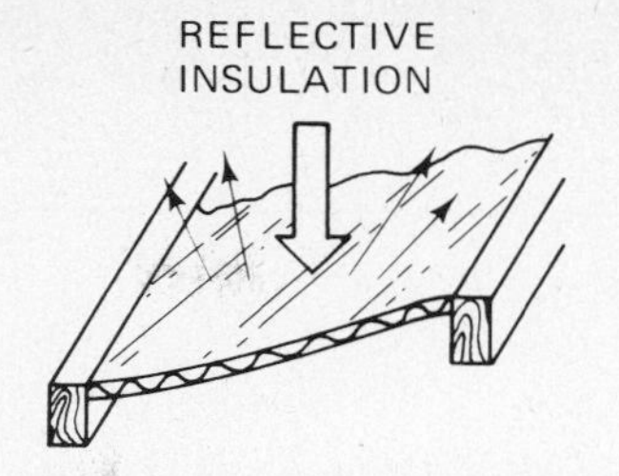

FIGURE 5-6 Reflective insulation.

Poly panels are also ideal for use in insulating the inside walls of crawl spaces (Fig. 5-8), or for insulating the perimeter foundation of slab-on-grade foundations and sometimes the below-grade portion of the outside walls of crawl spaces. They are impervious to moisture and ground acids, so decomposition will not occur. Poly panels have an approximate *R* value of 5.8 per inch of thickness.

Equally important to the *R* value of insulation is the manner or quality of application by the builder, contractor, or homeowner. Several excellent publications that detail specific recommendations are available from the National Mineral Wool Insulation Association, Inc., 211 East 51 Street, New York, New York 10222, or from the National Association of Home Builders Foundation, Inc., 627 Southlawn Lane, P. O. Box #1627, Rockville, Maryland 20850. There is also information available from the FHA in its *Minimum Property Standards for One and Two-Family Dwellings*, known as HUD/FHA Revision 1 to 4900.1, effective November 22, 1974. Also, HUD/FHA Minimum Property Standards for Multifamily Housing. The 1973 edition deals with the apartment and townhouse structure.

Insulating glass instead of single-pane glass is also a big help in reducing transmission loads. The heat flow factor of insulating glass that has a $\frac{1}{4}$-inch air space is exactly half of that for single pane windows. This is most important where large expanses of glass walls and windows are used, usually vacation and recreational homes. Storm windows with at least a 1-in. air space and very tightly sealed have about the same heat transfer resistance as double glass, but air-leakage factor can be a problem, depending on quality and fit.

One major consideration in reducing transmission loads is the handling of roof or attic ventilation (Fig. 5-9). In a flat roof situation, there is no opportunity for exhausting the hot air before it affects the room conditions; we must rely totally on the insulation. However, in a pitched roof situation *there is* an

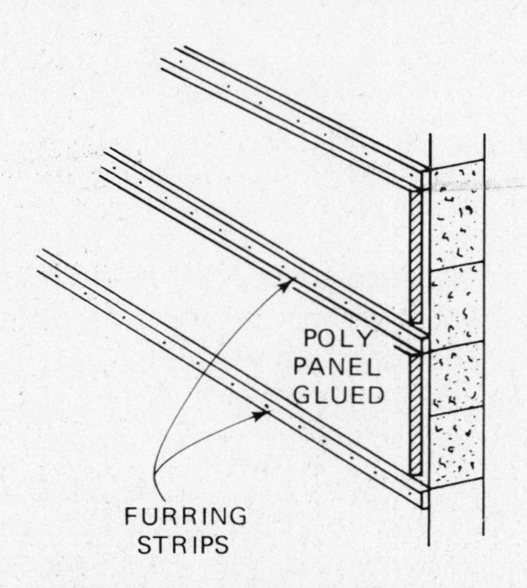

FIGURE 5-7 Rigid foam insulation panels.

opportunity to do something. Proper ventilation of the attic can reduce heat loads measurably. Attic temperatures can climb to 140° and above even if vented in accordance with HUD/FHA minimum recommended standards. These standards include proper treatment of gable ends, ridge vents, and eave openings, depending on the use of a vapor barrier at the ceiling. The flow of air must be created by natural convection currents, and the sizes of minimum ventilation openings are a function of the ratio of total square feet of ceiling to the square feet of ventilation opening.

More and more the use of mechanical exhaust fans (Fig. 5-10) is recommended to cut cooling loads where summer temperatures exceed 90°F. Several types of attic vent fans are available: roof-mounted or in the gable ends. The capacity of a fan should be large enough to change the attic air volume every 10 minutes, or six changes per hour. This should pull the attic temperature down to within a few degrees of the outdoor air. The heat penetration from the attic to the room below is a function of the temperature difference and R values of the ceiling and insulation (which remain constant). Therefore, a reduction of attic temperatures from 130° to about 100° means over a 25% reduction in ceiling heat load and a resulting lowering of air-conditioning operating costs. Such fans are usually not included by the builder, in order to keep the sale price of his houses down, and the home-owner is often unaware of the advantages of an exhaust fan.

Solar Effect

One of the greatest external sources of summer heat load is the sun. The sun's heat can get into a building in one of two ways: through glass and through the walls and roof (Fig. 5-11). Solar heat through glass is *absorbed* instantly into the room. This is in addition to the conducted heat passed by the glass. Methods of reducing solar radiation will be covered later.

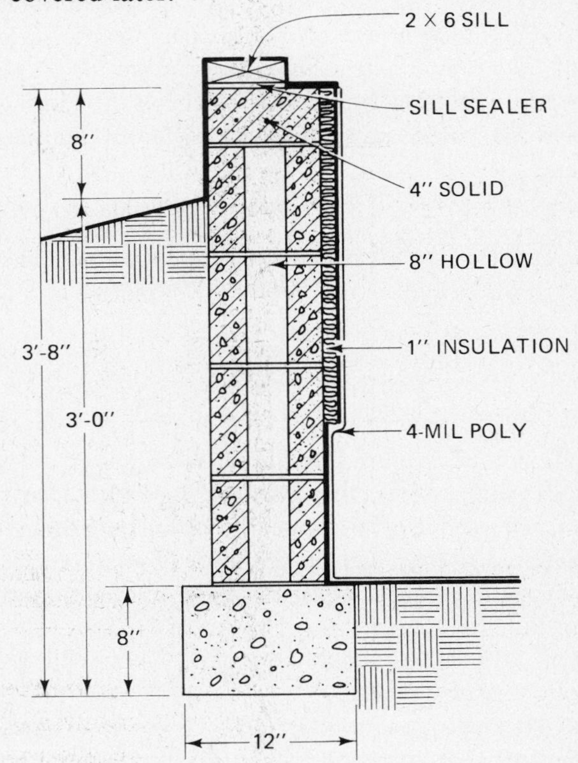

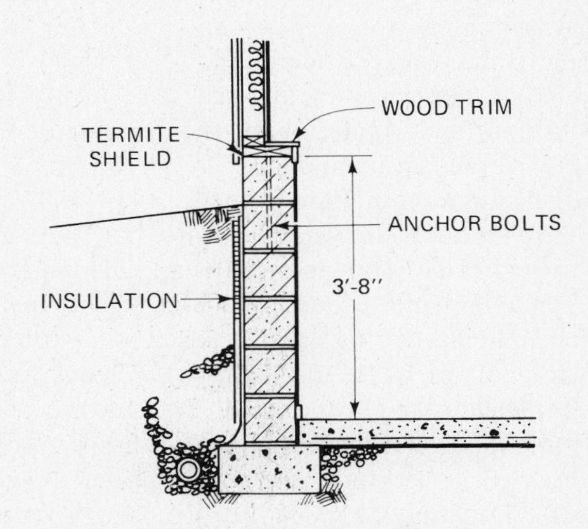

FIGURE 5-8 Poly panel insulation.

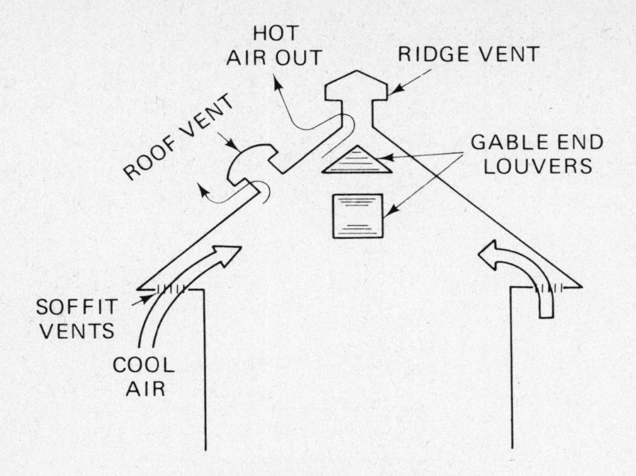

FIGURE 5-9 Roof ventilation.

In the case of the walls and roof, however, the sun heats up the outside surface and then the heat is *conducted* into the room. Depending on the type of construction and insulation, there is usually a time lag of from two to 10 hours before the heat reaches the room. This means heat may be pouring in *after* the sun has gone down.

The exact amount of heat gain from either source depends upon the area, the direction each wall faces, the shading, and the type and color of the surface exposed to the sun. The amount of heat also depends on the position of the sun; that is, the season of the year, the time of day, and the latitude at which the building is located. Figure 5-12 shows graphically the effect of the sun's heat through glass for a typical August day. Note the eastern exposure is at its high point at approximately 8:00 A.M. Southern exposure peaks at noon, and the western peak occurs at about 4:00 P.M. A northern exposure does get some reflected heat, but it's a very small amount compared to direct sunlight.

Thus the orientation of a structure is most important. For example, an automobile showroom facing east or west will have a terrific sun effect through the show window. The same is true for a residence with a large picture window glass exposure. When the direction of orientation cannot be changed, effective shading of the glass can result in an appreciable reduction in solar heat during the summer cooling period. (Obviously, solar heat during the winter is desirable to help reduce heating costs.)

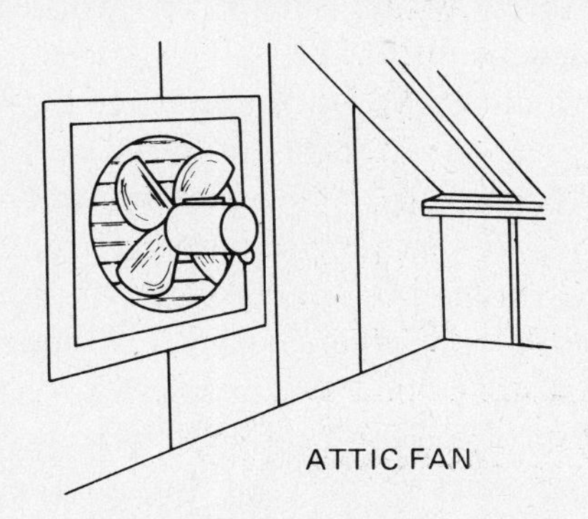

FIGURE 5-10 Attic vent fan.

The outside shading of glass (Fig. 5-13) may be achieved through the use of an overhang on the building or by outside awnings, a sun screen, or vertical blinds. These are about the most effective methods in turning away direct solar rays; they may reduce the solar heat about 75%. There will still be about 25% coming through because of diffused light.

Inside shading (Fig. 5-14) helps but is not as effective. Venetian blinds reflect only about 35% of the solar rays. Light drapery fabrics help to reflect outdoor light. Some fabrics are specifically recommended for minimum solar absorption. Insulated glass reduces the solar penetration about 10–20% over ordinary glass. Also available is special heat-

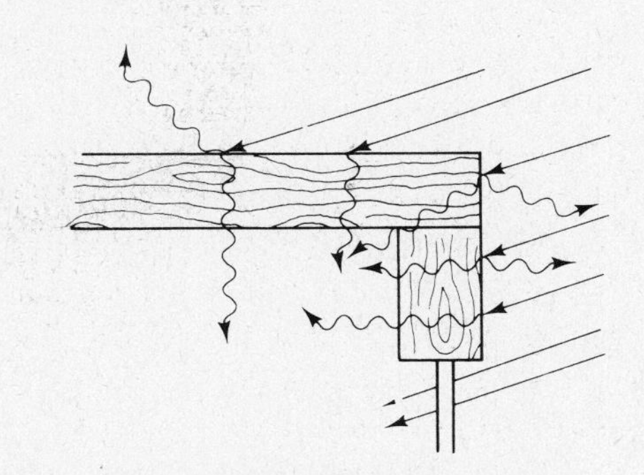

FIGURE 5-11 Solar penetration.

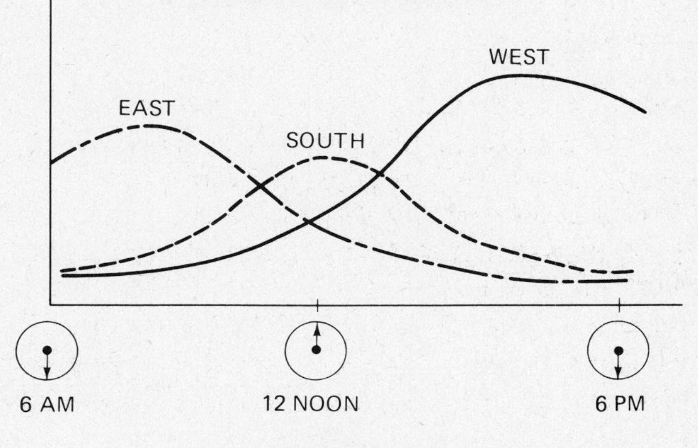

FIGURE 5-12.

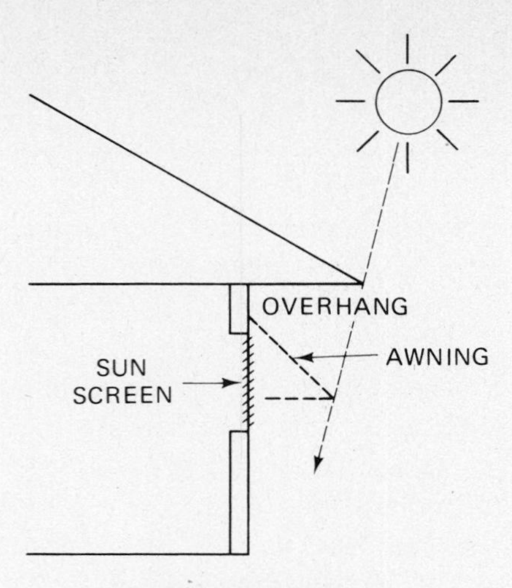

FIGURE 5-13.

reflecting glass, usually used in commercial buildings, which has screening and filtering agents built into the glass itself. We must keep in mind that the solar effect is becoming more and more important, since winter heating and devices that permanently block the sun's rays are apt to be reconsidered in view of their overall contribution to year-round operating cost. The ideal solution would be outside mechanically actuated shading devices that would react to the sun's location or to internal lighting and heat requirements and adjust themselves accordingly.

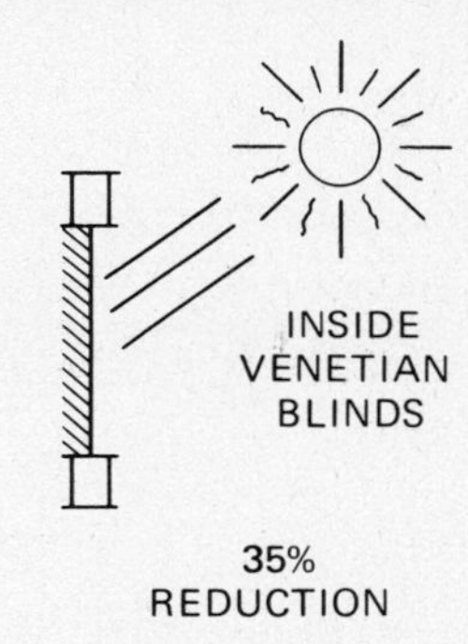

FIGURE 5-14.

FIGURE 5-15.

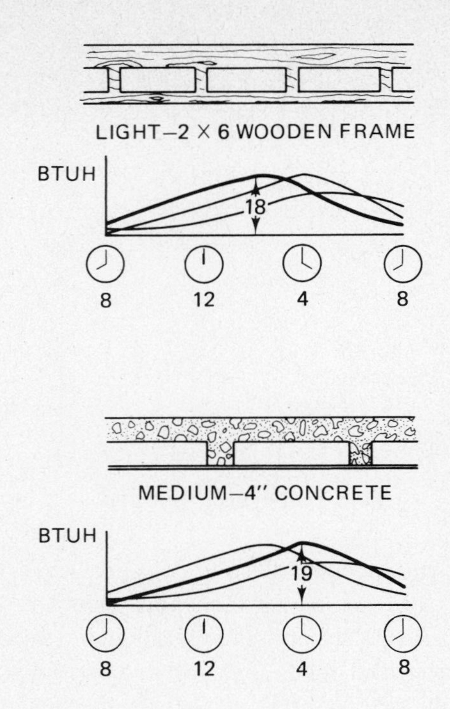

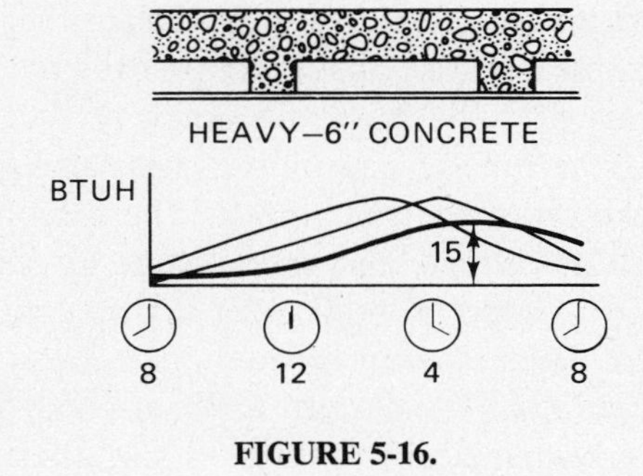

FIGURE 5-16.

Let's consider the sun's effect on walls and roof. In this case, the sun heats up the outside surface (Fig. 5-15), and then some of this heat gradually works its way into the building, while some is reflected and conducted back to the outside air. The amount of penetration depends on the wall or roof composition, insulation, and ventilation. Figure 5-16 shows a comparison of heat penetration and peak loads among a light wood frame roof (flat), a medium-thick concrete roof (4 in.), and a 6-in. heavy concrete roof. The less dense wood roof peaks at about 2 P.M., shortly after the high noon sun, whereas the 6 in. concrete roof doesn't peak till 6: 00 P.M. or later.

The effect of the sun beating down on the walls is similar to its effect on the roof, but another factor is involved—that of the wall facing. Walls and roof both cool at night. Roof heating starts almost immediately as the sun rises, but wall heating depends on when the sun rays actually start hitting the walls. Like glass, they too are subject to the time cycle. West walls tend to peak out at 6:00 P.M. or later, and because the ambient temperature is also at its peak, west walls generally have the most heat gain per square foot.

This delayed peaking and storage of solar heat in the walls and roof is referred to as the *flywheel effect* because it is not instantaneous as would be the case of glass. Therefore, equipment selection can be preplanned to take advantage of this situation by running it 24 hours a day and building an internal cooling storage effect ahead of peak outdoor conditions.

VENTILATION LOAD

No building is 100% air tight, and people must open doors to enter or leave as the case may be. Thus, a certain amount of outside air enters the building by this process; we call this *infiltration*. Additionally, as discussed in Chap. 2, outdoor air is desirable to replenish oxygen and to provide make-up air for appliances and exhaust systems. So it's common practice to provide the system with an outside air intake duct (Fig. 5-17). Properly screened against birds and insects, the outside air is then filtered and heated or cooled before entering the conditioned space. (Infiltration air is not filtered and thus adds to room contamination; nor is it preheated or precooled.) Mechanical ventilation has a plus factor in that it creates a slight positive internal air pressure and reduces untreated outdoor air infiltration.

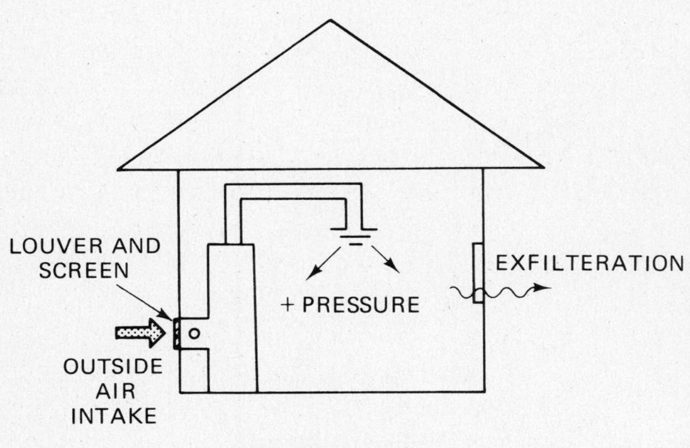

FIGURE 5-17.

In both cases outside air will add or subtract heat and humidity. In winter, it must be heated and should be humidified. In summer, heat and humidity must be removed to achieve comfort. Sensible and latent psychrometric processes are involved, and load calculation forms have precalculated factors to determine the loads. In homes, the usual load is 10 to 15 ft³/min per person with average smoking conditions. Commercial loads are much heavier.

In summary, outside or external load factors are made up of transmission, solar effect, and ventilation load sources that vary a great deal with weather, the sun, and the time of day and month; internal sources are not subject to the same variations.

INTERNAL LOADS

One of the most important sources of internal heat is people. People in a room give off both sensible and latent heat—even as you are doing right now. The exact amount is determined by the type activity (Fig. 5-18). People seated at rest in a theater give off about 350 Btu/h total heat per person, of which some 235 Btu are sensible heat and the balance of 115 Btu are latent. Compare that to people dancing where the total per person would be about 900 Btu/h, 325 would be sensible and 575 latent. Thus you can see the result of

FIGURE 5-18.

body perspiration and latent load with increased activity. Again the load calculation forms for residential and commercial situations provide us the proper factors. People-heat loads work for the system in winter but against the system in summer.

Lighting and appliances (Fig. 5-19) are other forms of internal heat sources. Incandescent lights give off 3.4 Btu/h per watt of electricity consumed. Fluorescent lights, however, require extra power—about 25% more. Thus a nominal 40-watt fluorescent lamp would consume 50 watts of current and produce 170 Btu/h of heat. Appliances like coffee makers produce over 1,000 Btu/h (770 sensible and 230 latent). Electric ranges with several burners and the oven on produce a large amount of heat, as do television sets and radios that have tubes rather than solid-state circuits. In the home, lighting and appliances are usually not on continuously, and it's impractical to assume 100% load. So, using test homes and samplings, an average hourly load has been estimated and included in the calculation form. On the other hand; commercial operations like fast-food outlets will have peaks during lunch or dinner hours where all equipment will be going. Values for commercial cooking apparatus (vented and unvented), meters, and machines are included on most calculation forms. The ASHRAE *Guide* or your local electric and gas utilities are additional sources of specific data on appliance loads where unusual circumstances exist. A case in point might be an electronic computer or a gas-burning manufacturing process, but these can be dealt with on an individual basis and are not considered necessary knowledge for the technician.

Bathing is another internal summer load contributor. People take more showers in the summer, and unless the bathroom is well ventilated (forced-fan-ducted to the outside), a great deal of moisture and sensible heat is added to the conditioned space.

Now, with an understanding of basic building materials and the factors that contribute to the heat load, we are ready to put them to work in doing load calculations, first for residential structures in Chap. 6 and then for light commercial structures in Chap. 7.

Throughout this book we have made repeated references to *load calculation forms* that are used to calculate loads. Most major manufacturers over the years have developed their own forms, and these are available to their dealer organizations. Naturally, each manufacturer has his own particular or proprietary methods, and frequently it becomes a matter of choice by the contractor, a choice that depends on which forms the contractor has learned first, or which may be the easiest for him to use.

For our purposes we have chosen to use the manuals prepared by NESCA, the National Environmental Systems Contractors Association, because they are perhaps the most easily available and widely used methods, and also because they have just been revised and reprinted in 1975.

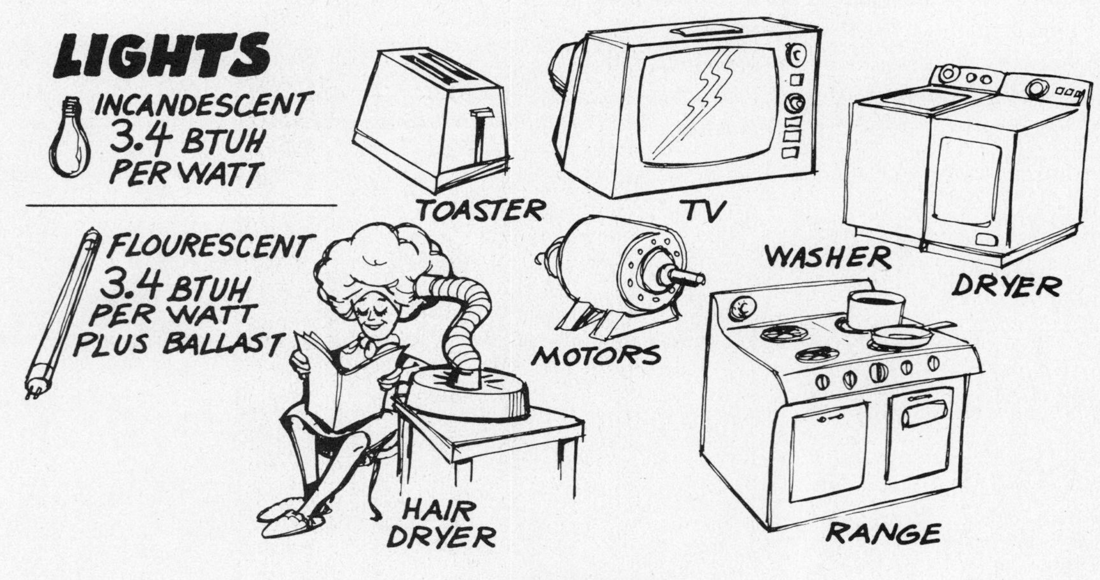

FIGURE 5-19.

PROBLEMS

1. Name the two factors that make up the air conditioning load.
2. Heat through walls, doors, ceilings, etc., is called ______________.
3. The term to define heat-flow resistance is ______________.
4. R values are expressed in ______________.
5. The quantity of heat flow is a function of ______________, ______________, and ______________.
6. Insulation has a high or low R value?
7. Bulk insulation for residential construction comes in three popular forms: ______________, ______________, and ______________.
8. Insulating glass reduces heat by 25, 50 or 75% over single glass?
9. Forced attic ______________ is effective in reducing summer cooling loads.
10. Solar heat gain can be reduced by ______________.
11. Internal heat loads are the result of ______________.
12. Incandescent lights give off ______________ Btu/h per watt of electricity.

6

LOAD CALCULATION—RESIDENTIAL

Manual J, *Load Calculation*, Fourth Edition, (copyright 1975) which will be reproduced here in full, supersedes all previous editions of Manual J published by the National Environmental Systems Contractors Association and its predecessor. *Manual J* contains accurate, practical procedures and data for calculating both heat loss and heat gain in residential structures. Much of the data in this manual is taken from the ASHRAE *Handbook of Fundamentals*. Load calculation is the essential first step in the design and installation of comfort-producing warm-air and air-conditioning systems.

The heat-gain calculation procedure contained in this manual is the result of an overall industry study of residential heat gains and calculation methods. It has been developed jointly and adopted by NESCA, ARI and the Hydronics Institute. The Manual contains design and installation procedures which are practial to use and which, when followed, will result in warm-air heating and air-conditioning systems that produce a high degree of comfort in the homes of the American public.

LOAD CALCULATION

TABLE OF CONTENTS

"This manual is published as a public service by the **national environmental systems contractors association.** The statements contained herein represent the consensus of members of the Association who participated in the preparation of the manual. The Association, however, does not warrant the applicability of the manual for any given situation."

SECTION 1 INTRODUCTION

1-1 Providing comfort conditioning for the occupants of the conditioned space is the objective of all comfort air conditioning systems. To provide this comfort during the winter and during the summer, the heat loss and heat gain of each room must be determined. Once this is done, equipment can be selected and the duct system can be designed, installed and adjusted to deliver the conditioned air where it is needed. Equipment capacity can be determined by calculating the load of the entire structure; heat loss and heat gain of **each room** must be determined to design the duct system.

1-2 Importance of accurate load calculation cannot be overemphasized. It is the essential first step to providing comfort in a structure. Accurate design data and logical, step-by-step calculation procedures are included in this manual.

1-3 EQUIPMENT CAPACITY Capacity of heating and cooling units selected should be as nearly equal to the calculated loads as possible. Equipment that is undersized does not have enough capacity to provide comfort at design conditions. On the other hand, oversized equipment may not provide comfort and may be uneconomical to operate. For detailed information on equipment selection refer to Manual K.

1-4 An "entire house" column is included on Worksheet J-1 so that the heat loss and heat gain of the structure, considered as a single room, can be more easily calculated. Other columns on this worksheet are to be used for calculating the heat loss and heat gain of each room or space to be conditioned.

1-5 Some of the terms and abbreviations used in the manual are explained in the following paragraphs.

1-6 Btu OR BRITISH THERMAL UNIT A unit used to express the amount of heat.

1-7 HEAT FLOW OR HEAT TRANSFER The movement of heat from one substance or region to another. Where one substance or region is at a higher temperature than an adjoining one, heat always flows from the higher temperature substance or region to the one at a lower temperature. (See Fig. 1-1.)

1-8 Btu OR BRITISH THERMAL UNITS PER HOUR A unit used to express the rate of heat flow. Heat loss and heat gain of a structure is expressed in Btuh. Capacity of heating equipment and cooling equipment is also expressed in Btuh —how many Btu can be supplied to or removed from a conditioned space each hour by the equipment.

1-9 Htm OR HEAT TRANSFER MULTIPLIER A term used to express the rate of heat flow (Btuh) through **one** square foot of a structural component at specific design conditions.

1-10 R-VALUE A number used to describe the resistance of a material to the flow of heat (thermal resistance). The larger the R-value of a material, the more difficult it is for heat to flow through that material. For example, it is more difficult for heat to flow through a well insulated wall than an uninsulated wall; therefore, the insulated wall has a larger R-value.

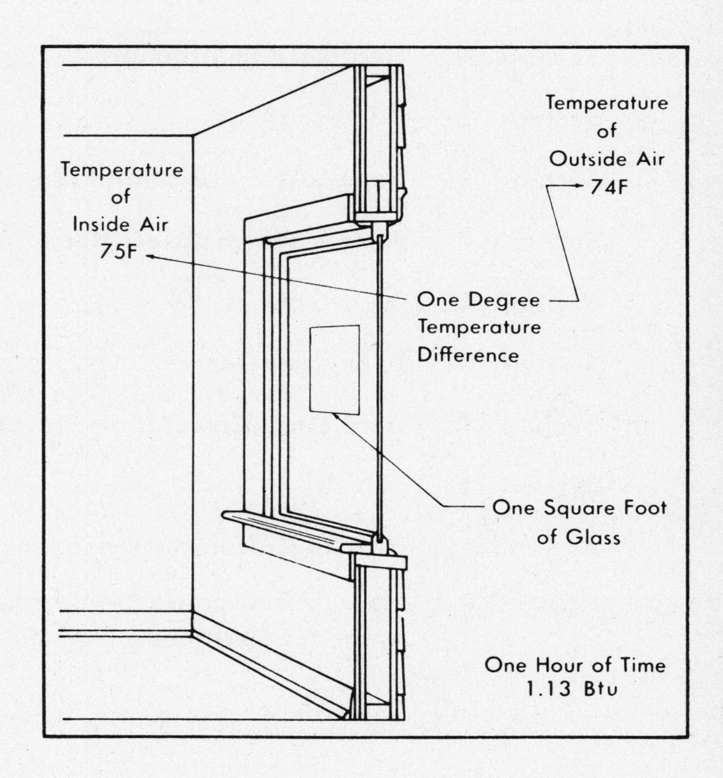

1-1 The heat transmission coefficient of single glass is 1.13 which means that this is the quantity of heat in Btu which the room air loses by transmission through ONE square foot of the glass during ONE hour of time when there is a temperature difference of ONE degree Fahrenheit between the room air and the outdoor air.

SECTION 2 MEASURING FOR HEAT TRANSMISSION

2-1 OBTAINING BUILDING DIMENSIONS Dimensions of a new structure are usually obtained from the building plans. At least one dimension for each direction should be checked on each sheet of the plans. If a dimension does not correspond to the scaled measurement, determine from the builder or owner which is correct. It is also wise to make sure that all construction changes are shown on the plans.

2-2 Plans for existing structures should be used only after they have been compared to the structure. In many cases, the measurements will have to be obtained from the building.

2-3 When measuring for heat transmission, dimensions and areas for closets and halls are usually included with those of adjoining rooms. Large closets or entrance halls should be considered as rooms and their dimensions obtained and recorded separately.

2-4 LENGTH OF EXPOSED WALL An exposed wall is one which faces the outside and may be either above or below grade level. In existing structures, the length of the wall is measured from inside surface to inside surface of the adjoining walls. Dimensions recorded on plans are usually from center line to center line of adjoining walls. Whether dimensions are obtained by actually measuring or from floor plans, they are recorded to the nearest foot on the worksheet. For example, if the distance is 14 ft 6 in., record 15 ft as the distance. If a room has two or more exposed walls, the total running feet of exposed wall is the sum of the individual wall lengths. (See Fig. 2-1.)

2-5 HEIGHT OF EXPOSED WALL Wall height is recorded to the nearest one-half foot. For example, if the measured height is 7 ft 9 in., record 8 ft on the worksheet. If more than one type of wall construction is used in a room, record the actual height of each construction. Record the average height for walls in a room when the height is not uniform. An example of this is a wall in a room with an open beam ceiling.

2-6 WALL AREA The **gross exposed wall area** of a room equals the length of exposed wall multiplied by its height. It is recorded to the nearest sq ft. Gross wall area includes wall, window, and door areas. To obtain the **net exposed wall area**, subtract the window and door areas from the

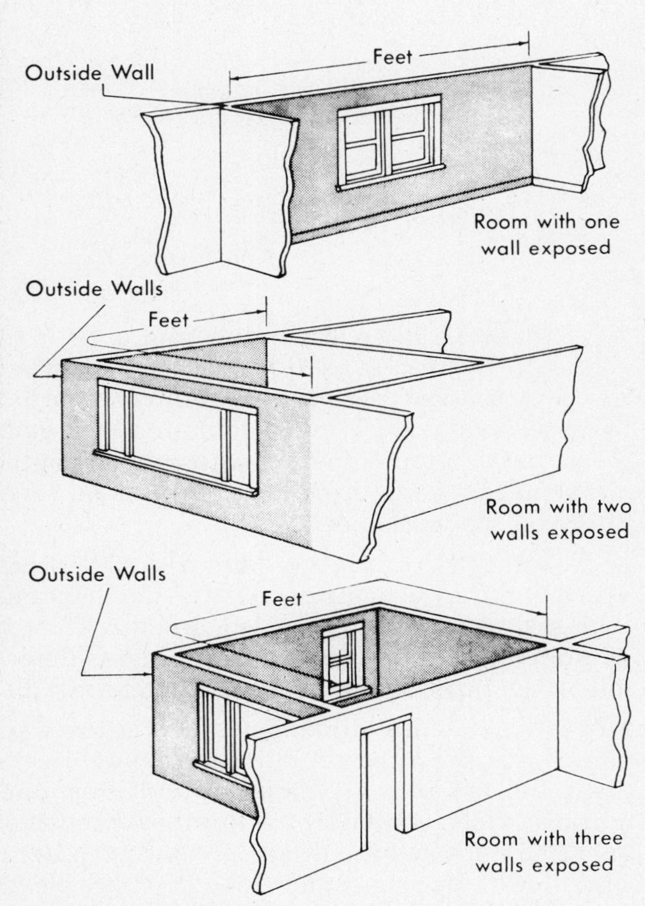

2-1 Floor plans showing running feet of exposed walls for rooms having 1, 2, and 3 wall exposures.

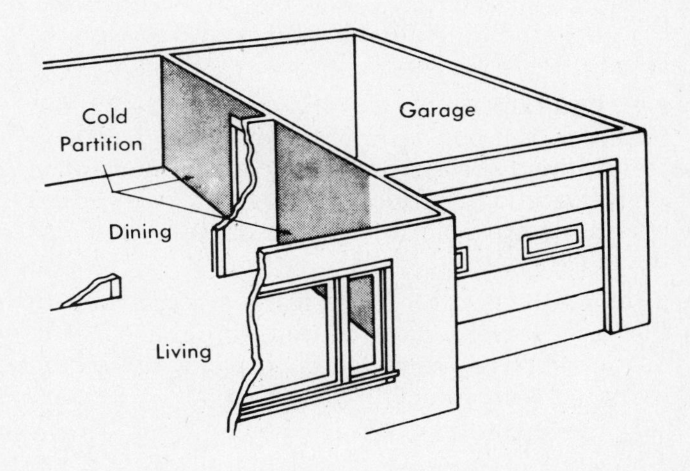

2-2 Cutaway showing cold partition wall.

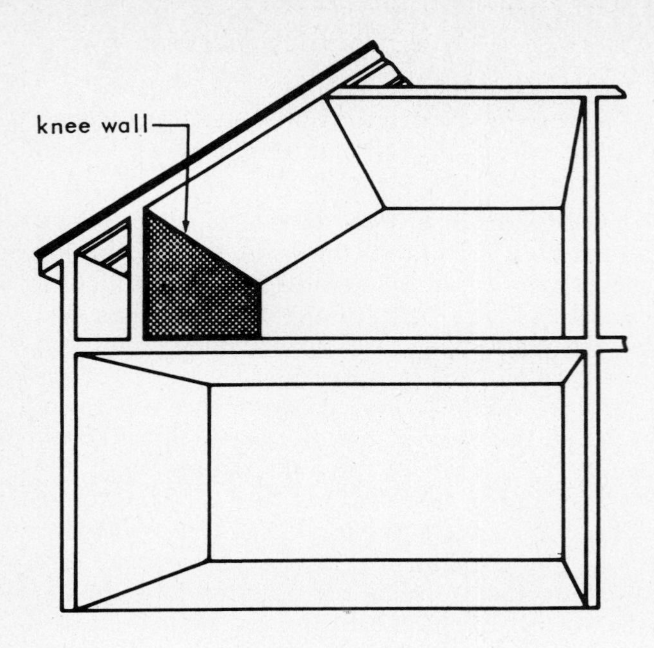

2-3 Cutaway showing knee wall.

gross exposed wall area. If more than one wall construction is used, be certain to subtract areas of windows and doors from the construction in which they are located.

2-7 When an open stairway is located adjacent to a cold outside wall, the exposed wall area up to the ceiling of the second story is included with that of the first story hall or room.

2-8 Wall areas above grade are usually recorded separately from wall areas below grade. If a wall extends less than 3 ft below grade, it is assumed to be entirely above grade; and the area is recorded as if the entire wall were above grade. When a wall extends more than 3 ft below grade, the above grade wall area and below grade wall area are recorded separately.

2-9 PARTITIONS A wall that is not directly exposed to outside temperature but separates a conditioned space from an unconditioned space is an exposed partition. Dimensions for exposed partitions are recorded in the same manner as those for exposed walls.

2-10 KNEE WALLS Knee walls are exposed to the same conditions as ceilings under an unconditioned space. Areas of knee walls should be included with the ceiling area.

2-11 WINDOWS AND DOORS Measurements of a window and door are illustrated in Fig. 2-4. Dimensions for windows and doors are measured to the nearest inch and are usually recorded to the nearest 0.1 ft. Dimensions in 0.1 ft which

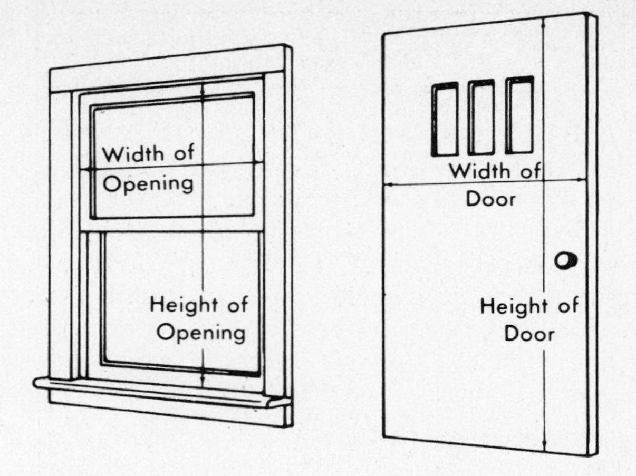

2-4 Dimensions needed for calculating window and door areas.

correspond to dimensions in inches are listed below:

Dimension Measured, in.	Dimension Recorded, ft
1	0.1
2	0.2
3	0.3
4	0.3
5	0.4
6	0.5
7	0.6
8	0.7
9	0.8
10	0.8
11	0.9

2-12 The area of each window and door is recorded to the nearest sq ft, and the **total area of each type of window and door** for each room is recorded to the nearest sq ft. When calculating heat gain, areas for each type of window (single or double glass) and for each direction are recorded separately on the worksheet.

2-13 Glass completely shaded by roof overhangs may be considered to be north facing. If glass is only partially shaded, and the effect of window shading on heat gain is to be considered; follow the procedure in Appendix B (page 426).

2-14 Full opening dimensions of windows are not always specified on builder's or architect's plans. Instead, they may be identified by a model number or designated by the number of lights or panes of glass and the dimensions of each light. Obtain full opening dimensions from the elevations or from the window manufacturer's catalog. In existing buildings, window opening can be measured at the same time that other dimensions are obtained.

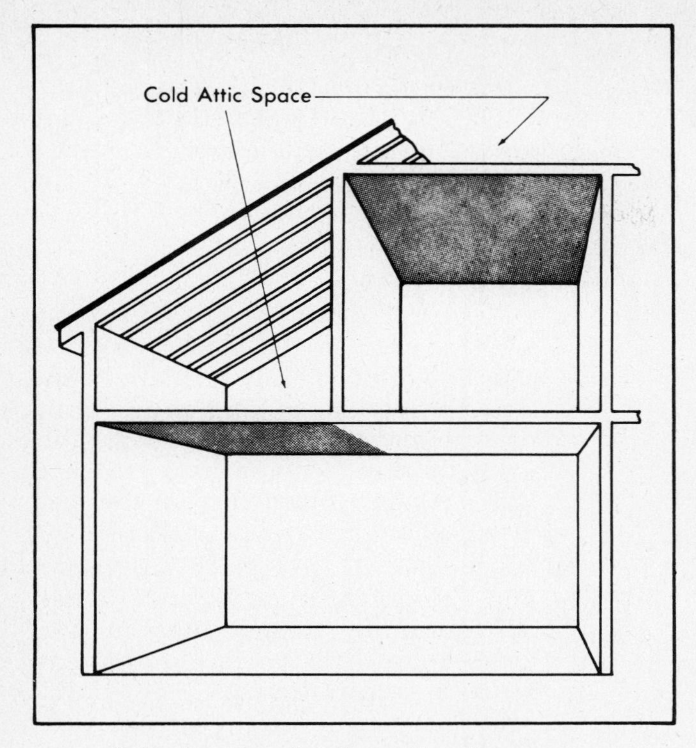

2-5 Elevation showing cold ceiling areas.

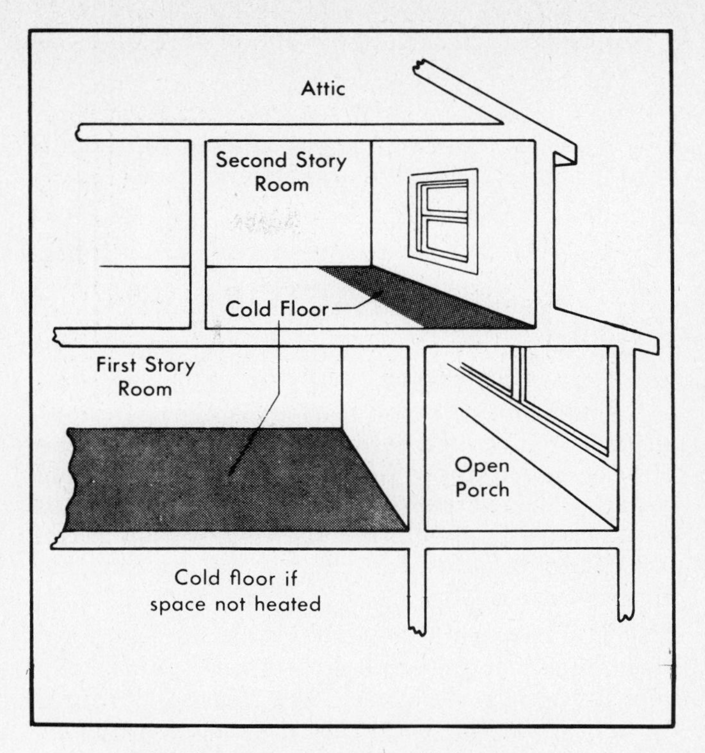

2-6 Elevation showing cold floor area.

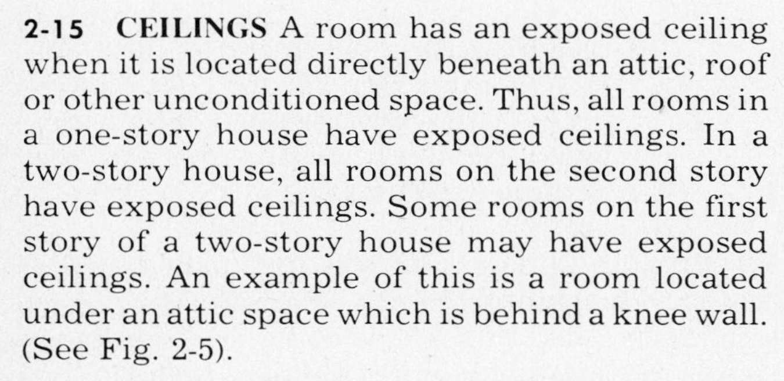

2-15 CEILINGS A room has an exposed ceiling when it is located directly beneath an attic, roof or other unconditioned space. Thus, all rooms in a one-story house have exposed ceilings. In a two-story house, all rooms on the second story have exposed ceilings. Some rooms on the first story of a two-story house may have exposed ceilings. An example of this is a room located under an attic space which is behind a knee wall. (See Fig. 2-5).

2-16 Determine the area of exposed ceiling for each room by multiplying the room width by its length. Width and legnth of the room are measured to the nearest ft and the area is recorded to the nearest sq ft. **NOTE: When only part of the ceiling of a room is exposed, determine the area for the exposed portion only.**

2-17 FLOORS When taking measurements for the purpose of determining **heat loss** through floors, all floors that are located over unconditioned spaces, exposed to outside temperature, or in contact with the ground must be considered. For determining **heat gain** from floors, only floors located over unconditioned spaces or those exposed to outside temperatures are considered. (Unconditioned spaces are rooms, basements, or enclosed **and** unvented crawl spaces that do not have conditioned air supplied to them.) If only part of a floor is exposed, then only that portion that is exposed is considered.

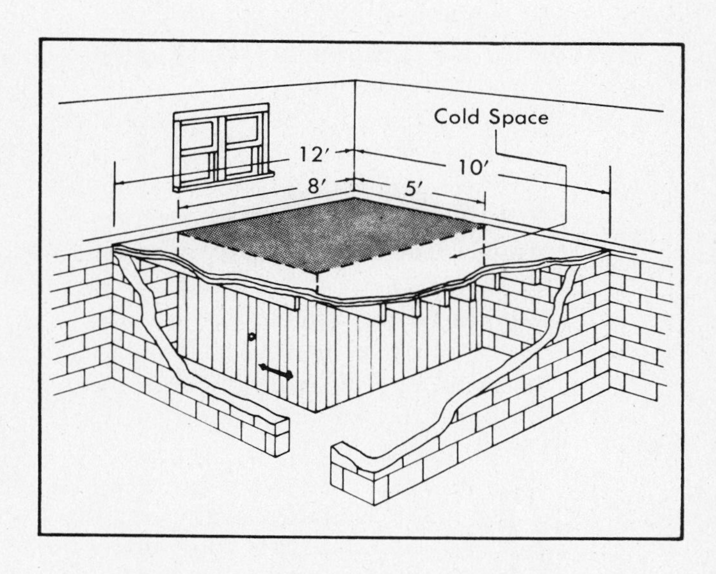

2-7 Cutaway showing cold floor above cold space.

An example of this is a room part of which is located over an open porch. (See Fig. 2-6.)

2-18 Since the rate of heat flow is uniform through floors located over unconditioned rooms, open or vented spaces, and basement floors; the heat transmission through those floors is based on their area. Measure the width and length of these floors to the nearest ft and record the area to the nearest sq ft.

2-19 Most of the heat loss from concrete slab floors and crawl space floors is near the edge of the floor. Heat loss for these floors is, therefore, based on the length of floor exposed to outside temperature. This is measured and recorded to the nearest ft.

2-20 Basement and concrete slab floors are not considered when obtaining measurements for the purpose of determining **heat gain.**

SECTION 3
FACTORS WHICH AFFECT HEAT LOSS

3-1 HOW A BUILDING LOSES HEAT The amount of heat a building loses is affected by the materials used in its construction. The architect and the builder can help provide for a more economical installation and more satisfactory operation of the system by specifying the use of insulation, tight fitting windows and doors, and double glazing or storm sash.

3-2 The construction and the design temperature difference determine the rate at which the heat is lost. To determine the heat loss from a room or building, obtain the area of each type of wall, ceiling, floor, window, and door having one side exposed to room air and the other side exposed to a lower temperature.

3-3 Ordinarily, the components which must be considered are the following:

A. **Walls** exposed to the outside air and walls below grade.

B. **Partitions** separating spaces which are at different temperatures.

C. **Windows and Doors** and any other glass surfaces.

D. **Ceilings** below spaces which are at lower temperatures.

E. **Floors** above spaces which are at lower temperatures or are in contact with the ground.

F. **Infiltration:** the heat lost due to infiltrated air is included in the window and door Heat Transfer Multipliers in Table 2.

G. **Ventilation** provides for the introduction of outside air under positive control. Ventilation requirements in residences are usually met by infiltration. However, if mechanical ventilation is provided, it imposes an additional load on the equipment and does not reduce infiltration. Strictly speaking ventilation is not a part of the building or space heat loss but does constitute part of the equipment load.

For one or two family residences, the intermittent use of kitchen and bathroom exhaust fans makes a negligible contribution to the heating load and can be safely omitted from the load calculation.

H. **Combustion Air:** gas and oil fired heating equipment must have sufficient air for combustion. In most houses, normal infiltration is adequate to provide combustion air. If the building is of unusually tight construction, air for combustion must be obtained from the outdoors or from spaces opening directly to the outdoors. If provided directly to the equipment area, combustion air does not contribute to the heating load. For a detailed treatment of the subject, see Section 8, Manual K.

3-4 In some homes there is unfinished space that is to be completed at a later date. Information as to how this space will be completed should be obtained so that the heat load can be calculated. If this space is not to be conditioned immediately, heat loss through constructions which separate the unfinished space from space that is to be conditioned should be calculated. This is necessary because it may be several years before the space is completed.

3-5 INSULATION AND STORM SASH It is good practice to advocate the use of insulation, weather stripping, storm windows, and caulking. The heat loss of the structure and the operating cost of the heating plant will be reduced by improved thermal performance, and the structure will be much easier to heat comfortably.

3-6 To avoid misunderstanding, the type of construction and design conditions assumed should be stated in writing. **Specify the wall construction, amount of insulation, types of windows, and storm sash on which calculations are based.** Space for indicating this information is included on Worksheet J-1.

3-1 Specify basis for load calculation.

3-7 Garages are seldom heated with a central warm air system that heats the living areas of the house. When garages are to be heated with one of these systems, however, the following precautions must be taken:

A. Return air **must not** be taken from the garage;
B. Outside air **must** be supplied to the return air side of the heating unit in a volume equivalent to the warm air delivered to the garage; and,
C. The heat loss of the garage should be calculated. (The inside design temperature for a garage is usually lower than that used for living areas of the house.)

Each supply outlet in a garage must be equipped with a backdraft damper so that air from the garage cannot enter the supply duct when the blower is not operating.

3-8 DESIGN TEMPERATURE To design a heating system, the equipment capacity requirements must be calculated. The **design temperatures** are used to determine this demand.

3-9 OUTSIDE DESIGN TEMPERATURE Recommended Outside Design Temperatures for many localities are listed in Table 1. The outside design temperature is **not** the lowest temperature recorded in each locality. In Urbana, Illinois, for example, the lowest temperature recorded over a period of about 50 years was 21 degrees below zero (—21 F). The temperature remained at this low point for less than two hours.

3-10 Heating systems in Urbana, Illinois, are not designed for requirements that occur only two hours out of 50 years. Instead, an outside design temperature of zero F is used. The experience in the Research Residences indicate that a system designed in this manner will operate satisfactorily during short periods at considerably lower temperatures.

3-11 INSIDE DESIGN TEMPERATURE The temperature to be maintained within the structure is the inside design temperature. Values commonly used as the inside design temperature are 70 F or 75 F. There are two basic considerations in establishing which temperature will be used in a particular structure. One of these is established either by local practice or code requirements. In many areas, the heating code specifies that the heating system must be adequate to heat the structure to at least 70 F. When the owner does not specifically request an inside design temperature higher than 70 F, this value is often used in the calculation. The other consideration in establishing the inside design temperature is the comfort requirements of the occupants of the structure. Some people prefer a higher temperature, about 75 F, in their homes.

3-12 Do not make fine distinctions of 2 or 3 degrees difference in design temperatures for different rooms in the same building. For example, do not specify design temperatures of 68 F in the hall, 70 F in bedrooms, and 72 F in the living room. **One design temperature for all rooms is sufficeint.** A warm air system can be adjusted so that the user can provide for small temperature variations after the system is installed, even when all rooms are designed for one temperature.

3-13 DESIGN TEMPERATURE DIFFERENCE The heat loss of the structure depends upon the difference between the inside design temperature and the outside design temperature. This is the design temperature difference.

3-14 If the outside design temperature is **warmer than 0 F,** subtract the outside design temperature from the inside design temperature to determine the design temperature difference. For example, if the outside design temperature is 20 F, and the inside design temperature is 70 F, the design temperature difference is 70 minus 20, or 50 degrees.

3-15 If the outside design temperature is colder than 0 F, add the outside design temperature to the inside design temperature to determine the design temperature difference. For example, if the outside design temperature is —20 F, and the inside design temperature is 70 F; the design temperature difference is 70 plus 20, or 90 degrees. (See Fig. 3-2.)

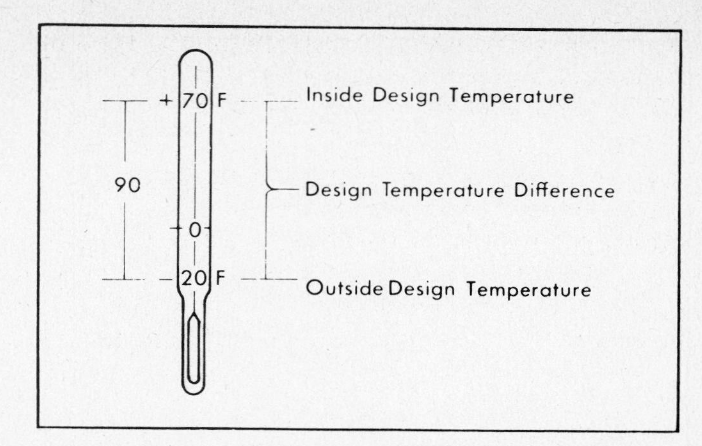

3-2 Design temperature difference for a —20 F temperature zone.

SECTION 4 HEAT LOSS OF A STRUCTURE

4-1 The design heat loss of the structure is that which occurs when the inside and outside temperatures are at design values and is used as a basis for equipment selection. When calculated on a room by room basis, the design heat loss also becomes the basis for system design.

4-2 HEAT TRANSFER MULTIPLIERS The heat loss through one square foot of a construction depends upon the insulation characteristics of the construction and upon the temperature difference. The **Heat Transfer Multipliers** of Table 2 give the design heat loss per square foot for most common constructions. Thus, the design heat loss through a wall equals the area of the wall multiplied by the Heat Transfer Multiplier tabulated for the design temperature difference.

4-3 Each construction in Table 2 is described in general terms; therefore, the Heat Transfer Multipliers represent average constructions. In most cases, these may be used satisfactorily. If more precise factors are required, use the procedure discussed in Appendix A to determine the Heat Transfer Multiplier.

4-4 WINDOWS AND DOORS Heat Transfer Multipliers for windows and doors (construction No. 1 through No. 9 in Table 2) include both transmission and infiltration heat loss. Therefore, the total heat loss through a window or door equals the **area** of the **window** or **door (sq ft)** multiplied by the **Heat Transfer Multiplier** specified for the type of construction and design temperature difference. For example, the Heat Transfer Multiplier for a double glass picture window (construction No. 4-b) is 70 Btuh per sq ft at a design temperature difference of 80 degrees. If the window area is 30 sq ft, the design heat loss is 2100 Btuh.

4-5 WALLS Heat loss through walls is obtained by multiplying the net wall area (sq ft) by the Heat Transfer Multiplier specified for the type of wall construction (constructions 10 through 13 in Table 2) and the design temperature difference. Such considerations as the variations of ground temperature have been considered in determining Heat Transfer Multipliers for walls below grade, and the higher temperatures which occur in crawl spaces used as a warm air plenum have been considered in the determination of Heat Transfer Multipliers for these walls. Factors included in Table 2 may therefore be used without any further correction.

4-6 CEILINGS AND ROOFS Heat Transfer Multipliers for determining heat loss through ceilings and roofs are included as constructions 14 through 16 in Table 2. To obtain the heat loss through ceilings and roofs multiply the ceiling area (sq ft) by the Heat Transfer Multiplier specified for the type of construction and the design temperature difference.

4-7 FLOORS Heat is lost from some floors at about the same rate throughout the entire area of the floor. To obtain heat loss for these floors the area (sq ft) is multiplied by the Heat Transfer Multiplier for the type of construction specified (constructions 17 through 19, Table 2) and the design temperature difference.

4-8 Other floors such as slabs on ground or heated crawl space floors have much greater heat loss near the edge of the floor than near the center. Heat Transfer Multipliers for these floors are based on the length of the exposed edge of the floor. To obtain the heat loss for these floors the length of the exposed edge (ft) is multiplied by the Heat Transfer Multiplier for the type of construction specified (constructions 20 through 23, Table 2) and the design temperature difference.

4-9 Included in the Heat Transfer Multipliers for basement floors and crawl space floors is an allowance for the variation between outside design temperature and ground temperature. Higher temperatures that occur in crawl spaces used as supply plenums have also been considered when Heat Transfer Multipliers for these floors were developed. Therefore, factors included in Table 2 may be used without further correction.

SECTION 5 DUCT HEAT LOSS

5-1 Heat is lost from a duct when the air in it is warmer than the space in which the duct is located. This loss decreases the amount of heat which is delivered to the conditioned space through the duct.

5-2 DUCTS LOCATED IN CONDITIONED SPACE Heat loss from ducts located in conditioned spaces helps heat space for which the heat loss has been calculated. Because it is alrady included in the heat loss of the structure, it is not an added load on the equipment. It is not considered as a load when determining the required heating capacity.

5-3 DUCTS LOCATED IN UNCONDITIONED SPACE When ducts are located in unconditioned spaces for which the heat loss has not been calculated, heat lost from them is also lost from the structure. This heat loss is therefore an added heating load and must be considered when determing required heating capacity.

5-4 Ducts installed in attics, unconditioned crawl spaces, attached garages, etc. should be insulated. Consequently, Heat Loss Multipliers indicated in Table 3 for ducts in these locations are for insulated ducts only. Since ducts located in basements are not normally insulated, even when the basement is not conditioned, the factors for basement ducts in Table 3 are for uninsulated ducts only. **Do not use** Heat Loss Multipliers for uninsulated ducts located in other unconditioned spaces.

5-5 Before the Duct Heat Loss Multiplier can be obtained from Table 3, the heat loss per sq ft of the living space must be determined. If the heat loss of the entire house has been determined, this figure can be obtained by dividing the heat loss of the living space (do not include heat loss of basement, crawl spaces, or garages) by the area of the living space. When heat loss has been calculated on a room by room basis, the sum of the heat losses of each room should be divided by the area of the living space.

5-6 The preferred method of obtaining duct heat loss is to multiply the heat loss of each room by the appropriate **duct heat loss multiplier** obtained from Table 3.

Duct Location and Insulation	Heat Loss of Living Space, Btuh/ft^2	Outside Design Temp	
		Below 0	0 to 15
Attic or Crawl Space; 1 in. Flexible Blanket Insulation [1], [2]	Less than 35	.25	.20
	35 to 45	.20	.15
	45 or more	.15	.10
Attic or Crawl Space; 2 in. Flexible or 1 in. Rigid Duct Insulation[3]	Less than 35	.20	.15
	35 to 45	.15	.10
	45 or more	.10	.05
Unconditioned Basement; No Duct Insulation	Less than 35	.20	.15
	35 to 45	.10	.10
	45 or more	.05	.05

5-1 Duct heat loss multiplier obtained from table 3 for duct wrapped with 2 in. insulation when calculated heat loss is 40 Btuh per sq ft.

5-7 The Outisde Design Temperature has the effect of defining the temperature of the space in which the duct is located. When the thickness of duct insulation has been determined, the Duct Heat Loss Multiplier can be selected from Table 3.

> Example: A residence is located in a city that has an Outside Design Temperature of 5 F. This residence has a calculated heat loss of 40 Btuh per sq ft; and the ducts, wrapped with 2 in. insulation, are installed in the attic. From Table 3, the Duct Heat Loss Multiplier selected for this residence is 0.10. (See Fig. 5-1).

5-8 DUCTS EMBEDDED IN SLAB Heat loss factors for concrete slabs with embedded ducts includes an allowance for duct heat loss; it is not necessary to calculate the duct heat loss for these systems.

SECTION 6
HEAT LOSS CALCULATION PROCEDURE

6-1 An example of a heat loss calculation for a one story house with a basement is included in this section. Since this is a well insulated house with storm windows and doors, the heat loss is only 62,308 Btuh. If this house were not insulated and had poorly fitted windows and no storms, the heat loss could be more than 150,000 Btuh.

6-2 STEP-BY-STEP PROCEDURE Study the heat loss calculation procedure below and refer to the paragraphs, figures and tables indicated for additional information required for each step in the example. Design conditions and type of construction assumed for this example are listed below, and a sample calculation on Worksheet Form J-1 is illustrated in Fig. 6-3.

6-3 The plan used and the constructions and design conditions assumed were chosen only to illustrate the use of data and procedures in the manual. They should not be interpreted as recommendations of construction features.

	ASSUMED DESIGN CONDITIONS AND CONSTRUCTION (Heating):	**From Table 2 Const. No.**	**HTM**
A.	Determine Outside Design Temperature		
B.	Select Inside Design Temperature		
C.	Design Temperature Difference: 80 Degrees		
D.	Windows: Living Room—Double Glass with ¼ in. Sealed Air Space	4(b)	70
	Basement—Uncertified Metal Casement Windows, No Storms	3(a)	205
	Others—Double-hung with Storms, Infiltration Less Than 0.75 cfm/ft @ 25 mph Certified by Test	2(c)	70
E.	Doors: Wood with Storm Doors	9(b)	195
F.	First Floor Walls: Basic Frame Construction with Insulation (R-11)	10(d)	6
	Basement Wall: 8 in. Concrete Block		
	Above Grade Height: 3 ft	13(a)	41
	Below Grade Height: 5 ft	13(f)	5
G.	Partitions: 8 in. Concrete Block, Furred with Insulation (R-5)	12(c)	10
H.	Ceiling: Basic Construction Under Vented Attic with Insulation (R-19)	14(c)	4
I.	Floor: Basement Floor, 4 in. Concrete	19	3
J.	Ducts: Located in Conditioned Space		

PROCEDURE FOR DETERMINING Btuh HEAT LOSS FROM ROOMS

PROCEDURE	**EXAMPLE**
A. Determine the **Outside Design Temperature** from Table 1. (See paragraphs 3-9 through 3-11.)	The outside design temperature for Atlanta, Georgia, is 20 F; the outside design temperature for Fort William, Ontario, is —25 F.
B. Select **Inside Design Temperature** for rooms to be heated. (See paragraphs 3-12 through 3-13.)	Use same temperature for all rooms.
C. The **Design Temperature Difference** is the difference in temperature between inside and outside at design conditions. (See paragraphs 3-13 through 3-15.) If the outside design temperature is above 0 F, the design temperature difference is equal to inside design temperature **minus** the outside design temperature.	For Atlanta, Georgia, the design temperature difference is 70 **minus** 20 or 50 degrees for all rooms to be heated.
If the outside design temperature is below 0 F, the design temperature difference is equal to the inside design temperature **plus** the number of degrees the outside design temperature is below 0 F.	For Fort William, Ontario, the design temperature difference is 70 **plus** 25 or 95 degrees for all rooms to be heated.

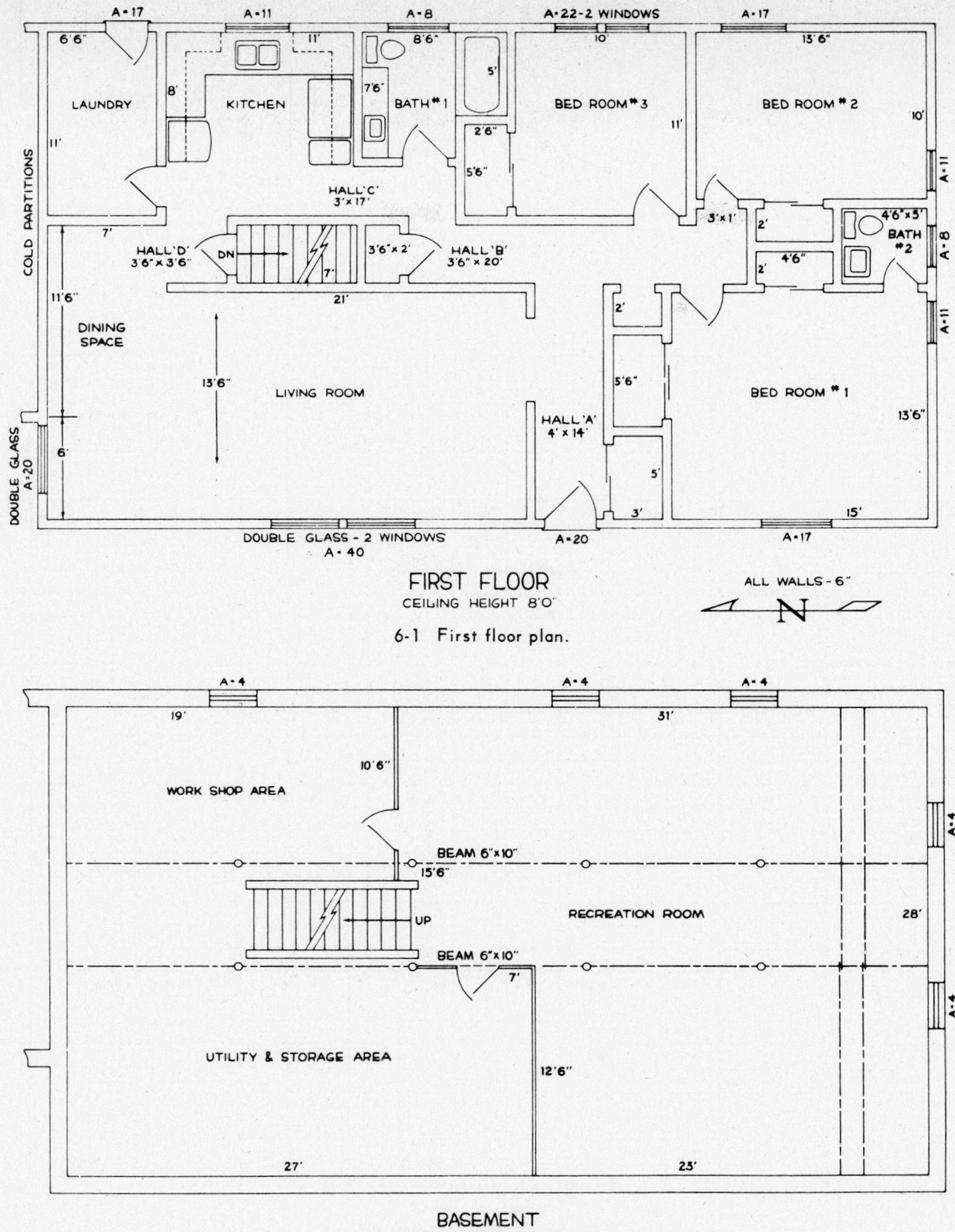

6-1 First floor plan.

6-2 Basement plan.

PROCEDURE	EXAMPLE
D. Determine all dimensions of each room to be conditioned and direction each outside wall faces. (See paragraphs 2-1 through 2-5.)	Complete lines 1 through 4 on Worksheet J-1 for each room. See Fig. 6-3.
E. In Table 2, locate description of exposed walls, windows, doors, cold partitions, cold ceilings, cold floors, and ventilation; and determine Construction Number.	Each Construction No. is entered in Construction No. column on Worksheet J-1. For example, a frame wall is described as Construction No. 10(a). This wall with insulation (R-11) would be described as Construction No. 10(d).

#	Type of Exposure		Const. No.	HTM Htg	HTM Clg	Entire House Area or Length	Entire House Btuh Htg	Entire House Btuh Clg	1 LIV Area or Length	1 LIV Btuh Htg	1 LIV Btuh Clg	2 HALL 'D' & DIN Area or Length	2 HALL 'D' & DIN Btuh Htg	2 HALL 'D' & DIN Btuh Clg	3 LAUN Area or Length	3 LAUN Btuh Htg	3 LAUN Btuh Clg	4 HALL 'C' & KIT Area or Length	4 HALL 'C' & KIT Btuh Htg	4 HALL 'C' & KIT Btuh Clg	5 BATH #1 Area or Length	5 BATH #1 Btuh Htg	5 BATH #1 Btuh Clg
1	Name of Room					Entire House			LIV			HALL 'D' & DIN			LAUN			HALL 'C' & KIT			BATH #1		
2	Running Ft Exposed Wall								21			12 & 13			11 & 7			11			9		
3	Room Dimensions, Ft								14 × 21			7×18 & 4×4			7×11			8×11 & 3×17			6×8 & 3×5		
4	Ceiling Ht, Ft		Directions Room Faces						8	W		8	N & W		8	N & E		8	E		8	E	
5	Gross Exposed Walls and Partitions	a	10(d)						168			104			56			88			72		
		b	12(c)									96			88								
		c	13(a)																				
		d	13(f)																				
6	Windows and Glass Doors (Htg)	a	2(c)	70														11	770		8	560	
		b	4(b)	70					40	2800		20	1400										
		c	3(a)	205																			
7	Windows and Glass Doors (Clg)		North																				
			E & W or NE & NW																				
			South or SE & SW																				
8	Other Doors		9(b)	195											17	3320							
9	Net Exposed Walls and Partitions	a	10(d)	6					128	768		84	504		39	234		77	462		64	384	
		b	12(c)	10								96	960		88	880							
		c	13(a)	41																			
		d	13(f)	5																			
10	Ceilings	a	14(c)	4					294	1176		142	568		77	308		139	556		63	252	
		b																					
11	Floors	a	19	3																			
		b																					
12	Ventilation																						
13	Sub Total Btuh Loss									4744			3432			4742			1788			1196	
14	Duct Btuh Loss																						
15	Total Btuh Loss						62,308																
16	People @ 300 and Appliances 1200																						
17	Sensible Btuh Gain (Structure)																						
18	Duct Btuh Gain																						
19	Sum of Lines 17 and 18 (Clg)																						
20	Total Btuh Gain (Line 19 x 1.3)				1.3																		
21	Btuh for Air Quantities																						

6–3 Example of heat loss calculation

6 BR #3			7 BR #2			8 BATH #2			9 BR #1			10 HALL 'A' & 'B'			11 REC RM			12 WORK & UTIL			1
10			24			5			29			7			82			74			2
10×11 & 3×6			10×14 & 2×5			5×5			14×15, 2×5, & 3×6			~~4×14, 3×5, 4×20~~ 2×4, 1×3, 2×3			23×28 & 8×16			19×28 & 8×12			3
8	E		8	S & E		8	S		8	S & W		8	W		3/5	E, S, & W		3/5	W, N, & E		4
Area or Length	Btuh		Area or Length	Btuh		Area or Length	Btuh		Area or Length	Btuh		Area or Length	Btuh		Area or Length	Btuh		Area or Length	Btuh		
	Htg	Clg		Htg	Clg		Htg	Clg		Htg	Clg		Htg	Clg		Htg	Clg		Htg	Clg	
80			192			40			232			56									5
															246			222			
															410			370			
22	1540		28	1960		8	560		28	1960											6
															16	3280		4	820		
																					7
												20	3900								8
58	348		164	984		32	192		204	1224		36	216								9
															230	9430		218	8940		
															410	2050		370	1850		
128	512		150	600		25	100		238	952		168	672		28	112					10
															772	2320		628	1884		11
																					12
	2400			3544			852			4136			4788			17,192			13,494		13
																					14
																					15
																					16
																					17
																					18
																					19
																					20
																					21

6–3 (cont.)

PROCEDURE	EXAMPLE
F. From Table 2, obtain heat transfer multipliers (HTM) determined for type of construction and design temperature difference.	Enter these HTM in the appropriate column opposite the corresponding Construction Number on Worksheet J-1. See Fig. 6-3.
G. Obtain area of exposed walls and cold partitions including windows and doors in sq ft. This is the **Gross Exposed Wall Area.** (See paragraphs 2-6 through 2-9.)	Enter in columns for "Area or Length" on Line 5, Worksheet J-1, for each room. See Fig. 6-3.
H. Obtain areas of windows and glass doors. (See paragraphs 2-11 through 2-14.)	Enter in columns for "Area or Length" on Line 6, Worksheet J-1. See Fig. 6-3.
I. Line 7 is not used for heat loss calculation.	Used for heat gain calculation only. See Fig. 10-1.
J. Obtain other door areas.	Enter in columns for "Area or Length" on Line 8, Worksheet J-1 for appropriate rooms. See Fig. 6-3.
K. Subtract areas of windows and doors in each room from gross exposed wall area in each room. This is the **Net Exposed Wall Area.** (See paragraph 2-6.)	Enter in columns for "Area or Length" on Line 9, Worksheet J-1. See Fig. 6-3.
L. Obtain for each room areas of cold ceilings and areas or length of perimeter of cold floors exposed to outside temperature or to cold spaces. (See paragraphs 2-15 through 2-19.)	Enter in columns for "Area or Length" on Lines 10 and 11, Worksheet J-1. See Fig. 6-3.
M. Determine air volume (cfm) to be provided for ventilation or makeup air. (See paragraph 3-3.)	Enter on Line 12, Worksheet J-1, in an "Area or Length" column that is not being used for an individual room. Ventilation or makeup air is calculated for the entire house. It is not included in the heat loss for an individual room. See Fig. 6-3.
N. Design Heat Loss equals (area) x (HTM) for walls, ceilings, doors, windows and some floors (see paragraph 2-18); (length of exposed perimeter) x (HTM) for other floors (see paragraph 2-19); (air volume) x (HTM) for ventilation.	Multiply each area, length, or air volume by the appropriate HTM and enter in "Btuh" columns on corresponding lines, Worksheet J-1. See Fig. 6-3. (The ceiling over the stairway is included in the recreation room.)
O. Obtain **Subtotal Btuh Loss** for each room by adding heat losses for windows, walls, cold partitions, cold floors, and cold ceilings. Repeat this procedure for each room (include all conditioned spaces).	Enter on Line 13, Worksheet J-1. See Fig. 6-3.
P. If ducts are located in unconditioned spaces, determine the **Duct Btuh Loss** for each room. (See Section 5.)	Since the ducts are located in conditioned spaces for this example, Line14 is not used.
Q. Obtain **Total Btuh Loss for each room** by adding Lines 13 and 14.	If there is no duct loss, Line 15 equals Line 13. See Fig. 6-3.
R. Obtain **Total Heat Loss for the entire house** by adding the heat loss calculated for ventilation air to the heat losses of the unconditioned rooms in the structure. This value is used to select the furnace size.	Selection of furnace size is discussed in Manual K.
S. Lines 16 through 21 are not used for heat loss calculation.	Used for heat gain calculation only. See Fig. 10-1.

SECTION 7
FACTORS WHICH AFFECT HEAT GAIN

7-1 For winter air conditioning, heat and moisture must be **added** to obtain comfort conditions. Conversely, heat and moisture must be **removed** to provide comfort conditions during the summer. Some of the factors which affect heat gain are discussed in this section.

7-2 Most of the heat gain of a residence comes from external sources such as outside air temperature and sun shining on exposed glass, roof, and walls. Heat gain from other sources, such as people and appliances, is relatively small in most residences; but it does increase the load, and it must be considered.

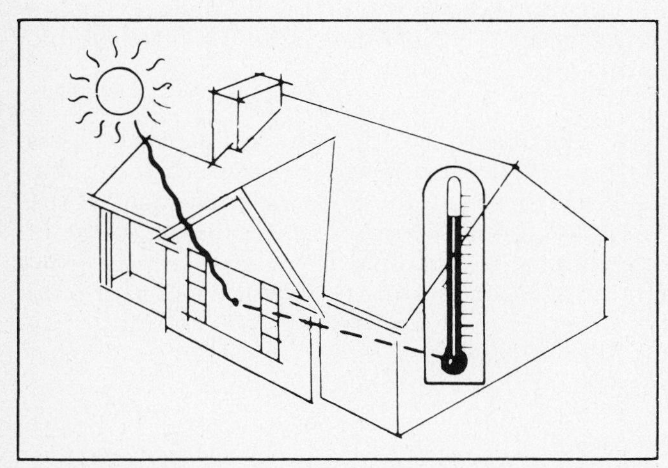

7-1 Sunshine on windows is a major source of residential heat gain.

7-3 OUTSIDE DESIGN TEMPERATURE Summer outside design temperatures lised in Table 1 are based on those normally exceeded for less than 75 hours during the four months June through September. Each occurence of higher temperature will usually last only an hour or two. It is unnecessary and uneconomical to calculate the heat gain for these occasional higher temperature conditions. If the locality for which you are calculating a load is not listed in Table 1, use the design temperature for the nearest locality listed or check with the nearest weather station.

7-4 OUTSIDE DAILY TEMPERATURE RANGE The number of degrees difference between the average maximum temperature and the average minimum temperature that occurs more than five percent of the time during the cooling season is the outside daily temperature range. Daily temperature range is identified as **low** when the difference is **less than 15 degrees, medium** when the difference is **15 to 25 degrees,** and **high** when the difference is **more than 25 degrees.**

7-5 To obtain the daily temperature range for a particular locality, see Table 1. If the locality is not listed in Table 1, use the value for the nearest listed locality. High, medium, and low daily temperature ranges shown in the table are average conditions, and local variations and preferences should be considered when determining the temperature range to be used for load calculation.

7-6 FLYWHEEL EFFECT Because the structural components of a house absorb some of the heat, peak outside temperatures do not affect the temperature inside the house immediately. This delayed response is referred to as the flywheel effect of the structure. Sun shining on glass does introduce heat instanteously, and unusually large unshaded glass areas tend to nullify the flywheel effect. To take full advantage of the flywheel effect, the equipment should operate with automatic temperature control set at the desired temperature 24 hours a day.

7-7 TEMPERATURE SWING Indoor temperature swing (cooling) is the difference between the maximum indoor temperature and the temperature set at the thermostat during a day on which the outdoor temperature reaches the design value. When the use of temperature swing results in the selection of lower capacity equipment, better humidity control but greater temperature variation will result on a design day. If the maximum desired indoor temperature is 78 F, and a 6 degree temperature swing is chosen, for example: The thermostat must be set at 72 F on a design day.

7-8 Factors to be used for maintaining inside temperatures within 3, 4½, and 6 degree temperature swings are shown in Table 7. Multiplying the total calculated heat gain by the Table 7 factor for the desired temperature swing and the applicable outside design temperature results in the equipment capacity necessary to maintain inside temperature limits as explained above.

7-9 Inside units of town or row housing have only two exposed walls. Except for units which have only north and south exposures, equipment oversizing can be prevented by using a 4½ or 6 degree temperature swing.

7-10 INSIDE DESIGN TEMPERATURE The heat transfer multipliers in this manual are to be used for the listed outside design conditions and 75 F inside design temperature. If an inside design temperature other than 75 F is desired, a correspondingly higher or lower outside design temperature must be used when selecting heat transfer multipliers from the tables.

7-11 Some people, particularly those in areas having moderate summer climates, operate their air conditioning equipment only part of the time. Others wish to air condition only certain rooms in their home. These practices **are not** recommended. Equipment operated on a part-time basis must have sufficient pickup to reduce temperatures rapidly and, therefore, must be oversized. Equipment operated so as to air condition only a few rooms when originally selected to air condition a whole house will likewise be oversized. As a result, in both cases, the equipment will cycle frequently and may not control humidity satisfactorily.

7-12 Localities having design temperatures lower than 85 F do have short periods of warmer weather. So that equipment will have enough capacity to control temperatures during these periods, calculate the heat gain using factors for an 85 F design temperature in low and medium daily range localities.

7-1 Compensate for heat gain caused by kitchen appliances.

7-13 INTERNAL HEAT GAIN Heat gain from people and appliances is relatively small, but it is a load on the equipment and must be considered. To compensate for the heat gain from people, add to the structure load 300 Btuh of sensible heat for each person.

7-14 If the number of occupants is unknown, a commonly used method of determining the number is to multiply the number of bedrooms by two. Never use a total of less than three people.

> Example: If a house has three bedrooms, the number of people who will reside in it is calculated 3 x 2 = 6. To determine the heat gain from the six people, multiply 300 by 6. The resulting 1800 Btuh of sensible heat must be removed by the air conditioning equipment.

7-15 The greatest amount of heat gain usually occurs when most of the people residing in the house are occupying the living areas. Therefore, the occupancy load should be divided among the living areas of the house, not the bedrooms, bathrooms, or the kitchen.

7-16 Although exceptions may occur, this method of determining and apportioning the load is sufficiently accurate for most cases. If the homeowner will do a considerable amount of entertaining, additional allowances may be desirable to assure comfort conditions. Care should be taken, however, not to overestimate the number of people since it could lead to oversizing the air conditioning equipment.

7-17 To compensate for heat gain from kitchen appliances and cooking, 1200 Btuh sensible heat is added to the kitchen heat gain.

7-18 Heat gain also comes from laundry appliances and bathing. In most installations, it is not practical to maintain controlled conditions in laundries and bathrooms during periods of maximum use. Exhaust fans are recommended to remove heat and moisture generated in these rooms. Do not increase the load to compensate these heat gains.

7-19 INFILTRATION AND MECHANICAL VENTILATION Natural infiltration is taken into account in the wall heat transfer multipliers for cooling in Table 5. Studies in the research residences at the University of Illinois have shown that mechanical ventilation imposes an additional load on the equipment and does not reduce the infiltration load. If mechanical ventilation is provided, this load must be calculated separately.

7-20 ATTICS HTM values for roofs, ceilings, duct heat gains, and duct heat losses listed in the tables in this manual are based on HUD-FHA minimum attic ventilation described as follows: *

Provide cross ventilation with vent openings protected against entrance of rain and snow. Diagrams below show **minimum total net free ventilating area required:**

Gable vents only without ceiling vapor barrier

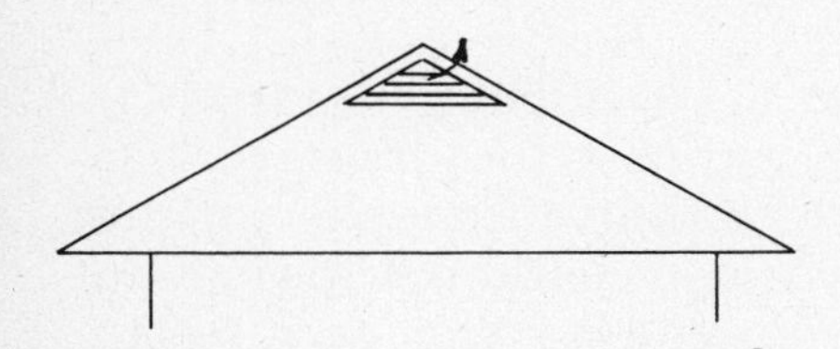

1 sq ft for each 150 sq ft of ceiling. Put half in each gable end. For ceiling area of 1,200 sq ft, 8 sq ft net ventilating area required, use 4 sq ft at each gable end.

Gable vents only with vapor barrier in ceiling

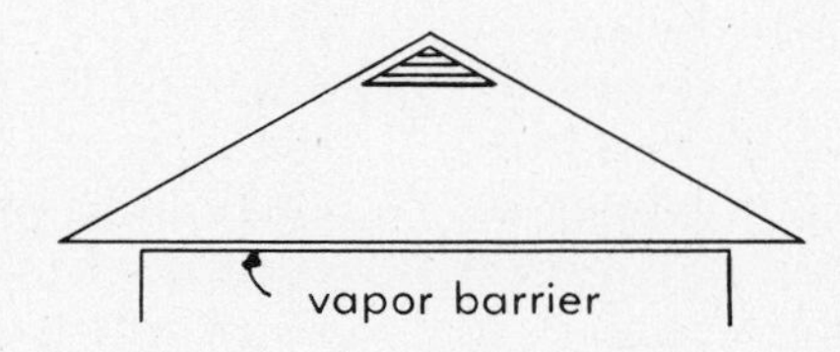

1 sq ft for each 300 sq ft of ceiling. Put half in each gable end. For ceiling area of 1,200 sq ft, 4 sq ft of net ventilating area required, use 2 sq ft at each gable end.

Combination of eave and roof vents without vapor barrier

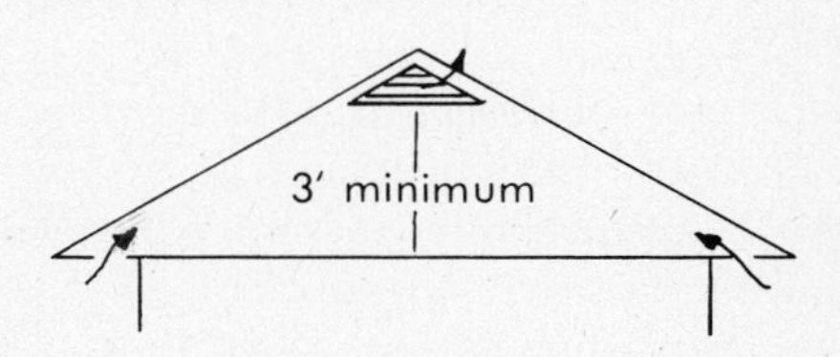

0.5 sq ft in each gable end or 1 sq ft at ridge for each 300 sq ft of ceiling **and** 0.5 sq ft of vent area in each soffit or at each eave for each 300 sq ft of ceiling, provided required gable end or ridge ventilating area is at least 3'-0" above eave or cornice vent. For ceiling area of 1,200 sq ft 4 sq ft of net ventilating area required, use 1 sq ft in each gable end and 1 sq ft at each eave.

Roof/Ceilings

Cathedral, flat, and mansard roof/ceilings **with** vapor barriers need vent area of 1 square foot per 300 square feet of ceiling. Cross ventilation may be achieved by placing half of required vent area at each soffit or eave. Experience has shown that if possible it is preferable, though, to provide half of the required area at the ridge and one quarter at each soffit or eave. **Without** a vapor barrier, the vent area should be doubled.

In flat or nearly flat roofs without a ridge, an additional vapor barrier may be used if vapor condensation problems are considered likely.

The above stated vent areas for attics and roof/ceilings refer to net free area of opening through which air can pass unobstructed. When screening, louvers, and rain/snow shields cover the vents, the area of the vent opening should be increased to offset the area of the obstructions. For convenience, the table below lists a recognized method of determining gross area of vent opening related to type of vent covering and required net free ventilating area.

Type of Covering	Size of Opening
¼" hardware cloth	1 x required net free area
¼" hardware cloth and rain louvers	2 x required net free area
⅛" mesh screen	1¼ x required net free area
⅛" mesh screen and rain louvers	2¼ x required net free area
1/16" mesh screen	2 x required net free area
1/16" mesh screen and rain louvers	3 x required net free area

*Data reprinted by permission from the Insulation Manual of the NAHB Research Foundation, Inc.

SECTION 8
HEAT GAIN OF THE STRUCTURE

8-1 Heat gain is calculated for all rooms to be cooled. See Section 2 for explanation of areas to be measured.

8-2 Sun shining through windows may constitute a large part of the heat gain of a structure. The cooling load due to transmitted and absorbed solar energy varies with the type of glass, type of shading and direction window faces. Usually, all windows in a house will be of one type—either single or double glass—and the type of inside shading will be the same. When entering window areas on Worksheet J-1, however, be certain to indicate the proper direction for each window. Glass heat gain factors are listed in Table 4.

8-3 When storm windows are used during the heating season, they should also be used during the summer for air conditioned homes. Their use will reduce the heat gain of the house and may reduce the capacity and operating cost of equipment.

8-4 **SHADED GLASS** When windows are shaded by a roof overhang, the shaded portion of the window is not affected by the direct rays of the sun and may be considered north facing. If the shading effect of roof overhangs is to be considered in the heat gain calculation, the shaded glass areas must be calculated using the data and procedures in Appendix B.

8-5 The possible need for calculating shaded glass areas should be considered for each installation. If a structure has large window areas facing either east or west, calculating shaded glass areas can reduce the design heat gain significantly. In general, if a house has roof overhangs 2 ft or more in width, large glass areas, and the direction which the house faces is known; the shading effect of the glass areas should be determined. When the effect of shading is considered, smaller equipment capacity can be specified than when it is not considered.

8-6 Areas of ceilings, walls, and floors used for calculating heat loss can be used for calculating heat gain except that heat gain for floors over a basement, enclosed and unvented crawl space, or concrete slab on ground is so small that it is not calculated. When a basement is conditioned, the heat gain for the basement floor and walls below grade is not calculated.

8-7 **TOTAL SENSIBLE HEAT GAIN** When the heat gain for an entire residence is calculated as one room the total sensible heat gain is the sum of all the Btuh heat gains from heat transmitted through the exposed walls, windows, doors, warm partitions, warm ceilings, and from the occupants. When the heat gain for each individual room is calculated, as required to size the distribution system, the total sensible heat gain is the sum of the heat gains of all the individual rooms. (See Section 9)

8-8 **LATENT HEAT LOAD** A latent heat load factor is included to estimate the moisture load. This load must be removed by the cooling equipment. To accurately determine the amount of moisture introduced by the various appliances and occupants, it would be necessary to calculate very accurately the grains of water emitted from these moisture sources. Experience has indicated, however, that the moisture load in a home depends directly on the amount of infiltration, occupancy loads, etc. This permits the use of a factor which is based upon the total sensible heat gain of the residence. The total sensible heat gain time 0.30 equals the allowance for the load caused by moisture in the conditioned area. This factor represents the moisture load for an average range of relative humidity conditions.

8-9 On Worksheet J-1, the total heat load is found by multiplying the sensible load by 1.3. This average factor is the performance expectancy designed into unitary equipment. Because the moisture load is due to both weather conditions and normal family living activities, the use of an average factor will cause negligible errors in most geographical locations.

8-1 Shaded glass area from roof overhang.

8-10 When field assembled equipment engineered specifically for the arid regions of the country (with oversize evaporator coils and air circulating equipment) is used, an appropriate engineering correction factor for the latent heat load can also be used in the load calculations.

8-11 TOTAL HEAT GAINS To obtain the total cooling equipment load, multiply the total sensible heat gain by 1.3. This is the entire load in Btuh that must be removed by the cooling equipment. To determine the size of the unit which should be used, **refer to your manufacturer's catalog** for the total capacity ratings in Btuh.

8-12 If for any reason the cooling unit to be installed has a cooling capacity different from the total shown under the column headed "Entire House," the volume of air circulated should be based on the actual **cooling capacity of the unit.** This is necessary to prevent frosting of the cooling coils. Therefore, the calculated heat gain of each room must be adjusted so that the sum of the room heat gains equals the capacity of the unit. The Adjusted Heat Gain for each room is determined in the following manner:

A. Obtain the sum of the calculated heat gains of all rooms to be conditioned (total of Line 20, Worksheet J-1).
B. Obtain Capacity Multiplier from Table 7 corresponding to the outside design conditions and the desired inside temperature swing.
C. Determine the minimum required equipment standard ARI capacity rating (Btuh) by multiplying the total calculated heat gain by the Capacity Multiplier.
D. It is recommended that the rated capacity of the equipment selected should be not less than 95 percent of the minimum required capacity determined above.
E. Divide the Btuh rating of the cooling unit selected by the total calculated heat gain to obtain the Adjustment Factor.
F. Obtain the Adjusted Heat Gain for each room by multiplying the calculated heat gain by the Adjustment Factor.

A. Total Btuh Gain, from Line 20	31,300
B. Capacity Multiplier, from Table 7	1.05
C. Equipment Standard ARI Capacity Rating, Minimum required, (Line A) x (Line B)	32,900
D. Capacity of Equipment Selected, not less than 95% of Line C, Enter in "Entire House" column, Line 21	36,000
E. Adjustment Factor (Line D) (Line A), Enter in "HTM" column, Line 21	1.15

8-2 Adjustment Factor determined using Table B, Form J-1

Table B, Form J-1, is used when determining the Adjustment Factor. This table is illustrated in Fig. 8-2.

LINE A
Enter the (Entire House) Total Btuh Gain obtained from Line 20, Worksheet J-1.

LINE B
Enter the Capacity Multiplier obtained from Table 7.

LINE C
Obtain Minimum Capacity Rating Required by multiplying (Line A) x Line B) and enter on Line C.

LINE D
Enter the capacity of equipment selected.

LINE E
Divide (Line D) by (Line A) to obtain Adjustment Factor.

8-13 Details concerning equipment selection are discussed in Manual K.

SECTION 9 DUCT HEAT GAIN

9-1 Duct heat gain occurs when air in a duct is cooler than the space in which the duct is located. This heat gain decreases the cooling capacity delivered to the conditioned space through the duct.

9-2 DUCT LOCATED IN CONDITIONED SPACE Heat gain to ducts located in conditioned spaces help cool space for which the heat gain has been calculated. Because it is already included in the heat gain of the structure, it is not an added load on the equipment. It is not considered a load when determining the required cooling capacity.

9-3 DUCTS LOCATED IN UNCONDITIONED SPACE When a duct system is located outside the conditioned space, the temperature rise of the air in the duct system must be considered. This air temperature rise is a result of heat gain to the duct, and the duct heat gain must be included with the heat gain of the residence when determining the required equipment capacity.

9-4 Unconditioned spaces in which ducts are commonly located include the following:

A. **Attic**—The space between a ceiling and a roof. Ducts should be insulated. Heat gain multipliers for ducts with insulation are shown in Table 6. Attic vent area is assumed to at least equal HUD-FHA minimum requirements.

B. **Above Grade Spaces**—A space such as a vented crawl space, an attached garage, or other unconditioned rooms above grade. Included are basement rooms having more than 25 percent of any wall above grade. Ducts should be insulated. Heat gain multipliers for ducts with insulation are shown in Table 6.

C. **Below Grade Spaces**—A space such as a basement below grade for which the heat gain has not been calculated. Multipliers for uninsulated duct are shown in Table 6.

D. **Perimeter-Loop in Slab**—Factors apply only to loop duct; feeder duct heat gains help cool the conditioned space. Heat gain multipliers in Table 6 are for systems in which slab edge is insulated, but ducts are not insulated.

9-5 To obtain duct heat gain, multiply the sensible heat gain by the appropriate factor from Table 6.

EXAMPLE: Duct system is installed in an unconditioned space above grade. Each duct is wrapped with 1 in. flexible duct insulation, and the duct heat gain multiplier is 0.05. (See Fig. 9-1.)

Duct Location and Insulation	Duct Heat Gain Multipliers
Attic; 1 in. Flexible Blanket Insulation	0.15
Attic; 2 in. Flexible or 1 in. Rigid Duct Insulation	0.10
Unconditioned Above Grade Space; 1 in. Flexible Duct Insulation	0.05
Unconditioned Below Grade Space; No Duct Insulation	0.05
Ducts in Slab; No Duct Insulation	0.05

9-1 Duct heat gain multiplier obtained from Table 6 for a duct installed in an unconditioned space above grade and wrapped with 1 in. flexible insulation.

SECTION 10
HEAT GAIN CALCULATION PROCEDURE

10-1 An example of heat gain calculation for a one story house with a basement is included in this section. This is the same structure used for the heat loss calculation example in Section 6.

10-2 Since the same residence is being used for both examples; the construction, amount of insulation, areas, windows, etc. are the same. For this example, however, the area calculations are repeated. If both the heat loss and heat gain were being calculated, it would be necessary to determine these areas only once because the method of calculating the cooling load parallels that of calculating the heating load.

10-3 **STEP-BY-STEP PROCEDURE** Study the heat gain calculation procedure below and refer to the paragraphs, figures, and tables indicated for additional information rquired for each step in the example. Design conditions and type of construction assumed for this example are listed below. A sample calculation on Worksheet Form J-1 is illustrated in Fig. 10-1.

10-4 The plan used and the constructions and design conditions assumed were chosen only to illustrate the use of data and procedures in the manual. They should not be interpreted as recommendations of construction features.

1 Name of Room				Entire House			1 LIV			2 HALL 'D' & DIN			3 LAUN			4 HALL 'C' & KIT			5 BATH # 1		
2 Running Ft Exposed Wall							21			12 & 13			11 & 7			11			9		
3 Room Dimensions, Ft							14 × 21			7 × 18 & 4 × 4			7 × 11			8 × 11 & 3 × 17			6 × 8 & 3 × 5		
4 Ceiling Ht, Ft	Directions Room Faces						8	W		8	N & W		8	N & E		8	E		8	E	
TYPE OF EXPOSURE	Const. No.	HTM Htg	HTM Clg	Area or Length	Btuh Htg	Btuh Clg	Area or Length	Btuh Htg	Btuh Clg	Area or Length	Btuh Htg	Btuh Clg	Area or Length	Btuh Htg	Btuh Clg	Area or Length	Btuh Htg	Btuh Clg	Area or Length	Btuh Htg	Btuh Clg
5 Gross Exposed Walls and Partitions a	10(d)						168			104			56			88			72		
b	12(c)									96			88								
c	13(a)																				
d	13(f)																				
6 Windows and Glass Doors (Htg) a																					
b																					
c																					
7 Windows and Glass Doors (Clg)	North		20							20		400									
	E &W or NE & NW		50 / 90				40		2000							11		550	8		400
	South or SE & SW		25 / 50																		
8 Other Doors	9(b)		13.0										17		221						
9 Net Exposed Walls and Partitions a	10(d)		3.0				128		384	84		252	39		117	77		231	64		192
b	12(c)		4.5							96		432	88		396						
c	13(a)		13.0																		
d	13(f)		—																		
10 Ceilings a	14(c)		2.0				294		588	142		284	77		154	139		278	63		126
b																					
11 Floors a	19		—																		
b																					
12 Ventilation																					
13 Sub Total Btuh Loss																					
14 Duct Btuh Loss																					
15 Total Btuh Loss																					
16 People @ 300 and Appliances 1200							3		900	3		900						1200			
17 Sensible Btuh Gain (Structure)									3872			2268			888			2259			718
18 Duct Btuh Gain																					
19 Sum of Lines 17 and 18 (Clg)						24,071															
20 Total Btuh Gain (Line 19 x 1.3)			1.3			31,300			5030			2950			1154			2940			933
21 Btuh for Air Quantities			1.15			36,000			5780			3390			1327			3380			1073

10-1 Example of heat gain calculation.

ASSUMED DESIGN CONDITIONS AND CONSTRUCTION (Cooling)

		From Tables 4 & 5 Const. No.	HTM
A.	Outside Design Temperature: Dry Bulb 95F		
B.	Daily Temperature Range: Medium		
C.	Inside Design Conditions: 75F, 3 Degree Temperature Swing		
D.	Types of Shading: Venetian Blinds on All First Floor Windows		
E.	Windows: All Double Glass on First Floor		
	North		20
	East or West		50
	South		25
	All Single Glass in Basement		
	East		90
	South		50
F.	Doors: Wood with Storm Doors	9	13.0
G.	First Floor Walls: Basic Frame Construction with Insulation (R-11)	10(d)	3.0
	Basement Wall: 8 in. Concrete Block, Above Grade: 3 ft	13(a)	13.0
	8 in. Concrete Block, Below Grade: 5 ft		
H.	Partition: 8 in. Concrete Block Furred, with Insulation (R-5)	12(c)	4.5
I.	Ceiling: Basic Construction Under Vented Attic with Insulation (R-19), Dark Roof	14(c)	2.0
J.	Occupants: 6 (Figured 2 per Bedroom, But Distributed 3 in Living, 3 in Dining)		
K.	Appliances: Add 1200 Btuh to Kitchen		
L.	Ducts: Located in Conditioned Space		
M.	Air Conditioning Unit: Air-Cooled		

Line	6 BR #3			7 BR #2			8 BATH #2			9 BR #1			10 HALL 'A' & 'B'			11 REC RM			12 WORK & UTIL		
2	10			24			5			29			7			82			74		
3	10 × 11 & 3 × 6			10 × 14 & 2 × 5			5 × 5			14 × 15, 2 × 5, & 3 × 6			4 × 14, 3 × 5, 4 × 20, 2 × 4, 1 × 3, 2 × 3			23 × 28 & 8 × 16			19 × 28 & 8 × 12		
4	8	E		8	S & E		8	S		8	S & W		8	W		3/5	E, S, & W		3/5	W, N, & E	
	Area or Length	Btuh Htg	Btuh Clg	Area or Length	Btuh Htg	Btuh Clg	Area or Length	Btuh Htg	Btuh Clg	Area or Length	Btuh Htg	Btuh Clg	Area or Length	Btuh Htg	Btuh Clg	Area or Length	Btuh Htg	Btuh Clg	Area or Length	Btuh Htg	Btuh Clg
5	80			192			40			232			56								
																246			222		
																410			370		
6																					
7																					
	22		1100	17		850				17		850				8		720	4		360
				11		275	8		200	11		275				8		400			
8													20		260						
9	58		174	164		492	32		96	204		612	36		108						
																230		2990	218		2830
																410		—	370		—
10	128		256	150		300	25		50	238		476	168		336	28		56			
11																772		—	628		—
12																					
13																					
14																					
15																					
16																					
17			1530			1917			346			2213			704			4166			3190
18																					
19																					
20			1989			2490			450			2870			915			5420			4150
21			2290			2860			518			3300			1052			6230			4770

PROCEDURE FOR DETERMINING Btuh HEAT GAIN TO ROOMS

PROCEDURE	EXAMPLE
A. Determine the **Outside Design Temperature** from Table 1. (See paragraph 7-3.)	The Outside Design Temperature for Nashville, Tennessee, is 95 F. The Outside Design Temperature for Windsor, Ontario, is 90 F.
B. Determine **Daily Temperature Range** from Table 1. (See paragraphs 7-4 and 7-5.)	The Daily Temperature Range in Salem, Oregon, is high and in Edmonton, Alberta, is medium.
C. Select **Inside Design Temperature** for rooms to be cooled. (See paragraph 7-10.)	Use a 75 F Inside Design Temperature unless customer insists on different design conditions.
D. Determine all dimensions of each room to be conditioned and direction each outside wall faces. If heat loss has been calculated, these items have already been obtained. (See paragraphs 2-1 through 2-5 and Fig. 6-3.)	If heat loss has not been calculated, complete lines 1 through 4 on Worksheet J-1 for each room. See Fig. 10-1.
E. In Table 5, locate the various descriptions of construction and ventilation used in the structure. If heat loss has been calculated, most of these construction numbers have already been obtained. Types of shade for windows must be determined. (See paragraphs 8-2 through 8-5.)	Each Construction Number is entered in Construction Number column on Worksheet J-1. Construction Numbers in Table 5 correspond to Construction Numbers in Table 2.

	PROCEDURE	EXAMPLE
F.	From Table 4, obtain heat transfer multipliers for each type of window and glass door facing each direction. (See paragraphs 8-2 and 8-3.)	Enter HTM in appropriate column on line corresponding to direction that windows and glass doors face on Line 7, Worksheet J-1. See Fig. 10-1.
G.	From Table 5, obtain heat transfer multipliers for other types of construction and ventilation.	Enter these HTM in the appropriate column opposite the corresponding Construction Number on Worksheet J-1. See Fig. 10-1.
H.	Obtain areas of exposed walls and partitions including windows and doors in sq ft. This is the **Gross Exposed Wall Area.** If heat loss has been calculated, these areas have already been obtained. (See paragraphs 2-6 through 2-9.)	Enter in columns for "Area or Length" on Line 5, Worksheet J-1, for each room. See Fig. 10-1.
I.	Line 6 is not used for heat gain calculation.	Used for heat loss calculation only. See Fig. 6-3.
J.	Obtain areas of windows and glass doors.	Enter in columns for "Area or Length" on line corresponding to direction they face on Line 7, Worksheet J-1, for each room. See Fig. 10-1.
K.	Obtain other door areas.	Enter in columns for "Area or Length" on Line 8, Worksheet J-1, for appropriate rooms. See Fig. 10-1.
L.	Subtract areas of windows and doors in each room from gross exposed wall area in each room. This is the **Net Exposed Wall Area.** If heat loss has been calculated, these areas have already been obtained. (See paragraph 2-6.)	Enter in columns for "Area or Length" on Line 9, Worksheet J-1. See Fig. 10-1.
M.	Obtain areas of warm ceilings and warm floors. If heat loss has been calculated, these areas have already been obtained. (See paragraphs 2-15 through 2-17 and 2-20.)	Enter in columns for "Area or Length" on Line 10 and 11 respectively, Worksheet J-1. See Fig. 10-1. (The ceiling over the stairway is included in the recreation room.)
N.	Determine air volume (cfm) to be provided for ventilation or makeup air.	Enter on Line 12, Worksheet J-1 in an "Area or Length" column that is not being used for an individual room. Ventilation or makeup air is calculated for the entire house. It is not included in the heat gain for an individual room. (See Fig. 10-1.)
O.	Lines 13 through 15 are not used for heat gain calculation.	Used for heat loss calculation only. See Fig. 6-3.
P.	Obtain number of persons occupying a home (see paragraphs 7-13 through 7-16) and add 1200 Btuh to the calculated heat gain for the kitchen. (See paragraph 7-17.)	See Worksheet Fig. 10-1, Line 16. Three bedrooms times two persons equals six. Distribute this load throughout the living area. Twelve hundred Btuh is the load allowance to compensate for heat gain from appliances.
Q.	The design heat gain for each window, door, warm ceiling, warm floor, ventilation, and people equals the area or number times the corresponding HTM.	Multiply each area or number by the appropriate HTM and enter in "Btuh" columns on corresponding lines, Worksheet J-1. See Fig. 10-1.

PROCEDURE	EXAMPLE
R. Obtain **Sensible Heat Gain** for each room by adding heat gain for windows, doors, wall, partitions, ceilings, floors, people, and appliances. Repeat the procedure for each room (include all conditioned spaces.)	Enter on Line 17, Worksheet J-1. See Fig. 10-1.
S. If ducts are located in unconditioned spaces, determine the **Duct Btuh Gain** for each room. (See Section 9.)	Since the ducts are located in conditioned spaces for this example, Line 18 is not used.
T. Add Lines 17 and 18.	Enter sum of Lines 17 and 18 on Line 19, Worksheet J-1, for each room.
U. Obtain Total Btuh Gain of each room by multiplying by 1.3. (See paragraphs 8-7 through 8-9.)	Enter total Btuh gain for each room on Line 20, Worksheet J-1. See Fig. 10-1.
V. Obtain the Total Btuh Gain for the entire house by adding all of the room heat gains from Line 20 to 1.3 times the ventilation Btuh from the Entire House column of Line 12.	Enter in "Entire House" column, Line 20, Worksheet J-1. See Fig. 10-1.
W. From Table 7, obtain minimum equipment capacity required and determine actual capacity of equipment to be used. (See paragraphs 7-15 through 7-18.)	Enter in "Entire House" column, Line 21, Worksheet J-1. See Fig. 10-1.
X. Obtain **Btuh for Air Quantities.** (See paragraph 8-12.)	Enter on Line 21, Worksheet J-1. See Fig. 10-1.

SECTION 11 TABLES

Table 1
OUTSIDE DESIGN CONDITIONS—UNITED STATES

State & City	Winter Dry-Bulb Temp[1] (F)	Summer Dry Bulb Temp[1] (F)	North Latitude (0)	Daily Range
ALABAMA				
Alexander City	20	95	35	M
Anniston	15	95	35	M
Auburn	25	95	30	M
Birmingham	20	95	35	M
Decatur	15	95	35	M
Dothan	25	95	30	M
Florence	15	95	35	M
Gadsden	20	95	35	M
Huntsville	15	95	35	M
Mobile ★ AP	25	95	30	M
Mobile ★ CO	30	95	30	M
Montgomery	25	95	30	M
Selma	25	95	30	M
Talladega	15	95	35	M
Tuscaloosa	20	95	35	M
ALASKA				
Anchorage	—20	70	60	M
Barrow	—45	55	70	L
Fairbanks	—50	80	65	M
Juneau	—5	70	60	M
Kodiak	5	65	55	L
Nome	—30	60	65	L
ARIZONA				
Douglas	20	100	30	H
Flagstaff	5	80	35	H
Fort Huachuca	25	95	30	H
Kingman	25	100	35	H
Nogales	20	100	30	H
Phoenix	30	105	35	H
Prescott	15	95	35	H
Tucson	30	100	35	H
Winslow	10	95	35	H
Yuma	40	110	30	H
ARKANSAS				
Blytheville	15	95	35	M
Camden	20	95	35	M
El Dorado	20	95	35	M
Fayetteville	10	95	35	M
Fort Smith	15	100	35	M
Hot Springs	20	95	35	M
Jonesboro	15	95	35	M
Little Rock	20	95	35	M
Pine Bluff	20	95	35	M
Texarkana	25	95	35	M
CALIFORNIA				
Bakersfield	30	100	35	H
Barstow	25	100	35	H
Blythe	35	110	35	H
Burbank	35	95	35	M
Chico	30	100	40	H
Concord	35	90	40	H
Covina	40	95	35	H
Crescent City	35	70	40	M
Downey	35	90	35	M
El Cajon	30	95	30	H
El Centro	35	110	35	H
Escondido	35	90	35	H
Eureka/Arcata	35	65	40	L
Fairfield	30	95	40	H
Fresno	30	100	35	H
Laguna Beach	35	80	35	M
Livermore	30	95	35	H
Lompoc	35	80	35	M
Long Beach	35	85	35	M
Los Angeles ★ AP	40	85	35	M
Los Angeles ★ CO	40	90	35	M
Merced	30	100	35	H
Modesto	35	100	35	H
Monterey	35	80	35	M
Napa	30	90	40	H
Needles	35	110	35	H
Oakland	35	80	35	M
Oceanside	40	80	35	L
Ontario	30	95	35	H
Oxnard	35	80	35	M
Palmdale	25	100	35	H
Palm Springs	35	110	35	H
Pasadena	35	95	35	H
Petaluma	30	90	40	H
Pomona	30	95	35	H
Redding	35	100	40	H
Redlands	35	95	35	H
Richmond	35	80	40	M
Riverside	30	95	35	H
Sacramento	30	95	40	H
Salinas	35	85	35	M
San Bernardino	30	100	35	H
San Diego	40	85	30	L
San Fernando	35	95	35	H
San Francisco ★ AP	35	80	35	M
San Francisco ★ CO	40	75	40	L
CALIFORNIA Cont.				
San Jose	35	90	35	H
San Luis Obispo	35	85	35	H
Santa Ana	35	90	35	H
Santa Barbara	35	85	35	M
Santa Cruz	30	85	35	H
Santa Maria	30	80	35	M
Santa Monica	45	75	35	M
Santa Paula	35	90	35	H
Santa Rosa	30	95	40	H
Stockton	30	100	40	H
Ukiah	30	95	40	H
Visalia	35	100	35	H
Yreka	15	95	40	H
Yuba City	30	100	40	H
COLORADO				
Alamosa	—15	80	35	H
Boulder	5	90	40	H
Colorado Springs	0	90	40	H
Denver	0	90	40	H
Durango	0	85	35	H
Fort Collins	—5	90	40	H
Grand Junction	10	95	40	H
Greeley	—5	90	40	H
La Junta	—5	95	40	H
Leadville	—5	75	40	H
Pueblo	—5	95	40	H
Sterling	—5	95	40	H
Trinidad	5	90	35	H
CONNECTICUT				
Bridgeport	5	90	40	M
Hartford	5	90	40	M
New Haven	5	85	40	M
New London	5	85	40	M
Norwalk	0	90	40	M
Norwich	0	85	40	M
Waterbury	0	90	40	M
Windsor Locks	0	90	40	M
DELAWARE				
Dover	15	90	40	M
Wilmington	15	90	40	M
DISTRICT OF COLUMBIA				
Washington	15	90	40	M
FLORIDA				
Belle Glade	35	90	25	M
Cape Kennedy	35	90	30	M
Daytona Beach	35	90	30	M
Fort Lauderdale	45	90	25	M
Fort Myers	40	90	25	M
Fort Pierce	40	90	25	M
Gainesville	30	95	30	M
Jacksonville	25	95	80	M
Key West	55	90	25	L
Lakeland	35	95	30	M
Miami	40	90	25	M
Miami Beach	45	90	25	L
Ocala	30	95	30	M
Orlando	35	95	30	M
Panama City	30	90	30	L
Pensacola	25	90	30	L
St. Augustine	30	90	30	M
St. Petersburg	35	90	30	M
Sanford	30	95	30	M
Sarasota	35	90	25	M
Tallahassee	25	95	30	M
Tampa	35	90	30	M
GEORGIA				
Albany	30	95	30	M
Americus	25	95	30	M
Athens	20	95	35	M
Atlanta	20	90	35	M
Augusta	20	95	35	M
Brunswick	25	95	30	M
Columbus	25	95	30	M
Dalton	15	95	35	M
Dublin	25	95	30	M
Gainesville	20	90	35	M
Griffin	20	95	35	M
La Grange	20	95	35	M
Macon	25	95	30	M
Marietta	20	95	35	M
Moultrie	30	95	30	M
Rome	20	95	35	M
Savannah	25	95	30	M
Valdosta	30	95	30	M
Waycross	25	95	30	M
HAWAII				
Hilo	55	85	20	M
Honolulu	60	85	20	L
HAWAII Cont.				
Kaneoke	60	85	20	L
Wakiawa	60	85	20	L
IDAHO				
Boise	10	95	45	H
Burley	5	95	40	H
Coeur d'Alene	5	90	50	H
Idaho Falls	—10	90	45	H
Lewiston	10	95	45	H
Moscow	0	90	45	H
Mountain Home	5	95	45	H
Pocatello	—5	90	45	H
Twin Falls	5	95	40	H
ILLINOIS				
Aurora	—5	90	40	M
Belleville	10	95	40	M
Bloomington	0	90	40	M
Carbondale	10	95	40	M
Champaign/Urbana	0	95	40	M
Chicago	0	90	40	M
Danville	0	95	40	M
Decatur	0	95	40	M
Dixon	—5	90	40	M
Elgin	—5	90	40	M
Freeport	—10	90	40	M
Galesburg	0	90	40	M
Greenville	5	95	40	M
Joliet	—5	90	40	M
Kankakee	0	90	40	M
LaSalle/Peru	0	95	40	M
Macomb	0	95	40	M
Moline	—5	90	40	M
Mt. Vernon	10	95	40	M
Peoria	0	90	40	M
Quincy	0	95	40	M
Rantoul	0	90	40	M
Rockford	—5	90	40	M
Springfield	0	90	40	M
Waukegan	—5	90	40	M
INDIANA				
Anderson	5	90	40	M
Bedford	5	95	40	M
Bloomington	5	90	40	M
Columbus	5	90	40	M
Crawfordsville	0	95	40	M
Evansville	10	95	40	M
Fort Wayne	5	90	40	M
Goshen	0	90	40	M
Hobart	0	90	40	M
Huntington	0	90	40	M
Indianapolis	0	90	40	M
Jeffersonville	10	95	40	M
Kokoma	0	90	40	M
Lafayette	0	90	40	M
La Porte	0	90	40	M
Marion	0	90	40	M
Muncie	0	90	40	M
Peru	0	90	40	M
Richmond	0	90	40	M
Shelbyville	5	90	40	M
South Bend	0	90	40	M
Terre Haute	5	95	40	M
Valparaiso	—5	90	40	M
Vincennes	5	95	40	M
IOWA				
Ames	—10	90	40	M
Burlington	0	90	40	M
Cedar Rapids	—5	90	40	M
Clinton	—5	90	40	M
Council Bluffs	—5	95	40	M
Des Moines	—5	90	40	M
Dubuque	—10	90	40	M
Fort Dodge	—10	90	40	M
Iowa City	—5	90	40	M
Keokuk	0	95	40	M
Marshalltown	—10	90	40	M
Mason City	—10	90	45	M
Newton	—5	95	40	M
Ottumwa	—5	95	40	M
Sioux City	—10	95	40	M
Waterloo	—10	90	40	M
KANSAS				
Atchison	0	95	40	M
Chanute	5	95	35	M
Dodge City	5	95	40	M
El Dorado	5	100	40	M
Emporia	5	95	40	M
Garden City	0	100	40	H
Goodland	0	95	40	H
Great Bend	5	100	40	H
Hutchinson	5	100	40	H
Liberal	5	100	35	H

★ AP - Airport
CO - City Office
[1] Temperatures are rounded off to permit use of precalculated HTM tables.

Table 1 (Cont'd)

State & City	Winter Dry-Bulb Temp[1] (F)	Summer Dry Bulb Temp[1] (F)	North Latitude (0)	Daily Range
KANSAS Cont.				
Manhattan	0	100	40	M
Parsons	5	95	35	M
Russell	0	100	40	H
Salina	5	100	40	H
Topeka	5	95	40	M
Wichita	5	100	35	M
KENTUCKY				
Ashland	10	90	40	M
Bowling Green	10	95	35	M
Corbin	5	90	35	M
Covington	5	90	40	M
Hopkinsville	10	95	35	M
Lexington	10	90	40	M
Louisville	10	95	40	M
Madisonville	10	95	35	M
Owensboro	10	95	40	M
Paducah	10	95	35	M
LOUISIANA				
Alexandria	25	95	30	M
Baton Rouge	30	95	30	M
Bogalusa	25	95	30	M
Houma	25	90	30	M
Lafayette	25	95	30	M
Lake Charles	30	95	30	M
Minden	25	95	30	M
Monroe	25	95	30	M
Natchitoches	25	95	30	M
New Orleans	35	90	30	M
Shreveport	25	95	30	M
MAINE				
Augusta	–5	85	45	M
Bangor	–5	85	45	M
Caribou	–15	80	45	M
Lewiston	–5	85	45	M
Millinocket	–15	85	45	M
Portland	0	85	45	M
Waterville	–5	85	45	M
MARYLAND				
Baltimore ★ AP	15	90	40	M
Baltimore ★ CO	20	90	40	M
Cumberland	5	90	40	M
Frederick	10	90	40	M
Hagerstown	10	90	40	M
Salisbury	15	90	40	M
MASSACHUSETTS				
Boston	10	90	40	M
Clinton	0	85	40	M
Fall River	5	85	40	M
Framingham	0	90	40	M
Gloucester	5	85	40	M
Greenfield	–5	85	40	M
Lawrence	0	90	40	M
Lowell	0	90	40	M
New Bedford	10	85	40	M
Pittsfield	–5	85	40	M
Springfield	0	90	40	M
Taunton	0	85	40	M
Worcester	0	85	40	M
MICHIGAN				
Adrian	0	90	40	M
Alpena	–5	85	45	H
Battle Creek	5	90	40	M
Benton Harbor	0	90	40	M
Detroit	5	90	40	M
Escanaba	–5	80	45	M
Flint	0	85	45	M
Grand Rapids	5	90	45	M
Holland	5	90	45	M
Jackson	0	90	40	M
Kalamazoo	5	90	40	M
Lansing	5	85	45	M
Marquette	–5	85	45	M
Mt. Pleasant	0	85	45	M
Muskegon	5	85	45	M
Pontiac	0	90	40	M
Port Huron	0	90	45	M
Saginaw	0	85	45	M
Sault Ste. Marie	–10	80	45	M
Traverse City	0	85	45	M
Ypsilanti	5	90	40	M
MINNESOTA				
Albert Lea	–10	90	45	M
Alexandria	–15	90	45	M
Bemidji	–30	85	45	M
Brainerd	–20	85	45	M
Duluth	–15	80	45	M
Faribault	–15	90	45	M
Fergus Falls	–20	90	45	M
MINNESOTA Cont.				
International Falls	–25	80	50	H
Mankato	–15	90	45	M
Minneapolis/ St. Paul	–10	90	45	M
Rochester	–15	90	45	M
St. Cloud	–20	90	45	M
Virginia	–25	85	45	M
Willmar	–15	90	45	M
Winona	–10	90	45	M
MISSISSIPPI				
Biloxi	30	90	30	M
Clarksdale	20	95	35	M
Columbus	15	95	35	M
Greenville	20	95	35	M
Greenwood	20	95	35	M
Hattiesburg	25	95	30	M
Jackson	20	95	30	M
Laurel	25	95	30	M
McComb	25	95	30	M
Meridian	20	95	30	M
Natchez	25	95	30	M
Tupelo	20	95	35	M
Vickburg	25	95	30	M
MISSOURI				
Cape Girardeau	10	95	35	M
Columbia	5	95	40	M
Farmington	5	95	40	M
Hannibal	0	95	40	M
Jefferson City	5	95	40	M
Joplin	10	95	35	M
Kansas City	5	95	40	M
Kirksville	–5	95	40	M
Mexico	0	95	40	M
Moberly	0	95	40	M
Poplar Bluff	10	95	35	M
Rolla	5	95	40	M
St. Joseph	0	95	40	M
St. Louis ★ AP	5	95	40	M
St. Louis ★ CO	10	95	40	M
Sedalia	5	95	40	M
Sikeston	10	95	35	M
Springfield	10	95	35	M
MONTANA				
Billings	–10	90	45	H
Bozeman	–15	85	45	H
Butte	–20	85	45	H
Cut Bank	–20	85	50	H
Glasgow	–20	95	50	H
Glendive	–20	95	45	H
Great Falls	–20	90	45	H
Havre	–15	85	50	H
Helena	–15	85	45	H
Kalispell	–5	85	50	H
Lewiston	–15	85	45	H
Livingston	–15	90	45	H
Miles City	–15	95	45	H
Missoula	–5	90	45	H
NEBRASKA				
Beatrice	0	95	40	M
Chadron	–10	95	45	H
Columbus	–5	95	40	M
Fremont	–5	95	40	M
Grand Island	–5	95	40	H
Hastings	0	95	40	H
Kearney	–5	95	40	H
Lincoln	0	95	40	M
McCook	0	95	40	H
Norfolk	–10	95	40	H
North Platte	–5	95	40	H
Omaha	–5	95	40	M
Scottsbluff	–5	95	40	H
Sidney	–5	90	40	H
NEVADA				
Carson City	5	90	40	H
Elko	–10	90	40	H
Ely	–5	90	40	H
Las Vegas	25	105	35	H
Lovelock	10	95	40	H
Reno ★ AP	5	90	40	H
Reno ★ CO	15	90	40	H
Tonopah	10	90	40	H
Winnemucca	5	95	40	H
NEW HAMPSHIRE				
Berlin	–15	85	45	M
Claremont	–10	85	45	M
Concord	–10	90	45	H
Keene	–10	90	45	M
Laconia	–10	85	45	M
Manchester	0	90	45	M
Portsmouth	0	85	45	M
NEW JERSEY				
Atlantic City	15	90	40	M
Long Branch	10	90	40	M
Newark	15	90	40	M
New Brunswick	10	90	40	M
Paterson	10	90	40	M
Phillipsburg	10	90	40	M
Trenton	15	90	40	M
Vineland	15	90	40	M
NEW MEXICO				
Alamagordo	20	100	35	H
Albuquerque	15	95	35	H
Artesia	15	100	35	H
Carlsbad	20	100	30	H
Clovis	15	95	35	H
Farmington	5	95	35	H
Gallup	–5	90	35	H
Grants	–5	90	35	H
Hobbs	15	100	30	H
Las Cruces	20	100	30	H
Los Alamos	5	85	35	H
Raton	0	90	35	H
Roswell	15	100	35	H
Santa Fe	10	90	35	H
Silver City	15	95	30	H
Socorro	15	95	35	H
Tucumcari	10	95	35	H
NEW YORK				
Albany ★ AP	0	90	45	M
Albany ★ CO	5	90	45	M
Auburn	0	85	45	M
Batavia	0	85	45	M
Binghampton	0	90	40	M
Buffalo	5	85	45	M
Cortland	–5	90	40	M
Dunkirk	5	85	40	M
Elmira	5	90	40	M
Geneva	0	90	45	M
Glens Falls	–10	85	45	M
Gloversville	–5	85	45	M
Hornell	–5	85	40	M
Ithaca	0	90	40	M
Jamestown	5	85	40	M
Kingston	0	90	40	M
Lockport	5	85	45	M
Massena	–15	85	45	M
Newburgh	5	90	40	M
New York City CO	15	90	40	M
(Kennedy) ★ AP	20	85	40	M
(LaGuardia) AP	15	90	40	M
Niagara Falls	5	85	45	M
Olean	–5	85	40	M
Oneonta	–5	85	40	M
Oswego	5	85	45	M
Plattsburgh	–10	85	45	M
Poughkeepsie	0	90	40	M
Rochester	5	90	45	M
Rome	–5	85	45	M
Schenectady	–5	90	45	M
Syracuse	0	85	34	M
Utica	–5	85	45	M
Watertown	–10	85	45	M
NORTH CAROLINA				
Asheville	15	90	35	M
Charlotte	20	95	35	M
Durham	15	90	35	M
Elizabeth City	20	90	35	M
Fayetteville	20	95	35	M
Goldsboro	20	90	35	M
Greensboro	15	90	35	M
Greenville	20	95	35	M
Henderson	15	90	35	M
Hickory	15	90	35	M
Jacksonville	20	90	35	M
Lumberton	20	95	35	M
New Bern	20	90	35	M
Raleigh	20	90	35	M
Rocky Mount	20	95	35	M
Wilmington	25	90	35	M
Winston-Salem	15	90	35	M
NORTH DAKOTA				
Bismarck	–20	90	45	H
Devil's Lake	–20	90	50	M
Dickinson	–20	95	45	M
Fargo	–20	90	45	M
Grand Forks	–25	85	50	M
Jamestown	–20	90	45	H
Minot	–20	90	50	M
Williston	–20	90	50	M
OHIO				
Akron	5	85	40	M
Ashtabula	5	85	40	M

★ AP - Airport
CO - City Office
[1]Temperatures are rounded off to permit use of precalculated HTM tables.

Table 1 (Cont'd)

State & City	Winter Dry-Bulb Temp[1] (F)	Summer Dry Bulb Temp[1] (F)	North Latitude (0)	Daily Range
OHIO Cont.				
Athens	5	90	40	M
Bowling Green	0	90	40	M
Cambridge	0	90	40	M
Chillicothe	5	90	40	M
Cincinnati	10	90	40	M
Cleveland	5	90	40	M
Columbus	5	90	40	M
Dayton	5	90	40	M
Defiance	0	90	40	M
Findlay	0	90	40	M
Fremont	0	90	40	M
Hamilton	5	90	40	M
Lancaster	5	90	40	M
Lima	0	90	40	M
Mansfield	0	90	40	M
Marion	5	90	40	M
Middletown	5	90	40	M
Newark	0	90	40	M
Norwalk	0	90	40	M
Portsmouth	5	90	40	M
Sandusky	5	90	40	M
Springfield	5	90	40	M
Steubenville	5	90	40	M
Toledo	5	90	40	M
Warren	0	90	40	M
Wooster	0	90	40	M
Youngstown	5	85	40	M
Zanesville	0	90	40	M
OKLAHOMA				
Ada	15	100	35	M
Altus	15	100	35	M
Ardmore	15	100	35	M
Bartlesville	5	100	35	M
Chickasha	15	100	35	M
Enid	10	100	35	M
Lawton	15	100	35	M
McAlester	15	100	35	M
Muskogee	15	100	35	M
Norman	15	100	35	M
Oklahoma City	15	95	35	M
Ponca City	10	100	35	M
Seminole	15	100	35	M
Stillwater	15	100	35	M
Tulsa	15	100	35	M
Woodward	5	100	35	H
OREGON				
Albany	25	90	45	H
Astoria	30	75	45	M
Baker	0	90	45	H
Bend	0	85	45	H
Corvallis	25	90	45	H
Eugene	25	90	45	H
Grants Pass	25	90	40	H
Klamath Falls	5	85	40	H
Medford	20	95	40	H
Pendleton	10	95	45	H
Portland ★ AP	20	85	45	M
Portland ★ CO	25	90	45	M
Roseburg	25	90	45	H
Salem	25	90	45	H
The Dalles	15	90	45	H
PENNSYLVANIA				
Allentown	5	90	40	M
Altoona	5	85	40	M
Butler	0	90	40	M
Chambersburg	5	90	40	M
Erie	10	85	40	M
Harrisburg	10	90	40	M
Johnstown	5	85	40	M
Lancaster	5	90	40	M
Meadville	0	85	40	M
New Castle	0	90	40	M
Philadelphia	15	90	40	M
Pittsburgh ★ AP	5	85	40	M
Pittsburgh ★ CO	10	90	40	M
Reading	5	90	40	M
Scranton/ Wilkes-Barre	5	85	40	M
State College	5	85	40	M
Sunbury	5	90	40	M
Uniontown	5	90	40	M
Warren	0	85	40	M
West Chester	10	90	40	M
Williamsport	5	90	40	M
York	5	90	40	M
RHODE ISLAND				
Newport	10	85	40	M
Providence	10	85	40	M
SOUTH CAROLINA				
Anderson	20	95	35	M
SOUTH CAROLINA Cont.				
Charleston ★ AP	25	90	35	M
Charleston ★ CO	30	95	35	L
Columbia	20	95	35	M
Florence	25	95	35	M
Georgetown	25	90	35	M
Greenville	20	95	35	M
Greenwood	20	95	35	M
Orangeburg	25	95	35	M
Rock Hill	20	95	35	M
Spartanburg	20	95	35	M
Sumter	25	95	35	M
SOUTH DAKOTA				
Aberdeen	—20	90	45	H
Brookings	—15	90	45	M
Huron	—15	95	45	H
Mitchell	—15	95	45	H
Pierre	—10	95	45	H
Rapid City	—10	95	45	H
Sioux Falls	—10	90	45	M
Watertown	—20	90	45	H
Yankton	—10	95	45	M
TENNESSEE				
Athens	15	95	35	M
Bristol	15	90	35	M
Chattanooga	15	95	35	M
Clarksville	15	95	35	M
Columbia	15	95	35	M
Dyersburg	15	95	35	M
Greenville	10	90	35	M
Jackson	15	95	35	M
Knoxville	15	90	35	M
Memphis	20	95	35	M
Murfreesboro	15	95	35	M
Nashville	15	95	35	M
Tullahoma	15	95	35	M
TEXAS				
Abilene	20	100	30	M
Alice	30	100	25	M
Amarillo	10	95	35	H
Austin	25	100	30	M
Bay City	30	95	30	M
Beaumont	25	95	30	M
Beeville	30	95	30	M
Big Spring	20	100	30	H
Brownsville	40	90	25	M
Brownwood	25	100	30	M
Bryan	30	100	30	M
Corpus Christi	35	95	30	M
Crosicana	25	100	30	M
Dallas	20	100	35	M
Del Rio	30	100	30	M
Denton	20	100	35	M
Eagle Pass	30	105	30	M
El Paso	25	100	30	H
Fort Worth	20	100	35	M
Galveston	35	90	30	L
Greenville	20	100	35	M
Harlingen	35	95	25	M
Houston	30	95	30	M
Huntsville	30	95	30	M
Killeen	25	100	30	M
Lamesa	15	100	35	H
Laredo	35	100	25	M
Longview	25	100	30	M
Lubbock	15	95	35	H
Lufkin	25	95	30	M
McAllen	35	100	25	M
Midland	20	100	30	H
Mineral Wells	20	100	35	M
Palestine	25	95	30	M
Pampa	10	100	35	H
Pecos	15	100	30	H
Plainview	10	100	35	H
Port Arthur	30	90	30	M
San Angelo	25	100	30	M
San Antonio	30	95	30	M
Sherman	20	100	35	M
Snyder	15	100	30	H
Temple	25	100	30	M
Tyler	20	95	30	M
Vernon	15	100	35	M
Victoria	30	95	30	M
Waco	25	100	30	M
Wichita Falls	15	100	35	M
UTAH				
Cedar City	5	90	35	H
Logan	5	90	40	H
Moab	15	100	40	H
Ogden	10	90	40	H
Price	5	90	40	H
Provo	5	95	40	H
Richfield	0	90	40	H
UTAH Cont.				
St. George	25	100	35	H
Salt Lake City	5	95	40	H
Vernal	—10	90	40	H
VERMONT				
Barre	—15	85	45	M
Burlington	—10	85	45	M
Rutland	—10	85	45	M
VIRGINIA				
Charlottsville	15	90	40	M
Danville	15	90	35	M
Fredericksburg	10	90	40	M
Harrisonburg	5	90	40	M
Lynchburg	15	90	35	M
Norfolk	20	90	35	M
Petersburg	15	95	35	M
Richmond	15	95	35	M
Roanoke	15	90	35	M
Staunton	10	90	40	M
Winchester	10	90	40	M
WASHINGTON				
Aberdeen	20	80	45	M
Bellingham	15	75	50	M
Bremerton	25	80	45	M
Ellensburg	5	90	45	H
Everett	20	80	50	M
Kennewick	15	95	45	H
Longview	20	85	45	H
Moses Lake	—5	95	45	H
Olympia	25	85	45	H
Port Angeles	25	75	50	M
Seattle ★ CO	20	80	45	M
(Boeing Field) AP	15	80	45	M
(Seattle-Tacoma	10	80	45	M
Spokane	0	90	45	H
Tacoma	20	80	45	M
Walla Walla	15	95	45	H
Wenatchee	5	90	45	H
Yakima	10	90	45	H
WEST VIRGINIA				
Beckley	5	90	40	M
Bluefield	10	85	35	M
Charleston	10	90	40	M
Clarksburg	5	90	40	M
Elkins	5	85	40	M
Huntington	10	95	40	M
Martinsburg	10	95	40	M
Morgantown	5	90	40	M
Parkersburg	10	90	40	M
Wheeling	5	90	40	M
WISCONSIN				
Appleton	—10	85	45	M
Ashland	—20	85	45	M
Beloit	—5	90	40	M
Eau Claire	—15	90	45	M
Fond du Lac	—10	85	45	M
Green Bay	—10	85	45	M
La Crosse	—10	90	45	M
Madison	—5	90	45	M
Manitowoc	—5	85	45	M
Marinette	—5	85	45	M
Milwaukee	—5	85	45	M
Racine	0	90	45	M
Sheboygan	0	85	45	M
Stevens Point	—15	85	45	M
Waukesha	—5	90	45	M
Wausau	—15	85	45	M
WYOMING				
Casper	—5	90	45	H
Cheyenne	—5	85	40	H
Cody	—10	85	45	H
Evanston	—10	80	40	H
Lander	—15	90	45	H
Laramie	—5	80	40	H
Newcastle	—5	90	45	H
Rawlins	—15	85	40	H
Rock Springs	—5	85	40	H
Sheridan	—10	90	45	H
Torrington	—10	90	40	H

★ AP - Airport
CO - City Office
[1]Temperatures are rounded off to permit use of precalculated HTM tables.

Table 1 (Cont'd)
OUTDOOR DESIGN CONDITIONS—CANADA

State & City	Winter Dry-Bulb Temp[1] (F)	Summer Dry Bulb Temp[1] (F)	North Latitude (0)	Daily Range
ALBERTA				
Calgary	—25	85	50	H
Edmonton	—30	85	55	M
Grande Prairie	—40	80	55	M
Lethbrige	—25	90	50	H
Medicine Hat	—30	95	50	H
BRITISH COLUMBIA				
Fort Nelson	—45	85	60	M
Penticton	0	90	50	H
Prince George	—35	80	55	H
Prince Rupert	15	70	55	L
Vancouver	15	80	50	M
Victoria	20	75	50	M
MANITOBA				
Brandon	—30	85	50	H
Churchill	—40	75	60	M
Dauphin	—30	85	50	M
The Pas	—35	80	55	M
Winnipeg	—25	85	50	M
NEW BRUNSWICK				
Campbellton	—15	85	50	M
Fredericton	—10	85	45	M
NEW BRUNSWICK Cont.				
Moncton	—10	85	45	M
Saint John	—10	80	45	M
NEW FOUNDLAND				
Corner Brook	—5	80	50	M
Gander	—5	80	50	M
St. John's	5	75	45	M
NORTHWEST TERRITORIES				
Frobisher Bay	—45	60	65	L
Resolute	—50	50	75	L
Yellowknife	—50	75	60	M
NOVA SCOTIA				
Halifax	0	80	45	M
Sydney	5	80	45	M
Yarmouth	5	75	45	M
ONTARIO				
Fort William	—25	90	50	M
Hamilton	0	90	45	M
Kapuskasing	—30	85	50	M
Kingston	—10	80	45	M
Kitchener	0	85	45	M
Tondon	0	90	45	M
North Bay	—20	85	45	M
Oshawa	—5	85	45	M
ONTARIO Cont.				
Ottawa	—15	85	45	M
Peterborough	—10	85	45	M
St. Catherines	5	90	45	M
Sudbury	—15	85	45	M
Toronto	0	85	45	M
Windsor	5	90	40	M
PRINCE EDWARD ISLAND				
Charlottetown	—5	80	45	M
QUEBEC				
Bayotville	—25	85	50	M
Montreal	—10	85	45	M
Quebec	—15	80	45	M
Sept. Iles	—25	80	50	M
Sherbrooke	—15	80	45	M
Trois Rivieres	—15	85	45	M
Val d'Or	—30	85	50	M
SASKATCHEWAN				
Prince Albert	—35	85	55	M
Regina	—30	90	50	H
Saskatoon	—30	85	50	M
Swift Current	—25	90	50	M
YUKON TERRITORY				
Whitehorse	—45	75	60	M

Table 2
HEAT TRANSFER MULTIPLIERS (Heating)

Item	Design Temperature Difference, Degrees														
	30	35	40	45	50	55	60	65	70	75	80	85	90	95	100
WINDOWS, WOOD OR METAL FRAME (Btuh per sq ft) Factors include heat loss for transmission and infiltration @ 15 mph															
No. 1 Double-Hung, Horizontal-Slide, Casement or Awning (Infiltration less than 0.50 cfm/ft @ 25 mph certified by test)															
(a) Single glass	40	50	55	60	70	75	80	90	95	105	110	115	125	130	135
(b) With double glass or insulating glass	25	30	35	40	45	45	50	55	60	65	70	75	75	80	85
(c) With storm sash	25	25	30	35	40	45	45	50	55	60	60	65	70	75	80
No. 2 Double-Hung, Horizontal-Slide, Casement or Awning (Infiltration less than 0.75 cfm/ft @ 25 mph certified by test)															
(a) Single glass	45	50	60	65	75	80	90	95	105	110	120	125	135	140	150
(b) With double glass or insulating glass	30	35	40	45	50	55	60	65	70	75	80	80	85	90	95
(c) With storm sash	25	30	35	40	45	50	55	60	60	65	70	75	80	85	90
No. 3 Other Double-Hung, Horizontal-Slide, Casement or Awning															
(a) Single glass	75	90	105	115	130	140	155	165	180	195	205	220	230	245	255
(b) With double glass or insulating glass	60	70	80	90	105	115	125	135	145	155	165	175	185	195	205
(c) With storm sash	35	40	45	55	60	65	70	75	85	90	95	100	105	110	120
No. 4 Fixed or Picture															
(a) Single glass	40	50	55	60	70	75	85	90	95	105	110	115	125	130	140
(b) Double glass or with storm sash	25	30	35	40	45	45	50	55	60	65	70	75	75	80	85
No. 5 Jalousie (Infiltration less than 1.5 cfm/sq ft @ 25 mph certified by test)															
(a) Single glass	60	70	80	90	100	105	115	125	135	145	155	165	175	185	195
(b) With storm sash	35	40	50	55	60	65	70	75	80	85	90	100	105	110	115
No. 6 Other Jalousie															
(a) Single glass	225	265	300	340	375	415	450	490	525	565	600	640	675	715	750
(b) With storm sash	65	75	90	100	110	120	135	145	155	165	175	190	200	210	220
DOORS (Btuh per sq ft) Factors include heat loss for transmission and infiltration @ 15 mph															
No. 7 Sliding Glass Doors (Infiltration less than 1.0 cfm/sq ft @ 25 mph certified by test)															
(a) Single glass	50	60	70	75	85	95	100	110	115	125	135	140	150	160	165
(b) Double glass	40	45	50	55	60	65	70	80	90	95	100	105	115	120	125
No. 8 Other Sliding Glass Doors															
(a) Single glass	75	85	100	115	125	140	150	165	175	190	200	210	225	240	250
(b) Double glass	60	70	80	90	100	110	120	130	140	150	160	170	180	190	200
No. 9 Other Doors															
(a) Weatherstripped **and** with storm door	40	45	55	60	65	75	80	85	90	100	105	110	120	125	130
(b) Weatherstripped **or** with storm door	70	85	95	110	120	135	145	155	170	180	195	205	215	230	240
(c) No weatherstripping or storm door	135	160	180	200	225	250	270	290	315	340	360	380	405	430	450
WALLS AND PARTITIONS (Btuh per sq ft)															
No. 10 Wood Frame with Sheathing and Siding, Veneer or Other Exterior Finish															
(a) No insulation	8	9	10	11	13	14	15	16	18	19	20	21	23	24	25
(b) Expanded polystyrene extruded board sheathing (R-5)	3	4	4	5	6	6	7	7	8	8	9	9	10	10	11
(c) R-7 batt insulation (2"-2¾")	3	4	4	5	5	6	6	7	7	8	8	9	9	10	10
(d) R-11 batt insulation (3"-3½")	2	2	3	3	4	4	4	5	5	5	6	6	6	7	7
(e) R-13 batt insulation (3½"-3⅝")	2	2	2	3	3	3	4	4	4	5	5	5	5	6	6
No. 11 Partition between conditioned and unconditioned spaces															
(a) Finished one side only, no insulation	17	19	22	25	28	30	33	36	39	41	44	47	49	52	55
(b) Finished both sides, no insulation	9	11	12	14	16	17	19	20	22	23	25	26	28	29	31

* Insulation is produced in different densities and materials such as mineral fibers, glass fibers, etc.; therefore, there is some variation in thickness for the same R-value among various manufacturers' products.

Table 2
HEAT TRANSFER MULTIPLIERS (Heating)

Item	Design Temperature Difference, Degrees														
No. 11 Cont'd	30	35	40	45	50	55	60	65	70	75	80	85	90	95	100
(c) Partition with 1" expanded polystyrene extruded board, R-5, finished both sides	4	4	5	5	6	7	7	8	8	9	10	10	11	11	12
(d) R-7 insulation (2"-2¾") finished both sides	3	4	4	5	5	6	6	7	7	8	8	9	9	10	10
(e) R-11 insulation (3"-3½") finished both sides	2	3	3	4	4	4	5	5	6	6	6	7	7	8	8
(f) R-13 insulation (3½"-3⅝") finished both sides	2	2	3	3	4	4	4	5	5	5	6	6	6	7	7
No. 12 Solid Masonry, Block or Brick															
(a) Plastered or plain	14	16	18	20	22	25	27	29	32	34	36	38	40	43	45
(b) Furred, no insulation	9	10	12	13	14	16	17	19	20	22	23	25	26	28	29
(c) Furred, with R-5 insulation (Nominal 1½")	4	5	5	6	6	7	8	8	9	10	10	11	12	12	13
No. 13 Basement or Crawl Space															
(a) Above grade, no insulation	15	18	20	23	26	28	31	33	36	38	41	43	46	48	51
(b) Wall of conditioned crawl space, R-3.57 insulation (molded bead bd.)	5	6	7	8	9	10	11	12	13	14	14	15	16	17	18
(c) Wall of conditioned crawl space, R-5 insulation (ext. polystyrene bd.)	4	5	6	6	7	8	9	9	10	11	12	12	13	14	14
(d) Wall of crawl space used as supply plenum, R-3.57 insulation (molded bead bd.)	11	12	13	14	15	16	16	17	18	19	20	21	22	23	24
(e) Wall of crawl space used as supply plenum, R-5 insulation (ext. polystyrene bd.)	9	10	10	11	12	12	13	14	15	15	16	17	17	18	19
(f) Below grade wall	2	2	2	3	3	3	4	4	4	5	5	5	5	6	6
CEILINGS AND ROOFS (Btuh per sq ft)															
No. 14 Ceiling Under Unconditioned Space or Vented Roof															
(a) No insulation	18	21	24	27	30	33	36	39	42	45	48	51	54	57	60
(b) R-11 insulation (3"-3¼")	2	3	3	4	4	4	5	5	6	6	6	7	7	8	8
(c) R-19 insulation (5¼"-6½")	2	2	2	2	2	3	3	3	4	4	4	4	4	5	5
(d) R-22 insulation (6"-7")	1	1	2	2	2	2	2	3	3	3	3	3	4	4	4
(e) Ceiling under unconditioned room	9	11	12	14	15	17	18	20	21	23	24	26	27	29	30
No. 15 Roof on Exposed Beams or Rafters															
(a) Roofing on 1½ in. wood decking, no insulation	10	12	14	15	17	19	20	22	24	26	27	29	31	32	34
(b) Roofing on 1½ in. wood decking, 1 in. insulation between roofing and decking	5	6	7	8	8	9	10	11	12	13	13	14	15	16	17
(c) Roofing on 1½ in. wood decking, 1½ in. insulation between roofing and decking	4	5	5	6	7	7	8	9	10	10	11	12	12	13	14
(d) Roofing on 2 in. coarse, shredded wood plank	6	7	8	9	10	11	12	14	15	16	17	18	19	20	21
(e) Roofing on 3 in. coarse, shredded wood plank	5	5	6	7	8	8	9	10	11	11	12	13	14	14	15
(f) Roofing on 1½ in. insulating fiberboard decking	6	7	8	9	10	10	11	12	13	14	15	16	17	18	19
(g) Roofing on 2 in. insulating fiberboard decking	4	5	6	7	8	8	9	10	11	11	12	13	14	14	15
(h) Roofing on 3 in. insulating fiberboard decking	3	4	4	5	6	6	7	7	8	8	9	9	10	10	11
No. 16 Roof-Ceiling Combination															
(a) No insulation	9	11	12	14	16	17	19	20	22	23	25	26	28	29	31
(b) R-11 insulation (3"-3½") between ceiling and decking	2	2	3	3	4	4	4	5	5	5	6	6	6	7	7
(c) R-19 insulation (5¼"-6½") between ceiling and decking	1	2	2	2	2	2	3	3	3	3	4	4	4	4	5
(d) R-22 insulation (6"-7")	1	1	2	2	2	2	2	3	3	3	3	3	4	4	4
FLOORS (Btuh per sq ft)															
No. 17 Floors Over Unconditioned Space															
(a) Over unconditioned room	4	5	6	6	7	8	8	9	10	11	11	12	13	13	14
(b) Over open or vented space or garage, no insulation	8	10	11	13	14	15	17	18	20	21	22	24	25	27	28
(c) Over open or vented space or garage, R-7 insulation (2"-2¾")	3	3	4	4	5	5	6	6	7	7	8	8	9	9	9
(d) Over open or vented space or garage, R-11 insulation (3"-3½")	2	2	3	3	4	4	4	5	5	5	6	6	6	7	7
(e) Over open or vented space or garage, R-19 insulation (5¼"-6½")	1	2	2	2	2	2	3	3	3	3	4	4	4	4	4
No. 18 Floor of Room Over Heated Basement or Crawl Space	0	0	0	0	0	0	0	0	0	0	0	0	0	0	0
No. 19 Basement Floor	1	1	1	1	1	2	2	2	2	2	3	3	3	3	3
FLOORS (Btuh per ft of Perimeter)															
No. 20 Concrete Slab Floor, Unheated															
(a) No edge insulation	35	40	40	45	45	50	50	55	55	60	60	65	65	70	75
(b) 1 in. edge insulation	25	30	30	35	35	40	40	45	45	50	50	55	55	60	60

Table 2
HEAT TRANSFER MULTIPLIERS (Heating)

Item	Design Temperature Difference, Degrees														
No. 20 cont'd	30	35	40	45	50	55	60	65	70	75	80	85	90	95	100
(c) 2 in. edge insulation	15	20	20	25	25	30	30	35	35	40	40	45	45	50	50
No. 21 Concrete Slab Floor with Perimeter System in Slab															
(a) No edge insulation	60	70	80	90	95	105	115	125	135	145	150	160	170	180	190
(b) 1 in. edge insulation	45	50	55	60	65	70	75	80	85	90	95	100	105	110	115
(c) 2 in. edge insulation	25	30	35	40	45	50	55	60	65	70	75	80	85	90	95
No. 22 Floor of Heated Crawl Space															
(a) Less than 18 in. below grade	35	40	40	45	45	50	50	55	55	60	60	65	65	70	75
(b) 18 in. or more below grade	15	20	20	25	25	30	30	35	35	40	40	45	45	50	50
No. 23 Floor of Crawl Space Used as Supply Plenum															
(a) Less than 18 in. below grade	45	50	50	60	60	65	65	70	70	80	80	85	85	90	95
(b) 18 in. or more below grade	20	25	25	30	30	40	40	45	45	50	50	60	60	65	65
OUTSIDE AIR (Btuh per cfm)															
No. 24 Ventilation or Makeup Air	30	40	45	50	55	60	65	70	75	80	85	90	95	105	110

ADDENDA TO TABLE 2, MANUAL J, 4th EDITION

Item	Design Temperature Difference, Degrees														
	30	35	40	45	50	55	60	65	70	75	80	85	90	95	100
WINDOWS (Btuh/ft²) Factors include transmission and infiltration															
No. 1 Double-Hung, Horizontal Slide, Casement, or Awning (Infiltration less than 0.50 cfm/ft @ 25 mph certified by test															
(d) Triple Glass	17	20	23	26	29	32	35	38	41	44	46	49	52	55	58
(e) Double Glass or insulating glass with storm sash	15	18	20	23	26	28	31	33	36	38	41	43	46	48	51
No. 2 Double-Hung, Horizontal Slide, Casement, or Awning (Infiltration less than 0.75 cfm/ft @ 25 mph certified by test)															
(d) Triple Glass	21	25	28	32	35	39	42	46	49	53	56	60	63	67	70
(e) Double Glass or insulating glass with storm sash	18	21	24	27	30	33	36	39	42	45	48	51	54	57	60
No. 3 Other Double-Hung, Horizontal Slide, Casement, or Awning															
(d) Triple Glass	30	35	40	45	50	55	60	65	70	75	80	85	90	95	100
(e) Double Glass or insulating glass with storm sash	27	32	36	41	45	50	54	59	63	58	72	77	81	86	90
No. 4 Fixed or Picture															
(c) Triple Glass or Double glass with storm sash	17	19	22	25	28	30	33	36	39	41	44	47	50	52	55
DOORS (Btuh/ft²) Factors include heat loss for transmission and infiltration @ 15 mph															
No. 9 Other Doors															
(d) Insulated (core, R-5) door weatherstripped *and* storm door	35	40	46	52	58	63	69	75	81	86	92	98	104	109	115
(e) Insulated (core) weatherstripped *or* storm door	68	79	90	101	113	124	135	146	158	169	180	191	203	214	225
(f) Insulated (core)	123	144	164	185	205	226	246	267	287	308	328	349	369	390	410
WALLS and PARTITIONS (Btuh/ft²)															
No. 10 Wood frame walls with sheathing and siding or veneer															
(f) Expanded polystyrene extruded board sheathing plus R-13 in cavity	1.5	1.8	2.0	2.3	2.5	2.8	3.0	3.3	3.5	3.8	4.0	4.3	4.5	4.8	5.0
(g) 6 in. framing with R-19 batt insulation	1.2	1.4	1.6	1.8	2.0	2.2	2.4	2.6	2.8	3.0	3.2	3.4	3.6	3.8	4.0
No. 11 Frame partition between cond. & unconditioned spaces															
(g) Expanded polystyrene extr. board sheathing plus R-13	1.5	1.8	2.0	2.3	2.5	2.8	3.0	3.3	3.5	3.8	4.0	4.3	4.5	4.8	5.0
(h) 6 in. framing with R-19	1.2	1.4	1.6	1.8	2.0	2.2	2.4	2.6	2.8	3.0	3.2	3.4	3.6	3.8	4.0

ADDENDA TO TABLE 2, MANUAL J, 4th EDITION (cont'd)

Item	Design Temperature Difference, Degrees														
	30	35	40	45	50	55	60	65	70	75	80	85	90	95	100
No. 13 Basement or Crawl Space Walls															
(g) Above grade with ¾ in. expanded polystyrene	5.4	6.3	7.2	8.1	9.0	9.9	10.8	11.7	12.6	13.5	14.4	15.3	16.2	17.1	18.0
(h) Above grade with ¾ in. expanded polyurethane or furred with R-5 batts	4.2	4.9	5.6	6.3	7.0	7.7	8.4	9.1	9.8	10.5	11.2	11.9	12.6	13.3	14.0
(i) Above grade with R-13 min. fiber insulation	2.1	2.5	2.8	3.2	3.5	3.9	4.2	4.6	4.9	5.3	5.6	6.0	6.3	6.7	7.0
(j) Below grade with R-5 min. fiber or ¾ in. expanded polystyrene or polyurethane	1.4	1.6	1.8	2.0	2.2	2.4	2.6	2.8	3.0	3.2	3.4	3.6	3.8	4.0	4.2
(k) Below grade with R-13 min. fiber	.8	.9	1.0	1.1	1.3	1.4	1.5	1.6	1.7	1.8	1.9	2.1	2.2	2.3	2.4
CEILING & ROOFS (Btuh/ft²)															
No. 14 Ceiling under unconditioned space or vented roof															
(f) Under vented roof R-26 insulation	1.1	1.3	1.4	1.6	1.8	2.0	2.2	2.3	2.5	2.7	2.9	3.1	3.2	3.4	3.6
(g) Under vented roof R-30 insulation	1.0	1.1	1.3	1.4	1.6	1.8	1.9	2.1	2.2	2.4	2.6	2.7	2.9	3.0	3.2
(h) Under vented roof R-33 insulation	.9	1.0	1.2	1.3	1.5	1.6	1.7	1.9	2.0	2.2	2.3	2.5	2.6	2.8	2.9
(i) Under vented roof R-38 insulation	.8	.9	1.0	1.1	1.3	1.4	1.5	1.6	1.8	1.9	2.0	2.1	2.3	2.4	2.5
(j) Under vented roof R-44 insulation	.7	.8	.9	1.0	1.1	1.2	1.3	1.4	1.5	1.7	1.8	1.9	2.0	2.1	2.2
No. 16 Roof-ceiling combination															
(e) R-26 insulation	1.0	1.2	1.4	1.6	1.8	1.9	2.1	2.3	2.5	2.6	2.8	3.0	3.2	3.3	3.5
(f) R-30 insulation	.9	1.1	1.2	1.4	1.5	1.7	1.8	2.0	2.1	2.3	2.4	2.6	2.7	2.9	3.0
FLOORS (Btuh/ft²)															
No. 17 Floor over uncond. space															
(e) Over open or vented space or garage, R-22 insulation	1.2	1.4	1.6	1.8	2.0	2.3	2.5	2.7	2.9	3.1	3.3	3.5	3.7	3.9	4.1

Table 3
DUCT HEAT LOSS MULTIPLIERS

Duct Location and Insulation	Heat Loss of Living Space, Btuh/ft²	Outside Design Temperature, F		
		Below 0	0 to 15	Above 15
Attic or Crawl Space; 1 in. Flexible Blanket Insulation [1], [2]	Less than 35	.25	.20	.15
	35 to 45	.20	.15	.10
	45 or more	.15	.10	.10
Attic or Crawl Space; 2 in. Flexible or 1 in. Rigid Duct Insulation[3]	Less than 35	.20	.15	.10
	35 to 45	.15	.10	.05
	45 or more	.10	.05	.05
Unconditioned Basement; No Duct Insulation	Less than 35	.20	.15	.10
	35 to 45	.10	.10	.05
	45 or more	.05	.05	.05

[1] Flexible-blanket duct insulation—density 0.6 to 0.75 lb per cu ft. Do not use insulation with density less than 0.6 lb per cu ft.

[2] Does not meet HUD-FHA Minimum Property Standard requirement in Paragraph 515-3.1.

[3] Rigid duct insulation—density minimum 2.5 to 4.0 lb per cu ft.

Table 4—GLASS HEAT TRANSFER MULTIPLIERS (Cooling)

Outside Design Temp	Single Glass 85	90	95	100	105	110	Double Glass 85	90	95	100	105	110
Direction Window Faces	**No Shading**											
N	25	30	30	35	40	45	20	20	25	25	30	30
NE and NW	55	60	65	70	70	80	50	50	50	55	55	60
E and W	80	85	90	95	95	105	70	70	75	75	80	80
SE and SW	70	75	80	85	85	90	60	60	65	65	70	70
S	40	45	50	55	55	60	35	35	40	40	45	45

Table 4—GLASS HEAT TRANSFER MULTIPLIERS (Cooling)

Outside Design Temp	Single Glass 85	90	95	100	105	110	Double Glass 85	90	95	100	105	110
Direction Window Faces	**Draperies or Venetian Blinds**											
N	15	20	25	30	30	35	15	15	20	20	20	25
NE and NW	35	35	40	45	50	55	30	30	35	35	35	40
E and W	50	55	55	60	65	70	45	45	50	50	50	55
SE and SW	40	45	50	55	55	60	35	40	40	45	45	45
S	25	30	30	35	40	45	20	25	25	30	30	30
	Roller Shades											
N	20	25	25	30	35	40	15	20	20	25	25	25
NE and NW	40	45	50	55	55	60	40	40	45	45	50	50
E and W	60	65	70	75	75	85	55	55	60	60	65	65
SE and SW	55	55	60	65	70	75	50	50	50	55	55	60
S	30	35	40	40	45	50	30	30	35	35	35	40
	Ventilated Awnings, Porches or Other Shade											
All Directions	20	25	30	35	35	45	15	15	20	20	25	25

NOTE: If heat absorbing double glass is used, obtain factors from ASHRAE Handbook.

ADDENDA TO TABLE 4, MANUAL J, 4th EDITION

PLAIN (CLEAR) GLASS

OUTSIDE DESIGN TEMPERATURE	TRIPLE GLASS 85	90	95	100	105	110
DIRECTION WINDOW FACES	**NO SHADING**					
N	15	16	18	19	20	21
NE and NW	37	38	40	41	42	44
E and W	55	56	58	59	60	62
SE and SW	47	49	51	52	53	54
S	26	27	29	31	32	33
	DRAPERIES OR VENETIAN BLINDS					
N	10	12	14	16	17	19
NE and NW	22	24	26	28	30	31
E and W	35	36	38	40	42	44
SE and SW	29	30	32	34	36	38
S	16	18	20	22	24	26
	ROLLER SHADES					
N	12	13	15	17	18	19
NE and NW	29	30	32	34	35	37
E and W	44	45	47	49	50	52
SE and SW	37	39	41	42	44	45
S	20	22	24	26	27	29
	AWNING, PORCHES OR OTHER SHADE					
ALL DIRECTIONS	15	17	19	21	23	25

ADDENDA TO TABLE 4, MANUAL J, 4th EDITION

TINTED (HEAT ABSORBING) GLASS

	Single Glass						Double Glass						Triple Glass					
OUTSIDE DESIGN TEMP.	85	90	95	100	105	110	85	90	95	100	105	110	85	90	95	100	105	110
DIRECTION WINDOW FACES	NO SHADING																	
N	18	22	26	30	33	37	12	14	17	19	21	23	9	10	12	13	14	16
NE and NW	37	41	45	49	53	56	27	29	32	34	36	38	20	21	23	25	26	27
E and W	56	60	64	68	72	76	42	44	47	49	51	53	32	33	35	36	38	39
SE and SW	47	51	55	59	63	67	35	37	40	42	44	46	26	28	30	31	32	33
S	27	31	35	39	42	46	19	21	24	26	28	30	14	15	17	19	20	21
	DRAPERIES OR VENETIAN BLINDS																	
N	13	17	21	24	27	30	9	11	14	16	18	20	6	8	10	12	13	15
NE and NW	29	33	37	40	43	46	20	22	25	27	29	31	13	15	17	19	20	22
E and W	44	48	52	55	58	61	30	32	35	37	39	41	20	21	23	25	27	28
SE and SW	35	39	43	46	49	52	24	26	29	31	33	35	16	18	20	21	23	24
S	22	26	30	33	36	39	15	17	20	22	24	26	10	12	14	15	17	18
	ROLLER SHADES																	
N	15	19	23	27	30	33	10	12	15	17	19	21	7	8	10	12	13	15
NE and NW	32	36	40	44	48	50	24	26	29	31	33	35	16	17	19	21	22	24
E and W	49	53	57	61	64	68	35	37	40	42	44	46	25	26	28	30	32	33
SE and SW	40	44	48	52	55	59	30	32	35	37	39	41	20	22	24	25	27	28
S	24	28	32	35	38	42	18	20	23	25	27	29	11	13	15	16	18	19
	AWNINGS, PORCHES, OR OTHER SHADE																	
ALL DIRECTIONS	18	22	26	30	33	37	12	14	17	19	21	23	9	10	12	13	14	16

ADDENDA TO TABLE 4, MANUAL J, 4th EDITION

REFLECTIVE COATED GLASS

	Single Glass						Double Glass						Triple Glass					
OUTSIDE DESIGN TEMP.	85	90	95	100	105	110	85	90	95	100	105	110	85	90	95	100	105	110
DIRECTION WINDOW FACES	NO SHADING																	
N	16	20	25	29	33	37	11	13	16	18	21	23	7	9	11	13	14	16
NE and NW	32	36	41	45	49	53	22	24	27	30	32	34	15	16	18	20	22	23
E and W	49	53	57	61	65	69	33	36	39	41	43	46	22	24	26	28	29	31
SE and SW	41	45	50	54	58	62	28	30	33	36	38	40	19	20	22	24	26	27
S	24	28	33	37	41	45	16	18	21	24	26	28	11	12	14	16	18	19
	DRAPERIES OR VENETIAN BLINDS																	
N	13	16	20	23	27	30	9	11	14	16	18	20	6	8	10	12	13	15
NE and NW	27	30	34	38	41	44	19	21	24	26	28	30	13	15	17	18	20	21
E and W	40	43	47	50	53	57	28	30	33	35	37	39	19	21	23	24	26	28
SE and SW	32	35	39	43	46	49	23	25	28	30	32	34	15	17	19	21	22	24
S	20	24	28	31	34	37	14	16	19	21	23	25	10	12	14	15	17	18
	ROLLER SHADES																	
N	14	17	22	25	29	33	9	11	14	16	19	21	6	8	10	12	13	15
NE and NW	29	32	37	41	44	48	20	22	25	27	29	31	13	15	17	18	20	21
E and W	44	47	51	55	58	62	30	32	35	37	39	42	20	22	24	25	27	29
SE and SW	36	39	44	48	51	55	25	27	30	32	34	36	16	18	20	22	23	25
S	21	25	30	33	37	40	14	16	19	22	24	26	10	11	13	15	17	18
	AWNINGS, PORCHES, OR OTHER SHADE																	
ALL DIRECTIONS	16	20	25	29	33	37	11	13	16	18	21	23	7	9	11	13	14	16

Table 5—HEAT TRANSFER MULTIPLIERS (Cooling)

Design Temperature	85	90	95		100		105	110
Daily Temperature Range	L or M	L or M	M	H	M	H	H	H
WINDOWS (Btuh per sq ft)								
No. 1 through No. 6—Obtain factors from Table 4								
DOORS (Btuh per sq ft) Factors include heat gain for transmission and infiltration @ 7.5 mph								
No. 7 & 8 Sliding Glass—Obtain factors from Table 4								
No. 9 Other Doors	7.5	10.0	13.0	10.5	16.0	13.5	16.0	19.0
WALLS AND PARTITIONS (Btuh per sq ft)* Factors include heat gain for transmission and infiltration @ 7.5 mph								
No. 10 Wood Frame with Sheathing and Siding, Veneer or Other Exterior Finish								
(a) No insulation	4.0	5.5	7.0	6.0	9.0	7.5	9.0	10.5
(b) Expanded polystyrene extruded board sheathing (R-5)	2.5	3.5	4.5	4.0	5.5	5.0	6.0	7.0
(c) R-7 batt insulation (2"-2¾")	2.0	3.0	3.5	3.0	4.5	4.0	5.0	5.5
(d) R-11 batt insulation (3"-3½")	1.5	2.5	3.0	3.0	4.0	3.5	4.0	5.0
(e) R-13 batt insulation (3½"-3⅝")	1.5	2.0	3.0	2.5	3.5	3.5	4.0	4.5
No. 11 Frame partitions between conditioned And unconditioned spaces								
(a) Finished one side only, no insulation	3.5	6.5	10.0	7.0	13.0	10.0	13.0	16.5
(b) Finished both sides, no insulation	2.5	4.5	6.0	4.5	8.0	6.5	8.5	10.5
(c) Finished both sides, R-5, 1" expanded polystyrene extruded board insulation	1.5	2.5	3.5	3.0	4.5	4.0	5.0	6.0
(d) R-7 insulation (2"-2¾") finished both sides	1.0	2.0	3.0	2.5	4.0	3.5	4.0	5.0
(e) R-11 insulation (3"-3½") finished both sides	1.0	2.0	2.5	2.5	3.5	3.0	4.0	4.5
(f) R-13 insulation (3½"-3⅝") finished both sides	1.0	2.0	2.5	2.0	3.5	3.0	3.5	4.5
WALLS AND PARTITIONS (Btuh per sq ft)—cont'd								
No. 12 Solid Masonry, Block or Brick								
(a) Plastered or plain	3.5	6.0	8.5	6.5	11.5	9.0	11.5	14.0
(b) Furred, no insulation	2.5	4.5	6.0	4.3	8.0	6.5	8.0	10.0
(c) Furred, with R-5 insulation (nominal 1½")	1.5	2.5	3.5	3.0	4.5	4.0	5.0	6.0
No. 13 Basement or Crawl Space								
(a) Above grade, no insulation	7.5	10.0	13.0	10.5	16.0	13.5	16.0	19.0
(b) Wall of conditioned crawl space, R-3.57 insul. (molded bead board)	3.0	4.5	5.5	5.0	7.0	6.0	7.5	8.5
(c) Wall of conditioned crawl space, R-5 insul. (extruded polystyrene board)	2.5	3.5	5.0	4.0	6.0	5.0	6.0	7.5
(d) Wall of crawl space used as supply plenum, R-3.57 insulation (molded bead board)	7.0	8.0	9.5	8.5	10.5	9.5	11.0	12.0
(e) Wall of crawl space used as supply plenum, R-5 insulation (extruded polystyrene board)	5.5	6.5	7.5	7.0	8.5	8.0	9.0	10.0
(f) Below grade wall	0	0	0	0	0	0	0	0
CEILINGS AND ROOFS (Btuh per sq ft)								
No. 14 Ceiling Under Unconditioned Space or Vented Roof								
(a) No insulation dk	8.0	9.0	10.0	9.0	11.5	10.0	11.5	12.5
lt	6.0	7.0	8.5	7.0	9.5	8.5	9.5	10.5
(b) R-11 insulation (3"-3½") dk	2.0	2.5	3.0	2.5	3.0	3.0	3.0	3.5
lt	1.5	2.0	2.5	2.0	2.5	2.5	2.5	3.0
(c) R-19 insulation (5¼"-6½") dk	1.5	1.5	2.0	1.5	2.0	2.0	2.0	2.5
lt	1.0	1.5	1.5	1.5	1.5	1.5	1.5	2.0
(d) R-22 insulation (6"-7") dk	1.0	1.5	1.5	1.5	1.5	1.5	1.5	2.0
lt	1.0	1.0	1.5	1.0	1.5	1.5	1.5	1.5
(e) Ceiling under unconditioned room	2.5	3.5	3.5	2.5	4.5	3.5	4.5	5.5

* Insulation is produced in different densities and materials such as mineral fibers, glass fibers, etc; therefore, there is some variation in thickness for the same R-value among various manufacturers' products.

Table 5 (cont'd)

	Design Temperature		85	90	95		100		105	110
	Daily Temperature Range		L or M	L or M	M	H	M	H	H	H
	No. 15 Roof on Exposed Beams or Rafters									
(a)	Roofing on 1½ in. wood decking, no insulation	dk	10.0	11.5	13.0	11.5	14.5	13.0	14.5	16.0
		lt	7.5	9.0	10.5	9.0	12.0	10.5	12.0	13.5
(b)	Roofing on 1½ in. wood decking. 1 in. insul. between roofing and decking	dk	5.5	6.5	7.0	6.5	8.0	7.0	8.0	9.0
		lt	4.0	5.0	6.0	5.0	6.5	6.0	6.5	7.5
(c)	Roofing on 1½ in. wood decking, 1½ in. insul. between roofing and decking	dk	4.5	5.0	6.0	5.0	6.5	6.0	6.5	7.0
		lt	3.5	4.0	5.0	4.0	5.5	5.0	5.5	6.0
(d)	Roofing on 2 in. coarse, shredded wood plank	dk	6.5	7.5	8.5	7.5	9.5	8.5	9.5	10.5
		lt	5.0	6.0	7.0	6.0	8.0	7.0	8.0	9.0
(e)	Roofing on 3 in. coarse, shredded wood plank	dk	5.0	5.5	6.5	5.5	7.0	6.5	7.0	8.0
		lt	4.0	4.5	5.0	4.5	6.0	5.0	6.0	6.5
(f)	Roofing on 1½ in. insulating fiberboard decking	dk	6.0	7.0	7.5	7.0	8.5	7.5	8.5	9.5
		lt	4.5	5.5	6.5	5.5	7.0	6.5	7.0	8.0
(g)	Roofing on 2 in. insulating fiberboard decking	dk	5.0	5.5	6.0	5.5	7.0	6.0	7.0	7.5
		lt	3.5	4.5	5.0	4.5	5.5	5.0	5.5	6.5
(h)	Roofing on 3 in. insulating fiberboard decking	dk	3.5	4.0	4.5	4.0	5.0	4.5	5.0	5.5
		lt	2.5	3.0	3.5	3.0	4.0	3.5	4.0	4.5
	No. 16 Roof-Ceiling Combination									
(a)	No insulation	dk	10.0	11.5	13.0	11.5	14.5	13.0	14.5	16.0
		lt	7.5	9.0	10.5	9.0	12.0	10.5	12.0	13.5
(b)	R-11 insulation (3"-3½") between ceiling and decking	dk	2.5	2.5	3.0	2.5	3.5	3.0	3.5	3.5
		lt	2.0	2.0	2.5	2.0	3.0	2.5	3.0	3.0
(c)	R-19 insulation (5¼"-6½") between ceiling and decking	dk	1.5	2.0	2.0	2.0	2.5	2.0	2.5	2.5
		lt	1.0	1.5	1.5	1.5	2.0	1.5	2.0	2.0
(d)	R-22 insulation (6"-7") between ceiling and decking	dk	1.5	1.5	1.5	1.5	2.0	1.5	2.0	2.0
		lt	1.0	1.0	1.5	1.0	1.5	1.5	1.5	2.0
FLOORS (Btuh per sq ft)										
	No. 17 Floors over unconditioned space									
(a)	Over unconditioned room		1.5	2.5	3.5	2.5	5.0	3.5	5.0	6.0
(b)	Over open or vented space or garage, no insulation		1.5	3.5	5.0	3.5	7.0	5.0	7.0	8.5
(c)	Over open or vented space or garage, R-7 insulation (2"-2¾")		0.5	1.0	1.5	1.0	2.0	1.5	2.0	2.5
(d)	Over open or vented space or garage, R-11 insulation (3"-3½")		0.5	0.5	1.0	0.5	1.5	1.0	1.5	2.0
(e)	Over open or vented space or garage, R-19 insulation (5¼"-6½")		0.5	0.5	1.0	0.5	1.0	1.0	1.0	1.5
	No. 18 through No. 23 Over basement or enclosed crawl space (not vented); concrete slab on ground; basement or crawl space floor		0	0	0	0	0	0	0	0
OUTSIDE AIR (Btuh per cfm)										
	No. 24 Ventilation or Makeup Air		11.0	16.0	22.0	22.0	27.0	27.0	32.0	38.0

ADDENDA TO TABLE 5, MANUAL J, 4th EDITION

DESIGN TEMPERATURE		85	90	95		100		105	110
DAILY TEMPERATURE RANGE		L or M	L or M	M	H	M	H	H	H
DOORS (Btuh/ft²) factors include heat gain for transmission and infiltration @ 7.5 mph									
No. 9 Other doors									
(d) Insulated (core, R-5) door weatherstripped *and* storm door		2.3	3.3	4.2	3.5	5.2	4.4	5.4	6.3
(e) Insulated (core) weatherstripped		2.8	4.0	5.0	4.2	6.2	5.3	6.4	7.5
OR storm door		2.5	3.6	4.5	3.8	5.6	4.8	5.8	6.8
(f) Insulated (core)		3.0	4.3	5.5	4.7	6.8	5.9	7.1	8.3
WALL and PARTITIONS (Btuh/ft²) factors include transmission and infiltration heat grain @ 7.5 mph									
No. 10 Wood frame walls with sheathing and siding or veneer									
(f) Expanded polystyrene extruded board sheathing plus R-13 in cavity		1.4	2.0	2.7	2.4	3.3	3.1	3.6	4.3
(g) 6 in. framing with R-19 batt insul.		1.2	1.8	2.4	2.2	3.0	2.8	3.3	3.9
No. 11 Frame partition between cond. & unconditioned spaces									
(g) Expanded polystyrene extruded board sheathing plus R-13		1.0	1.6	2.3	2.0	2.9	2.7	3.2	3.9
(h) 6 in. framing with R-19		.9	1.5	2.1	1.9	2.7	2.5	3.0	3.6
No. 13 Basement or crawl space walls									
(g) Above grade with ¾ in. expanded polystyrene		1.6	2.8	4.1	3.2	5.3	4.5	5.6	6.9
(h) Above grade with ¾ in. expanded polyurethane *OR* furred with R-5 batts		1.4	2.5	3.6	2.9	4.7	4.0	5.0	6.1
(i) Above grade with R-13 mineral fiber insulation		1.1	1.8	2.6	2.2	3.3	3.0	3.6	4.4
(j) Below grade with R-5 min. fiber *OR* ¾ in. expanded polystyrene or polyurethane		0	0	0	0	0	0	0	0
(k) Below grade with R-13 min. fiber		0	0	0	0	0	0	0	0
CEILINGS & ROOFS (Btuh/ft²)									
No. 14 Ceiling under unconditioned space or vented roof									
(f) Under vented roof, R-26 insulation	dk	1.1	1.3	1.5	1.3	1.6	1.5	1.6	1.8
	lt	.9	1.0	1.2	1.0	1.4	1.2	1.4	1.5
(g) Under vented roof, R-30 insulation	dk	1.0	1.1	1.3	1.1	1.4	1.3	1.4	1.6
	lt	.8	.9	1.0	.9	1.2	1.0	1.2	1.3
(h) Under vented roof, R-33 insulation	dk	.9	1.1	1.2	1.1	1.3	1.2	1.3	1.5
	lt	.7	.8	1.0	.8	1.1	1.0	1.1	1.2
(i) Under vented roof, R-38 insulation	dk	.8	.9	1.1	.9	1.2	1.1	1.2	1.3
	lt	.6	.7	.9	.7	1.0	.9	1.0	1.1
(j) Under vented roof, R-44 insulation	dk	.7	.8	.9	.8	1.0	.9	1.0	1.1
	lt	.5	.7	.8	.7	.9	.8	.9	1.0
No. 16 Roof-ceiling combination									
(e) R-26 insulation	dk	1.2	1.4	1.5	1.4	1.7	1.5	1.7	1.9
	lt	.9	1.1	1.3	1.1	1.4	1.3	1.4	1.6
(f) R-30 insulation	dk	1.0	1.2	1.4	1.2	1.5	1.4	1.5	1.7
	lt	.8	1.0	1.1	1.0	1.3	1.1	1.3	1.4
FLOORS (Btuh/ft²)									
No. 17 Floors over uncond. space									
(f) Over open or vented space or garage, R-22 insulation		.2	.4	.6	.4	.8	.6	.8	1.0

Table 6—DUCT HEAT GAIN MULTIPLIERS[1]

Duct Location and Insulation	Duct Heat Gain Multipliers
Attic; 1 in. Flexible Blanket Insulation[2,3]	0.15
Attic; 2 in. Flexible or 1 in. Rigid Duct Insulation[4]	0.10
Unconditioned Above Grade Space; 1 in. Flexible Duct Insulation	0.05
Unconditioned Below Grade Space; No Duct Insulation	0.05
Ducts in Slab; No Duct Insulation	0.05

[1] Insulation must be covered with a vapor barrier and all joints must be sealed against vapor penetration.
[2] Flexible-blanket duct insulation—density 0.6 to 0.75 lb per cu ft. Do not use insulation with density less than 0.6 lb per cu ft.
[3] Does not meet HUD-FHA Minimum Property Standard requirement in Paragraph 515-4.1.
[4] Rigid duct insulation—density minimum 2.5 to 4.0 lb per cu ft.

Table 7—CAPACITY MULTIPLIER FOR SELECTION OF AIR CONDITIONING UNITS

Outside Design Temperature, F	Desired Inside Temperature Swing, Degrees[1]		
	6	4½	3
AIR-COOLED UNITS			
85-90	0.70	0.85	1.00
95	0.75	0.90	1.05
100	0.80	0.95	1.10
105	0.85	1.00	1.15
110	0.90	1.05	1.20

WATER-COOLED UNITS

Outside Design Temperature, F	Leaving Water Temperature, F														
	90	95	100	105	110	90	95	100	105	110	90	95	100	105	110
80-90	0.74	0.77	0.81	0.83	0.86	0.88	0.93	0.97	1.00	1.03	1.03	1.08	1.12	1.16	1.20
95	0.78	0.82	0.85	0.88	0.91	0.93	0.97	1.01	1.05	1.08	1.07	1.12	1.17	1.21	1.25
100	0.82	0.87	0.91	0.94	0.97	0.97	1.02	1.06	1.09	1.12	1.10	1.16	1.20	1.24	1.28
105	0.86	0.90	0.94	0.97	1.00	1.00	1.05	1.10	1.14	1.17	1.15	1.20	1.25	1.30	1.34
110	0.89	0.94	0.98	1.01	1.04	1.04	1.09	1.14	1.18	1.22	1.18	1.24	1.30	1.34	1.38

This table to be used according to the relationship:
(Calculated Heat Gain) X (Capacity Multiplier) = (Equipment Standard ARI Capacity Rating.)

[1]Thermostat must be set number of degrees below 78F as degree swing desired. (See paragraph 7-7.)

APPENDIX A
HOW TO CALCULATE HEAT TRANSFER MULTIPLIERS

A-1 Heat Transfer Multipliers for most types of construction are listed in Table 2 (for heating) and Table 5 (for cooling). Occasionally, constructions are encountered that are not included in these general descriptions. A procedure for determining Heat Transfer Multipliers for these constructions is outlined in this appendix.

A-2 CONDUCTANCE AND RESISTANCE There are two ways in which the heat transmission characteristics of a construction can be evaluated. Thermal conductance is the most commonly used concept. It expresses the amount of heat which flows each hour through one square foot of a construction for each degree Fahrenheit of temperature difference.

A-3 The conductance of a construction can be calculated directly. The arithmetic involved is not complicated, but it can be confusing. Thermal resistance is a concept which is more easily applied. Conductance and resistance are directly related in that the U-value of a construction equals one divided by the thermal resistance of the construction. $(U=\frac{1}{R})$

A-4 Thermal resistance is the temperature drop in degrees Fahrenheit througn a heat barrier when heat is passing through the barrier at the rate of 1.0 Btu per sq ft of surface per hour. Thus, the thermal resistance is expressed in degrees per Btuh per sq ft. Larger numerical values of thermal resistance indicate better heat barriers. Because the units of thermal resistance are unwieldy, the term resistance units is more commonly used. This is abbreviated ru.

A-5 RESISTANCE COMPONENTS The thermal resistance concept is easy to apply because the total thermal resistance of a wall or ceiling equals the sum of the thermal resistances of each component of the construction. To calculate the thermal resistance:

A. Identify each component—see paragraph A-6,

B. Determine thermal resistance of each component from Table A-1, and

C. Add these thermal resistances to obtain the total thermal resistance of the construction.

A-6 Components that are to be considered when determining the thermal resistance of a construction are:

A. **Air Surfaces,** inside and outside

B. **Air Spaces**—at least ¾ in. thick—between construction materials

C. **Building Materials,** including interior and exterior finishing materials and insulation when used.

A-7 AIR SURFACES The air surface resistance is dependent upon the thickness of the air film which is adjacent to the surface. The air film thickness depends upon the air movement near the surface and the direction in which heat is flowing. For example, the air film adjacent to an outside surface is being continuously wiped off by the wind. Consequently, it is not as thick as the air film which is adjacent to the inside surface since the room air is not moving rapidly and there is little or no wiping action. Natural air currents also affect the air film thickness and create a thicker film when heat flows downward to or from a horizontal surface than when heat flows upward.

A-8 Thermal characteristics of surface materials also affect the air surface resistance. Reflective or shiny surfaces have higher thermal resistance than nonreflective surfaces. The term nonreflective describes wood, plaster, brick, or other common building materials. For example, Table A-1, Number 1, includes a horizontal surface which has a thermal resistance of 0.61 ru when the heat flow is up and the surface is nonreflective. A similar surface has a thermal resistance of 1.32 ru, however, when the heat flow is up and the surface is bright aluminum foil. A horizontal surface with the heat flowing up will usually be a surface such as a ceiling. An aluminum foil surface which faces an attic space is considered, in this manual, to be nonreflective. It is suggested that this recommendation be followed when using the information in Table A-1.

A-9 The air surface resistance is 0.17 ru for a surface exposed to a 15 mph wind. This value applies regardless of the direction of heat flow, the direction in which the surface faces, or the type of surface material.

A-10 AIR SPACES An air space is totally enclosed and is of limited width and thickness. For example, a space formed by two studs and an inside and outside wall is considered to be an air space. An air space less than ¾ in. thick has practically no insulating value and is not considered

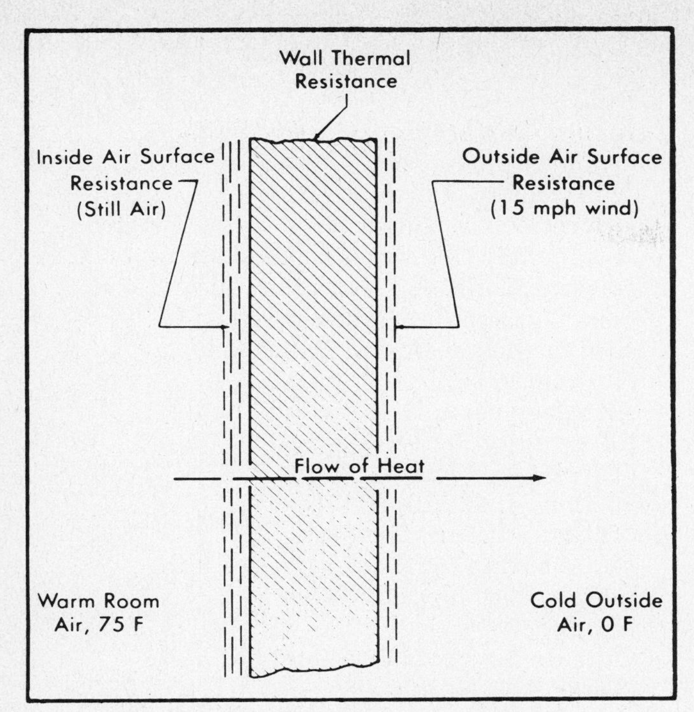

A-1 Thermal resistance to heat flow.

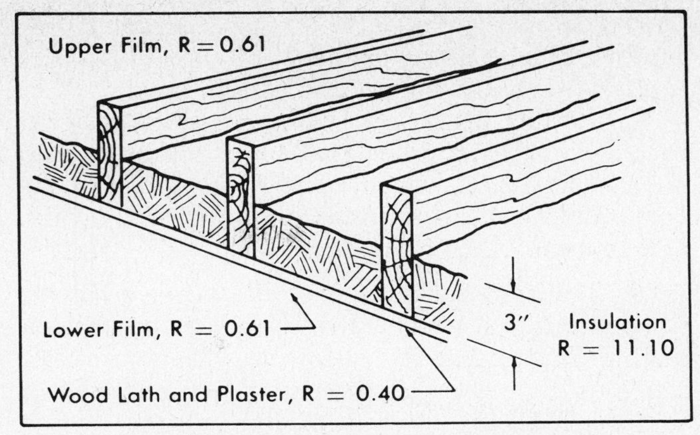

A-2 Thermal resistance for insulated ceiling.

when determining the thermal resistance of a construction. Thus if 3 in. of insulation were installed in a 3⅝ in. stud space, the air space remaining would be less than ¾ in. thick and would not be considered in the thermal resistance calculation.

A-11 The thermal resistance of an air space is affected by the direction of heat flow and the type of material facing the air space. For example, the thermal resistance of a vertical air space with a horizontal heat flow is 0.97 ru when it is faced with nonreflective material. If this were a horizontal space with heat flowing in an upward direction the thermal resistance would be 0.85 ru.

A-12 Values for different types of surfaces facing an air space are included in Table A-1, Number 2. These surfaces are bright aluminum foil, aluminum painted or coated paper, and nonreflective materials. If at least one surface of an air space is faced with bright aluminum foil, the air space is considered to be reflective, and the highest resistance value is used. An example is an air space lined on one side by foil-backed gypsum board or gypsum lath. Similarly, if one side of an air space is enclosed by a surface composed of aluminum coated or painted paper, the "Aluminum Painted Paper" resistance value is used. The nonreflective resistance values apply to air spaces completely enclosed by ordinary construction materials such as gypsum board of gypsum lath, wood, masonry materials, etc.

A-13 THERMAL RESISTANCE OF BUILDING MATERIALS The thermal resistance of a wall, ceiling, or floor is the sum of the thermal resistances of the air surfaces, the air spaces, and the materials of construction. The thermal resistance of each construction component depends upon the following:

1. The thickness of the material and
2. The ability of the material to resist the flow of heat.

A-14 Some types of building material can be used in any thickness and are uniform throughout this thickness such as a poured concrete wall or floor. Thus, the total thermal resistance of concrete is the resistance per inch of thickness multiplied by the thickness of the wall in inches.

A-15 Other construction materials are commonly manufactured and sold in several standard thicknessess (such as wood, fibrous insulation, concrete, and insulating board). For example, fibrous insulation is often purchased in blanket or batt form which is available in the thicknesses shown in Table 2. Similarly, wood is usually obtained as 1, 2, 4, 6 in. etc. boards. However, wood board dimensions are nominal, and a 1 in. board, for example, is about 25/32 in. thick. Similarly, a 2 in. board is 1-⅝ in. thick, a 4 in. board is 3-⅝ in. thick, etc.

A-16 A third type of building material is manufactured in standard sizes. Brick, concrete block, and hollow clay tile are examples of this type of material. Because these materials often have holes cast into them, it is not possible to show their thermal resistances on a per inch of thickness basis. Consequently, thermal resistance values of the product as manufactured in standard sizes are indicated.

A-17 The last several sections of Table A-1 include the thermal resistance of many construction materials. Some of these are listed on a per inch of thickness basis, and others are listed for specific thicknesses. For example, in Number 4, the thermal resistance of asbestos-cement board is listed as 0.25 ru per in. of thickness. One of the commonly used thicknesses, however, is ⅛ in., and its resistance is listed as 0.03 ru.

A-18 Thermal resistance of materials such as fibrous mineral wool is listed only on a per inch of thickness basis. To obtain the thermal resistance of this material, the resistance per inch of thickness, 3.70 ru, is multiplied by the thickness in inches. Concrete, such as that composed of Portland cement and sand and gravel or sand aggregate which has not been oven dried, has a thermal resistance of 0.08 ru per inch of thickness. The thermal resistance of the concrete in an 8 in. poured wall is 8 x 0.08 or 0.64 ru.

A-19 CALCULATION OF THERMAL RESISTANCE OF WALLS As an example of the calculation of the thermal resistance of a frame wall, consider a wall which has a wood siding exterior, 25/32 in. insulating sheathing, and a ⅜ in. gypsum lath and a ½ in. plaster interior. (See Figure A-3.) Table A-1, shows the following resistances for this wall:

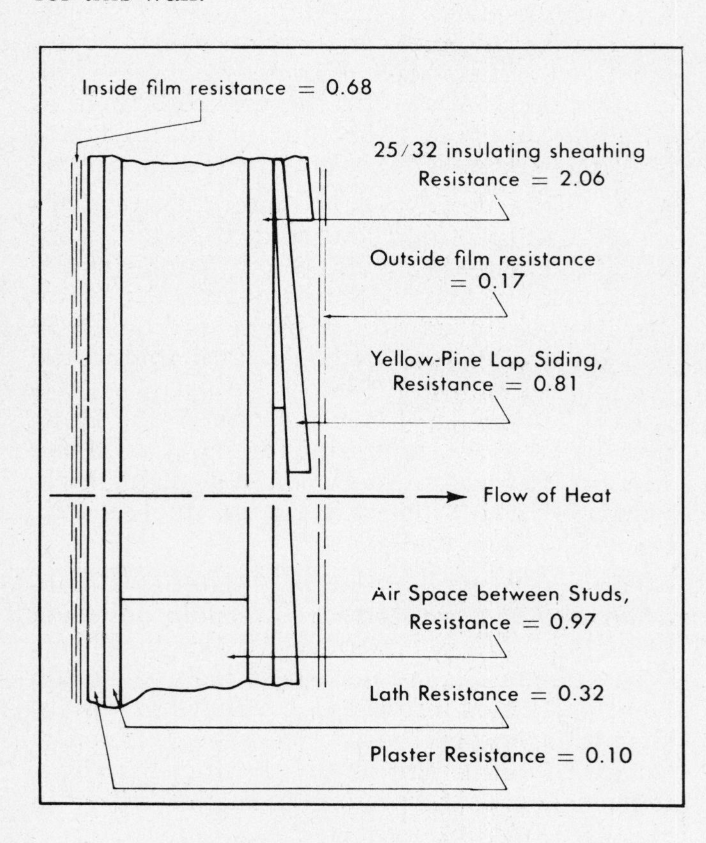

A-3 Thermal resistance for wall construction described in paragraph A-19.

	Resistance Units, Ru
Inside air surface resistance (Table A-1, Number 1, Vertical Wall, Horizontal Heat Flow, Nonreflective Surfaces)	0.68
½ in. cement plaster with sand aggregate (Table A-1, Number 13)	0.10
⅜ in. Gypsum lath (shown as gypsum or plaster board, Table A-1, Number 4)	0.32
Air space vertical, horizontal heat flow, nonreflective surfaces (Table A-1, Number 2)	0.97
25/32 in. insulating sheathing (Table A-1, Number 8)	2.06
Wood siding (entered as bevel ½ in. x 8 in. lapped, Table A-1, Number 17)	0.81
Outside air surface resistance, (Table A-1, Number 3)	0.17
Total	5.11

A-20 For this example, the U-value equals 0.196 (1/5.11). Or when rounded off to two decimal places, it will equal 0.20. If 2 in. insulation (R-7) is added to this construction, the thermal resistance is 12.11 ru (5.11+ 7.00) and the U-value is 0.083 or 0.08.

A-21 Thermal resistances of ceilings and floors are calculated in a similar manner, but resistance values of the components vary because of heat flow direction and other factors.

A-22 Once the U-value has been determined, the Heat Transfer Multiplier for heating is obtained by multiplying the U-value by the Design Temperature Difference. To obtain the Heat Transfer Multiplier for cooling, the U-value is multiplied by the Equivalent Temperature Difference.

A-23 Equivalent Temperature Difference values for average building constructions and normal design conditions are included in Table A-2. Infiltration Factors that must be considered when calculating heat gain through doors and walls are included in Table A-3.

Table A-1

RESISTANCE VALUES FOR COMMON BUILDING AND INSULATING MATERIALS

(Used with permission of The American Society of Heating, Refrigerating and Air-Conditioning Engineers)

Position of Air Space	Direction of Heat Flow	Air Space Thickness, in.	Thermal Resistance (R)		
			Bright Aluminum Foil	Aluminum Painted Paper	Non-Reflective
No. 1 Still Air Surfaces					
(a) Horizontal	Up		1.32	1.10	0.61
(b) 45° Slope	Up		1.37	1.14	0.62
(c) Vertical	Horiz.		1.70	1.35	0.68
(d) 45° Slope	Down		2.22	1.67	0.76
(e) Horizontal	Down		4.55	2.70	0.92
No. 2 Air Spaces					
(a) Horizontal	Up (Winter)	¾ to 4	2.06	1.62	0.85
(b) Horizontal	Up (Summer)	¾ to 4	2.75	1.87	0.80
(c) 45° Slope	Up (Winter)	¾ to 4	2.22	1.71	0.88
(d) Vertical	Horiz. (Winter)	¾ to 4	2.62	1.94	0.94
(e) Vertical	Horiz. (Summer)	¾ to 4	3.44	2.16	0.91
(f) 45° Slope	Down (Summer)	¾ to 4	4.36	2.50	0.90
(g) Horizontal	Down (Winter)	¾	3.55	2.39	1.02
(h) Horizontal	Down (Summer)	¾	3.25	2.08	0.84
(i) Horizontal	Down (Winter)	1½	5.74	3.21	1.14
(j) Horizontal	Down (Summer)	1½	5.24	2.76	0.93
(k) Horizontal	Down (Winter)	4	8.94	4.02	1.23
(l) Horizontal	Down (Summer)	4	8.08	3.38	0.99
No. 3 Moving Air Surfaces (Any Position or Direction)					
(a) 15 mph Wind (Winter)					0.17
(b) 7½ mph Wind (Summer)					0.25

Description	Density lb per cu ft	Thermal Resistance (R)	
		Per Inch of Thickness	For Thickness Listed
No. 4 Building Board, Boards, Panels, Sheathing, etc.			
(a) Asbestos-cement board	120	0.25	
(b) Asbestos-cement board—⅛ in.	120		0.03
(c) Gypsum or plaster board—⅜ in.	50		0.32
(d) Gypsum or plaster board—½ in.	50		0.45
(e) Plywood	34	1.25	
(f) Plywood—¼ in.	34		0.31
(g) Plywood—⅜ in.	34		0.47
(h) Plywood—½ in.	34		0.63
(i) Plywood—⅝ in.	34		0.78
(j) Plywood or wood panels—¾ in.			0.94
(k) Wood fiber board, laminated or homogeneous	26	2.38	
	31	2.00	
	33	1.82	
(l) Wood fiber, hardboard type	65	0.72	
(m) Wood fiber, hardboard type—¼ in.	65		0.18
(n) Wood, fir or pine sheathing—25/32 in.			0.98
(o) Wood, fir or pine—1⅝ in.			2.03

Table A-1 (cont'd)

Description		Density lb per cu ft	Thermal Resistance (R)	
			Per Inch of Thickness	For Thickness Listed
No. 5 Building Paper				
(a) Vapor-permeable felt				0.06
(b) Vapor-seal, two layers or mopped 15 lb felt				0.12
(c) Vapor-seal, plastic film				Negl
No. 6 Flooring Materials				
(a) Carpet and fibrous pad				2.08
(b) Carpet and rubber pad				1.23
(c) Cork tile ⅛ in.				0.28
(d) Floor tile or lineoleum—average value—⅛ in.				0.05
(e) Terrazzo—1 in.				0.08
(f) Wood subfloor—25/32 in.				0.98
(g) Wood, hardwood finish—¾ in.				0.68
No. 7 Insulating Materials, Blanket and Batt				
(a) Cotton fiber		0.8 - 2.0	3.85	
(b) Mineral wool, fibrous form,	(2"-2¾")	1.5 - 4.0		7
processed from rock, slag	(3"-3½")			11
or glass	(3½"-3⅝")			13
	(5¼"-6½")			19
	(6"-7")			22
(c) Wood fiber		3.2 - 3.6	4.0	
No. 8 Insulating Materials, Board				
(a) Glass fiber		9.5 - 11.0	4.00	
(b) Wood or cane fiber Acoustical tile—½ in.				1.19
(c) Wood or cane fiber Acoustical tile—¾ in.				1.78
(d) Wood or cane fiber Interior finish, (plank, tile, lath)		15.0	2.86	
(e) Wood or cane fiber Interior finish, (plank, tile, lath)—½ in.		15.0		1.43
(f) Roof deck slab, approximately—1½ in.				4.17
(g) Roof deck slab, approximately—2 in.				5.56
(h) Roof deck slab, approximately—3 in.				8.33
(i) Sheathing, impregnated or coated		20.0	2.63	
(j) Sheathing, impregnated or coated—½ in.		20.0		1.32
(k) Sheathing, impregnated or coated—25/32 in.		20.0		2.06
No. 9 Insulating Materials, Board and Slabs				
(a) Cellular glass		9.0	2.50	
(b) Expanded Urethane			5.88	
(c) Expanded rubber		4.5	4.55	
(d) Hog hair, with asphalt binder		8.5	3.00	
(e) Expanded Polystyrene, molded beads		1.0	3.57	
(f) Expanded Polystyrene extruded		2.2	5.00	
(g) Wood shredded, cemented in preformed slabs		22.0	1.82	
(h) Mineral wool with resin binder		15.0	3.45	
(i) Mineral wool with asphalt binder		15.0	3.22	
No. 10 Insulating Materials, Loose Fill				
(a) Macerated paper or pulp products		2.5-3.5	3.57	
(b) Mineral wool, glass, slag, or rock		2.0-5.0	3.33	
(c) Sawdust or shavings		8.0-15.0	2.22	
(d) Silica Aerogel		7.6	5.88	
(e) Vermiculite, expanded		7.0-8.2	2.08	
(f) Wood fiber, redwood, hemlock, or fir		2.0-3.5	3.33	
(g) Wood fiber, redwood bark		3.0	3.22	
(h) Wood fiber, redwood bark		4.0	3.57	
(i) Wood fiber, redwood bark		4.5	3.84	

Table A-1 (cont'd)

	Description	Density lb per cu ft	Thermal Resistance (R) Per Inch of Thickness	Thermal Resistance (R) For Thickness Listed
	No. 11 Roof Insulation, Preformed, for Use Above Deck			
(a)	Approximately—½ in.			1.39
(b)	Approximately—1 in.			2.78
(c)	Approximately—1½ in.			4.17
(d)	Approximately—2 in.			5.56
(e)	Approximately—2½ in.			6.67
(f)	Approximately—3 in.			8.33
	No. 12 Masonry Materials, Concretes			
(a)	Cement mortar	116	0.20	
(b)	Gypsum-fiber concrete, 87½% gypsum, 12½% wood chips	51	0.60	
(c)	Lightweight aggregates including	120	0.19	
		100	0.28	
	expanded shale, clay or slate; expanded	80	0.40	
	slags; cinders; pumice; perlite; vermicu-	60	0.59	
	lite; also cellular concretes	40	0.86	
		30	1.11	
		20	1.43	
(d)	Sand and gravel or stone aggregate, oven dried	140	0.11	
(e)	Sand and gravel or stone aggregate, not dried	140	0.08	
(f)	Stucco	116	0.20	
	No. 13 Plastering Materials			
(a)	Cement plaster, sand aggregate	116	0.20	
(b)	Cement plaster, sand aggregate—½ in.			0.10
(c)	Cement plaster, sand aggregate—¾ in.			0.15
(d)	Gypsum plaster, lightweight aggregate—½ in.	45		0.32
(e)	Gypsum plaster, lightweight aggregate—⅝ in.	45		0.39
(f)	Gypsum plaster, lightweight aggregate on metal lath—¾ in.			0.47
(g)	Gypsum plaster, perlite aggregate	45	0.67	
(h)	Gypsum plaster, sand aggregate	105	0.18	
(i)	Gypsum plaster, sand aggregate—½ in.	105		0.09
(j)	Gypsum plaster, sand aggregate—⅝ in.	105		0.11
(k)	Gypsum plaster, sand aggregate on metal lath—¾ in.			0.10
(l)	Gypsum plaster, sand aggregate on wood lath			0.40
(m)	Gypsum plaster, vermiculite aggregate	45	0.59	
	No. 14 Masonry Units			
(a)	Brick, common	120	0.20	
(b)	Brick, face	130	0.11	
(c)	Hollow clay tile, one cell deep—3 in.			0.80
(d)	Hollow clay tile, one cell deep—4 in.			1.11
(e)	Hollow clay tile, two cells deep—6 in.			1.52
(f)	Hollow clay tile, two cells deep— 8 in.			1.85
(g)	Hollow clay tile, two cells deep—10 in.			2.22
(h)	Hollow clay tile, three cells deep—12 in.			2.50
(i)	Stone, lime or sand		0.08	
(j)	Gypsum partition tile, 3 in. x 12 in. x 30 in.—solid			1.26
(k)	Gypsum partition tile, 3 in. x 12 in. x 30 in.—4 cell			1.35
(l)	Gypsum partition tile, 4 in. x 12 in. x 30 in.—3 cell			1.67

Table A-1 (cont'd)

	Description	Density lb per cu ft	Thermal Resistance (R)	
			Per Inch of Thickness	For Thickness Listed
	No. 15 Concrete Blocks			
(a)	Sand and gravel aggregate, three oval core—4 in.			0.71
(b)	Same as (15a) but—8 in.			1.11
(c)	Same as (15a) but—12 in.			1.28
(d)	Cinder aggregate, three oval core—3 in.			0.86
(e)	Same as (15d) but—4 in.			1.11
(f)	Same as (15d) but—8 in.			1.72
(g)	Same as (15d) but—12 in.			1.89
(h)	Sand and gravel aggregate, two core—8 in., 36 lb			1.04
(i)	Same as (15h) but—with filled cores			1.93
(j)	Lightweight aggregate, expanded shale, clay, slate or slag; pumice—3 in.			1.27
(k)	Same as (15j) but—4 in.			1.50
(l)	Same as (15j) but—8 in.			2.00
(m)	Same as (15j) but—12 in.			2.27
(n)	Lightweight aggregate, expanded shale, clay, slate or slag, pumice—2-core, 8 in., 24 lb			2.18
(o)	Same as (15n) but—with filled cores			5.03
(p)	Same as (15n) but—3 core, 6 in., 19 lb			1.65
(q)	Same as (15p) but—with filled cores			2.99
(r)	Same as (15n) but—3 core, 12 in., 38 lb			2.48
(s)	Same as (15r) but—with filled cores			5.82
	No. 16 Roofing			
(a)	Asbestos-cement shingles	120		0.21
(b)	Asphalt roll roofing	70		0.15
(c)	Asphalt shingles	70		0.44
(d)	Built-up roofing—⅜ in.	70		0.33
(e)	Slate—½ in.			0.05
(f)	Sheet Metal		Negl	
(g)	Wood shingles			0.94
	No. 17 Siding Materials (On Flat Surface)			
(a)	Wood shingles, 16 in., 7½ in. exposure			0.87
(b)	Wood shingles, double, 16 in., 12 in. exposure			1.19
(c)	Wood shingles, plus insulation 5/16 in. backer board			1.40
(d)	Asbestos-cement siding, ¼ in., lapped or shingles			0.21
(e)	Asphalt roll siding			0.15
(f)	Asphalt insulating siding (½ in. board)			1.46
(g)	Wood siding, drop, 1 in. x 8 in.			0.79
(h)	Wood siding, bevel, ½ in. x 8 in., lapped			0.81
(i)	Wood siding, bevel, ¾ in. x 10 in., lapped			1.05
(j)	Wood siding, plywood, ⅜ in., lapped			0.59
(k)	Structural glass			0.10
	No. 18 Woods			
(a)	Maple, oak, and similar hardwoods	45	0.91	
(b)	Fir, pine, and similar soft woods	32	1.25	
(c)	Fir, pine, and similar soft woods—25/32 in.	32		0.98
(d)	Fir, pine, and similar soft woods—1⅝ in.	32		2.03
(e)	Fir, pine, and similar soft woods—2⅝ in.	32		3.28
(f)	Fir, pine, and similar soft woods—3⅝ in.	32		4.55

Table A-2
EQUIVALENT TEMPERATURE DIFFERENCES*

	Design Temperature, F	85	90	95		100		105	110
	Daily Temperature Range	L or M	L or M	M	H	M	H	H	H
A.	WALLS AND DOORS**								
	1. Wood Frame Walls and Doors	13.6	18.6	23.6	18.6	28.6	23.6	28.6	33.6
	2. Solid Masonry, Block or Brick Walls	6.3	11.3	16.3	11.3	21.3	16.3	21.3	26.3
	3. Partitions	5.0	10.0	15.0	10.0	20.0	15.0	20.0	25.0
B.	CEILINGS AND ROOFS***								
	a. Dark Exterior	34.0	39.0	44.0	39.0	49.0	44.0	49.0	54.0
	2. Light Exterior	26.0	31.0	36.0	31.0	41.0	36.0	41.0	46.0
C.	FLOORS								
	1. Over Unconditioned, Vented or Open Space	5.0	10.0	15.0	10.0	20.0	15.0	20.0	25.0
	2. Over Conditioned Space or On or Below Grade	0	0	0	0	0	0	0	0

* Equivalent Temperature Differences in this Table are for the Outside Design Conditions specified and an Inside Design Temperature of 75 F.

** HTM for Walls and Doors equals ("U" value times Equivalent Temperature Difference) plus the appropriate infiltration factor obtained from Table A-3.

*** Ceilings and Roofs: Factors for light-colored roofs are calculated using footnote for Table 9, 1959 Guide, Chapter 13. For roofs in shade, eight hour average = 11 degrees temperature difference. At 90 F design temperature and medium daily range, equivalent temperature difference for light-colored roof equals $11 + (0.71)(39 - 11) = 31$.

Table A-3
INFILTRATION FACTORS*

Outside Design Temperature, F	85	90	95	100	105	110
INFILTRATION Factors, (Btuh/ft^2)**	0.7	1.1	1.5	1.9	2.2	2.6

* Infiltration Factors in this Table are for the Outside Design Temperature specified and an Inside Design Temperature of 75 F.

** HTM for Walls and Doors equals the appropriate infiltration factor plus (the "U" value times the Equivalent Temperature Difference obtained from Table A-2).

APPENDIX B

HOW TO DETERMINE SHADED AND UNSHADED GLASS AREAS FOR HEAT GAIN CALCULATION

B-1 For most houses it is not necessary to consider the effect of roof overhangs when calculating the heat gain for windows. When a house has an unusually large amount of glass area or when a very precise load calculation is required, the effect of a roof overhang should be considered. Glass shaded by roof overhangs is not affected by sun position and is considered as north facing glass for purposes of calculating heat gain.

B-2 Shaded areas are not calculated for windows that face north, northeast, or northwest. When glass areas facing other directions are partially shaded by roof overhangs; either the shaded area must be determined, or the glass area must all be considered as sunlit glass.

B-3 Table A, Form J-1, is used when determining the shaded and unshaded glass areas beneath permanent shading devices such as roof overhangs. A separate calculation is made for each window. This table is illustrated in Fig. B-2. In many instances, there is not enough shade from roof overhangs to have appreciable effect upon the glass sun gain. In these cases, this table need not be used for the heat gain calculation.

LINE 1

On line 1, enter the compass direction for each window shaded by a roof overhang.

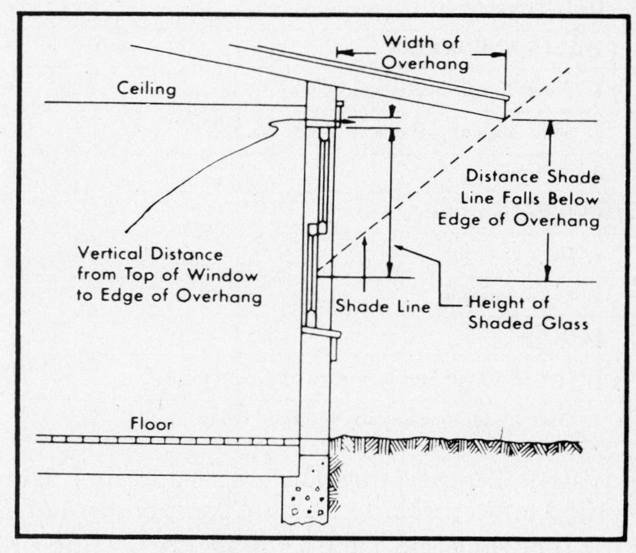

B-1 Location of shade line from overhang

Table B-1
SHADED GLASS AREAS

Window Width, Ft	Less than 2.0			2.1 to 2.5			2.6 to 3.0			3.1 to 3.5			3.6 to 4.0			4.1 to 4.5			4.6 to 5.0			5.1 to 5.5			5.6 to 6.0		
Direction Window Faces	E W	SE SW	S	E W	SE SW	S	E W	SE SW	S	E W	SE SW	S	E W	SE SW	S	E W	SE SW	S	E W	SE SW	S	E W	SE SW	S	E W	SE SW	S
Latitude Degrees	Area Of Shaded Glass Per Foot Overhang, Sq Ft																										
25	0	2	17	1	3	22	1	4	26	1	5	31	1	5	36	1	6	41	1	7	46	2	7	50	2	8	55
30	0	2	9	1	2	11	1	3	14	1	4	16	1	4	18	1	5	21	1	5	23	2	6	26	2	6	28
35	0	2	5	1	2	7	1	2	9	1	3	10	1	3	12	1	4	13	1	4	15	2	5	16	2	5	18
40	0	1	4	1	2	5	1	2	6	1	3	7	1	3	8	1	3	9	1	4	10	2	4	11	2	5	12
45	0	1	3	1	1	3	1	2	4	1	2	5	1	2	6	1	3	6	1	3	7	2	3	8	2	3	9
50	0	1	2	1	1	3	1	1	3	1	2	4	1	2	5	1	2	5	1	2	6	2	3	6	2	3	7
55	0	1	2	1	1	2	1	1	2	1	1	3	1	2	3	1	2	4	1	2	4	2	2	5	2	2	5

NOTE: If top of window is more than 1 ft below edge of overhang the wall area between top of window and overhang must be subtracted from shaded area obtained from Table B-1. Shaded areas are not calculated for NE, N, or NW windows. Use glass heat transfer multiplier indicated in Table 4 for entire glass area of window facing these directions.

LINE 2

On line 2, enter the total area of each window to the nearest sq ft.

LINE 3

On line 3, enter the width of each window to the nearest 0.1 ft.

LINE 4

Obtain the shaded area per ft of overhang from Table B-1, and enter on Line 4. This area depends upon the direction which the window faces, the width of the window and the degrees of north latitude of the city in which the structure is located. The degrees of north latitude can be obtained from Table 1.

LINE 5

On line 5, enter the width of the roof overhang to the nearest one-half ft. For example, a 3 ft, 4 in. roof overhang would be entered as having a width of 3.5 ft.

LINE 6

Calculate the total area of shaded glass to the nearest sq ft and enter on Line 6. This area equals the shaded area per ft of overhang (Line 4) multiplied by the width of the overhang (Line 5). If the area obtained by multiplying Line 4 by Line 5 is greater than the entire window area (Line 2), enter the actual window area.

LINE 7

On line 7, enter the total area of unshaded glass to the nearest sq ft. The area of unshaded glass is equal to the total glass area entered on Line 2 minus the shaded glass area entered on Line 6.

1. Direction which window faces.	W	W	E	E	E	E	E	S	S	S	W
2. Total window area, sq ft.	20	20	11	8	11	11	17	11	8	11	17
3. Width of window, ft.	4.0	4.0	3.5	2.5	2.5	2.5	3.7	2.5	2.5	2.5	3.7
4. Shaded area per foot of overhang from Table B−1, sq ft.	1	1	1	1	1	1	1	5	5	5	1
5. Width of overhang, ft.	2	2	2	2	2	2	2	2	2	2	2
6. Total area of shaded glass, sq ft. (Line 4) x (Line 5).	2	2	2	2	2	2	2	10	8	10	2
7. Total area of unshaded glass, sq ft. (Line 2) − (Line 6).	18	18	9	6	9	9	15	1	0	1	15

B-2 Calculation of shaded areas of windows using Table A, Form J-1

FORM J—1

Copyright by the
Air Conditioning Contractors
of America, Inc.
1228 – 17TH Street, N.W.
Washington, D.C. 20036
Printed in U.S.A.

January, 1968

Plan No. ____________
Date ____________
Calculated by ____________

WORKSHEET FOR MANUAL J

LOAD CALCULATIONS FOR RESIDENTIAL AIR CONDITIONING

For: Name ____________
Address ____________
City and State or Province ____________
By: Contractor ____________
Address ____________
City ____________

Winter Design Conditions

Outside ______ F Inside ______ F Temperature Difference ______ Degrees

(**Insert data below only after all heat loss calculations have been completed**)

Total Heat Loss (Btuh) ______ (From Line No. 15) Model No. ______
Serial No. ______ Manufactured by ______
Rating Data: Input ______ Btuh Output at Bonnet ______ Btuh
Description of Controls ______

Summer Design Conditions

Outside ______ F Inside ______ F
North Latitude ______ Degrees Daily Range ______

(**Insert data below only after all heat gain calculations have been completed**)

Total Heat Gain (Btuh) ______ (From Line No. 20 or 21, if used)
Equipment Capacity Multiplier ______ Model No. ______
Serial No. ______ Manufactured by ______
Rating Data: Cooling Capacity ______ Btuh Air Volume ______ Cfm
Description of Controls ______

Winter Construction Data (**See Table 2**)

Walls and Partitions ______

Windows and Doors ______

Ceilings ______

Floors ______

Summer Construction Data (**See Table 5**)

Direction House Faces ______

Windows and Doors ______

Walls and Partitions ______

Ceilings ______

Floors ______

FILE

Table A

(USE TO CALCULATE SHADED AND UNSHADED GLASS AREAS)

1. Direction which window faces.															
2. Total window area, sq ft.															
3. Width of window, ft.															
4. Shaded area per foot of overhang from Table B–1, sq ft.															
5. Width of overhang, ft.															
6. Total area of shaded glass, sq ft. (Line 4) x (Line 5).															
7. Total area of unshaded glass, sq ft. (Line 2) – (Line 6).															

Table B

(Use to Determine Adjustment Factor)

A. Total Btuh Gain, from Line 20	
B. Capacity Multiplier, from Table 7	
C. Equipment Standard ARI Capacity Rating, Minimum required, (Line A) x (Line B)	
D. Capacity of Equipment Selected, not less than Line C, Enter in "Entire House" column, Line 21	
E. Adjustment Factor (Line D) ÷ (Line A), Enter in "HTM" column, Line 21	

DO NOT WRITE IN SHADED BLOCKS

1 Name of Room					Entire House			1			2			3			4			5			6		
2 Running Ft Exposed Wall																									
3 Room Dimensions, Ft																									
4 Ceiling Ht, Ft		Directions Room Faces																							
TYPE OF EXPOSURE		Const. No.	HTM		Area or Length	Btuh		Area or Length	Btuh		Area or Length	Btuh		Area or Length	Btuh		Area or Length	Btuh		Area or Length	Btuh		Area or Length	Btuh	
			Htg	Clg		Htg	Clg		Htg	Clg		Htg	Clg		Htg	Clg		Htg	Clg		Htg	Clg		Htg	Clg
5 Gross	a																								
Exposed	b																								
Walls and	c																								
Partitions	d																								
6 Windows	a																								
and Glass	b																								
Doors (Htg)	c																								
7 Windows	North																								
and Glass	E &W or NE & NW																								
Doors (Clg)	South or SE & SW																								
8 Other Doors																									
9 Net	a																								
Exposed	b																								
Walls and	c																								
Partitions	d																								
10 Ceilings	a																								
	b																								
11 Floors	a																								
	b																								
12 Ventilation																									
13 Sub Total Btuh Loss																									
14 Duct Btuh Loss																									
15 Total Btuh Loss																									
16 People @ 300 and Appliances 1200																									
17 Sensible Btuh Gain (Structure)																									
18 Duct Btuh Gain																									
19 Sum of Lines 17 and 18 (Clg)																									
20 Total Btuh Gain (Line 19 x 1.3)				1.3																					
21 Btuh for Air Quantities																									

DO NOT WRITE IN SHADED BLOCKS

						7			8			9			10			11			12			
1	Name of Room																							1
2	Running Ft Exposed Wall																							2
3	Room Dimensions, Ft																							3
4	Ceiling Ht, Ft	Directions Room Faces																						4
	TYPE OF EXPOSURE		Const. No.	HTM Htg	HTM Clg	Area or Length	Btuh Htg	Btuh Clg	Area or Length	Btuh Htg	Btuh Clg	Area or Length	Btuh Htg	Btuh Clg	Area or Length	Btuh Htg	Btuh Clg	Area or Length	Btuh Htg	Btuh Clg	Area or Length	Btuh Htg	Btuh Clg	
5	Gross	a																						5
	Exposed	b																						
	Walls and	c																						
	Partitions	d																						
6	Windows	a																						6
	and Glass	b																						
	Doors (Htg)	c																						
7	Windows	North																						7
	and Glass	E & W or NE & NW																						
	Doors (Clg)	South or SE & SW																						
8	Other Doors																							8
9	Net	a																						9
	Exposed	b																						
	Walls and	c																						
	Partitions	d																						
10	Ceilings	a																						10
		b																						
11	Floors	a																						11
		b																						
12	Ventilation																							12
13	Sub Total Btuh Loss																							13
14	Duct Btuh Loss																							14
15	Total Btuh Loss																							15
16	People @ 300 and Appliances 1200																							16
17	Sensible Btuh Gain (Structure)																							17
18	Duct Btuh Gain																							18
19	Sum of Lines 17 and 18 (Clg)																							19
20	Total Btuh Gain (Line 19 x 1.3)				1.3																			20
21	Btuh for Air Quantities																							21

7

LOAD CALCULATION—COMMERCIAL

Manual N, *Load Calculation* for commercial summer and winter air conditioning, which will be reproduced here in full, was published in 1976 by ACCA. It supersedes all previous ACCA publications dealing with commercial load calculations. The purpose of this manual is to provide practical data and procedures for calculating both heat loss and heat gain in commercial structures using unitary air-conditioning equipment and systems.

Like *Manual J* for residential load calculation, *Manual N* is also the result of joint committee efforts on the part of ACCA and ARI to present the latest industry thinking on the art of commercial load calculation. Much of the data was obtained from, or based on, the latest edition of the ASHRAE *Handbook of Fundamentals*.

The table of contents of *Manual N* reflects an orderly development of instructions to understand and use its procedures properly.

LOAD CALCULATION FOR COMMERCIAL SUMMER AND WINTER AIR CONDITIONING

(Using Unitary Equipment)

TABLE OF CONTENTS

The purpose of this manual is to provide simplified, practical data and procedures for calculating both heat loss and heat gain in commercial structures using unitary air conditioning equipment and systems.

The development and publication of this manual was accomplished by the Joint Committee on System Design and Installation of ARI (Air Conditioning and Refrigeration Institute) and ACCA. Much of the data in this manual is secured from, or based on, the latest edition of the ASHRAE Handbook of Fundamentals.

Grateful acknowledgement is made of those members of both ARI and ACCA who contributed to the preparation of this manual.

This manual is published as a public service by the Air Conditioning Contractors of America, Inc. The statements contained herein represent the consensus of members of the Association who participated in the preparation of the manual. The Association, however, does not warrant the applicability of the manual for any given situation.

SECTION 1

INTRODUCTION

1.1 **GENERAL.** Air conditioning application engineering is as much an art as it is a science. Science has evaluated all the factors required to determine a heating or cooling load through years of experimentation, tests, and analysis. It is in the application of these factors in determining the building or space load that much care and judgement must be exercised.

1.2 It is the purpose of this manual to provide data and procedures for load calculations for summer and winter air conditioning in commercial and public assembly applications. It is not intended to be a comprehensive application manual for equipment selection and system design, subjects which are covered in other ACCA manuals.

1.3 To arrive at accurate cooling and heating load calculations, a thorough and accurate field survey must be made, and/or a detailed set of architectural drawings must be studied. Consideration should be given to the need for zoning to accommodate the various exposures and occupancies occurring.

1.4 For heating load calculations, it is usually sufficient to select the proper heat transmission factors to be applied to the various elements of the building construction.

1.5 For cooling, there are other factors to be considered in order to arrive at an accurate load calculation. Building orientation—the directions in which the several walls of the space to be cooled face—must be considered. Time of day at which the peak load will occur must be determined.

1.5.1 To illustrate how building orientation and how time of day the peak load occurs can affect the cooling load, consider the following examples:

> An office on the lower floor of a building has a more or less uniform internal load of lights and people. If a wall with a high percentage of glass faces east, the peak load will occur in the morning. If the wall faces west, the peak load will occur late in the afternoon.
>
> A cafeteria might experience its greatest cooling load at noon; a restaurant in the evening; a night club at night, depending upon the changes in occupancy and ventilation requirements.

1.5.2 There are other factors that may affect the cooling load such as the thickness of insulation and glass shading.

> For example, the cooling load for a space with an uninsulated roof might have its peak at 3:00 pm, regardless of internal load variation.
>
> The effect of building orientation on the cooling load can be reduced or minimized through the use of double glass, tinted glass, heat absorbing glass, and draped or outside-shaded windows.

1.6 Because of the various factors involved in determining a peak load time, it is advisable to calculate the load for more than one time of day whenever there is doubt.

1.7 Time of year can also affect the cooling load in certain commercial buildings such as retail establishments which are crowded during the Easter and Christmas seasons. Obviously, one would not consider the maximum number of customers the store expects during these periods when calculating a cooling load for the hot summer months.

1.7.1 This same type of occupancy analysis must be made when calculating cooling loads for other buildings such as churches, auditoriums, and other places of public gathering where there may be a decrease in occupancy during the summer months.

1.8 To arrive at an accurate cooling load, internal heat sources must also be considered as well as related heat transmission through ceilings, floors, and partitions separating cooled spaces from kitchens, boiler rooms, or other hot spaces. Internal heat sources are discussed in Section 10, page 441.

1.9 Much good judgement must be used to determine the amount of ventilation required. Outside air introduced for any reason—including makeup air to equal that being exhausted—must be accounted for in arriving at the total outside air quantity.

1.9.1 Here the length of time air is being exhausted and the time(s) of day or night it is being exhausted relative to other peak load components should be given careful consideration. Many heating or cooling installations prove inadequate because the outside air loads are not properly calculated.

1.10 Occupancy of the building and the type of activity the occupants will be engaged in are usually relatively easy to establish. It is important to be certain that the total number of occupants is included in the cooling load calculations.

1.11 When a building is to be divided into zones, peak loads must be calculated for the overall space or building and for each zone so that equipment may be properly selected, ducts and piping properly sized, and proper controls selected.

1.12 When all the factors entering into a cooling load calculation are considered and the judgement involved in these considerations is recognized, it can readily be seen that air conditioning application engineering is an art as well as a science. Constant study, diligent application, good judgement, and analysis are required to become an expert.

1.13 **PURPOSE.** The purpose of the manual is to establish for commercial comfort air conditioning the following:

a) Definitions of air conditioning terms

b) Outside and inside design conditions

c) Ventilation requirements

d) Factors and calculation procedures for heating and cooling loads

1.13.1 This manual is to be construed as defining recommended practices rather than hindering progress or preventing the use of other manuals where such is justified by the economics of a situation and where departures from conventional practices are made known and clear to the buyer.

1.14 **SCOPE.** This manual is intended primarily for summer, winter, and year-round conditioning systems employing unitary equipment (packaged or split system, including heat pumps) of all types installed in commercial buildings. This manual does not apply to large buildings employing central systems using applied or central station equipment for air conditioning purposes.

1.14.1 For residential applications, see ACCA's Manual J, "Load Calculation for Residential Winter and Summer Air Conditioning."

1.15 **SPECIAL APPLICATIONS** for applications where life, safety, or specific processes depend on maintaining specified indoor design conditions, a consulting engineer competent in the particular application should be employed.

SECTION 2

DEFINITIONS

2.1 **COMMERCIAL COMFORT AIR CONDITIONING.** Commercial comfort air conditioning is the process of treating air so as to control simultaneously its temperature, humidity, cleanliness, and distribution to meet the comfort requirements of the occupants of the commercial space.

2.2 **DESIGN LOAD.** For the practical purpose of this manual, the design load (heating or cooling) is the load imposed on the conditioning equipment as the equipment maintains inside design conditions when outside design conditions of temperature and humidity prevail and when all sources of load are taken at the maximum that will occur coincidentally.

2.3 **SENSIBLE HEAT.** If the addition of heat to or the removal of heat from a substance can be measured with a dry-bulb thermometer, that heat is called sensible heat. An example of sensible heat is the heat given off by a light bulb.

2.4 **LATENT HEAT.** If the addition of heat to or the removal of heat from a substance does not cause a change in temperature but results in a change or state—from a liquid to a vapor, for example—that heat is known as latent heat. An example of latent heat addition is the evaporation of water at 212 F and sea level atmospheric pressure wherein the liquid water changes to water vapor, or steam, without a change in temperature.

2.5 **INSIDE DESIGN CONDITIONS.** Inside design conditions are the indoor dry-bulb (DB) temperature and relative humidity (RH) specified for design load calculation.

2.6 **OUTSIDE DESIGN CONDITIONS.** Outside design conditions are the outdoor dry-bulb (DB) and wet-bulb (WB) temperatures specified for design load calculation.

SECTION 3

FACTORS WHICH AFFECT HEAT LOSS AND GAIN

3.1 In order to perform an accurate estimate of heating and cooling loads, an accurate survey of the load components of the space to be conditioned must be made. Following are the factors which affect the heat loss and heat gain of a space:

1. Construction materials
2. Physical dimensions
3. Orientation of building or space
4. Occupancy
5. Lighting
6. Appliances and equipment
7. Infiltration
8. Ventilation
9. Schedule of use
10. Outside design conditions
11. Inside design conditions

SECTION 4

TIME OF DAY OF PEAK COOLING LOAD

4.1 The time of day of peak space cooling load is not always immediately apparent. This is because the major load components making up the cooling load may not reach individual peaks at the same time. While peak outside temperatures occur about 3:00 pm (Standard Time) in the summer months, peak solar heat gains through windows may occur anytime from 7:00 am to 5:00 pm depending on glass exposure. Furthermore, internal heat gains may reach a peak at any time. It may, therefore, be necessary to make total heat gain calculations for more than one time of day to determine the peak cooling load.

4.1.1 Note that the tables in this manual are based on Standard Time. It is important to correct for Daylight Saving Time by adding one hour to Standard Time to arrive at the correct Daylight Saving Time. For example, 3 o'clock Standard Time becomes 4 o'clock Daylight Saving Time.

4.1.2 An important aspect of calculating cooling loads at times other than 3:00 pm that is sometimes overlooked is correction of the outdoor design dry-bulb temperatures for the time of day. Obviously, if the peak outdoor dry-bulb temperature reaches the daily maximum at 3:00 pm, it must be lower at all other times. Consequently, the difference between the outdoor and indoor dry-bulb temperatures at times other than 3:00 pm will be less than at 3:00 pm design conditions. Table 6 (page 476) provides factors for making the necessary corrections, and note #2 under the table shows an example of how to make a typical calculation. The correction applies not only to the indoor-outdoor temperature difference, but also to total equivalent temperature differentials for walls (Table 8, page 480) and roofs (Table 9, page 481).

SECTION 5

OUTSIDE DESIGN CONDITIONS

5.1 **COOLING.** The outside design dry-bulb and wet-bulb temperatures for calculating cooling loads should not be lower than those shown in Table 1 (page 470) for the cities listed and geographic locales that are the same climatically.

5.2 **HEATING.** The outside design dry-bulb temperatures for calculating heating loads should not be higher than those shown in Table 1 for the cities listed and geographic locales that are the same climatically.

SECTION 6

INSIDE DESIGN CONDITIONS

6.1 **HEATING.** The inside design dry-bulb temperature should be determined by the use of the space.

6.1.1 Where controlled humidification is provided, in order to avoid condensation on windows, the inside design relative humidity should not exceed that shown in Table 21 (page 492) for the various outside temperatures.

6.2 **COOLING.** The inside design conditions should be determined by the use of the space.

6.3 **COMFORT.** ASHRAE Standard 55-74 defines the comfort zone or envelope as a quadrilateral on the psychrometric chart. When the mean radiant temperature equals the dry-bulb temperature, the corners of the quadrilateral are located approximately at the following four points: 72 F DB and 71% RH (65 F WB), 78 F DB and 58% RH (67 F WB), 73 F DB and 24% RH (53 F WB), and 80 F DB and 19% RH (56 F WB). See Figure 1.

Table 16b(page 488) shows data based on indoor conditions of 80 F DB and 50% RH, clearly outside the comfort zone, although not by a great amount. Similarly, Table 22a (page 493) gives data based on indoor conditions of 68 F DB and RH as low as 5%, again clearly outside the comfort zone. The purpose in presenting data based on marginal comfort conditions is to enable the user of this manual to satisfy the requirements of jobs which place priority on energy conservation.

6.3.1 The range of conditions under which most people are comfortable is from 71.5 F and 71% RH to 79.7 F and 20% RH. Generally speaking, to maintain comfort, the higher the dry-bulb temperature, the lower the relative humidity must be, and vice versa.

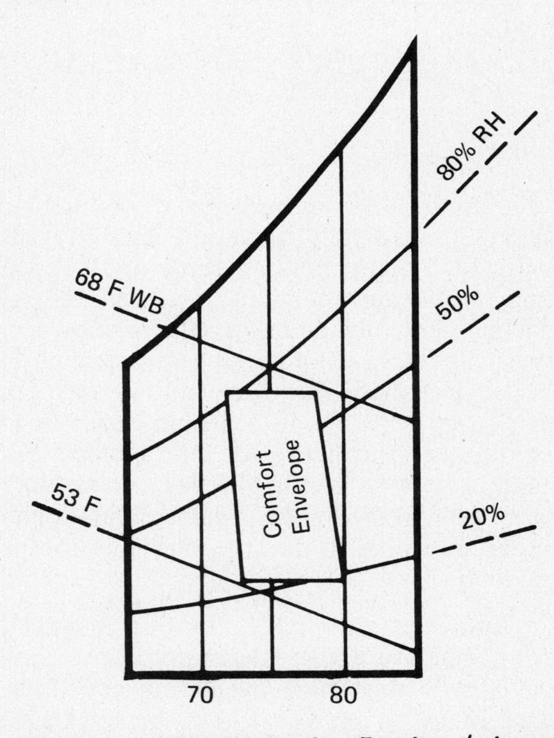

Figure 1 — ASHRAE Comfort Envelope (schematic)

SECTION 7

SOLAR RADITION HEAT GAIN THROUGH GLASS

7.1 Solar radiation heat gain values for clear glass are shown in Table 2 (page 475). Correction factors for other glass and various shading devices are shown in Table 3 (page 475).

7.1.1 **WINDOW SHADING FROM OVERHANGS.** When a window is shaded by an overhang, the portion of the glass that is shaded is subject to the same solar radiation heat gain as glass facing the North.

To compute the shaded and sunlit areas of a window partially shaded by an overhang, find the distance the shade line falls below the bottom edge of the overhang by multiplying the length of the overhang by the applicable factor from Table 4 (page 476). Next, subtract the vertical distance from the bottom edge of the overhang to the top of the window from the distance the shade line falls below the overhang. The result is the height of the shaded portion of the window, and multiplying it by the width of the window determines the area of the shaded portion of the window which is treated as north glass. The remaining area of the window is sunlit and subject to the solar radiation heat gain for whatever direction it faces as shown in Table 2.

Example: A building located at 40° N latitude has an office wall facing the Southwest and containing a 6 ft x 6 ft clear glass window. The roof overhang, 1 ft above the top of the window, is 4 ft deep. What is the solar radiation heat gain through the window at 3:00 pm in summer?

Solution: (see Figure 2)

a) Length of overhang (4 ft) times factor from Table 4 (.86) equals 3.44 ft, distance shade line falls below bottom edge of overhang.

b) Distance shade line falls below overhang (3.44 ft) minus distance from bottom of overhang to top of window (1 ft) equals 2.44 ft, height of shaded portion of window.

c) Width of window (6 ft) times height of shaded portion (2.44 ft) equals area of shaded portion (14.64 sq ft).

d) Area of whole window (36 sq ft) minus area of shaded portion (14.64 sq ft) equals area of sunlit portion of window (21.36 sq ft).

e) Shaded area (14.64 sq ft) times solar gain for north glass in Table 2 (30 Btu/h per sq ft) equals shaded solar gain of 439 Btu/h.

f) Sunlit area (21.36 sq ft) times solar gain for southwest glass in Table 2 (187) equals sunlit gain of 3,990 Btu/h.

g) Total solar gain of window is 4,429 Btu/h. (Without shading, gain would be 6,732 Btu/h.)

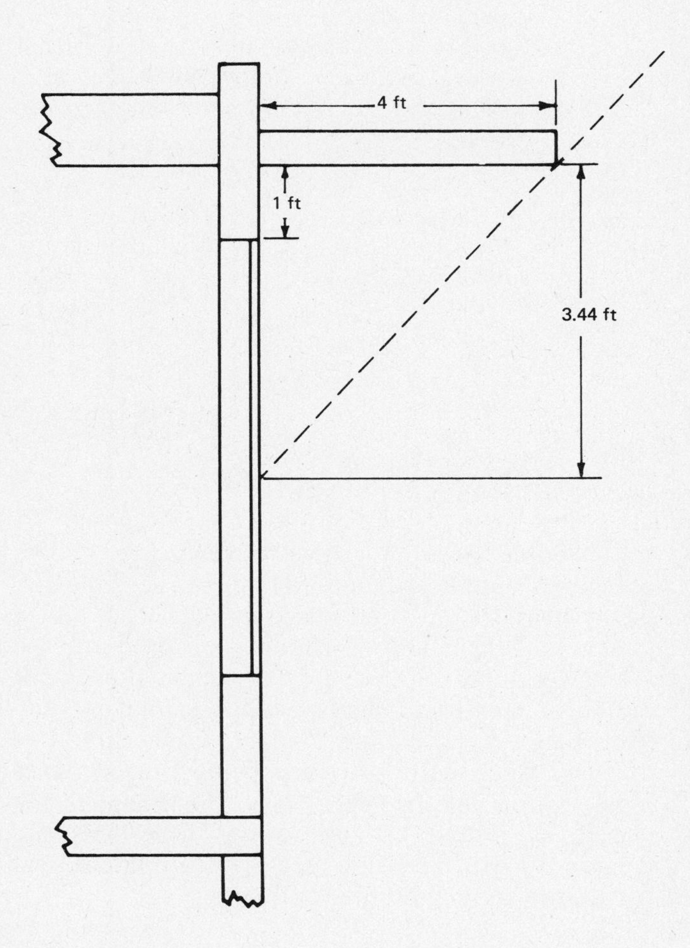

Figure 2 – Typical overhang

SECTION 8

HEAT TRANSMISSION

THROUGH STRUCTURAL COMPONENTS

8.1 Transmission of heat through structural components takes place in accordance with the following formula:

$$Q = U \times A \times T.D.$$

where

Q = rate of heat flow through the component in Btu/h

U = overall heat transmission factor for the structural component in Btu per hour per square foot per degree F temperature difference between inside and outside [Btu/(hr)(ft^2)(deg F)]

A = area of structural component surface exposed to inside and outside temperatures

T.D. = difference between indoor and outdoor temperatures, degrees F

Table 7 (page 477) shows the U-values for commonly used wall, roof-ceiling, and floor constructions. Table 5 (page 476) shows U-values for glass.

8.2 **EQUIVALENT TEMPERATURE DIFFERENCE.** Equivalent temperature difference applied to walls and roofs combines the effects of solar radiation, time lag or storage effect, and air temperature difference. It is that temperature difference which would alone produce the same rate of heat flow into a space without the radiation effect as actually occurs as a result of simultaneous radiation and convection. Tables 8 and 9 (pages 480 and 481) show equivalent temperature differences for various constructions at various times of day for walls and roofs respectively.

SECTION 9

DESIGN OCCUPANCY

9.1 People occupying the space to be cooled contribute significant amounts of both sensible and latent heat to the total space cooling load. Occupancy load calculations should be based on the average number of people in the space during the period of maximum design cooling load. Contributions to the total cooling load by people should be in accordance with activities being carried on as indicated in Table 10 (page 482). The heat gains for people in restaurant applications have been increased 30 Btu/h sensible and 30 Btu/h latent heat per person to include the food served.

Table 15 (page 486) shows estimates of occupancy in square feet per person to be used when more accurate data are not available.

9.2 Commercial and public assembly applications may or may not require operation of equipment 24 hours per day. In some, the equipment is turned off at the close of the business day and turned on again at or before the start of the next business day. In others, the equipment may be operated continuously to take full advantage of the flywheel effect of the structure and its contents. In special applications, including churches, occupancy may be only a few hours per week, and the space can be precooled, possibly below indoor design temperature, to better use the flywheel effect. Judgement is required in determining how design occupancy affects the equipment and system capacities.

SECTION 10

HEAT GAIN FROM OTHER INSIDE SOURCES

10.1 Among the sources of heat gain within the space to be cooled are lights, office machines, appliances, and electric motors. Table 12 (page 483) shows the heat gains for various restaurant appliances and other miscellaneous appliances, and Table 11 (page 482) shows heat gains from motors with alternative locations of motors and driven machinery.

10.1.1 Since much electric motor driven machinery is seldom in continuous use, it will usually be necessary to apply a use factor to the Btu/h heat gains shown in Table 11. If a motor runs approximately 3/4 of the time, the use factor is .75; if a motor runs half the time, the use factor is .50; etc.

10.1.2 The heat gains shown in Table 12 are for appliances located in conditioned spaces and are already corrected for average usage. No usage factor should be applied to values in Table 12.

10.1.3 When heat producing equipment such as steam tables, ranges, fryers, and the like are hooded, only a percentage of the heat output becomes a load in the space; however, the total load of the makeup air to offset the air exhausted through the hood must be calculated.

10.1.4 Heat given off by hot water or steam pipes serving heat producing machinery also must not be overlooked.

10.2 Usually the nameplate on the appliance, machine, or equipment will provide information necessary to approximate the heat gain from the apparatus. Where input wattage is given on the nameplate, treat it the same as incandescent lights by multiplying the watts by 3.4 to find the Btu/h. If motor horsepower is given on the nameplate, use Table 11 to find the Btu/h. If only motor voltage and current in amperes are shown on the nameplate, multiply the volts times the amperes times a reasonable power factor (from 0.6 for small motors to 0.9 for large motors) times 3.4 to find the Btu/h. In all cases, apply an approximate usage factor.

SECTION 11

INFILTRATION AND VENTILATION

11.1 **INFILTRATION** is the inward leakage of outside air to a space through cracks and joints around windows and doors and through floors and walls. The amount of infiltration depends on the type of construction and condition of a building.

11.1.1 Table 13 (page 484) shows values for infiltration estimates based on the air change method.

11.2 **VENTILATION** is the intentional displacement of inside air with outside air with or without the aid of a fan or fans. Usually, ventilation air is taken directly from the outdoors, mixed with return air, and passed through the heating or cooling equipment before being delivered to the conditioned space. The temperature of the mixture of outside and return air entering the heating equipment must be determined to establish the temperature of the air leaving the heating equipment for use in estimating duct heat loss.

11.2.1 Summer and winter requirements for ventilation may be specified by local codes and ordinances. In all cases, the requirements must be met. In the absence of such requirements, the values shown in Table 14 (page 485) should be used.

11.2.2 There are available filtering and odor removal devices, the use of which may greatly reduce ventilation requirements. Their use is recommended in the interest of energy conservation so long as applicable code and ordinance requirements are met.

11.3 Both infiltration and ventilation add significantly to the heating load and cooling load.

11.3.1 In accounting for the addition to the heating or cooling load resulting from infiltration and ventilation, use only the larger of the two loads, not both.

SECTION 12

DEHUMIDIFICATION AND HUMIDIFICATION

12.1 **DEHUMIDIFICATION.** The removal of moisture from air introduced into a conditioned space in summer is accomplished by the cooling coil in an air-conditioning unit. Dehumidification by this means is accomplished when the air is cooled to or below its dewpoint temperature thus condensing and draining off a portion of the water vapor in the air. Condensation of water vapor requires the removal of (latent) heat at the rate of approximately 1,070 Btu per pound of water condensed.

Table 16a(page487) shows the rate of heat removal per 100 ft^3/min of outdoor air required to reach indoor conditions of 75 F dry-bulb temperature and 50% relative humidity, and Table 16b (page 488) shows the heat removal rate per 100 ft^3/min to reach 80 F dry-bulb and 50% RH. Interpolation for room dry-bulb temperatures between 75 F and 80 F will yield results sufficiently accurate for ordinary load calculations.

Example: Assume that the outdoor design dry-bulb temperature is 94 F and the wet-bulb temperature is 78 F, the desired indoor conditions are 75 F dry-bulb temperature and 50% RH, and the ventilation rate is 1,200 ft^3/min. Find the dehumidification (or latent) load.

Solution: In Table 16a find 3,766 Btu/h for 94 F DB and 78 F WB. Since the 3,766 Btu/h is for 100 ft^3/min, multiply by 12 to arrive at 45,200 Btu/h dehumidification load.

12.2 **HUMIDIFICATION.** Although indoor relative humidity above 20% is indicated for maximum human comfort and for certain materials and processes, humidification in existing commercial buildings should be approached with caution. The humidity level that a building can tolerate without concealed condensation may be much lower than that indicated by visible condensation on window glass.

Table 21 (page 492) shows maximum indoor relative humidities to avoid condensation on window glass at various outdoor temperatures for both single and double glass. Humidification of buildings which have inadequate vapor barriers and insulation is not recommended because of possible structural damage resulting from concealed condensation.

12.2.1 The addition of humidity to outdoor air introduced into a conditioned space in winter requires the expenditure of energy in the form of heat. The heat required to evaporate water in a humidifer amounts to approximately 1,070 Btu per pound of water evaporated.

Tables 22a, 22b, and 22c (page 493, 494, and 495) show the rate at which water, in gallons per day per 100 ft^3/min, must be evaporated and supplied to the conditioned space to reach the desired indoor relative humidity level. Also shown in Tables 22a, 22b, and 22c are the required Btu/h heating rates for each humidification value.

Example: Assume that the outdoor design dry-bulb temperature is −10 F, that the ventilation air flow rate is 1,200 ft^3/min and that the desired indoor conditions are 72 F dry-bulb temperature and 20% relative humidity. Find the required humidification rate and humidification heating load.

Solution: In Table 22b, find 3.8 gallons/day and 1,410 Btu/h per 100 ft^3/min. Since there are 1,200 ft^3/min of outdoor air, multiply each of the values from the table by 12. The required design humidification rate is, therefore, 12 x 3.8 or 45.6 gallons per day; and the humidification heat load is 12 x 1,410 or 16,920 Btu/h.

SECTION 13

DUCT HEAT GAIN AND HEAT LOSS

13.1 Heat gain or loss through ducts located in unconditioned spaces can be appreciable. Neither can be accurately calculated until the duct system is designed, but it must be estimated and included in the heating and cooling load calculations. After the duct system has been designed, the duct heat gains and losses may be rechecked and adjustments made if necessary.

13.1.1 Chart 17a (page 489) shows the % of Subtotal Sensible Load that must be added for various lengths of duct and various Subtotal Sensible Loads to account for the heat gain to the supply air through the duct walls. The curves shown are applicable only to the conditions listed in the legend. Similarly, Chart 23a (page 496) shows the % of Subtotal Heating Load that must be added for various lengths of duct and various Subtotal Heating Loads to account for the heat loss from the supply air through the duct walls. The legend defines the conditions to which the curves apply.

13.1.2 For conditions other than those listed in the legends in Charts 17a and 23a, multipliers from the charts and tables following each Chart must be selected and applied to the % Addition found in the Charts. Discussions and examples showing how to select and apply the multipliers appear in steps (7) and (15) in Section 15, Calculation of Summer and Winter Design Loads (page 444 and 448, respectively).

SECTION 14

EQUIPMENT SELECTION (COOLING)

14.1 For a normal cooling application, the equipment capacity should match the cooling loads (sensible and latent) as closely as possible. Manufacturers' published ratings show sensible and latent heat capacities at various air flow rates (ft^3/min).

14.1.1 In order to properly select equipment, the total sensible and latent heat loads as well as the required ft^3/min air flow must be known.

14.2 Because most commercial applications involve intermittent operation of equipment as opposed to continuous (or 24-hour) operation on residential applications, equipment is not normally selected on the basis of an expected temperature swing as recommended in Manual J (Residential Load Calculation).

14.3 Many commercial applications require zoning because of varying occupancies and exposures. In using unitary equipment, the use of multiple units to serve the different zones will usually result in the maintenance of more acceptable room conditions.

14.4 Applications in geographic locations of high humidity also benefit from the use of multiple units as opposed to a single cooling unit. The use of a single cooling unit may result in satisfactory humidity control, however, if the unit is equipped with capacity control.

14.5 Multiple units may be superior to a single unit (not equipped with capacity control) in maintaining satisfactory indoor conditions under part load operation. Since outdoor design conditions occur for relatively few hours during a cooling season, equipment operates most of the time under part load conditions. Under part load conditions with multiple units, only as many units will operate as are necessary to carry the existing load. With a single unit under part load conditions, the unit will cycle on and off thus losing control of humidity in the conditioned space.

SECTION 15

CALCULATION OF SUMMER AND WINTER DESIGN LOADS

15.1 An air conditioning load calculation form is shown on pages 458 thru 461. The form is designed to make the load calculations easier to perform in a systematic manner. It may be used for all normal applications. For special or unusual applications, or where a more detailed analysis of the loads is desired, tables and methods contained in the *ASHRAE Handbook of Fundamentals* should be used.

15.1.1 The form is divided into two sections: one for cooling load calculations, the other for heating load calculations. For the usual heating application, no credit is taken for heat gains from people, lights, motors, appliances, or solar radiation since heat from such sources is not always available at the time of peak heat loss. However, these sources are important and should be considered for zoning purposes.

15.2 **COOLING LOAD.** The numbers of the following instructions correspond to the numbers of the several load components in the cooling load section of the form. If the total space is to be zoned, separate calculations should be made for each zone.

(1) **DESIGN CONDITIONS**: fill in the time of day, daily range, inside design DB and RH, and outside design DB and WB. Subtract the inside DB from the outside DB to find the DB temperature difference. See Table 1 (page 471) for outside design conditions.

(2) **SOLAR RADIATION HEAT GAIN THROUGH GLASS**: fill in the square feet of glass for each exposure, and multiply by the appropriate factors from Table 2 and Table 3 (page 475). If appreciable shading from overhang is present, use Table 4 (page 476) and the procedure described in Section 7 (page 439) to calculate solar loads.

(3) **TRANSMISSION GAIN**: fill in the total square feet of glass, the net area of walls, and the areas of partitions, roofs and/or ceilings, and floors. Select the proper heat transmission factors from Tables 5 and 7 (pages 476 and 477), and multiply by the applicable equivalent (TD) or dry-bulb temperature differences. Tables 8 and 9 (pages 480 and 481) show equivalent TD.

(4) INTERNAL HEAT GAINS:

a. Occupants: fill in the number of people expected to occupy the space, and multiply by the applicable sensible and latent heat quantities from Table 10 (page 482).

b. Lights and Other: fill in the total wattage for incandescent lights and fluorescent lights, and multiply by the factors shown on the form.

Find in Table 11 (page 482) the horsepower of electric motors in use during peak load conditions in the conditioned space, and enter on the form the appropriate heat gains shown in Table 11. Multiply by an appropriate use factor.

Fill in the appropriate sensible and latent loads from Table 12 (page 483) for appliances in use during the time of peak load.

Add any other coincidental loads not already accounted for.

(5) **INFILTRATION OR VENTILATION**: determine the ft^3/min of infiltration and ventilation air from Tables 13 and 14 (pages 484 and 485), and enter the larger of the two on the form. Enter the dry-bulb temperature difference in the sensible load line and perform the indicated multiplication. In Table 16a or 16b (pages 487 and 488), find the latent heat load corresponding to the indoor and outdoor conditions, and enter it in the latent load line. Perform the indicated computation.

(6) **SUBTOTAL SENSIBLE LOAD**: the purpose of subtotalling the sensible loads at this point is to provide a basis for using Chart 17a to estimate the duct heat gain as required in step (7).

(7) **DUCT HEAT GAIN**: (see Section 13, page 443). Following is the procedure to be used in estimating the duct heat gain:

A. Find the subtotal sensible heat load on the scale at the bottom of Chart 17a. Follow a vertical line from the sensible heat load to its intersection with the applicable duct length curve. If the duct length does not correspond with any of the curves shown on the chart, estimate the location of the applicable length curve.

B. From the intersection of the vertical cooling load line with the duct length curve, follow a horizontal line to the scale on the left side of the chart. Read the % Addition to be made to the Subtotal Sensible Load.

C. Enter the % Addition figure in the space provided in #7 on the load calculation form.

NOTE: It is seldom that the conditions noted in the legend in Chart 17a correspond to the conditions of the job for which the cooling load is being calculated. The variables likely to be encountered are the following:

1. velocity of the air in the duct
2. temperature of the air surrounding the duct
3. temperature of the air entering the duct
4. design dry-bulb temperature of the room or space to be conditioned
5. aspect ratio of the duct (or round versus rectangular duct)
6. insulation applied to the duct

The variables noted above must be fixed by calculation, measurement, or assumption in order to arrive at a useful estimate of the duct heat gain.

Following are the considerations that relate to fixing the variables:

1. **Air Velocity**—On the one hand, to minimize duct heat gain and duct sizes, maximum air velocities are indicated. On the other hand, to minimize noise generation and fan horsepower, minimum duct velocities are indicated. Reasonable compromises that are known to have given satisfactory results are indicated in Table 18, page 491. Chart 17b shows multipliers to be applied to the % Addition for velocities other than 1500 ft/min.

2. **Surrounding Air Temperature**—Table 19 on page 491 will serve as a guide in estimating the temperature of air surrounding ducts in various environments. Measurement of actual temperatures in environments similar to that in which the ducts are to be located is recommended. Chart 17c shows multipliers to be applied to the % Addition for various surrounding air temperatures.

3. **Entering Air Temperature** –
Except for air entering a return air duct or a branch supply duct, the temperature of the air entering a supply duct is the same as the temperature of the air leaving the air conditioning unit. This latter air temperature depends on the temperature of the air entering the equipment (usually a mixture of outside and return air), the operating characteristics designed into the unit by the manufacturer, and the ratio of the room sensible heat to the total (sensible plus latent) heat. Normally, the temperature of the supply air entering a duct from an air conditioning unit will be from 50 F to 60 F. Neglecting to accurately estimate the temperature of the air entering the duct directly from the unit will ordinarily result in an error of not more than 2 or 3% in the supply duct heat gain estimate, well within the range of probable error in other estimates used in making a cooling load calculation. Consequently, no multiplier chart or table is provided in conjunction with Chart 17a to correct the duct heat gain % Addition for various entering air temperatures in supply ducts.

4. **Room Temperature**—The curves in Chart 17a are based on a room design temperature of 78F. For other room design temperatures, find the corresponding multipliers in Chart 17d on page 490. Apply these multipliers to the *supply duct* % Addition as indicated on the load calculation form.

   The temperature of room air entering *return ducts* may be from 75F to 80F. Such temperatures are significantly greater than the 55F air temperature on which the curves in Chart 17a are based, and a multiplier must be applied to the % Addition from Chart 17a. Multipliers for various return air temperatures are shown in Chart 17f on page 491.

5. **Aspect Ratio**—The aspect ratio of a rectangular duct is the ratio of its long side to its short side. The rate at which heat is transferred through the walls of a duct is proportional to the perimeter of the duct. A square duct (aspect ratio = 1) will have a lower heat gain than any rectangular duct of the same cross-sectional area, all other factors being equal. A round duct will have a lower heat gain than a square duct (all other factors equal) as well as any rectangular duct.

   Chart 17e shows multipliers for round, square, and rectangular ducts with aspect ratios up to 8. As in the case of the other multipliers associated with Chart 17a, aspect ratio multipliers are to be applied to the % Addition found in Chart 17a when a duct with an aspect ratio other than 2 is chosen. Generally, economy of materials and minimizing duct heat loss dictate the use of as low an aspect ratio as possible (round duct preferred), but space limitations often require the use of rectangular duct, sometimes with a rather large aspect ratio. The use of an aspect ratio larger than 8 is not recommended.

6. **Duct Insulation**—The duct heat gain curves in Chart 17a are based on sheet metal ducts with various thicknesses of flexible blanket composed of fibers processed from rock, slag, or glass *or equivalent* insulating material to provide an appropriate overall heat transmission coefficient (U-value). In those cases where duct heat gain will be significant, it is recommended that no less than 2 inches *or equivalent* of insulation be used.

   For load estimation purposes, fibrous blanket insulation of 2 inch thickness *(or equivalent)* may be safely assumed. If final duct heat gain/loss calculations are deemed worthwhile after the duct system is designed, appropriate corrections can be made for other combinations of duct material and insulation.

EXAMPLE—Duct Heat Gain

Given: Industrial Building

Inside design dry-bulb temperature, 80F

Outside design dry-bulb temperature, 95F

Subtotal sensible cooling load, 90,000 Btu/h

Supply duct length (main trunk), exposed to well ventilated warehouse space, 60 ft

Return duct length, also exposed to well ventilated warehouse space, 50 ft

Duct insulation, 2 in. glass fiber blanket

Selected aspect ratio of duct, round duct

Find: Duct heat gain, supply and return

SOLUTION:

A. In Chart 17a (page 489) locate the Subtotal Sensible Load of 90,000 Btu/h on the scale at the bottom of the chart. Follow a vertical line upward to its intersection with the 50 ft and 70 ft duct length curves. Locate a point on the vertical line halfway between the two curves.

   From the point thus located, follow a horizontal line to the scale on the left side of the chart, and read the % Addition of 1.7%. Enter the 1.7 on the load calculation form (see page 459).

B. Turn to Table 18 (page 491) and select a duct velocity compatible with the job requirements. The application is in an industrial building where, in this example, it is assumed that noise is not a factor but sheetmetal cost is. A velocity of 2,000 ft/min is selected.

   In Chart 17b for Velocity Multipliers, find 2,000 ft/min on the bottom scale. Follow a vertical line upward to its intersection with the curve. From the intersection with the curve, follow a horizontal line to the lefthand scale, and read the multiplier .86. Enter the .86 on the load calculation form.

C. Turn to Table 19 (page 491) and select a temperature for the air surrounding the duct. It would be preferable to measure the air temperature in the warehouse space. In this example, it is assumed that it is not possible to measure the temperature, and an air temperature of 95 F is selected.

In Table 17c for Surrounding Air Multipliers, estimate (interpolate) the 1.15 multiplier for 95 F surrounding air. Enter the 1.15 on the load calculation form.

D. In Chart 17d for Room Temperature Multipliers, select the .96 multiplier corresponding to 80 F. Enter the .96 on the load calculation form.

E. From Chart 17e for Aspect Ratio Multipliers, read the .84 multiplier for round duct. Enter the .84 on the load calculation form.

F. Apply the four multipliers to the 1.7% factor taken from Chart 17a as indicated on the load calculation form as follows: 1.7% x .86 x 1.15 x .96 x .84 = 1.36%, the % Addition to be made to the Subtotal Sensible Load to account for the supply duct heat gain.

G. Return duct heat gain is estimated in a manner entirely similar to the method used in estimating supply duct heat gain, with one exception: the temperature of the air leaving the conditioned space and entering the return duct is appreciably higher than the supply air temperature of 55 F entering the supply duct on which the curves in Chart 17a are based. Consequently, in addition to the multipliers for velocity, surrounding air temperatures, and aspect ratio, a multiplier for the temperature of the air entering the return duct must be applied to the % Addition from Chart 17a. The applicable multipliers are found in the table of Return Air Multipliers, Chart 17f.

The complete return duct heat gain calculation is as follows:

1) Chart 17a, read the % Addition of 1.4% corresponding to the 90,000 Btu/h Subtotal Sensible Load and 50 ft length.

2) Apply the velocity multiplier .86 as found in Step B above for the supply duct.

3) Apply the surrounding air multiplier 1.15 found in Step C above for the supply duct.

4) Apply the .84 aspect ratio multiplier for round duct found in Step E above for the supply duct.

5) Select and apply the .29 multiplier corresponding to the 80 F inside design temperature from the table of Return Air Multipliers, Chart 17f.

6) The complete return air duct heat gain calculation as provided for on the load calculation form (page 461) is as follows:

   1.4% x .86 x 1.15 x .84 x .29 = .34%, the % Addition to be made to the Subtotal Sensible Load to account for the return duct heat gain.

H. The total duct heat gain to be entered on the load calculation form is obtained by adding the supply duct Adjusted % Addition to the return duct Adjusted % Addition, dividing the sum by 100, and multiplying by the Subtotal Sensible Load, as indicated on the form. In this example the calculation is as follows:

$$(90{,}000 \div 100) \times 1.70 = 1530 \text{ Btu/h}$$

(8) **TOTAL SENSIBLE AND LATENT LOADS**: add the subtotal sensible load to the duct heat gain to obtain the total sensible load, and total the latent loads.

(9) **TOTAL COOLING LOAD**: add the total sensible and total latent loads to find the Total Cooling Load.

15.3 HEATING LOAD. The numbers of the following instructions correspond to the numbers of the several load components in the heating load section of the form. If the total space is to be zoned, separate calculations should be made for each zone.

(10) **DESIGN CONDITIONS**: fill in the inside and outside design dry-bulb temperatures and subtract to arrive at the design difference. See Table 1 (page 470) for outside design temperatures.

(11) **TRANSMISSION LOSSES**: fill in areas (in square feet) of windows, walls, roofs, floors, and partitions; use the heat transmission factors from Tables 5 and 7 (pages 476 and 477) and the temperature difference from #10

above. Perform the multiplications indicated to arrive at the transmission heat losses of the several structural components.

(12) **INFILTRATION OR VENTILATION**: determine the infiltration and ventilation air quantities from Tables 13 and 14 (pages 484 and 485). Fill in the larger of the two quantities, and multiply by the design temperature difference and the factor shown on the form to obtain the heating load.

(13) **HUMIDIFICATION**: if the space is to be positively humidified during the heating season, find in Table 21 (page 492) the maximum relative humidity to avoid condensation on windows. Find in Table 22a or 22b (pages 493 and 494) the maximum relative humidity, or the next lower one shown in the tables, and find the humidification heating load opposite the outdoor design dry-bulb temperature. Divide the same ft^3/min as used in #12 above by 100, and multiply by the Btu/h found in Table 22a or 22b to find the total humidification heating load.

Enter the required gallons of water/day on the form for future reference in selecting a humidifier.

(14) **SUBTOTAL HEATING LOAD**: add all of the individual heat losses calculated thus far, and enter the total on the form.

(15) **DUCT HEAT LOSS**: (see Section 13, page 443.)

A. **If a cooling load has been calculated**, use the following procedure:

1. Calculate the *air flow* rate (ft^3/min) through the heating unit by assuming 400 ft^3/min per 12,000 Btu/h of Total Cooling Load.

2. Finding the *temperature rise* through the heating unit by performing the following steps:

(a) Establish the temperature of the mixture of return air and outside air entering the heating unit using the formula

$$t_M = \frac{(A_V \times t_V) + (A_R \times t_R)}{A_T}$$

where

t_M = temperature of mixture of outside and return air (deg F)

A_V = ventilation air flow rate (ft^3/min)

t_V = temperature of ventilation air (deg F, outside design temperature)

A_R = return air flow rate (ft^3/min) ($A_R = A_T - A_V$) see A_T below

t_R = temperature of return air (deg F)

A_T = total air flow rate through heating unit (ft^3/min from step 1 above)

(b) Make a tentative selection of the heating unit from a manufacturer's catalog. Choose a unit that is capable of providing the necessary air flow rate (A_T) and from 110% to 120% of the Subtotal Heating Load.

(c) Calculate the temperature rise through the heating unit with the formula

$$TR = \frac{Q}{1.1 \times A_T}$$

where

TR = temperature rise (deg F)

Q = heating unit capacity (Btu/h)

A_T = total air flow rate through heating unit (ft^3/min)

3. Find the *temperature of the air leaving* the heating unit, or entering the supply duct, as follows:

$$t_L = t_M + TR$$

where

t_L = temperature of air entering the duct (deg F)

t_M = temperature of air entering heating unit (deg F)

TR = temperature rise through heating unit (deg F from step 2 above)

B. **If a cooling load has not been calculated,** use the following procedure:

1. Make a *tentative selection* of the heating unit from a manufacturer's catalog. Choose a unit capable of providing from 110% to 120% of the Subtotal Heating Load.

2. Assume an air *temperature rise* through the heating unit. A temperature rise of 50 to 100 degrees (F) should be acceptable. A check of acceptability can be made later.

3. Calculate the *air flow* rate through the heating unit using the formula

$$A_T = \frac{Q}{1.1 \times TR}$$

where

A_T = air flow rate through heating unit (ft^3/min)
Q = output capacity of heating unit (Btu/h)
TR = temperature rise of air through heating unit (deg F from step 2 above)

4. Find the *temperature of the mixture* of ventilation air and return air entering the heating unit by means of the formula

$$t_M = \frac{(A_V \times t_V) + (A_R \times t_R)}{A_T}$$

where

t_M = temperature of mixture of outside (ventilation) air and return air (deg F)
A_V = ventilation air flow rate (ft^3/min)
t_V = temperature of ventilation air (deg F, outside design temperature)
A_R = return air flow rate (ft^3/min) ($A_R = A_T - A_V$) see A_T below
t_R = temperature of return air (deg F)
A_T = total air flow rate through heating unit (ft^3/min from step 3 above)

5. Find the *temperature of the air leaving* the heating unit, or entering the supply duct, with the formula

$$t_L = t_M + TR$$

where

t_L = temperature of air leaving heating unit (deg F)
t_M = temperature of air entering heating unit (deg F from step 4 above)
TR = temperature rise through heating unit (deg F from step 2 above)

Check the acceptability of the TR assumed in step 2 by subtracting the inside design temperature from the leaving air temperature found in step 5. If the result is between 35 and 80 degrees (F), the assumed TR is acceptable. If the result is appreciably less than 35 degrees, the room may be drafty. If the result is much over 80 degrees, the room may be stuffy. To reduce the risk of draftiness or stuffiness, adjust the TR assumed in step 2, and repeat steps 3 through 5.

C. Now that the temperature of the air entering the supply duct has been determined by following the procedure under either A or B above, it can be used later in finding the duct heat loss with the following procedure:

1. In Chart 23a, page 496, locate the Subtotal Heating Load on the scale at the bottom of the chart. Follow a vertical line from the heating load to its intersection with the applicable duct length curve. If the duct length does not correspond with any of the curves shown on the chart, estimate (interpolate) the location of the applicable curve.

2. From the intersection of the vertical heating load line with the duct length curve, follow a horizontal line to the scale on the left side of the chart. Read the % Addition to be made to the Subtotal Heating Load.

NOTE: As in calculating duct heat gain for cooling load calculations, it is seldom that the conditions shown in the legend of Chart 23a will correspond to those of the job at hand. The same variables listed as applying to duct heat gain (page 443) apply also to duct heat loss.

Following are the considerations related to fixing the variables:

(a) **Air Velocity**—To minimize duct heat loss and duct sizes, maximum air velocities are required. To minimize noise and fan horsepower, however, minimum velocities are required. Table 18 (page 491) shows velocities for various applications that experience has shown give satisfactory results.

If a cooling load has been calculated, use the same velocity for heating. For air velocities other than 1500 ft/min, select a multiplier from Chart 23b.

(b) **Surrounding Air Temperature**—Table 24 on (page 448) will serve as a guide in estimating the temperature of the air surrounding ducts in various environments. Measurement of actual temperatures in environments similar to that in which the duct will be located is recommended.

Chart 23c, shows multipliers for various surrounding air temperatures. Select, directly or by interpolation, a multiplier corresponding with the expected surrounding air temperature.

(c) **Room Temperature**—The curves in Chart 23a are based on a room design temperature of 70F. For other room design temperatures, find the corresponding multipliers in Chart 23d on (page 497). Apply these multipliers to the *supply duct* % Addition as indicated on the load calculation form.

(d) **Entering Air Temperature**—The temperature of air entering a supply duct as it leaves a heating unit varies much more widely from job to job than the temperature of air leaving a cooling unit. Heating air temperatures range from 90 F to 150 F or more, while cooling air temperatures usually fall in the range of 50 F to 60 F.

Consequently, for air temperatures determined in either A or B above that are more than a few degrees from 120 F (the entering air temperature on which the curves in Chart 23a are based), a multiplier should be selected from Chart 23e and applied to the % Addition.

The temperature of air entering a *return* duct is usually the same as the room air temperature. Table 23g (page 498) lists multipliers to be applied to the % Addition for various return air temperatures. See step H4 (page 451) for example.

(e) **Aspect Ratio**—The remarks that appear in the first paragraph under the heading of aspect ratio on page 14 apply to duct heat loss as well as duct heat gain with the substitution of the term "heat loss" for "heat gain" wherever it appears. Chart 23f shows multipliers to be applied to ducts with aspect ratios other than 2.

(f) **Duct Insulation**—The Remarks under the heading of duct insulation on page 446 also apply to duct heat loss.

3. Enter the supply and return duct heat loss % Addition figures from Chart 23a along with each of the required multipliers, perform the indicated multiplications, and add the results to arrive at the combined supply and return Adjusted % Addition. Apply the latter total to the Subtotal Heating Load as indicated on the form.

EXAMPLE—Duct Heat Loss

Given: Same industrial building as in example for duct heat gain, page 446

Inside design temperature, 68F

Outside design temperature, 15F

Ventilation air flow, 500 ft^3/min

Subtot. Heating Load, 200,000 Btu/h

Supply duct length (main trunk), exposed to warehouse space heated to 55F, 60 ft

Return duct length, same exposure as supply, 50 ft

Duct insulation, 2 in. glass fiber blanket

Selected aspect ratio of duct, round duct

Find: Duct heat loss, supply and return

SOLUTION:

A. In Chart 23a (page 496), locate the Subtotal Heating Load of 200,000 Btu/h on the scale at the bottom of the chart.

Follow a vertical line upward to its intersection with the 50 ft and 70 ft duct length curves. Locate a point on the vertical line halfway between the two curves. From the point thus located, follow a horizontal line to the scale on the left side of the chart, and read the % Addition factor of 1.5. Enter the 1.5 on the load calculation form (see page 459).

B. In this example, a cooling load has been calculated, and the same 2000 ft/min velocity is used. See step B, page 448 and step (a), page 448. Enter the same .86 multiplier on the load form.

C. In this example, the temperature of the air surrounding the duct is assumed to be 55 F because the duct will be located in the warehouse space which will be heated to 55 F in the winter.

In the table for Surrounding Air Multipliers, Chart 23c, interpolate the .93 multiplier for 55 F surrounding air temperature. Enter the .93 on the load calculation form.

D. The room design temperature, in this example, is given as 68F. In Chart 23d, find the .98 multiplier corresponding to 68F, and enter the .98 on the load calculation form.

E. In this example, the cooling load has been calculated. It will be necessary, therefore, to use the procedure described under step A (page 448) to determine the temperature of the air entering the supply duct (leaving the heating unit). The calculations are performed as follows:

1. Find the total air flow through the heating unit. For this example, total cooling load is assumed to be 120,000 Btu/h. The air flow rate through the heating unit is 400 ft^3/min per 12,000 Btu/h of Total Cooling Load as indicated in step 1, page 448. The total air flow through the heating unit is calculated as follows:

$$\frac{400\ ft^3/min}{12,000\ Btu/h} \times 120,000\ Btu/h = 4,000\ ft^3/min$$

2. Find the temperature of the mixture of ventilation air (outside air) and return air entering the heating unit. Use the formula given in step 2(a) on page 448:

$$t_M = \frac{(A_V \times t_V) + (A_R \times t_R)}{A_T}$$

In this example A_V was given as 500 ft^3/min, t_V = 15 F, t_R = 68 F, and A_T = 4,000 ft^3/min (from step 1 immediately above).

$$A_R = A_T - A_V \text{ or}$$
$$A_R = 4,000\ ft^3/min - 500\ ft^3/min$$
$$A_R = 3,500\ ft^3/min$$

then

$$t_M = \frac{(500 \times 15) + (3,500 \times 68)}{4,000} = 61F$$

3. Find the temperature rise through the heating unit. Assume the selection of a heating unit having an output capacity of 230,000 Btu/h (115% of the Subtotal Heating Load) at the required 4,000 ft^3/min air flow. The Q = 230,000 Btu/h, and the temperature rise through the heating unit is calculated in accordance with step 2(c) on page 448 as follows:

$$TR = \frac{Q}{1.1 \times A_T}$$

$$= \frac{230,000\ Btu/h}{1.1\ \frac{Btu/h}{deg\ F - ft^3/min} \times 4,000\ ft^3/min}$$

$$= 52\ deg\ F$$

4. The temperature of the air leaving the heating unit is calculated, using the formula given in step 3 on page 448, as follows:

$$t_L = t_M + TR$$
$$= 61 + 52 = 113\ F$$

In Chart 23e read the .97 multiplier for 113F entering air temperature, and enter it on the load calculation form.

F. On Chart 23f, *Aspect Ratio* Multipliers, read the multiplier for round duct (.84), and enter it on the load calculation form.

G. Perform the multiplication indicated on the load calculation form, and enter the result (.96) in the space provided for *Adjusted % Addition*.

H. Return duct heat loss is estimated in a similar manner as follows:

1. In Chart 23a, read the *% Addition* of 1.3 corresponding to the Subtotal Heating Load of 200,000 Btu/h and the duct length of 50 ft.

2. Enter on the form the same .86 *velocity* multiplier as found in step B above.

3. Enter on the form the .93 *surrounding air* multiplier found in step C above.

4. In Chart 23g, *Return Air* Multipliers, find the .25 multiplier corresponding to 68 F, and enter the .25 on the load calculation form.

5. Enter on the form the *aspect ratio* multiplier, .84, found in step F above.

6. Perform the multiplication indicated on the form, and enter the result (.22) in the space provided for *Adjusted % Addition*.

I. *Total* the Adjusted % Additions for *supply and return* ducts, and enter the resulting 1.18% in the indicated space. Perform the required multiplication by the 200,000 Btu/h Subtotal Heating Load and division by 100, and enter the result (2360 Btu/h) in the Heating Load column.

(16) **TOTAL HEATING LOAD**: add the Duct Heat Loss (2360) to the Subtotal Heating Load (200,000) to arrive at the Total Heating Load of 202,360 Btu/h.

SECTION 16

LOAD CALCULATION EXAMPLES

16.1 Following are two sample heating and cooling load calculations, one for a drugstore and one for a furniture store.

EXAMPLE 1

16.2 The Rex drugstore is located in Atchison, Kansas. The floor plan is shown in Figure 3.

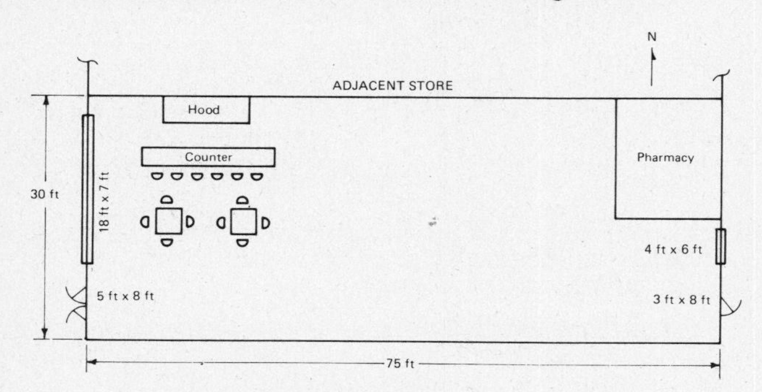

Figure 3 – Rex Drugstore

Details of construction are as follows:

Walls: 4 in. brick veneer, frame construction, ½ in. gypsum wallboard inside finish, R-7 insulation; dark color.

Roof-ceiling: built-up, flat, dark color roof on 1 in. wood deck with 2 in. insulation. Ceiling height, 10 ft.

Floor: wood frame construction with 2½ in. batt insulation over 8 ft basement.

Windows: single plate glass. West window protected by ventilated awning. East window has inside venetian blind.

Doors: single glass, tempered plate. East door, normally closed.

Other design details are as follows:

Lighting: 6,000 watts fluorescent.

Appliances: 1 gas-fired coffee brewer-warmer; 1 gas-fired steam table, 8 ft^2; hood over both appliances, 600 ft^3/min exhaust; 1 refrigerator, 1/4 hp and 1/8 hp electric motors.

Occupancy: 20 people at noon (including 1 person in pharmacy). Average time of occupancy, 1 hr.

Inside Design Conditions: winter, 72 F and maximum RH to avoid condensation on windows; summer, 75 F DB and 50% RH.

Calculate the cooling load and the heating load. The basement is to be heated in the winter but not cooled in the summer.

The following steps correspond to the numbers on the load calculation form shown on pages 26 and 27.

16.2.1 COOLING LOAD

(1) *Design conditions:* see Table 1 (page 471) for the latitude, the daily range, and the design outdoor dry-bulb and wet-bulb temperatures for Atchison, Kansas: 40° latitude, 23 degree daily range, 95F DB, and 78F WB. Enter the values thus found and the inside design conditions in the appropriate spaces, and also fill in the dry-bulb temperature difference. (Selection of time of day is discussed in the following step.)

(2) *Solar radiation heat gain through glass:*

(a) Enter the area of the east window (4 x 6 = 24 ft^2) on the form.

(b) Enter the area of the east door (3 x 8 = 24 ft^2) on the form.

(c) Enter the area of the west door (5 x 8 = 40 ft^2).

(d) Enter the area of the west window (18 x 7 = 126 ft^2) on the form. *It is considered a north facing window because of the ventilated awning which shields the window from direct solar radiation.*

(e) See Table 2 (page 475) to find the solar radiation heat gain factors for the glass.

The question of time of day of the peak cooling load now arises. The peak load might occur at noon because of maximum occupancy and appliance use, or it might occur at 3:00 pm because of peak outside temperatures and solar effects. The only way to answer the question with certainty is to calculate the loads for both times. For this example, only the noon load calculation will be explained in detail. The results of the 3:00 pm load calculation are shown on the load calculation form on pages 475 and 476.

The noon factors under 40° latitude from Table 2 and the correction factors from Table 3 (page 475) are as follows:

(window)	(E)	68 and .64 for shading correction for plate glass
(door)	(E)	68 and .95 for plate glass
(door)	(W)	36 and .95 for plate glass
(west window)	(N)	34 and .95 for plate glass

Multiplication of the areas of the windows by the radiation and correction factors yields the following loads:

(window)	(E)	24 x 68 x .64 = 1,044 Btu/h
(door)	(E)	24 x 68 x .95 = 1,550 Btu/h
(door)	(W)	40 x 36 x .95 = 1,368 Btu/h
(west window)	(N)	126 x 34 x .95 = 4,070 Btu/h

(3) *Transmission gains:*

(a) Glass:

Total the glass areas without inside shading (126 + 24 + 40 = 190 ft^2), and enter on the form. Enter the area of the shaded east window (24 ft^2) on the form. In Table 476 (page 44) find the U-values for the glass: 1.06 for the unshaded and .81 for the shaded. Enter the U-values and the dry-bulb temperature difference for noon.

To find the dry-bulb temperature difference for noon, first find the dry-bulb temperature at noon in accordance with Table 476 (page 44), and subtract from it the indoor design dry-bulb temperature. Calculate as follows: 95 – (23 x .23) = 95 – 5.29 = 89.71 or 90 F. Outdoor temperature at noon minus room temperature = dry-bulb temperature difference for noon, 90 – 75 = 15 F. Enter the 15 F temperature difference on the form.

Make the indicated multiplications to arrive at the 3,020 Btu/h unshaded window load and the 292 Btu/h shaded window load.

(b) Walls:

Enter on the form the net wall area for each of the three exposures: east, 300 – 48 = 252 ft^2; south, 750 ft^2; west, 300 – 166 = 134 ft^2. In Table 7 (page 477) find the U-value for the walls, .09. In Table 8 (page 480), for Light Construction and dark color, find the equivalent temperature differential for each wall exposure at noon: east, 38; south, 27; and west, 24. Because the indoor-outdoor temperature difference is only 15 deg F as determined in step (3) above, subtract 5 degrees from each of the factors from Table 8: east, 38 – 5 = 33; south, 27 – 5 = 22; and west, 24 – 5 = 19. Enter the latter values on the form. Multiply as follows to arrive at the three wall loads.

East:	252 x .09 x 33 = 748 Btu/h
South:	750 x .09 x 22 = 1,485 Btu/h
West:	134 x .09 x 19 = 229 Btu/h

(c) Roof/Ceiling:

Enter the area of the roof/ceiling (2,250 ft^2). Find the U-value (.11) for the roof/ceiling in Table 7 (page 477), and enter on the form. Find in Table 9 (page 481) the equivalent temperature difference under Dk (dark) at 12 noon. The equivalent TD is 65 deg which must be corrected in accordance with note #1 for the noon 15 deg F outdoor-indoor temperature difference as follows: 65 – 5 = 60 deg.

Enter the 60 on the form, and multiply 2,250 x .11 x 60 to arrive at the load of 14,850 Btu/h.

(d) Floor:

Enter the area of the floor (2,250 ft², same as ceiling) on the form. Find the U-value (.09) in Table 7 (page 477), and enter on the form. Enter the dry-bulb temperature difference (15 degrees) determined in step (3a, 2nd ¶) above, and perform the indicated multiplication to find the 3,038 Btu/h heat gain. (It is assumed, in this example, that the air in the basement will be outside air because of ventilation.)

(4) *Internal heat gains—people:* See Table 10 (page 482) for the heat gains from occupants. Assume, since the time is noon, that 14 of the 20 people are eating and the remaining six are standing or walking around the store. The sensible and latent heat values to enter on the form are 275 and 250 sensible and 275 and 250 latent, respectively.

Multiply the entries as follows to find the loads:

Sensible, eating = 14 x 275 = 3,850 Btu/h
walking = 6 x 250 = 1,500 Btu/h
Latent, eating = 14 x 275 = 3,850 Btu/h
walking = 6 x 250 = 1,500 Btu/h

Internal heat gains—lights and others:

(a) Enter on the form 6,000 watts for fluorescent lights, and multiply by 4.1 to find the lighting load of 24,600 Btu/h.

(b) In Table 11 (page 482), find the Btu/h values for the 1/4 and 1/8 hp refrigerator motors, and enter them on the form (1,000 and 580 respectively). The usage factor choice is a matter of judgement. In this case a value of .75 seems reasonable during the noon hour. Multiply the figures as follows:

1/4 hp: 1,000 x .75 = 750 Btu/h
1/8 hp: 580 x .75 = 435 Btu/h

(c) Appliances: In Table 12 (page 483), find the heat gains from the coffee maker and the steam table. Since the appliances are served by an exhaust hood, the loads will be sensible only. For the coffee maker, the gain is 500 Btu/h. For the steam table, the gain is calculated as follows: 8 x 250 = 2,000 Btu/h.

(5) *Infiltration or ventilation:*

Infiltration:

See Table 13 (page 484) for air changes per hour for infiltration. The value shown is 1.2; therefore, the ft³/min for infiltration is calculated as follows:

$$\frac{10 \times 75 \times 30 \times 1.2}{60} = 450 \text{ ft}^3/\text{min}$$

Add the infiltration for the west door as indicated in Table 13 as follows:

$$\text{TR} = \frac{20 \text{ persons}}{2 \text{ doors} \times 1 \text{ hr (avg. stay)}} = 10$$

TD = [95 – (23 x .23)] – 75 = 15. (See Table 6, page 44.) (Use 20.)
Inf = 2 doors x 8 ft³/(min)(door)
= 16 ft³/min

The total infiltration is 450 + 16 = 466 ft³/min.

Ventilation:

In Table 14 (page 485) find the ventilation requirements for occupants as follows:

Pharmacist:	1 x 20 =	20
Sales area:	5 x 7 =	35
Eating area:	14 x 30 =	420
Total		475 ft³/min

The exhaust hood requires 600 ft³/min of makeup air which is more than adequate for the ventilation requirements.

Since ventilation exceeds infiltration, enter 600 ft³/min and 15 F dry-bulb temperature (corrected for time-of-day per Table 6) difference in the sensible load line, and multiply as indicated (600 x 15 x 1.1 = 9,900 Btu/h) to find the outside air sensible load.

Find the 4,220 Btu/h value for 90 F DB (noon temperature) and 78 F WB in Table 16a (page 55), and enter on the latent line. Multiply and divide as indicated to find the latent load of 25,320 Btu/h.

(6) *Subtotal Sensible Load:* add the outside air sensible load and the page 1 subtotal (65,329 Btu/h) to arrive at the 75,229 Btu/h Subtotal Sensible Load.

(7) *Duct heat gain:*

(a) A judgement must be made as to whether or not any ductwork outside the conditioned space will be exposed to a surrounding air temperature higher than that of the supply air inside the ducts. In this example, it is reasonable to assume that the ducts will be located in the unconditioned, well ventilated basement.

(b) With the above assumption made, it is necessary to choose a tentative location of equipment and layout of ductwork. In this example, it would be advantageous to choose two systems for better partial load operation. A possible layout of the systems is shown in Figure 4.

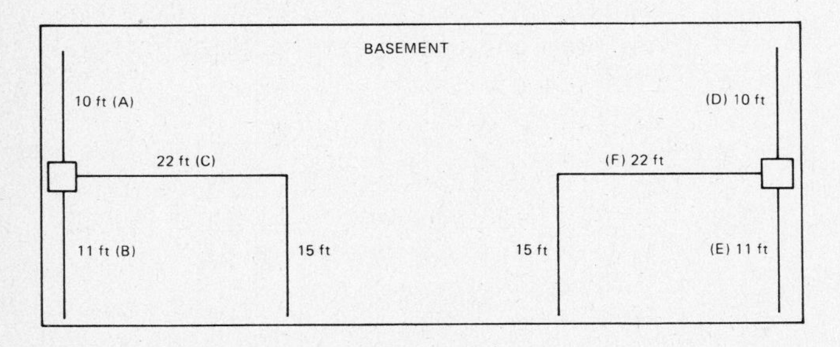

Figure 4 — Possible Duct Layout

The return grilles are assumed to be at or very close to the units. Duct insulation is assumed to be 2 in. of glass fiber blanket.

(c) To perform the calculations for duct heat gain, follow the procedure described in step (7), Section 15 (page 444), with the additional considerations indicated below:

1. Set up a schedule as shown in Figure 5, and fill in the indicated information for each duct shown in Figure 4. Notice that reasonable assumptions must be made, namely, that the load will be divided equally between the two systems and that the load on each system will be divided equally among the three ducts.

Duct	Sensible Load	Length	% Addition	Multipliers				Adjusted % Addition	Duct Heat Gain
				Velocity	Surrounding Air	Room Temperature	Aspect Ratio		
A	16,100	10 ft	0.7	1.3	---	1.08	.84	.83	134
B	↓	11 ft	0.7	↓		↓	↓	.83	134
C		37 ft	2.5					2.95	475
D		10 ft	0.7					.83	134
E		11 ft	0.7					.83	134
F		37 ft	2.5					2.95	475
								Total	1,486

Figure 5 — Duct Heat Gain Schedule

2. With these assumptions made, the % Addition for each duct can be read in Chart 17a and entered on the schedule.

3. Turn to Table 18 (page 491), and select a duct velocity. The choice of velocity is a matter of judgement that should be based on considerations of noise generation (the higher the velocity, the greater the potential for noise), ductwork costs (the higher the velocity, the smaller the ducts), fan horsepower (the higher the velocity, the greater the horsepower required), and duct heat gain (the higher the velocity, the lower the heat gain), all with due consideration of the owner's objectives and wishes. In this example, let us assume that, after due consideration of the factors noted above, an average velocity of 900 ft/min is chosen.

   Read in Chart 17b, the 1.3 multiplier for 900 ft/min and enter it on the schedule in the space provided.

4. The temperature of the air surrounding the ducts must now be determined to see whether a multiplier must be selected and applied to the % Addition for each duct. In Table 19 (page 491), the duct location described as an "unconditioned room, well ventilated" appears to be the description best fitting the basement. The corresponding temperature is the outdoor dry-bulb temperature at noon (90 F) found in step (3)(a) on page 21.

   Since the curves in Chart 17a are based on a surrounding air temperature of 90 F, no multiplier need to be used for surrounding air.

5. Since the inside design dry-bulb temperature for the drugstore in this example is 75 F while the curves in Chart 17a are based on a room design temperature of 78 F, a multiplier must be selected and applied to the % Addition. In Table 17d, find the 1.08 multiplier corresponding to 75 F, and enter it on the schedule.

6. It is now appropriate to consider the configuration of the ducts (round,

square, or rectangular) to judge whether an aspect ratio multiplier should be applied to the % Additions.

There is apparently enough headroom in the basement to accommodate a round, square, or rectangular duct plus the necessary clearance for insulation. To minimize duct cost and duct heat gain, round ducts are selected in this example.

Turn to Chart 17e and find the multiplier for round duct. Enter the .84 on the schedule.

7. Apply the multipliers to the % Addition for each duct to arrive at the Adjusted % Additions.
8. Multiply the sensible load for each duct by its Adjusted % Addition divided by 100 to arrive at the individual duct heat gains shown in the schedule.
9. Total the individual duct heat gains on the schedule.
10. Since it is assumed in this example that the return air grilles are located close enough to the units to result in a negligible return duct heat gain, no calculations need be made.
11. Enter the total duct heat gain from the schedule in the space provided on the load calculation form.

(8) *Total Sensible and Latent Loads:* add the duct heat gain to the Subtotal Sensible Load (1,328 + 75,229 = 76,557 Btu/h), and the latent load (25,320 Btu/h) to the page 1 subtotal latent load (5,350 Btu/h).

(9) *Total Cooling Load:* add the total sensible and latent loads to arrive at the total cooling load (76,557 + 30,670 = 107,227).

16.2.2 COOLING LOAD FOR 3:00 PM

Rather than a detailed description of the step-by-step procedure for a 3:00 pm load calculation, a completed load form is shown on pages 475 and 476.

Note the differences in factors and quantities for the two calculations. There is a difference in the total cooling loads, over 3,000 Btu/h, though not significant in itself. More significant are the differences in total sensible and latent loads. It appears that the choice of calculating the load at both noon and 3:00 pm was a good one. It is important in this case to choose equipment capable of handling the higher sensible load expected at 3:00 pm as well as the higher latent load expected at noon. For a further discussion of the subject of equipment selection, see NESCA Manual Q.

16.2.3 HEATING LOAD

(10) *Design conditions:* the winter design dry-bulb temperature for Atchison, Kansas, is shown in Table 1 (page 38) as 2 F. Enter 72 F in (a), 2 F in (b), and the difference of 70 F in (c).

(11) *Transmission losses:*

(a) Total the areas of the windows and doors from step #2 above, and enter the total on the form:

West window	126
West door	40
East window	24
East door	24
Total	214 ft^2

(b) Total the areas of exposed walls from step #3 above, and enter the total on the form:

West wall	252
South wall	750
East wall	134
Total	1,136 ft^2

(c) Calculate the wall area of the basement, 8 x (30 + 75 + 30) = 1,080 ft^2, and enter on the form.

(d) Enter the area of the roof/ceiling (2,250 ft^2) from step #3 on the form.

(e) Enter the floor area of the basement (same as ceiling area) on the form.

(f) Find the U-values for the walls and roof/ceiling in Table 7 (page 477) and for the windows and doors in Table 5 (page 476), and enter them on the form as follows:

Windows (and doors)	1.13
Walls, store	.09
Roof/ceiling	.12

(g) Find in Table 20 (page 492) the heat loss factors for the basement floor and walls, and enter them on the form (floor, 2.0; walls, 4.0).

(h) Enter the dry-bulb temperature difference (70 F) from step #10 above on the form, and multiply to find the heating loads as follows:

Windows	214 x 1.13 x 70 = 16,927 Btu/h
Walls, store	1,136 x .09 x 70 = 7,157 Btu/h
Walls, basement	1,080 x 4.00 = 4,320 Btu/h
Roof/ceiling	2,250 x .12 x 70 = 18,900 Btu/h
Floor (basement)	2,250 x 2.00 = 4,500 Btu/h

(12) *Infiltration or Ventilation:* for infiltration, calculate the volume of the store, 10 ft x 75 ft x 30 ft = 22,500 ft^3. Multiply the volume by the number of air changes per hour (2.0) from Table 13 (page 484), and divide by 60 to get 750 ft^3/min for the sales space of the store.

Even though the basement has some windows below grade, because there will be no infiltration through doors or walls, it can be considered a room with no windows or doors in using Table 13; and the air change factor (.25) under "Weatherstripping or Storm Sash" may be used. Infiltration for the basement is calculated as follows:

$$\frac{8 \text{ ft} \times 75 \text{ ft} \times 30 \text{ ft} \times .25 \text{ AC/h}}{60 \text{ min/h}} = 75 \text{ ft}^3\text{/min}$$

To the window and wall infiltration, add the door infiltration in a manner similar to that used in step #5 above (page 454):

$$TR = \frac{20 \text{ persons}}{2 \text{ doors} \times 1 \text{ hr}} = 10$$

TD = 70 (interpolate to find 28 ft^3/min)

Inf = 2 doors x 28 ft^3/(min)(door)
= 56 ft^3/min

Total infiltration for the store is 750 + 75 + 56 = 881 ft^3/min. Since infiltration exceeds ventilation (600 ft^3/min for the hood), enter the 881 ft^3/min in the space provided on the form, and multiply by the dry-bulb temperature difference (70 deg) and the 1.1 factor to obtain the infiltration heating load (67,837 Btu/h).

(13) *Humidification load:* see Table 21 (page 492) for the maximum relative humidity, 11%, and enter on the form. See Table 22b (page 494) for the required humidifier capacity (at 10% indoor RH and 0 degrees F outdoor temperature), 1.3 gallons/day per 100 ft^3/min, and the corresponding heating load, 480 Btu/h per 100 ft^3/min. Enter 881 ft^3/min and 480 Btu/h on the form, and complete the indicated calculations to find the humidification heat load of 4,229 Btu/h. Similarly find the 11.5 gallons/day humidifier capacity.

(14) *Subtotal heating Load:* total all the preceding component loads.

(15) *Duct heat loss:* since the ductwork will be located in the heated basement, the heat lost by the ducts will contribute to the heating of the basement and will not be lost outside the conditioned space; therefore, it need not be calculated.

(16) *Total heating load:* simply bring down the subtotal heating load of 123,870 Btu/h.

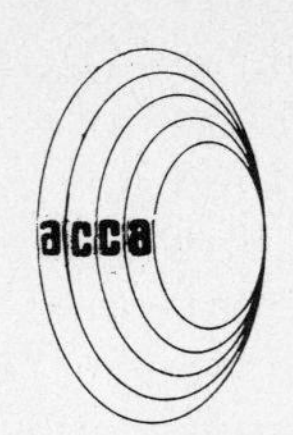

COMMERCIAL LOAD CALCULATIONS

* Reference Building & Systems DATA
Form 1 File No. ....

FORM N-1 1175

1228 - 17TH Street, N.W. Washington, D.C. 20036

For: Name REX DRUGSTORE Phone 743-2678
Address 421 Bellevue City Atchison State KS Zip
By: Contractor ABC Co. Phone 629-4711
Address 1700 Commerce City Atchison State KS Zip

COOLING LOAD

1. DESIGN CONDITIONS (Time of Day NOON) (Daily Range 23) (Latitude 40°)
a) Inside DB 75 RH 50% b) Outside DB 95 WB 78
Outside DB 95 (-) minus Inside DB 75 = Difference 20 (corr. to 15 for noon)

2. SOLAR RADIATION HEAT GAIN THROUGH GLASS — COOLING LOAD

Exposure	Sq Ft		Solar Factor		Shading and/or Glass Factor		Sensible	NOTES
E	24	X	68	X	.64	=	1,044	
E	24	X	68	X	.95	=	1,550	door
W	40	X	36	X	.95	=	1,368	door
W(N)	126	X	34	X	.95	=	4,070	shaded
		X		X		=		
		X		X		=		
		X		X		=		
		X		X		=		

3. TRANSMISSION GAINS

	Exposure	Sq Ft		Factor		Equivalent or Dry-Bulb Temp Diff		Sensible	Notes
Glass		190	X	1.06	X	15	=	3,020	unshaded
		24	X	.81	X	15	=	292	shaded
			X		X		=		
			X		X		=		
Walls	E	252	X	.09	X	33	=	748	
	S	750	X	.09	X	22	=	1,485	
	W	134	X	.09	X	19	=	229	
			X		X		=		
Doors			X		X		=		
			X		X		=		
Partitions			X		X		=		
			X		X		=		
Roof/Ceiling		2,250	X	.11	X	60	=	14,850	
Floors		2,250	X	.09	X	15	=	3,038	
			X		X		=		

4. INTERNAL HEAT GAIN

a. Occupants

Number		Sensible		Latent		Sensible	Latent
14	X	275			=	3,850	
6	X	250			=	1,500	
14			X	275	=		3,850
6			X	250	=		1,500

b. Lights & Others

	Watts			Sensible	Latent
Incandescent Lights		X 3.4	=		
Flourescent Lights	6,000	X 4.1	=	24,600	

Motors HP	Btu/h		Usage Factor	Sensible	Latent
1/4	1,000	X	.75	750	
1/8	580	X	.75	435	
		X			

		Sensible	Latent
Appliances	Coffee brewer-warmer	500	—
	Steam table	2,000	—
Other			

* This form designed to be used with NESCA Manual N

Page 1 Subtotal | 65,329 | 5,350

Page 1 Subtotal	65,329	5,350
5. INFILTRATION or VENTILATION	Sensible	Latent
ft³/min 600 X Dry-Bulb Temp Diff 15 X 1.1 =	9,900	
ft³/min 600 ÷ 100 X Btu/h 4,220 =		25,320
6. SUBTOTAL SENSIBLE LOAD	75,229	
7. DUCT HEAT GAIN		NOTES
a. Supply Ducts: % Addition * X Velocity * X Surr'dg. Air * X Room Temp * X Aspect Ratio * = Adjusted % Add *		* see attached schedule (Fig. 5)
b. Return Ducts: % Addition X Velocity X Surr'dg. Air X Return Air X Aspect Ratio =		
Subtotal Sensible Load ÷ 100 X (Total) Adj % Add * =	1,328	
8. TOTAL SENSIBLE LOAD	76,557	Latent
TOTAL LATENT LOAD		30,670
9. TOTAL COOLING LOAD	107,227	

HEATING LOAD

10. DESIGN CONDITIONS

Inside DB 72 (–) minus Outside DB 2 = Difference 70

11. TRANSMISSION LOSSES

	Exposure	Sq Ft		Factor		Dry-Bulb Temp Diff		Heating Load	NOTES
Windows		214	X	1.13	X	70	=	16,927	
			X		X		=		
			X		X		=		
Walls		1,136	X	.09	X	70	=	7,157	store
		1,080	X	4.0	X	—	=	4,320	basement
			X		X		=		
			X		X		=		
Roof/ceiling		2,250	X	.12	X	70	=	18,900	
			X		X		=		
			X		X		=		
Floor		2,250	X	2.0	X	—	=	4,500	
Other			X		X		=		
			X		X		=		

	HEATING LOAD
12. INFILTRATION or VENTILATION ft³/min 881 X Dry-Bulb Temp Diff 70 X 1.1 =	67,837
13. HUMIDIFICATION LOAD Inside RH (Desired 20%)(Max 11%) ft³/min 881 ÷ 100 X Btu/hr 480 = (water) gal/day 1.3 X (air) ft³/min 881 ÷ 100 = 11.5	4,229
14. SUBTOTAL HEATING LOAD	123,870
15. DUCT HEAT LOSS a. Supply Ducts: % Addition X Velocity X Surr'dg Air X Room Temp X Temp Ent Duct X Aspect Ratio X = Adjusted % Add; b. Return Ducts: % Addition X Velocity X Surr'dg Air X Return Air X Aspect Ratio = ; Subtotal Heating Load ÷ 100 X (Total) Adj % Add =	—
16. TOTAL HEATING LOAD	123,870

NESCA Form N-1 1175

COMMERCIAL LOAD CALCULATIONS

* Reference Building & Systems DATA
Form 1 File No. . . R-17 . . .

FORM N-1 1175

1228 - 17TH Street, N.W. Washington, D.C. 20036

For: Name . . . REX DRUGSTORE . . . Phone . . .
Address . . . City . . . State . . . Zip . . .
By: Contractor . . ABC Co. . . . Phone . . .
Address . . . City . . . State . . . Zip . . .

COOLING LOAD

1. DESIGN CONDITIONS (Time of Day 3 pm.) (Daily Range 23 . . .) (Latitude . . . 40° . . .)
a) Inside DB . . 75 RH . . 50% . . . b) Outside DB . . 95 . . WB . . . . 78 . . . .
Outside DB . . . 95 . . . . (-) minus Inside DB . . . 75 . . . . = Difference . . . 20 . . . .

2. SOLAR RADIATION HEAT GAIN THROUGH GLASS

Exposure	Sq Ft		Solar Factor		Shading and/or Glass Factor		COOLING LOAD Sensible	NOTES
E	24	X	30	X	.64	=	461	
E	24	X	30	X	.95	=	684	
W	40	X	165	X	.95	=	6,270	
W(N)	126	X	30	X	.95	=	3,591	
		X		X		=		
		X		X		=		
		X		X		=		
		X		X		=		

3. TRANSMISSION GAINS

	Exposure	Sq Ft		Factor		Equivalent or Dry-Bulb Temp Diff		Sensible
Glass		190	X	1.06	X	20	=	4,028
		24	X	.81	X	20	=	389
			X		X		=	
			X		X		=	
Walls	E	252	X	.09	X	29	=	658
	S	750	X	.09	X	32	=	2,160
	W	134	X	.09	X	34	=	410
			X		X		=	
Doors			X		X		=	
			X		X		=	
Partitions			X		X		=	
			X		X		=	
Roof/Ceiling		2,250	X	.11	X	88	=	21,780
Floors		2,250	X	.09	X	20	=	4,050
			X		X		=	

4. INTERNAL HEAT GAIN

a. Occupants

Number		Sensible		Latent	=	Sensible	Latent
8	X	250			=	2,000	
	X				=		
8			X	250	=		2,000
			X		=		

b. Lights & Others

	Watts		Sensible	Latent
Incandescent Lights		X 3.4 =		
Flourescent Lights	6,000	X 4.1 =	24,600	

	HP	Btu/h		Usage Factor	Sensible	Latent
Motors	1/4	1,000	X	.50	500	
	1/8	580	X	.50	290	
			X			

		Sensible	Latent
Appliances	Coffee brewer-warmer	500	
	Steam table		
Other			

* This form designed to be used with NESCA Manual N

Page 1 Subtotal . . . 72,371 . . 2,000

	Sensible	Latent
Page 1 Subtotal	72,371	2,000

5. INFILTRATION or VENTILATION

	Sensible	Latent
ft^3/min 600 X Dry-Bulb Temp Diff 20 X 1.1 =	13,200	
ft^3/min 600 ÷ 100 X Btu/h 3,649 =		21,894

6. SUBTOTAL SENSIBLE LOAD

85,571

7. DUCT HEAT GAIN

a. Supply Ducts:

% Addition		Velocity		Surr'dg. Air		Room Temp		Aspect Ratio		Adjusted % Add
*	X	*	X	*	X	*	X	*	=	*

b. Return Ducts:

% Addition		Velocity		Surr'dg. Air		Return Air		Aspect Ratio		
......	X	......	X	......	X	......	X	....	=	....

Subtotal Sensible Load .... ÷ 100 X (Total) Adj % Add * = 1,380

NOTES

* see attached schedule

8. TOTAL SENSIBLE LOAD

86,951

TOTAL LATENT LOAD

Latent 23,894

9. TOTAL COOLING LOAD

110,845

HEATING LOAD

10. DESIGN CONDITIONS

Inside DB .... (–) minus Outside DB ...... = Difference ....

11. TRANSMISSION LOSSES

HEATING LOAD

	Exposure	Sq Ft		Factor		Dry-Bulb Temp Diff		Heating Load
Windows	.....	....	X	....	X	....	=	....
	.....	....	X	....	X	....	=	....
	.....	....	X	....	X	....	=	....
Walls	.....	....	X	....	X	....	=	....
	.....	....	X	....	X	....	=	....
	.....	....	X	....	X	....	=	....
	.....	....	X	....	X	....	=	....
Roof/ceiling	.....	....	X	....	X	....	=	....
	.....	....	X	....	X	....	=	....
	.....	....	X	....	X	....	=	....
Floor	.....	....	X	....	X	....	=	....
Other		....	X	....	X	....	=	....
		....	X	....	X	....	=	....

NOTES

12. INFILTRATION or VENTILATION

ft^3/min .... X Dry-Bulb Temp Diff .... X 1.1 = ...

13. HUMIDIFICATION LOAD Inside RH (Desired .....)(Max .....)

ft^3/min .... ÷ 100 X Btu/hr .... = ...

(water) gal/day .... X (air) ft^3/ min .... ÷ 100 = ...

14. SUBTOTAL HEATING LOAD

....

15. DUCT HEAT LOSS

a. Supply Ducts:

% Addition		Velocity		Surr'dg Air		Room Temp		Temp Ent Duct		Aspect Ratio		Adjusted % Add
......	X	......	X	......	X	......	X	......	X	.....	X =	......

b. Return Ducts:

% Addition		Velocity		Surr'dg Air		Return Air		Aspect Ratio		
......	X	......	X	......	X	......	X	....	=	....

Subtotal Heating Load .... ÷ 100 X (Total) Adj % Add .... = ...

16. TOTAL HEATING LOAD

....

NESCA Form N-1 1175

EXAMPLE 2

16.3 The Fox furniture store is a one story building located in Meridian, Mississippi. Figures 6 and 7 show plan and elevation details of the store.

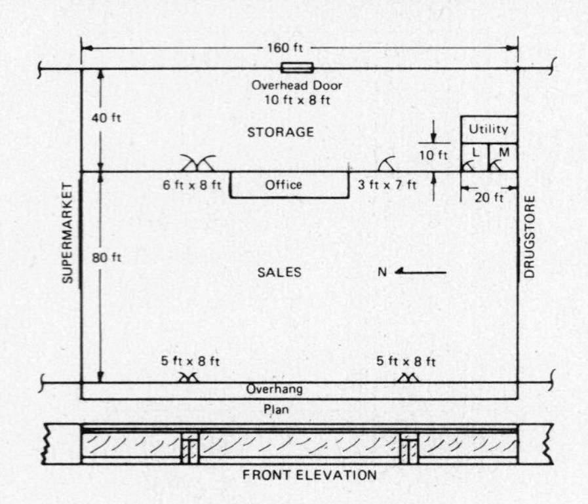

Figure 6 — Fox Furniture Store

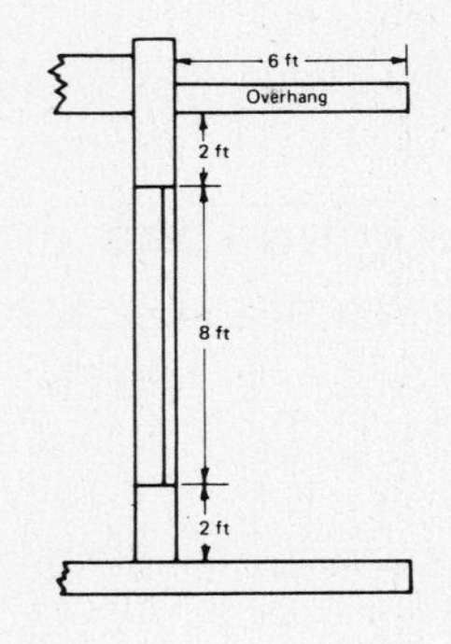

Figure 7 — Details of front windows and overhang

Construction details are as follows:

Windows: single plate glass.

Walls:

front: 4 in. face brick, 8 in. cinder block, 1 in. extruded polystyrene insulation board, and ½ in. gypsum wallboard, dark color.

rear: 8 in. cinder block, 1 in. extruded polystyrene insulation board, and ½ in. gypsum wallboard.

Partitions: 4 in. cinder block.

Ceiling height: 12 ft.

Doors: front, tempered plate glass; partition and overhead, 1 in. wood (overhead, motor-driven).

Roof/Ceiling: light color, built-up roofing; steel deck; 2 in. insulation; ½ in. gypsum wallboard.

Floor: concrete slab on grade with 1 in. edge insulation.

Other design details are as follows:

Lighting:

fluorescent: 320 40-watt tubes in sales area.

incandescent: 20 100-watt lamps in storage area.

Appliances: seven 1/8 hp motors in office equipment.

Occupancy: (peak) 50 customers, 7 employees. Customer avg. occupancy, ½ hr.

Indoor design conditions:

winter: 72 F DB; maximum RH to avoid window condensation.

summer: 80 F DB; 50% RH.

Storage room to be heated but not cooled.

16.3.1 COOLING LOAD: The following numbered steps correspond to the numbered items on the load calculation form shown on pages 467 and 468.

(1) *Design conditions:* in Table 1 (page 470) find and fill in on the form the daily range (22), the latitude (32°), and outdoor design conditions (95 F DB and 79 F WB). Fill in the indoor design conditions and the DB temperature difference.

(2) *Solar radiation heat gain through glass:* Because of the 6 ft overhang partially shading the west glass at the front of the building and because 3:00 pm is the logical time of day to expect the peak load, the calculation of sun radiation load on the glass is not a single step calculation. There are three steps: first, the calculation of the area of the glass that is shaded by the overhang; second, the calculation of the area of the glass exposed directly to the sun; and third, the calculation of the two radiation heat gains using the proper heat gain and correction factors. See Section 7, page 439, for detailed procedures.

(a) Multiply the length of the overhang (6 ft) by the factor (.97) from Table 4 (page 476) for a west exposure at 3:00 pm and 32° latitude to find the vertical distance that the shade line falls below the bottom of the overhang (6 ft x .97 = 5.82 ft). Subtract *the distance from the overhang to the top of the window from the distance the shade line falls below the overhang*. The result (5.82 ft – 2 ft = 3.82 ft) is the height of the shaded area of the glass. The height times the length equals the shaded area of the glass (3.82 ft x 160 ft = 611 ft^2). Enter the 611 on the form.

(b) To find the area of sunlit glass, subtract the area of shaded glass from the total glass area (1,300 – 611 = 689 ft^2). Enter the 689 on the form.

(c) From Table 2 (page 475) select the factor (32) for north glass under 32° and 3:00 pm for the shaded area, and enter on the form.

From Table 2, select the factor (168) for west glass under 32° and 3:00 pm for the sunlit glass area, and enter on the form. Enter also the correction factor (.95) for plate glass from Table 3.

Perform the indicated multiplications to find the cooling loads as follows:

611 x 32 x .95 = 18,574 Btu/h

689 x 168 x .95 = 109,964 Btu/h

It is interesting to note that without the overhang, the solar radiation heat gain for the glass would be 207,480 Btu/h.

(3) *Transmission gains:*

Windows:

Enter the total area of glass (1,300 ft^2) on the form. Find the U-value (1.06) for glass in Table 5 (page 476), and enter on the form. Enter the dry-bulb temperature difference, and multiply as follows to find the transmission heat gain:

1,300 x 1.06 x 15 = 20,670 Btu/h

Walls:

In finding the transmission gain for the walls, note that the portion of the front wall above the glass will be shaded by the overhang at 3:00 pm and, therefore, should be considered facing north. The portion of the wall below the glass will be sunlit at 3:00 pm and should be considered facing west.

Calculate the area of the front wall above the glass (160 x 2 = 320 ft^2), and enter on the form. Find in Table 7 (page 477) the U-value for the wall (.12), and enter it on the form. From Table 8 (page 480), select the equivalent temperature difference (12) for a dark north wall of heavy medium construction (nearest construction in note #5 is 4 in. face brick + 8 in. clay tile + 1 in. insulation) at 3:00 pm, and correct it according to note #2 as follows: 12 – (20 – 15) = 12 – 5 = 7 F. Enter the 7, and perform the indicated multiplication as follows to find the load:

320 x .12 x 7 = 269 Btu/h

Enter the area of the sunlit portion of the front wall (320 – 20 = 300 ft^2) on the form. Enter the U-value (.12) on the form, find the equivalent temperature difference (17) in Table 8, and correct per note #2 (17 – 5 = 12). Enter the 12 on the form, and perform the multiplication to find the load:

300 x .12 x 12 = 432 Btu/h

Doors:

Enter the total area (48 + 21 = 69 ft^2) of the doors in the partition between the sales and storage areas on the form, and find the U-value (.61) in Table 7 (page 477). Use a temperature difference of 10 F in accordance with note #3, Table 8 (page 480). Multiply as follows to find the load:

69 x .61 x 10 = 421 Btu/h

Partitions:

Find the area of the partitions between the storage area and the sales area, the storage area and the restroom, and the utility room and restrooms, (140 + 10 + 20) x 12 = 2,040 ft^2; and subtract the door areas (69 ft^2) to find the net area of the partitions: 2,040 – 69 = 1,971 ft^2. Enter the net area on the form, and find the U-value (.40) in Table 7 (page 477). Enter .40 on the form with the same TD (10) used for the partition doors. Multiply (1,971 x .40 x 10 = 7,880 Btu/h) to find the load.

Ceiling:

Calculate the ceiling areas of the conditioned spaces, (160 x 80) + (10 x 20) = 13,000 ft^2, and enter on the form. Find the U-value (.17) in Table 7 (page 477) and enter on the form. In Table 9 (page 481), find the equivalent temperature difference, 50 – (20 – 15) = 50 – 5 = 45. Enter the 45 on the form, and multiply to find the load as follows:

13,000 x .17 x 45 = 99,450 Btu/h

Floors:

There is no floor heat gain for a concrete slab on grade.

(4) *Internal heat gain—occupants:* assume that all 57 people are in the conditioned areas at the time of peak load. Find the sensible and latent components in Table 10 (page 482). Enter the proper values on the form and multiply to find the loads as follows:

Sensible:	57 x 250 = 14,250 Btu/h
Latent:	57 x 250 = 14,250 Btu/h

Lights and Others:

Enter total fluorescent wattage in the line for fluorescent lights, 320 x 40 = 12,800 watts. Multiply to find the cooling load: 12,800 x 4.1 = 52,480 Btu/h.

Find in Table 11 (page 482) the Btu/h gain (580) for a 1/8 hp motor (motor and driven machine in space), and multiply by 7 to find the total load, 4,060 Btu/h. Enter a reasonable usage factor such as .25. Multiply to find the cooling load, 1,015 Btu/h.

(5) *Infiltration or Ventilation:*

Infiltration:

The infiltration for summer conditions should be calculated only for the sales area because the storage room is not conditioned. The calculations based on the factors in Table 13 (page 484) are as follows:

$$\frac{160 \text{ ft} \times 80 \text{ ft} \times 12 \text{ ft} \times .60}{60} = 1{,}536 \text{ ft}^3/\text{min}$$

Door infiltration is estimated as follows:

$$TR = \frac{50}{.5 \times 4} = 25 \text{ (use 20)}$$

TD = 15 (use 20)

Inf = 4 doors x 16 ft^3/(min)(door)

= 64 ft^3/min

Total infiltration:

1,536 + 64 = 1,600 ft^3/min

Ventilation:

Table 14 (page 485) shows ventilation requirements as follows:

Sales:	7 ft^3/min per person
Restrooms:	15 ft^3/min per person

Ventilation calculations are as follows:

Sales	55 x 7 = 385
Restrooms	2 x 15 = 30
Total	415 ft^2/min

Infiltration far exceeds ventilation requirements; therefore, enter 1,600 ft^3/min on the form, and multiply by the 15 F temperature difference and 1.1 to find the sensible load. Divide by 100 and multiply by the 3,291 factor from Table 16b (page 488) to find the latent load. The load calculations follow:

1,600 x 15 x 1.1 = 26,400 Btu/h, sensible

$$\frac{1{,}600}{100} \times 3{,}291 = 52{,}656 \text{ Btu/h, latent}$$

(6) *Subtotal sensible load:* add the outside air sensible load to the page 1 Subtotal.

(7) *Duct heat gain:* a reasonable choice of equipment for the store might be rooftop heating/cooling units with integral supply/returns located directly underneath the units. Unit heaters might be used in the storage area. With this choice, there would be no exposed ducts of appreciable length and, therefore, no duct heat gains or losses. If the choice were for an alternate system requiring some exposed ductwork, duct heat gains and losses would have to be calculated.

(8) *Total sensible and latent loads:* bring down the Subtotal Sensible Load, and total all latent loads.

(9) *Total cooling load:* add the total sensible and latent loads to arrive at the total cooling load.

16.3.2 HEATING LOAD:

(10) *Design conditions:* enter the indoor design DB temperature (72 F). Find the outdoor winter design DB temperature for Meridian, Mississippi, in Table 1 (page 38), and enter (24 F). Enter the difference (72 – 24 = 48 F).

(11) *Transmission losses:*

(a) Windows:

Enter the total square feet of area of windows and glass doors, 611 + 689 = 1,300 ft^2, on the form. Calculate the area of the overhead door (10 x 8 = 80 ft^2), and enter on the form. Find the U-value (1.13) for the glass in Table 5 page (476), and enter on the form. Find the U-value (.64) for the wooden overhead door in Table 7 (page 477) and enter on the form. Enter the dry-bulb temperature difference (48 F), and perform the multiplications to find the losses as follows:

1,300 x 1.13 x 48 = 70,512 Btu/h

80 x .64 x 48 = 2,458 Btu/h

(b) Walls:

Calculate the gross areas of outside walls as follows:

front:	160 x 12 = 1,920 ft^2
rear:	160 x 12 = 1,920 ft^2

Subtract the window and door areas to obtain net wall areas:

front:	1,920 – 1,300 = 620 ft^2
rear:	1,920 – 80 = 1,840 ft^2

Enter the net wall areas on the form. In Table 7 (page 477), find the U-value for the walls (front, .12; rear, .12) and enter them on the form. Finally, enter the temperature difference and multiply to find the losses as follows:

front:	620 x .12 x 48 = 3,571 Btu/h
rear:	1,840 x .12 x 48 = 10,598 Btu/h

(c) Roof/Ceiling:

Calculate the area of the roof/ceiling (120 x 160 = 19,200 ft^2) and enter on the form. Find the U-value (.13) for the roof/ceiling in Table 7 (page 477) and enter on the form. Enter the TD, and multiply as follows to arrive at the loss:

19,200 x .13 x 48 = 119,808 Btu/h

(d) Floor:

Total the linear feet of exposed slab floor (160 + 160 = 320 ft) and enter on the form. In Table 7 (page 477), find the edge loss factor (30) and enter on the form. Multiply edge length by edge loss factor to find the loss (320 x 30 = 9,600 Btu/h).

(12) *Infiltration or Ventilation:*

(a) Infiltration:

Calculate the total volume of the store (sales and storage):

sales:	160 x 80 x 12 = 153,600 ft^3
storage:	160 x 40 x 12 = 76,800 ft^3

Multiply the volumes by the air change/hour factor (1.00) from Table 13 (page 484) and divide by 60:

sales: $\frac{153{,}600 \times 1.0}{60} = 2{,}560\ ft^3/min$

storage: $\frac{76{,}800 \times 1.0}{60} = 1{,}280\ ft^3/min$

3,840 ft^3/min

In accordance with the note at the bottom of Table 13, divide the total ft^3/min found by 2 to find the infiltration rate for the whole building, 1,920 ft^3/min.

Add door infiltration as follows:

Front doors:

$TR = \frac{50}{.5 \times 4} = 25$ (use 20)

TD = 48 (use 40)

Inf = 4 doors x 846 ft^3/(min)(door)
= 3,384 ft^3/min

Rear door: *careful and reasonable consideration based on information from store management personnel is recommended.* Possible estimate might be as follows:

Worst condition:

TR = 400 (equiv.) (door stays open)

TD = 48 (use 40)

Inf = 4 doors (equiv.) x 1,750 ft^3/(min) (door)
= 7,000 ft^3/min

Best condition:

TR = 0 (no deliveries or shipments)

TD = 48 (use 40)

Inf = 4 x 746 = 2,984 ft^3/min

Likely condition: By closing the motor-driven door as much as practical during loading and unloading, infiltration on a design day might be an average of the two estimates made above:

$\frac{7{,}000 + 2{,}984}{2} = 4{,}992$ or 5,000 ft^3/min

Total door infiltration is then 3,384 + 5,000 = 8,384 or 8,000 ft^3/min.

Total infiltration is 1,920 + 8,000 = 9,920 ft^3/min

(b) Ventilation:

See Table 14 (page 485) for minimum ventilation rates. The rates shown for the various spaces are as follows:

sales	7 ft^3/min per person
storage	5 ft^3/min per person
offices	15 ft^3/min per person
restrooms	15 ft^3/min per person

A reasonable estimate of the number of people occupying each space might be as follows:

sales	46
storage	2
offices	5
restrooms	4
	57

Multiplying the occupancy values by the ft^3/min rates yields the following:

sales	46 x 7 = 322
storage	2 x 5 = 10
offices	5 x 15 = 75
restrooms	4 x 15 = 60
Total	467 ft^3/min

Since the infiltration rate (3,360 ft^3/min) exceeds the ventilation rate, enter the infiltration rate in the form along with the TD (48 F), and multiply to find the loss (3,360 x 48 x 1.1 = 177,408 Btu/h).

(13) *Humidification:*

See Table 21 (page 492) for the maximum RH to avoid condensation on windows. Interpolating between 20 F and 30 F for 24 F outdoor design temperature and 72 F indoor design temperature indicates a maximum indoor RH for single glass of 26.6%. Enter Table 22b (page 494) at 25% indoor RH to be on the safe side with respect to condensation, and read, opposite 25 F outdoor dry-bulb temperature, 2.5 gallons/day per 100 ft^3/min and 920 Btu/h per 100 ft^3/min. Enter on the form the 920 Btu/h from the table and the 3,360 ft^3/min from Step 12 above, and perform the indicated calculation to find the humidification load.

(3,360 ÷ 100 x 920 = 30,912 Btu/h)

(14) *Subtotal heating load:* total the individual component loads.

(15) *Duct heat loss:* in accordance with the discussion in Step 7 (page 444), there will be no duct heat loss.

(16) *Total heating load:* bring down the subtotal heating load of 424,867 Btu/h.

COMMERCIAL LOAD CALCULATIONS

* Reference Building & Systems DATA
Form 1 File No. 164C

FORM N-1 1175

1228 - 17TH STREET, N.W. WASHINGTON, D.C. 20036

For: Name Fox Furniture — Phone 639-7856
Address 1310 W. Lambard — City Meridian — State MS — Zip
By: Contractor XYZ Co — Phone 549-6841
Address 816 N. Fourth — City Meridian — State MS — Zip

COOLING LOAD

1. DESIGN CONDITIONS (Time of Day 3 pm.) (Daily Range 22) (Latitude 32°)
a) Inside DB 80 RH 50% b) Outside DB 95 WB 79
Outside DB 95 (-) minus Inside DB 80 = Difference 15

2. SOLAR RADIATION HEAT GAIN THROUGH GLASS — **COOLING LOAD**

Exposure	Sq Ft		Solar Factor		Shading and/or Glass Factor		Sensible	NOTES
W(N)	611	X	32	X	.95	=	18,574	shaded
W	689	X	168	X	.95	=	109,964	unshaded
		X		X		=		
		X		X		=		
		X		X		=		
		X		X		=		
		X		X		=		
		X		X		=		

3. TRANSMISSION GAINS

	Exposure	Sq Ft		Factor		Equivalent or Dry-Bulb Temp Diff		Sensible	Notes
Glass		1,300	X	1.06	X	15	=	20,670	
			X		X		=		
			X		X		=		
			X		X		=		
Walls	W(N)	320	X	.12	X	7	=	269	
	W	300	X	.12	X	12	=	432	
			X		X		=		
			X		X		=		
Doors		69	X	.61	X	10	=	421	partition
			X		X		=		
Partitions		1,971	X	.40	X	10	=	7,880	
			X		X		=		
Roof/Ceiling		13,000	X	.17	X	45	=	99,450	
Floors			X		X		=		
			X		X		=		

4. INTERNAL HEAT GAIN

a. Occupants

Number		Sensible		Latent		Sensible	Latent
57	X	250			=	14,250	
	X				=		
57			X	250	=		14,250
			X		=		

b. Lights & Others

	Watts		Sensible
Incandescent Lights		X 3.4 =	
Flourescent Lights	12,800	X 4.1 =	52,480

Motors HP		Btu/h		Usage Factor	Sensible
			X		
1/8	(7×580)	4,060	X	.25	1,015
			X		

Appliances

Other

* This form designed to be used with NESCA Manual N

Page 1 Subtotal 325,405 | 14,250

	Page 1 Subtotal	325,405	14,250

5. INFILTRATION or VENTILATION — Sensible / Latent

ft³/min 1,600 X Dry-Bulb Temp Diff 15 X 1.1 = 26,400

ft³/min 1,600 ÷ 100 X Btu/h 3,291 = 52,656

6. SUBTOTAL SENSIBLE LOAD 351,805

7. DUCT HEAT GAIN — NOTES

a. Supply Ducts:

% Addition		Velocity		Surr'dg. Air		Room Temp		Aspect Ratio		Adjusted % Add
......	X	......	X	......	X	......	X	....	=	....

b. Return Ducts:

% Addition		Velocity		Surr'dg. Air		Return Air		Aspect Ratio		
......	X	......	X	......	X	......	X	....	=	....

Subtotal Sensible Load .... ÷ 100 X (Total) Adj % Add .... = —

8. TOTAL SENSIBLE LOAD 351,805

TOTAL LATENT LOAD Latent 66,906

9. TOTAL COOLING LOAD 418,711

HEATING LOAD

10. DESIGN CONDITIONS

Inside DB 72 (−) minus Outside DB 24 = Difference 48

11. TRANSMISSION LOSSES — HEATING LOAD

	Exposure	Sq Ft		Factor		Dry-Bulb Temp Diff		Heating Load	NOTES
Windows		1,300	X	1.13	X	48	=	70,512	
			X		X		=		
			X		X		=		
Walls		620	X	.12	X	48	=	3,571	front
		1,840	X	.12	X	48	=	10,598	rear
			X		X		=		
			X		X		=		
Roof/ceiling		19,200	X	.13	X	48	=	119,808	
			X		X		=		
			X		X		=		
Floor		*320	X	30	X	—	=	9,600	* linear feet
Other	Door	800	X	.64	X	48	=	2,458	
			X		X		=		

12. INFILTRATION or VENTILATION

ft³/min 3,360 X Dry-Bulb Temp Diff 48 X 1.1 = 177,408

13. HUMIDIFICATION LOAD Inside RH (Desired .....)(Max 25%)

ft³/min 3,360 ÷ 100 X Btu/hr 920 = 30,912

(water) gal/day 2.5 X (air) ft³/ min 3,360 ÷ 100 = 84

14. SUBTOTAL HEATING LOAD 424,867

15. DUCT HEAT LOSS

a. Supply Ducts:

% Addition		Velocity		Surr'dg Air		Room Temp		Temp Ent Duct		Aspect Ratio		Adjusted % Add
......	X	......	X	......	X	......	X	......	X	.....	X =	......

b. Return Ducts:

% Addition		Velocity		Surr'dg Air		Return Air		Aspect Ratio		
......	X	......	X	......	X	......	X	....	=	....

Subtotal Heating Load .... ÷ 100 X (Total) Adj % Add .... = —

16. TOTAL HEATING LOAD 424,867

NESCA Form N-1 1175

APPENDIX

Tables and Charts

Table 1

OUTDOOR DESIGN CONDITIONS FOR UNITED STATES AND CANADA [1,2]

Location[4]	Latitude[3]	Winter	Summer		
		Design Dry-Bulb	Design Dry-Bulb	Design Wet-Bulb	Daily Range
UNITED STATES					
ALABAMA					
Alexander City	33	20	94	78	21
Anniston AP	34	19	94	78	21
Auburn	33	25	96	79	21
Birmingham AP	34	22	94	78	21
Decatur	35	19	95	78	22
Dothan AP	31	27	95	80	20
Florence AP	35	17	95	78	22
Gadsden	34	20	94	77	22
Huntsville AP	35	17	95	77	23
Mobile AP	31	29	93	79	18
Mobile CO	31	32	94	79	16
Montgomery AP	32	26	95	79	21
Selma-Craig AFB	32	27	96	80	21
Talladega	34	19	95	78	21
Tuscaloosa AP	33	23	96	80	22
ALASKA					
Anchorage AP	61	–20	70	61	15
Barrow	71	–42	54	51	12
Fairbanks AP	65	–50	78	63	24
Juneau AP	58	– 4	71	64	15
Kodiak	58	12	66	60	10
Nome AP	64	–28	62	56	10
ARIZONA					
Douglas AP	32	22	98	69	31
Flagstaff AP	35	5	82	60	31
Fort Huachuca AP	32	28	93	68	27
Kingman AP	35	29	100	69	30
Nogales	31	24	98	71	31
Phoenix AP	34	34	106	76	27
Prescott AP	35	19	94	66	30
Tucson AP	33	32	102	73	26
Winslow AP	35	13	95	65	32
Yuma AP	33	40	109	78	27
ARKANSAS					
Blytheville AFB	36	17	96	79	21
Camden	34	23	97	80	21
El Dorado AP	33	23	96	80	21
Fayetteville AP	36	13	95	76	23
Fort Smith AP	35	19	99	78	24
Hot Springs Nat. Pk.	34	22	97	78	22
Jonesboro	36	18	96	79	21
Little Rock AP	35	23	96	79	22
Pine Bluff AP	34	24	96	80	22
Texarkana AP	34	26	97	79	21
CALIFORNIA					
Bakersfield AP	35	33	101	71	32
Barstow AP	35	28	102	72	37
Blythe AP	34	35	109	77	28
Burbank AP	34	38	94	70	25
Chico	40	33	100	70	36
Concord	38	36	92	67	32
Covina	34	41	97	72	31
Crescent City AP	42	36	69	60	18
Downey	34	38	90	71	22
El Cajon	33	34	95	73	30
El Centro AP	33	35	109	80	34
Escondido	33	36	92	72	30
Eureka/Arcata AP	41	35	65	59	11
Fairfield-Travis AFB	38	34	94	69	34
Fresno AP	37	31	99	72	34
Hamilton AFB	38	35	85	68	28
Laguna Beach	34	39	80	68	18
Livermore	38	30	97	69	24
Lompoc, Vandenburg AFB	35	38	79	63	20
Long Beach AP	34	38	84	70	22
Los Angeles AP	34	43	83	68	15
Los Angeles CO	34	44	90	70	20
Merced-Castle AFB	37	32	99	72	36
Modesto	38	36	98	71	36
Monterey	37	37	79	63	20
Napa	38	34	92	68	30
Needles AP	35	37	110	75	27
Oakland AP	38	37	81	63	19
Oceanside	33	40	81	68	13
Ontario	34	34	97	71	36
Oxnard AFB	34	37	80	69	19
Palmdale AP	35	27	101	68	35
Palm Springs	34	36	108	78	35
Pasadena	34	39	93	70	29
Petaluma	38	32	90	68	31
Pomona CO	34	34	96	72	36
Redding AP	40	35	101	69	32
Redlands	34	37	96	71	33
Richmond	38	38	81	64	17
Riverside-March AFB	34	34	96	71	37
Sacramento AP	38	32	97	70	36
Salinas AP	37	35	85	65	24
San Bernardino, Norton AFB	34	33	98	73	38
San Diego AP	33	44	83	70	12
San Fernando	34	37	97	72	38
San Francisco AP	38	37	79	63	20
San Francisco CO	38	44	77	62	14
San Jose AP	37	36	88	67	26
San Luis Obispo	35	37	85	64	26
Santa Ana AP	34	36	89	71	28
CALIFORNIA (continued)					
Santa Barbara CO	34	36	84	66	24
Santa Cruz	37	34	84	65	28
Santa Maria AP	35	34	82	64	23
Santa Monica CO	34	45	77	68	16
Santa Paula	34	36	89	71	36
Santa Rosa	38	32	93	68	34
Stockton AP	38	34	98	70	37
Ukiah	39	30	96	69	40
Visalia	36	36	100	72	38
Yreka	42	17	94	66	38
Yuba City	39	34	100	70	36
COLORADO					
Alamosa AP	38	–13	82	61	35
Boulder	40	8	90	63	27
Colorado Springs AP	39	4	88	62	30
Denver AP	40	3	90	64	28
Durango	37	4	86	63	30
Fort Collins	41	– 5	89	62	28
Grand Junction AP	39	11	94	63	29
Greeley	40	– 5	92	64	29
La Junta AP	38	– 2	95	71	31
Leadville	39	– 4	73	55	30
Pueblo AP	38	– 1	94	67	31
Sterling	41	– 2	93	66	30
Trinidad AP	37	5	91	65	32
CONNECTICUT					
Bridgeport AP	41	8	88	76	18
Hartford, Brainard Field	42	5	88	76	22
New Haven AP	41	9	86	76	17
New London	41	8	86	75	16
Norwalk	41	4	89	76	19
Norwich	42	2	86	76	18
Waterbury	42	4	88	76	21
Windsor Locks, Bradley Field	42	2	88	75	22
DELAWARE					
Dover AFB	39	15	90	78	18
Wilmington AP	40	15	90	77	20
DISTRICT OF COLUMBIA					
Andrews AFB	39	16	91	77	18
Washington National AP	39	19	92	77	18
FLORIDA					
Belle Glade	27	39	91	79	16
Cape Kennedy AP	28	40	89	80	15
Daytona Beach AP	29	36	92	80	15
Fort Lauderdale	26	45	90	80	15
Fort Myers AP	27	42	92	80	18
Fort Pierce	28	41	91	80	15
Gainesville AP	30	32	94	79	18
Jacksonville AP	30	32	94	79	19
Key West AP	24	58	89	79	9
Lakeland CO	28	39	93	79	17
Miami AP	26	47	90	79	15
Miami Beach CO	26	48	89	79	10
Ocala	29	33	94	79	18
Orlando AP	28	37	94	79	17
Panama City, Tyndall AFB	30	35	91	80	14
Pensacola CO	30	32	90	81	14
St. Augustine	30	35	92	80	16
St. Petersburg	28	42	91	80	16
Sanford	29	37	93	79	17
Sarasota	27	39	91	80	17
Tallahassee AP	30	29	94	79	19
Tampa AP	28	39	91	80	17
West Palm Beach AP	27	44	91	80	16
GEORGIA					
Albany, Turner AFB	32	30	96	79	20
Americus	32	25	96	79	20
Athens	34	21	94	77	21
Atlanta AP	34	23	92	77	19
Augusta AP	33	23	95	79	19
Brunswick	31	31	95	80	18
Columbus, Lawson AFB	32	26	96	79	21
Dalton	35	19	95	77	22
Dublin	32	25	96	79	20
Gainesville	34	20	92	77	21
Griffin	33	22	93	78	21
La Grange	33	20	94	78	21
Macon AP	33	27	96	79	22
Marietta, Dobbins AFB	34	21	93	77	21
Moultrie	31	30	95	79	20
Rome AP	34	20	95	77	23
Savannah-Travis AP	32	27	94	80	20
Valdosta-Moody AFB	31	31	94	79	20
Waycross	31	28	95	79	20
HAWAII					
Hilo AP	20	61	83	73	15
Honolulu AP	21	62	85	74	12
Kaneohe	21	61	83	73	12
Wahiawa	22	61	84	74	14
IDAHO					
Boise AP	44	10	93	66	31
Burley	42	8	93	66	35
Coeur d'Alene AP	48	7	91	65	31
Idaho Falls AP	44	– 6	88	64	38
Lewiston AP	46	12	96	66	32
Moscow	47	1	89	63	32
Mountain Home AFB	43	9	96	66	36
Pocatello AP	43	– 2	91	63	35
Twin Falls AP	42	8	94	64	34

See footnotes at end of table.

Table 1 (continued)

Location[4]	Latitude[3]	Winter	Summer		
		Design Dry-Bulb	Design Dry-Bulb	Design Wet-Bulb	Daily Range
ILLINOIS					
Aurora	42	−3	91	77	20
Belleville, Scott AFB	38	10	95	78	21
Bloomington	40	3	92	78	21
Carbondale	38	11	96	79	21
Champaign/Urbana	40	4	94	78	21
Chicago, Midway AP	42	1	92	76	20
Chicago, O'Hare AP	42	0	90	75	20
Chicago CO	42	1	91	76	15
Danville	40	4	94	78	21
Decatur	40	4	93	78	21
Dixon	42	−3	91	77	23
Elgin	42	−4	90	76	21
Freeport	42	−6	90	77	24
Galesburg	41	0	92	78	22
Greenville	39	7	94	78	21
Joliet AP	42	−1	92	77	20
Kankakee	41	1	92	77	21
La Salle/Peru	41	1	93	77	22
Macomb	41	1	93	78	22
Moline AP	40	−3	91	77	23
Mt. Vernon	38	10	95	78	21
Peoria AP	41	2	92	77	22
Quincy AP	40	2	95	79	22
Rantoul, Chanute AFB	40	3	92	77	21
Rockford	42	−3	90	76	24
Springfield AP	40	4	92	78	21
Waukegan	42	−1	90	76	21
INDIANA					
Anderson	40	5	91	77	22
Bedford	39	7	93	78	22
Bloomington	39	7	92	78	22
Columbus, Bakalar AFB	39	7	92	78	22
Crawfordsville	40	2	93	77	22
Evansville AP	38	10	94	78	22
Fort Wayne AP	41	5	91	76	24
Goshen AP	42	0	90	76	23
Hobart	42	0	91	76	21
Huntington	41	2	92	76	23
Indianapolis AP	40	4	91	77	22
Jeffersonville	38	13	94	78	23
Kokomo	40	4	92	76	22
Lafayette	40	3	92	77	22
La Porte	42	0	91	76	22
Marion	40	2	91	76	23
Muncie	40	2	91	77	22
Peru, Bunker Hill AFB	41	1	89	76	22
Richmond AP	40	3	91	77	22
Shelbyville	40	6	92	77	22
South Bend AP	42	3	89	76	22
Terre Haute AP	40	7	93	78	22
Valparaiso	41	−2	90	76	22
Vincennes	39	9	94	78	22
IOWA					
Ames	42	−7	92	78	23
Burlington AP	41	0	92	78	22
Cedar Rapids AP	42	−4	90	76	23
Clinton	42	−3	90	77	23
Council Bluffs	41	−3	94	78	22
Des Moines AP	42	−3	92	77	23
Dubuque	42	−7	90	76	22
Fort Dodge	42	−8	92	77	23
Iowa City	42	−4	91	77	22
Keokuk	40	1	93	78	22
Marshalltown	42	−6	91	77	23
Mason City AP	43	−9	88	75	24
Newton	42	−5	93	77	23
Ottumwa AP	41	−2	93	78	22
Sioux City AP	42	−6	93	77	24
Waterloo	42	−8	89	76	23
KANSAS					
Atchison	40	2	95	78	23
Chanute AP	38	7	97	78	23
Dodge City AP	38	7	97	73	25
El Dorado	38	8	99	77	24
Emporia	38	7	97	77	25
Garden City AP	38	3	98	73	28
Goodland AP	39	4	96	70	31
Great Bend	38	6	99	76	28
Hutchinson AP	38	6	99	76	28
Liberal	37	8	100	73	28
Manhattan, Fort Riley	39	4	98	78	24
Parsons	37	9	97	78	23
Russell AP	39	4	100	76	29
Salina	39	7	99	76	26
Topeka AP	39	6	96	78	24
Wichita AP	38	9	99	76	23
KENTUCKY					
Ashland	38	10	92	76	22
Bowling Green AP	37	11	95	78	21
Corbin AP	37	9	91	77	23
Covington AP	39	8	90	76	22
Hopkinsville, Campbell AFB	37	14	95	78	21
Lexington AP	38	10	92	77	22
Louisville AP	38	12	93	78	23
Madisonville	37	11	94	78	22
Owensboro	38	10	94	78	23
Paducah AP	37	14	95	79	20

Location[4]	Latitude[3]	Winter	Summer		
		Design Dry-Bulb	Design Dry-Bulb	Design Wet-Bulb	Daily Range
LOUISIANA					
Alexandria AP	31	29	95	80	20
Baton Rouge AP	30	30	94	80	19
Bogalusa	31	28	94	79	19
Houma	30	33	92	80	15
Lafayette AP	30	32	93	81	18
Lake Charles AP	30	33	93	79	17
Minden	33	26	96	80	20
Monroe AP	32	27	96	81	20
Natchitoches	32	26	97	80	20
New Orleans AP	30	35	91	80	16
Shreveport AP	32	26	96	80	20
MAINE					
Augusta AP	44	−3	86	73	22
Bangor, Dow AFB	45	−4	85	73	22
Caribou AP	47	−14	81	70	21
Lewiston	44	−4	86	73	22
Millinocket AP	46	−12	85	72	22
Portland AP	44	0	85	73	22
Waterville	44	−5	86	73	22
MARYLAND					
Baltimore AP	39	15	91	78	21
Baltimore CO	39	20	92	78	17
Cumberland	40	9	92	75	22
Frederick AP	39	11	92	77	22
Hagerstown	40	10	92	76	22
Salisbury	38	18	90	78	18
MASSACHUSETTS					
Boston AP	42	10	88	74	16
Clinton	42	2	85	74	17
Fall River	42	9	86	74	18
Framingham	42	3	89	74	17
Gloucester	42	6	84	73	15
Greenfield	42	−2	87	74	23
Lawrence	43	1	88	74	22
Lowell	42	3	89	74	21
New Bedford	42	13	84	73	19
Pittsfield AP	42	−1	84	72	23
Springfield, Westover AFB	42	2	88	74	19
Taunton	42	0	86	75	18
Worcester AP	42	1	87	73	18
MICHIGAN					
Adrian	42	4	91	75	23
Alpena AP	45	−1	85	73	27
Battle Creek AP	42	5	89	74	23
Benton Harbor AP	42	3	88	74	20
Detroit Met. CAP	42	8	88	75	20
Escanaba	46	−3	80	71	17
Flint AP	43	3	87	75	25
Grand Rapids AP	43	6	89	74	24
Holland	43	6	88	74	22
Jackson AP	42	4	89	75	23
Kalamazoo	42	5	89	75	23
Lansing AP	43	6	87	75	24
Marquette CO	46	−4	86	71	18
Mt. Pleasant	44	1	87	74	24
Muskegon AP	43	8	85	74	21
Pontiac	43	4	88	75	21
Port Huron	43	3	88	74	21
Saginaw AP	44	3	86	75	23
Sault Ste. Marie AP	46	−8	81	71	23
Traverse City AP	45	4	86	73	22
Ypsilanti	42	5	89	74	22
MINNESOTA					
Albert Lea	44	−10	89	76	24
Alexandria AP	46	−15	88	74	24
Bemidji AP	48	−28	84	72	24
Brainerd	46	−20	85	73	24
Duluth AP	47	−15	82	71	22
Faribault	44	−12	88	75	24
Fergus Falls	46	−17	89	74	24
International Falls AP	48	−24	82	69	26
Mankato	44	−12	89	75	24
Minneapolis/St. Paul AP	46	−10	89	75	22
Rochester AP	44	−13	88	75	24
St. Cloud AP	46	−16	88	75	24
Virginia	48	−21	83	71	23
Willmar	45	−14	88	75	24
Winona	44	−8	89	76	24
MISSISSIPPI					
Biloxi, Keesler AFB	30	32	92	81	16
Clarksdale	34	24	96	80	21
Columbus AFB	34	22	95	79	22
Greenville AFB	34	24	96	80	21
Greenwood	34	23	96	80	21
Hattiesburg	31	26	95	79	21
Jackson AP	32	24	96	78	21
Laurel	32	26	95	79	21
McComb AP	31	26	94	79	18
Meridian AP	32	24	95	79	22
Natchez	32	26	94	80	21
Tupelo	34	22	96	79	22
Vicksburg CO	32	26	95	80	21
MISSOURI					
Cape Girardeau	37	12	96	79	21
Columbia AP	39	6	95	78	22
Farmington AP	38	8	95	78	22
Hannibal	40	4	94	78	22
Jefferson City	39	6	95	78	23

See footnotes at end of table.

Table 1 (continued)

Location[4]	Latitude[3]	Winter	Summer		
		Design Dry-Bulb	Design Dry-Bulb	Design Wet-Bulb	Daily Range
MISSOURI (continued)					
Joplin AP	37	11	95	78	24
Kansas City AP	39	8	97	77	20
Kirksville AP	40	−3	94	78	24
Mexico	39	3	94	78	22
Moberly	40	2	94	78	23
Poplar Bluff	37	13	96	79	22
Rolla	38	7	95	78	22
St. Joseph AP	40	3	95	78	23
St. Louis AP	39	8	95	78	21
St. Louis CO	39	11	94	78	18
Sedalia, Whiteman AFB	39	9	94	77	22
Sikeston	37	14	96	79	21
Springfield AP	37	10	94	77	23
MONTANA					
Billings AP	46	−6	91	66	31
Bozeman	46	−11	85	60	32
Butte AP	46	−16	83	59	35
Cut Bank AP	49	−17	86	63	35
Glasgow AP	48	−20	93	67	29
Glendive	47	−16	93	69	29
Great Falls AP	48	−16	88	63	28
Havre	48	−15	87	64	33
Helena AP	47	−13	87	63	32
Kalispell AP	48	−3	84	63	34
Lewiston AP	47	−14	86	63	30
Livingston AP	46	−13	88	62	32
Miles City AP	46	−15	94	69	30
Missoula AP	47	−3	89	63	36
NEBRASKA					
Beatrice	40	1	97	77	24
Chadron AP	43	−9	95	70	30
Columbus	42	−3	96	76	25
Fremont	42	−3	97	77	22
Grand Island AP	41	−2	95	75	28
Hastings	41	1	96	75	27
Kearney	41	−2	95	75	28
Lincoln CO	41	0	96	77	24
McCook	40	0	97	72	28
Norfolk	42	−7	95	76	30
North Platte AP	41	−2	94	73	28
Omaha AP	41	−1	94	78	22
Scottsbluff AP	42	−4	94	69	31
Sidney AP	41	−2	92	69	31
NEVADA†					
Carson City	39	7	91	61	42
Elko AP	41	−7	92	62	42
Ely AP	39	−2	88	59	39
Las Vegas AP	36	26	106	71	30
Lovelock AP	40	11	96	64	42
Reno AP	40	7	92	62	45
Reno CO	40	17	92	62	45
Tonopah AP	38	13	92	63	40
Winnemucca AP	41	5	95	62	42
NEW HAMPSHIRE					
Berlin	44	−15	85	71	22
Claremont	43	−9	87	73	24
Concord AP	43	−7	88	73	26
Keene	43	−8	88	73	24
Laconia	44	−12	87	73	25
Manchester, Grenier AFB	43	1	89	74	24
Portsmouth, Pease AFB	43	3	86	73	22
NEW JERSEY					
Atlantic City CO	40	18	88	77	18
Long Branch	40	13	91	76	18
Newark AP	41	15	91	76	20
New Brunswick	40	12	89	76	19
Paterson	40	12	91	76	21
Phillipsburg	41	10	91	76	21
Trenton CO	40	16	90	77	19
Vineland	40	16	90	77	19
NEW MEXICO					
Alamagordo, Holloman AFB	33	22	98	69	30
Albuquerque AP	35	17	94	65	27
Artesia	33	19	99	70	30
Carlsbad AP	32	21	99	71	28
Clovis AP	34	17	97	69	28
Farmington AP	37	9	93	65	30
Gallup	36	−1	90	63	32
Grants	35	−3	89	63	32
Hobbs AP	33	19	99	71	29
Las Cruces	32	23	100	69	30
Los Alamos	36	9	86	63	32
Raton AP	37	2	90	65	34
Roswell, Walker AFB	33	19	99	70	33
Santa Fe CO	36	11	88	63	28
Silver City AP	33	18	93	67	30
Socorro AP	34	17	97	66	30
Tucumcari AP	35	13	97	70	28
NEW YORK					
Albany AP	43	0	88	74	23
Albany CO	43	5	89	74	20
Auburn	43	2	87	73	22
Batavia	43	3	87	74	22
Binghamton CO	42	2	89	72	20
Buffalo AP	43	6	86	73	21
Cortland	45	−1	88	73	23
Dunkirk	42	8	86	74	18
Elmira AP	42	5	90	73	24
Geneva	43	2	89	73	22
NEW YORK (continued)					
Glens Falls	43	−7	86	72	23
Gloversville	43	−2	87	73	23
Hornell	42	−5	85	72	24
Ithaca	42	0	88	73	24
Jamestown	42	5	86	73	20
Kingston	42	2	90	74	22
Lockport	43	6	85	74	21
Massena AP	45	−12	84	74	20
Newburgh-Stewart AFB	42	6	89	76	21
NYC-Central Park	41	15	91	76	17
NYC-Kennedy AP	41	21	87	76	16
NYC-LaGuardia AP	41	16	90	76	16
Niagara Falls AP	43	7	86	74	20
Olean	42	−3	85	72	23
Oneonta	42	−3	87	72	24
Oswego CO	44	6	84	74	20
Plattsburg AFB	45	−6	84	73	22
Poughkeepsie	42	3	90	75	21
Rochester AP	43	5	88	74	22
Rome-Griffiss AFB	43	−3	87	74	22
Schenectady	43	−1	88	73	22
Suffolk County AFB	41	13	84	75	16
Syracuse AP	43	2	87	74	20
Utica	43	−2	87	73	22
Watertown	44	−10	84	74	20
NORTH CAROLINA					
Asheville AP	36	17	88	74	21
Charlotte AP	35	22	94	77	20
Durham	36	19	92	77	20
Elizabeth City AP	36	22	91	79	18
Fayetteville, Pope AFB	35	20	94	79	20
Goldsboro, Seymour-Johnson AFB	35	21	92	79	18
Greensboro AP	36	17	91	76	21
Greenville	36	22	93	80	19
Henderson	36	16	92	78	20
Hickory	36	18	91	76	21
Jacksonville	35	25	92	80	18
Lumberton	35	22	93	80	20
New Bern AP	35	22	92	80	18
Raleigh/Durham AP	36	20	92	78	20
Rocky Mount	36	20	93	79	19
Wilmington AP	34	27	91	81	18
Winston-Salem AP	36	17	91	76	20
NORTH DAKOTA					
Bismarck AP	47	−19	91	72	27
Devel's Lake	48	−19	89	71	25
Dickinson AP	47	−19	93	70	25
Fargo AP	47	−17	88	74	25
Grand Forks AP	48	−23	87	72	25
Jamestown AP	47	−18	91	73	26
Minot AP	48	−20	88	70	25
Williston	48	−17	90	69	25
OHIO					
Akron/Canton AP	41	6	87	73	21
Ashtabula	42	7	87	75	18
Athens	39	7	91	76	22
Bowling Green	42	3	91	75	23
Cambridge	40	4	89	76	23
Chillicothe	39	9	91	76	22
Cincinnati CO	39	12	92	77	21
Cleveland AP	41	7	89	75	22
Columbus AP	40	7	88	76	24
Dayton AP	40	6	90	75	20
Defiance	41	1	91	76	24
Findlay AP	41	4	90	76	24
Fremont	41	3	90	75	24
Hamilton	39	8	92	77	22
Lancaster	40	5	91	76	23
Lima	41	4	91	76	24
Mansfield AP	41	3	89	75	22
Marion	41	6	91	76	23
Middletown	40	7	91	76	22
Newark	40	3	90	76	23
Norwalk	41	3	90	75	22
Portsmouth	39	9	92	76	22
Sandusky CO	42	8	90	75	21
Springfield	40	7	90	76	21
Steubenville	40	9	89	75	22
Toledo AP	42	5	90	75	25
Warren	41	4	88	74	23
Wooster	41	3	88	75	22
Youngstown AP	41	6	86	74	23
Zanesville AP	40	3	89	76	23
OKLAHOMA					
Ada	35	16	100	78	23
Altus AFB	35	18	101	76	25
Ardmore	34	19	101	78	23
Bartlesville	37	9	99	78	23
Chickasha	35	16	101	76	24
Enid-Vance AFB	36	14	100	77	24
Lawton AP	34	16	101	77	24
McAlester	35	17	100	78	23
Muskogee AP	36	16	99	78	23
Norman	35	15	99	77	24
Oklahoma City AP	35	15	97	77	23

Table 1 (continued)

Location[4]	Latitude[3]	Winter	Summer		
		Design Dry-Bulb	Design Dry-Bulb	Design Wet-Bulb	Daily Range
OKLAHOMA (continued)					
Ponca City	37	12	100	77	24
Seminole	35	16	100	77	23
Stillwater	36	13	99	77	24
Tulsa AP	36	16	99	78	22
Woodward	36	8	101	74	26
OREGON					
Albany	45	27	88	67	31
Astoria AP	46	30	76	60	16
Baker AP	45	1	92	65	30
Bend	44	4	87	62	33
Corvallis	44	27	88	67	31
Eugene AP	44	26	88	67	31
Grants Pass	42	26	92	66	33
Klamath Falls AP	42	5	87	62	36
Medford AP	42	23	94	68	35
Pendleton AP	46	10	94	65	29
Portland AP	46	24	85	67	23
Portland CO	46	29	88	68	21
Roseburg AP	43	29	91	67	30
Salem AP	45	25	88	67	31
The Dalles	46	17	91	68	28
PENNSYLVANIA					
Allentown AP	41	5	90	75	22
Altoona CO	40	5	87	73	23
Butler	41	2	89	74	22
Chambersburg	40	9	92	75	23
Erie AP	42	11	85	74	18
Harrisburg AP	40	13	89	75	21
Johnstown	40	5	87	73	23
Lancaster	40	6	90	76	22
Meadville	42	4	86	73	21
New Castle	41	4	89	74	23
Philadelphia AP	40	15	90	77	21
Pittsburgh AP	40	9	87	74	22
Pittsburgh CO	40	11	88	74	19
Reading CO	40	9	90	76	19
Scranton/Wilkes-Barre	41	6	87	74	19
State College	41	6	87	73	23
Sunbury	41	7	89	75	22
Uniontown	40	8	88	74	22
Warren	42	1	87	73	24
West Chester	40	13	90	76	20
Williamsport AP	41	5	89	75	23
York	40	8	91	76	22
RHODE ISLAND					
Newport	42	11	84	74	16
Providence AP	42	10	86	75	19
SOUTH CAROLINA					
Anderson	34	22	94	76	21
Charleston AFB	33	27	92	80	18
Charleston CO	33	30	93	80	13
Columbia AP	34	23	96	79	22
Florence AP	34	25	94	79	21
Georgetown	33	26	91	80	18
Greenville AP	35	23	93	76	21
Greenwood	34	23	95	77	21
Orangeburg	34	25	95	79	20
Rock Hill	35	21	95	77	20
Spartanburg AP	35	22	93	76	20
Sumter-Shaw AFB	34	26	94	79	21
SOUTH DAKOTA					
Aberdeen AP	46	−18	92	75	27
Brookings	44	−15	90	75	25
Huron AP	44	−12	93	75	28
Mitchell	44	−11	94	76	28
Pierre AP	44	− 9	96	74	29
Rapid City AP	44	− 6	94	71	28
Sioux Falls AP	44	−10	92	75	24
Watertown AP	45	−16	90	74	26
Yankton	43	− 7	94	76	25
TENNESSEE					
Athens	34	18	94	76	22
Bristol-Tri City AP	36	16	90	75	22
Chattanooga AP	35	19	94	78	22
Clarksville	37	16	96	78	21
Columbia	36	17	95	78	21
Dyersburg	36	17	96	79	21
Greenville	36	14	91	75	22
Jackson AP	36	17	95	79	21
Knoxville AP	36	17	92	76	21
Memphis AP	35	21	96	79	21
Murfreesboro	36	17	94	78	22
Nashville AP	36	16	95	78	21
Tullahoma	35	17	94	78	22
TEXAS					
Abilene AP	32	21	99	75	22
Alice AP	28	34	99	80	20
Amarillo AP	35	12	96	71	26
Austin AP	30	29	98	78	22
Bay City	29	33	93	80	16
Beaumont	30	33	94	80	19
Beeville	28	32	97	80	18
Big Spring AP	32	22	98	73	26
Brownsville AP	26	40	92	80	18
Brownwood	32	25	100	75	22
Bryan AP	31	31	98	78	20
Corpus Christi AP	28	36	93	80	19
Corsicana	32	25	100	78	21
TEXAS (continued)					
Dallas AP	33	24	99	78	20
Del Rio, Laughlin AFB	29	31	99	77	24
Denton	33	22	100	78	22
Eagle Pass	29	31	104	79	24
El Paso AP	32	25	98	69	27
Fort Worth AP	33	24	100	78	22
Galveston AP	29	36	89	81	10
Greenville	33	24	99	78	21
Harlingen	26	38	95	80	19
Houston AP	30	32	94	80	18
Houston CO	30	33	94	80	18
Huntsville	31	31	97	79	20
Killeen-Gray AFB	31	26	99	77	22
Lamesa	33	18	98	73	26
Laredo AFB	28	36	101	78	23
Longview	32	25	98	80	20
Lubbock AP	34	15	97	72	26
Lufkin AP	31	28	96	80	20
McAllen	26	38	100	79	21
Midland AP	32	23	98	73	26
Mineral Wells AP	33	22	100	77	22
Palestine CO	33	25	97	79	20
Pampa	36	11	98	72	26
Pecos	31	19	100	71	27
Plainview	34	14	98	72	26
Port Arthur AP	30	33	92	80	19
San Angelo, Goodfellow AFB	31	25	99	75	24
San Antonio AP	30	30	97	77	19
Sherman-Perrin AFB	34	23	99	78	22
Snyder	33	19	100	74	26
Temple	31	27	99	78	22
Tyler AP	32	24	97	79	21
Vernon	34	18	101	76	24
Victoria AP	29	32	96	79	18
Waco AP	32	26	99	78	22
Wichita Falls AP	34	19	100	76	24
UTAH					
Cedar City AP	38	6	91	64	32
Logan	42	7	91	65	33
Moab	39	16	98	65	30
Ogden CO	41	11	92	65	33
Price	40	7	91	64	33
Provo	40	6	93	66	32
Richfield	39	3	92	65	34
St. George CO	37	26	102	70	33
Salt Lake City AP	41	9	94	66	32
Vernal AP	40	− 6	88	63	32
VERMONT					
Barre	44	−13	84	72	23
Burlington AP	44	− 7	85	73	23
Rutland	44	− 8	85	73	23
VIRGINIA					
Charlottesville	38	15	90	77	23
Danville AP	36	17	92	77	21
Fredericksburg	38	14	92	78	21
Harrisonburg	38	9	90	77	23
Lynchburg AP	37	19	92	76	21
Norfolk AP	37	23	91	78	18
Petersburg	37	18	94	79	20
Richmond AP	38	18	93	78	21
Roanoke AP	37	18	91	75	23
Staunton	38	12	90	77	23
Winchester	39	10	92	76	21
WASHINGTON					
Aberdeen	47	27	80	61	16
Bellingham AP	49	18	74	65	19
Bremerton	48	29	81	66	20
Ellensburg AP	47	6	89	65	34
Everett-Paine AFB	48	24	78	65	20
Kennewick	46	15	96	68	30
Longview	46	24	86	66	30
Moses Lake, Larson AFB	47	− 1	93	66	32
Olympia AP	47	25	83	65	32
Port Angeles	48	29	73	58	18
Seattle-Boeing Fld	48	27	80	65	24
Seattle CO	48	32	79	65	19
Seattle-Tacoma AP	48	24	81	64	22
Spokane AP	48	4	90	64	28
Tacoma-McChord AFB	47	24	81	66	22
Walla Walla AP	46	16	96	68	27
Wenatchee	47	9	92	66	32
Yakima AP	46	10	92	67	36
WEST VIRGINIA					
Beckley	38	6	88	73	22
Bluefield AP	37	10	86	73	22
Charleston AP	38	14	90	75	20
Clarksburg	39	7	90	75	21
Elkins AP	39	5	84	73	22
Huntington CO	38	14	93	76	22
Martinsburg AP	39	10	94	77	21
Morgantown AP	40	7	88	74	22
Parkersburg CO	39	12	91	76	21
Wheeling	40	9	89	75	21
WISCONSIN					
Appleton	44	− 6	87	74	23
Ashland	46	−17	83	71	23
Beloit	42	− 3	90	76	24

See footnotes at end of table.

Table 1 (continued)

Location[4]	Latitude[3]	Winter	Summer		
		Design Dry-Bulb	Design Dry-Bulb	Design Wet-Bulb	Daily Range
WISCONSIN (continued)					
Eau Claire AP	45	−11	88	74	23
Fond du Lac	44	− 7	87	74	23
Green Bay AP	44	− 7	85	73	23
La Crosse AP	44	− 8	88	76	22
Madison AP	43	− 5	88	75	22
Manitowoc	44	− 1	86	74	21
Marinette	45	− 4	86	72	20
Milwaukee AP	43	− 2	87	75	21
Racine	43	0	88	75	21
Sheboygan	44	0	87	74	20
Stevens Point	44	−12	87	73	23
Waukesha	43	− 2	89	75	22
Wausau AP	45	−14	86	72	23
WYOMING					
Casper AP	43	− 5	90	62	31
Cheyenne AP	41	− 2	86	62	30
Cody AP	44	− 9	87	60	32
Evanston	41	− 8	82	57	32
Lander AP	43	−12	90	62	32
Laramie AP	41	− 2	80	59	28
Newcastle	44	− 5	89	67	30
Rawlins	42	−11	84	61	40
Rock Springs AP	42	− 1	84	57	32
Sheridan AP	45	− 7	92	65	32
Torrington	42	− 7	92	67	30
CANADA					
ALBERTA					
Calgary AP	51	−25	85	64	26
Edmonton AP	54	−26	83	67	23
Grande Prairie AP	55	−37	81	64	23
Jasper CO	53	−28	84	64	28
Lethbridge AP	50	−24	88	66	28
McMurray AP	57	−39	84	67	28
Medicine Hat AP	50	−26	93	69	28
Red Deer AP	52	−28	86	65	25
BRITISH COLUMBIA					
Dawson Creek	56	−35	81	64	25
Fort Nelson AP	59	−41	84	64	23
Kamloops CO	51	−10	94	69	31
Nanaimo CO	49	20	78	64	20
New Westminster CO	49	19	84	66	20
Penticton AP	50	3	91	69	31
Prince George AP	54	−31	82	65	26
Prince Rupert CO	54	15	71	60	13
Trail	49	3	91	68	30
Vancouver AP	49	19	78	66	17
Victoria CO	48	23	76	62	16
MANITOBA					
Brandon CO	50	−26	87	73	26
Churchill AP	59	−38	75	66	18
Dauphin AP	51	−26	86	72	24
Flin Flon CO	55	−36	81	69	19
Portage la Prairie AP	50	−22	87	74	22
The Pas AP	54	−32	81	71	20
Winnipeg AP	50	−25	87	74	23
NEW BRUNSWICK					
Campbellton CO	48	−14	84	71	20
Chatham AP	47	−10	87	71	22
Edmundston CO	47	−16	81	72	21
Fredericton AP	46	−10	86	70	23
Moncton AP	46	− 7	85	71	21
Saint John AP	45	− 7	79	68	18
NEWFOUNDLAND					
Corner Brook CO	49	− 5	81	68	18
Gander AP	49	− 1	82	68	20
Goose Bay AP	53	−25	81	67	18
St. John's AP	48	6	77	68	17
Stephenville	48	− 1	76	68	13

Location[4]	Latitude[3]	Winter	Summer		
		Design Dry-Bulb	Design Dry-Bulb	Design Wet-Bulb	Daily Range
NORTHWEST TERRITORIES					
Fort Smith AP	60	−46	83	65	25
Frobisher Bay AP	64	−42	59		14
Inuvik	68	−48	77	61	23
Resolute AP	75	−47	51		10
Yellowknife AP	62	−47	76	63	17
NOVA SCOTIA					
Amherst	46	− 5	82	70	21
Halifax AP	45	4	80	68	16
Kentville CO	45	0	83	70	23
New Glasgow	46	− 5	81	70	21
Sydney AP	46	5	82	70	20
Truro CO	45	− 7	81	70	22
Yarmouth AP	44	9	73	68	15
ONTARIO					
Belleville CO	44	− 7	86	75	21
Chatham CO	42	6	90	75	20
Cornwall	45	− 9	86	75	23
Thunder Bay AP	48	−23	83	70	23
Hamilton	43	3	88	75	21
Kapuskasing AP	50	−28	84	71	23
Kenora AP	50	−28	83	73	20
Kingston CO	44	− 7	82	75	20
Kitchener	44	1	85	75	24
London AP	43	3	88	75	22
North Bay AP	46	−17	84	70	18
Oshawa	44	− 2	87	75	21
Ottawa AP	45	−13	87	74	21
Owen Sound	44	− 1	84	72	21
Peterborough CO	44	− 9	87	74	22
St. Catharines CO	43	5	88	75	20
Sarnia	43	6	90	74	19
Sault Ste. Marie CO	46	−15	85	70	22
Sudbury	46	−15	86	70	25
Timmins CO	48	−28	87	71	24
Toronto AP	44	1	87	75	22
Windsor AP	42	7	90	75	20
PRINCE EDWARD ISLAND					
Charlottetown AP	46	− 3	81	70	16
Summerside AP	46	− 3	81	70	16
QUEBEC					
Bagotville	48	−22	84	71	20
Chicoutimi CO	48	−20	83	71	20
Drummondville CO	46	−13	85	74	22
Granby	45	−12	84	74	21
Hull	46	−13	87	74	21
Mégantic AP	46	−16	81	73	19
Montréal AP	46	−10	86	74	18
Québec AP	47	−13	82	73	21
Rimouski	48	−12	74	69	18
St. Jean	45	−10	85	74	20
St. Jérôme	46	−13	84	74	23
Sept Îles AP	50	−22	78	64	17
Shawinigan	46	−15	85	74	21
Sherbrooke CO	45	−13	84	73	20
Thetford Mines	46	−14	83	73	22
Trois Rivières CO	46	−13	85	74	23
Val d'Or AP	48	−27	85	71	22
Valleyfield	45	− 9	85	74	21
SASKATCHEWAN					
Estevan AP	49	−25	89	73	25
Moose Jaw AP	50	−27	89	71	27
North Battleford AP	53	−29	86	69	25
Prince Albert AP	53	−35	84	70	25
Regina AP	50	−29	88	71	27
Saskatoon AP	52	−30	86	69	25
Swift Current AP	50	−25	89	70	24
Yorkton AP	51	−28	85	72	23
YUKON TERRITORY					
Whitehorse AP	61	−42	75	60	22

[1] Data extracted from *ASHRAE Handbook of Fundamentals–1972*. Winter design dry-bulb temperature shown is for 97½% incidence; summer dry-bulb and wet-bulb temperatures represent the highest 2½% of all the hours during the summer months.

[2] More detailed data for Arizona, California, and Nevada have been published by the Golden Gate Chapter of ASHRAE and for Southern California, Arizona, and Nevada by the Southern California Chapter of ASHRAE.

[3] Latitude shown to nearest whole degree.

[4] Weather station locations are indicated as follows: AP = airport, AFB = Air Force Base, CO = urban location. No designation indicates semirural location.

Table 2*
SOLAR RADIATION HEAT GAIN THROUGH GLASS
Btu/h per ft²

Latitude	24°				32°				40°				48°				56°			
St'd. Time	9 am	Noon	3 pm	6 pm	9 am	Noon	3 pm	6 pm	9 am	Noon	3 pm	6 pm	9 am	Noon	3 pm	6 pm	9 am	Noon	3 pm	6 pm
(Facing) N (or shaded)	28	37	33	12	27	36	32	14	25	34	30	15	23	32	28	15	22	29	26	15
NE	135	47	32	9	118	41	31	9	101	36	30	10	145	33	28	11	69	30	26	11
E	202	71	32	9	202	70	31	9	199	68	30	10	195	65	28	11	188	61	26	11
SE	153	83	33	9	168	107	45	9	181	131	34	10	191	151	35	11	197	167	38	11
S	32	68	48	9	41	104	71	11	59	141	98	14	76	171	123	18	91	194	143	22
SW	26	50	147	65	25	64	168	98	24	84	187	90	23	98	200	100	22	112	210	108
W	26	39	171	100	25	38	168	115	24	36	165	129	23	33	161	139	21	30	154	145
NW	26	38	153	76	25	36	82	85	24	34	65	92	23	32	50	96	21	29	39	97
Horiz.	153	267	215	39	150	256	207	41	142	239	194	42	131	215	176	43	116	187	154	42

*Based on data in *ASHRAE Handbook of Fundamentals—1972.* Values shown are for August 21st, double strength (1/8 in.) sheet glass.

Table 3
FACTORS FOR SOLAR RADIATION HEAT GAIN THROUGH GLASS

Type of Glass	No Shading	Shading (Closed Venetian Blind, Lined Drapery, or Roller Shade)
Sheet Glass (1/8 in.)	1.00	.64
Plate Glass (1/4 in.)	.95	.64
Heat Absorbing or Tinted (3/16 in.)	.72*	.57*
Reflective (1/4 in.)	.30–.60*	.25–.50*
Reflective Film Applied to Inside of Clear Glass	.25–.45*	.21–.35*
Double Glass		
Clear Sheets (1/8 in.)	.90	.57
Clear Plates (1/4 in.)	.83	.57
Heat Absorbing Plate out; Clear Plate in	.56	.39

*Refer to manufacturer's ratings for exact values.

Table 4
SHADING FACTORS FOR OVERHANGS[1]

Latitude	24°				32°				40°				48°				56°			
St'd. Time	9 am	Noon	3 pm	6 pm	9 am	Noon	3 pm	6 pm	9 am	Noon	3 pm	6 pm	9 am	Noon	3 pm	6 pm	9 am	Noon	3 pm	6 pm
(Facing) N	––	––	––	.58	––	––	––	.63	––	––	––	.83	––	––	––	1.37	––	––	––	1.61
NE	1.89	––	––	––	2.17	––	––	––	2.13	––	––	––	3.03	––	––	––	3.45	––	––	––
E	1.00	––	––	––	.97	––	––	––	.89	––	––	––	.83	––	––	––	.74	––	––	––
SE	.93	4.55	––	––	1.00	3.33	––	––	.86	2.33	––	––	.73	1.67	––	––	.61	1.33	––	––
S	4.35	3.57	4.35	––	2.63	2.38	2.63	––	1.85	1.59	1.85	––	1.33	1.19	1.33	––	1.08	.93	1.08	––
SW	––	4.55	.93	––	––	3.33	1.00	––	––	2.33	.86	––	––	1.67	.73	––	––	1.33	.61	––
W	––	––	1.00	*	––	––	.97	*	––	––	.89	*	––	––	.83	*	––	––	.74	*
NW	––	––	1.89	*	––	––	2.17	*	––	––	2.13	*	––	––	3.03	*	––	––	3.45	*

[1]Extracted from the *ASHRAE Handbook of Fundamentals—1972.*

*Excessive length of overhang required.

(——)Glass is entirely shaded.

Table 5
HEAT TRANSMISSION FACTORS FOR GLASS*

Type of Glass	U-Value*		
	Summer		Winter
	No Shading	Shading**	
Single Glass	1.06	.81	1.13
Double Glass (1/4 in. air space)	.61	.52	.65
Prime Window + Storm Window	.54	.47	.56

*Btu/(hr)(ft^2) (deg F temp diff)

**Values apply to tightly closed Venetian blind, lined drapery, or roller shade.

Table 6
TEMPERATURE CORRECTION FOR TIME-OF-DAY[1]

Standard Time	9 am	Noon	3 pm	6 pm
Daily Range Factor[2]	.71	.23	0	.21

[1]Extracted from *ASHRAE Handbook of Fundamentals*—1972.

[2]Example:

Design dry-bulb temperature = 95 F
Daily range = 25 deg F
Find temperature at 6 pm

Temperature at 6 pm = Design DB – (Daily range x factor)
= 95 – (25 x .21)
= 95 – 5.25 = 89.75 or 90 F

Table 7
HEAT TRANSMISSION FACTORS
(U-values) *

Construction	U-value	
	Summer	Winter
WALLS		
Frame with wood siding, sheathing, and inside finish		
No insulation	.22	.23
R-7 insulation (2 in.—2¾ in.)	.09	.09
R-11 insulation (3 in.—3½ in.)	.07	.07
Frame with 4 in. brick or stone veneer, sheathing, and inside finish		
No insulation	.24	.24
R-7 insulation	.09	.09
R-11 insulation	.07	.07
Frame with 1 in. stucco, sheathing, and inside finish		
No insulation	.29	.29
R-7 insulation	.10	.10
R-11 insulation	.07	.07
Masonry:		
8 in. concrete block, no finish	.49	.51
12 in. concrete block, no finish	.45	.47
Masonry (8 in. concrete block):		
Inside finish:		
furred gypsum wallboard (½ in.); no insulation	.29	.30
furred, foil-backed gypsum wallboard (½ in.); no insulation	.29	.30
1 in. polystyrene insulation board (R-5), and ½ in. gypsum wallboard	.13	.13
Masonry (8 in. cinder block or hollow clay tile):		
Inside finish:		
furred gypsum wallboard (½ in.); no insulation	.25	.25
furred, foil-backed gypsum wallboard (½ in.); no insulation	.17	.17
1 in. polystyrene insulation board (R-5), and ½ in. gypsum wallboard	.12	.12
Masonry (4 in. face brick and 8 in. cinder block or 8 in. hollow clay tile):		
Inside finish:		
furred gypsum wallboard (½ in.); no insulation	.22	.22
furred, foil-backed gypsum wallboard (½ in.); no insulation	.15	.16
1 in. polystyrene insulation board (R-5), and ½ in. gypsum wallboard	.12	.12
Masonry (12 in. hollow clay tile or 12 in. cinder block):		
Inside finish:		
furred gypsum wallboard (½ in.); no insulation	.24	.24
furred, foil-backed gypsum wallboard (½ in.); no insulation	.16	.17
1 in. polystyrene insulation board (R-5), and ½ in. gypsum wallboard	.12	.12
Masonry (4 in. face brick, 4 in. common brick):		
Inside finish:		
furred gypsum wallboard (½ in.); no insulation	.28	.28
furred, foil-backed gypsum wallboard (½ in.); no insulation	.18	.19
1 in. polystyrene insulation board (R-5), and ½ in. gypsum wallboard	.13	.13
Masonry (8 in. concrete or 8 in. stone):		
Inside finish:		
furred gypsum wallboard (½ in.); no insulation	.33	.34
furred, foil-backed gypsum wallboard (½ in.); no insulation	.21	.21
1 in. polystyrene insulation board (R-5), and ½ in. gypsum wallboard	.14	.14
Metal with vinyl inside finish, R-7 (3 in. glass fiber batt)	.14	.14

*Btu per hour, sq ft, deg temp difference. $\left[\frac{\text{Btu/h}}{(\text{ft}^2) \times (\text{TD, deg F})}\right]$

Table 7 (continued)

Construction	U-value	
	Summer	Winter
PARTITIONS		
Frame (½ in. gypsum wallboard one side only):		
No insulation	.55	.55
Frame (½ in. gypsum wallboard each side):		
No insulation	.31	.31
R-11 insulation	.08	.08
Masonry (4 in. cinder block):		
No insulation, no finish	.40	.40
No insulation, one side furred gypsum wallboard (½ in.)	.26	.26
No insulation, both sides furred gypsum wallboard (½ in.)	.19	.19
One side 1 in. polystyrene insulation board (R-5), and ½ in. gypsum wallboard	.13	.13
CEILING-FLOOR		
Frame (asphalt tile floor, 5/8 in. plywood, 25/32 in. wood subfloor, finished ceiling):		
Heat flow up	.23	.23
Heat flow down	.20	.19
Concrete (asphalt tile floor, 4 in. concrete deck, air space, finished ceiling):		
Heat flow up	.34	.33
Heat flow down	.26	.25
ROOF (flat roof, no finished ceiling)		
Steel deck:		
No insulation	.64	.86
1 in. insulation (R-2.78)	.23	.25
2 in. insulation (R-5.56)	.15	.16
1 in. Wood deck:		
No insulation	.40	.48
1 in. insulation (R-2.78)	.19	.21
2 in. insulation (R-5.56)	.12	.13
2.5 in. Wood deck:		
No insulation	.25	.28
1 in. insulation (R-2.78)	.15	.16
2 in. insulation (R-5.56)	.10	.11
4 in. Wood deck:		
No insulation	.17	.18
1 in. insulation (R-2.78)	.12	.12
2 in. insulation (R-5.56)	.09	.09
ROOF-CEILING (flat roof, finished ceiling)		
Steel deck:		
No insulation	.33	.40
1 in. insulation (R-2.78)	.17	.19
2 in. insulation (R-5.56)	.12	.13
1 in. Wood deck:		
No insulation	.26	.29
1 in. insulation (R-2.78)	.15	.16
2 in. insulation (R-5.56)	.11	.11
2.5 in. Wood deck:		
No insulation	.18	.20
1 in. insulation (R-2.78)	.12	.13
2 in. insulation (R-5.56)	.09	.10
4 in. Wood deck:		
No insulation	.14	.15
1 in. insulation (R-2.78)	.10	.10
2 in. insulation (R-5.56)	.08	.08
4 in. Light weight concrete deck:		
No insulation	.14	.15

Table 7 (continued)

Construction	U-value	
	Summer	Winter
ROOF-CEILING (continued)		
6 in. Light weight concrete deck:		
No insulation	.10	.11
8 in. Light weight concrete deck:		
No insulation	.08	.09
2 in. Heavy weight concrete deck:		
No insulation	.32	.38
1 in. insulation (R-2.78)	.17	.19
2 in. insulation (R-5.56)	.11	.12
4 in. Heavy weight concrete deck:		
No insulation	.30	.36
1 in. insulation (R-2.78)	.16	.18
2 in. insulation (R-5.56)	.11	.12
6 in. Heavy weight concrete deck:		
No insulation	.28	.33
1 in. insulation (R-2.78)	.16	.17
2 in. insulation (R-5.56)	.11	.12
ROOF-CEILING (wood frame pitched roof, finished ceiling on rafters)		
No insulation	.28	.29
R-19 insulation (5¼ in.—6½ in.)	.05	.05
ROOF-ATTIC-CEILING (attic with natural ventilation)		
No insulation	.15	.29
R-19 insulation (5¼ in.—6½ in.)	.04	.05
FLOORS		
Concrete slab on grade:		
No insulation	0	50*
1 in. expanded polystyrene board, 2 ft deep or 2 ft wide	0	30*
Floor over unconditioned space, no ceiling		
Wood frame:		
No insulation	.33	.27
R-7 insulation (2 in.—2¾ in.)	.09	.08
Concrete deck:		
No insulation	.59	.43
R-7 insulation	.10	.09
DOORS		
Solid wood:		
1 in. thick	.61	.64
1½ in. thick	.47	.49
2 in. thick	.42	.43
Steel:		
1¾ in. thick, mineral fiber core	.58	.59
1¾ in. thick, polystyrene core	.46	.47
1¾ in. thick, urethane foam core	.39	.40

*Btu/h per liner foot of exposed edge.

Table 8
EQUIVALENT TEMPERATURE DIFFERENCES FOR SUNLIT AND SHADED WALLS
(degrees F)

Wall* Construction	Standard Time	NE		E		SE		S		SW		W		NW		(Shade) N	
		Dk	Lt	Dk	Lt	Dk	Lt	Dk	Lt	Dk	Lt	Dk	Lt	Dk	Lt	Dk	Lt
LIGHT CONSTRUCTION	9 am	28	17	35	20	29	17	16	10	18	12	18	12	15	10	14	9
	Noon	27	17	38	22	38	23	27	17	24	15	24	15	20	14	17	12
	3 pm	24	17	29	20	31	21	32	21	37	24	34	22	26	18	20	15
	6 pm	23	17	26	19	26	18	26	18	41	25	47	30	37	24	21	16
LIGHT MEDIUM CONSTRUCTION	9 am	12	8	14	9	11	7	6	4	8	5	9	6	7	5	7	5
	Noon	25	14	34	18	27	15	11	7	9	7	9	6	9	4	10	6
	3 pm	29	18	35	23	39	22	26	16	21	16	18	12	15	11	16	11
	6 pm	30	20	37	24	39	25	36	24	41	24	38	25	29	20	22	17
HEAVY MEDIUM CONSTRUCTION	9 am	14	11	17	13	16	12	14	11	18	12	20	16	17	11	12	10
	Noon	17	11	21	14	19	12	13	9	15	10	16	11	14	10	11	8
	3 pm	21	14	28	19	25	15	16	11	14	11	17	11	14	10	12	9
	6 pm	25	16	32	19	30	18	23	15	23	15	22	15	18	12	15	11
HEAVY CONSTRUCTION	9 am	20	14	26	16	23	15	20	14	24	16	26	17	21	15	15	11
	Noon	19	13	24	15	22	14	19	13	24	15	24	16	20	14	14	11
	3 pm	20	13	24	16	22	15	18	13	22	14	23	15	19	13	14	10
	6 pm	20	14	26	16	25	16	19	13	22	14	23	15	18	13	14	11

*For examples of each type of wall construction, see note 5 below.

NOTES:

1. Based on outdoor design temperature of 95 F, indoor design temperature of 75 F.
2. When indoor-outdoor temperature difference is greater (or less) than 20 degrees, add the excess to (or subtract the deficiency from) the values in the table.
3. To calculate the heat loss or gain through a partition separating a conditioned space from an unconditioned space, use a temperature difference 5 degrees less than the design temperature difference unless another temperature is expected to prevail.
4. All walls include an inside finish of 3/4 in. plaster or gypsum board.
5. Wall construction details:

LIGHT CONSTRUCTION
1 in. stucco + frame
4 in. light weight concrete block + airspace

LIGHT MEDIUM CONSTRUCTION
1 in. stucco + 4 in. common brick
or + 4 in. heavy weight concrete block with or without 2 in. insulation
or + 8 in. heavy weight concrete block with or without 1 in. insulation
or + 2 in. insulation + 4 in. heavy weight concrete block
4 in. face brick + 4 in. light weight concrete block with or without 1 in. insulation

HEAVY MEDIUM CONSTRUCTION
4 in. face brick + 4 in. common brick
or + 2 in. insulation + 4 in. concrete block or common brick or 8 in. heavy weight concrete block
or + 8 in. clay tile + 1 in. insulation
or + 8 in. common brick
or + airspace + 4 in. heavy weight concrete block
1 in. stucco + 8 in. clay tile + 1 in. insulation or airspace
or + 2 in. insulation + 4 in. common brick
or + 12 in. heavy weight concrete block

HEAVY CONSTRUCTION
4 in. face brick + 8 in. common brick + 1 in. insulation
or + 2 in. insulation or airspace + 8 in. clay tile or 4 in. common brick or 4 in. heavy weight concrete block or 8 in. heavy weight concrete block
or + 2 in. insulation + 8 in. common brick or 8 in. heavy weight concrete
or + 8 in. clay tile + airspace
or + airspace + 8 in. common brick or 12 in. heavy weight concrete
or + 2 in. insulation + 12 in. heavy weight concrete
1 in. stucco + 2 in. insulation + 8 in. common brick or 12 in. heavy weight concrete

Table 9
EQUIVALENT TEMPERATURE DIFFERENCES FOR HEAT GAIN THROUGH FLAT ROOFS[1]
(degrees F)

Roof Construction[2]	STANDARD TIME							
	am				pm			
	9		12		3		6	
	Dk	Lt	Dk	Lt	Dk	Lt	Dk	Lt
LIGHT CONSTRUCTION								
Steel deck with 1 in.–2 in. insulation	34	14	81	42	90	50	56	34
1 in. Wood with 1 in.–2 in. insulation	19	6	65	32	88	48	70	40
2.5 in. Wood with 1 in.–2 in. insulation	7	−1	38	17	68	35	73	40
MEDIUM CONSTRUCTION								
4 in. Wood with 1 in.–2 in. insulation	8	1	21	8	44	19	60	32
4 in. Light Weight concrete (no insulation) or 2 in. Heavy Weight concrete with 1 in.–2 in. insulation	8	1	40	17	70	36	75	41
6 in. or 8 in. Light Weight concrete (no insulation)	4	−1	16	6	41	19	62	32
HEAVY CONSTRUCTION								
4 in. Heavy Weight concrete with 1 in.–2 in. insulation	11	3	21	8	39	19	53	28
6 in. Heavy Weight concrete with 1 in.–2 in. insulation	18	9	21	9	33	15	44	22
ROOFS IN SHADE								
Light	3		11		18		17	
Medium	2		7		15		17	
Heavy	3		5		11		15	

[1]Values in this table are based on 20 degree design difference between outdoor and indoor temperature. When the design difference is greater (or less) than 20 degrees, add the excess to (or subtract the deficiency from) the values in the table.

[2]Includes ½ in. slag, a membrane, and 3/8 in. felt on top.

Table 10
HEAT GAIN FROM OCCUPANTS[1]

Degree of Activity	Typical Application	Total Heat Btu/h	Sensible Heat Btu/h	Latent Heat Btu/h
Seated, at rest	Theater/Matinee, Grade School Classroom	330	225	105
	Theater/Evening	350	245	105
Seated, very light work	Office, Hotel, Apartment, High School Classroom	400	245	155
Moderately active office work	Office, Hotel, Apartment, College Classroom	450	250	200
Standing, light work; walking slowly	Drug Store, Bank	500	250	250
Sedentary work	Restaurant[2]	550	275	275
Light bench work	Factory	750	275	475
Moderate dancing	Dance Hall	850	305	545
Walking, 3 mph; Moderately heavy work	Factory	1,000	375	625
Bowling[3] Heavy work	Bowling alley, Factory	1,450	580	870

[1]This table was extracted from the *1972 ASHRAE Handbook of Fundamentals*.

[2]The adjusted total heat value for sedentary work, restaurant, includes 60 Btu/h for food per individual (30 Btu/h sensible and 30 Btu/h latent).

[3]For bowling, figure one person per alley actually bowling and all others sitting (400 Btu/h) or standing (550 Btu/h).

NOTE: The above values are based on 75 F room dry-bulb temperature. For 80 F room dry-bulb temperature, the total heat gain remains the same; but the sensible heat values should be decreased by 20 percent and this amount of heat load added to the latent heat value to arrive at the same total heat.

Table 11
HEAT GAIN FROM ELECTRIC MOTORS
(Continuous Operation)*
(Btu/h)

Motor Horsepower	Location of Equipment with Respect to Air Stream or Conditioned Space		
	Motor and Driven Machine In	Motor Out Driven Machine In	Motor In Driven Machine Out
1/8	580	320	260
1/6	710	430	280
1/4	1,000	640	360
1/3	1,290	850	440
1/2	1,820	1,280	540
3/4	2,680	1,930	750
1	3,220	2,540	680
1-1/2	4,770	3,820	950
2	6,380	5,100	1,280
3	9,450	7,650	1,800
5	15,600	12,800	2,800
7-1/2	22,500	19,100	3,400
10	30,000	25,500	4,500
15	44,500	38,200	6,300
20	58,500	51,000	7,500
25	72,400	63,600	8,800

*From Table 30, p. 417, Chapter 22, *ASHRAE Handbook of Fundamentals—1972.*

Table 12

HEAT GAIN FROM APPLIANCES

(Btu/h)[1]

TYPE OF APPLIANCE	ELECTRIC				GAS				STEAM			
	Without Hood			Hood[2]	Without Hood			Hood[2]	Without Hood			Hood[2]
	Sensible	Latent	Total	All Sensible	Sensible	Latent	Total	All Sensible	Sensible	Latent	Total	All Sensible
Broiler-Griddle 31 in. x 20 in. x 18 in.					11,700	6,300	18,000	3,600				
Coffee brewer/warmer					1,750	750	2,500	500				
per burner	770	230	1,000	340								
per warmer	230	70	300	90								
Coffee urn:												
3 gallon	2,550	850	3,400	1,000	3,500	1,500	5,000	1,000	2,180	1,120	3,300	1,000
5 gallon	3,850	1,250	5,100	1,600	5,250	2,250	7,500	1,500	3,300	1,700	5,000	1,600
8 gallon (twin)	5,200	1,600	6,800	2,100	7,000	3,000	10,000	2,000	4,350	2,250	6,600	2,100
Deep fat fryer:												
15# fat	2,800	6,600	9,400	3,000	7,500	7,500	15,000	3,000				
21# fat	4,100	9,600	13,700	4,300								
Dry food warmer per sq ft top	320	80	400	130	560	140	700	140				
Griddle, frying per sq ft top	3,000	1,600	4,600	1,500	4,900	2,600	7,500	1,500				
Hot plate (two heating units)					5,300	3,600	8,900	2,800				
Short order stove (open grates) per burner					3,200	1,800	5,000	1,000				
Steam table per sq ft top					750	500	1,250	250	500	325	825	260
Toaster:												
Continuous—												
360 slices per hour	1,960	1,740	3,700	1,200	3,600	2,400	6,000	1,200				
720 slices per hour	2,700	2,400	5,100	1,600	6,000	4,000	10,000	2,000				
Pop-up (4 slice)	2,230	1,970	4,200	1,300								
Waffle iron 18 in. x 20 in. x 13 in. (2 grids)	1,680	1,120	2,800	900								
Hair dryer:												
Blower type	2,300	400	2,700									
Helmet type	1,870	330	2,200									
Lab burners:												
Bunsen					1,680	420	2,100					
Fishtail					2,800	700	3,500					
Meeker					3,360	840	4,200					
Neon sign per foot of tube	60		60									
Sterilizer	650	1,200	1,850									
Vending machines:												
Hot drink			1,200									
Cold drink			625									

[1] Average operation for one hour.

[2] Does not include the latent and sensible heat loads imposed by the outdoor makeup air exhausted through the hood. Includes only the sensible heat load resulting from direct radiation from the appliance.

Table 13
INFILTRATION*
(ft³/min)

KIND OF ROOM OR BUILDING	AIR CHANGES PER HOUR			
	Summer		Winter	
	Ordinary	Weatherstripping or Storm Sash	Ordinary	Weatherstripping or Storm Sash
No Windows or Outside Doors	0.30	0.15	0.50	0.25
Entrance Halls	1.20 to 1.80	0.60 to 0.90	2.00 to 3.00	1.00 to 1.50
Reception Halls	1.20	0.60	2.00	1.00
Bath Rooms	1.20	0.60	2.00	1.00
Infiltration through Windows:				
Rooms, 1 side exposed	0.60	0.30	1.00	0.50
Rooms, 2 sides exposed	0.90	0.45	1.50	0.75
Rooms, 3 sides exposed	1.20	0.60	2.00	1.00
Rooms, 4 sides exposed	1.20	0.60	2.00	1.00

*The air quantity is computed as follows (applicable to buildings up to 3 stories in height):

$$\frac{(H) \times (L) \times (W) \times (AC)}{60} = \text{ft}^3/\text{min}$$

Where: H = room height, ft
L = room length, ft
W = room width, ft
AC = air changes per hour

NOTE: The total simultaneous infiltration for an entire building will be approximately 50 percent of the sum of the infiltration allowances of individual rooms.

DOOR INFILTRATION*,[1,6]
(ft³/min)
Without Vestibule[3]

Temperature Difference[4] (TD) *(degrees F)*	Traffic Rate (TR)[5] (Traffic Rate = Persons/hr for EACH DOOR)							
	10	20	40	60	80	100	200	400
10	4	8	16	24	32	40	80	160
20	8	16	32	48	64	80	160	320
40	16	32	64	96	128	160	320	640
60	24	48	96	144	192	240	480	960
80	32	64	128	192	256	320	640	1280
100	40	80	160	240	320	400	800	1600

*Based on data in ASHRAE *Handbook of Fundamentals* (1972), especially Chapter 19, pages 339-342. Table values are based on the assumption that during severe weather, doors other than main entrance doors *in one wall only* will be kept closed.

[1]Values in table are for a one-story building. For a two-story building, multiply table value by 1.50; for a three-story building, multiply table value by 1.75.

[2]Door is 7 ft x 3 ft hinged door. Infiltration for a door other than 3 ft x 7 ft is proportional to the perimeter of the door.

[3]*With vestibule*, multiply table value by 0.60.

[4]*Difference* between indoor temperature and outdoor temperature at time for which load is being calculated. See footnote 2, Table 6 (page 44), for time of day correction. Nearest TD is sufficiently accurate for most estimating purposes.

[5]Determine *traffic rate* by actual count or by dividing the expected number of persons in the conditioned space by the average time (hours) of occupancy and by the number of doors.

[6]Calculate *total door infiltration* by multiplying the number of doors by the table value corresponding to the applicable TR and TD.

Table 14
VENTILATION REQUIREMENTS[1]

Application	Ft^3/min per Person
Banking space	7
Barber shop	7
Beauty parlor	25
Bowling alley	15
Cocktail lounge, bar	30
Department store	
Retail shop	7
Storage area serving sales area	5
Drug store	
Pharmacists' workroom	20
Sales area	7
Factory[2,3]	10–35
Garage	
Parking	1.5[4]
Repair[5]	1.5[4]
Hospital[3]	
Single or double room	10
Ward	10
Corridor	20
Operating room[6]	20
Food service center	35
Hotel	
Bedroom	7
Living room (suite)	10
Bath	20
Corridor	5
Lobby	7
Conference room (small)	20
Assembly room (large)	15
Public restroom	15
Laboratory	15
Office	
General	15
Conference room	25
Waiting room	10
Pool or Billiard room	20
Restaurant	
Dining room	10
Kitchen	30
Cafeteria, short order, drive-in	30
School[3]	
Classroom	10
Laboratory	10
Shop	10
Auditorium	5
Gymnasium	20
Library	7
Office	7
Lavatory	15
Locker room[7]	30
Dining room	10
Corridor	15
Dormitory bedroom	7
Theater	
Lobby	20
Auditorium	
Smoking	10
No smoking	5
Restroom	15

[1] Data extracted from *ASHRAE Standard 62-73*. Minimum values used.

[2] Special contaminant control systems may be required.

[3] State or local codes are usually determining factor.

[4] Ft^3/min per square foot of floor area.

[5] Where engines are run, positive engine exhaust withdrawal system must be used.

[6] All outside air often required to avoid hazard of anesthetic explosion.

[7] Special exhaust systems required.

Table 15
OCCUPANCY ESTIMATES[1]

Application	Square Feet Floor Area per Person
Assembly hall, church or school auditorium, funeral parlor, theater	7
Barber shop	40
Beauty parlor	20
Bowling alley	(seating capacity plus 6 persons per alley)
Classrooms	20
Conference room	14-17
Department stores, retail shops:	
Basement and first floor	33
Other floors	50
Dormitory	50
Food services:	
Dining room	14
Cafeteria, short order, drive-in	10
Kitchen	50
Laboratory	20
Library	50
Office, general	100
Recreation facilities:	
Ballroom	10
Pool room	40
Restrooms, public	10
Tavern, bar, cocktail lounge:	
Stand-up crowd	7
Average crowd	10

[1]Code requirements should be followed whenever they apply.

Table 16a

DEHUMIDIFICATION (75 F)

Latent Heat Load to Dehumidify Outdoor Air to 50% RH
at *75 F* Indoors—Btu/h per 100 ft^3/min

Outdoor Dry-bulb Temperature *(degrees F)*	Outdoor Wet-bulb Temperature *(degrees F)*																	
	64	65	66	67	68	69	70	71	72	73	74	75	76	77	78	79	80	81
110											227	663	1074					
109											339	774	1186	1621	2062	2526	2977	3470
108											450	886	1302	1733	2173	2643	3093	3586
107											566	1002	1413	1844	2289	2754	3204	3698
106											678	1113	1525	1960	2401	2865	3320	3814
105											789	1225	1641	2072	2517	2981	3436	3930
104											900	1336	1752	2183	2628	3098	3553	4046
103											1016	1447	1868	2294	2744	3209	3664	4162
102									329	731	1128	1563	1980	2405	2856	3320	3780	4274
101								68	440	842	1239	1675	2091	2522	2972	3436	3896	4390
100								179	557	953	1350	1786	2207	2633	3083	3553	4008	4506
99								290	668	1070	1467	1897	2318	2744	3194	3664	4124	4617
98							19	402	779	1181	1578	2009	2430	2856	3311	3775	4240	4734
97							131	513	891	1292	1689	2125	2546	2967	3422	3891	4351	4850
96							242	629	1007	1404	1800	2236	2657	3078	3538	4008	4467	4966
95							358	741	1118	1515	1917	2347	2768	3194	3649	4119	4583	5082
94						102	469	852	1229	1631	2028	2459	2885	3306	3766	4230	4695	5193
93						213	581	963	1346	1742	2139	2570	2996	3417	3877	4346	4811	5309
92						319	692	1074	1457	1854	2251	2686	3107	3528	3993	4462	4927	5426
91						431	803	1186	1568	1965	2367	2798	3223	3640	4104	4574	5043	5537
90					169	542	920	1297	1684	2076	2478	2909	3335	3751	4220	4685	5155	5653
89					290	649	1031	1408	1796	2193	2589	3020	3451	3867	4332	4801	5271	5769
88				39	407	760	1142	1520	1907	2304	2701	3131	3562	3978	4443			
87				150	523	871	1254	1631	2023	2415	2817	3248	3674	4090	4559			
86				261	644	983	1365	1742	2134	2526	2928	3359	3790	4201				
85			24	378	765	1089	1481	1859	2246	2638	3040	3470	3901					
84			136	489	881	1200	1592	1970	2357	2754	3151	3582	4012					
83			247	600	997	1312	1704	2081	2473	2865	3262	3693	4129					
82		29	358	716	1118	1418	1815	2193	2585	2977	3378	3809	4240					
81		140	469	828	1239	1529	1926	2309	2696	3088	3490							
80		252	581	939	1355	1641	2042	2420	2807									
79	19	363	692	1050	1471	1752												
78	131	474	803	1162	1592	1859												
77	242	586	915	1278	1713	1970												
76	353	697	1026	1389	1830	2081												
75	465	808	1137	1500	1946	2188												
74	576	920	1249	1617	2067	2299												
73	687	1031	1360	1728	2188													
72	799	1142	1471	1839	2304													
71	910	1254																
	64	65	66	67	68	69	70	71	72	73	74	75	76	77	78	79	80	81

Table 16b

DEHUMIDIFICATION (80 F)

Latent Heat Load to Dehumidify Outdoor Air to 50% RH
at *80 F* Indoors—Btu/h per 100 ft^3/min

Outdoor Dry-bulb Temperature *(degrees F)*	Outdoor Wet-bulb Temperature *(degrees F)*																	
	64	65	66	67	68	69	70	71	72	73	74	75	76	77	78	79	80	81
110													247					
109													358	794	1234	1699	2149	2643
108												58	474	905	1346	1815	2265	2759
107												174	586	1016	1462	1926	2376	2870
106												286	697	1133	1573	2038	2493	2986
105												397	813	1244	1689	2154	2609	3102
104											73	508	924	1355	1800	2270	2725	3219
103											189	620	1041	1467	1917	2381	2836	3335
102											300	736	1152	1578	2028	2493	2952	3446
101										15	411	847	1263	1689	2144	2609	3069	3562
100										126	523	958	1379	1805	2255	2725	3180	3678
99										242	639	1070	1491	1917	2367	2836	3296	3790
98										353	750	1181	1602	2028	2483	2948	3412	3906
97									63	465	862	1297	1718	2139	2594	3064	3524	4022
96									179	576	973	1408	1830	2251	2710	3180	3640	4138
95									290	687	1089	1520	1941	2362	2822	3291	3756	4254
94								24	402	803	1200	1631	2057	2478	2938	3403	3867	4366
93								136	518	915	1312	1742	2168	2589	3049	3519	3983	4482
92								247	629	1026	1423	1859	2280	2701	3165	3635	4099	4598
91								358	741	1137	1539	1970	2396	2812	3277	3746	4216	4709
90							92	469	857	1249	1650	2081	2507	2923	3393	3857	4327	4825
89							203	581	968	1365	1762	2193	2623	3035	3504	3974	4443	4942
88							315	692	1079	1476	1873	2304	2735	3151	3615			
87						44	426	803	1195	1588	1989	2420	2846	3262	3732			
86						155	537	915	1307	1699	2101	2531	2962	3373				
85						261	653	1031	1418	1810	2212	2643	3073	3485				
84					53	373	765	1142	1529	1926	2323	2754	3185					
83					169	484	876	1254	1646	2038	2435	2865	3301					
82					290	590	987	1365	1757	2149	2551	2981	3412					
81					411	702	1099	1481	1868	2260	2662							
80				111	528	813	1215	1592	1980									
79				223	644	924												
78				334	765	1031												
77			87	450	886	1142												
76			198	561	1002	1254												
75			310	673	1118	1360												
74		92	421	789	1239	1471												
73		203	532	900	1360													
72		315	644	1012	1476													
71	82	426																
	64	65	66	67	68	69	70	71	72	73	74	75	76	77	78	79	80	81

Chart 17a
DUCT HEAT GAIN

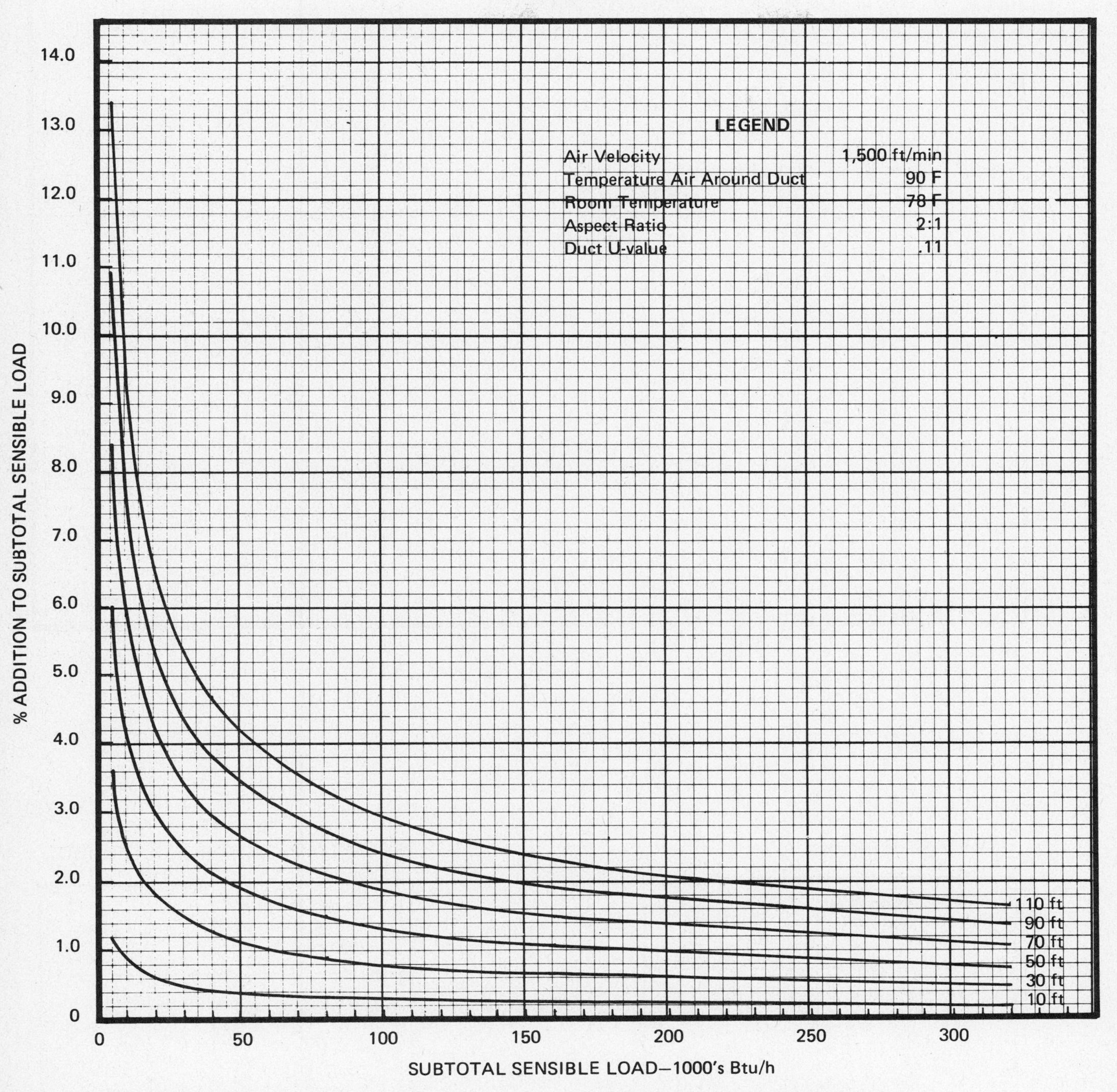

For estimating purposes, it is sufficiently accurate to use the total length of the exposed portions of *all* ducts, trunk and/or branch and/or radial.

Chart 17b
VELOCITY MULTIPLIERS

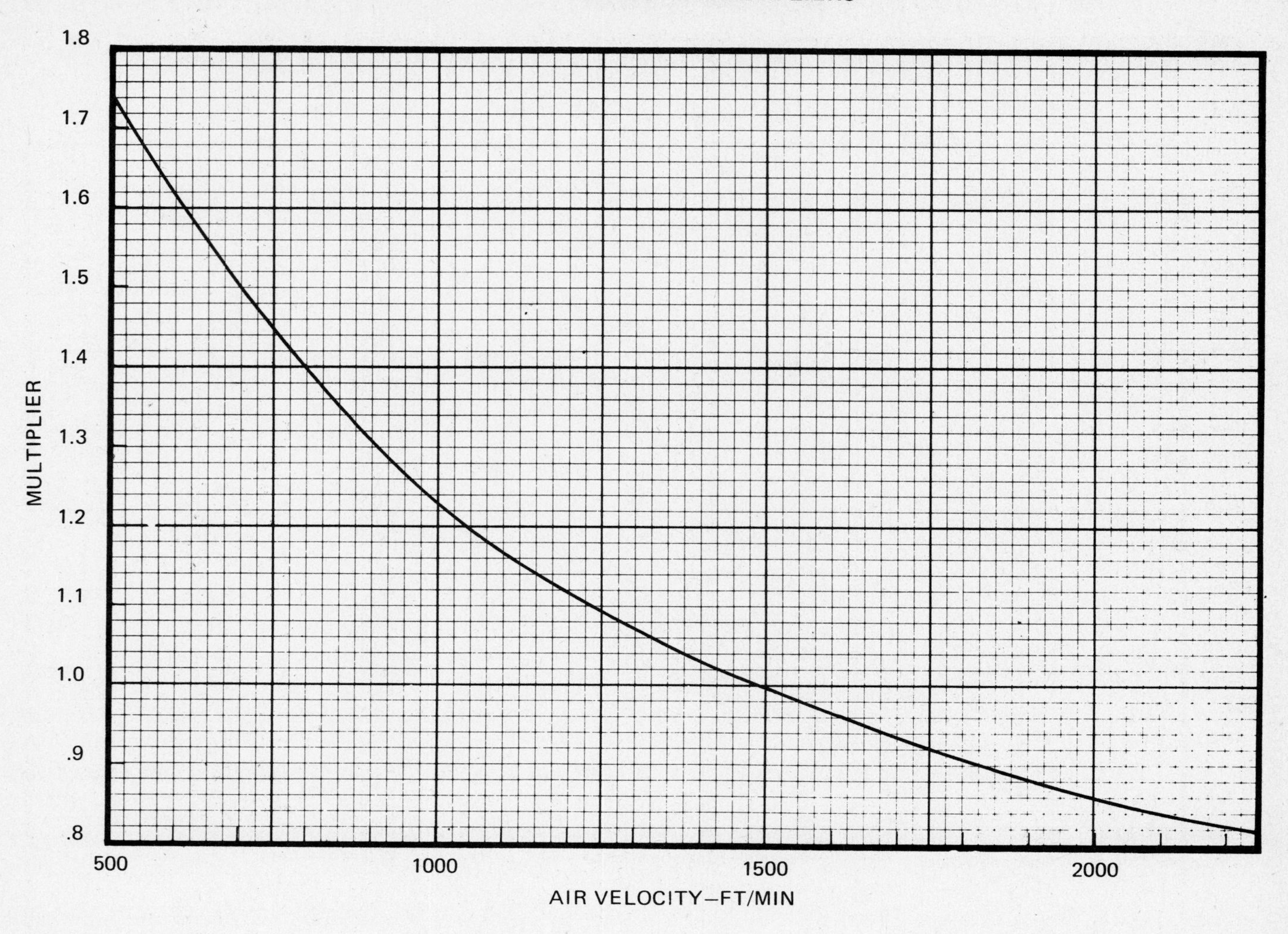

Chart 17c
SURROUNDING AIR MULTIPLIERS

Temperature of Air Surrounding Duct *(degrees F)*	Multiplier
150	2.8
140	2.5
130	2.2
120	1.9
110	1.6
100	1.3
90	1.0
80	.7
75	.6

Chart 17d
ROOM TEMPERATURE MULTIPLIERS

Room Design Temperature *(degrees F)*	Multiplier
75	1.08
76	1.05
77	1.02
78	1.00
79	.98
80	.96

Chart 17e
ASPECT RATIO MULTIPLIERS

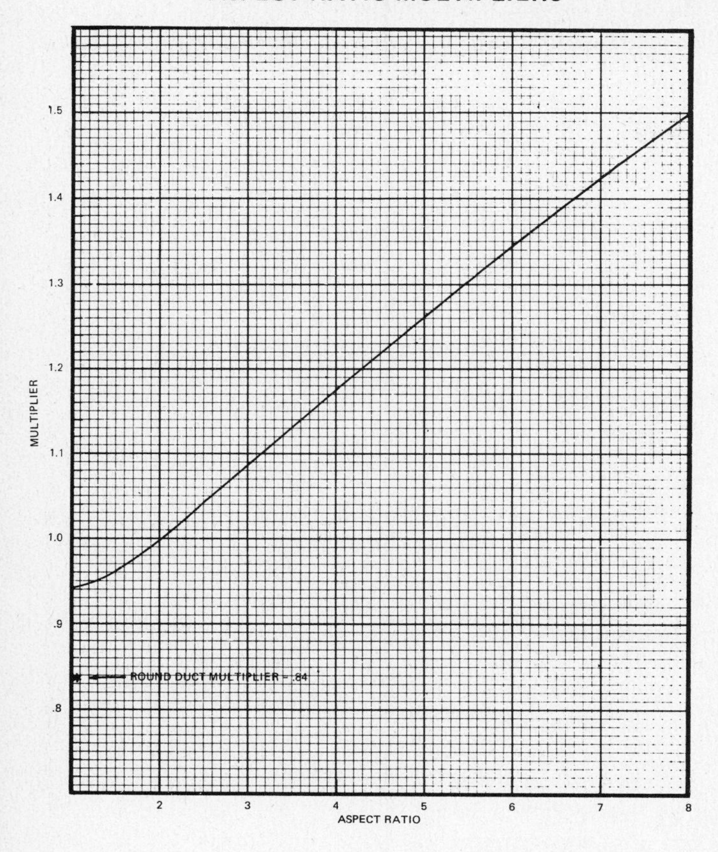

Chart 17f
RETURN AIR MULTIPLIERS
(Summer Conditions)

Return Air Temperature *(degrees F)*	Multiplier
80	.29
79	.31
78	.34
77	.37
76	.40
75	.43

Table 18*
DUCT VELOCITIES

Type Duct	Schools, Theaters, Public Buildings	Industrial Buildings
	Recommended Velocities (ft/min)	
Main trunk	1000–1300	1200–1800
Branch	600– 900	800–1000
Riser	600– 700	800
	Maximum Velocities	
Main trunk	1100–1600	1300–2200
Branch	800–1300	1000–1800
Riser	800–1200	1000–1600

*Excerpted from Table 6, page 481, *ASHRAE Handbook of Fundamentals (1972).*

Table 19
AIR TEMPERATURE SURROUNDING DUCTS*
(Summer Conditions)

Duct Location	Approximate temperature of air surrounding the duct
Unconditioned room, not well ventilated	5 degrees F below outside dry-bulb temperature
Unconditioned room, well ventilated	Outside dry-bulb temperature
Attic space, minimum ventilation	Outside dry-bulb temperature plus 25 to 50 degrees
Attic space, well ventilated	Outside design dry-bulb temperature plus 10 to 20 degrees F
Other unconditioned spaces (boiler rooms, kitchens, etc.)	Base on dry-bulb temperature measurements in similar spaces.
Exposed directly to outdoor conditions	Outdoor dry-bulb temperature plus 5 to 20 degrees (depending on exposure to sunlight)

*Data based on research conducted at the University of Illinois (for example, see Transactions, The American Society of Heating and Ventilating Engineers, Volume 59, 1953, No. 1477, *Cooling a Small Residence with a Two-Horsepower Mechanical Condensing Unit*, by H. T. Gilkey, et al.) and other empirical data, as well as theoretical considerations.

It is recommended that actual temperatures be measured whenever possible.

Table 20
HEAT LOSSES FOR BELOW GRADE BASEMENT FLOORS AND WALLS*

Outdoor DB Design Temperature	Floor Loss Btu/h per sq ft	Wall Loss Btu/h per sq ft
Below 0 F	3.0	6.0
0 to 25 F	2.0	4.0
Above 25 F	1.0	2.0

*Adapted from data in *ASHRAE Handbook of Fundamentals*—1972.

Table 21
MAXIMUM (WINTER) INSIDE RELATIVE HUMIDITY (% RH)

Outside Design Dry-bulb Temperature *(degrees F)*	Inside Dry-bulb Temperature					
	68 F		72 F		75 F	
	Single Glass	Double Glass	Single Glass	Double Glass	Single Glass	Double Glass
−30	3	18	3	18	3	18
−20	5	23	5	22	5	21
−10	8	27	7	26	6	25
0	12	33	11	31	10	29
10	17	39	16	37	15	35
20	24	46	23	44	22	42
20	34	55	32	52	30	49

Table 22a

HUMIDIFICATION HEATING LOAD (68 F)

(Btu/h per 100 ft^3/min of outside air flow for Inside Design Conditions of *68 F* and desired RH *and* gallons of water/day required)

Outdoor Dry-bulb Temperature *(degrees F)* (80% RH)	Indoor Relative Humidity (% at 68 F Dry-bulb Temperature)								
	40	35	30	25	20	15	10	5	
40	*2.1* 775	*1.2* 440	*0.2* 80						*Gallons* Btu/h
35	*3.1* 1150	*2.1* 775	*1.2* 440	*0.2* 80					*Gallons* Btu/h
30	*3.9* 1445	*3.0* 1110	*2.0* 735	*1.1* 400					*Gallons* Btu/h
25	*4.7* 1755	*3.7* 1370	*2.8* 1035	*1.8* 665					*Gallons* Btu/h
20	*5.3* 1975	*4.3* 1610	*3.4* 1260	*2.5* 920	*1.5* 555	*0.6* 215			*Gallons* Btu/h
15	*5.8* 2145	*4.8* 1795	*3.8* 1410	*2.9* 1075	*2.0* 735	*1.0* 365			*Gallons* Btu/h
10	*6.2* 2300	*5.2* 1940	*4.2* 1570	*3.3* 1225	*2.4* 885	*1.4* 520	*0.5* 180		*Gallons* Btu/h
5	*6.5* 2410	*5.5* 2045	*4.6* 1715	*3.6* 1330	*2.7* 995	*1.7* 625	*0.8* 285		*Gallons* Btu/h
0	*6.7* 2490	*5.8* 2145	*4.8* 1795	*3.9* 1445	*2.9* 1075	*2.0* 735	*1.0* 365		*Gallons* Btu/h
−5	*6.9* 2560	*5.9* 2185	*5.0* 1865	*4.0* 1495	*3.1* 1150	*2.2* 810	*1.2* 440	*0.3* 115	*Gallons* Btu/h
−10	*7.0* 2600	*6.0* 2225	*5.1* 1905	*4.1* 1535	*3.2* 1190	*2.3* 850	*1.3* 480	*0.4* 145	*Gallons* Btu/h
−15	*7.1* 2640	*6.1* 2265	*5.2* 1940	*4.2* 1570	*3.3* 1225	*2.4* 885	*1.4* 520	*0.5* 180	*Gallons* Btu/h
−20	*7.2* 2680	*6.2* 2300	*5.3* 1975	*4.3* 1610	*3.4* 1260	*2.5* 920	*1.5* 555	*0.6* 215	*Gallons* Btu/h
−25	*7.3* 2715	*6.3* 2340	*5.4* 2010	*4.4* 1645	*3.5* 1295	*2.6* 960	*1.6* 590	*0.7* 250	*Gallons* Btu/h
−30	*7.4* 2750	*6.4* 2375	*5.5* 2045	*4.5* 1680	*3.6* 1330	*2.7* 995	*1.7* 625	*0.8* 285	*Gallons* Btu/h

Table 22b
HUMIDIFICATION HEATING LOAD (72 F)
(Btu/h per 100 ft^3/min outdoor air flow for Indoor Conditions of *72 F* and desired RH *and* gallons of water/day required)

Outdoor Dry-bulb Temperature *(degrees F)* (80% RH)	Indoor Relative Humidity (% at 72 F Dry-bulb Temperature)								
	40	35	30	25	20	15	10	5	
40	*3.2* 1210	*2.2* 810	*1.0* 365						*Gallons* Btu/h
35	*4.2* 1570	*3.1* 1150	*2.0* 735	*0.9* 325					*Gallons* Btu/h
30	*5.1* 1905	*4.0* 1495	*2.9* 1075	*1.8* 665	*0.7* 250				*Gallons* Btu/h
25	*5.8* 2145	*4.7* 1755	*3.6* 1330	*2.5* 920	*1.5* 555	*0.4* 145			*Gallons* Btu/h
20	*6.4* 2375	*5.3* 1975	*4.2* 1570	*3.2* 1190	*2.1* 775	*1.0* 365			*Gallons* Btu/h
15	*6.9* 2560	*5.8* 2145	*4.7* 1755	*3.6* 1330	*2.5* 920	*1.4* 520	*0.4* 145		*Gallons* Btu/h
10	*7.3* 2715	*6.2* 2300	*5.1* 1905	*4.0* 1495	*2.9* 1075	*1.8* 665	*0.8* 285		*Gallons* Btu/h
5	*7.6* 2830	*6.5* 2410	*5.4* 2010	*4.3* 1610	*3.2* 1190	*2.1* 775	*1.1* 400		*Gallons* Btu/h
0	*7.8* 2910	*6.7* 2490	*5.6* 2080	*4.6* 1715	*3.5* 1295	*2.4* 885	*1.3* 480	*0.3* 115	*Gallons* Btu/h
−5	*8.0* 2980	*6.9* 2560	*5.8* 2145	*4.8* 1795	*3.7* 1370	*2.6* 960	*1.5* 555	*0.4* 145	*Gallons* Btu/h
−10	*8.1* 3020	*7.0* 2600	*5.9* 2185	*4.9* 1830	*3.8* 1410	*2.7* 995	*1.6* 590	*0.5* 180	*Gallons* Btu/h
−15	*8.2* 3060	*7.1* 2640	*6.0* 2225	*5.0* 1865	*3.9* 1445	*2.8* 1035	*1.7* 625	*0.6* 215	*Gallons* Btu/h
−20	*8.3* 3100	*7.2* 2680	*6.1* 2265	*5.1* 1905	*4.0* 1495	*2.9* 1075	*1.8* 665	*0.7* 250	*Gallons* Btu/h
−25	*8.4* 3135	*7.3* 2715	*6.2* 2300	*5.2* 1940	*4.1* 1535	*3.0* 1110	*1.9* 700	*0.8* 285	*Gallons* Btu/h
−30	*8.5* 3170	*7.4* 2750	*6.3* 2340	*5.2* 1940	*4.1* 1535	*3.1* 1150	*2.0* 735	*0.9* 325	*Gallons* Btu/h

Table 22c
HUMIDIFICATION HEATING LOAD (75 F)

(Btu/h per 100 ft^3/min outdoor air flow for Indoor Conditions of *75 F* and desired RH *and* gallons of water/day required)

Outdoor Dry-bulb Temperature *(degrees F)* (80% RH)	Indoor Relative Humidity (% at 75 F Dry-bulb Temperature)								
	40	35	30	25	20	15	10	5	
40	*4.2*	*2.9*	*1.7*						*Gallons*
	1570	1075	625						Btu/h
35	*5.1*	*3.8*	*2.7*	*1.4*	*0.2*				*Gallons*
	1905	1410	995	520	80				Btu/h
30	*6.0*	*4.7*	*3.5*	*2.3*	*1.1*				*Gallons*
	2225	1755	1295	850	400				Btu/h
25	*6.7*	*5.4*	*4.3*	*3.1*	*1.8*	*0.7*			*Gallons*
	2490	2010	1610	1150	665	250			Btu/h
20	*7.3*	*6.1*	*4.9*	*3.7*	*2.4*	*1.3*	*0.1*		*Gallons*
	2715	2265	1830	1370	885	480	35		Btu/h
15	*7.8*	*6.5*	*5.4*	*4.1*	*2.9*	*1.7*	*0.6*		*Gallons*
	2910	2410	2010	1535	1075	625	215		Btu/h
10	*8.2*	*6.9*	*5.8*	*4.5*	*3.3*	*2.1*	*1.0*		*Gallons*
	3060	2560	2145	1680	1225	775	365		Btu/h
5	*8.5*	*7.2*	*6.1*	*4.8*	*3.6*	*2.4*	*1.3*	*0.1*	*Gallons*
	3170	2680	2265	1795	1330	885	480	35	Btu/h
0	*8.8*	*7.5*	*6.3*	*5.1*	*3.9*	*2.7*	*1.5*	*0.3*	*Gallons*
	3255	2795	2340	1905	1445	995	555	115	Btu/h
−5	*9.0*	*7.7*	*6.5*	*5.3*	*4.1*	*2.9*	*1.7*	*0.5*	*Gallons*
	3330	2870	2410	1975	1535	1075	625	180	Btu/h
−10	*9.1*	*7.8*	*6.7*	*5.4*	*4.2*	*3.0*	*1.9*	*0.7*	*Gallons*
	3365	2910	2490	2010	1570	1110	700	250	Btu/h
−15	*9.2*	*7.9*	*6.8*	*5.5*	*4.3*	*3.1*	*2.0*	*0.8*	*Gallons*
	3405	2925	2525	2045	1610	1150	735	285	Btu/h
−20	*9.3*	*8.0*	*6.9*	*5.6*	*4.4*	*3.2*	*2.1*	*0.9*	*Gallons*
	3440	2980	2560	2080	1645	1210	775	325	Btu/h
−25	*9.4*	*8.1*	*7.0*	*5.7*	*4.5*	*3.3*	*2.2*	*1.0*	*Gallons*
	3480	3020	2600	2120	1680	1225	810	365	Btu/h
−30	*9.4*	*8.1*	*7.0*	*5.7*	*4.5*	*3.3*	*2.2*	*1.0*	*Gallons*
	3480	3020	2600	2120	1680	1225	810	365	Btu/h

Chart 23a
DUCT HEAT LOSS

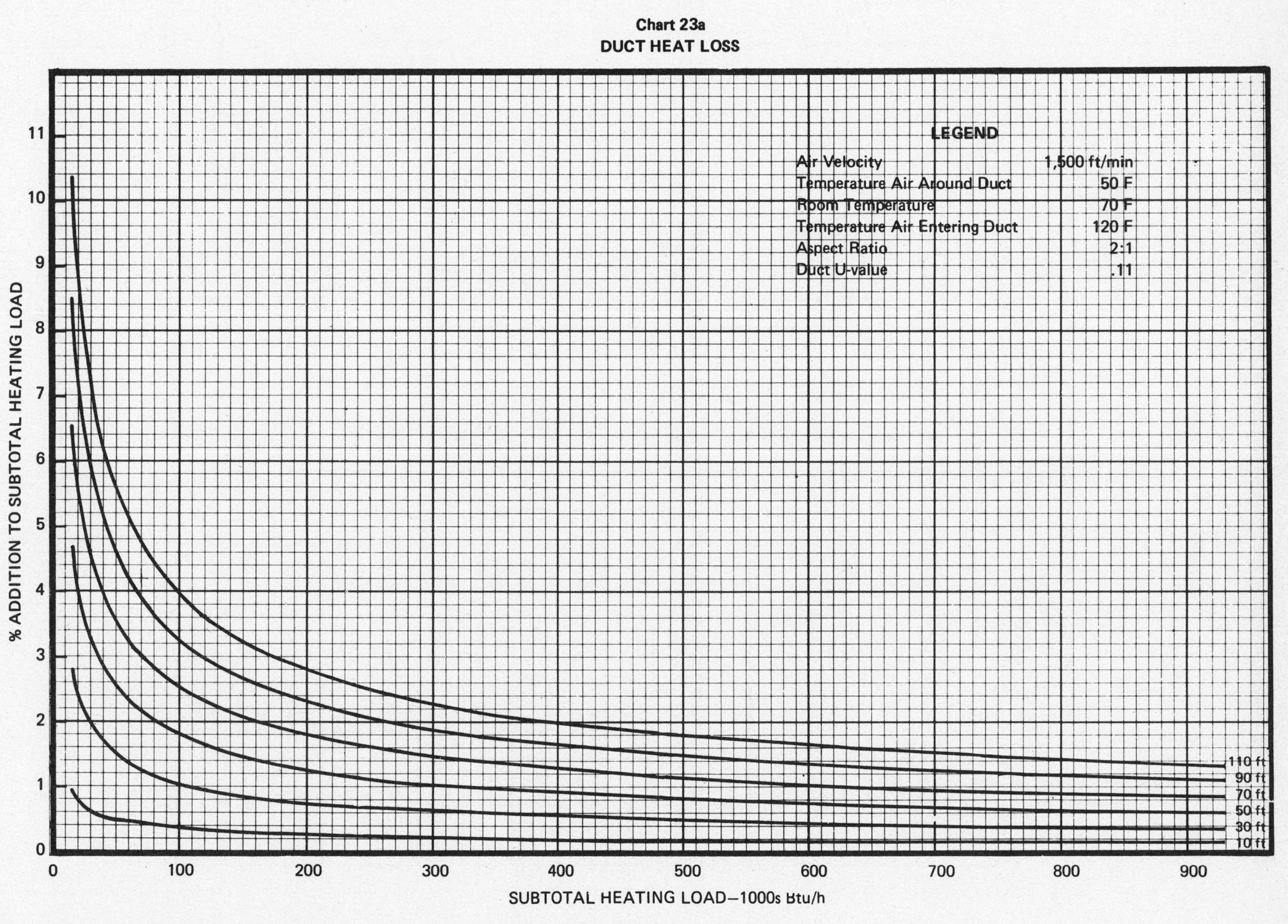

For estimating purposes, it is sufficiently accurate to use the total length of the exposed portions of *all* ducts, trunk and/or branch and/or radial.

Chart 23b
VELOCITY MULTIPLIERS

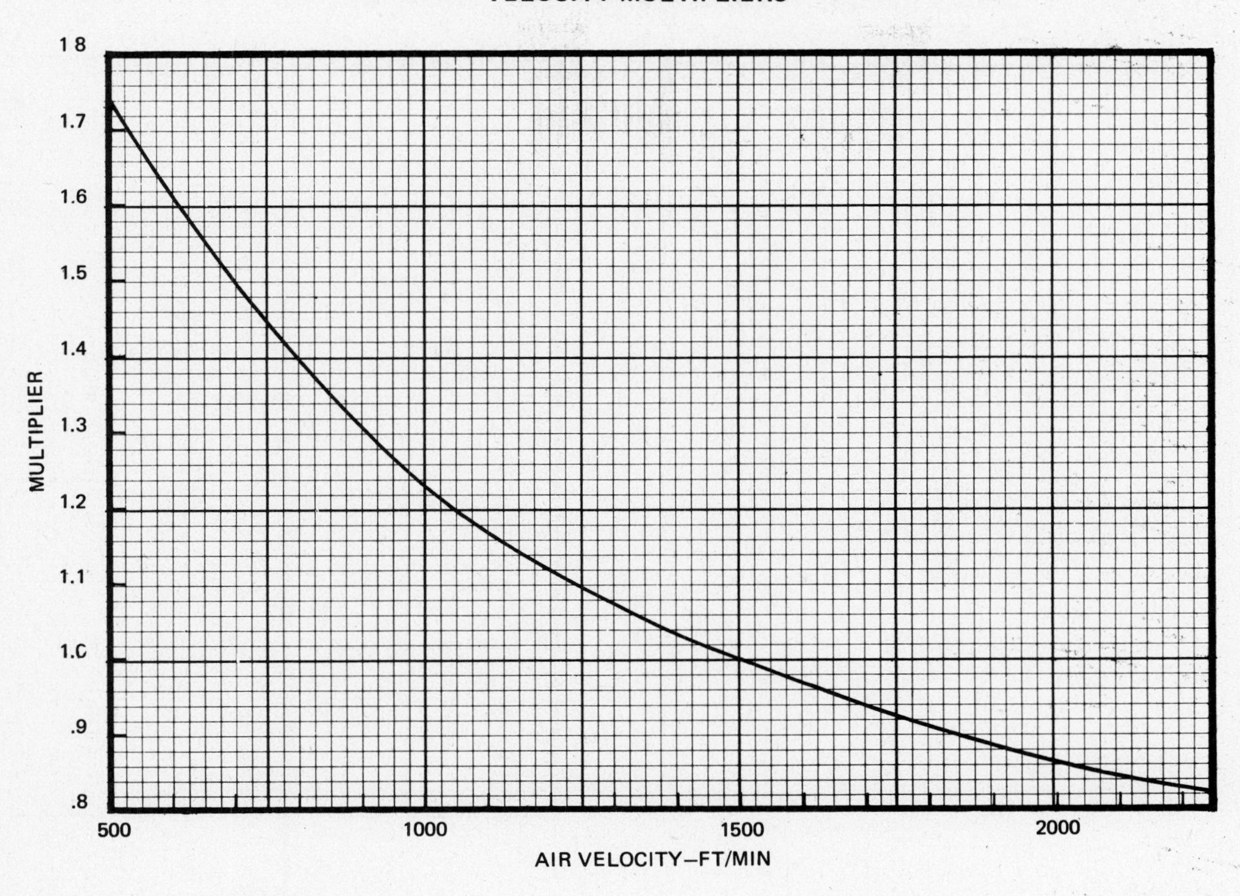

Chart 23c
SURROUNDING AIR MULTIPLIERS

Temperature of Air Surrounding Duct *(degrees F)*	Multiplier
70	.71
60	.86
50	1.00
40	1.14
30	1.29
20	1.43
10	1.57
0	1.71
−10	1.86
−20	2.00
−30	2.14

Chart 23d
ROOM TEMPERATURE MULTIPLIERS

Room Design Temperature *(degrees F)*	Multiplier
75	1.05
74	1.04
73	1.03
72	1.02
71	1.01
70	1.00
69	.99
68	.98

Chart 23e
ENTERING TEMPERATURE MULTIPLIERS

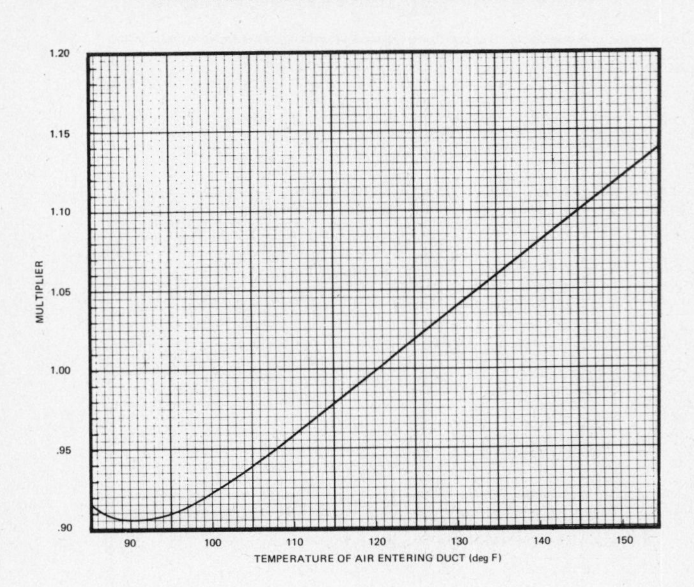

Chart 23f
ASPECT RATIO MULTIPLIERS

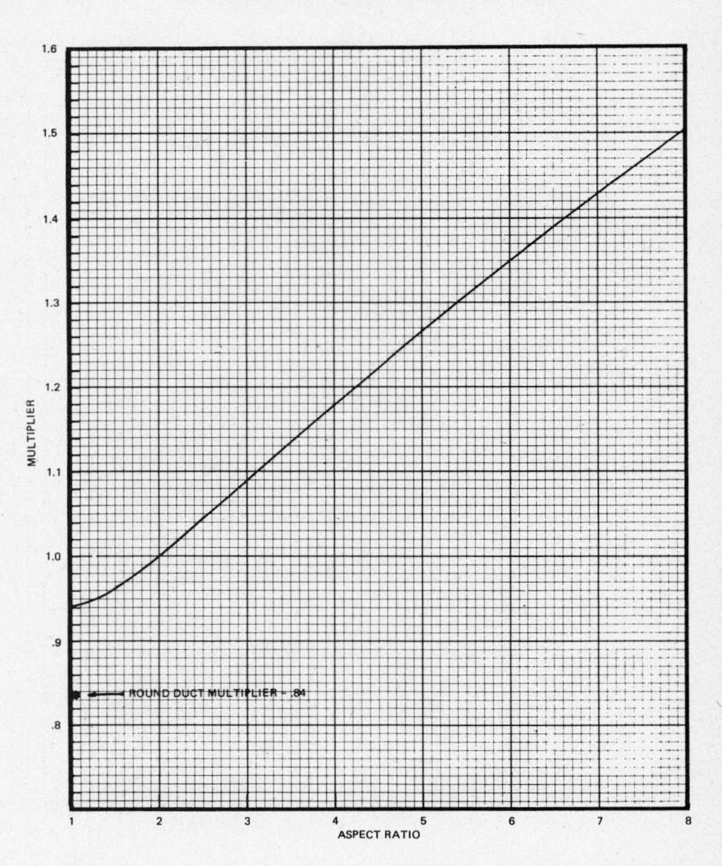

Chart 23g
RETURN AIR MULTIPLIERS

Temperature of Return Air Entering Duct *(degrees F)*	Multiplier
75	.38
74	.36
73	.34
72	.32
71	.30
70	.29
69	.27
68	.25

Table 24
AIR TEMPERATURE SURROUNDING DUCTS*
(Winter Conditions)

Duct Location	Approximate temperature of air surrounding the duct
Unheated room, not well ventilated	20 to 50 degrees above outside temperature
Unheated room, well ventilated	10 to 30 degrees above outside temperature
Attic space, minimum ventilation	5 to 20 degrees above outside temperature
Attic space, well ventilated	Outside temperature
Exposed directly to outdoor conditions	5 to 15 degrees below outside temperature to compensate for wind exposure

*Data based on research conducted at the University of Illinois (for example, see Transactions, The American Society of Heating and Ventilating Engineers, Volume 59, 1953, No. 1477, *Cooling a Small Residence with a Two-Horsepower Mechanical Condensing Unit*, by H. T. Gilkey, et al.) and other empirical data, as well as theoretical considerations.

It is recommended that actual temperatures be measured whenever possible.

8

WINTER COMFORT

FUELS

Heating requires the expenditure of some form of energy to raise the temperature of water or air, depending on the type of equipment used. Discounting solar heating, one can classify four types of energy sources (fuels) as the principal means of domestic heating; coal, oil, gas, and electricity. The choice of fuel is usually based on supply (availability), economy, operating requirements, dependability, cleanliness, and control. Major attention is now being focused on the cost of operation.

Coal

Coal is a solid fuel consisting of carbon, hydrogen, oxygen, nitrogen, sulfur, and ash, and, depending on the proportions of each, we can classify two types—hard and soft (Fig. 8-1). Anthracite (hard) is a clean, dense, hard coal which creates little dust. It's hard to ignite but burns uniformly and with little smoke. Soft coal is classified as bituminous; it ignites easily and burns freely with a long flame, producing much smoke and soot if improperly fired. Both varieties have a heat content of around 14,000 Btu per pound but do differ in carbon and oxygen content, which accounts for the difference in burning characteristics.

This description is an oversimplification of the coal classifications, but, because coal is no longer a major source of fuel for heating homes or commercial buildings, we need not go into greater detail on the subject. Coal, however, is still a vital energy source for power company generating plants and certain industrial applications. And with the focus on energy conservation and the high cost of fuels, it may well be that coal will return to the heating scene with new technologies and combustion equipment.

Oils

Fuel oil is a major source of heating energy for homes and commercial buildings. It is very popular in the northern and eastern sections of the country and is found in many rural areas where the availability or desirability of gas or electrical energy are not alternate options. Fuel oils are mixtures of hydro-

FIGURE 8-1 Coal samples.

carbons derived from crude petroleum by various refining processes. They are classified by dividing them into grades according to their characteristics, mainly viscosity; however, other properties like flash point, pour point, water and sediment content, carbon residue, and ash, are important in the storage, handling, and types of burning equipment for oil. The viscosity determines whether the fuel oil can flow or be pumped through lines or if it can be atomized into small droplets.

For comfort heating applications, we are primarily interested in two grades of fuel oil; No. 1 and No. 2 (Fig. 8-2), which contain 84 to 86% carbon, up to 1% sulfur, and the remainder is mostly hydrogen. The heavier grades, No. 4 and No. 5, have even higher carbon contents, but also considerably more sulfur than is permissible in domestic grades. Number 1 grade fuel oil is considered premium quality and priced accordingly, and is used in room-type space heaters, which do not use high-pressure burners and depend on gravity flow; thus the need for the lower viscosity of No. 1. Number 2 grade is the standard heating oil sold by most oil supply firms. It weighs between 7.296 and 6.870 pounds per gallon as compared to water at 8.34 pounds per gallon. Number 2 oil is used in equipment that has pressure-type atomizing, which covers most forced-warm-air furnaces and boilers. The heating value is approximately 135,000 to 142,000 Btu per gallon. Thus the hourly heating effect is found by multiplying the gallons burned by the heat value per gallon.

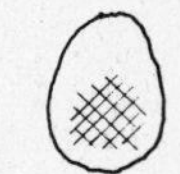

FIGURE 8-2 Fuel oil.

Gases

Fuel gases are employed for various heating and air-conditioning (cooling) processes and fall into three broad categories; natural, manufactured, and liquefied petroleum. Stories about the discovery and use of gas date back as far as 2,000 B.C. History notes that the Chinese piped gas from shallow wells through bamboo poles and boiled sea water to obtain salt. Early explorers of America reported seeing "burning springs," which were probably jets of gas escaping from holes in the earth. Settlers often struck gas when drilling water wells.

Today gas has a thousand and one uses. In addition to the heating industry, the power generation industry, iron and steel mills, oil refineries, food service industries, and the glass, cement, and pottery industries are large users of gas. In 1973 AGA (American Gas Association) reported 38 million homes had gas service.

Natural gas comes from wells. It's the product of the organic material of plants and animals that over millions of years were chemically converted into gas or oil. Most reserves of gas in the United States are are not dissolved in or in contact with oil. Natural gas (Fig. 8-3) is mostly *methane*, which consists of

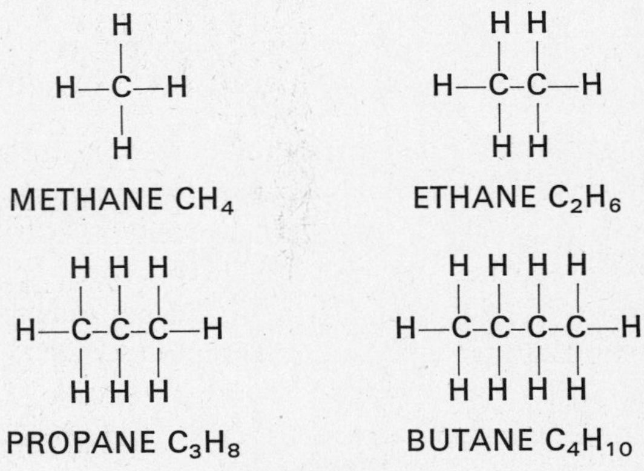

FIGURE 8-3.

one carbon atom linked to four hydrogen atoms, and *ethane*, which consists of 2 carbons and 6 hydrogen atoms. Liquid petroleum (LP) gases are *propane* and *butane* or a mixture of the two, and these fuel gases are obtained from natural gas or as a by-product of refining oil. From Fig. 8-3 you will note they contain more carbon and hydrogen atoms than natural gases, are thus heavier, and, as a result, have more heating value per cubic foot.

Manufactured gas, as the name states, is man made as a by-product from other manufacturing operations. For example, in ironmaking large amounts of gases are produced that can be used as fuel gases. The use of manufactured fuel gases has declined greatly in the United States. Today over 99% of sales by gas distribution and transmission companies is natural gas. However, the man-made kind is still a popular fuel in Europe.

Mixed gas, as the name implies, is also a man-made mixture of gases. A common example is a mixture of natural gas and manufactured gas.

It is important to know the specific gravity of a gas. Compared to standard air (Fig. 8-4), which has a

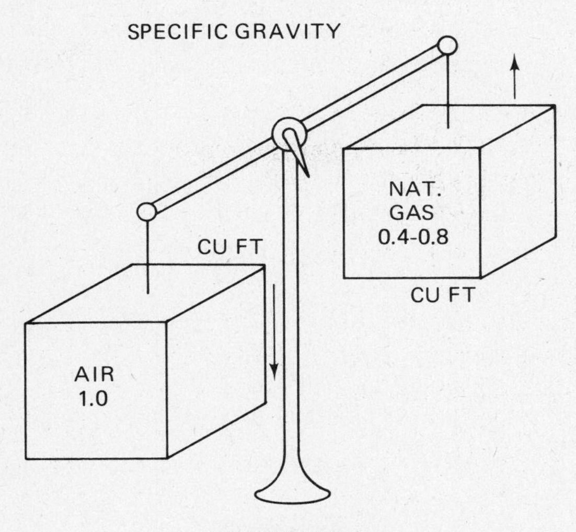

FIGURE 8-4.

specific gravity (spgr) of 1.0, natural gas ranges from 0.4–0.08 and thus is lighter than air. On the other hand, of the liquid petroleum gases, the spgr of propane is 1.5 and that of butane is 2.0, meaning they are heavier than air. The density of the gases is important because it affects the flow of the gas through orifices (small holes) and then to the burner. Should a leak develop in a gas pipe, natural gas will rise while LP gases will not and may drift to low spots and collect in pools, creating a hazard if open flames are present. Specific gravity also affects gas flow in supply pipes and the pressure needed to move the gas.

The *heating value* (or *heat value*) of a gas is the amount of heat released when one cubic foot of the gas is completely burned (Fig. 8-5). Natural gas (which is largely methane) has a heating value of about 950 to 1,150 Btu per cubic foot. Propane has a heat value of approximately 2,500 Btu/ft³ and butane, about 3,200 Btu/ft³. The hourly rate of a heating value is thus the number of cubic feet burned times the heating value. For example: If a furnace burned 75 ft³ of natural gas in one hour and the heat value is

HEAT CONTENT WHEN BURNED

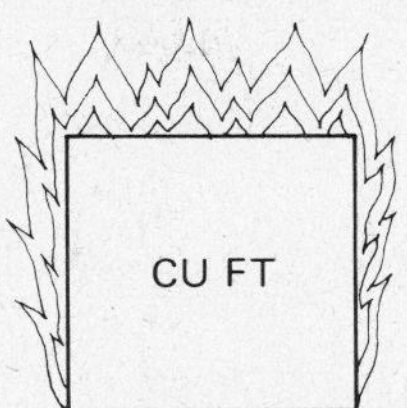

NATURAL GAS	950 TO 1150 BTU/FT^3
PROPANE	2,500 BTU/FT^3
BUTANE	3,200 BTU/FT^3

FIGURE 8-5.

1000 Btu/ft³, the total heating *input* is $75 \times 1{,}000 =$ 75,000 Btu/hr. The actual output depends on the operating efficiency of the furnace. The exact heating value of gases in your local area can be obtained from your gas company or LP gas distributor. Propane is the LP gas most used for domestic heating. Butane has more agricultural and industrial applications.

Combustion and ventilation are most important elements and should be understood by the technician. Combustion (Fig. 8-6) takes place when fuel bases are burned in the presence of air. Methane gas combines with the oxygen and nitrogen present in the air and the resulting combustion reaction produces heat, with by-products of carbon dioxide, water vapor, and nitrogen. *For each one cubic foot of methane gas, ten cubic feet of air are needed for complete combustion* (Fig. 8-7). Although natural gas requires a 10-to-1 ratio of air, LP fuels require much more, due to the concentration of carbon and hydrogen atoms. Liquid petroleum combustion must have more than 24 ft³ of air per cubic foot of gas to support proper combustion.

The by-products called *flue products* are vented to the outside. Insufficient make-up air can produce

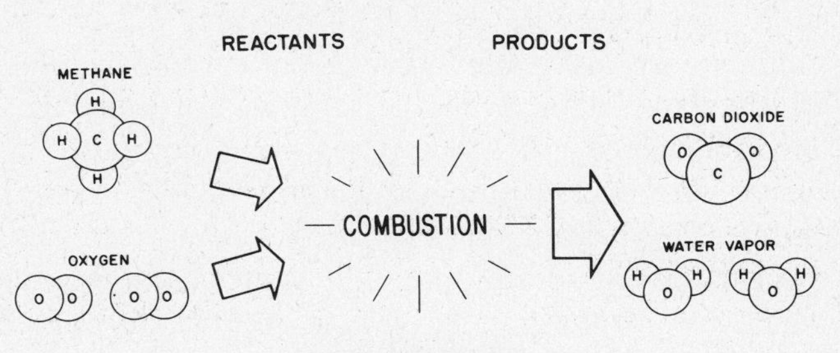

FIGURE 8-6.

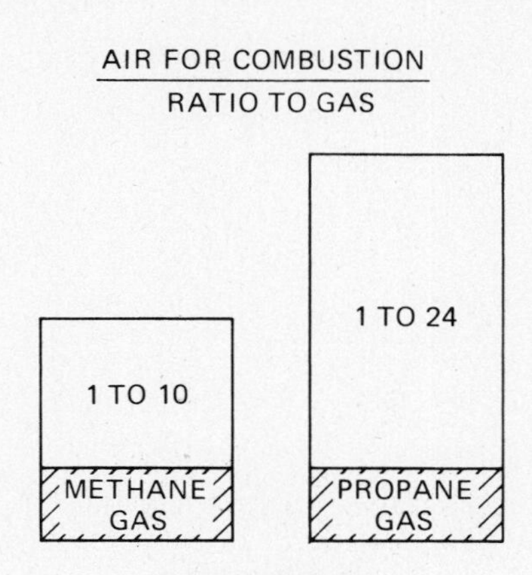

FIGURE 8-7.

hazardous results. If too little oxygen is supplied, part of the by-products will be dangerous carbon monoxide gas (CO) rather than harmless carbon dioxide gas (CO_2). Secondly, lack of make-up air causes poor flue action and spillage of combustion products into the room, and carbon monoxide in a living area is a serious problem. Specific recommendations on venting and make-up air will be covered later, but remember, combustion heating ventilation is most important.

There is an excellent manual entitled *Fundamentals of Gas Combustion* prepared by AGA (American Gas Association), 1515 Wilson Boulevard, Arlington, Virginia 22209, which covers combustion and the design of burners, their operation, and their maintenance. This author highly recommends it for all who wish to specialize in gas heating.

Electrical Energy

The growth of electric heating came immediately after World War II and found its main use in areas of cheap public power such as the TVA (Tennessee Valley Authority) and in the Pacific Northwest. In both situations, the cost of power and the winter climate were favorable to the use of electric heating. In the late 1950s, when it became apparent that investor-owned private utilities were going to be faced with heavy summer cooling loads, electric utilities and heating manufacturers began promoting electric heating to build up their winter load. Special rates and promotion programs like "Live Better Electrically," "Total Electric Home," etc., were introduced in the sixties and the market developed rapidly. The situation in the mid 1970s changed somewhat due to higher and higher power-generating costs, which are being passed on to the consumer. Also, the federal government directed that the use of special promotional programs be stopped. However, even with these current obstacles, the use of electric heating continues to grow, due to availability, convenience, and other considerations. With more and more nuclear generating stations being built, the cost of power may be stabilized and may even start to decline in the future.

The heating value of electric resistance heat is easy to remember and calculate (Fig. 8-8). For each

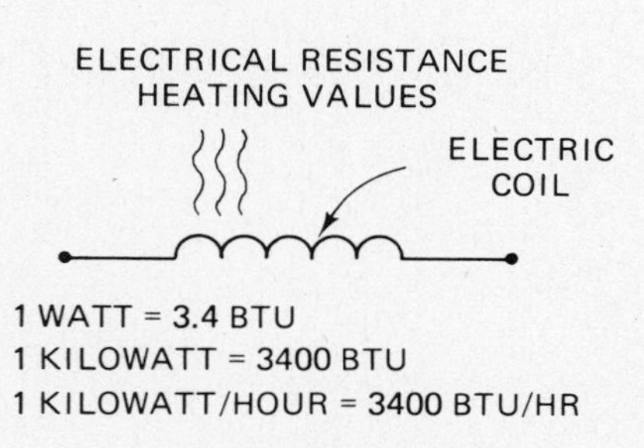

FIGURE 8-8.

watt of power consumed, 3.4 Btu of heat output will be generated. Resistance heating is 100% efficient; there are no losses such as those experienced with oil and gas combustion processes. Electric rates are measured in kilowatts (one kilowatt is 1,000 watts). Therefore, consuming one kilowatt in one hour is called a kilowatt/hour. And if the heat ratio is 3.4 Btu/watt, then a kilowatt/hour of electric resistance heating will produce 3,400 Btu/hour.

This discussion of electric resistance heating does not include the all-electric heat pump. Its unique reverse refrigerant cycle can produce heating with efficiency ratios of 1 input to 3 output under ideal conditions, and 1 to 2 or 2.5 under normal operation. That means for each unit of electricity put in, 2 to 2.5 units of heat are produced, as compared to the 1-to-1 ratio of straight resistance heat. The operation of heat pumps will be explained more fully under subsequent chapters, but because of their ability to produce all-electric heating and cooling, they are a vital factor in future electric energy developments.

Without engaging in a "battle of the fuels," we will cover the most common types of heating equipment that use gas, oil, and electric energy and attempt to establish the advantages and disadvantages as noted by manufacturers and other experts in the field.

Common to all types of heating apparatus sold domestically is the need to be tested, certified, or listed by the proper regulatory agencies.

The AGA (the American Gas Association) establishes the minimum construction safety and performance standards for gas heating equipment. The AGA maintains laboratories to examine and test furnaces, and also maintains a field inspection service. Furnaces submitted and found to be in compliance are listed in the AGA *Directory*; they also bear the Blue Star Seal of Certification (Fig. 8-9).

FIGURE 8-9 AGA seal (Courtesy, American Gas Association).

Oil heating equipment and electric heating products are subject to examination, testing, and approval by UL (Underwriters Laboratories, Inc.). Most of us are familiar with the UL stamp (Fig. 8-10)

FIGURE 8-10 UL seal (Courtesy, Underwriters Laboratories, Inc.).

appearing on everything from toasters to electric blankets, but UL also gets deeply involved in the approval and listing of heating and air-conditioning equipment, even dealing with large centrifugal machines one hundred tons and over. Local city codes and inspectors are guided by UL standards, and failure to comply with them may be costly to the manufacturer and installer. Underwriters Laboratories maintains testing laboratories for certain types of products; however, they often perform the necessary tests at the manufacturer's plant.

Manufacturers who actively sell their products in Canada seek CSA (Canadian Standards Association) approvals as well. CSA is the Canadian counterpart to our UL. So if you see the CSA symbol (Fig. 8-11) on a product, you will recognize its meaning.

FIGURE 8-11 CSA seal (Courtesy, Canadian Standards Association).

SPACE HEATERS

Gas and Oil

Space heater is a general term applied to room heaters, wall furnaces, and floor furnaces that are fired by gas or oil combustion fuels.

A room heater (Fig. 8-12) is a self-contained, free-standing, nonrecessed, gas or oil-burning, air-heating appliance intended for installation in the space being heated and not intended for duct connection. A room heater may be of the gravity or mechanical air circulation type, vented or unvented. If unvented, the input for gas should not exceed 50,000 Btu/hr. Because unvented room heaters discharge their products of combustion into the space being heated and get their combustion air from the same space, it is essential that they be used in well-ventilated rooms. AGA and many local codes prohibit their use in public spaces like hotels, motels, or institutions such as nursing homes and sanitariums. This form of heating is low cost, but leaves much to be desired in comfort control.

FIGURE 8-12 Vented room heater (Courtesy, Dearborn Stove Company).

FIGURE 8-13 Wall insert heater (Courtesy, Gould, Inc.).

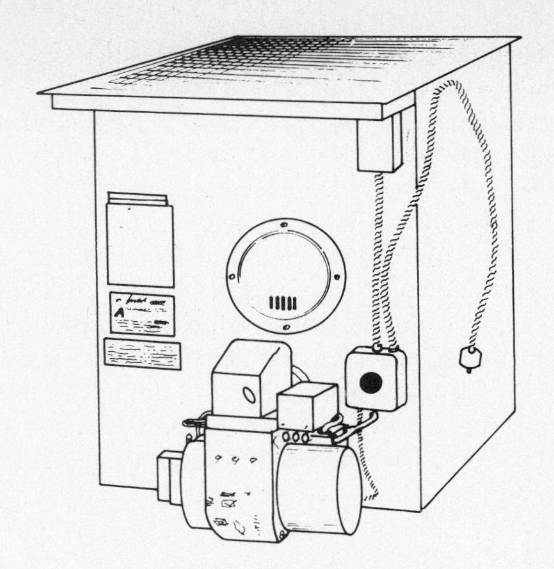

FIGURE 8-14 Floor furnace.

Wall furnaces (Fig. 8-13) are also self-contained vented appliances complete with grilles or their equivalent, designed for incorporation into, or permanently attached to, a wall or partition. They furnish heated air circulated by gravity or by fan directly into the space to be heated. Venting is usually accomplished by an approved integral venting system to the outside wall, although some have optional vertical venting through the roof.

Floor furnaces (Fig. 8-14) are also completely self-contained units designed to be installed on the floor of the space being heated, taking combustion air from outside the living space. The user is able to observe the unit and its lighting; it also can be serviced from this same outside space. Floor furnaces have been very popular in the south for low-cost housing; the crawl space of the house is used as the outside air and service access source. The air flow may be by either gravity or fan-forced circulation into the room. Their width is designed so as to enable them to fit between joist spaces, with enough clearance to provide protection from combustible products. Venting of combustion products must be by proper flue arrangements to outdoors—*not* through the crawl space. Floor furnaces are an improvement over in-space heaters in that they don't take up usable space; however, the circulation of warm air with this type of furnace leaves much to be desired.

The selection and application of oil and gas space heaters is not really a technical art, and in most cases the choice of heater is the result of a recommendation by a furniture or hardware dealer, or a retail store. The operation and service of space heaters is relatively simple; so a technician who understands the rest of this book, dealing with more complicated products, can readily handle this type of equipment.

Electric Space Heating

Radiant heaters for residential use are available in several forms. Most popular is the wall-mounted variety (Fig. 8-15) that may be flush mounted (recessed) or surface mounted. Only about two inches deep, they have an enclosed heating element with a highly polished heat reflector behind. A built-in thermostat automatically controls temperature. These heaters are excellent for bathrooms and spot heating for workrooms, playrooms, etc. Sizes range from 500 to 1,000 watts (1,700 to 3,400 Btu/h) and come suitable for 120, 208, or 240-volt operation. Service is easy; it is usually a matter of changing a defective thermostat or heating element.

Another variety of heater is the cove radiant heater (Fig. 8-16), which takes advantage of the convective air movement that is produced as room temperatures warm from the radiant effect. Wattages range from 450 to 900 watts, with a choice of voltage.

Ceiling panel radiant heaters (Fig. 8-17) are equally useful in residential or commercial buildings. They attach easily to ceiling joists or hang in the T-bar suspended ceiling. Panel sizes conform to ceiling tile dimensions so they are easy to fit into standard building dimensions.

Another form of ceiling-mounted radiant heater is the tiered round grille design (Fig. 8-18). Frequently, this type is used in conjunction with a wall switch timer or wall-mounted thermostat. It adapts to a standard fixture outlet box.

Commercially, there are quite a range of infrared lamp heaters for spot heating for indoor or outdoor application (Fig. 8-19). Typically, these are used in industrial plants, aircraft hangars, garages, etc., where convective heating of the air may not be feasible. They may also be found under covered entrance ways to hospitals, stores, etc., to melt ice and snow. Control is by manual switch operation rather than by a thermostatic device.

FIGURE 8-15 Radiant wall heater (Courtesy, Singer Company).

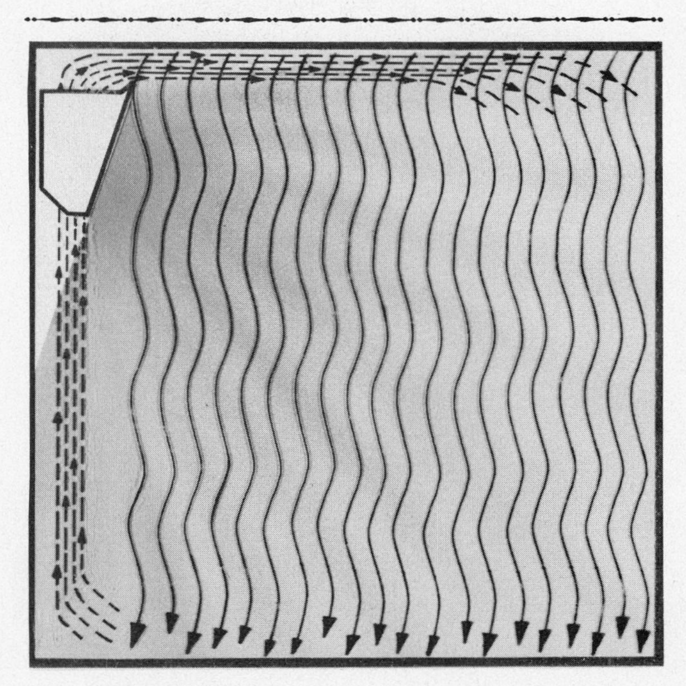

FIGURE 8-16 Ceiling cove radiant heater (Courtesy, Federal Pacific Electric Company).

Forced-air heaters add a measure of comfort to electric space heating. In homes and offices the recessed models (Fig. 8-20) combine style with the forced air circulation, resulting in more uniform room temperature control. Most are wall mounted but models are also available for recessing into the ceiling if wall installation isn't practical. There are even space heaters for use under kitchen counters. Wall-mounted units go up to 3,000 watts (10,240 Btu/h), which can warm larger rooms or offices.

Suspended force-fan unit heaters (Fig. 8-21) are most functional, being used in homes (garages, work-

SURFACE MOUNTED

JUNCTION BOX
INSTALLATION
CEILING SURFACE
TOGGLE BOLT
WOOD SCREW

RECESS MOUNTED

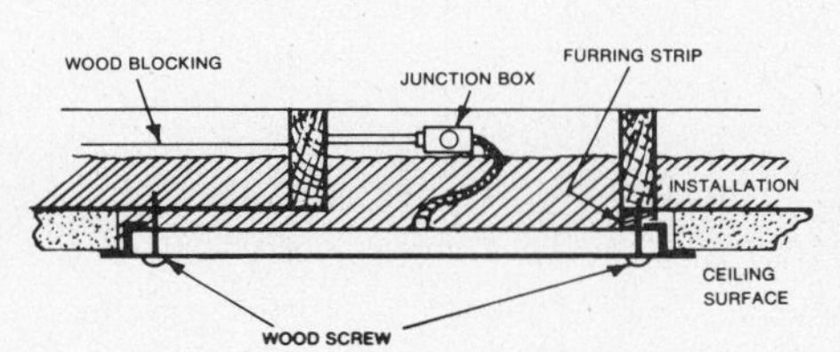

T-BAR CEILING MOUNTED

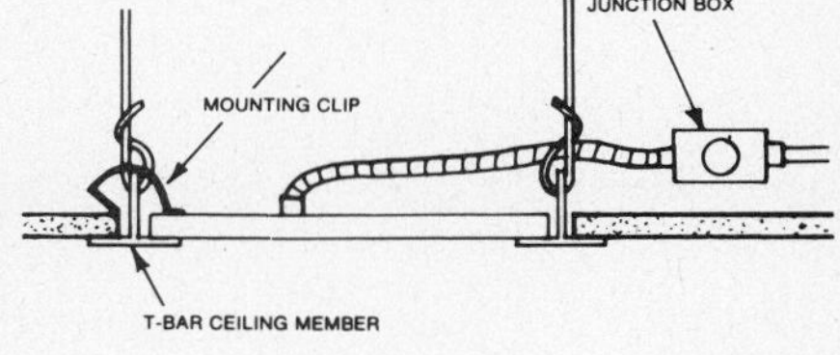

FIGURE 8-17 Ceiling panel radiant heaters (Courtesy, Federal Pacific Electric Company).

FIGURE 8-18 Ceiling heater (Courtesy, Singer Company).

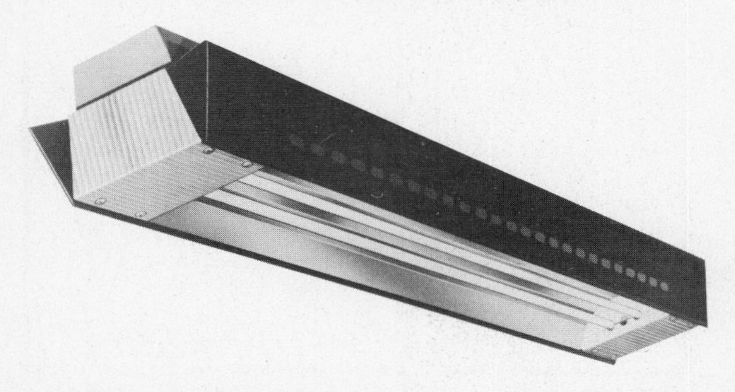

FIGURE 8-19 Infrared lamp heater (Courtesy, Federal Pacific Electric Company).

rooms, playrooms, etc.,) where large capacity and positive air circulation are important, as well as in commercial and industrial establishments of all kinds. They may be suspended in vertical or horizontal fashion with a number of control options such as wall switch only, thermostats, timers, and even night setback operation. Capacities generally range from 3 kW to 12 kW and above. Discharge louvers may be set to regulate air motion patterns. The heavy-duty construction of this heater category is geared for commercial applications.

Cabinet-type space heaters (Fig. 8-22) are found in classrooms, corridors, foyers, and similar areas of commercial and institutional buildings where cooling is not a consideration. Heating elements go up to 24 kW, and the fans are usually the centrifugal type for quiet but high volume air flow. Cabinets have provision for introducing outside air where use and/ or local codes require minimum ventilation air. Outdoor air intake damper operation can be manual or automatic. Control systems can be a simple one-unit operation or multiunit zone control from a central control station.

The most frequently used electric space heater is the convective baseboard model (Fig. 8-23a) that has

FIGURE 8-20 Wall forced-air heater (Courtesy, Federal Pacific Electric Company).

FIGURE 8-21 Suspended forced-air heater (Courtesy, Federal Pacific Electric Company).

FIGURE 8-22 Cabinet space heater (Courtesy, Federal Pacific Electric Company).

been installed in millions of American homes as well as in countless commercial buildings. A cross section through the heater shows the convective air flow across the finned tube heating element. The contour of the casing provides warm air motion away from the walls thus keeping them cooler and cleaner.

It is important that drapes do not block the air flow and cause overheating. Likewise, it is important that carpeting does not block inlet air. Most baseboard units have a built-in linear type of thermal protection that prevents overheating. If blockage

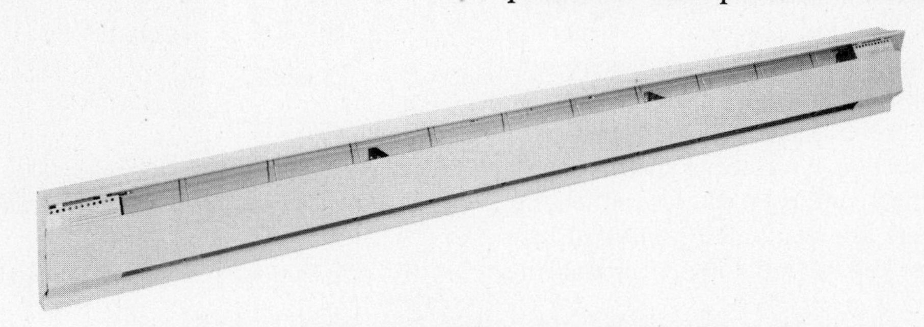

FIGURE 8-23(a) Baseboard heater (Courtesy, Federal Pacific Electric Company).

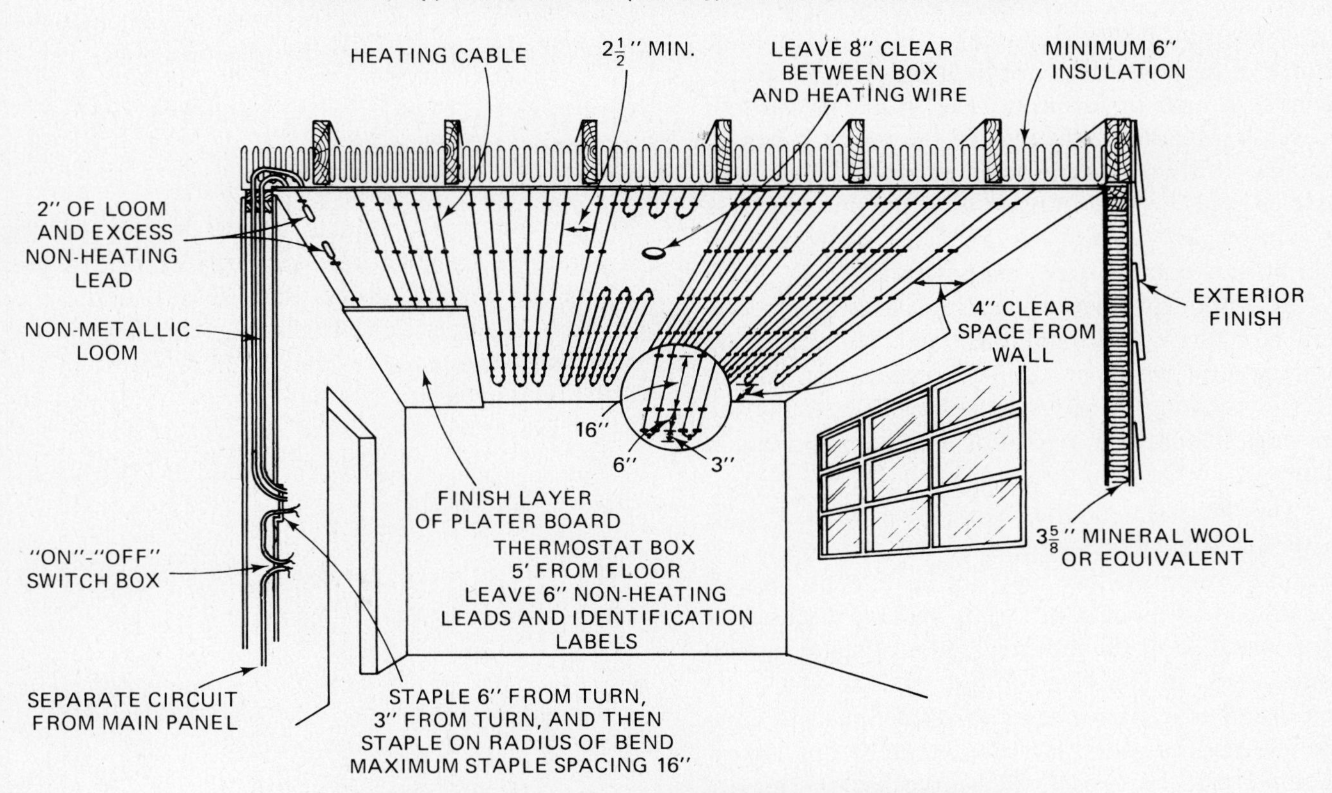

FIGURE 8-23(b) Electric heating cable (Courtesy, Singer Company).

occurs, the safety limit stops the flow of electrical current. It cycles off and on until the blockage is removed. Baseboard heaters come in lengths of 2 feet to 10 feet in the nominal 120, 208, 240, and 277 volt rating. Standard wattage per foot is 250 at the rated voltage. But lower heat output can be obtained by applying heaters on lower voltage. For example, a four-foot heater rated at 1,000 watts on 240 volts, when applied on 208 volts, will produce about 750 watts. Most manufacturers offer these reduced output (low-density) models by simply changing the heating element to 187 watts per foot. Low density is generally preferred by engineers and utilities and is recommended for greatest comfort. Controls may be incorporated in the baseboard or through the use of wall-mounted thermostats. Accessories such as corner boxes, electrical outlets, and even plug-in outlets for room air conditioners are available. Baseboards are finished painted, but can be repainted to match the room decor.

Electric radiant heating cable (Fig. 8-23b) was one of the most popular methods of heating early in the development of this technology. It is an invisible source of heat and does not interfere with the placement of drapes or furniture. The source of warmth is evenly spread over the area of the room. It can be installed within either plaster or dry-wall ceilings. Staples hold the wire in place until the finished layer of plaster or dry-wall is applied. Wall-mounted thermostats control space temperature.

An important electric heating industry association is NEMA (National Electrical Manufacturers Association), which not only publishes manuals on electric heating load calculations (Fig. 8-24) but provides standards of operation for baseboard equipment.

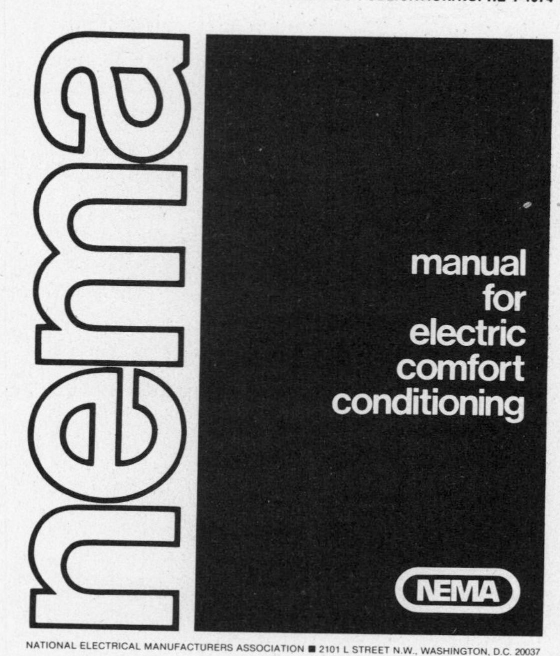

FIGURE 8-24 (Courtesy, National Electrical Manufacturers Association).

This discussion on the room type of space heating covers a broad cross section of this type of winter heating equipment. We may not have mentioned all the individual models, but what we have discussed is representative of the industry. Sales of space heaters total many millions of dollars and as such, are important to certain phases of residential and commercial requirements. However, there is one important point to be considered—space heaters are not designed for the addition of cooling, and at best offer little or no filtering or humidity control and only limited air circulation. Therefore, in terms of their contribution to total comfort, we must conclude they provide a minimum objective: heating only.

WINTER HEATING—GAS FORCED-AIR FURNACE SYSTEM

By far the most popular central system heating apparatus is the gas forced-air furnace; in 1974 over 1.5 million units were sold. Because of the difference in housing types, construction methods, and air distribution systems, several different forms of this system are needed.

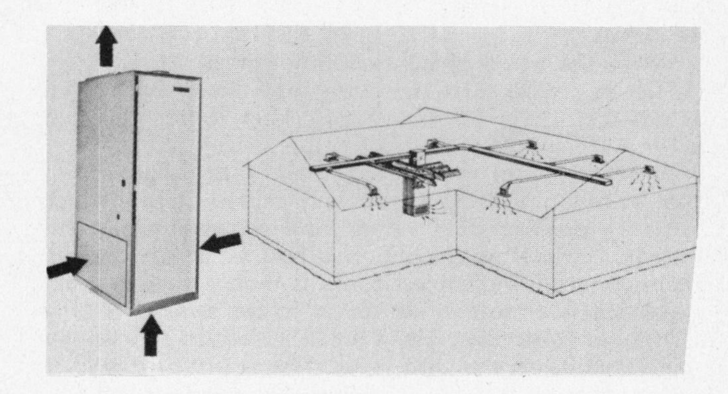

FIGURE 8-25 Upflow high boy furnace (Courtesy, Westinghouse Electric Corporation).

Types of Furnaces

Furnaces can be categorized into four styles: Upflow highboy, lowboy, counterflow, and horizontal.

The upflow highboy (Fig. 8-25) is most popular. Its narrow width and depth allows for location in first-floor closets and/or utility rooms. However, it can still be used in most basement applications for heating only or with cooling coils where head room space permits. Blowers are usually direct drive multispeed. Air intake can be either from the sides or from the bottom.

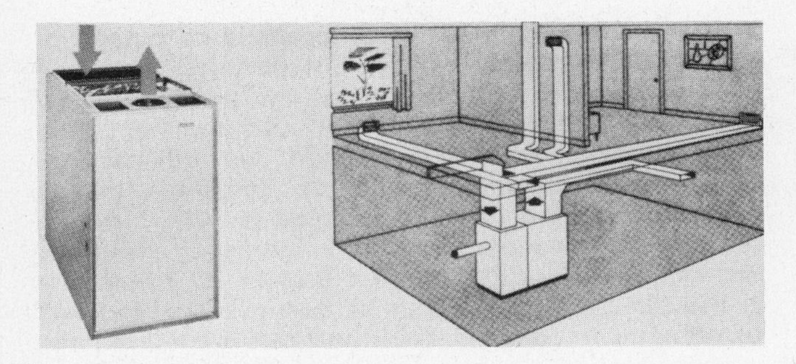

FIGURE 8-26 Low boy furnace (Courtesy, Westinghouse Electric Corporation).

Lowboy furnaces (Fig. 8-26) are built low in height and are the necessary choice where head room is minimum. These furnaces are a little over four feet high, providing easy installation in a basement. Supply and return ducts are on top for easy attachment. Blowers are most commonly belt driven. Many are sold as replacement equipment in older homes.

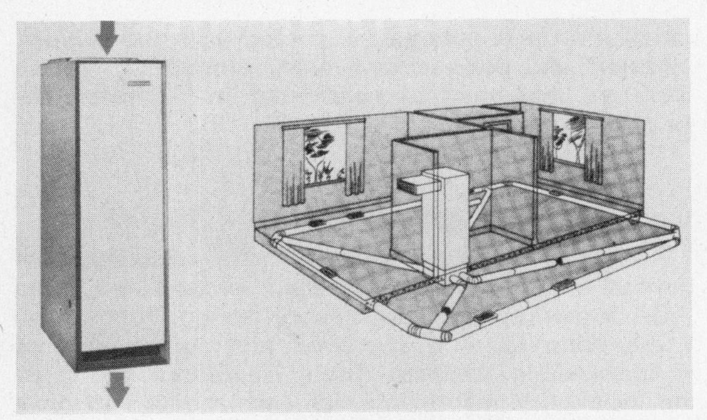

FIGURE 8-27 Counterflow furnace (Courtesy, Westinghouse Electric Corporation).

The counterflow or downflow (Fig. 8-27) is similar in design and style to the highboy, except that the air intake and fan are at the top and the discharge is at the bottom. These are widely used where duct systems are set in concrete or in a crawl space beneath the floor. When mounted on a combustible floor, accessory bases are required. An extra safety limit control is also used.

The fourth type is the horizontal furnace (Fig. 8-28a), which is adaptable to installation in crawl spaces, attics, or basements due to its low height. It requires no floor space. Intake air is at one end and discharged out the other. Burners are usually field changeable for left or righthand application. Figure 8-28b is a composite drawing of typical furnace installation.

Furnace Ratings

Gas furnaces and even gas-burning space heaters are rated on input Btu/h capacity by AGA. The output is a function of the individual furnace design, but it's common practice to assume 80% efficiency. Therefore, a furnace rated at 100,000 Btu/h input will have 80,000 Btu/h output.

Naturally, in selecting a furnace to satisfy the heating load as calculated in Chap. 6, we would base our recommendation on the output as stated in the manufacturer's specifications. This is true for all gases; natural, manufactured, mixed, and LP. Oil furnaces, on the other hand, are tested and rated by UL for output.

Basic Furnace Design

In its simplest form the cross section of a gas upflow warm-air furnace (Fig. 8-29) consists of a cabinet housing, a heat exchanger, a burner assembly, a blower to move air over the heat exchanger, an air filter, a flue vent system, and the necessary safety and operation controls. Let's examine each.

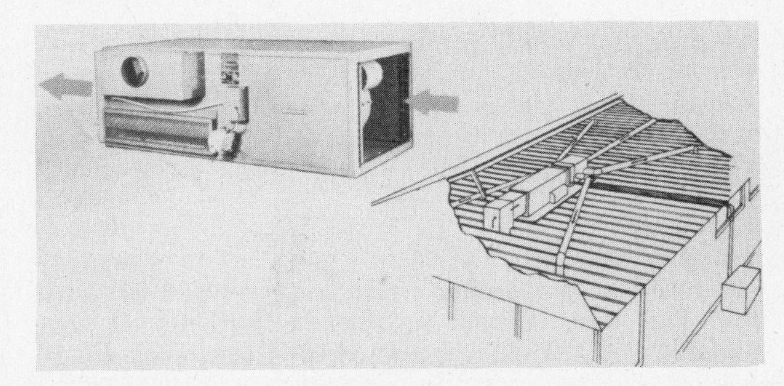

FIGURE 8-28(a) Horizontal furnace (Courtesy, Westinghouse Electric Corporation).

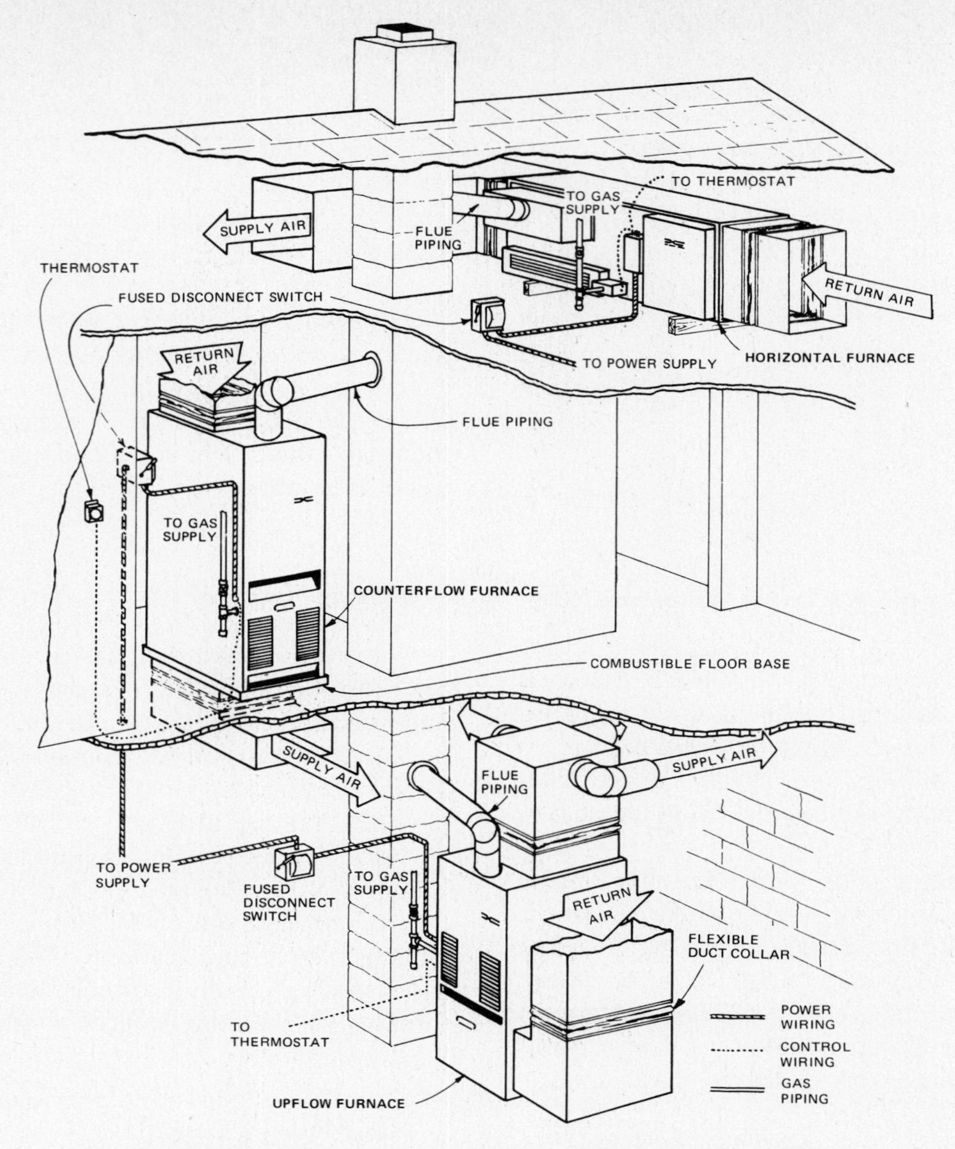

FIGURE 8-28(b) (Courtesy, York, Div. of Borg-Warner Corporation).

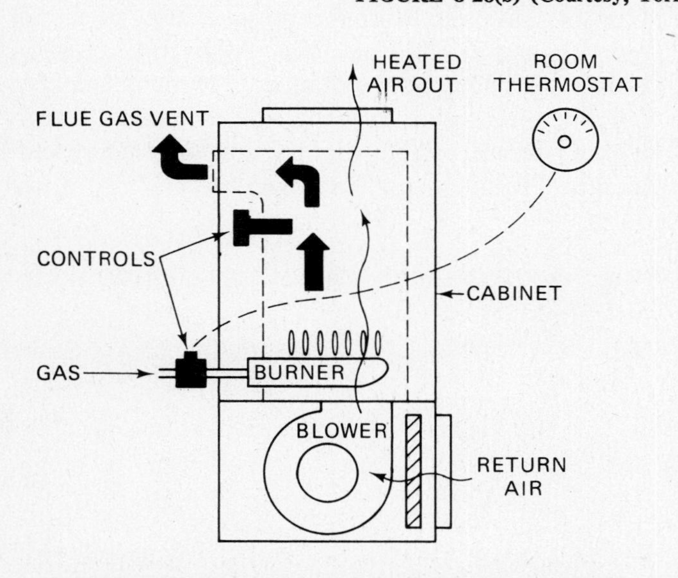

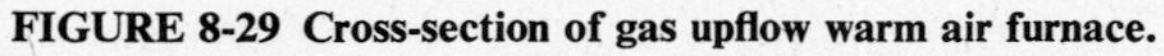

FIGURE 8-29 Cross-section of gas upflow warm air furnace.

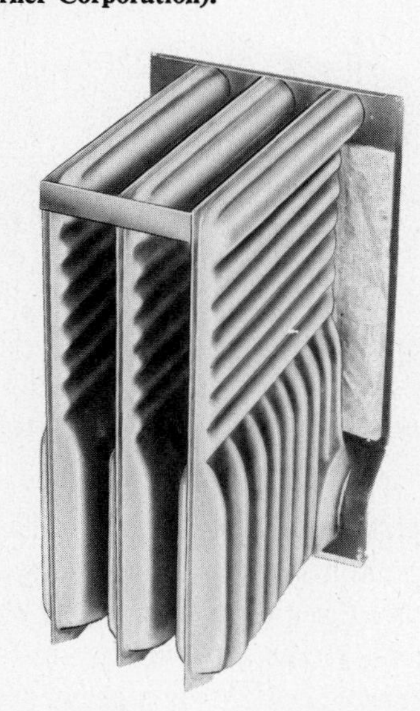

FIGURE 8-30 Heat exchanger (Courtesy, Airtemp Corporation).

Heat Exchanger: The typical gas furnace heat exchanger (Fig. 8-30) is made from two stamped or formed steel sheets which are then seam welded to make a section. These sections are then headered at top and bottom and welded into fixed position. The number of sections depends on the ultimate capacity. Most manufacturers estimate about 25,000 to 40,000 Btu/h per section. Thus, a 100,000 Btu/h furnace will have four sections and so on. The heat exchanger is of heavy gauge steel and is then finished with protective coatings to help prevent rusting from condensation that occurs in the normal combustion process. Note the contours in the surface of the sections. This is to create air turbulence for better heat transfer, inside and outside.

Burner: The burner assembly that fits into the lower openings of the heat exchanger is sized and designed to produce the right amount of heat, flame pattern, and flue effect for efficient gas combustion. Figure 8-31 is a cross section of a typical atmospheric burner.

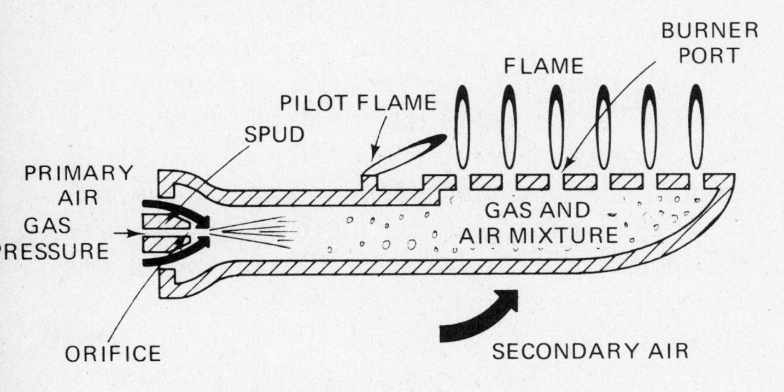

FIGURE 8-31 Burner cross-section.

(Power burners are not used on residential work.) Gas is piped from the meter to the furnace and enters the burner cavity through the spud, which has a precisely drilled orifice to meter the correct amount of gas. This gas ejects itself into the burner venturi throat, and from basic physics we learn that a negative pressure or vacuum is produced behind the gas jet. *Primary air* is thus drawn into the tube where it mixes with gas and forms a positive pressure mixture ready for exhausting through the burner ports. As the mixture comes out of the ports, it is ignited by a pilot device and a flame is established. In a previous discussion we noted the importance of air for combustion. About half of the air is needed for primary air, but in order to expell the combustion byproducts, a *secondary air* source is needed to help create a good flue effect. This secondary air through the burner pouch opening is controlled by restrictor baffles designed into the heat exchanger. The primary air is controlled by air shutters (Fig. 8-32), as would be seen from the burner end, if this was possible. Turning the shutters opens or closes the venturi and thus the amount of induced primary air. With a fixed gas jet, we can then vary the flame characteristics for most efficient burning.

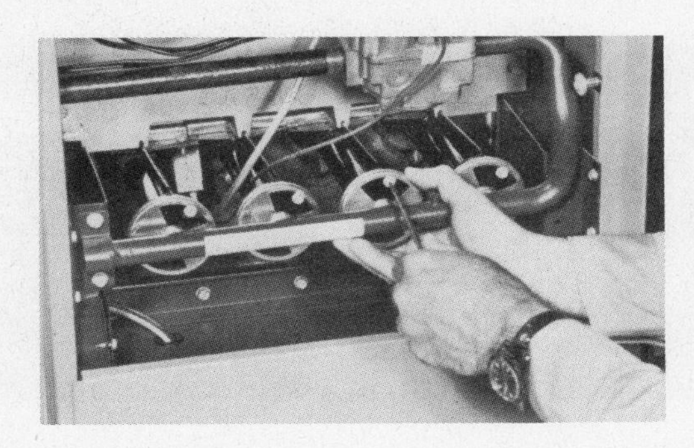

FIGURE 8-32 Adjusting primary air (Courtesy, Westinghouse Electric Corporation).

The exact flame characteristics depend on the burner, but it's recognized that a soft, clear blue flame is the most desirable. Closing the air shutters and reducing primary air will cause the flame to go yellow, and soot will result from incomplete combustion. Fully opened shutters will allow too much primary air, and the flame will raise off the ports and tend to become "blowing" and unstable. All the heat content may not be released, but carbon monoxide might. Blowing flames are also noisy.

Pilot: Notice in Fig. 8-33 the existence of a pilot runner attached to the main burner tube. Its function is to carry burning gas to each of the main burners from the one which has the pilot lighter attached to it. The alignment of these runners is most important

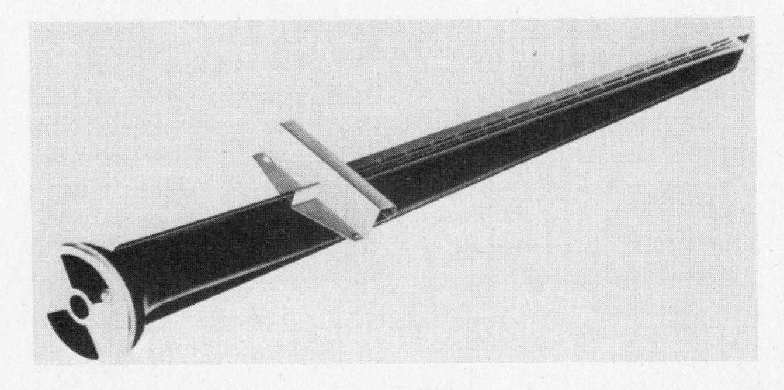

FIGURE 8-33 Slotted port burner (Courtesy, Westinghouse Electric Corporation).

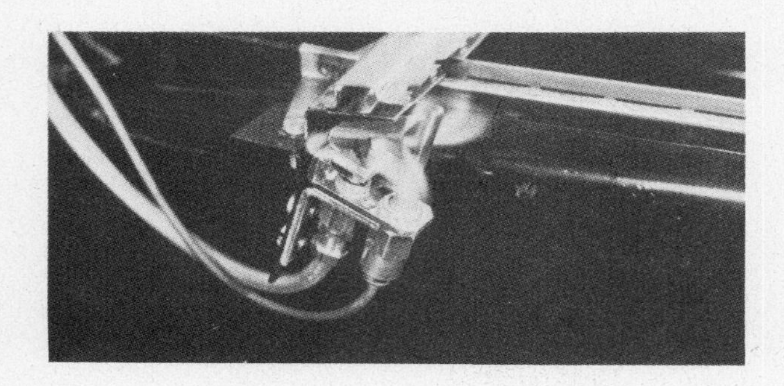

FIGURE 8-34 Pilot flame (Courtesy, Westinghouse Electric Corporation).

for smooth, positive ignition. If it is out of line, a delayed ignition can create noise, possibly with enough force to blow out the pilot.

The pilot lighter and flame (Fig. 8-34) are fed gas from the main gas valve (to be discussed later) by a small $\frac{1}{4}$-in. line. Attached to the bracket is a thermocouple sensor that, when heated, produces a small current of electricity (millivolts), which is directed back to the gas valve to actuate a magnetic operator. The flame adjustment should give a soft steady flame enveloping $\frac{3}{8}$ to $\frac{1}{2}$ in. of the tip of the thermocouple, which will glow a dull red. The flame should be correctly positioned to supply the pilot runners across all burners.

Main Gas Valve: Although control of the main flame is accomplished by adjusting the burner shutters, the gas supply or flow is controlled in the gas manifold section through the use of a combination gas valve (Fig. 8-35) that performs many functions. Valves from various manufacturers may differ, but they all accomplish essentially the same functions:

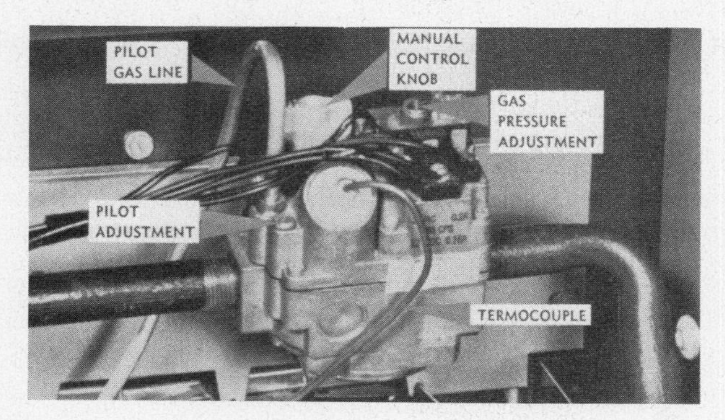

FIGURE 8-35 Combination gas valve (Courtesy, Westinghouse Electric Corporation).

(1) Manual control for ignition and normal operation
(2) Pilot supply, adjustment, and safety shut-off
(3) Pressure regulation of burner gas feed
(4) On-off electric solenoid valve controlled by the room thermostat

Figure 8-36 shows a basic valve sketch of these functions for a natural gas valve. The schematic drawing of the valve shows the main diaphragm valve in the open condition that occurs during heat demand. It is assumed that the main diaphragm open schematic applies to a gas heating appliance with the pilot flame burning, that a thermocouple is connected to the automatic pilot magnetic operator, and that the lighting operation was previously performed to open the automatic pilot valve. It is also assumed that the main gas cock has been turned to the ON position after the pilot is lit. In the application, the 24-volt transformer circuit includes the 24-volt operator and room thermostat in series. Closure of the room thermostat switch on heat demand has energized the 24-volt operator, which causes the armature to be attached to the pole face of the magnet, resulting in a clockwise rotation of the armature, as indicated by the arrows at the end of the armature. This rotation has overcome the valve spring and has pulled the valve stem of the dual operator valve downward, allowing the diaphragm above it to seat on the valve seat encircling the valve stem. The seating of this diaphragm shuts off the bypass porting. Bleed gas can enter the actuator cavity only through the bleed orifice. Bleed gas is allowed to flow from the actuator cavity through the dual valve operator and the regulator to the outlet through the outlet pressure sensing port. (The valve stem has a square cross section in a circular guide, allowing gas to pass between the stem and the guide.) The resultant drop in pressure within the actuator cavity and through the main valve operator porting to beneath the main diaphragm allows the inlet pressure above the main diaphragm to open the main diaphragm valve against the force of the main diaphragm valve spring.

After opening, as shown in Fig. 8-36 (main diaphragm valve open), *straight line* pressure regulation is secured by the feedback of outlet pressure through the outlet pressure-sensing port to the pressure regulator in the bleed line. A rise in pressure at the control outlet above the set pressure causes a proportional closure of the regulator valve in the bleed line. The proportional closure of the regulator valve causes a proportional rise in pressure in the

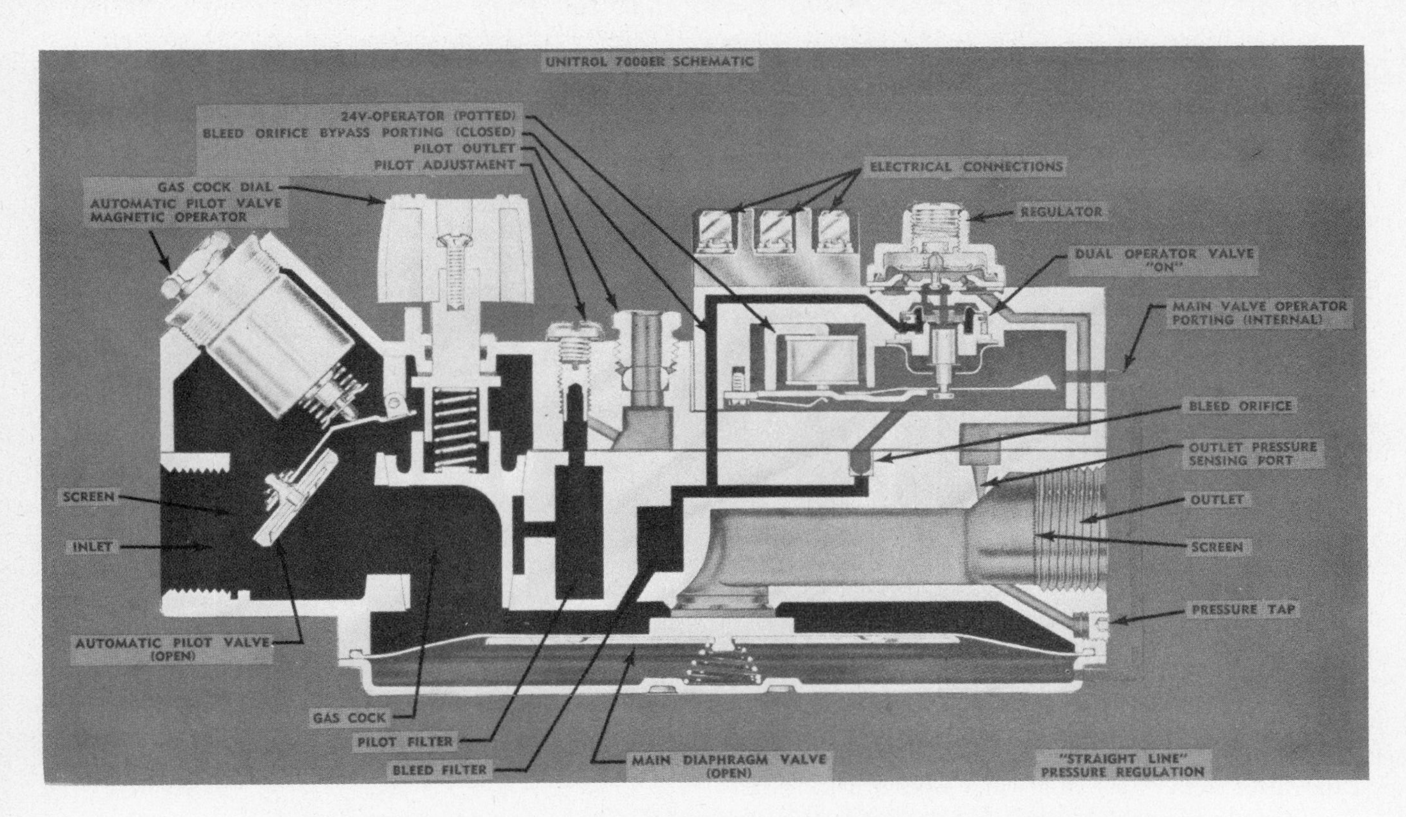

FIGURE 8-36 Combination gas valve heating demand (Courtesy, Robertshaw Controls Company).

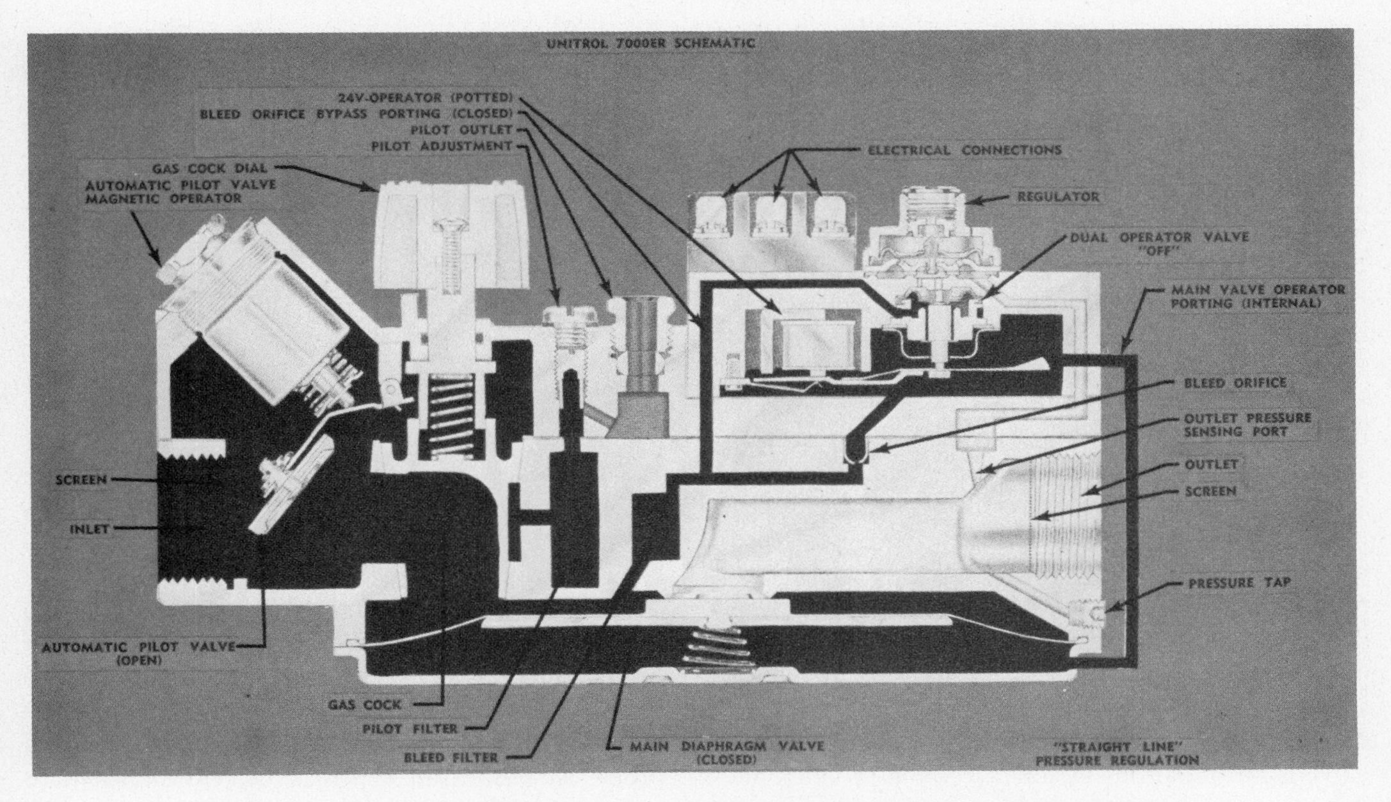

FIGURE 8-37 Combination gas valve heating satisfaction (Courtesy, Robertshaw Controls Company).

bleed line ahead of the regulator. This rise in pressure in the actuator cavity acts through the main valve operator port to increase the pressure beneath the main diaphragm and causes a partial closure of the main diaphragm valve sufficient to lower the control outlet pressure down to the pressure setting. Upon a drop in control outlet pressure, a corresponding decrease in bleed-line pressure transmitted to the underside of the main diaphragm through the action of the bleed-line pressure regulator causes a proportional increase in the main valve opening to bring the delivered outlet pressure back up to the set pressure.

The schematic of the closed main diaphragm (Fig. 8-37) shows the action of the bypass valve in the dual valve operator during a fast "off" response independent of the bleed orifice. Upon heating "satisfaction," resulting in the opening of the thermostat switch, the 24-volt operator is deenergized. A loss of magnetic attraction allows the return spring on the valve stem of the dual valve operator to force the stem upward, closing the center port of the small diaphragm above it and shutting off bleed gas to the bleed regulator and outlet sensing port. The resultant counterclockwise rotation of the armature is indicated by the arrows at the ends of the armature. As the small diaphragm above the valve stem is forced upward, the bleed orifice bypass porting is opened by the diaphragm, leaving the valve seat encircling the valve stem. The actuator cavity, main valve operator port, and the cavity beneath the diaphragm are rapidly exposed to full inlet pressure, which acts to close the main diaphragm valve. The bleed orifice bypass port allows the pressure above and below the mean diaphragm to be equalized rapidly, independently of the bleed orifice. When this pressure is equalized, the main valve is closed by the main valve spring.

The dual operator valve that provides the bleed orifice bypass porting control for the fast "off" response is an important feature. It helps to prevent flash-back conditions. Flash back can cause pilot outage problems as well as burner and appliance sooting. Relatively slow closing has been a problem on larger capacity applications. The dual operator valve is an important factor in helping to eliminate this application problem.

The valve has built-in protection against gas line contaminants through the use of highly effective inlet and outlet screens for the main gas passages, a pilot filter for the pilot line, and a bleed filter for the bleed control line. These means of self-protection are highly effective and protect the control from malfunction due to such contaminants entering the control passages. In the highly unlikely case that the bleed orifice restrictor will become clogged with the main valve in the open or "on" position, in spite of the bleed filter protection, the dual operator valve will enable the control to operate to shut off the main gas valve when signaled to do so by the room thermostat or limit control.

On an LP furnace, the gas valves do not have the pressure regulation function, but there is still 100% pilot shut off in the event of pilot outage, which is most important. Liquid petroleum gas is heavier than air, and over a prolonged period enough pilot gas would collect to be a hazard.

The above description is of a method of pilot ignition and gas valve control that is still the most common in residential equipment use. However, the industry has developed and applied electronic ignition devices in roof-top commercial equipment, and we are just beginning to see this technique being offered in residential furnaces.

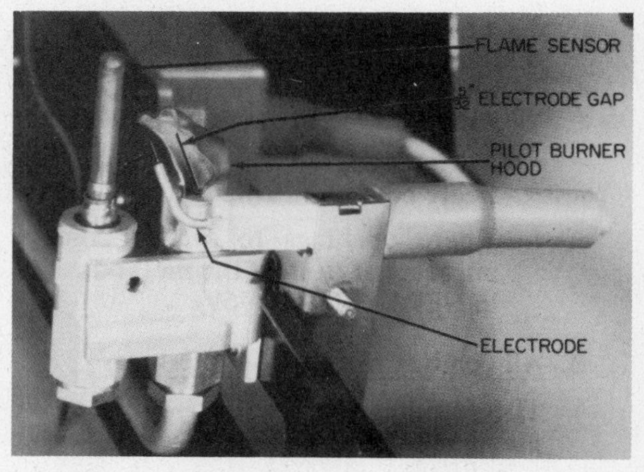

FIGURE 8-38 Electronic ignition system (Courtesy, Carrier-Air Conditioning Company).

Figure 8-38 is a picture of an electronic ignition system. Pilot gas is fed to the assembly and burner orifice. The spark electrode is so positioned as to ignite the gas upon a signal from the room thermostat. With the pilot ignited and burning, a sensing probe establishes an electrical current sufficient to energize a relay that opens the main gas valve by closing normally open electrical contacts. At the same time, the relay's normally closed electrical contacts in the spark ignition circuit open, terminating the spark. As long as the sensing probe recognizes the pilot flame, the relay coil will be energized and normally closed contacts in the spark ignition system will remain open.

Venting: Another important application consideration is the matter of proper venting of flue gases. The basic function of the vent to which a gas furnace is connected is two-fold:

(1) To provide a safe and effective means for moving the products of combustion from the draft hood to the outside atmosphere without contaminating the room air.

(2) To provide the mechanics for producing and maintaining a draft, which induces a supply of replacement air into the room where the furnace is situated.

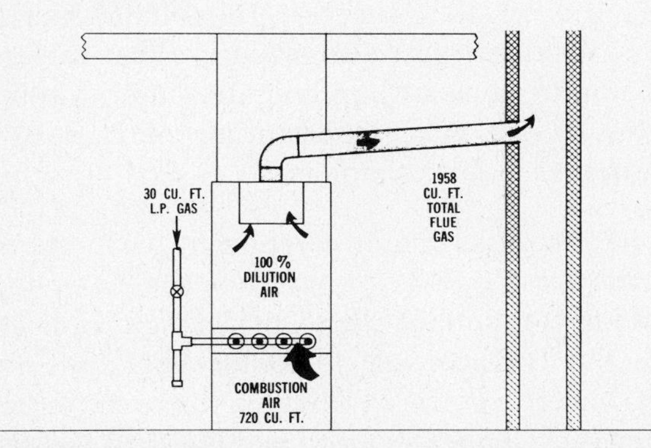

FIGURE 8-39 75,000 Btu hr L.P. furnace (Courtesy, Westinghouse Electric Corporation).

For example, a 75,000 Btu LP furnace (Fig. 8-39) burns 30 ft^3 of gas per hour. Remembering that LP also requires approximately 24 ft^3 of air for each cubic foot of gas to support combustion, we then add 720 ft^3 of air. When these two quantities are burned, the resultant release of combustion byproducts, plus the 100% dilution of air entering the draft diverter brings the total vent gases to 1,440 ft^3—more than 60 times the input gas capacity. A natural gas furnace would have about half that much, but this is why supply air and venting is so important on all gas appliances.

The draft hood on a furnace (Fig. 8-40) is the place where flue gases and dilution air are mixed. Under normal operation, furnace flue gases entering the draft hood are mixed with dilution air entering the relief opening and then flow out of the draft hood exit at a lower temperature than incoming flue gases.

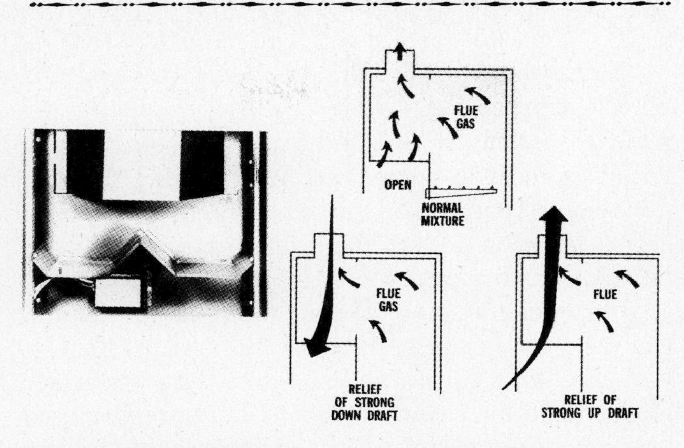

FIGURE 8-40 Draft diverter action (Courtesy, Westinghouse Electric Corporation).

Under strong updraft conditions, such as may be caused by vent terminal wind conditions, the low resistance to the flow of dilution air through the relief opening of a properly designed draft hood tends to neutralize the effect of excess stack action on the furnace.

If a down draft occurs, the relief opening also provides a discharge opening for flue gases and down-flowing air so that pilot outage and other possible effects are minimized.

Controls: Another set of components in the furnace is the controls. For a simple highboy "heating only" system, the illustration in Fig. 8-41 shows the basic wiring of the furnace.

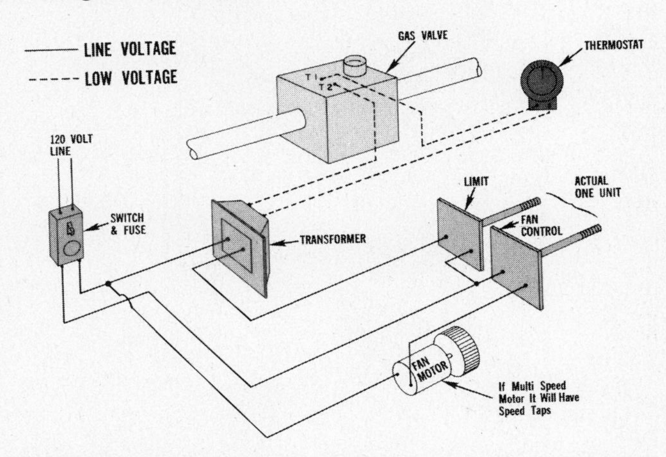

FIGURE 8-41 Typical furnace wiring (Courtesy, Westinghouse Electric Corporation).

Line voltage to the furnace should be fused and wired according to local codes. (Check the unit's data plate to determine full-load amps and the sizes of wire and fuses in compliance with codes.) The line voltage then passes through the normally closed limit switch before it goes to the primary tap on the step-down transformer, where it provides a 24-volt secondary circuit. Note that the room thermostat and gas valve are in series in the 24-volt system. Thus, if the safety limit opens, all voltage to the transformer is killed and the gas valve goes to its normally closed position. Counterflow furnaces require a second limit, which must be manually reset.

The blower motor and fan control are also in series so that the fan can be set to cycle off and on as desired. Actually, the limit and fan control are physically in the same casing, with temperature-sensing elements projecting into the furnace air stream.

The combination fan and limit switch is illustrated in Fig. 8-42 and has a dual function. First, as a safety limit as mentioned above, some have fixed limit temperature settings; others are adjustable (approximately 180°F to 200°F is the usual range). This allows a 50° to 60°F rise above normal operation before it opens. The fan control switch is also a temperature-sensing device that is set to turn on the fan after the furnace has warmed up at least 10 to 15°F above room conditions, so that cold drafts are not experienced. It also stops the blower after the burner cuts off, so again there are no uncomfortable drafts. Note: we must also recognize that some systems employ constant fan operation and thus override this switch. CAC (constant air circulation) will be discussed in later chapters.

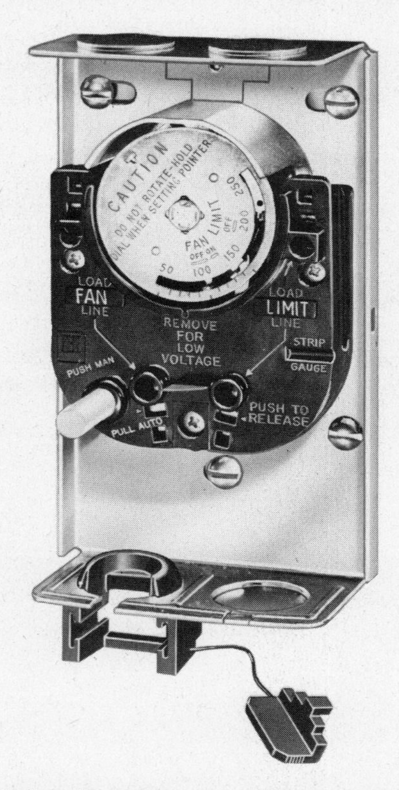

FIGURE 8-42 Combination fan and limit control (Courtesy, Honeywell, Residential Division).

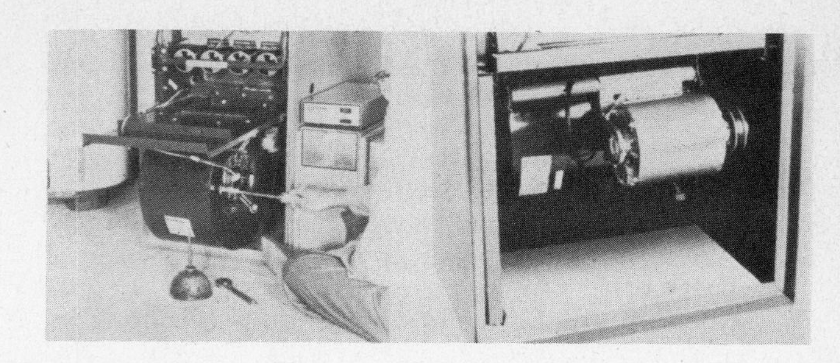

FIGURE 8-43 Blower (Courtesy, Westinghouse Electric Corporation).

Blower: Another major component of our furnace is the blower (Fig. 8-43), which provides the energy to distribute conditioned air to the living space. Most residential furnaces use a double inlet centrifugal blower. Direct drives are possible where the motor is mounted inside the fan scroll and the shaft is directly connected to the fan wheel. Fan speed then is the same as the motor rpm. Most direct-drive blowers have multispeed motors for ft^3/min adjustment. Larger furnace models often employ belt drives where the ft^3/min adjustment can be varied over a wide range due to pulley diameter ratios.

The multispeed direct-drive motors are wound with speed taps so that the rpm can be changed either at the motor itself or at a remote location. Some manufacturers use solid-state control devices to vary the ft^3/min, depending on winter or summer needs or to compensate for additional duct pressures when cooling, humidifiers, air cleaners, and the like are added to the system. These solid-state devices vary motor speed by changing the electrical frequency sine waves of power input.

Cabinet: The final element of our built-up furnace is the cabinet itself (Fig. 8-44). Its primary purpose,

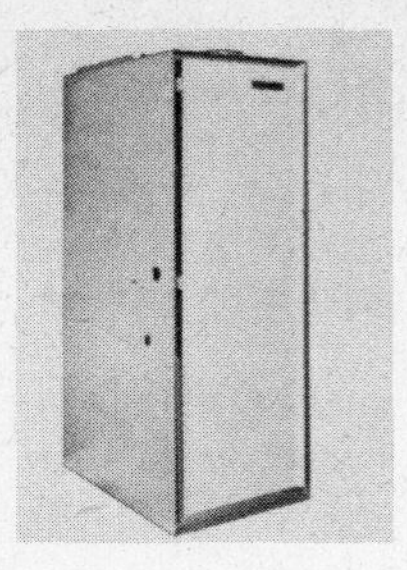

FIGURE 8-44 Cabinet (Courtesy, Westinghouse Electric Corporation).

of course, is to support all of the internal components rigidly; but it must also provide a reasonably attractive appearance. Many furnaces are installed in living areas, closets, utility rooms, etc., where an attractive appearance and sound control are desirable. The cabinet must also incorporate important safety features; i.e., to keep children away from the combustion area, to present a surface temperature suitable to the touch, and to provide proper clearances from combustible surfaces. The hot air passing over the outside heat exchanger shell surface is naturally in contact with the cabinet. The heated air, plus radiant heat, produce a surface temperature that must be considered in the safety aspects of a furnace.

This then leads to discussion of clearances. When gas furnaces are approved and listed by AGA (American Gas Association) testing laboratories, they are done so with approved minimum clearances from combustible surfaces. These clearances are listed in the installation instructions for each model and on a decal in the furnace cabinet. Figure 8-45 is a sample of the required clearances for an upflow highboy. The clearances shown in the table below are from any surface of the enclosure:

	Clearance
From top of plenum	1 inch
From top of duct within 6 feet of unit	1 inch
From front of unit	5 inches
Horizontally from flue or draft hood	6 inches
Vertically from flue or draft hood	6 inches
From rear of unit	1 inch
From left side	1 inch
From right side	1 inch
From side of plenum	1 inch

A 24-inch service access must be provided at the front of the unit. It is necessary to comply with these minimum clearance standards. If they are not met, the risk of fire and voided insurance coverage may become liabilities for the dealer.

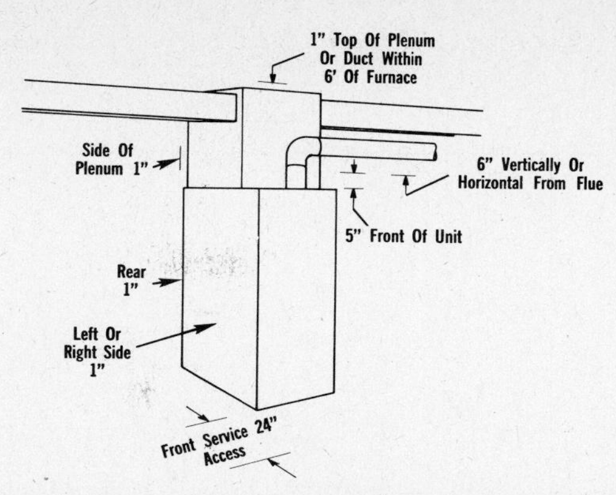

FIGURE 8-45 Typical furnace clearances (Courtesy, Westinghouse Electric Corporation).

As previously mentioned, when a counterflow furnace is installed on a combustible floor, AGA requires the use of an approved floor base (Fig. 8-46). Sufficient clearance space is allowed for the passage of the plenum through the floor opening in the event of an overheating problem.

With a basic understanding of the furnace construction and the function of internal components, let's look at some related application data; for the furnace can be no better than the installation. Lack of knowledge or attention to these elements can cause faulty or possibly hazardous operation.

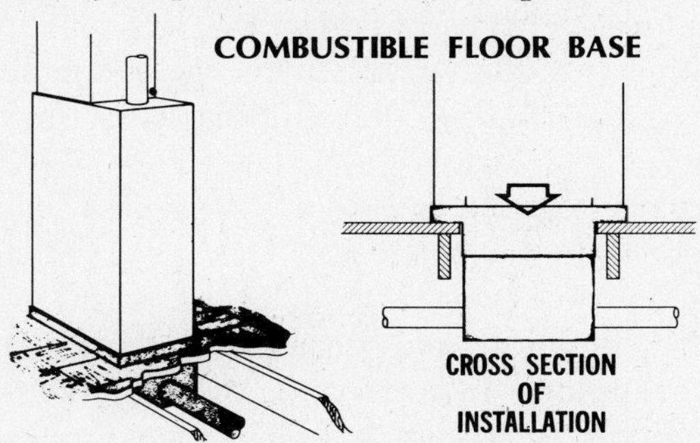

FIGURE 8-46 Counterflow furnace (Courtesy, Westinghouse Electric Corporation).

Flue/Vents: The vent connector pipe between the furnace outlet and the vent or chimney is illustrated in Fig. 8-47. A large portion of the resistance to the flow of flue gases occurs in this vent connection. It is important to employ a minimum run and elbows and use a pipe at least equal in size to the draft hood outlet. Long horizontal runs and fittings increase flow resistance and also lower the flue gas temperature before it reaches the vertical vent. Always slope the vent connector up toward the chimney connector. The vent connector is then inserted into, but not beyond, the inside wall of the chimney flue liner.

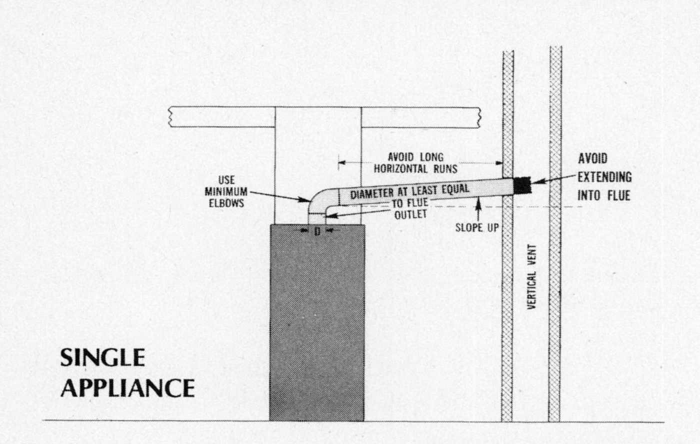

FIGURE 8-47 Vent connector (Courtesy, Westinghouse Electric Corporation).

Where two or more appliances (furnace and water heater) are vented into the same flue (Fig. 8-48), the connector should be at least the size of the largest flue, plus 50% of the other appliance. The chart shown gives venting combinations, and it is usually in the furnace installation bulletin for quick reference. The critical point occurs when one appliance is not operating and dilution air enters the common vent, lowering the draft-producing flue gas of the other appliance. For example, a water heater with a 3-in. vent and a furnace with 5-in. vent would require a common 6-in. flue connector.

The vertical vent (chimney or flue stack) must be properly designed and installed to create proper draft conditions. In some cases the heating and air-conditioning technician has little control over these conditions; in others, the installation of prefabricated vents may be in the heating contract from the builder or owner.

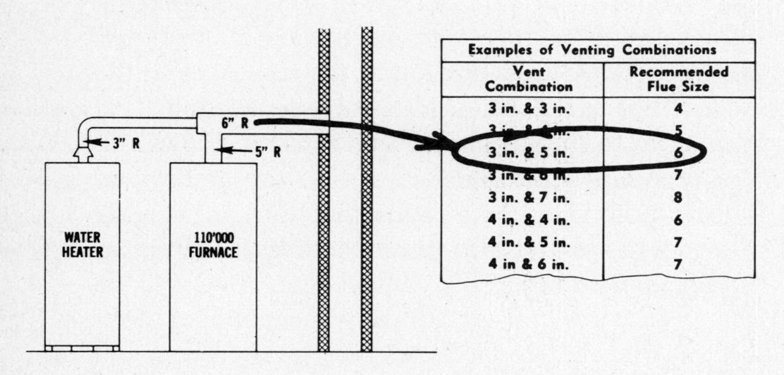

Examples of Venting Combinations	
Vent Combination	Recommended Flue Size
3 in. & 3 in.	4
[illegible]	5
3 in. & 5 in.	6
[illegible]	7
3 in. & 7 in.	8
4 in. & 4 in.	6
4 in. & 5 in.	7
4 in & 6 in.	7

FIGURE 8-48 Vent connector multiple appliances (Courtesy, Westinghouse Electric Corporation).

Masonry chimneys (Fig. 8-49) are field constructed and should be built in accordance with local and/or nationally recognized codes or standards. The size of the flue liner, offsets, terminal vent and the chimney height can materially affect the draw or draft.

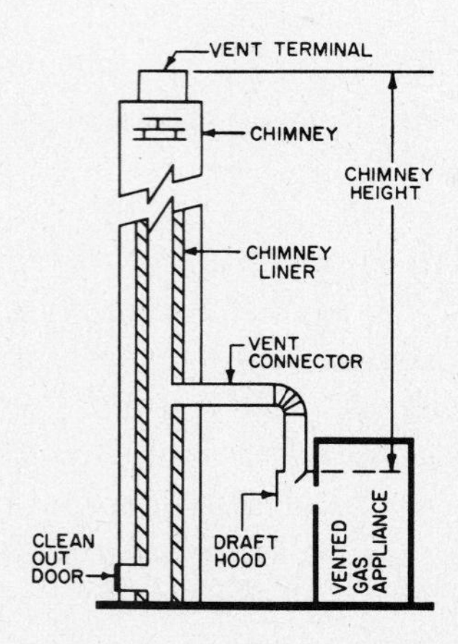

FIGURE 8-49 Vertical vents (masonry chimney) (Courtesy, Westinghouse Electric Corporation).

The chimney in Fig. 8-50 is factory built and listed by a nationally recognized agency. It is made of metal of an adequate thickness and is insulated, galvanized, and properly welded or riveted. It is commonly called a *Type B gas vent* and is double walled. The double wall helps to conserve heat in the flue gas and thus promotes better draft as well as producing a lower surface temperature on the outer surface; it also requires less clearance than the single-walled type.

Single-walled metal vents are not generally considered suitable for venting residential heating equipment and are prohibited by many local codes because they cannot be used in concealed spaces.

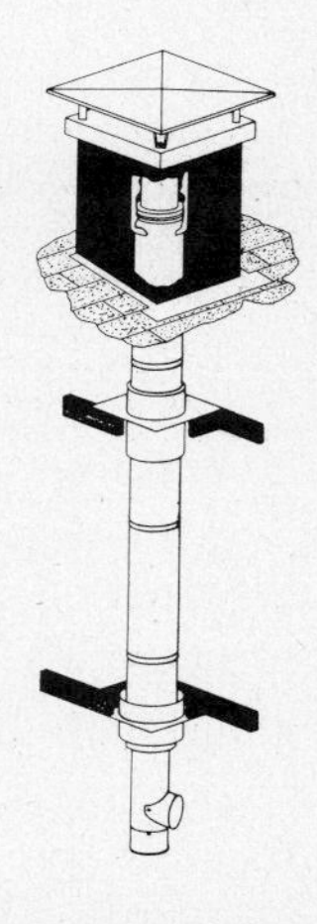

FIGURE 8-50 Vertical vents (prefabricated chimney).

As mentioned above, although you, as a service technician, in most cases will not have a part in the type or design of the vertical flue, you can observe those poor practices that may reflect on the performance and operation of equipment or lead to serious service problems. This is important when you are replacing a furnace; you then must check the venting arrangement.

Most city venting codes are based on the standards published by AGA, the *National Fire Protection Association Bulletin*, or by the National Board of Fire Underwriters (NBFU) and Gas Vent Institute. Your local building inspector and gas company keep these publications on file.

AGA Report No. 1319 provides tables for the proper sizing of both masonry and metal flues based on heat input, the height of the chimney, horizontal run of connector, etc. The table shown in Fig. 8-51 for masonry chimneys gives the maximum allowance of heat input based on the chimney height, vent connector size, and distance of horizontal run. For example, a chimney height of 15 feet with a 6-in. vent connection of 5 feet of horizontal run will handle 163,000 Btu per hour.

Type B double wall gas vents are figured in a similar manner. Manufacturers' specifications will list the capacity based on height, horizontal run, and vent size. For example, in Fig. 8-52 an individual appliance with vent height of 20 feet, 6 inches round size and a 10-foot horizontal run will carry 228,000 Btu. This would be typical of a ranch-house application, with the furnace in the basement and a short stack to the roof. The specifications also list the sizes for multiple appliance application (furnace and water heater).

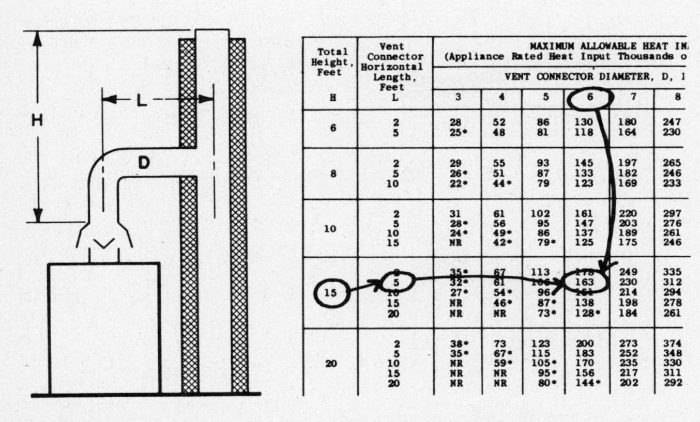

Total Height, Feet H	Vent Connector Horizontal Length, Feet L	MAXIMUM ALLOWABLE HEAT IN (Appliance Rated Heat Input Thousands o					
		VENT CONNECTOR DIAMETER, D, I					
		3	4	5	6	7	8
6	2	28	52	86	130	180	247
	5	25*	48	81	118	164	230
8	2	29	55	93	145	197	265
	5	26*	51	87	133	182	246
	10	22*	44*	79	123	169	233
10	2	31	61	102	161	220	297
	5	28*	56	95	147	203	276
	10	24*	49*	86	137	189	261
	15	NR	42*	79*	125	175	246
15	[illegible]	35*	67	113	[illegible]	249	335
	5	32*	61	[illegible]	163	230	312
	[illegible]	27*	54*	96*	[illegible]	214	294
	15	NR	46*	87*	138	198	278
	20	NR	NR	73*	128*	184	261
20	2	38*	73	123	200	273	374
	5	35*	67*	115	183	252	348
	10	NR	59*	105*	170	235	330
	15	NR	NR	95*	156	217	311
	20	NR	NR	80*	144*	202	292

FIGURE 8-51 Masonry chimney with single vent connector (Courtesy, Westinghouse Electric Corporation).

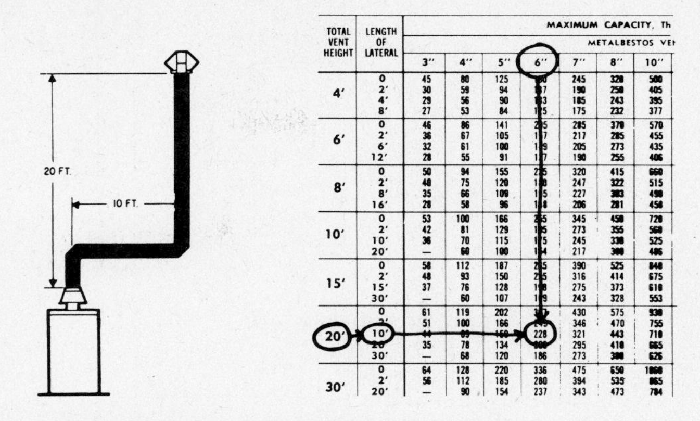

TOTAL VENT HEIGHT	LENGTH OF LATERAL	MAXIMUM CAPACITY, Th						
		METALBESTOS VE						
		3″	4″	5″	6″	7″	8″	10″
4′	0	45	80	125	[illegible]	245	320	500
	2′	30	59	94	[illegible]	190	250	405
	4′	29	56	90	[illegible]	185	243	395
	8′	27	53	84	[illegible]	175	232	377
6′	0	46	86	141	[illegible]	285	370	570
	2′	36	67	105	[illegible]	217	285	455
	6′	32	61	100	[illegible]	205	273	435
	12′	28	55	91	[illegible]	190	255	406
8′	0	50	94	155	[illegible]	320	415	660
	2′	40	75	120	[illegible]	247	322	515
	8′	35	66	109	[illegible]	227	303	490
	16′	28	58	96	[illegible]	206	281	450
10′	0	53	100	166	[illegible]	345	450	720
	2′	42	81	129	[illegible]	273	355	560
	10′	36	70	115	[illegible]	245	330	525
	20′	—	60	100	[illegible]	217	300	486
15′	0	58	112	187	[illegible]	390	525	840
	2′	48	93	150	[illegible]	316	414	675
	15′	37	76	128	[illegible]	275	373	610
	30′	—	60	107	[illegible]	243	328	553
20′	0	61	119	202	[illegible]	430	575	930
	[illegible]	51	100	166	[illegible]	346	470	755
	10′	[illegible]	[illegible]	[illegible]	228	321	443	710
	[illegible]	35	78	134	[illegible]	295	410	665
	30′	—	68	120	186	273	300	626
30′	0	64	128	220	336	475	650	1060
	2′	56	112	185	280	394	535	865
	20′	—	90	154	237	343	473	784

FIGURE 8-52 Type B gas vent selection table (Courtesy, Westinghouse Electric Corporation).

In addition to the type and size of vertical flue, the terminal point in respect to the roof is also important (Fig. 8-53). Two specific rules apply:

(1) If the vent terminates without a listed cap or similar device and depends on natural draft, it must extend at least two feet above the point where it passes through the roof, and be at least two feet higher than any portion of a building that is within 10 feet of the vent.

(2) If the vent terminates with a listed or approved top, it must terminate in accordance with the terms under which such listing was approved.

Additionally, Type B vents must be equipped with an approved cap having a capacity at least as great as the vent stack. Also, this cap must be at least five feet above the highest connected appliance draft hood.

Altitude also has a pronounced effect on combustion air and draft, and above 2,000 feet the AGA calls for reducing the furnace input 4% for each 1,000 feet above sea level.

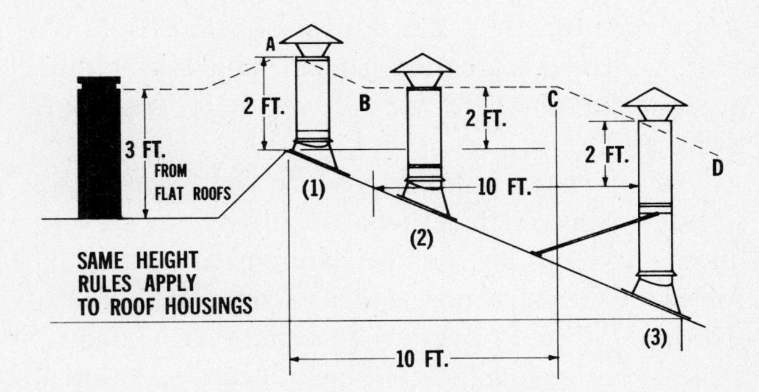

FIGURE 8-53 Vertical vent roof termination (Courtesy, Westinghouse Electric Corporation).

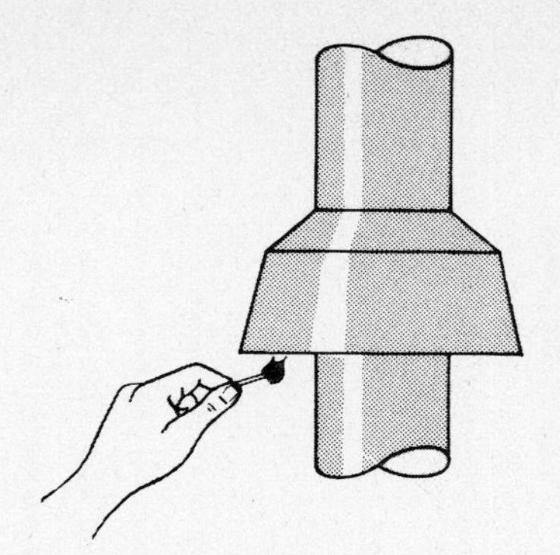

FIGURE 8-54 Vent test.

METHOD A

AIR SUPPLY OPENINGS

ALL AIR FROM INSIDE BUILDING

NOTE: Each opening shall have a free area of not less than one square inch per 1,000 BTU per hour of the total input rating of all appliances in the enclosure.

FIGURE 8-55 (Courtesy, American Gas Association).

Outside masonry chimneys or vents lose heat faster and take longer to heat up than do their indoor counterparts; thus prolonged priming and spilling of flue products from the draft hood relief will occur when the vented appliances are put into operation.

A simple test for proper vent operation can be made at the draft hood of the appliance (Fig. 8-54). Light a match near the relief opening and move it around the entire perimeter of the hood. If vent gases are spilling out, the match will be extinguished because of a lack of oxygen, not because of the velocity of the air. The test should be made under actual (or even extreme) operating conditions.

Combustion Air

Equally important as venting is the matter of sufficient supply air to maintain proper combustion air for diluting the flue gas. As previously pointed out, a 75,000 Btu LP furnace requires 1,440 ft^3 of air per hour. Any other gas-burning appliances within the structure add to the problem. With the tight construction of modern homes, equipped with storm doors and windows, normal infiltration cannot be depended on to provide such a volume of air, particularly if there is a kitchen ventilating system, a clothes dryer, or a fireplace competing for air to perform its function. The danger of a short supply of combustion air is greatest in small, single-floor houses or apartments where the furnace is located in the living area.

As a furnace installer, you do share the responsibility to insure that there are provisions to introduce make-up air to the equipment room. The following illustrstions taken from the NFPA's (National Fire Protection Association) Manual 54 indicate various methods of introducing make-up air.

Method A (Fig. 8-55) is the most critical and least desirable because it depends on the infiltration

METHOD B

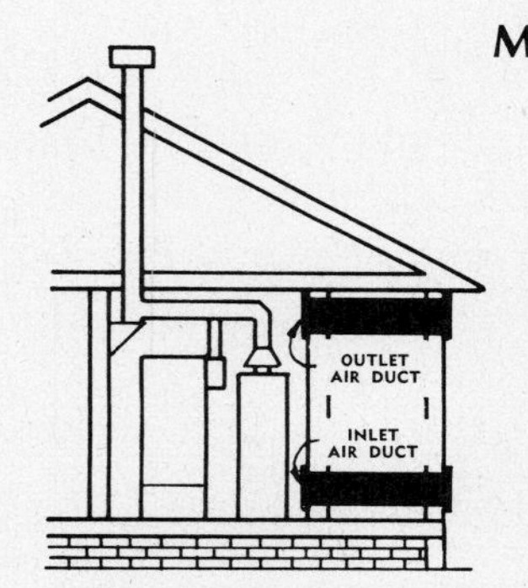

ALL AIR FROM OUTDOORS

NOTE: Each air duct opening shall have a free area of not less than one square inch per 2,000 BTU per hour of the total input rating of all appliances in the enclosure.

FIGURE 8-56 (Courtesy, American Gas Association).

METHOD C

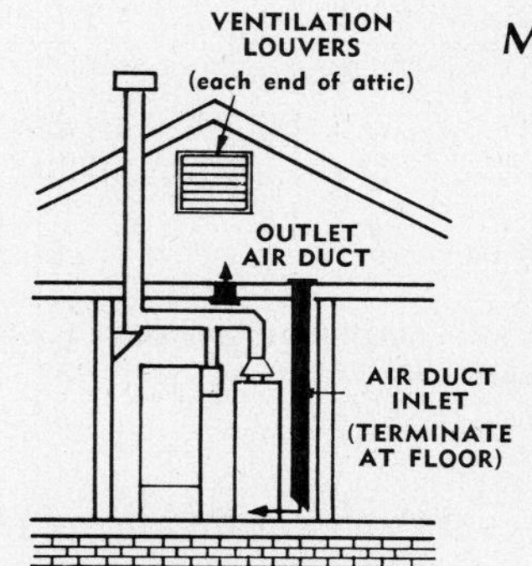

ALL AIR FROM VENTILATED ATTIC

NOTE: The inlet and outlet air openings shall each have a free area of not less than one square inch per 4,000 BTU per hour of the total input rating of all appliances in the enclosure.

FIGURE 8-57 (Courtesy, American Gas Association).

METHOD D

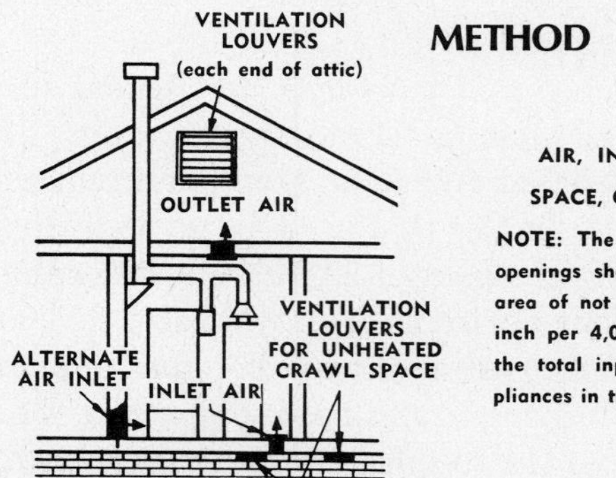

AIR, IN FROM CRAWL SPACE, OUT INTO ATTIC

NOTE: The inlet and outlet air openings shall each have a free area of not less than one square inch per 4,000 BTU per hour of the total input rating of all appliances in the enclosure.

FIGURE 8-58 (Courtesy, American Gas Association).

air getting into the house and then through the supply air openings. This is typical of closet applications. The grilles are sized by dividing the total input Btu (of all appliances) by 1,000. Remember this is free area in square inches, and the particular grille must be selected from the manufacturer's tables to be at least that.

Method B (Fig. 8-56) provides all air from the outdoors.

Method C (Fig. 8-57) is good for single-floor residences where air may be taken from a ventilated attic. Caution: Make sure louvers in the sttic are adequate and open in winter and, of course, there should be no attic exhaust fan.

Finally, Method D (Fig. 8-58) may be used where there is a crawl space. Again vent louvers must be large enough and open during the winter.

Basement installations can usually use infiltration air from window openings, garage door leakage, etc., particulary if there is an abundance of open space.

Over the years closet installations have proven to be particularly dangerous where installers tried to use the equipment room as a plenum chamber for return air (Fig. 8-59). Even though the installer calculated the theoretical air opening size or used a louvered door, there were, and still are, too many deaths reported from carbon monoxide poisoning due to homeowners intentionally or unintentionally blocking openings and grilles and creating a negative pressure. Lack of fresh air causes poor combustion and produces excessive carbon monoxide, which is then introduced from the draft hood into the return air and circulated to the living area. *Always connect the furnace return air to a return duct from the living space.*

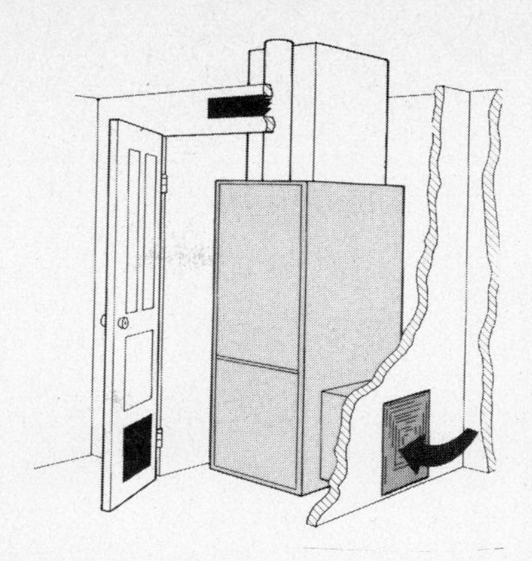

FIGURE 8-59 Closet installation (Courtesy, Westinghouse Electric Corporation).

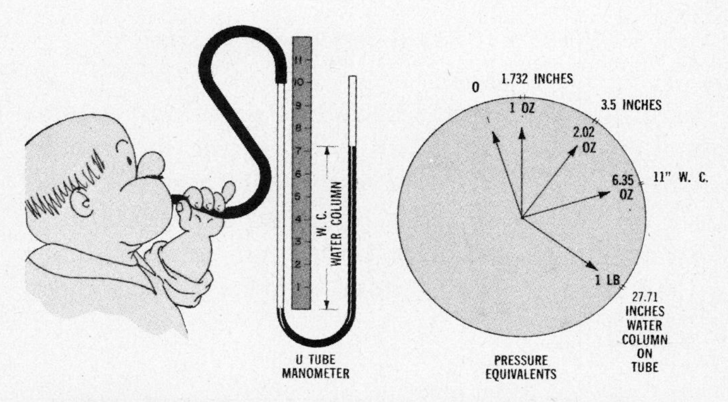

FIGURE 8-60 Measuring pressure (Courtesy, Westinghouse Electric Corporation).

Pressure Measurements

Before we go any further, it would be helpful to pause and make sure the reader understands the subject of pressure equivalents, since it becomes necessary to change from one unit to another. In heating work, in measuring gas pressure, draft, or air pressure in ducts, we use units of less than a pound pressure. Thus, we introduce a measurement tool called the *manometer* (Fig. 8-60), which is simply a U-tube of glass, or preferably plastic, with a sliding inch scale. Fill the tube with water as shown. By connecting a rubber tube to one side, we can blow into the tube and cause the water to rise and measure the pressure we exert. If we exert one pound of pressure as seen on dial, the water would rise 27.71 inches, assuming the glass tube was that long; Hence the term *inches of water column* (W. C.). One ounce of pressure equals 1.732 inches of water column. So when we talk later about adjusting the natural gas pressure to 3.5 inches of water column, we are talking about 2.02 ounces of pressure; we adjust the LP gas so that 11 inches of water column equals 6.35 ounces.

On the left side of Fig. 8-61 is a typical gas pressure manometer that measures up to 15 inches W. C. Later, in dealing with duct work and positive air pressure, we will work in fractions of an inch W. C. and use an inclined type of manometer, as shown on the right. It is also possible to measure negative or suction pressure on the return air duct.

Gas Piping—Natural Gas

Another area of application is the gas piping, which must be installed in accordance with local regulations. We will consider this topic in terms of

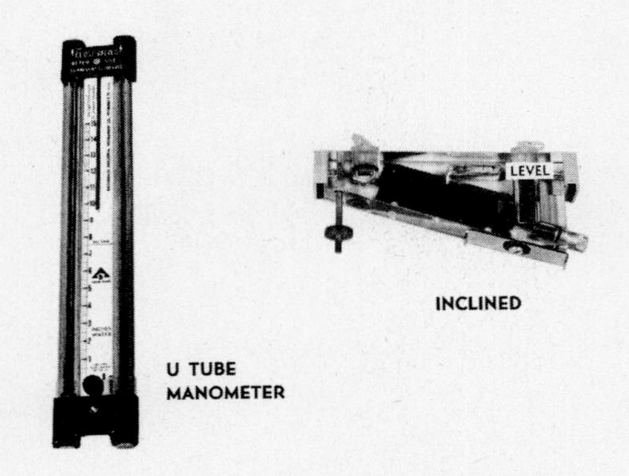

FIGURE 8-61 Instruments for measuring pressure.

natural (including mixed) and LP gases. For natural and mixed gas distributed by city mains, the amount of gas required can be estimated from the table shown in Fig. 8-62. For example, a 100,000 Btu/h input rating will require 1.67 ft³/min of gas rated at 1,000 Btu/ft³. Hourly, this would be 1.67×60, or 100.2 ft³/h.

The next table (Fig. 8-63) gives the carrying capacity of nominal pipe sizes versus length of run (pressure drop not to exceed 0.3 in. W. C.). Our example requires 100 ft³/h and assumes a run of 60 feet; $\frac{3}{4}$ in. pipe can handle 80 ft³/h, but a 1 in. line will do the job. Remember though, we assumed only a furnace; if there is also gas water heating, plus gas cooking, the main would have to include all the appliances and be sized accordingly. Branches would be sized for the individual appliance.

With the main and branch lines sized correctly, the next step is the furnace. Provide a manual shut-off valve that feeds the furnace (Fig. 8-64). Check your codes on allowable heights above the floor. On the lower tee outlet provide a drip trap consisting of a length of pipe and a cap. This is to catch dirt or scale that may come from the supply lines. Next, connect the tee outlet to the furnace gas manifold assembly and provide a union between the tee and gas valve in case removal of the manifold is required. When connecting the pipe to the gas valve, apply only a moderate amount of pipe dope and leave two end threads bare.

Also, in tightening the pipe into the valve, be sure and use the wrench on the valve as shown in Fig.

	BTU CONTENT OF GAS							
	550		604		1,000		1,100	
Input Rating BTU/hr.	Cu. Ft. per Minute	Sec. per Cu. Ft.	Cu. Ft. per Minute	Sec. per Cu. Ft.	Cu. Ft. per Minute	Sec. per Cu. Ft.	Cu. Ft. per Minute	Sec. per Cu. Ft
60,000	1.82	33	1.65	36.4	1.00	60	.91	66
65,000	1.97	30.3	1.79	33.5	1.08	55.5	.985	61
70,000	2.12	28.2	1.93	31	1.17	51	1.06	56.5
75,000	2.28	26.4	2.06	29.10	1.25	48	1.135	53
80,000	2.42	24.8	2.20	27.30	1.33	45	1.21	49.6
85,000	2.58	23.2	2.34	25.60	1.42	42	1.29	46.5
90,000	2.72	00.0	2.40	24.20	1.50	40	1.36	44
95,000	3.00	20.8	2.62	22.9	1.59	38	1.44	41.7
100,000	3.04	19.5	2.76	21.8	1.67	36	1.52	39.4
105,000	3.18	18.85	2.90	20.7	1.62	34.3	1.59	37.8
110,000	3.34	17.95	3.04	19.7	1.83	32.8	1.67	35.9
115,000	3.50	17.12	3.18	18.85	1.92	31.1	1.74	34.5
120,000	3.64	16.50	3.32	18.10	2.00	30.0	1.82	33.0
125,000	3.79	15.80	3.46	17.30	2.08	28.8	1.89	31.7
130,000	3.95	15.20	3.60	16.65	2.17	27.6	1.97	30.5
135,000	4.10	14.62	3.74	16	2.25	26.7	2.04	29.4
140,000	4.25	14.10	3.88	15.45	2.33	25.7	2.12	28.3
145,000	4.40	13.60	4.02	14.9	2.42	24.8	2.20	27.2
150,000	4.55	13.20	4.16	14.4	2.50	24.0	2.27	26.4
155,000	4.70	12.75	4.30	13.9	2.58	23.2	2.34	25.6
160,000	4.85	12.35	4.44	13.5	2.67	22.5	2.42	24.8
165,000	5.00	12.00	4.58	13.1	2.75	21.8	2.50	24.0

FIGURE 8-62 Gas consumption rate (Courtesy, Westinghouse Electric Corporation).

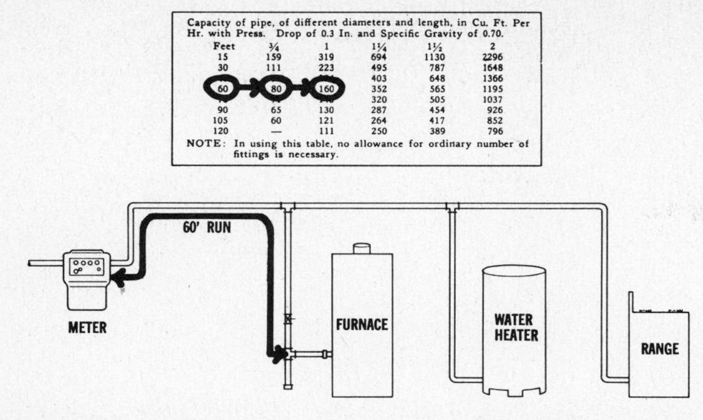
Capacity of pipe, of different diameters and length, in Cu. Ft. Per Hr. with Press. Drop of 0.3 In. and Specific Gravity of 0.70.

Feet	¾	1	1¼	1½	2
15	159	319	694	1130	2296
30	111	223	495	787	1648
[illegible]	[illegible]	[illegible]	403	648	1366
60	80	160	352	565	1195
[illegible]	[illegible]	[illegible]	320	505	1037
90	65	130	287	454	926
105	60	121	264	417	852
120	—	111	250	389	796

NOTE: In using this table, no allowance for ordinary number of fittings is necessary.

FIGURE 8-63 **City gas piping (Courtesy, Westinghouse Electric Corporation).**

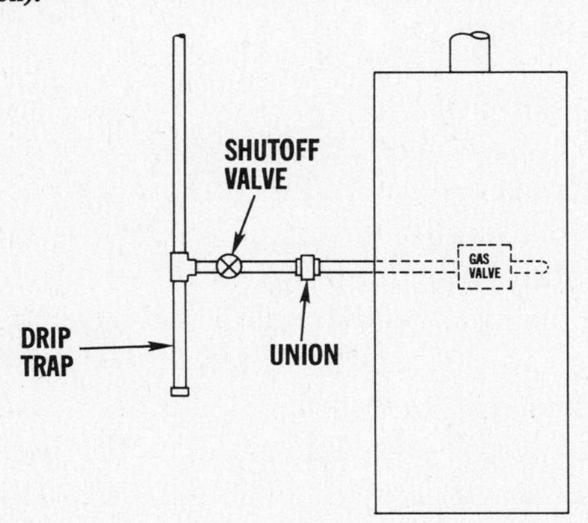

FIGURE 8-64 **Gas piping at furnace (Courtesy, Westinghouse Electric Corporation).**

FIGURE 8-65 **Gas piping at furnace (Courtesy, Westinghouse Electric Corporation).**

8-65. Otherwise, you may cause distortion and damage to the operator section. Codes vary on the exact details of this entire assembly, so be sure to become familiar with your specific regulations.

LP System

Let's consider the LP gas supply system (Fig. 8-66). Actually the LP supplier will provide the tank installation, pressure regulating devices, and piping

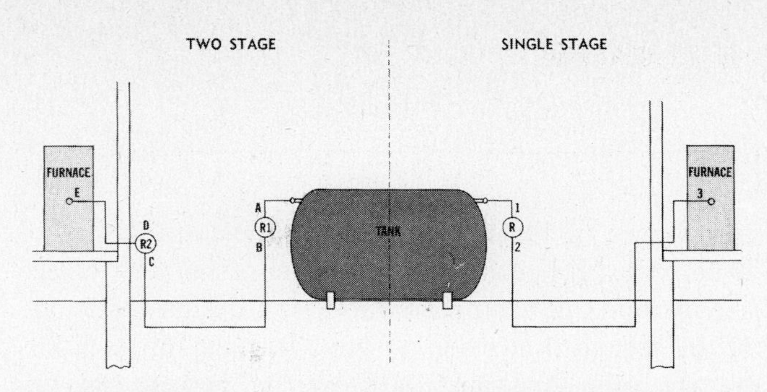

FIGURE 8-66 **L.P. Gas supply system (Courtesy, Westinghouse Electric Corporation).**

to the furnace; but you should be familiar with the hook-up procedure. There are two basic systems, as illustrated. The single system on the right, which is most frequently used in residential work, uses only one pressure regulator located at the tank. The two-stage system is used mainly in commercial work where the number of appliances and volume of gas must be greater. Therefore, the outlet pressure at point *B* is 15 lb, whereas at point *2* on the single-stage system, it is down to approximately 11 inches of water column.

Let's be more specific about the *single stage* system. Pressure at point *1* will be direct tank pressure, and this will vary with fuel capacity and temperature. Pressure at point *2*, the outlet of the single stage regulator, should not be less than 9.6 in. of W. C., even with the lowest tank pressure and a full load. Ideally it should control at 11 in. of W. C. or slightly above. Pressure at point *3*, the furnace manifold, should never be less than 9 in. of W. C., regardless of temperature and load. Maximum pressure drop in the connecting line should not be more than 0.6 in. of W. C. If it is more, the pipe is too small. The lock-up pressure at point *3*, with no load on the system, should not be more than 13 in. of W. C. when tank pressure is at its highest. As we will discuss later, the firing rate on the furnace will be set at 11 in. of W. C. Pipe sizing for LP gas is much like that for natural gas. The LP distributor has tables relating the gas flow and pressure drop for both copper and iron pipe.

Blower Performance

Of major importance to the application of the heating system is the ability of the blower to provide the required air movement against the resistances within the system. First, there are the internal resistances of the filter, losses in the fan itself, the heat exchanger, and flue collector. Fortunately, these are included in each manufacturer's design selection of the fan, so that when we get a furnace in the field, we are only concerned with external resistances, or more correctly *static pressure losses*, which can be both positive and negative.

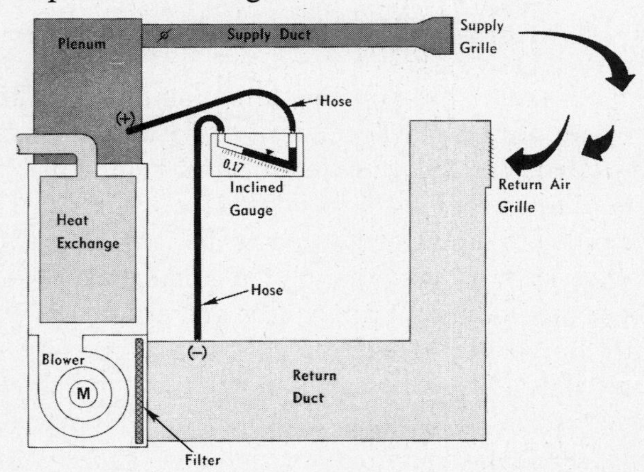

FIGURE 8-67 Heating system (Courtesy, Westinghouse Electric Corporation).

In a simple heating system (Fig. 8-67) the principal losses come from the supply air duct system and its related fittings, dampers, registers, etc., *plus the return air duct and its fittings and grille.* It is common practice to design the total system not to exceed 0.2 in. of W. C. as measured on an inclined manometer. Pressure from the discharge duct will be positive, and on the return negative. The arithmetic sum will be the total. Connecting two rubber hoses, we can measure this in one operation. Assuming this is a 110,000 Btu heating-only furnace and we measured 0.17 in. of W. C., it is only necessary to check the furnace performance data to see if the furnace size selected can produce about 950 ft³/min at 0.2 in. of W. C., for example. Normally, this furnace will handle the requirement without difficulty.

However, if this were to be a total comfort application (Fig. 8-68) with a cooling coil and electronic air cleaner, we must reconsider the total static very carefully.

Assuming a three-ton cooling requirement, we must now increase the air flow to around 1,200 ft³/min. The three-ton cooling coil (from the catalog

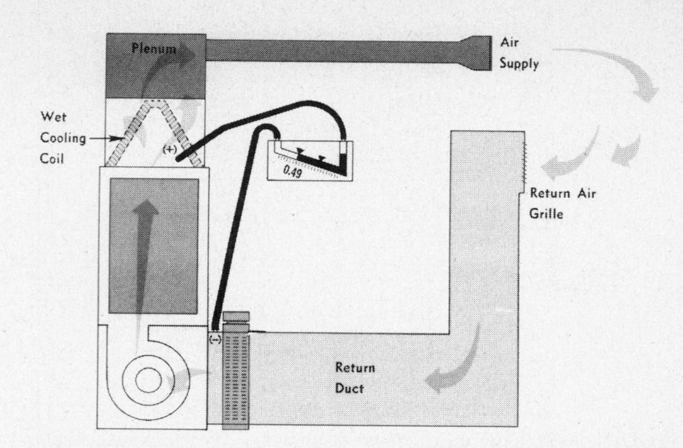

FIGURE 8-68 Total comfort system (Courtesy, Westinghouse Electric Corporation).

data) has a wet coil pressure drop of 0.28 in. of W. C. at about 1,200 ft³/min. Next, we check the electronic air cleaner curves and find its resistance to be 0.03 in. of W. C. Naturally, when the air flow is increased, the duct static will go up slightly. Now if we take our pressure reading, we will find a total loss of 0.49 in. of W. C.

Chances are the furnace selection would now require a model with increased motor horsepower and/or a larger blower. The heating Btu would remain the same.

The example above was used to demonstrate that when you are adding cooling to either a new or existing furnace, you must assume the total static will approach 0.5 in. of W. C. This is why furnaces that are approved have noted on the nameplate the maximum ft³/min that can be expected at 0.5 in. of W. C. static.

Air Rise Considerations

In addition to furnaces being approved for static pressures, they are also checked for maximum and minimum air rise measured between return air and outlet air temperature. This is important because of the relation of cooling air requirement versus normal heating operation. If too large a cooling ft³/min is required, the heating air temperature rise may be too low, and improper heating operation will result. Continuing our example of a 110,000 Btu input furnace with an 88,000 Btu output, three tons of cooling at 1,200 ft³/min will produce a 68°F air temperature rise with about 138°F air coming off the bonnet. (See the chart in Fig. 8-69 and the intersection of the plotted lines.)

This condition falls within the 45°F to 75°F AGA air rise noted on the rating table and also on the data

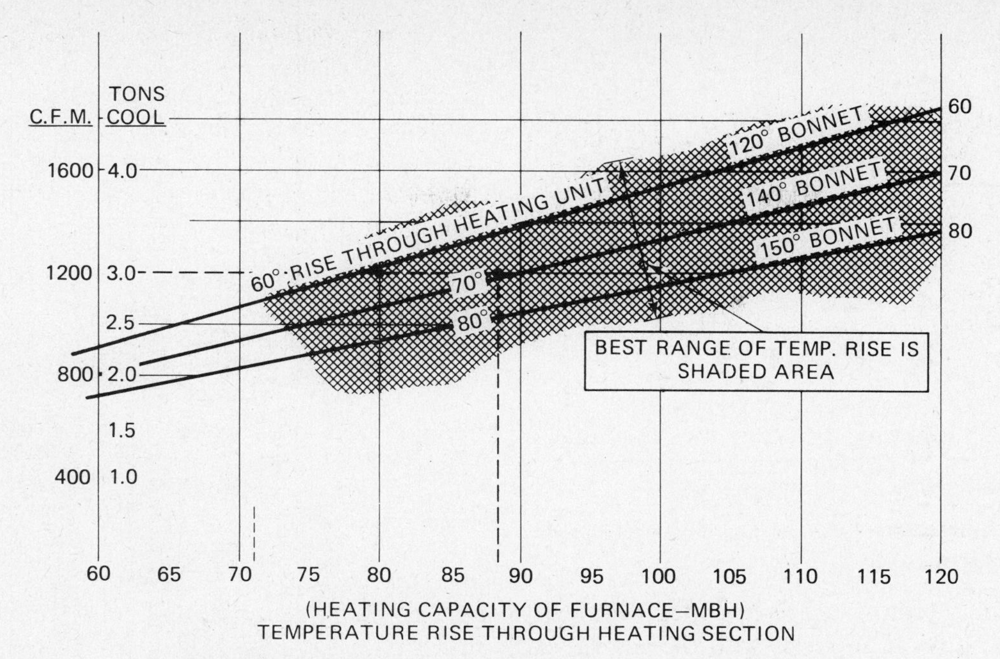

FIGURE 8-69 (Courtesy, Westinghouse Electric Corporation).

plate. This chart can also be used in reverse for checking ft³/min under actual operating conditions, and it will be used for start-up and servicing in later chapters. Air rise temperatures below 45°F will produce too much condensation within the heat exchanger and cause excessive rusting. If the heating air rise is too low, the fan speed must be lowered in winter, thus lowering the ft³/min and thereby raising the air rise. This is the function of motor speed controls or changing pulley arrangements on belt-driven models.

Room Thermostat Installation

Many customer complaints stem from improper location and installation of the room thermostat (Fig. 8-70). The following are six points that will help you choose a proper thermostat location:

(1) Take the initiative and select two or more good locations for the thermostat, and then let the customer choose between them.

(2) Locate the thermostat four to five feet above the floor and in the most lived-in room. Avoid such rooms as the kitchen, bathroom, or long hallway. The thermostat should be mounted on an inside or partitioning wall where it will respond to normal room temperature rather than the cold of an outside wall. Seal the hole in the plaster where wires come through to shield the thermostat from cold drafts.

(3) The thermostat should be open to natural air circulation, without interference from furniture or other obstructions. Behind a door, in an alcove, or above a piece of furniture are poor locations.

(4) Locate the thermostat so it will not be subject to heat or cold other than the normal circulation of warm air. To be avoided are locations subject to heat from sunlight, lamps, fireplace, registers, electric appliances, and concealed hot water pipes. The thermostat should be protected from cold coming from windows, outside doors, concealed cold water pipes, and cold air ducts.

(5) Before installing the thermostat, remove any pins or other devices used to protect delicate parts during shipment. Remember that a thermostat is a precision instrument and handle it with care. Level the instrument according to instructions.

(6) Be careful with the wiring. Do a neat job, use new cable, avoid splices, and follow the manufacturer's wiring instructions. Be sure and check the number of conductors needed, particularly if cooling and humidification are involved. Also, the incorrect gauge wire and distance can cause voltage drop and erratic operation. Below are recommended minimums for a 24-volt system.

Wire Size	*Maximum Length*
14 Ga	100 ft
16 Ga	64 ft
18 Ga	40 ft
20 Ga	25 ft

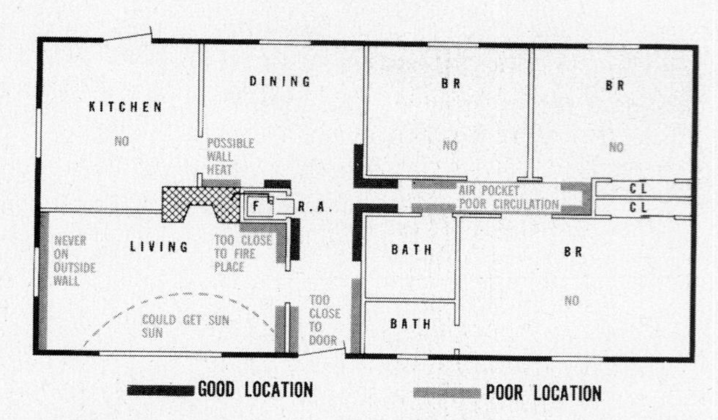

FIGURE 8-70 Room thermostat locations (Courtesy, Westinghouse Electric Corporation).

Most heating thermostats are equipped with heat anticipators to be wired in series with the gas valve. The heat anticipator is actually an electrical resistance that produces a small amount of heat inside the thermostat cover. This heat will cause the switch contacts to open a little sooner and shut off the burner shortly before the desired room temperature is reached. The stored heat in the heat exchanger will then be enough to bring the temperature up to normal. The result is a much more even cycle of the burner and room temperature.

Wrong or incorrect heat anticipation can cause discomfort, so be sure you have made the right selection and match it to the total amps through the thermostat. Some stats are fixed; some have adjustable anticipation.

Recalibration of thermostats is seldom necessary. Calibration should be checked only if performance proves to be unsatisfactory. Follow the manufacturer's instructions for checking and changing calibration. Controls for heating and cooling will be covered in greater detail in Chaps. 17 and 18.

WINTER HEATING—OIL FORCED-AIR HEATING

As previously mentioned, oil has been a popular fuel in the northeast part of the United States for some time. It is also widely used in rural areas where gas mains don't exist and/or where the use of electricity for heating is not practical. Vacation or second homes frequently use oil for primary heating.

Types of Furnaces

Like its counterpart the gas furnace, oil forced-air furnaces are available in upflow (Fig. 8-71), lowboy, downflow (counterflow), and horizontal models. The type of general application, ductwork, and air distribution into the conditioned space are essentially the same. There are, however, differences in the internal combustion chamber, burners, controls, clearances, and flue venting requirements.

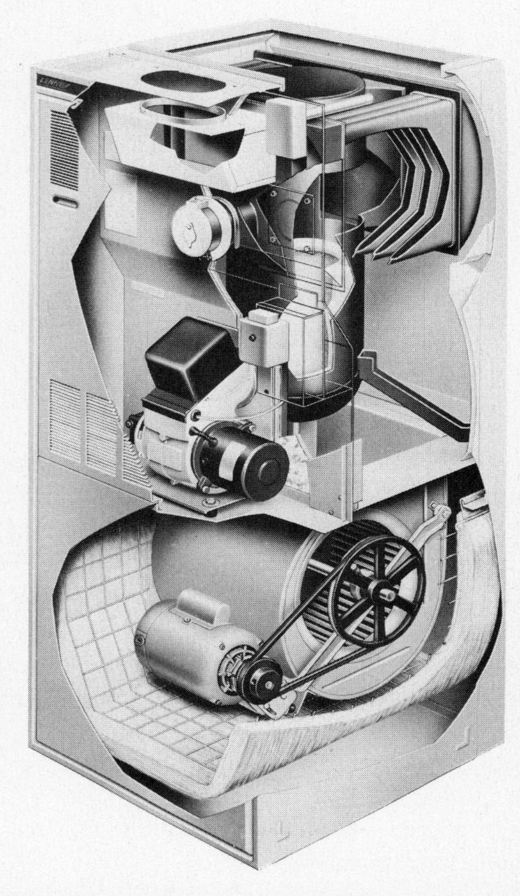

FIGURE 8-71 Oil upflow furnace (Courtesy, Lennox Industries).

Heat Exchanger

The typical oil-fired heat exchanger (Fig. 8-72) is a cylindrical shell of heavy-gauge steel where combustion takes place; it offers additional surfaces for heat transfer from the products of combustion to the air over the outside of this heat exchanger. This type of heat exchanger is called a *drum and radiator*. The portion containing the flame is called the *primary surface*, and the additional section or sections are called the *secondary surface*.

FIGURE 8-72 Oil-fired heat exchanger (Courtesy, Airtemp Corporation).

Some manufacturers add baffles, flanges, fins, or ribs to the surfaces to provide faster heat transfer to the air passing over the surfaces.

The burner assembly is bolted to the heat exchanger with its firing assembly and blast tube extending into the primary surface in correct relationship to the combustion change or refractory. A flame inspection port is provided just above the upper edge of the refractory. This is essential to observe proper ignition, flame, and for measurement of over-fire draft for startup and service operations.

Refractory

To produce maximum burning efficiency of the oil/air mixture, high temperatures are required in the flame area of the heat exchanger. To provide this increase in temperature, a reflective material is

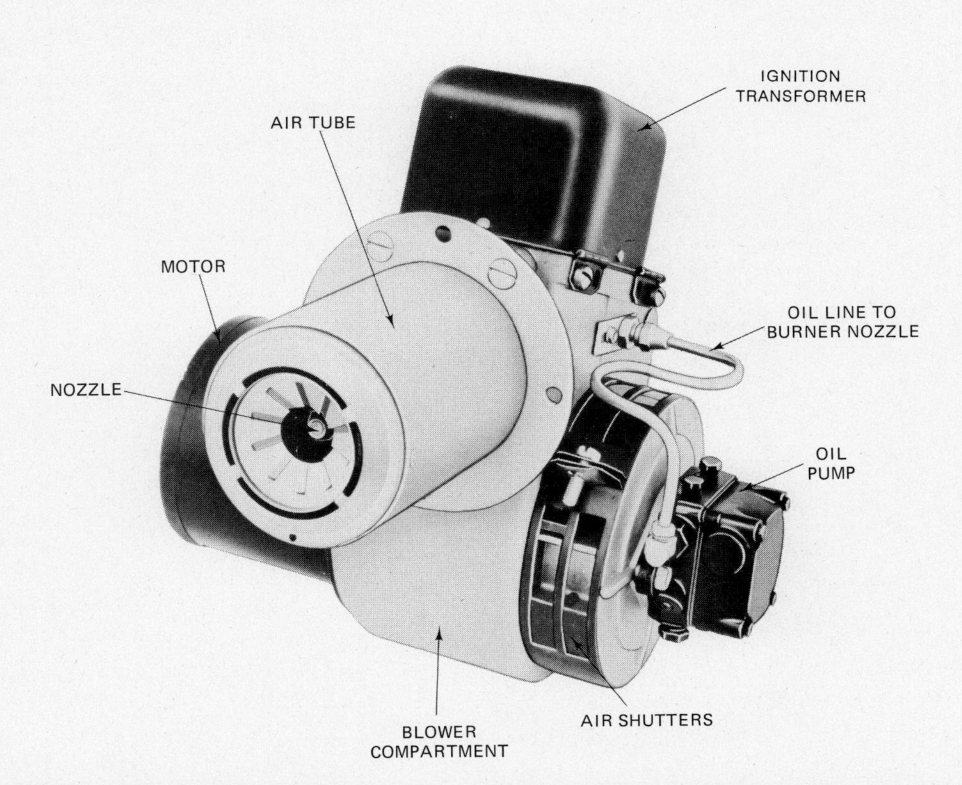

FIGURE 8-73 High pressure atomizing gun type burner (Courtesy, Wayne Home Equipment Company).

installed around the combustion area. This insulation type of material, called the *refractory*, is designed to obtain white-heat surface temperatures very quickly and to withstand high temperatures with minimum deterioration.

Burner

The high-pressure atomizing gun burner (Fig. 8-73) is the type used on most residential and small commercial oil-fired heating systems. Viewed from the exterior, it has several functional components:

(a) The oil pump
(b) The air blower
(c) Electric motor
(d) Control compartment
(e) The blast tube with included nozzles and ignition System

A cross section of the burner (Fig. 8-74) reveals a more detailed view of the operation. The pump and blower are driven from a common motor shaft. The oil pump is a gear arrangement, and by changing a bypass plug in the housing, the pump can operate as a single-stage or a two-stage system. Single-stage operation is used where the oil tank is above the burner and gravity oil feed to the burner is permitted. Two-stage operation is needed where the oil tank is below the

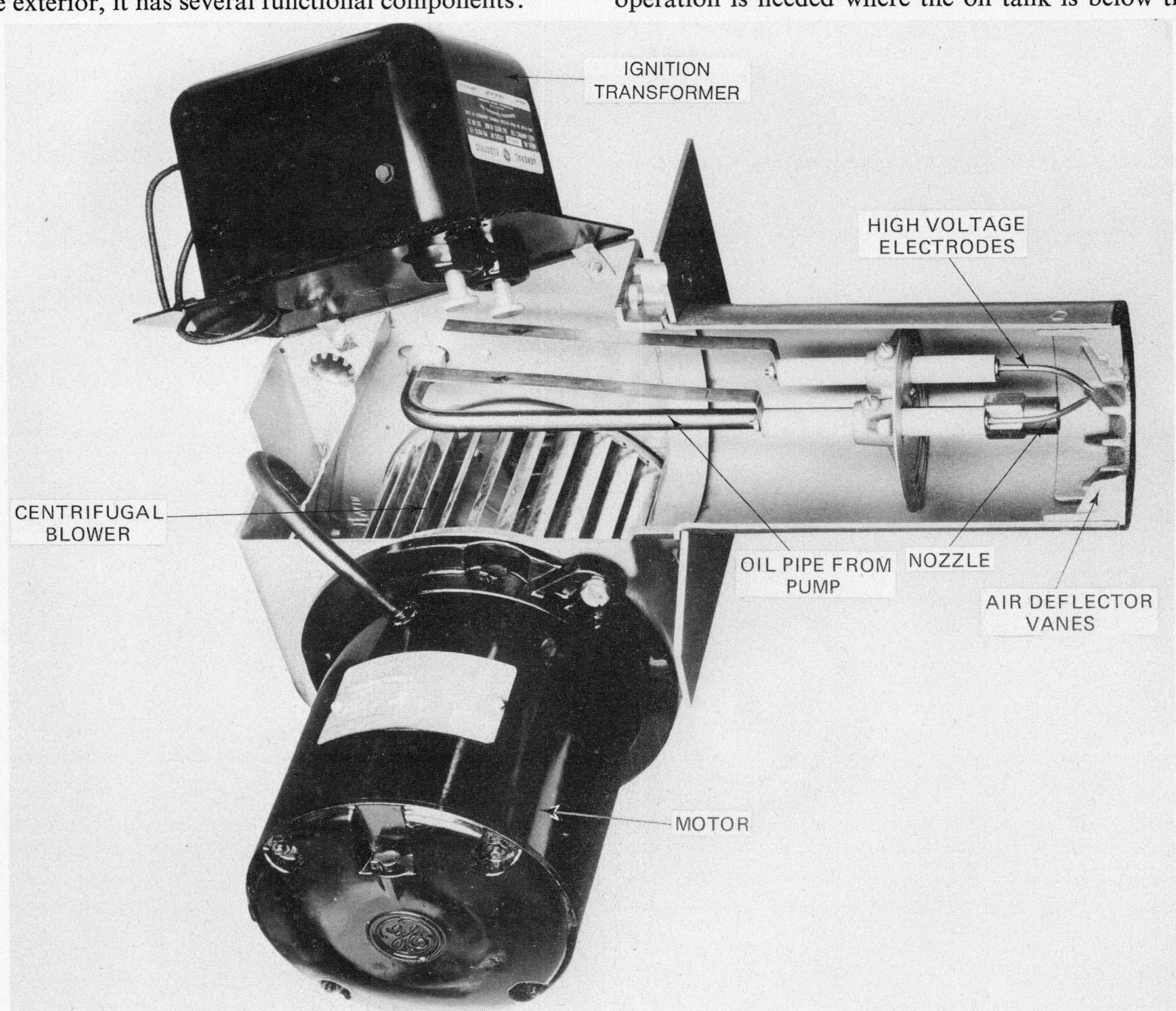

FIGURE 8-74 **Cross section of burner (Courtesy, Wayne Home Equipment Company).**

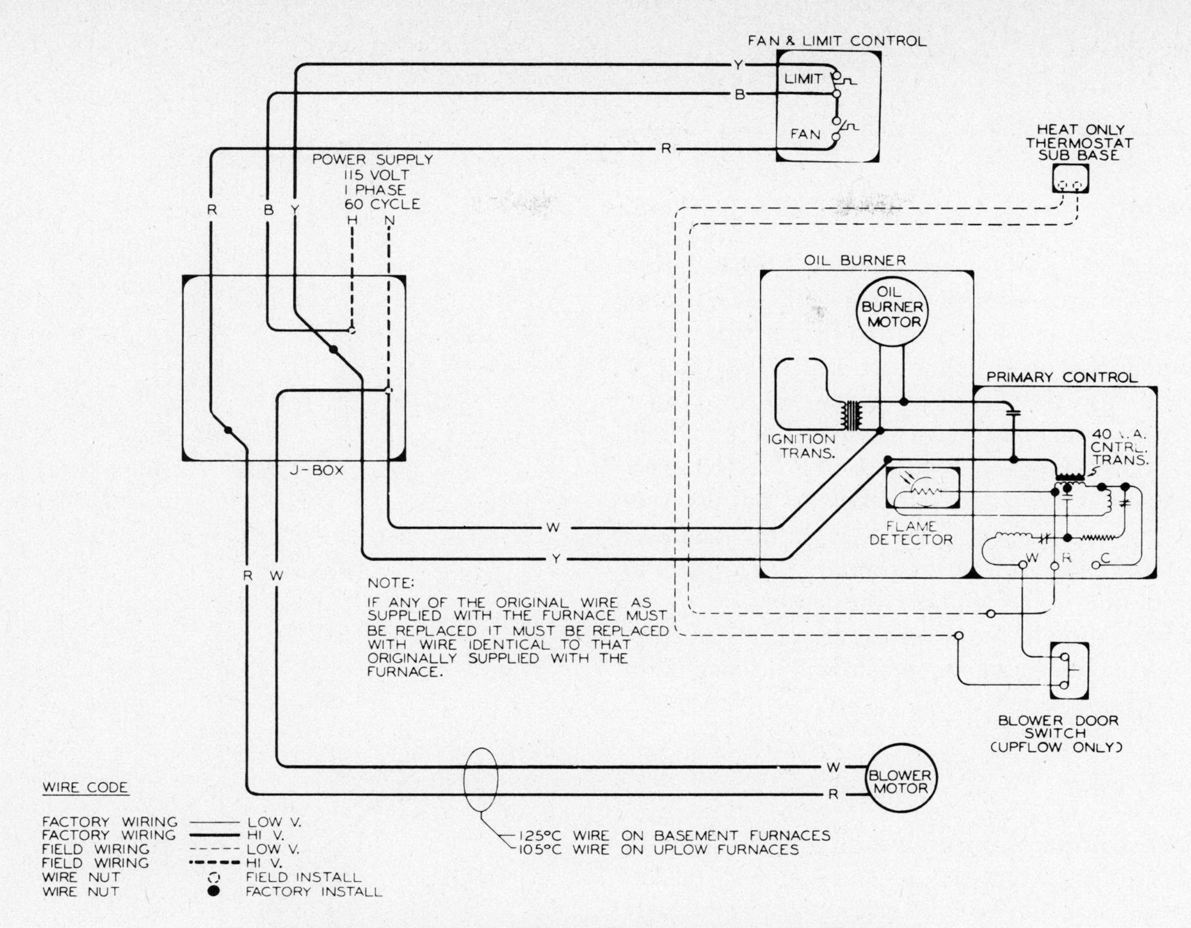

FIGURE 8-75 Typical oil furnace control (Courtesy, Luxaire, Inc.).

burner and the pump must lift the oil from the tank as well as furnish pressure to the nozzles. The pump supplies oil to the nozzle at 100 to 300 psi.

The burner blower is a centrifugal wheel also mounted on the common motor shaft. It furnishes air through the blast tube and turbulator for proper combustion. The amount of air is controlled by a rotating shutter band on the blower housing section.

The nozzle is mounted in an adapter and receives oil by means of a pipe from the pump. The orifice opening in the nozzle is factory bored to produce the correct firing rate. As a rule of thumb, the firing rate for No. 2 grade fuel oil will be approximately 0.8 gal/h for each 100,000 Btu/h output of the furnace. *Do not* attempt to change the firing rate by changing orifices or drilling an enlarged opening. Under high pressure the oil is atomized into fine droplets and mixes with primary air.

The air deflector vane ring (turbulator) serves an important function in getting the mixture of atomized oil and air to rotate or spin into the cavity of the heat exchanger. Ignition is established by a high-voltage electric spark, which may be continuous (on when the motor is on) or interrupted (on only to start combustion). The electric current is provided by an ignition transformer located in the burner control compartment.

Also located in the burner control compartment (not shown) is a flame-detecting device called a *cad cell* that is light sensitive. It has a hermetically-sealed light-actuated cell that has low electrical resistance in the presence of light rays from the flame, thus permitting current flow, and high electrical resistance in the absence of light rays from the flame, thus prohibiting current flow. Therefore if there is a flame failure for whatever reason, the cad cell will stop the burner motor so that oil cannot flow into the heat exchanger. Some furnaces have a lockout system requiring the control to be reset manually before the burner can run again.

Controls

External to the burner controls, the oil-fired forced-air furnace control system (Fig. 8-75) must also have a fan and limit control that performs the same function as on a gas furnace. The safety limit, if open, will stop the electrical current to the burner before overheating becomes excessive. The fan switch is set to cycle the furnace blower as desired. The space thermostat (24 volts) feeds directly to burner low-voltage secondary control circuit, actuating the flame detector and relay to feed the primary line voltage to the burner motor and ignition transformer.

The operation of the controls, as well as startup, adjusting, and servicing the system will be covered in subsequent chapters.

Flue Venting

Oil-fired furnaces must have an ample supply of make-up air for combustion, and the methods of introducing make-up air for a gas furnace will also be adequate for oil equipment.

Masonry chimneys used for oil-fired furnaces should be constructed as specified in the National Building Code of the National Board of Fire Underwriters. A masonry chimney shall have a minimum cross-sectional area equivalent to either an 8-in. round or an 8 × 8-in. square; it should never be less than the flue outlet of the furnace. Prefabricated light-weight metal chimneys (Fig. 8-76) are also available for use with oil. The double-wall type is filled with insulation. These vents are class A flues. Class B flues, made specifically for gas-fired equipment, are not suitable for use with solid or liquid fuels. Be sure the class rating states "all fuels."

The terminal vent heights above the roof recommendations as previously outlined for gas should also be observed for oil.

Oil-fired furnaces operate on positive pressure from the burner blower, and it is most important to have a chimhey that will develop a minimum draft of 0.01 to 0.02 in. of W. C. as measured at the burner flame inspection port. Consistency or stability of draft is also more critical, and the use of a barometric damper (Fig. 8-77) is required. The damper is usually installed in the horizontal vent pipe between the furnace and chimney. (Note: Some manufacturers attach them directly to the furnace flue outlet.) The damper has a movable weight so that it can be set to counterbalance the suction and to maintain reasonably constant flue operation. It's adjusted while the

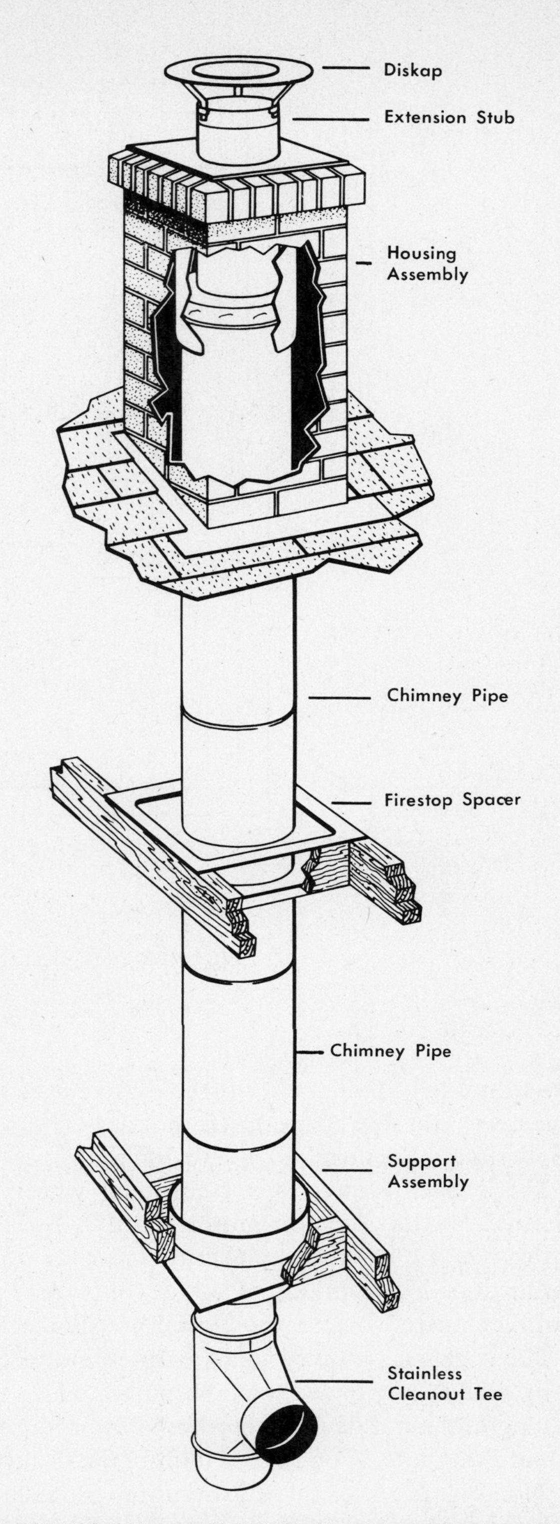

FIGURE 8-76 Factory-built "all fuel" chimney (Courtesy, Metalbestos Systems, Wallace Murray Corporation).

FIGURE 8-77 Barometric damper (Courtesy, Field Control, Div. of Conco, Inc.).

furnace is in operation and the chimney is hot. The overfire draft at the burner should be in accordance with the equipment installation instructions.

Clearances

Cabinet temperatures and flue pipe temperatures run warmer for oil-burning equipment, and the clearances from combustible material should be adjusted accordingly. One-inch clearances are common for the sides and rear of the cabinet, as opposed to zero clearances for many gas units. Flue pipe clearances of 9 inches or more are needed, whereas only 6 in. of clearance is needed for gas.

Front clearance is generally determined by the space needed to remove the burner assembly. Combustible floor bases for oil are generally increased as compared to gas. Horizontal oil furnaces are particularly important in respect to attic installations and the installer must check the recommendations carefully.

Oil furnaces are rated and listed by Underwriters Laboratories, and clearances are a vital part of this inspection and compliance procedure along with may other safety considerations.

Oil Storage

The installation of the fuel oil tank and connecting piping must conform to the standards of the National Fire Protection Association (Standard No. 31) and/or local code requirements. Regulations and space permitting, the oil tank can be located indoors as shown in Fig. 8-78. Note the proper use of shut-off valves and an effective filter to catch impurities. If the oil tank is placed outdoors above ground, firm footings must be provided. Exposed tanks and piping above ground are subject to more condensation of water vapor and the possibility of freezing during extremely low temperatures.

If a large tank is installed below ground (Fig. 8-79), it is important to keep it well filled with oil during periods of high water level (i.e., spring rains); otherwise ground water may force it to float upwards. Extra concrete for added weight over the tank is advised. Flexible copper lines are recommended so as to give with ground movement. Notice the use of suction and return lines; as mentioned previously, it is normal practice to use a two-stage fuel pump with a two-pipe system whenever it is necessary to lift oil from a tank that's below the level of the burner. The oil burner pump suction is measured in terms of inches of mercury vacuum. A two-stage pump should never exceed a 15-inch vacuum. Generally, there is one inch of vacuum for each foot of vertical oil lift and one inch of vacuum for each 10 feet of horizontal run of supply piping.

WINTER HEATING—ELECTRIC FORCED-AIR

At the end of 1973 the Electric Energy Association reported that over 5 million American residential dwellings were heated electrically. The growth rate is approximately 800,000 per year. Some of these systems are combination heating and cooling units (heat pumps or packaged terminal units), but the majority of installations (47.6% in 1973) were forced-air

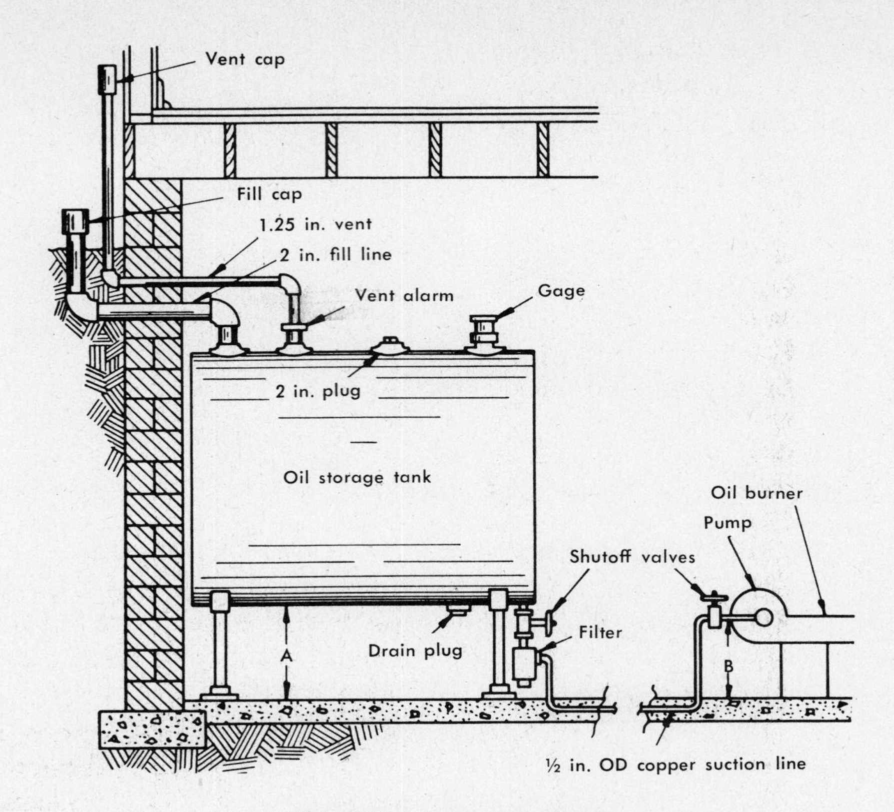

FIGURE 8-78 Indoor oil storage tank.

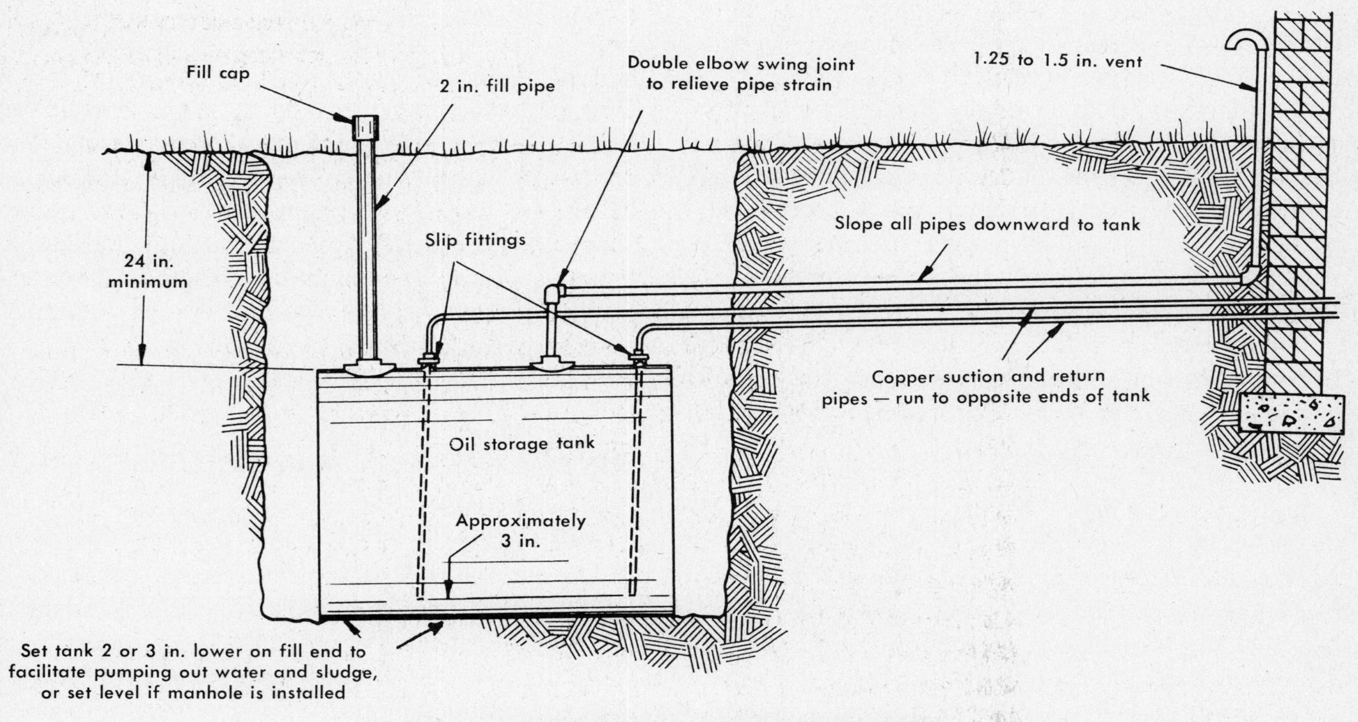

FIGURE 8-79 Outside oil storage tank two-pipe details.

electric furnaces and all other forms of ducted resistance heating, and industry forecasts project this trend will continue.

Electric Furnace

The typical electrical furnace (Fig. 8-80) is a most flexible and compact heating unit. It consists of a cabinet, blower compartment, filter, and resistance heating section. The cabinet size is generally more compact than the equivalent gas or oil furnaces, and due to cooler surface temperature, most units enjoy "zero" clearances all around from combustible materials. Thus, they may be located in very small

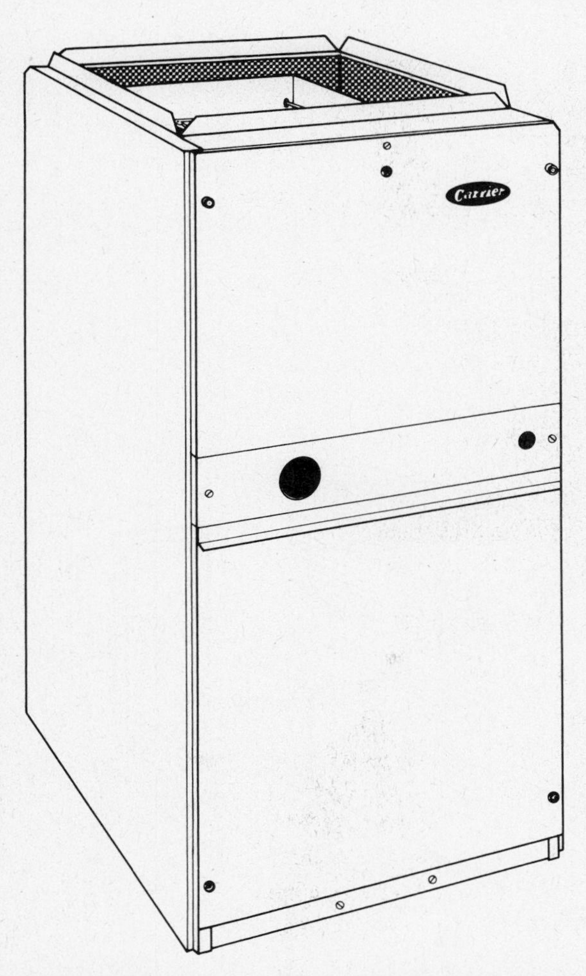

FIGURE 8-80 (Courtesy, Carrier Air-Conditioning Company).

closets. Additionally, since there is no combustion process involved, there is no requirement for venting pipes, chimney, or make-up air, thus simplifying installation and reducing building costs. Service access is the only dimensional consideration. Also, the absence of combustion permits mounting the units for up, down, or horizontal air flow application (Fig. 8-81). Some manufacturers provide space within the furnace cabinet for a cooling coil.

With the panel removed (Fig. 8-82) observe that the blower compartment usually houses a centrifugal, multispeed direct-drive fan or one of a belt-driven design where larger units are involved. Air flow through an electric furnace has less resistance and fan performance is more efficient. Mechanical filters of either the throw-away or cleanable variety are available.

The heating section consists of banks of resistance heater coils wound from nickel-chrome wire held in place by ceramic spacers. The heater resistance is designed to operate on 208 to 240-volt power with output according to the actual voltage used. The amount of heat per bank is a function of the amperage draw and/or staging. National and local electrical codes control the amount of current that can be put on the line in one surge, so it's necessary for manufacturers to limit this impact by the size (kW) of the heaters in a bank and the sequence of operation. Total furnace output capacities range from a low of 5 kW (17,000 Btu) to 35 kW (119,400) Btu/h. At approximately 35 kW the amperage draw on 240 volts approaches 150 amps and, considering that 200 amps is the total service to a residence, this only leaves some 50 amps for other electrical uses. Seldom are all on continuously, but codes must assume so for safety purposes. Note, however, that a well-insulated, electrically heated home requiring 35 kW will have considerable square feet. Zoning too is easier with two electric furnaces since location is not critical.

Electrical heating elements are protected from any overheating that may be caused by fan failure or a blocked filter. These are high-limit switch devices that sense air temperature and when overheated "open" the electrical circuit. Some furnaces also employ fusible links wired in series with the heater; these links melt at approximately 300°F and open the circuit. This is an auxiliary back-up for the high-limit switches.

Built-in internal fusing or circuit breakers are provided by some manufacturers and are divided to comply with the National Electric Code. This can be

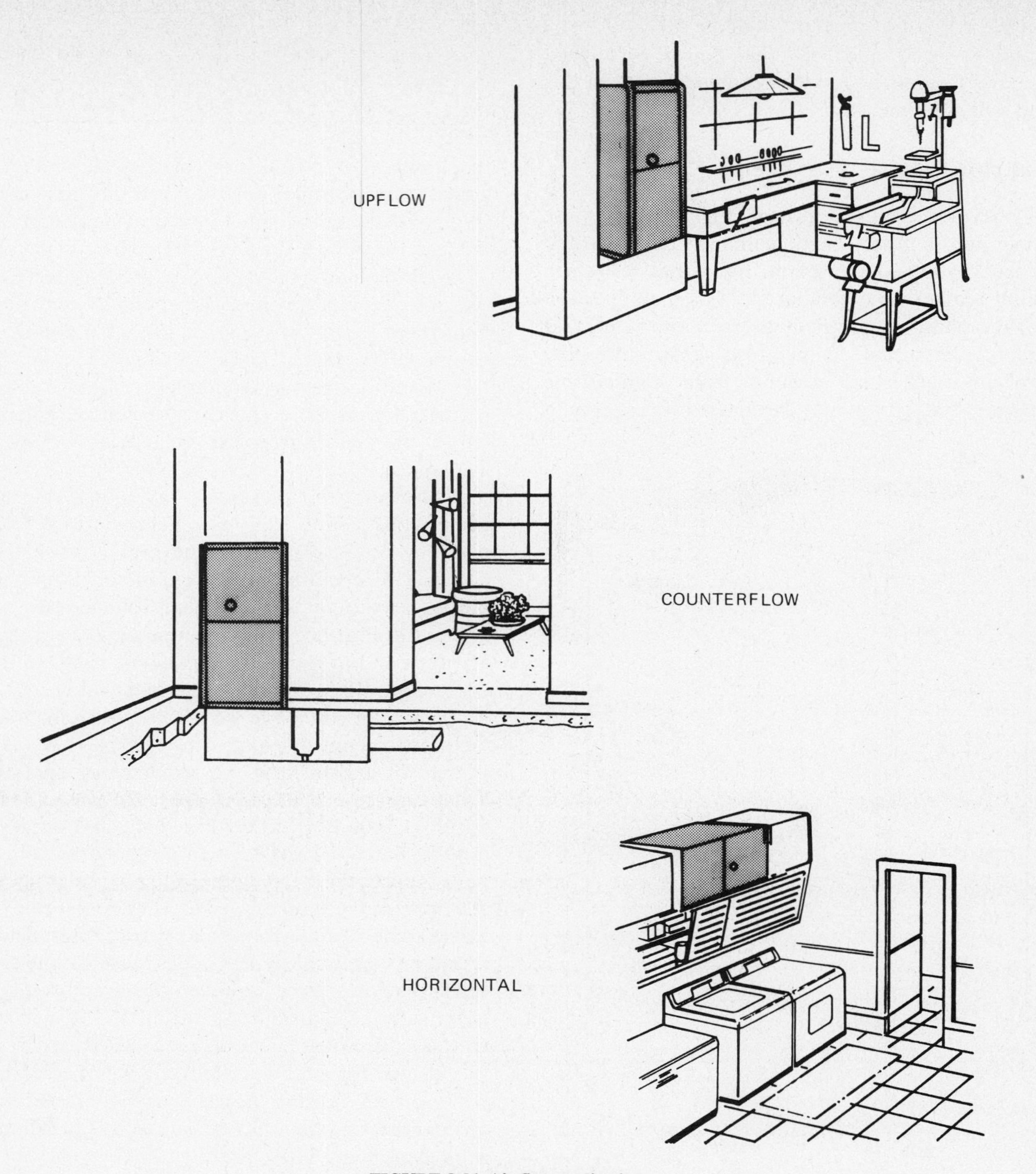

FIGURE 8-81 Air flow applications.

convenient and may also mean installation savings for the contractor; external fuse boxes and associated on-site labor are eliminated.

The sequence control system is an important operation of the electric furnace. On a single-stage thermostat system there is an *electric sequencer control*, which contains a bimetal strip. On a call for heat, the thermostat closes a circuit which applies 24 volts across the heater terminals of the sequencer. As the bimetal element heats up, the blower and first heater come on. The operation characteristics of the sequencer are such that there is a time delay before each additional heater stage is energized. When the thermostat is satisfied, the sequencer is deenergized

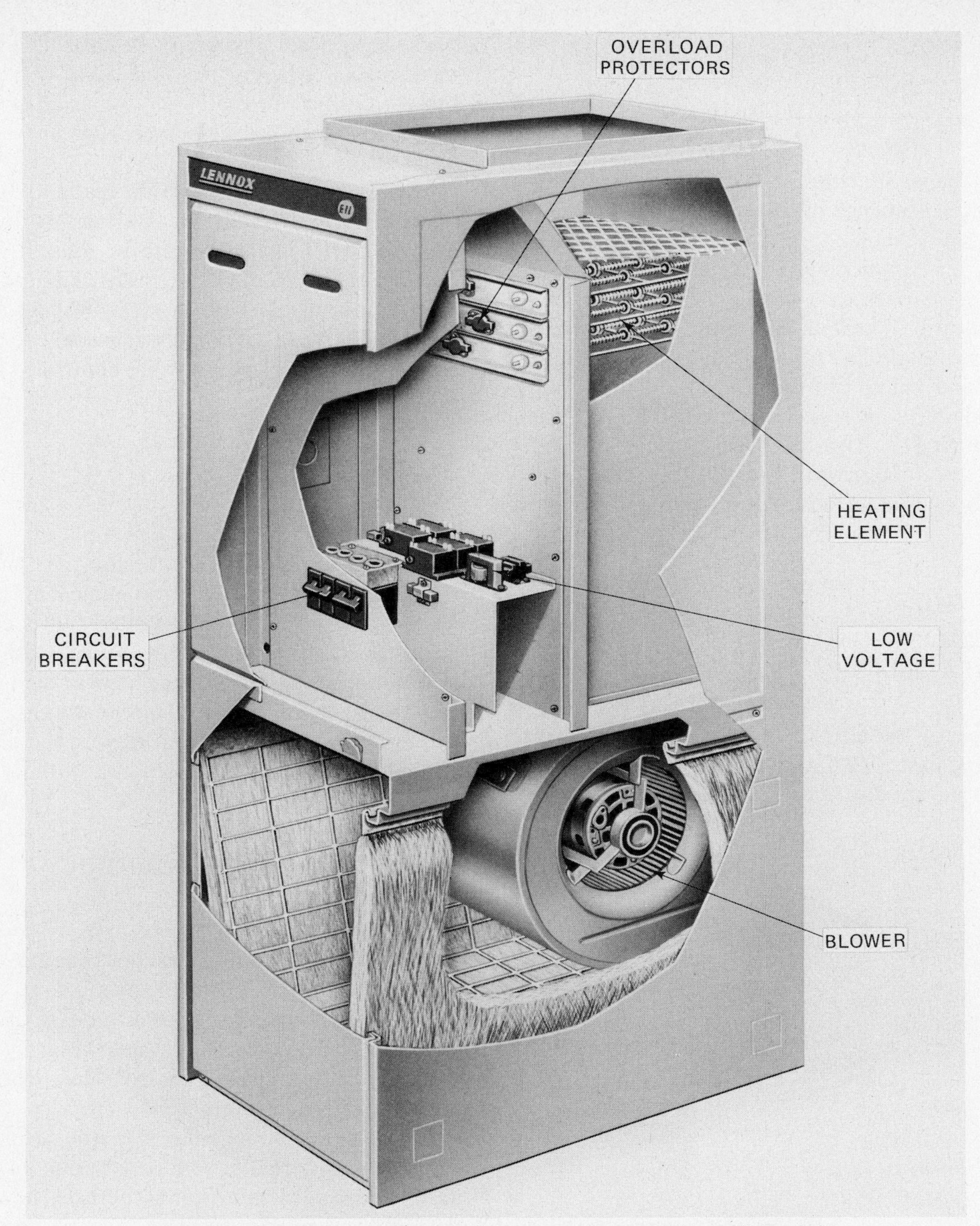

FIGURE 8-82 Lennox electric E-II upflow furnace (Courtesy, Lennox Industries).

and the elements are turned off. The time delay (in seconds) is adequate to stagger power inrush and minimize the shock on the power system.

Where larger furnaces are installed, it is common to use two-stage thermostats in connection with the sequencer control. The first stage would operate as previously described and would bring on at least 50% of total capacity. The second stage of the thermostat would respond only when full heating capacity is needed. With this added control, wide variations of indoor temperature are avoided.

Previously the text underlined *electric* sequencer control as opposed to a *motor-driven* variety that may be used on furnaces having four to six elements. It fulfills the same function and, by a series of cams, times out the sequence of bringing heaters "on" and "off."

The air distribution system for an electric furnace should receive extra care due to the normally lower air temperature of heated air coming off the furnace, as compared to gas and oil equipment. Temperatures of 120°F and below can create drafts if improperly introduced into the space. Additional air diffusers are recommended. Also, duct loss through unconditioned areas can be critical, so *well-insulated ducts are a must*, to maintain comfort and reduce operating cost.

AIR-HANDLING UNITS WITH DUCT HEATERS

A variation of the electric furnace is the use of an air-handling unit with duct-type electric heaters (Fig. 8-83). The air handler consists of a blower housed in

FIGURE 8-83 Electric duct heater (Courtesy, Airtemp Corporation).

an insulated cabinet with openings for connections to supply and return ducts. Electric resistance heaters are installed either in the primary supply trunk or in branch ducts leading from the main trunk to the rooms in the dwelling. In terms of sales this system has not been as popular as the complete furnace package concept, primarily because it complicates installation requirements and adds cost. There is more comfort flexibility by zone control when heaters are installed in branch runs; rooms can be individually controlled.

Duct heaters (Fig. 8-84) are made to fit standard duct sizes and contain overheating protection. Electric duct heaters can also be utilized with other types of ducted heating systems to add heat in remote duct runs or to beef up the system if the house has been expanded. They may be interlocked to come on with the furnace blower and are controlled by a room thermostat.

SUMMARY

Forced-air heating systems, be they gas, oil, or electric, offer the advantage of mechanical air movement whereby the air is not only heated but is cleaned by filtration, humidified, and "freshened" with an intake of outdoor air, which is circulated to the living area. Thus four of the five elements of total comfort air conditioning can be performed, with the fifth option of cooling as an easy addition.

WINTER HEATING—HYDRONIC SYSTEMS

Another form of heating, and one that has existed for many years, is the hydronic method of conveying heat energy to the point of use by means of hot water or steam. The early steam system (Fig. 8-85) used a boiler partially filled with water. Heat from the fuel converts the water into steam (vapor). The steam passes through piping mains to the terminal units (radiators), where it is condensed upon giving up its heat. The condensate (liquid) returns to the boiler through the piping system, thus maintaining the proper water level within the boiler. The flow of heat is by pressure; no mechanical force is used.

In contrast to the partially filled boiler of the steam-heating system, the boiler, piping, and terminal units of a hot-water heating system (Fig. 8-86) are entirely filled with water. Heat produced by the fuel, be it gas, oil, or electric, is transferred to the water within the boiler or heat exchanger. The heated water is circulated through piping and terminal units of the system. The heat given off as the water passes through

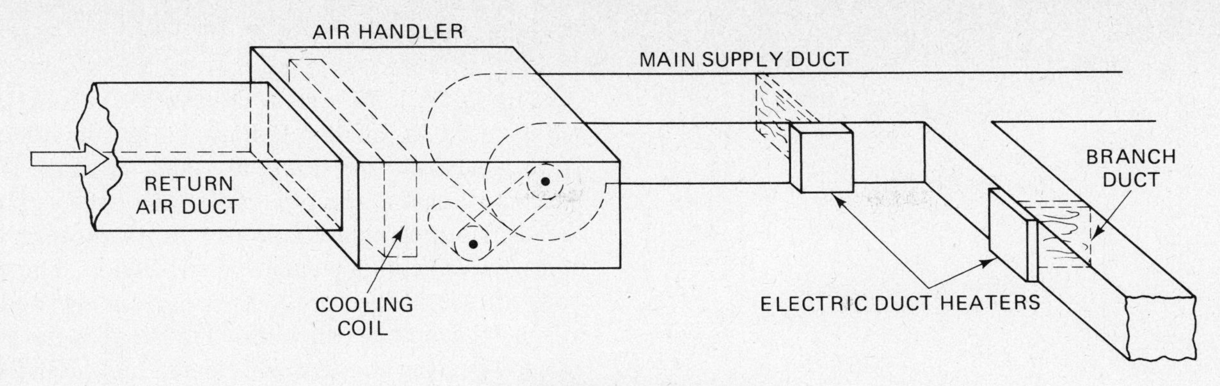

FIGURE 8-84 Duct heaters.

the terminal units causes the water to cool. The cooled water returns to the boiler and is reheated and recirculated. (Note: A similar but reverse process can be used for cooling by circulating chilled water; this technique will be covered later.)

By definition, hydronics includes all three systems: steam, hot water, and chilled water. However, since steam-heating systems are not frequently installed in residences or small commercial buildings today, this discussion will be confined to hot-water heating.

Hot-water Heating

In Fig. 8-86 the flow of water is dependent on gravity circulation due to the difference in the density of water in the supply and return sides. Like steam heating, gravity-flow hot-water systems are pretty much obsolete; much more popular today is the forced circulation system using a pump to circulate the water from the boiler to terminal units and back again. Pipe sizes can be smaller and longer, and terminal units can be smaller and may be located below the boiler if needed. Forced circulation insures better distribution control.

Hot-water systems may be classified according to operating temperatures and pressures:

	Boiler Pressure	*Hot Water Maximum Temperature*
LTW—Low Temp	30 psi	250°F
MTW—Med Temp	150 psi	350°F
HTW—High Temp	300 psi	400–500°F

Medium and high-temperature systems are limited to large installations and are beyond the scope of this text.

Low-temperature hot-water heating systems are classified according to the piping layout. Four common types of systems are:

(1) Series loop
(2) One pipe
(3) Two pipes
(4) Hot-water panel

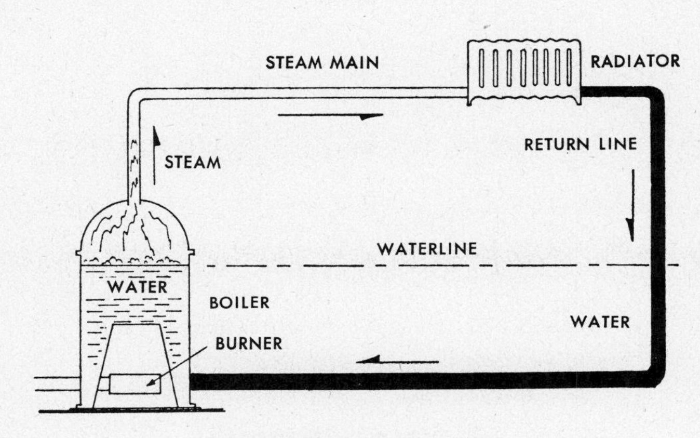

FIGURE 8-85 Steam system.

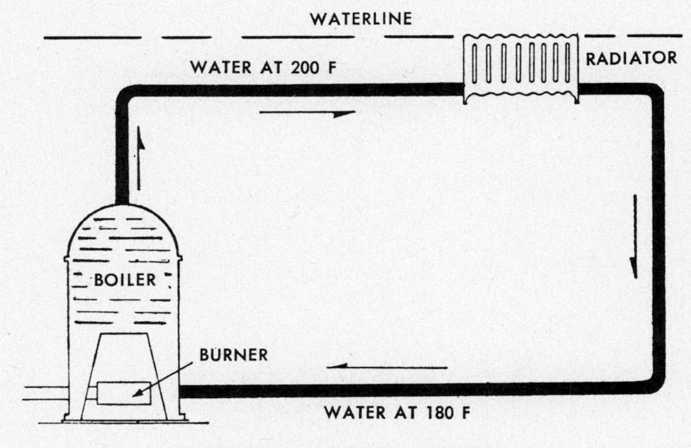

FIGURE 8-86 Hot water system.

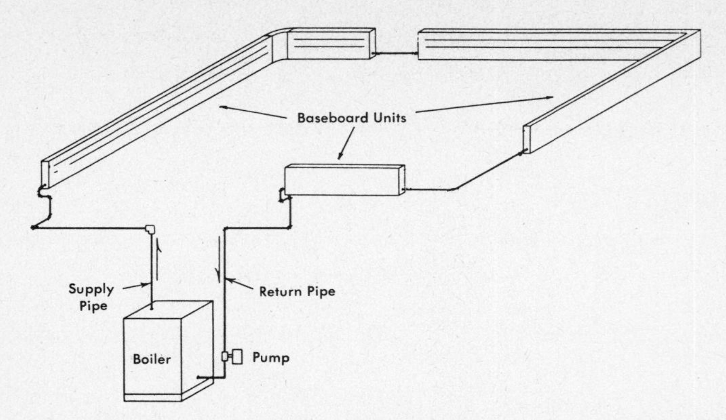

FIGURE 8-87 Series loop baseboard system (single circuit).

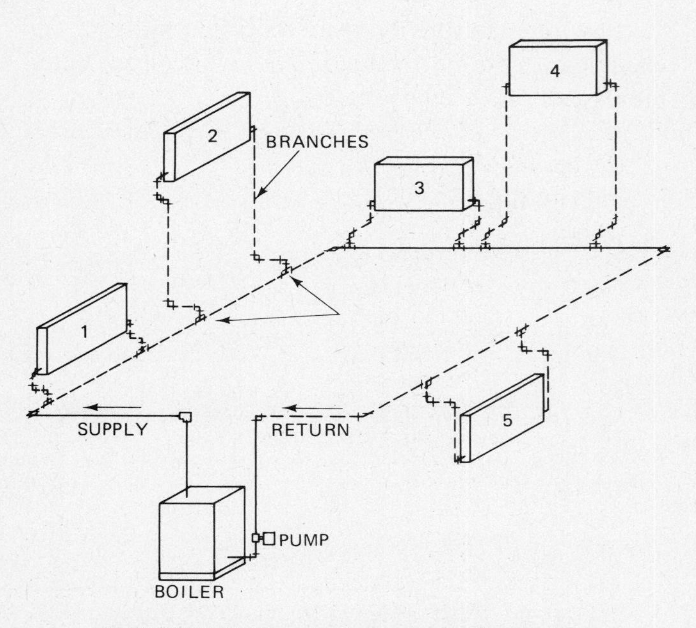

FIGURE 8-88 One pipe hot water heating system.

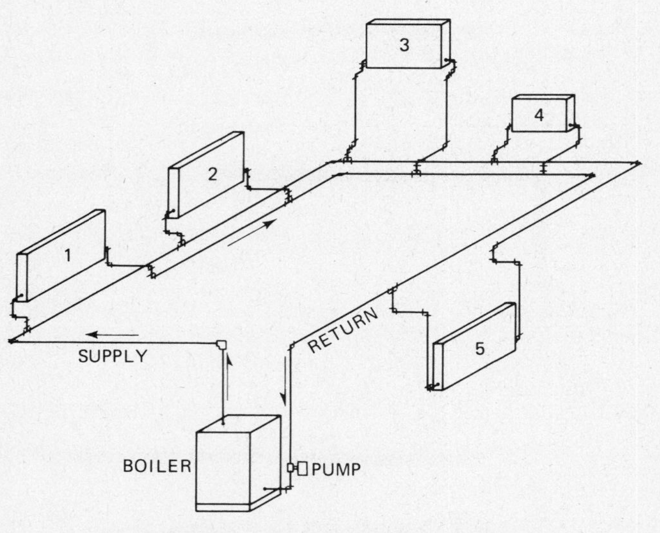

FIGURE 8-89 Two pipe reverse hot water heating system.

Piping Systems

The series loop system (Fig. 8-87) is most commonly used in small buildings or in subcircuits of large systems. The terminal units, usually of the baseboard or fin tube type radiation, serve as part of the distribution system. Water flows through each consecutive heating element in succession. The water temperature, therefore, is progressively reduced around the circuit, which may require that the length of each heating element be selected and adjusted for its actual operating temperature. The series loop system has the advantage of lower installed costs. Also, interconnecting piping can be above the floor, eliminating the need for pipe trenches, insulation, and furring. Its disadvantage lies in the fact that the water temperature to each unit cannot be regulated, and thus some form of manual air damper on the terminal is desirable. Also, circuit capacity is limited by the size of the tube(s) in the terminal element(s).

The one-pipe system (Fig. 8-88) is a variation of the loop, except that the terminals are connected to the pipe loop by means of branches. Individual terminals can be controlled, but special tees are needed, and these fittings are available only from a few manufacturers. A one-pipe system will generally cost more to install than a comparable series-loop system, because of the extra branch pipe, fittings, and special fittings themselves. The one-pipe system also shows a progressive temperature drop around the circuit with resulting limitations.

Two-pipe systems (Fig. 8-89) use one pipe (supply) to carry hot water to the terminal and a second to return cool water to the boiler. Known as a *two-pipe reverse return*, it is set up so that while terminal 1 is closest to the boiler on the hot water supply, it's the farthest on the return main. The reverse is true for terminal 5; it is farthest away on the hot water supply and closest on the return main. This equilization of the distance the water travels through each unit provides even distribution of water throughout the system—and better control. Zones or several circuits may be employed with equal success. The two-pipe system is usually preferred when low pressure-drop terminal units like baseboard are used.

Panel systems (Fig. 8-90) are installed in floors or ceilings and utilize a supply or return header where all of the coils can be brought together. Balancing valves and vents on each of the coils can then be made accessible in a pit or in the ceiling of a closet.

The preceding discussions and illustrations were of single-circuit systems. Any of these hydronic sys-

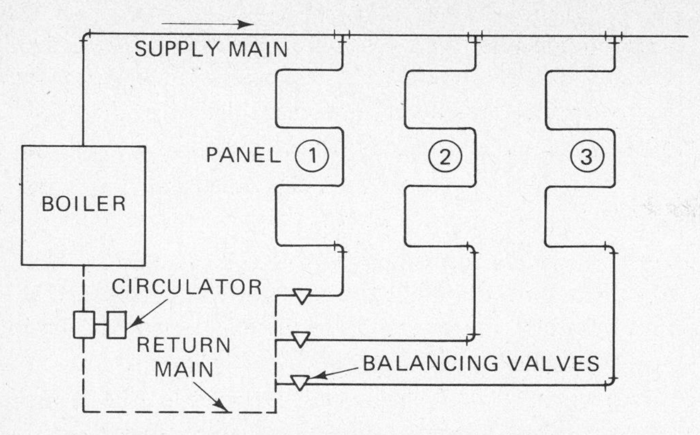

FIGURE 8-90 Forced circulation hot water panel system.

tems may also be installed as multiple circuits where a main or mains form two or more parallel loops between the boiler supply and return.

Multiple circuits are employed for several reasons; to reduce the total length of circuits, to reduce the number of terminal units on one circuit, to reduce pipe size or the quantity of water circulated through a circuit, and for simplification of pipe design in certain types of buildings.

Boilers

Steam boilers were in existence prior to the mid-1800s, and by the 1870s, hot water was beginning to replace steam. Modern boiler design includes the use of cast iron, steel, and copper materials.

Cast-iron boilers are assembled from cast-iron sections (Fig. 8-91). They can burn gas, oil, or coal.

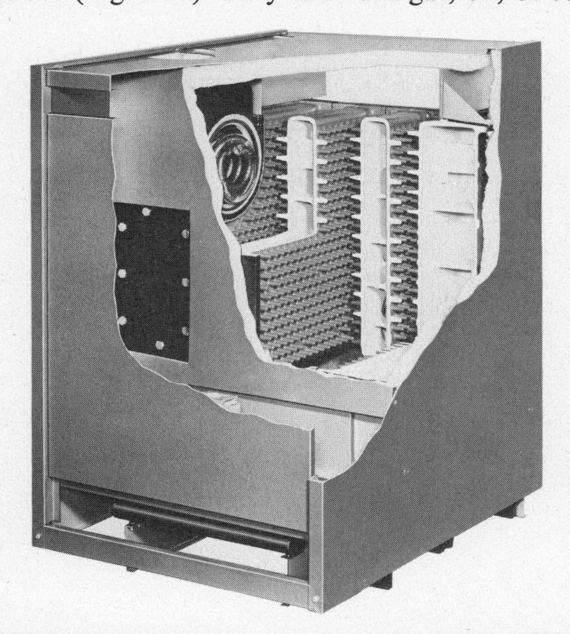

FIGURE 8-91 Gas boiler (Courtesy, Weil-McLain, Div. of Wylain, Inc.).

Steel boilers are designed with tubes inside a shell. Usually, the system water is in the tubes and the fire and hot gases surround the tubes and thus heat the water. This is called a *water tube boiler* (Fig. 8-92).

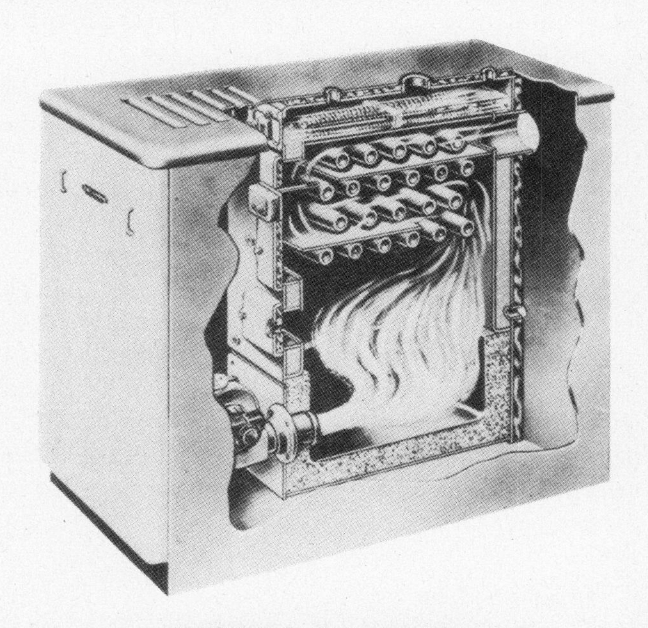

FIGURE 8-92 Water tube boiler.

FIGURE 8-93 Wall type electric boiler (Courtesy, Weil-McLain, Div. of Wylain, Inc.).

However, some boilers function in just the reverse fashion—the fire and hot gases pass through the tubes and heat the water in the shell. These are called *fire-tube boilers*. Residential-size steel boilers are sold as packages with pump, burner, and controls mounted and wired.

Flash boilers for gas or oil burning consist of copper coils which contain the system water surrounded by the fire and hot gases. These are very compact, and some models are designed to be used outdoors.

An electric boiler is essentially a conventional boiler with a large internal volume and electric heaters directly immersed in water. There are also instantaneous types of electric boilers small enough to "hang on the wall" (Fig. 8-93), which are fitted with either immersed electric heaters or electrical heating elements bonded to the outside of the water vessel. The pump and expansion tank are enclosed in the cabinet.

Typical System

A typical low-temperature hot-water installation (Fig. 8-94) consists of:

(1) The boiler, be it gas, oil, or electric, as described above and operated at 30 psi, with heating water from 180 to 220°F for normal supply.

(2) A boiler water temperature control to cycle the burner or heaters to provide control of hot water temperature.

(3) A circulating water pump large enough to allow the flow of sufficient water to heat the space. The control of the pump is from a space or zone thermostat.

(4) A relief valve that will "open" if the boiler pressure exceeds 30 psi.

(5) Pressure-reducing and check valves and an automatic water make-up valve that supplies water to the boiler for initial fill and also if pressure gets too low.

(6) A closed air-cushion tank, which, on initial fill, has a pocket of trapped air within the tank. When water is heated it expands and compresses the trapped air as a cushion, thus providing space for extra water without creating excessive pressure.

(7) Air vents at the terminals or high point of the piping system to purge unwanted trapped air.

(8) A flow-control valve (check valve) to prevent gravity circulation of water when the pump is not in operation.

(9) The necessary cocks, purge valves, and drain cocks for balancing, service, and maintenance adjustment.

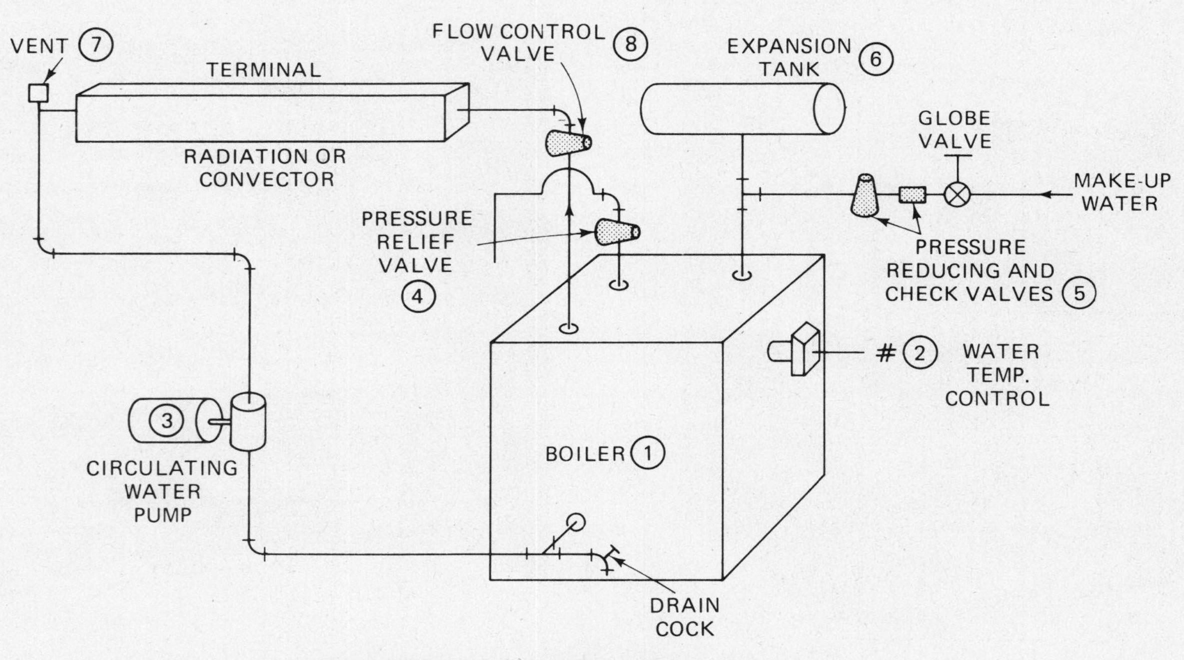

FIGURE 8-94 Typical residential hot water system.

Terminal units vary from cast-iron radiators (seldom used today) to finned-tube residential baseboard radiation (Fig. 8-95), to cabinet-type convectors (Fig. 8-96), to hot-water coils for air-handling units or duct mounting (Fig. 8-97). The selection of the type of unit depends on whether it is used in residential, commercial, or institutional buildings, etc.

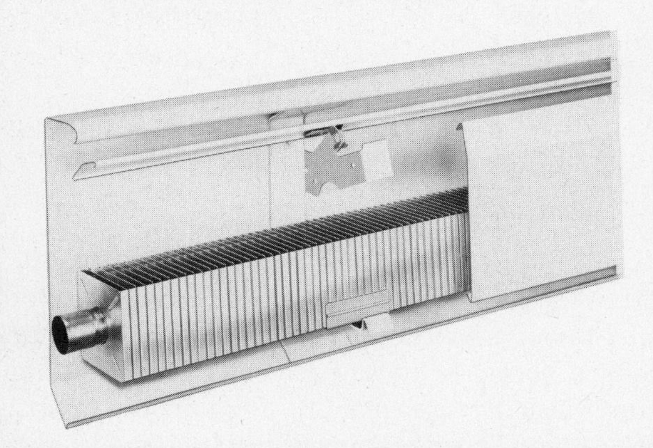

FIGURE 8-95 Finned tube baseboard radiation (Courtesy, Weil-McLain, Div. of Wylain, Inc.).

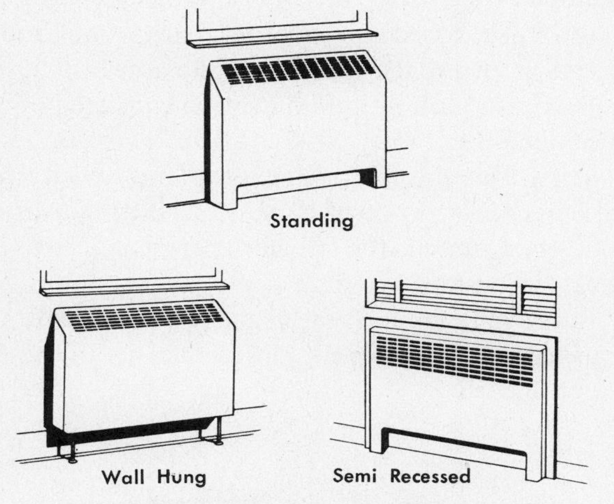

FIGURE 8-96 Convector radiation.

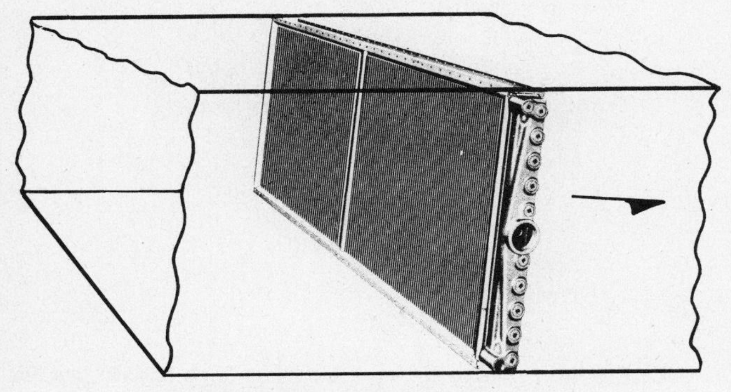

FIGURE 8-97 Hot water coil in duct.

The selection of the terminal size is essentially a function of the temperature of the water, the flow rate, the surface area of radiation, and the type of space air circulation (natural or fan-forced). Manufacturers' catalogs provide rating tables which prescribe the proper selection.

The overall design of hydronic heating systems is in itself a special art, and we would suggest that those who become involved in this area of heating take advantage of the information available through the Hydronic Institute, 35 Russo Place, Berkeley Heights, New Jersey 07922, and the North American Heating and Air-Conditioning Wholesalers Association, 1661 West Henderson Road, Columbus, Ohio 43220. Both of these associations offer excellent training data on the design, installation, and service of hydronic systems.

WINTER HEATING—HUMIDIFICATION

In Chap. 2 the importance of maintaining proper humidity was discussed, and some of the harmful effects of poor humidity control on people and materials.

Room-type humidifiers are available that add moisture to the air, but these require constant attention in terms of manually filling with water. Their evaporation capacity is limited, and at best they are useful only for bedside medical purposes.

As previously mentioned, central forced-air heating and air-conditioning systems offer the ideal solution of adding moisture to the air at the furnace location and circulating it to the total living area, not to just one room.

How much moisture should be added to the air in a residence? This, of course, is subject to the outside and inside temperature and moisture conditions, all of which are constantly changing. We could go through a rather lengthy discussion involving vapor pressure and moisture content, permeability of building materials, etc., but the answer would still only represent one sample and not be useful to practical application.

Fortunately, ARI developed a certification program known as *Standard 610*, open to all manufacturers, whereby a humidifier "size" is rated in terms of capacity—a measure in gallons of the amount of water the equipment can evaporate into the heated home in 24 hours. The size of the home and its air tightness are two important factors in determining the needed capacity required in a humidifier. Figure 8-98 is a sample chart published by one manufacturer of a specific model showing maximum moisture requirements. It is based on design conditions of: an outdoor dry-bulb temperature of 20°F, 80% relative humidity, an indoor dry-bulb temperature of 70°F and 40% relative humidity, and minimum moisture production from residential operations for an absolute humidity difference of 0.0049 lb/h. The definition of a tight house is one that is well insulated,

VOLUME OF RESIDENCE (FT^3)	TIGHT HOUSE		AVERAGE HOUSE	
	Pounds Per Day	*Gallons Per Day*	*Pounds Per Day*	*Gallons Per Day*
8,000	1.76	5.09	3.52	10.17
10,000	2.21	6.35	4.41	12.72
12,000	2.64	7.63	5.29	15.26
14,000	3.09	8.91	5.92	17.08
16,000	3.53	10.18	7.06	20.35
18,000	3.97	11.45	7.94	22.89
20,000	4.41	12.72	8.82	25.44
22,000	4.85	13.99	9.71	27.98
24,000	5.29	15.27	10.59	30.52
26,000	5.74	16.54	11.47	33.07
28,000	6.18	17.81	12.35	36.51
30,000	6.62	19.08	13.24	38.16

FIGURE 8-98 Maximum moisture requirements.

has vapor barriers, tight-fitting storm doors and windows with weather stripping, and its fireplace will be dampered. An average house will be insulated, have vapor barriers, loose storm doors and windows, and a dampered fireplace.

With the volume of the residence known (in ft^3—length × width × ceiling height) one can quickly determine how many pounds of water per hour or gallons per day will be needed to maintain the above conditions. This is an average guide but will apply to

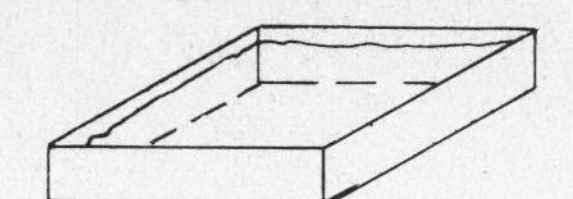

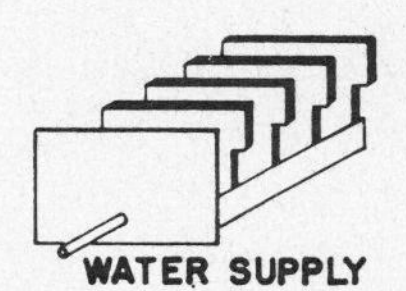

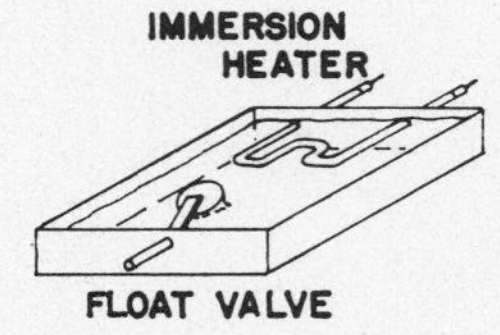

FIGURE 8-99 Pan type humidifiers.

most of the country. Remember, the *average* outdoor winter temperature doesn't stay below 20°F for any extended period of time. Each ARI certified manufacturer will have similar information on capacity and/or home dimensions relative to particular units.

With a method of determining how much water to evaporate, the next question is what kind(s) of humidifiers are available. This, too, is a choice or compromise based on need, capacity, method of operation, price, and ease of maintenance.

Early attempts at humidification consisted of pan-type humidifiers (Fig. 8-99). These were limited in capacity, fraught with scale problems when solids were precipitated out of hard water during evaporation, or required too much energy, such as the immersed electric heater.

Improvements have been made to the plate type by using rotating discs as represented in Fig. 8-100.

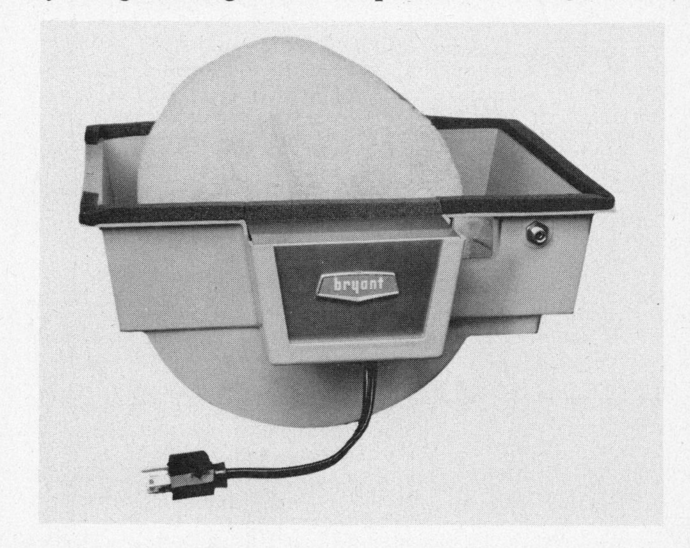

FIGURE 8-100 Rotating disc-type humidifier (Courtesy, Bryant Air-Conditioning Company).

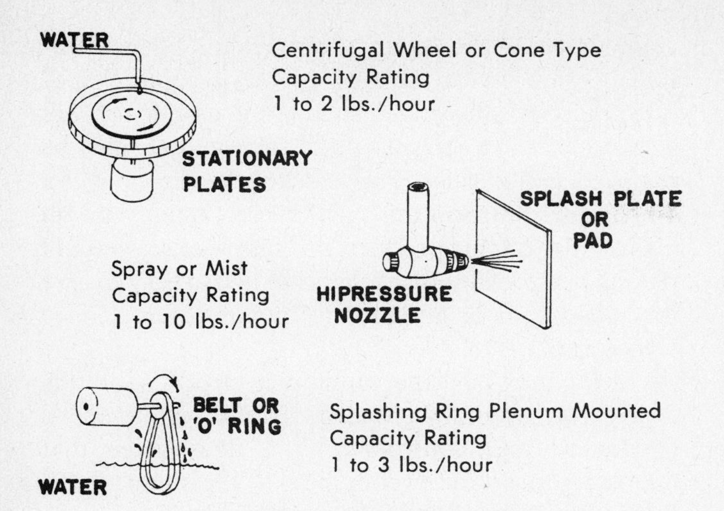

FIGURE 8-101 Atomizing humidifiers.

The discs are made of a noncorrosive material and are rotated by a small motor (similar to the way a barbeque spit works). The disc constantly presents a wetted surface to the air. This type of unit is mounted under a warm-air duct and can evaporate upwards of 25 gallons of water per day. It is also relatively inexpensive to install and operate.

Another version is the atomizing humidifier (Fig. 8-101), which mechanically breaks the water into fine droplets to accelerate evaporation. Several methods are shown. Seldom is the high-pressure nozzle used in residential application. They are, however, used extensively in spray coil commercial systems that need large volumes of water. The centrifugal wheel design (Fig. 8-102) has been a popular unit for mounting on the furnace return plenum or suspended under a return air duct. A motor-driven centrifugal wheel throws the water against a set of slotted plates that break the water into a fine mist, which is then blown into the conditioned air. The limited capacity of this system is a disadvantage for larger homes.

The third and most effective type is called a *wetted element humidifier* (Fig. 8-103), which combines the advantages of a positive air stream passing over a large, wetted surface. The plenum power type of humidifier (Fig. 8-104) uses a small internal fan to draw the air through the wetted evaporating pad and then return it to the furnace air stream. Its capacity is high, but a significant amount of electrical power is consumed by the humidifier fan.

The bypass humidifier (Fig. 8-105) does not have a fan but depends on the air pressure differential between the warm-air plenum (+pressure) and the cold-air return (−pressure). Its capacity rating is

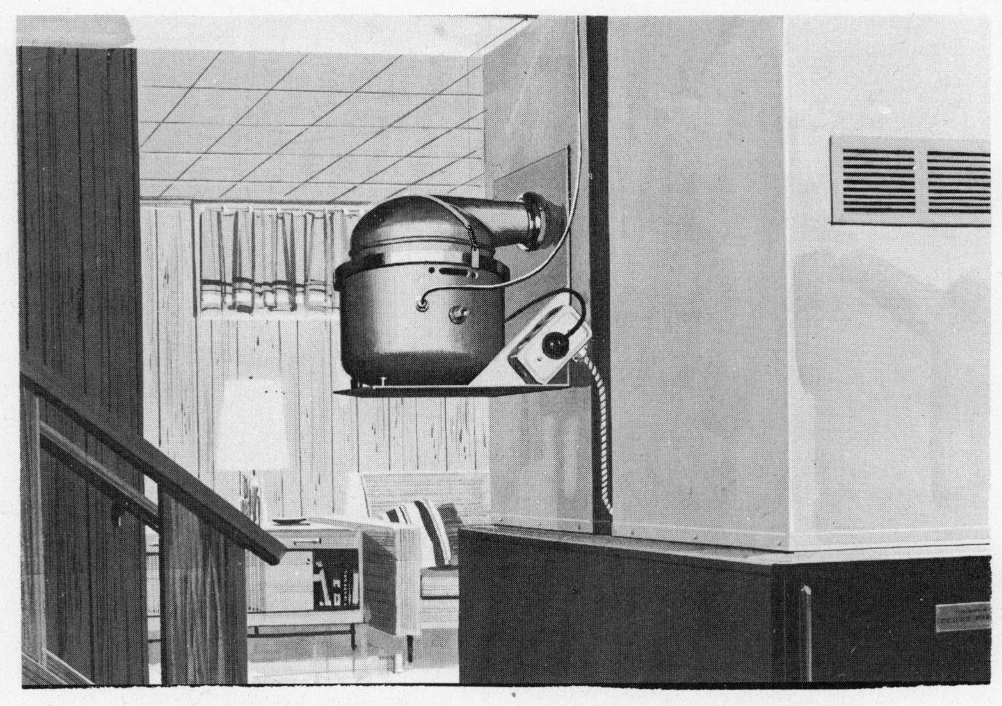

FIGURE 8-102 Centrifugal wheel-type humidifier (Courtesy, Walton Laboratories).

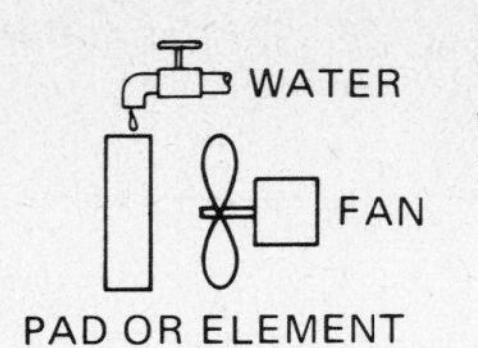

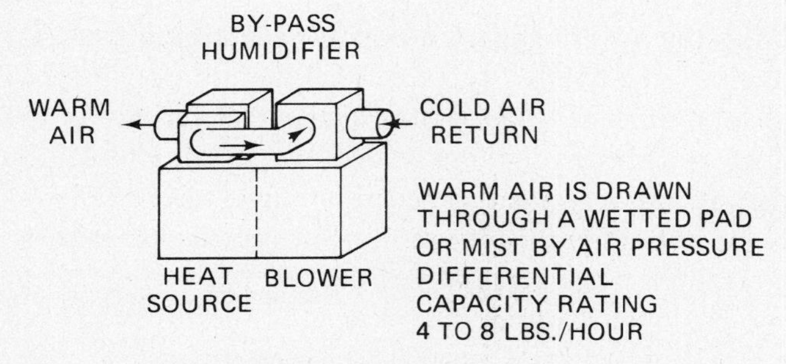

FIGURE 8-103 Wetted element humidifiers.

slightly less than that of the fan model above, but no added electrical cost is introduced.

To recap, the decision as to which method, model, or brand to install is a function of need, application, price, and maintenance. And on the last point, maintenance is the key to effective humidifier operation. Regardless of the type, water is being evaporated, and any suspended solids such as calcium will collect and leave a crust on the surface that must be removed periodically. The question is, can or will a homeowner perform this task or must he depend on his installer? Some products are simple to service and require no special tools; others are not so easy to care for and should be attended to by competent service technicians.

The control of the humidifier depends on the degree of accuracy needed. Water make-up is introduced and controlled by a float valve that maintains a certain level or flow to the sump. If the humidifier has no internal power fan, such as in the bypass model, it will function as the furnace fan cycles on and off. But as long as the furnace fan runs, moisture will be introduced, so in order not to over-humidify, a shut-off water solenoid valve is employed.

And where close control of humidity is desired, a humidistat may be used to energize or deenergize the water solenoid and/or furnace fan relay. (This will be

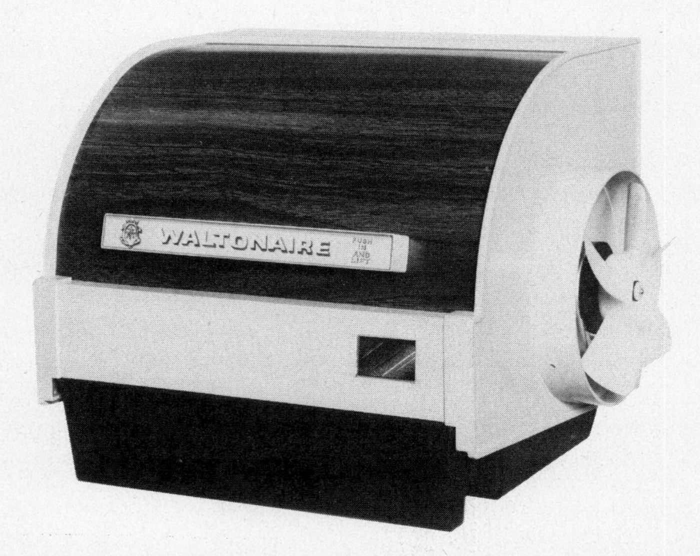

FIGURE 8-104 Plenum power type humidifier (Courtesy, Walton Laboratories).

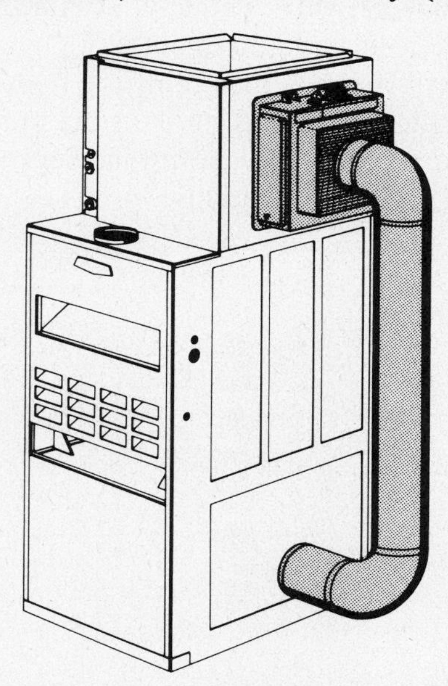

FIGURE 8-105 By-pass humidifier (Courtesy, Bryant Air-Conditioning Company).

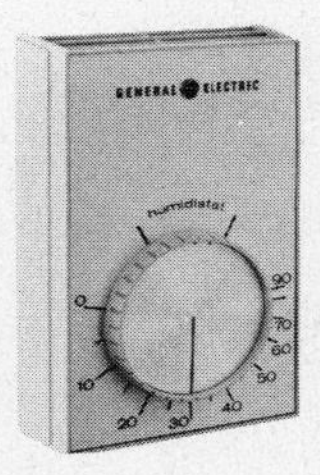

FIGURE 8-106 Humidistat (Courtesy, General Electric Air-Conditioning, Business Division).

reviewed later under electrical controls systems.) A humidistat (Fig. 8-106) is a switch mechanism actuated by changes in moisture. The sensing or hygroscopic element of a typical residential humidistat is composed of tightly strung human hair or a Teflon strip that will contract or expand under varying relative humidity conditions. This movement is sufficient to open or close an electrical contact. Some are wall mounted, but others are made for duct mounting and may be equipped with a sail switch to sense air movement in the duct.

PROBLEMS

1. Name four types of common fuels.
2. What grades of fuel oil are used for residential heating? The approximate heating value is ______________.
3. Natural gas is mainly made up of ______________.
4. LP fuel stands for ______________. Name the two most common types.
5. Is LP gas lighter or heavier than air?
6. How much combustion air (ratio) is needed to burn natural gas?
7. In electric heating, 1 kilowatt equals ______________ Btu.
8. What is the most frequently used electric heating product?
9. Name four types of forced-air furnaces.
10. 100% pilot shut-off is not needed on LP furnaces: true or false?
11. Furnace clearances are distances from ______________.
12. Flue venting methods are not important to the furnace installer: true or false?
13. Air rise through a furnace is a measurement of ______________.
14. Should the choice of the room thermostat location be left to the homeowner?
15. Oil burners for residential forced-air furnaces are of the ______________ type.
16. Are oil furnace vents and clearances more or less critical than gas?
17. Is hydronic heating mainly steam or hot water?
18. Electric forced-air furnaces use a ______________ to stagger the heating elements.

9

MECHANICAL AND ELECTRONIC FILTRATION

In earlier discussions on forced-air heating equipment, mechanical filters were included as standard equipment. The term *mechanical* describes a typical fibrous material filter (Fig. 9-1) commonly called a *throwaway* or replacement variety. It consists of a cardboard frame with a metal grille to hold the filter material in place. The filter media material consists of continuous glass fibers packed and loosely woven that face the entering air side and more dense packing and weave on the leaving air side. Air direction is clearly marked. Filter thickness is usually one inch for residential equipment. Filters come in a wide variety of standard sizes (length and width). The recommended maximum air velocity across the face should be around 300 ft/min, with a maximum allowable pressure drop of about 0.5 in. W.C. Some makes also have the fiber media coated with an adhesive substance in order to attract and hold dust and dirt.

Many manufacturers and filter suppliers also offer what is known as a *permanent* or *washable* filter (Fig. 9-2). It consists of a metal frame with a washable viscous-impingement type of air-filter material supported by metal baffles with graduated openings or air passages. The filter material is coated with a thin spray of oil or adhesive in order to help the filtering effect. When needed, these filters may be removed and cleaned with detergent, dried, and recoated.

Another variation of dry filter is a pad or roll of material supported by a metal mesh frame of some type as shown in Figure 9-3a and 9-3b.

In commercial and industrial work there are variations of wet and dry filters for special use where larger volumes of air are involved, or the filtration of chemical substances such as in the removal of paint in spray booths, commercial laundries, hospital operating rooms, etc., but this is a matter for experts to handle.

The efficiency of standard residential and light commercial mechanical filters can vary from 25% for a typical window air conditioner to perhaps 75 to 80% for the better heating and central air-conditioning equipment. This means they are effective in removing lint, hair, large dust particles, and somewhat effective on common ragweed pollen. They are relatively ineffective on smoke and staining particles. So for average use they do a creditable job. However, getting the homeowner to keep his filters clean is not easy, and dirty filters are probably the main contributor to malfunctioning equipment. As the mechanical filter clogs with dirt, it can reduce air flow to a point where the cooling coil will freeze and perhaps lead to a compressor failure. Or, in a heating system, it can cause overheating and reduce the life of the heat exchanger, or it can cause nuisance tripping of the limit switch. Dirty filters increase operating costs as system efficiency goes down.

Where a high degree of filtering efficiency is desired (above 80%), the use of electronic air cleaners is highly recommended. Chapter 2 mentioned the many advantages to health, cleanliness, and savings resulting from having clean air. Industry product development has put these devices within the price range of most homeowners.

The principle of electrostatic precipitation in filtering air is not a recent discovery and has been used in commercial and industrial work such as smoke stacks, clean rooms, etc. The adaptation to residential use, made possible by mass production and effective marketing, is more recent, within the last 10 to 15 years. There are two types of electrostatic cleaners: the ionizing type and the charged media.

In ionizing cleaners (Fig. 9-4), positive ions are generated in the ionizing section by the high-potential ionizer wires. The ions flow across the air stream, striking and adhering to any dust particles carried by the air stream. These particles then pass into the collecting cell, a system of charged and grounded plates. They are driven to the collecting plates by the force exerted by the electrical field on the charges they carry. The dust particles which reach the plates are thus removed from the air stream and held there until washed away. A prefilter/screen is usually employed to catch large particles such as lint and hair.

In a typical ionizing electrostatic air cleaner design (Fig. 9-5), a dc current potential of 12,000 volts or more is used to create the ionizing field, and a similar or slightly lower voltage is maintained between the plates of the collector cell. These voltages necessitate safety precautions, which are designed into the equipment. The power pack boosts the 110-volt supply to 12,000 volts by means of a high-voltage transformer and converts ac current to dc, using solid-state rectifiers that contain silicon or

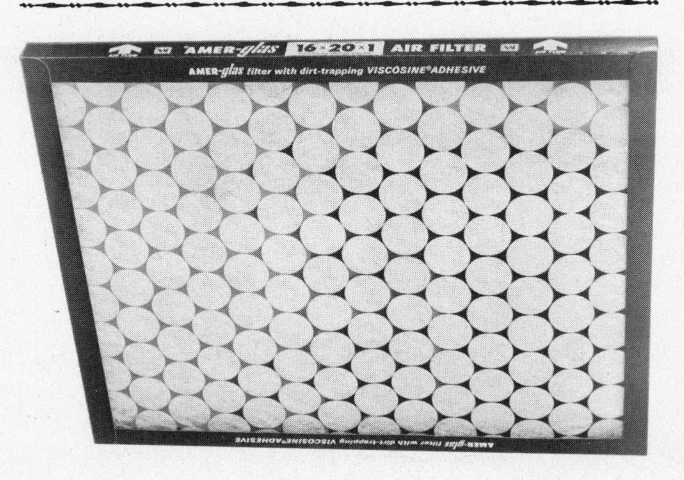

FIGURE 9-1 Fibrous material throw-away type filter (Courtesy, American Air Filter Company).

FIGURE 9-2 Washable filter (Courtesy, American Air Filter Company).

selenium. Power consumption is relatively low; 50 watts or less. The collector cell (Fig. 9-6), made of nonrusting aluminum plates, requires periodical removal for cleaning by soaking in a tub or any suitable container in detergent or liquid soap and then washing down with a hose. Some cells are small enough to be cleaned in an automatic dishwasher.

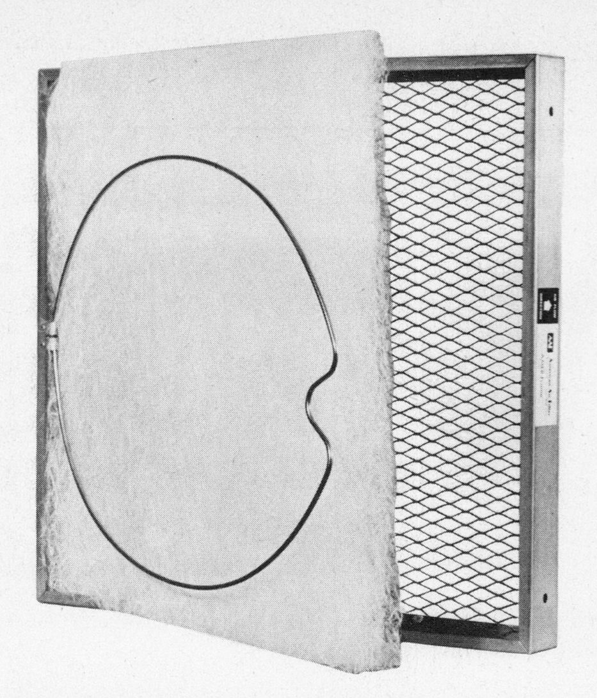

FIGURE 9-3(a) Dry filter pad (Courtesy, American Air Filter Company).

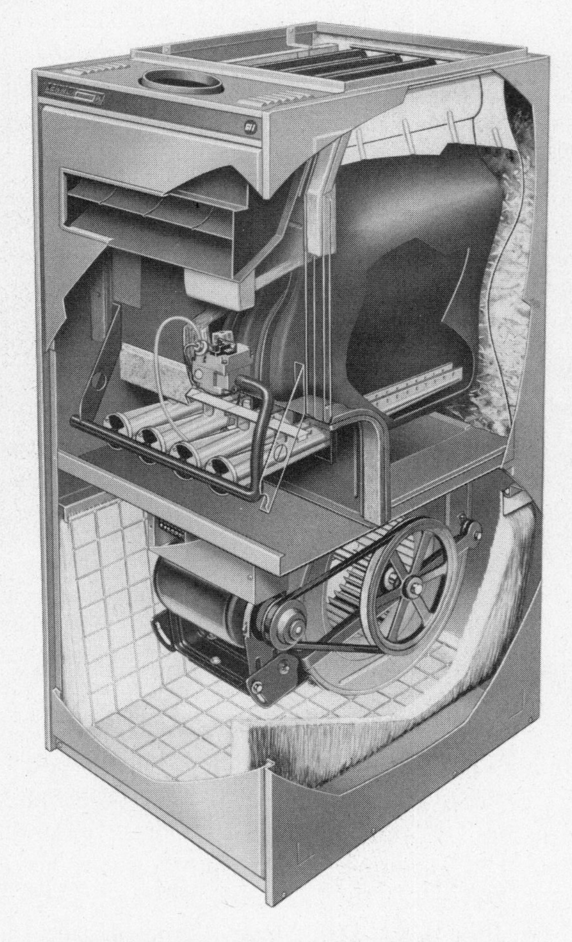

FIGURE 9-3(b) (Courtesy, Lennox Industries).

The *charged-media type of air cleaner* has certain characteristics of both dry filters and electrostatic cleaners. It consists of a dielectric filtering medium, usually arranged in pleats or pads. Ionization is not used. The dielectric medium is in contact with a

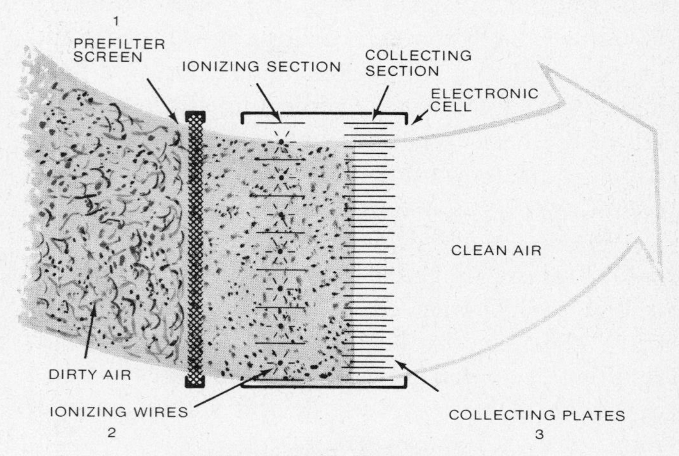

FIGURE 9-4 Ionizing type cleaner.

FIGURE 9-5 Electronic air cleaner (Courtesy, Airtemp Corporation).

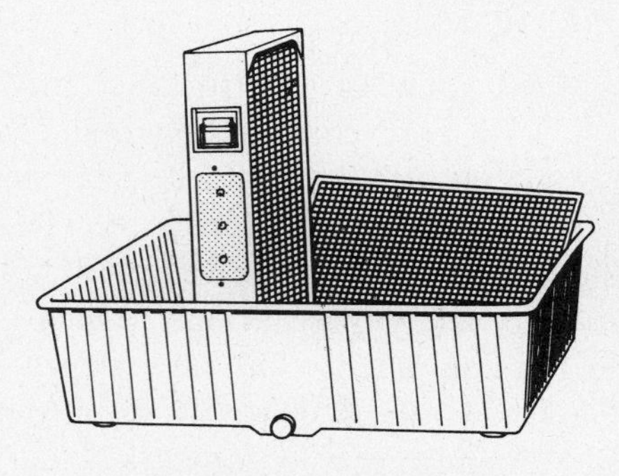

FIGURE 9-6 Collector cell.

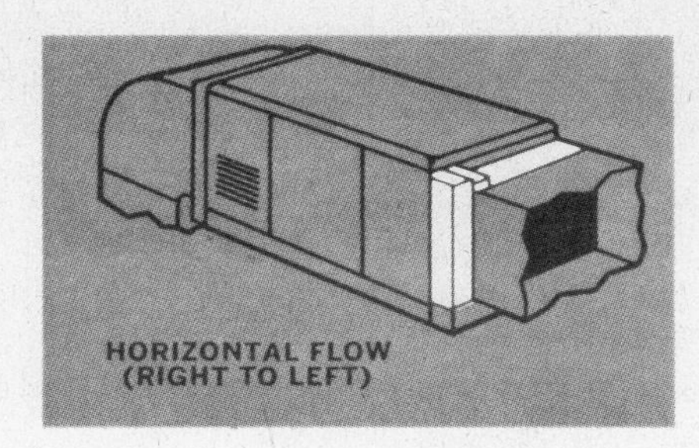

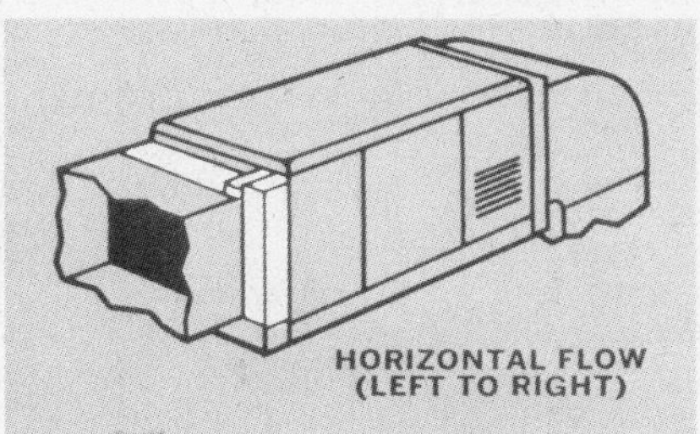

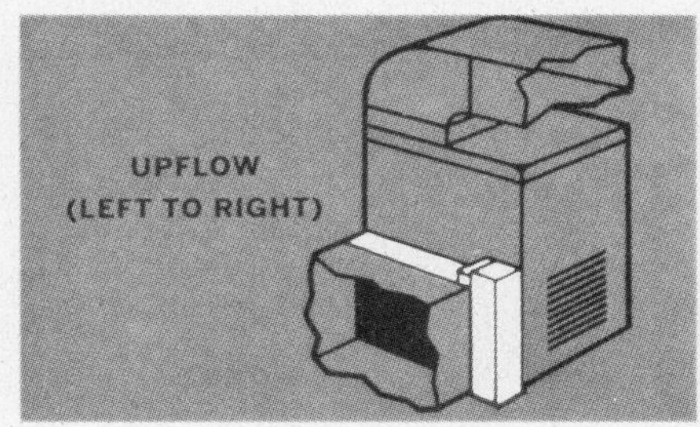

FIGURE 9-7 Air cleaner installations.

gridwork consisting of alternately charged members held at very high voltage potential. An intense electrostatic field is thus created, and airborne particles approaching the field are polarized and drawn into the filaments or fibers of the media. This type of cleaner offers much higher resistance to air flow, and the efficiency of the media is impaired by high humidity. At this time, charged-media air cleaners are not sold in any appreciable volume for residential application.

How effective are electronic air cleaners? Most are capable of filtering airborne particles of dust, smoke, pollen, and bacteria ranging in size from diameters of 50 microns down to 0.03 micron. The micron is a unit of measure (length) equal to one millionth of a meter or 1/25,400 of an inch. The dot made by a sharp pencil is about 200 microns in diameter. It takes special electronic microscopes to examine particles of this size.

The operating efficiency of electronic air cleaners is a function of air flow. With a fixed number of ionizer wires and collector plates, a particular model may be used over a range of air-flow quantities (ft^3/min). It will be rated at a nominal ft^3/min, and then for other efficiency ratings for the recommended span of operation. The higher the ft^3/min of air and the resultant velocity, the lower the efficiency. Residential electrostatic air cleaners come in sizes of 800 ft^3 to 2,000 ft^3. The number of models depends on the manufacturer. Most offer two sizes: 800 to 1,400 ft^3 and 1,400 to 2,000 ft^3. Static pressure drop will run about 0.20 in. of W.C. at the nominal air-flow rating point. Ratings are published in accordance with National Bureau of Standards Dust Spot Test and/or are certified under ARI standards.

The application of an air cleaner is most flexible (Fig. 9-7). When mounted at the furnace or air handler, it is installed in the return air duct for upflow, downflow, or horizontal air flow. Position does not affect its operation or performance. (Note: Some units are equipped with built-in washing, and, obviously, these must be installed vertically.)

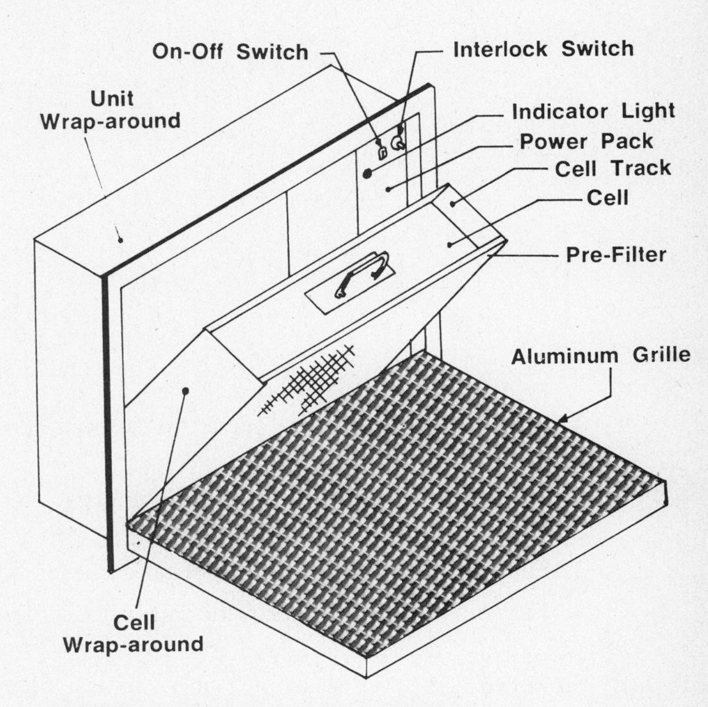

FIGURE 9-8 Wall type air cleaner installation.

Where furnaces are installed in very small closets or utility rooms and no duct space is available, some manufacturers offer a model (Fig. 9-8) which will recess into a wall and actually provide a decorative return air louver that faces the living area. Service access is through the louver.

Some units, as shown in Fig. 9-9 have built-in washing without removing the cells. On an automatic control, washing action sprays water over the plates. The runoff is collected in the bottom where it drains out.

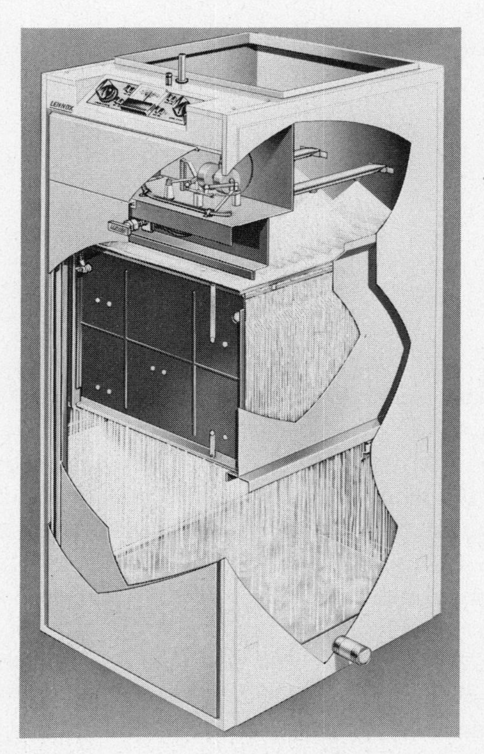

FIGURE 9-9 Electronic air cleaner (Courtesy, Lennox Industries).

PROBLEMS

1. What is the common name for a noncleanable mechanical filter?
2. Mechanical filters are generally effective up to ______________%.
3. Filter maintenance is not important: true or false?
4. The most popular electronic air cleaner is the ______________.
5. Voltage in an electronic can run as high as ______________.
6. Electronic air cleaners are rated in ______________.
7. What is a micron?
8. The power consumption of a typical electronic air cleaner is ______________ watts.
9. The position of the electronic air cleaner is not critical: true or false?
10. The collector cell is usually cleaned by ______________.

10

UNITARY PACKAGED COOLING

The text on *Refrigeration* mentioned the basics of a mechanical refrigeration cycle and reviewed each component in terms of its function in the system and its operating characteristics. This and subsequent chapters will build on that background and discuss the products that make up the broad range of cooling and combination heating and cooling systems that contain a mechanical refrigeration cycle. Chapter 10 deals with products essentially designed to provide cooling only, although it is sometimes difficult to make such a distinction because of the similar characteristics of and crossover between certain systems.

ROOM AIR CONDITIONERS

With annual sales of over 5 million units per year, the room unit (Fig. 10-1) obviously is meeting a need. It is installed in the window of a room and is meant to provide cooling for that immediate area, although in actual practice, the cooling effects may extend to other areas if there is good circulation. For example, a unit installed in the living room may offer some relief to bedrooms or kitchen. Not all units are window mounted. Models are available for permanent installation through the wall. This is common in low cost installations in motels, small offices, and apartments, but such units are not often mounted in the walls of single-family residential dwellings.

Room units have been used extensively in homes employing nonducted baseboard or ceiling cable heating systems.

The capacities of room units range from 5,000 Btu/h to over 30,000 Btu/h. The larger units are more frequently associated with new residential construction, where the necessary electrical service and structural support is preplanned, and also for commercial applications. Most average bedrooms use a $\frac{1}{2}$ to $\frac{3}{4}$-ton unit. For existing homes the industry has developed what is known as a $7\frac{1}{2}$-amp unit that will plug into an ordinary 15-amp lighting circuit. Codes assume that the air conditioner will not exceed 50% of the circuit rating, so that breakers or fuses won't open when the compressor starts up. The maximum size of $7\frac{1}{2}$-amp models are about 9,000 Btu/h. Larger units are wired

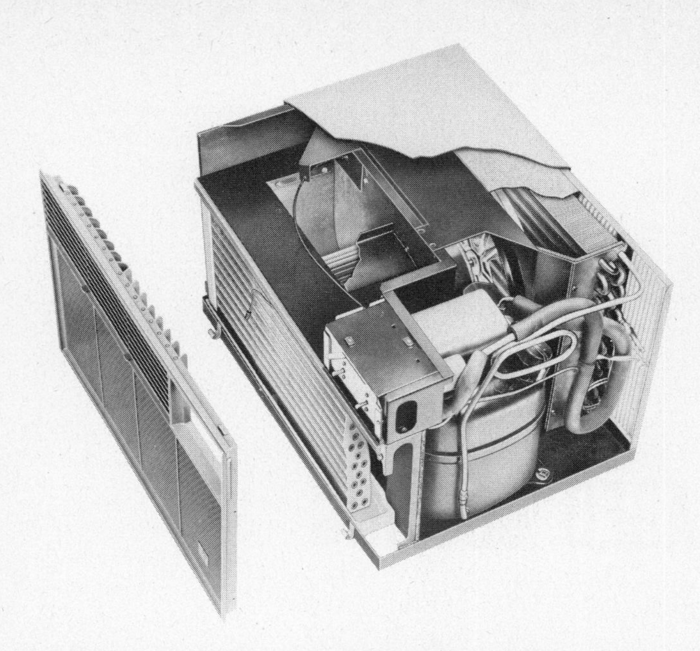

FIGURE 10-1 Room air-conditioning unit (Courtesy, Airtemp Corporation).

independently of the lighting system for 208-230 volt operation.

The sales and distribution of room units is predominately through appliance and chain store marketing channels rather than through professional air-conditioning contractors. Many installations are do-it-yourself situations. Capacity ratings and other technical standards are subject to the Association of Home Appliance Manufacturers (AHAM) rather than to the ARI for all other cooling apparatus. Also, in terms of warranty and service, the room unit more closely follows the appliance industry warranties.

In terms of operation there is little or nothing that can be changed to control comfort except perhaps directional louvers on the room air outlet grille to minimize drafts. Filter replacement is the essential service requirement.

In summary, the room unit serves a definite need; spot cooling at a minimum installed cost and mobility of location, both of which preclude the consideration of central system cooling units.

CENTRAL SYSTEM COOLING UNITS

Packaged Air-Cooled Units

FIGURE 10-2 Two-ton packaged air cooled unit (Courtesy, Airtemp Corporation).

Basically developed as a residential central system cooling unit, the packaged air-cooled unit (Fig. 10-2) has also been broadly adapted to light commercial application and mobile home conditioning. The usual design configuration is a rectangular box with supply and return air connections on the front and provisions for the condensing section air (in and out) on the back and sides. The inside arrangement is relatively simple. Return air is drawn across the fin-tubed evaporator by a centrifugal blower and discharged out the front. On smaller sizes the blower is of the direct-drive variety. Larger units have adjustable belt drives. A condensate pan beneath the evaporator collects moisture runoff; it is connected to a permanent drain. The evaporator compartment is amply insulated to prevent heat gain and sweating. The filter is usually located in the return air duct. Separating the evaporator compartment and the condensing compartment is a barrier wall, which isolates air movements and insulates for minimum thermal and sound transmission to the conditioned air. The compressor and condenser coil form the *high-side* refrigerant circuit. Condenser air is drawn in from the sides and discharged through the condenser coil. This is called a *blow-through* arrangement. Some units use a *draw-through* arrangement and exhaust through the sides. The condenser fan is most frequently a propeller type. It moves large volumes of air where there is little resistance. Propeller fans are not intended for ducting. Refrigerant R-22 is almost universally used by all manufacturers, as is capillary expansion on smaller tonnage equipment.

The control box (Fig. 10-3) includes capacitors for fan motors and/or compressors, starting relays, and the terminal connection for a remote thermostat.

The capacities of packaged air-cooled cooling units range from $1\frac{1}{2}$ tons to $7\frac{1}{2}$ tons for residential

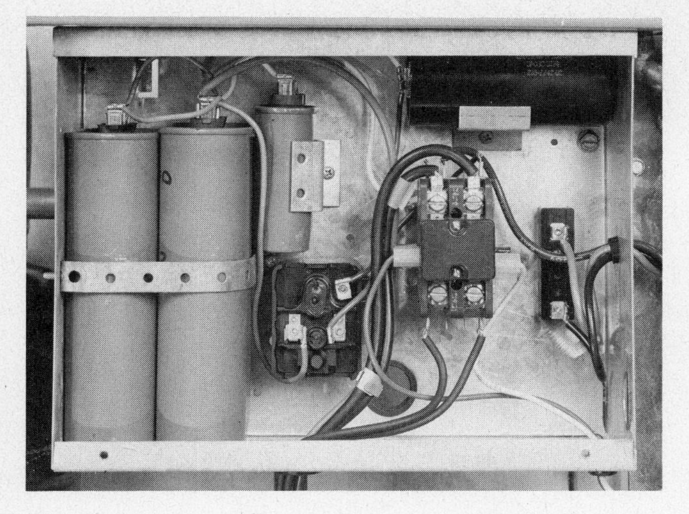

FIGURE 10-3 Three-ton control box (Courtesy, Airtemp Corporation).

use and upwards of 30 tons in commercial work. Most units are rated and certified in accordance with ARI Standard 210 rating conditions, which are 80°F dry-bulb and 67°F wet-bulb temperature of the entering air to the indoor evaporator coil and 95°F D.B. of the entering air to the outdoor condensing coil. Included in ARI Standard 210 is also the requirement that the unit must be capable of operating up to 115°F outdoor ambient without cutting off on high pressure or causing compressor overload cycling.

Diagrammatically, the system operating at ARI conditions has the characteristics shown in Fig. 10-4; at a dry-bulb temperature of 80°, return air from the conditioned space at the rate of 400 to 450 ft³/min per ton goes through the filter and then through the evaporator where it is cooled and dehumidified. Air coming off the coil will be about 58° to 60° D.B. Thus, there is approximately a 20–22°F D.B. temperature drop across the coil. The ratio of sensible cooling to total cooling will be about 0.75. Suction pressure coming from R-22 off the coil would then be about 73 to 76 psig. The conditioned air is at 60°, and, assuming it picks up a small amount of heat in the ducts, it will enter the room at 62° to 65° dry-bulb (15° to 18° temperature difference, T.D.), which is a desirable temperature.

On the refrigerant high side, outdoor air for condensing will be introduced at 95°F across the coil. Depending on the coil design, air flow over the condenser coil will be a nominal 800 ft³/min per ton. The resulting pumping head of the compressor on R-22 will be in the range of 295 psig discharge pressure. The average condenser temperature will be 130°F, with approximately 16°F liquid subcooling to a liquid "off" temperature of 114°F.

In a single package air-cooled unit as described above, it is assumed there will be a supply and return duct system. In a residence the unit can be mounted on a concrete slab at ground level (Fig. 10-5), with ducts into the basement or crawl space. This is a common method of adding cooling to an existing home that has hydronic or electric baseboard heating. Some units can be mounted through a concrete block foundation wall.

Attic, garage, or carport installations may be found in areas like Florida, where homes have con-

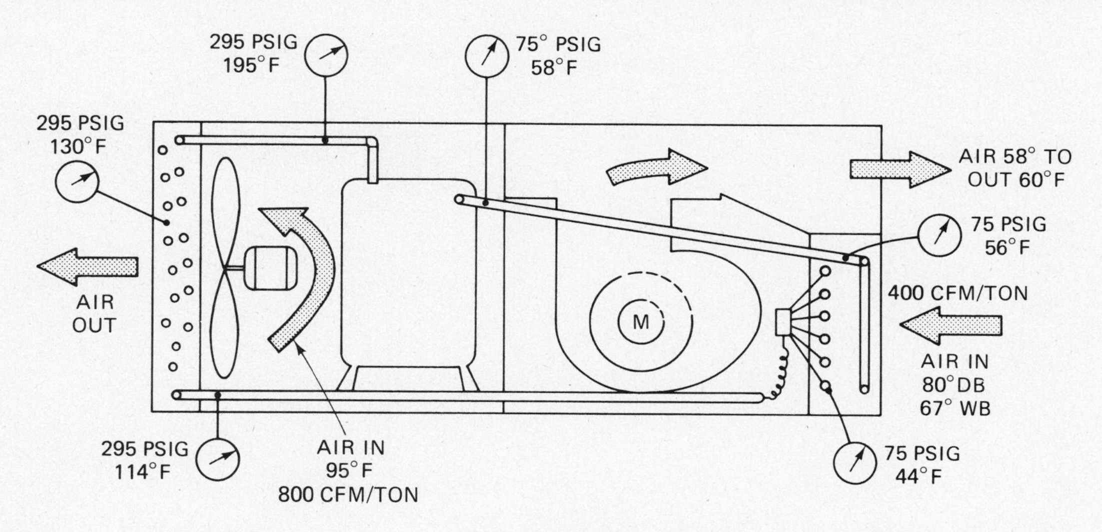

FIGURE 10-4.

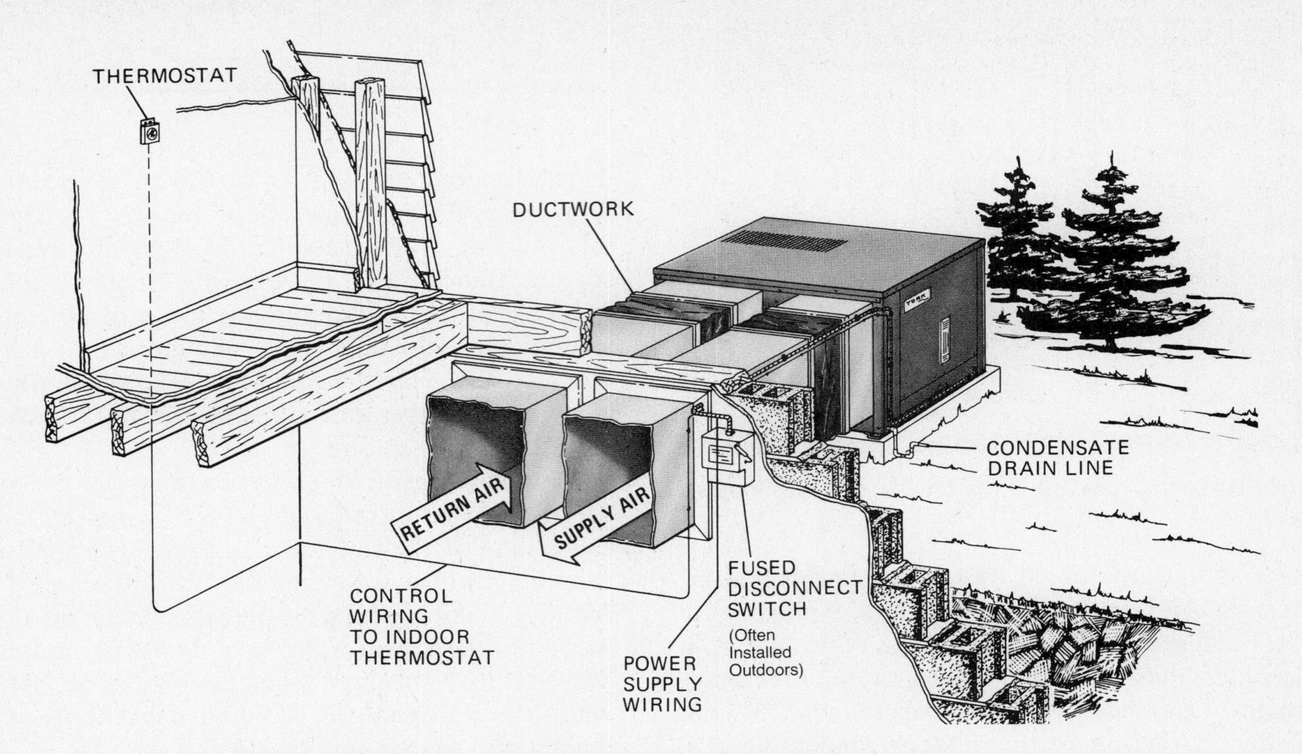

FIGURE 10-5 Typical ground-level installation (Courtesy, York, Div. of Borg-Warner Corporation).

crete slab floors (Fig. 10-6). Structural support, adequate condensing air intake, and well-insulated ducts are important for proper operation.

In light commercial applications (Fig. 10-7), the ability to mount the unit on a roof, with short supply and return ducts, allows for simple and minimum cost installations. The installer must make sure the roof is strong enough to support the weight and that a proper water seal exists around the ducts. On larger units, roof curbs are available to support the equipment.

Another commercial application approach is to install the unit through an outside wall (Fig. 10-8) and provide a simple air distribution plenum on the

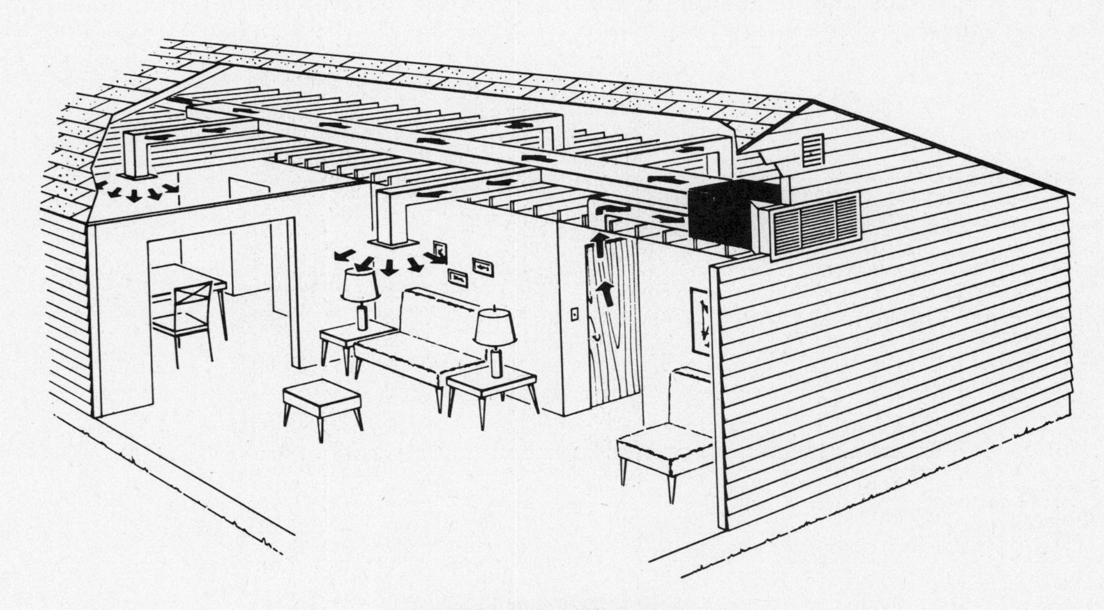

FIGURE 10-6 Attic installation.

On the roof

Rooftop installations are practical and economical. For a single room, one central air diffuser is adequate. A multiple duct installation is best for several rooms or spaces without an open floor plan.

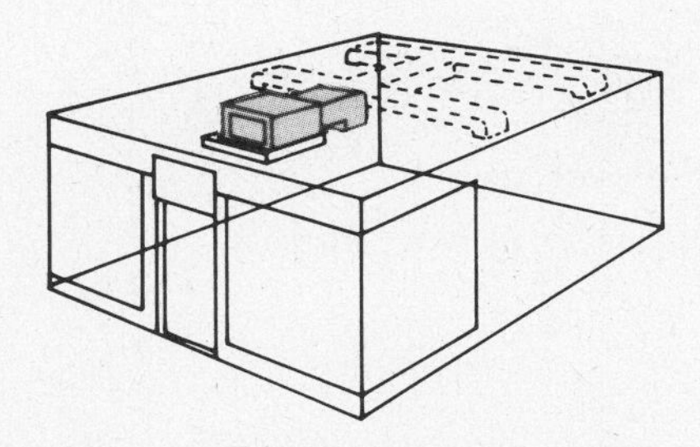

FIGURE 10-7 Roof top installation.

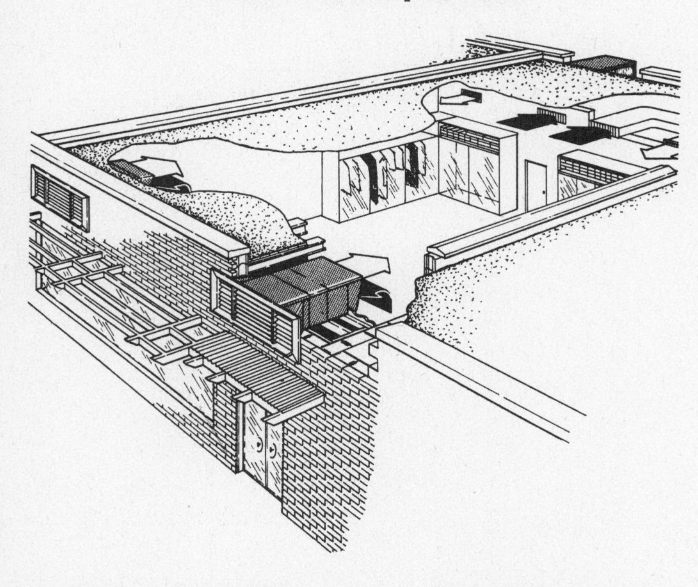

FIGURE 10-8 Commercial outside wall installation (Courtesy, Westinghouse Electric Corporation).

FIGURE 10-9 Ducted resistance heater (Courtesy, Lennox Industries).

front of the unit. Conditioned air is free-blown into the space out of the top grille and returned through the bottom louvers; ductwork is thus eliminated.

As mentioned at the beginning of Chap. 10, the classification of a unit as "cooling only" is not really accurate, because most manufacturers have provided internal space for the option of adding electric heating resistance coils to provide winter heating. These coils function much as noted for an electric furnace. Other manufacturers offer ducted resistance heaters (Fig. 10-9) that are installed in the main trunk duct and/or in branch runs. Either method will convert the basic cooling only unit to a full year-round comfort system for either residential or commercial use.

Many manufacturers have also adapted the basic packaged air-cooled unit to mobile home air conditioning (Fig. 10-10). An adapter box is placed on the front section. Flexible supply and return ductwork feeds under the structure to floor outlets and a return grille. The air conditioner mounts on a slab on the paved surface provided in a mobile home park. The system can be disconnected and transported to a new location without too much difficulty.

In commercial air-conditioning it is quite common that internal heat loads (due to lights, people, cooling, etc.), require cooling even when outdoor temperatures are at 35°F or lower. This presents problems for the standard refrigeration cycle. As outdoor temperatures fall, so does the refrigerant condensing pressure, until there may not be sufficient pressure drop across the expansion device (capillary tube or expansion valve) to feed the evaporator properly. The coil can freeze up, presenting a danger to the system and to compressor operation. The solution is to raise the head pressure. Several methods are used. Where only one condenser fan is present, its speed may be reduced by electrical speed controls, and thus air flow across the condenser coil is also reduced. The compressor head pressure rises as the condenser air quantity is reduced. Solid-state control devices are common.

Another popular method of low ambient control is the use of two or more condenser fans. Then it's only a matter of cutting off fan motors in stages until proper pressures are established. Many units have multiple condenser fans and can be successfully operated down to 0°F and below outdoor ambient.

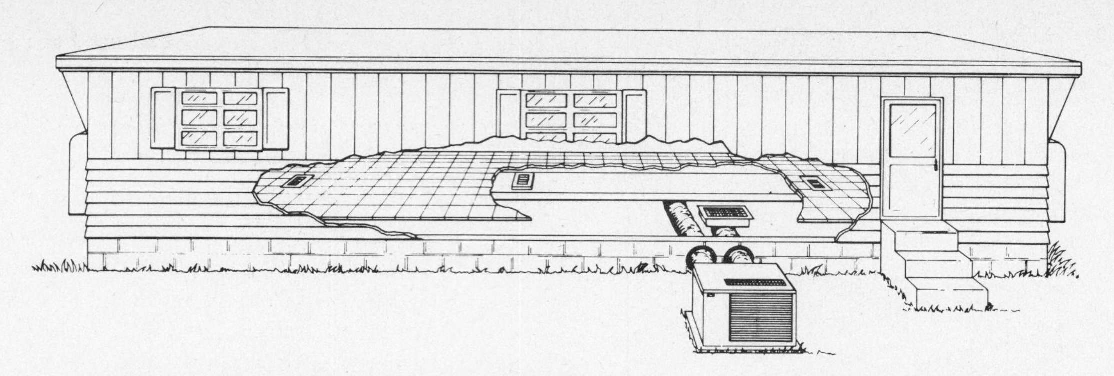

FIGURE 10-10 Mobile home air-conditioning (Courtesy, York, Div. of Borg-Warner Corporation).

(Note: It is assumed there is enough internal heat to keep evaporator coils above freezing.)

A third method of head-pressure control is to place dampers on the condenser air circuit, and so throttle the air flow across the coil. This system too is essentially automatically preset to maintain a minimum head pressure.

No matter which technique is used, it is important to understand that there is a difference between mild weather control (35°F and above) and true low ambient operation down to perhaps −20°F and below. Mild weather operation is standard on many units, while low ambient control is usually an optional extra.

The configuration of a horizontal rectangular cabinet is the most conventional design; however, there are variations that include vertical arrangements. Figure 10-11 depicts an air-cooled self-contained unit specifically designed for use in modernizing older high-rise buildings. The unit has been carefully designed to fit into elevators, to go through existing doors, and to be placed against outside window openings for the intake and exhausting of condenser air. With these limiting specifications, sizes range from 7½ to 20 tons; but when used in multiples, these units can effectively condition large areas.

Whether horizontal or vertical, the operation is essentially the same, and packaged air-cooled units continue to offer many advantages. They are completely factory assembled and tested, are relatively easy to install with minimum electrical and plumbing work, and need have little or no ductwork for simple air distribution.

FIGURE 10-11 Vertical air-cooled self contained unit (Courtesy, York, Div. of Borg-Warner Corporation).

Packaged Water-Cooled Units

The self-contained vertical water-cooled packaged cooling unit (Fig. 10-12) is one of the earliest approaches to commercial air conditioning and is still offered by several major manufacturers. Capacities range in sizes from 3 to 60 tons, with the range of 3 to 15 tons for in-space, free-standing application utilizing discharge and return grilles. Larger units are generally located out of the conditioned space with necessary ducting for air distribution.

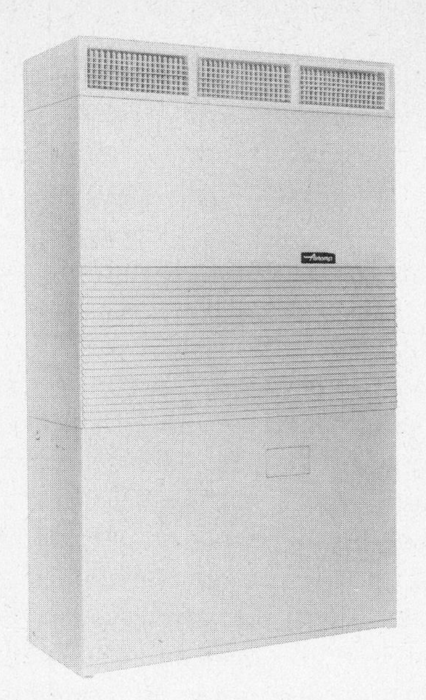

FIGURE 10-12 Packaged water-cooled unit (Courtesy, Airtemp Corporation).

FIGURE 10-13 Packaged water-cooled unit (Courtesy, Airtemp Corporation).

The original concept of operation (Fig. 10-13) is a water-cooled circuit, with the compressor and water-cooled condenser located in the bottom compartment. The center section houses filters and the fin-tubed cooling coil. (Note: Some units also allow space for the installation of a combination steam and hot-water coil.) The upper portion contains centrifugal blower(s) motor and the motor drive. Some units offer an optional fan discharge arrangement, depending on the application and space considerations.

The watei-cooled condenser may be piped for series flow city-water circuit, controlled by an automatic head-pressure water valve, or piped for parallel flow water tower operation. The temperature rise and the pressure drop through the condenser vary with the flow arrangement. City water supply can mean water from the city mains or from a well or lake source. The automatic water valve is set to regulate the proper condensing pressure. Water tower operation (Fig. 10-14), as reviewed in *Refrigeration*, is a function of tower selection, wet-bulb temperature, and flow rate. The normal design is to get water from the tower and to the condenser at about $7\frac{1}{2}$°F approach to the wet bulb. So, if the design wet-bulb temperature was 78°F, the cold water to the condenser would be 85°. The water out of the condenser is usually based on a 10° rise and thus would be 95°F. The gal/min flow rate at these conditions is determined from the tower performance data. Pump selection is a function of the water flow (gal/min) and the resistance head (expressed in feet) for the piping, condenser, and tower.

Whether on city water or water tower operation, the efficiency of the condenser operation is affected by water scale and other contamination. Tube scaling reduces the heat flow between water and refrigerant, and reduces the effective pipe diameter as well cutting water flow and increasing the pumping head. Towers are also affected by dust, dirt, and algae and slime formations. Proper water treatment is necessary and regular maintenance is a must. Cleaning procedures and chemicals are suggested by

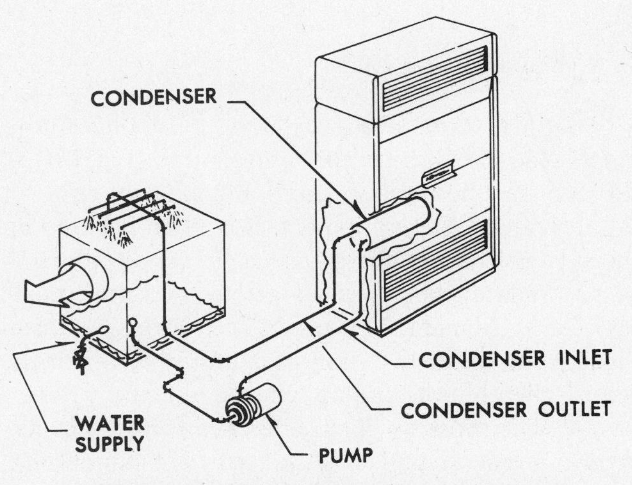

FIGURE 10-14 Typical water tower application.

the air-conditioning equipment manufacturers, as well as by water treatment companies.

Some localities do not allow the use of city water for air-conditioning; also, wastewater can be expensive. In addition, some users don't wish to have the service problems associated with water towers; however, they do like the design of a vertical self-contained air conditioner. As a result, manufacturers offer models without the water-cooled condenser, permitting application of a remote air-cooled condenser (Fig. 10-15). The compressor stays with the air conditioner. Liquid and discharge lines are run between indoor and outdoor units. This section functions like an air-cooled system, but the installer must be skilled in running extended refrigeration lines, installing proper oil traps and pipe insulation, and dealing with noise considerations.

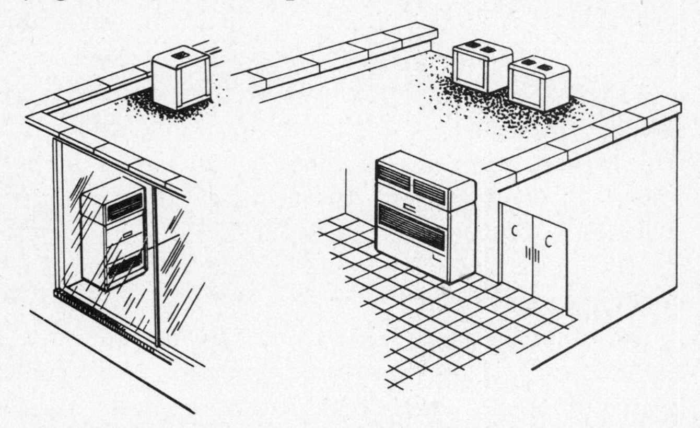

FIGURE 10-15 Typical air-cooled application (Courtesy, Westinghouse Electric Corporation).

Computer Room Unit

Another form of single package cooling equipment is the so-called computer room unit (Fig. 10-16), for use in the precise environmental control of sophisticated EDP equipment. Its design is based on the fact that computer rooms have raised floors to accommodate the maze of electronic connections between equipment. This raised floor forms a natural plenum for the distribution of conditioned air or the return thereof, depending on the specific product design. The most obvious difference from other air-conditioning equipment is the attractive cabinet that must relate to the highly styled surrounding computer equipment. High-efficiency filters and built-in humidification are most important for the clean air and constant humidity conditions required. Equipment sizes range from 3 to 15 tons. At $7\frac{1}{2}$ tons and above, the units generally employ multiple compressor circuits for higher capacity reduction and more accurate control.

Computer rooms must frequently operate on a 24-hour schedule all year long, and thus the air-conditioning equipment must also operate under all weather conditions. Thus the selection and operation of the condensing function is carefully planned. Some units use a remote air-cooled refrigerant condenser (as shown) that employs low ambient fan speed modulation for head-pressure control, along with the staging of multicircuit compressor operation. Units are also available with conventional water-cooled condensers where city water or tower operation can be used without danger of freezing. Where freezing is a problem, some manufacturers offer a form of remote air-cooled condenser that uses a closed loop system and circulates antifreeze liquid (glycol) instead of water. It is pumped to the computer unit condenser.

Since reliability and service is mandatory to minimize down time of EDP computers, these units are top shelf in terms of quality and ease of access. Operational and warning lights permit constant monitoring of the system.

Unitary Split-System Cooling

The development of split-system air-conditioning for both residential and commercial applications is related to certain limitations of the packaged units. In residential work there are millions of American homes that have existing furnaces and duct systems suitable for the addition of cooling, but to tie in a packaged cooling unit and to use the common ducts presents difficult and expensive dampering and control problems. This is particularly true where the furnace is centrally located in a closet or utility room.

In many apartments, it's very difficult to find outdoor space to put packaged units except on the roof; running ducts down several stories is not a practical and economical consideration when compared to running refrigerant lines. The same situation is true for many multifloor commercial buildings. So, to meet the many variations in construction techniques and to make use of existing ductwork, the

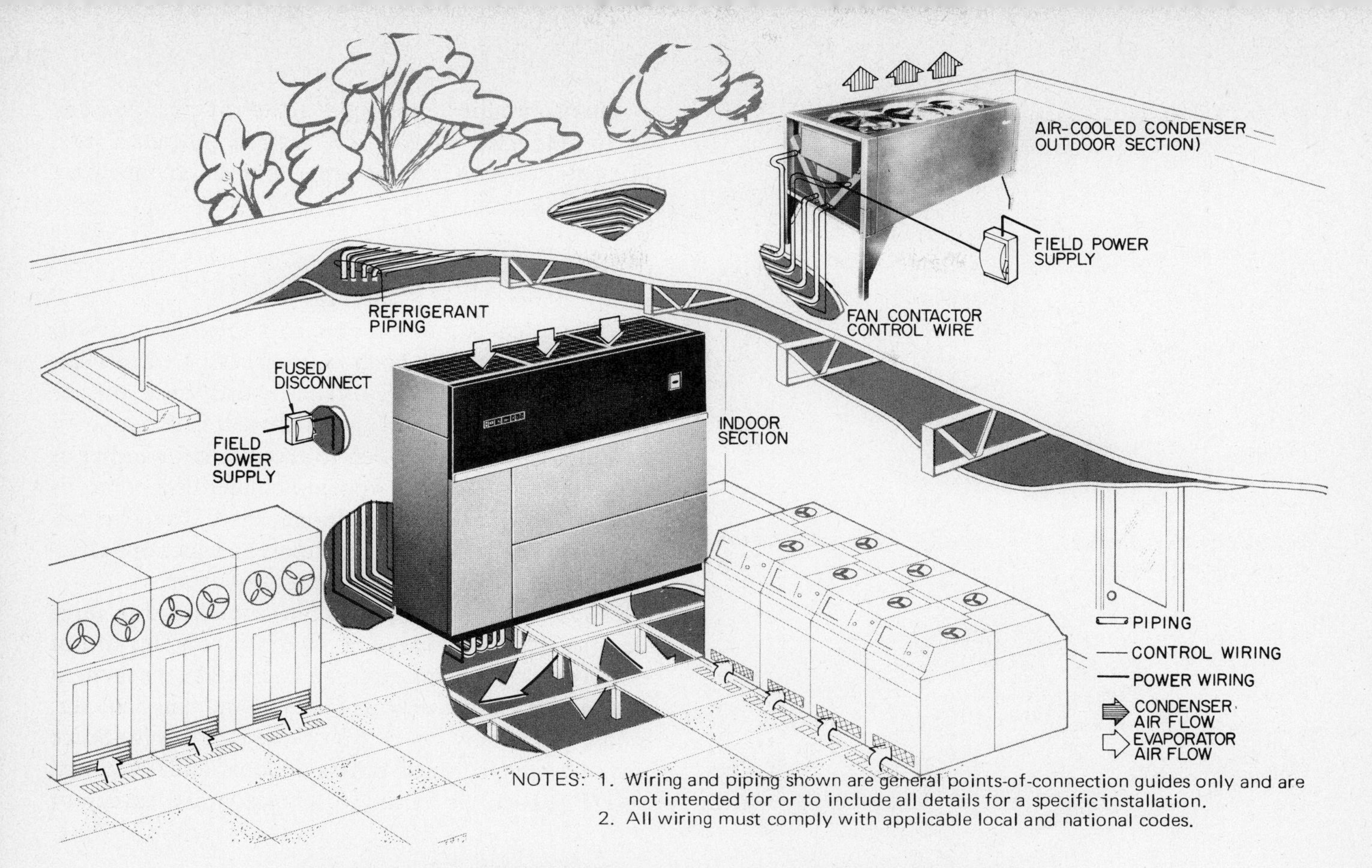

FIGURE 10-16 Computer room unit (Courtesy, Carrier Air-Conditioning Company).

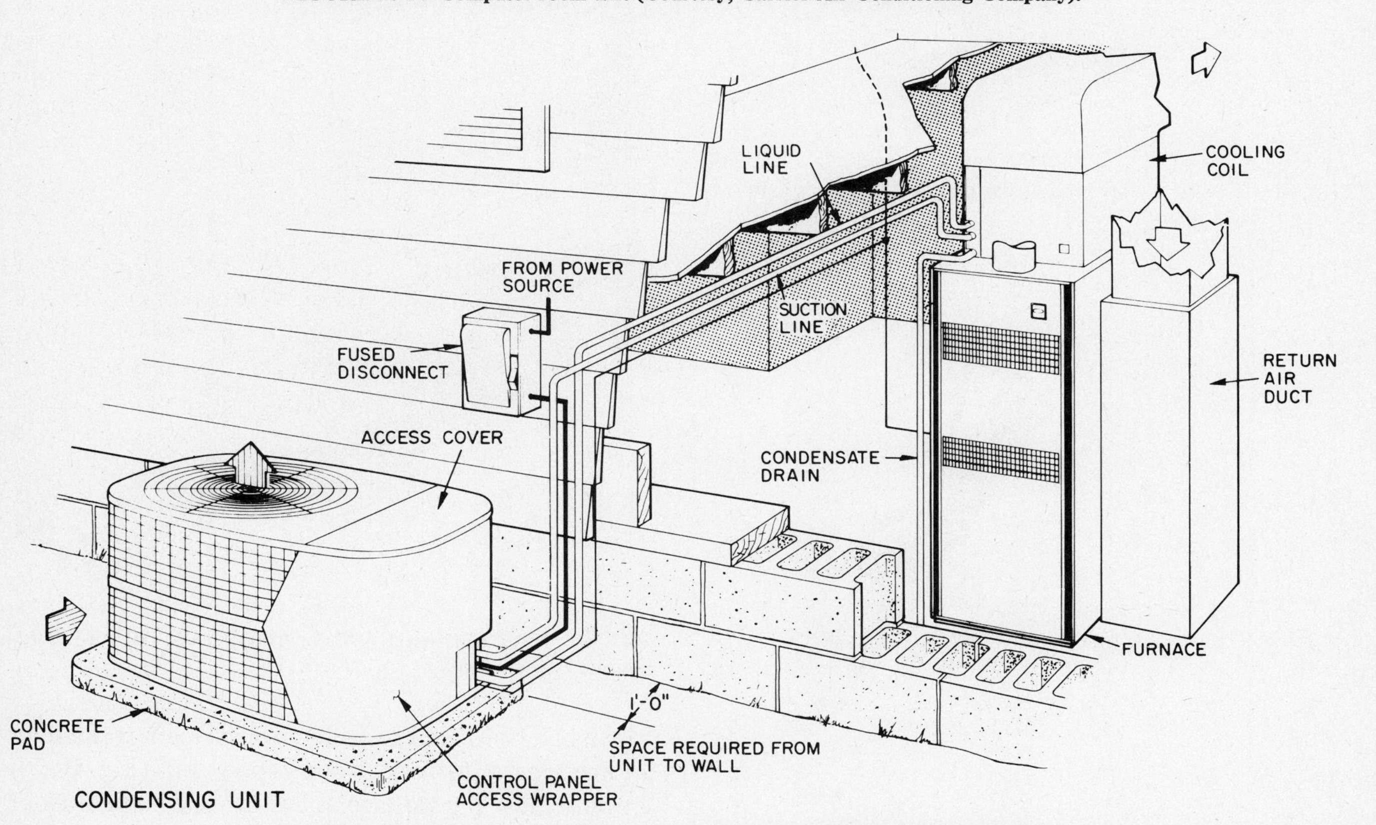

FIGURE 10-17 Add-on cooling system (Courtesy, Carrier Air-Conditioning Company).

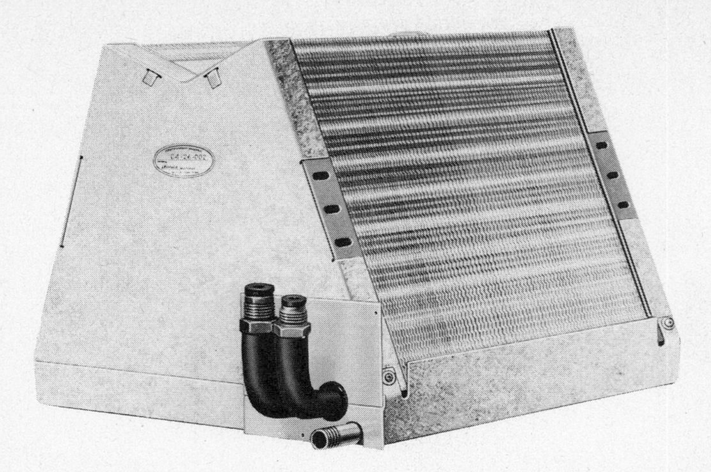

FIGURE 10-18(a) Lennox C-4 series upflow cooling coil (Courtesy, Lennox Industries).

FIGURE 10-18(b) Upflow coil (Courtesy, York, Div. of Borg-Warner Corporation).

FIGURE 10-18(c) Typical two-ton slant coil (Courtesy, Airtemp Corporation).

industry developed the split system, which consists of an indoor cooling section (with or without a fan), an outdoor condensing unit, and interconnecting refrigeration piping.

Residential Split Systems

The so-called add-on cooling system (Fig. 10-17) accounts for a huge potential market in modernization work. Where the furnace exists (gas, oil, or electric) and the size of the ductwork is adequate, a cooling coil may be added to the discharge outlet of the furnace. An air-cooled condensing unit is located outdoors on a suitable base. The two are connected by properly sized liquid and suction refrigerant lines. Let's examine each component.

Add-on cooling coils (no fan) are offered in three designs. Upflow coils (Fig. 10-18) mount on top of upflow or lowboy furnaces, and most take the form of the letter **A**; however, there are flat and diagonal versions, which may cut down on the coil height. Some are cased in a cabinet; others are not. The horizontal coil (Fig. 10-19) is for use with horizontal furnaces, or

FIGURE 10-19 Horizontal coil (Courtesy, York, Div. of Borg-Warner Corporation).

else it can be installed in a horizontal main trunk duct of any type of furnace. Counterflow coils (Fig. 10-20) install below the furnace and must offer a cabinet of sufficient strength to support the furnace. The coun-

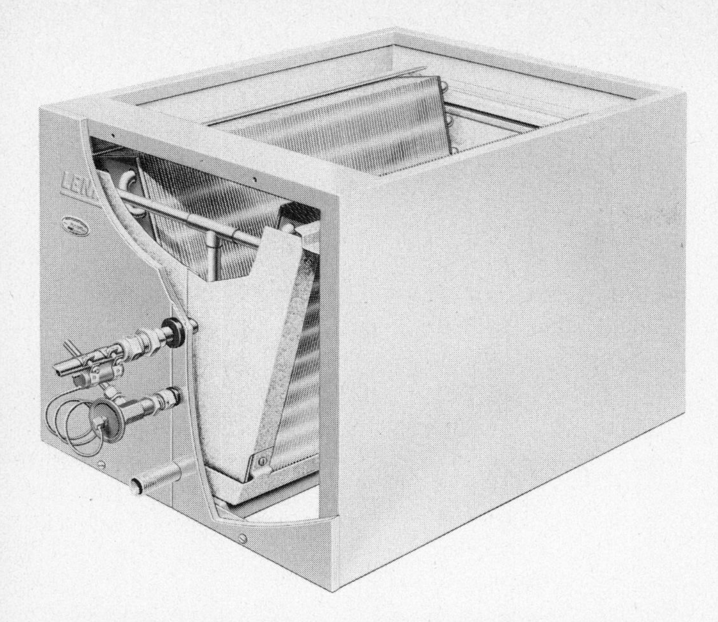

FIGURE 10-20 Counterflow coil (Courtesy, Lennox Industries).

terflow coil casing must also have proper clearances from combustible surfaces, for it is used for heating air distribution as well.

In all three instances the coils of a particular manufacturer are usually designed to match their furnace outlets so as to simplify installation and eliminate costly ductwork transitions.

Coil cabinets are insulated to prevent sweating, and all models have adequate drain pans for collecting condensate runoff. Low-cost plastic hose may be used to pipe condensate drain water to the nearest sink or open plumbing drain in the house. If no nearby drain is available, a small condensate pump may be installed to pump the water to a drain or perhaps even to an outdoor disposal arrangement.

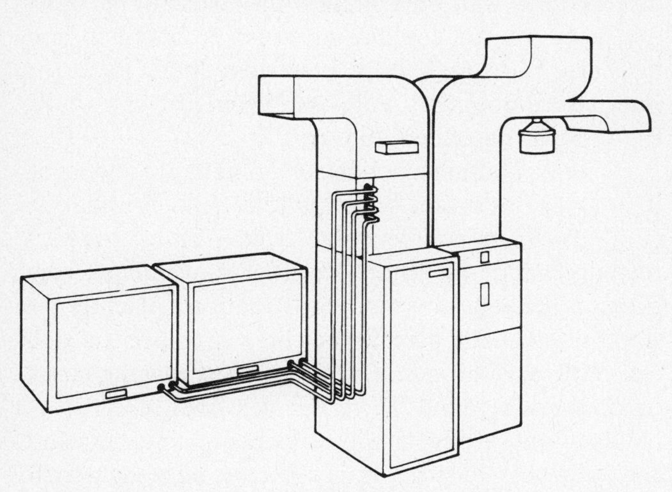

FIGURE 10-21 Split circuit coil (Courtesy, Westinghouse Electric Corporation).

An important consideration in applying an add-on cooling coil to an existing furnace is the air resistance it adds to the furnace fan. During the heating cycle, the coil is inactive and thus is *dry*. In summer, when the unit is cooling and dehumidifying, the coil is termed *wet*. On the average, the wet coil will add 0.20 in. to 0.30 in. of water column of static pressure loss. This added resistance will be sufficient to require a change in furnace fan speed; also, an increase in motor horsepower will probably be needed. If the installation is new, then the air conditioning contractor will select a furnace fan capacity capable of producing sufficient external static.

A variation of the add-on cooling coil offered by a few manufacturers is the split-circuit coil (Fig. 10-21), where two condensing units are employed and controlled by a two-stage thermostat for capacity control. The evaporator coil is row split rather than face split, and so when operating on only one unit all the air is conditioned. Dual circuit systems are considered a more deluxe application and are most frequently used in the 4 and 5-ton range for larger houses.

Continuing with the add-on concept, we next examine the outdoor air-cooled condensing unit (Fig. 10-22). It consists of a compressor, condenser

FIGURE 10-22 Outdoor air cooled condensing unit (Courtesy, Lennox Industries).

coil, a condenser fan, and the necessary electrical control box assembly. On residential condensing units, the full hermetic compressor is universally used. It is sealed from dust and dirt and requires little ventilation. The condenser coil is a finned-tube arrangement. The rows of depth and face area are a function of the particular manufacturer's design. On the condenser a large area is desirable, and thus you will see many units that offer almost a complete wraparound coil arrangement to gain maximum surface area.

The configuration of the condenser fan also depends on each manufacturer's design. Most utilize a propeller fan, because it can move large volumes of air where little resistance is offered. This is another reason why the condenser coil tube depth is limited so as not to create added resistance. Few air-cooled condensing units are designed for ducting, but some are available and these use centrifugal blowers. (Later we will see that these designs are meant to be used with heat pumps that require more positive air flow on defrosting.)

Air-flow direction becomes a function of the cabinet and coil arrangement, and there is no one best arrangement. Most units, however, do use the draw-through operation over the condenser coil. Outlet

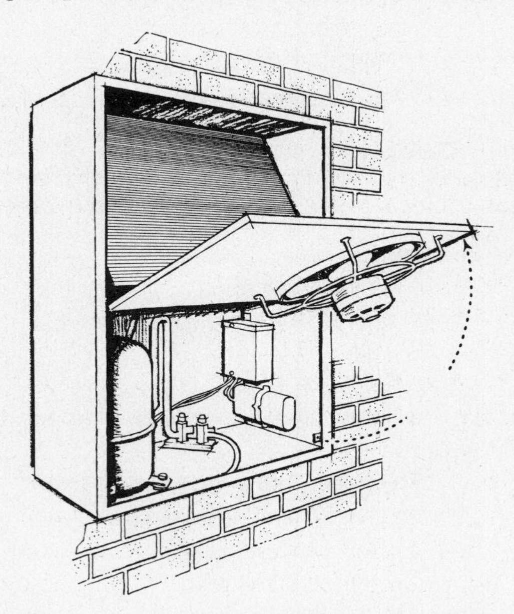

FIGURE 10-23 In-wall residential condensing unit (Courtesy, Tappan Air-Conditioning Company).

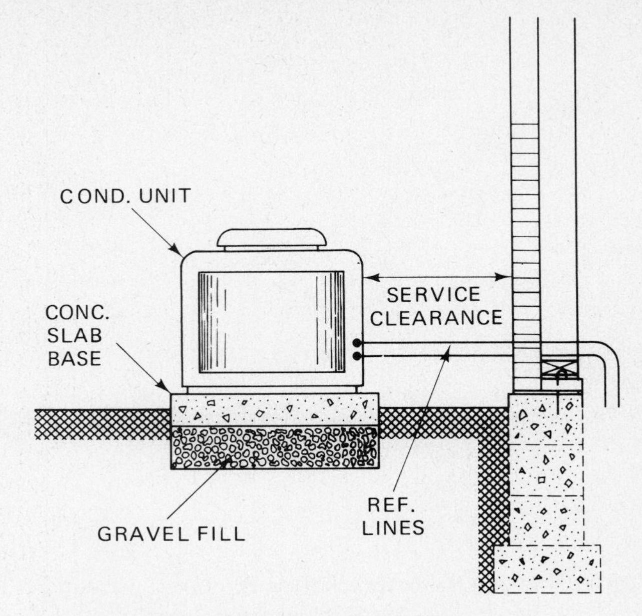

FIGURE 10-24 Condensing unit at ground level.

air, in its direction and velocity, can have an effect on surrounding plant life. Top discharge is the most common arrangement. The fan motors are sealed or covered with suitable rain shields. Fan blades must be shielded so as to offer protection for prying hands and fingers.

The foregoing description of an add-on condensing unit applies to a free standing configuration, where there is walk space around the unit. But we must pause to acknowledge that there are residential condensing units (Fig. 10-23) that are primarily designed for in-wall mounting in new construction. An apartment may have a condensing unit installed in a balcony wall or a small utility closet, and obviously the air for condensing must be taken in and discharged through only one side. Otherwise, the internal components and operation are essentially the same as described above.

When installing a condensing unit at ground level, it is most important to provide a solid foundation (Fig. 10-24). A concrete slab over a fill of gravel is recommended to minimize the movement from ground heaving. In new construction, placing the slab on new back fill may not be wise, since the slab can settle several inches or more, thus placing stress on or even breaking the refrigerant lines.

Roof mounting small condensing units doesn't generally present structural problems, since the weight per square foot is relatively low. Larger commercial units will, however, require consideration of the

FIGURE 10-25 Outdoor condensing unit.

support, and it may be necessary to provide steel rails spanning several bar joists.

Location of the outdoor condensing unit (Fig. 10-25) is a compromise among several factors; the actual available space, the length of run for refrigerant lines, the aesthetic effects on the home or landscaping, and sound considerations as they relate to the home or to neighbors. In general, avoid placement of the unit directly underneath a window or immediately adjacent to an outdoor patio. Do try to locate it where plants and shrubs may screen the unit visually, as well as provide some sound-absorbing assistance.

Noise pollution, along with air pollution, is gaining widespread attention. Outdoor air-cooled air conditioning is a mechanical device and does produce sound levels that cannot be ignored. As a matter of fact, in the 1960s certain localities, such as Coral Gables, Florida, enacted local codes to help curb the compounding noise pollution of air conditioners, particularly from window units. Sound is a difficult subject to understand. It is hard to measure and even harder to evaluate and is frequently misunderstood by local code makers.

However, these early attempts to create local ordinances prompted action on the part of the ARI to establish an industry method of equipment sound rating and application standards (which local communities could adopt), setting forth acceptable sound standards.

Thus ARI Standard 270 was created in 1971 with the publication of its first directory of sound-rated outdoor unitary equipment. Under the program, all participating manufacturers are required to rate the sound power levels of their outdoor units in accordance with rigid engineering specifications. They must

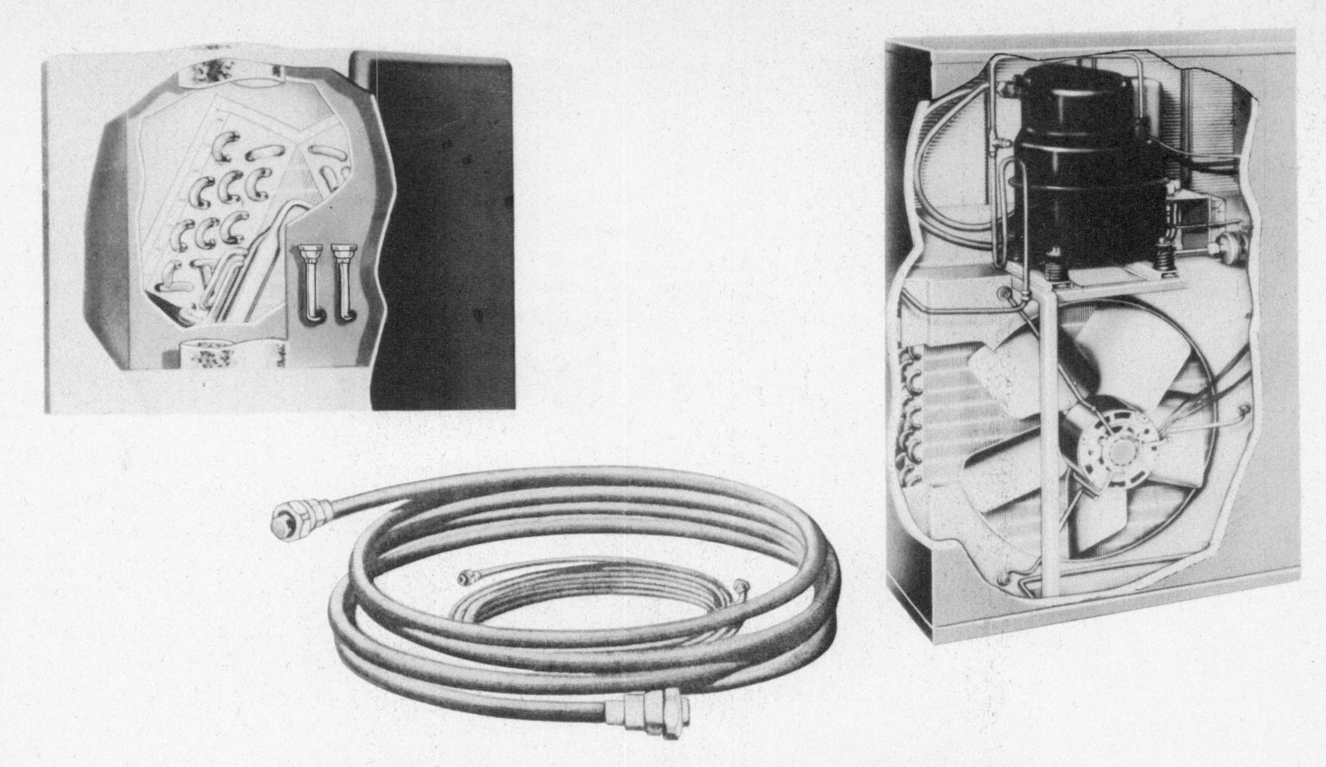

FIGURE 10-26 Precharged refrigerant lines (Courtesy, Aeroquip Corporation).

submit test results on all their units to ARI engineers for review and evalution and make any of their units available for sound testing by an independent laboratory. Units are sound-rated with a single number—the Sound Rating Number (SRN). Most units will be rated between 14 to 24 on the ARI sound-rating scale. The ARI anticipates that the sound-rating program will guide and encourage manufacturers to produce quieter units in the future.

Along with equipment ratings, ARI Standard 275 sets forth recommended application procedures for using a sound-rated unit SRN to predict and, through the design of the system, control the sound level at a given point. The application procedures are worked out so that installation contractors can use them to properly predict and control the sound level. A technician involved in selling and installing outdoor units must become familiar with these standards and use the recommendations accordingly. These standards and recommendations apply to both packaged and split-system outdoor equipment.

Refrigerant piping between the indoor coil and the outdoor condensing unit can take several forms. On basically residential-type equipment, non-soldered methods are used. As mentioned in Chap.1, the development of flexible, quick-connect, precharged refrigerant lines (Fig. 10-26) played a major role in the growth and reliability of home air conditioning. The liquid and suction lines are made of bendable copper or convoluted steel tubing. Suction lines are factory covered with a foam-rubber insulation material. Each end of the tubing is fitted with a coupling half that mates with a coupling half on the equipment (Fig. 10-27). Both coupling halves have diaphragms that provide a seal, which prevents refrigerant loss before connection. The male half (on the equipment) contains a cutter blade, the metal refrigerant sealing diaphragm, and an intermediate synthetic rubber seal to prevent a loss of refrigerant while the coupling is being connected. The female half (on the tubing) contains a metal diaphragm, which is a leak-proof metal closure. Tightening the

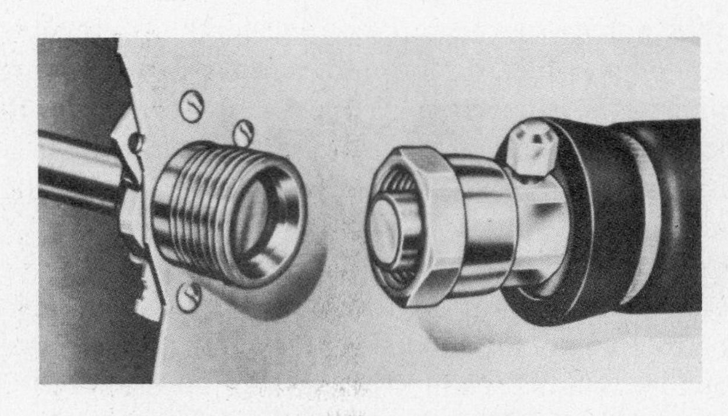

FIGURE 10-27 (Courtesy, Aeroquip Corporation).

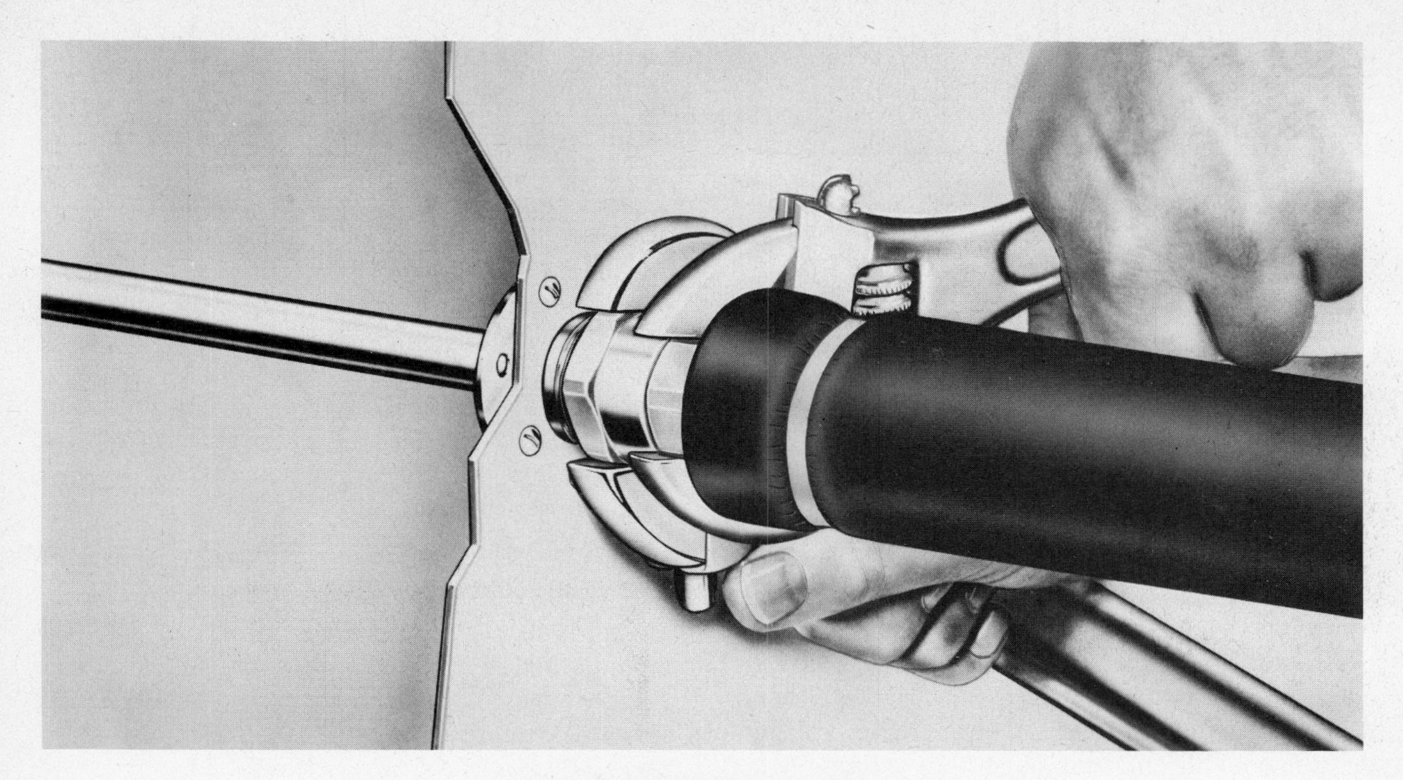

FIGURE 10-28(a) (Courtesy, Aeroquip).

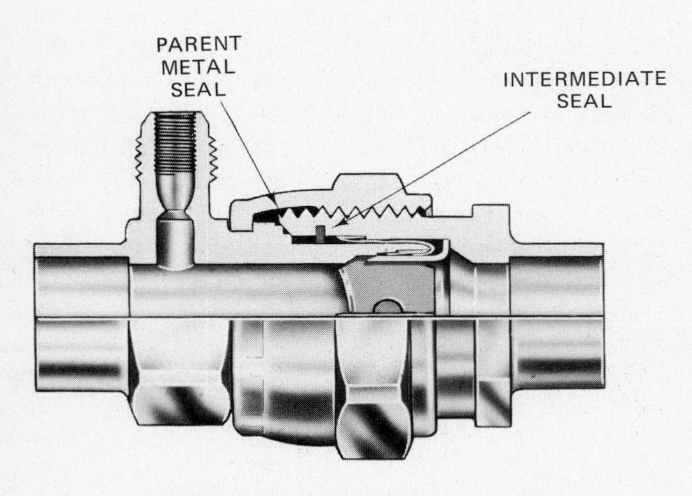

FIGURE 10-28(b) (Courtesy, Aeroquip).

union nut draws the coupling halves together (Fig. 10-28a), piercing and folding both metal diaphragms back, and opening the fluid passage. When fully coupled (Fig. 10-28b) a parent metal seal forms a permanent leak-proof joint between the two coupling halves. Note the service port for checking refrigerant pressure. The port is equipped with a valve (not shown), similar to a tire valve, that opens when depressed by gauge lines. When installing quick-connect precharged refrigerant lines, follow the manufacturer's installation directions on the radius of the bend for the size of tubing, lubrication of the coupling, and proper torque values when couplings are joined together. If excess tubing is present, form a loop or coil in a flat, horizontal manner; do not form vertical loops, which will create an oil trap. Precharged lines are manufactured in various sizes and lengths from 10 to 50 feet. It's necessary to plan the installation in order to have sufficient but not excessive tubing, which adds cost and pressure drop to the system. The coupling described above is called a *permanent, one-shot connection*, since the coupling cannot be undone and reused. There are, however, more expensive spring-loaded designs that do permit uncoupling, reuse, and self-sealing for service reasons.

Another method of mechanically sealing refrigerant tubing is the compression fitting technique (Fig. 10-29). The refrigerant tube is first cut square and deburred. A coupling nut is slipped on the tubing, followed by a compression **O** ring. When mated with the male adapter, the **O** ring is compressed to form a tight and leak-proof seal. The male adapter on the air-conditioning unit is also part of a service valve assembly. Refrigerant tubing is shipped in coils that are clean and dehydrated. Removal of the seals and piping connection should be made as quickly as possible to minimize moisture condensation. Piping connections are first made hand-tight. By cracking

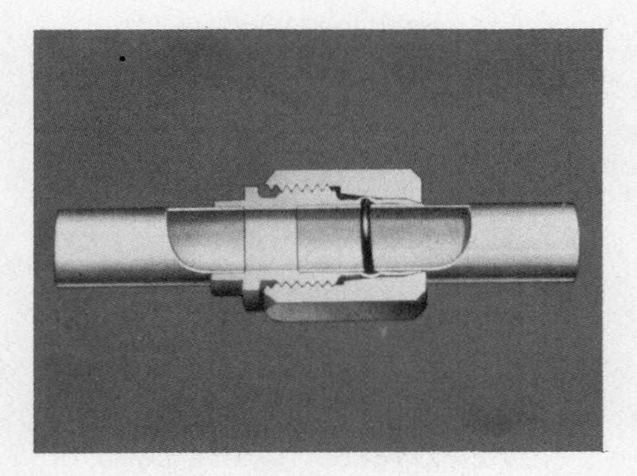

FIGURE 10-29 Compression fitting (Courtesy, Carrier Air-Conditioning Company).

open the service valves, refrigerant pressure is allowed to pressurize the tubing, thus purging air and moisture out through the hand-tight fittings. After purging, the compression fittings are securely tightened to form a leak-proof joint.

The third form of mechanical coupling is the *flare connection*, which has been widely used on refrigeration systems for years. Flare connections are generally limited to tubing sizes of about $\frac{3}{4}$ in. outside diameter, and the installer must be skilled in making the flare. The tubing kit is factory-cleaned and capped at both ends to minimize dirt and moisture penetration, but it is not precharged with refrigerant. In the process of making the connections, if the system is open to the atmosphere for more than five minutes, a complete procedure of purging and leak testing and full evacuation and recharging will be needed.

In the case of quick-connect couplings, compression fittings, and flared fittings, no field soldering is needed, and pumpdown, evacuation, and refrigerant charging is not usually required. The condensing unit and evaporators are factory-sealed. The condensing unit is also factory-charged with refrigerant for the complete system up to about 25 feet. On-site labor, tools, time, and mechanical skills are greatly reduced. All, however, have the common disadvantage of not being practical or economical in large pipe sizes. The limiting size of soft-drawn copper tubing that can be formed or bent is about $\frac{7}{8}$ in. O.D. Thus, single compressor precharged systems stop at the nominal 5 to $7\frac{1}{2}$ ton range. Unless dual circuits are employed the large systems must go to the field-fabricated and soldered lines.

In addition to the physical limitations of handling larger pipe sizes and because of refrigerant characteristics, split-system piping has limitations in both the length of line and the height of refrigerant lift, which must be observed by the technician (Fig. 10-30). When the evaporator is located above the condensing unit, there is a pressure loss in the liquid line due to the weight of the column of liquid and the friction of the tubing walls. The friction and weight of the refrigerant column may be interpreted in terms of pressure loss in pounds per square inch (psi), as shown in the following table for Refrigerant 22.

Vertical Lift in Feet	5	10	15	20	25	30
Static Pressure Loss PSI	$2\frac{1}{2}$	5	$7\frac{1}{2}$	10	$12\frac{1}{2}$	15

Thirty feet of vertical lift is considered the maximum limitation in normal use. If the pressure loss is great enough, vapor will form in the liquid line prior to normal vaporization in the evaporator, and operation and capacity are adversely affected.

Line lengths and the separation between a condenser and an evaporator using standard precharged refrigerant lines is 50 feet, or less if so noted by the manufacturer. Seventy-five feet is considered the maximum separation and usually requires the addition of refrigerant. Where the evaporator is above the condenser, pitch the suction line toward the condenser at least 1 inch in 10 feet of run to insure proper gravity oil return. (More information on refrigerant line application will be covered when we discuss commercial split system cooling.)

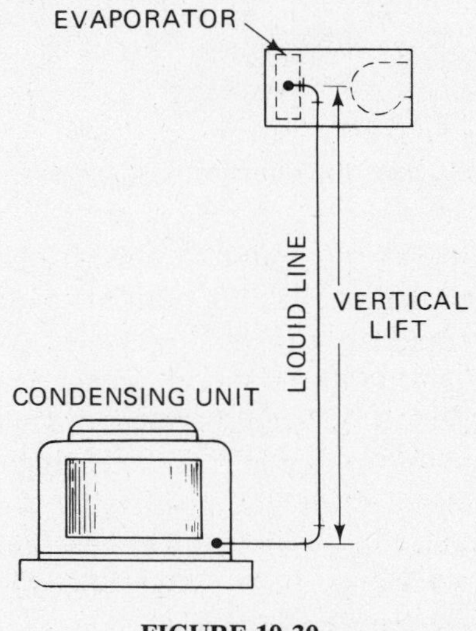

FIGURE 10-30.

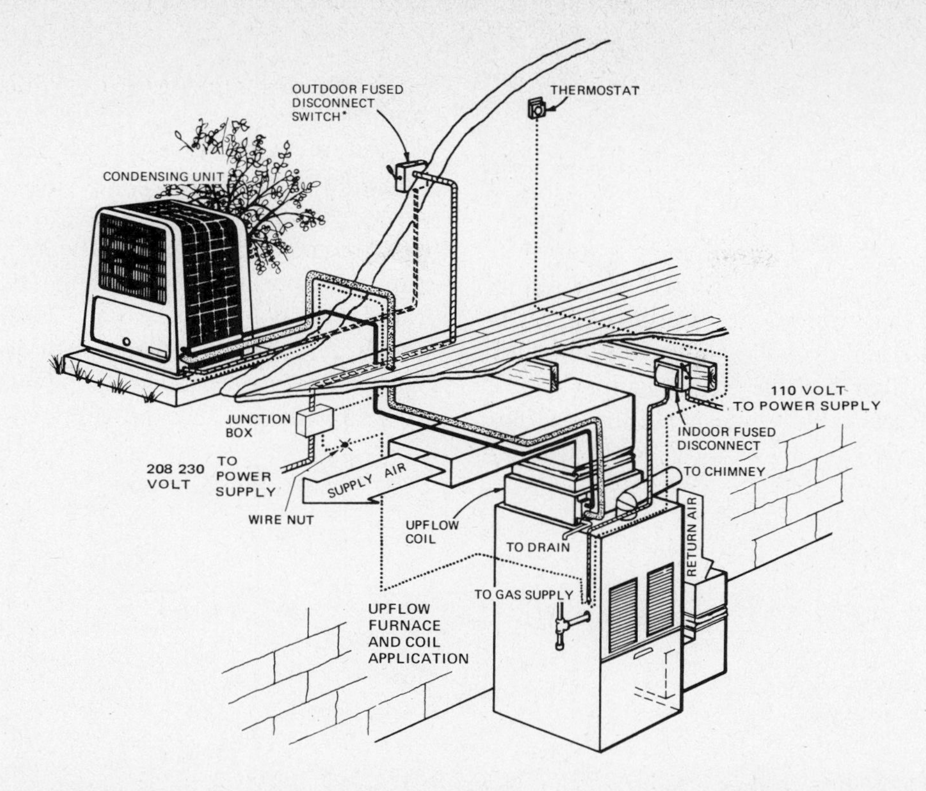

FIGURE 10-31 Single circuit add-on split system (Courtesy, York, Div. of Borg-Warner Corporation).

Wiring and piping a typical single circuit add-on split system is relatively simple, as shown diagrammatically in (Fig. 10-31). The electrical wiring consists of a line voltage (208/240 volt) to the outdoor condensing unit through a properly fused outside disconnect switch. The line voltage (usually 110 volts) is supplied to the furnace, which furnishes power to the fan as well as for the low-voltage (24-volt) secondary "off" transformer. A 24-volt combination heating and cooling room thermostat controls the "on-off" of cooling or heating through interlocking relays. The practice of using permanent split capacitor motors in small hermetic compressors for residential work is included in the electrical review. It is assumed that refrigeration pressures equalize during the "off" cycle, and thus the compressor is not required to start against a load. The auxiliary winding remains energized at all times during motor operation. The running capacitor is added to provide additional torque both during starting and while the unit is running, but in cases where low-voltage fluctuation exists, it may be necessary to add a starting capacitor for "extra kick." Hard-start kits, which essentially consist of a starting capacitor to be field applied, are offered to overcome low-voltage starting problems and minimize light flicker.

The sizes of residential add-on split systems run from 1 ton to 7½ tons, with the bulk of sales in the lower capacities, from 2 to 3 tons. Models are usually available in elements of 6,000 Btu/h; i.e., 12,000 Btu/h, 18,000, 24,000, 30,000, and so on. Most manufacturers rate and certify their equipment in accordance with ARI Standard 210, which is based on mating specific sizes condensing units and cooling coils. This combination rating is published in the ARI directory. Some manufacturers do publish mix-match ratings, which means selecting a different combination of condensing unit and cooling coil. This most notably occurs in the southwest part of the country, which has low humidity but high temperatures and thus needs a higher air flow and sensible cooling ratio, even though the condensing unit total capacity is not changed. These mix-match combinations may or may not be ARI-certified.

Fan-Coil Units

Not all residential split systems are of the add-on variety for existing homes. Many are installed in new home construction. Many are also planned for apartments and motels, where the indoor element is a fan-coil unit rather than a furnace. Figure 10-32 is a ceiling evaporator blower designed to be installed in a dropped-ceiling application in furred areas above closets, hallways, bathrooms, etc., where room air is returned through a ceiling grille and conditioned air is discharged through a grille into a single room. These systems are basically not designed for ducting. Complete cabinets are eliminated to reduce weight and cost.

As noted previously, it is common to refer to a "cooling system" but then acknowledge it can also be equipped for heating. Ceiling evaporator blowers in apartments and motels can be so supplied with electric resistance heaters. Cooling sizes of ceiling evaporator blowers range from 12,000 Btu/h to 30,000 Btu/h and are installed with quick-connect, precharged refrigerant lines. Condensate drain lines are treated with care, because an overflow could result in

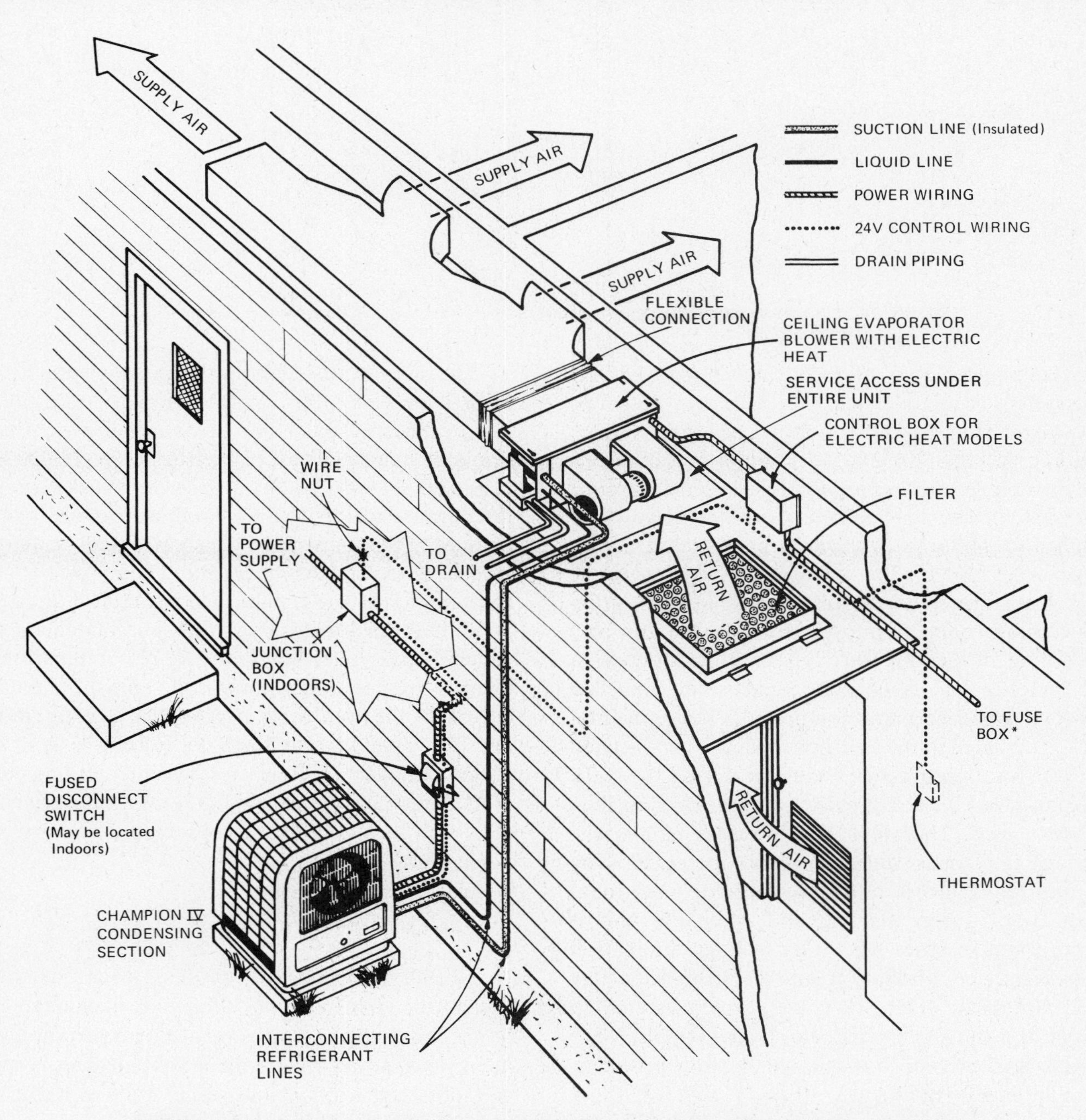

FIGURE 10-32 Ceiling evaporator blower system (Courtesy, York, Div. of Borg-Warner Corporation).

FIGURE 10-33 Evaporator blower air-conditioning unit (Courtesy, Airtemp Corporation).

damage to the ceiling. Safety is recommended and required in some locations.

Larger evaporator blowers for homes and commercial use (Fig. 10-33) are available in sizes from $1\frac{1}{2}$ tons to $7\frac{1}{2}$ tons and are complete with total cabinet enclosure. These are also commonly called *air-handling units*. Models for horizontal application are suspended in a basement or floor-mounted in an attic. Vertical models for upflow or counterflow are easily adapted to closet installation. As mentioned above, these also can be equipped with electric resistance heaters for winter heating. The centrifugal blowers provide ample external static pressure for a complete air distribution system; optional motor speed control permits varying the air flow. Frequently these units are used in small commercial applications with ductwork or with an accessory discharge plenum for free-blow air distribution.

UNITARY SPLIT-SYSTEM COOLING

Commercial Split Systems

Up to $7\frac{1}{2}$ tons there is no distinct rule to classify a piece of equipment as commercial or residential, for there is a wide degree of application in both markets, using the same products. However, above $7\frac{1}{2}$ tons the application becomes distinctly commercial, and the product designs use different component standards.

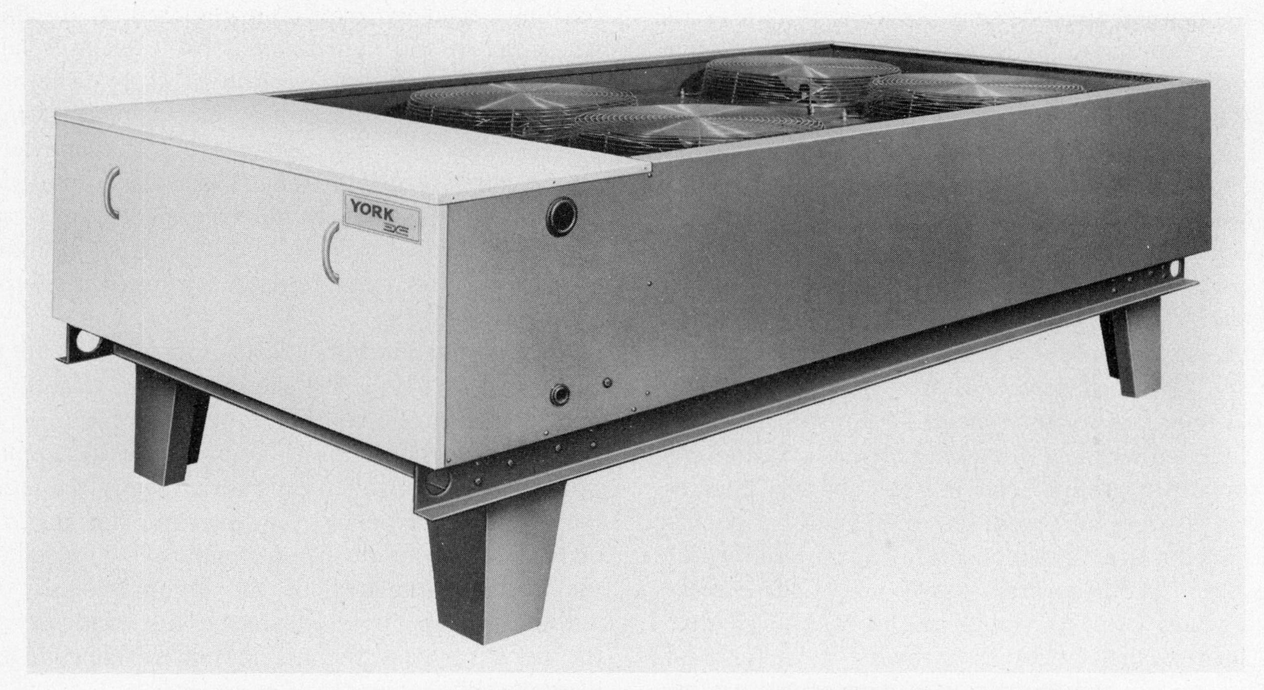

FIGURE 10-34 Typical condensing unit design (Courtesy, York, Div. of Borg-Warner Corporation).

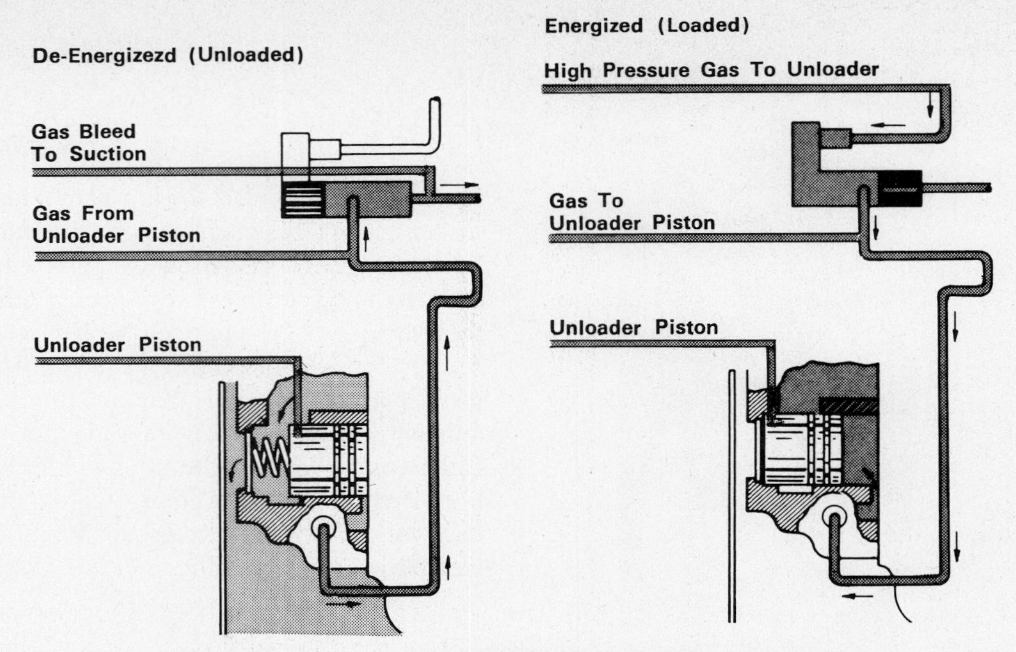

FIGURE 10-35 Cylinder unloader (Courtesy, Westinghouse Electric Corporation).

The first notable difference other than size and capacity is the use of soldered refrigerant lines and expansion valves rather than capillary tubes. Capacity reduction methods and low ambient operation are improved by virtue of multiple compressors and condenser fans. Structural design becomes somewhat more functional and heavier, with less emphasis on appearance. Compressors all have three-phase electrical characteristics.

Commercial split systems range from $7\frac{1}{2}$ tons to 100 tons and above, and the condensing unit design most often used is similar to that in Fig. 10-34, where air discharge is vertical and inlet air is from the sides or bottom. Vertical discharge has several advantages. First, these units move large volumes of air and the best way to avoid blowing against obstructions or creating a disturbance is to blow straight up. Second, the horizontal arrangement of fan blades is least affected by windmilling if the unit is off and then tries to start with its blades rotating in the wrong direction. The use of direct-drive multiple propeller fans is almost universal in this type of equipment. Ducting is not needed when equipment is normally located on the roof or at ground level in open space. Capacity control is necessary because commercial air-conditioning loads are rarely constant. Capacity reduction is accomplished in several ways. When multiple compressors are used, as would be the case for higher tonnages, it then becomes a matter of staging or sequencing the operation by cutting compressors on and off to match the load. Where single compressors are used (and even in multiple arrangements), they can be equipped with cylinder unloaders to vary the pumping capacity down 25% or less. In other words, the machine starts almost unloaded, and the cylinders are cut in as the heat load demands. Unloaded starting reduces power demand.

An example of a cylinder unloader (Fig. 10-35) consists of a gas bypass valve in the cylinder head that is opened by a spring and closed by discharge gas pressure, and that is actuated by solenoid valves that are responsive to suction pressure. The compressor is always unloaded at start-up for maximum power economy. The decrease in horsepower is almost directly proportional to capacity reduction. By the use of this method, the capacity in a compressor is limited only by the number of cylinders. Other unloaders use oil pressure to open the bypass valve.

Another method of capacity control is the hot gas bypass (Fig. 10-36). The hot gas bypass valve in the bypass line can be controlled by temperature or gas pressure, depending upon the nature of the application. As the controller calls for capacity reduction, the bypass valve opens, allowing some hot gas to go directly to the coil distributor manifold. This reduces the effective capacity of the compressor by the amount of gas bypassed around the condenser. In the hot gas bypass, there is no appreciable change in horsepower.

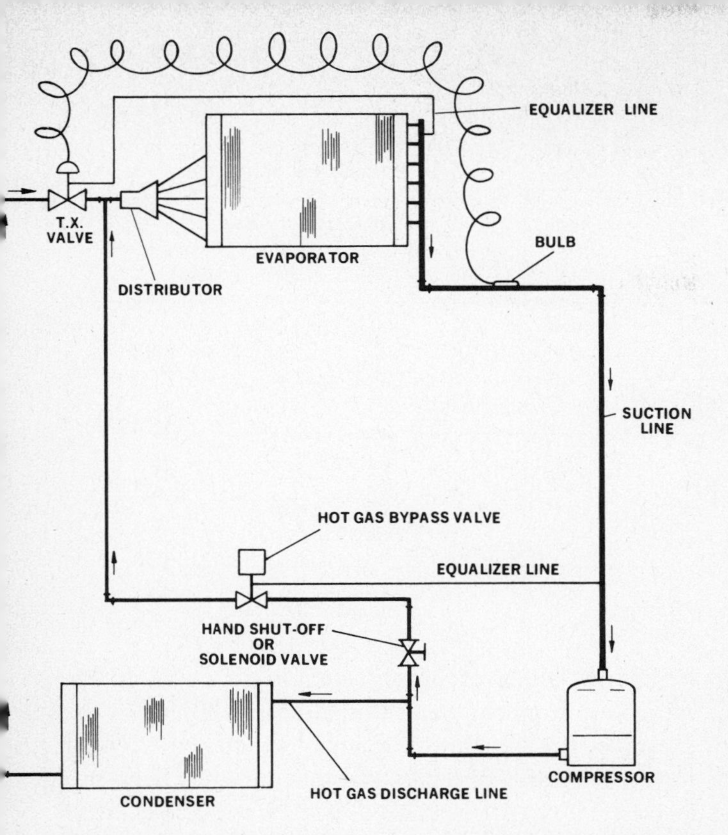

FIGURE 10-36 Hot-gas bypass (Courtesy, Westinghouse Electric Corporation).

Low ambient control is most frequently provided by cycling the condenser fans as previously explained, except that the number of steps is increased by virtue of the multiple fans; closer control is thus achieved. There are other techniques, which include dampers to throttle the condenser air (Fig. 10-37). Choice is determined by the manufacturer's design. And since commercial equipment equipped for both mild weather and low ambient temperatures must operate under cold conditions, crankcase heaters in the compressor are needed to keep the lubrication oil warm. A reverse-acting switch will energize the heater as soon as the compressor is "off." Crankcase heaters also vaporize liquid refrigerant in the crankcase, so that on cold weather starts liquid slugging doesn't harm the compressor valves.

Control systems that are more sophisticated include an oil-pressure failure switch, high and low-pressure refrigerant switches, lookout protector relays, time-delay relays between compressor starts and anticycling timers to prevent rapid or short-cycling of the compressors. Timing ranges from 3 to 5 minutes. The indoor air-handling equipment takes two main forms, as shown in Fig. 10-38. Horizontal

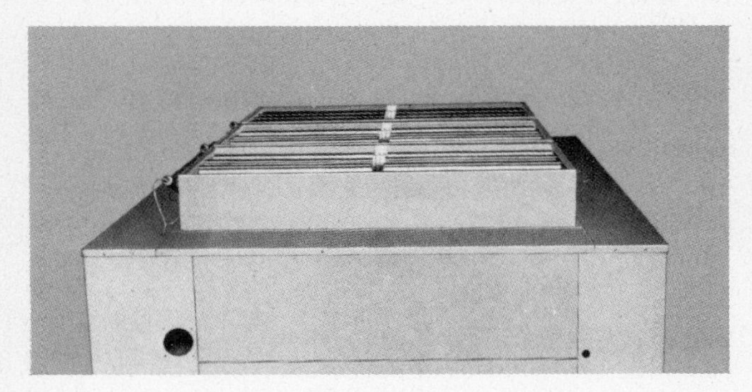

FIGURE 10-37 Dampers to throttle the condenser air (Courtesy, Fedders).

FIGURE 10-38(a) Horizontal model, evaporator blower section

FIGURE 10-38(b) Vertical model, evaporator blower section (Courtesy, York, Div. of Borg-Warner Corporation).

Cooling capacity—95°F DB air on condenser—MBH

System		Air on Evaporator Ft³/min	Air on Evaporator WB Temp. °F	Total Capacity Mbh	WB Temp. off Evap. °F	Comp. and Cond. Fan Input kW	Sensible Capacity-MBH, Dry-Bulb Temp. on Evaporator—°F 70	75	80	85	90
CA 91	(EB92-B) (EBV92-B) (C90UX[4]) (92DX)	3300	72	101	65.9	10.4	—	40	56	69	89
			67	93	60.6	9.9	43	57	73	88	91
			62	86	55.4	9.4	57	73	84	86	86
			57	78	50.2	9.0	67	72	78	78	78
CA 91	(EB122-B) (EBV122-B) (122DX)	4400	72	—	—	—	—	—	—	—	—
			67	98	60.6	10.2	43	59	77	93	98
			62	91	55.4	9.7	61	79	87	91	91
			57	84	50.2	9.3	83	84	84	84	84

FIGURE 10-39 Ratings.

evaporator blower sections are suspended from a ceiling either for in-space free-blow application with accessory plenum and discharge grille, or are remotely located with conditioned air ducted into the comfort space. Vertical models are free standing and can also be free blow or ducted. The choice of free blow or ducting is a function of cost, space, appearance, and type of facility. Large, open-space stores are typical applications for free blow, while an office with many partitions must be ducted.

The centrifugal blower must deliver a higher air flow and static pressure than are called for on residential work, and thus a **V**-belt drive with an optional variable pitch motor pulley and motor horsepower is needed. The fan scroll can frequently be rotated within the cabinet to obtain alternate discharge arrangements for greater application flexibility.

The evaporator coil(s) will be circuited to match the number of compressors. Also, space is generally provided for the inclusion of either nonfreezable steam coils or hot water coils, for winter heating.

Ratings of *matched* condensing units with specific air handlers are published by the manufacturer and are certified under ARI Standard 210. An example of a specification is shown in Fig. 10-39. For a system combination (condensing unit and air handler) at a rated ft³/min (usually 400 ft³/min/ton), 95°F outdoor air on the condenser, and an indoor entering 67° wet-bulb temperature to the evaporator coil it lists

(a) The total capacity in Mbh (93)
(b) Leaving WB temperature off the evaporator (60.6°F)
(c) Total power input in kW (9.9)
(d) The sensible capacity in Mbh

based on whatever entering dry-bulb exists in the return air from the conditioned space (80°F). In this example the sensible capacity would be 73 Mbh for an S/T ratio of 0.785. All this data has been precalculated, so the selection of a standard combination is very easy, and the alert application technician will always consider using standard equipment to keep cost and delivery time down. There are occasions when nonstandard combinations are desired, and so the art of selecting the air-handling equipment first and then cross plotting the coil capacity against the condensing unit is necessary, in order to determine the correct system balance in terms of refrigerant operating temperature. This procedure is presented by the manufacturer and varies with individual methods for system performance.

A knowledge of refrigerant piping for commercial split systems is important for the installation and servicing technician. The soldering skill needed to run hard-drawn copper piping has been discussed elsewhere; however, here we are concerned with system design in terms of proper oil return to the compressor, pressure loss in the lines, and the velocity of gas flow. Pressure drop in the suction line is more critical than on the liquid line, since it means a loss in system capacity, because it forces the compressor to operate at a lower suction pressure to maintain the desired evaporator temperature. Capacity falls off while the horsepower actually increases. The larger the suction line, the less the pressure drop, but oversizing also has its problems. First, the velocity of gas flow may not be sufficient to return entrained oil back to the compressor, and second, the cost of larger pipe becomes an economic factor. Therefore, as a technical and economical practice, suction lines are sized with nominal pressure drop corresponding to 2°F, or approximately 3 psi for R-22. Liquid-line sizes have nominal pressure drops corresponding to 1°F, or approximately 2 psi for R-22.

The size of the liquid and suction line connections on the equipment usually represents a compromise between maximum and minimum velocities at maximum and minimum loads with nominal pressure drops in pipes and fittings. It is impossible to specify one line size to cover all conditions; however, based on the volume above, standard line sizes on the equipment should fall within the suction lift of 30 feet, and the length of the liquid and suction lines should be from 50 to 75 feet. System applications exceeding these values should be checked by calculating the actual piping and fitting loss. Below is a procedure to follow.

First, determine the total pressure loss from pipe and fittings. This is done by the equivalent length method. Since every valve, fitting, and bend in a refrigerant line offers resistance to the smooth flow of refrigerant, each one also increases the pressure drop. To avoid the necessity of calculating the pressure drop of each individual fitting, an equivalent length of straight tube, or pipe, has been established for each fitting, as shown in Fig. 10-40. The line sizing tables shown are set up on the basis of the pressure drop per 100 feet of straight pipe or tubing. This permits consideration of the entire length of line, including all fittings, as an equivalent length of straight pipe and allows the data shown to be used directly.

OD in. Line Size	*Globe Valve*	*Angle Valve*	*90° Elbow*	*45° Elbow*	*Tee Line*	*Tee Branch*
$\frac{1}{2}$	9	5	0.9	0.4	0.6	2.0
$\frac{5}{8}$	12	6	1.0	0.5	0.8	2.5
$\frac{7}{8}$	15	8	1.5	0.7	1.0	3.5
$1\frac{1}{8}$	22	12	1.8	0.9	1.5	4.5
$1\frac{3}{8}$	28	15	2.4	1.2	1.8	6.0
$1\frac{5}{8}$	35	17	2.8	1.4	2.0	7.0
$2\frac{1}{8}$	45	22	3.9	1.8	3.0	10.0
$2\frac{5}{8}$	51	26	4.6	2.2	3.5	12.0
$3\frac{1}{8}$	65	34	5.5	2.7	4.5	15.0
$3\frac{5}{8}$	80	40	6.5	3.0	5.0	17.0

NOTE: For complete information on equivalent lengths, refer to ASHRAE Data.

FIGURE 10-40 Equivalent length in feet of straight pipe for various valves and fittings.

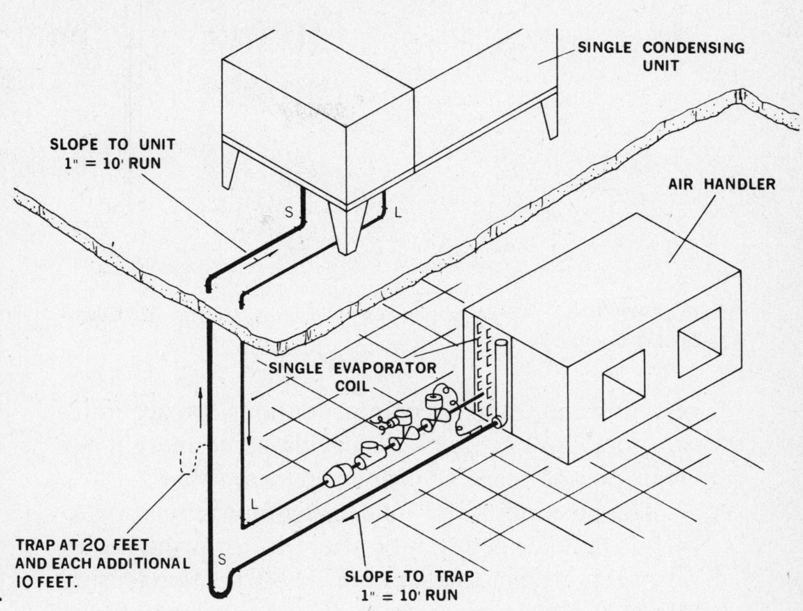

FIGURE 10-41 One condensing unit with one single circuit evaporator coil (Courtesy, Westinghouse Electric Corporation).

Assume you are calculating the suction pressure loss for the 10-ton system shown in Fig. 10-41. The actual separation between the condensing unit and the air handler is 75 feet for combined vertical and horizontal distances. The suction line layout includes an angle valve and six 90° elbows. From the manufacturer's catalog or by actually checking the line size on the equipment itself you determine the standard suction line connection is $1\frac{3}{8}$ in. OD copper. Next, determine the fitting loss from Fig. 10-40.

(1) $1\frac{3}{8}$ in. OD Angle valve = 15.0 ft equivalent length

(2) $1\frac{3}{8}$ in. OD
90° elbows @ 2.4 = 14.4 ft equivalent length

Total Fitting Loss = 29.4 ft equivalent length

Add this to the 75 feet of actual straight pipe for a total system value of 104.4 feet, or roughly 105 feet.

Referring to the manufacturer's catalog we find $1\frac{3}{8}$ in. OD tubing has a friction loss of 2.8 psi per 100 feet for R-22 at a 10-ton load, and, since we have 105 feet, the actual suction pressure loss would be $2.8 \times 1.05 =$ **2.95** psi, which is just marginal in terms of the stated 2°F or approximately 3 psi recommended maximum. The liquid lines are calculated in the same way.

Most manufacturers publish line size tables for various lengths of run including the average number

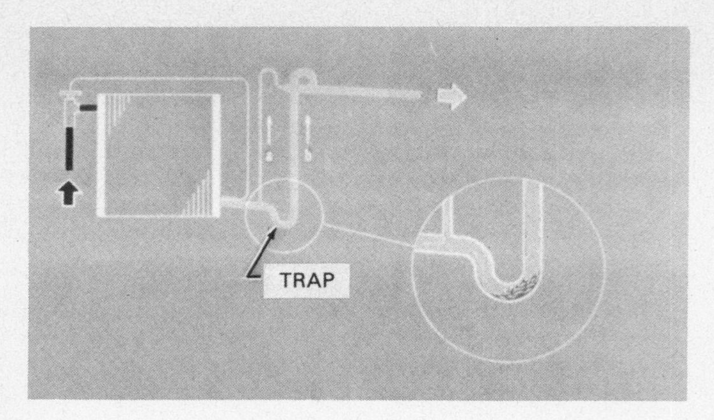

FIGURE 10-42 Double suction riser (Courtesy, Carrier Air-Conditioning Company).

of fittings, and their recommendations are generally safe to follow in terms of leaving some margin for error; do not intentionally exceed these values.

The position of the condensing unit in relation to the air handler determines where to form the necessary suction line traps. In Fig. 10-41 the condensing unit is above the evaporator. In the suction riser, oil is carried upward by the refrigerant gas. A minimum gas velocity must be maintained to entrain the oil. The sump or trap at the bottom of the riser promotes free drainage of liquid refrigerant away from the evaporator.

Where system capacity is variable, through compressor capacity control or some similar arrangement, a short riser will usually be sized smaller than the remainder of the suction line to insure oil return up the riser. Some manufacturers also recommend a trap in the vertical riser at approximately 20 feet and at each additional ten feet thereafter.

Where the system capacity is variable over a wide range, it may not be possible to find a pipe size for a single suction riser that will insure oil return under minimum load conditions and still have a reasonable pressure drop during maximum conditions. A double suction riser as shown in Fig. 10-42 should be considered. In a double riser, the small pipe, *A*, on the lift is sized to return oil when the compressor is "unloaded" to minimum capacity. The second pipe, *B*, which is usually larger but not necessarily so, is sized so that the pressure drop through both pipes under maximum load is satisfactory. A trap is located between the two risers. During partial load operation the trap will fill up with oil until riser *E* is sealed off. The gas then travels up riser *A* only and has sufficient velocity to carry oil along with it back into the horizontal suction main. Both risers loop at the top before entering the horizontal suction line. This is to prevent oil drainage into an idle riser during partial load operation.

In Fig. 10-41 note the recommended slope of horizontal suction lines to either the traps or the condensing unit. Figure 10-43 shows how to form a line trap from standard fittings. Figure 10-44 illustrates the correct position for clamping the expansion valve bulb to the suction line. Between the nine o'clock and the three o'clock position the bulb feels only superheated gas temperature but will detect any excess liquid surge.

The choice of controls and coil circuiting determines the headers and the number of solenoid valves used, but in all cases it is recommended practice to install a filter/drier and sight glass in the main liquid line. Evaporator coils may be single circuit, dual circuit, face split, row split, or combinations thereof, and the proper mainfold piping is usually recommended by the manufacturer.

Although the liquid-line pressure loss from refrigerant flow is not as critical as the suction, the vertical lift is an important consideration, as explained earlier. (Figure 10-30 presented the static pressure loss for the vertical lift of R-22 refrigerant.) Remember, lifts over 30 feet may cause "flash gas," possibly

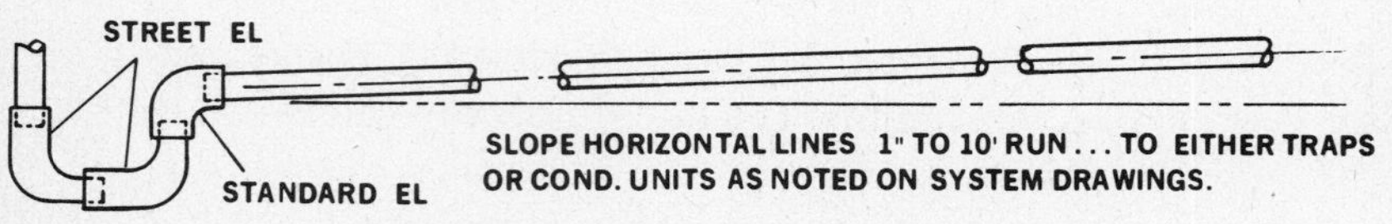

FIGURE 10-43 Typical line trap.

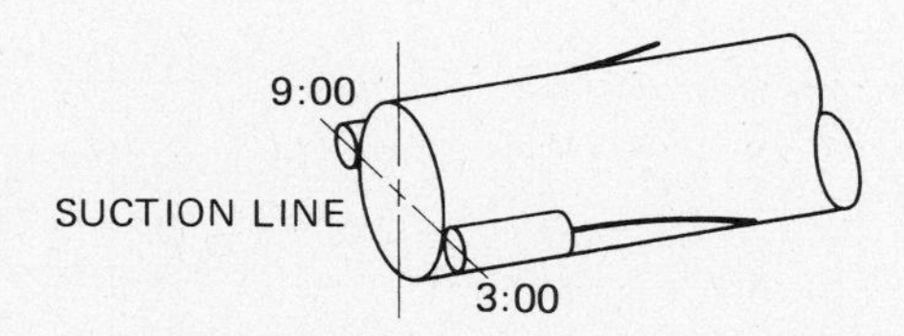

FIGURE 10-44 Typical expansion valve bulb locations.

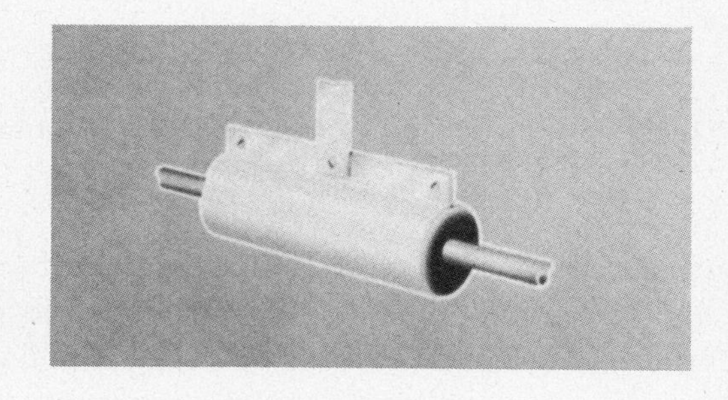

FIGURE 10-45 Pipe support.

resulting in damage to the expansion valve or at least erratic control of the evaporator.

Structural support of the refrigerant piping is important to hold the weight and also to isolate vibration and minimize sound transmission. Figure 10-45 illustrates typical hanging methods. Supports should provide insulation between themselves and the pipe to absorb vibration, but it should not be tight enough to cut regular pipe insulation.

Thermal insulation of suction lines is necessary to prevent sweating and heat gain. The thickness of insulation will depend on the type of application and be subject to the specific manufacturer's recommendations. The correct thickness must raise the surface temperature above the surrounding dewpoint temperature. When exposed to the outdoors the insulation must also have an adequate vapor barrier. Some insulations, by virtue of their material substance, have a built-in or integral vapor barrier.

Evaporators are equipped with condensate drain connections, and the field-fabricated drain lines should remain full size (not reduced). A line trap similar to the one shown in Fig. 10-46 *must* be provided close to the unit to form an air lock. Otherwise, negative suction pressure from the fan on a draw-through arrangement could create improper draining.

Drain lines must also be protected from freezing; they must empty into an open drain and not to a closed sanitary plumbing connection where sewer gas could be drawn into the air handler in the event the unit water trap went dry on the winter cycle.

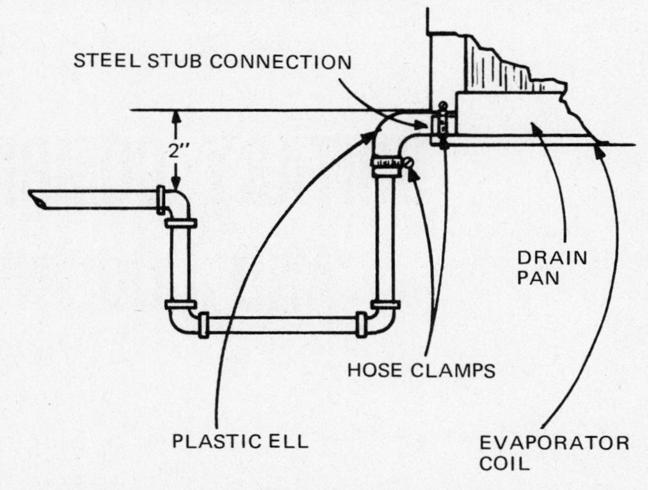

FIGURE 10-46 Condensing drain piping (Courtesy, York, Div. of Borg-Warner Corporation).

PROBLEMS

1. What is the ARI standard operating capacity for air-cooled packaged units in terms of indoor and outdoor conditions?
2. Conventional S/T ratios for residential cooling units are approximately ______________.
3. ARI-listed air-cooled units must operate to ______________ °F outdoor ambient.
4. Operating a cooling unit down to 55°F outdoor ambient is termed ______________.
5. Operating a cooling unit to 0° outdoor ambient is termed ______________.
6. The most frequent method of low ambient control is ______________.
7. Water-cooled conditioners for water tower operation will have condenser piping arranged for ______________ flow.
8. Computer room units will frequently use ______________ solution instead of water for year-round operation.
9. Name three types of add-on evaporators.
10. Name three types of mechanical seals for joining refrigerant piping.
11. Pressure loss in a liquid line because of height difference is called ______________.
12. What main difference in equipment specifications occurs in the crossover from residential to commercial use?
13. Capacity control can be accomplished by several means; name two.
14. Traps in the refrigerant lines are primarily to ______________.
15. Evaporator drain lines should have water traps to form an ______________.

11

UNITARY COMBINATION HEATING AND COOLING EQUIPMENT

The overall scope of heating and cooling products as previously mentioned is not clearly defined. Many packaged and split-system cooling products may offer heating as an option to provide year-round comfort. There are, however, three categories of unitary products designed to perform both heating and cooling, and in this chapter we will review their aspects.

PACKAGED TERMINAL UNIT

Figure 11-1 portrays what are sometimes called *incremental systems*, which have shown rapid market growth in the past ten years. This type of unit is geared to the new construction in motels, hotels, schools, offices, apartments, hospitals, nursing homes, etc., where the nature of the construction and the need are based on conditioning one room or one space. Ducted systems cannot provide individual control, and in some cases the mixing of room air such as in a hospital or nursing home is not desirable. Also, in an office building, each tenant can be on a separate meter. And finally, the very nature of the product lends itself to simple installation and ease of service. One of the largest installations of packaged terminal units is in the U. S. Department of Transportation building in Washington, D. C., with approximately 2,000 units installed. Where great quantities of units are involved, it's common practice to have spares, so when major service is needed, it's done by exchanging the entire unit with minimum labor and disruption; this is one of the major design considerations. Packaged terminal units are not the same as through-the-wall window units. The former offer higher quality construction, reliability, and much broader application capability.

The typical design of a packaged terminal unit is usually in several components (Fig. 11-2). First a wall sleeve is permanently inserted and attached to the outside wall of the building. The wall sleeve also includes the outside condenser air grille. It must be attractive as well as functional, in terms of directing air intake and discharge, shedding rain water, and above all it must be noncorrosive.

FIGURE 11-1 Packaged terminal unit, incremental system (Courtesy, Singer Company).

The refrigeration chassis is a complete assembly that slides into the wall sleeve. It includes the entire cooling cycle and also space for electric resistance heaters or, in some cases, steam or hot water coils. Ninety percent of all systems use electric heating, but there are still areas where wet heat is available and is so specified. Within the refrigeration chassis components such as the fan deck, control console, etc., are the plug-in type and may easily be removed for quick service.

Facing the conditioned space is the room cabinet, which provides the discharge and air return functions. The air outlets are at the top and have directional control vanes to throw and spread the air pattern as needed. Return air comes back at the floor level and enters the unit through a concealed opening. The room cabinet must be decorative and sturdy. Projection into the room is a function of the individual manufacturer's model. Naturally the least possible extension is a desirable feature. Filter(s) are also included in the room cabinet and are easily accessible for routine replacement.

Several operational factors must be noted for this type of equipment. First, since all mechanical operation is contained in a comparatively small enclosure, accoustical control becomes vital to reduce the room-side sound level. Fan-speed control also helps to reduce air noise. Second, the condensate disposal system is accomplished without plumbing. Condensate water must be mechanically vaporized and exhausted in the condenser air stream. This also includes rain water should it collect during a storm; it cannot be allowed to drip down the exterior wall. Third, the wall-sleeve installation and cabinet mounting system must be properly sealed to prevent

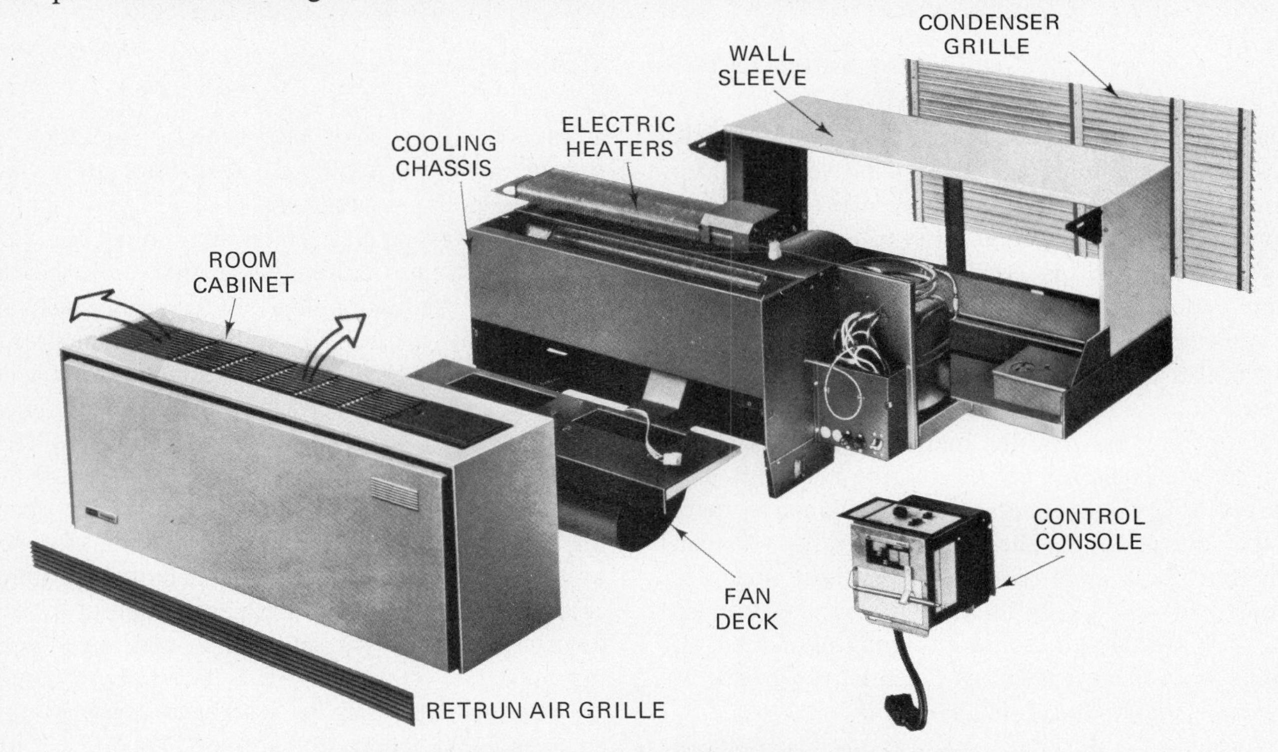

FIGURE 11-2 Packaged terminal unit (Courtesy, Singer Company).

FIGURE 11-3 Hurricane test (Courtesy, Westinghouse Electric Corporation).

moisture or air infiltration. These units are frequently used in high-rise buildings where wind velocity, even in normal weather, can be high. Because of this, some manufacturers subject their units to a *hurricane test* (Fig. 11-3), which is conducted under conditions where water and air are blowing at 75 mph against the outside grille.

Another operating provision is the introduction of outside air. This is done through a small, dampered opening between the indoor and outdoor section. The air is filtered before entering the room. The damper is normally operated by a manual control on the control panel. The amount of outside air can sometimes be a code requirement if these units are installed in schools, hospitals, and nursing homes, etc.; 25% outside air would be a minimum need. In motels and offices it is left to personal preference to overcome smoke and stale air.

Since these units must operate on a year-round basis, it is entirely possible that mild weather or low ambient cooling may be needed. Therefore, the system must be capable of operating safely under low head pressure situations.

Control of packaged terminal units can range from a simple, manually-operated fan, heating, cooling, and ventilation selection to very sophisticated central-point control for multiple unit operation. An example of the latter would be a motel where the desk clerk could activate a guestroom at check-in time. Offices too can be "zone controlled" from the building maintenance office. A night set-back arrangement can also be included to reduce operating time during no-occupancy hours. Motorized outside air dampers can be controlled or programmed as a function of outdoor ambient temperature or in connection with night set-back to minimize operating costs.

Cooling capacities of packaged terminal units go from 7,000, 9,000, 12,000 to 15,000 Btu/h, with a few up to the 18,000 Btu/h. Electric heating capacity usually ranges from 4,000 Btu to 16,000 Btu. Most manufacturers use two-stage heating arrangements to split electrical circuits and to offer closer room control without overheating.

The electrical installation is simple. 208/240-volt units may be permanently wired or equipped for plug-in to wall-mounted receptacles; 15, 20, or 30-amp, depending on the requirements. A power cord, or *pig tail*, with the proper amp rating is prewired at the factory. The 277-volt operation common to many offices is another consideration. Underwriters Laboratories requires that 277-volt electrical installations be protected by a raceway, and so the use of a subbase with a built-in receptacle (Fig. 11-4) is

FIGURE 11-4 Sub-base with built-in receptacle (Courtesy, Westinghouse Electric Corporation).

needed. The subbase also serves another function, and that is to support the weight of the unit and wall sleeve when the unit is installed in a curtain wall. Most curtain-wall construction has no load-bearing capability and cannot be used for that purpose.

Although a packaged terminal unit is basically designed to serve one room, the use of a duct kit (Fig. 11-5) allows conditioned air to be discharged into an adjoining room. Such might be the case in a hospital or nursing home room with an adjoining bathroom. Return air may not be brought back to the unit, and outside air is used to compensate for the loss.

Packaged terminal unit servicing is about half-way between that of a refrigerator and a packaged air-cooled central unit. All units use welded hermetic compressors and thus no on-the-spot compressor repairs can be made. Most units do not have service valves for checking operating pressures, and they rely on temperature measurements to monitor performance. Fan motors (evaporator and condenser) are direct drive and require little or no oiling. If a fan

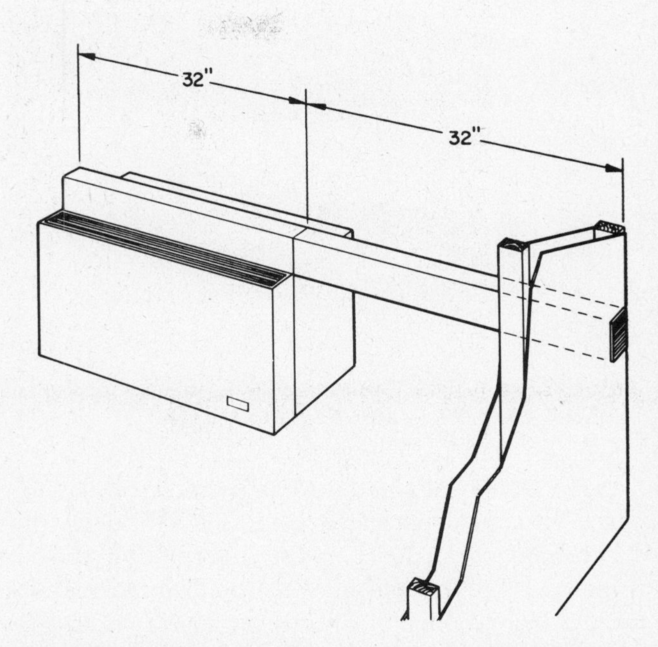

FIGURE 11-5 Duct kit (Courtesy, Carrier Air-Conditioning Company).

motor failure occurs, the motor simply unplugs for quick removal. Filter maintenance is a key to successful operation. Space simply doesn't permit large filters, and so frequent replacement or cleaning is a must; fortunately, in larger, multiunit installations, this is a regularly scheduled activity. Control consoles are designed for quick access or complete removal by plug-in arrangement. Much repair and maintenance can be done on a bench in the maintenance shop, and because of the availability of spare replacement units, down time and disruption are kept to a minimum.

Packaged terminal units are not usually used in single-family residences, where ducted systems offer a better approach for whole-house conditioning.

Rooftop Units

Combination heating and cooling equipment for ducted installation fits into the description of a one-piece *rooftop unit* (Fig. 11-6). This unit is ideal for low-rise shopping centers and commercial and industrial buildings. This particular category of year-round air-conditioning has experienced the fastest growth rate among all those reported to the ARI.

FIGURE 11-6 Rooftop combination gas heating and cooling unit (Courtesy, York, Div. of Borg-Warner Corporation).

The design of this unit has features similar to a large air-cooled packaged cooling conditioner, plus a heating section and an air distribution section. Figure 11-7 represents an oversimplified cross section of these components. Return air is drawn into the unit over filters and through the cooling coil by the centrifugal fan. It is then forced through the heat exchanger and then out through a supply duct to the conditioned space. The air-cooled condenser portion functions as described in previous products. Due to the weight involved and the need to provide a base that can be built in during building construction, a *roof curb* is needed; it actually becomes part of the roof itself. A cutaway picture of a single-zone unit is shown in Fig. 11-8, and the major components can be identified. This illustration depicts a gas heating system; however, the use of electric resistance heating is also very popular. Some models also burn oil; but such units are in the minority. The cooling capacities of standard models begin at around $1\frac{1}{2}$ tons and go up to 60 tons. Some manufacturers even offer custom-type units of up to 100 tons. The range of gas heating to go with standard cooling products runs from 45,000 Btu/h to over a million Btu/h in very large units. The Btu capacity ratio of gas heat to cooling is about 2 or 2.5 to 1. The ratio of electric heating is a bit lower, because of the assumption that tighter building construction, insulation, and internal heat loads will require less heating input.

The description of a rooftop unit really doesn't do justice to the versatility of this product line; it can also be installed at ground level with ductwork through the wall to the conditioned space (Fig. 11-9). This application is very popular in the smaller sizes for use in residential or light commercial work, because it is a complete factory-tested unit in one cabinet; it requires no chimney or flue; all mechanical sound is outside; and service is made easy by walk-around space at ground level.

Let's look more closely at certain features of this unit. *Single-zone units* as shown in Figs. 11-7, 8, and 9 direct all of the conditioned air into one trunk duct, which then distributes air to branch runs as required. *Multizone units* (Fig. 11-10a) differ in that the air is divided inside the unit and then is directed to several zone ducts that serve various parts of a building. Zone thermostats signal the proper mode (heating

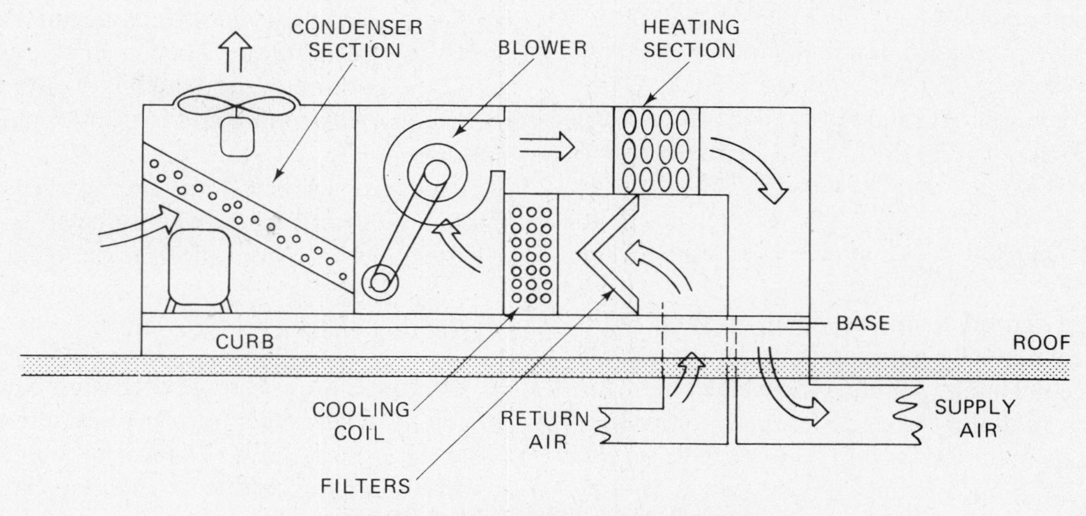

FIGURE 11-7 Cross-section of rooftop unit.

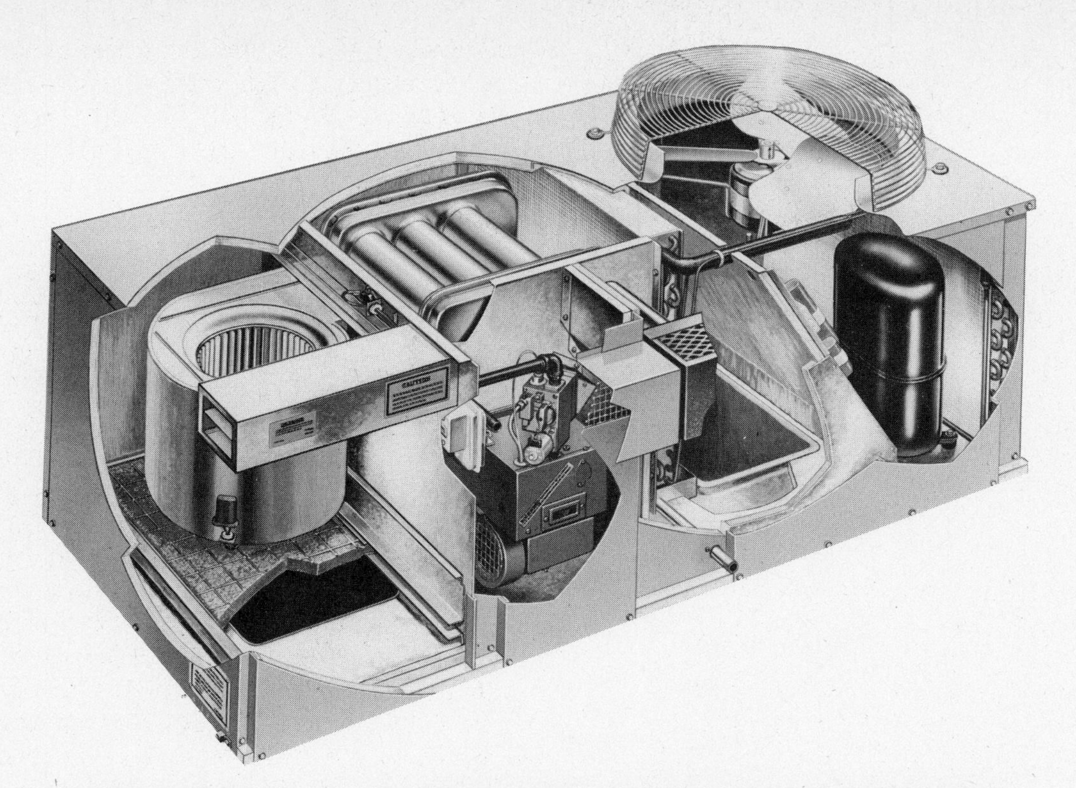

FIGURE 11-8 Single package gas heating and electric cooling unit (Courtesy, Lennox Industries).

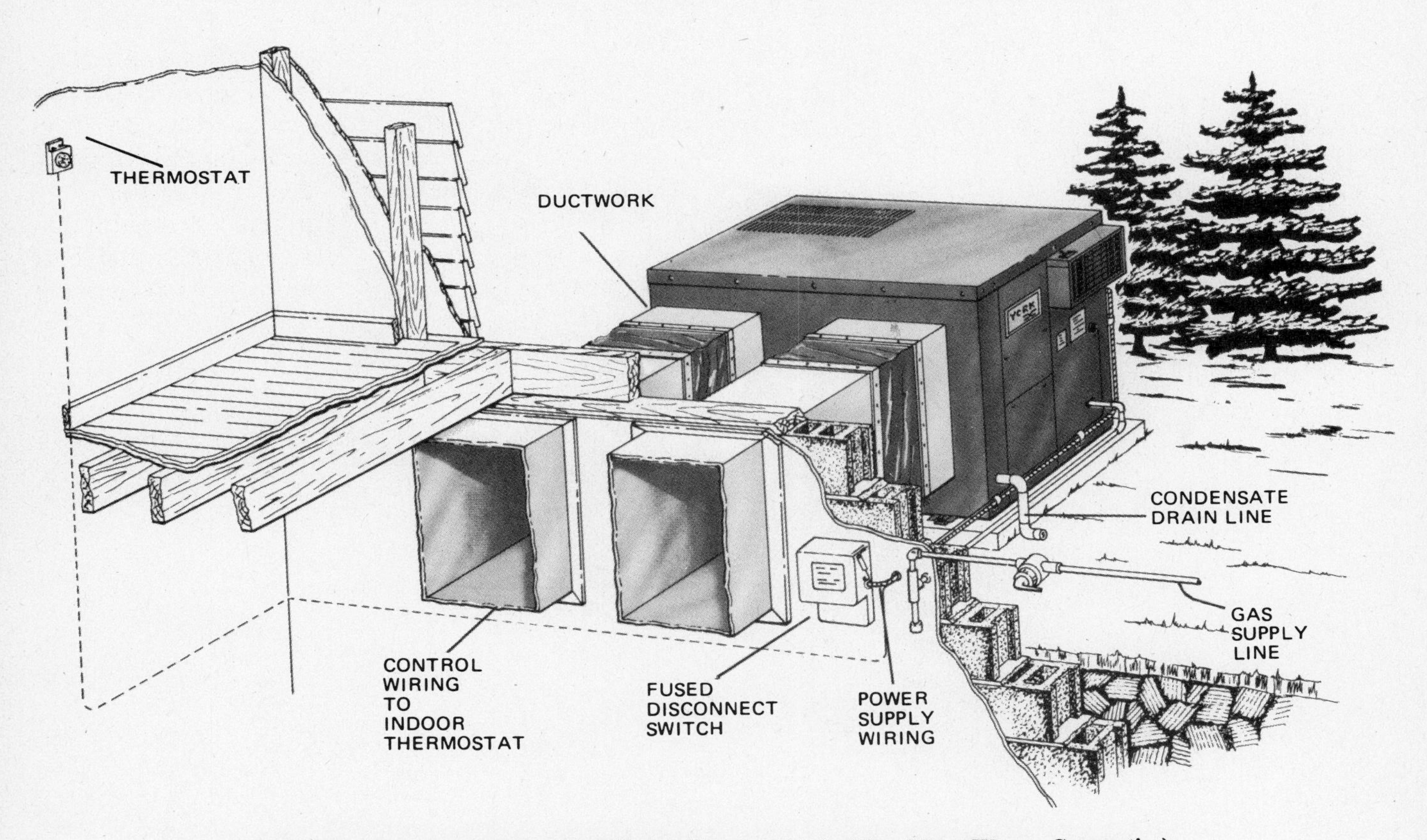

FIGURE 11-9 Typical ground-level installation (Courtesy, York, Div. of Borg-Warner Corporation).

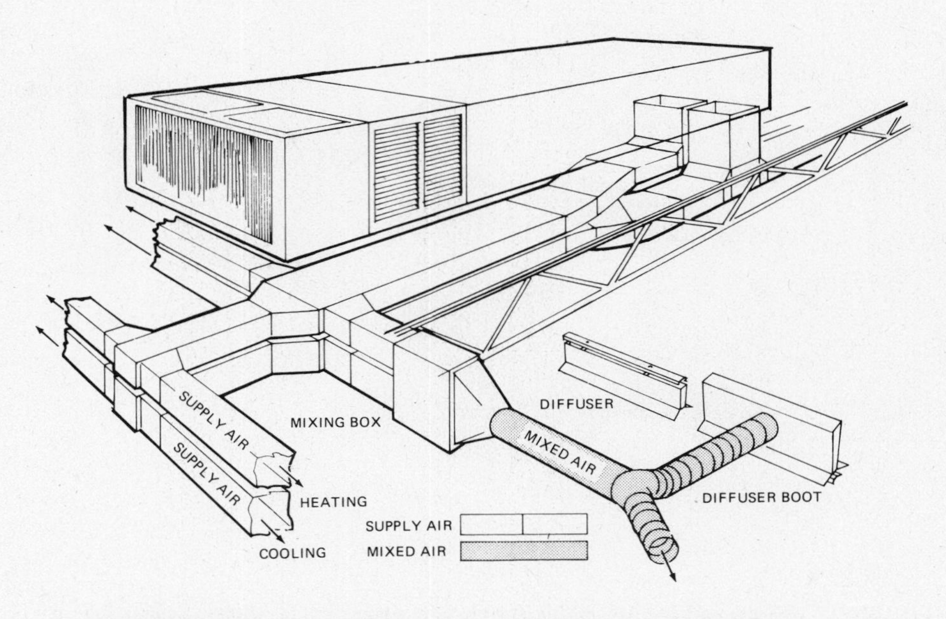

FIGURE 11-10(a) Direct multizone duct system (Courtesy, Lennox Industries).

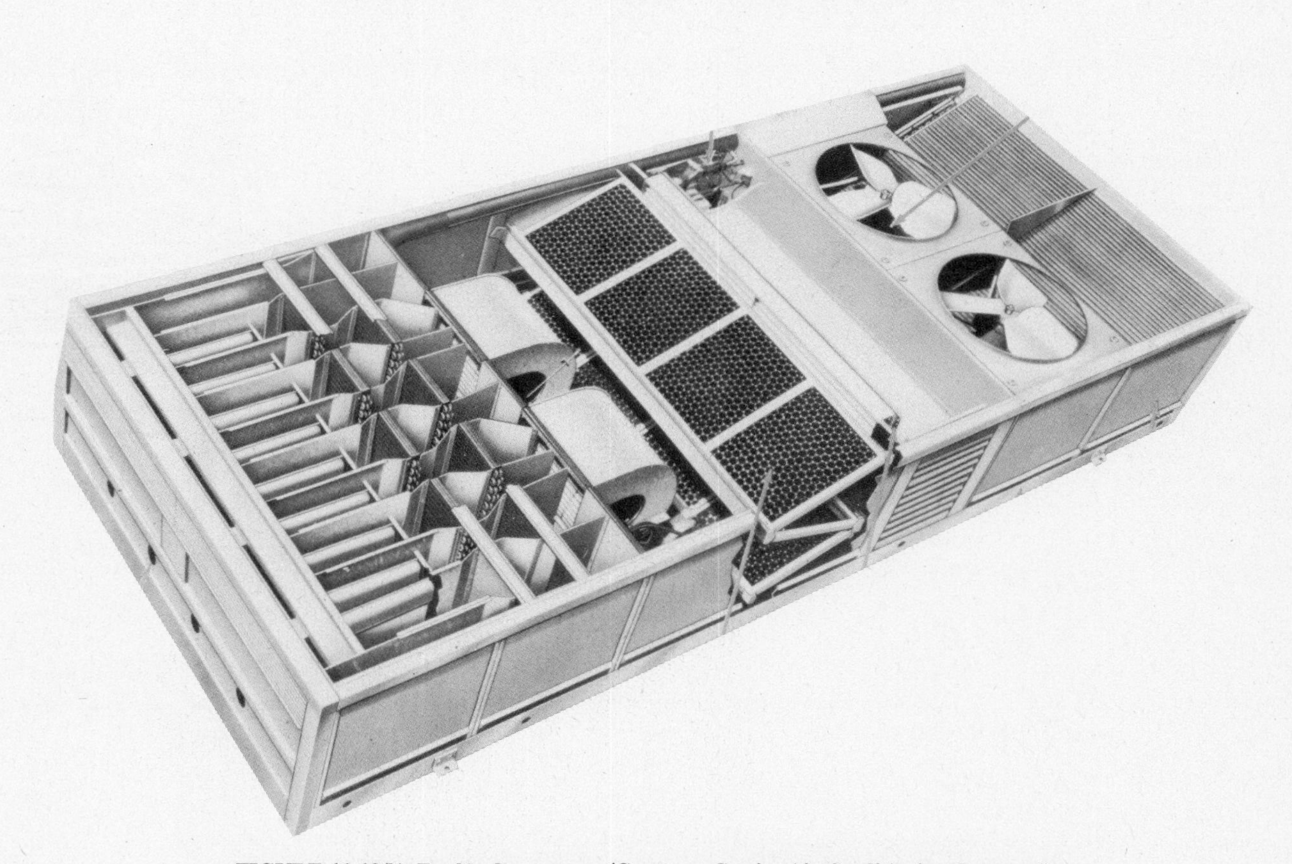

FIGURE 11-10(b) Double duct system (Courtesy, Carrier Air-Conditioning Company).

or cooling) in response to comfort needs. Multizone units do differ in their design concepts. Some manufacturers use constant air circulation and temper the heating, cooling, and humidity control, while others use a damper arrangement and premix the air to each zone. Still others use a double duct system as shown in Fig. 11-10b (hot and cold), and mix the air out in the conditioned space. In terms of sales, single-zone units are by far the most widely used, because they are the least complicated to apply, to install, and to control for the average need.

One important feature of rooftop units is the so-called economizer cycle, which uses filtered outdoor air for cooling rather than mechanical refrigeration when the outdoor ambient temperature is less than the indoor comfort zone temperatures. This is accomplished in the return air chamber by means of dampers on the return air and on an outside air inlet opening (Fig. 11-11). These dampers

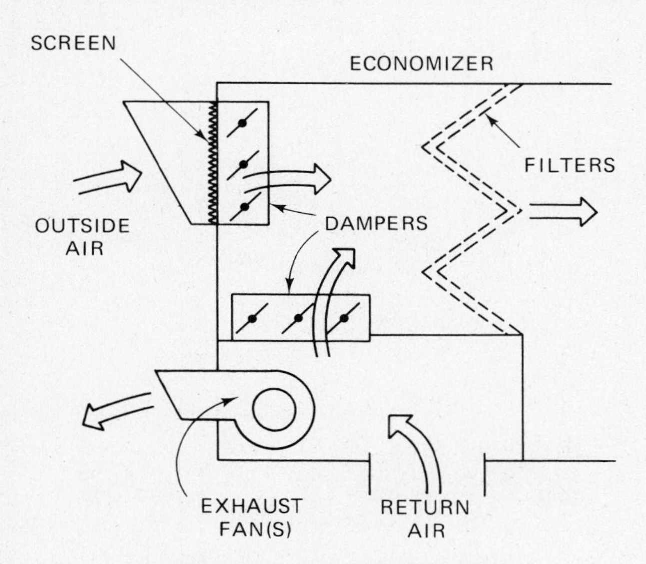

FIGURE 11-11 Economizer cycle.

work in reverse, and as the outdoor air opens the return closes, mixing the two. Some units can take in 100% outside air, which means no return air comes back. This creates excessive building pressures, which must be exhausted somewhere to relieve pressure. Some units are equipped with built-in power exhaust fans as shown in Fig. 11-11. Others depend on remotely located exhaust fans that are energized by the economizer control. The resultant air mixture of outdoor and return air is controlled so as to provide a ducted air temperature to the conditioned space of not less than about 55°F. Below this the economizer will modulate to a closed or minimum outside air position; thereafter, heating will be needed.

In mild weather, low ambient operation of the cooling cycle is needed and is accomplished by the techniques of speed control, staging, or sequencing "off" the condenser fans. Some units also offer compressor unloading for capacity control. These devices must be programmed to work in conjunction with the economizer system.

The gas-fired heating system has several notable features. The most obvious is that there is no extended flue stack to create the needed draft for exhausting combustion products. Small units have short terminal vents. However, on larger units power venters (centrifugal fans, Fig. 11-12) are needed to provide mechanical draft through the heat exchanger and to discharge combustion products into the atmosphere. Because of this forced-air movement, heat exchanger design includes tubular steel, as well as the conventional sectionalized type. Ignition is provided by a solid-state electrical spark. If the pilot light goes out, it will relight automatically. Pilot outage could be a real nuisance on rooftop equipment, but the spark method has solved this problem.

Due to the quantity of heat associated with large stores, the commercially-sized rooftop equipment usually has a two-stage furnace operation controlled by a two-stage heating thermostat so as to match load variations more closely. Natural gas and propane LP fuels are the standard gas-heating options.

Electric heating (Fig. 11-13) is accomplished by banks of nicrome wire elements supported by ceramic insulators. As previously described, they are pro-

FIGURE 11-12 Power ventor (Courtesy, Fedders).

FIGURE 11-13 Electric heating (Courtesy, Fedders).

tected from overheating by thermal cut-out switches and are backed up by fusible links. Some manufacturers install the heaters at the factory; others offer them as field-installed kits to be placed in the discharge air stream. Prewired fusing and contactors are contained in the equipment panel to simplify and reduce on-site electrical work.

Frequently you will hear the expression "single-point electrical hook-up." This means that the total power requirement for the unit is brought to one point and then is split internally into its various functions. This is in contrast to individual external wiring to cooling, heating, fans, etc. Obviously this, too, reduces on-site labor and materials and simplifies the building electrical layout.

The roof curb (Fig. 11-14) varies for different

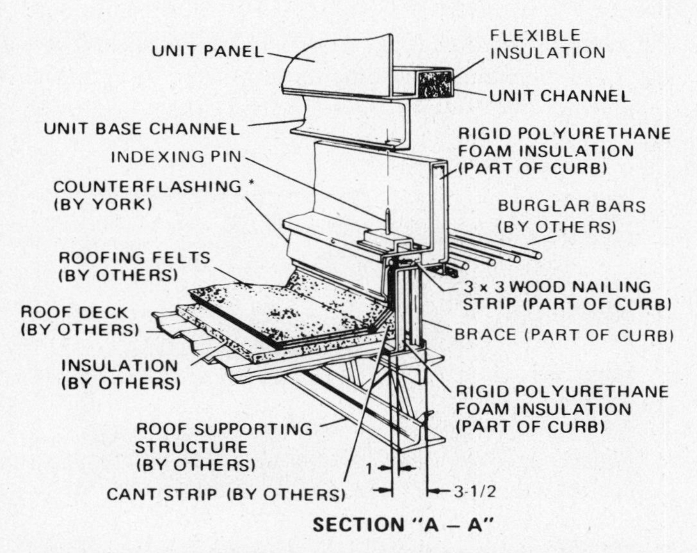

FIGURE 11-14 Roof curb mounted directly on roof supports (Courtesy, York, Div. of Borg-Warner Corporation).

manufacturers, but essentially it is a rigid galvanized steel base that is attached to the roof joists prior to the roof installation. The positioning of the curb depends on the unit location. Leveling the curb is important to assure that the unit drains properly. Wood nailing strips are bolted to the curb and are used by the roofer to nail his flashing and counter-flashing. Insulation around the entire curb perimeter is needed to prevent sweating. When the unit is set over the curb, a gasket between the two prevents air leaks, and it's common practice to let the unit channels overlap the curb to form a rain shield. Roof leakage was a common problem in the early stages of this equipment's development, but that has been overcome with better equipment design and experienced installers.

Lifting the unit onto the base is an interesting event. Large cranes with extension booms are usually used to spot the equipment on small buildings. But when many units are involved, it is faster to use a helicoptor airlift as shown in Fig. 11-15. The tech-

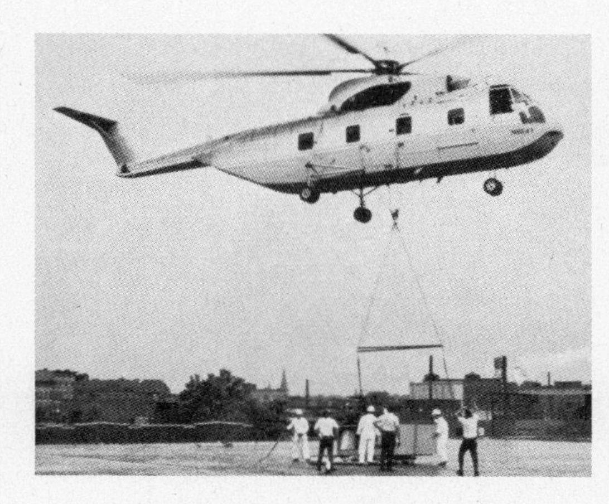

FIGURE 11-15 Helicopter air lift.

nique of maneuvering this size of equipment and setting it down on the curb indeed takes an expert pilot.

Two methods of air distribution are used with single-zone units when mounted on the roof of a commercial job. First, conventional trunk duct work (supply and return) go down into the drop ceiling of the building as shown in Fig. 11-8. It then is routed to branch ducts and air diffusers as needed. The second method is to use a concentric duct connection (Fig. 11-16), consisting of a supply and return to a single ceiling diffuser below. This arrangement depends on the feasibility of spotting multiple units over the area so that complete air coverage is achieved.

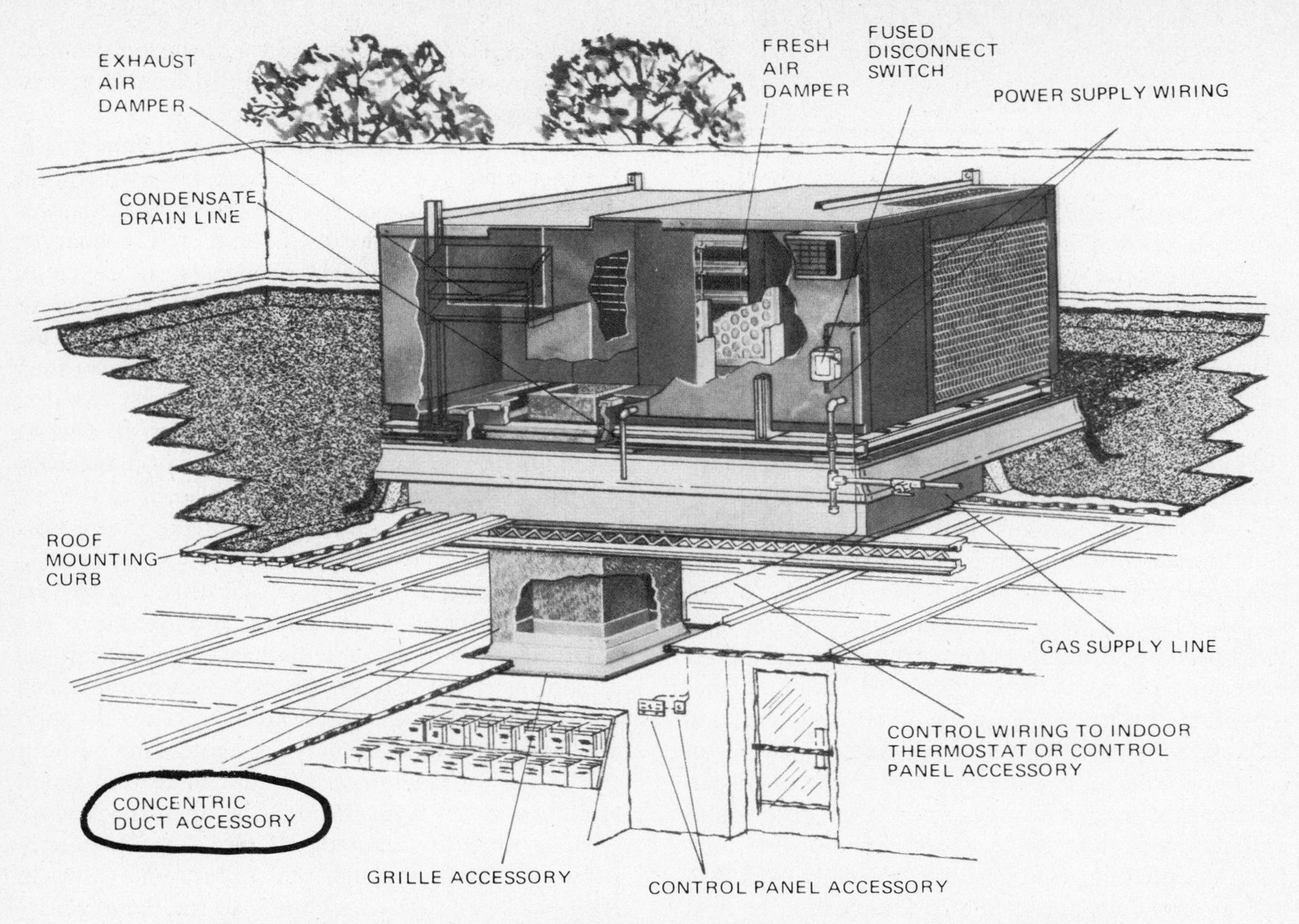

FIGURE 11-16 Typical rooftop installation with accessories (Courtesy, York, Div. of Borg-Warner Corporation).

But when this method can be used, considerable savings in ductwork and installation cost are achieved.

The indoor centrifugal fans on large equipment are belt driven and must be capable of producing much higher duct static pressure than is necessary for residential equipment. Pressures of $1\frac{1}{2}$ in., 2 in., and almost 3 in. of W.C. are needed, depending on the type of air distribution. Pulley and motor options provide a wide selection of speeds.

FIGURE 11-17 In-space control panel (Courtesy, York, Div. of Borg-Warner Corporation).

Since rooftop equipment is remotely located and not readily visible, the need for an in-space control panel (Fig. 11-17) is essential to permit the operator to select whichever mode (heat, cool, or ventilation) the system should be on. Panel lights may also be included that signal dirty filters or warn of malfunctions.

HEAT PUMP UNITS

A third and distinct form of combination heating and cooling unit is the *heat pump*, also generally known as a *reverse cycle heating and cooling unit*. Both mean the same thing: A system in which refrigeration equipment is used in such a way that heat is taken from a heat source and given up to the conditioned space when heating is wanted; or where heat is removed from the space and discharged when cooling is desired.

The heat pump is certainly not a new idea. The principle was first introduced in 1852 by Lord Kelvin. He proposed the use of a compressor as a "warming engine" and as a means of heating buildings to replace the equipment that burned fuels directly. The idea remained primarily of laboratory interest until the last 40 years. The first actual residential heat pump was probably installed in Scotland in 1927. Most early heat pumps used water as a heat source, and these became known as *water-to-air* units. Later on, systems were built up with large coils buried in the ground as the heat source, and these became known as *ground-to-air* units.

After World War II, air-conditioning installations became so numerous that the electrical power load due to summer air conditioning became higher than that due to winter uses. When this situation became apparent, power companies, because of economics, began to actively promote the use of all-electric heat pumps in order to equalize their winter and summer loads. Since water of the proper temperature, quantity, and cost is not usually available, and ground coils are too bulky and costly, the next practical medium was air, and air has become the chief source of heat. This development paralleled that of air-cooled equipment, and these heat pumps are known as *air-to-air*.

Let's look at an air-to-air heat pump cycle. First, if we consider the conventional refrigeration cooling cycle (Fig. 11-18), we know that heat is absorbed by the indoor evaporator and discharged by the outdoor air-cooled condenser. If we could physically reverse these components and absorb heat from the outdoor air and, by means of the refrigerant, discharge it into the indoor space, then we have created heating. This is exactly what a reverse cycle heat pump does—except it doesn't actually physically reverse the evaporator and condenser. By means of a reversing valve, it can direct the refrigerant flow alternately to make the process heat or cool.

Thus the heat pump cycle looks like Fig. 11-19a. We relabel the coils as indoor and outdoor. The reversing valve directs the discharge and suction gas as shown by the arrows. Check valves are installed in the outlets of both coils. Carefully follow the solid arrows for cooling. Discharge gas enters the outdoor coil in the usual manner. It's condensed to a liquid and flows to the right through the open check valve (No. 1), since it is not possible to force high-temperature liquid through an expansion valve in reverse. Check valve 2 is closed, so the liquid refrigerant goes to the indoor expansion device (TVX

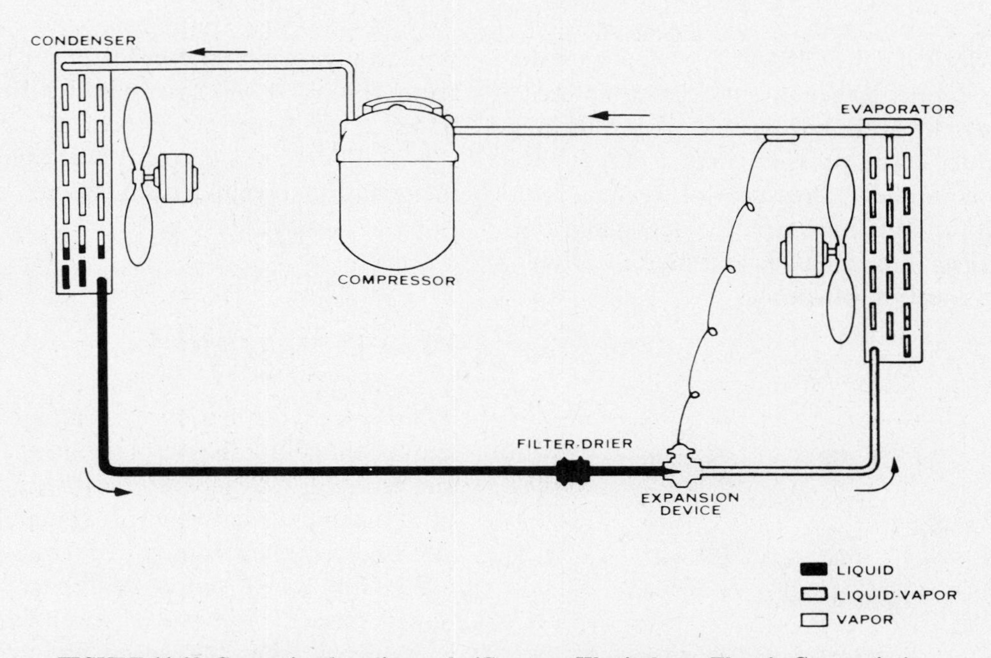

FIGURE 11-18 Conventional cooling cycle (Courtesy, Westinghouse Electric Corporation).

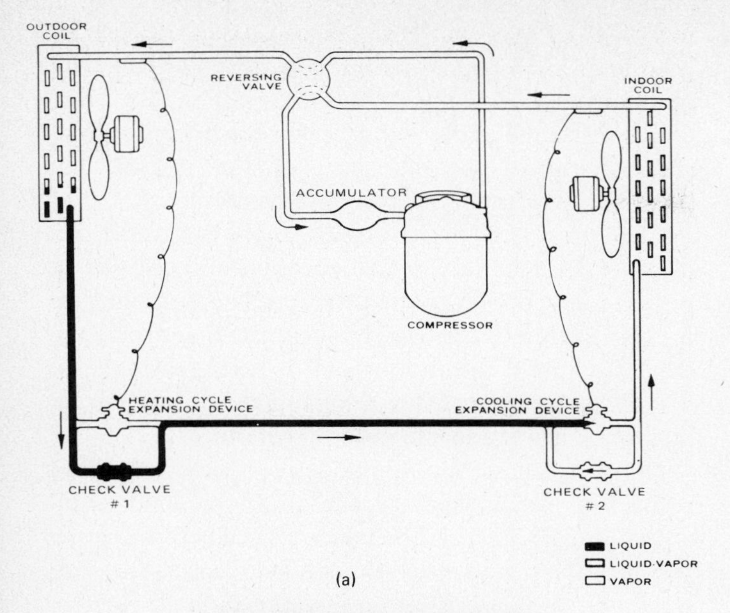

FIGURE 11-19(a) Cooling phase.

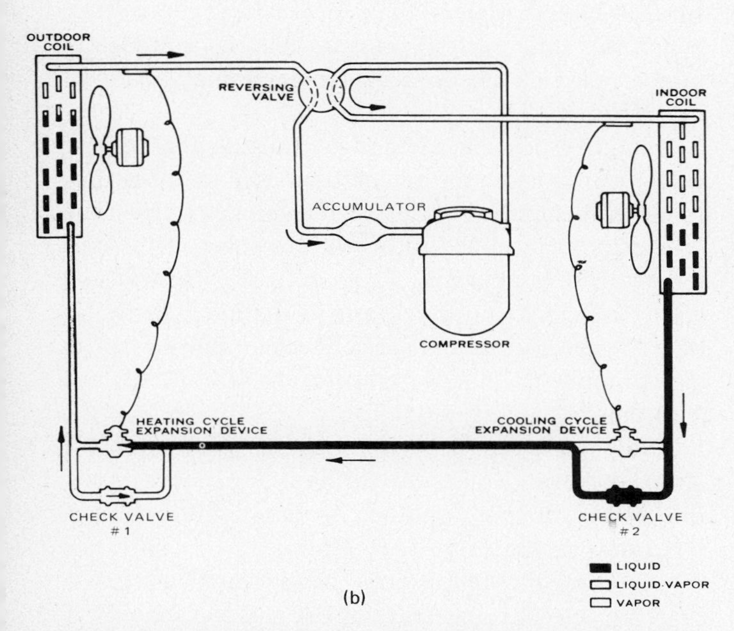

FIGURE 11-19(b) Heating phase (Courtesy, Westinghouse Electric Corporation).

valve or capillary tube) and evaporates in a normal manner. Suction gas comes off the top of the indoor coil and is routed back to the compressor through the reversing valve.

Upon heating, the system reverses as shown in Fig. 11-19b. The compressor discharge gas goes to the indoor coil where it gives up its heat and the refrigerant is condensed to a liquid. Check valve 2 is open to flow and permits liquid to travel to the left. Check valve 1 is now closed to flow and the liquid goes to the expansion valve for the outdoor coil where it's allowed to evaporate and absorb heat.

The one additional element of a heat pump refrigerant cycle that is most important is a suction line accumulator that protects the compressor from refrigerant floodback during the changeover cycles (heating to cooling and vice versa) and during the defrost cycle. The physical volume of this element depends on the capacity of the unit. Its shape is a function of the brand.

The preceding description is illustrative of how the heat pump system operates. In actual products from different manufacturers there are variations in system design or components, but the end result is essentially the same. If he is involved in the installation and service of a heat pump, the air-conditioning technician will become familiar with a specific brand and learn its specifications and operational characteristics.

Physically, a typical reversing valve looks like Fig. 11-20 on the outside. Internally (Fig. 11-21), it is composed of two pistons connected to a sliding block or cylinder with two openings. The illustration shows heating and cooling modes. The four-way valve is actuated by a solenoid valve that uses high-

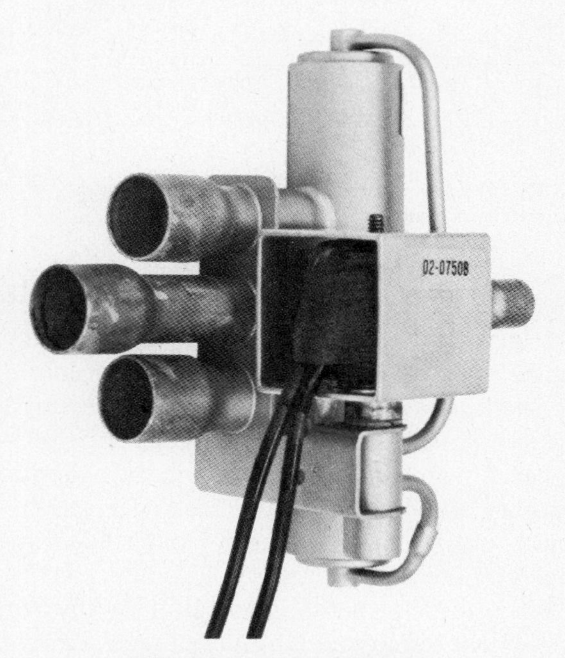

FIGURE 11-20 Reversing valve.

FIGURE 11-21(a) Reversing valve—cooling.

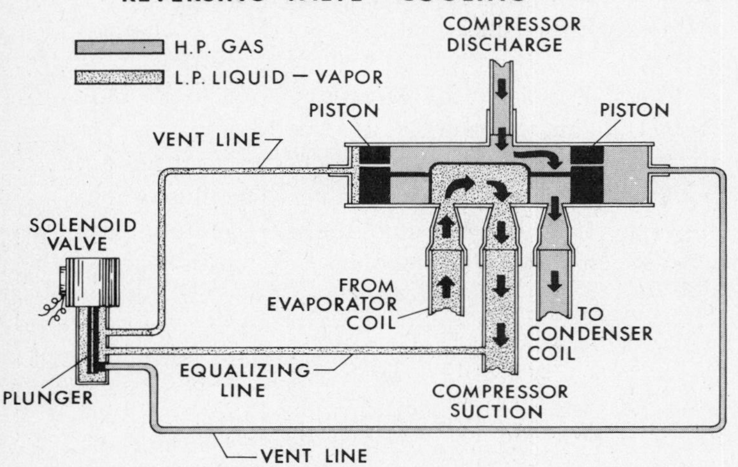

FIGURE 11-21(b) Reversing valve—heating (Courtesy, Westinghouse Electric Corporation).

pressure compressor gas to move the piston left or right depending on which mode is needed. The plunger position changes the vent lines accordingly. The small white lines in the pistons represent small diameter passages through which the high-pressure gas can slowly find its way, causing the action to be smoother and quieter. In the crossover action, too rapid a change in pressures could result in system shock and excessive noise.

When the heat pump is operating on the heating cycle so that refrigerant is evaporating in the outdoor coil and the temperature of the coil surface falls below 32°F, frost will begin to appear on the coil. If frosting is allowed to continue, the deposit of ice will gradually build up until the flow of air through the coil is restricted. This will decrease heat transfer and seriously affect the efficiency of the system. So periodic defrosting is needed to remove frost.

The process of defrosting is simply to reverse the cycle and direct hot gas to the outdoor coil for a period long enough to melt the ice. The outdoor fan is "off" during this action. No heat is being produced inside (actually the indoor coil is refrigerating), so it's necessary to provide supplementary heat from resistance heaters in the indoor section.

Starting and stopping the defrost cycle is accomplished by one of several techniques. One method of automatically starting the defrost cycle is to measure the air pressure across the outside coil. As frost builds on the coil, it becomes more difficult for the air to find its way through the coil, so upstream pressure becomes higher than downstream pressure. When this difference in pressure reaches a predetermined point on the control, the reversing valve will be activated.

Another popular method of starting the defrost cycle uses a combination of time and temperature. A clock mechanism is set to reverse the refrigerant flow at predetermined intervals. However, if the coil temperature is above 32°F the defrost cycle will not start. The clock continues, and at the next scheduled time the temperature is again checked and defrost is started only if outside temperature conditions are sufficient to cause frost to form.

Now let's see how the defrost cycle may be stopped. Several different methods have been used; one based on time, another on temperature, and a third on pressure.

A *timing mechanism* may be activated when the defrost cycle starts so that defrost will continue for a preset period. At the end of the time period the reversing valve again turns back to the heating cycle. The time period can be adjusted depending on the severity of winter conditions and an estimate of the maximum amount of frost that might be formed. It is sometimes difficult to set the timer to cover all frosting conditions and still keep the defrost cycle within reasonable limits.

A more positive method of terminating the defrost cycle is by *temperature*. While the frost is melting, the temperature in the vicinity of the outside coil will remain fairly constant—in the neighborhood of 32°F. When all the frost is gone, the temperature will begin to rise. This temperature change can be used to reverse the flow of refrigerant. If a temperature device is used to start the defrosting, the same thermocouple or thermal bulb can be used to stop the defrost cycle.

The *pressure* of the refrigerant in the outside coil can also be used to end the defrost cycle. As with temperature, the pressure of the refrigerant will stay nearly constant while the defrosting operation is going on. When the frost is all gone, the pressure of the refrigerant in the coil will rise and this change can be used to signal the four-way valve to reverse.

The *air pressure differential* means of terminating the defrost cycle is, of course, used in conjunction with the air-pressure control that initiated the action. When the restriction is removed, the control will signal the reversing valve to return to heating.

Now with an understanding of how a heat pump functions, the reader might very well ask "How do heat pumps save on electric heating costs?"

In previous discussions, we referred to the fact that a heat pump is 2.5 times more efficient than straight electric resistance heating. For each unit of electricity put in, resistance heaters produce one unit of heat output; we call this a 1-to-1 ratio. But for each unit of electricity put into a heat pump, under ideal circumstances more than 2.5 units of heat output will be produced, or a 2.5-to-1 ratio. This ratio of heat output is known as the COP, or the coefficient of performance. This is the amount of heat moved, divided by the energy needed to move it. One may still ask, "How is this 2.5-to-1 COP possible?"

Even at 0°F outdoor temperature there is heat in the outdoor air—approximately 0.835 Btu/lb of dry air. The expanding refrigerant in the outdoor coil, at 10° to 15° below the heat source, picks up this heat and sends it to the compressor by means of the suction line. The compressor raises its temperature and pressure and pumps it to the indoor coil at 10° to 20°F higher than the heating medium (indoor air). Without going into a detailed discussion of the Carnot or Rankin cycle, it suffices to say that the work load needed to raise the temperature and pressure of the refrigerant from about −15°F to 110°F is considerable.

The modern compressor-motor combination (hermetic compressor) is about 85 to 90% efficient, and the resulting heat generated during compression (like an air pump) is added to the refrigerant. In addition there is the heat of friction developed between the moving parts in the compressor and the heat generated in the motor windings as it drives the pistons, both of which are added to the refrigerant gas. The sum of all three of these heats (less a 10 to 15% loss through the shell) is discharged to the indoor coil.

The electrical energy input to the compressor motor when compared to the same amount of heat produced by electrical resistance heaters is approximately half as much; or, putting it another way, the heat pump is twice as efficient. This example was at 0°F. This is why the heat pump is an important energy conservation unit.

However, the COP is variable because the capacity of the heat pump on the heating cycle becomes less as outdoor temperatures become lower. Figure 11-22 represents the effect of outdoor temperature on the heating COP of an air-to-air heat pump.

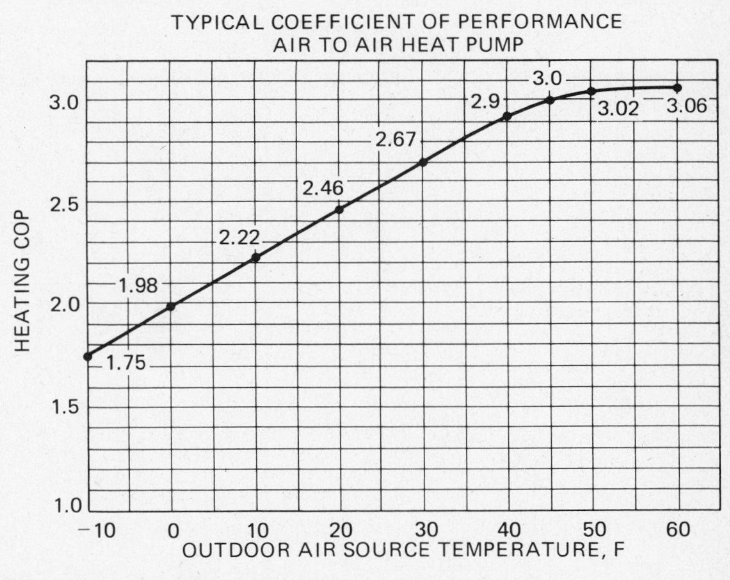

FIGURE 11-22 Typical heat pump COP (Courtesy, Carrier Air-Conditioning Company).

Note that a 3.0 COP occurs at approximately 45°F, but even at 0°F it is approximately 2.0, which is still most favorable. An example is the heating season in Chicago (Fig. 11-23), which represents a northern application. Outdoor temperature is shown at the bottom, reading from 60°F on left to −20°F on right. The hours of occurrence are shown vertically. The curve is a historical weather pattern.

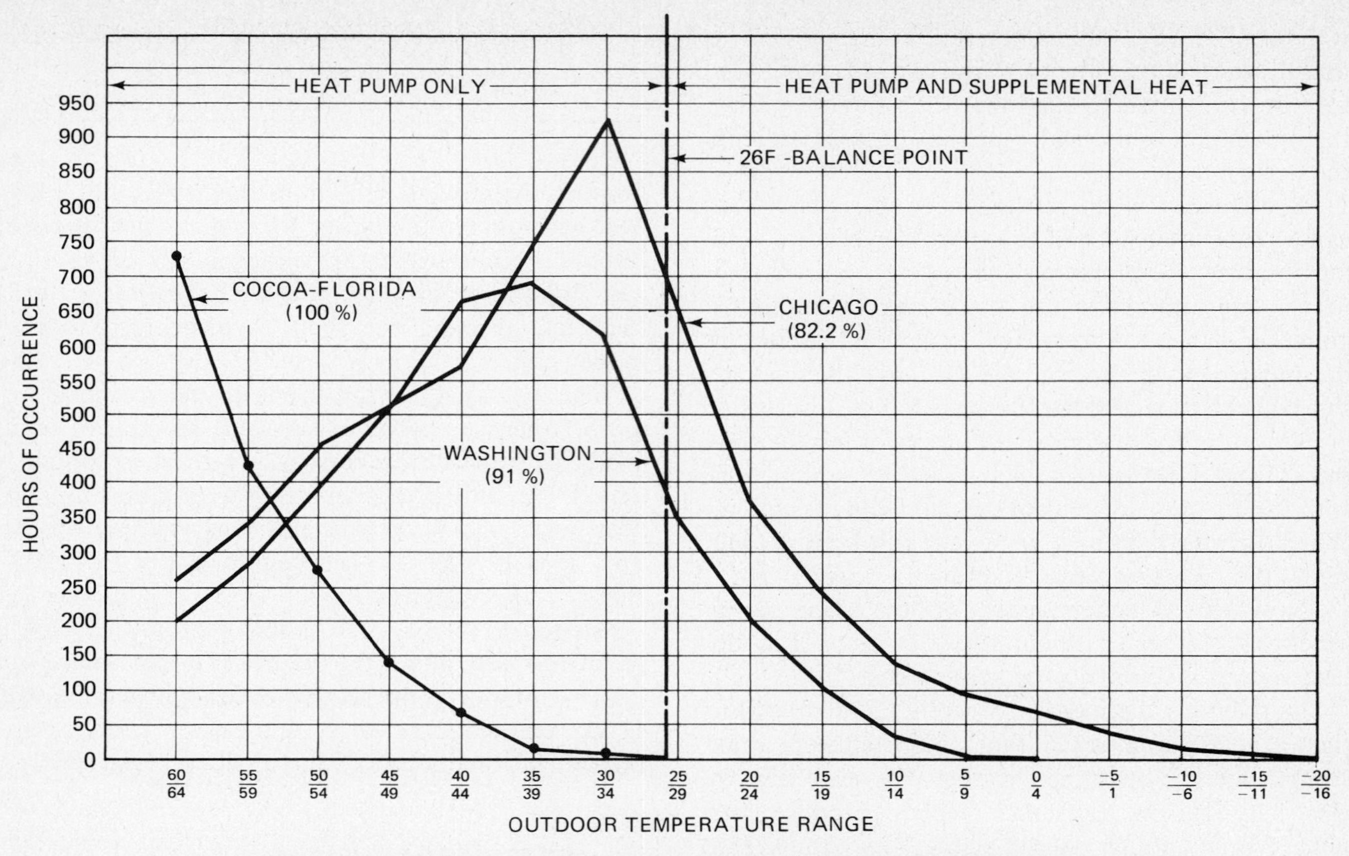

FIGURE 11-23 Outdoor temperature range (Courtesy, Carrier Air-Conditioning Company).

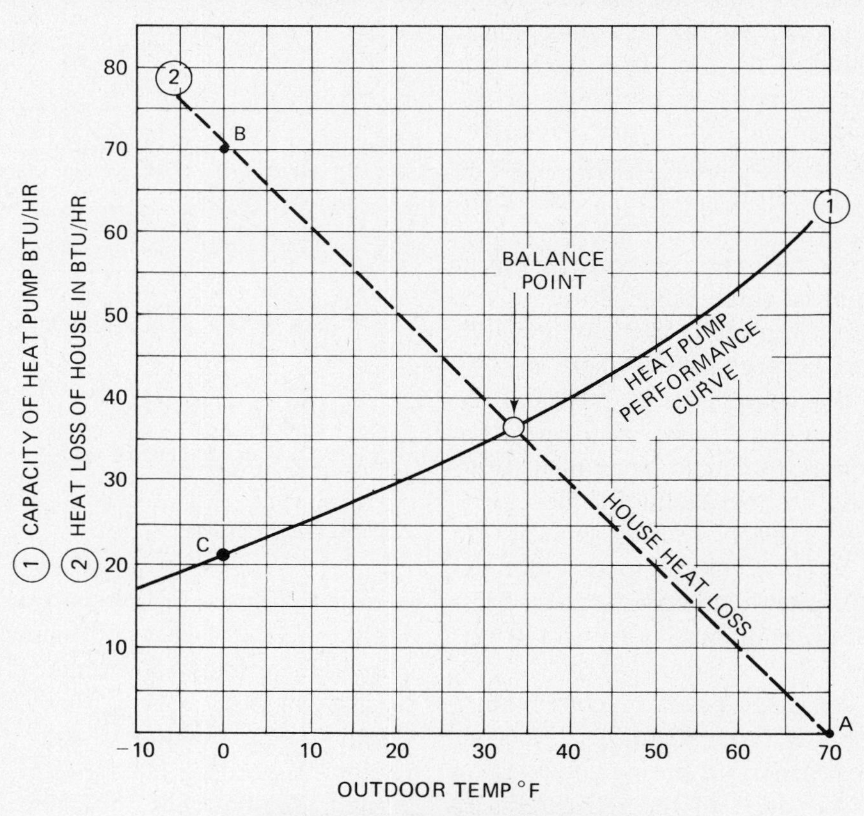

FIGURE 11-24 Heat pump balance point.

Note some 82% of the heating hours is above 25°F, and the COP of the example heat pump shown in Fig. 11-24 at 25°F is 2.5. Thus considerable savings in operating costs can be achieved over straight resistance heating. Naturally the farther south, the better the efficiency, as illustrated for Washington, D. C. and Cocoa Beach, Florida.

Can a heat pump supply all the heat a home will need without booster heaters? The answer is generally *no*, except in extreme southern Florida. A heat pump is basically a cooling unit and should be selected for that use. Oversizing the cooling to gain heating efficiency will result in poor summer operation. The unit will short cycle, causing temperature and humidity control to be erratic—not to mention the added cost of unneeded equipment. Undersized cooling can penalize both winter and summer operation. So select the cooling size just as for any air-conditioning system.

Once the cooling capacity is determined you can plot the resultant heating operation in a home. Figure 11-24 represents the heat loss from a typical house. At point *A* (70°F) no heat is needed. At point *B* (0°F) it requires 70,000 Btu/h, and you can determine the heating requirement at any temperature condition in between by drawing a straight line between *A* and *B*.

On this same chart you can plot the heating capacity of the selected heat pump (data obtained from the manufacturer's performance tables) at the various outdoor temperatures. The point at which these two lines cross is called the *balance point*. In this example, the balance point is about 33°F. Below 33°F the heat pump cannot provide enough heat to keep indoor temperature at 70°F. Thus supplemental electric heaters are needed to support the reverse cycle operation. They are either built into the indoor unit or mounted in the supply air duct. These same heaters are also used during the defrost cycle. The number or capacity of supplemental heaters must at least equal the difference of heat needed between point *B* and point *C* at 0°F design conditions (approximately 50,000 Btu/h, or around 15 kW).

The range of balance points obviously shifts with the geographic conditions. An identical house built in Chicago, Ill. and Mobile, Ala. will have the same cooling selection based on 95°F summer design, but the winter design temperature is −4°F and +28°F respectively. Thus the balance point in Mobile will be considerably higher, but there will still be the need for some supplementary heating and for defrosting.

Air-to-air heat pump product designs are very similar to the unitary air-cooled systems previously reviewed. The self-contained horizontal packaged heat pump (Fig. 11-25) is designed to be installed at the ground level of a residential application, with ductwork extending into the basement or crawl

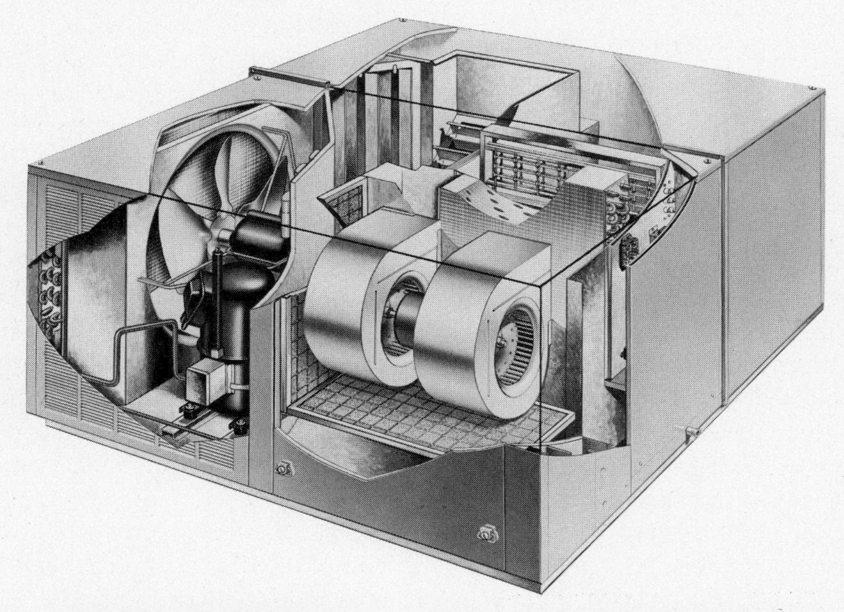

FIGURE 11-25 Horizontal package heat pump (Courtesy, Lennox Industries).

FIGURE 11-26 Vertical wall model (Courtesy, Westinghouse Electric Corporation).

space. Ducts could also go into the attic of a slab-type home as shown in Fig. 11-25. This same design can also be mounted on the roof of a commercial building. Sizes begin at the nominal 2-ton cooling capacity and go up to 15 tons. The larger commercial pumps from some manufacturers also contain the economizer system, which provides free cooling by using outdoor air during intermediate seasons.

In addition to the familiar horizontal configuration there are vertical models which can be mounted (hung) on an exterior wall (Fig. 11-26). These are more frequently associated with apartments, mobile structures, and special applications rather than conventional whole-house systems.

Split-system heat pumps (Fig. 11-27), like their counterparts in cooling systems, are the most flexible over a broad range of applications from apartments to residential to commercial usage. The outdoor unit and indoor unit are connected by refrigerant lines, except that in heat-pump technology they are not called liquid and suction; they are liquid and vapor lines. The vapor line can be for suction or discharge gas depending on which mode the unit is operating. Tubing connections are of the precharged, quick-connect, compression, flared, or soldered type, depending on size and brand. Also, in heat pump work the precautions against moisture entering the system during hookup are most critical. The hazard of refrigerant flow restriction due to ice is far greater where the outdoor unit is functioning as the evaporator with refrigerant temperatures at 0°F or below. For this reason, the quick-connect precharged methods are more popular and reliable on the smaller residential unit. It's common practice to provide a liquid-line drier either factory installed or field applied.

The discussion about proper concrete foundations and gravel fill also applied to heat pumps installed at ground level. But there is an additional requirement peculiar to heat pumps. During winter there will be drainage from defrosting and the possibility of ice accumulation around the equipment. Its elevation above ground should be sufficient so there is good water runoff and a minimum chance of blockage.

Air-to-air split-system heat pumps range from small 18,000 Btu ($1\frac{1}{2}$ tons) to approximately 50 tons. The indoor unit will have supplementary heaters installed within the cabinet or as an accessory section on the fan discharge outlet. Indoor units for residential application are available for upflow, downflow, and horizontal air flow. For commercial application, the indoor unit may be equipped with an accessory discharge grille and plenum.

Most heat pumps (packaged and split) are capacity tested, rated, and certified against ARI Standard 240—on cooling, 95°F DB outside air; 80°F DB and 67°F WB indoor air, and on heating, 45°F DB and 43°F WB outdoor air and 70°F DB indoor. Heating capacity is for the compressor operation only and does not include any supplementary heating. Residential heat pumps are also considered for rating and certification in compliance with ARI Sound Rating Standard 270.

The control system of a heat pump is a bit more complicated than a conventional cooling system. The indoor thermostat will be a 24-volt variety as shown in Fig. 11-28, with either manual or automatic

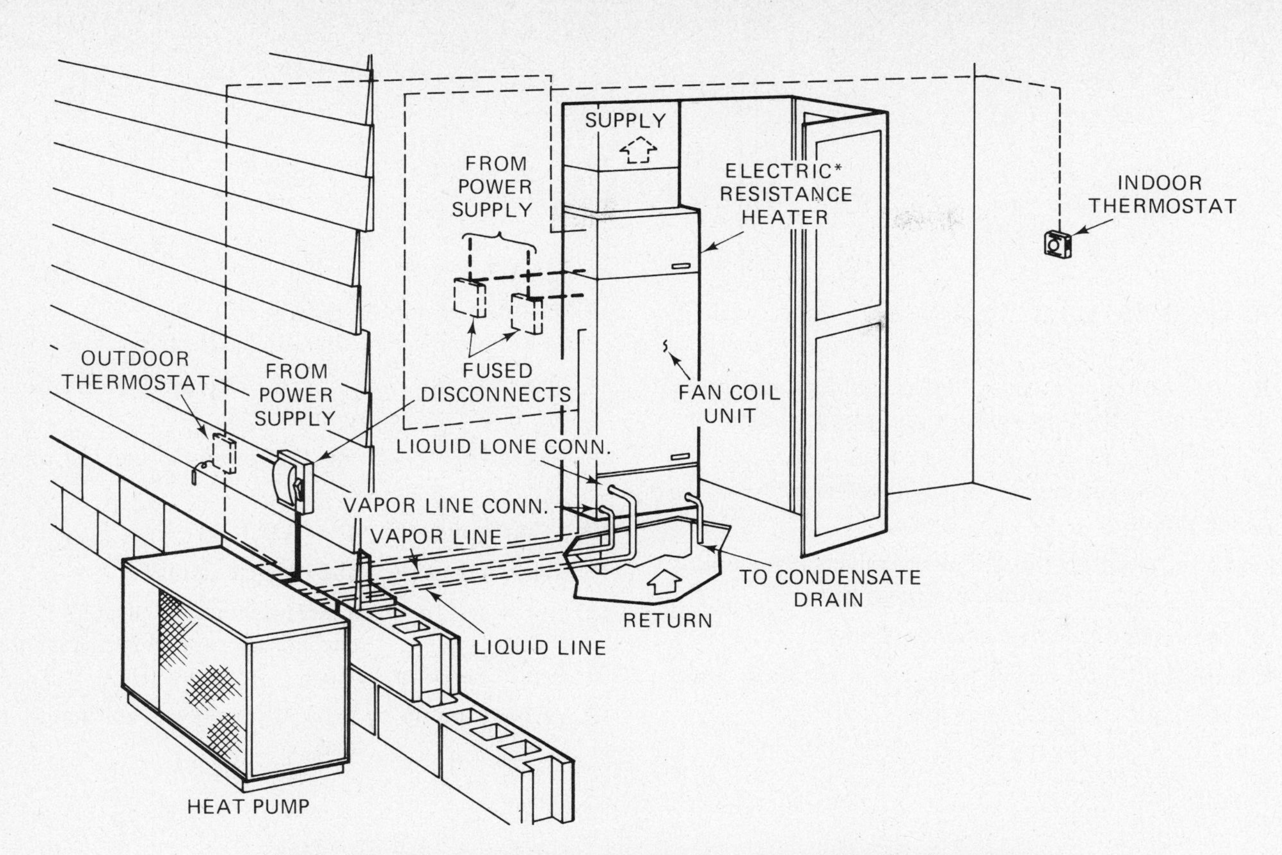

FIGURE 11-27 Split system heat pump (Courtesy, Carrier Air-Conditioning Company).

changeover from heating to cooling. If an automatic model is installed, there will be about a 4° difference between heating and cooling settings. The indoor fan may be operated continuously (ON) or set to automatic, which will cycle it off and on with the compressor. Some thermostats are equipped with an emergency heat switch that permits the homeowner to turn on full supplementary heat in the event there is no heat from the refrigerant cycle, for whatever reason. A red warning light will stay on until the switch is returned to normal after the malfunction is corrected.

The heating portion of the thermostat is a two-stage operation. The first stage brings on the compressor. If this is not sufficient, the second stage will automatically bring on the first element of electric heaters. The second stage of the heating thermostat is also wired in series with the outdoor thermostat control circuit, so it can bring on more supplementary heat if necessary.

Outdoor thermostats are used to control resistance supplementary heaters so that they operate only when required for maximum economy. They are actuated by outside air temperature and function during the normal heating cycle, when the system switch is set at automatic to quickly sense sudden drops in outside temperatures and bring in additional stages of electric heat before the inside temperature becomes uncomfortable. Each outdoor thermostat has an adjustable temperature range and is set in conjunction with the system balance point.

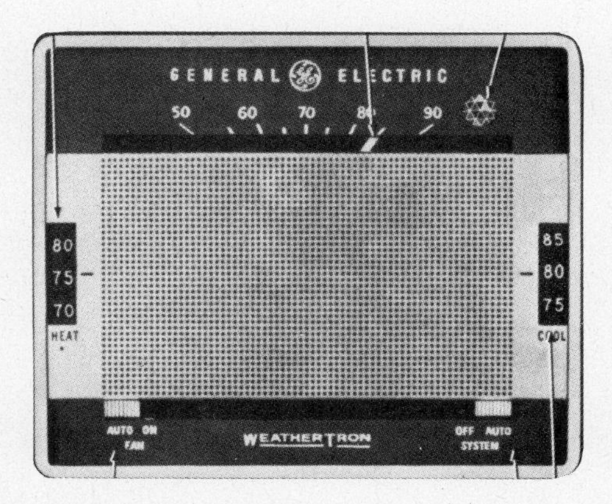

FIGURE 11-28 Indoor thermostat (Courtesy, General Electric Corporation).

PROBLEMS

1. Packaged terminal units are also called ______________.
2. Rooftop combination heating and cooling units also divide into two other descriptions based on air flow. Name them ______________, ______________.
3. The use of outdoor air for mild-weather cooling is called the ______________ cycle.
4. Gas fired rooftop units almost universally employ ______________ ignition and reignition.
5. What is meant by *single point hook-up*?
6. Concentric duct connection on a rooftop unit permits ______________.
7. The heat pump in general will be ______________ to-1 more efficient than direct electric heating.
8. The ______________ valve changes the direction of gas flow in a heat-pump cycle.
9. Removing ice from the outdoor coil is termed ______________.
10. What is meant by the COP?
11. What is meant by the balance point?
12. Booster or supplementary heaters are ______________ needed to supply supplementary heating and for defrost.
13. A heat pump is basically selected to match the ______________ load.

12

CENTRAL STATION SYSTEMS

The name *central station system* or *equipment* is also commonly called an *applied system*, *applied equipment*, *applied machinery*, and *engineered system*. Essentially these are names adopted by the various companies who manufacture the equipment as a means of identifying the division or departments within their organizations. The end results are essentially the same.

Central station equipment is associated with installations where the cooling plant is located in the basement or in a penthouse on the roof of multistory buildings. It serves air-handling equipment and air distribution systems throughout the building. Although size is not necessarily the crossover point between unitary and central station, it is usually acknowledged that central station equipment starts at 25 to 50 tons and extends upward to the multi-thousand-ton systems. Unitary equipment taper off at the 50 to 75-ton range.

Another distinguishing difference is that central station systems use the medium of liquid (mostly water) to transfer heating and cooling to a space air terminal, while unitary systems are based on distributing conditioned air directly to the conditioned space. Unitary equipment makes use of factory packaged, balanced, and tested equipment, which requires a minimum of on-site labor and material to be operational. Central station systems are made up of separate components such as chillers, air handlers, water towers, controls, etc., which can become quite complex in terms of on-site installation and labor and related trades and crafts. Central station equipment is closely associated with the *plan and spec* engineered systems put out by consulting engineering firms. Equipment is then selected and/or built to order in compliance with these specifications. Delivery time can vary from a few months to even a year or more on very large apparatus.

The art of selecting and matching the components is done by skilled professional engineers, and the servicing technician need not be concerned with that function. The service technician is, however, concerned with the system's installation, operation, and maintenance and, therefore, must learn the basics of the equipment being utilized and the system application. So we will now examine the scope of the

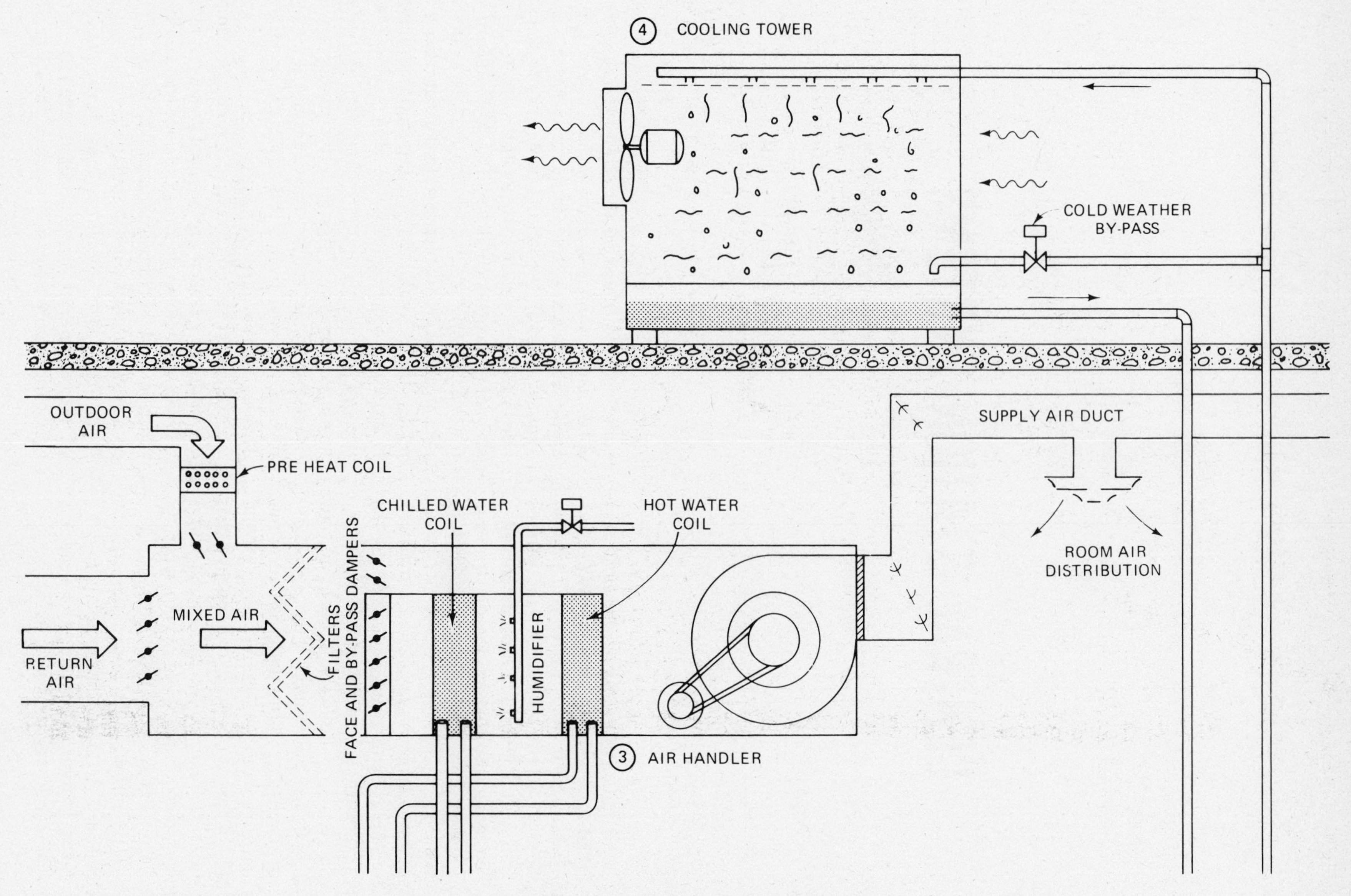

FIGURE 12-1 Conventional central station system.

systems available, beginning with an orientation sketch of a conventional central system (Fig. 12-1). First, identify the major components:

(1) Water chiller

(2) Boiler

(3) Air-handling unit

(4) Water-cooling tower

(5) Control system

The water chiller will produce 40°–45°F cold water, and by means of a pump circulate it to the cold-water coil in the air handler. Water off the coil will generally return at a 10° rise.

Similarly in winter the boiler will produce hot water at 180° to 200°F and pump it to the hot-water coil. Note that it is possible to have the boiler and chiller operating at the same time, because in large buildings there may be need for cooling and heating in different zones.

Condenser water off the chiller (95°F) is pumped to cooling tower spray nozzles where it is cooled within the tower to around 85°F and then is returned to the condenser. The tower bypass permits regulating return water temperature as a function of ambient temperature changes, but it is never less than 70°F.

The air-handling unit or units, depending on the number of floors or zones, generally contain:

- Chilled-water coils
- Main hot-water coil (can be steam)
- Humidifier
- Filters
- Face and bypass dampers
- Dampers for mixing return air and outside air
- Blower and motor

A preheat coil is frequently required where large amounts of outside air at or below 32°F are needed either by code or because of special application. It is installed in the outside duct and is controlled as a function of entering air temperature to the main heating coil.

The face and bypass dampers either permit all the air to go through the humidifiers and the heating and cooling coils or allow some of it to be bypassed, depending on the particular situation. All air to the conditioned space is always filtered or cleaned.

The control system must direct the operation of these elements and do it automatically. In large installations the control system is usually a separate installation job from the air-conditioning equipment. Controls may be electric, electronic, or pneumatic (air), or a combination of all three and must be included in the initial construction stages. Although it takes a thorough knowledge of controls to plan and install the system, it isn't so complicated when the elements are studied individually. Basically the water chiller operates by sensing the temperature of chilled water and controls it accordingly. The tower fan and condenser water pump come on when the chiller operates. The boiler is activated by hot-water temperature. The tower bypass valve controls the water by temperature. Space thermostats and humidistats control the functions within the air handler. Chilled-water and hot-water circulating pumps are a function of summer/winter operation and generally operate on a continuous basis.

Naturally, this is an oversimplification of the control system, and there are many details a technician must learn, but these come with exposure and experience and are also a function of the type of system. This illustration is only one approach; others may include air-cooled condensing.

With the system components identified let's review the equipment hardware.

WATER-CHILLING EQUIPMENT

Packaged Chillers

The design and range of water chillers is somewhat related to the compressor and condensing medium (air or water). Starting at the lower capacity, the so-called packaged water chiller uses a reciprocating compressor (or multiples thereof). Figure 12-2 represents a small water-cooled chiller in the 20-ton range that has reciprocating compressors, liquid chillers and water-cooled condensers, compressor starters, controls, and refrigerant and oil pressure gauges neatly enclosed in a cabinet. This style of unit is ideal for motels, small office buildings, etc.

FIGURE 12-2 Small package chiller (Courtesy, York, Div. of Borg-Warner Corporation).

FIGURE 12-3 Reciprocating large package chiller (Courtesy, York, Div. of Borg-Warner Corporation).

Because it is packaged it can be factory operated and tested to job specifications. It must be used with a remote water tower or other sources of water for condensing.

Continuing with the reciprocating compressor, water-cooled models (Fig. 12-3), the larger packaged chillers range upward from 40 to 200 tons. These are characteristically multiple compressor models that permit close control of capacity; they have, as well, standby compressor protection in the event of a malfunction. Starting inrush current is reduced. Where single compressors are used they are equipped with cylinder unloading to allow capacity reduction and minimum starting power requirements.

The basis of heat rejection for both units described above is a shell-and-tube condenser (Fig. 12-4).

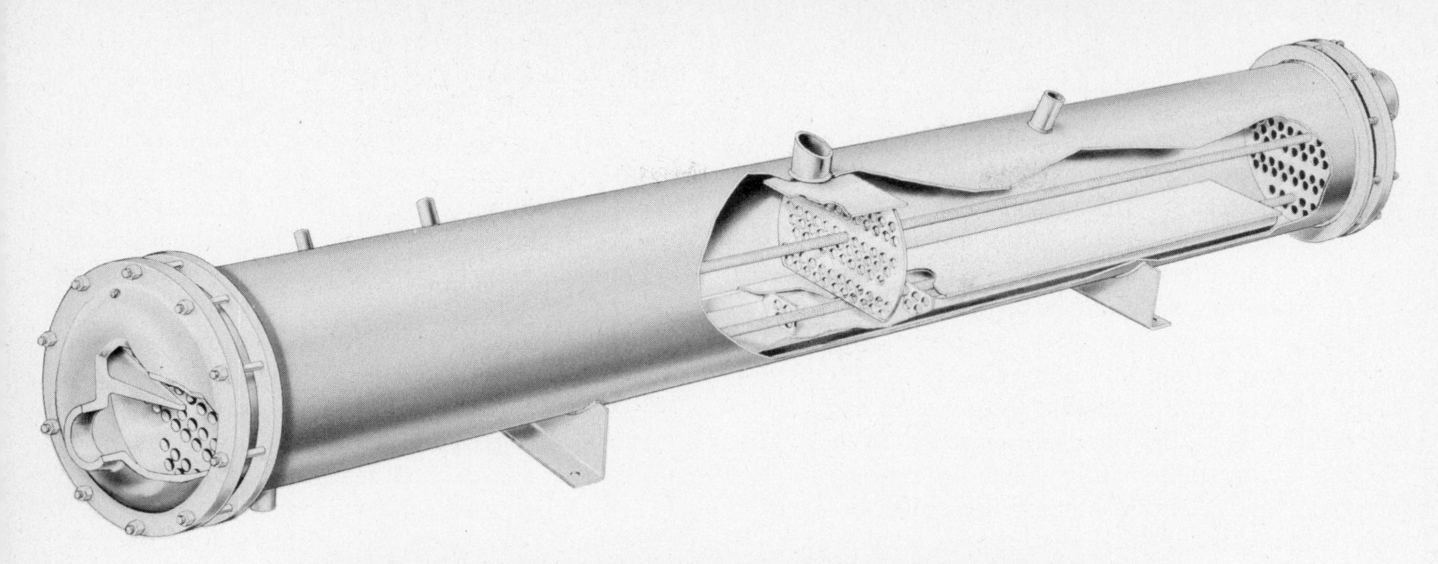

FIGURE 12-4 Shell and tube condenser (Courtesy, York, Div. of Borg-Warner Corporation).

Water flows through the tubes, and refrigerant vapor fills the shell, condensing to a liquid that is collected at the bottom and is subcooled 10° to 15°F for greater cooling capacity. In the water circuit, the coldest condenser water enters the lower part of the shell and circulates through the tubes. The water will make two or three passes through the shell before it is discharged. This is arranged by circuit baffles in the condenser heads. The higher the number of passes the greater the pressure drop or pressure required to produce the required flow rate. There are also cross baffles within the shell which serve to hold the tube bundles but also help to spread the refrigerant gas over the entire length of the shell. Condenser capacity is normally based on 85°F entering water temperature with a 10° rise. Calculated into the performance rating is a *fouling factor*, which assumes there will be scale deposits collecting inside the water tubes. This amounts to about a 4% reduction from the performance of a clean condenser Cleaning water-cooled condensers is influenced by the local water hardness conditions in addition to atmospheric contaminants that collect in the water tower operation. Chemical water treatment is a necessary routine maintenance activity, but eventually the condenser will need attention. Removable heads permit the mechanical cleaning of the tubes by reaming out the scale.

Most manufacturers offer a special sea-water condenser for marine duty where salt water is used as the cooling medium. These tubes are made of cupro-nickel steel to withstand the salt corrosion effects.

In the chiller size ranges associated with R-22 reciprocating compressors, the chiller (cooler), shown in Fig. 12-5, is of the direct expansion design.

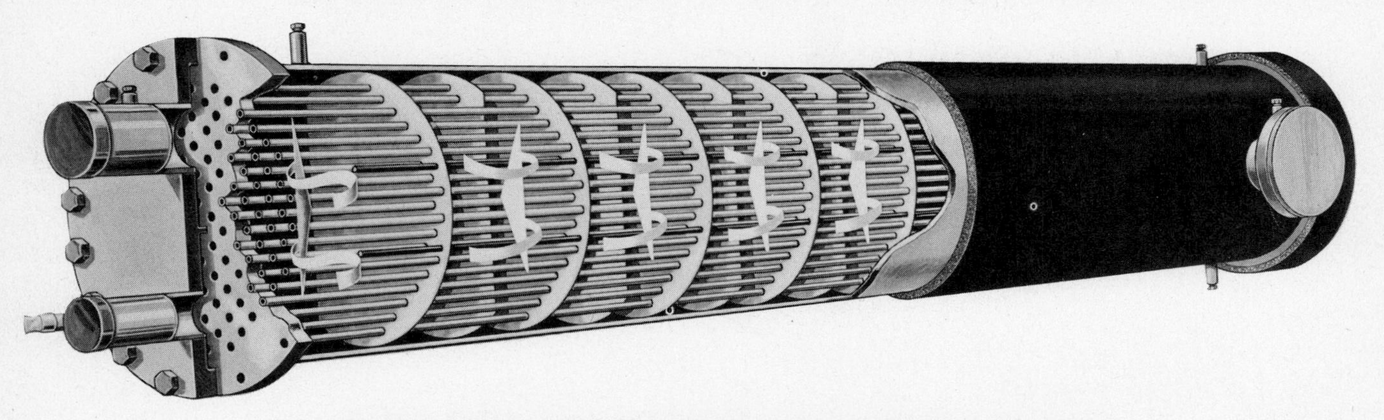

FIGURE 12-5 Chiller (Courtesy, York, Div. of Borg-Warner Corporation).

Refrigerant flows through the tubes and will generally make two passes for standard operation, which gives a counterflow arrangement versus the water flow pattern. Water connections are on the sides instead of the ends. Cross baffles direct the chilled water back and forth over the refrigerant-filled tubes at the optimum velocity for best heat transfer. The water system within the building is usually a closed loop. That is, it's not open to evaporation and contamination as is the condenser circuit. Even so a fouling factor is assumed in the chiller ratings.

The cooler shell and suction lines must be properly insulated to prevent sweating; this is done at the factory with a layer of closed-cell foam insulation prior to painting.

Both condenser and cooler shells must comply with ANS B9.1 and the applicable ASME safety codes for pressure vessels.

Standard chiller rating points are based on ARI Standard 590: 44°F leaving water temperature off the cooler at 105°F and 120°F condensing temperatures; also, 95°F leaving water off the condenser with a 10° rise. The 95°F condenser water rating will produce a condensing temperature near the 105°F figure. The rating point of 120°F condensing temperature is used for applications where remote air-cooled condensers (Fig. 12-6) are used instead of the water cooled type.

A more popular method of offering air-cooled water-chiller operation is the complete package (Fig. 12-7), which mounts on the roof and is not unlike a large air-cooled condensing unit, except that the cooler shell is suspended beneath the condenser coil and fan section. The cooler internal construction is as previously described, except that it must be protected against freezing. Electric heating elements are wrapped around the shell and then covered with a thick layer of insulation. Some manufacturers also add a final protective metal jacket that doubles as a good vapor barrier.

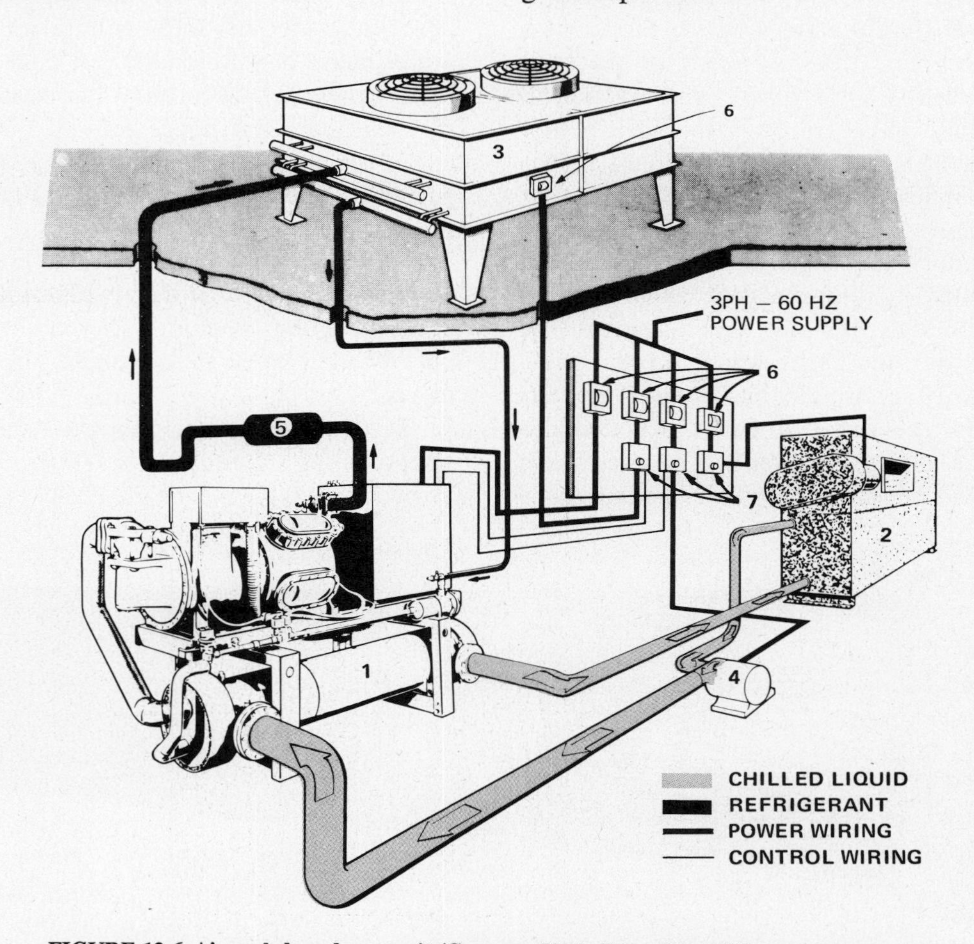

FIGURE 12-6 Air-cooled condenser unit (Courtesy, York, Div. of Borg-Warner Corporation).

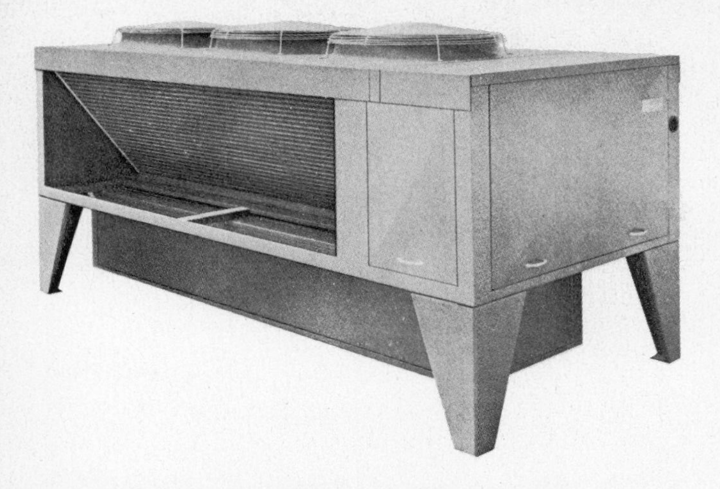

FIGURE 12-7 Packaged air-cooled water chiller (Courtesy, Westinghouse Electric Corporation).

The sizes of reciprocating compressor air-cooled packaged water chillers range from 10 to over 100 tons. These are also rated in accordance with ARI Standard 590 at 44°F leaving chilled water temperature and 95°F DB condenser entering air temperature.

In addition to its space-saving advantages and the elimination of the water cooled condenser and tower problems, the air-cooled packaged chiller has the ability to furnish chilled water at mild weather and low ambient conditions where water-cooled condensing methods would be subject to freezing problems. The techniques for low ambient control are as previously described for air-cooled condensing units and/or rooftop equipment.

CENTRIFUGAL CHILLERS (HERMETIC)

For very large installations the industry offers a range of hermetic centrifugal compressor water chillers of up to 1,300 tons in a single assembly; when used in multiples they handle applications of huge magnitude such as sports arenas, airports and high-rise office buildings. Interestingly, however, the comparatively new small centrifugal chillers, down to 100 tons and below, are beginning to challenge the popularity of the reciprocating machines at lower tonnages.

Hermetic centrifugal compressors (Fig. 12-8) vary in design and refrigerant use. Some are single stage, others multistage. Some are direct drive while

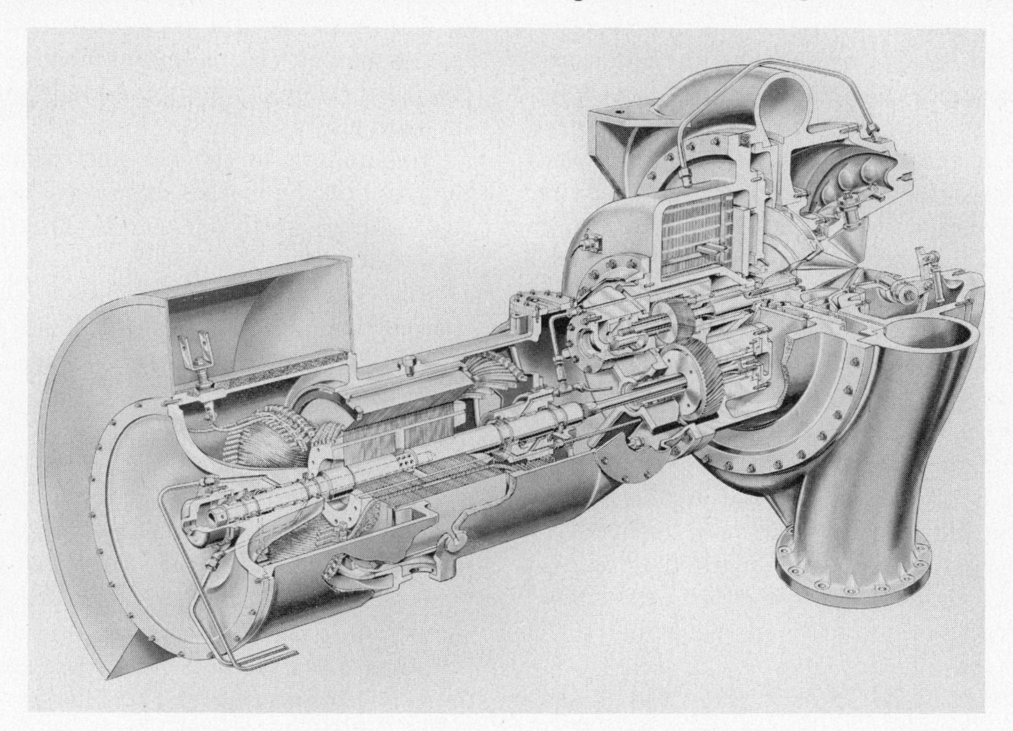

FIGURE 12-8 (Courtesy, York, Div. of Borg-Warner Corporation).

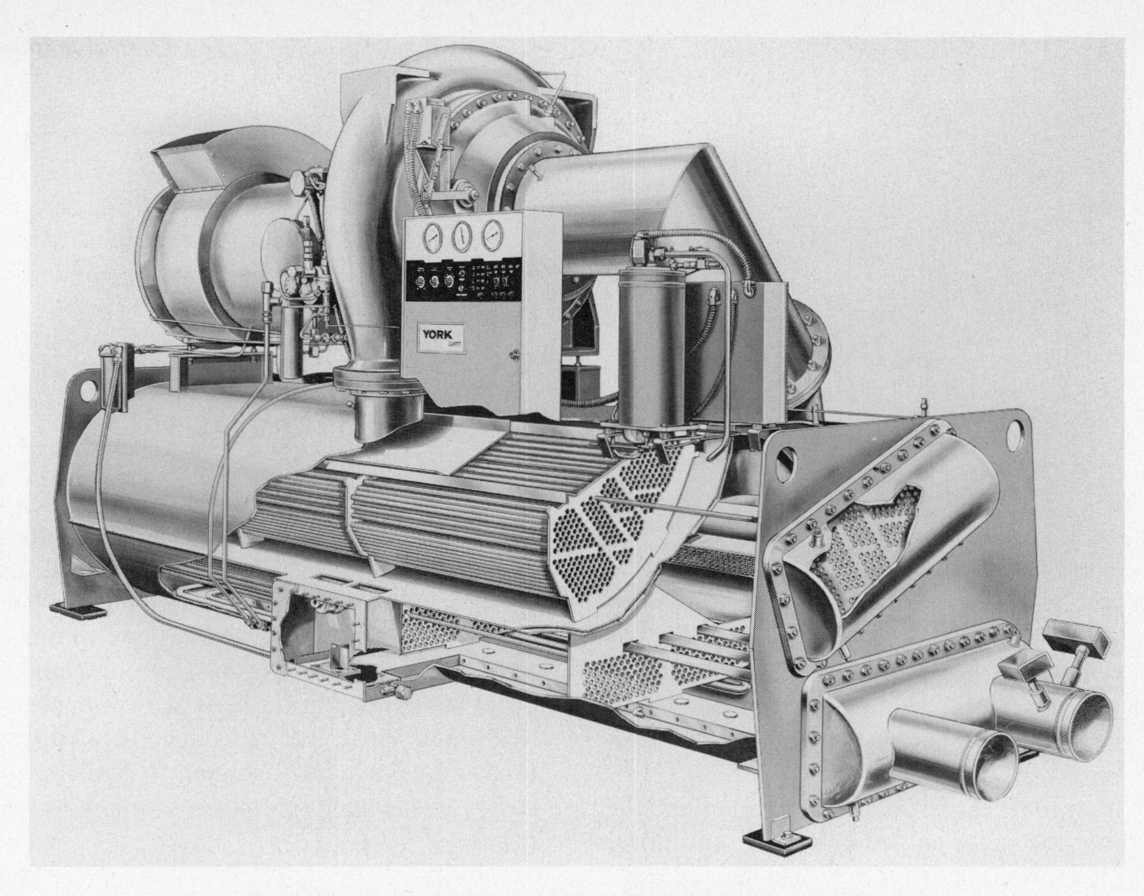

FIGURE 12-9 (Courtesy, York, Div. of Borg-Warner Corporation).

others are gear driven. However, as reviewed in *Refrigeration*, the basic principle of all hermetic compressors is the same—and that is to use a rotating impeller to draw suction gas from the chiller (cooler) and compress it through a voluted discharge passage into the condenser (water or air). The speed of the impeller is a function of the design, but some gear-driven units go up to 20,000 to 25,000 rpm. Capacity control is accomplished by a set of inlet vanes that throttle the suction gas to load or unload the impeller wheel.

A complete water-cooled hermetic centrifugal chiller assembly is illustrated in Fig. 12-9. This particular unit utilizes a combination chiller and condenser in one shell, although they are separated internally according to the respective function of each. Chillers of this size and design do not use expansion valves as such; they use either a float control or a metering orifice to flood the chiller shell with liquid refrigerant.

Hermetic centrifugal chillers are also being assembled in complete air-cooled units for roof mounting. Current product offerings range from 130 to 320 tons. The size and weight of these units are factors in their future development, as is the energy efficiency rating of air-cooled condensing methods in increasing tonnages.

In addition to the hermetic centrifugal design, there are on the market open drive centrifugal units that produce up to 5,000 tons. The choice of drive can be a compatible gas, steam, diesel, or electric motor. This category of centrifugal unit is somewhat specialized and not as easily designed, selected, and installed as are the complete packaged units.

Screw-Compressor Chillers

An important type of central station chiller is the screw-compressor design (Figure 12-10), which utilizes mating sets of gear lobes that rotate; during rotation the space or mesh between the lobes becomes smaller and smaller so that suction gas entering from the front is compressed and discharged at the end of the screw arrangement. The illustrated compressor is not hermetic but is driven by an external motor. The condenser and chiller (cooler) design is similar

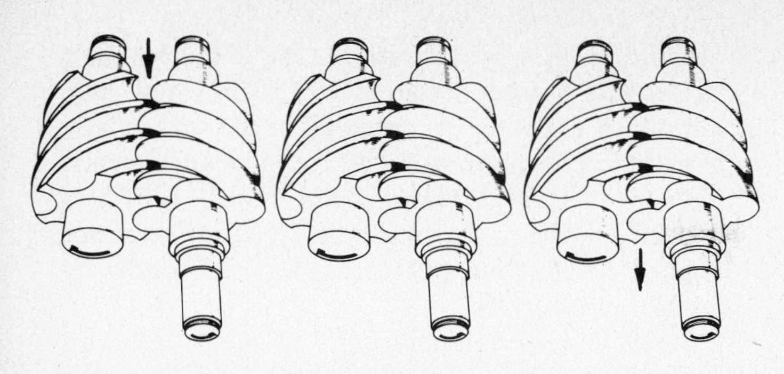

FIGURE 12-10(a).

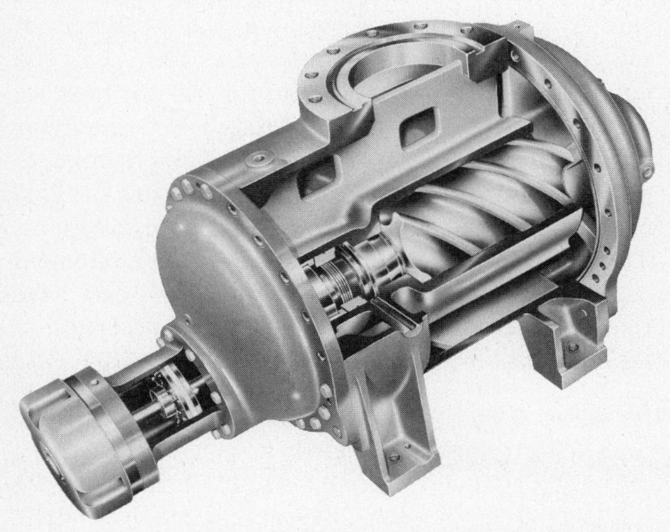

FIGURE 12-10(b) Screw compressor chiller (Courtesy, York, Div. of Borg-Warner Corporation).

to the shell-and-tube system already reviewed. The development of screw-compressor chillers has been mostly useful for the refrigeration needs of the food and chemical industries, but recently this type of compressor is becoming popular in the comfort air-conditioning market.

Absorption Chillers

Unlike the conventional mechanical compression refrigeration cycle used in all the other equipment discussed, an absorption chiller (Fig. 12-11) uses either steam or hot water as an energy source to produce a pressure differential in a generator section. The absorption cycle substitutes physico-chemical processes for the purely mechanical processes of the compression cycle. A complete thermodynamic analysis of an absorption cycle is relatively complex and is beyond the scope of this book. Also, it is a specialized field that should be considered by those who wish to become involved in that type of work. For further information, consult the ASHRAE *Handbook of Fundamentals and Equipment*. For our purposes it is sufficient to note that these units are available from 100 tons to over 1,500 tons. The weight and size of an absorption chiller is a major consideration in the selection of a system, its application, and its installation.

In the review of chilling equipment, we have worked with water as the fluid medium to be cooled and circulated, and so it is for most comfort air-conditioning applications. However, these machines will provide liquid cooling for other purposes, such as the circulation of low-temperature brine or glycol solutions for refrigerating in ice skating rinks, freezing plants, and the chemical, drug, and petrochemical industries.

With source of chilled water established, the next step is to review the terminal equipment that is used to transfer the cooling effect from the water to the conditioned space.

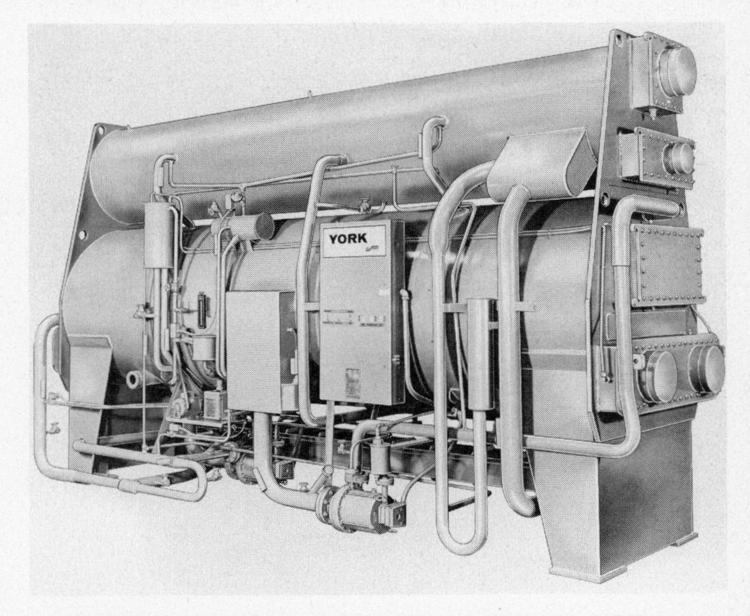

FIGURE 12-11 Absorption chiller (Courtesy, York, Div. of Borg-Warner Corporation).

AIR DISTRIBUTION EQUIPMENT

Fan-Coil Units

Fan-coil units for combination heating and cooling come in a variety of designs. Perhaps the most familiar is the individual room conditioner (Fig. 12-12) so frequently seen in office buildings, apartments, dormitories, motels, and hotels. It has an attractive cabinet enclosure with an air return on the bottom and a set of discharge directional louvers on top. Inside it consists of a filter, direct-driven centrifugal fan(s) and a coil suitable for handling chilled or hot water. The size of the unit is based on its cooling ability. Heating then is usually more than adequate. For use with chilled water and hot water, a variety of water-flow control packages are available for manual, semiautomatic, or fully automatic motorized or solenoid valve operation. Air flow is controlled by fan speed adjustment (manual or optional automatic). Outside air is introduced through a dampered opening to the outside wall. The size of the cabinet is designated in ft^3 of air flow and ranges from 200 to 1,200 ft^3/min. Coils are then picked to match cooling conditions based on the number of rows of coil depth.

These units may be installed on 2-pipe, 3-pipe, or 4-pipe system designs as illustrated and described in Fig. 12-13.

With slight modifications in cabinetry, the same components are assembled in a horizontal ceiling-mounted version (Fig. 12-14). Where hot-water heating is not desirable, electric resistance heaters may be installed in the cabinet on the leaving air side of the cooling coil.

FIGURE 12-12 Fan-coil unit (Courtesy, Carrier Air-Conditioning Company).

Another version of the room type of fan-coil unit is a vertical column design (Fig. 12-15), which may be installed exposed or concealed in the wall. These are placed in common walls between two apartments, motel rooms, etc. Water-piping risers are also included in the same wall cavity. They are not designed for ducting but can serve two rooms by adding supply grilles.

Small ducted fan-coil units (Fig. 12-16) with water coils may be installed in a drop ceiling or a closet, with ducts running to individual rooms in an apartment. These range in sizes from 800 to 2,000 ft^3. The unit's cooling capacity is selected based on the desired ft^3/min and the number of rows of coil depth needed.

Larger fan-coil units (Fig. 12-17), used to condition offices, stores, etc., where common air distribution is feasible, are somewhat like the cabinets of self-contained store conditioners. They can be equipped with supply and return grilles for in-space application, or they may be remotely located for ducting. Sizes range from about 800 ft^3/min to 15,000 ft^3/min. In larger sizes, cabinet and fan discharge arrangements permit flexible installations. These units approach the next category of equipment called *air handlers*, but in general, they don't have the size and functions available in central station air handlers.

Central Station Air Handlers

The air handler illustrated in Fig. 12-1 is a single-zone, low-pressure, draw-through system, which is typical of a small chiller application or where multiple air handlers are used to provide zone control However, where large volumes of air are to be distributed and/or where multizone or double-duct applications are needed, the blow-through type unit as shown in Fig. 12-18 is used. First, the unit is made up of dimensionally matching sections. Each section can then be selected for its own performance requirement.

Note the combination mixing box and filter section. The size of the outside air intake damper is such that 100% outside air may be taken in during intermediate seasons, just as was described for the economizer cycle of rooftop units. Filters may be of the throw-away or permanent (washable) type.

The blow-through fan section utilizes a large centrifugal fan, which is classified for either low-

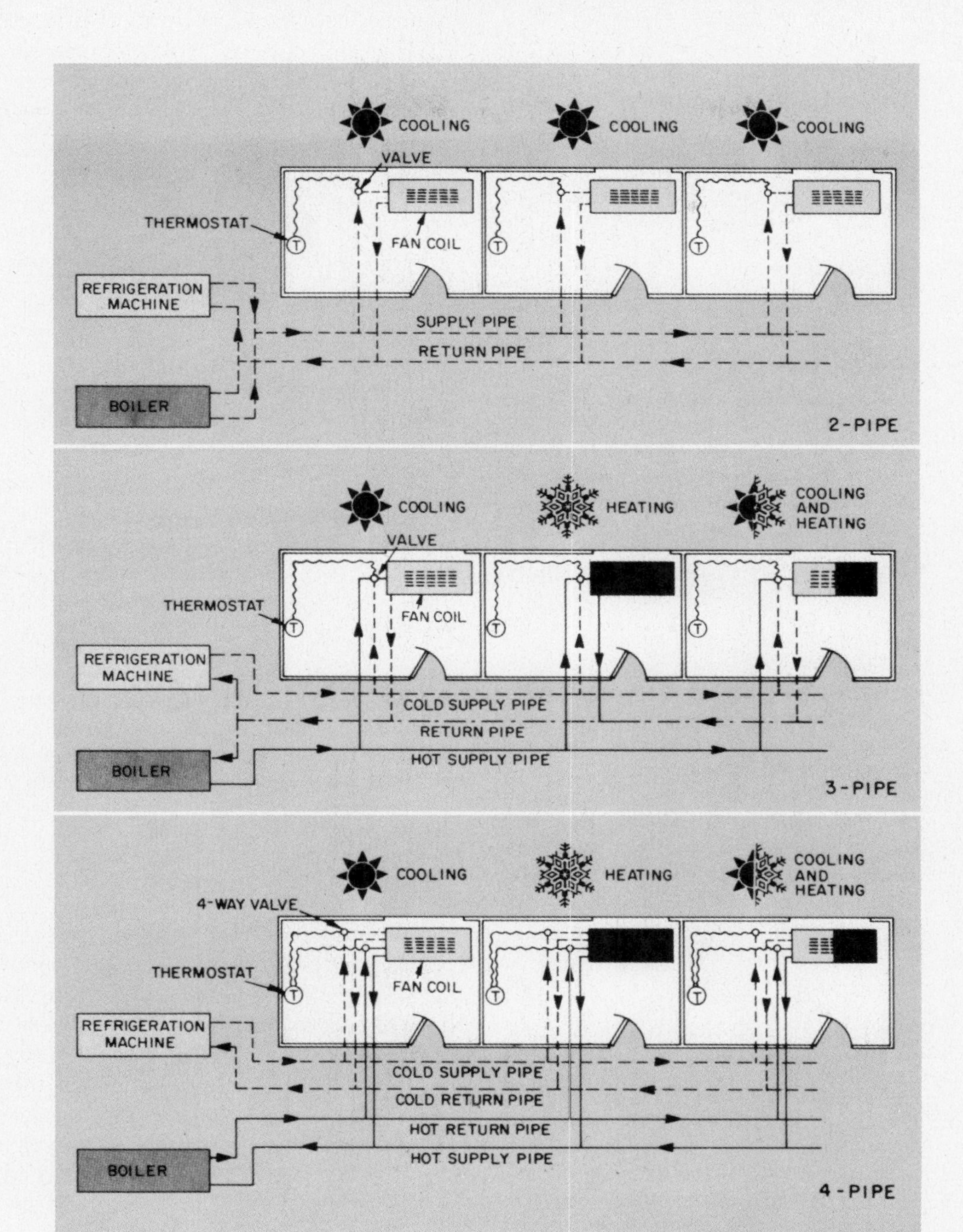

2-pipe system

Either hot or chilled water is piped throughout the building to a number of fan-coil units. One pipe supplies water and the other returns it. Cooling operation is illustrated here.

3-pipe system

Two supply pipes, one carrying hot and the other chilled water, make both heating and cooling available at any time needed. One common return pipe serves all fan-coil units.

4-pipe system

Two separate piping circuits — one for hot and one for chilled water. Modified fan-coil unit has a double or split coil. Part of this heats only; part cools only.

FIGURE 12-13 Fan-coil units (Courtesy, Carrier Air-Conditioning Company).

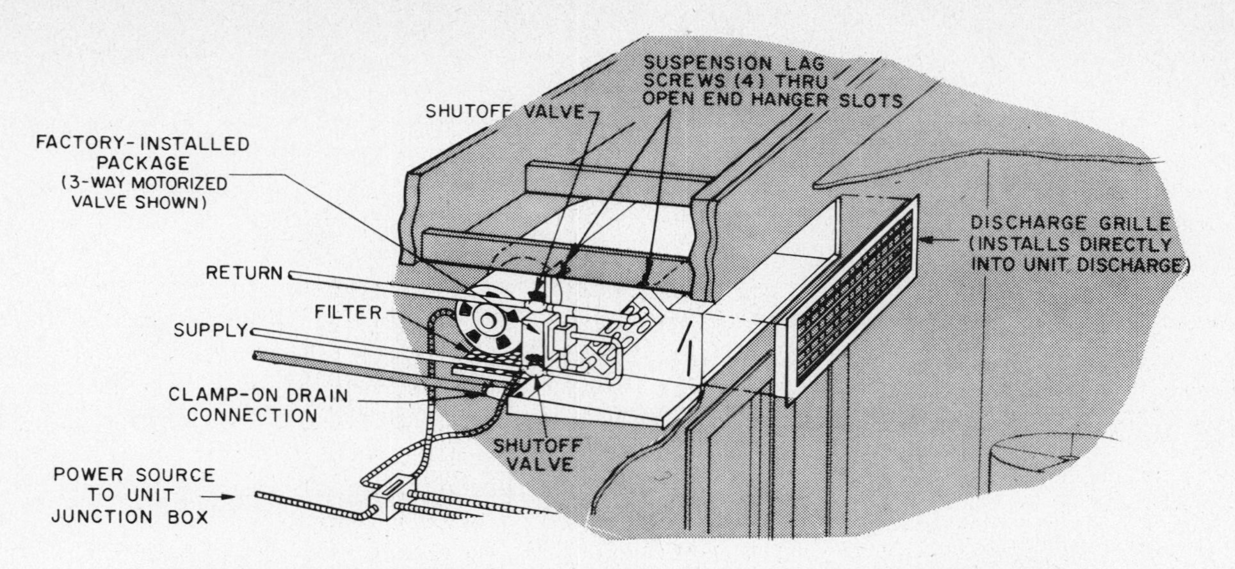

FIGURE 12-14 Ceiling mounted fan-coil unit (Courtesy, Carrier Air-Conditioning Company).

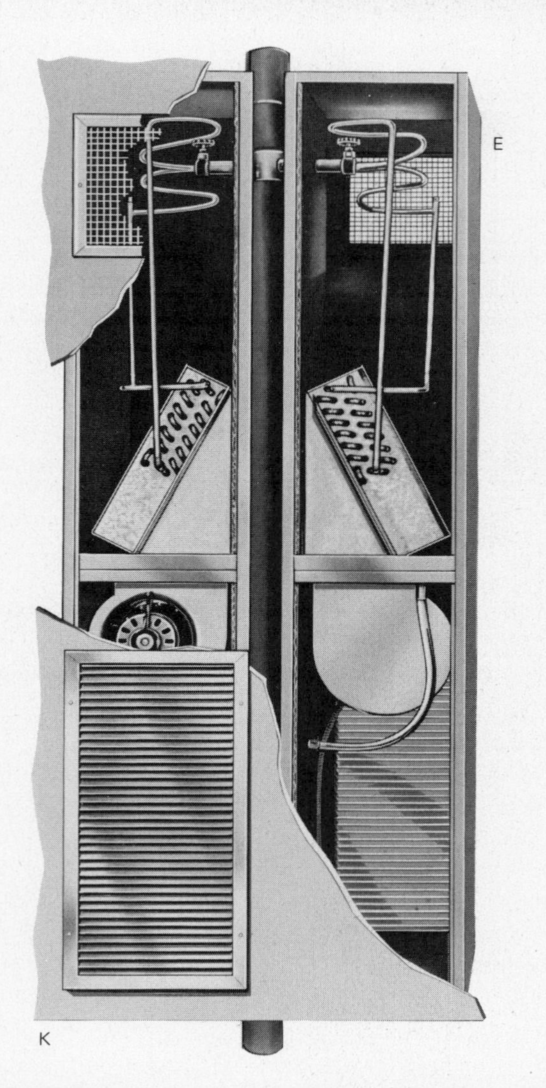

FIGURE 12-15 Room type vertical column fan-coil unit (Courtesy, York, Div. of Borg-Warner Corporation).

FIGURE 12-16 Ducted vertical fan-coil unit (Courtesy, York, Div. of Borg-Warner Corporation).

pressure or medium-pressure performance. Low-pressure wheels produce external static pressures from 0 to 3 in. of W.C. and usually have forward-curved blades. For medium static pressure duty from 3 in. of W. C. to 6 in. of W. C., units may use the air-

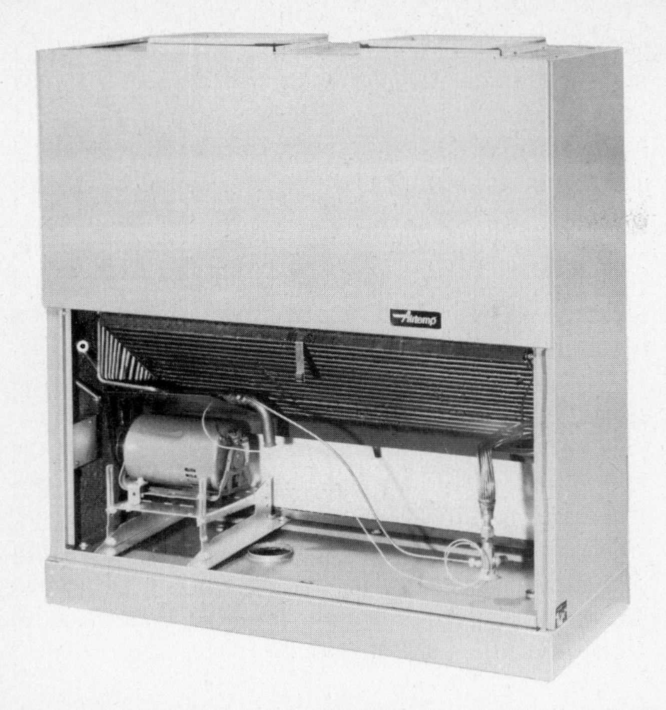

FIGURE 12-17 Blower fan-coil unit (Courtesy, Airtemp Corporation).

foil type of wheels as illustrated in Fig. 12-18. The larger fans may also be equipped with variable inlet vanes (Fig. 12-19) to control air flow performance. (Chapter 13 contains more information on basic fan designs and application.)

The coil section consists of a cooling coil and heating coil. Finned-tubed water coils for cooling (Fig. 12-20) are selected on the basis of the square feet of face area needed for the most efficient use of space; the rows and fin spacing necessary to meet performance standards at the lowest cost; and the proper circuiting for the best transfer, within pressure-drop limitations. Actually there are many possible options, and many manufacturers offer a computerized selection service to help the engineer in making the optimum selection.

Heating coils using hot water look identical to those shown for cooling, except for the face area and the number of rows of depth, which rarely exceeds three, since the temperature difference between the 180°F hot water and the 75°F conditioned air is so great, compared to the 44°F chilled water and 75°F air. Steam coils are also available and would be positioned in the area noted for the hot-water coil.

Note the grid diffuser (Fig. 12-18), which spreads the fan discharge air over the entire coil face. The position and separation of the two coils form a cold deck and a hot deck, so that air may be blended to

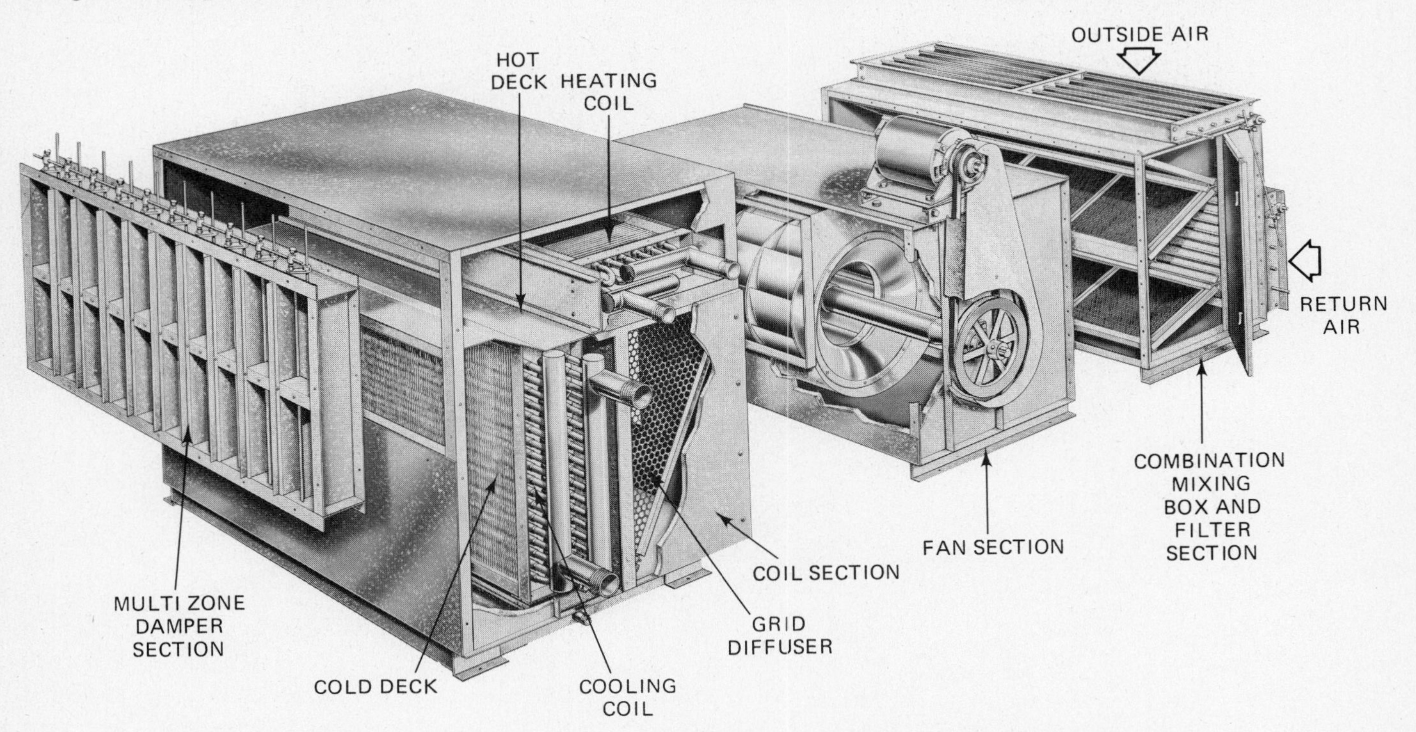

FIGURE 12-18 Blow-through central station air handler (Courtesy, York, Div. of Borg-Warner Corporation).

FIGURE 12-19 Variable inlet vane dampers (Courtesy, York, Div. of Borg-Warner Corporation).

provide the individual zone requirement. In many large buildings the perimeter (outside) zones may need heating while the interior core, because of lights, people, and equipment, may call for cooling. The multizone damper section will individually control each zone requirement.

Not shown in the air handler illustrated in Fig. 12-1 are humidifiers that can be used to add moisture to the air. This would be done in the hot deck as a function of heating need. Several types are used (Fig. 12-21), depending on need and application. Water-spray types are used with hot water heating and provide optimum performance in applications where the humidity level is fairly low and the most precise control is not required.

The steam pan type is used when the introduction of steam directly into the air stream is undesirable. The vaporization of water from the pan provides moisture to the conditioned air.

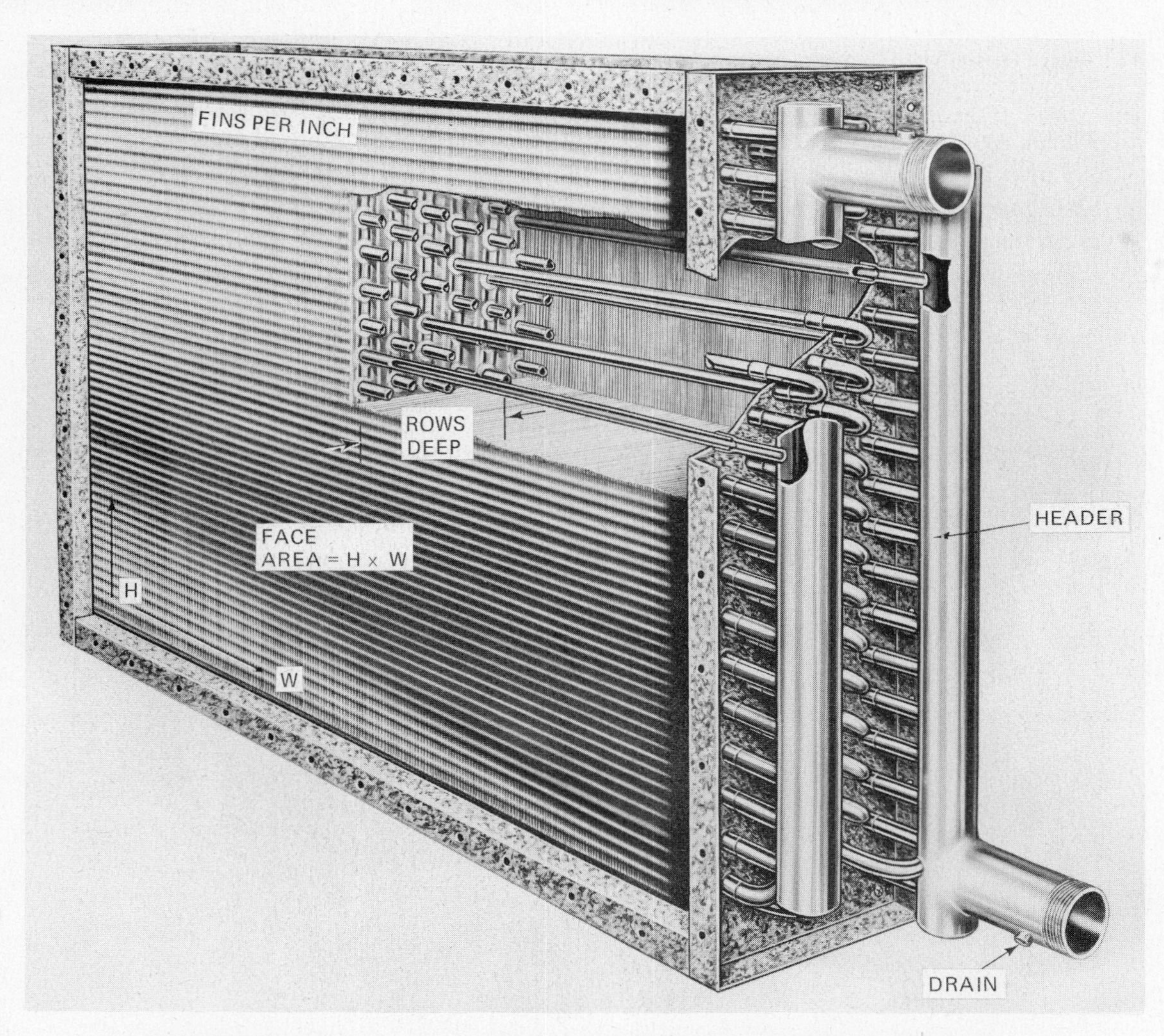

FIGURE 12-20 Finned tube water coils (Courtesy, York, Div. of Borg-Warner Corporation).

The steam grid type is highly recommended because it offers simplicity in construction and operation, and humidification can be closely controlled. Obviously a source of steam must be available.

The duct systems into the conditioned space vary with the application and type of air handler. Some are relatively simple; others are complex due to the nature of the need. Later chapters will discuss commercial duct systems, induction boxes, space terminals, etc., that will be common to both central systems and some unitary equipment.

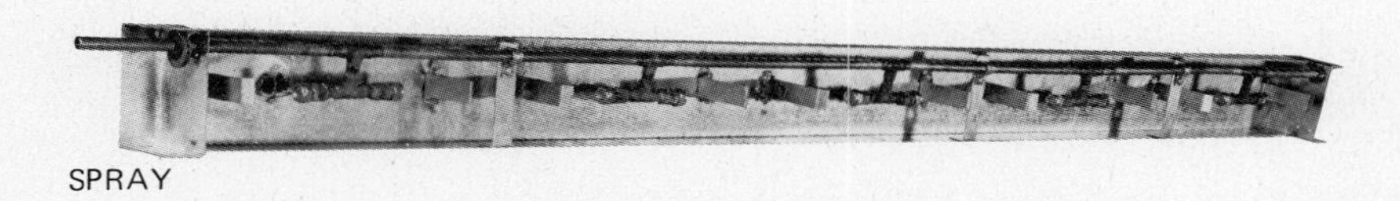

FIGURE 12-21(a) Water-spray type.

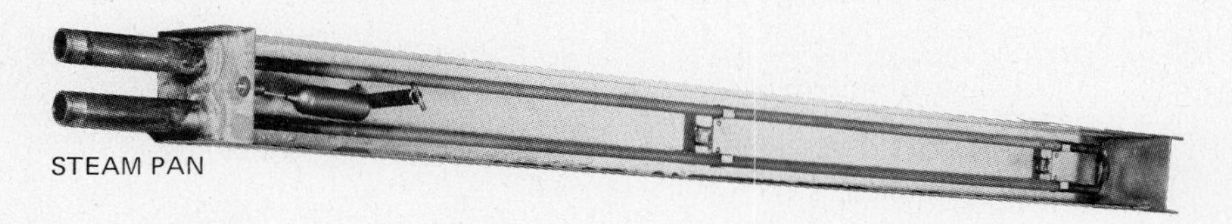

FIGURE 12-21(b) Steam pan type.

FIGURE 12-21(c) Steam grid type.

PROBLEMS

1. Central station equipment is also called ______________.
2. Central station systems are ______________ fabricated.
3. Heat transfer is by ______________ to ______________ to ______________.
4. Chilled water plants for comfort application produce ______________ to ______________ °F cold water.
5. What is the function of face and bypass dampers?
6. Chillers up to a nominal 100 tons are mostly ______________.
7. What is meant by "number of passes" in a chiller or condenser?
8. What is meant by "fouling factor" of a condenser?
9. Centrifugal compressors may be ______________ driven or ______________ driven.
10. Central station air handlers are fabricated and assembled in ______________ to match specific job requirements.

13

BASIC AIRFLOW PRINCIPLES

In previous discussions virtually every air-conditioning product or system depended on the flow or movement of air to perform its function. Determining the size of ductwork or the performance of the fan involves some understanding of basic fluid motion. The energy or force to set it in motion, the container (duct or pipe) to direct its flow, the resistance to motion because of friction and constriction, the effect of velocity on noise considerations, the methods of air distribution and types of fans, and finally the use of instruments to measure the performance and balance the system are subjects to which the next several chapters will address themselves. First, here are some of the fundamentals a technician should know.

FLUID FLOW

The flow of a fluid (Fig. 13-1) is caused either by a pressure difference due to an increase of pressure at some point in the path of flow because of a mechanical device like an air pump, or by a change in the density of the fluid, due to a temperature difference such as in an oil lamp. For purposes of this discussion, thermal motion is not involved. However, in precise system designs the engineers do consider the temperature of the fluid (air) because its density per ft^3 is a function of temperature and varies accordingly.

If the mechanical movement is further refined and a fan is substituted for the piston (Fig. 13-2), air can be set into motion. If the air is allowed to free-blow into the atmosphere, there is no resistance to flow, nor is there any direction or control. If a container (duct) is attached to provide a sense of direction and distribution, then the energy input required by the fan increases due to the added resistance of the ducting system.

RESISTANCE PRESSURES

Resistance to air movement in a duct system (Fig. 13-3) has several causes. First, friction is created by air moving against the duct wall surface, even in

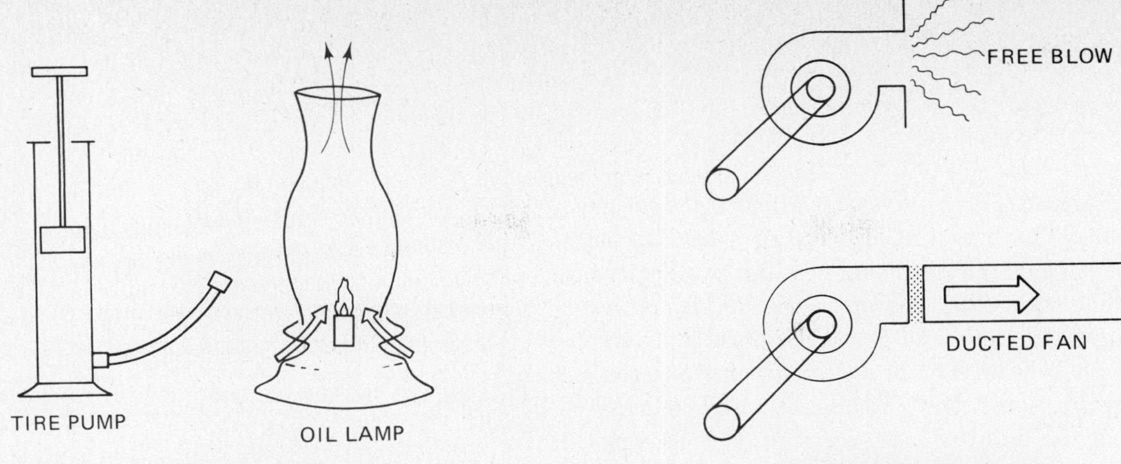

FIGURE 13-1 Tire pump and an oil lamp.

FIGURE 13-2 Ducted fan.

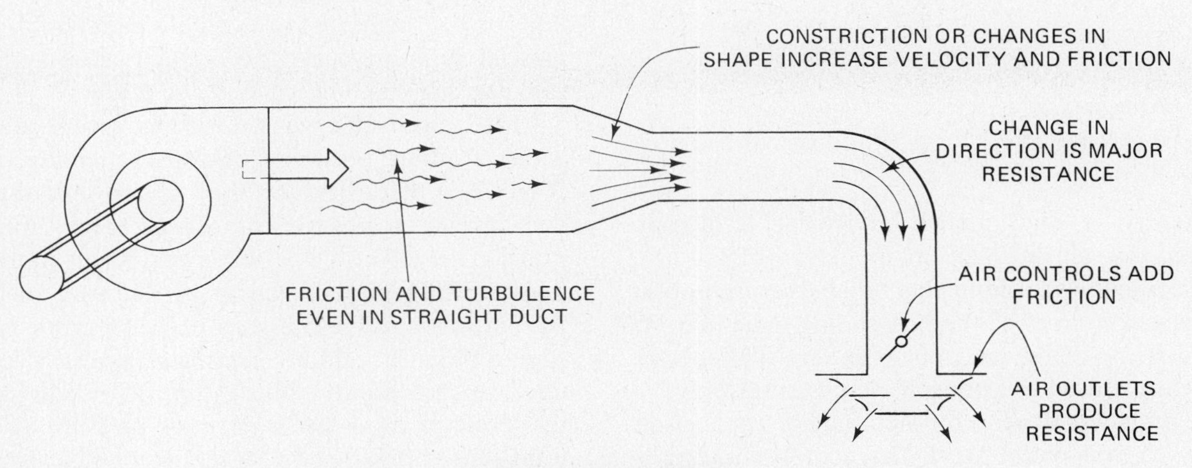

FIGURE 13-3 Duct system.

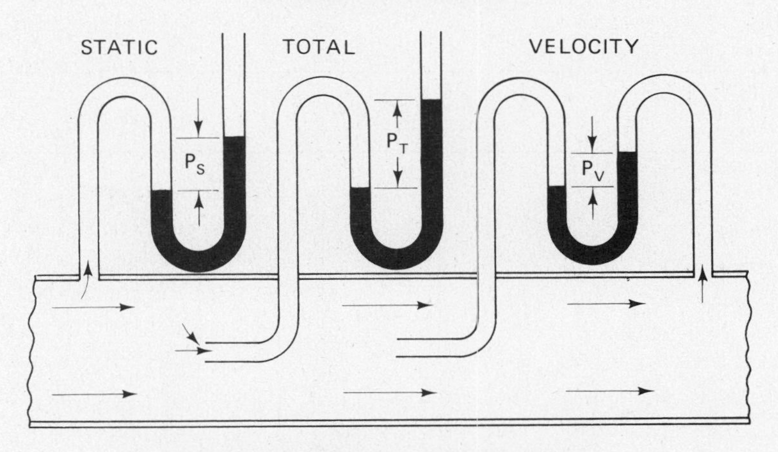

FIGURE 13-4 Duct pressure measurements.

straight pipe. Air does not move along in a nice, placid stream. Rather, it moves in what engineers call a *turbulent flow*, constantly churning and mixing. Metal, fiberglass, and flexible ducts each have a somewhat different friction effect. Constrictions or changes in shape require more pressure to speed up the velocity so that the volume of air flow is unchanged. An additional push is needed to make air change direction, such as a 90° turn. Restrictions such as filters, coils, dampers, grilles, and registers all add to the pressure requirement.

The total pressure (P_t, Fig. 13-4) required to move a desired amount of air through a duct is composed of two elements. Static pressure (P_s) is the pressure exerted against the side wall of the duct in all directions. Think of it as bursting pressure. It can be positive (+) on the push side of a fan or negative (−) on the suction side. Velocity pressure (P_v) is the pressure in the direction of flow. Think of it as the impact or push needed to bring the air up to speed. Total pressure is the sum of the static and velocity pressures at the point of measurement. For illustrative purposes the use of a **U**-tube manometer is shown.

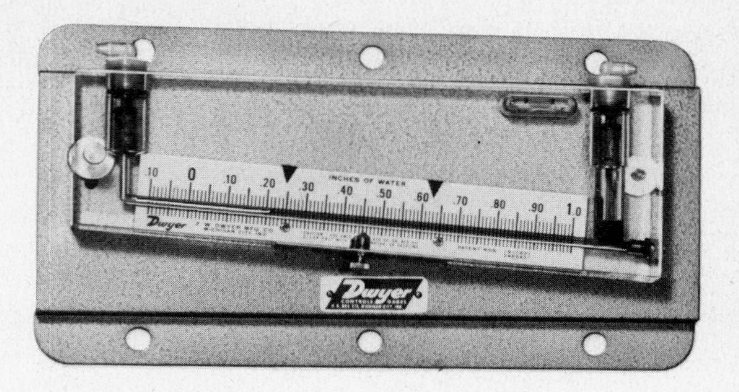

FIGURE 13-5 Inclined manometer (Courtesy, Dwyer Instruments, Inc.).

AIRFLOW MEASURING INSTRUMENTS

Pressures in a duct system are most frequently measured by an inclined manometer (Fig. 13-5), which is filled with a fluid that will be set in motion by pressures from either end. A sliding scale permits the calibration of the instrument at zero (0) pressure when it is exposed to atmospheric pressures only.

The scale will depend on the range of application needed. In residential work the common range is 0.10 to 1.0 inches of water column (W.C.). Commercial high-pressure duct systems may require a range as high as 6 in. of W.C. or more.

The manometer can be used in several different ways. Putting positive pressure at connection *A* (Fig. 13-6) will cause the fluid to move to the right and register a positive numerical value. Similarly, putting negative or suction pressure at connection port *B* will register negative pressure on the scale. By putting positive pressure on both ports, a pressure difference can be measured across a coil or filter, as shown in the illustration. By placing a positive pressure at *A* and a negative pressure at *B*, the total static pressure of a system can be measured.

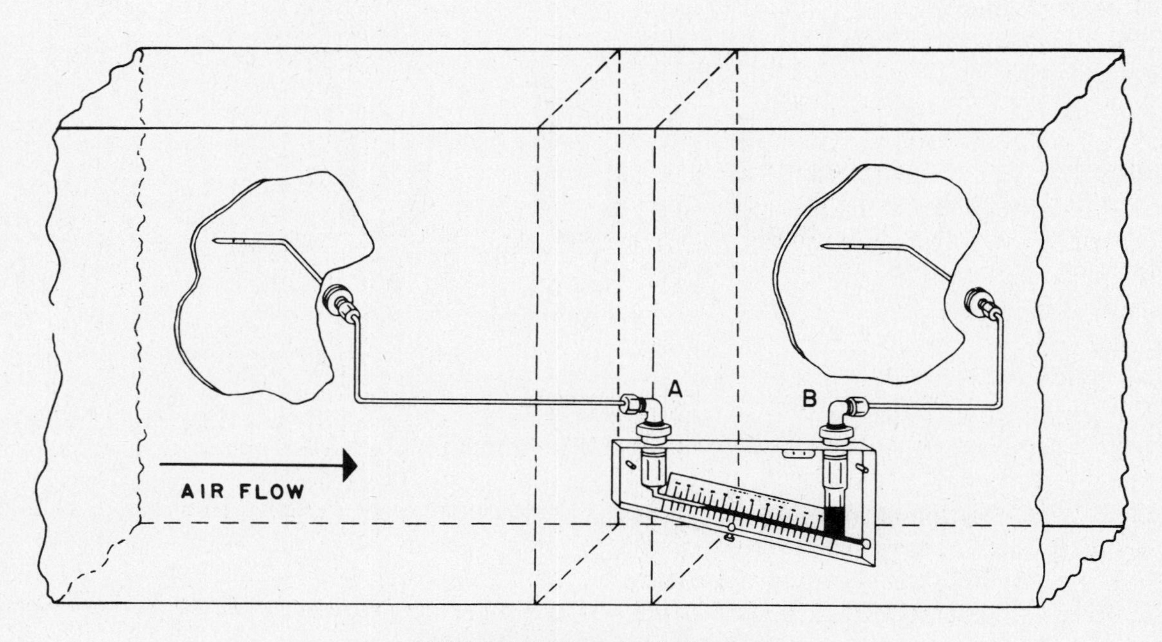

FIGURE 13-6 Measuring duct pressure.

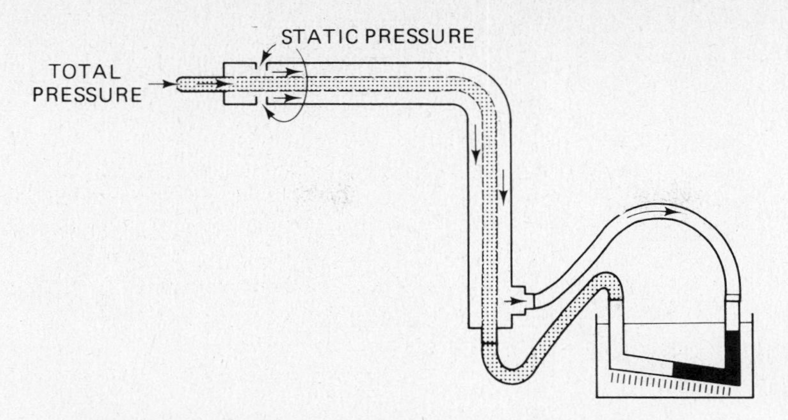

FIGURE 13-7 Pitot tube and manometer measuring velocity pressure.

Going back to the pressures in a duct, the manometer can record static, velocity, and total pressures, as indicated by Fig. 13-4. In actual laboratory work the use of a pitot tube (Fig. 13-7) at the end of the air probe allows the engineer to measure total pressure and static pressure with only one probe. When connected as shown, it will indicate the difference, which is the velocity pressure.

In normal air-conditioning work the measurement of velocity pressure is not generally needed. Velocity in terms of speed or motion in feet per minute (ft/min) is important, since it is related to noise considerations and to balancing the systems.

Instrument manufacturers have modified the basic inclined manometer into a more compact, hand-held field instrument such as illustrated in Fig. 13-8a. An identical model can be used with a Pitot tube, and the scale will read velocity in feet per minute rather than in inches of water (Fig. 13-8b).

Another instrument used to measure the velocity of air from or into a grille is the *anemometer* (Fig. 13-9). It consists of propellers on a shaft that revolve when held in an air stream. In the center is a dial that reads velocity in ft/min (feet per minute). A stop watch is used to clock the instrument at whatever time period is desired. Some situations might require the anemometer to run longer than one minute. Longer time measurements will provide averages that are more accurate. With the velocity known, the ft^3/min can be determined by multiplying the grille's net free area in square feet times the velocity in ft/min. As an example, assume the grille in Fig. 13-9 had 2 ft^2 of net free area (data obtained from the manufacturer's catalog), and the velocity was recorded at 500 ft/min. The ft^3/min would thus be 500 ft/min times 2 ft^2 = 1,000 ft^3/min.

The understanding of pressures, velocity, and the use of instruments can be valuable in designing, balancing, and servicing air distribution systems. The design of a system starts with the understanding of friction loss in straight ducts.

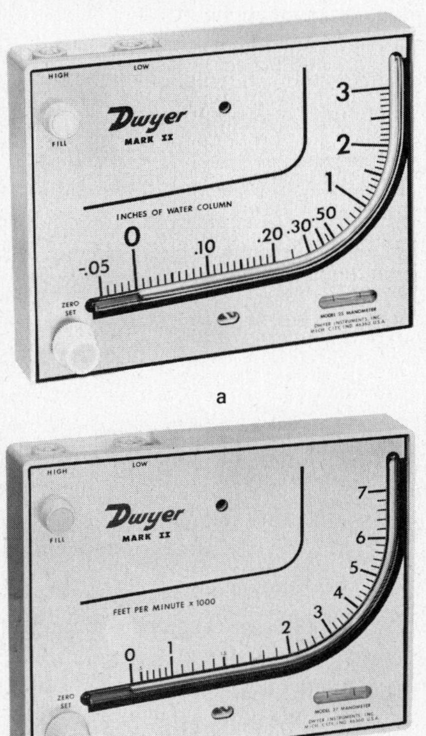

FIGURE 13-8 (a) Manometer to measure inches of water; (b) Manometer to measure velocity in feet per minute (Courtesy, Dwyer Instruments, Inc.).

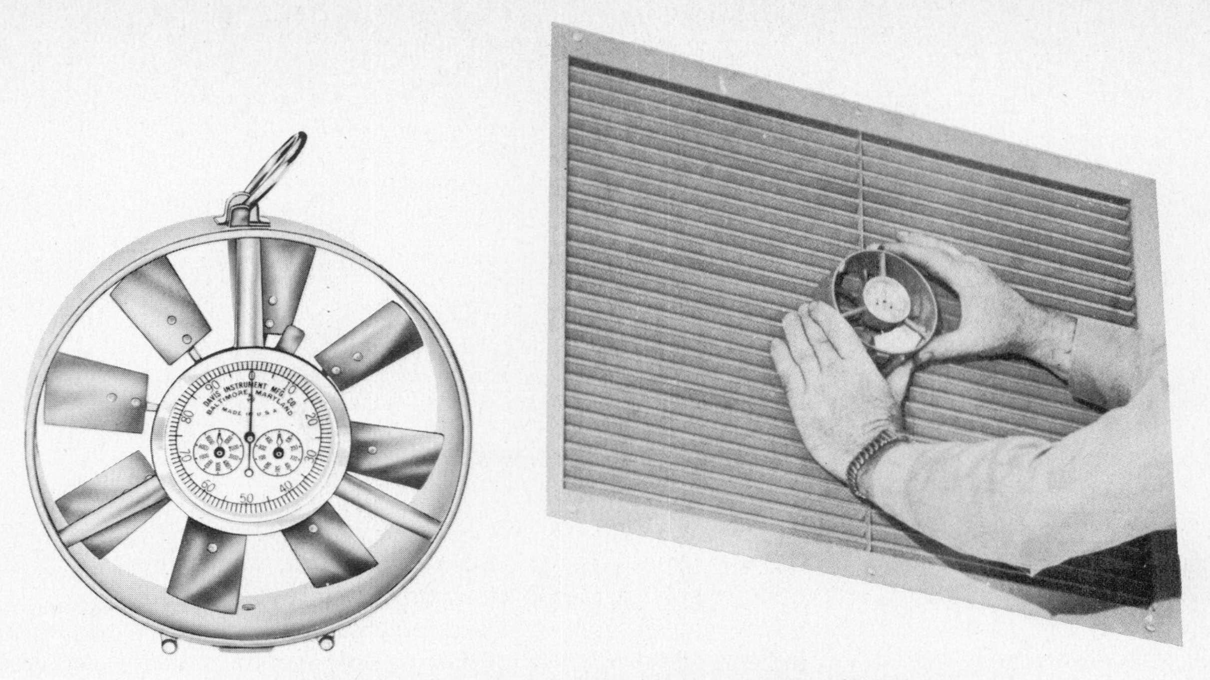

FIGURE 13-9 **Anemometer (Courtesy, Davis Instruments Manufacturing Company).**

AIR FRICTION CHARTS

The ASHRAE air friction chart (Fig. 13-10) is nothing more nor less than a graph on which the coordinates are the friction loss in inches of water (W.C.) for each 100 feet of equivalent length of duct and the cubic feet of air per minute (ft^3/min) carried through the duct. As a result of laboratory studies and calculations, a separate line for each size of duct is plotted on the graph. Calculated duct air velocities are also plotted. Logarithmic graph paper is used in plotting the graph so that the lines will be almost straight. This makes the coordinates look a bit unusual, but the air friction chart is much easier to use. The chart illustrated covers up to 2,000 ft^3/min air volume for residential and light commercial work up to 5 tons. Figure 13-11 is for volumes up to 100,000 ft^3/min. Both charts are reprints from the ASHRAE *Guide and Data Book* and are based on standard air conditions flowing through average, clean, round, galvanized metal ducts having approximately 40 joints per 100 ft.

An example of using Fig. 13-10 would be to assume a 6-in. round duct has 200 ft^3/min flowing. We wish to find its velocity and friction loss. Follow across from 200 ft^3/min on the left vertical scale to the intersection of the 6-in. diagonal line marked *A*. Read down to the base line and note a loss of 0.3 inches of water per 100 ft of length. Read the velocity from the other set of diagonal lines at just over 1,000 ft/min. If the actual duct is only 75 feet long, the friction loss is calculated as $0.3 \times 0.75 = 0.225$ in. of W.C. friction loss.

If any two coordinates are known, the other values can be determined. Assuming that a maximum of 500 ft/min velocity is the important element to consider and the duct size cannot exceed 8 in., what would be the resultant ft^3/min and friction loss? Plotting from the 500 ft/min diagonal line to the intersection with the 8-in. round duct (point *B*), read the volume as 175 ft^3/min and the friction loss as 0.058 in. of W.C. per 100 ft of duct.

The significance of plotting friction loss per 100 ft of equivalent length cannot be overemphasized. The relationship between the friction loss and the design pressure must be understood if the friction loss chart is to be used correctly. If, for example, the total equivalent of a duct is 200 ft and the system design pressure is to be 0.20 in. of W.C., the total friction loss is 0.20 in. of W.C. For each 100 ft of equivalent length the friction loss cannot exceed 0.10 in. of W.C. In other words, if the duct were shortened to 100 ft of equivalent length, the same ft^3/min of air would be delivered if the pressure were decreased from 0.20 to 0.10 in. of W.C.

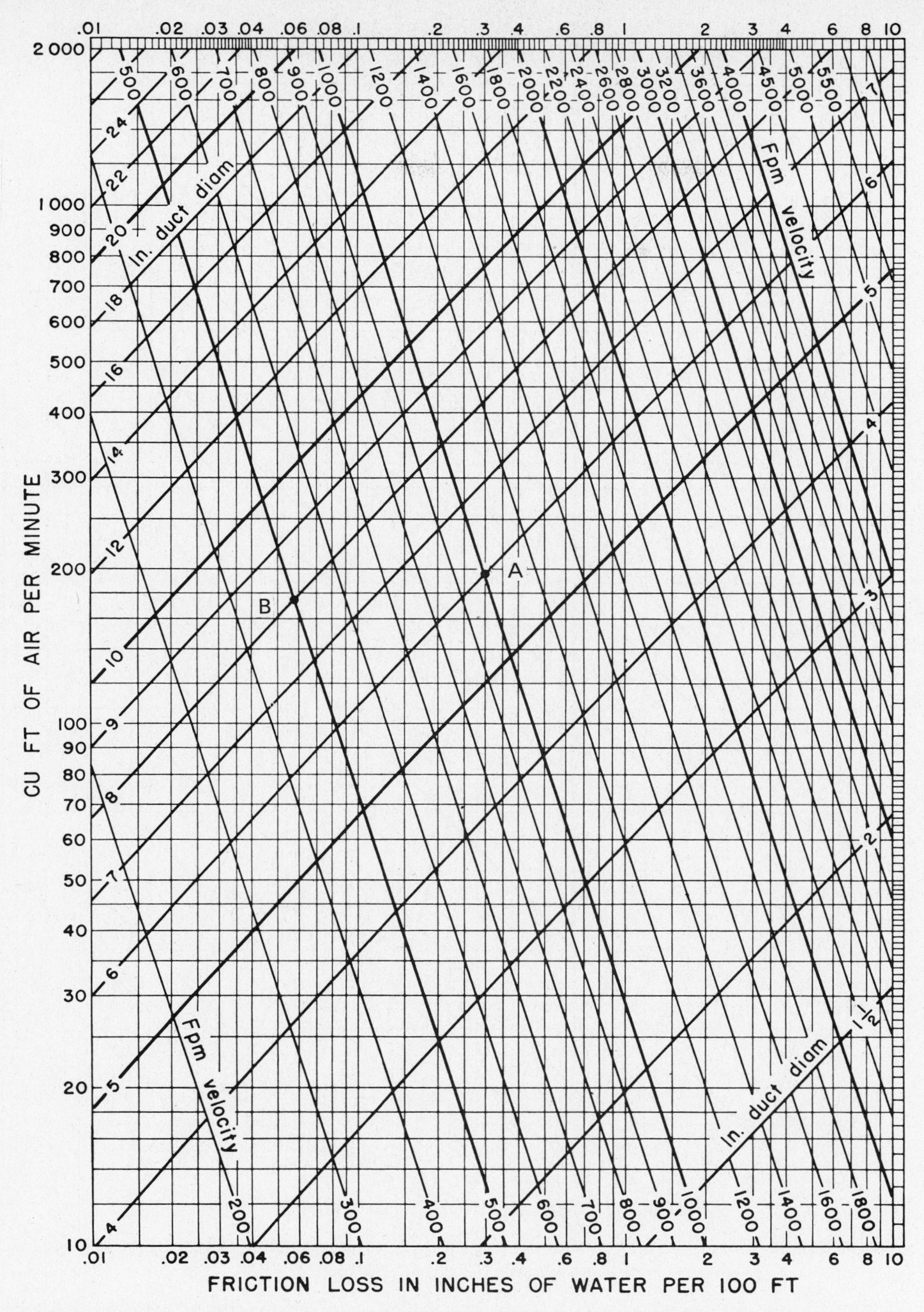

(Based on Standard Air of 0.075 lb per cu ft density flowing through average, clean, round, galvanized metal ducts having approximately 40 joints per 100 ft.) Caution: Do not extrapolate below chart.

FIGURE 13-10 Friction of air in straight ducts for volumes of 10 to 2000 cfm., (Reprinted by by permission from ASHRAE Guide and Data Book).

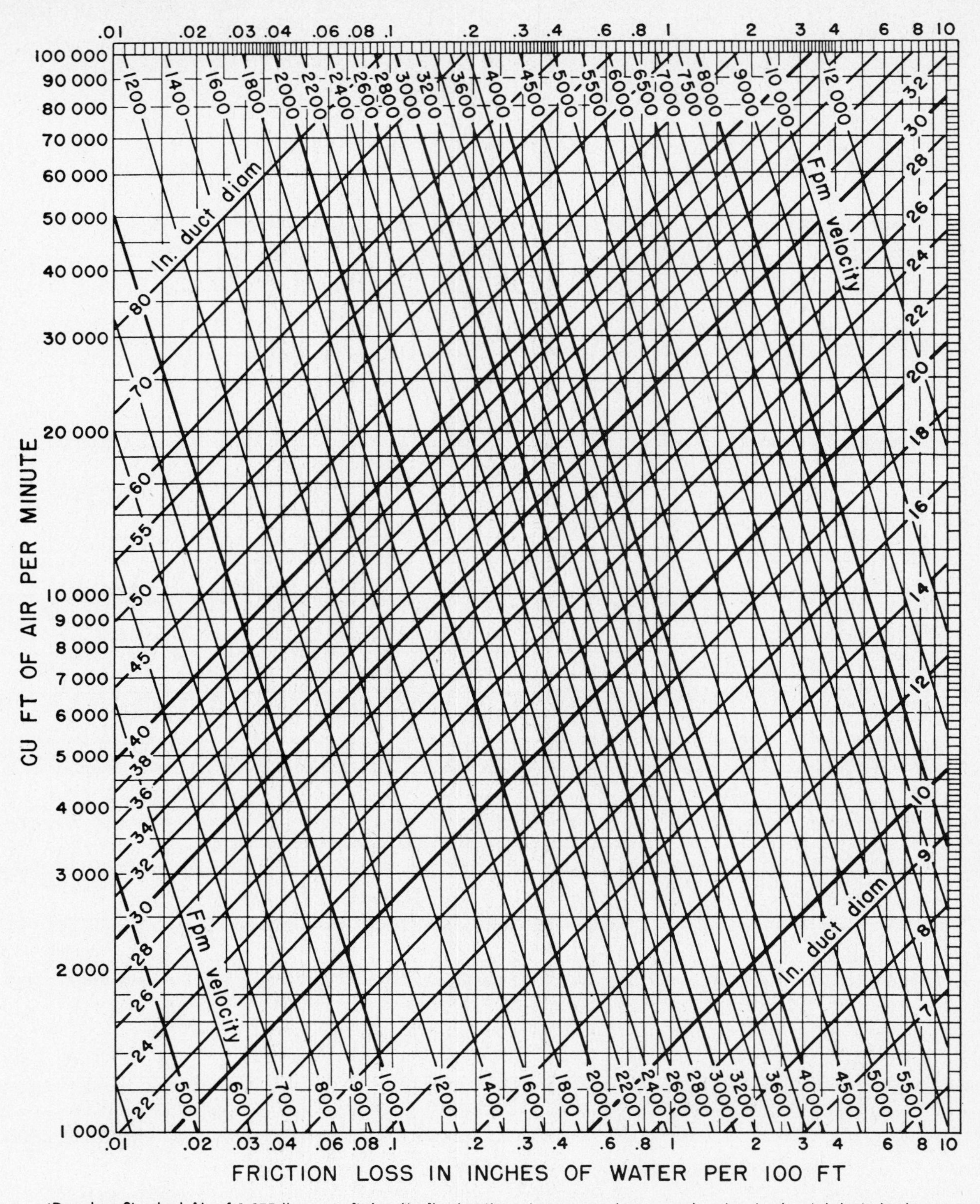

(Based on Standard Air of 0.075 lb per cu ft density flowing through average, clean, round, galvanized metal ducts having approximately 40 joints per 100 ft.)

FIGURE 13-11 Friction of air in straight ducts for volumes of 1000 to 100,000 cfm., (Reprinted by permission from ASHRAE Guide and Data Book).

The friction charts are based on the diameters of round ducts; to express the equivalent cross section in square inches of rectangular dimensions, refer to Fig. 13-12. Note the large numbers overprinted on the table. These are called *duct class numbers*, and are numerical representations of the initial cost of the ductwork. The larger the duct class the more expensive the duct. Always use the smaller duct

FIGURE 13-12 Duct dimensions, section area, circular equivalent diameter, and duct class (Courtesy, ASHRAE).

SIDE	6		8		10		12		14		16		18		20		22	
	Area sq ft	Diam in.	Area sq ft	Diam in.	Area sq ft	Diam in.	Area sq ft	Diam in.	Area sq ft	Diam in.	Area sq ft	Diam in.	Area sq ft	Diam in.	Area sq ft	Diam in.	Area sq ft	Diam in.
10	.39	8.4	.52	9.8	.65	10.9												
12	.45	9.1	.62	10.7	.77	11.9	.94	13.1										
14	.52	9.8	.72	11.5	.91	12.9	1.09	14.2	1.28	15.3								
16	.59	10.4	.81	12.2	1.02	13.7	1.24	[illegible]	1.45	16.3	1.67	17.5						
18	.66	11.0	.91	12.9	1.15	14.5	1.40	[illegible]	1.63	17.3	1.87	18.5	2.12	19.7				
20	.72	11.5	.99	13.5	1.26	15.2	1.54	[illegible]	1.81	18.2	2.07	19.5	2.34	20.7	2.61	21.9		
22	.78	12.0	1.08	14.1	1.38	15.9	1.69	17.6	1.99	19.1	2.27	20.4	2.57	21.7	2.86	22.9	3.17	24.1
24	.84	12.4	1.16	14.6	1.50	16.6	1.83	18.3	2.14	19.8	2.47	21.3	2.78	22.6	3.11	23.9	3.43	25.1
26	.89	12.8	1.26	15.2	1.61	17.2	1.97	19.0	2.31	20.6	2.66	22.1	3.01	23.5	3.35	24.8	3.71	26.1
28	.95	13.2	1.33	15.6	1.71	17.7	2.09	19.6	2.47	21.3	2.86	22.9	3.25	24.4	3.60	25.7	4.00	27.1
30	1.01	13.6	1.41	16.1	1.82	18.3	2.22	20.2	2.64	22.0	3.06	23.7	3.46	25.2	3.89	26.7	4.27	28.0
32	1.07	14.0	1.48	16.5	1.93	18.8	2.36	20.8	2.81	22.7	3.25	24.4	3.68	26.0	4.12	27.5	4.55	28.9
34	1.13	14.4	1.58	17.0	2.03	19.3	2.49	21.4	2.96	23.3	3.43	25.1	3.89	26.7	4.37	28.3	4.81	29.7
36	1.18	14.7	1.65	17.4	2.14	19.8	2.61	21.9	3.11	23.9	3.63	25.8	4.09	27.4	4.58	29.0	5.07	30.5
38	1.23	15.0	1.73	17.8	2.25	20.3	2.76	22.5	3.27	24.5	3.80	26.4	4.30	28.1	4.84	29.8	5.37	31.4
40	1.28	15.3	1.81	18.2	2.33	20.7	2.88	23.0	3.43	25.1	3.97	27.0	4.52	28.8	5.07	30.5	5.62	32.1
42	1.33	15.6	1.86	18.5	2.43	21.1	2.98	23.4	3.57	25.6	4.15	27.6	4.71	29.4	5.31	31.2	5.86	32.8
44	1.38	15.9	1.95	18.9	2.52	21.5	3.11	23.9	3.71	26.1	4.33	28.2	4.90	30.0	5.55	31.9	6.12	33.5
46	1.43	16.2	2.01	19.2	2.61	21.9	3.22	24.3	3.88	26.7	4.49	28.7	5.10	30.6	5.76	32.5	6.37	34.2
48	1.48	16.5	2.09	19.6	2.71	22.3	3.35	24.8	4.03	27.2	4.65	29.2	5.30	31.2	5.97	33.1	6.64	34.9
50			2.16	19.9	2.81	22.7	3.46	25.2	4.15	27.6	4.84	29.8	5.51	31.8	6.19	33.7	6.87	35.5
52			2.22	20.2	2.91	23.1	3.57	25.6	4.30	28.1	5.00	30.3	5.72	32.4	6.41	34.3	7.14	36.0
54			2.29	20.5	2.98	23.4	3.71	26.1	4.43	28.5	5.17	30.8	5.90	32.9	6.64	34.9	7.38	36.8
56			2.38	20.9	3.09	23.8	3.83	26.5	4.55	28.9	5.31	31.2	6.08	33.4	6.87	35.5	7.62	37.4
58			2.43	21.1	3.19	24.2	3.94	26.9	4.68	29.3	5.48	31.7	6.26	33.9	7.06	36.0	7.87	38.0
60			2.50	21.4	3.27	24.5	4.06	27.3	4.84	29.8	5.65	32.2	6.50	34.5	7.26	36.5	8.12	38.6
64			2.64	22.0	3.46	25.2	4.24	27.9	5.10	30.6	5.91	33.1	6.87	35.5	7.71	37.6	8.59	39.7
68					3.63	25.8	4.49	28.7	5.37	31.4	6.26	33.9	7.18	36.3	8.12	38.6	9.03	40.7
72					3.83	26.5	4.71	29.4	5.69	32.3	6.60	34.8	7.54	37.2	8.50	39.5	9.52	41.8
76					4.09	27.4	4.91	30.0	5.86	32.8	6.83	35.4	7.95	38.2	8.90	40.4	9.98	42.8
80					4.15	27.6	5.17	30.8	6.15	33.6	7.22	36.4	8.29	39.0	9.21	[illegible]	10.4	43.8
84							5.41	31.5	6.41	34.5	7.54	37.2	8.55	39.6	9.75	[illegible]	10.8	44.6
88							5.58	32.0	6.64	34.9	7.87	38.0	8.94	40.5	10.1	[illegible]	11.2	45.4
92							5.79	32.6	6.91	35.6	8.12	38.6	9.39	41.5	10.4	43.8	11.7	46.3
96							5.90	33.0	7.14	36.2	8.40	39.2	9.70	42.1	10.8	44.5	12.1	47.2
100									7.40	36.9	8.50	39.5	9.80	42.5	11.3	45.5	12.3	47.6
104									7.60	37.4	8.90	40.5	10.3	43.5	11.6	46.2	13.0	48.8
108									7.90	38.0	9.20	41.2	10.6	44.0	12.0	47.0	13.4	49.6
112									8.10	38.6	9.50	41.8	10.9	44.7	12.3	47.5	13.8	50.3
116											9.80	42.4	11.3	[illegible]	12.6	48.1	14.3	51.3
120											10.0	42.8	11.5	46.0	13.1	49.1	14.4	51.5
124											10.3	43.5	11.9	46.7	13.4	49.6	15.0	52.4
128											10.6	44.1	12.1	47.1	13.8	50.4	15.5	53.3
132													12.5	47.9	14.1	50.9	15.8	53.9
136													12.8	48.5	14.5	51.6	16.2	54.5
140													13.0	48.8	14.7	52.0	16.5	55.0
144													13.3	49.4	15.2	52.9	16.8	55.6

class wherever possible. The cross-sectional areas for equivalent round and rectangular ducts never agree exactly, so in using the chart go to the next higher rectangular value. For example, if you wish to convert a nominal 10-inch round duct to a rectangular duct with one side 8 in. long, you would select a 12 × 8-in. equivalent, using the chart. The 10 × 8-in. size would be slightly undersized.

With ductwork made of other materials than galvanized metal or aluminum, it is recommended that the manufacturer's application data be consulted for similar friction-loss tables of conversion factors that can be applied to the metal duct information. Fiberglass ducts and flexible ducting have somewhat different airflow characteristics and must be treated accordingly. Also, if metal ducts are lined with thermal or sound absorbing insulation, they not only have different friction values but also are smaller in the inside dimension by the thickness of the insulation, so don't forget that in the original size determination.

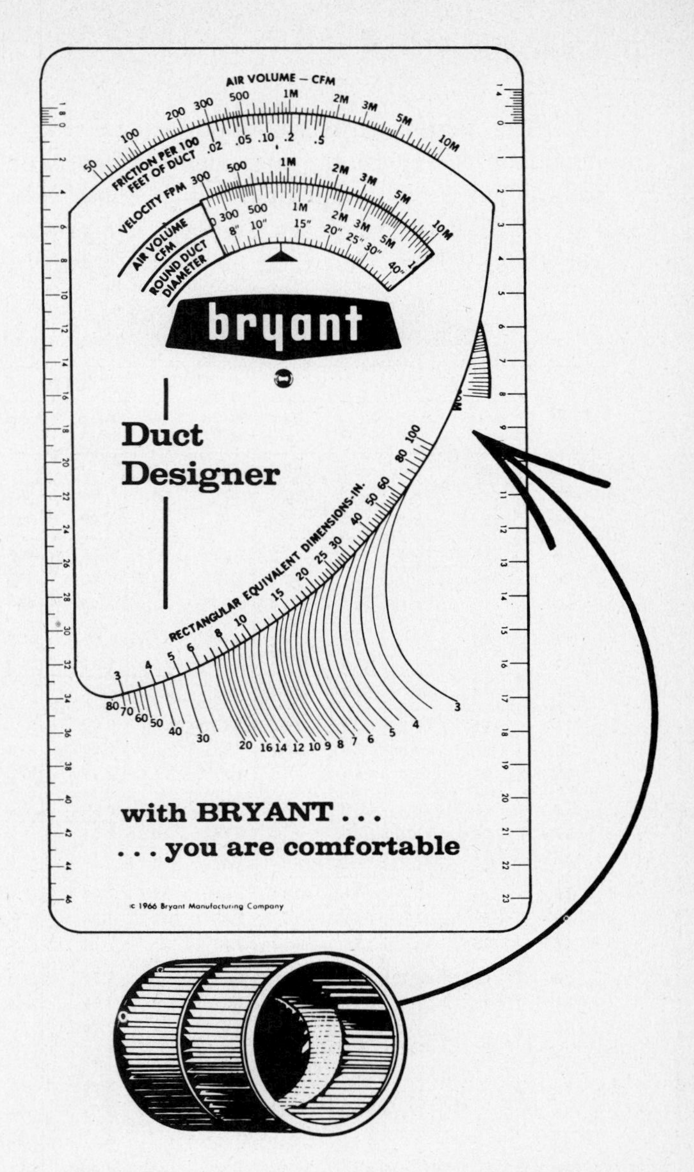

FIGURE 13-13 Duct calculator (Courtesy, Bryant Air-Conditioning Company).

DUCT CALCULATORS

For those who frequently work with duct design there are handy duct calculators (Fig. 13-13), which are based on the friction charts and provide the same information on ft³/min. velocity, static pressure, etc. The setting also provides an instant conversion from round duct to its rectangular equivalent.

DUCT FITTINGS

Fittings is the name given to the various parts of a duct system such as turning elbows, trunk duct take-offs, boot connections, etc. These offer resistance to air flow and represent a major portion of duct loss; they also affect the sound characteristics of the system. The ASHRAE *Guide*, NESCA manuals, and duct manufacturers list conventional fittings (Fig. 13-14) and express the air losses in terms of equivalent length of straight duct. For example a type *B*, 90° elbow from Group 2, with a trunk width of 12 to 21 inches will have the same or equivalent loss as 15 feet of the same size of straight duct. A glance at the chart will quickly reveal the fact that it doesn't take many fittings to add an appreciable resistance.

Figure 13-15 is an example of several types of trunk duct take-offs. Note method *G*, a round duct butt joint that has a 35-ft equivalent length as compared to *B*, which is a transition collar and is only 15 feet. Top takeoffs like *A*, *C*, and *F* are even higher in resistance because they actually require two 90°

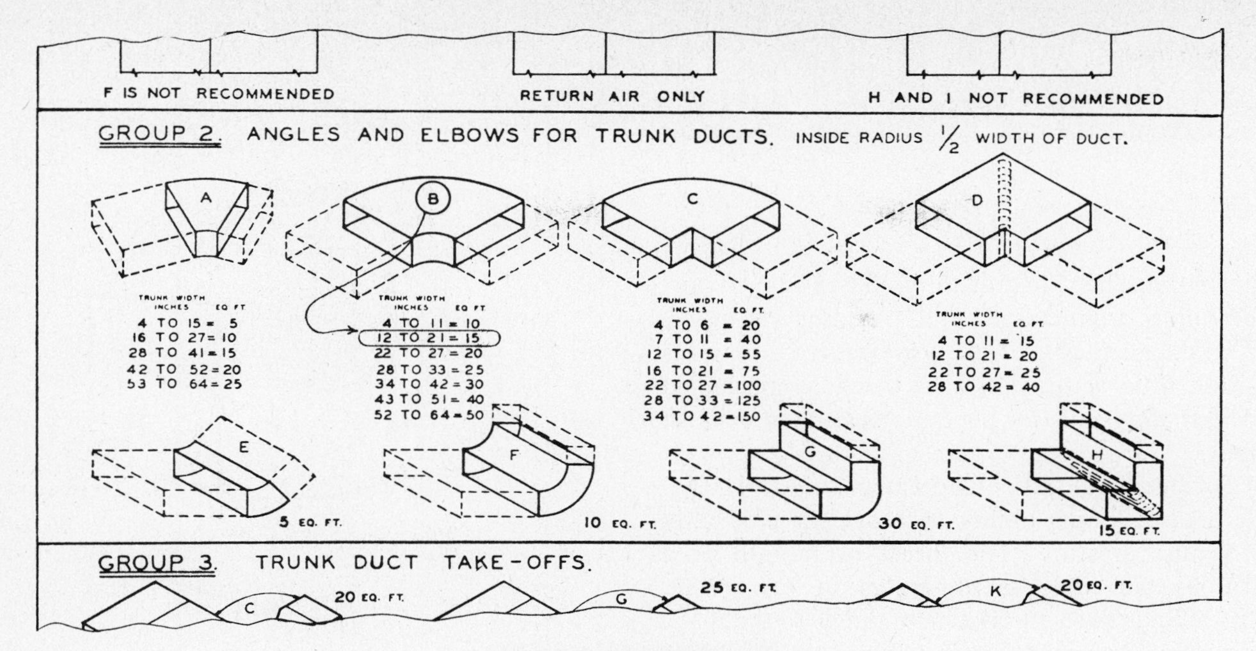

FIGURE 13-14 Duct fittings.

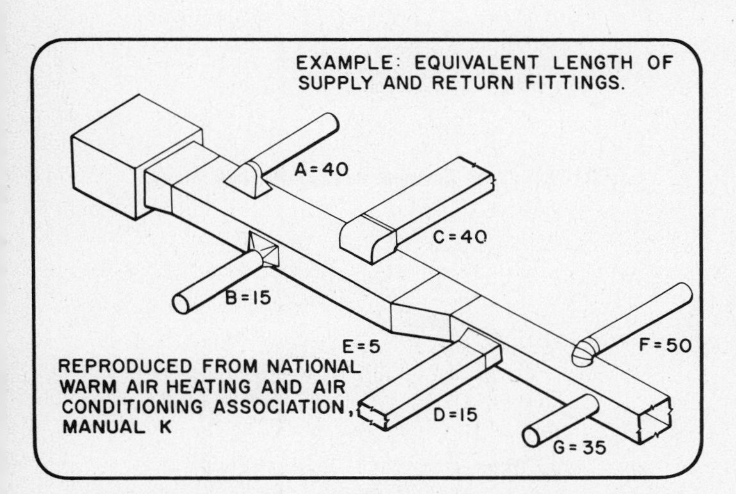

FIGURE 13-15 Trunk duct take-offs (Courtesy, Bryant Air-Conditioning Company).

turns of the air as opposed to only one for a side outlet.

Filters, if installed in duct work rather than in the rest of the equipment, plus air outlet and return air grilles, also present restrictions to airflow that must be considered in the overall system. The resistance of filters and grilles are given directly in terms of inches of water column static pressure taken from the manufacturer's catalogs. They vary widely, from low-pressure residential systems to higher-pressure commercial applications. The use of this information will be more fully covered under system designs.

SOUND LEVEL

The noise or sound level of a duct system is the function of velocity, and the permissible limits for acceptable levels are tabulated in Fig. 13-16. These should not be exceeded, and if sound is a major consideration use branch duct values for the entire system design. Although most duct design is initially determined by the static loss method a careful check of velocity is advised so as to minimize the possibility of undesirable noise.

We have discussed the basic restrictions to airflow and velocity recommendations in a duct system, but what about the "push" needed to generate the movement: the fan or blower?

RECOMMENDED MAXIMUM DUCT VELOCITIES, FPM

APPLICATION	MAIN DUCTS*		BRANCH DUCTS	
	SUPPLY	RETURN	SUPPLY	RETURN
Apartments	1000	800	600	600
Auditoriums	1300	1100	1000	800
Banks	2000	1500	1600	1200
Hospital Rooms	1500	1300	1200	1000
Hotel Rooms	1500	1300	1200	1000
Industrial	3000	1800	2200	1500
Libraries	2000	1500	1600	1200
Meeting Rooms	2000	1500	1600	1200
Offices	2000	1500	1600	1200
Residences	1000	800	600	600
Restaurants	2000	1500	1600	1200
Retail Stores	2000	1500	1600	1200
Theaters	1300	1100	1000	800

* When noise control is critical use branch duct values.

FIGURE 13-16 (Courtesy, Bryant Air-Conditioning Company).

FANS

Propeller fans, as illustrated in the review of air-conditioning products, are not used for air movement in ducts; they are fine, however, for situations requiring little static resistance, such as air-cooled condenser duty and for separate free-blow exhaust systems.

The centrifugal fan is used extensively in HVAC work because of its ability to move air efficiently and against pressure. The thrust or motion of a centrifugal fan wheel is illustrated in Fig. 13-17. The rotation is fixed and is at a 90° angle to the shaft. The motion imparted by the shape of the blade varies considerably depending on the design. A straight radial blade as illustrated in Fig. 13-17 is seldom used in HVAC work. It is efficient for handling abrasives and dust, and for exhausting fumes containing dirt, grease, or acids.

The typical heating and air-conditioning blower (Fig. 13-18) has a forward-curved multiblade wheel, with the resultant force as shown in the sketch. It moves large masses of air at low rpm and at medium pressure applications, which covers most residential and light commercial work.

Forward-curved blade fans are smaller and are considered relatively quiet and thus are universally used in the unitary type of heating and cooling equipment.

The backward-inclined wheel (Fig. 13-19) has blades inclined toward the rear, away from the fan

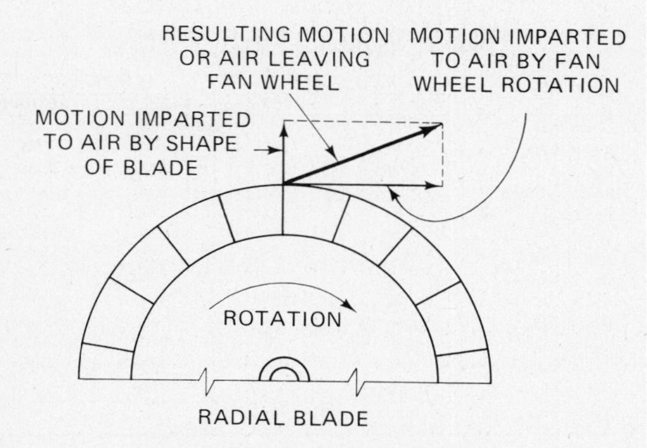

FIGURE 13-17 Radial blade fan.

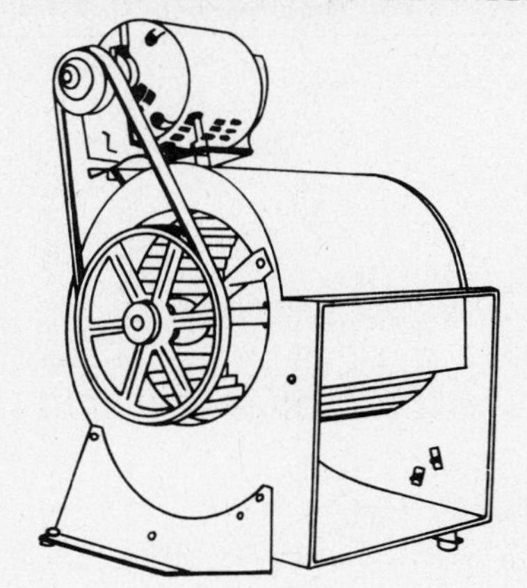

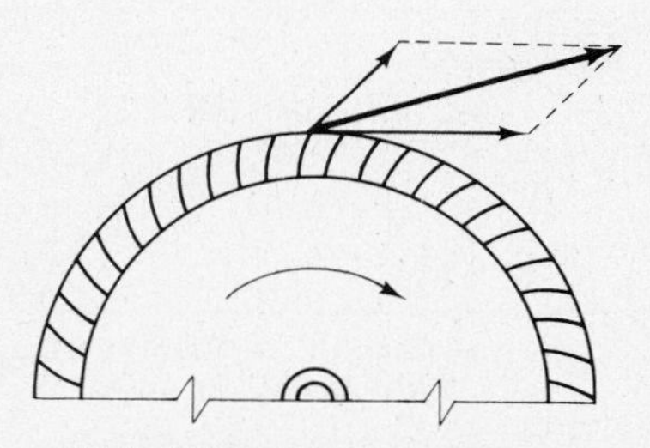

FIGURE 13-18 Forward curve multi-blade wheel.

BACKWARD INCLINED
BLADE TYPE

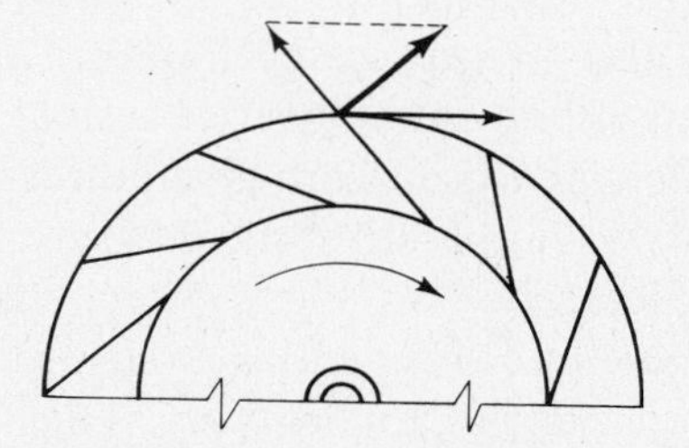

FIGURE 13-19 Backward inclined blade wheel.

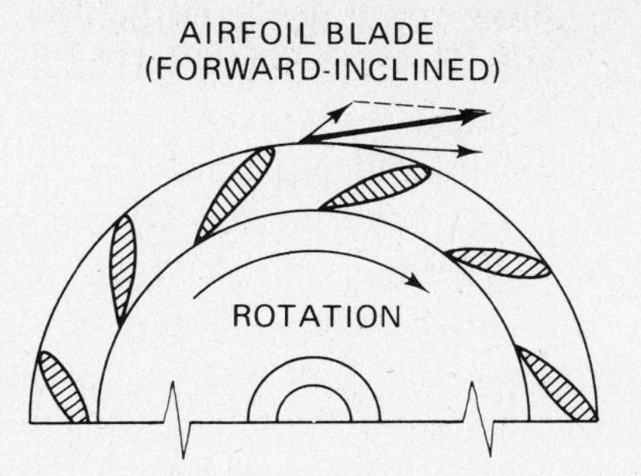

FIGURE 13-20 Air foil blade wheel.

wheel rotation. They may also be curved. It is sturdy, high speed, and has a low noise level with nonoverload horsepower characteristics. Its mechanical efficiency is high and it requires little power.

A variation of both the forward and backward-inclined blade is the airfoil wheel (Fig. 13-20) with its fewer but airfoil shaped blades. Note that the force vector is more forward, which increases its static pressure capability for use in high-pressure systems but maintains a relatively low speed and power requirement. The airfoil blade and wheels are more expensive to manufacture and thus are only used on larger equipment.

A comparison of performance curves for backward and forward curved fans (Fig. 13-21) shows efficiency and brake horsepower compared to ft^3/min at a constant rpm. One significant point can be observed from these curves. For the backward-curved fan, the horsepower rises to a maximum and then drops off and is said to have a *nonoverloading characteristic*. In contrast, the forward-curved fan horsepower continues to rise, and this type, therefore, is considered to be *overloading*. Selection is made within the range shown. Ideally it would be at the point of highest efficiency and minimum horsepower. When the efficiency of these two types of fans are compared, it can be seen that the backward-curved fan has a higher peak efficiency but it also has a wider efficiency variation. Thus for general purpose application and size, the forward-curved multiblade wheel provides a better selection.

Of interest is the axial type of fan (Fig. 13-22), which has a tube configuration with blades or vanes on a center shaft, much like the aircraft engine turbine. It's versatile, simple in design, and relatively cheap for industrial ventilation and exhaust duty. Its high speed and the resulting noise preclude its use in ducted comfort air conditioning.

Fans are generally rated by the AMCA (Air-Moving and Conditioning Association) by class according to the operating limits of static pressure.

Class I Fans— $3\frac{3}{4}$ in. maximum total pressure
II Fans— $6\frac{3}{4}$ in. maximum total pressure
III Fans—$12\frac{1}{4}$ in. maximum total pressure
IV Fans—Greater than $12\frac{1}{4}$ in. total pressure

Most normal air-conditioning requirements fall in the Class I or II rating.

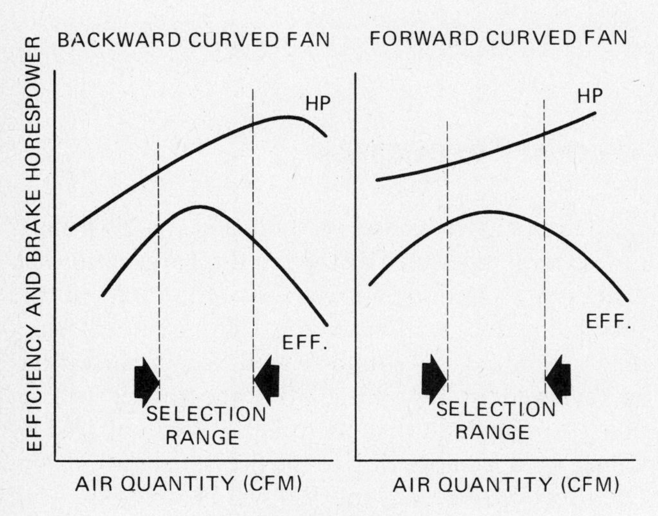

FIGURE 13-21.

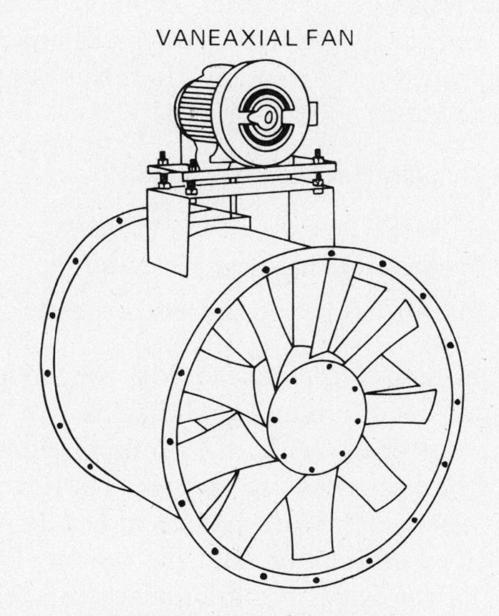

FIGURE 13-22 Axial type fan.

AIR DUCT DESIGN METHODS

The basic function of a duct system is to transmit air from the air-handling unit to the individual spaces to be conditioned. The designer must take into consideration the space available, sound levels, friction loss, initial cost, and heat gain or leakage factors. The prime objective is to consider all of them and determine which system can most nearly satisfy the collective requirements.

A simple duct system (Fig. 13-23) consists of a supply fan, ductwork, transition fitting, discharge grilles, and a return air duct from a room. The fan picks up the air from the room at an air velocity of about 25 feet per minute, and after passing through the return air system, the fan discharges the air at a velocity of say 1,500 feet per minute. In this process, from point 1 to point 2 four things have been done to the air:

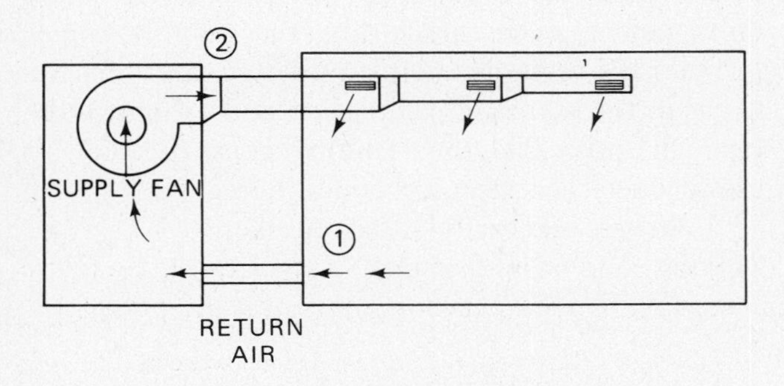

FIGURE 13-23 Duct system diagram.

(1) Velocity has been increased

(2) Static pressure has been increased

(3) Temperature has been increased

(4) The air has been compressed slightly

These four items represent the work the fan has done.

On air-conditioning applications we do not concern ourselves with the compression of air because the pressures are so low. The temperature increase due to fan work has been accounted for in the heat load calculation. But the velocity and static pressure of the fan at point 2 are the two factors we must concern ourselves with in duct design.

Static pressure in the duct and behind the grille forces air through the grille (Fig. 13-24). The selection of these grilles is based on a given quantity of air at a given outlet velocity and static pressure. If the static pressure from point 2 to point 3 were constant, we would be able to set the vanes of each outlet in the same way and get the same amount of air and the same performance. But if we were to take static pressure readings as the air flowed down the duct, we would read perhaps 1.0 in. at point *A* and 0.9 in. at point *B*. So a loss of 0.1 in. of W.C. in static pressure has occurred between these two points. The outlets at the far end of the duct will deliver less air than those in the first part. We must be able to predict the losses in a duct system in order to design a system that will deliver the proper air quantities to all conditioned spaces.

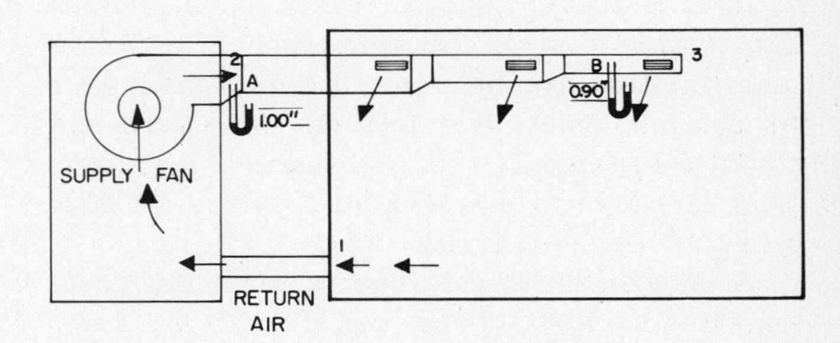

FIGURE 13-24.

There are three methods of sizing and predicting duct performance:

(1) Velocity reduction

(2) Equal friction

(3) Static regain

VELOCITY REDUCTION

In the velocity reduction method (Fig. 13-25) a starting velocity is selected at the fan discharge, and then arbitrary reductions in velocity are made in the succeeding duct sections or runs. The assumed velocities are based on sound-related recommendations as reviewed in Fig. 13-16. But the reader will be quick to conclude that there is no constant way of telling what is happening to the pressure relationship. Each section, V1, V2, and V3, will be different, both in velocity and pressure. Therefore, since velocity

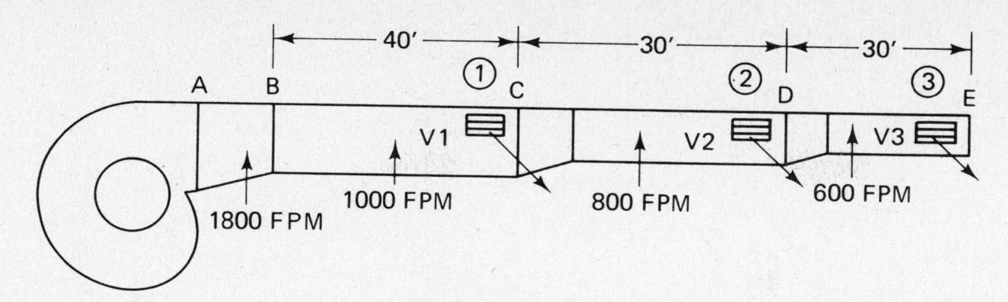

FIGURE 13-25.

reduction is an arbitrary method, it is recommended only for very sophisticated and experienced duct designers who have developed a "feel" for required velocity reductions versus the length of the duct section, the duct velocity, pressures, etc.

EQUAL FRICTION METHOD

In the equal friction method each section of duct (Fig. 13-26) is designed to have the same friction loss per foot of duct. It matters not how long sections P_1, P_2, P_3, etc. actually are since their rate of loss per foot is constant. This method of sizing is the most popular today and is used to size both the supply and return duct systems. It is far superior to the velocity reduction method, since it requires less balancing for symmetrical duct layouts and results in more economical duct sizes.

STATIC REGAIN

With duct systems designed on an equal friction basis, the calculated friction loss from one end to the other is not generally as great as calculated. As the air proceeds down the duct, the air velocity is reduced. This velocity represents energy, and, since energy cannot be created or destroyed, the same total energy must still be available. Actually, every time the velocity is reduced there is a conversion of velocity pressure to static pressure. This pressure will partially offset the friction in the next duct section.

This is similar to what happens to a skier on a hill (Fig. 13-27). At the top of the slope all of the available energy is potential. Progressing down the hill the skier picks up speed, and at the bottom where he really speeds along, all energy is motion. This energy is sufficient to carry him part way up the next slope, but friction and wind resistance prevent him from reaching the top. This is comparable to what happens in a duct. Static pressure is the potential energy. If air progresses down a duct with increasing velocity pressure, there would be a decrease in static pressure. Likewise, decreasing the velocity of air going through a duct increases the static pressure.

This carryover effect is called *static regain* in ducts. The regain efficiency is about 60%, and in high-velocity systems the use of this energy can be significant in reducing duct sizes and, therefore, initial costs. But for low-pressure duct systems as used in residential or light commercial work, the effect is not enough to be included in the design and its procedures.

With this review of the basic principles of air movement, fans, and duct design methods, we are ready to discuss the actual design of a typical residential system.

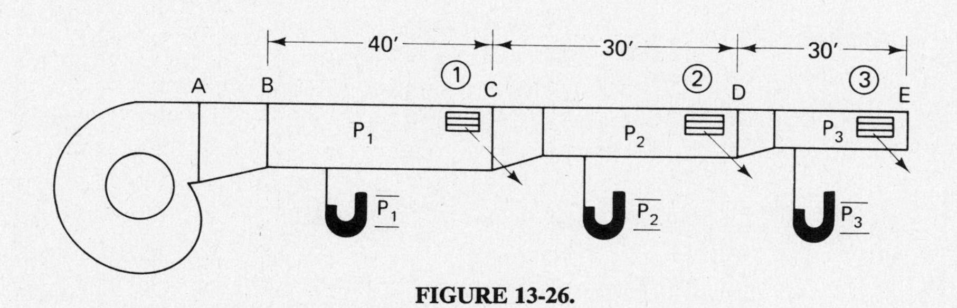

FIGURE 13-26.

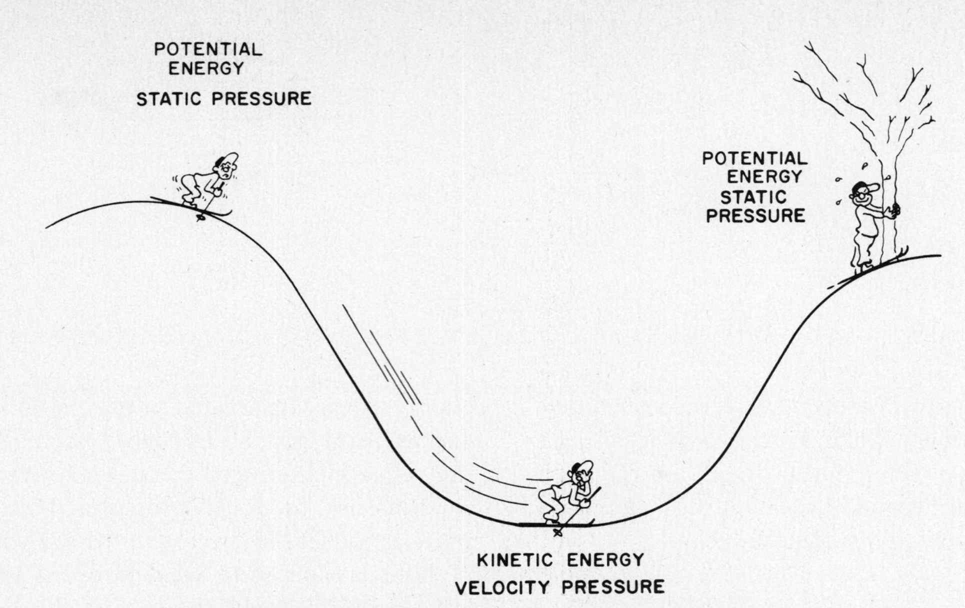

FIGURE 13-27 (Courtesy, Carrier Air-Conditioning Company).

PROBLEMS

1. Fluid flow may be caused by changes in ______________.

2. Air friction in ducts is caused by ______________.

3. Total pressure is made up of ______________ and ______________.

4. An instrument to measure air pressure is called a ______________.

5. An instrument to measure air velocity is called a ______________.

6. Using the friction chart in this chapter, determine the size of a duct to carry 1,000 ft^3/min so as not to exceed 0.1 in. of W.C. friction loss per 100 feet.

7. In Prob. 6, what is the velocity?

8. Convert the round duct to one with a height of 12 in. What is the width?

9. What are the recommended main and branch duct velocities for residential systems?

10. Most packaged air conditioners use a ______________ fan wheel.

14

RESIDENTIAL FORCED-AIR DUCT SYSTEMS

There are many types of forced-air duct installations used to heat and air condition a residence. The type used depends on a number of considerations, including geography, type of house, and economic considerations. This chapter will illustrate the most commonly used applications.

First, let's examine systems based on type of house, construction considerations, and equipment variations.

TYPES OF SYSTEMS

Figure 14-1 shows a one-story house with basement; it has a typical furnace installation in the basement with individual runs to the various outlets. This system is very popular for smaller homes where the individual runs are not excessively long. Notice also that there is a central return duct. This is a perimeter design, because the outlets are all on the outside (perimeter) wall, usually under the windows. Either a highboy or lowboy furnace may be used depending on the head room.

Figure 14-2 shows a basement installation in a two-story home with extra runs to the second floor. This could also apply to a larger single-story home. The difference between Figs. 14-1 and 14-2 is the use of an extended plenum. The extended plenum is a supply air trunk duct leading from the furnace plenum as noted in the equipment sketch. It carries sufficient air for several rooms. Individual round duct run-outs are then taken off to the various air outlets. A return air trunk duct comes from both floors. The location of the air outlet is also shown on the outside wall.

Figure 14-3 illustrates a crawl-space installation with a horizontal furnace. Again the extended plenum is used. For air conditioning, a horizontal cooling coil would be installed in the furnace supply air outlet duct. Although the furnace is installed in the crawl space, this home construction also lends itself to a packaged heating cooling unit outside, mounted on a slab with ducts coming into the crawl

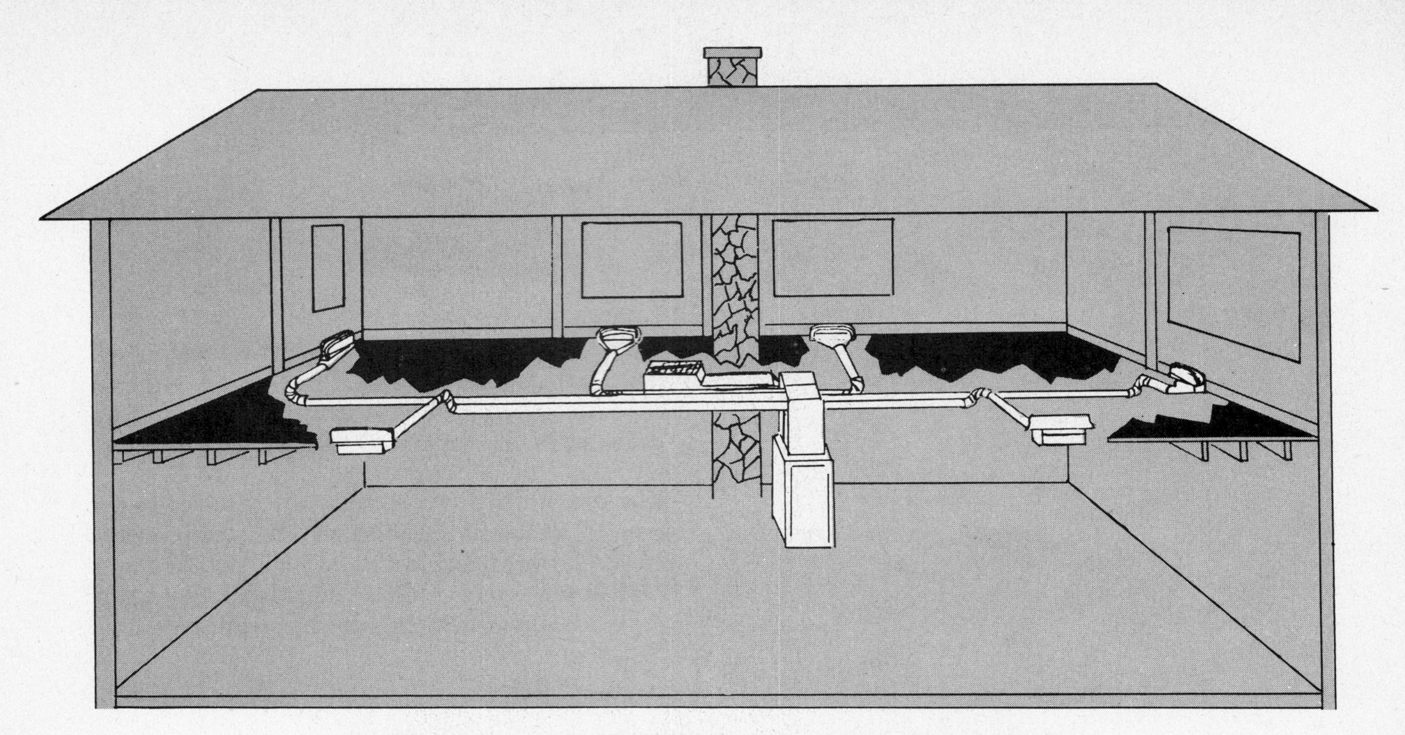

FIGURE 14-1 One-story house with basement duct system (Courtesy, NESCA).

FIGURE 14-2 Two-story house with basement extended plenum system (Courtesy, NESCA).

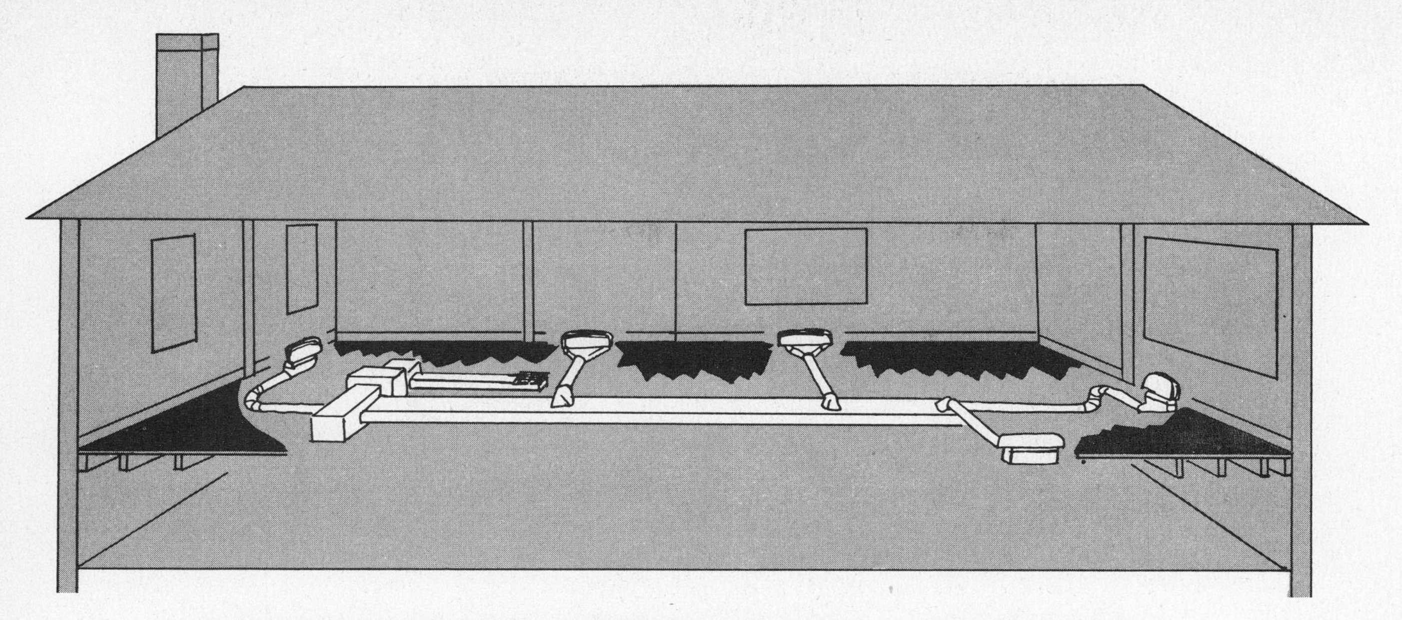

FIGURE 14-3 One-story house with extended plenum system in crawl space (Courtesy, NESCA).

space. The duct distribution system would be the same.

Figure 14-4 is also of a crawl-space home using a counterflow furnace in the living area, but with ductwork under the floor. Again an extended plenum duct design is used, but it could very well have individual radial ducts to each room air outlet. A central return is common practice.

Attic air distribution systems are very popular in the south where slab floors are used. They may be handled in several ways. Figures 14-5 and 14-5a illustrate a conventional upflow furnace and cooling coil installed on the first floor, but the supply air plenum goes into the attic with an insulated duct system above that serves ceiling outlets. The supply ducts can be an extended plenum as shown or else radial ducts. The return is usually a single low outlet in a hallway. Note that when single returns are used, doors are undercut to allow air passage even when door is closed.

Another variation of the attic installation (Fig. 14-6) shows a horizontal furnace with radial feeder ducts to ceiling outlets. The single return also frequently uses a combination return air/filter grille so that filter maintenance can be done from the living area. Supply ducts should be well insulated for heat transfer and have a moisture vapor barrier. Any heat loss in an attic is of no value and is costly, while in

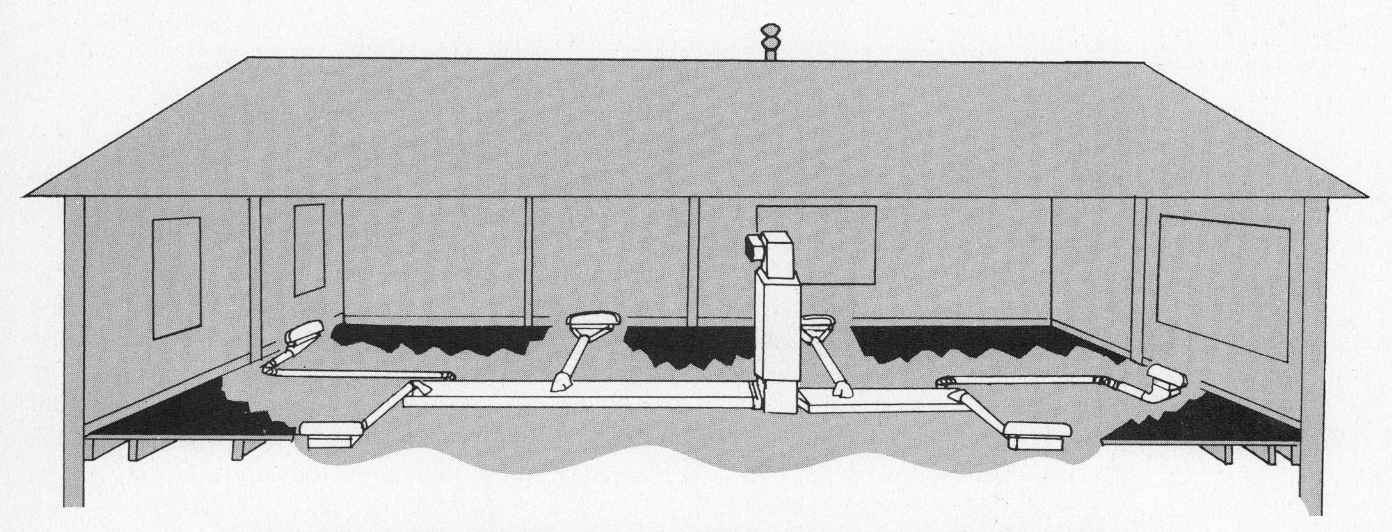

FIGURE 14-4 Utility counterflow furnace with duct system in crawl space (Courtesy, NESCA).

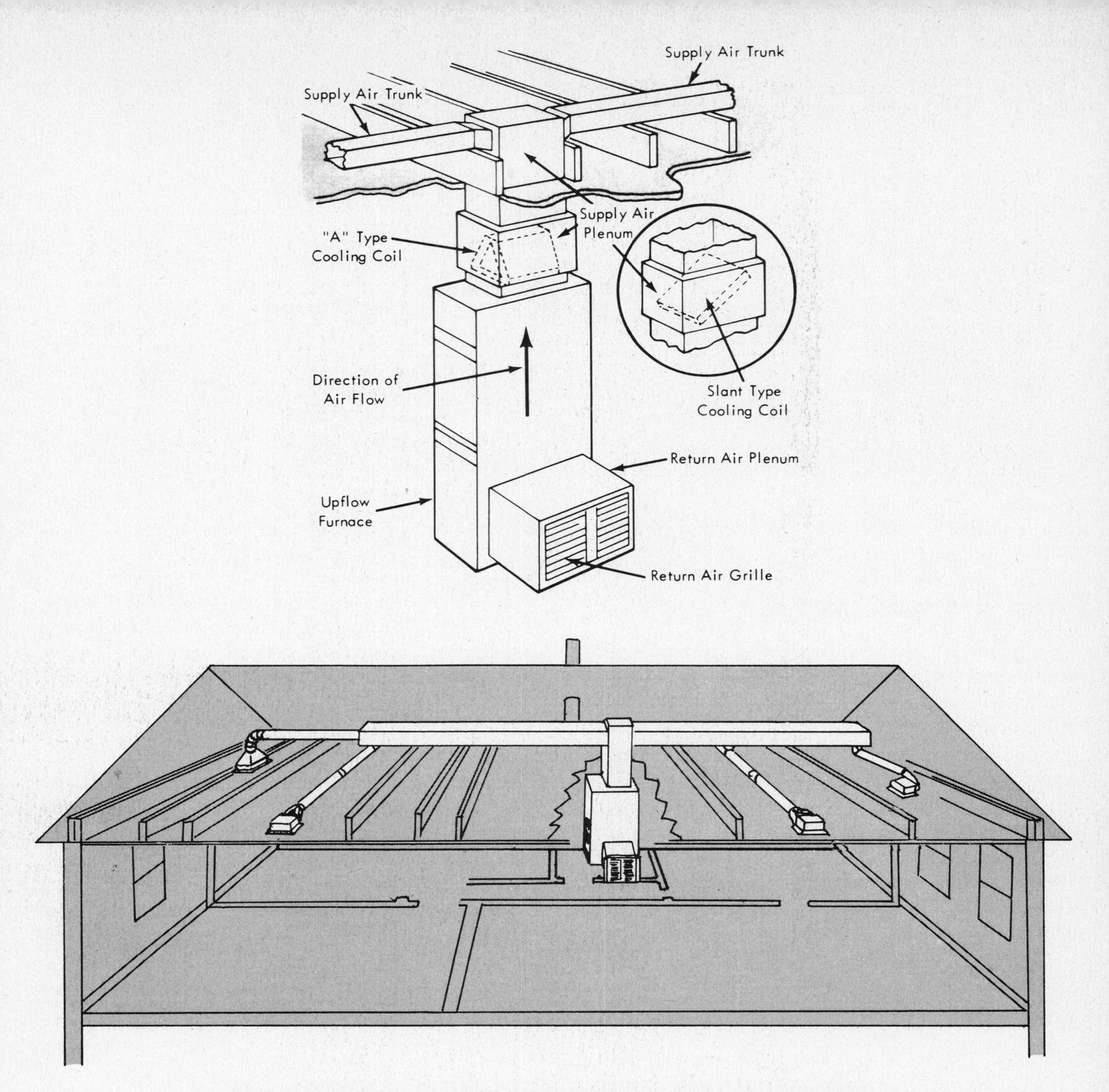

FIGURE 14-5 High-boy furnace with extended plenum system in attic (Courtesy, NESCA).

a basement or crawl space it has some floor-warming effect.

In some areas of the country, air-conditioning equipment may be mounted on the roof (Fig. 14-7) with proper support. Ducts penetrate the roof and go to ceiling outlets, usually in a radial fashion. A ceiling return (with filter) is common.

The last type of house construction is the concrete slab floor. It can be handled with a counterflow furnace with a radial duct design as shown in Fig. 14-8, but heating the floor at the outside wall is marginal. In a slab-floor structure, the loop perimeter system (Fig. 14-9) is generally preferred for heating and should be used wherever practicable, because it produces more uniform floor temperatures. The loop, which is buried in the concrete, keeps a continuous supply of warm air circulating around the slab edge, and outlets are stubbed up into the rooms as needed. Feeder ducts are also buried in the slab and planned so as to provide equal air distribution.

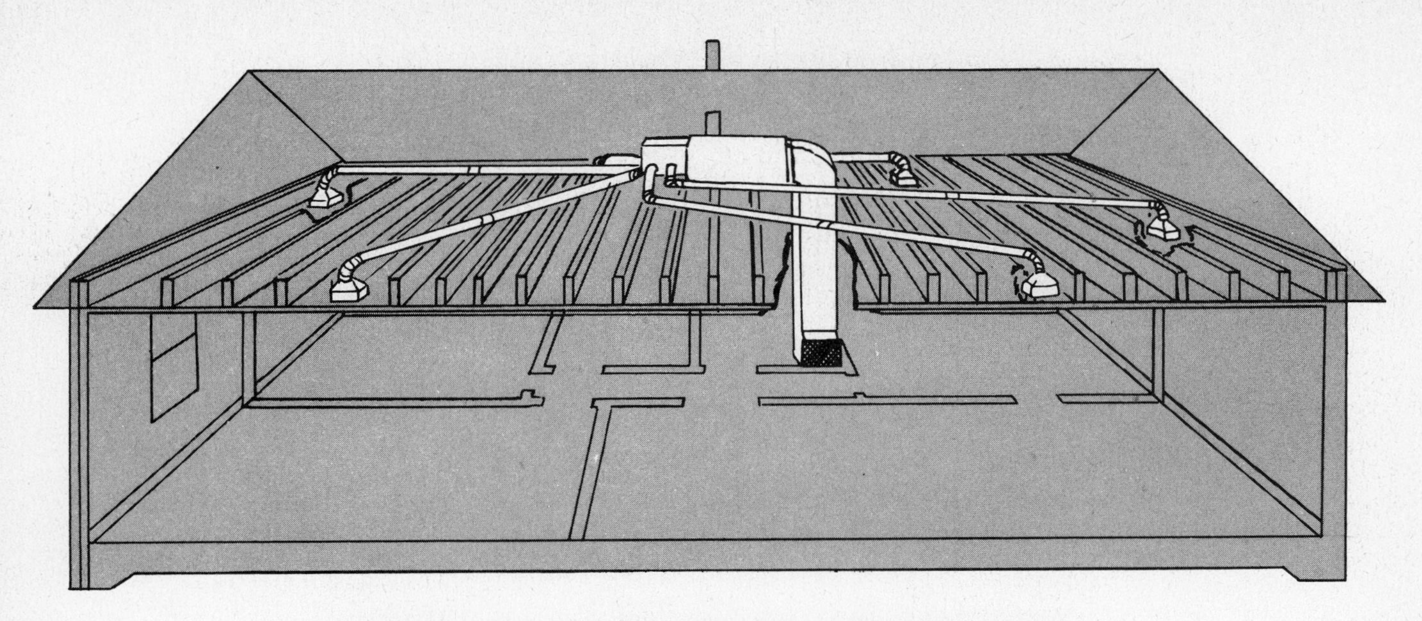

FIGURE 14-6 Horizontal furnace attic installation (Courtesy, NESCA).

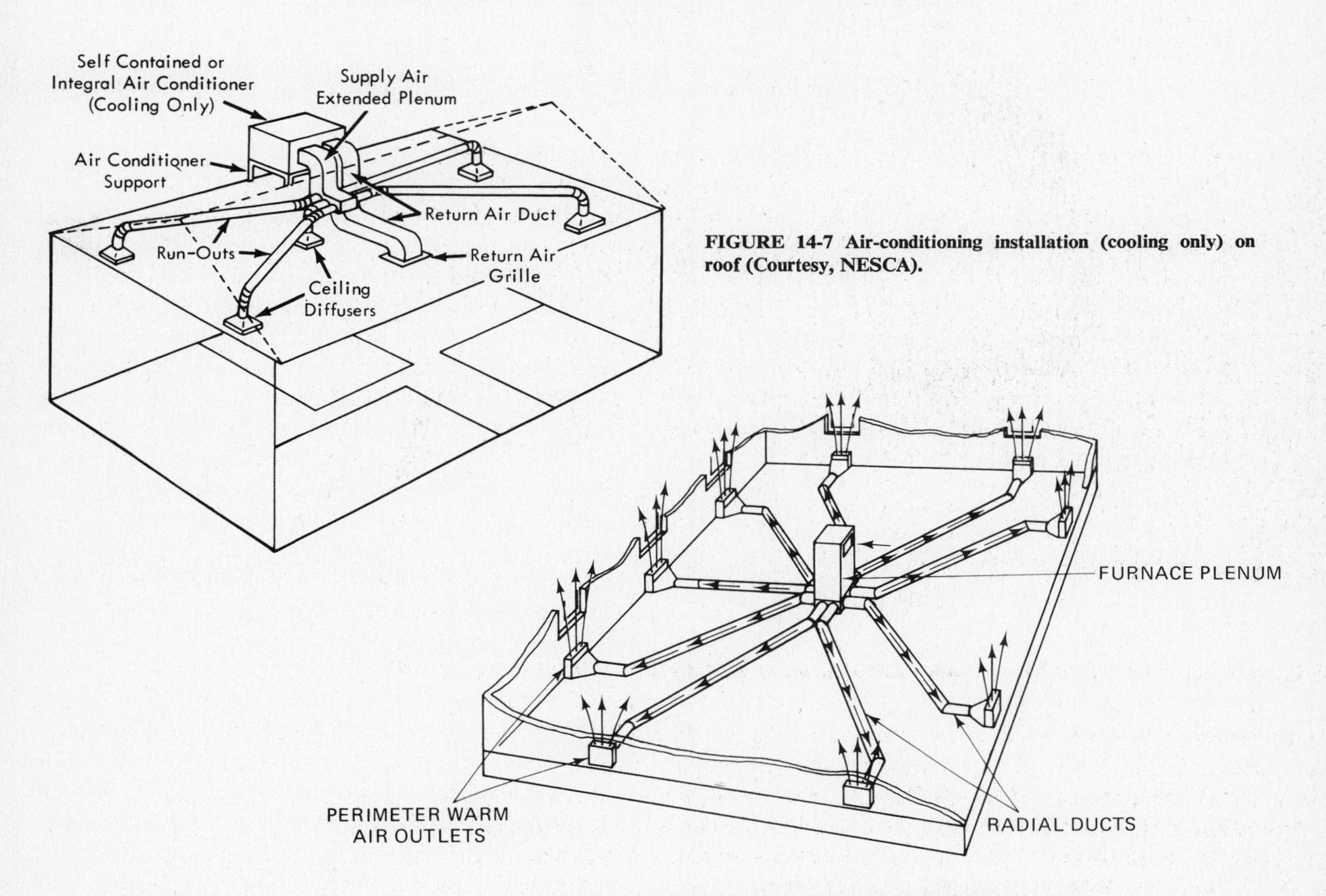

FIGURE 14-7 Air-conditioning installation (cooling only) on roof (Courtesy, NESCA).

LESS EXPENSIVE THAN A PERIMETER LOOP SYSTEM, THE RADIAL DUCT SYSTEM PROVIDES FOR WARM FLOORS AND IS APPLICABLE BOTH IN CRAWL SPACE CONTRUCTION AND CONCRETE SLABS'

FIGURE 14-8 (Courtesy, NESCA).

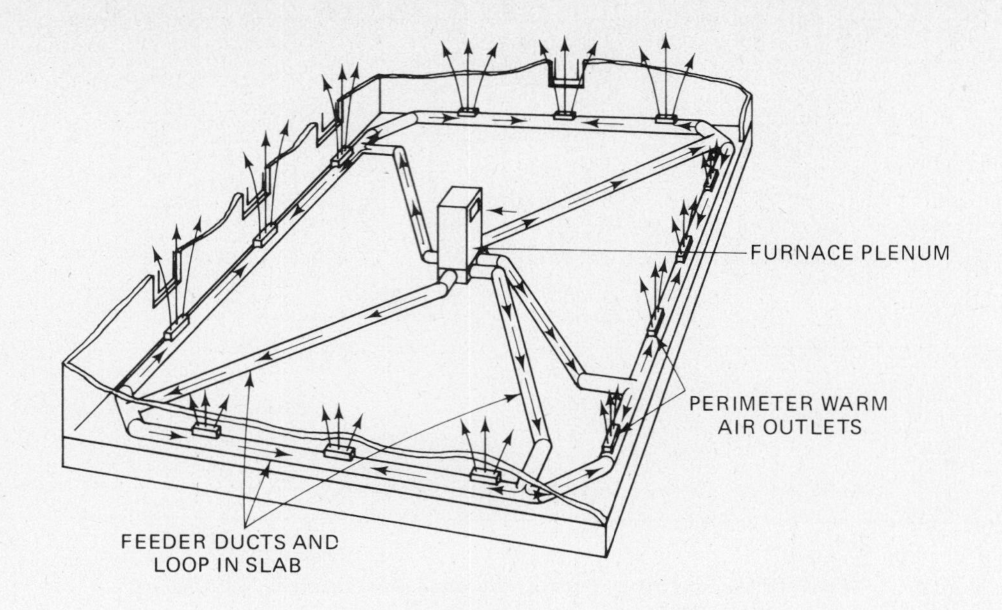

THE PERIMETER LOOP SYSTEM IS PARTICULARLY EFFECTIVE IN COLDER CLIMATES. IT HAS EXCELLENT UTILITY IN HOUSES BUILT ON SLABS IN THAT GOOD HEAT DISTRIBUTION IS INSURED AND FLOORS ARE KEPT WARM ALSO.

FIGURE 14-9 (Courtesy, NESCA).

Heating is a combination of warm air and the radiant effect from the warm floor.

This discussion on the loop perimeter applies to geographic zones where heating is an important consideration. In the south and southwest many buildings have slab floors but utilize one of the other duct systems, because in these cases cooling is the prime comfort consideration and air outlets are located in the ceiling or high on the side wall.

CLIMATE EFFECT ON THE DISTRIBUTION SYSTEM

The effect of climate is a prime factor in the proper selection of an air distribution system. In northern areas, it is important that the system provide warm floors in the winter. For a similar home in an area where it gets very hot, the ability to cool the home is most important. Thus, in northern areas a perimeter system is best for year-round air conditioning, while in the deep south an overhead system is best.

To help determine which type of system should be installed, climate zones, defined in terms of the number of degree days, can be established. Degree days are a measure of the average winter temperature. For any given day, when the mean temperature is less than 65°F, there are as many degree days as there are Fahrenheit degrees difference in the temperature of the day and 65°F. If the maximum temperature for the day is 54°F and the minimum is 32°F, the mean temperature, which is the average of the maximum and minimum temperatures, will be 43°F. There are then 65 — 43, or 22 degree days for that day. Figure 14-10 is a map of the United States divided into three areas:

Zone A—more than 3,500 degree days

Zone B—2,000 to 3,500 degree days

Zone C—Less than 2,000 degree days

Climatic conditions within zones B and C are fairly similar, but in zone A there can be a considerable difference in temperature characteristics. For example Dallas, Texas, and Baltimore, Md., are both in zone A. Both have the same winter design temperature (10°F), but Baltimore has twice the number of degree days in a normal heating season. Prolonged low temperatures result in colder surfaces of exposed components. Therefore, it is important to have a perimeter system in Baltimore, while in Dallas, which

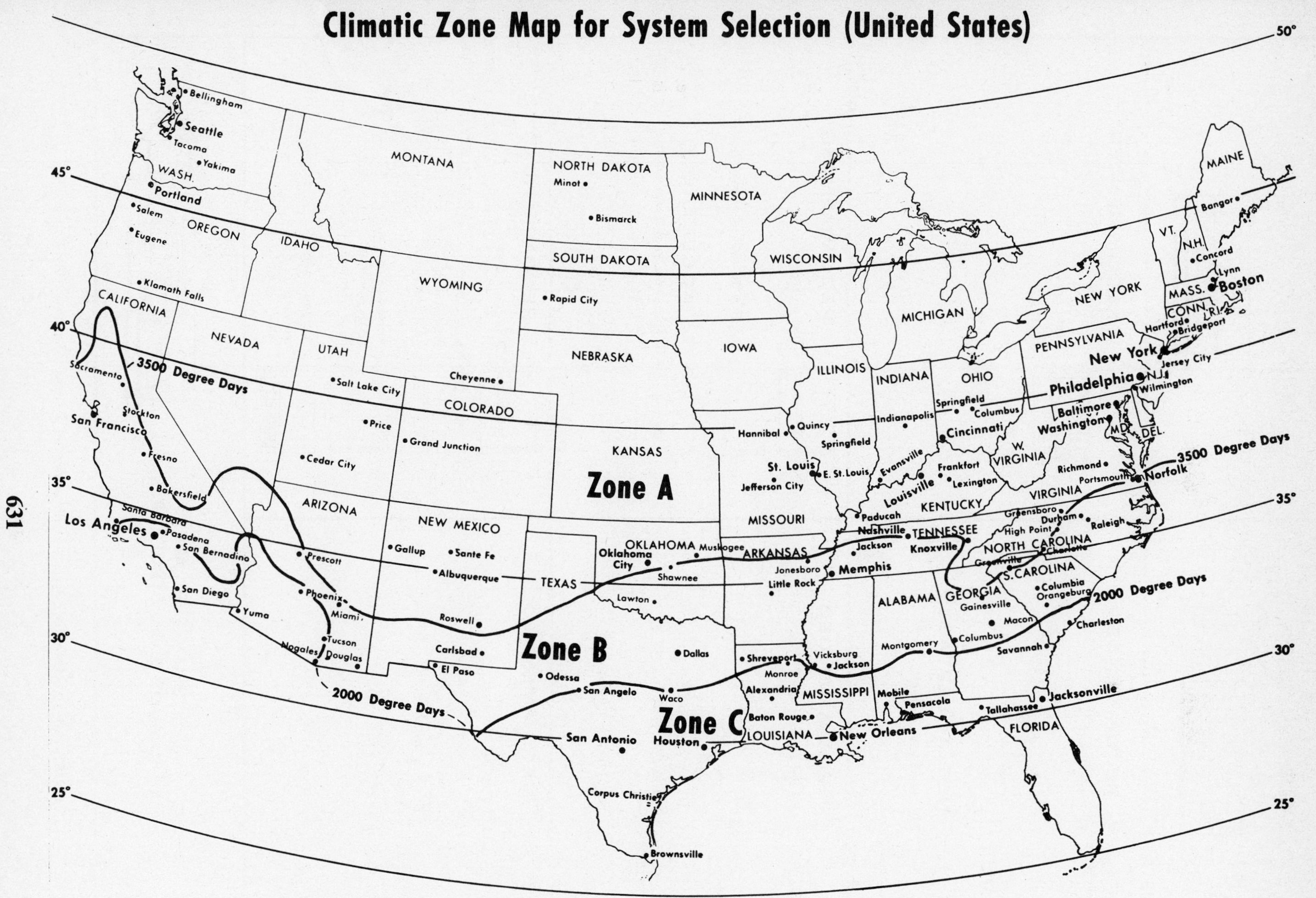

Climatic zones.

FIGURE 14-10 Climatic zone map for system selection (United States) (Courtesy, NESCA).

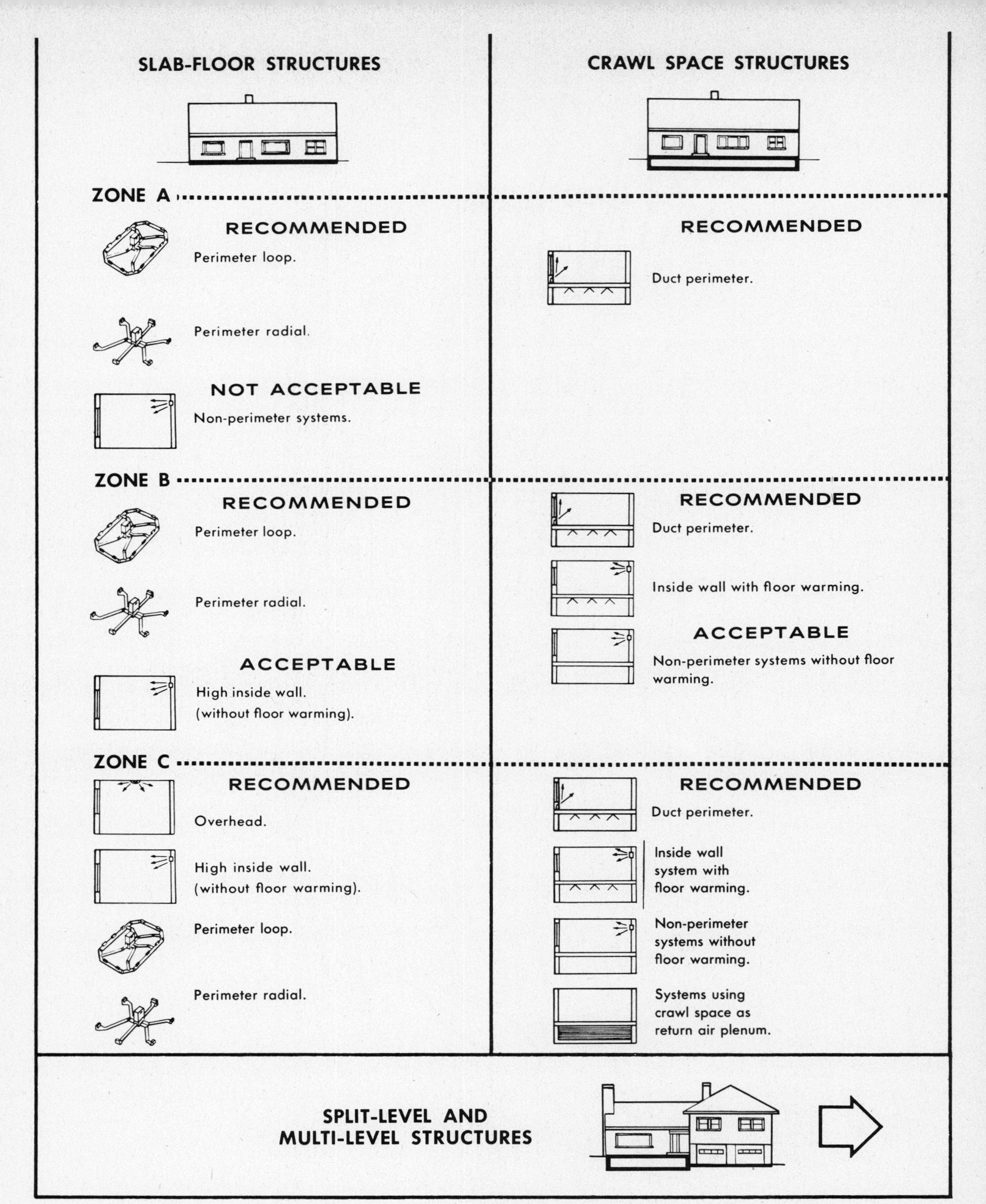

FIGURE 14-11 Guide for the selection of air-distribution systems (Courtesy, NESCA).

BASEMENT STRUCTURES	MULTIPLE DWELLING STRUCTURES
ZONE A	
RECOMMENDED	**RECOMMENDED**
Perimeter (for upstairs, heated basement). Use perimeter system in basement too, if rooms are to be used as living quarters.	Perimeter system on all floors.
ACCEPTABLE	**ACCEPTABLE**
Inside wall systems with floor warming.	Overhead system for intermediate stories.
ZONE B	
RECOMMENDED	**RECOMMENDED**
Perimeter (for upstairs; heated basement).	Perimeter system on all floors.
Inside wall system with floor warming.	
ACCEPTABLE	**ACCEPTABLE**
Inside wall system without floor warming.	Overhead system for intermediate stories.
ZONE C	
RECOMMENDED	**RECOMMENDED**
Perimeter. (With or without floor warming).	Overhead systems for all stories. (Perimeter systems may be desirable for improved air distribution and greater comfort.)
Inside wall. (With or without floor warming.)	

Select distribution system to match sections of structure. For example, if room is built over basement, use system for basement structures. If room is located over slab floor, use slab floor practices, etc.

FIGURE 14-11 (Cont'd).

has comparatively short periods of severe cold and long periods of hot weather, the use of an overhead system would be acceptable.

Figure 14-11 is a guide to help select the proper system; the guide is a composite of recommendations by manufacturers, contractors, and others actively engaged in the industry. The category "Recommended" is the system which the majority of the experts felt would give the optimum degree of comfort for the installed cost. It does not mean a deluxe system with frills or accessories. "Acceptable" means the system is a type that has been used for many years and produces a degree of comfort which would be classified as fair. "Not acceptable" means the system does not provide an adequate degree of comfort.

The chart also includes a reference to multiple dwelling structures such as apartments or townhouses, where the effect of conditioning on one floor affects the one above or below.

DUCT DESIGN

The actual design of the duct sizes is a function of the type of system and the heating/cooling air volumes needed for each room. NESCA Manual K, *Equipment Selection and System Design Procedure*, gives detailed instructions on the methods and calculations needed for each system. We suggest it as additional reference.

EQUAL FRICTION METHOD

As previously mentioned, there are several acceptable techniques for duct design. We, however, have elected to use the *equal friction* method, because it lends itself to a greater variety of applications. The success of the equal friction method of duct design lies in the fact that the same static pressure is available to push air through each duct in the system whether it is 5 ft or 25 ft long; whether it is to deliver 10 ft^3/min or 100 ft^3/min, or whether it contains one fitting or three friction-creating fittings.

First, it is necessary to determine the equivalent length of the supply run that offers the greatest resistance to air flow. If the duct run with the greatest resistance is not readily apparent, it may be necessary to calculate the resistance of several or all of the runs in order to be certain that the longest equivalent length is used to establish a supply system pressure loss. It is also necessary to apply the same logic to the return system.

The total equivalent length of a run of ductwork is the sum of the measured horizontal and vertical lengths of the air path plus the equivalent length of the fittings in the air path. Figure 14-12 illustrates many of the ductwork components used in indoor comfort systems. In practice, no one installation uses all the parts shown. With experience the many parts become familiar. The equivalent lengths of most fittings, expressed in feet of straight duct that would have the same resistance to air flow, can be determined from the accompanying drawings in Fig. 14-13, Group 1 through 7. These have been reproduced from *Manual K.* For each duct fitting, the capital letter identifies the type of fitting. The numerals indicate the equivalent length in feet of straight duct for that fitting.

When designing a duct system, no uniformly sized section of an extended plenum should be longer than 24 feet. When the total length of a trunk duct exceeds 24 feet, it is recommended that the size of each duct be reduced every 15 to 20 feet in the manner illustrated in Fig. 14-14.

SYSTEM PRESSURE

Referring to the manufacturer's product data sheet, which pertains to the equipment selected, will indicate the pressure available to overcome the total pressure losses external to the air-handling equipment. It should be noted whether losses through filters, coils, and heat exchangers have been considered in determining the external static pressure available. This pressure must be capable of overcoming the losses of the supply and return duct system, the diffusers, the return grille, the cooling coil (if it has not been included as an integral part of the air-handling equipment), and any other external equipment that is placed in the air path. This remaining pressure must be available under conditions compatible with the air volume that is necessary to handle the heating and/or cooling load. The total available pressure can be divided between the supply and return duct system in a manner that will aid in

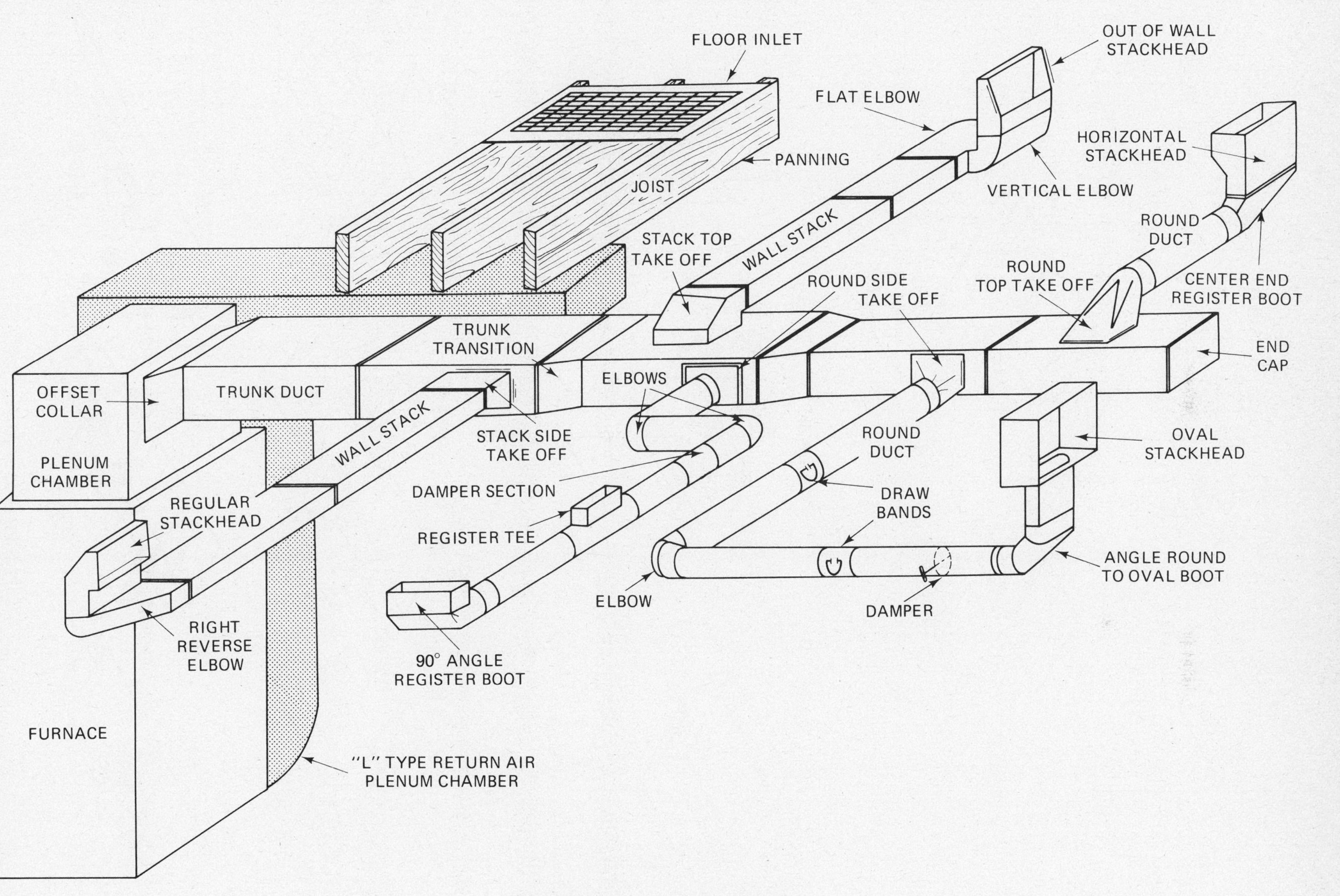

FIGURE 14-12 Duct work components.

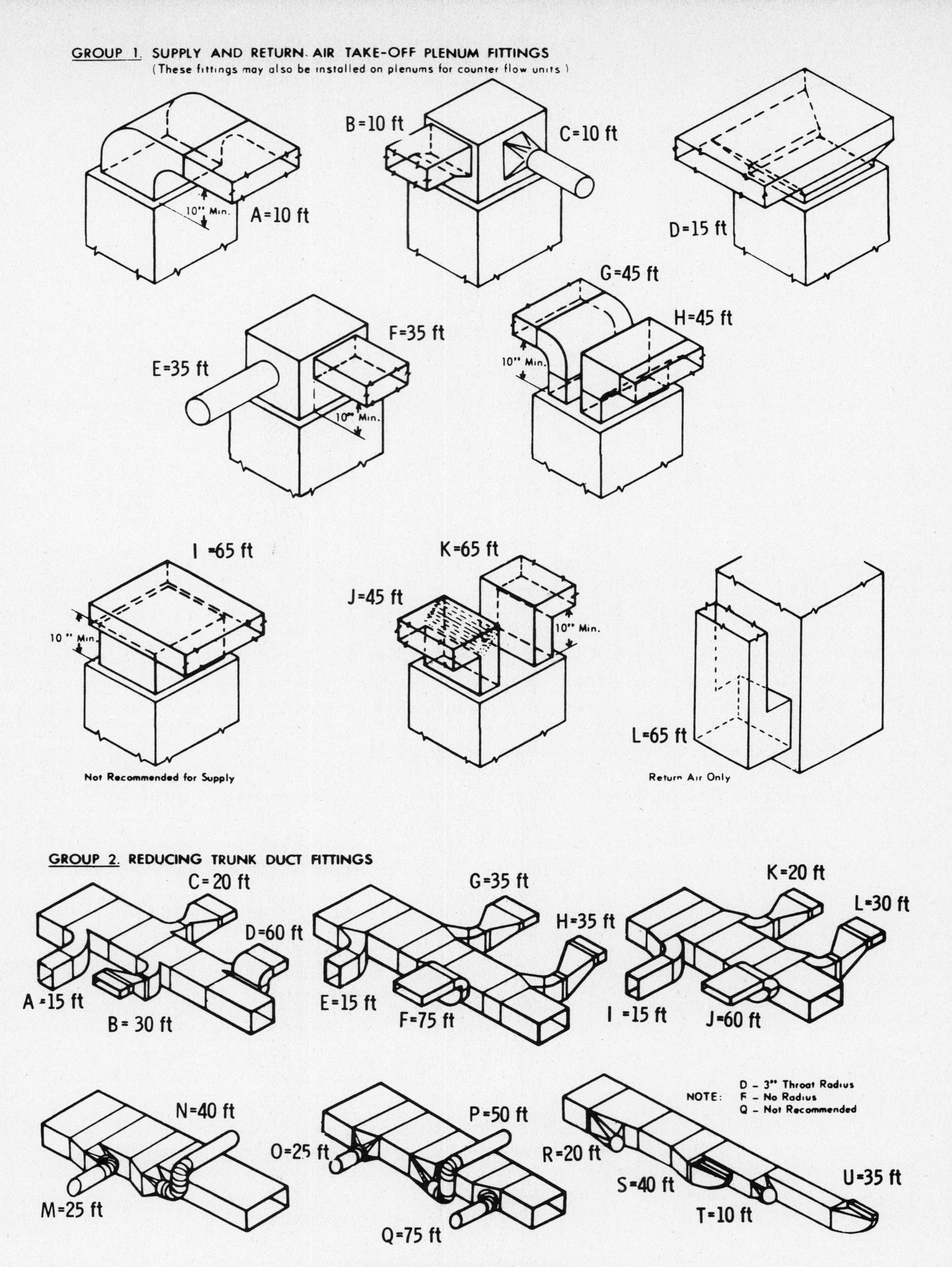

FIGURE 14-13 (Courtesy, NESCA).

GROUP 3. EXTENDED PLENUM FITTINGS

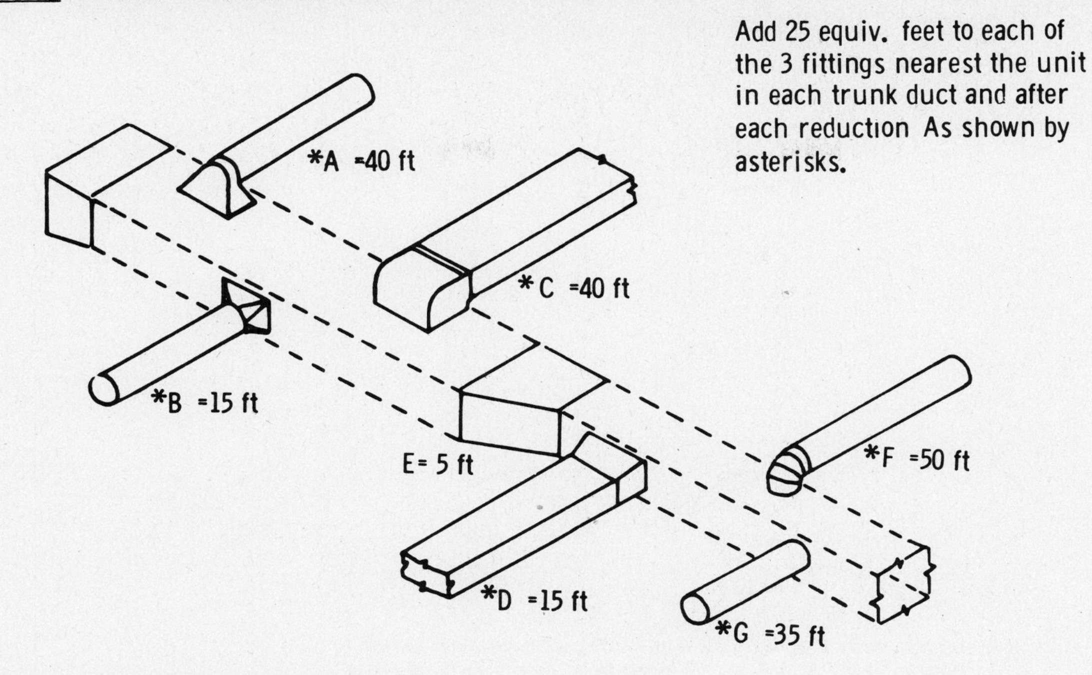

GROUP 4 ROUND TRUNK DUCT FITTINGS
(Add 25 equivalent feet to each of the 3 fittings nearest the unit in each Trunk Duct)

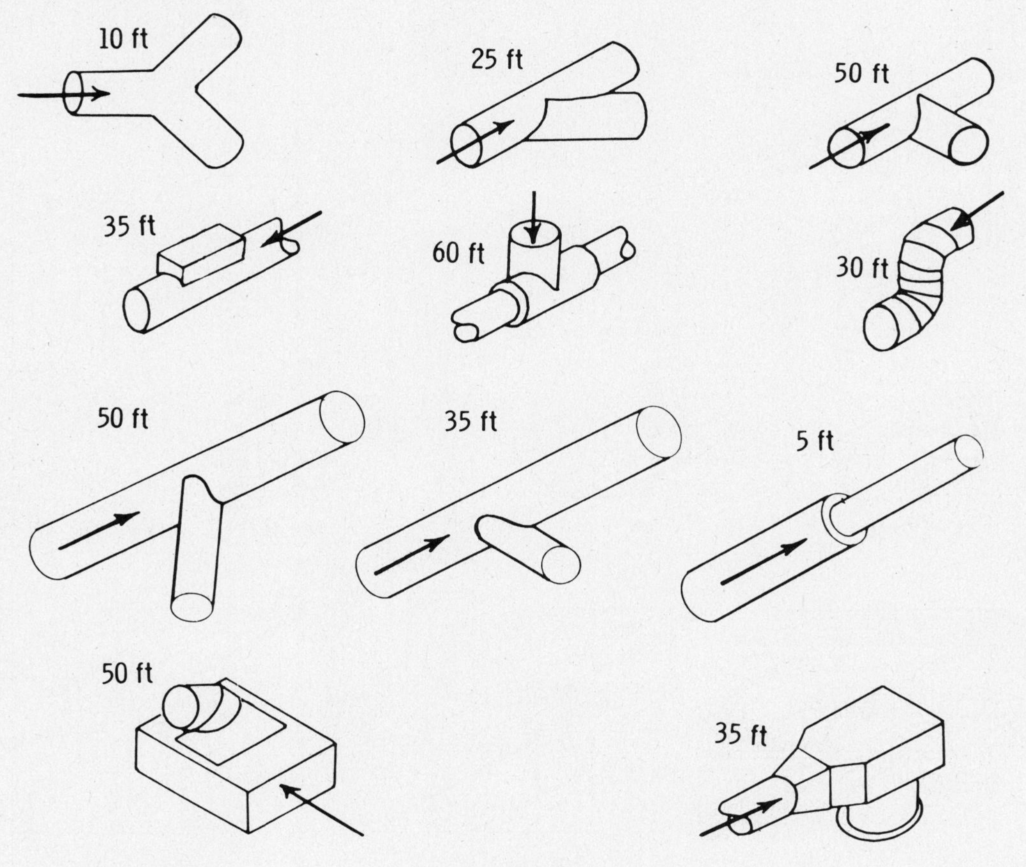

FIGURE 14-13 (Cont'd).

GROUP 5. ANGLES AND ELBOWS FOR TRUNK DUCTS
(Inside Radius = ½ Width of Duct)

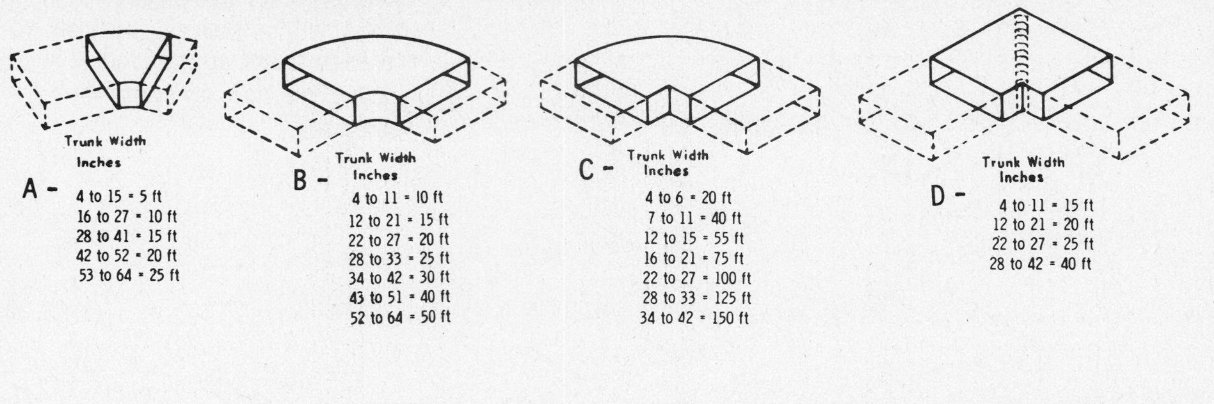

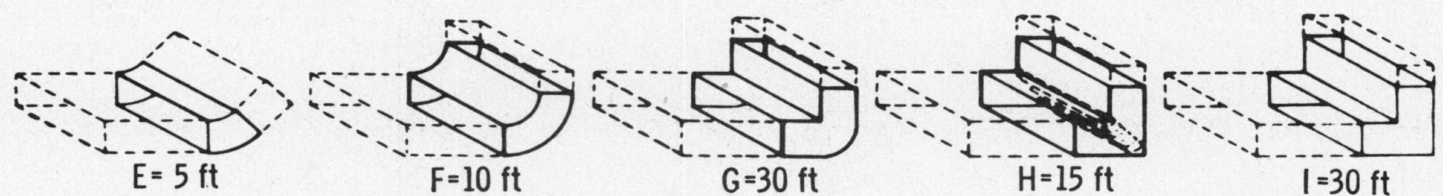

GROUP 6 ANGLES AND ELBOWS FOR INDIVIDUAL AND BRANCH DUCTS
(Inside Radius for "A" and "B" = 3 in. and for "F" and "G" = 5 in.)

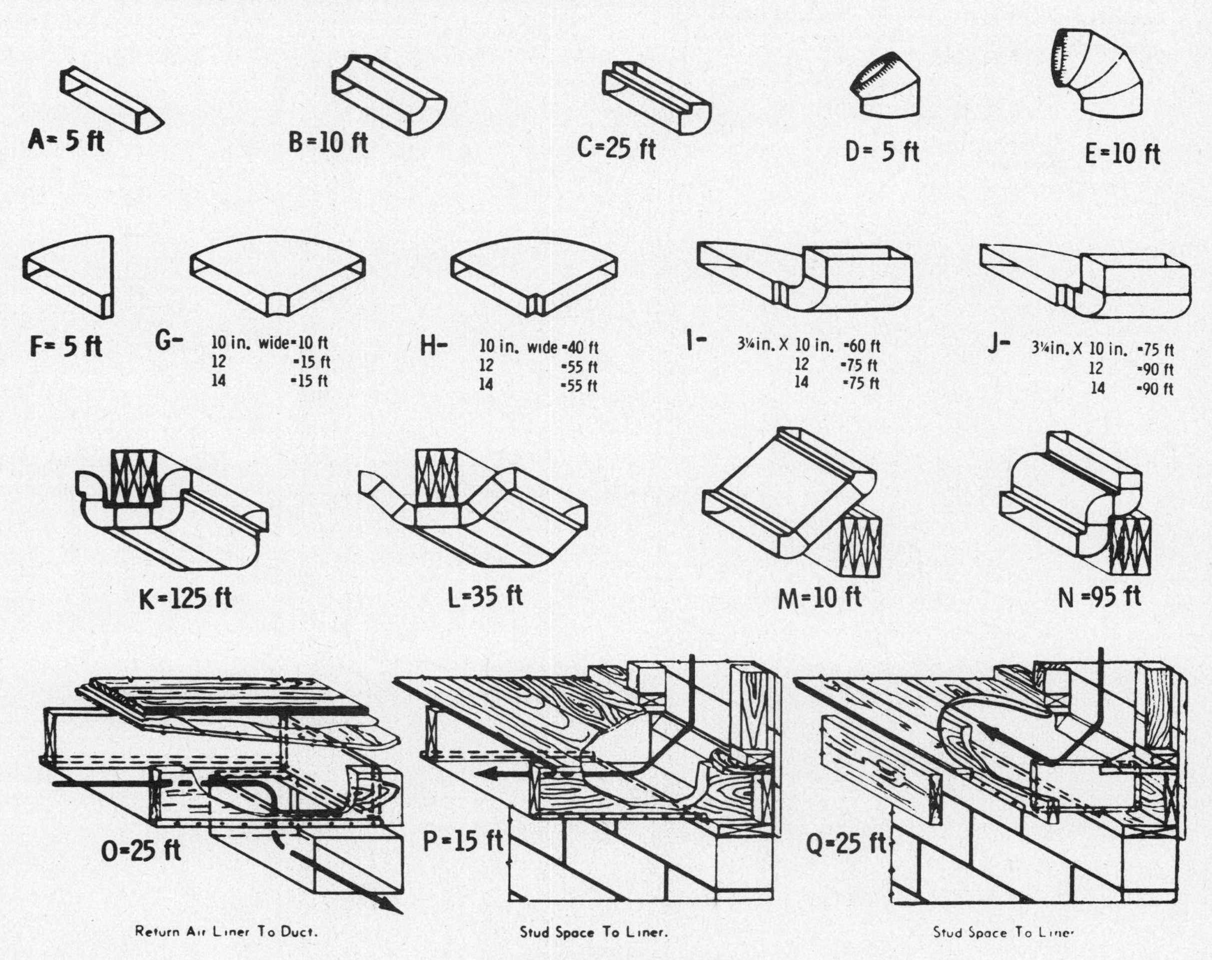

FIGURE 14-13 (Cont'd).

GROUP 7. BOOT FITTINGS
(These values may also be used for floor Diffuser Boxes)

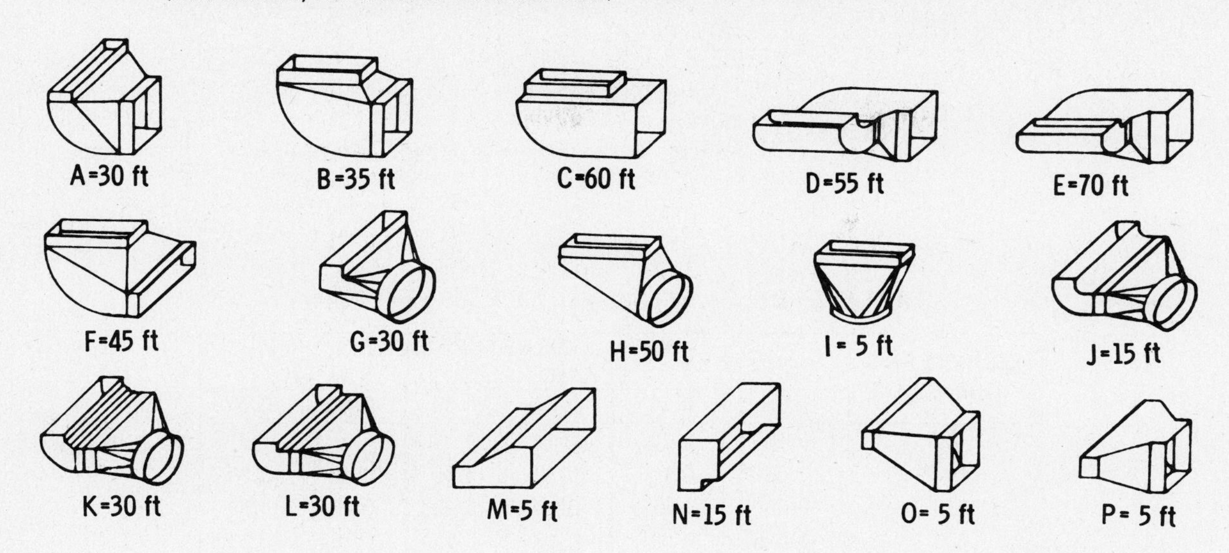

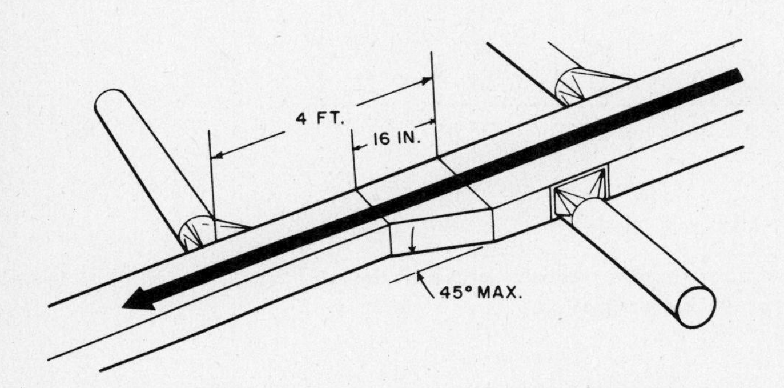

FIGURE 14-14 Transition filling for reducing plenum (Courtesy, Bryant Air-Conditioning Company).

economical duct design. Figure 14-15 will be of help in selecting and dividing the total available static pressures between the supply and return duct systems.

Where a quiet system is desired, a low static pressure will permit a lower fan speed and quieter equipment performance. A lower pressure loss will also reduce duct and grille velocity—two possible noise sources.

DUCT SIZING

After the total static pressure has been determined and divided between the supply and return duct systems, with the necessary deductions made for the coil, diffuser, and all other applicable items in the air path, then Fig. 14-16 can be used to determine the design friction per 100 equivalent feet of duct.

Suggested Division of Static Pressure (inches of water column) between Supply and Return Systems

SYSTEMS WITH MULTIPLE RETURNS (One for each room or area)											
Total available pressure	.1	.12	.14	.16	.18	.2	.4	.6	.8	1.0	1.2
Supply system pressure	.05	.06	.07	.08	.09	.1	.2	.3	.4	.5	.6
Return system pressure	.05	.06	.07	.08	.09	.1	.2	.3	.4	.5	.6
SYSTEMS WITH RETURNS IN SEVERAL AREAS											
Total available pressure	.1	.12	.14	.16	.18	.2	.4	.6	.8	1.0	1.2
Supply system pressure	.07	.08	.09	.10	.12	.13	.28	.42	.56	.7	.8
Return system pressure	.03	.04	.05	.06	.06	.07	.12	.18	.24	.3	.4
SYSTEMS WITH SIMPLE (SINGLE) RETURNS											
Total available pressure	.1	.12	.14	.16	.18	.2	.4	.6	.8	1.0	1.2
Supply system pressure	.08	.09	.10	.12	.13	.15	.3	.45	.6	.8	.9
Return system pressure	.02	.03	.04	.04	.05	.05	.1	.15	.2	.2	.3

NOTE: This table shows only suggested total pressure and its division into supply and return system pressures. The selection and division may be altered to meet the requirements of the job.

FIGURE 14-15 (Courtesy, Bryant Air-Conditioning Company).

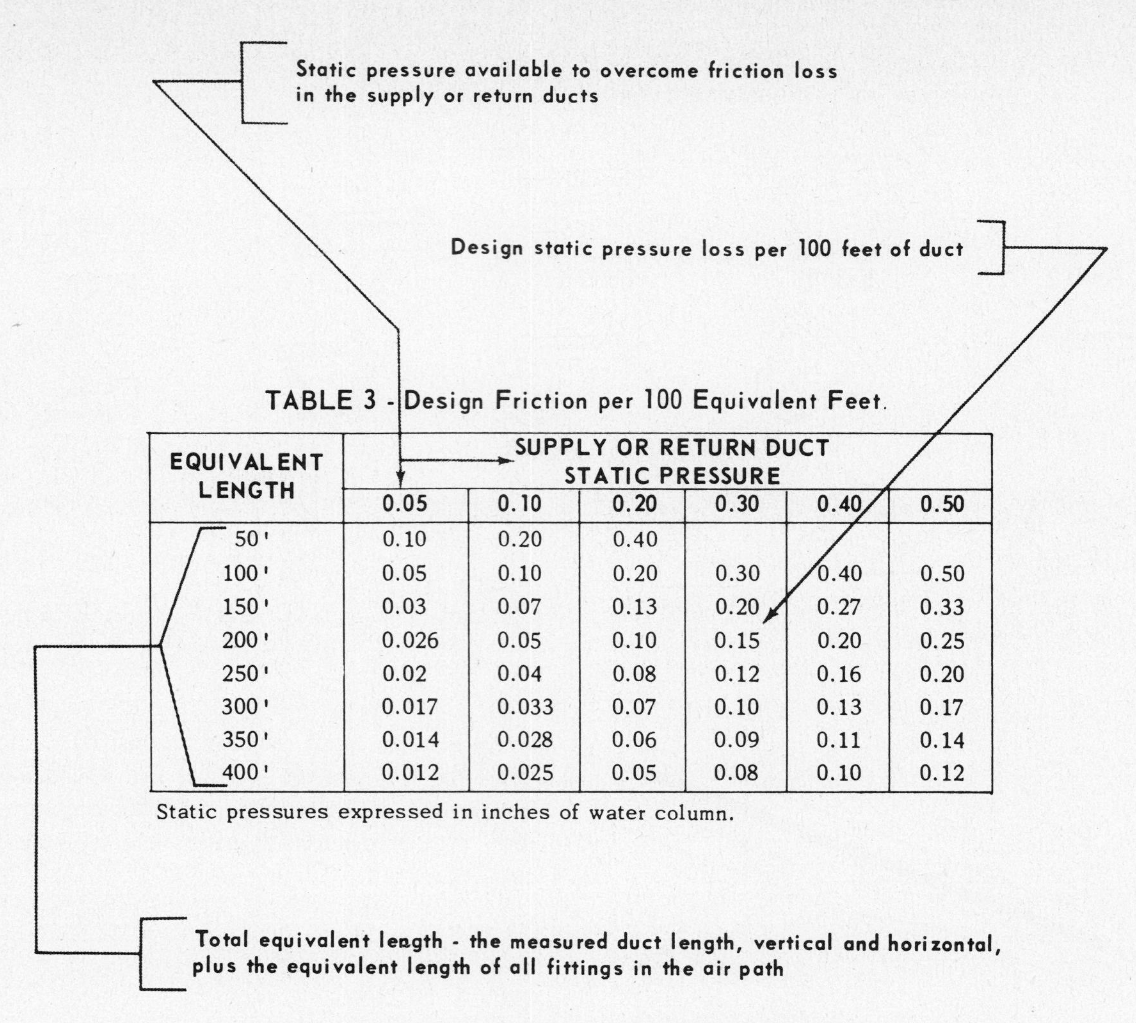

TABLE 3 - Design Friction per 100 Equivalent Feet.

EQUIVALENT LENGTH	SUPPLY OR RETURN DUCT STATIC PRESSURE					
	0.05	0.10	0.20	0.30	0.40	0.50
50'	0.10	0.20	0.40			
100'	0.05	0.10	0.20	0.30	0.40	0.50
150'	0.03	0.07	0.13	0.20	0.27	0.33
200'	0.026	0.05	0.10	0.15	0.20	0.25
250'	0.02	0.04	0.08	0.12	0.16	0.20
300'	0.017	0.033	0.07	0.10	0.13	0.17
350'	0.014	0.028	0.06	0.09	0.11	0.14
400'	0.012	0.025	0.05	0.08	0.10	0.12

Static pressures expressed in inches of water column.

FIGURE 14-16 (Courtesy, Bryant Air-Conditioning Company).

When the allowable design friction loss per 100 equivalent feet of duct has been established, the friction loss charts and conversion tables from Chap. 13 can be used to size properly all of the round and rectangular ducts in the system. Air velocities must not exceed those recommended in Chap. 13.

SAMPLE HOUSE PROBLEM

Step 1—Calculating Loads

First, calculate the heat loss and/or heat gain of the whole house plan as illustrated in Fig. 14-17 and presented under Chapter 6, using *Manual J* procedures. Assume this was done; the block table represents the resulting room and the total heat gain and loss for the particular geographic area.

Step 2—Selecting the Equipment

Select the heating and/or cooling unit from the appropriate manufacturer's product data sheet. The Btu/h output capacity of the heating unit should be equal to, or 15% greater than, the calculated heat load if the heating unit or duct system is outside the heated space. (Output capacity depends on the type of fuel. Remember, gas furnaces are rated on input, so be sure to check the manufacturer's specifications on whatever product is used.) The Btu/h output of the cooling unit should not be less than 90% or more than 110% of the minimum required capacity. These selection factors are based on the assumption that the equipment is operating under command of the room thermostat—greater capacity will be needed when the owner turns the equipment off for long periods of time and then desires a rapid return to a comfort condition in a short period of time. If the ducts are installed outside the conditioned space, they should be properly insulated. In the following example, the ducts are in an unconditioned basement, so the 15% safety factor is used:

Heating Unit = 81,000 × 1.15 = 93,150 Btu/h

A cooling unit with a nominal 36,000 Btu/h rating is the recommended choice.

Step 3—Locate the Outlets

Indicate the outlets for each room or space to be conditioned on the floor plan. Use at least one outlet for each 4,000 Btu/h heat gain, or 8,000 Btu/h heat loss, whichever is greater, in any room or area. Since, in this example, the cooling requirement is more than the heating requirement, we will size our system on the cooling requirements. Number each outlet S-1, S-2, S-3, etc., for supply, and R-1 and R-2 for returns (see Fig. 14-17).

Step 4—Layout of the Distribution System

Make a line drawing of the proposed basement duct system (Fig. 14-18), and use the same numbering system for supply diffusers and return air grilles that were used on the floor plan.

Step 5—Determine Air Quantities

Determine the necessary air quantities for the supply outlet and the return air grille and tabulate as noted in Fig. 14-19.

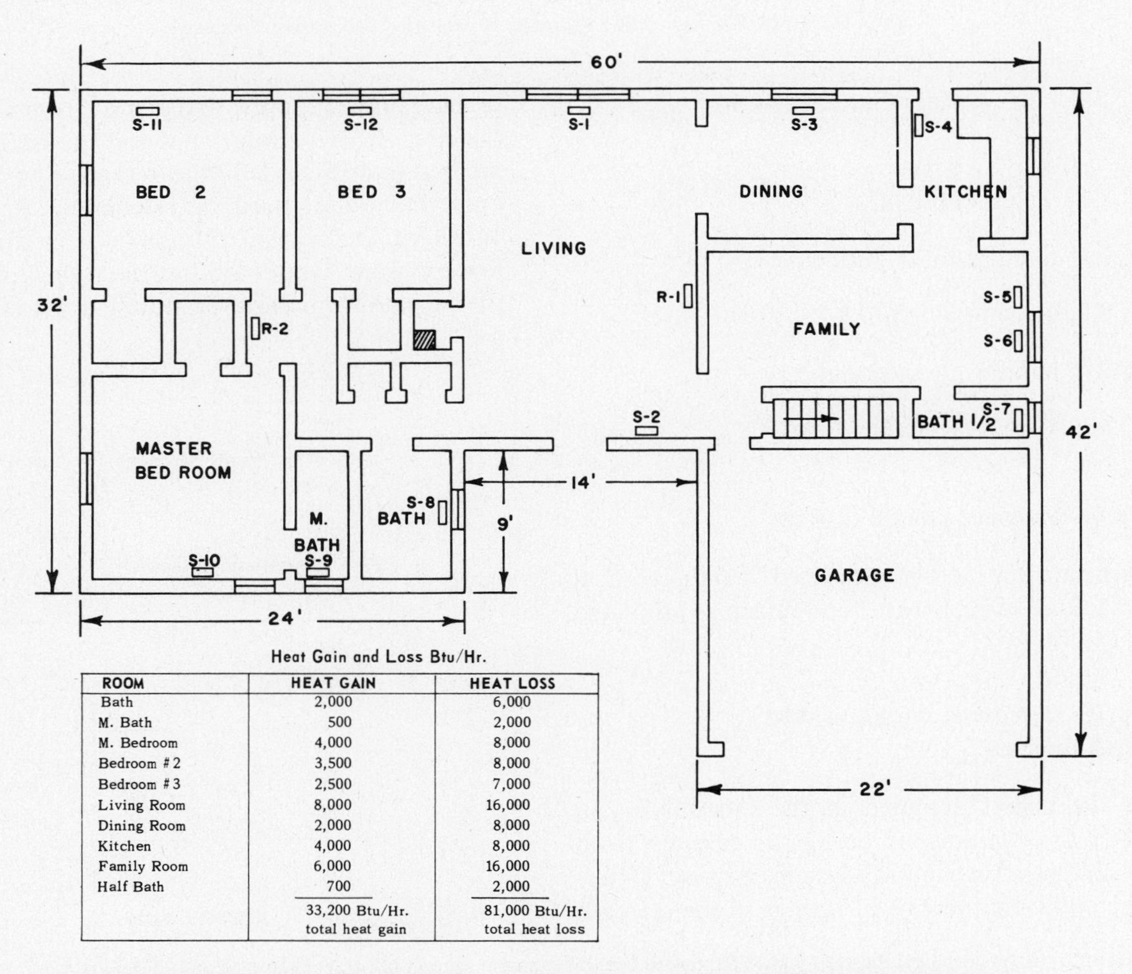

Heat Gain and Loss Btu/Hr.

ROOM	HEAT GAIN	HEAT LOSS
Bath	2,000	6,000
M. Bath	500	2,000
M. Bedroom	4,000	8,000
Bedroom #2	3,500	8,000
Bedroom #3	2,500	7,000
Living Room	8,000	16,000
Dining Room	2,000	8,000
Kitchen	4,000	8,000
Family Room	6,000	16,000
Half Bath	700	2,000
	33,200 Btu/Hr. total heat gain	81,000 Btu/Hr. total heat loss

FIGURE 14-17 Floor plan showing register locations (Courtesy, Bryant Air-Conditioning Company).

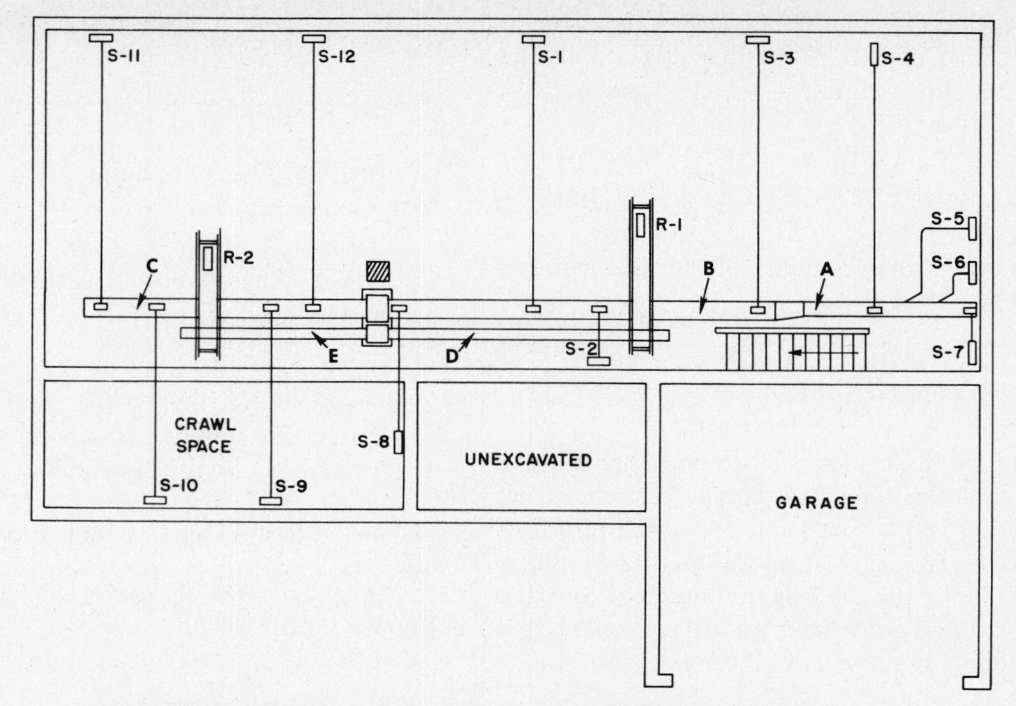

FIGURE 14-18 Basement plan (Courtesy, Bryant Air-Conditioning Company).

$$\text{Total ft}^3/\text{min} = \frac{\text{Heat Gain in Btu/h}}{12{,}000} \times 400^*$$

$$= \frac{36{,}000}{12{,}000} \times 400 = 1200 \text{ ft}^3/\text{min}$$

Return R-1 handles these outlets:

S-1, S-2, S-3, S-4, S-5, S-6, S-7

Return R-2 handles these outlets:

S-8, S-9, S-10, S-11, S-12

Step 6—Measure Length of Runs

Determine the actual measured length, vertical and horizontal, of each outlet and return as tabulated in Fig. 14-20.

Step 7—Determine the Equivalent Length of Runs

Use the values assigned in the illustrations in Fig. 14-13 for the various fittings as selected from Fig. 14-22, the air distribution system drawing. Determine the equivalent length of fittings in the air path to each outlet, supply, and return from the outlet to the unit. If, by careful analysis, the outlet with the greatest equivalent length can readily be determined, only that outlet need be calculated. Refer to Fig. 14-22 for the type of fittings used in this problem. For example: Outlet S-5 has the following fittings in the air path from the air-handling unit to the diffuser:

Outlet	*Ft³/min Required*
S-1	145
S-2	145
S-3	65
S-4	135
S-5	110
S-6	110
S-7	25
S-8	70
S-9	30

FIGURE 14-19.

* Will vary with geographical location from 350 to 450. For this example, we will use 400 as the factor, and since a 3-ton unit is selected, 36,000 Btu will be used as the heat gain.

Outlet	*Ft³/min Required*
S-10	145
S-11	120
S-12	100
Total	1,200
Return R-1	735
R-2	465
Total	1,200

FIGURE 14-19 Cont'd.

Outlet	*Measured Length Ft*
S-1	29
S-2	19
S-3	43
S-4	49
S-5	42
S-6	42
S-7	42
S-8	12
S-9	21
S-10	28
S-11	37
S-12	23
Return R-1	24
R-2	18

FIGURE 14-20.

1	G-30 floor diffuser box	30 ft
3	E-10 90° adjustable elbow	30 ft
1	B-15 side take-off	15 ft
1	E-15 reduction	5 ft
	For one of first three take-off fittings after a trunk reduction	25 ft
1	B-10 plenum take-off	10 ft
	TOTAL	115 ft

Fig. 14-21 is a tabulation of the equivalent lengths of fittings for each run of our sample house problem:

Outlet	*Equivalent Length of Fittings*
S-1	105
S-2	105
S-3	80
S-4	130
S-5	115
S-6	105
S-7	105
S-8	125
S-9	105
S-10	105
S-11	80
S-12	105
Return R-1	140
R-2	140

FIGURE 14-21.

Step 8—Determine the Total Equivalent Length of the Longest Run

Select the supply and return runs that have the longest total equivalent length. The total equivalent length is the sum of the measured length and the equivalent length of fittings.

The supply run S-4 has a measured length of 49 ft and an equivalent length of fittings of 130 ft; therefore, the total equivalent length of the longest supply run is 179 ft. Return run R-1 has a measured length of 24 ft and an equivalent length of fittings of 140 ft. Therefore, the total equivalent length of the longest return run is 164 ft.

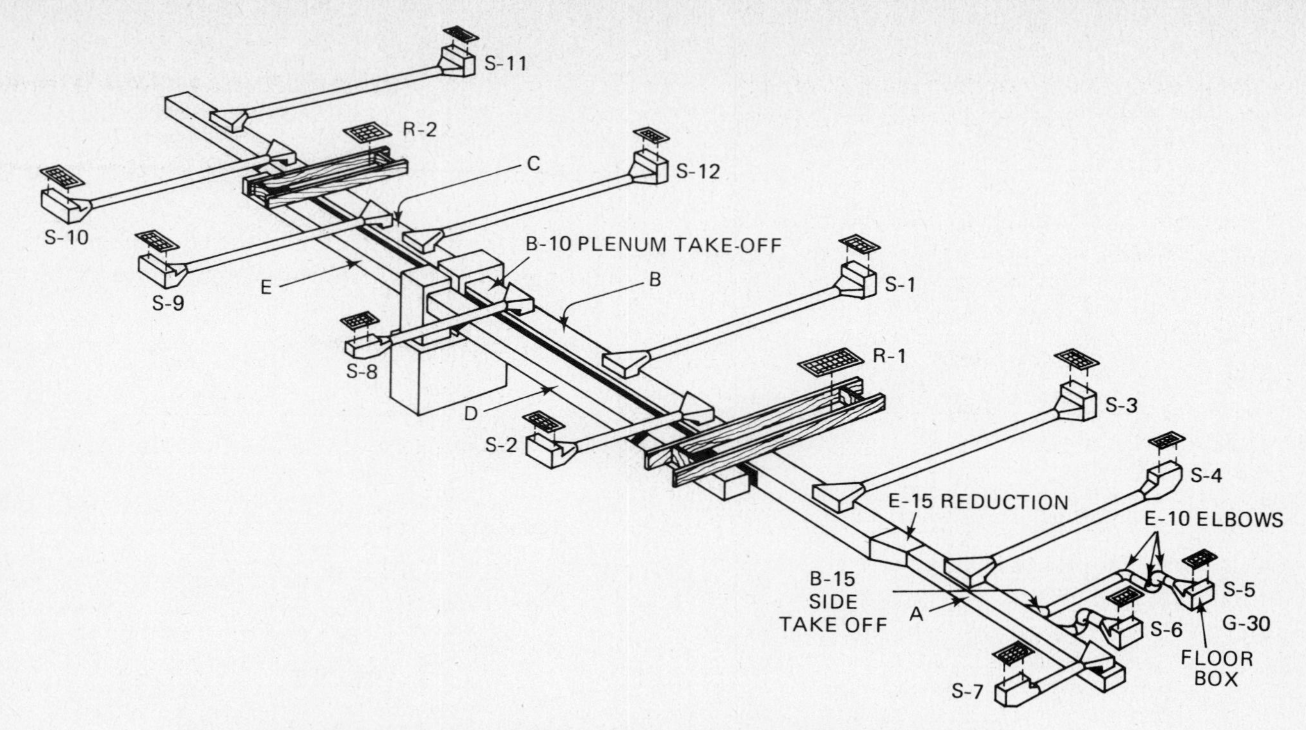

FIGURE 14-22 Air distribution system (Courtesy, Bryant Air-Conditioning Company).

From the furnace product data sheet (Fig. 14-23) select the total ft³/min requirement nearest the actual calculated amount. In this case 1,200 ft³/min is matched with the standard fan motor on high speed to produce required available external static pressure.

Step 9—Friction Loss of the Cooling Coil

Subtract the friction loss of the cooling coil from the available static pressure. The friction loss of the coil is obtained from the product data sheet (Fig. 14-24). For a wet coil at 1,200 ft³/min it is 0.24 in. of W.C. External static is therefore

$$0.68 - 0.24 = 0.44$$

static pressure available for the air distribution system.

Step 10—Compute the Supply and Return Static Pressures

Divide the available external static pressure between the supply and return duct systems in accordance with the table in Fig. 14-15 of suggested static pressures. In this example use returns from several areas.

$$0.44 \text{ in.} = 0.14 \text{ in. return} + 0.30 \text{ supply}$$

Subtract the loss of the supply diffuser with the greatest pressure drop (obtained from the manufac-

Air Delivery CFM	EXTERNAL STATIC PRESSURE AVAILABLE							
	Type L Motor				Type E Motor			
	Hi	Med	Med-Lo	Lo	Hi	Med	Med-Lo	Lo
900			.48				.74	
1000		.75	.06			.78	.62	
1100	.77	.59				.67	.15	
1200	.68	.43			.76	.52		
1300	.57	.18			.67	.34		
1400	.39				.56			
1500	.20				.44			
1600					.30			
1700					.16			

FIGURE 14-23.

turer's data) from the available supply static pressure (0.30 — 0.09* = 0.21) to determine the permissible loss in the supply duct.

Subtract the pressure loss of the return grille with the greatest pressure drop (obtained from the manufacturer's data) from the available return static pressure (0.14 — 0.01* = 0.13) to determine the permissible loss in the return duct.

Ft³/min	*Friction Inches W.C.*
1000	0.17
1100	0.20
1200	0.24
1300	0.28
1400	0.32

FIGURE 14-24 Wet coil air friction.

Step 11—Select Design Static Pressure

From Fig. 14-16 select the design air friction per 100 equivalent feet. For the supply system, enter the table at 0.20 in. of static pressure and 200 ft of equivalent length (the closest numbers to the 0.21 supply pressure and the 179 ft equivalent length calculated in the problem), and read 0.10 in. of design friction loss per 100 equivalent ft of duct. For the return system enter the table at 0.10 in. of static pressure and 150 ft of equivalent length (the closest number to 164 ft obtained in the problem), and read 0.07 in. of design friction loss per 100 equivalent ft of run.

Step 12—Size the Branch Ducts

Using the air friction chart (Fig. 13-13), plot the ft³/min requirement for each supply duct at 0.10 friction per 100 feet of duct, and read the round duct size from diagonal lines. Tabulate each outlet as illustrated in Fig. 14-25. Note that S-1, S-2, and S-10 indicate the use of a 7-in. round duct, but since standard sizes are in even numbers, 8-in. ducts would be used. Likewise, a 6-in. duct would be used on S-8.

* Note: Assumed for purposes of example problem.

Outlet	*Ft³/min Required*	*Round Pipe Size Inches*
S-1	145	7
S-2	145	7
S-3	65	6
S-4	135	6
S-5	110	6
S-6	110	6
S-7	25	4
S-8	70	5
S-9	30	4
S-10	145	7
S-11	120	6
S-12	100	6

FIGURE 14-25.

Step 13—Size the Main Supply Trunk

Size the rectangular trunk in the same manner as the round branch ducts, based on the ft³/min flowing through each section of the trunk.

Use the rectangular equivalent scale from Fig. 13-12 and tabulate the results as shown in Fig. 14-26.

Supply Trunk Section	*Ft³/min*	*Size*
A	380	8 × 12
B	805	8 × 26
C	395	8 × 12

FIGURE 14-26.

Step 14—Size the Return Trunk

To size the rectangular return air trunk, use 0.07 in. of friction per 100 feet of duct and the appropriate ft³/min, and read the round duct size. Convert to rectangular size as shown in Fig. 14-27. Since the

Return Trunk Section	Ft^3/min	Size
D	777	8 × 22
E	469	8 × 16

FIGURE 14-27.

floor joist spaces are 12 × 14 in., one space is adequate for each of the two return air inlets.

Step 15—Balance the System

Place volume dampers in each supply run for final air balance, to assure that the proper ft^3/min is delivered through each supply grille. Otherwise, the low-friction runs will deliver more than their share of air, and the longest run will deliver less than desired.

PROBLEMS

1. There are several types of residential duct systems. Name them.
2. What is the term used to define winter heating requirements?
3. There are several methods for designing duct systems. Name two.
4. High inside wall air distribution would be _____________ for a slab floor structure.
5. *Plenum chamber* is the name for _____________.
6. In Fig. 14-13, the friction loss is expressed in _____________.
7. In Group 2, Fig. 14-13, would take-off *Q* be recommended over *O*?
8. Assume you have a system with a single return, and the total available pressure is 0.4 in. W.C., what division of supply and return pressure is recommended?
9. Assume that your example above the highest duct run had 200 feet equivalent, what supply duct static pressure would be used to design the duct?
10. Is the static pressure drop across an evaporator coil expressed as wet or dry when figuring cooling?

15

RESIDENTIAL AIR DISTRIBUTION

The design of an air distribution system to a great extent depends on the type and location of supply and return ductwork, as explained in Chap. 14. Room air grilles are selected on the basis of application requirements, which are related to building construction, the primary use of the grille (for heating or cooling or both), and certain performance standards or air patterns to be achieved.

Improper room air distribution is frequently the cause of customer complaints such as "there's not enough heat," or, "it's too hot" or "too cold." In fact, it may not be the whole system but simply poor room distribution causing drafts, cold floors, hot spots, etc. How and where the air is admitted (Fig. 15-1), the temperature differential between the supply air and the room air, temperature variations in the room, the velocity and pattern of distribution, and of course the size of the room and the number of outlets are factors to be considered.

If a system is to be used for both winter and summer air-conditioning (Fig. 15-2), the air distribution system must be able to supply the air volume required for cooling, since the cooling air requirements are usually greater. In some cases, a different air volume will be used in the winter time, usually by utilizing a different blower speed. This is determined from the load calculation procedure as presented in Chap. 6, where air quantities for both heating and cooling are calculated for the entire structure and for each room or area to be conditioned.

AIR VELOCITY

Even though a system delivers the required amount of conditioned air to a space, discomfort may result if the air is not correctly distributed. One of the most critical factors affecting room comfort is room air velocity. Velocities of less than 15 ft/min may cause a feeling of air stagnation, and velocities higher than 65 ft/min could result in a sensation of drafts.

Air velocities of 25 to 35 ft/min in the occupied zone are considered to be the most satisfactory (Fig. 15-3). The air supplied to the outlet is at a greatly different temperature and velocity than the air in

FIGURE 15-1 (Courtesy, Bryant Air-Conditioning Company).

FIGURE 15-2 (Courtesy, Bryant Air-Conditioning Company).

FIGURE 15-3 (Courtesy, Bryant Air-Conditioning Company).

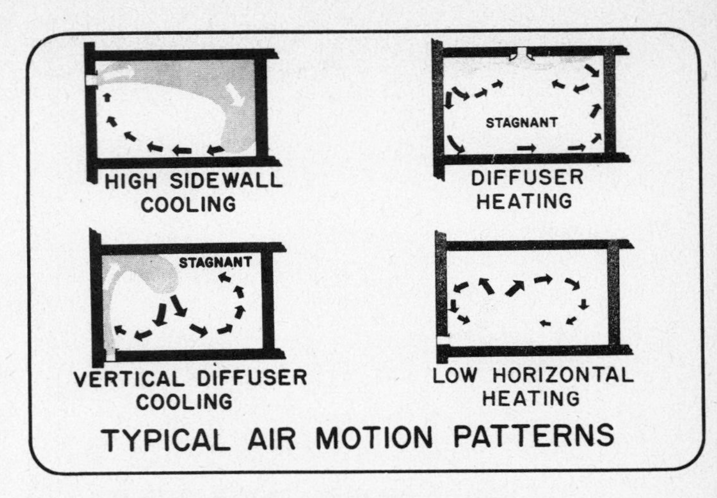

FIGURE 15-4 (Courtesy, Bryant Air-Conditioning Company).

the occupied zone. Therefore, proper air distribution requires that air leaving the diffusers and registers entrain and mix with the room air before entering that part of the room occupied by people. The air furnished by supply outlets must also counteract the detrimental natural convection and radiation effects within occupied areas.

The diagrams illustrated in Fig. 15-4 are air motion patterns encountered in the everyday residential applications of air distribution systems—for both heating and cooling, employing both high and low-designed diffusers (outlets).

TYPES OF OUTLETS

Horizontal discharge outlets (Fig. 15-5), such as high side-wall grilles and side-wall diffusers furnish air to the occupied zone, so that it drops into the zone at some distance from the outlet. The distance is dependent on air quantity, supply velocity, and temperature differential between supply and room air, along with the deflection setting, ceiling effect, and the types of loading within the occupied zone. Cooling with this type of outlet holds temperature variations within the room to a minimum, and there is practically no stagnant area.

The use of the high side-wall discharge outlets (Fig. 15-6) for heating is somewhat restricted. Warm air tends to rise, and it occupies the region near the ceiling. A large stagnant layer results between the floor and some level above the floor; also, a large temperature gradient tends to exist between the floor

FIGURE 15-5 (Courtesy, Bryant Air-Conditioning Company).

and the stagnant layer of air. With this type of distribution, additional radiation at floor level to take care of perimeter exposures is usually beneficial.

Ceiling diffusers (Fig. 15-7) and linear ceiling diffusers (extra long diffusers) furnish air to the occupied area in much the same pattern as high side-wall grilles or diffusers. They are an acceptable method of controlling temperature variations within the room, with little or no stagnant area. However, with both horizontal and ceiling diffusers, overblow —terminal velocity is higher than usual—may cause a region of high velocity accompanied by uncomfortable temperatures.

The use of overhead discharge outlets (Fig. 15-8) in the form of ceiling or linear ceiling diffusers for heating is limited. Since warm air, as it rises, occupies the region near the ceiling, a stagnant layer is formed

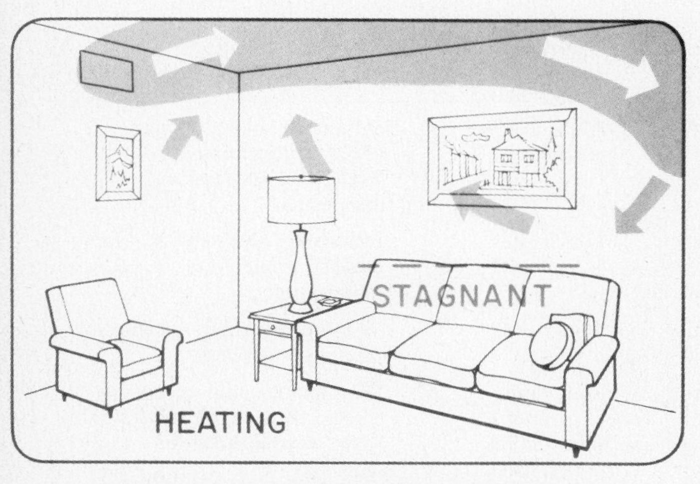

FIGURE 15-6 (Courtesy, Bryant Air-Conditioning Company).

FIGURE 15-7 (Courtesy, Bryant Air-Conditioning Company).

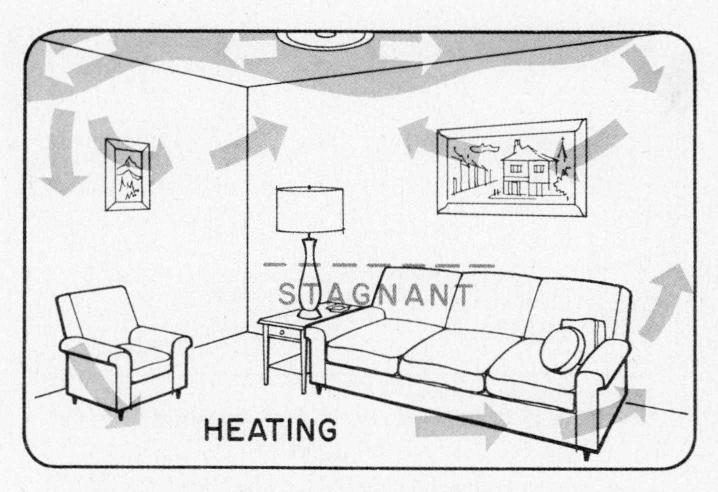

FIGURE 15-8 (Courtesy, Bryant Air-Conditioning Company).

between the floor and some level above it, along with temperature differences.

Vertical discharge from floor registers (Fig. 15-9), baseboard units, low side-wall units, and linear grilles in the floor, all with no deflecting vanes, offer good air distribution for cooling. The air rises from these units, and its velocity injects and mixes it with room air, which strikes the ceiling and fans out in all directions. It follows the ceiling for a distance and drops toward the occupied zone. A stagnant layer is formed outside the occupied zone of the room. Below the stagnant layer, the air temperature is uniform, and complete cooling is effected.

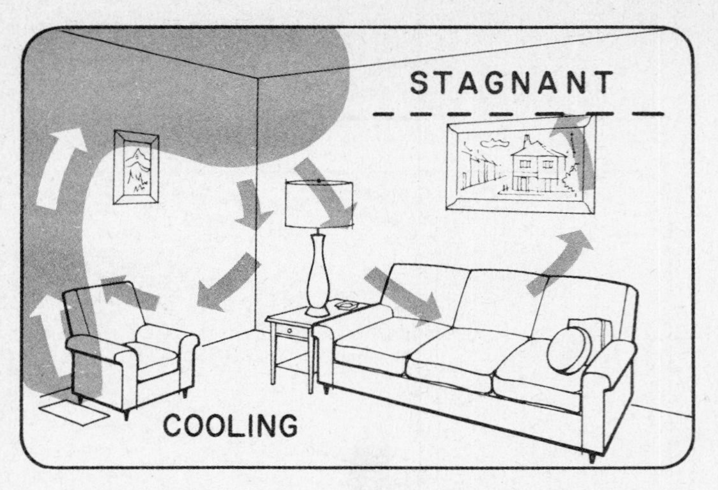

FIGURE 15-9 (Courtesy, Bryant Air-Conditioning Company).

FIGURE 15-11 (Courtesy, Bryant Air-Conditioning Company).

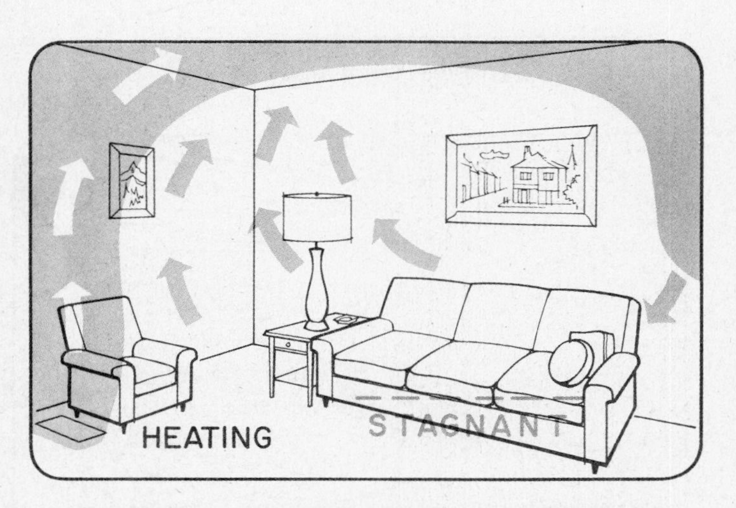

FIGURE 15-10 (Courtesy, Bryant Air-Conditioning Company).

FIGURE 15-12 (Courtesy, Bryant Air-Conditioning Company).

Group	*Outlet Flow Pattern*	*Outlet Type*	*Most Effective Application*	*Preferred Location*	*Size Determined by*
A	Vertical Spreading	Floor diffusers, baseboard, and low sidewall	heating and cooling	Along exposed perimeter	Minimum supply velocity—differs with type and acceptable temperature differential
B	Vertical, Nonspreading	Floor registers, baseboard, and low sidewall	cooling and heating	Not critical	Maximum acceptable heating temperature differential
C	High Horizontal	Ceiling and high sidewall	cooling	Not critical	Major application—whether heating or cooling
D	Low Horizontal	Baseboard and low sidewall	heating only	Long outlet at perimeter short outlet, not critical	Maximum supply velocity should be less than 300 ft/min. (Not recommended for cooling.)

FIGURE 15-13 General characteristics of outlets.

When used for heating, the low, nonspreading type of outlets (Fig. 15-10) form a small stagnant layer near the floor, but a fair temperature gradient is achieved in the occupied area. Low outlets of this type with deflecting vanes achieve a similar cooling pattern, except the unoccupied stagnant zone is somewhat larger. When used for heating, the pattern is similar to the nonspreading type, with the exception that the stagnant layer is smaller.

Low side-wall and similar horizontal diffusers (Fig. 15-11) discharge air in single or multiple streams. Since the air is discharged horizontally across the floor, the air remains near the floor during cooling, and a large stagnant layer is found in the upper region of the room. When this type of discharge is used for heating (Fig. 15-12), because of the natural buoyant effect of heated air, the air rises toward the ceiling and uniform temperature variations are established.

The characteristics of outlets may be summarized as to their outlet pattern, type, and most effective application, as noted in Fig. 15-13. Physically, these are grouped as shown in Fig. 15-14. Style and designs vary depending on the manufacturer, but essentially their performance and application are very similar.

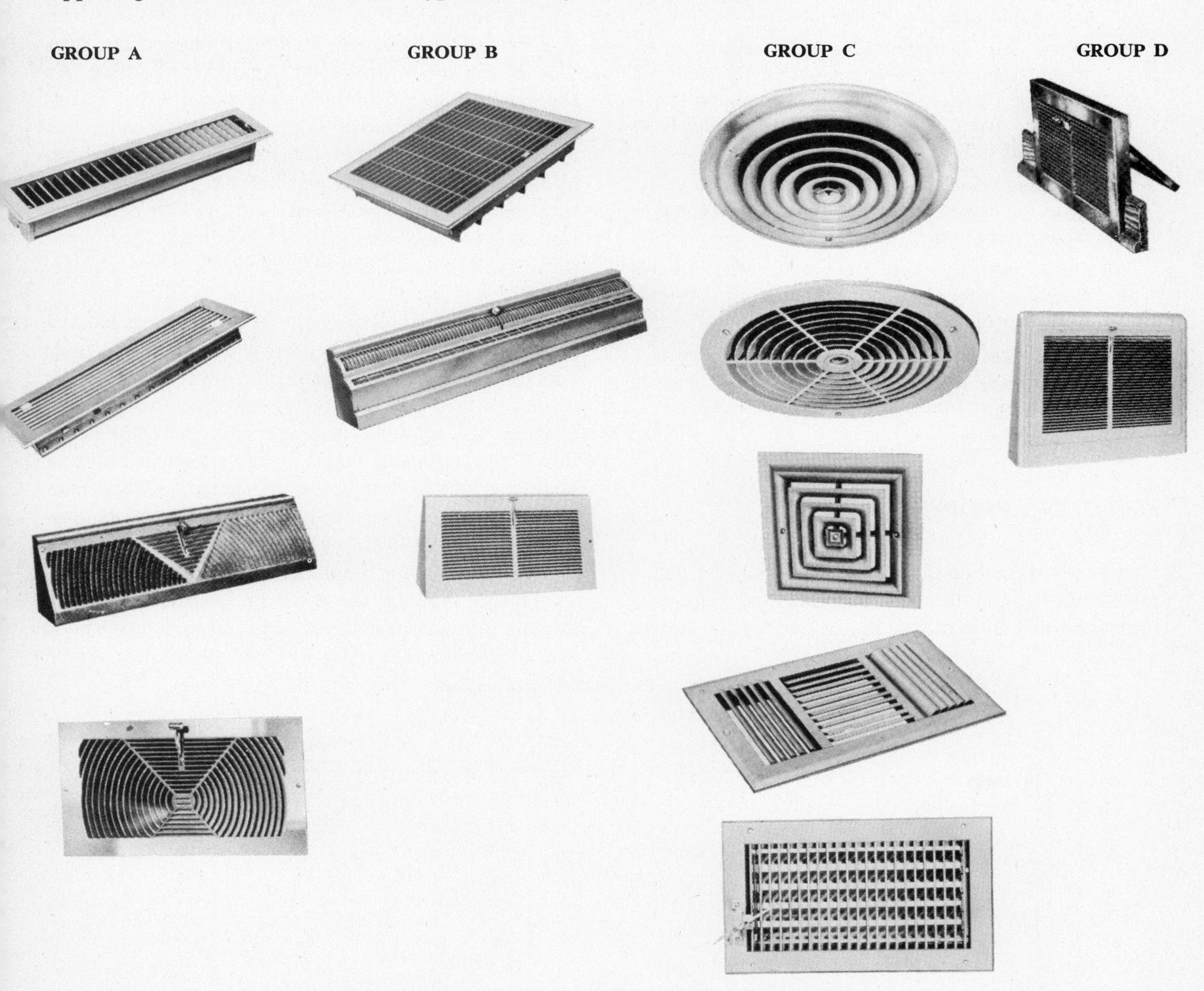

FIGURE 15-14 (Courtesy, Hart and Cooley Manufacturing Company).

SPECIFICATIONS OF OUTLETS

Specifications of outlets are stated in terms of (see Fig. 15-15).

(1) Size—physical dimensions such as 8 × 14 in., 8-in. round, etc.

(2) Capacity—as measured in ft³/min.

(3) Spread—the maximum total width in feet of the air pattern at the point of terminal velocity. For ceiling diffusers, it's the radius of diffusion.

(4) Throw—the distance measured in feet that an air stream travels from the outlet to the point of terminal velocity. The throw is measured horizontally from a register and vertically from perimeter diffusers.

(5) Terminal velocity—the average air stream velocity at the end of the throw, generally accepted as 50 feet per minute.

(6) Outlet velocity—the average velocity of air emerging from the outlet measured in the plane of the opening.

(7) Drop—the vertical distance the lower edge of a horizontally projected air stream drops between the outlet and the end of its throw.

SELECTING PROPER OUTLETS

Neither the type nor the size of outlet can be determined until the system has been selected. The only outlet characteristic of real importance in the system design is the pressure loss; for, so far as the duct system is concerned, the outlet is just another fitting. Outlet characteristics like *throw*, *spread*, and *face velocity* are important factors in room air distribution.

When the perimeter distribution system is to be used, the type of outlet will usually be a floor diffuser, baseboard, or low side wall, and will deliver the air in a vertical spreading pattern. These outlets with a wide spread are recommended for heating. Warm air from the outlet blankets the cold surface, traveling upward along the exposed wall, across the ceiling, and down the far wall to form a relatively small stratified zone. Temperature variations throughout the occupied area are small.

The vertical throw from the diffuser should be approximately four feet. Perimeter outlets that provide a long throw are recommended for cooling. As the horizontal spread increases, the upward throw is decreased. Therefore, higher supply air velocities are needed to project the total air toward the ceiling, allowing it to mix with warmer air before it drops into the occuped zone. Since heating with perimeter outlets is accomplished with little difficulty, perimeter diffusers are usually selected, based on cooling requirements for year-round applications.

Vertical nonspreading outlets are registers located in or near the floor. When air is delivered in a nonspreading jet directed vertically upward, it hits the ceiling surface and fans out in all directions. During heating, the total air moves across the ceiling and down the opposite wall of the room, forming a slightly larger stratified zone than occurs with perimeter diffusers. Furthermore, the stratification boundary is somewhat farther above the floor because the exposed wall is not blanketed with warm air.

During cooling, the total air is delivered to the ceiling, follows along the ceiling, and then drops back

FIGURE 15-15 Specification of outlets.

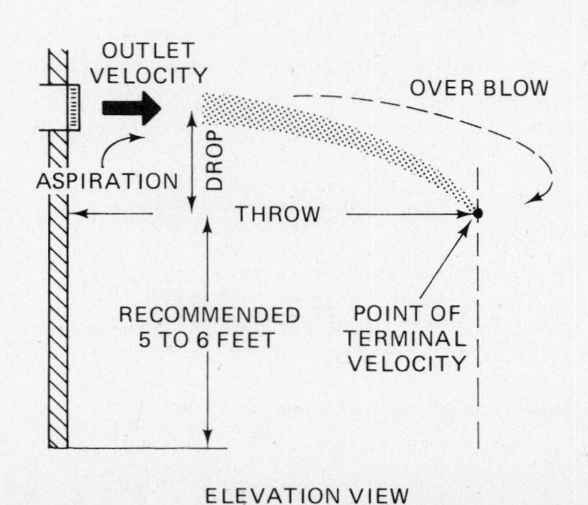

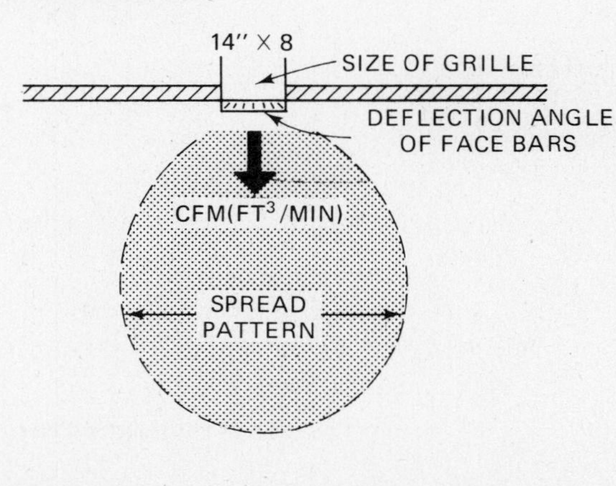

into the room in a semicircular zone some distance from the outlet. The nonspreading outlets are better for cooling than perimeter diffusers, because the projected air does not fold back upon itself before it hits the ceiling, causing the natural currents to form a smaller stratified zone of warm air near the ceiling and away from the occupied area.

Most ceiling outlets deliver the air high in the room in a horizontal pattern and also provide some downward air motion. Therefore, both throw (radius of diffusion) and mounting height are important. During cooling, the total air drops into the occupied zone near a wall; there is practically no stratified zone at all. During heating, the total air remains near the ceiling, causing the stratification boundary to form over the whole room at some distance above the floor. This results in a large stratified zone and relatively great temperature differentials near the floor level.

Ceiling diffusers of two different types are sometimes used. One type delivers the air with a 360-degree spread and is usually placed near the center of the room. The other type is often placed near an inside wall and directs the supply air toward the opposite wall of the room. In some instances, the throw will be only in one direction with little spread; in other cases, the spread may be semicircular.

Ceiling diffusers that deliver air with a 360-degree spread have a greater rate of induction of room air into the projected air stream, and the throw from the diffuser is shorter than for a wall outlet delivering the same air quantity at the same outlet velocity. Because of the shorter throw, ceiling diffusers can be selected that use higher air quantities and velocities than wall outlets and may be sized smaller to handle the same air volume. When ceiling outlets are installed on the bottom of a duct, the throw will be decreased approximately 40%, due to the lack of ceiling effect.

The first step in selecting a ceiling outlet is to determine the required throw (radius of diffusion). In the case of diffusers that have a wide angle of spread, the throw should be equal to the distance from the diffuser to the nearest wall. For a diffuser that has little spread of its discharge pattern, the throw should be equal to approximately three-fourths of this distance. When two diffusers of the same type are located so they are delivering air toward each other, the throw of each diffuser should equal half the distance between them.

Wall registers located high on the inside wall are generally sized so that the throw from the register projects about three-quarters of the distance across the room. In this way, air will not strike the opposite wall and bounce into the occupied zone. If it appears that the drop is likely to be a problem because a long throw is required, the solution is to use two or more small outlets, each having the required throw. The reason for this is that the total volume of air delivered from an outlet has been found to have a greater effect toward causing drop than velocity has on reducing drop. The use of a greater number of outlets, each delivering a small volume of air, therefore will reduce the drop.

Outlets should be located high enough for the air stream to be six feet above the floor level at the termination of the throw; or, in other words, the drop should not be more than the difference between the mounting height and the zone of occupancy. The maximum throw (if overblow is not used) for a given ceiling height may be obtained by locating the outlet lower on the wall, arching the throw, and sweeping the air across the flat area. In some installations, due to limited ceiling heights or structural design, the outlet must be installed low on the wall.

High side-wall registers are excellent for cooling and for year-round conditioning in mild climate

Application	*Recommended Velometer Velocities*	*Application*	*Recommended Velometer Velocities*
Broadcasting studios	500 FPM	Motion picture theatres	1,000 to 1,250 FPM
Residences	500 to 750 FPM	Private office, not treated	1,000 to 1,250 FPM
Apartments	500 to 750 FPM	General offices	1,250 to 1,500 FPM
Churches	500 to 750 FPM	Stores, upper floors	1,500 FPM
Hotel bedrooms	500 to 750 FPM	Stores, main floors	1,500 FPM
Legitimate theatres	500 to 1,000 FPM	Industrial buildings	1,500 to 2,000 FPM
Private offices			
Acoustically treated	500 to 1,000 FPM		

areas. However, their use is not recommended for heating in cold climate areas where warmed floors are essential.

The sound caused by an air outlet in operation is directly proportional to the velocity of the air passing through it. By selecting outlets of the proper size, air velocities can be controlled within safe sound limits. The recommended outlet velocities are within safe sound limits for most applications.

SIZING OF OUTLETS

Figure 15-16 is a typical manufacturer's selection chart for a side-wall outlet. For a given ft³/min of air quantity there is a choice of sizes, based on the application need in terms of pressure loss, horizontal spread, vertical throw, and average face velocity. Assume that a residential outlet application requires 150 ft³/min. Note that the spread and throw are not too much different for the three sizes shown, but there is an important difference in pressure drop and velocity.

From the table above, we see that for residences the maximum recommended velocity is 750 ft/min and a 10 × 6-in. outlet would not be acceptable at 830 ft/min. A 12 × 6-in. outlet would be the better choice. However, its pressure loss must also be in keeping with the overall system design. Most grille and register manufacturers have engineering data as a part of their catalogs to assist in selecting the proper outlets depending on the application requirements.

One additional term that needs explanation is *overblow* (Fig. 15-15), or its opposite, *underblow*. The throw of a horizontally discharged outlet must be sufficient to produce satisfactory conditions throughout the occupied space. Underblowing may cause the heated air to rise to the ceiling before it has a chance to mix with the rest of the air in the room, thus creating excessive stratification. In cooling, underblow may cause cold air to drop into the

ENGINEERING DATA FOR NO. 401 SIDEWALL "DIFFUSAIRE"®

Diffuser Size Free Area		CFM (Cubic Feet per Minute)									
		50	70	90	110	130	150	170	190	210	230
10 x 6	Pressure Loss (In. W.G.)	.007	.015	.025	.038	.055	.074	.095	.120	.150	.198
	Horiz Spread (Ft.)	10	11	13	14	15	16	17	18	19	20
26 sq. in.	Vertical Throw (Ft.)	5.5	6	6.5	7	7.5	8	8.5	9	9.5	10
	Avg. Face Velocity (FPM)	280	390	500	610	720	830	940	1050	1160	1270
12 x 6	Pressure Loss (In. W.G.)	.005	.010	.017	.025	.035	.046	.060	.075	.092	.110
	Horiz Spread (Ft.)	9	10	12	13	14	15	16	17	18	19
33 sq. in.	Vertical Throw (Ft.)	4	5	5.5	6	6.5	7	7.5	8	8.5	9
	Avg. Face Velocity (FPM)	220	310	390	480	570	650	740	830	920	1000
14 x 6	Pressure Loss (In. W.G.)	.004	.007	.012	.017	.024	.032	.041	.054	.062	.075
	Horiz Spread (Ft.)	7	9	11	12	13	14	15	16	17	18
40 sq. in.	Vertical Throw (Ft.)	3	4	5	5.5	6	6.5	7	7.5	8	8.5
	Avg. Face Velocity (FPM)	180	250	320	400	470	540	610	680	760	830

PRESSURE LOSS is total pressure (static plus velocity) for diffuser and stackhead only. **SPREAD** is horizontal distance of the air pattern measured in feet to a terminal velocity of 50 fpm. **THROW** is the vertical distance of the air pattern measured in feet to a terminal velocity of 50 fpm. **VELOCITY** is the average calculated face velocity in fpm.

FIGURE 15-16 (Courtesy, Hart and Cooley Manufacturing Company).

occupied zone before supply and room air are mixed sufficiently, thereby creating drafts.

On the other hand, the improper use of overblow can result in objectionable downdrafts from any surface the projected air stream may strike. However, the proper use of overblow is gaining wide acceptance in modern room air distribution systems. Effective overblow occurs when total air is knowingly directed against a surface, such as ceiling, wall, or floor, to fan out and improve air distribution in the space.

maximum inlet face velocity. So in selecting a return grille, be sure the effective free area is large enough to provide the correct velocity. The catalog data will state the ft^3/min, the area, and the velocity.

Where larger central return air openings are desired (for example, in a hallway) it's common to use a combination return air hinged filter grille (Fig. 15-18), which is a frame to hold the air filter. Face velocity should be dropped to the 400 ft/min range so as to provide good filter performance.

Outlet Group	*Heating Intake Location*	*Cooling Intake Location*	*Year-Round Intake Location*
A—Spreading Vertical Jet	No Preference	Near Ceiling*	Near Ceiling*
B—Non-spreading Vertical Jet	Near Floor	No Preference	Near Floor
C—Horizontal Discharge, High	Near Floor	No Preference	Near Floor
D—Horizontal Discharge, Low	No Preference	(System not recommended for cooling)	(System not recommended for year-round conditioning)

* Preferred location is near ceiling on opposite wall from outlet. However, the location is not critical.

FIGURE 15-17 Preferred location of return intakes (Courtesy, Hart and Cooley Manufacturing Company).

The discussion so far has been directed at the supply air outlets. The return intake affects only the air motion within its immediate vicinity. Even the natural convection currents possess enough energy to overcome the draw of the intake. This does not mean that the location of the return intake is not important—only that such location has little effect on the room air motion. The return generally should be located inside the stratified zone. In this location, the warmest air is returned during cooling, and the coolest air during heating. Where the conditioning system is to be used for year-round applications, the return should be located according to the relative size of the heating and cooling stratified zones for the type of outlet selected. For example, with horizontal projection at high levels, no appreciable stratified zone exists during cooling, but a large stratified zone might exist during heating. Thus, the return should be located near the floor to favor the heating performance. Figure 15-17 is a table of preferred locations for return air intakes based on the selection of the supply outlet.

Noise in return air grilles is as important as it is for the outlets, and a good guide is 400 to 600 ft/min

Although the information in this discussion is geared toward residential applications, many of the principles apply to occupied spaces in commercial work.

For additional reading on room air distribution, *Manual E*, by NESCA, is a highly recommended reference; information is also available from air-conditioning equipment manufacturers and grille manufacturers.

FIGURE 15-18 Return air hinged filter grille (Courtesy, Hart and Cooley Manufacturing Company).

PROBLEMS

1. Room air distribution has little effect on the total system. True or false?
2. The recommended air velocity in an occupied space is ______________.
3. There are several types of air outlets used in residential systems; name three.
4. The capacity of an outlet or return is expressed in ______________.
5. Define *spread.*
6. Define *throw.*
7. Define *terminal velocity.*
8. Define *drop.*
9. Is ceiling distribution recommended for heating systems?
10. From Fig. 15-16, select the proper size of outlet to handle 100 ft^3/min with a 6-foot throw and pressure loss not to exceed 0.025 in. of W.C.

16

COMMERCIAL AIR DISTRIBUTION

In terms of the principles of air flow, velocity, static pressure, etc., the air distribution systems for commercial work use the same basic design information as previously presented for residential work. The major differences lie in the fact that houses, their construction, and occupancy factors are really all quite similar to each other when compared to the huge variety of applications found in commercial and industrial comfort conditioning. The requirements for air conditioning a broadcasting studio are quite different than the air distribution in a textile plant. So to a great extent the application determines the design conditions.

In a high-rise office building the need to provide perimeter air distribution is a function of the external heat gain or loss through walls and windows; at the same time the center core air distribution is a function of the heat load from people, lighting, and machines. The perimeter loads are functions of time and would be affected by the sun and weather conditions. The core loads are a function of whether the building is occupied or not. So the air distribution systems of most commercial applications have to cope with wide variations in demand over a 24-hour period. Thus we find the use of systems that are able to vary temperature and air volume to the space, and some even have provisions to transfer heat from where it's not needed to areas of need, thus conserving energy.

TYPES OF SYSTEMS

For simplicity, types of commercial air distribution may be classified as low, medium, or high-velocity systems.

Low-velocity systems are those associated with the application of smaller unitary packaged and split-system units as illustrated in Fig. 16-1. The use of limited ductwork or none at all (free blow) is typical of this classification. Where ductwork is used, the external static pressure is held down to the 0.25 in. of W.C. to 0.50 in. of W.C. range. The use of concentric supply and return ducts, as shown in the illustration of the supermarket, is a common application. Duct and air outlet velocities and thus resultant noise are

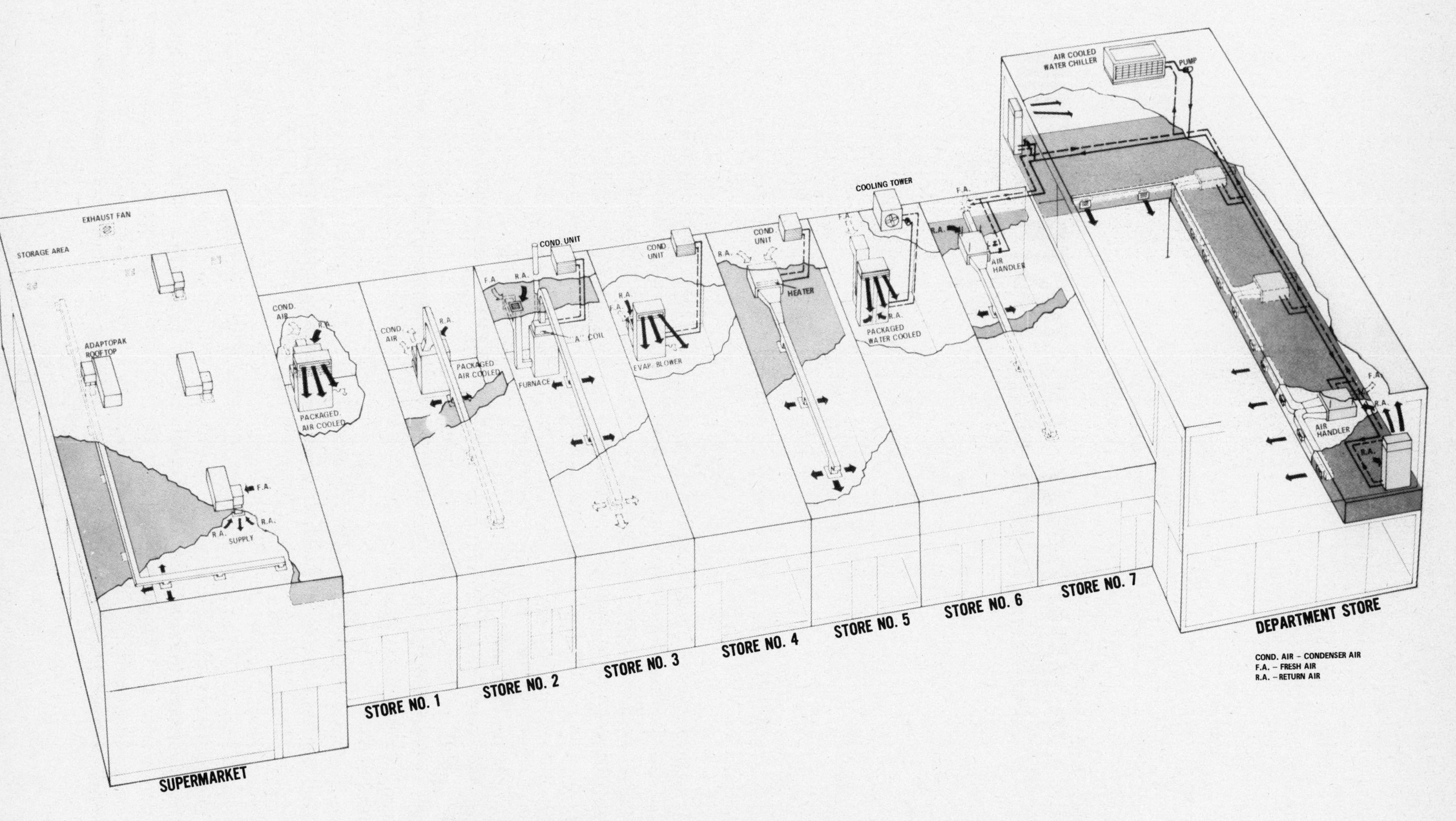

FIGURE 16-1 Shopping center air-conditioning installations (Courtesy, Fedders Company).

somewhat higher than that of residential work; they should be kept within the recommendation noted in Chap. 13. The type of duct design is the same as that of the equal friction method, which permits the prediction or control of total static.

Medium velocity systems are those associated with conditioners or air handlers that provide larger air volumes at external static pressures of up to 2.0 to 3.0 in. of W.C. Typical of this category are the larger rooftop and air-handling units for single and multi-zone application. It is in this category that the use of different air distribution techniques becomes necessary, depending on the application and economic considerations. The following discussions are presented to acquaint the reader with general systems, not necessarily the specific design illustrations.

SYSTEM APPLICATIONS

Application in core areas may be handled in several ways.

(1) Single-zone heating and cooling. If the interior core of a building is a large open area, a single-zone, constant-volume system (Fig. 16-2) can be used at a reasonable initial cost. A single heat/cool thermostat provides year-round environmental control with automatic changeover; top floor and ground floor systems will have heating for morning warm-up. Intermediate floor systems usually omit the heat source. The design of the

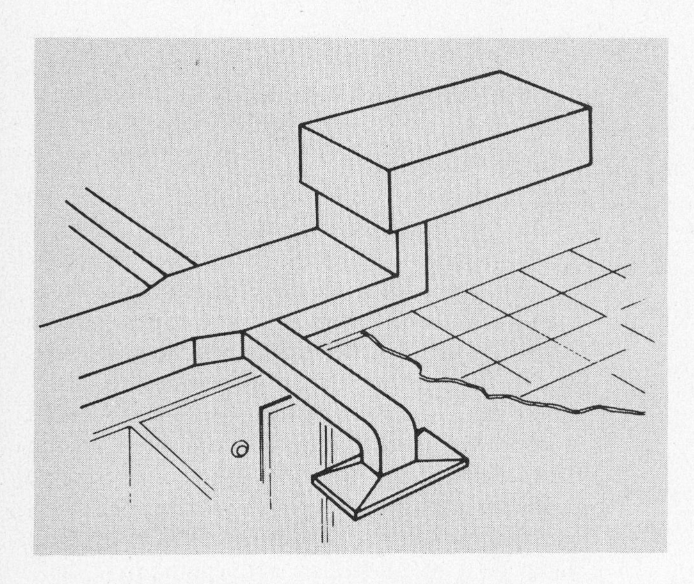

FIGURE 16-2 Single-zone heating/cooling constant volume.

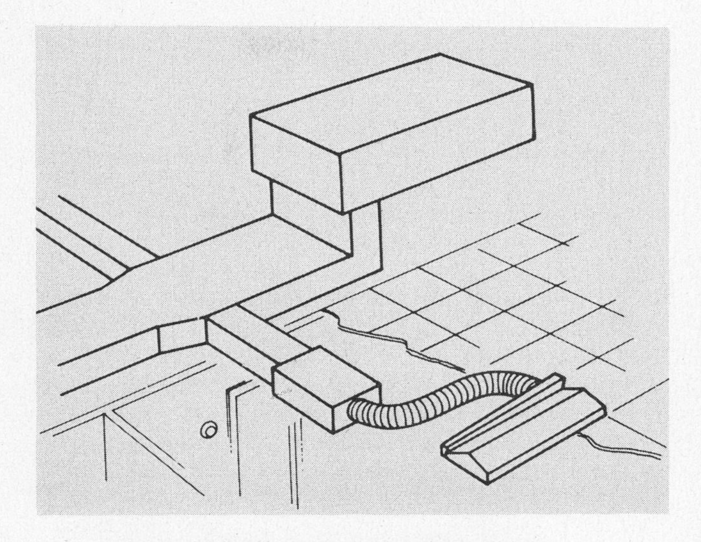

FIGURE 16-3 Single-zone variable volume.

ducts follows conventional practices, and room air distribution will usually be from the ceiling.

(2) For an area composed of numerous spaces subject to variations in lighting loads or people density, the variable-volume, single-zone system (Fig. 16-3) is a low-cost answer. Instead of delivering a constant flow of conditioned air to each space, the air quantity is thermostatically varied for each space to match fluctuations in space loads exactly. Under partial load conditions, excess air is simply bypassed back to the rooftop unit. The variable volume control terminal (Fig. 16-4) consists of a main duct inlet and one or more branch outlets that feed the supply air diffusers. A damper operator modulates the air to each branch, and relieves or bypasses excess air to the ceiling plenum or the return air duct as well. Air diffusers are designed to fit into the **T**-bar suspended ceiling. Flexible round ducts feed the control terminal and air diffusers.

Variable-volume systems can be either low velocity or high velocity, depending on the type of controller and air diffusers used. The volume of primary air is automatically adjusted to the total cooling load of the building by pressure-controlled inlet vanes on both the supply and return air fans.

Heating and cooling requirements for various perimeter areas can easily be met with cooling-only single-zone equipment used in conjunction

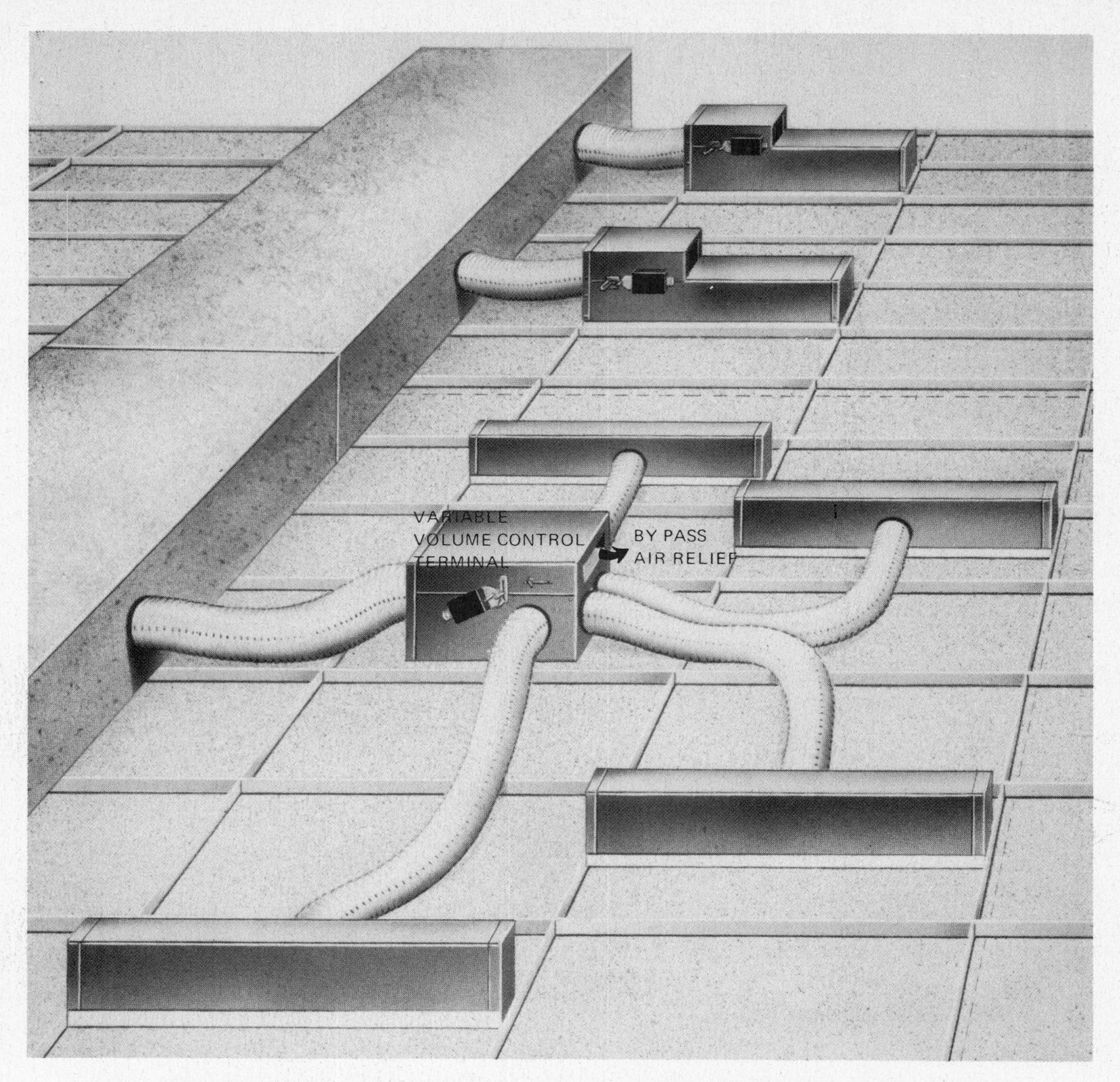

FIGURE 16-4 Variable-volume control terminal (Courtesy, York, Div. of Borg-Warner Corporation).

with individual, thermostatically-controlled electric heaters or hot-water coils in the branch duct to the occupied space (Fig. 16-5). The duct coils provide tempering reheat in spring and fall and full heat in winter.

The single-zone, variable-volume system may also be applied to the perimeter areas, but air is introduced to the space at floor level on the outside wall rather than through ceiling outlets. Branch ducts are installed in the ceiling plenum of the floor below, and air diffusers are placed in the floor along outside walls under window exposures.

(3) As a straightforward solution to perimeter air-conditioning problems, *multizone* heating and cooling systems are usually the best. In single-zone systems, a single main duct supplies all spaces. In the multizone system (Fig. 16-6), a small branch duct connects the rooftop unit or air handler with each space directly. A thermostat in each space modulates mixing dampers at the discharge of the multizone unit. This provides exactly the right supply air temperature to satisfy cooling and heating requirements. Since there is a separate thermostat, duct run, and zone mixing damper for each space, one area can receive heating while another receives cooling. Because a constant quantity of air is delivered to each space under all load conditions, better humidity, filtration, and odor control are possible.

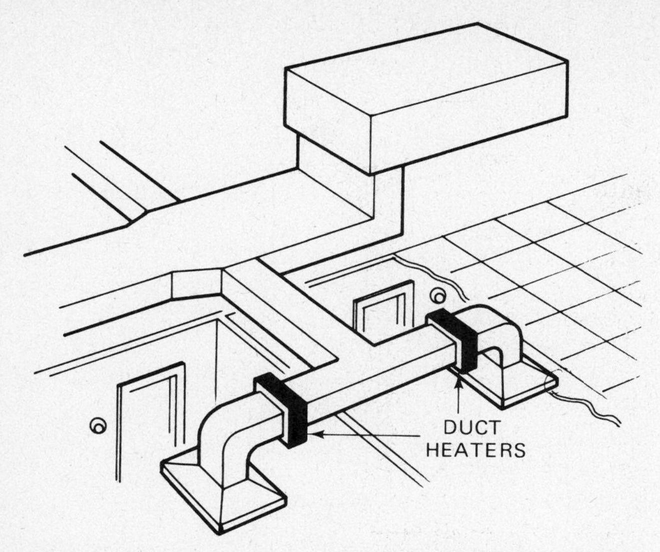

FIGURE 16-5 Single-zone remote duct heaters constant volume.

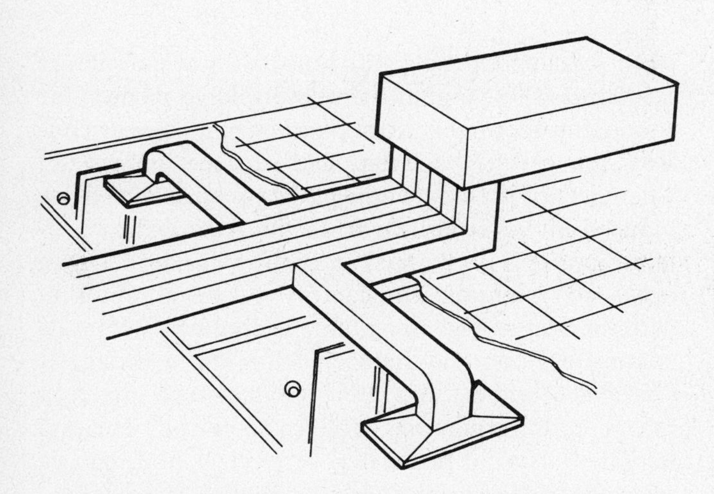

FIGURE 16-6 Multizone heating/cooling constant volume.

(4) Double-duct heating/cooling systems (Fig. 16-7) have the highest initial cost of all systems, but they offer distinct advantages for certain building requirements. The main benefit is maximum flexibility. This type of system is also advantageous where there are a large number of areas requiring less than 400 ft³/min. The duct system size is double that of a single-zone system because, as the name implies, two duct systems run to each space—one for warm air, the other for cool air. Both ducts connect to a mixing box serving each space, which is controlled by an individual thermostat; therefore, the supply air mixture temperature to the space is maintained at exactly the right level.

The double-duct system is a high-velocity and high-pressure type. The maximum velocity in the trunk supply ducts will range from 2,500 ft/min to 6,000 ft/min, depending on the air quantity. Static pressures can run in excess of 4 in. of W.C. Branch ducts feeding the mixing boxes will then be sized to help slow the air down; and when it enters the mixing box its velocities are further reduced, so the low-pressure ducts leaving the mixing box can be sized as any other conventional ductwork. The low-pressure ducts feeding the air diffusers are also lined with accoustical material to assure a quiet installation.

Return air ducts in a double-duct system and any other system should be sized at low velocity as in any conventional system. Attempts to pressurize the return duct have not as yet proved to be entirely successful.

When to use a combination of these systems is a function of building size and economics. Small buildings frequently cannot justify one system for the core and another for the perimeter. In these instances, any one of the three perimeter systems can be used for the entire building.

The systems are all of the type that use primary (supply) air from the air conditioner, and the maximum quantity needed is based on the maximum load conditions of the area. Primary air temperatures will be approximately 20 to 22°F below the room air

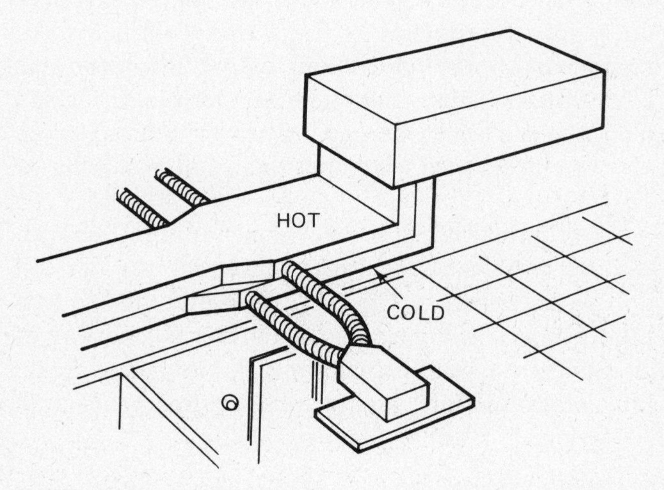

FIGURE 16-7 Double-duct heating/cooling variable volume.

temperature. So if room air is held at 78°F, the entering primary air is at approximately 56°F. The quantity of air is a function of the sensible heat load and will be about 400 ft³/min per ton for cooling.

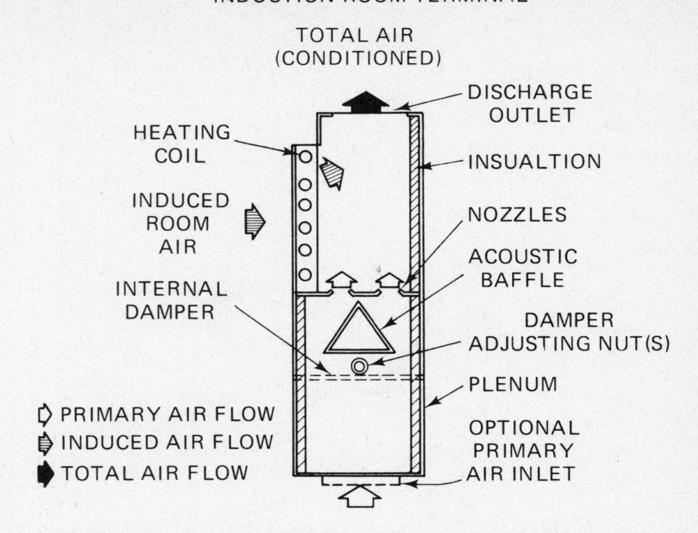

FIGURE 16-8 Induction room terminal (Courtesy, Carrier Air-Conditioning Company).

INDUCTION AIR DISTRIBUTION SYSTEMS

Since the cost of ductwork is a function of size, the smaller the ductwork the lower the initial cost and the less labor is needed to install it. A reduction in the size of the ductwork means a reduction in the quantity of primary air, and to reduce the quantity of air it is necessary to lower the temperature so that the required sensible cooling capacity remains the same. This can be seen in the formula:

$$\text{LAT* (Leaving air temperature)} = \text{room temp} - \frac{\text{room sensible heat capacity (RSHC)}}{1.1 \times \text{ft}^3/\text{min}}$$

If the room temperature is held constant (for example, 78°F DB) and the RSHC also remains constant, it becomes apparent that a reduction in ft³/min will require a reduction in the LAT or the primary air temperature. Thus, if the primary air is reduced to 38 to 40°F, there can be a considerable savings in the supply air ductwork. But supply air cannot be introduced into the room at 38°F, for this is 40° below the room conditions, which will cause uncomfortable cold drafts. Thus, induction air terminals, which blend cold primary air with warmer induced room air, were developed to temper the resultant discharge air.

An induction type of room terminal has an external physical appearance much like the fan-coil unit as described in Chap. 12; however, inside it is totally different. Figure 16-8 is an internal cutaway of an induction type of room terminal. Note there are no fans. High-pressure, high-velocity, primary air at 38 to 40°F is introduced at the bottom (or at the side) at 1,500 to 2,000 ft/min into a well-insulated plenum for sound and thermal control. An internal damper controls the primary air that feeds to the air nozzles where jets of air create a negative pressure and cause induced air to be drawn from the room. The resultant mixture is discharged at about 55 to 60°F DB in order to maintain satisfactory room air motion without drafts. The air volume is constant. Note the heating coil on the induced room-air opening. It provides total heat in winter; it can also be used as a reheat coil during the cooling operation. Heating may be by steam, hot water, or electric, and may be selected on the basis of utility rates.

Room terminal induction units are most often used in schools, laboratories, hospitals, and single-floor or low-rise office buildings. The typical piping system is represented in Fig. 16-9. This system is also a high-velocity method. The main trunk duct will be sized for velocities of 2,000 to 5,000 ft/min. Branch take-offs to terminals should be a maximum of 2,000 ft/min. Round ducts are preferred to rectangular because of their greater rigidity. The ducts are sealed to prevent leakage of air, which may cause objectionable noise. A number of variations in equipment design are offered, including horizontal models that may be suspended in a dropped ceiling. Primary air at 38 to 40°F must be supplied from a chilled-

* LAT is the leaving air temperature off the coil and is assumed to be the same temperature as the primary ducted air without duct loss.

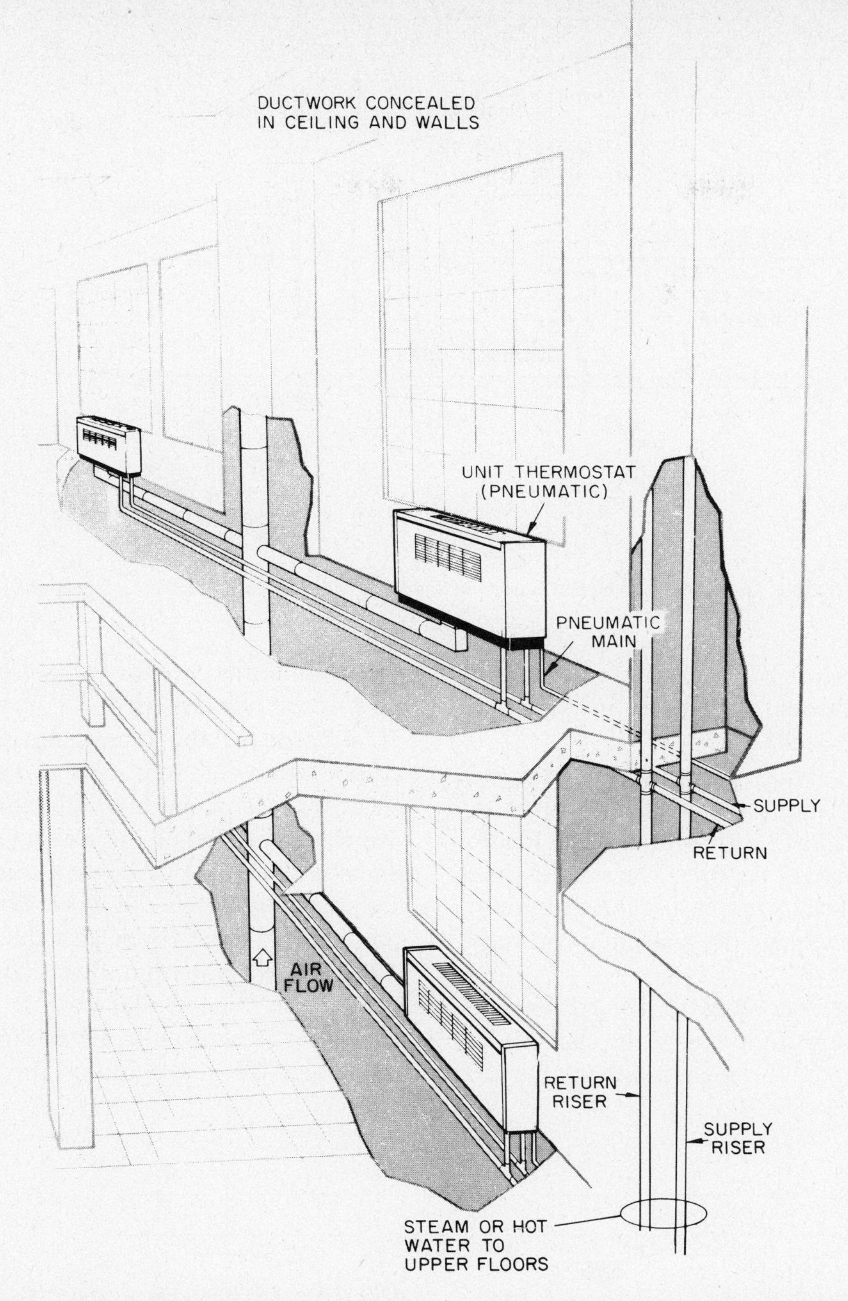

FIGURE 16-9 Piping system of room terminal induction unit (Courtesy, Carrier Air-Conditioning Company).

water coil included in a central station air-handling apparatus that circulates a brine-cooling solution. Room-type induction units, even though they operate at high velocity, are considered medium-pressure applications, since the inlet static pressure to the terminal is normally 0.5 to 1.0 in. of W.C. at rated air flow; the maximum would be 2.0 in. of W.C. In the selection of room induction units, two parameters must be satisfied: The unit must supply air at an acceptable sound power level, and it must have enough unit capacity to maintain the proper room temperatures.

Ceiling induction systems use induction units installed in the primary supply duct and the feeding air diffusers. Figure 16-10 shows a single induction unit that draws air from the room and blends it with the cold primary air. The high-velocity primary air is supplied to induction boxes in each zone in a variable volume at a constant temperature. The induction boxes are designed for both interior zones and perime-

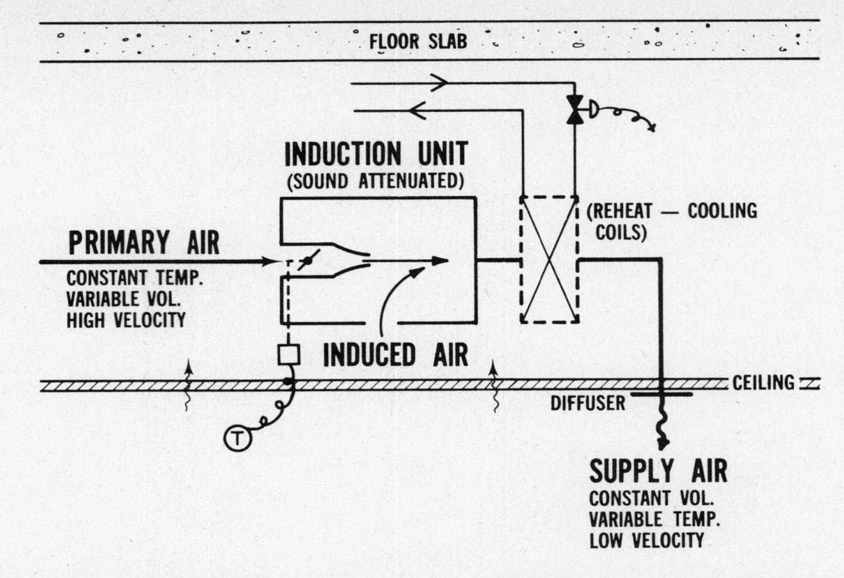

FIGURE 16-10 Single induction terminal (Courtesy, Westinghouse Electric Corporation).

ter zones. The temperature of the primary air can be regulated to meet seasonal variations in the total load, especially at the perimeter.

The low-velocity room supply air is discharged into a zone from an air induction box that induces air from the zone and mixes it with primary air. Room air is supplied at a constant volume. The temperature of the supply air is raised by increasing the amount of air induced and by decreasing the amount of primary air at the same time.

Additional heating or cooling may be required to cope with local variations that exceed the capacity of the system. The limits of the system are established by the minimum primary air setting and the temperature of the primary air.

Because of the increasing emphasis on energy conservation, as well as the desire to reduce duct sizes, the use of double induction is becoming more popular. The double induction principle (Fig. 16-11) extends the range of control over space conditions. Double induction makes it possible to induce air from two sources—the room and the plenum—each at a different heat potential, thus taking advantage of the heat in the plenum that was created by the lighting. In winter the plenum air damper is open to take advantage of recirculating the heat given off by

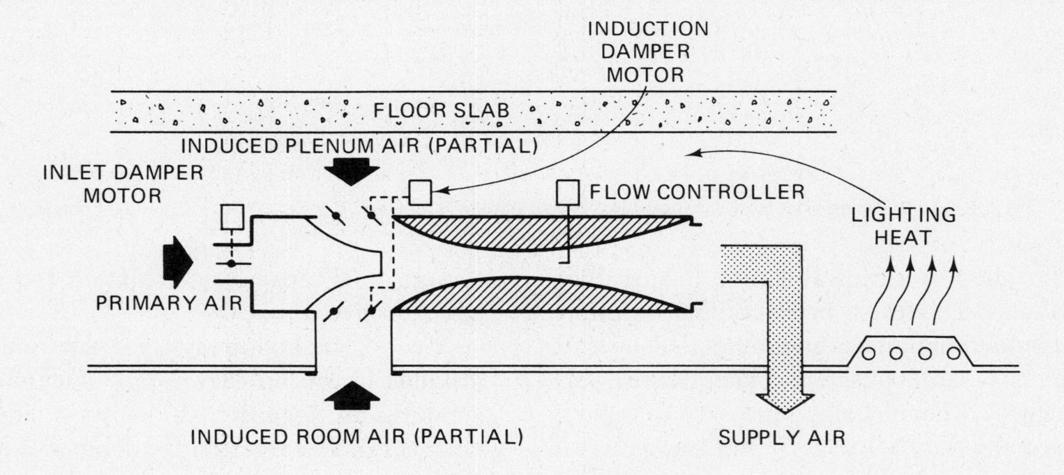

FIGURE 16-11 Double induction terminal (Courtesy, Westinghouse Electric Corporation).

electric lighting. In summer it is mostly induced room air that is mixed with the cold primary air to maintain zone control. The positions of all the dampers are determined by automatically modulating operators in response to the zone controller.

The main return air ductwork is extended into the ceiling plenum from service ducts or shafts at appropriate locations to collect the return air from the plenum and transfer it back to the central plant. In most buildings the ceiling plenum is used as a void. Air is drawn from the space below into the plenum through grilles, registers, or slots in the lighting fixtures.

The double-induction method offers a more closely controlled room temperature; it reduces the total ft^3/min delivery, the fan horsepower, and the air-handling apparatus size, and it results in lower building and operating costs. But, although it can use some of the heat produced by lighting and thus reduce the ft^3/min requirement, it cannot reduce the refrigeration requirement.

Recently this problem has been solved by the use of water-cooled lighting fixtures, or *luminaires* (Fig. 16-12). The object of all luminaire designs that remove light heat by return air is to minimize the amount of heat that becomes a space-conditioning load. However, the lighting heat (during the summer) must be removed from the return air before it is recirculated through the building. This means that the refrigeration plant must be large enough to cool the total heat generated by the light fixtures.

Therein lies the advantage in using water-cooled luminaires; they capture the maximum amount of lighting heat in a medium (water) that permits the heat to be either redistributed or rejected from the building. A water-cooled luminaire is a simple heat exchanger in the shape of a standard fluorescent luminaire fixture. Tubing is an integral part of the fixture housing. Water circulated through this tubing absorbs most of the lighting heat (60 to 80%) before it can increase the plenum temperatures and the heat regain in the area. The size of the central refrigeration plant can thus be reduced by the amount of heat removed from the fixture. The resulting hot water may be rejected by a cooling tower or used in the building domestic hot-water system. Lighting efficiency is improved by some 20% as a result of lower fixture temperatures. Ceiling induction air terminals are used as previously explained to blend the primary, room, and plenum air for space temperature control. However, the size of the terminal can be affected by the reduction in the heat removal requirement.

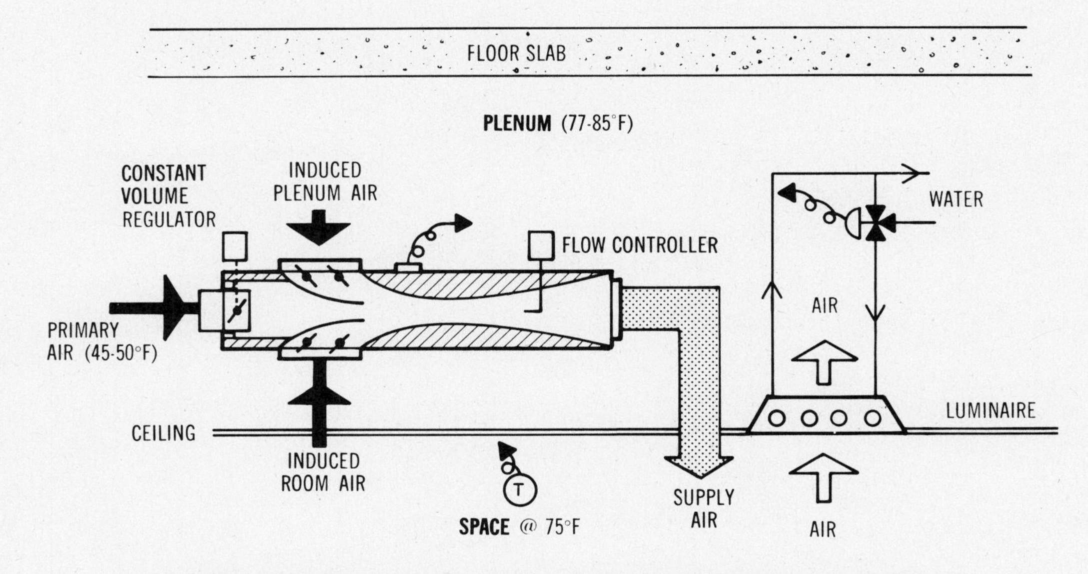

FIGURE 16-12 Luminaire design (Courtesy, Westinghouse Electric Corporation).

PROBLEMS

1. Commercial air distribution systems are generally classified as ______________, ______________, and ______________ pressure.
2. In Fig. 16-1 stores number 1, 4, and 6 would have what kind of distribution?
3. Commercial duct systems take several forms; name two.
4. Do induction room terminals require a fan for air movement?
5. Commerical high-velocity systems have duct velocities in the range of ______________.
6. High-velocity primary air to room terminals will have a temperature range of ______________.
7. Ceiling induction terminals may be classified as ______________ or ______________ volume units.
8. Ceiling terminals can also be classified as ______________ or ______________ induction units.
9. Ceiling terminals may take advantage of the heat of electrical lighting fixtures to temper primary air. True or false?
10. Water-cooled lighting fixtures (luminaires) reduce the heat load of an area in summer. True or false?

17

CONTROLS

The development of controls and control systems has gone hand-in-hand with the development of overall heating and air-conditioning equipment. Particular controls may have answered the need to improve operation, safety, personal convenience, economy, or a combination of several factors. The next several chapters will review the scope of these factors and the hardware which performs the action.

Very simply, a control system checks or regulates within prescribed limits. Such a system consists of two elements (1) a controlling device, and (2) a controlled device.

A controller (controlling device) measures some variable conditions that must be held constant, for example room temperature, and then activates the controlled device, such as a gas valve, to adjust heat input to maintain the room air between prescribed temperature limits. A medium regulated by the controlled device (gas flow, in this instance) is called the *control agent*, to distinguish it from the controlled variable (room air) sensed by the controller. A controller is made up of two recognizable components—its sensing element and its transducer. The sensing element actually measures any change in the variable that is being held constant. The transducer converts the effect of the sensor into a more useful, perhaps more powerful, action to motivate the controlled device, which may be located some distance from the controller.

For example, in an ordinary controller, a bimetal element that expands with temperature is the sensing element. The switch attached to the bimetal that closes an electric circuit is the transducer.

In a similar manner, a controlled device can be viewed as having components—its operator and its final control element.

In a solenoid valve, the magnetic coil that converts electrical energy into the physical movement of a valve stem is the operator. The final control element is the valve itself. Together the components form what we call the *controlled device.*

Safety is the third ingredient. In addition to a controller and a controlled device, every control system contains one or more safety devices that stop the action, or take a new action, whenever a dangerous condition develops.

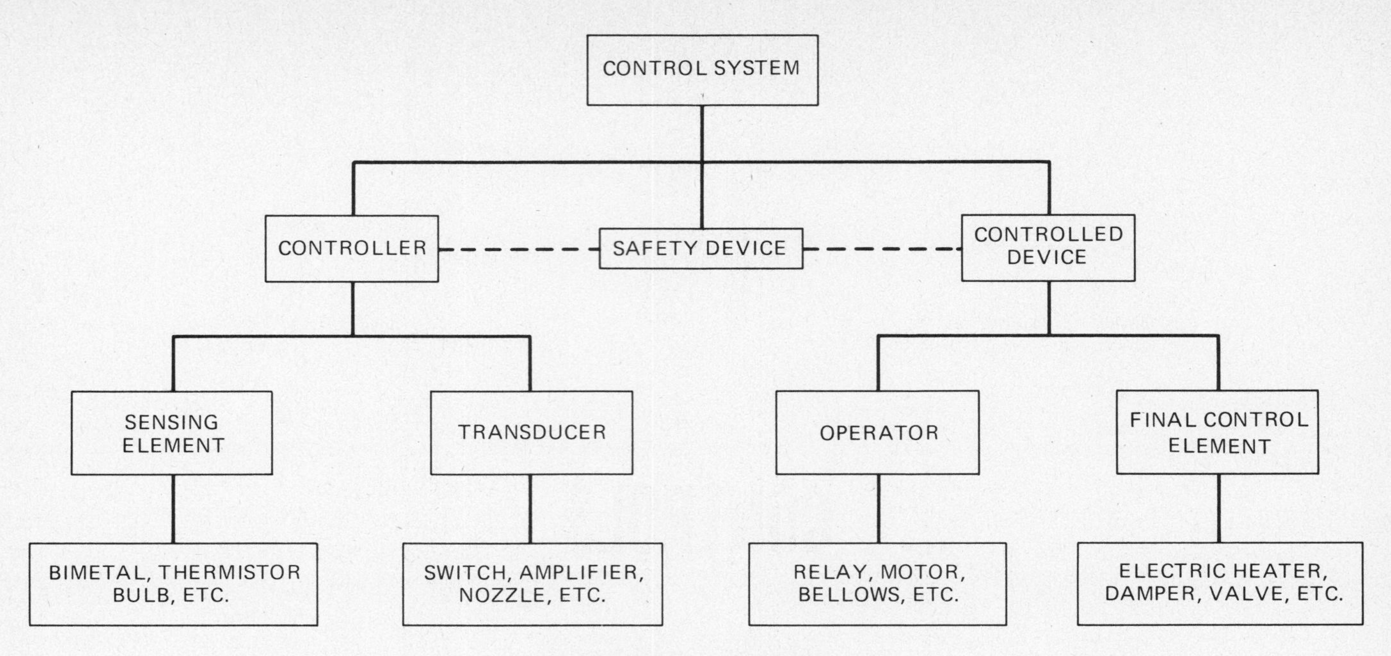

FIGURE 17-1 Control system.

The family of devices that perform one or more of these functions—controller, controlled devices, or safety device—are collectively referred to as the *controls of a control system* and are graphically illustrated in Fig. 17-1.

POWER SOURCE

Obviously, there must be a source of power to operate a control system. Automatic controls for residential and light commercial heating and air conditioning are powered by electricity. The controls in engineered systems can be electrical, electronic, pneumatic (compressed air), or a combination of all three. Here we will concentrate on the electrical type of controls for residential and light commercial application.

We will first consider the power source; heating and cooling control circuits can be designed to operate on either line voltage (115–120 volts) or low voltage, which is designated as a 24-volt system. A low-voltage control circuit is superior to a line voltage circuit because first, the wiring is simplified and safer, and second, low-voltage thermostats provide closer temperature control than line-voltage thermostats.

Essential to the development of low-voltage controls is the economical ac voltage transformer (Fig. 17-2). A stepdown or low-voltage transformer is used in heating and air-conditioning control systems to reduce line voltage to operate the control components. Inside, a simple stepdown transformer consists of two unconnected coils of insulated wire wound around a common iron core (Fig. 17-3). To go from 120 volts (primary) to 24 volts (secondary), there are 5 primary turns to 1 secondary turn. For a 240-volt primary the ratio will be 10 to 1, and so on. Thus the induction ratio is a direct proportion. Stepup transformers would be just the reverse.

In reducing the voltage, was any energy lost? Not really, because the value of the current on the secondary side would be five times higher than on the primary side, so that the power on both sides of the transformer remains the same—assuming a trans-

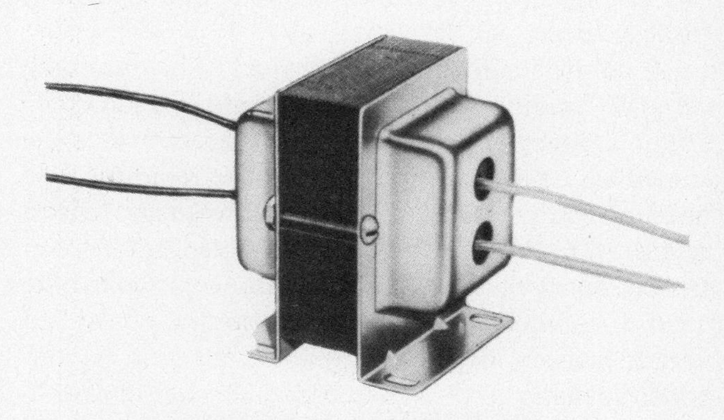

FIGURE 17-2 Voltage transformer.

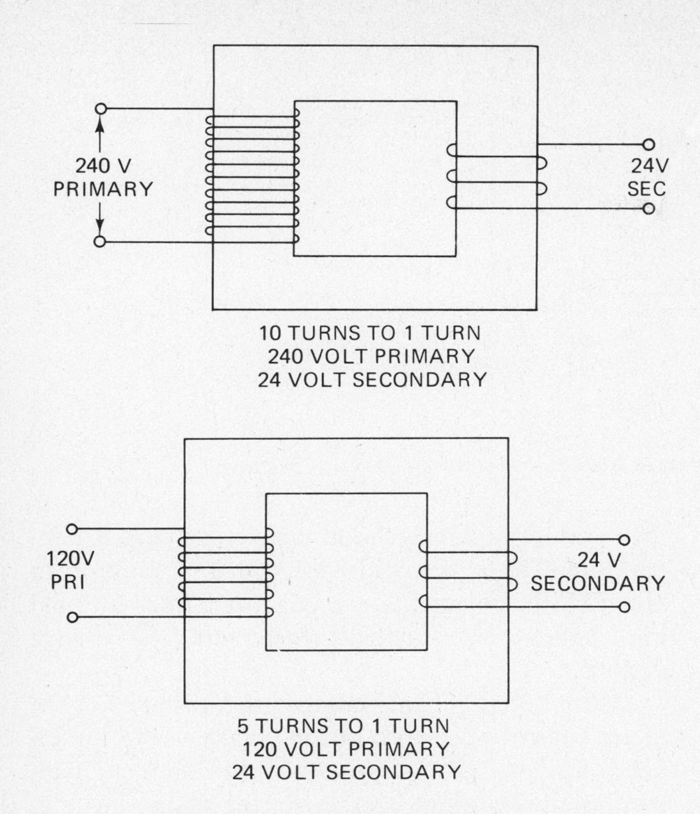

FIGURE 17-3 Stepdown transformer.

former that is 100% efficient. Thus, the ratio can be expressed by a formula:

$$\underset{\text{(Primary)}}{\text{Volts} \times \text{Amps}} = \underset{\text{(Secondary)}}{\text{Volts} \times \text{Amps}}$$

So, if we wanted a 40-volt-amp capacity rating on the secondary, it would take a 0.334-amp input on the primary at 120 volts to produce 1.67 amps on the 24-volt secondary side:

$$120 \times .0334 = 24 \times 1.67 = 40 \text{ volt-amps}$$

Transformers are available in a variety of voltages and capacities. The capacity refers to the amount of electrical current expressed in volt-amperes; a transformer for a control circuit must have a capacity rating sufficient to handle the current (amperage) requirements of the loads connected to the secondary. Forty volt-amp ratings are used only for heating forced-air furnaces. Heavier volt-amp ratings are needed for air-conditioning duty, because electrical devices containing a coil and iron, such as solenoid valves and relays, have a power factor of approximately 50%—thus, for secondary circuits with such controls the capacity of a transformer must be equal to or greater than twice the total nameplate wattages of the connected loads.

The proper transformer rating for the electrical control circuits will have been selected by the equipment manufacturer. However, if accessory equipment is added, the additional power draw must be considered and might possibly result in an increased rating. This situation is common, for example when cooling is added to an existing furnace and where the original transformer is too small. Also, when replacing a defective transformer, make sure the rating is equal to or greater than the original equipment.

With a source of power established, let's turn our attention to the most familiar controller—the room thermostat. As previously stated, the low-voltage thermostat is much more accurate than line-voltage stats, but a number of line voltage stats are used in direct electrical resistant heaters—wall, baseboard, room air conditioners, etc.—and in order to appreciate the differences, let's first examine the line-voltage type.

THERMOSTATS

Early typical line-voltage stats, sometimes called *snap action thermostats*, consisted of bimetal temperature sensing elements (Fig. 17-4) made of two or more metallic alloys welded together, each having different coefficients of expansion when exposed to heat. One metal will expand more rapidly than the other and thus cause the bimetal to change its curvature when it "feels" a change in temperature. The bimetal may be a straight form (cantilever), U-shaped, or spiral. Line voltage usually flows through the bimetal, which has a moving contact that closes against a fixed con-

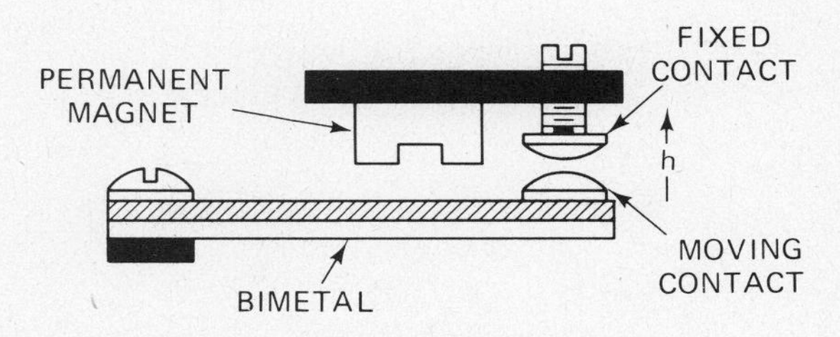

FIGURE 17-4 Snap action thermostat.

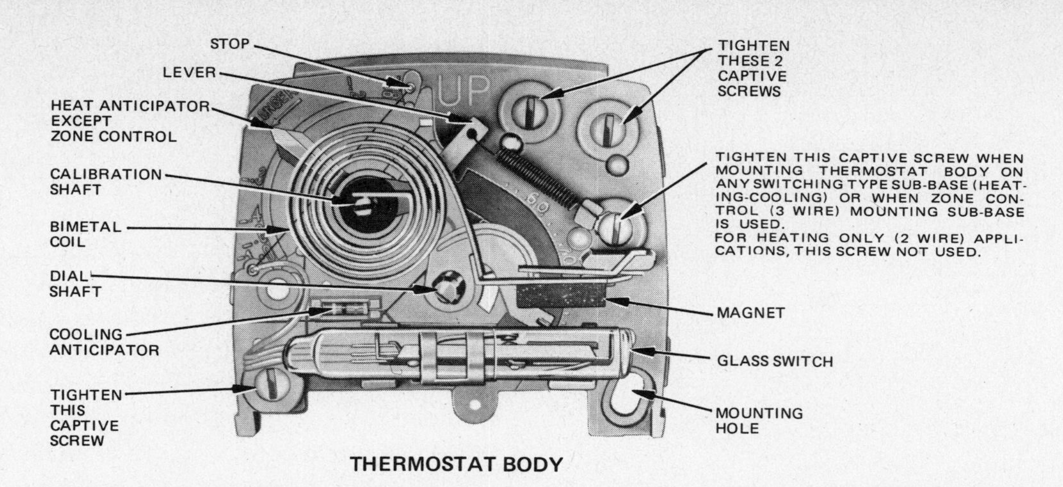

FIGURE 17-5 Low-voltage thermostat with cover removed (Courtesy, Robertshaw Controls Company).

tact point. The distance (h) the moving contact travels is the differential range between *off* and *on*, and in a simple, low-cost line stat this distance can be 3 degrees or more. If a knob is attached to the fixed contact screw the temperature sensing range can be adjusted within a fairly narrow range. The function of the permanent magnet is to snap the contacts closed when the moving contact comes within range, in order to help reduce the arcing of contacts. This type of control can be used for heating or cooling, depending on whether the contact is made on a rise or a drop in room temperature. Dual contact models can thus perform both functions.

Snap-action line-voltage thermostats are satisfactory for limited control situations where close temperature swings are not economically justified. Also, line-voltage stats can suffer from dirty contacts or pitting of the contacts due to arcing when making or breaking contact. The control of a snap-action stat also is a delayed reaction. This means that by the time the bimetal has traveled its range and the heating or cooling equipment is shut down, the actual room temperature will overshoot the desired space temperature. The anticipation method is a solution to above (to be covered later), but it is harder to build into high-voltage controls than into low-voltage designs.

More recently, line-voltage thermostats for the direct control of electric heating make use of liquid-filled elements that respond to both ambient temperature and radiant heat. Also, the design includes slower moving elements, which can handle direct resistance loads of up to 5,000 watts at 240 volts ac. The differential between the set point and room conditions is reported to be less than that of the snap-action variety.

From an installation viewpoint, wall-mounted line-voltage thermostats must be served by heavy-duty wiring. On commercial work, codes may require that they be run in conduit. Also where many wires are involved, identification coding is difficult. In conclusion, therefore, line-voltage stats are almost totally limited to space heaters and the room and store type of self-contained air-conditioning products.

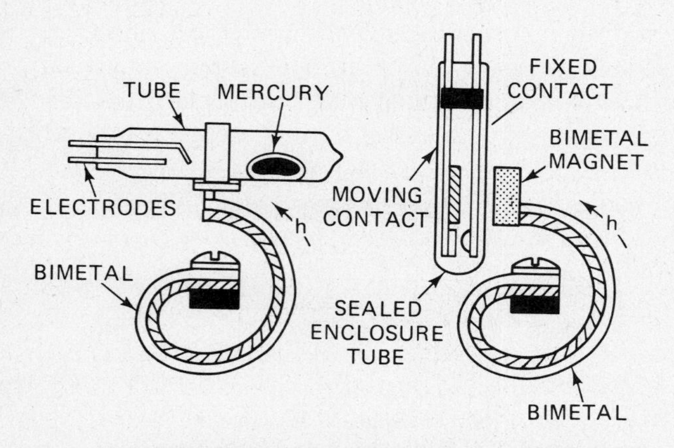

FIGURE 17-6 Single-action mercury bulb thermostat.

FIGURE 17-7 Thermostat subbase (Courtesy, ITT Nesbitt).

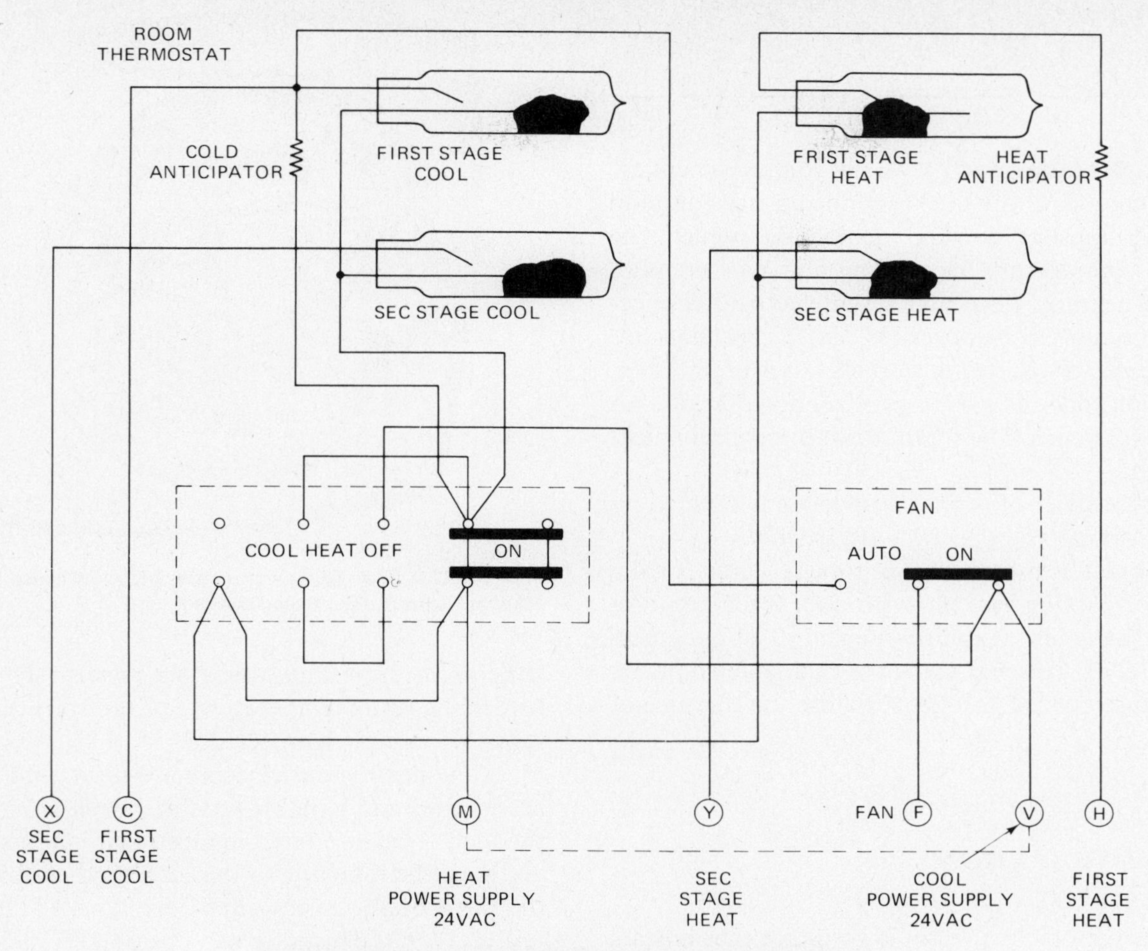

FIGURE 17-8 Two-stage heating and cooling wiring diagram.

The low-voltage thermostat (Fig. 17-5) overcomes the limitations of line-voltage models and is almost universally used in modern central system control circuits. First, the use of spiral-shaped, light-weight bimetal increases the effective length and thus the sensitivity to temperature change. Second, the use of sealed contacts virtually eliminates the problem of dirt and dust. Although there are minor variations among different manufacturers, the contacts are always sealed in a glass tube. Figure 17-6 (on the left) represents a single-action mercury bulb design. As the bimetal expands and curves the mercury fluid moves to the left, completing an electrical circuit between the two electrodes, which carry only 24 volts. The differential gap between OFF and ON is very small—$\frac{3}{4}$ to 1°F from the set point. On the right is a sealed tube that has a metal-to-metal contact. The magnet provides the force that closes the contacts.

Combination heating and cooling thermostats may have two sets of contacts—two single-pole, single-throw mercury tubes, or one single-pole, double-throw mercury tube—to provide control of both the heating and the cooling system.

Room thermostats are designed to control room temperatures over a fairly wide range—usually about 50°F to 90°F; normal settings are between 68° to 80°F. A dial or arm permits the homeowner to select the desired conditions.

The thermostat subbase (Fig. 17-7) not only provides a mounting base for the thermostat, but it is also used to control the system operation through a series of electrical switches. The system (SYS) switch selects COOL, OFF, or HEAT. The blower operation is controlled by the fan switch, which is a simple two-position switch. When set in automatic, the fan will cycle with the furance on heating. If in the ON position, the fan will run continuously.

Somewhat more sophisticated thermostats contain two dials that establish different control points for heating and cooling, with an automatic changeover from heating to cooling. Figure 17-8 represents a two-stage heating and two-stage cooling schematic wiring diagram. Two-stage thermostats are common in heat pumps or rooftop equipment, which use multiple compressors for cooling and two or more stages of heating. Note that seven electrical connections are required; however, *M* and *V* terminals are the same power source, and so six wires are needed. With color-coded low-voltage wire, however, it's no problem to connect the thermostat to the mechanical equipment.

The sensitivity of room thermostats is affected by both *system lag* and *operating differential.*

System lag is the amount of time required for the heating or cooling system to produce a temperature change that is felt at the thermostat. The operating differential of a thermostat is the change in room air temperature needed to open or close the thermostat contacts.

HEAT ANTICIPATION

Heat anticipators are used in low-voltage thermostats to reduce the effects of the system-lag operating differential. They are nothing more than small electrical resistors that generate a small amount of heat when current flows through it. If the thermostatic bimetal is exposed to a small amount of "artificial" heat, the sensitivity of the thermostat will increase, because the bimetal reacts to changes in surrounding air temperature and will *lead* actual changes in room temperature. For example, if the room air temperature increases a certain number of degrees in five minutes, the anticipated bimetal will feel this change in only three minutes, while the unanticipated bimetal will not feel the change in air temperature until six minutes have passed. Thus, the anticipated bimetal will shut the furnace burner off before the actual room temperature reaches the thermostat set point. Anticipation results in closer room temperature control and less overshooting of heating due to residual heat in the furnace heat exchanger that must be dissipated.

A similar technique is used for cold anticipation, except that the cold anticipator is energized only when the cooling is OFF. The bimetal senses the

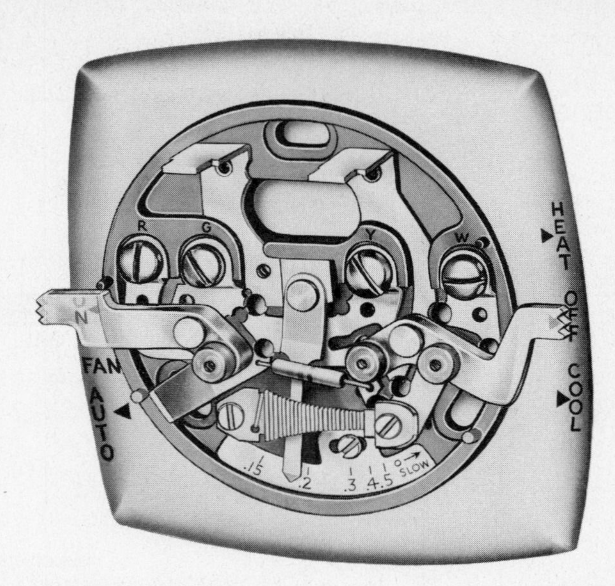

FIGURE 17-9 Low-voltage stat showing heat anticipator (Courtesy, Johnson Controls, Inc., Penn Division).

anticipator heat and turns the system ON again before the room temperature gets too warm . . . thus, it *leads* room air temperature.

Some heating anticipators are fixed, while others are wire-wound variable resistors *wired in series* with the load. They are rated in amps or fractions thereof and must be matched to the total load controlled by the thermostat. On the adjustable type, as illustrated in Fig. 17-9, the installer will position the sliding arm to the proper load rating. If ON-OFF cycles are too long or too short, the system operation can be changed to give a faster or slower response. Fixed anticipators obviously cannot do this, and are thus usually used on economy type heating applications.

Cooling anticipators are fixed resistors *wired in parallel* with the thermostat contacts; they do not have the capacity to energize the relays, solenoid valves, or contactors normally found in cooling products. These anticipators are factory selected and should not be changed.

HUMIDISTAT

Another popular controller found in residential total comfort applications is the wall-mounted humidistat (Fig. 17-10). It is very similar to the low-voltage thermostat, and it contains both a sensing element and a low-voltage electric switch. The sensing element consists of either an exceptionally thin moisture-sensitive nylon ribbon or strands of human hair

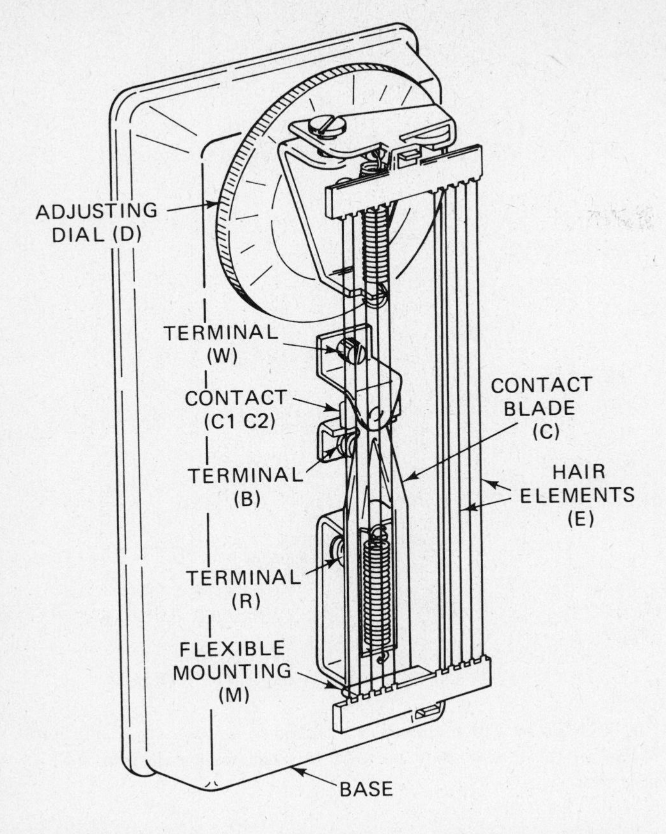

FIGURE 17-10 Construction details of a three-wire wall-mounted humidistat.

that react to changes in humidity. The movement of the sensing element is sufficient to make and break electrical contacts directly or when coupled with a mercury-type switch. Humidistats for mounting on ductwork are also available.

Figure 17-11 is a simple schematic wiring arrangement for year-round control of humidity. In winter the elements contract and close the contact to the humidifier. In summer the humidstat will expand in

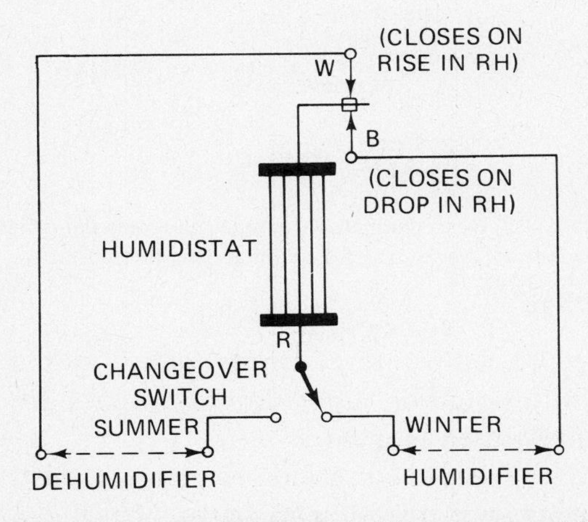

FIGURE 17-11 Schematic diagram of wiring arrangement for year-round control of humidity.

response to a rise in the relative humidity and close the contacts to the dehumidifier (cooling unit) circuit. A summer/winter changeover switch selects the appropriate operation.

The humidistat is not quite as precise a controller as a standard low-voltage room thermostat. Normally, a change of 5% relative humidity is required to actuate switching action. But this is acceptable for normal comfort applications, since people cannot detect, nor do they react to, changes in relative humidity with the same sensitivity as room temperature.

The reader should be aware, however, that more precise controls are available for specialized applications, such as computer rooms, libraries, printing plants, etc., where controllers such as *wet-bulb* and *dew-point* thermostats would be used.

SENSING DEVICES

Before proceeding with other controls, let us examine other types of sensing elements that react to temperature and/or pressure. Bimetal temperature controllers are available in various forms as illustrated in Fig. 17-12. There are metal actuators that produce rotary movement, elongation, warping, bending, or snap action when exposed to heat. The mechanical movement is relatively large in response to small changes in temperature.

Bellows and diaphragms (Fig. 17-13) react to changes in pressure. Because of a relatively large internal volume, the bellows will produce a greater mechanical movement, but the diaphragm is more accurate; the choice will depend on the amount of movement or sensitivity and the working pressures that must be contained.

The bellows and/or diaphragm can also be adapted to control temperature by the addition of a remote-bulb sensing element (Fig. 17-14) filled with a vapor, a vapor/liquid, or a solid/liquid. Heat will cause the fill to expand into the bellows or diaphragm cavity, thus creating mechanical movement.

The Bourdon tube (Fig. 17-15) is another sensing element that reacts to changes in pressure. It's an elliptical tube that is sealed and linked to a trans-

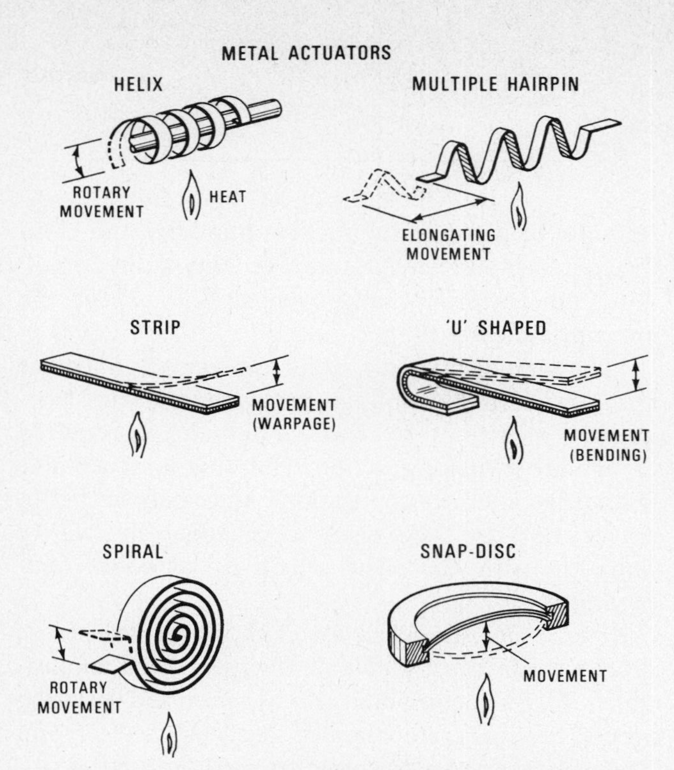

FIGURE 17-12 Common bimetal configurations used in thermostats, warm-air controllers and some water-immersion controls.

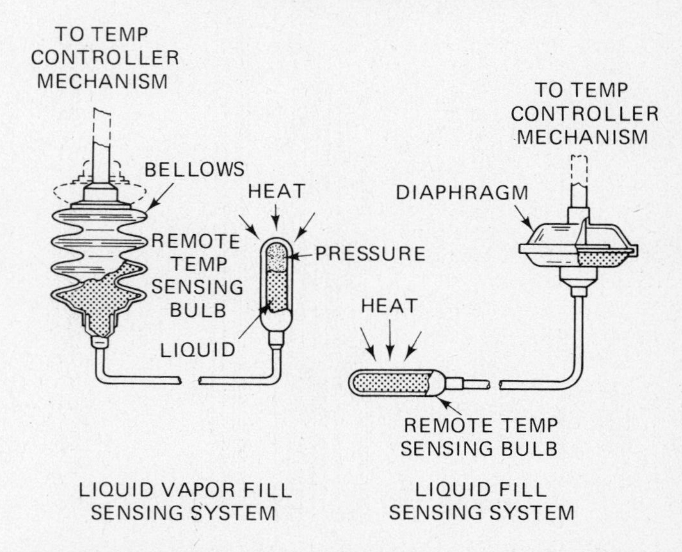

FIGURE 17-14 Filled actuators attached to bellows and diaphragms to permit these devices to be used to sense temperature as well as pressure.

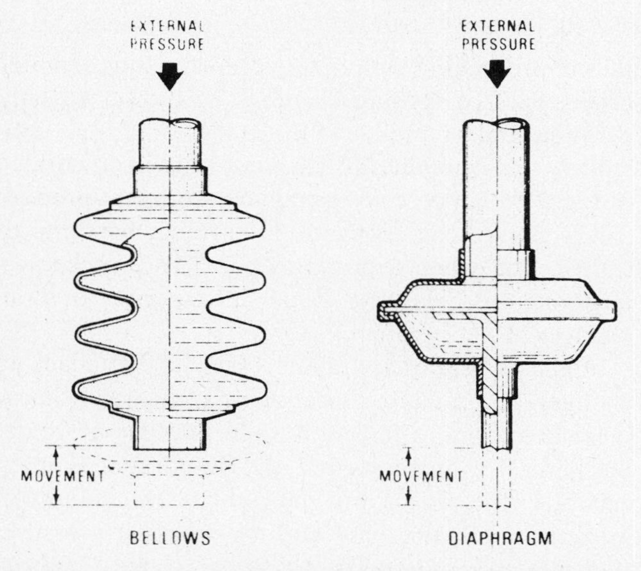

FIGURE 17-13 Bellows and diaphragms react to pressure changes and are often used in pressure controllers.

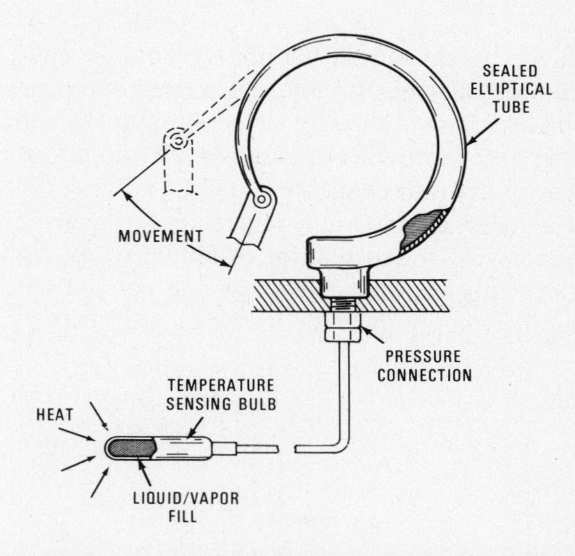

FIGURE 17-15 Bourdon tube used in many gauges is shown here used as a temperature sensor, say for an immersion thermostat.

ducer. As pressure is applied to the pressure connection, the tube will tend to straighten and thus create a mechanical movement. Pressure can come from refrigerant, water, oil, steam, or any fluid that can be pressurized. The Bourdon tube may also be adapted to temperature sensing with a closed bulb filled with liquid or vapor. Bourdon tubes are most extensively used in gauges.

The rod-and-tube sensing element is constructed by placing an inert rod in an active tube (Fig. 17-16).

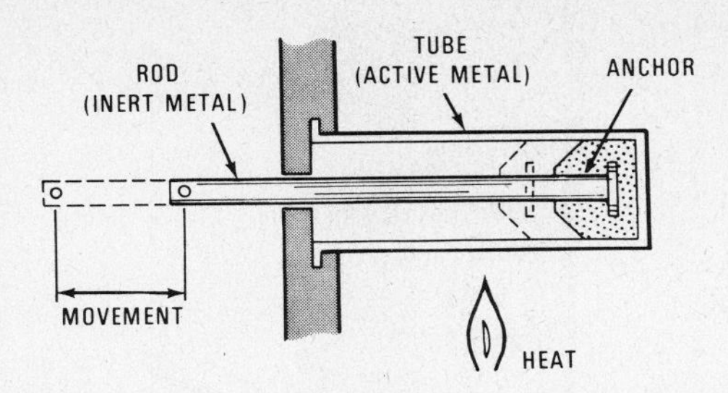

FIGURE 17-16 Rod and tube sensors are used in some controllers.

Since the rod is anchored to one end of the tube, a temperature change will cause a pushing or pulling movement that can be made to operate a switch.

Resistance elements are sensing elements designed to vary their electrical resistance when exposed to changes in temperature or light intensity. Thermistors, photoresistance cells, and photo cells are typical. A thermistor is an electrical device that will increase its resistance to the flow of current as its temperature decreases (Fig. 17-17). Changes in current flow can be used to activate or deactivate a remote control for various uses.

The cadmium sulfide cell (CdS cells Fig. 17-18), or photoresistance cell, is a resistance element that reacts to changes in light intensity and is highly resistive to the passage of electrical current when its sensing base is in complete darkness. When exposed to light, the electrical resistance declines proportionally to the intensity of the light.

Some of the more common controls that use the principle of the sensing elements just explained are discussed in the section on residential and light commercial equipment.

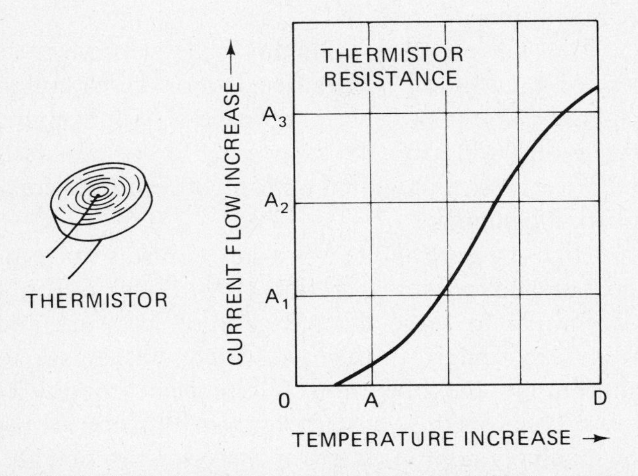

FIGURE 17-17 Thermistor resistance changes with temperature (decreases as temperature increases). Thus current flow in a circuit would increase with a temperature rise.

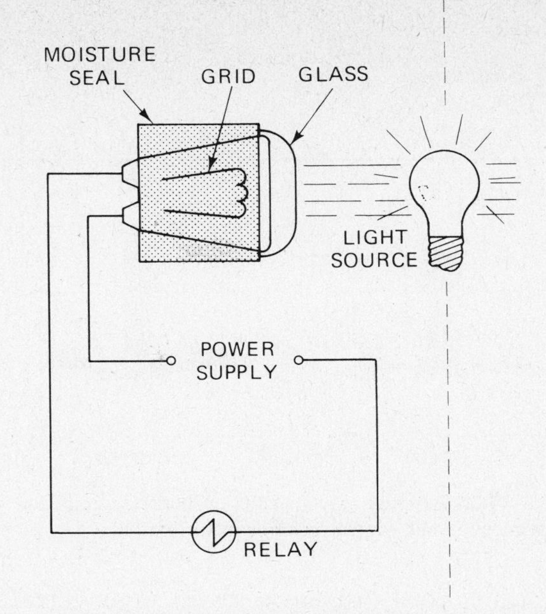

FIGURE 17-18 CdS cell.

In the area of safety the high-limit thermal cutout control is widely used (Fig. 17-19). It employs a snap-disc bimetal to open an electrical circuit. These controls are common in electrical space heaters, electric duct heaters, electric furnaces, heat pumps, and any application where the small, quarter-sized control can be mounted and exposed to heat sources.

The inherent protector used in hermetic compressors is an excellent example of bimetal action. The overload (Fig. 17-20) is a current and heat-sensing device and consists of a bimetal contact in series with the motor contactor coil control circuit. Whenever the compressor motor is overheated, overloaded, or stalled, the heavy continuous current through the heater warps the bimetal contact until it opens the motor contactor low-voltage relay coil circuit and stops the flow of current to the compressor.

Another form of high-limit cutout, previously noted in the review of baseboard electric heaters, is

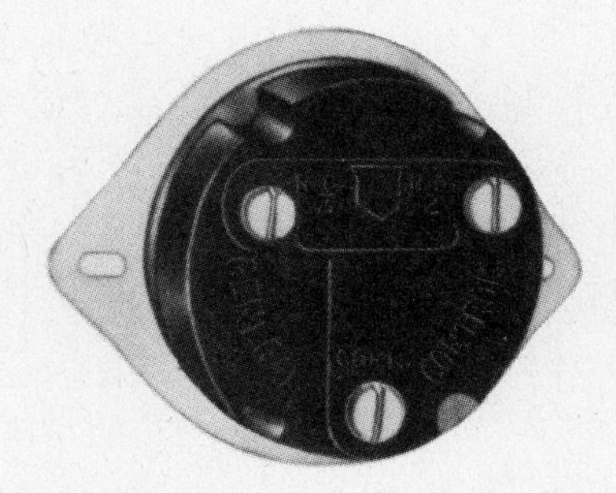

FIGURE 17-19 High-limit thermal cutout control (Courtesy, General Controls Company).

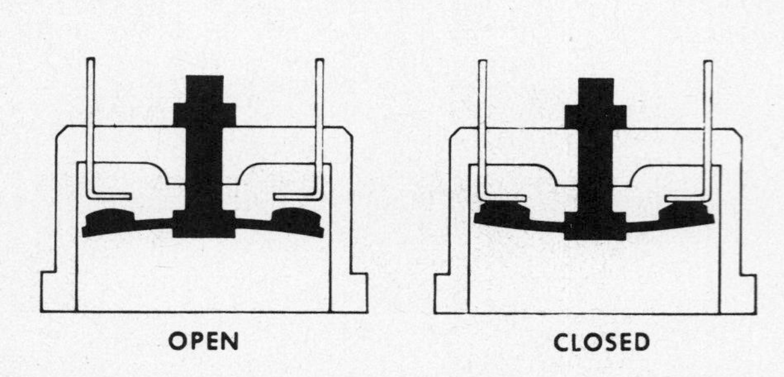

FIGURE 17-20 External type (mounts on housing) snap disc line break overload. Some designs fit inside motor winding.

the lineal control that has a sealed vapor/liquid fill tube extending the entire length of the heater. If the heater is blocked by drapes, etc., the safety limit will open.

One type of combination of fan and limit control for a typical gas or oil warm-air furnace (Fig. 17-21) uses the power created by the rotary movement of a helix bimetal. The rotating cam makes or breaks the separate fan and limit electrical contacts. Other models may use spiral or flat bimetal or even liquid-filled elements. Some forms of duct-mounted limit controls use a rod and tube or a liquid-filled bulb to sense the air conditions.

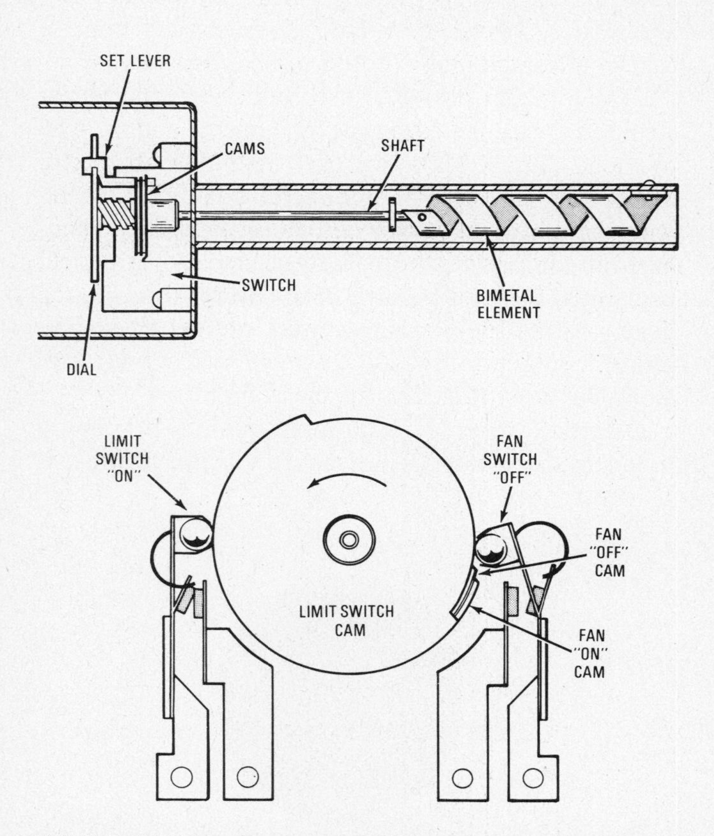

FIGURE 17-21 Details of one combination fan-limit control show how twisting motion of heated helix turns cam to activate fan and limit switches. Arrow on cam indicates counter clockwise rotation on temperature rise.

FIGURE 17-22 Immersion controller (Courtesy, Honeywell Residential Division).

A common control in hydronic heating is the immersion controller, which uses a liquid-filled bulb that operates a snap-acting switch (Fig. 17-22). Immersion controls are inserted directly into the boiler to detect water temperature and can be used as high-limit cutouts, and low-limit and circulator pump controls.

A prime example of the thermistor sensing device is the high-temperature cutout on a hermetic compressor (Fig. 17-23). The thermistor, not much larger than an aspirin tablet, is imbedded deep in the motor windings. It senses overheating and will signal a transducer to shut down the compressor until normal temperatures are restored.

The use of the cadmium cell principle was reviewed under oil burner operation. The cadmium cell reacts to the presence or absence of light from the flame and will stop the oil burner if ignition fails before excessive unignited oil can collect in the combustion chamber.

High-pressure and low-pressure controls for compressor and system operation (Fig. 17-24) utilize a diaphragm to sense the refrigeration pressure and react accordingly through a switch mechanism to shut down the operation if abnormal conditions exist. High-pressure cutout controls frequently must be manually reset in order for the system to operate. This prevents rapid short cycling, which can eventually cause compressor overheating and possible damage.

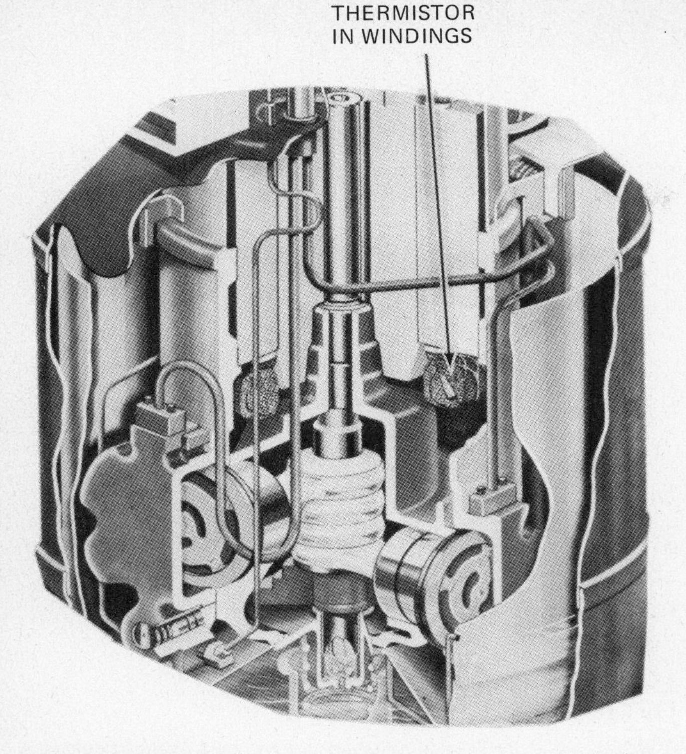

FIGURE 17-23 Thermistor sensing device (Courtesy, Westinghouse Electric Corporation).

Pressure controls can also be used on large compressors to detect the buildup of oil pressure. They are equipped with a time-delay mechanism (usually one minute or less), and if normal oil pressure is not achieved within the time period, the controller will stop the compressor before damage can occur. It can also be connected to an alarm indicator.

The diaphragm type of air pressure switch (Fig. 17-25) is used in heat pumps to start and stop the defrost cycle, as previously reviewed. In that case they are used for measuring the differential pressure across the outdoor coil. These devices can also be used to measure positive pressure as in a central system air duct and energize an alarm circuit when a change in air pressure is sensed.

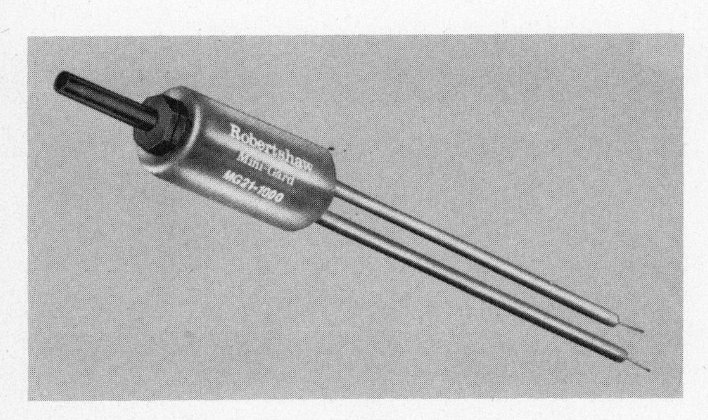

FIGURE 17-24 Pressure switch for excessive high- or low-pressure protection (Courtesy, Robertshaw Controls Company).

FIGURE 17-25 Diaphragm-type air pressure switch.

Static pressure regulators (diaphragm device) are also air controllers that are used in larger systems to measure static pressure and then, through an electric circuit to a damper motor, to regulate the air damper to maintain a constant static pressure in the system, as illustrated in Fig. 17-26. A static pressure regulator can also be connected to a control that varies the central fan inlet air vanes to maintain a constant fan performance.

A different form of sensing element not previously described but frequently used in residential and light commercial work is the sail switch (Fig. 17-27). As the name implies, it detects the movement of air by the use of a sail or paddle attached to an arm; the arm, in turn, is attached to an electrical switch. The illustration is of a combination duct-mounted humidistat and sail switch. The humidistat mounts in the return air duct of a forced-air system. The function of the sail switch is to shut down the humidifier when the supply air is cut off.

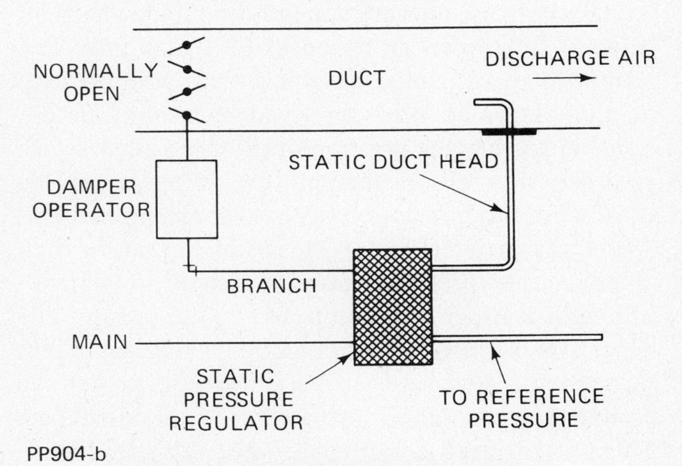

FIGURE 17-26 Static pressure regulator.

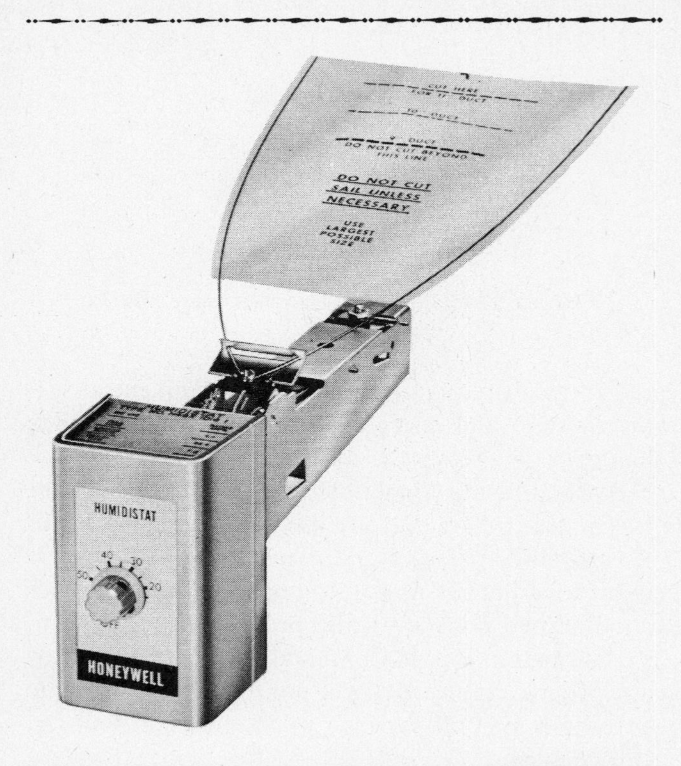

FIGURE 17-27 Sail switch (Courtesy, Honeywell Residential Division).

Sail switches are also used as safety devices to detect air flow or the absence thereof and cut off heating and/or cooling before abnormal system operation occurs.

Most of the controllers reviewed above are of the two-position variety—that is, they are OFF or ON, OPEN or CLOSED—however you wish to describe the action. But, even within the scope of residential and light commercial work, there is a need for modulating controls to vary an action somewhere in between fully open or closed or OFF and ON. The proportional controller does just that. Proportional or modulating action is achieved by using a potentiometer to vary the position of the controlled device in proportion to any temperature (it could also be pressure or flow) variation felt by the sensing element. A potentiometer (Fig. 17-28) is nothing more than an adjustable three-wire variable resistor in conjunction with a wiper arm connected to the sensor. The wiper arm is the third wire contact. As the arm moves through the complete stroke (throttling range) the current flow changes in proportion to the resistance. This current flow is sent to a motor relay that actu-

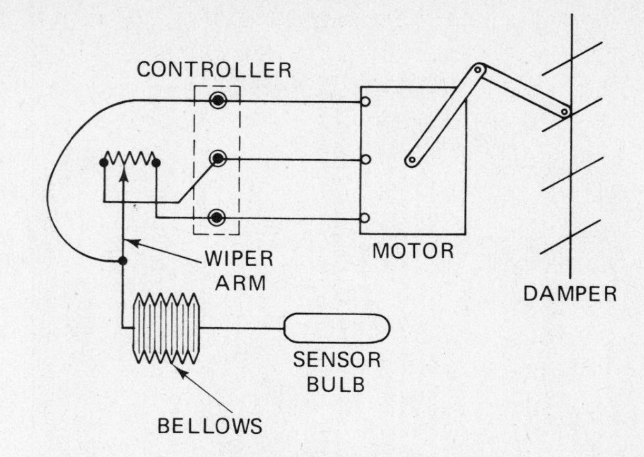

FIGURE 17-28 Potentiometer.

ates a damper motor, motorized valve, etc., to bring about the required temperature change in the controlled space.

OPERATORS

Operators and final control elements, as mentioned at the beginning of this chapter, are on the receiving end of the controller signal. The operator can be thought of as a secondary action to cause another action, while the final element is the last action.

The electric relay (Fig. 17-29) is the best example of an operator. Its function is to take a low-voltage signal from a controller and, through a magnetic coil, close a set of higher voltage contacts, which in turn will start a fan or compressor or perhaps only supply high voltage to another control circuit. The

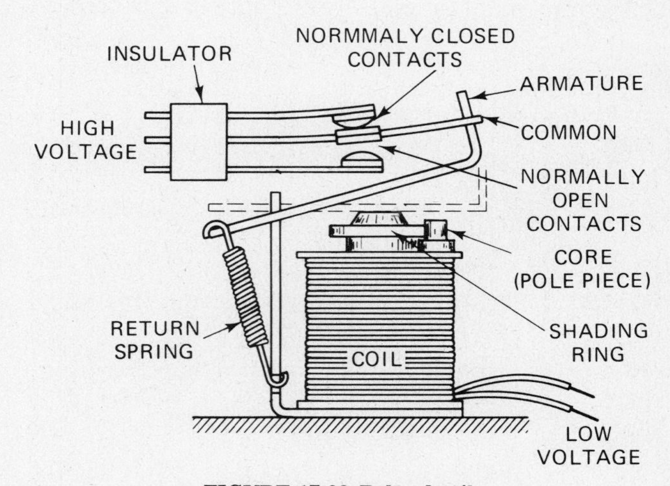

FIGURE 17-29 Relay details.

coil which sets up the magnetic field is installed around a pole piece. The armature moves in response to coil energization to open or close the electrical contact. Return springs are used to pull the armature to the deenergized position.

Relays are used to control electrical loads that draw less than 20 amps, such as small motors, valves, ignition transformers, and small compressors. A *contactor* is a heavy-duty relay designed to handle over 20 amps, and in most cases is used on compressor and pump motors. The volt-amp draw of the commonly used relay and contactor coils is low; therefore, low-voltage transformers can supply enough power to operate these devices. (NOTE: A *starter* is a form of contactor or relay that contains overload protection.)

To complete the relay story, it is important to mention briefly a second type of electric relay that has come into use comparatively recently—the so-called thermal relay, which uses a resistance heater and a bimetal instead of a magnetic coil and switch. When the low-voltage circuit is energized, perhaps by thermostatic action, the low-voltage heater warms the bimetal, which warps and closes a switch on the high voltage side. The chief advantage of this relay is its quiet action, and it is therefore well suited to low-capacity residential type of switching needs.

Sealed plug-in relays have been used extensively in commercial work and are becoming popular in residential use to simplify the addition of humidity control and electronic air cleaning. They look much like a square radio tube and plug into a control panel arrangement. Since the electrical contacts are not exposed to dust and dirt, reliability is greatly improved. Also, the plug-in arrangement permits quick replacement in the event of trouble.

Another form of operator is a motor (Fig. 17-30) that responds to a command of a controller and mechanically operates a damper, valve, step controller, etc., that produce a final action. The crank arm is connected to the final apparatus by means of linkage rods. Motors are available for two-position response and may swing through a 160 to 180° arc from start to stop. Others respond to a modulating controller and operate in an intermediate position to bring the controlled temperature, pressure, etc., back to the set point. Motors vary in arc rotation from 15 seconds to 4 or 5 minutes depending on the application. They are also rated according to the torque force the arm can exert (in inch-pounds or the size in square feet of the damper it can handle). Some models have a spring return to some normal setting in the event of power failure. Auxiliary switches can be included to perform a variety of other automatic switching combinations for secondary equipment.

FIGURE 17-30 Damper motor (Courtesy, Johnson Controls, Inc., Penn Division).

SOLID-STATE CONTROLS

Solid-state controls are finding their way into the HVAC industry just as they have in the space program, computers, and TV industry. Thimble-sized circuits in heating and air-conditioning equipment are not as important as are the features of flexibility, faster service and maintenance, and most importantly the development of new functional controls. The use of the solid-state thermistor to detect motor overheating or the use of cadmium cells in the oil burner are practical examples.

Another use of solid-state components is the fan-motor speed control so frequently mentioned in our equipment discussions. The SCR (silicon-controlled rectifier) can be considered an electronic switch: it has no moving parts but is capable of turning current off and on. When applied to a fan motor the SCR can "chop off" part of the time that power is supplied to the motor or part of each cycle of ac power and, as a result, reduce and regulate the speed. This process of changing the wave shape of the voltage and current to the motor is termed *phase modulation.*

Solid-state electronics is in itself a complete field and it is recommended the reader do more outside research on this subject prior to doing installation and service work on solid-state circuits.

This review of the more common controls found in basic residential and light commercial work is only an example of a rather extensive control industry, which in itself is a major portion of the overall heating and air conditioning industry. Many controls are preselected by the equipment manufacturer and only require that the technician know how to adjust and maintain their operation properly. Other controls are field applied and do require a careful examination of the function, application, and installation requirements prior to selection.

PROBLEMS

1. In a control system, the two main elements are ______________ and ______________.

2. Give a common example of a controller.

3. Give a common example of a controlled device.

4. Low-voltage control (24 volts) is produced from a ______________.

5. A 24-volt 70 volt-amp secondary output transformer requires ______________ amps on a 120-volt primary.

6. What is the *bimetal principle*?

7. Low-voltage thermostats are considered more accurate than line-voltage types. True or false?

8. Heat anticipators are actually small ______________.

9. Heat anticipators may be ______________ or ______________.

10. Humidistats measure and control ______________.

11. Humidistat sensing elements are made of ______________.

12. A thermistor changes its ______________ with changes in temperature.

18

TYPICAL RESIDENTIAL CONTROL SYSTEMS

With a basic understanding of the mechanical construction of residential heating and air-conditioning as presented in Chap. 8 through 11 and a review of some of the more common electrical controls as discussed in Chap. 17, let's now examine some typical heating and cooling circuits and become familiar with the function and operation of the component parts.

Different manufacturers will have variations in the physical wiring, but the function remains the same in all systems. With a firm understanding of the construction and use of schematic wiring diagrams, the student should be able to wire any manufacturer's product. The industry uses standardized electrical symbols, as shown in Fig. 18-1, to help the technician identify common controls. Additionally, the legend designations, as presented in Fig. 18-2, are related alphabetically to the devices they represent. For example *R* is a general relay, while *CR* is a cooling relay, *HR* is a heating relay, *DR* is a defrost relay, etc. Legends may vary with the particular wiring diagram or the manufacturer's methods of labeling.

Study the symbols and legend designations carefully, for they are all-important to the following discussion, and they will also serve later as a reference.

SERVICE ENTRANCE

The starting point of the electrical system is the service entrance. Electricity is distributed through high-power lines by the power companies, and these are reduced down into local residential lines that run on every street. A transformer on the pole will serve several residences, reducing down the high-voltage in the transmission lines to normal house entrance voltage. Entrance service will be 120/240 volts single phase, and modern houses will have anywhere from 100–200-amp service to a circuit breaker or fuse board at the central distribution point. Three wires are brought in by the power company to the house, two of which are the hot wires and one of which is

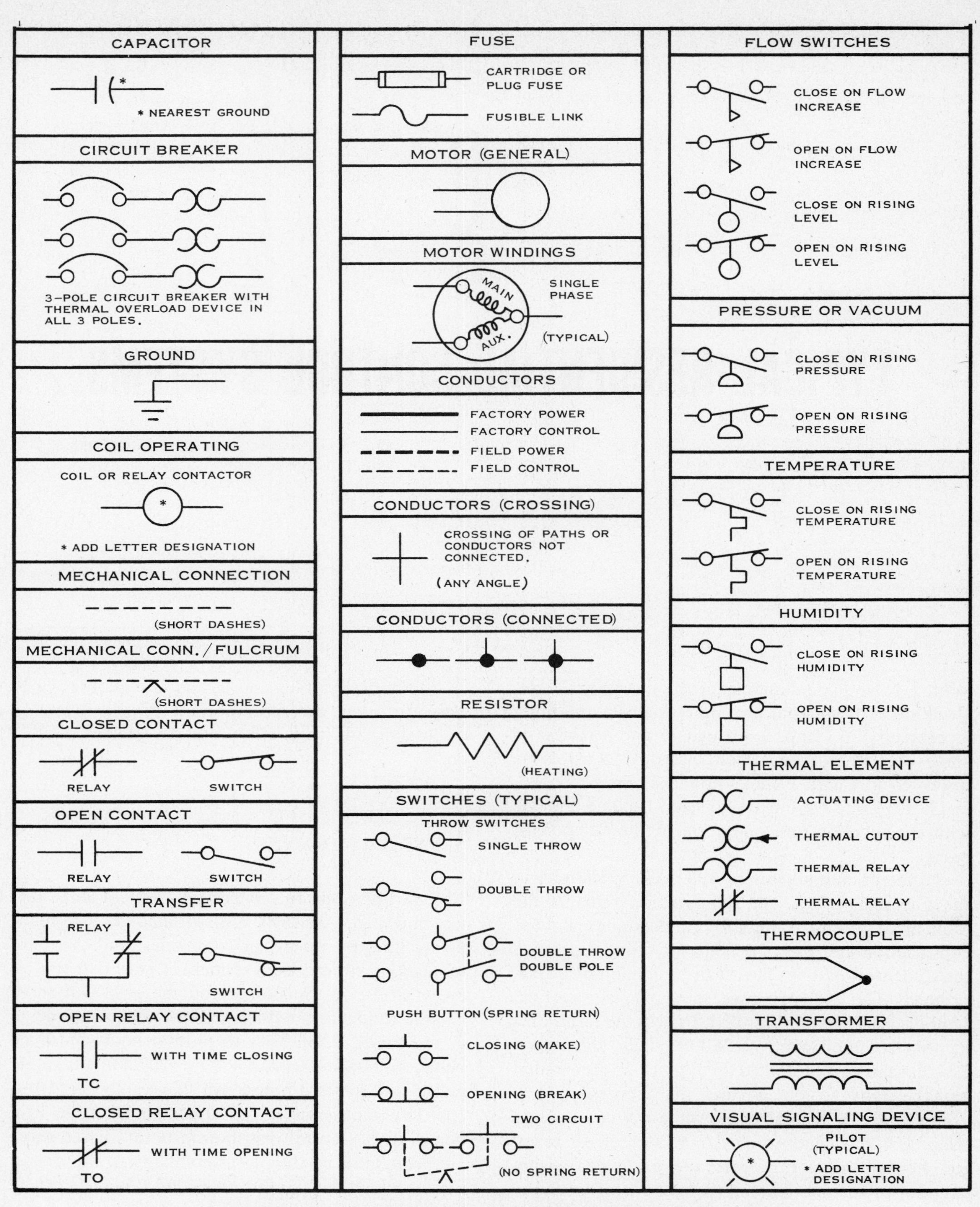

FIGURE 18-1 Electrical symbols (Courtesy, Westinghouse Electric Corporation).

grounded to the board. The serviceman has no control over the entering house voltage and, therefore, if this is not correct, his only recourse is to call the power company. *But*—always check the actual voltage value.

At the house service entrance, power is supplied to various branch circuits to be distributed throughout the house. These branch circuits will be 120-V, single-phase, two-wire circuits for use in lighting, receptacles, appliances, and a separate circuit for the furnace. The furnace should be the only appliance on that circuit, and its normal fusing is 15 amps.

Appliances drawing higher currents such as electric ranges, clothes dryers, and air conditioners will have 240-V service, which is obtained by tapping off each hot side of the entrance circuit. The fuse or circuit-breaker size for these circuits will be in accordance with the current draw of the given appliance; in the case of an air conditioner this might be 30–40 amps per leg. The circuit breakers for each leg on a 240-V circuit are mechanically interlocked, so that if high current is experienced in one leg, the breaker will trip both legs of the supply service.

WIRE SIZE

Wire sizes and fusing for independent appliances will be determined by the manufacturers and specified on their wiring diagrams. Local codes should be examined to determine whether or not conduit is

LEGEND DESIGNATIONS

RELAYS

R - Relay General
CR - Cooling Relay
DR - Defrost Relay
FR - Fan Relay
IFR - Indoor Fan Relay
OFR - Outdoor Fan Relay
GR - Guardistor Relay
HR - Heating Relay
LR - Locking Relay (Lock-in or Lockout)
PR - Protection Relay (Relay in series with protective devices)
VR - Voltage Relay
TD - Time Delay Device
THR - Thermal Relay (type)
M - Contactor
MA - Auxiliary Contact

SOLENOIDS

S - Solenoid, General
CS - Capacity Solenoid
GS - Gas Solenoid
RS - Reversing Solenoid

SWITCHES

DI - Defrost Initiation
DT - Defrost Termination
DIT - Defrost Initiation-Defrost Termination (dual function device)

GP - Gas Pressure
HP - High Pressure
LP - Low Pressure
HLP - Combination High-Low Pressure
OP - Oil Pressure

RM - Reset, Manual

FS - Fan Switch
SS - System Switch
HS - Humidity Switch (Humidistat)

TA - Thermostat, Ambient
TC - Thermostat, Cooling
TH - Thermostat, Heating
TMA - Thermostat, Mixed Air

CT - Thermostat, Compressor Motor
HT - High Temperature
LT - Low Temperature
RT - Refrigerant Temperature
WT - Water Temperature

MISCELLANEOUS

C-HTR - Crankcase Heater
RES - Resistor
HTR - Heater
PC - Program Control
OL - Overload
L - Indicating Lamp
⊕ - Manual Reset Device
+ - Automatic Reset Device

FIGURE 18-2 Legend designations (Courtesy, Westinghouse Electric Corporation).

required in bringing the wire from the main circuit breaker to the furnace or air conditioner. Some codes require conduit for parts of the circuit even on 120 V. Specification sheets will give the minimum wire and fuse sizes that will meet the requirements of the National Electric Code, and this wire size is based on 125% of the full-load current rating. The length of the wire run will also determine the wire size, and Fig. 18-3 shows the maximum length of two wire runs for various wire sizes. This table is based on holding the voltage drop to 3% at 240 V. For other voltages, the multipliers at the bottom should be used. If the required run is nearly equal to or somewhat more than the maximum run shown by the calculation of the table, then the next larger wire size should be used as a safety factor. It is always possible to use larger wire than called for on the specifications, but a smaller wire size should never be used, as this will cause nuisance trips, affect the efficiency of operation, and create a safety hazard. All replacement wire should be the same size and type as the original wire and should be rated at 90–105°C.

HEATING CIRCUIT

Let's take a quick glance at the simple schematic diagram we will soon be building (Fig. 18-4). It is for a typical gas-fired upflow air furnace. The wire conductors are represented by lines and the other components by means of symbols and letter designations. Notice the very important LEGEND, which identifies the various types of wiring, the symbols, and the letter designations. Although other types of diagrams will be discussed in later discussions, we are interested now in the schematic, since this is the diagram used by service personnel to tell how the system works and why.

There are only five major electrical devices in our diagram: the fan motor, which circulates the heated air; the 24-volt automatic gas solenoid valve to control the flow of gas to the burner; the combination blower and limit control, which controls fan opera-

MAXIMUM LENGTH OF TWO-WIRE RUN*

WIRE SIZE	AMPERES												
	5	10	15	20	25	30	35	40	45	50	55	70	80
14	274	137	91										
12		218	145	109									
10			230	173	138	115							
8					220	182	156	138					
6								219	193	175	159		
4									309	278	253	199	
3										350	319	250	219
2											402	316	276
1												399	349
0												502	439
00													560

*To limit voltage drop to 3% at 240V. For other voltages use the following multipliers:

110V	0.458
115V	0.479
120V	0.50
125V	0.521
220V	0.917
230V	0.966
250V	1.042

EXAMPLE: Find maximum run for #10 wire carrying 30 amp at 120V...... 115 x 0.5 = 57 ft.

NOTE: If the required length of run is nearly equal to, or even somewhat more than, the maximum run shown for a given wire size - select the next larger wire. This will provide a margin of safety. The recommended limit on voltage drop is 3 per cent. Something less than the maximum is preferable.

FIGURE 18-3.

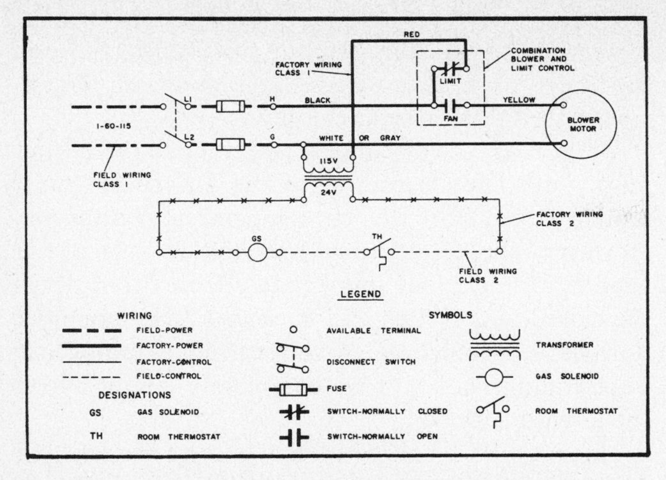

FIGURE 18-4 Heating circuit diagram for gas-fired upflow air furnace (Courtesy, Westinghouse Electric Corporation).

tion and governs the flow of current to the low-voltage control circuit; the step-down transformer, which provides 24-volt current to operate the automatic gas valve; and the room thermostat, which directly controls the opening and closing of the automatic gas valve.

Now, we'll build this same schematic diagram a step at a time, beginning with the power supply and adding each component as we go along. The wires for the power supply are usually run by the installing technician. They must carry line voltage (single phase, 60 hertz, 115 volts) to operate the blower motor, so we will represent them as heavy broken lines to indicate field wiring (Fig. 18-5).

As a protection to our circuit, the power supply must be run through a fused disconnect switch. The disconnect switch is the main switch to the entire system. During normal operation, it must be closed to energize the circuit, but it may be opened or closed manually as necessary for troubleshooting and servicing electrical components. The fuses must be properly sized; large enough to carry the normal flow

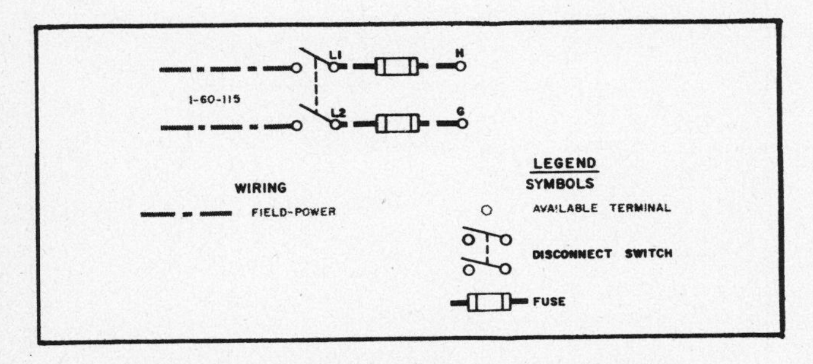

FIGURE 18-5 (Courtesy, Westinghouse Electric Corporation).

of current, but small enough to "blow" in case of an overload that might damage the circuit. If the fused disconnect switch is located outside of the building, it must be installed in a weatherproof enclosure.

In a gas-fired heating system, the fan motor is the heaviest electrical load, so it will be added to the diagram next and connected to the power supply (Fig. 18-6). Since the fan motor is clearly identified on the diagram, we will not add it to the legend. Internal windings are not shown.

Now, close the disconnect switch and see what happens. The fan motor runs, since we have completed a simple circuit from *L1* through the windings of the blower motor and back to neutral. But the automatic gas valve is not connected, and we have no means of controlling its operation automatically, so let's keep the disconnect switch open and continue with our schematic diagram.

Since we must have automatic control of both the fan motor and the automatic gas solenoid valve, we will begin to connect the combination fan and limit control (Fig. 18-7). This control, as previously discussed, consists of a sensing element, which feels the temperature of the heated air inside the furnace,

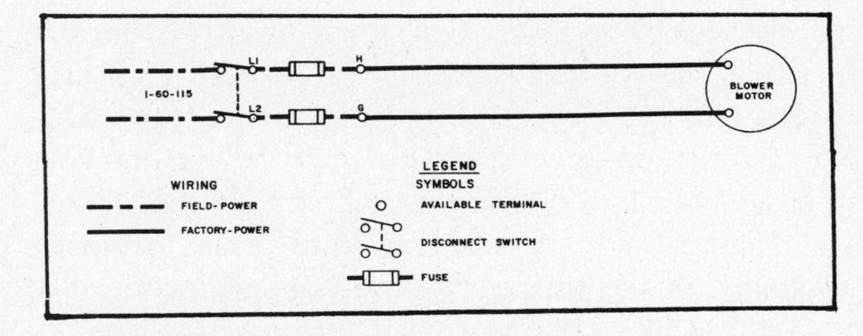

FIGURE 18-6 (Courtesy, Westinghouse Electric Corporation).

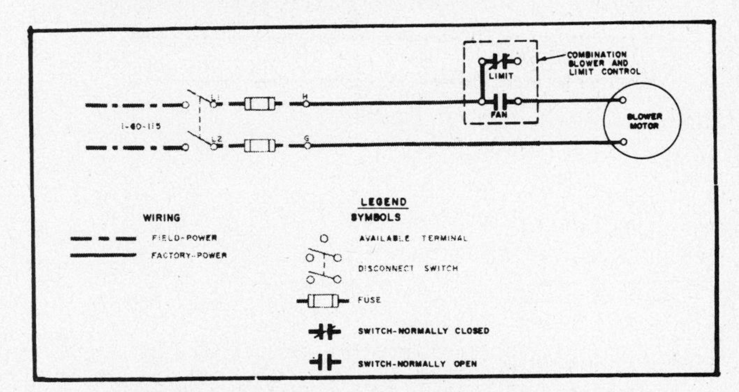

FIGURE 18-7 (Courtesy, Westinghouse Electric Corporation).

and two adjustable switches—a fan switch and a limit switch.

The fan switch controls the starting and stopping of the blower motor. When the burner starts, the fan switch will not close to start the blower until the air temperature inside the furnace plenum chamber has warmed up to the FAN ON setting. When the burner stops, the fan switch remains closed, keeping the blower in operation until the air temperature inside the furnace cools down to the FAN OFF setting.

The limit switch is a safety control; it prevents the furnace from overheating. As long as the plenum chamber temperature is below the setting of the limit switch, the switch remains closed. If the plenum chamber temperature rises to the switch setting, it opens to deenergize the entire 24-volt circuit and close the automatic gas valve.

As illustrated, the combination fan and limit control is partially connected. The fan switch is wired into *L1* in series with the blower motor windings. Even if we close the disconnect switch now, our fan motor will not run because the fan switch in the fan motor circuit is open, and it will remain open until we can warm up the air temperature in the furnace to the switch setting.

We learned earlier that small electrical devices, such as relays and solenoid valves, do very little work and therefore require little current for operation, so a step-down transformer to reduce line voltage to 24 volts for the control circuit is needed.

Figure 18-8 shows the transformer properly installed in the wiring diagram—one side connected to neutral and the other side connected to *L1* through the limit switch. This places the limit switch in control of the entire 24-volt circuit.

If we close the disconnect switch now, the blower motor still cannot run, since the fan switch must remain open until the air temperature warms up. However, the transformer and the 24-volt power source are energized immediately from *L1* through the closed limit switch and back to *L2*. Although we now have a source of 24-volt current, nothing can be accomplished until we complete a circuit across the 24-volt supply.

Now, with a source of low-voltage current, we are ready to connect the automatic gas solenoid valve and the room thermostat. Essentially, the automatic gas valve is a closed solenoid valve with an adjustable bypass for the pilot flame. During normal operation, the pilot burns continuously to ignite the gas admitted to the burner when the solenoid is energized to open the main gas valve. A warm-air heating system must be properly controlled to maintain a comfortable temperature in the heated space. For this purpose, an adjustable low-voltage room thermostat, which opens on an increase in temperature, is used as the primary operating control. It senses the room temperature and energizes or deenergizes the automatic gas solenoid valve as necessary to maintain the desired temperature.

Now that we have reviewed the functions of the automatic gas solenoid valve and the heating thermostat, let's wire them into our circuit (Fig. 18-9). Since the limit switch is a safety control and not an operating control, the thermostat must control the opening and closing of the gas solenoid, so we will wire them in series across the 24-volt circuit. The gas valve is factory-wired, so we will use a light, unbroken line to indicate factory control wiring to identify

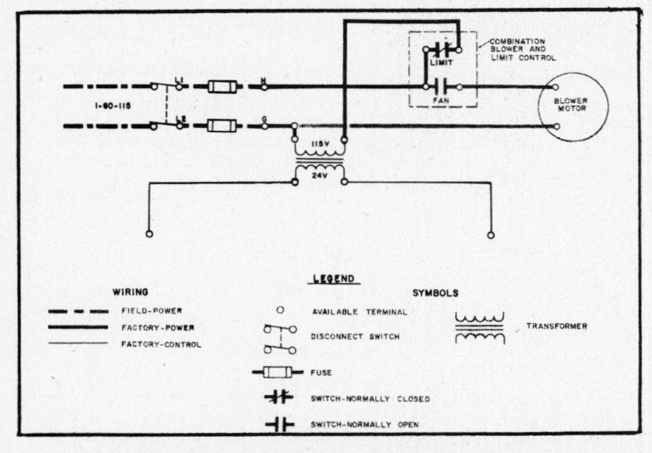

FIGURE 18-8 (Courtesy, Westinghouse Electric Corporation).

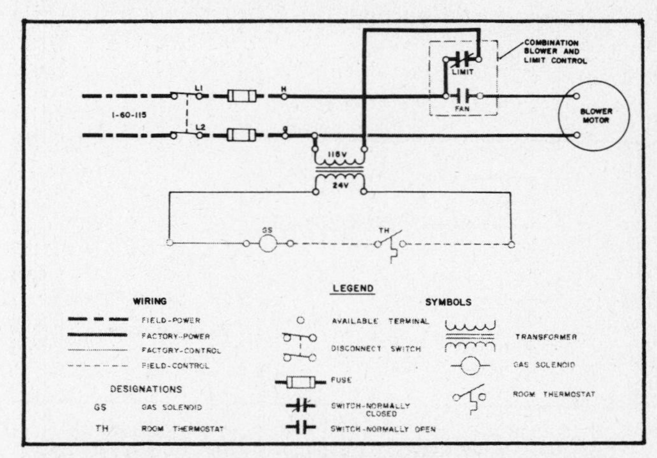

FIGURE 18-9 (Courtesy, Westinghouse Electric Corporation).

terminals. Normally, the thermostat is located in a room away from the furnace and must be field-wired, as indicated by the light broken lines used to represent control circuit field wiring.

The circuit is now completely wired, so let's close the disconnect switch and see if our system performs as it should. The limit switch should be closed; the thermostat is open. The transformer and low-voltage circuit are immediately energized from *L1* through the closed limit switch back to neutral. The thermostat is open, preventing current from flowing to energize the automatic gas valve, which remains closed. The fan motor cannot operate because the fan switch is open, and it will remain open until the air temperature in the furnace warms up to the FAN ON setting.

When the room becomes cool enough to close the thermostat contacts, the automatic gas valve is energized (Fig. 18-10). This opens the valve to admit gas to the burner, and we now begin to heat the air in the plenum chamber of the furnace. The fan switch would still be open, so the fan motor cannot run, but as soon as the air temperature in the furnace plenum chamber warms up to the FAN ON setting, the fan switch closes to start the fan motor and circulate warm air to the heated space. The blower will continue to operate as long as the heated air temperature remains above the FAN OFF setting.

When the room temperature warms up to the setting of the thermostat, the thermostat contacts open the 24-volt circuit to the automatic gas valve. This closes the valve and extinguishes the burner. However, the fan switch is still closed and the blower is still circulating warm air. With the burner extinguished and the blower running, the air in the furnace plenum chamber becomes cooler and cooler. When the air temperature falls to the FAN OFF setting, the fan switch opens, the blower motor is deenergized and will remain stopped until the air temperature in the furnace plenum is again warmed up to the FAN ON setting.

So far, we have paid very little attention to the all-important limit switch, but it is the overriding system safety control. If, at any time during furnace operation, the plenum chamber tends to overheat, the air temperature soon reaches the setting of the limit switch. This opens the limit switch contacts and immediately deenergizes the entire 24-volt circuit. When this happens, the automatic gas valve closes and the burner is extinguished, even though the room thermostat may still be closed and calling for more heat. Although the limit switch can override the room thermostat, it has no effect on fan operation. The fan switch remains closed and the blower continues to run as long as the heated air temperature remains above the FAN OFF setting.

The operating sequence just completed proves that the wiring diagram is correct. The disconnect switch controls the entire electrical system; the blower motor operates independently of the burner, under the control of the fan switch, and the automatic gas valve is cycled on and off by the room thermostat, under the control of the limit switch. However, we have only been concerned with lines and symbols so far. Actually the finished diagram (Fig. 18-11) contains the complete story on the electrical system just reviewed; the LEGEND clearly identifies all symbol and letter designations used in the diagram, and the NOTES call attention

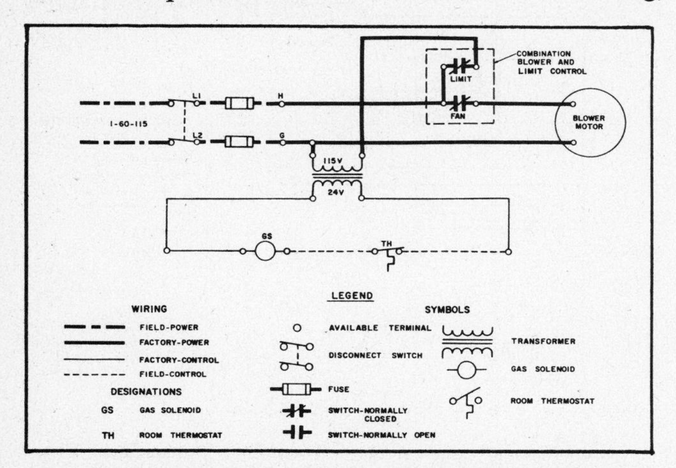

FIGURE 18-10 (Courtesy, Westinghouse Electric Corporation).

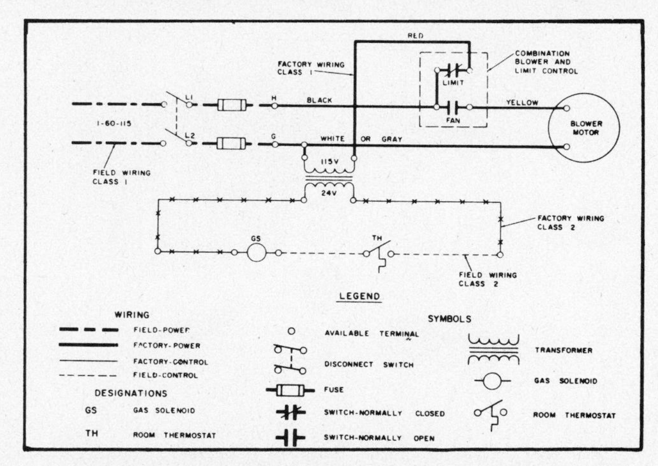

FIGURE 18-11 (Courtesy, Westinghouse Electric Corporation).

to important facts relating to the circuit. When you have a thorough knowledge of the simple principles just outlined, the functions of the controls and their relationship to each other in the electrical circuit, you have taken a big step toward becoming a professional in installation, trouble-shooting, and servicing the electrical circuit of a gas-fired heating system—or other heating apparatus.

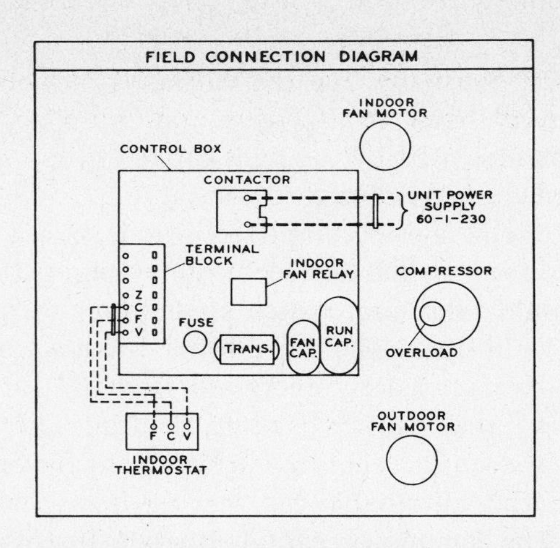

FIGURE 18-13 Field connection diagram (Courtesy, Westinghouse Electric Corporation).

COOLING CIRCUIT

At one time or another, any one just starting a career in air conditioning probably has regarded the formidable cooling system wiring diagram with a feeling of awe and suspicion. We looked at the maze of lines and the dozens of symbols (Fig. 18-12) and wondered what it all meant. Actually, it really isn't that bad. When you understand the meaning of a few symbols and their relationship to each other in the electrical circuit, you can tell exactly what is happening in an electrical system: how it happens, when it happens, and why. The ability to read and understand the wiring diagram often makes the difference between an amateur and a well-trained, professional service technician.

Wiring diagrams are made for different purposes, and it is important to recognize each type and know the intended purpose of each. There are *field connection diagrams* and *schematic diagrams.*

FIGURE 18-12 Cooling circuit diagram (Courtesy, Westinghouse Electric Corporation).

The *field connection diagram* (Fig. 18-13) identifies the various electrical controls in the control box and indicates the necessary field wiring by means of broken lines and shaded areas. The *field connection diagram* is designed to instruct the installing electrician or technician in how to run the proper power supply to the unit and in the correct wiring between sections if the unit consists of more than one section. Internal wiring is not shown, since we are only interested in installation at this time.

Now let's look at Fig. 18-14, a typical *schematic diagram* for an air-cooled packaged residential air-conditioning unit. One glance tells us that it contains more information than the field connection diagram. This is the diagram used by service personnel that shows how the system works and why. Most manufacturers include a copy of the diagram in the control cover or access panel of the unit. The load devices (motors) and the controls are represented by symbols, the wires by lines. Notice that some lines

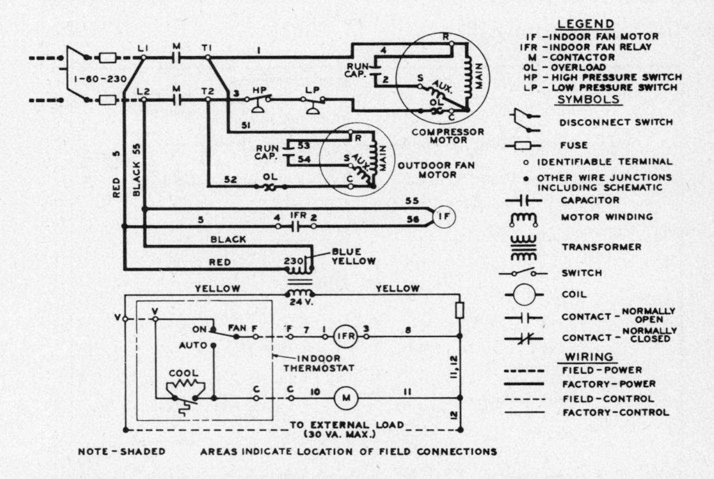

FIGURE 18-14 Schematic diagram (Courtesy, Westinghouse Electric Corporation).

are heavy while others are light by comparison. The heavy lines represent wires in the *power circuit*, which supplies higher line-voltage current to the heavy electrical loads, such as motors. The light lines represent the *control circuit*, which supplies low voltage through a step-down transformer to the light load devices, like relay coils that are used only to actuate switches.

Now, start at the beginning and build this cooling electrical circuit step by step, component by component, just as was done for heating. First, the power supply to the unit (Fig. 18-15). Since these wires must be run by the installing electrician or technician in the field, we represent them by heavy broken lines. To protect the circuit, our supply must pass through a properly sized fused disconnect switch. Proper fuse sizes are found in the equipment installation manual. If the fused disconnect switch is located outdoors, it must be installed in a weatherproof enclosure. The power supply is shown as single-phase, 60-hertz, 230-volt current, and the incoming wires are designated as *L1* and *L2*. Whenever we complete a path between *L1* and *L2* and install a load in the path, we will have a circuit; current will flow and work will be accomplished.

Since the compressor motor is the heaviest load, it will be placed in the diagram next and connected to the power supply. If we closed the disconnect switch, there would be a completed circuit from *L1* through the compressor motor windings back to *L2* and the compressor would run, as a result of manual switch operation. But the system must run automatically, so we will keep the disconnect open and continue with our diagram.

Before the simplest air-cooled air-conditioning system is complete, we need an outdoor fan and an

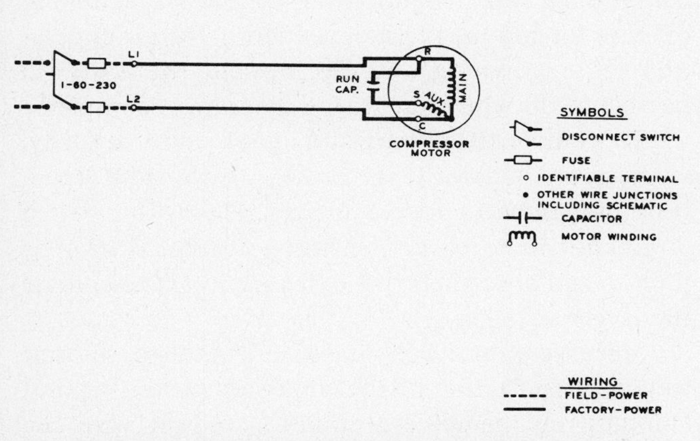

FIGURE 18-15 (Courtesy, Westinghouse Electric Corporation).

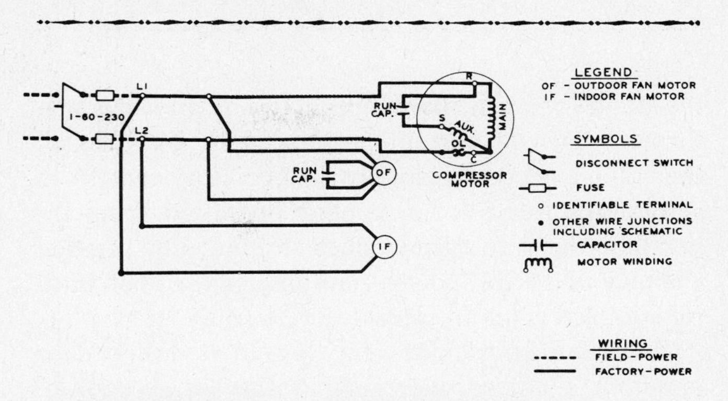

FIGURE 18-16 (Courtesy, Westinghouse Electric Corporation).

indoor fan. So let's put them into our diagram next (Fig. 18-16). It is easy to see that both fans are connected across *L1* and *L2*, and all three motors are in individual circuits of their own. At this point we have a complete wiring diagram for a simple, manually controlled air-conditioning system. If we closed the disconnect switch, the compressor motor would run, the outdoor fan would operate, and the indoor fan would circulate the cooled air to the conditioned space; but it still needs controls to make the system operate automatically—and safely.

Earlier we learned that electrical controls that do very little work require very little current; therefore we again use a low-voltage (24 volt) control circuit. This permits the use of much smaller wire, makes the circuit much safer for residential use and, most important, it provides much closer control of system operation. So, we need a source of low-voltage current for our controls and we get this by means of the small step-down transformer. Figure 18-17 shows

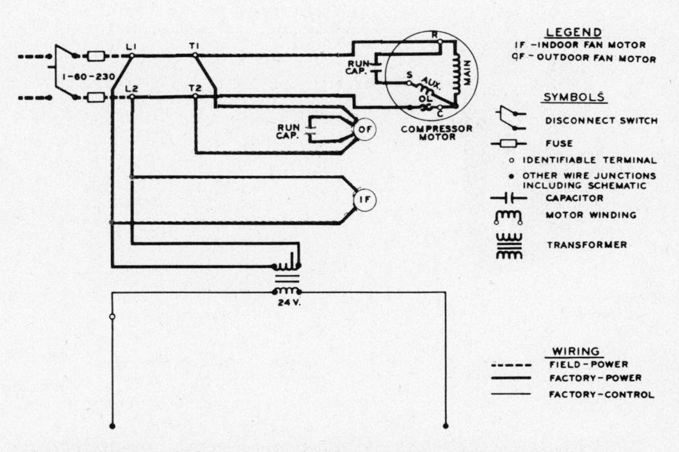

FIGURE 18-17 (Courtesy, Westinghouse Electric Corporation).

the transformer in the wiring diagram. In the transformer, 230-volt current is reduced to 24 volts, all the voltage needed to supply the necessary current to actuate the relays for automatic operation. Remember that the transformer does nothing but provide a source of 24-volt power—nothing can happen until we complete a circuit across the 24-volt source.

Since an air-conditioning system is designed to maintain a comfortable temperature in the conditioned space, a low-voltage adjustable indoor cooling thermostat is used as the primary system control (Fig. 18-18). It senses the air temperature and tells the secondary controls to start or stop the compressor and fans as necessary. Although there are many variations of indoor thermostats, the one illustrated has two slide switches—a fan switch with two positions, ON and AUTO, and a cooling control switch with positions marked OFF and COOL.

Internally, the thermostat consists of two sets of switches. The fan switch controls the indoor fan only. In the ON position, the fan will operate continously; in AUTO, fan operation is actually controlled by the position of the cooling control switch. The contacts of the cooling switch are actually sealed inside a tilting bulb partially filled with mercury. When the bulb is tilted so that mercury covers only one contact, the switch is open. When the bulb is positioned so that mercury covers both contacts, the switch is closed. The bulb is tilted by means of a bimetal element that warps as it feels a change in temperature, to either close or open the mercury

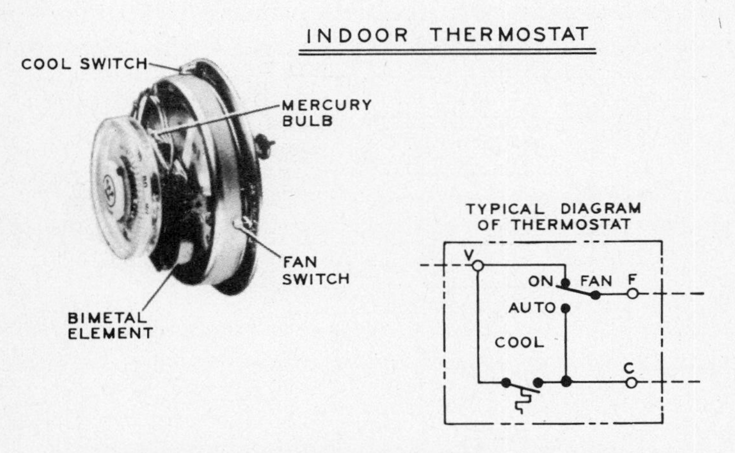

FIGURE 18-18 (Courtesy, Westinghouse Electric Corporation).

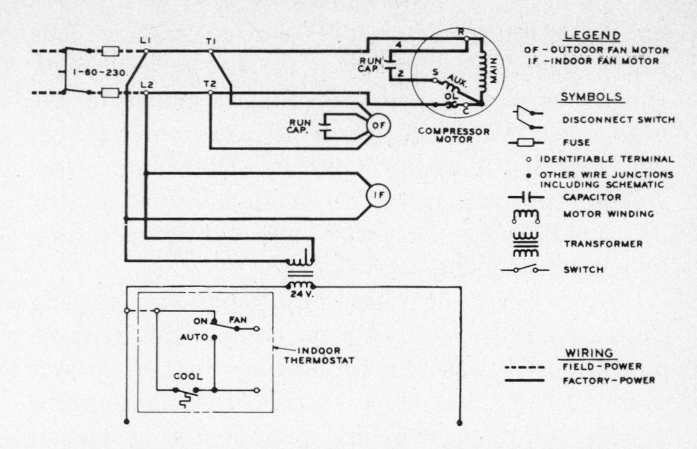

FIGURE 18-19 (Courtesy, Westinghouse Electric Corporation).

switch. The symbol *V* is the common low-voltage terminal feeding through the fan switch to terminal *F* and through the mercury bulb to terminal *C*.

We have just seen how our thermostat works mechanically, so let's put it into our control circuit and see what happens electrically (Fig. 18-19). We have connected our common terminal *V* to the *L1* side of the low-voltage supply, but nothing new can happen electrically until we complete a circuit to the other side of the 24-volt supply.

If we close the main disconnect switch, all three motors will immediately start and continue to operate until the disconnect switch is manually opened. The transformer will be energized and the low-voltage power supply will be available to the thermostat, but no low-voltage circuit is made; we still need a few more controls for automatic operation.

For comfort cooling, an air-cooled residential cooling system must be wired so that both the indoor and outdoor fans operate all the time the compressor is running, the outdoor fan to condense the refrigerant vapor and maintain a safe discharge pressure, and the indoor fan to circulate the cooled air to the conditioned space. In addition, we should be able to operate the indoor fan independently for air circulation without cooling, and this is exactly what we will be able to do when our wiring diagram is complete.

To control the compressor motor automatically, we need a contactor (Fig. 18-20), which, we learned, is nothing more than an oversized relay with a switch large enough to carry the heavy current drawn by motors and a magnetic coil strong enough to actuate the larger switches.

Since we want the cooling switch in the thermostat to govern the action of the contactor in starting and stopping the compressor motor, we will wire the

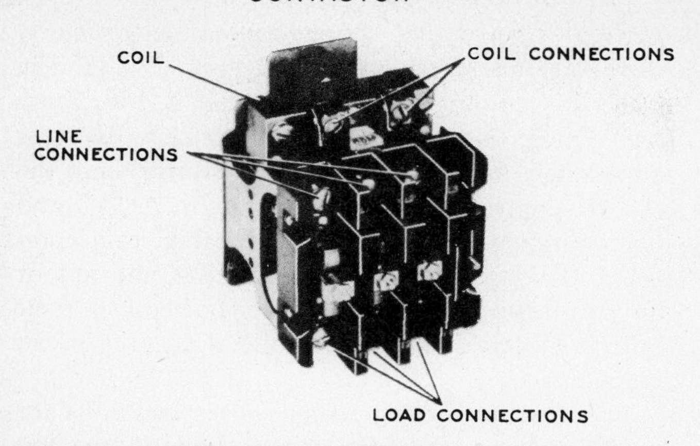

FIGURE 18-20 (Courtesy, Westinghouse Electric Corporation).

contactor coil *M* in series with the cooling switch of the thermostat (Fig. 18-21), but we must connect the contactor switches in the power circuit so that they control both the compressor motor and the outdoor fan motor, but not the indoor fan; it must be able to operate independently. Notice that the contactor coil and both the starter switches are marked *M* for identification. This simply means that coil *M* actuates the two switches marked *M*. Since both switches are normally open, they will close when the coil is energized and open when it is deenergized.

What happens now, when we close our main disconnect switch? Current flows to the indoor fan motor and the 24-volt transformer immediately, but the compressor motor and outdoor fan motor cannot start. The thermostat cooling switch is open, so no current can flow to contactor coil *M* to close its switches marked *M* in the power circuit to the compressor and outdoor fan motors.

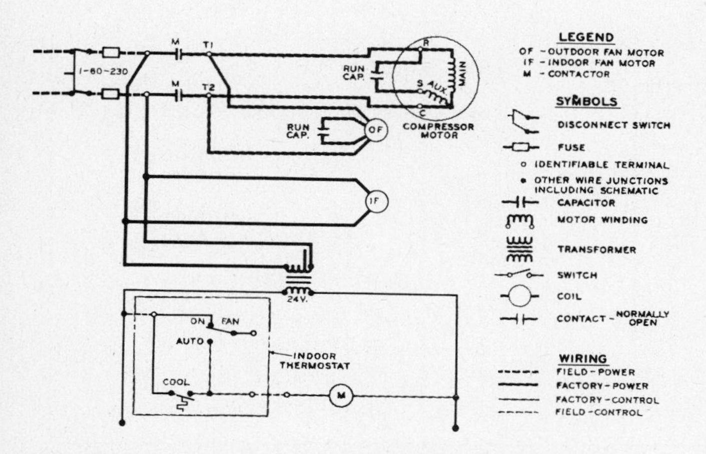

FIGURE 18-21 (Courtesy, Westinghouse Electric Corporation).

What happens when the thermostat cooling switch closes and calls for cooling (Fig. 18-22)? Current immediately flows through starter coil *M*, and the two switches marked *M* in the power circuit close at the same time. This energizes and starts both the compressor motor and the outdoor fan motor at the same time. When the thermostat is satisfied, the cooling switch again moves to the open position. The compressor and outdoor fan motors stop, but our indoor fan motor continues to run, and it will remain in operation as long as the disconnect is closed, regardless of the position of the fan switch.

It seems we still need another control for our indoor fan motor, *IF*. This time we'll use a relay with a coil and a normally open switch, both marked *IFR* for indoor fan relay (Fig. 18-23). Notice that

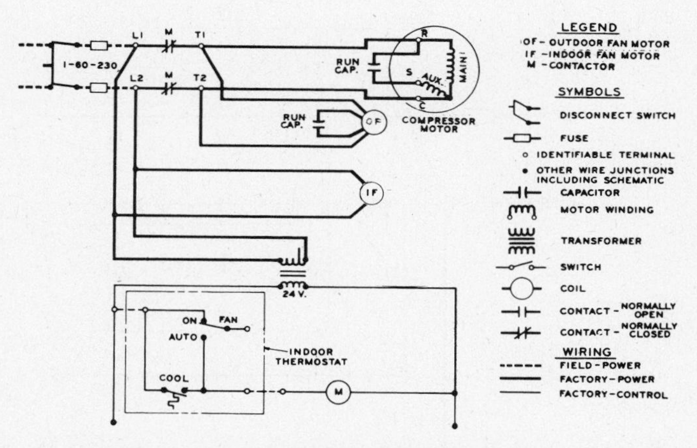

FIGURE 18-22 (Courtesy, Westinghouse Electric Corporation).

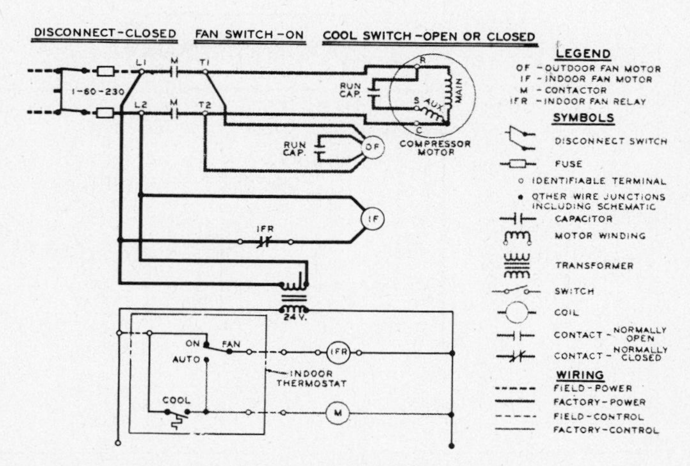

FIGURE 18-23 (Courtesy, Westinghouse Electric Corporation).

the coil is wired in series with the fan switch in the thermostat, and its switch is wired into the indoor fan motor circuit.

With the fan switch in the ON position, low-voltage current now flows through the fan switch to energize relay coil *IFR* and close its switch in the indoor fan power circuit. The indoor fan is now running, and it will continue to run as long as the fan switch remains in the ON position, regardless of the position of the cooling switch. The cooling system—the compressor and the outdoor fan—is free to cycle on and off as the cooling switch dictates with no effect on the indoor fan.

We'll change our wiring diagram once more. Move the fan switch to AUTO (Fig. 18-24). The thermostat cooling switch is open; it is not calling for cooling. Now, close the disconnect. Current immediately flows through the transformer into the 24-volt supply, but the fan switch is in AUTO and the cooling switch is open, so we can't get a completed 24-volt circuit. Contactor coil *M* and indoor fan relay coil *IFR* cannot be energized, so their switches all remain open, and none of the motors can operate.

Assume the cooling switch closes and calls for cooling without changing the position of the fan switch or the disconnect—disconnect closed; fan on AUTO (Fig. 18-25). Current immediately flows through the cooling switch and through both the contactor coil *M* and the indoor fan relay coil *IFR* at the same time. This closes both switches marked

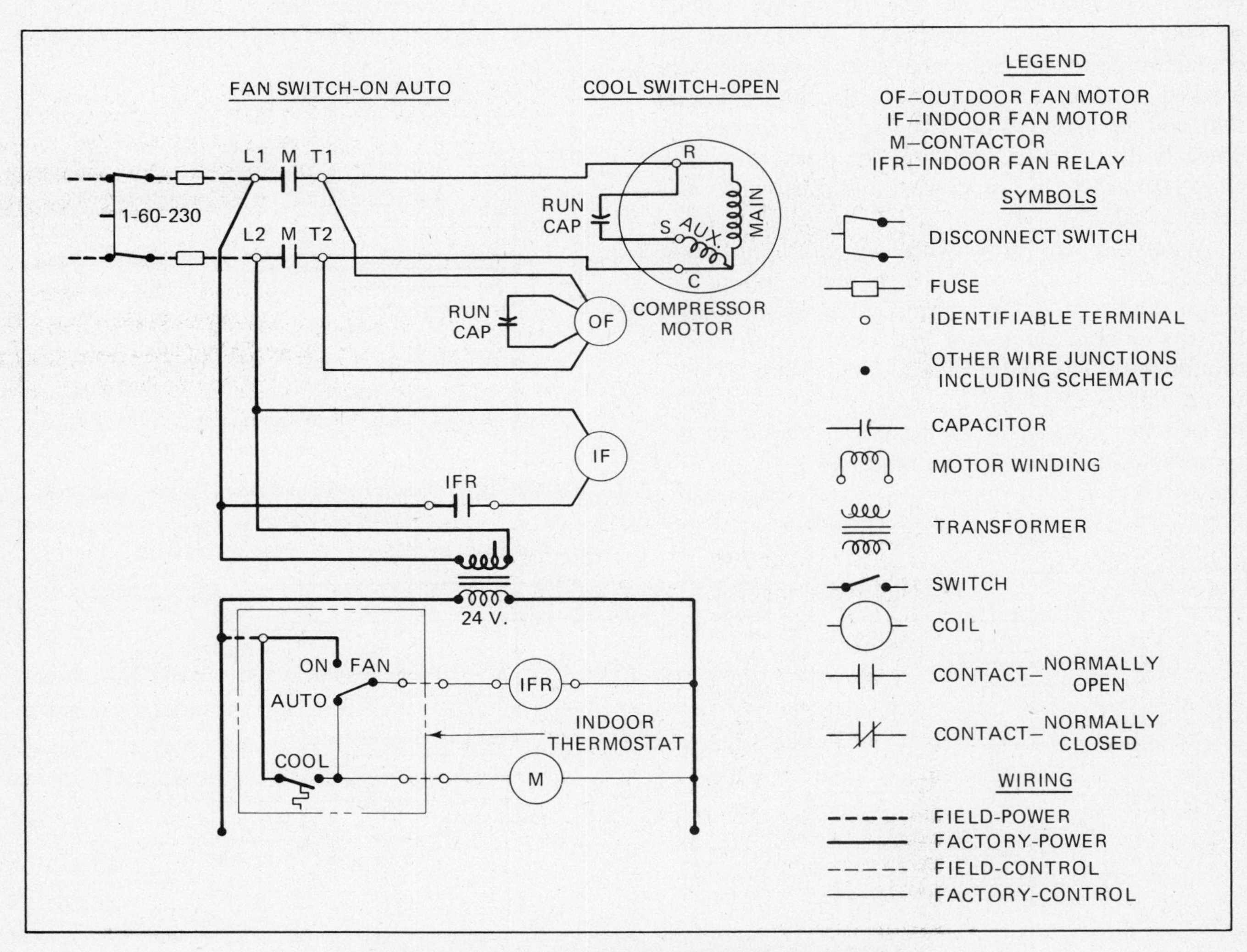

FIGURE 18-24 (Courtesy, Westinghouse Electric Corporation).

M in the power circuit to the compressor and outdoor fan motors, as well as the switch marked *IFR* in the power circuit to the indoor fan motor.

The compressor and outdoor fan motors are energized and started through closed switches *M*; at the same time, the indoor fan motor is also started through closed switch *IFR*. The entire system is now energized and all motors are running automatically.

When the thermostat is satisfied, the cooling switch opens automatically, and we're back where we started when we closed the disconnect switch; the cooling switch opens the circuit to both the contactor and indoor fan relay coils, their respective switches immediately snap to their normally open positions, and all motors stop—compressor, outdoor fan, and indoor fan.

Now that we know our wiring diagram is correct from the standpoint of operation, we will add safety controls to protect our system and improve its operating performance (Fig. 18-26).

Overloads *OL* have been added to protect the compressor and outdoor fan motor windings against excessive current flow. Overloads may be installed inside the motor housing as shown for the compressor motor or externally as shown for the outdoor fan motor. Although the windings are not shown for the indoor fan motor, all small single-phase motors are equipped with some type of protection.

As an added precaution against compressor operation under abnormal conditions such as excessive discharge pressure or dangerously low suction

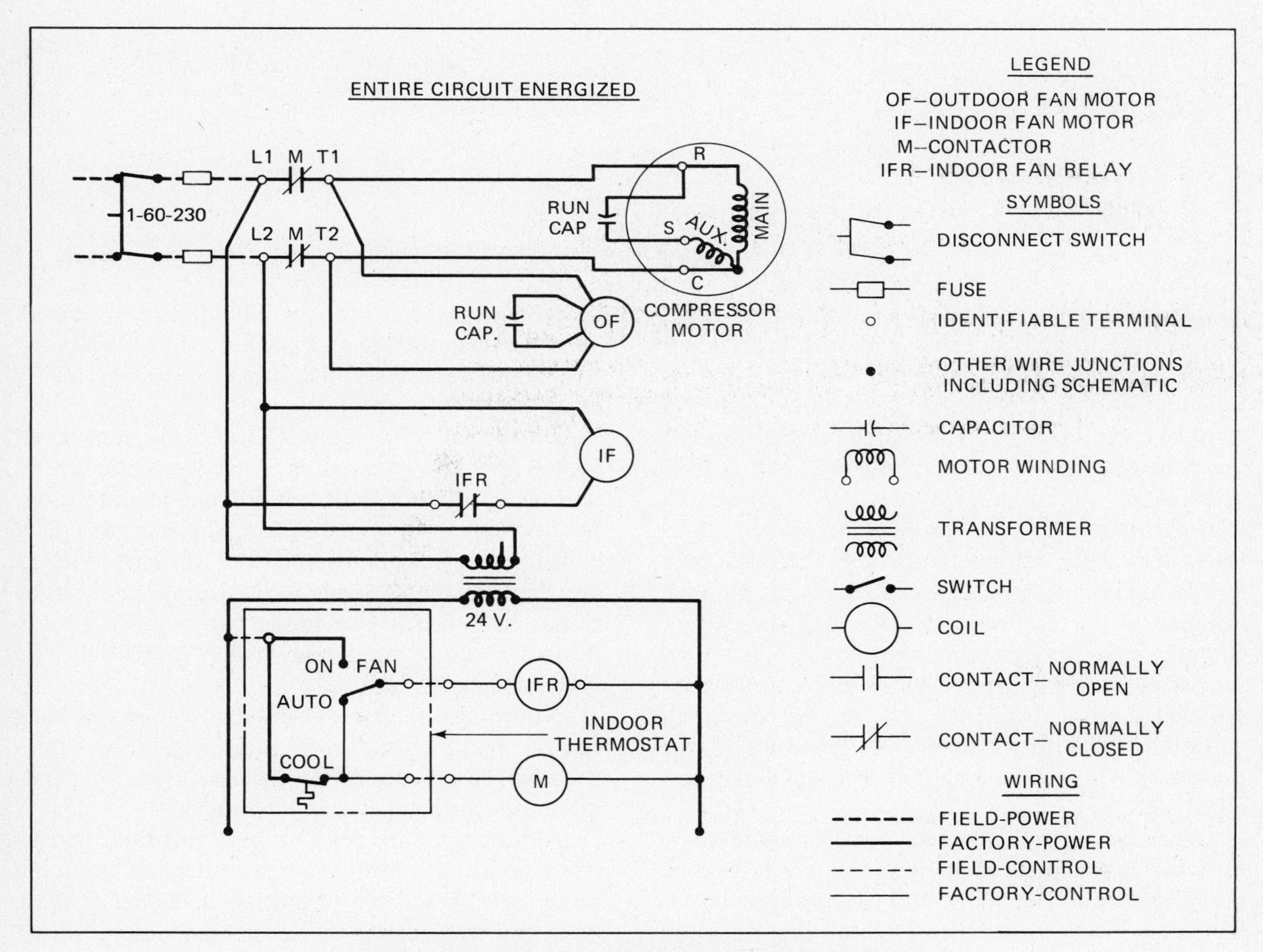

FIGURE 18-25 (Courtesy, Westinghouse Electric Corporation).

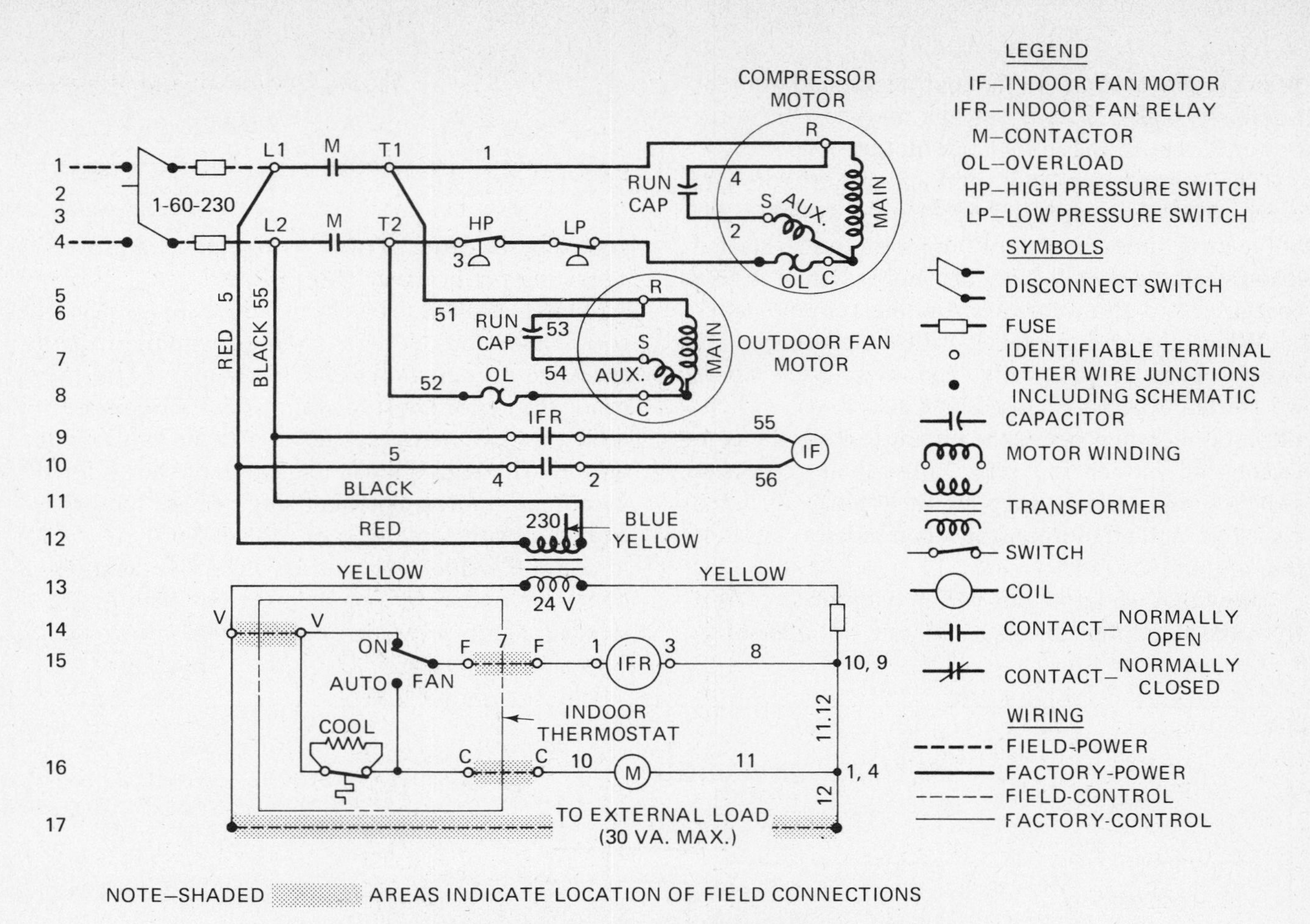

FIGURE 18-26 (Courtesy, Westinghouse Electric Corporation).

pressure, we have wired a high-pressure switch and a low-pressure switch in series with the compressor motor windings. The high-pressure switch, *HP*, is actuated by system discharge pressure to open and stop the compressor if the discharge pressure exceeds the switch setting. The low-pressure switch, *LP*, is actuated by system suction pressure to open and stop the compressor if the suction pressure falls below the switch setting. To protect the relay coils and wiring in the low-voltage control circuit, we have installed a fuse properly sized to open the 24-volt circuit in case of excessive current flow.

For closer control of air temperature in the conditioned space as explained in Chap. 17, an anticipator has been wired across the terminals of the thermostat cooling switch. It is a fixed, nonadjustable type.

When we discussed the relay, we learned that one relay coil may actuate as many switches as it is possible to build into one relay, and each switch may be located in a different section of the wiring diagram. As the diagram becomes larger and more relays are used, these switches become more difficult to locate. To solve this problem and make it easy to locate the switches of each relay, the *ladder diagram* system is often used.

On the ladder diagram (Fig. 18-26), each horizontal line is numbered in sequence from top to bottom on the left-hand side for line identification. On the right side of the diagram, opposite each relay coil, are shown the line numbers in which its switches are located. For example, our diagram shows that contactor coil *M* in line 16 has two switches located in lines 1 and 4, while relay coil *IFR* has only one switch located in line 10.

Notice also the factory-installed wiring has been color-coded for ease of identification. (Sometimes there will be a numerical designation instead, depending on the equipment manufacturer.)

Although the diagram we have just built is relatively simple, all wiring diagrams are based upon the same logic. When the technician understands these basic principles, even the most complicated wiring diagram is simplified and much easier to read.

It is impossible to review all wiring diagrams that cover the broad range of residential heating and cooling products, along with humidity control and electronic air cleaning that provide a total comfort system. And each manufacturer has his own techniques for combining the various functions and methods of presentation. Most publish diagrams or at least make application recommendations to assist the electrician or the installer in hooking up the systems.

For simplicity of demonstration, this text will review a total comfort system (TCS) using an upflow gas furnace with single speed fan operation, a split-system air-cooled condensing unit (with indoor coil), a combination heat/cool thermostat, a humidistat, and an electronic air cleaner.

First, the physical wiring and point-to-point connections are illustrated in (Fig. 18-27). This is what the electrical installer will use as a guide during the installation. Note the need for two junction boxes to make the required interlocks. The low-voltage wiring is also detailed as to the number of conductors needed and the terminal or junction connections. Power wiring is black (*B*) and white (*W*), and ground conductors are sized according to the load and to meet local codes requirements. The internal wiring of the furnace, condensing unit, etc., is not shown on the installation diagram. Study the layout for the components and the connections involved.

Figure 18-28 is the schematic ladder diagram of this same system. First locate and relate the major components to the previous drawing; the furnace blower motor, the humidifier, the electronic air cleaner, the condensing unit, etc. Note that the combination thermostat has a HEAT-OFF-COOL system selection and an ON-AUTO fan selection. Anticipators are not shown. The system diagram is shown on the heating mode.

Closing the disconnect switch (line 4) will produce a 115-volt line-voltage current flow to the transformer. Trace the black wire through the closed

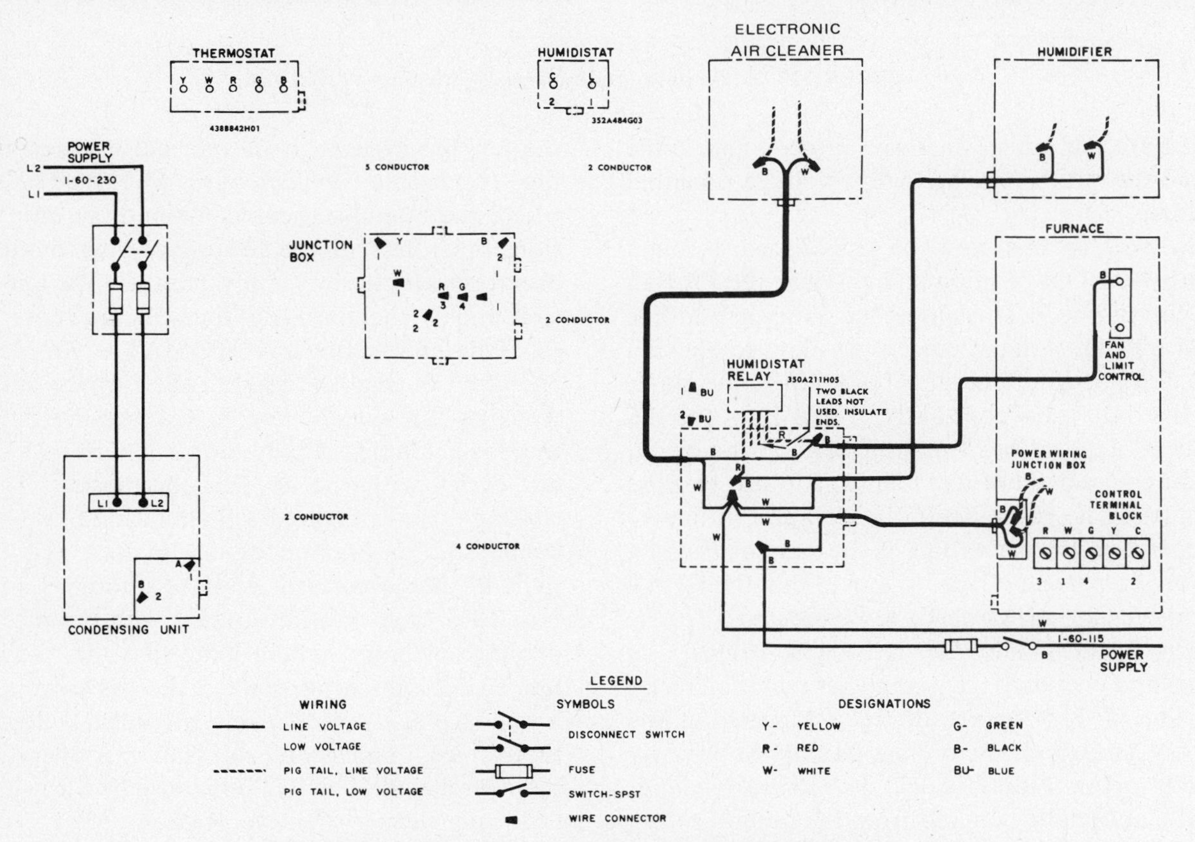

FIGURE 18-27 (Courtesy, Westinghouse Electric Corporation).

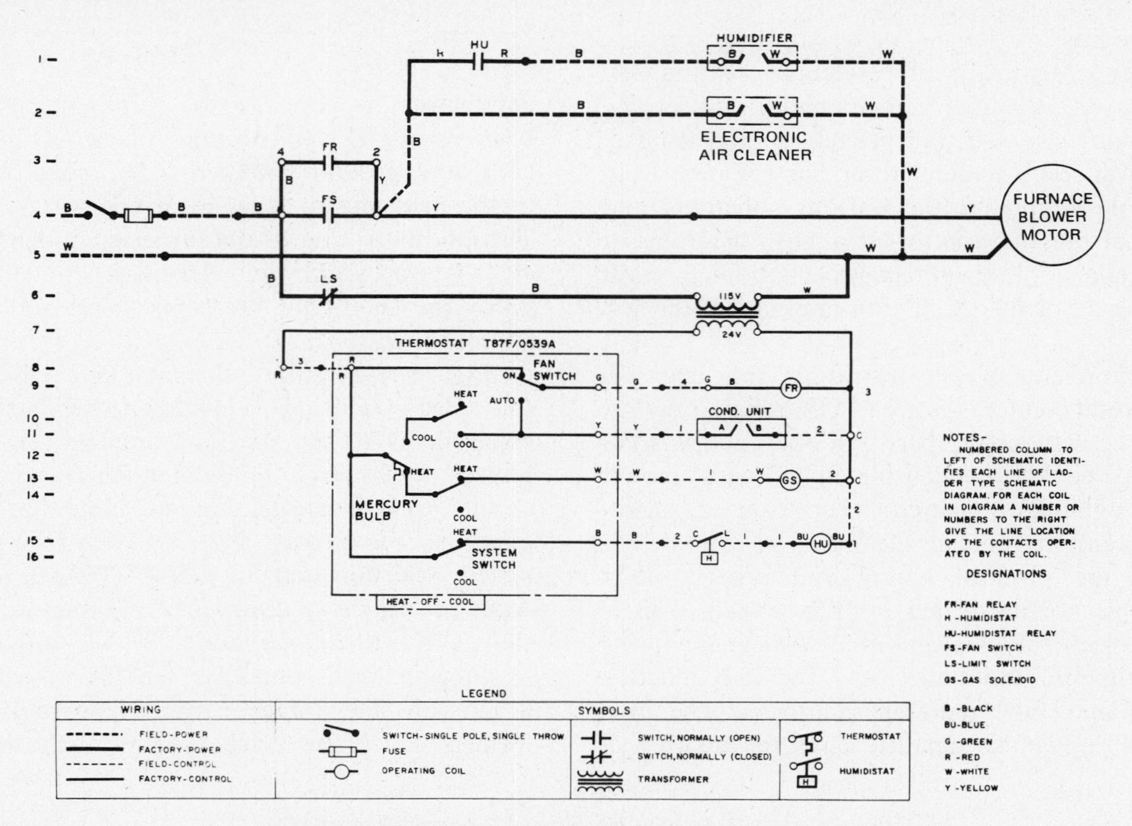

FIGURE 18-28 (Courtesy, Westinghouse Electric Corporation).

limit (*LS*) to the transformer primary coil and to the white power return (line 5). A low-voltage potential is created.

Low voltage can now be established through the fan switch ON position and energize the *FR* (fan relay) coil in line 9. This closes the *FR* switch in line 3, allowing line-voltage current to flow to the fan motor and to the humidifier relay (*HU*) in line 1. Thus the fan runs constantly. Note that the *FS* (furnace switch) in line 4 may or may not be closed depending on whether there is sufficient heat to close its contacts. The gas valve (*GS*) operation in line 13 will depend on the activation of the mercury bulb by the room temperature. Note in line 15 that the HEAT position selector switch feeds low voltage to *H* (the humidistat), and if humidity is needed, it closes its contacts and energizes the humidistat relay coil (*HU*) also in line 15. The contacts of *HU* are located in line 1 and will activate the humidifier (fan motor or water solenoid valve). This is called *permissive humidification* by automatic demand from the humidistat on the heating cycle with the normal cycle of furnace fan. (*Command humidification* would be a system where the humidistat can command fan-only operation, regardless of the heating/cooling mode.) The electronic air cleaner simply parallels the fan operation and is energized any time the fan runs.

If the fan selection is AUTOMATIC, *FR* cannot be energized during heating, but the furnace fan cycles from its fan switch, line 4. On cooling, however, there is a contact to the automatic terminal and the fan cycles with the cooling thermostat. Also, on cooling there is a completed circuit on line 11 to the condensing unit contactor terminals *A* and *B*. This pulls in the compressor and the outdoor condenser fan. They then cycle off and on with the room temperature mercury switch. Note that the middle system switch on cool drops out the gas valve and the lower switch drops out the humidistat. Note also that on cooling, the furnace fan can operate from the *FR* relay (ON or AUTO) and thus furnish power to the air cleaner whenever it runs.

This is how a simple but common TCS system is installed and wired schematically. The introduction of furnace fan speed control, two-stage cooling and heating thermostats, command humidification, etc., add sophistication and enlarge the diagram, but these items do not necessarily add to its complexity if the schematic is analyzed step by step as to function and operation.

HEAT PUMP CIRCUIT

Figure 18-29 is a schematic diagram for a 4-ton packaged air-cooled heat pump using a single-stage cooling and a two-stage heating thermostat. Again for simplicity we have omitted the wiring detail to supplemental electric heaters, but we will show how they are energized. First observe that the thermostat has a system OFF and AUTO position. On AUTO it switches from heat to cool (automatically) by virtue of the position of the mercury contacts responding to room temperature. There is a dead spot between the changeover so as to prevent too-rapid cycling. Indoor fan control may be ON (continuous) or AUTO (cycling with the compressor).

In the cooling mode this circuit functions much as the one in our previous discussion. The cooling thermostat contact in line 33 furnishes low-voltage power to the indoor fan relay (*IFR*), if it is set in AUTO, and also energizes the low-voltage cooling relay, *CR*. The contacts of *CR* are in line 22, which closes and energizes *M*, the compressor contactor coil. Power contacts for *M* are located in lines 11, 14, and 17.

Terminals *T1* and *T3* furnish power to the compressor, while *T2* and *T3* furnish power to the outdoor fan. Note that terminal *C* on the thermostat also feeds power to the outdoor fan relay, *OFR*, so it can cycle with the defrost relay *DR* (the defrost relay switch is in line 37 and is normally closed during cooling).

Note the use of a motor winding thermostat, *MTW*, in line 22 and also a high pressure cutout (*HIP*) in series with the cooling relay. Either can open the circuit in the event of trouble.

In the heating mode, low voltage from terminal *V* feeds both heating stages. Stage 1 closes contacts to terminal *Z* and *C* and power to the indoor fan (if it is set in automatic). *C* terminal energizes the *CR* relay and will start the compressor, but simultaneously the reversing valve *RVS*, in line 37, is also energized, changing the reversing valve as discussed in Chap. 11 to reverse the flow of refrigerant so that the compressor is now heating the indoor coil.

If more heat is needed than the compressor can furnish, the second-stage heating contacts on line 43 close, providing a low-voltage circuit to terminal *H2* and the separate electric heaters. Although they are not shown, it's normal to bring on at least one heater with the closing of the second stage of heat, and the balance of heaters are staged ON and OFF by outdoor thermostats. The low-voltage circuit that does this is between *H2* and 1, and is represented by the dotted line. Note that the maximum current draw would be 36 volt-amps to operate the heater relay coil.

What happens when the outdoor coil frosts up? Defrost relay *DR* in line 39 is activated by one of the defrost methods as explained in Chap. 11. As a result, *DR* contacts in line 37 open, dropping out the outdoor fan relay, *OFR*, and deenergizing the reversing valve *RVS*, which then mechanically goes back into the cool position, thus sending condensing heat to the outdoor coil. In order to furnish heat indoors, another set of *DR* contacts close in line 36, providing power to terminal *H2* and the electric resistance heaters. Usually this brings on enough heat to overcome the effects of the compressor during the defrost cycle. Once defrost is terminated, defrost relay *DR* is deactivated, which, through normally closed contacts, switches the *RVS* (*reversing valve*) back to the heating mode again.

From this basic description, one can trace and understand the functions involved in most heat pump circuits. Where multiple compressors are involved, two-stage cooling contacts are included. Emergency electric heat control and indicator lights are optional refinements of this system. Some systems have multispeed indoor fans. Humidification and electronic air cleaning can also be included in the circuits. In commercial work, an economizer cycle might even be involved to reduce operating costs. However, these are specific systems to be reviewed as the need arises.

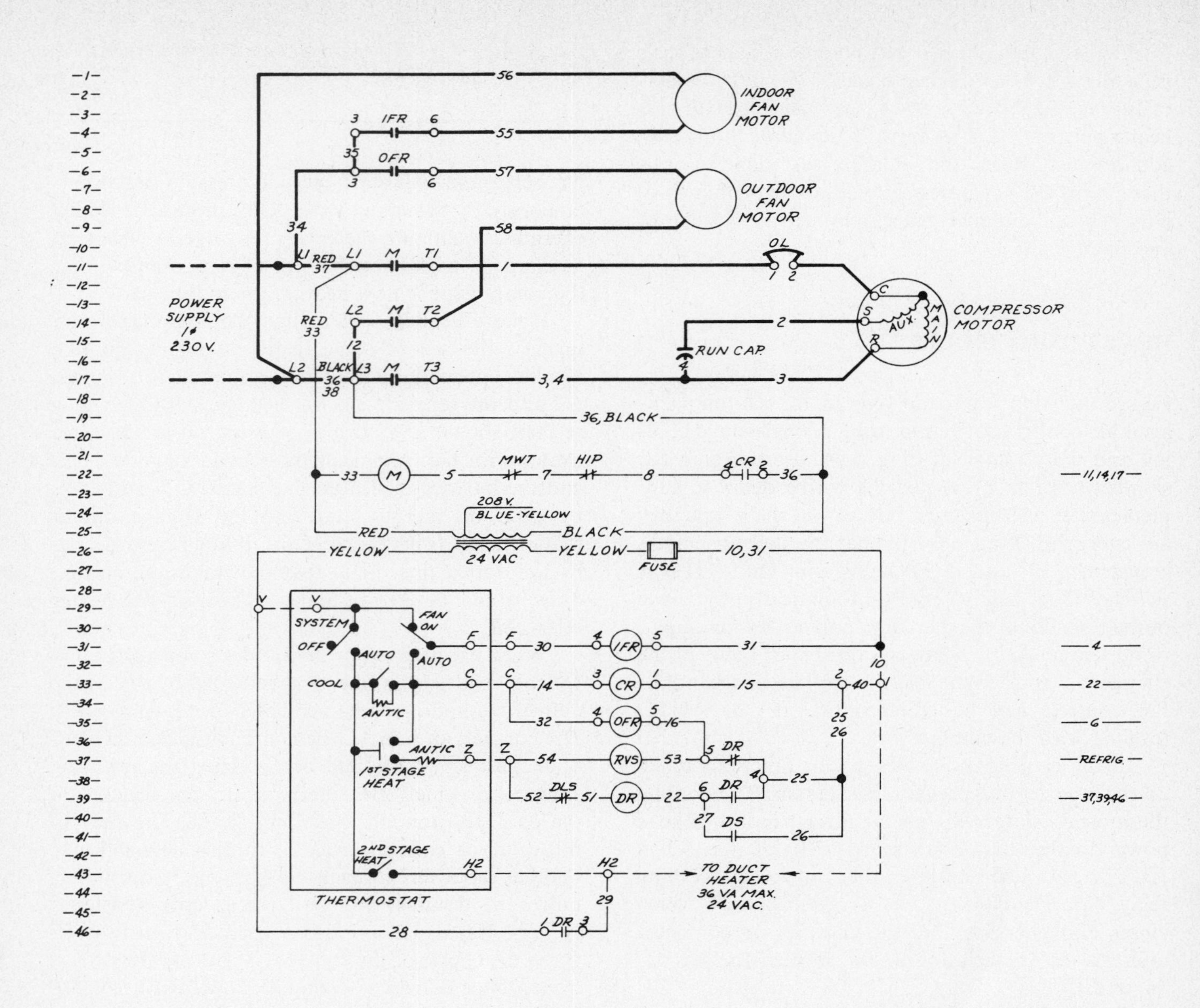

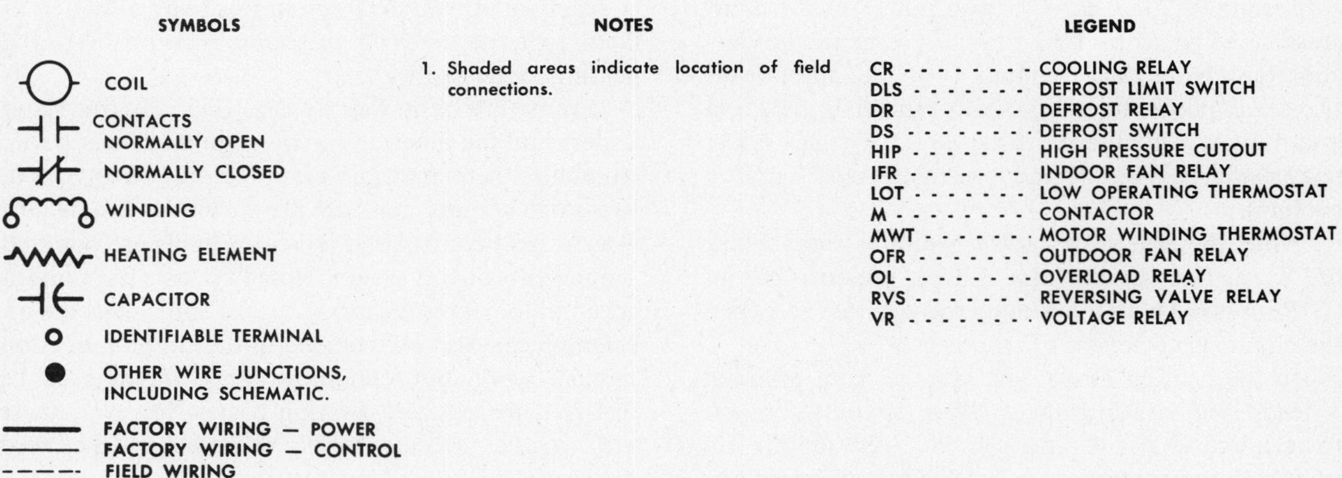

FIGURE 18-29 (Courtesy, Westinghouse Electric Corporation).

PROBLEMS

1. The electrical service entrance for residential power will be ______________ volts, ______________ phase, ______________ hertz.
2. Main breakers (or fuses) range from ______________ to ______________ amps.
3. The voltage drop in selecting wire sizes should be limited to ______________%.
4. Wiring diagrams usually require ______________ and ______________ for proper understanding.
5. Single-phase power is usually designated as two letters and color codes. Name them.
6. Which is the hot wire and which is neutral?
7. Manufacturers usually furnish two wiring diagrams. Name them.
8. What is the function of the limit switch in a furnace?
9. The furnace fan switch does what?
10. The term *ladder diagram* refers to what?
11. In Fig. 18-26 the coil *IFR* in line 15 closes what?
12. In the heat pump wiring diagram (Fig. 18-29), the reversing valve relay is located in line ______________.
13. Field wiring is shown by ______________ lines.

19

COMMERCIAL AND ENGINEERED CONTROL SYSTEMS

The differences between residential control systems and those for commercial uses and engineered systems lie in the increased equipment size, more sophisticated application requirements (low ambient control and economizer cycles, etc.), and in the nature of the space to be controlled.

The unitary type of central heating and air-conditioning (up to 5 tons), which is used in residential work, is all single-phase power, and this same kind of product may also be employed in small commercial situations with no essential changes. In residential work, there is usually only one space thermostat (an exception would be a zone control with two units). Single-point space thermostat control is also common in small and light commercial work where a single zone duct system is used.

It is at the $7\frac{1}{2}$-ton size that the electrical characteristics of the equipment really become three-phase service, although some manufacturers do offer three-phase units in smaller sizes to meet special applications. Above $7\frac{1}{2}$ tons the need for more heavy-duty compressor starting and protection becomes apparent. And in the large tonnage systems, special equipment, such as reduced-voltage starters, open and closed transition starters, etc., become important considerations.

The shift from a single-zone duct system and single-zone space thermostat to the multizone duct systems, induction, and double-duct application, etc., means also a shift to multispace control. That constitutes a change in the point of control for the air-conditioning equipment itself. And as the need for the separation of equipment and space control becomes more pronounced, *control centers* with monitors must be employed to manage the operation, to avoid having personnel running around the entire building checking conditions.

Small and medium commercial systems using packaged equipment usually utilize preengineered electrical controls that require a minimum of on-site electrical work. This reduces installation costs, promotes better reliability, and simplifies the problems of maintenance for the operating and service personnel.

On the other hand, *engineered control systems* are almost always tailored to a particular application, and the design is included in the HVAC *plans and specifications* prepared by a *consulting engineer*. It is in this area where the use of pneumatic and electronic type of controls, in conjunction with regular electrical controls, becomes more prevalent.

The installation of *energized control systems* is also a specialized field, and within the HVAC work this job is separately done by control contractors who not only install the systems but do the start-up, balancing, and service. This is a specialized field, and one which offers challenging employment opportunities to the qualified technician.

COMMERCIAL ELECTRICAL CONTROL CIRCUIT

With this brief introduction of the commercial and engineered system, let's look at the electrical system of a typical commercial rooftop unit with natural gas heating, low ambient control, and an economizer cycle. It is common for the manufacturer to present the information in three parts: (1) the basic refrigeration power circuit, (2) the particular control needed for heating gas, oil, electric (in this example, natural gas heating is used), and (3) the necessary legend to identify symbols and other items.

Figure 19-1 shows the basic power wiring for compressors and fan motor circuits (note that this is a dual compressor system). Figure 19-2 shows the low-voltage control circuit, and Fig. 19-3 is the legend.

Referring to Fig. 19-1, the power supply is assumed to be 208/230 volts, three-phase, 60 hertz, for purposes of sizing fuses and disconnect switch. From the manufacturer's electrical table (Fig. 19-4) on fuses and wire sizes, the model shown at 208/230 volts will require a 100-amp disconnect and a maximum of 70 amp dual element fuses. The minimum power supply wire size is a function of the distance from the switch to the panel terminals *LL1*, *LL2*, and *LL3* on the conditioner. Assume the distance is 125 feet one way. The minimum wire size is No. 4 AWG 60°C wire.

Compressor motors are three phase, with the external overloads (*OL*) in the motor winding circuit. The evaporator motor is also three phase, but it has inherent motor protection. All three motors are started by the contactors marked *1M*, *2M*, and *3M*, respectively.

The condenser fan motors and the balance of controls must be single phase so as to place elements in series circuits; this is not possible with three-phase operation. Single-phase line voltage is connected at terminals 1 and 3 on line 10. Ignore the transformer (see the note); its only function is to reduce power from 460 volts to 230 volts, and in this example it is not used. Note the added fuse protection *1 FU* and *2 FU* in the single-phase hook-up. Drop down to lines 11 and 12 and note the compressor crankcase heaters (*1 HTR* and *2 HTR*) that come on when the crankcase thermostat closes in response to oil temperature. They can operate as long as power is ON, regardless of the system operation. Therefore, it's important that *the power not be turned off in cold weather*.

Both condenser fans have single-phase motors with external running capacitors (*1 RC* and *2 RC*) and inherent overload protection. As stated, this is a dual circuit unit equipped with low ambient control. Note that in line 13 the No. 2 condenser fan motor is energized by the cooling relay *IR-1* to start and run, but it can also be dropped out by the first low ambient thermostat *1 TH*.

If it were not for the economizer cycle, the No. 1 condenser fan motor would be wired in as shown for the No. 2 fan. But the economizer system works within the same temperature range as the first-stage low ambient, and the two must be electrically interlocked as shown from line 15 to line 26.

IR-2 in line 26 is the No. 2 cooling relay and the key to energizing the circuit that feeds terminals 2 and 6 and the No. 1 condenser fan motor; trace out that circuit. But when the low ambient accessory kit is used, the jumper between terminals 2 and 6 is removed and the circuit, shown in dotted lines, is connected. Following line 26, the relay switch *IR-2* power flows to *6R* coil and on to *2R* (cooling contacts) and to the contactor coil *2M*, or it closes the No. 2 compressor contactor. *2R* functions from the second stage of the cooling thermostat. The *6R* coil is the key to furnishing power to line 25 through contacts *6R*. Line 25 has a low-pressure cutout (*2LP*), high-pressure cutout (*2HP*), electrical overload switches *40L* and *50L*, and thermal cutout *60L* . . . all of which protect the No. 2 compressor.

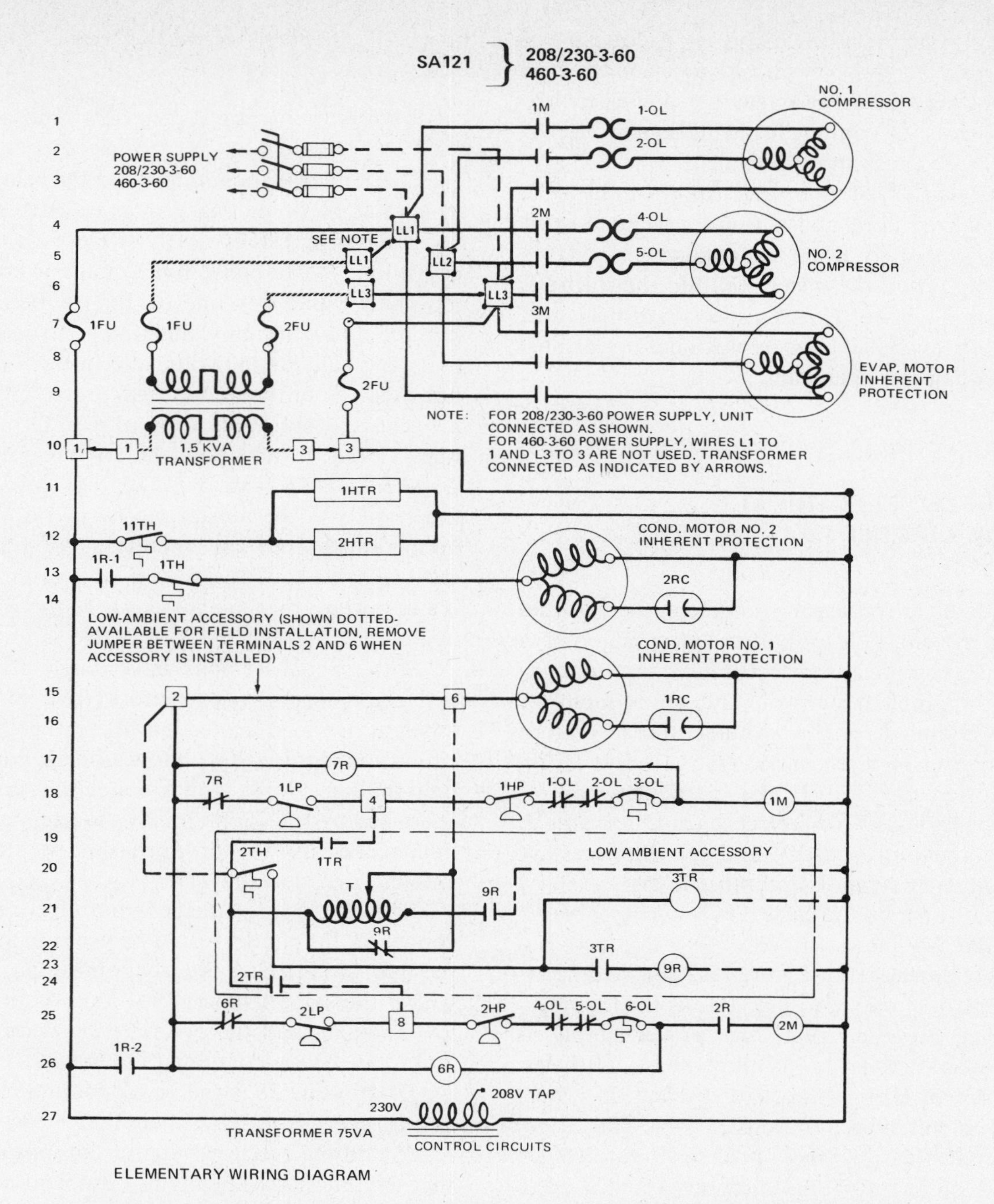

FIGURE 19-1 High voltage wiring diagram (Courtesy, York, Div. of Borg-Warner Corporation).

Note that line 18 is a similar protective circuit for the No. 1 compressor. The two lockout relays *6R* and *7R* are normally closed.

Without a complete detailed explanation of the low ambient accessory, it can be noted that *2 TH*, line 20 (low ambient thermostat) is the key to controlling the No. 1 condenser fan motor through a system of the time delay relays *1 TR*, *2 TR*, and *3 TR*, eventually feeding to terminal 6 and the condenser fan motor. *T* (line 20) is an auto transformer fan speed control that modulates the fan motor rpm instead of providing a straight OFF-ON. Note that it is put into the circuit when *9R* switches reverse on response to the need for low ambient operation. Otherwise, the fan speed is at full rpm when the current goes around *T*, because *9R* on line 21 is open.

Line 27 is the primary line-voltage side of the low-voltage control transformer. It is rated at 75 volt-amps.

Now go to Fig. 19-2, the low-voltage diagram. Line 1 is the 24-volt secondary with a fused protection based on the amp draw. The space thermostat

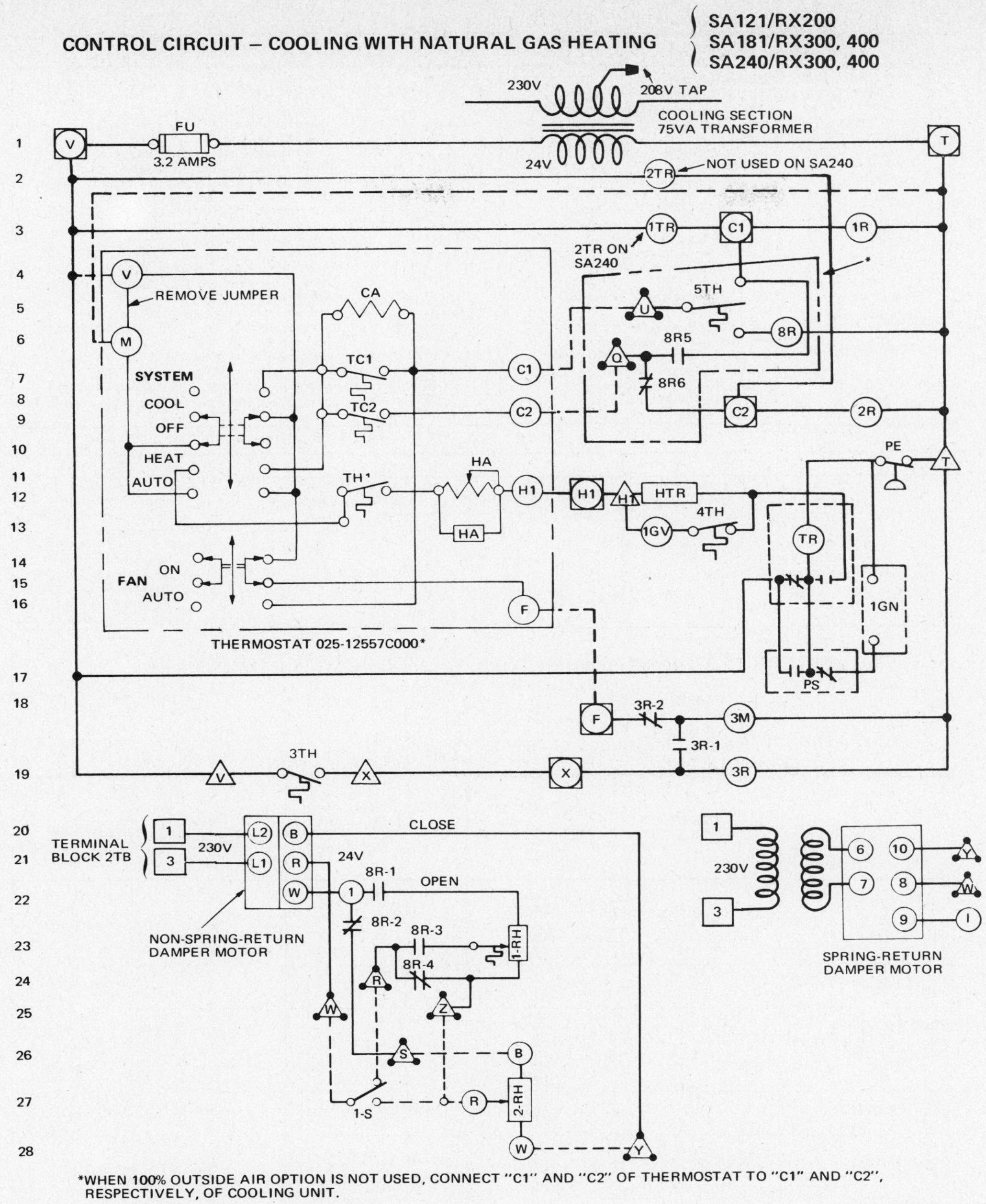

FIGURE 19-2 Low-voltage control circuit (Courtesy, York, Div. of Borg-Warner Corporation).

has a COOL-OFF-HEAT-AUTO system selection. In automatic, the system will bring on heating, cooling, or the economizer system without manual selection. *TC1* and *TC2* are first- and second-stage cooling stats. *TH1* is the single-stage heating stat. The indoor fan selection is ON (continuous) or AUTO to cycle with the *3 TH* blower control.

The economizer control thermostat *5 TH* in the mixing box is tied into the *CI* first-stage cooling circuit, so that the No. 1 compressor cannot operate as long as it is open, meaning the temperature of the outside air is sufficient to provide cool air without refrigeration. But when the outside ambient is high enough to call for cooling, *5 TH* closes, furnishing power to relay *8R* and, through contacts *8R5* and *8R6*, allows both cooling relays *R1* and *R2* to close on command of the cooling stats. *1 TR* and *2 TR* are time-delay relays in the cooling circuits that prevent both compressors from coming on at the same time.

On the heating mode, the economizer system is bypassed. Current from the *TH-1* stat (line 12) goes

	COMMON LEGEND FOR SA121, SA181, SA240 ELEMENTARY WIRING DIAGRAMS		
CA	Anticipator, Cooling	1RC,2RC, 3RC,4RC	Running Capacitor
HA	Anticipator, Heating	S	Switch, Oil Pressure
T	Auto-Transformer, Speed Controller	1-S	Switch, Bypass
FU,1FU,2FU	Fuse	TC1	Thermostat, Cooling 1st Stage
1M,2M	Contactor, Compressor	TC2	Thermostat, Cooling 2nd Stage
3M	Contactor, Blower Motor	TH1	Thermostat, Heating 1st Stage
1GV	Gas Valve	TH2	Thermostat, Heating 2nd Stage
2GV	Gas Valve, Second Stage	1TH,2TH	Thermostat, Low Ambient Control
HTR	Heater in 3TH Fan Thermostat	3TH	Thermostat, Blower (Heat)
1HTR,2HTR	Heater, Compressor Crankcase	4TH	Thermostat, Limit (Heat)
1HP,2HP	High Pressure Cutout, Refrig.	5TH	Thermostat, Mixing Box
IGN	Ignition Trans. (For Pilot Relighter)	11TH	Thermostat, Crankcase Heater
PE	Low Gas Pressure Switch	TR	Time Delay Relay, Pilot Ignition
1LP,2LP	Low Pressure Cutout, Refrig.	VFS	Venter Fan Sail Switch
1-OL,2-OL 3-OL,4-OL 5-OL,6-OL	Overload Protectors, Compressor	10R,11R	Venter Motor Relays
		1RH	Control Damper, Mixed Air
PS	Pilot Safety Switch	2RH	Control Damper Controller, Min. Position
R	Second Stage Gas Valve Relay		Terminal Block, 1TB
1R,2R	Relay, Control Cooling		Terminal Block, 2TB
3R	Low Voltage Control Relay		24-Volt Terminal Block, 3TB
4R,5R	Relay, Control Electric Heat		Identified Connection in Heating Section
8R	Relay, Control, Mixing Box		0-100% Outside Air Terminal Block, 6TB (SA121); 7TB (SA181); 4TB (SA240)
9R	Relay, Low Ambient Control		

SPECIFIC LEGEND FOR SA121, SA181, SA240 ELEMENTARY WIRING DIAGRAM

Model	SA121		SA181		SA240
6R	Relay, Lockout No. 2 System	6R	Relay, Lockout No. 2 System	6R	Relay, Lockout
7R	Relay, Lockout No. 1 System	7R	Relay, Lockout No. 1 System	MP	Compr. Protection
1TR,2TR	Time Delay Relay (Low Ambient Accessory)	1TR,2TR	Time Delay Relay, Low Ambient Control	1TR	Time Delay Relay, Part Winding Start
3TR	Time Delay Relay, Cond. Fan			2TR	Time Delay Relay, Low Ambient Control
T	Auto-Transformer (Low Ambient Accessory)			3TR	Time Delay Relay, Oil Pressure Switch
				1-SOL	Solenoid, Compr. Unloader
				2-SOL	Solenoid, Evap. Unloader

FIGURE 19-3 Legend (Courtesy, York, Div. of Borg-Warner Corporation).

–WIRE, FUSE AND DISCONNECT SWITCH SIZES
COOLING ONLY AND GAS-FIRED HEAT MODELS SA91, SA121, SA181, SA240, SA360

Model	Power Supply	Length Circuit One Way, Ft. Up To	Min. Wire Size Copper AWG 60 C 2% Voltage Drop	Max. Fuse Size Dual Element	Disconnect Switch Size, Amps
SA91-25A	A	100 175 200 250	6 4 3 2	55	60
SA91-45A	A	175 250	10 8	30	30
208/230 VOLT SA121-25B	A	125 175 200 250	4 3 2 1	70	100
SA121-46B	A	200 250	8 6	40	60
SA181-25A	A	150 200 250	1 0 000	100	100
SA181-45A	A	250	4	60	60
SA240-25C	A	150 200 250	00 000 250MCM	175	200

FIGURE 19-4 (Courtesy, York, Div. of Borg-Warner Corporation).

through the heat anticipator *HA* to terminal *H1*. From *H1* it feeds the gas valve *1GV*, provided that the limit control contact *4 TH* is closed. Another circuit on line 17 provides an alternate flow of current to the pilot safety switch, *PS*, and through its contacts feeds the *1GN* pilot ignition transformer. But ignition of the pilot cannot take place unless there is sufficient gas pressure to close the *PE* (pressure/electric) switch (line 11). At the same time the *TR* (time-delay relay) is also energized, the contacts below it are reversed, and the gas valve opens. All this happens instantly.

If the fan selector is at the ON position, current from terminal *F* goes through normally closed *3R*-2 switch and energizes *3M* (the contactor coil) for the indoor or evaporator fan. In the AUTO position, that circuit is essentially bypassed. Current then (on line 19) flows through *3 TH*, the blower control thermostat (if heat is present to close the switch), and to the *3R* coil, which then closes *3R-1* and opens *3R-2*. Thus *3M* is again energized and the indoor blower comes on. It will remain on until *3 TH* opens to reflect on heat in the plenum.

The power (230 volts on lines 20 and 21) to operate the economizer outside air damper is picked up at terminal blocks 1 and 2 from line 10 on Fig. 19-1. Two diagrams are shown. One is of a damper motor with a nonspring return; the other has a spring return. With the spring return the motor will close completely in the event of a power failure in the damper motor system. If there is no spring return, the damper motor will close to some minimum position based on the minimum amount of outside air needed. Note that relay *8R*, coil line 6, is the key to all the switch actions; *8R-1*, *8R-2*, *8R-3*, and *8R-4* to position the control dampers *1-RH* and *2-RH*.

This review may seem complicated to the beginner, but with a little concentration, it will be seen that its circuits are not that much different than those in residential systems. The difference is the addition of crankcase heaters, low ambient control, economizer dampers, electronic ignition, and dual compressor operation.

Other options that may be found in similar products are two-stage heating, flue venting motors and their control relays, and a ventor fan sail switch to verify air flow.

The control system for this same basic cooling unit but with electric heating instead of gas will be quite different in the low-voltage heating arrangement. The base drawing shown in Fig. 19-1 essentially does not change.

ENGINEERED PNEUMATIC CONTROL SYSTEMS

Pneumatic control systems use compressed air to supply energy for the operation of valves, motors, relays, and other pneumatic control equipment. Consequently, the circuits consist of air piping, valves, orifices, and similar mechanical devices.

Pneumatic control systems are reported to offer a number of distinct advantages, especially in commercial and industrial applications:

- Pneumatic equipment is inherently adaptable to modulating operation, yet two-position or positive operation can easily be provided.
- A great variety of control sequences and combinations are available by using relatively simple equipment.
- Pneumatic equipment is said to be relatively free of operating difficulties.
- It is most suitable for controlling explosion hazards.
- Costs may be less than electrical controls if building codes require the total use of electrical conduit.

Pneumatic control systems are made up of the following elements:

(1) A source of clean, dry compressed air to provide the operating energy.

(2) Air lines, usually copper or plastic, from the air supply to the controlling devices. These are called *mains*.

(3) Controlling devices such as thermostats, humidistats, humidity controllers, relays, and switches. These are the *controllers*.

(4) Air lines leading from the controlling devices to the controlled devices. These are called *branch lines*.

(5) Controlled devices such as valves and motors. These are called *operators* or *actuators*.

The air source is an electrically driven compressor (Fig. 19-5), which is connected to the storage tank in which the pressure is maintained between fixed limits (usually between 20 and 35 PSI for low-pressure systems). Air leaving the tank is filtered to remove oil and dust, and in many installations a small refrigeration unit is included to condense out any entrained moisture. Pressure-reducing valves control the air pressure to the main that feeds the controller (thermostat).

If suitable compressor air already exists in the building for other purposes, the compressor can be eliminated and a reducing valve station may be installed to clean and reduce the air pressure to required conditions.

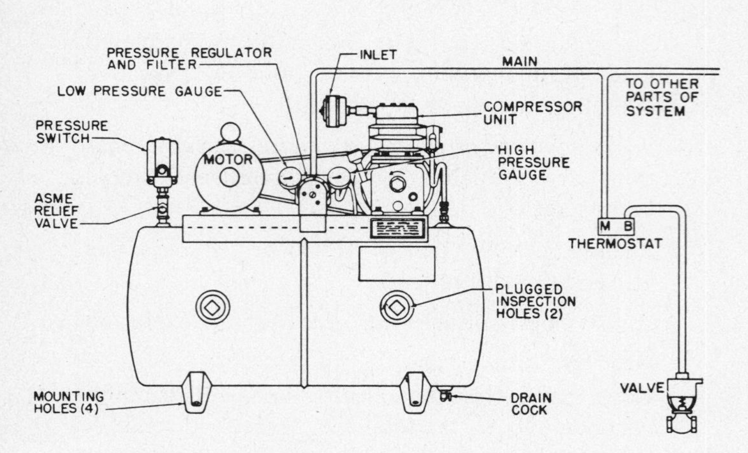

FIGURE 19-5 Diagram of basic pneumatic control system.

The controller function is to regulate the positioning of the controlled device. It does this by taking air from the supply main at a constant pressure and delivering it through the branch line to the controlled device at a pressure that is varied according to the change in the measured conditions.

A bleed type of thermostat control is illustrated in Fig. 19-6. The bimetal element, which reacts to temperature, positions the vane near or away from the air nozzle, and thus the pressure in the branch line is relative to how much air is bled off. Bleed controls used directly do not have a wide range of control, so they are frequently coupled to a relay that is separately furnished with air for activating purposes, and the bleed thermostat simply controls the relay action. Bleed controls naturally cause a constant drain on the compressed air source.

Nonbleed type controllers use air only when the branch line pressure is being increased. The air

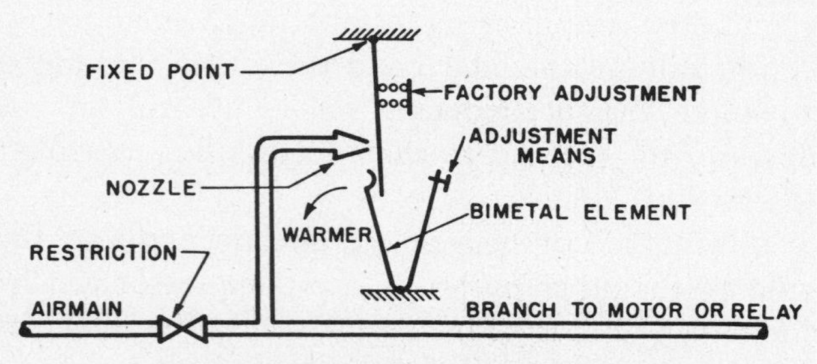

FIGURE 19-6 Diagram of bleed thermostat.

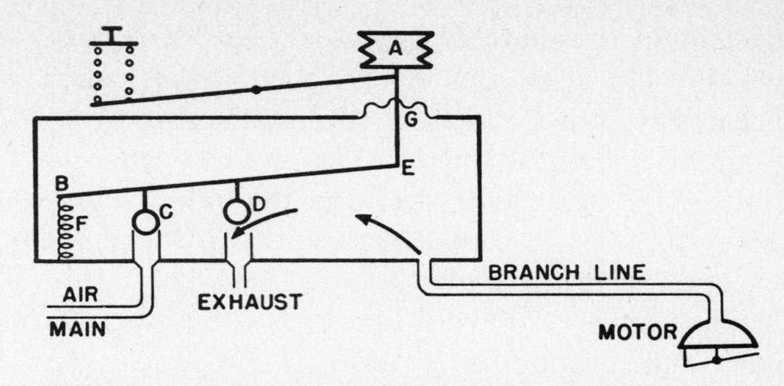

FIGURE 19-7 Diagram of a nonbleed stat on a decrease in the measured condition.

pressure is regulated by a system of valves (Fig. 19-7), which eliminates the constant bleeding of air that is present in the bleed thermostat. Valves *C* and *D* are controlled by the action of the bellows (*A*) resulting from changes in room temperature. Although the exhaust is in a sense a bleeding action, it is relatively small and occurs only on a pressure increase.

The controlled devices, actuators or operators, are mostly pneumatic damper motors and valves. The principle of operation is the same for both. Figure 19-8 is a diagram of a typical motor. The movement of the bellows as branch-line pressure changes actuates the lever arm or valve stem. The

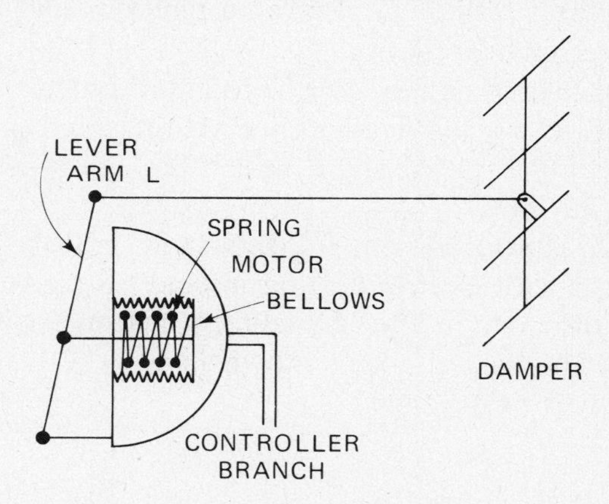

FIGURE 19-8 Pneumatic actuator and normally open damper.

spring exerts an opposing force so that a balanced, controlled position can be stabilized. The motor arm *L* can be linked to a variety of functions.

Figure 19-9 is a pictorial review of some of the functions that can be involved in a pneumatic engineered control system. Note that there will always be some crossover between the air devices and the electrical system, and the device most widely used is a pneumatic/electric relay.

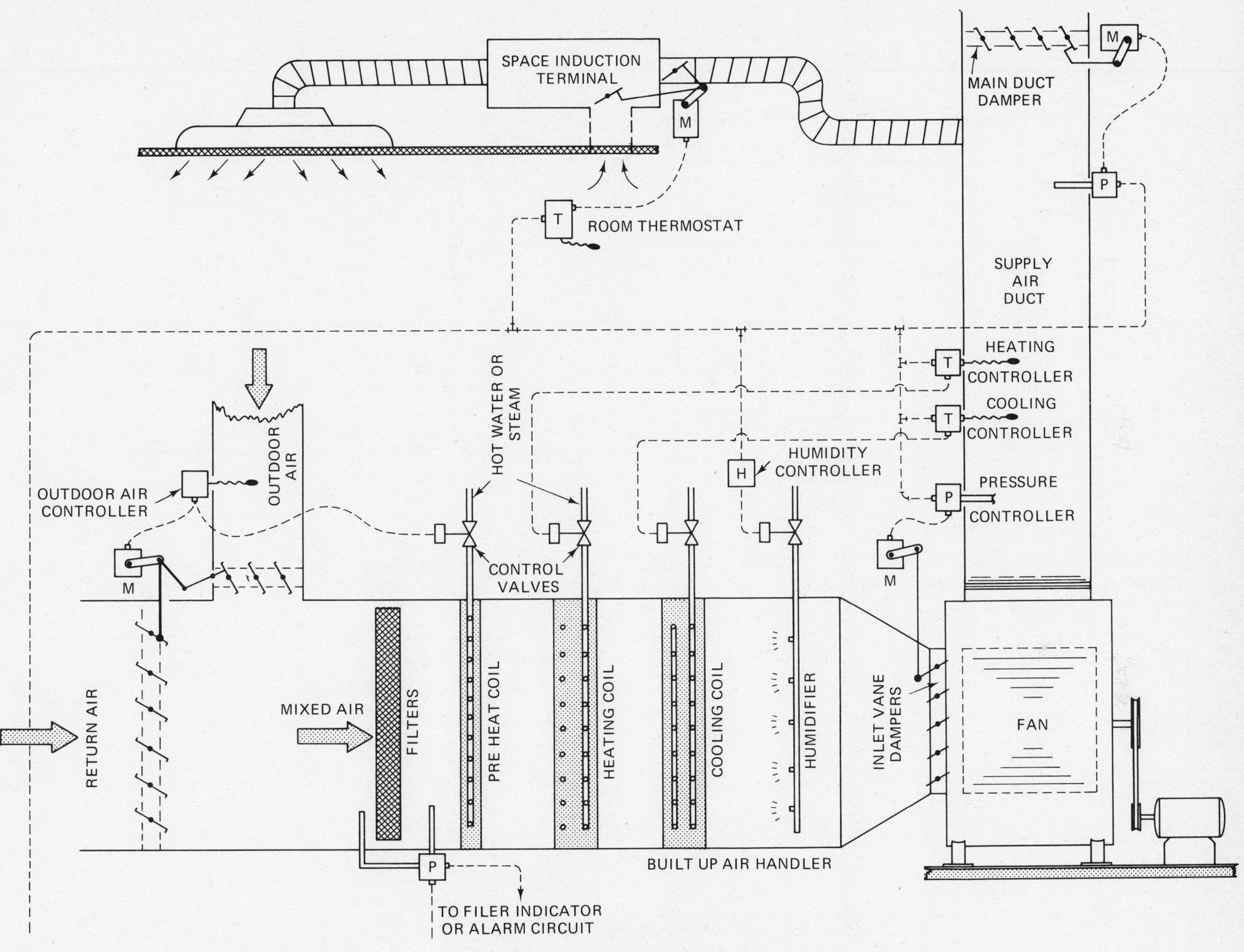

FIGURE 19-9 Pneumatic engineered control system.

FIGURE 19-9 Cont'd.

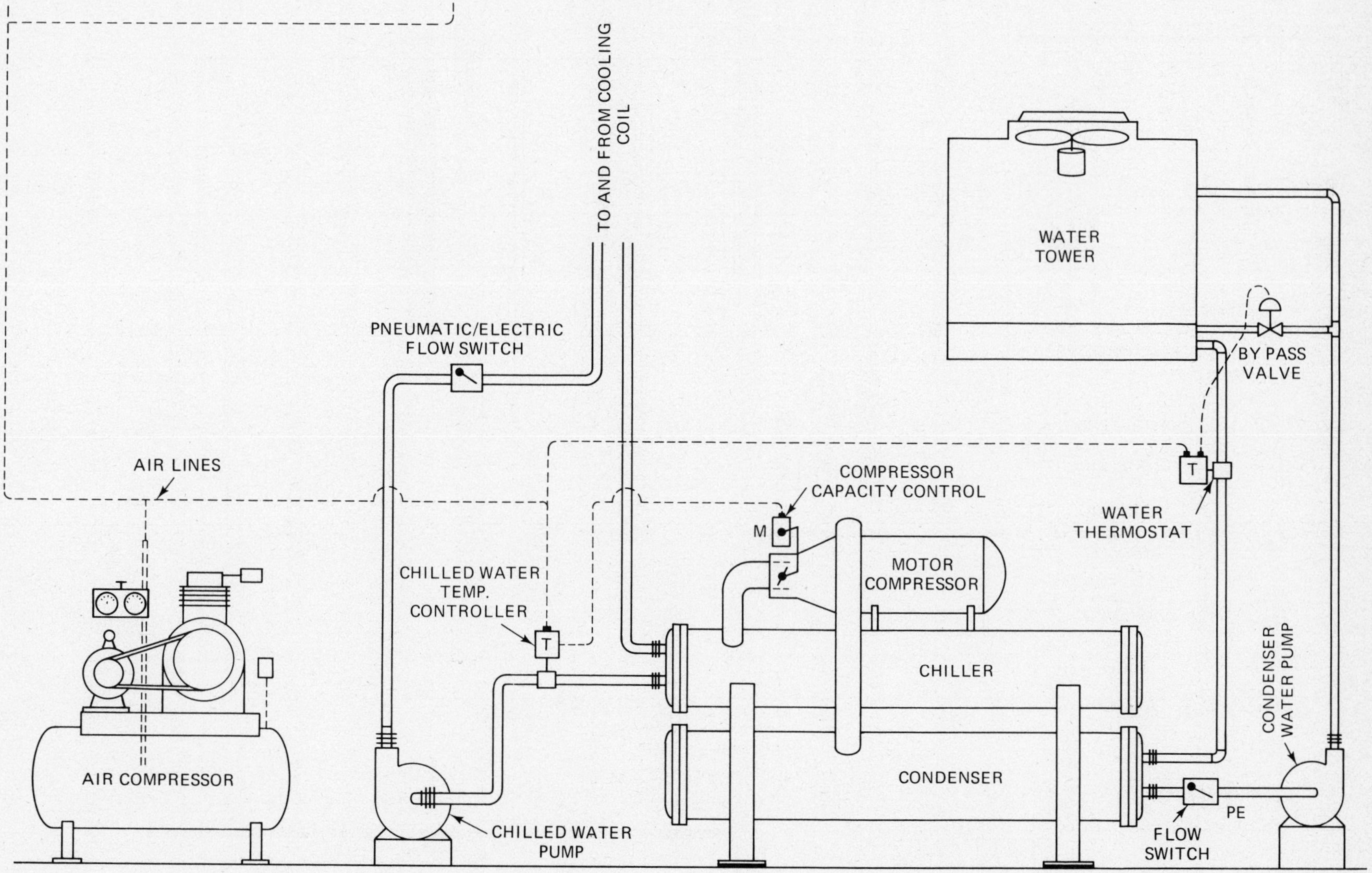

Electronic control systems may also be used effectively in the control of HVAC systems in commercial and industrial work and may become more popular for residential use. Electronic automatic control is the newest of the three types of controls (electric, electronic, and pneumatic). It offers a number of advantages. The sensing elements of electronic controllers are constructed simply; there are no moving parts to interfere with dependable operation. Response is fast. The regulatory element of the controller is usually some distance from the sensing element, and this offers several benefits: (1) all adjustments can be made from a central location, (2) the central area may be cleaner than the place where the sensing element is located, and (3) temperature averaging can be more easily accomplished.

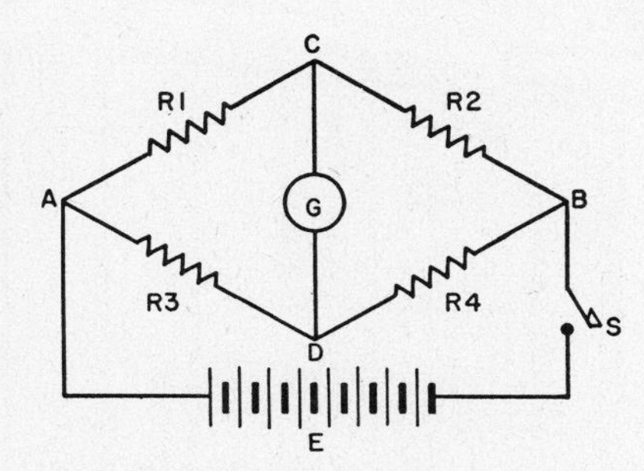

FIGURE 19-10 Wheatstone bridge.

Only simple low-voltage connections are needed between the sensing element and the electric circuit. Flexibility is an important plus, since electronic circuits can be combined with both electric and pneumatic circuits to provide results that ordinarily cannot be achieved separately. Electronic circuits can coordinate temperature changes from several sources (room, outdoor air, fan discharge air, etc.) and program actions accordingly. Another example is a sensing element of perhaps 25 feet in length in a duct to average temperatures as compared to a single-point bulb location.

Electronic controls are based on the Wheatstone bridge concept (Fig. 19-10), which is composed of two sets of two series resistors (*R1* and *R2*; *R3* and *R4*) connected in parallel across a dc voltage source. A galvanometer, *G* (a sensitive indicator of electrical current), is connected across the parallel branches at junctions *C* and *D* between the series resistors. If switch *S* is closed, voltage *E* (dc battery) is impressed across both branches. If the potential at point *C* equals that at *D*, the potential difference is zero and the galvanometer will indicated zero current. When this condition exists, the bridge is said to be *balanced.*

But if the resistance in any one leg is changed, the galvanometer will register a current flow, indicating the existence of a potential difference between points *C* and *D*. The bridge is now *unbalanced.*

If that resistance value were changed as a result of temperature reaction, we now have an electronic method of measuring the current flow in relation to the temperature change. With a few minor changes we can create an electronic main bridge circuit such as in Fig. 19-11. A 15-volt ac circuit replaces the dc

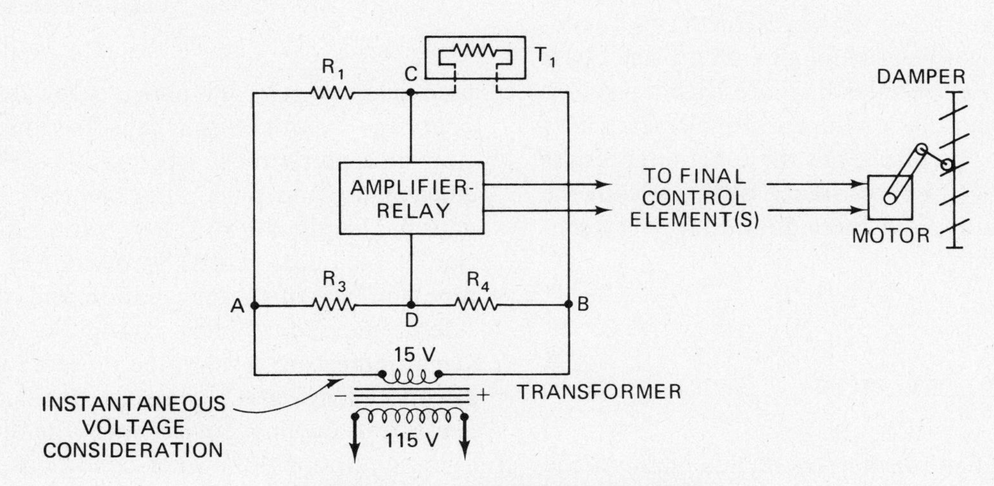

FIGURE 19-11 Electric main bridge circuit.

battery. The galvanometer is replaced by a combination voltage amplifier-phase discriminator-switching relay unit. Resistor *R2* is replaced by a sensing element *T* of an electronic controller.

The purpose of the voltage amplifier is to take the small voltage from the bridge and increase its magnitude by stage amplification to do work. Phase discrimination simply means determining the sensor action. In an electric bimetal thermostat the mechanical movement is directly related to temperature changes. However, the electronic sensing element is a nonmoving part, and the phase discriminator determines whether the signal will indicate a rise or fall in temperature. The relay then operates the final element action. Phase discrimination can be two-position or, with certain modifications, can be converted to a modulating system.

The crossover point from electronic to electric occurs at the output of the amplifier and relay signal. The motor is a conventional ON-OFF, or a proportional (modulating) electric motorized valve, damper actuator, etc.

Electronic temperature-sensing elements are room thermostats, outdoor thermostats, insertion thermostats for ducts (from several inches to 25 feet or over), insertion thermostats for liquid lines, etc. The typical room thermostat is a coil of fine wire wound on a bobbin. The resistance of the wire varies directly with temperature changes.

Obviously, electronic controls do not sense pressures, but they can detect and control humidity within narrow limits. The sensing element of a humidity controller is gold leaf embossed on a plastic base and coated with a special salt. Its resistance value is much greater than that of the sensing elements used in temperature controllers, but the functional results are the same. Electrical resistance will vary with the change in humidity, and these devices are very sensitive and accurate.

CONTROL AUTOMATION

As mentioned at the beginning of this chapter, the complexity of functions and variables to be controlled and monitored in engineered systems have given rise to the use of *building automation techniques* using data control centers (Fig. 19-12). In large buildings, or in a complex of buildings such as a college or university, it is impossible to have enough operating and maintenance personnel to stop, start and watch each system or component during a 24-hour day.

The control center collects key operating data from the heating, ventilating, and air-conditioning system and incorporates remote-control devices to supervise the system's operation. Although it can be used in preventive maintenance by detecting faulty operation before it can cause serious trouble, the control center's most important contribution is its efficient use and scheduling of personnel to investigate and handle the overall operation. The sophistication of such data centers is, of course, related to the type and size of the installation and economic considerations. Some have continuous scanners with alarm indicators to monitor refrigeration machines, oil and refrigeration pressures, chilled water temperatures, cooling tower water temperatures, air filter conditions, low water conditions in the boiler, etc., as well as the conventional space temperature and humidity conditions in each zone.

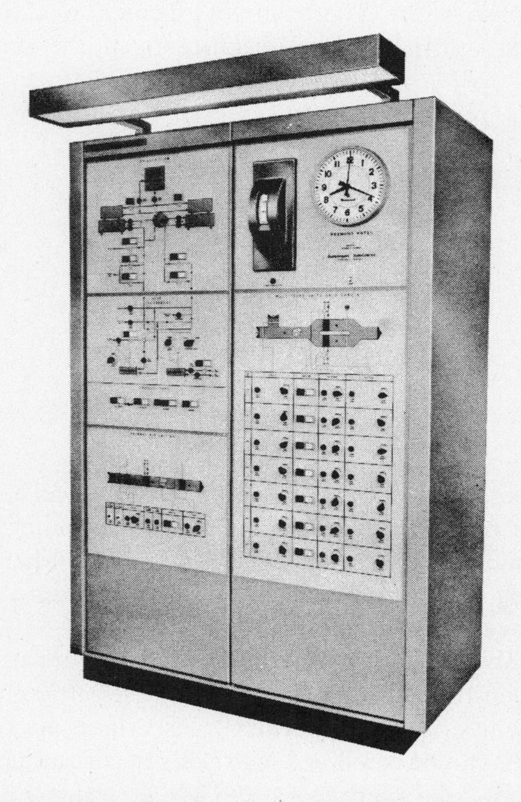

FIGURE 19-12 (Courtesy, Honeywell Residential Division).

PROBLEMS

1. What is the main electrical distinction between residential and commercial equipment?
2. Commercial systems wiring diagrams are usually prepared by _____________.
3. Engineered control systems are usually prepared by _____________.
4. Name three types of commercial and engineered control systems.
5. In Fig. 19-1, the low-voltage transformer capacity rating is _____________.
6. In Fig. 19-2, the heating anticipator is located in line _____________. Is it shown as a fixed type?
7. Electronic control circuits are based on what concept?
8. Name two types of pneumatic controllers.
9. What is the normal air pressure range in the compressed air storage tank?
10. The centralized control system is also referred to as _____________.

20

HEATING STARTUP AND OPERATION

With a background of heating application practices from Chap. 8 and a general review of residential controls and circuits from Chap. 18, we are now ready to become familiar with the startup and operation of a conventional upflow gas-fired, forced-air furnace. We recognize that there will be many other forms of heating apparatus the technician will incounter in the field, but we cannot cover them all here. Instead, we will present a technique of check, test, and start procedures on a natural gas furnace, and if the technician will read the equipment manufacturer's installation and operation manuals thoroughly and then follow a similar approach to other systems, he will be prepared to tackle all types of heating equipment.

Before attempting to check the furnace operation, or for that matter any equipment, the technician should have the proper tools and know how to use them. Figure 20-1 represents a group of instruments for checking the basic furnace operation and consists of:

(1) a 15 in. W.C. U-tube manometer

(2) a CO_2 indicator with aspirator

(3) a draft gauge to at least 0.15 in. of W.C.

(4) a 0–1000°F stack thermometer with a $5\frac{1}{2}$-in. stem

(5) a fire efficiency slide rule

These items usually comprise a kit that may be purchased from several suppliers.

Next, you should have a clamp-on ammeter, a voltage and ohm resistance meter or meters, plus several stem or dial pocket thermometers reading up to approximately 220°F, as illustrated in Fig. 20-2.

During installation and before operating any new installation, a thorough check should be made of the furnace casing for any signs of visible external damage that would indicate rough or abnormal treatment in shipping or handling. Check the fit of the door for proper air clearance. Proceed then to check the other elements.

Remove the door and examine the draft diverter for tightness. Also check the lower baffle (Fig. 20-3). After installation, carefully check the flue pipe connection from the furnace to the vent opening; it must be rigid and tight.

With the door removed, check the AGA data-rating plate (Fig. 20-4) to make sure you have the

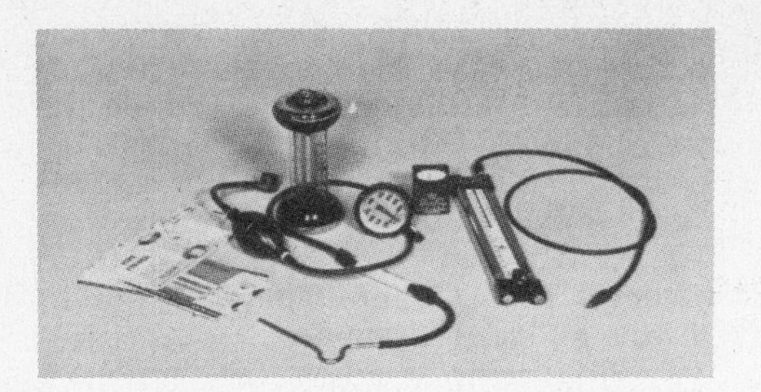

FIGURE 20-1 Recommended service tools (Courtesy, Westinghouse Electric Corporation).

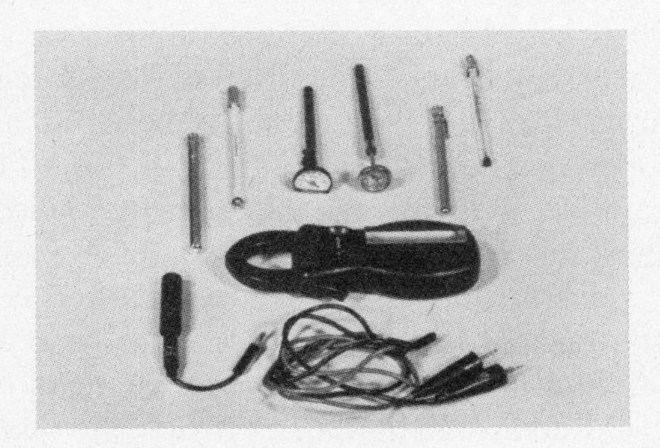

FIGURE 20-2 Recommended service tools (Courtesy, Westinghouse Electric Corporation).

correct furnace. Our example is a forced-air furnace for natural gas; 110,000 Btu/h input; 88,000 Btu/h bonnet capacity, temperature rise from 70°F to 100°F. The ft^3/min rating should be from 815 to 1160 ft^3/min at 0.5 in. of WC external static pressure. The overall electrical rating is 7.0 amps for 115-volt, 60-cycle operation.

The model and serial numbers are important for warranty and parts reference. Check also to assure that the homeowner's operating manual is present. *Do not throw it away*. Make sure it stays with the furnace for later review by the user. Included are instructions on lighting, field wiring, oiling the fan motor, filters, etc. (Fig. 20-5). These instructions vary of course among the different manufacturers' equipment and recommendations. So *do not assume* that all gas furnaces are alike; read the instructions carefully.

Note also the Installation Clearances decal, as well as the electrical wiring diagram (Fig. 20-6).

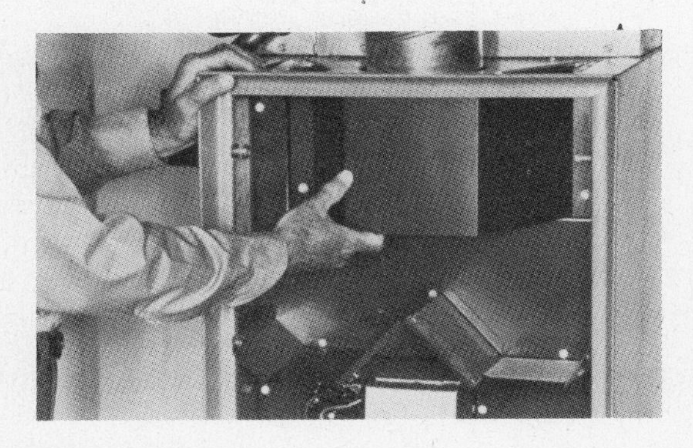

FIGURE 20-3 Inspect draft diverter (Courtesy, Westinghouse Electric Corporation).

FIGURE 20-5 Check instruction decals (Courtesy, Westinghouse Electric Corporation).

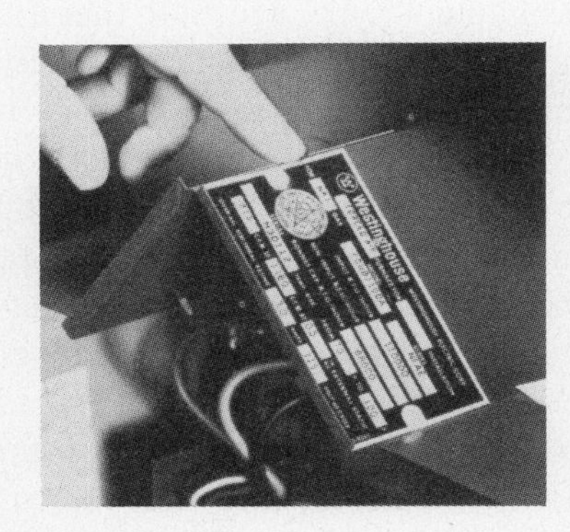

FIGURE 20-4 Check furnace nameplate data (Courtesy, Westinghouse Electric Corporation).

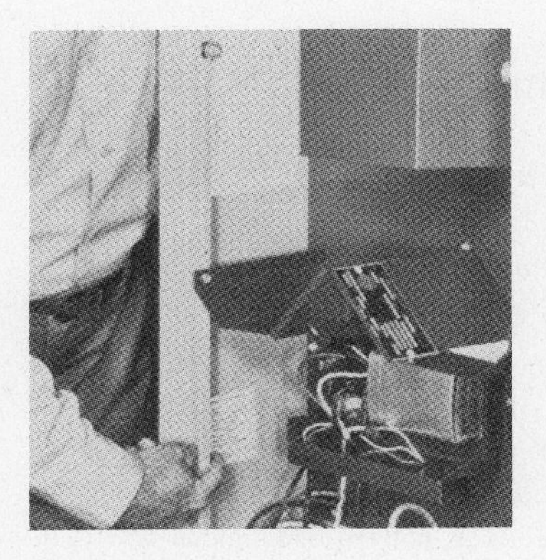

FIGURE 20-6 Wiring diagram and clearance instructions (Courtesy, Westinghouse Electric Corporation).

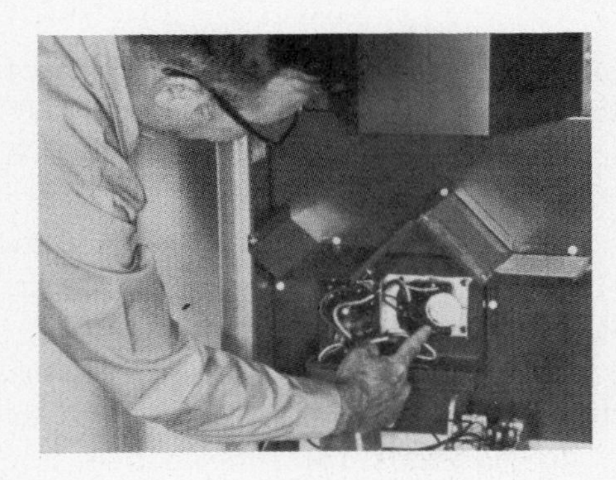

FIGURE 20-7 Check wiring controls (Courtesy, Westinghouse Electric Corporation).

Proper clearances should have been allowed during installation, but it's possible that some other tradesman (carpenter, etc.) might have placed a combustible material too close to the flue or plenum. Always double check.

Move down into the burner compartment (Fig. 20-7). Remove the fan limit control cover. Check the ON-OFF settings of the fan to the cycle, based on the manufacturer's recommendation. The high limit on the furnace may be factory preset, and if so, it will be so noted on the control cover. If the high limit is adjustable, set it at the manufacturer's recommended limit. Also test the tightness of all electrical connections to the control and fan relay terminals. If the furnace is a multispeed, direct-drive model, observe which fan speed lead is connected—high, medium, or low. For initial startup it is recommended the fan be set for medium speed, subject to later verification.

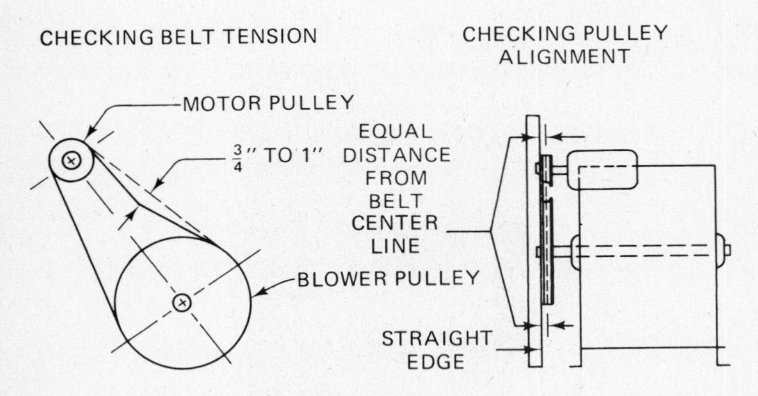

FIGURE 20-8 Align pulleys and tighten belt (Courtesy, Westinghouse Electric Corporation).

If you have a belt-driven model (Fig. 20-8) make sure that the belt tension is correct to avoid slippage, but not too tight to cause excessive wear. About $\frac{3}{4}$ to 1 inch of play should be allowed for each 12 inches of distance between the motor and the blower shaft. Also check the pulley alignment by using a straight edge (yardstick). Misalignment of pulleys will cause noise and excessive vibration and wear on both the blower and the motor bearings.

One of the important prechecks is the positioning of the burners (Fig. 20-9). See that they are positioned in the baffle slots but are free to permit some in and out expansion as well as rotational movement. Check the pilot assembly to see that it is securely fastened to the burner. Set the primary air shutters at a one-third open position as a preliminary setting. Loosen the manifold pressure tap on the burner pipe and retighten; it will be used later.

If you are using the standard air filter shipped with the furnace, see that it fits in the channels or brackets and that the airflow arrow direction is correct. If in the installation an electronic air cleaner is going to be used, the filter may be omitted, subject to the furnace instructions. If the standard filter is a washable permanent type, see that there is a sticky film on the surface, . . . if not, apply the recommended film.

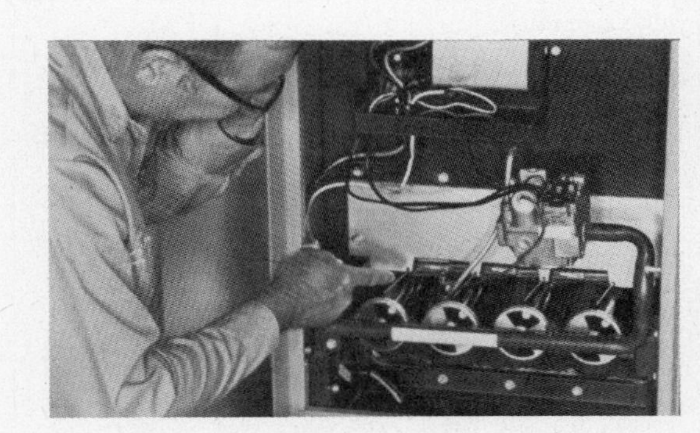

FIGURE 20-9 Check burners (Courtesy, Westinghouse Electric Corporation).

Finally and most important, grasp the blower assembly to see that it is secure in the mounts (Fig. 20-10). Spin the blower wheel to test for free movement. Check the motor mounts. Check the electrical leads to the motor and capacitor. Make sure the motor oil cups are in an upright position; add oil if it is recommended. Some fan bearings and motors are permanently lubricated and won't normally need such attention.

FIGURE 20-10 Check fan assembly (Courtesy, Westinghouse Electric Corporation).

Assuming the furnace has now been prechecked, set it in place and connect it to the gas piping, the flue vent, and the supply and return air ductwork as previously discussed.

Provide the proper power wiring to a fused switch on or near the furnace. The fuse rating and wire sizes must conform to national and/or local codes. The nameplate data calls for 7 amp normal current load. A conventional 15-amp combination fuse and switch would be satisfactory.

Inside the furnace junction box, make up the power wiring to the black and white leads provided. Wire nuts are recommended.

Bring the low-voltage thermostat leads to the furnace and apply them to the correct terminals (Fig. 20-11). Follow the heating wiring diagram provided with the equipment or a separate electrical diagram from the manufacturer if it is to be a total comfort system. It is becoming more common to provide an electrical terminal strip and even plug-in relays because of the a variety of equipment options offered; these items simplify the technician's job.

The following is a suggested startup technique on initial firing: With the manual electrical switch in the OFF position set the room thermostat in the HEATING position, above the room temperature, so that the thermostat will make contact and call for heat. Set the heat anticipator amp rating to match the required volt-amp draw. Then return to the furnace.

With the power OFF, open the main gas valve on the incoming line (Fig. 20-12). Check the gas line for leaks, using a soap bubble solution (not matches). It may be necessary to purge the gas line of trapped air by cracking open the union. Tighten and recheck for leaks.

Turn the control knob on the gas valve to the PILOT position (Fig. 20-13). Then depress the knob that allows gas to the pilot orifice. With a match or burning straw, light the pilot while you continue to depress the knob. In approximately 60 seconds, the heat will generate sufficient current in the thermocouple to hold the port valve open. Release the

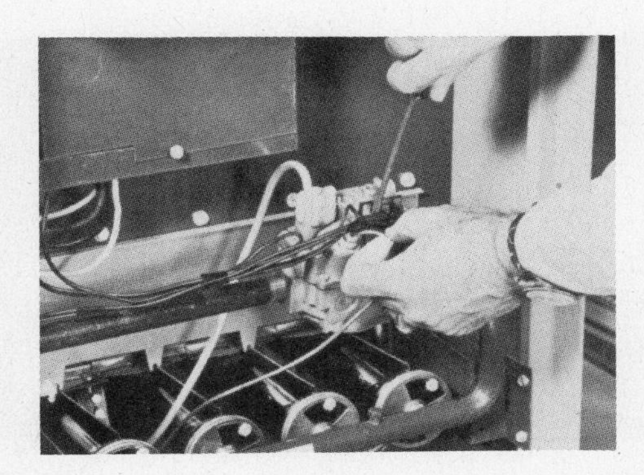

FIGURE 20-11 Low-voltage connections (Courtesy, Westinghouse Electric Corporation).

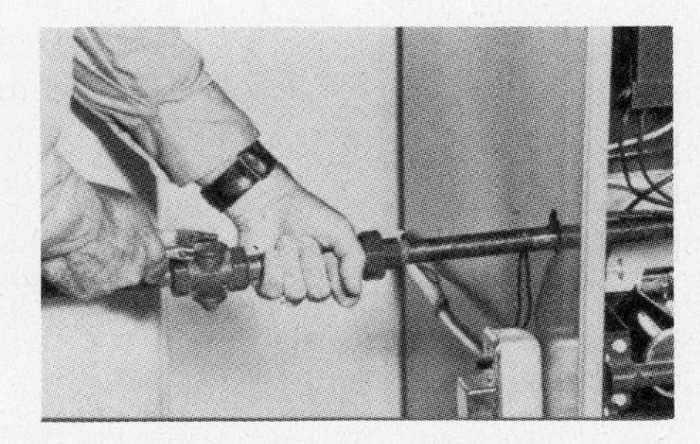

FIGURE 20-12 Open gas valve and check for leaks (Courtesy, Westinghouse Electric Corporation).

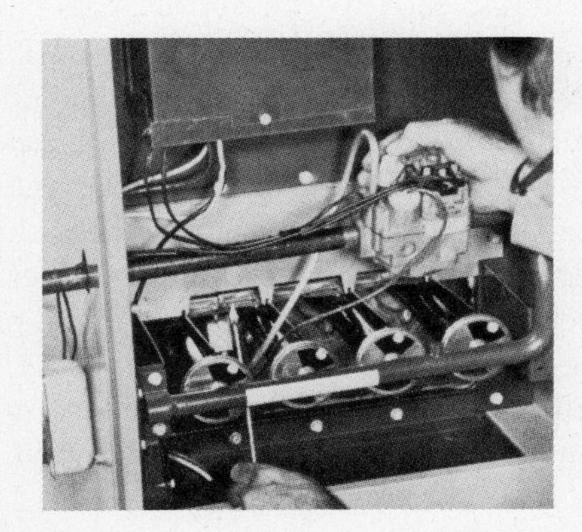

FIGURE 20-13 Light the pilot (Courtesy, Westinghouse Electric Corporation).

pressure on the knob. Note: where electrical ignition is used, follow the manufacturer's recommendations.

Observe the pilot flame. It should be soft (not noisy) and cover $\frac{3}{8}$ in. to $\frac{1}{2}$ in. of the thermocouple tip, which should glow dull red.

If the flame goes out, wavers too much, or is too strong, remove the pilot adjustment cap on the gas valve. With a small screwdriver, adjust to proper height and color. This should be done with other gas-burning appliances in operation. Failure to achieve a proper flame may suggest an improper main gas line supply pressure. With the proper pilot established, turn the control knob to the ON position. This allows the gas to go to the pressure regulation chamber; but with power OFF and the electric solenoid closed, no gas flow will take place.

Turn on the electric power. Listen for the immediate opening of the 24-volt gas valve solenoid and the "light off" of the burners. Burner ignition should take place smoothly and with minimum noise. Make sure all burners are firing. If any abnormal operation seems apparent, shut off the electrical power switch. Don't worry about flame adjustment (primary air) at this point. After a short period, the fan should cycle on (again, don't worry about the exact temperature at this point). When the fan does cycle on, listen for normal operating sound. If the sound is normal, let the furnace run for five to 10 minutes. There will probably be some smoke generated in the living area because some of the paint from the outside surface on the heat exchanger will burn. Open the windows to vent this smoke if it is too objectionable. Also, be certain that a positive up-draft is being produced in the flue vent.

The gas meter (Fig. 20-14) is the beginning step in the adjustment operation. Most meters have a convenient two-cubic-foot dial for computing gas flow.

With all other gas-burning appliances shut off, it is necessary to determine the amount of fuel input to the furnace by timing the rate of flow. You must also know the heat content of the gas being burned. Check with the gas company for the exact rating. For this example, we will assume a heat content of 1,000 Btu per cubic foot.

In this example, it takes 60 seconds for the two-cubic-foot dial to make one revolution. This means

FIGURE 20-14 Timing gas flow (Courtesy, Westinghouse Electric Corporation).

the furnace is consuming 2 cubic feet of gas per minute.

From the gas input chart in Fig. 20-15 (frequently included in the furnace instructions), under the 1,000 Btu column, it can be seen that gas flowing at 2 cubic feet per minute would equal a 120,000 Btu/h input. Our nominal 110,000 Btu/h furnace is then being overfired, and its capacity should be reduced. Remember, an overfire can cause the life of the heat exchanger to be shortened, plus the nuisance tripping of the limit control as the air filter becomes dirty.

If you don't have a chart, use this formula.

$$\text{Input Btu/h} = \underset{\text{(Burned)}}{\text{ft}^3\text{/h}} \times \underset{\text{(Heat Content)}}{\text{Gas Btu/ft}^3}$$

EXAMPLE:

$$\underset{\text{(Burned)}}{120\text{ft}^3\text{/hr}} \times \underset{\text{(Heat Content)}}{1{,}000\text{ Btu/ft}^3} = 120{,}000\text{Btu/h input rate}$$

In our example the actual input is 120,000 Btu/h and the desired input rating is 110,000 Btu/h. Turn the furnace gas valve control knob back to the PILOT position. Remove the manifold pressure tap.

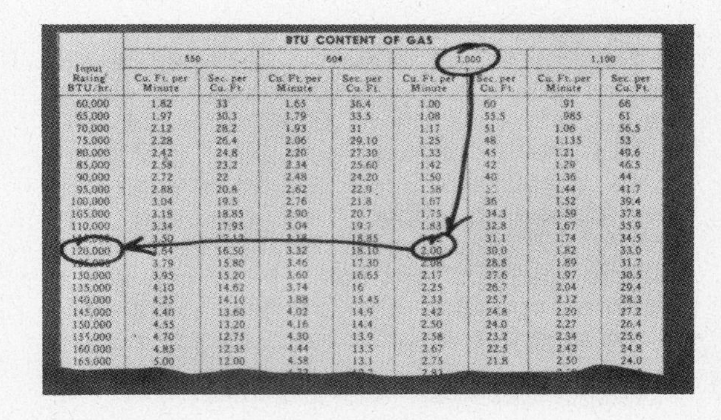

Input Rating BTU/hr.	BTU CONTENT OF GAS: 550		604		1,000		1,100	
	Cu. Ft. per Minute	Sec. per Cu. Ft.	Cu. Ft. per Minute	Sec. per Cu. Ft.	Cu. Ft. per Minute	Sec. per Cu. Ft.	Cu. Ft. per Minute	Sec. per Cu. Ft.
60,000	1.82	33	1.65	36.4	1.00	60	.91	66
65,000	1.97	30.3	1.79	33.5	1.08	55.5	.985	61
70,000	2.12	28.2	1.93	31	1.17	51	1.06	56.5
75,000	2.28	26.4	2.06	29.10	1.25	48	1.135	53
80,000	2.42	24.8	2.20	27.30	1.33	45	1.21	49.6
85,000	2.58	23.2	2.34	25.60	1.42	42	1.29	46.5
90,000	2.72	22	2.48	24.20	1.50	40	1.36	44
95,000	2.88	20.8	2.62	22.9	1.58	[illegible]	1.44	41.7
100,000	3.04	19.5	2.76	21.8	1.67	36	1.52	39.4
105,000	3.18	18.85	2.90	20.7	1.75	34.3	1.59	37.8
110,000	3.34	17.95	3.04	19.7	1.83	32.8	1.67	35.9
[illegible]	3.50	[illegible]	[illegible]	18.85	[illegible]	31.1	1.74	34.5
120,000	[illegible]	16.50	3.32	18.10	2.00	30.0	1.82	33.0
[illegible]	3.79	15.80	3.46	17.30	2.08	28.8	1.89	31.7
130,000	3.95	15.20	3.60	16.65	2.17	27.6	1.97	30.5
135,000	4.10	14.62	3.74	16	2.25	26.7	2.04	29.4
140,000	4.25	14.10	3.88	15.45	2.33	25.7	2.12	28.3
145,000	4.40	13.60	4.02	14.9	2.42	24.8	2.20	27.2
150,000	4.55	13.20	4.16	14.4	2.50	24.0	2.27	26.4
155,000	4.70	12.75	4.30	13.9	2.58	23.2	2.34	25.6
160,000	4.85	12.35	4.44	13.5	2.67	22.5	2.42	24.8
165,000	5.00	12.00	4.58	13.1	2.75	21.8	2.50	24.0

FIGURE 20-15 Determine input from chart (Courtesy, Westinghouse Electric Corporation).

Fasten or hang the U-tube manometer (Fig. 20-16) to the furnace so it can be easily read. Insert the manometer hose adapter into the pipe tap. Open the thumb wheel valves on top of the manometer. Adjust the sliding scale to the point where the level line on the scale matches the actual liquid level. You are now ready to take a reading. Turn on the other gas appliances. Set the furnace gas valve knob back to ON, permit the furnace to fire a minute or so, and then observe the stabilized pressure.

In this case, the level line was at $7\frac{1}{2}$ in. and the vertical column now reads nearly $11\frac{1}{2}$ in. Thus, subtracting the difference, we have about 4.0 in. of WC manifold pressure, which would verify an overfire situation. (Remember that 3 to 3.5 in. of WC is normal for natural gas.)

To adjust the manifold pressure, remove the cap on the gas valve and, with a screw driver, gradually turn the screw to lower the pressure to approximately $10\frac{3}{4}$ in. of WC pressure on the vertical scale.

With the manometer still connected and the other gas-burning appliances again turned off, return to the gas meter and recheck the input. This time we record 66 seconds for a revolution of the two-cubic-foot dial, which equals 33 seconds per cubic foot. From the input chart (Fig. 20-15) we determine that the input rating of 33 seconds/ft^3 is slightly under 110,000 Btu/h.

Return to the furnace. Turn the control knob to PILOT. Disconnect the manometer and replace the pressure tap on the manifold piping.

With the pilot, gas pressure, and input rating established, adjust the primary air shutters to obtain a visual flame check (Fig. 20-17). Close the shutters until the yellow tips just disappear and a blue flame is established.

Some installers also like to set the primary air adjustment by checking the CO_2 content in the flue gas, using the gas analyzer. Follow the manufacturer's directions.

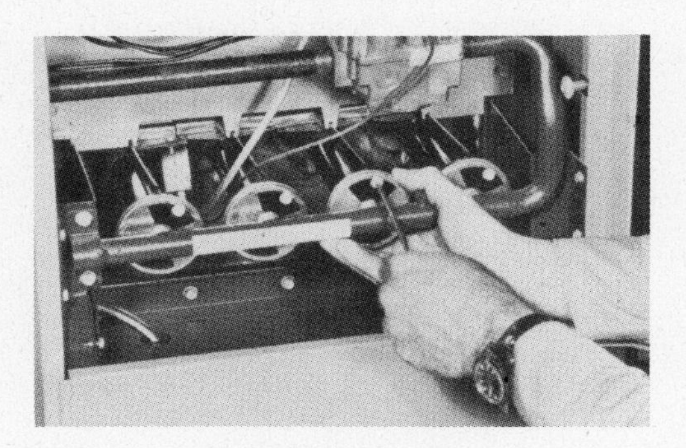

FIGURE 20-17 Adjust primary air (Courtesy, Westinghouse Electric Corporation).

Check on the flue draft with the small draft gauge (Fig. 20-18). Insert the tube into the hole in the flue. Placing a finger over the hole in the rear of the case, obtain the 0 setting. Remove your finger and measure the draft. In normal operation there will be a steady updraft (negative reading). Any prolonged down-draft (positive reading) would indicate problems with the terminal vent on the roof caused by wind currents.

With the furnace firing normally, now turn to the problem of moving air. Assume this is a total comfort system with a $2\frac{1}{2}$ ton cooling coil, so we would want to obtain a flow of approximately 900–1,000 ft^3/min. 800 ft^3/min would be the minimum, to prevent the coil from "icing."

With all dampers, grilles, etc, in the full OPEN position, and the direct-drive blower operating on the medium speed tap, measure the supply and return air temperatures (Fig. 20-19). In this system

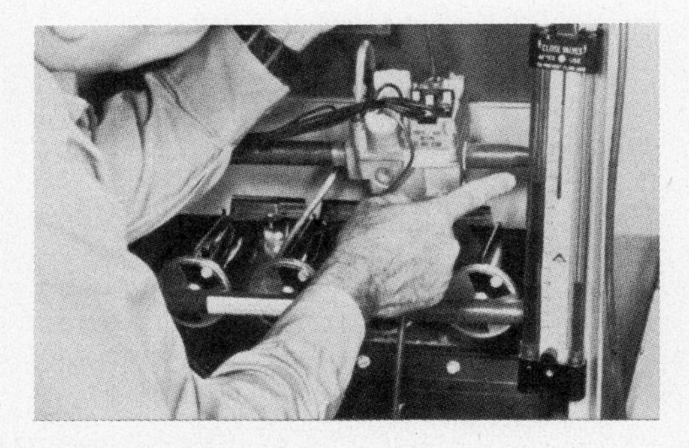

FIGURE 20-16 Adjust manifold pressure (Courtesy, Westinghouse Electric Corporation).

FIGURE 20-18 Measure draft conditions (Courtesy, Westinghouse Electric Corporation).

FIGURE 20-19 Measure air rise (Courtesy, Westinghouse Electric Corporation).

we recorded 70°F on the return and 152°F on the supply. Thus, there is an air temperature rise of 82°F (152 − 70).

Plot the 82°F air rise on the ft³/min air rise chart (Fig. 20-20) against an 88,000 Btu/h furnace output. Find 88,000 Btu/h on lower scale. Follow it up to intersect the diagonal line at 82° (80° is the closest). From that point follow across to read slightly less than 1,000 ft³/min is being circulated. This is within the desired air range. Should the ft³/min be too high or low, the motor speed could be changed accordingly to low or high speed taps. The formula for this is:

$$\text{ft}^3/\text{min} = \frac{\text{Btu/h Output}}{1.08 \times \text{Temperature Rise}} = \frac{88{,}000}{1.08 \times 82}$$
$$= 994\ \text{ft}^3/\text{min}$$

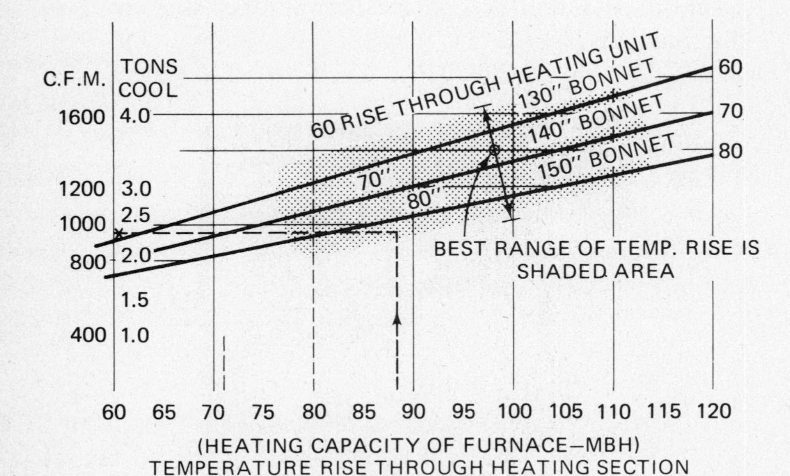

FIGURE 20-20 Determine CFM from air rise (Courtesy, Westinghouse Electric Corporation).

Another method is to determine the total resistance of the system (Fig. 20-21). Place an inclined static pressure gauge in a convenient location. The operation of this instrument was covered in Chap. 13. Adjust the liquid level and sliding scale to 0. Connect the positive hose (right tap) to the hole just above the furnace fan outlet. In this case it was in the **V** below our cooling coil. The negative pressure hose (left tap) is connected to the return air duct. Read the total static for this example as slightly below 0.5 in. of WC. Checking the installation manual fan performance table for medium speed, we note that the model used will deliver approximately 935 to 940 ft³/min between 0.40 in. and 0.50 in. Thus, we assume from our two checks that the air flow is adjusted correctly.

Another method of air flow verification is to check the fan motor amps by placing a clamp from an ammeter on the fan leads from the relay terminal (Fig. 20-22). Compare the reading obtained with the full load amp rating on the unit nameplate. In this example the reading should be slightly under the 7-amp rating. Checking the fan amps is also the procedure to follow to determine that the fan motor and the fan are running free with no bearing problems.

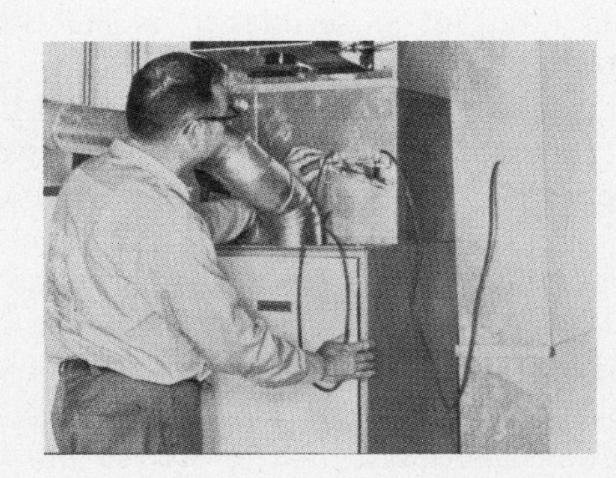

FIGURE 20-21 Measure total static (Courtesy, Westinghouse Electric Corporation).

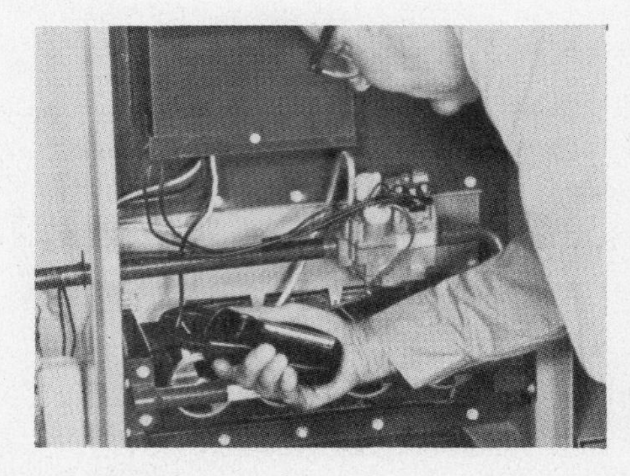

FIGURE 20-22 Check fan amps (Courtesy, Westinghouse Electric Corporation).

Continuing at the furnace, check the operation of the high-limit control. This may be done by temporarily removing one of the fan leads on the fan so the blower motor can't run. The gas valve remains energized, and thus the heat exchanger will build up heat rapidly and open the normally closed high-limit switch, which will close the gas valve and shut off the gas supply to the burners. As soon as the high-limit switch functions, reconnect the fan lead immediately to reduce the heat exchanger temperature. Another method is to block the return air by placing a piece of cardboard against the filter. Here again the temperature in the heat exchanger will rise and the high-limit control will open, which will close the gas valve and shut off the gas supply to the burners.

The other adjustment to the blower is the fan OFF and ON temperature. In the early days of gas heating, contractors would normally set the fan to cycle ON or OFF at 165°F bonnet temperature. This meant the blower would operate intermittently even when the outdoor air temperature dropped to design conditions. Wide variations in room temperatures resulted. Furnaces frequently overheated and wasted fuel. On the OFF cycle, the rooms furthest from the furnace suffered, while on start-up, those nearer overheated.

The unsatisfactory temperatures which resulted led the industry to use the method of CAC (continuous air circulation), or nearly continuous air flow. Briefly, the rules are:

(1) Select the correct size furnace and adjust the burner input to the lowest value that will handle the load at the design temperature.

(2) Adjust blower ft^3/min for 100°F rise (or less if cooling is applied)

(3) Adjust the fan switch to cycle the blower ON at 125°F and to cycle OFF at approximately 90°F, or as low as possible to prevent drafts.

These rules apply to heating-only applications; otherwise, the fan may be in constant operation where electronic air cleaning is used or where command humidification can cycle the fan ON even though heating is not required. There are variations, controlled by the thermostat selector switching and/or humidistats, which were previously discussed.

Referring to rule 3, these settings must be high enough to prevent drafts in the conditioned space. With your pocket thermometer measure the actual air temperature at the discharge grille farthest from the furnace on both START and STOP of blower

FIGURE 20-23 Measure supply air temperature (Courtesy, Westinghouse Electric Corporation).

(Fig. 20-23). Due to excessive runs or high heat loss in ducts, the discharge air temperature may be too low and cause drafts. If so, increase the values of rule 3 by 5° to 10°F. This will naturally raise the air temperature of the grilles closest to furnace; thus, their volume of air must be controlled to prevent overheating. With your fan switch setting determined, now proceed to balance the air distribution in each room.

Most installers balance residential systems by using common pocket thermometers placed about 30 to 36 inches above the floor (Fig. 20-24), where air can circulate around them, but away from direct heat sources such as lamps and sunshine. Open all dampers and grilles and allow the temperature to stabilize. Increase or decrease the air volume by opening or closing the branch dampers, if the duct system is so equipped, or the grilles, if dampers are not installed.

FIGURE 20-24 Measure room air conditions (Courtesy, Westinghouse Electric Corporation).

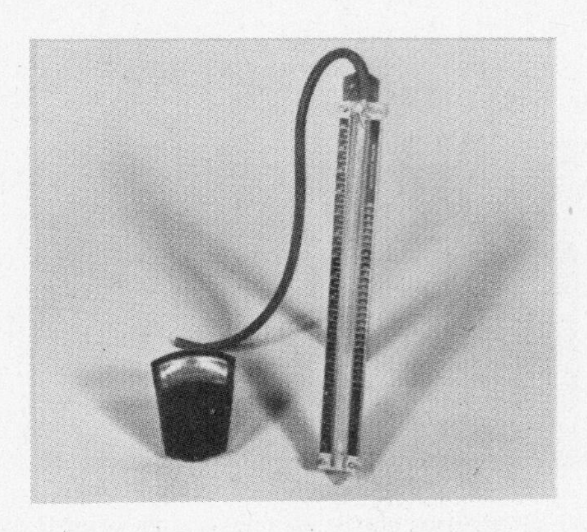

FIGURE 20-25 Air meters (Courtesy, Schilling Chilling Air Conditioning Company).

Some installers use air meters such as the ones shown in Fig. 20-25 to adjust the room ft³/min as calculated by the heat loss methods in Chap. 6. Both instruments measure the air velocity, and the one on the right can also measure pressure. They are both relatively inexpensive.

Take several readings over the face of the air outlet and average the results to determine the average velocity (Fig. 20-26). With the free area of the grille provided by manufacturer's catalog, simply multiply the velocity (ft/min) by the grille area (in square feet) to obtain the ft³/min. Compare this to calculated values. Increase or decrease the air flow as needed by adjusting the damper on the duct or at the individual registers. Admittedly this method requires some practice and also the specifications of the grille used, which may not be readily available to the technician at time of startup.

FIGURE 20-26 Taking velocity reading at grille (Courtesy, Westinghouse Electric Corporation).

During the room air-balancing period, also place a thermometer alongside the room thermostat and observe the differential in cycles. Close control will indicate good heat anticipation and proper air flow conditions. Erratic or wide cycles would indicate bad anticipation, stagnated air, or a thermostat that is not level or is out of calibration. Consult the thermostat instructions for changing calibration. Also compare the thermostat thermometer reading with your pocket thermometer. A faulty thermostat may be detrimental to the homeowner.

The final step of operation should be to go over user operations with the purchaser (Fig. 20-27). Review with him the thermostat operation, how to turn the gas off and on, lighting the pilot, normal CAC operation (if the system is so designed), *filter maintenance*, and any necessary safety precautions. Also be sure to explain the equipment warranty, and who to call for service. Chapter 24 covers customer service in more detail.

FIGURE 20-27 Explaining home owners manual (Courtesy, Westinghouse Electric Corporation).

We have just discussed the installation and operation of a conventional upflow natural gas furnace, which, in terms of units sold, is the most popular forced air furnace. There are important differences on counterflow and horizontal models. For example, the counterflow models require an auxiliary limit control mounted in the fan compartment as well as in the discharge air outlet. Clearances for horizontal and counterflow furnaces are frequently more critical than for upflow due to the nature of air movement.

Liquid petroleum gas furnaces in all cabinet styles must be treated with extreme care, because of the hazardous nature of LP fuels. Complete pilot shutoff is mandatory. Should pilot outage occur, the heavier-than-air LP gas would collect in the heat exchanger and furnace room, and upon reignition could cause a major explosion. Thus complete shutdown is absolutely necessary. Leak testing of all pipe joints and fittings must be done with great care.

Liquid petroleum gas furnaces are fired at a different rate than natural gas. The gas manifold pressure runs approximately 11 in. of WC as compared to the normal 3.5 in. of WC for natural gas. No gas meters are used, so the installer must determine the input capacity from the manifold pressure setting.

Space does not permit a discussion of all variations in heating apparatus, so it's important for the technician to follow the manufacturer's installation instructions and related training material covering the specific product in question.

PROBLEMS

1. What instrument is needed to measure gas manifold pressure?

2. The AGA data-rating plate on a furnace lists what important data?

3. Do furnace clearances refer to service space requirements only?

4. Belt tension should be approximately ______________.

5. The pilot flame should cover about ______________ of the thermocouple tip.

6. A furnace using 100 ft^3 of gas per hour and rated at 950 Btu/ft^3 has an input capacity of ______________.

7. If the two-cubic-foot gas meter dial makes two revolutions in one minute, the rate of flow is ______________ ft^3/h.

8. Natural gas furnaces will normally fire at ______________ in. of WC manifold pressure.

9. Liquid petroleum gas furnaces will normally fire at about ______________ in. of WC manifold pressure.

10. What does *CAC* mean?

21

HEATING SERVICE AND TROUBLE ANALYSIS

This chapter will acquaint the reader with some of the more common service and maintenance problems that do arise in connection with residential gas heating; we will again use a typical upflow gas-fired furnace as our example. Once a problem does develop it will usually be the job of the technician to handle the complaint, first to diagnose the cause and second to take the necessary remedial action to cure the problem.

FILTER MAINTENANCE

To start with, the most frequent cause of problems is the filter (Fig. 21-1). Again we stress the importance of showing the homeowner how to remove and replace dirty filters, with particular attention to replacing them with others of the same size and with the proper direction of air flow.

On occasion the homeowner might remove the fiber filters and replace them with a so-called cleanable metal filter. First of all, metal filters have a higher resistance to air flow. Also, when washed, particularly if they are sprayed with a high-velocity hose, they will tend to pack and further restrict flow. Also, wet lint is practically impossible to remove from the core, but when it is dry, it will find its way into the living area.

Strange as it seems, many homeowners will remove dirty filters and never replace them. Dirt and grease will adhere to the fan wheels, motors, and most importantly, clog a cooling coil in a hurry. The same damaging effect can also be caused by the homeowner turning off an electronic air cleaner. Again we cannot overemphasize the importance of proper filter maintenance.

PILOT PROBLEMS

Pilot problems are probably the second most frequent cause of heating complaints.

Pilot Outage

This probem can be traced to several possible causes:

FIGURE 21-1 Filter maintenance (Courtesy, Westinghouse Electric Corporation).

(1) Flashback, delayed ignition, flame rollout, or whatever else this phenomenon may be called, can quickly extinguish the pilot and cause a loud noise, and also alarm the homeowner.

(2) Excessive draft on the pilot can cause a floating flame on the thermocouple and disrupt the current long enough to close the safety valve. The draft could be due to flue problems, opening a garage door during strong winds, or on occasion, someone removing the burner baffle plate.

(3) Momentary low gas pressure, caused by too many other appliances cycling on at the same instant, particularly with undersized lines and a high gas consumption drain on the total utility system. Main line pressure should not drop below 7 in. of W.C. for a natural gas system.

(4) A clogged orifice due to scale, rust, or corrosion from inside the gas lines or, externally, scale particles falling from the heat exchanger surface. The latter condition usually occurs after summer operation with high humidity conditions.

(5) A defective thermocouple—the safety valve won't stay open.

If the pilot runners (the metal channels on top of each burner that carry the flame, as explained in Chap. 8 on the general construction of a furnace) get out of alignment, one or more burners may fail to ignite until too much gas has accumulated in the heat exchanger cavity. Finally, when the burners do ignite (Fig. 21-2) a small explosion takes place, and the resultant expansion creates a flashback through the secondary air channel strong enough to blow out the pilot, in addition to creating a loud noise. The force has even been known to blow off the furnace door.

FIGURE 21-2 Flashback delayed ignition flame rollout (Courtesy, Westinghouse Electric Corporation).

To check both the pilot runners and burners, remove the baffle plate (Fig. 21-3). After attempting to align the burners and pilot runners by visual sighting, it may be necessary to use a hand level to check them. For purposes of illustration, the burners have been removed from our furnace (Fig. 21-4). First, place a level across two or more burners to check the horizontal level plane. Then place the level on top of each burner and measure the level from front to back.

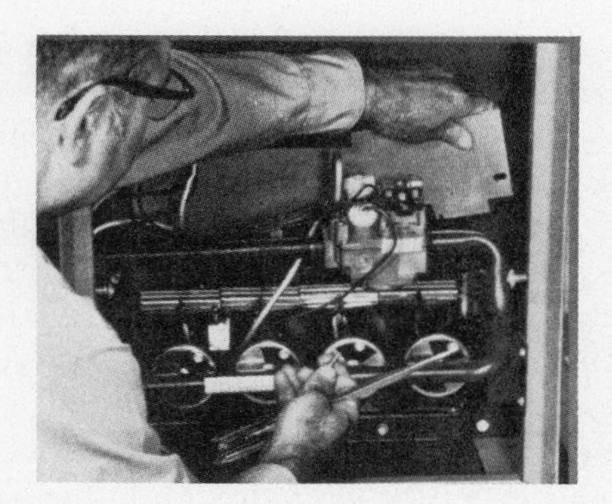

FIGURE 21-3 Remove baffle plate (Courtesy, Westinghouse Electric Corporation).

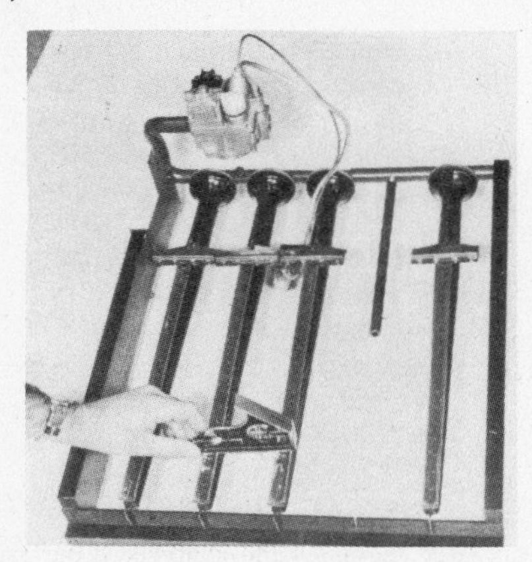

FIGURE 21-4 Checking burner level (Courtesy, Westinghouse Electric Corporation).

Although it occurs very infrequently, it is possible that the pipe tap hole for the gas spuds may be drilled wrong, so that it doesn't align with the burner tube. This would cause the gas to enter the tube at an angle and possibly result in poor mixing in the venturi. Delayed ignition or premature burning could result. To check this, remove the gas spud. Screw a 12-inch length of 1/4-inch threaded pipe into the manifold opening and, with your level, check the level plane from front to back. If the port is drilled wrong, it will be necessary to replace the manifold piping.

Referring back to item 5 in our list of pilot outage causes, *erratic thermocouple operation* may be traced to several sources. An example would be dirty contacts at the thermocouple connection to the gas valve. Loosen the nut with a wrench. (Fig. 21-5). Examine the surface contacts and clean them. Tighten the thermocouple nut first by hand, then with a wrench. An additional quarter turn will be sufficient to seat the lock washer. Overtightening may cause damage to the thermocouple or magnet.

If cleaning the contacts fails to cure the problem, then look for a defective thermocouple lead. Some installers check a thermocouple by substituting a new one, and if normal operation is restored, they discard the original equipment; however this is not a generally recommended procedure, for you may not always have a proper replacement unit at hand.

Instead you can test a thermocouple lead by disconnecting it from the gas valve. Hold the gas valve control knob down, light the pilot, and, with a millivolt meter, check the current flow (Fig. 21-6). If

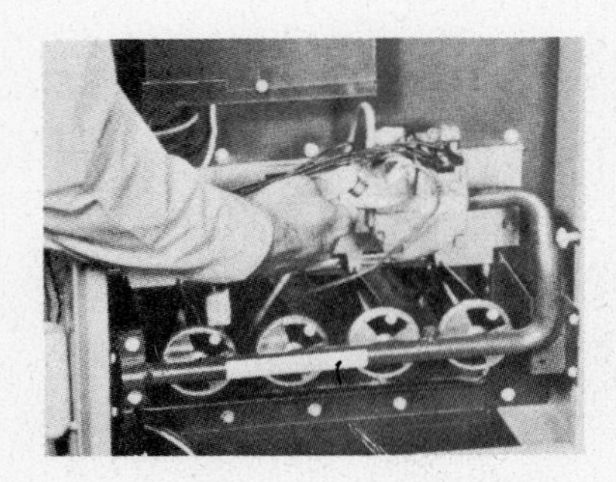

FIGURE 21-5 Check thermocouple contacts (Courtesy, Westinghouse Electric Corporation).

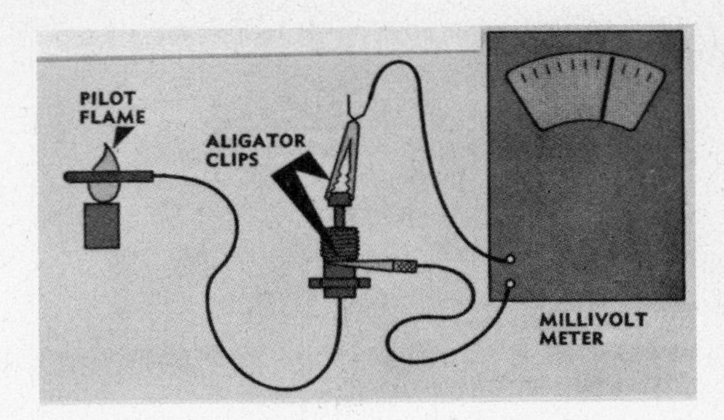

FIGURE 21-6 Checking thermocouple millivolt output (Courtesy, Westinghouse Electric Corporation).

it is less than 8 to 10 millivolts, you can assume there is a defective lead and replace it.

On rare occasions there might be a failure in the automatic pilot magnet assembly. To replace it, shut off the gas, remove the thermocouple, and with a socket wrench unscrew the magnet assembly. Replace the assembly with a new unit and apply thread lubricant. Tighten sufficiently and check for gas leakage. Reassemble the thermocouple and the magnet assembly and test for normal operation. NOTE: This discussion applies to the specific gas valve being reviewed. Other make valves may not use a magnet operator, and the recommendations for each particular valve design should be followed.

Intermittent Pilot Outage (incomplete outage)

This condition can also cause nuisance complaints. Here are some associated problems.

(1) Wrong or improper mounting position. The pilot may have been twisted, sprung, or loosened in transit. The flame may miss the thermocouple or fail to line up with the pilot runner. The adjustment may be up, down, or rotational.

(2) Low gas pressure, which will cause a soft and wavering flame as previously described.

(3) High gas pressure, which will cause the flame to change direction and also to miss the thermocouple.

(4) Wrong flame adjustment at the gas valve.

To readjust the pilot, remove the adjustment cap and, with screw driver, reestablish the flame characteristics described in Chap. 20.

Assuming that you have established normal pilot and thermocouple operation, let's now consider some problems associated with an inoperative gas valve.

GAS VALVE PROBLEMS

If you have an *inoperative valve* (and the pilot is burning normally) look for these conditions:

(1) No secondary voltage at the valve. If, with a test meter, you fail to get a 24-volt supply then suspect:

(a) a burned-out transformer. Check the secondary taps of the transformer; and if there is no voltage, go to the primary side. If you have a 115-volt input, replace the transformer.

(b) the thermostat contacts are dirty or not closed. Clean the contacts with a business card or other unprinted, hard-surface paper.

(c) open circuit in low-voltage wiring, a loose connection, or a short circuit.

(2) A voltage at the gas valve. You should suspect a sticking valve solenoid or clogged internal gas ports. To check the solenoid, remove one 24-volt wire at the valve terminal, touch it to the terminal, and listen for a "click." If there is no click, replace the entire valve.
If the solenoid clicks and the valve does not open, the valve is sticking or its internal ports are clogged. To clean the main valve, carefully follow the service instructions provided by the valve manufacturer.

(3) No primary voltage. Obviously, if there is no primary voltage to the transformer, there will not be any 24-volt secondary to the gas valve. Check for a blown fuse or loose connections, and also for a defective high-limit control. If the high-limit opens, all power to the transformer will be cut off. Disconnect the power wires to the switch and, with an ohmmeter, check the resistance. If there is no resistance, the switch has failed to open.

In addition to the problems associated with the pilot and gas valve, there are other types of service situations to be examined.

BURNER PROBLEMS

Flashback (where the flame burns in the venturi)

The situation depicted in Figure 21-7 can result in noisy and inefficient burner operation. This condition can be the result of insufficient primary air and improper mixing. It is also caused by too low a manifold pressure, which won't induce enough primary air into the tube for proper burning. The velocity of the mixture should be high enough to carry outside the burner inserts.

Damaged or clogged burners will result in the same condition. Inspect the ports and repair them if damaged. Sometimes, when a furnace is located near a clothes dryer, lint will be pulled into the secondary air stream and clog the ports. Heat exchanger rust scale can also sometimes clog burner slots.

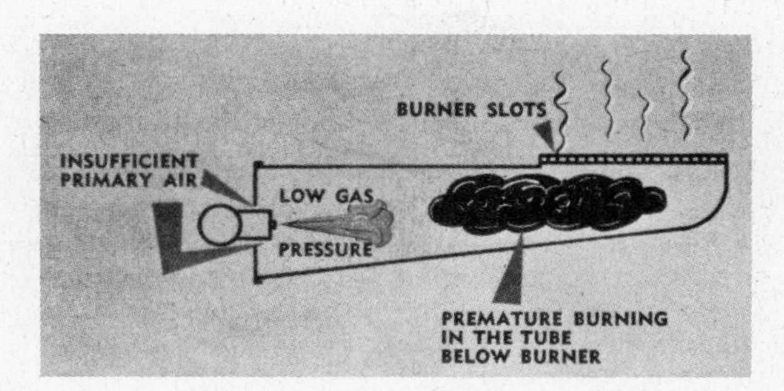

FIGURE 21-7 Flashback in burner (Courtesy, Westinghouse Electric Corporation).

To check the burner ports, remove the baffle plate as previously shown and then extract each burner. With a wire brush or other suitable tool, carefully clean the slots or ports and remove all foreign matter. Straighten any bent openings.

INSUFFICIENT HEAT COMPLAINT

If the homeowner's complaint is *not enough heat*, look for these possibilities:

- ***Undersized Furnace*** It is possible that the original application selection was in error, or it was assumed the homeowner would provide additional insulation, storm doors, and windows. Perhaps the owner is trying to heat areas such as basements or garages that were not considered in the original load calculation survey. Be very careful in dealing with the customer on this point.

- ***Furnace Underfiring*** Can be checked at the meter. If so, raise the manifold pressure, taking care not to exceed the furnace's rated input capacity. Gas lines that are too small and a falling main line pressure can also cause reduced heat during design temperature conditions.
- ***Not Enough Heat*** Could be caused by the *thermostat* being set too low, out of calibration, or affected by heat sources such as fireplaces, lamps, TV, or sunshine. These conditions could cause erratic cycles.
- ***A Malfunctioning High-Limit Control*** The control could be out of calibration and cause burner cycling.
- ***Low Air Flow*** is generally the most common cause of unsatisfactory operation of a ducted heating system. It can cause insufficient heat transfer, cycling of the high-limit control, burned-out heat exchangers, and excessive temperature rise through the heat exchanger. 100°F rise is the maximum AGA rating condition. Temperature rises in excess of 100°F could be caused by (a) dirty filters, (b) undersized ducts or restricted air flow at registers or grilles, or (c) blower running too slow. If the blower is belt-driven and running too slow, check the belt for slippage; also check the motor variable-speed pulley set screw. If direct-drive, check the speed tap selection. In either case, when you increase the air flow, check the motor amp draw against the nameplate rating. DO NOT EXCEED allowable amps.
- ***Furnace Sooted*** Caused by carbon buildup on the inside of the heat exchanger, this condition can create a heat transfer barrier that is allowing too much heat to be exhausted up the stack and not enough heat transfer through the heat exchanger to the conditioned air. The main cause of soot is insufficient primary air. A yellow flame will produce soot and fumes. The incorrect adjustment of the primary air shutters, excessive manifold pressure, an insufficient stack draft, or a combination of all three will create a sooting problem.

Sometimes the homeowner's complaint will be *too much heat*. Look for the possibility that the thermostat is set too high or is out of calibration; more frequently, the thermostat is being influenced by cold air drafts created by opening doors, poor weatherstripping, fireplace draft, etc. Check also the possibility that the burners keep operating when the thermostat is satisfied. This could be caused by a short in the thermostat wiring. Remove one wire at the valve; if the valve fails to close, there is a short. Also, if the valve does not close, it may be sticking open mechanically, and the furnace will operate continuously, resulting in excessive heating in the house.

If the homeowner reports that the furnace is cycling too frequently, check for improper thermostat location (too near a door or where rapid changes in air flow may exist). Recheck the heat anticipator setting of the thermostat to see that it matches the gas valve amperage.

The most dangerous situation of rapid cycling would be on the high-limit control due to low air flow and other causes. DO NOT increase the high-limit setting beyond the manufacturer's factory setting. Instead, look for the causes of the high-limit cycling and correct them. An oversized furnace will produce overheating and inefficient operation.

If you have furnace set for almost complete CAC (constant air circulation), there will be more rapid cycling of the burner and fan. This is normal and should be explained to the homeowner along with other general operating instructions.

When the complaint is that there are too frequent or too long OFF cycles and excessive swings in temperature, the thermostat may be out of calibration or may be improperly located and not sensing temperature changes, particularly if the air is stagnant. Also, the heat anticipation could be set for too long a cycle. Dirty thermostat contacts, if they are of the open type, might produce erratic and long cycles.

An oversized furnace and a high fan ON setting will definitely produce overheating and long OFF periods, because of the residual heat in the heat exchanger. Temperature swings will be increased. It may be necessary to derate the furnace according to the manufacturer's recommendations and lower the blower ON setting.

HEAT EXCHANGER BURNOUT

A burned-out heat exchanger (Fig. 21-8) can be caused by excessive overfiring, coupled with low air flow and poor draft conditions. Its life can also be

FIGURE 21-8 Burned-out heat exchanger.

shortened by water condensate dripping from a cooling coil during summer operation. There may also be internal rust formed by the combination of low-temperature inlet air and too low a temperature rise through the furnace. One of the major reasons for heat exchanger failure is that the combustion air contains corrosive elements.

A burned-out heat exchanger can sometimes be detected by the presence of strong fumes (odor) in the conditioned air stream, or by observing the burner flame pattern before and after the fan starts. If the flame waves when the blower starts, it is a strong indication that there is a break in the heat exchanger. Some service personnel also use a smoke device to check leakage into the air stream.

FAN MAINTENANCE

Most direct-drive furnaces are designed so that the direct-drive blower can be serviced by first loosening the sheet-metal screws that hold the deck assembly to the furnace (Fig. 21-9). Pull forward on the fan deck while at the same time feeding the electrical leads through the grometted hole. When the deck is fully extended, you have access to the motor mounts and the capacitor. Oil the fan motor per instructions on the furnace, and make sure the oil cups are pointed in an upward direction. Techniques for testing the fan motor and capacitor are covered under the electrical instructions in the installation bulletin.

As mentioned in the startup, if you have a belt-driven model, make sure that the belt tension is correct to avoid slippage but not too tight to cause excessive wear; about $\frac{3}{4}$ to 1 inch of play should be allowed. Also check the pulleys alignment by using a straight edge. Misalignment of pulleys will cause excessive vibration and wear on both the blower and motor bearings.

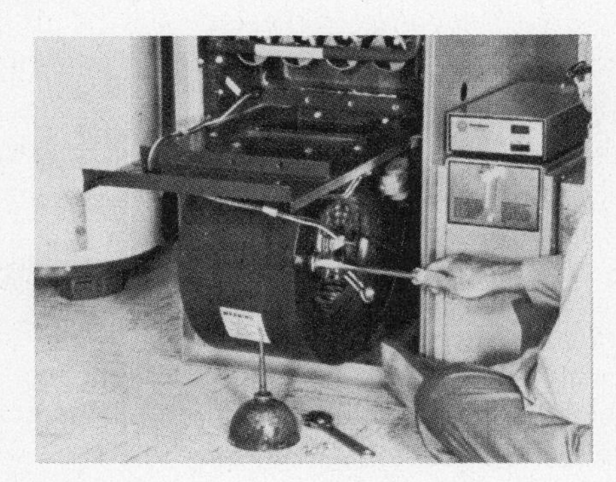

FIGURE 21-9 Servicing fan assembly (Courtesy, Westinghouse Electric Corporation).

NOISE COMPLAINTS

A noisy system operation will also be a source of customer complaints. Here are some common problems related to ductwork (Fig. 21-10, 11, and 12).

- Ducts are too small, and the velocity exceeds the recommendations for residential application. The main ducts should be 700–900 ft/min; the branch ducts, 600 ft/min; and the branch riser 500 ft/min.
- Grilles and resisters also have critical velocities and can cause "hissing" or "whistling" if the velocity is too high. With all the different types of supply outlets, it's difficult to set exact standards; however, a good rule of thumb is not to exceed 400 ft/min on return grilles.

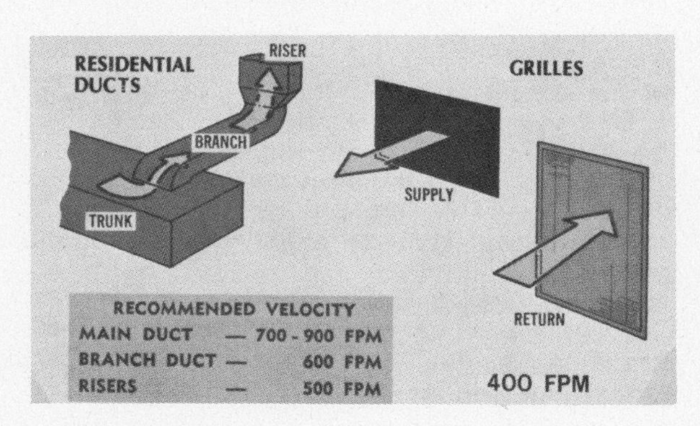

FIGURE 21-10 Air noise (Courtesy, Westinghouse Electric Corporation).

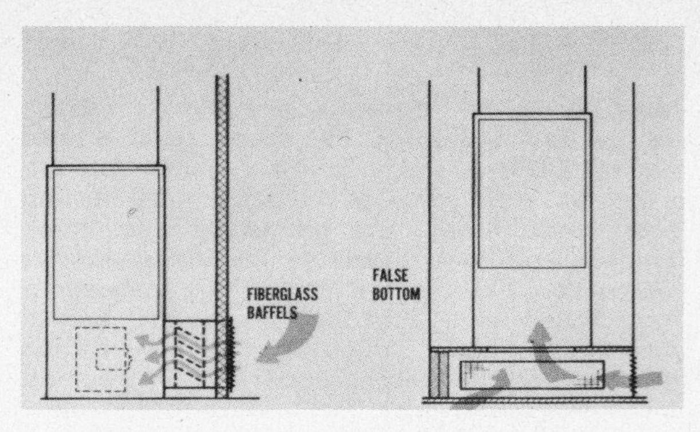

FIGURE 21-11 Ducting practices (Courtesy, Westinghouse Electric Corporation).

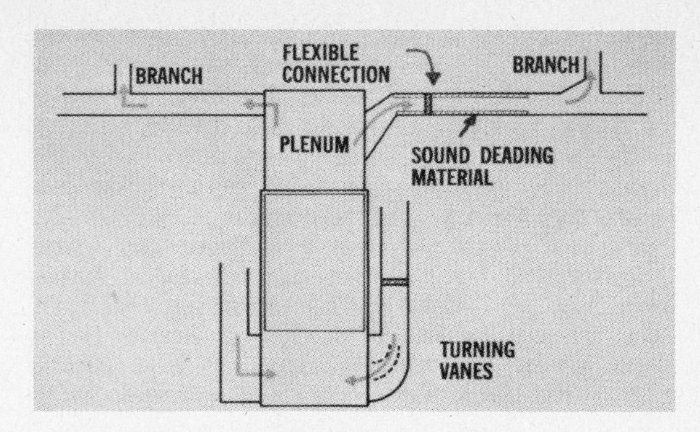

FIGURE 21-12 Short returns (Courtesy, Westinghouse Electric Corporation).

- Vibration and noise transmission. Although most modern furnaces have reasonably low operating noise and vibration levels, it is good practice to provide flexible connections on both the supply and return trunk connections; metal ducts transmit sound rapidly. Also, the flexible connections help reduce expansion noises on startup or shutdown. Nonmetal ductwork (fiberglass) systems have excellent sound-absorbing qualities and provide thermal insulation as well.
- Poorly designed ductwork, such as a 90° turn at the plenum, can create unnecessary pressure and noise; a gradual transition is recommended. This also holds true for branch runs; use formed take-offs. On short returns, turning vanes will reduce static and possible noise.
- The use of sound-deadening fiber material in a short section of supply and return duct can often do much to quiet air noise. This is particularly true of a closet installation that has only a short connection for return air. Here it is recommended that fiberglass be used on insulated baffles, which are staggered so that motor and fan noise cannot be directly transmitted to the living space.
- Also, in closet applications that use a bottom return with a raised platform, make sure that the raised base is well supported and doesn't act as a drum. Central returns handle a lot of air, so be liberal in the size of the return openings and the grilles.

The discussion in this chapter is directed toward technical causes of customer complaints. Frequently the complaint is more from a lack of understanding by the homeowner of what to expect and how to operate the equipment properly. For example, many homeowners feel that the thermostat is like the accelerator in a car; the higher it is turned the faster the house will get warm. And so they become impatient.

So, in answering a customer complaint, see if there is a technical problem, and solve it first. Then take time to explain your actions and what the homeowner might have done to minimize problems. Good customer relations and product service are vital keys to your business image in the community. Therefore, in Chap. 24 we will address ourselves to this problem of human communications.

TRUE OR FALSE

1. Throw-away filters, furnished with the furnace, should be thrown away and replaced with the metal permanent type. ______________
2. A defective thermocouple will generally cause delayed ignition. ______________
3. Improperly aligned pilot flame runners and burners can cause delayed ignition and flame rollout. ______________
4. Gas manifold assembly and/or gas burners do not have to be level for proper burner operation. ______________

5. Never attempt to redrill gas orifices. ______________

6. Thermocouples can not be checked—they can only be replaced. ______________

7. No 24-volt power at the gas valve always indicates a defective transformer.______________

8. 24-volt power to the gas valve indicates proper operation of the valve. ______________

9. An "open" high-limit switch would terminate power to the gas valve but not to the transformer. ______________

10. A burned-out heat exchanger could be caused by over-firing, coupled with low air flow and poor draft conditions. ______________

PROBLEMS

1. What is the cause of most common service problems?

2. Pilot outage can be caused by several problems. Name one.

3. Delayed ignition can be the result of misaligned ______________ runners.

4. The function of the thermocouple is to generate what?

5. Flashback is a condition where the gas burns inside the ______________.

6. Flashback can result from several causes. Name two.

7. Too high an air rise could be the result of several factors. Name two.

8. Long cycles and excessive temperature swings could be caused by the improper setting of thermostat ______________.

9. High air flow can cause tripping of high limit control. True or false?

10. A yellow gas flame will produce ______________.

22

COOLING SYSTEM—STARTUP AND OPERATION

This chapter is a continuation of the check-test and startup methods that would be applied to the cooling components of a residential heating and air-conditioning system. We will assume the indoor furnace equipment, cooling coil, and ductwork were installed in keeping with the general recommendations given in previous chapters, as well as the specific recommendations offered by the manufacturer.

Figure 22-1 represents the installation of the outdoor air-cooled condensing unit, connecting wiring, and refrigeration lines. The physical location of the condensing unit is ideally the shortest distance from the evaporator, but this choice must sometimes be compromised because of a possible noise problem, and the necessity of free air movement into and from the unit itself. Observe the manufacturer's recommendations on minimum clearances; see Fig. 22-1, which shows the proper setback from walls or overhead obstructions as well as minimum service considerations.

A poured concrete slab or precast concrete slab over gravel fill should be level and extend 3 inches above the ground, and it should be at least 3 inches larger in each direction than the unit itself. Do not tie the poured slab into the building.

The hole in the wall should be large enough to accommodate the insulated suction line and the liquid line. Route the suction and liquid line from the evaporator to the condensing unit. If excess refrigerant line(s) are present, coil the excess in *a horizontal* fashion as illustrated, and, if needed, support both lines with sheet metal hangers. Next, make the proper refrigerant connections, depending on which form of seal is used (precharged quick-connect couplings, compression fittings, or flares). Here the method of assembly purging and leak testing of refrigerant lines must follow the exact equipment specifications. Seal the exterior wall opening with Permagum or its equivalent.

For exterior power to the condensing unit, install a weatherproof fused disconnect switch located in a low position so that it can be easily reached during any service operation. Wiring should also be in a weather-tight flexible conduit and naturally must meet or comply with national and local electrical codes. Power wire sizes and fuses are determined from the equipment installation bulletin. Low-voltage wiring is usually No. 16 wire, depending on the distance between the thermostat and the condensing unit. Electrical point-to-point hook-up should follow a similar pattern to that explained in Chap. 18. Be sure to identify the correct terminals and field splices properly before making connections.

Not illustrated in Fig. 21-1 is the installation of a proper condensate drain line for the evaporator coil fan OFF. As previously noted, plastic tubing is most commonly used for residential work. See that it terminates in an approved open drain.

With these installation recommendations completed, proceed to make a series of visual electrical and mechanical checks of the system as follows:

(1) *Keep the power off* by *not closing* the disconnect switch. But *do check*, with a volt meter, the actual incoming voltage against the unit's nameplate rating. Most single-phase residential units are dual rated at 208/230 volts, and the maximum permitted voltage variations can run from −5% (197 volts) on the low side to +10% on the high side (253 volts). Outside these limits the equipment will not perform properly and may be subject to failures. Should failure occur, it is doubtful that the warranty would be in effect. We will assume the voltage is correct.

(2) Verify the proper fuse ratings in the disconnect switch.

(3) With the condensing unit service panels removed:

(a) Spin the condenser fan blades by hand to make sure the blades don't hit the shroud and also to assure that the motor bearings are free. Oil the fan motor in accor-

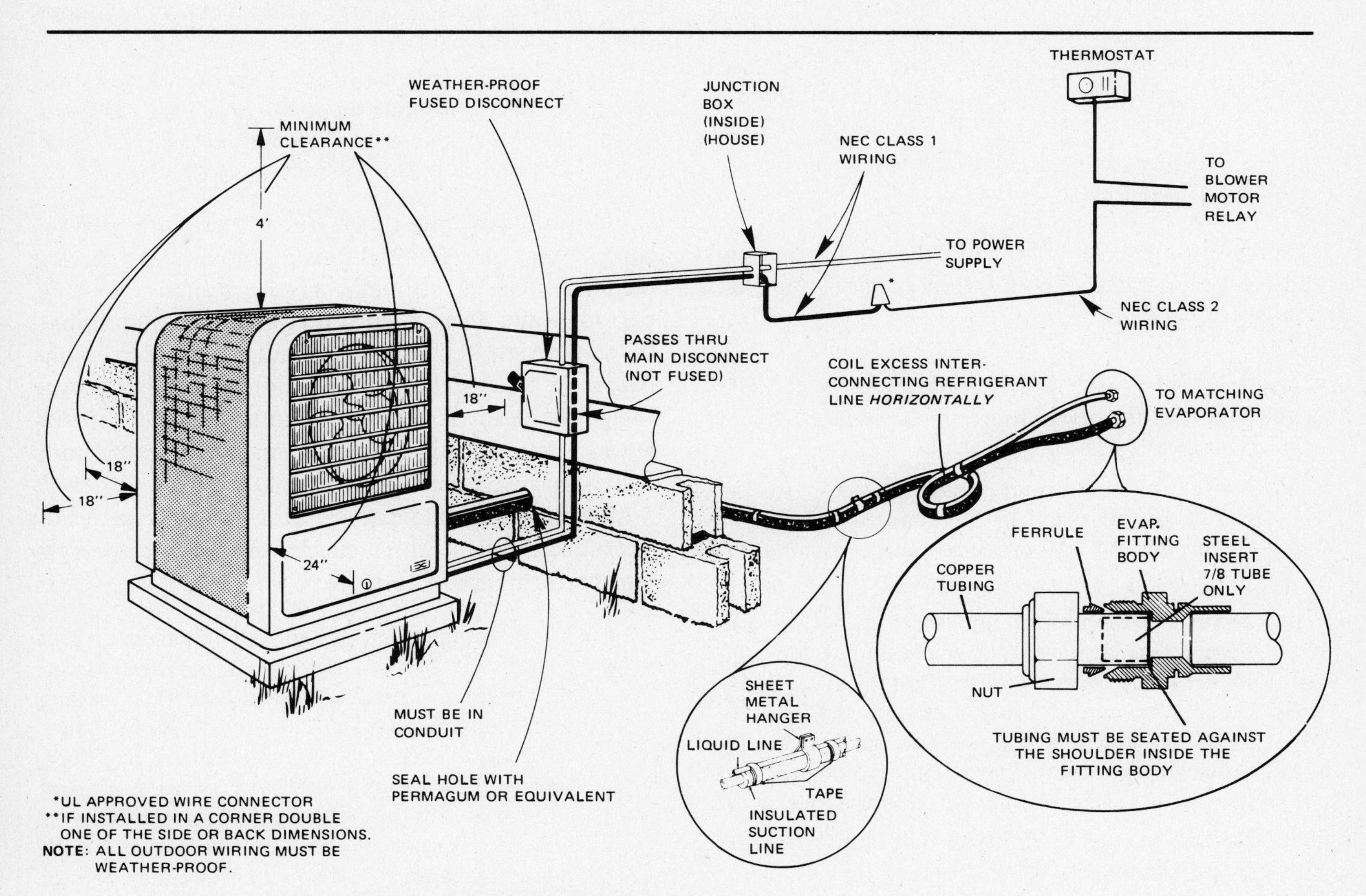

FIGURE 22-1 Installation of outdoor air-cooled condensing unit (Courtesy, York, Div. of Borg-Warner Corporation).

dance with the instructions. (Some are permanently lubricated).

(b) Check the electrical connections at all terminals, even those made at the factory, to ensure a tight fit.

(c) If required, loosen the compressor hold-down bolts to permit floating action.

(d) If refrigerant valves are furnished, open (back-seat) first the suction and then the liquid valves.

(4) Replace the service panels.

(5) Indoors,

(a) Recheck the low-voltage thermostat to see that it is level, and place the system selector in the OFF position.

(b) Check the air filter to make sure it is clean and that the direction of the air flow is correct.

(c) Check the cooling coil service cover for tight fit.

With these checks completed the system is now ready to be operated. Follow these general steps for startup:

(1) Keep the outside power disconnect switch OFF.

(2) Turn ON the power source to the indoor fan and control circuit.

(3) Move the thermostat system selector to the COOL position, and set the temperature selector to the desired indoor temperature (assume 75° to 80°F).

(4) Move the thermostat fan selector to ON, and observe the furnace fan, which should come on immediately with the air flow established. Verify the ft^3/min, as previously presented in Chap. 20, by either the temperature rise method or by measuring the total external static pressures and comparing them to the fan performance data.

(5) Return to the condensing unit and close the fused disconnect switch. Observe the condenser fan start and the cooling relay close to start the compressor. Listen for any abnormal sounds (this will come with experience).

(6) Let the system operate for several minutes to stabilize the refrigerant flow, and then observe the following indications of proper operation:

- The condenser fan is running with warm air coming off the condensing coil.
- The indoor blower is circulating cool air to the supply ducts.
- The suction line at the evaporator coil is cold to the touch.
- The liquid line at the condensing unit is warm to the touch.

(7) Return to the indoor thermostat and set the fan selector at AUTOMATIC; the system should continue to operate. Remove the cover and slowly move the cooling temperature selector toward a warmer position until it breaks contact. Immediately check to see that the condensing unit and the indoor fan have gone off. *Wait five minutes* for the refrigerant pressures to equalize, and then slowly return the cooling selector to start the system again. Observe both indoor and outdoor units come on.

(8) If the outdoor weather is hot it will take the residence an hour or more to pull the temperature down so that the cooling unit can cycle automatically from the thermostat.

During this pull-down time there are several things the professional installation and service technician can do to make this a productive waiting period. First, from a customer relations standpoint he can review the owner's manual with the homeowner to be sure that he or she understands the system. This point is very important, and in Chap. 25 we will develop this concept from both product and communication considerations.

Second, the technician can run through a series of operational verifications checks to measure the system performance. These are:

(1) Using an ammeter, measure the condensing unit power draw off the *L1* electrical leg at the disconnect switch (Fig. 22-2). Compare this reading with the nameplate data or product performance information for normal operation. Remember, this is total current for both compressor and condenser fan motors. If the current draw is higher than normal (10% or more), check each motor separately. Also recheck the voltage while the unit is operating.

(2) Now that the system is under full operating pressures recheck all the refrigerant piping connects for leaks. When detecting leaks outdoors

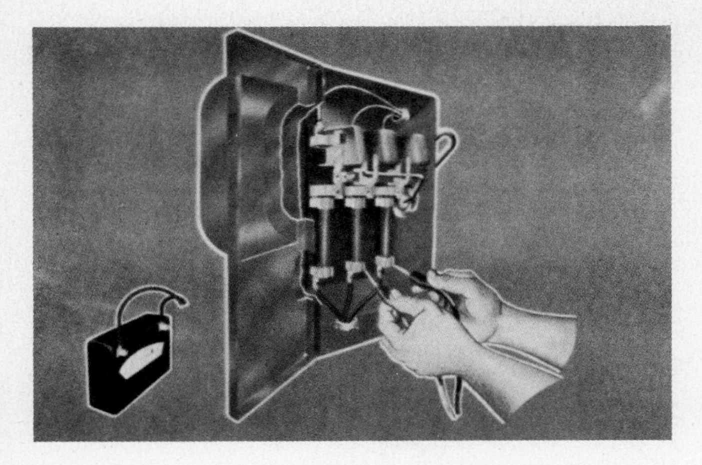

FIGURE 22-2 (Courtesy, Carrier Air- Conditioning Company).

around the condensing unit there will be considerable air movement so go slowly—be sure!

(3) Using dial or stem thermometers read the return air temperature coming to the coil and the air off the coil (Fig. 22-3). For the average residential system (50% relative humidity and 400 ft³/min per ton of air flow), the air temperature drop will be in the 18° to 20°F range. For high air-flow systems as used in the southwest where humidity is low, the temperature drop may be in the 15°F range. In this chapter we will assume normal operation. Chapter 23 will examine abnormal operation.

(4) An experienced technician will also examine the operating refrigeration pressures and compare them with the manufacturer's curves for normal operation. To do so he will attach his gauge manifold. Assume the unit has service valves and is operating.

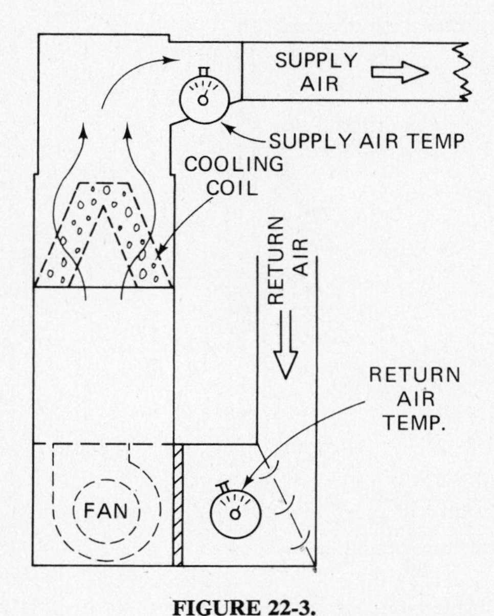

FIGURE 22-3.

The first step is to purge the gauge manifold of contaminants (Fig. 22-4) before connecting it to the system. Remove the valve stem caps from the service valves and check that both service valves are back seated, or that gauge ports are closed. Remove the gauge port covers from both service valves. Connect the center hose from the gauge manifold to a refrigerant cylinder, obviously using the same type of refrigerant as the system. Open the valve on the refrigerant cylinder for about two seconds; then close it. This will purge contaminants from the gauge manifold and hoses. Next, connect the gauge manifold hoses to the gauge ports (Fig. 22-5), the low-pressure gauge (compound) to the suction service valve, and the high-pressure gauge to the liquid service valve. Front-seat or close both valves on the gauge manifold. Next crack (turn clockwise) both air-conditioning unit service valves one turn off the back seat. The system pressure is thus allowed to register on each gauge.

If the Schraeder or drill type of valves are used (Fig. 22-6), the gauge hose coupling must be equipped with valve core depressors, or the special Schraeder adaptor must be used to allow the system pressure to register on the gauge. Either method will let you observe the suction and liquid (head) pressures during operation. But are they normal? Most manufacturers offer an operating chart, or charging chart, similar to that in Figure 22-7, which is based on the comparison of liquid and suction pressures with indoor air wet-bulb temperature (to the coil) and outdoor condensing air temperature to the condenser. Figure 22-7 is a text illustration only and not to be used in actual practice. Assume a 95°F outdoor air temperature and a normal 67°F WB indoor air temperature. The intersection of these points (*A*) would indicate that for R-22 refrigerant the pressures should be about 272 psi for the head pressure at the liquid valve and 77 psi suction pressure at the suction valve. A comparison of actual pressures against the chart will indicate the degree of departure from ideal operation. A wide difference does not necessarily indicate faulty equipment, but is simply another indication that additional corrective action is needed. Corrective action will be discussed in Chap. 23. Let's assume the system is running normally, but we will leave the gauges connected.

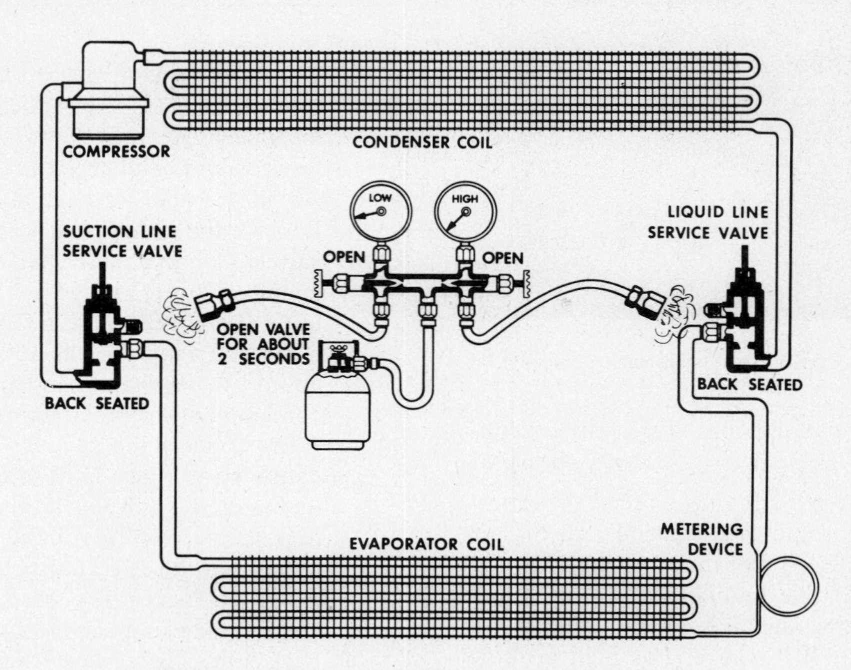

FIGURE 22-4 (Courtesy, Bryant Air-Conditioning Company).

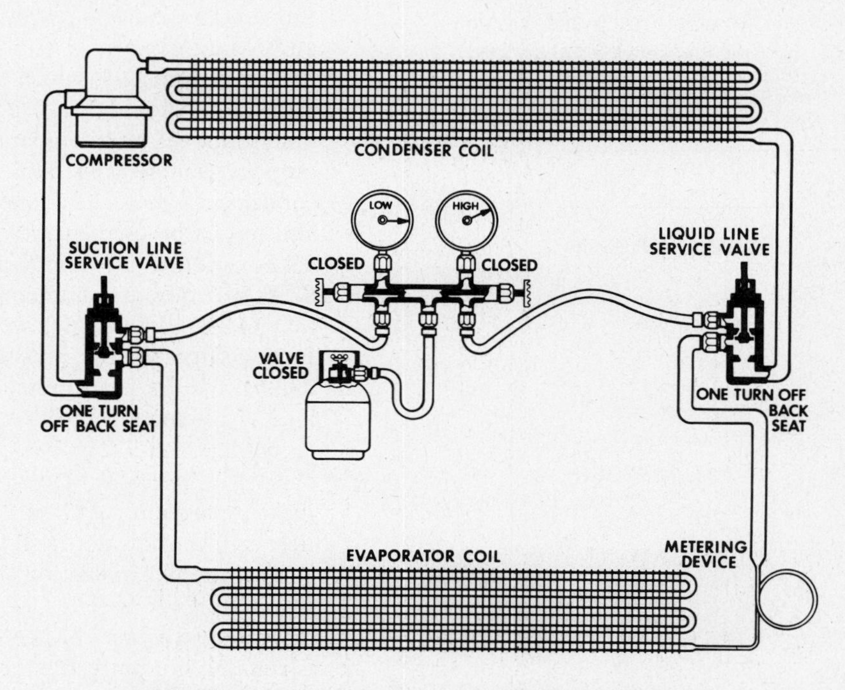

FIGURE 22-5 (Courtesy, Bryant Air-Conditioning Company).

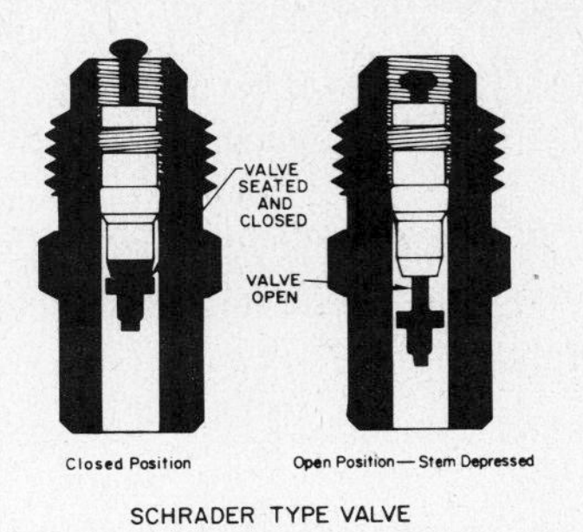

FIGURE 22-6 (Courtesy, Bryant Air-Conditioning Company).

(5) If the condensing unit is equipped with high and low-pressure safety cutouts, these may be checked to insure proper operation. First, test the low-pressure safety control by blocking the indoor air flow. Place cardboard or paper over the air inlet side of the filter. Observe the low-pressure refrigerant gauge as suction pressure falls until it reaches the cutout setting. The switch should then operate. Remove the air blockage and permit the system to resume normal operating pressures.
To check the high-pressure cutout, block the inlet air to the condenser; however, this may be a bit difficult to do on some round unit configurations. Cover the inlet air with cardboard, paper, or a dropcloth. Observe the buildup in head pressure and the resulting trip-off of the high-pressure switch. Remember, the compressor motor will probably also have thermal and current overload protection, but normally these would operate beyond the pressure conditions needed to trip off the high limit; thus a second safety backup also assumes that the HP switch opens.
Remove the air blockage from the condenser air inlet and let the refrigerant pressure return to normal. If the high-pressure cutout has an automatic reset, the unit will come on once safe conditions are reached. If cutout has manual reset, wait until the pressure falls to near normal conditions (see Fig. 22-7) and then push the button; the compressor should start.

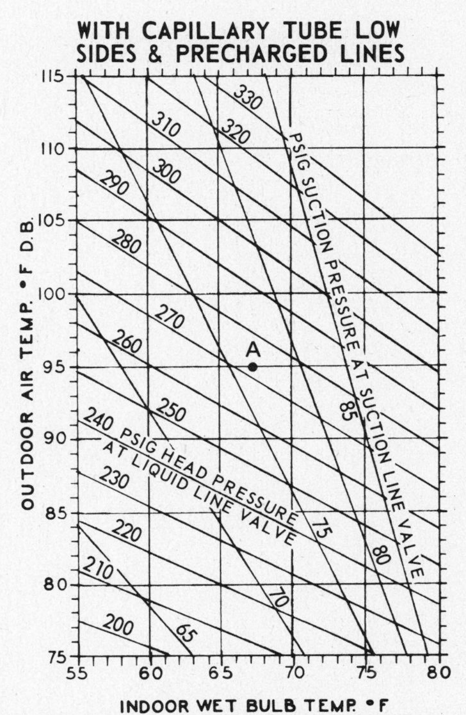

FIGURE 22-7 Operating or charging chart.

(6) Check the evaporator coil condensate drain line to ensure it is running free with no restriction.

(7) By the time the technician has made these observations, the house conditions should be approaching normal temperature, at least at the central control point (thermostat). But this doesn't mean each room is comfortable; one may be too hot, another too cold. So once again the air distribution system may have to be balanced. This is easy to understand. For example, a room facing north will need lots of warm air for heating but not too much cool air on cooling, etc.

But don't touch anything! That is, until you have marked the heating position of any main trunk dampers or branch dampers. Do this with a permanent marker. If no duct dampers are used, adjust the supply air outlet volume controls to allow more or less air as needed for cooling. *But*, also show the homeowner how to make these seasonal adjustments.

Fortunately, the modern residence is usually well insulated, and the use of shading devices, indoors and outdoors, minimizes the temperature differences between rooms. Hopefully then a compromise air quantity can be set for year-round operation.

The above description of startup and operating procedures is realistic in terms of the reliability of industry products to perform the first time the switch is pulled, provided the installing technician has done his job properly.

However, as with all mechanical equipment, there will be some equipment and installation problems that will require the skills of a trained technician to diagnose and correct. Chapter 23 will examine some of the more common causes, symptoms, and corrective actions.

PROBLEMS

1. Outdoor equipment at ground level should have a firm ______________.
2. Excess refrigerant tubing should be coiled in a vertical plane. True or false?
3. The outdoor disconnect switch should be located ______________.
4. Outdoor wiring in accordance with National Electrical Code should be ______________.
5. An outdoor condensing unit must have the necessary clearance from the wall for ______________ and ______________.
6. What are the permissible voltage variations on single-phase 208/230-volt hermetic compressors?
7. What is the normal air temperature reduction across a cooling coil?
8. Refrigerant pressures are taken by means of connecting ______________.
9. Referring to Fig. 22-7, if the outdoor temperature is 95°F and your gauges read 260 psig head pressure and about 72 psig suction pressure, what condition would you suspect?
10. Should the evaporator condensate drain into closed sanitary plumbing?

23

TROUBLE-SHOOTING AIR-CONDITIONING SYSTEMS—GENERAL OBSERVATIONS

When an air-conditioning system fails to perform properly, the underlying cause will usually fall into one of four categories: parts failure, improper adjustment, poor installation, or poor design.

(1) Parts failure

- Defective in manufacture
- Subjected to conditions beyond rated capacity
- Not properly maintained
- Worn out

(2) Improper adjustment

- Excess ventilation air
- Dirt, scale, rust
- Poor control settings
- Seasonal damper imbalance

(3) Poor installation

- Equipment inaccessible or noisy
- Tower or air-cooled condenser annoying
- Piping faulty
- Ductwork faulty

(4) Poor design

- Unit undersized
- Duct system undersized
- Drafty air distribution
- Poor choice of controls
- Changes in operating conditions

PARTS FAILURE

A parts failure is, perhaps, the easiest malfunction for the service technician to correct since, once detected, a simple replacement puts the system back into satisfactory operation. But why did the part fail? Many parts failures are obvious as, for example, a broken belt, a worn-out bearing, or a blown fuse.

Some, however, require considerable skill to detect. A broken compressor valve is just such a part. In between these extremes there are failures such as a leaking refrigerant pipe joint, a burned-out motor capacitor, or a defective control. These failures may require that a process-of-elimination procedure be followed in order to pin-point the trouble.

Parts fail for several reasons, all of which can be summarized as follows:

(a) defective manufacture

(b) subjected to conditions beyond their rated capacity

(c) not properly maintained

(d) worn out

Modern factory quality control methods used by the leading manufacturers prevent most defective parts from leaving the factory. To protect the owner from the few that do, one to five-year guarantees are offered. Any part that functions satisfactorily for a year has a good chance of remaining in good operating condition for many more years of service.

Parts that fail because they are used beyond their rated capacity are usually electrical items such as motors and controls. It is possible to overload both fan and compressor motors by imposing conditions in the field that are beyond their intended service. A fan motor can easily be overloaded by abnormal increases in fan speed, since brake horsepower requirements increase as the cube of the speed ratio. Thus, for example, a fan requiring 1 hp at 500 rpm would need 8 hp at 1,000 rpm.

Electrical controls are all rated by their respective manufacturers for maximum ampere draw at each operating voltage. When these conditions are exceeded, the life of the electrical components in the control is shortened considerably.

Parts that fail due to improper maintenance can be either mechanical or electrical. Any system whose components are exposed to the outside weather is subject to rust and other types of deterioration due to rain and prolonged sunlight. The latter is particularly hard on parts made of plastics and on electrical insulation.

Any part that requires periodic lubrication is subject to premature failure when regular lubrication maintenance is ignored. Motor and fan bearings fall into this classification. The compressor itself can also be included when it is an open type where oil can leak from the system.

The fourth classification of parts failures is wear and tear. Every system will be subject to this type of failure. The most vulnerable items here are electrical components. However, mechanical components with moving parts, such as bearings and pistons, are not far behind. Scheduled lubrication will prolong the life of the latter.

IMPROPER ADJUSTMENTS

A second major reason why an air-conditioning system fails to perform properly is, as noted previously, improper adjustment. Compared to a system that does not function at all due to a parts failure, a system that is not in proper adjustment may cause the owner even greater dissatisfaction. The reason is that in the former case he is sure something is wrong, so he calls the serviceman and has the condition corrected. When a system is out of adjustment, however, the condition may develop so gradually that the owner may not be certain that something is wrong until he has put up with considerable annoyance.

A system that is out of adjustment may result in an owner complaint because:

(a) Cooling capacity seems to be decreasing.

(b) Cooling is uneven and drafty.

(c) Operating cost is increasing.

(d) Noise level is rising.

What are the possible changes in adjustments that can affect cooling capacity? One is the relative amount of outside ventilation air drawn into the system as compared with the total air circulated. If this percentage should be accidentally increased, say by a dislodged damper, its greatest effect will be felt on a peak-load day, when so much of the cooling unit capacity must be used to lower the temperature and humidity of the outside air. There is then insufficient capacity remaining to handle the principal load imposed by the conditioned space.

A change that takes place over a period of time is the reduction in conditioned air due to dirt accumulation on the air filter. Every 5% reduction in the airflow rate results in a capacity reduction of about 1%. Dirt that is allowed to accumulate on the cooling coil or blades of the fan or dampers and restrict air flow unnecessarily, or rugs or furniture placed so that air cannot move freely through a supply outlet or return grille, will have the same effect upon capacity as the dirty filter.

Another change in conditions that reduces system capacity has to do with the condenser. If it is water-cooled, it can accumulate scale on the water side of the tubes, which cuts down the transfer of heat from refrigerant to water. The result is that the compressor must operate at a higher discharge temperature and pressure with reduced capacity. The condenser can be restored to its original condition by chemical treatment to remove the scale. Proper water conditioning will lengthen the periods between cleaning.

The high head-pressure operation described above can also be experienced with an air-cooled condenser if its efficiency is allowed to decrease due to dirt accumulation, recirculation of warm air, or fan-belt slippage. Over a period of time, it is even possible for fans shrouds to rust at the fasteners and fall away, allowing some air to bypass the coil.

Even though the air-conditioning unit itself seems to be operating properly, it is quite possible that the owner will complain of improper operation because of poor air distribution. This is particularly a problem where a single air-handling system serves for year round conditioning. A system cannot always be designed with outlets located so they may serve for cooling and heating with a single setting. For this reason, compromise outlet locations must be fitted with registers that have the necessary flexibility to be adjusted seasonally so as to throw the conditioned air in a pattern that will not cause discomfort.

When system operating costs increase and cause an owner complaint, the serviceman should check out the refrigerant charge. If it is low, the system capacity will be low and the compressor operating time will be extended. He should also check the system head pressure. If it is high, due to an inefficient condenser, the serviceman will have found another reason for high cost resulting from extended operating time due to low system capacity.

NOTE:

An increasing operating cost is not always the result of a malfunctioning system. Sometimes it is due to a change in the owner's operating schedule or in the amount of ventilation air he uses.

Complaints about noise result from adjustments to the air flow, from changes in damper positions, from worn-out equipment vibration isolators, from worn-out bearings and belts, and from loosened fastenings. Generally, a noise that develops after the system has been used is not as difficult to correct as an original noise.

POOR DESIGN/INSTALLATION

So far we have discussed poor performance of an air-conditioning system from the standpoints of parts failure and improper adjustment. We shall now discuss briefly the effects of poor installation and poor design on performance. The responsibilities of the installer and designer are closely related. Theoretically, the designer should take enough time to plan and specify every part of the system down to the location and size of the last screw. If he could constantly supervise to see that his ideas were carried out to the letter, it would be practically impossible for any installation faults to develop. In actual practice, however, the designer passes a good deal of the design responsibility to the man who does the installation work on the premise that he is a skilled mechanic who also recognizes and practices the type of standard, detailed work that makes for good installation. The designer usually takes the responsibility for selecting the proper size of the cooling plant, for sizing the ductwork, for selecting and locating supply registers and return grilles, and for designing the control system. The installer assumes all other work in connection with the system. Any trouble that cannot be traced to a defective part, maladjustment, or installation fault may ultimately be the responsibility of the designer. No field correction will add capacity to a unit that was selected with insufficient capacity in the first place. However, it is sometimes possible to reduce the cooling load by the use of simple measures such as shading devices for glass areas or added insulation.

Duct systems found to be too small can sometimes be corrected by adding a branch or two. Distribution that is not satisfactory because of drafts or uneven cooling can sometimes be improved by relocating a supply register or return grille.

Controls sometimes fail to function properly due to poor choice or location, or lack of owner understanding of their purpose. These are design faults that can usually be corrected.

Modern air-conditioning equipment is manufactured within very narrow operation tolerances. There are many models, types, and sizes of units to meet practically every application requirement. However, the serviceman cannot always assume that the proper equipment was selected to fit the application. Neither can he assume that the operating conditions as he views the installation are the same as the original design conditions.

It is important that the customer be consulted regarding a change of operating conditions, and that he understands the effects of these changes on the operation of the equipment.

Manufacturers have added safety devices to protect equipment operating beyond its design limitations or when a component fails. Since the prime source of power is electricity, safety devices are incorporated in the electrical circuits to interrupt the flow of current should danger threaten the heart of the system—the hermetic motor compressor. Fortunately, this does happen when operating limitations are exceeded. Eventually, an electrical device will stop the unit (low or high-pressure control, fuse, or thermal overload). However, an electrical device may also break down because its limits have been exceeded.

Operating limitations may be exceeded for four basic reasons:

(a) The operator of the unit is trying to make it do something beyond its design limits. After a unit has been installed and operating for a season or two, the owner might add a room or an enclosed porch which he expects his unit to cool. He may have planted shrubs in front of the condenser, reducing the air required for cooling.

(b) The owner is operating his unit when outside temperatures are below the minimum (usually 69°F.), unless low ambient protection has been provided (usually in commercial installations). If the unit does not have special devices to permit low ambient operation, the system is in trouble, and you may get a service call. Usually by the time the serviceman arrives, the outside temperature has risen above the minimum and the unit is operating normally. Be aware of these possibilities. Consider the changes that can take place between the time the customer reports trouble and the time you arrive on the job.

(c) Improper selection; not the right piece of equipment for the job. Sometimes this is due to poor sales engineering; a salesman who was just a little careless in his zeal to make a sale. Sometimes the requirements change. This most frequently happens when the ownership changes hands. As an example, an owner had an air conditioner that worked very well and was quite satisfactory. He sold his little store, and the new owner wanted to store pharmaceutical supplies at a temperature of 65°F. The unit just could not do it—it was not selected to do it, it could not be expected to. In another case, a small store in a shopping center complex was converted to a short-order food outlet with added cooking appliances and ventilation requirements that far exceeded the original cooling unit capacity.

(d) Poor installation is "built-in" trouble; it is frustrating to go out on a job, find that the installation is the cause of the trouble and try to do a good and inexpensive job of correcting it. Too often, back-alley mechanics undercut the legitimate contractor and do a poor job of installation, which gives the owner nothing but grief. He finally calls in someone else to correct his troubles and is not too happy when he is told that the installation is at fault. A remote unit, where the condensing unit is too far from the evaporator, the wrong line size, a poor location for sufficient, nonrecirculated outside air for the condenser, no provision for the oil return, the location of the thermostat, and inadequate ductwork or supply registers and return grilles are just some parts of the original installation that can interfere with the proper operation of an air conditioner.

It is usually easy to determine what has caused a unit to stop or perform unsatisfactorily, but *why* is the difficult question the serviceman must answer.

Replacing a blown fuse, resetting a low-pressure cutout, or adding refrigerant to a system that is "short" may get the unit started—but that is like pumping up a flat tire without repairing the puncture first.

Something has caused a fuse to blow, a low-pressure cutout to open, a system to be short of refrigerant. In the case of the fuse, the cause may be a temporary one, such as a power surge in the line, and it may not happen again for the remainder of the season, but this is rare. If a fuse is blown, the serviceman should do some thorough checking.

If a system needs refrigerant (a much-used "cure-all"), it must have lost some of its original charge, which means a leak—find it and fix it. And just because the low-pressure cutout is open does not mean the unit needs refrigerant. Most of the time it is due to an air restriction over the evaporator coil. Replacement of dirty filters or a blower adjustment may solve the problem, where adding refrigerant would only create more problems.

The foregoing is brought to your attention to help you in your approach to a trouble job, which should be like a large floodlight at first, taking in the entire scene. Sometimes a bad condition will show up before you open your tool box; it will give you a clue that will save you time and trouble. If the problem is a little deeper, the beam of light is narrowed to focus on one part at a time.

Experienced servicemen develop a *check-list* procedure; that is, a sequence of actions to perform once they arrive at a trouble-shooting job. Helpful guides are included in this chapter. However, more detailed guides are included in the "On-the-Job Servicing Guides," Chap. 24.

After all, what are the usual owner complaints? Three major and two minor complaints are:

(1) Unit will not start.

(2) Unit starts but soon stops.

(3) Unit runs but does not cool.

(4) Unit is noisy.

(5) Unit runs continuously, too cool.

Sometimes the complaint is that some rooms do not cool as well as others. This is not a unit fault, but it is a legitimate complaint, for the owner expects and deserves complete air-conditioning. This then is a problem of balancing the air distribution system. Adjustment of supply grilles and dampers may solve the problem, but sometimes a more costly cure is necessary, such as replacing existing grilles with the damper type, installing a damper in the ductwork, adding or rerouting ducts, or insulating ceiling of rooms under the roof.

If you replace a component, from fuse to compressor, very carefully study the possible causes of the failure of the component in order to prevent reoccurrence of failure.

REFRIGERANT LEAKS

Refrigerant leaks are probably the largest single cause for system failure. In spite of the extreme care and ultrasensitive leak-detecting devices used in production, leaks do occur and probably always will.

Soft copper tubing is often used for the interconnecting lines of a refrigerant system. It is easy to form, but once bent it becomes "hardened." The more it is bent, the harder it becomes, until it is so brittle it will break like glass. Sometimes vibrations may cause a slow "conditioning" of the tubing, making it more and more brittle until it cracks and splits, causing a leak. In production most discharge and suction lines are formed to a shape and size that will minimize the effects of vibration. It is not uncommon for breaks to occur near brazed joints, because of the hardening effect of heat on copper. This is especially true if too much heat is used during brazing, or if a joint has been opened and rebrazed several times. Thus it is not recommended that a tubing crack be repaired by simply brazing over the area. The entire section of tubing between the existing brazed joints should be replaced.

The brazed joint that requires the most skill and care is on small-diameter copper lines such as restrictor (capillary) tubes. Any repair that requires brazing the end of a restrictor tube to a fitting must be done very carefully. Enter the tube far enough into the mating piece to prevent the brazing compound from flowing across the open end; yet the end of the tube must not be so close to the strainer or the wall of the fitting that refrigerant flow is restricted. After positioning, the restrictor must be held in a manner that will prevent movement while brazing. Always use nitrogen for any brazing operation.

There are four reasons why leaks occur in a brazed joint:

- uneven or insufficient heat
- oxidized or dirty copper
- loose-fitted joint
- improper use of flux

When a leaking brazed joint is found, it should be pulled apart, cleaned, and inspected before rebrazing. It is sometimes hard to tell if the brazing material is flowing properly into the joint. It may appear to form a neat seal around the outer edge of a joint, which may hold temporarily, but in a good brazing job the brazing material flows deeply and evenly throughout the entire contact of the two pieces of tubing. This is assured by the proper application of heat during the brazing operation. The material flows toward the heat, so by placing the flame of the torch in the area beyond the end of the tube you will cause the flow to go where it is needed.

TROUBLE-SHOOTING GUIDE

Unit Will Not Start:

Check source of power
Check circuit breaker (right ampere rating?)
Check fuses (right size?). If blown, find cause
Check thermostat. Contacts made? Loose wire?
Check safety pressure cutouts

Low-Pressure Cutout Open:

restriction in suction or liquid line
low on refrigerant
poor air return over evaporator coil
stuck thermostatic expansion valve or restricted capillary tube

High-Pressure Cutout Open:

poor condenser air (or water)
excessive suction pressure
dirty condenser
overcharge

Faulty Controls:

Check the control circuit transformer.
Contacts made? Stuck open?
Check connections of terminals and contactor

Unit Starts But Shortly Stops:

Check low and high-pressure safety cutouts

Low-Pressure Cutouts:

restriction in suction or liquid line
poor air over evaporator
low on refrigerant

High-Pressure Cutouts:

poor condenser air (or water)
dirty condenser
overcharge
excessive suction pressure
noncondensables

Check Thermal Overload:

Though this control requires time to reset, it may cause the unit to stop a short time after restarting.

Check Amperage Draw: (see unit nameplate)

High-Amperage Draw May Be Due To:

excessive discharge pressure
defective start relay
defective capacitor
high or low voltage
incorrect wiring
tight compressor
burned contacts

Unit Runs But Does Not Cool:

First ask the owner if unit ever did cool to his satisfaction. Many times an owner is "oversold" or does not know what he can expect of his system. Unreasonable temperature requirements are sometimes not within the capabilities of the equipment installed.

Check the TD (temperature drop) across the evaporator coil. An 18°F to 22°F split indicates the unit is doing its job at the coil. The fact that it is not cooling the house or store is a different problem.

If an owner is suspected of being unreasonable or ignorant of the limitations of his unit, report this to the service or sales manager of your company.

Sometimes an owner reports a "no-cool" condition when it is really only one or two rooms he is concerned about. This falls into the realm of system balancing by the adjustment of supply grilles, dampers, etc. Rooms under a roof may need more adequate insulation—not less than four inches, and preferably six inches.

If a unit has cooled satisfactorily at one time, look for a physical change in the living pattern of the household or store:

Room addition?

Greater internal loads? (Motors, computer equipment, etc.)

Doors opening more often?

Look for something that could increase the load beyond the limits of the unit. Do not overlook the "psychological" changes. After visiting a neighbor's air-conditioned house (which has a bigger unit, or better air distribution), he becomes dissatisfied with the operation of his unit, which may be actually adequate for comfort conditions. Without doubt, the psychological area is a difficult one and should be referred to someone in your organization with more experience in problems of this kind.

Checking Ducts:

By far one of the most overlooked and difficult areas of trouble-shooting, poor or inadequate ducts are very often the underlying cause of poor operation complaints, especially so when an air-conditioning system is an "add-on" to an existing warm-air system.

Duct sizes for a warm-air system in the milder climates are rarely adequate for cool air distribution, especially in multistory dwellings. Warm air is light and tends to rise, while cool air is heavy and needs more power to force it through the ducts to upper areas. To make matters worse, the upper floors usually need the most cooling.

The biggest problem seems to be inadequate return air. A common solution is to provide additional returns; however, providing an extra return or two may not solve the air distribution problem.

First look at the supply grilles, a good installer will always make sure the supply grilles are of the correct type and are adequate in number and size to assure successful operation. If, however, this has been omitted or overlooked (to cut corners on price), replacement with proper and adequate grilles will greatly improve the system. They should, of course, be the multivane type, adjustable for vertical and horizontal directional air flow.

Supply grilles may not be the cure-all in themselves. The entire ductwork system must be adequate to distribute the heavier air without creating too much resistance for the blower and motor to overcome. NOTE: A blower motor of higher horsepower does not deliver more air. It turns the blower at the same speed as a motor of lower horsepower. It will, however, maintain that speed under a higher load than will the smaller hp motor. A larger motor is almost always required when cooling is added to an existing heating system in the warmer areas. A two-speed motor is often used—the slower speed for heating and the higher speed for cooling.

Air in Refrigeration System:

Air remains in the system after incomplete evacuation or purging or flushing. Air in the system is not desirable, because it is noncondensible and increases condenser pressure. Excessive air in the system may increase the discharge pressure as much as 30 to 40 lb.

Unit Too Noisy:

Though it may be considered a minor complaint, an unusual mechanical noise that is brought to the attention of the serviceman may be a forewarning of impending serious mechanical trouble, which can be averted if dealt with in time.

Usually a noise complaint is received from an oversensitive owner or neighbor. And here again we enter the realm of psychological problems.

Owners of warm-air systems are already adjusted to blower noise. However, the normal operation of an air-cooled condenser that is new to them may be annoying at first. Sometimes a serviceman goes through the motions of doing something about the noise, while he is actually making a routine inspection.

The physical location of the condensing unit, keeping in mind the direction of the discharge air flow, will greatly influence noticeable operating noise. For instance, a vertical air discharge condenser, placed too near the house under a wide roof overhang, may trap sound; a corner formed by a house and another wall may create a "megaphone" effect, amplifying operating noise to the annoyance of a neighbor several houses away.

Then there are the not unusual rattles, squeaks, hums, and thumps that may take a little or a lot of time to correct and may warn the alert serviceman of impending trouble.

Air rush and whistling noises may be due to inadequate ductwork, which causes a high-velocity air flow. Enlarged supply and return grilles can do much to reduce air velocity noises created this way.

Unit Runs But Gets Too Cold:

This is the rarest of complaints, but it deserves mention here if only to complete our list.

The most obvious place to look is the thermostat, the location of which is important, for it must be able to "feel" the average temperature in the space. Too, the blower should be set for constant operation during the cooling season so as to circulate the cooled air and prevent stratification.

An "out-of-adjustment" or stuck thermostat is one of the most common causes for this condition. A thermostat that does not have an anticipator for cooling should be replaced with one that has far closer temperature control.

DIAGNOSING PROBLEMS IS NO EASY TASK

The serviceman who has a comprehensive view of the air-conditioning system troubles as we have discussed should be better equipped to diagnose poor performance than the man whose view is limited only to the strictly technical details.

PROBLEMS

1. Air-conditioning systems service complaints may generally be grouped into four categories. Name them.
2. Dirty filters cause low air flow. Each 5% reduction in air flow results in a ____________% reduction in capacity.
3. Water-cooled condenser scale creates ____________ head pressure and loss of ____________.
4. Poor design and installation are easy to correct. True or false?
5. A refrigerant leak in a length of copper tubing should be repaired by brazing over the leak. True or false?
6. The safety pressure cutout has little to do with the failure of a unit to start. True or false?
7. High current draw can be due to excessive discharge pressure. True or false?
8. Air in the refrigeration system is not desirable because it ____________.
9. The condition that does not cause a low pressure cutout to open is: **(a)** dirty filters, **(b)** low refrigerant, **(c)** blower speed too slow, or **(d)** blower speed too fast.
10. The technique that would not cause refrigerant leaks in a brazed joint is: **(a)** uneven heat, **(b)** oxidizing the copper, **(c)** improper use of flux, and **(d)** use of nitrogen.

24

ON THE JOB SERVICING

Now that you have gained some basic knowledge of air-conditioning trouble-shooting, you are probably anxious to begin applying that knowledge to specific on-the-job situations. In order to guide you in quickly diagnosing and correcting malfunctions, this chapter presents a sample industry guide for air-conditioning service as published by the Air-Conditioning Division of Tappan, Elyria, Ohio. This guide relates to Tappan equipment, and the recommendations may vary slightly from other manufacturer's products and procedures. It is called the *Trouble Tracer* and is based on four common categories of service problems:

Figure 24-1 Unit will not start

Figure 24-2 Unit starts and then locks out

Figure 24-3 Unit cycles

Figure 24-4 Unit runs, but there is inadequate cooling

Before proceeding to the use of the *Trouble Tracer* charts let us first review the fact that proper servicing equipment is needed before tackling any situation. Experience has indicated that most service requirements are related to electrical problems. However, many problems occur in the refrigeration system and others in the air distribution system; still others result from lack of proper instructions to the owner or his lack of understanding of what he should expect from his system.

SERVICE TOOLS NEEDED

There are certain basic tools and instruments required in servicing an air-conditioning system. Special tools for certain models may also be required as specified by manufacturers. In addition, there are other instruments available that are perhaps not absolutely essential but can be extremely helpful in diagnosing problems.

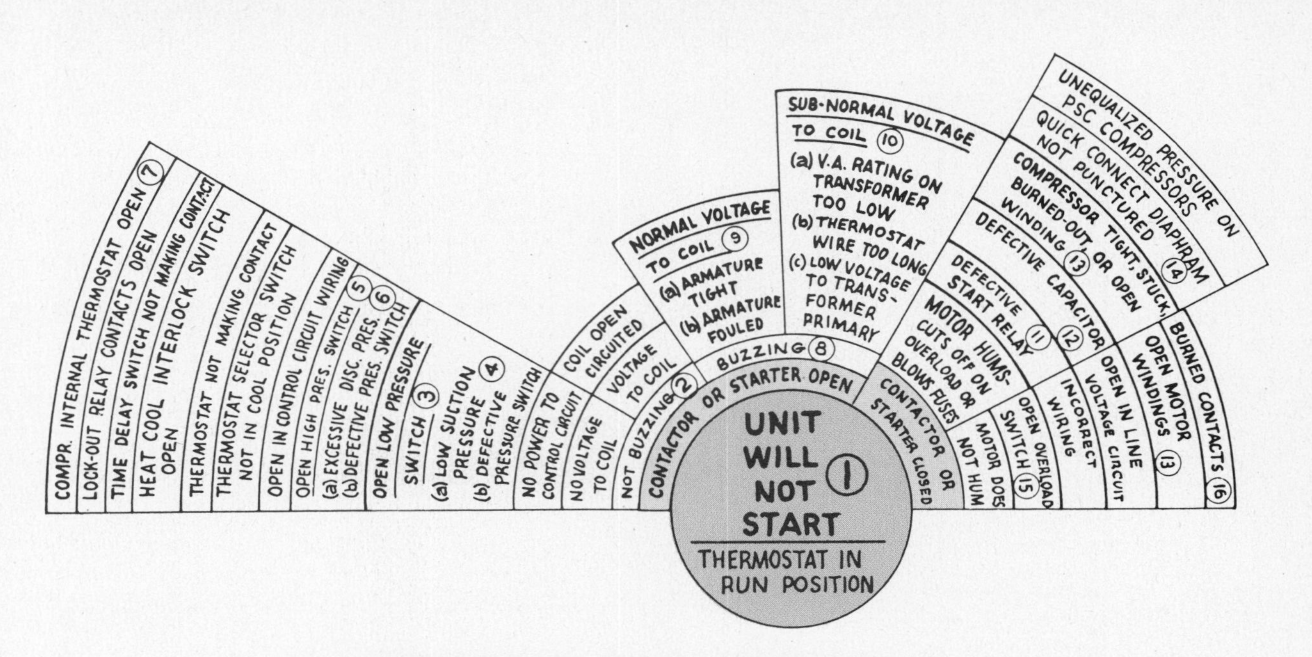

FIGURE 24-1 (Courtesy, Tappan Air-Conditioning Company).

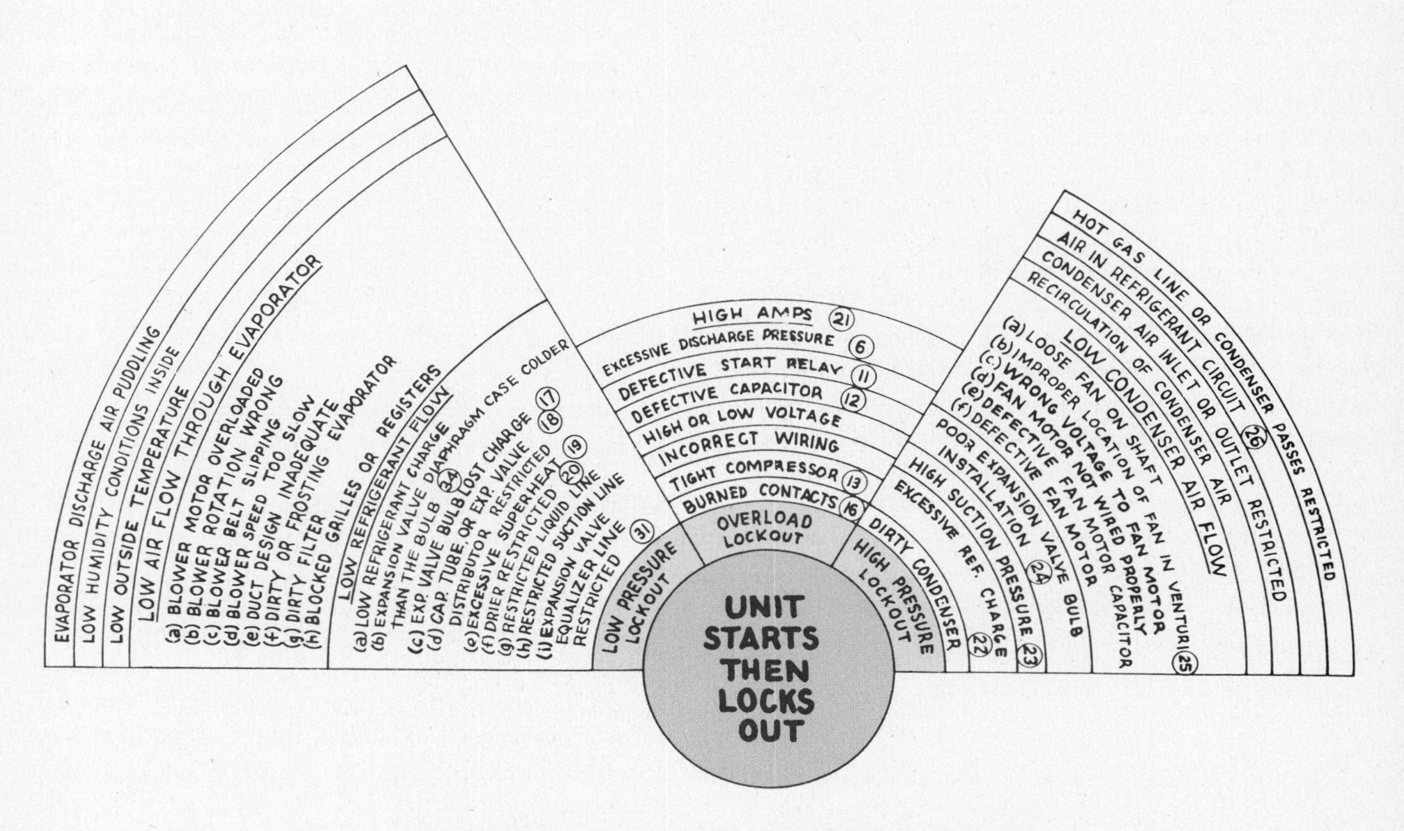

FIGURE 24-2 (Courtesy, Tappan Air-Conditioning Company).

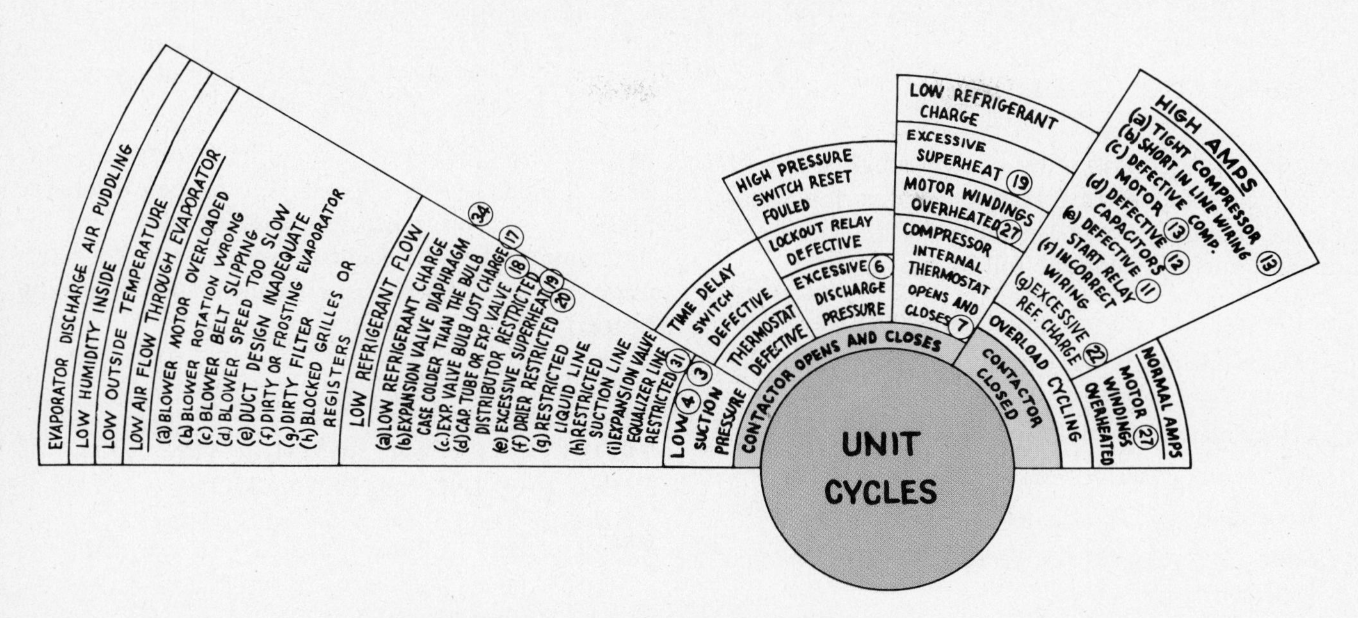

FIGURE 24-3 (Courtesy, Tappan Air-Conditioning Company).

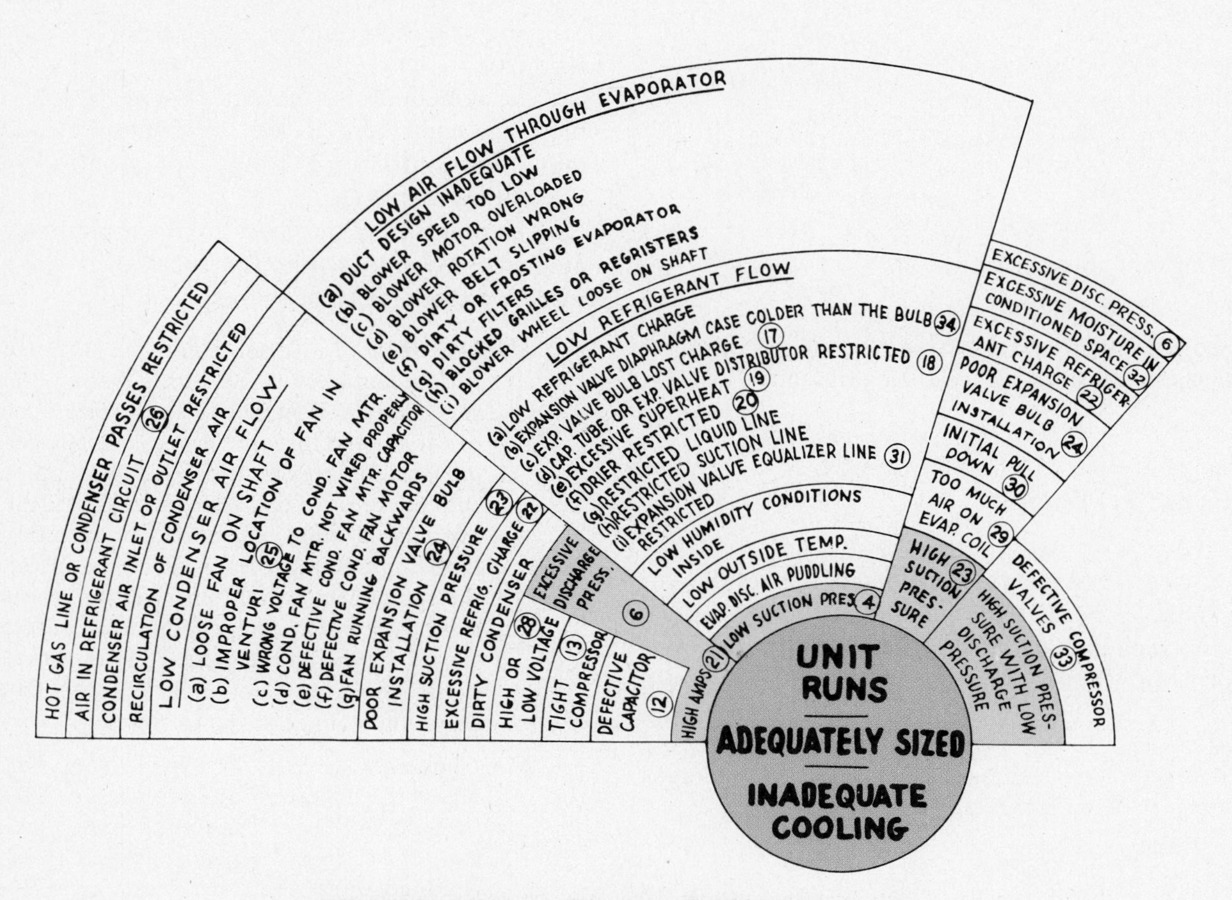

FIGURE 24-4 (Courtesy, Tappan Air-Conditioning Company).

REFRIGERATION TOOLS

Because an air-conditioning system consists of both mechanical and electrical components, various types of tools are needed. Some examples of the basic tools and equipment include the following:

Head-pressure gauge registering 0-400 psig

Compound pressure gauge registering from 30 inches vacuum to 150 psig

Gauge manifold with flexible lines

Refrigerant charging hose

Refrigerant service cylinder

Scales for weighing refrigerant charge or *charging cylinder*

Valve stem wrenches (do not use pliers)

Leak detector (halide torch or electronic type)

Vacuum pump

Brazing torch outfit

Tube-cleaning kit for brazing

Tube cutters

Tube benders

Tube flaring and swadging tools

Refrigerant grade thermometers

Sling psychrometer

Assorted screwdrivers, wrenches, pliers, chisels, punches, hacksaws, files, electric drills, etc.

ELECTRICAL TOOLS

The screwdrivers, pliers, etc., listed above should include those needed in making various electrical tests and repairs, such as wirecutters and insulation strippers. In addition, the following should be provided:

Combination volt-ohm-ammeter (measures voltage, amperage, and ohms)

Continuity test lamp

Neon test lamp

Capacitor tester

Jumper wires and insulated alligator clips

A velometer used to measure air velocity is extremely useful in checking air delivery and balancing a system, and a manometer is frequently used to check static pressures within the ductwork.

Satisfactory operation of an adequately sized air-conditioning system depends on proper discharge pressure, suction pressure, current draw and air distribution. The serviceman must be able to determine these in order to diagnose the system operation properly.

TROUBLE-TRACER: SUMMER AIR-CONDITIONING—EXPLANATION AND GUIDANCE OF FIGURES 24-1, 2, 3, AND 4

The center of the chart states the basic trouble; begin there. Adjacent partial circles state specific troubles that could cause the basic trouble. Determine which of these specific troubles exists before proceeding further.

The remaining partial circles state various troubles that contribute to the specific trouble. Determine the contributing trouble or troubles before attempting correction.

The numbers in the chart are for explanation and guidance. Where no numbers are shown, the statement is considered to be self-explanatory.

(1) Unit Will Not Start: Some units are equipped with a contactor, some with a starter; they are very similar. However, a starter has a built-in overload and usually has a manual reset to prevent compressor cycling if the compressor cuts off because of overload. A contactor does not have a built-in overload or manual reset. If the unit will not start, check the fuses, close the disconnect switch, and set the controls in run position. Observe whether the contactor or starter is open or closed. If it is not closed, press the starter reset. If it still doesn't close, check the voltage at the starter coil.

(2) Contactor or Starter Open, Not Buzzing: When the contactor or starter is open but not buzzing, it is an indication of no voltage to its coil, or voltage to its coil but the coil is open circuited. Press the starter reset and the high-

pressure control reset. If the unit does not start, check the voltage at the coil with a voltmeter. If there is voltage, the coil is open circuited. Replace the contactor or the starter coil. If there is no voltage, check the power to the control circuit. If there is voltage on the control circuit, use your VOM and check, one at a time, the high-pressure control; the low-pressure control; the time-delay switch; the lockout relay; and the thermostat. Also check the control wiring for a loose connection or open wiring.

(3) ***Open Low-Pressure Switch:*** This switch is usually factory set to open contacts at approximately 25 psi, with a 35 to 40 psi differential. It usually resets automatically.

(4) ***Low Suction Pressure:*** Suction pressure under normal operating conditions should not be lower than approximately 65 psi with R-22.

(5) ***Open High-Pressure Switch:*** It is usually factory set to open contacts at approximately 415 psi, with approximately 65 psi differential. Some require manual resetting; others are automatically reset.

(6) ***Excessive Discharge Pressure:*** The discharge pressure should generally not exceed the pressure corresponding to the ambient temperature (air-cooled units) plus 30° to 32°F for R-22 when the suction pressure is approximately 70 psi.

(7) ***Compressor Internal Thermostat Opens and Closes:*** Some compressors are equipped with internal thermostats. The thermostat may be 24 volts or line voltage. Its purpose is to sense excessive compressor winding temperatures quickly and open the power circuit to prevent a burnout of the motor. This high temperature may be caused by excessive current draw, or the high temperature of the suction gas, making it inadequate to cool the motor. Warm suction gas may be due to an undercharge, too much superheat, a restriction in the liquid or suction line, or a restriction in the drier. When the internal thermostat opens, it may take from 5 to 30 minutes for it to cool sufficiently to close. To check a 24-volt internal thermostat for an open circuit, temporarily jump the internal thermostat terminals. If the unit starts, the thermostat contacts are open. Wait until the compressor cools before attempting to start it again. In the meantime, check for conditions that cause the trouble. To check the line voltage internal thermostat for open circuit, disconnect the power from the circuit, and carefully remove all wires from the compressor terminals. Be sure not to loosen the studs where they come through the compressor housing, as this may result in refrigerant leak. Run a continuity check between all terminals with ohmmeter. If infinite resistance results, either the thermostat or the winding may be open. Let the compressor cool and check again.

(8) ***Contactor or Starter Open and Buzzing:*** When the contactor or starter is open but buzzing, it is an indication that its coil is energized but the contactor or starter is unable to close. See Items 9 and 10 for the cause.

(9) ***Normal Voltage To Coil:*** Check the voltage to the coil. It should not be less than 10% below the rated voltage as the contactor or starter tries to close: that is, 21.6 volts for a 24-volt coil. If the voltage is normal, the mechanism may be tight or fouled. Remove and inspect the mechanism, and clean it if necessary. If it is too sluggish, replace the contactor or starter mechanism.

(10) ***Subnormal Voltage To Coil:*** Check the voltage to the coil. If it is lower than 10% below rated voltage, it may be because: (a) volt-ampere rating on the transformer is too low, or (b) transformer primary voltage is too low.

(11) ***Defective Start Relay:*** A start relay is used only on single-phase capacitor-start, capacitor-run (CSR) compressors. This type of compressor is required to start "loaded", that is, when the discharge pressure is higher than the suction pressure. Starting relay contacts should be in a closed position when the compressor is idle and should open when the compressor is at approximately 85% of running speed. Its purpose is to remove the start capacitor from the motor starting winding as the motor approaches running speed. Sometimes the contacts may be stuck open or may be closed but not conducting electrically. If the discharge pressure and the suction pressure are equal, the compressor may start and continue to run. To test for this condition, start the unit, stop it, then restart it before pressures have a chance to equalize. If the relay contacts are open or not conducting electrically, the compressor will not start, and

the line fuse will probably blow. Replace the relay. Sometimes the contacts will fuse together and not open. In this case the compressor will cut off by overload. Replace the relay.

(12) ***Defective Capacitor:*** The purpose of the start capacitor is to provide starting torque on a single-phase, capacitor-start, capacitor-run (CSR) compressor. The purpose of the run capacitor is to improve the power factor, thus reducing run current draw on single-phase, capacitor-start, capacitor-run (CSR) compressors. Its purpose on permanent split capacitor (PSC) compressors is to provide some starting torque and reduce the run current draw. Shorted start or run capacitors will blow the line fuse or cycle the compressor by the overload. An open start capacitor will cause a higher than normal run current on CSR compressors and may cycle the compressor by the overload. An open run capacitor on PSC compressors will immediately cycle the compressor by the overload.

(13) ***Compressor Tight, Stuck, Burned Out, or Open Winding:*** Before concluding that this condition exists, be sure to make a continuity check on the compressor motor. Carefully remove all wires from the compressor terminals. Be sure not to loosen the studs where they come through the compressor housing, as this may result in refrigerant leak. If no burnout or open winding is indicated, check all other components in the compressor motor circuit, such as the overload stuck in an open position, before removing the compressor. A tight or stuck compressor is commonly due to a lack of oil returning to the compressor. Be sure to check the suction line for the oil return whenever a compressor is tight or stuck. Whenever the suction line rises on field-fabricated lines, there should be a trap where it exits from the evaporator. If the compressor is three phase, interchange any two lines to the motor. This will reverse the direction of rotation and may free the compressor. If an open motor winding is indicated:

(a) release refrigerant

(b) remove the compressor and seal its opening

(c) replace the compressor

(d) install a new drier

(e) pressurize the circuit to 75 psi and check for leaks

(f) evacuate the circuit three times to 28 inches of mercury, breaking the vacuum each time with refrigerant vapor and raising the pressure to 5 psi (do not use the circuit compressor to evacuate)

(g) recharge the circuit

(14) ***Burned-Out Compressor:*** Whenever a compressor burnout occurs, the circuit immediately becomes badly contaminated and must be thoroughly cleaned to prevent repeat burnouts. Proceed as follows:

(a) attach $\frac{1}{4}$-inch copper tubing with a shutoff valve in it to the liquid-line valve or discharge-service valve. Release refrigerant into a covered container to prevent splatter and acid burns. The chemical action caused by the burnout produces acid in the system, which can be harmful. Release the refrigerant rapidly to pull some of the contaminant out of the system

(b) remove the compressor and seal its openings

(c) remove the expansion valve and drier or capillary tube and strainer, and put a spacer in their place

(d) plug the expansion valve equalizer connection on the suction line

(e) make an opening in the liquid line and attach the $\frac{1}{4}$-inch tubing to the condenser side of the liquid line

(f) close the shut-off valve in the tubing

(g) connect the R-22 drum to the discharge line (formerly connected to the compressor), invert the drum, and charge at least 3 lb of liquid refrigerant into the high side; wait 10 minutes

(h) release refrigerant into the container through the shutoff valve

(i) clean the low side in similar manner by charging into the liquid line and out the suction

(j) clean the expansion valve and reinstall it, or replace the capillary tube and strainer with new one

(k) install a replacement compressor

(l) install a suction line filter-drier of adequate size with a new core in the suction line; install a new liquid-line drier

(m) pressurize the circuit to 75 psi and check for leaks

(n) triple-evacuate the refrigerant circuit with the vacuum pump and refrigerant-charging cylinder. Do not use the circuit compressor to evacuate

(o) charge the circuit with the correct amount by weight

(p) operate for 48 hours

(q) release the refrigerant through the suction valve, or pump it down if the circuit is provided with liquid and suction line valves. To pump down, close the liquid-line valve and temporarily jump the low pressure switch. Do not pump into the vacuum. Close the suction line valve

(r) remove the filter-drier, rejoin the suction line, and install a new liquid-line drier

(s) pressurize the circuit to 75 psi and check for leaks

(t) reevacuate the lines or the entire circuit and recharge; or open the liquid and suction-line valves if the circuit was pumped down

(15) Quick-Connect Diaphragms Not Punctured: If the diaphragms are not punctured, the discharge and suction pressures do not equalize rapidly; therefore the compressor cuts off on overload or blows a fuse.

(16) Open Overload Switch: Sometimes overloads will fail with contacts in the open position, or the contacts may be closed but not conducting electrically. To check this, disconnect the power circuit and, with an ohmmeter, check the resistance to determine if the switch is open or closed. Zero ohms will indicate an open position or no path of flow for the current.

(17) Burned Contacts: Sometimes contacts will close mechanically but will not conduct electrically. To check for this, disconnect the power circuit and measure the contact resistance with an ohmmeter. The meter should read zero ohms; if not, replace the contactor or contact sets.

(18) Expansion Valve Bulb Lost Charge: If the bulb of the expansion valve loses its charge, there will be no pressure to open the valve, thus causing low suction pressure. To check this, remove the expansion valve bulb from the suction line and hold it in your hand. If the suction pressure does not increase in a few minutes and there are no restrictions in the refrigerant circuit, it is an indication that the bulb has lost its charge. Replace the bulb element or expansion valve.

(19) Capillary Tube or Expansion Valve Distributor Restricted: To check this, check the suction pressure (very low suction pressure is an indication of restriction or excessive undercharge), temporarily cut off air to the evaporator, and allow the condensing unit to operate. If there is a partial restriction or excessive undercharge, frost will occur at that point. If there is no restriction, the evaporator coil will frost uniformly. If there is a total restriction anywhere in the refrigerant circuit from the condenser through the evaporator and back to the compressor, there will be no frost, the suction pressure may go into vacuum, and the discharge pressure will correspond to approximately ambient temperature because there will be no vapor to compress.

(20) Excessive Superheat: Superheat is the temperature of the refrigerant vapor above the temperature corresponding to the vapor pressure. Excessive superheat is an indication that the evaporator coil is "starved"; that is, there is not enough liquid refrigerant in the coil. To check for superheat,

(a) operate the unit for 20 minutes

(b) check the suction pressure

(c) check the temperature of the suction gas as it exits the evaporator

(d) compare the temperature of the suction gas with the temperature corresponding to the suction pressure

The temperature of the suction gas above the temperature corresponding to the suction pressure is the amount of superheat. Excessive superheat may be due to undercharge, a restriction in the refrigerant circuit, a low discharge pressure, lost expansion valve bulb charge, too much load on the evaporator, or refrigerant flashing ahead of the expansion valve or capillary tube due to pressure drop, as in the case of the evaporator being far enough above the condensing unit to cause more pressure drop than is offset by the amount of subcooling. This pressure drop is due to the weight of the liquid. Each foot of lift of R-22 liquid results in approximately $\frac{1}{2}$ psi pressure drop. Some expansion valves have external superheat adjustment, others have internal adjustment. Some have the internal adjustment on the inlet side, others on the outlet side. Turning the adjustment screw to the left lowers superheat; to the right, it raises superheat. For those with internal adjustment, pump the circuit down. Do not pump into the vacuum. Disconnect the expansion valve and turn the adjustment one turn. Be sure to purge both sides of the expansion valve thoroughly before tightening the connections when replacing it. Check for leaks. Operate the unit for 20 minutes and recheck superheat.

(21) ***Drier Restricted:*** To check this, feel the drier at the inlet and outlet. If the temperature at the inlet is higher than at the outlet, the drier is restricted; replace it. If the circuit is equipped with liquid-line and suction-line valves:

(a) pump it down by closing the liquid-line valve and temporarily jumping the low-pressure switch. Do not pump into the vacuum

(b) close the suction-line valve

(c) remove the drier

(d) pressurize the lines removed from the drier through the manifold, but close the manifold just before the pressure reaches zero psi

(e) open the suction-line and liquid-line valves. If the circuit is not equipped with liquid-line and suction-line valves, it will be necessary to remove the refrigerant from the circuit

(i) release the refrigerant slowly to prevent the removal of oil

(ii) install a new drier

(iii) pressurize and check for leaks

(iv) release the pressure and evacuate the circuit three times, breaking the vacuum each time with refrigerant vapor and raising the pressure to 5 psi. Do not use the circuit compressor to evacuate

(v) recharge the circuit

(22) ***High Amperes:*** Refer to the nameplate on the unit. The actual amperes should not exceed the rating more than 10%.

(23) ***Excessive Refrigerant Charge:*** Operate the unit for 20 minutes. Check the subcooling by checking the discharge pressure and the temperature of the liquid line as it exits the condensing unit. Compare the temperature of the liquid line with the temperature corresponding to the discharge pressure and refer to the manufacturer's operating data to determine if the subcooling is abnormal. If it is excessive, release some refrigerant slowly to prevent the removal of oil.

(24) ***High Suction Pressure:*** The suction pressure under normal operating conditions should not be higher than approximately 80 psi with R-22.

(25) ***Poor Expansion Valve Bulb Installation:*** The expansion valve bulb should be securely mounted on a horizontal straight piece of clean pipe, parallel to the pipe, with a firm metal-to-metal contact, and upstream from any suction line oil trap. It is usually good practice to mount the bulb above the center of the pipe.

(26) ***Improper Location of Fan Blade in Venturi:*** Check the blade to see if it is in center of the venturi opening. The distance of the blade from the condenser must be as recommended by the manufacturer of the condensing unit.

(27) ***Air in Refrigerant Circuit:*** If all other symptoms of high pressure have been checked and the pressure remains excessive, there may be air in the circuit.

(a) release the refrigerant slowly to prevent removal of oil

(b) evacuate the circuit three times to 28 inches mercury, breaking the vacuum each time with refrigerant vapor and raising the pressure to 5 psi. Do not use the circuit compressor to evacuate. Recharge the circuit

(28) ***Motor Windings Overheated:*** When the compressor is drawing normal amperes and becomes overheated and cycles by the overload or internal thermostat, it is because the temperature of the suction gas is too high to remove the heat from the compressor motor. This in turn is due to an undercharge, superheat that is too high, liquid and suction lines wrapped together, a restriction in the refrigerant circuit, or the compressor located in an excessively high ambient.

(29) ***High or Low Voltage:*** Check the nameplate on the unit for the voltage rating. Check the voltage at the contactor or starter while the unit is operating. For a hermetic compressor this voltage should not vary by more than 10% above the unit-rated voltage, or 5% below the unit-rated voltage.

(30) ***Too Much Air on Evaporator Coil:*** Too much air on the coil results in high suction pressure and reduces moisture removal; thus causing high humidity in the conditioned area. To reduce the air, open the blower motor variable-speed pulley one-half turn at a time. For direct-drive blowers, restrict the return air or change the speed tap.

(31) ***Initial Pulldown:*** High suction pressure is a characteristic of startup and initial pulldown. Wait until the conditioned space is down to approximately 80°F before checking the suction pressure.

(32) ***Expansion Valve Equalizer Line Restricted:*** To check this, pump the circuit down. Do not pump into a vacuum. Remove the expansion valve. Check the equalizer line and the suction line connections for restrictions. Be sure to purge both sides of the expansion valve thoroughly before tightening the connections when replacing it.

(33) ***Excessive Moisture in Conditioned Space:*** The most practical way to control moisture in the conditioned area is to prevent it from entering the space or remove it at its source. If any part of the conditioned space is over a crawl space, make sure that there is a vapor barrier over the earth in the crawl space. Exhaust fans installed in the kitchen and bathrooms will help remove water vapor generated from cooking or bathing.

(34) ***Defective Compressor Valves:*** To check this, make sure that the circuit is adequately charged and there are no restrictions in the circuit. Cover the condenser and observe whether the discharge pressure increases rapidly. If the discharge pressure does not increase rapidly, it is an indication of defective valves. If the valves are defective on a sealed hermetic compressor, replace the compressor.

(a) Release the refrigerant

(b) Remove the compressor and seal its opening

(c) Replace the compressor

(d) Install a new drier

(e) Pressurize the circuit to 75 psi and check for leaks

(f) Evacuate the circuit three times, breaking the vacuum each time with refrigerant vapor and raising the pressure to 5 psi. Do not use the circuit compressor to evacuate

(g) Recharge the circuit

(35) ***Repair Parts:*** Whenever repair parts are required, furnish the serial number and model number of the condensing unit to the distributor or parts source.

The preceding explanations and guidance cover the most common situations that face the service technician on basic cooling problems. Naturally, there will be slight differences in equipment and specific techniques of corrective action depending on the particular manufacturer, but in general the problems are all about the same.

Heat pump service is a little more complicated or involved, due to the reverse cycle operation and also because of the differences in methods and defrost controls used. The problems on the cooling cycle are not too much different than those above, but on the heating cycle the effects of undercharge and low indoor air circulation can be most critical to the operation. It is recommended that the specific manufacturer's instructions be followed closely.

In addition to the use of tools and the methods of troubleshooting, it is of paramount importance that the service technician be familiar with safety standards and the handling of injuries on the job site.

The greatest burden in improving the safety record lies with each individual. All sorts of protective devices can be used, laws can be passed, safety campaigns can be waged; but if the technician is careless, he can get hurt. He must be constantly on the lookout to protect himself and others. He must think and pay attention to what he is doing. *There is no substitute for vigilance*, and safe workmanship saves time, material, money, and suffering.

Accidents in the refrigeration industry can be divided into six general classes according to the cause and nature of the injury.

(1) physical injuries from mechanical causes

(2) electrical injuries

(3) injuries due to high pressure

(4) burns and scalds

(5) injuries due to explosions

(6) breathing toxic gases

In addition to information from the manufacturers, suppliers, and governmental agencies, excellent data and information are available from trade associations like the RSES (Refrigeration Service Engineers Society) that covers specific causes of such accidents, the precautions that will prevent them, and what to do if an accident does occur.

PROBLEMS

1. Proper service tools are important to the service technician. True or false?

2. If an ohmmeter reads zero when checking for an open circuit, it indicates that the circuit is closed. True of false?

3. Low head pressure and high suction pressure can be caused by a faulty compressor valve. True or false?

4. The appearance of a capacitor will not indicate its operating condition. True or false?

5. A stuck compressor may be due to

(**a**) insufficient oil return

(**b**) excessive voltage

(**c**) high ambient temperature

(**d**) restricted capillary tube

6. If the expansion valve bulb should lose its charge, it will cause

(**a**) low suction pressure

(**b**) high suction pressure

(**c**) increased amp draw

(**d**) high discharge pressure

7. Excessive superheat is an indication that the evaporator coil is ______________.

8. A restricted drier will have a higher outlet temperature than inlet temperature. True or false?

9. Excessive subcooling is an indication of refrigerant ______________.

10. Heat pump service is much the same as for a cooling unit.

25

CUSTOMER SERVICE

There are two phases to good customer service; first, the attention to technical factors involving the product, its application, and its maintenance, and second, the person-to-person human relations between the servicing agent and his customer. Usually both are involved and the professional dealer/contractor and service technician will be wise to recognize the importance of the latter. Its not enough to know only how to take care of mechanical problems.

The situations in which an air-conditioning technician finds himself will determine the necessary course of action or actions to follow. For example, he may be assigned to install air-conditioning in a new home or apartment while it is under construction, and he may never actually see or meet the eventual owner or occupant. Here technical skills plus a respect for the other person's property are all he needs.

Another situation may find him adding cooling to an existing furnace/duct system where installation, startup, and customer relations are involved. This is a happy event (Fig. 25-1); the homeowner is proud of his new investment and looks forward to the added comfort and to showing the results to his friends and neighbors.

But not all home situations are pleasant events (Fig. 25-2), as, for example, the case of a system problem during a 95°F-plus heat wave, particularly if there was an extended delay in getting a new part.

Another situation confronting the technician who becomes involved in commercial work is the handling of *maintenance contracts* (Fig. 25-3). Here he may become personally involved with the building owner or manager or the maintenance superintendent. Although they may be fine people, their job is to protect their investment at the least possible cost. It's a business relationship, and many times these people must be sold on the idea of spending additional money just to make sure the system is inspected on a routine basis. But when the system does break down, they are the first to want immediate attention because it could mean lost dollars in sales or production. So an air-conditioning technician should have a bit of salesmanship and persuasive skills in his makeup.

Speaking of salesmanship, the serviceman can

FIGURE 25-1.

frequently bring in new business to his firm by recommending additional products to his customer. A classic example could be when he is on a routine home air-conditioning service call and, in conversation with the homeowner, learns that one of the children has an allergy to dust or pollen. The recommendation of an electronic air cleaner can bring

FIGURE 25-2.

welcome relief to the child and represents new dollars in business for the air-conditioning firm.

Although not all situations can be covered, let's present several and develop some habits to follow.

First we will dicuss the pleasant situation of adding cooling to an existing home furnace/duct system. Naturally the salesman has processed the order and now the homeowner anxiously awaits the arrival of the equipment—and it all starts there.

We will assume that the installer has checked with the salesman to make sure he has the right equipment, and also to anticipate what changes are needed in wiring, ductwork, etc., so he can bring along the proper accessory items.

Then the homeowner should be notified by telephone several days in advance to confirm a date to start the installation. It is a good idea to suggest that the homeowner advise the neighbors who might

FIGURE 25-3.

become suspicious of a strange truck parked in the driveway. This also alerts the neighbors to your customer's new improvement, which perhaps will result in other sales.

The moment the installer drives up in his truck, the air-conditioning company's image starts to emerge. Although the selection of the truck may not be up to the technician, he can suggest these ideas to the company owner:

- Is the truck adequate to contain the equipment, tools, etc. without a lot of things hanging on the sides, top, etc?

- Is it well identified so as to be conspicious to the neighbors—to show that air conditioning is being installed?
- Is it clean? Maybe this is your job.
- And does it get the proper maintenance so breakdowns don't occur just when it's really needed? Again perhaps this is your job. If so, see to it, for time is money.

Now what about you and your helper? Are you neat and clean? Are you wearing a uniform identifying your company and/or the product brand you represent? A namepatch on the shirt or jacket is a courtesy to the homeowner so he or she can remember your name. Introduce yourself and your helper.

Before unloading any equipment or tools ask to look at the home and the furnace/duct system. Wipe your feet as you go in. Make a visual survey of the existing indoor equipment, electrical power source, room thermostat, room air distribution grilles, routing for refrigerant lines, etc. Next go outdoors to reverify the placement of the condensing unit with the homeowner in regard to proximity to shrubs, windows, patio, etc. After making the survey you should be in a position to know if you have all the equipment needed. If not, call the office and start the action to get things arranged.

Proceed with the installation as recommended by the manufacturer or as discussed earlier in this book. *But remember*! when you are inside the house take precautions to protect the rugs, etc., and clean your shoes. Use drop cloths in the area where mechanical work is being done. And when jobs are finished clean up any debris that is left. Vacuum the area if necessary—the homeowner will be glad to let you use his or her equipment.

Assuming the installation is completed, started, and is functioning normally, now is the time to instruct the homeowners, both husband and wife if possible, on the proper care and operation of their system.

Most manufacturers include with their equipment an envelope containing

(1) the installation instructions (which you have used)

(2) a homeowner's guide book

(3) a warranty statement

Start with the *Homeowner's Guide* (Fig. 25-4). The professional service technician will always take

FIGURE 25-4.

time to review this important document. First, identify the components by name, the coil, the condensing unit, the thermostat, etc., and briefly explain to the new owners that it is much like a large refrigerator system; it cools and dehumidifies in one operation. Review the thermostat operation; discuss what the system selector means and why a homeowner may or may not want constant or automatic fan operation. Show them how to select the indoor temperature, and in view of the rising cost of energy discuss the proper optimum temperature for comfort versus cost of operation. Once a temperature is decided upon, point out the advantage of setting the thermostat and leaving it alone. It is paradoxical that an owner-user will purchase equipment with an excellent automatic control system and then frustrate its performance by continually overriding the thermostat. Also, point out to the owner *not to move the temperature selector rapidly* since the system must have time to equalize refrigerant pressures.

Tell the homeowner what to expect from the system after a prolonged shutdown. For example, in a heating-cooling installation the heating section of the equipment can often bring the area up to satisfactory temperature in 15–30 minutes on a cold day. The homeowner should be instructed to understand why the cooling section cannot bring the area temperature *down* satisfactorily in the same time period on a hot day. Point out that *both the temperature and humidity* must be lowered before the area is comfortable, and this process takes more time, and that this is normal.

Should the air distribution system require summer-winter adjustment this may include resetting duct dampers or grille damper positions. Indoor fan speed changes are usually automatic on new equipment, but converting an older existing furnace blower may require a manual adjustment of belt drives or changing a two-speed switch.

Review the maintenance checks and operations a homeowner should do; generally they are simple but important:

(1) *Most Important—Keep Indoor Filters Clean*! Show the owners how to change or clean them. If a "filter flag" is installed instruct them how to read the gauge.

(2) Show them the fuses or circuit breakers that furnish power to both the heating and cooling equipment. Recommend they check these first if the equipment doesn't function.

(3) Keep the outdoor condenser coil clean and free of leaves or other debris that may restrict air flow.

(4) Caution children not to poke sticks or fingers into the condenser fan blades.

(5) Check the cooling coil drain to see that it runs free.

(6) Periodically wash the outdoor condenser cabinet, and apply a coat of wax to prolong its finish.

If the heating installation is also involved, the instructions will include how to turn on the gas, light the pilot, etc.

The degree of cooperation and understanding you will get from the homeowner depends on the individuals involved. Some men have mechanical skills and will follow the instructions above; others will not. If the system is installed in a home occupied by a widow or elderly people it's recommended that the air-conditioning contractor work out a maintenance inspection and startup service agreement to insure that the system works properly at each seasonal changeover.

After reviewing the operation and maintenance of the particular system, point out some of the following tips to improve the comfort and economy of the cooling system.

(1) Keep the sun from adding unnecessary heat to the building by shading the exposed windows. Draw shades, blinds or drapery on the east windows in the morning and on the west windows in the afternoon. The sun will pour heat through unshaded windows; heat will be absorbed by the furniture, walls, and other materials in the building. Operating costs can be reduced and comfort increased by eliminating some of this unnecessary heat. Better yet, provide exterior shading, such as awnings or shade screens or trees.

(2) Closed windows and doors keep out the hot, moisture-laden outside air. Excessive amounts of outside air only add to the total load that the cooling system must work against. Closed doors and windows also keep out street noise, dust, and dirt. Even basement windows should be kept closed so as to eliminate condensation in basements and the extra load on the floor of the conditioned area.
The normal activity in a house includes frequent opening and closing of doors, so an adequate supply of air leaks into the average home each day. This outside air leakage is usually the equivalent of changing all of the air in a home approximately six times a day and is sufficient for ordinary ventilation needs.

(3) Operate the system continuously. Part-time cooling is usually poor economy. If the cooling unit is turned off during morning hours, the building has a chance to soak up heat all morning long. Then, the cooling equipment must not only remove this excess heat, but also the normal hourly heat gain of the building in the hot afternoon. A properly sized cooling unit has a capacity equal to a maximum heat gain for the building and no more. This means that a properly sized unit will have very little reserve capacity to handle the extra load of the stored-up morning heat. For this reason, it is usually false economy to cool part time. The cooling unit will run only when cooling is needed. Therefore, the unit usually will be off during the early morning hours. It should run only enough to take care of any excess heat buildup in the building. As the afternoon heat gain load reaches its peak, the cooling unit will have its full capacity available to take care of only the heat gain on the building at that peak time of day.

(4) Avoid frequent thermostat resetting. Once you have found the temperature that you prefer, it is

best to leave the thermostat at that setting. A good thermostat is a precision instrument that is far more sensitive to temperature changes than a person's body. It will respond to the slightest change in air temperature and will turn on the cooling section even before the average person is aware that cooling is needed.

(5) Turn on the kitchen exhaust fan as soon as you start cooking or baking. It is also a good idea to open a kitchen window slightly to provide for the amount of air being exhausted from that room. The exhaust fan will remove not only excess heat, but humidity and cooking odors as well. It will remove this excess heat from the house economically and pull a slight amount of cool air from other portions of the house as well which helps to offset any additional load on the kitchen during that period of time.
The average four-burner gas stove and oven, with all the burners in use for a period of one full hour, can add over four tons of cooling load to a house. This is about twice the cooling capacity of a unit used for the average 1,000 ft^2 home.
Always be sure to turn off the kitchen exhaust fan and close the window when all the appliances that give off excess heat and moisture are turned off.

(6) Do laundry during the early morning or evening hours on extremely hot days. Washing and drying clothes, as well as bathing and floor mopping, add heat and moisture to the home. If at all practical, do these jobs when the demand on the cooling system is the lowest; this usually means the early morning or evening hours.

(7) Vent the clothes dryer to the outside. The average clothes dryer is capable of removing approximately a gallon of moisture per load. If this moisture is added to the home through an unvented clothes dryer, the cooling unit must work that much harder, and the cost of operation will go up. The small additional cost of the vent will quickly be repaid through the savings in cooling.

(8) *Miscellaneous:* In general, avoid any activity that will add heat or moisture to the building during hot afternoon hours. Use a plastic shower curtain that easily sheds water, instead of a cloth curtain that will absorb and hold water for a long time.

A final item in the homeowner's checkout is the warranty. The equipment manufacturer will include this as part of the owner's manual or installation bulletin; sometimes it is a separate document. There may or may not be a registration form to be returned to the factory (Fig. 25-5). If so, help the homeowner to complete the information, which includes the model and serial number of the equipment, the date of purchase, and the name of the installing contractor. Sometimes it will call for other data such as "Why did you select this equipment brand?" or "Why did you buy from this particular contractor?" This is really marketing information not pertinent to the registration of the warranty, but it is information useful to the manufacturer in improving his image.

The record of the date of purchase is important and should be retained by the homeowner and also by the installing contractor. A cancelled check, paid invoice, or other evidence of purchase should be filed with the other materials.

The standard industry equipment warranty on residential products is usually one year on all parts, plus extended coverage on compressors and the furnace heat exchanger. Hermetic compressors up to 5-ton capacity usually carry a four-year extended coverage. It is also standard practice for the coverage of the first-year warranty to start with the installation date (if known) or 18 months from date of shipment

FIGURE 25-5.

from the factory, whichever comes first. The latter allows six months of inventory by the wholesaler or contractor.

The equipment manufacturer's first-year warranty (Fig. 25-5) implies the replacement of parts at no charge if they are found to be defective during the period of warranty. Labor allowances may or may not be included. Extended coverage warranties on compressors generally do not include labor allowances. Furnace heat exchangers, for example, may be warrantied for as much as 20 years, but the replacement cost is prorated against the actual years of use. Be sure to spell out the specific conditions of parts and labor included in your installation.

In addition to the manufacturer's warranty, the contractor may choose to supplement this with his owner *system warranty*, which may include labor allowances and seasonal system startup and adjustments. Naturally, these are specifically related to the individual situation.

Before terminating the homeowner's checkout, encourage the owner *to file all* the material, for it may be very useful in future service situations. If the home is sold this material should be turned over to the new owner.

The professional air-conditioning technician will finish the installation by recording the telephone number to call for prompt service or maintenace assistance. Many contractors have pressure-sensitive decals (Fig. 25-6) that are prominantly placed on the equipment, listing the number to call.

FIGURE 25-6.

The above situation represents the pleasant event—assuming that all goes well.

But what about the call that comes into the office from another customer who has an important dinner party planned at 6:00 PM; it's already 95°F at noon time, and the cooling is *off*? Words can't describe the emotional trauma that may be reflected in the voice of the housewife—and you receive the call. The first rule is—don't loose your cool! The second rule is—be honest! Don't be evasive or unsure about answering the complaint because of other scheduled service, thus leaving her undecided as to what to do about dinner.

If the call can be answered that day give her an approximate time you expect to be there. Then proceed to ask some necessary information such as:

- Complete address and telephone number.
- Type and brand name of equipment installed, and the model number, if known.
- The age of the installation.
- Did you buy the system from us (meaning that your company might still have the job records)?
- When did the system go off?
- Does it try to start and then stop?
- Did you hear or smell anything unusual?
- Is there any air coming out of the grilles?
- Do you know how to check the fuses or circuit breaker?
- Did the power go off or was there a thunderstorm in the area?
- Have you had other complaints? When?
- Who did the previous service work?
- Is the thermostat switch on COOL?
- Is the temperature set at normal conditions?

Don't expect to diagnose the problem over the phone, but a checklist of information may save time in correcting the problem. If you know the type of equipment, a copy of service instructions or a wiring diagram can be pulled from the service manual or job file. Don't assume that the customer has retained that information. Check your compressor replacement stock—just in case a new compressor may be needed. On the way to the job, review the troubleshooting suggestions that may be included with the service manual and plan your approach so as not to waste time. From your personal telephone call you may conclude that the homeowner has inadvertently done something wrong or the fuse(s) have blown, rather than a more serious problem.

Upon arrival try to reassure the homeowner that you will have the system operational as soon as possible—but do point out that air-conditioning, like any mechanical equipment, will occasionally *malfunction* and need adjustment or replacement. Do not use the term *failure*; it has a negative sound to someone who is obviously concerned.

In the process of trouble-shooting the professional service technician never criticizes the equipment design or the work of a previous serviceman or competitor. He realizes that throwing mud dirties his own hands.

Don't imply that the owner should not have bought that brand, or that you have had all sorts of problems with this particular model. Should a part need replacement—fan motor, compressor, or whatever, and you don't have that particular part with you or even in stock, call the office and try to get a date established when one will be available. Explain to the homeowner what has happened and that you will call as soon as the replacement part is available.

If the system has become inoperative because of a leak or restriction or something that can be field repaired and made operational again, do so with reasonable haste .... but don't compromise quality of workmanship for speed. Explain the problem to the homeowner, and whether it is covered under the warranty. In either case, in warranty or not, get the customer to sign your work order and explain the company policy on billing and collections, so there are no surprises later.

The professional concludes the service call by asking, "Do you have any questions?" He makes certain the owner-user or operator feels that he has made an excellent choice of equipment and is aware of the fact that no matter how well a product has been designed and tested before leaving the manufacturer, it may require mechanical adjustments. He emphasizes that abuse and neglect of the equipment can only bring dissatisfaction with its performance and the expense of unnecessary repairs.

Nor does the professional leave without telling the homeowner what to expect in terms of how soon there will be normal temperature. If it is very hot and humid and the system has been off for any length of time, make sure the owner understands the needed recovery time for pulldown.

Don't leave with the expression, "I think it's going to be OK, but if not, call us and we will try again." Say, "I'm confident everything is again normal and we were pleased to have served you."

So much for the human relations side of a service call. But what about the business administration detail? The customer work order may state what was done, but chances are the owner-contractor will have a service report form that details labor, mileage, parts used, etc. He needs this for billing purposes, employment compensation, inventory control, and tax records. Complete this form accurately and promptly.

If an in-warranty part has been replaced it will be necessary to fill out a manufacturer's service report (Fig. 25-7) and attach this information to the defective material when it's returned to the factory or wholesaler/jobber for credit. We can't overemphasize the importance of supplying the needed information accurately, because it may provide a clue to the manufacturer as to why the equipment failed. He can then make corrections in design or manufacture to preclude future problems. This is particularly true of compressors, and too often the report will merely say "stuck," "won't start," "bad," etc.; when the compressor is examined at the factory it is often found to be fully operative. The fault lies elsewhere, but the service technician jumped to conclusions and didn't do a thorough job of trouble-shooting.

The service report will actually provide a convenient and organized approach to recording model numbers, serial numbers, voltage, operating pressures, etc., and hopefully will lead the technician to ask himself, "Why did the part fail in the first place?"

FIGURE 25-7.

CONDENSING SECTION
CHAMPION IV SPLIT-SYSTEM AIR CONDITIONERS

RENEWAL PARTS

Supersedes: Preliminary Form 550.08-RP1 Coded 274 | 1174 | Form 550.08-RP1

MODELS MC12, MC18, MC24, MC30, MC36, MC42, MC48, AND MC60

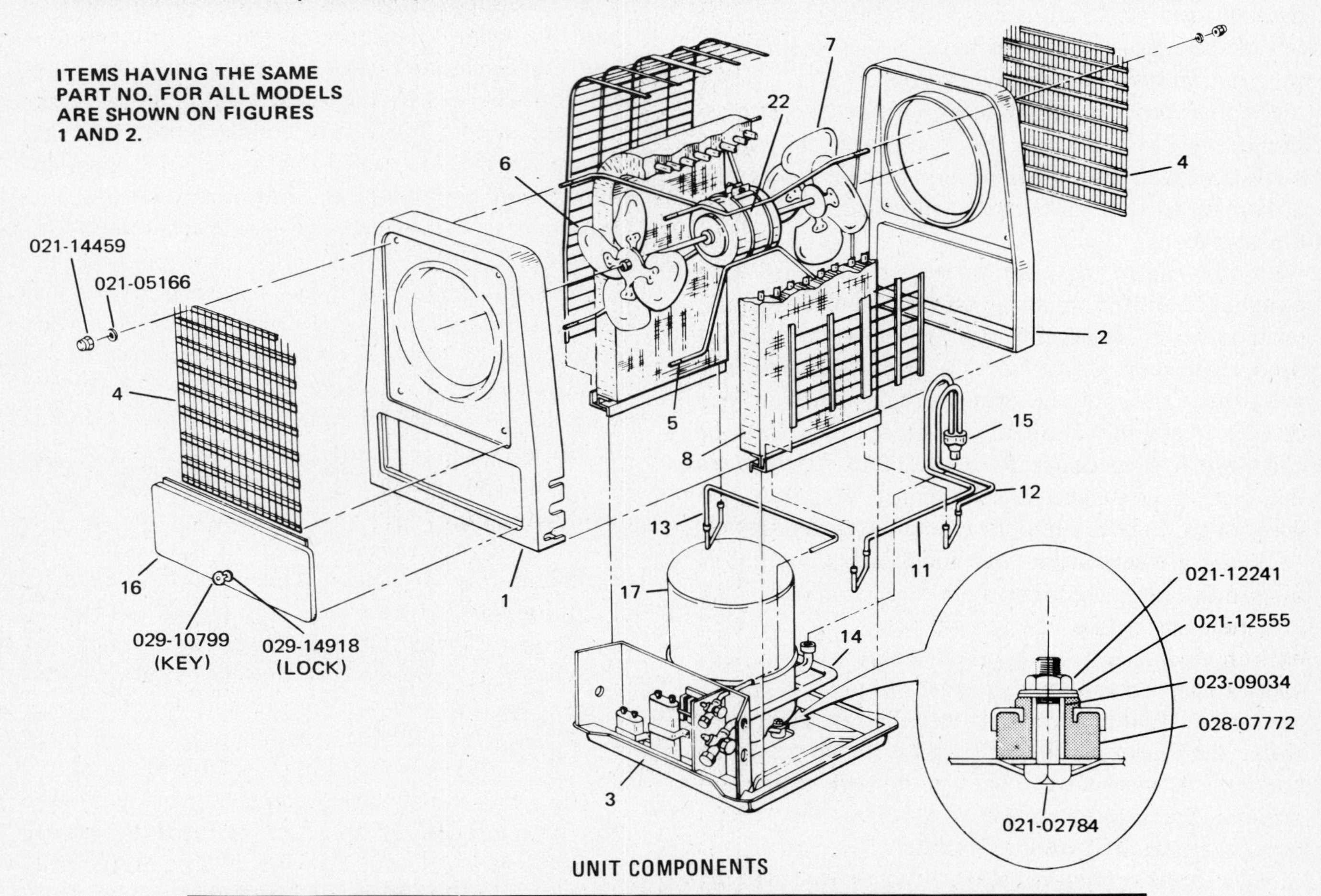

UNIT COMPONENTS

Item No.	Description	MC12	MC18	MC24	MC30	MC36	MC42	MC48	MC60
		Part Numbers							
1	Front Panel	063-54507	063-54507	063-54507	063-54508	063-54508	063-54509	063-54509	063-54509
2	Rear Panel	063-54510	063-54510	063-54510	063-54511	063-54511	063-54513	063-54513	063-54513
3	Base Pan	063-54514	063-54514	063-54514	063-54514	063-54515	063-54515	063-54515	063-54515
4	Fan Grille	026-19364	026-19364	026-19364	026-19365	026-19365	026-19366	026-19366	026-19366
5	Fan Motor Mtg. Bracket	026-19373	026-19373	026-19373	026-19373	026-19373	026-19374	026-19374	026-19374
6	Front Fan Blade	026-19367	026-19367	026-19367	026-19371	026-19371	026-19369	026-19369	026-19369
7	Rear Fan Blade	026-19368	026-19368	026-19368	026-19372	026-19372	026-19370	026-19370	026-19370
8	Condenser Coil								
	Style A (Copper)	363-54065	363-54066	363-54067	363-54068	363-54068	363-54070	363-54072	363-54072
	Style B (Alum.)	363-57659	363-57660	363-57661	363-57662	363-57662	363-57663	363-57664	363-57664
9*	Liquid Valve (w/conn.)	363-54581	363-54581	363-54582	363-54582	363-54582	363-54580	363-54580	363-54580
10*	Suction Valve (w/conn.)	363-54575	363-54575	363-54577	363-54578	363-54579	363-55058	363-54576	363-54576
11	Discharge Connection	063-54562	063-54559	063-54564	063-54563	063-54565	063-54525	063-55059	063-55059
12	Discharge Connection	---	---	---	063-55053	063-55054	063-54528	---	---
13	Liquid Connection								
	Style A (Copper)	363-54557	363-54568	363-54567	363-54556	363-54556	363-54593	363-54594	363-54594
	Style B (Alum.)	363-54557	363-57660	363-57962	363-57966	363-57966	363-54593	363-54594	363-54594
14	Suction Connection	063-54583	063-54586	063-54584	063-54585	063-54561	063-54573	063-54566	063-54566
15	Distributor	---	---	---	026-19363	026-19363	026-19363	---	---
16	Access Door	363-54503	363-54503	363-54503	363-54503	363-54503	363-54504	363-54504	363-54504

*See Figure 2.

FIGURE 25-8 Renewal parts data sheet (Courtesy, York, Div. of Borg-Warner Corporation).

"Will it happen again?" and, "Is there a system deficiency that may cause repeated failures?" Don't assume that the new part will correct everything.

And, speaking of renewal parts, this is one area of business administration where a service technician can be most helpful to his supplier when ordering parts. Most manufacturers publish renewal parts data sheets (Fig. 25-8), which offer a picture or drawing of the product and then itemize each component part by number or letter. The items are then tabulated by description and assigned a multidigit part number.

If anything drives a supplier "up the wall," it is a frantic call such as "I need an expansion valve for a 3-ton brand *X* unit." Too often the stock clerk on the other end is not a skilled service technician and can't be expected to know off the top of his head what to send. Instead he needs to know the product model, the part description, and the part number, so he can cross-check against the manufacturer's data. If you don't know parts data, at least give the clerk the right product model and an accurate part description; he then can usually find the part number. Try not to waste his time or yours.

This last discussion has emphasized residential air conditioning, where the system is relatively simple and inexpensive, and the service technician is dealing with the owner who is also the operator. It's a rather straightforward relationship.

Commercial customer service relationships on the other hand are more complicated, depending on the type of application or customer. The technical art of rendering the actual service is subject to the type of equipment itself—from a small rooftop to a thousand-ton centrifugal unit, not to mention commercial refrigeration, which is a whole field in itself. Obviously, the technical knowledge is assumed in order for a firm to be engaged in this work.

Certain large commercial customers, which we will call national syndicate buyers, such as Sears, J.C. Penney, Wards, and chain grocery stores may have their own inhouse service. Their crews provide the routine inspections, as well as general service and repair, unless a specific product requires attention by factory-authorized personnel.

Some large industrial customers like chemical plants, pharmaceutical manufacturers, may also fall into this category of inhouse maintenance. Due to the broad scope of mechanical work they can afford to hire and train their own people.

On the other hand, there are many large office buildings, locally owned retail stores, hotels, motels, and a host of commercial/industrial customers who own and operate air-conditioning plants vital to their existence, but who cannot economically justify a full-time highly skilled refrigeration/air-conditioning technician. Instead these customers will use general maintenance personnel merely to operate the heating and air-conditioning, as well as the elevators, lighting, and other forms of electrical and or mechanical equipment. Thus, when specific service on the HVAC equipment is needed, the outside experts are called in. This class of customer is the basis for the very important *service contract* business.

SERVICE CONTRACTS

Frequently known as a *service contract*, it may also be called a *maintenance contract* or *preventive maintenance contract*; sometimes the word *agreement* is used instead of contract.

There are four main types of refrigeration and air-conditioning service and maintenance contract programs being offered today, and the one selected by the user will depend on his particular situation.

Type I

This consists of an arrangement whereby the equipment covered by such a contract is usually checked monthly, and sometimes bimonthly. The following is the type of service included in these checkup calls:

- oil all open motors
- check all temperatures, and adjust controls or valves where necessary
- check belt tension and adjust pulleys if needed
- check refrigerant charge and for leaks
- clean air-cooled condensers
- clean spray nozzles
- perform air-filter service
- purge air and noncondensibles from refrigerant system

Any parts or supplies used are billed at regular rates, and any labor or extra service calls over and above that specified in the contract are charged.

Thus a Type I service contract is fundamentally an inspection, cleaning, and adjustment program, and it should be understood that it is nothing more.

Type II

This contract provides the same periodic inspection as in Type I. In addition, all labor for any and all necessary repairs are included at no additional charge. This includes extra service calls at any time required. The customer is still charged for all parts and supplies used.

With this plan the customer pays a fixed, known fee for the labor involved and takes his chances with the parts. Its main feature is a kind of insurance against excessive labor costs in the event of an unusual run of trouble that requires repeated service calls.

Type III

This program provides the protection of Type II, as well as all parts and supplies except motors and compressors. The cost must be higher not only to make provision for replacement of an estimated average number of parts and supplies but to build a fund to take care of those systems where more than average replacements are necessary.

This plan is a compromise between the cost of the partial protection provided by Type II program and the higher costs of *absolute protection*. It offers the benefit of the stabilization of costs. And for the service organization it is a safer plan than absolute coverage.

Type IV

This program provides *complete* or *absolute* coverage for all labor and material. To the customer such a program offers the insurance of a guaranteed fixed charge for maintenance and repair costs. However, it must be remembered that insurance is only as dependable as the organization backing it. Such programs are usually offered only by well-established service organizations or national suppliers of refrigeration/air-conditioning equipment. A solid financial foundation is a must. Also, the organization should have a broad range of contracts so as to average the risks.

Which program is best for the user will depend upon the size of the plant and the amount of "insurance" that the user decides to buy. For quite small systems a Type I plan is often selected, partially because it is cheapest, and partially because even service companies believe it is best for the small system.

For those users who want to consider the insurance type of programs (II, III, or IV) the question is: how will such a plan affect costs, and how hard would a bad breakdown hit the company financially. It is a problem for management to work out, after getting all the available information from the service company.

What are some of the realized customer benefits of a proper maintenance agreement?

- It maintains his system at its peak performance—"It's like having a new system every year."
- Greater safety, because all inspections include a safety check.
- It cuts down time lost and protects user investment in terms of sales, production, or personnel productivity.
- Lowers ownership and operating costs.
- Fixed annual charges (depending on plan selected).
- Reduces administrative problems for the customer.
- Reduces customer investment in parts and personnel.
- And lastly, provides customer management with an updated written report of his system conditions.

Benefits to the service organization include:

- Profitable business on a year-round basis.
- Better utilization and scheduling of manpower.
- Exposure to new equipment sales opportunities.

Manufacturers and the industry in general benefit from service/maintenance contracts because they obviously improve the reliability of operation and reduce customer dissatisfaction, not to mention the increased employment opportunities for competent service personnel.

Designers and consulting engineers are beginning to include the recognition of service and maintenance in the initial design specifications, rather than making it an afterthought to be dealt with after the manufacturer/contractor installation warranties expire. It's part of the consulting or design engineer's responsibility to offer an owner some assistance in evaluating an effective preventive maintenance program.

There is no set rule governing who will be the service technician's personal contacts on a commercial job. It may be the owner, the store manager, the building superintendent, the maintenance supervisor, the operating engineer, etc., but it's important to recognize who has the decision-making responsibility. When working with subordinates be careful not to engage in discussions that might be critical of customer management or methods of handling the equipment maintenance. Also, avoid situations which might place you in the position of going over the head of a subordinate, thus creating ill will.

CONCLUSION

Because of the complexity of modern heating and cooling equipment, the professionsl service technician must have a working knowledge of electrical circuits and components, compression and expansion of liquids and vapors, heat transfer methods, air treatment and distribution and, by all means, public relations.

After any necessary mechanical repairs have been made he must then repair or restore the owner-user's opinion of and confidence in the equipment. This process begins the moment the serviceman arrives on the scene. It is best conveyed by the manner in which he conducts himself. The professional technician is courteous, neat in appearance and workmanship, thorough, and businesslike.

The professional knows that "a workman is only as good as his tools." He equips himself with good, up-to-date tools and information and takes care of them so they will provide him with years of accurate and useful support.

The professional is prompt for his appointments. He does not smoke on the customer's premises without asking permission. He protects the premises from the debris and dirt resulting from a repair. He cleans the equipment and the work area before he leaves the job. Such conduct leaves a lasting impression. It eases any inconvenience and the customer is left with the impression that the equipment and the people involved with it are competent and of good quality. The introduction to air-conditioning starts with a salesman or consulting engineer, but lasting customer satisfaction rests in the hands of the installing and servicing technician who is usually the final communication channel with the purchaser. It's a responsibility to live up to—and of which the service technician can be proud.

PROBLEMS

1. Customer service requires only technical skills. True or false?
2. Service technicians should be good diplomats. True or false?
3. Personal appearances are not important. True or false?
4. Respect for other persons' property is not really important to a service technician. True or false?
5. The installation instructions and homeowner's guide should be discarded when the installation is complete. True or false?
6. What is the most important user responsibility in maintaining his system?
7. Warranty records are an unnecessary nuisance. True or false?
8. Warranty responsibility is not important. True or false?
9. Always criticize the equipment as being the problem. True or false?
10. Always knock competitive products. True or false?
11. Service contracts can be valuable to the user and the servicing agency. True or false?
12. Service contracts vary widely and must be chosen according to the particular need. True or false?

APPENDICES

TABLE 3 . . . PROPERTIES OF LIQUID AND SATURATED VAPOR (continued)

TEMP F	PRESSURE lb per sq in		VOLUME cu ft per lb		DENSITY lb per cu ft		ENTHALPY ** Btu per lb			ENTROPY ** Btu per (lb) (°R)		TEMP F
t	Absolute P	Gage p	Liquid v_f	Vapor v_g	Liquid $1/v_f$	Vapor $1/v_g$	Liquid h_f	Latent h_{fg}	Vapor h_g	Liquid s_f	Vapor s_g	t
-25	13.556	2.320*	0.010730	2.7295	93.197	0.36636	3.1724	71.391	74.563	0.007407	0.17164	-25
-24	13.886	1.649*	.010741	2.6691	93.098	0.37466	3.3848	71.288	74.673	.007894	.17151	-24
-23	14.222	0.966*	.010753	2.6102	92.999	0.38311	3.5973	71.185	74.782	.008379	.17139	-23
-22	14.564	0.270*	.010764	2.5529	92.899	0.39171	3.8100	71.081	74.891	.008864	.17126	-22
-21	14.912	0.216	.010776	2.4972	92.799	0.40045	4.0228	70.978	75.001	.009348	.17114	-21
-20	15.267	0.571	0.010788	2.4429	92.699	0.40934	4.2357	70.874	75.110	0.009831	0.17102	-20
-19	15.628	0.932	.010799	2.3901	92.599	0.41839	4.4487	70.770	75.219	.010314	.17090	-19
-18	15.996	1.300	.010811	2.3387	92.499	0.42758	4.6618	70.666	75.328	.010795	.17078	-18
-17	16.371	1.675	.010823	2.2886	92.399	0.43694	4.8751	70.561	75.436	.011276	.17066	-17
-16	16.753	2.057	.010834	2.2399	92.298	0.44645	5.0885	70.456	75.545	.011755	.17055	-16
-15	17.141	2.445	0.010846	2.1924	92.197	0.45612	5.3020	70.352	75.654	0.012234	0.17043	-15
-14	17.536	2.840	.010858	2.1461	92.096	0.46595	5.5157	70.246	75.762	.012712	.17032	-14
-13	17.939	3.243	.010870	2.1011	91.995	0.47595	5.7295	70.141	75.871	.013190	.17021	-13
-12	18.348	3.652	.010882	2.0572	91.893	0.48611	5.9434	70.036	75.979	.013666	.17010	-12
-11	18.765	4.069	.010894	2.0144	91.791	0.49643	6.1574	69.930	76.087	.014142	.16999	-11
-10	19.189	4.493	0.010906	1.9727	91.689	0.50693	6.3716	69.824	76.196	0.014617	0.16989	-10
- 9	19.621	4.925	.010919	1.9320	91.587	0.51759	6.5859	69.718	76.304	.015091	.16978	- 9
- 8	20.059	5.363	.010931	1.8924	91.485	0.52843	6.8003	69.611	76.411	.015564	.16967	- 8
- 7	20.506	5.810	.010943	1.8538	91.382	0.53944	7.0149	69.505	76.520	.016037	.16957	- 7
- 6	20.960	6.264	.010955	1.8161	91.280	0.55063	7.2296	69.397	76.627	.016508	.16947	- 6
- 5	21.422	6.726	0.010968	1.7794	91.177	0.56199	7.4444	69.291	76.735	0.016979	0.16937	- 5
- 4	21.891	7.195	.010980	1.7436	91.074	0.57354	7.6594	69.183	76.842	.017449	.16927	- 4
- 3	22.369	7.673	.010993	1.7086	90.970	0.58526	7.8745	69.075	76.950	.017919	.16917	- 3
- 2	22.854	8.158	.011005	1.6745	90.867	0.59718	8.0898	68.967	77.057	.018388	.16907	- 2
- 1	23.348	8.652	.011018	1.6413	90.763	0.60927	8.3052	68.859	77.164	.018855	.16897	- 1
0	23.849	9.153	0.011030	1.6089	90.659	0.62156	8.5207	68.750	77.271	0.019323	0.16888	0
1	24.359	9.663	.011043	1.5772	90.554	0.63404	8.7364	68.642	77.378	.019789	.16878	1
2	24.878	10.182	.011056	1.5463	90.450	0.64670	8.9522	68.533	77.485	.020255	.16869	2
3	25.404	10.708	.011069	1.5161	90.345	0.65957	9.1682	68.424	77.592	.020719	.16860	3
4	25.939	11.243	.011082	1.4867	90.240	0.67263	9.3843	68.314	77.698	.021184	.16851	4
5	26.483	11.787	0.011094	1.4580	90.135	0.68588	9.6005	68.204	77.805	0.021647	0.16842	5
6	27.036	12.340	.011107	1.4299	90.030	0.69934	9.8169	68.094	77.911	.022110	.16833	6
7	27.597	12.901	.011121	1.4025	89.924	0.71300	10.033	67.984	78.017	.022572	.16824	7
8	28.167	13.471	.011134	1.3758	89.818	0.72687	10.250	67.873	78.123	.023033	.16815	8
9	28.747	14.051	.011147	1.3496	89.712	0.74094	10.467	67.762	78.229	.023494	.16807	9
10	29.335	14.639	0.011160	1.3241	89.606	0.75523	10.684	67.651	78.335	0.023954	0.16798	10
11	29.932	15.236	.011173	1.2992	89.499	0.76972	10.901	67.539	78.440	.024413	.16790	11
12	30.539	15.843	.011187	1.2748	89.392	0.78443	11.118	67.428	78.546	.024871	.16782	12
13	31.155	16.459	.011200	1.2510	89.285	0.79935	11.336	67.315	78.651	.025329	.16774	13
14	31.780	17.084	.011214	1.2278	89.178	0.81449	11.554	67.203	78.757	.025786	.16765	14
15	32.415	17.719	0.011227	1.2050	89.070	0.82986	11.771	67.090	78.861	0.026243	0.16758	15
16	33.060	18.364	.011241	1.1828	88.962	0.84544	11.989	66.977	78.966	.026699	.16750	16
17	33.714	19.018	.011254	1.1611	88.854	0.86125	12.207	66.864	79.071	.027154	.16742	17
18	34.378	19.682	.011268	1.1399	88.746	0.87729	12.426	66.750	79.176	.027608	.16734	18
19	35.052	20.356	.011282	1.1191	88.637	0.89356	12.644	66.636	79.280	.028062	.16727	19
20	35.736	21.040	0.011296	1.0988	88.529	0.91006	12.863	66.522	79.385	0.028515	0.16719	20
21	36.430	21.734	.011310	1.0790	88.419	0.92679	13.081	66.407	79.488	.028968	.16712	21
22	37.135	22.439	.011324	1.0596	88.310	0.94377	13.300	66.293	79.593	.029420	.16704	22
23	37.849	23.153	.011338	1.0406	88.201	0.96098	13.520	66.177	79.697	.029871	.16697	23
24	38.574	23.878	.011352	1.0220	88.091	0.97843	13.739	66.061	79.800	.030322	.16690	24
25	39.310	24.614	0.011366	1.0039	87.981	0.99613	13.958	65.946	79.904	0.030772	0.16683	25
26	40.056	25.360	.011380	0.98612	87.870	1.0141	14.178	65.829	80.007	.031221	.16676	26
27	40.813	26.117	.011395	0.96874	87.760	1.0323	14.398	65.713	80.111	.031670	.16669	27
28	41.580	26.884	.011409	0.95173	87.649	1.0507	14.618	65.596	80.214	.032118	.16662	28
29	42.359	27.663	.011424	0.93509	87.537	1.0694	14.838	65.478	80.316	.032566	.16655	29
30	43.148	28.452	0.011438	0.91880	87.426	1.0884	15.058	65.361	80.419	0.033013	0.16648	30
31	43.948	29.252	.011453	0.90286	87.314	1.1076	15.279	65.243	80.522	.033460	.16642	31
32	44.760	30.064	.011468	0.88725	87.202	1.1271	15.500	65.124	80.624	.033905	.16635	32
33	45.583	30.887	.011482	0.87197	87.090	1.1468	15.720	65.006	80.726	.034351	.16629	33
34	46.417	31.721	.011497	0.85702	86.977	1.1668	15.942	64.886	80.828	.034796	.16622	34
35	47.263	32.567	0.011512	0.84237	86.865	1.1871	16.163	64.767	80.930	0.035240	0.16616	35
36	48.120	33.424	.011527	0.82803	86.751	1.2077	16.384	64.647	81.031	.035683	.16610	36
37	48.989	34.293	.011542	0.81399	86.638	1.2285	16.606	64.527	81.133	.036126	.16604	37
38	49.870	35.174	.011557	0.80023	86.524	1.2496	16.828	64.406	81.234	.036569	.16598	38
39	50.763	36.067	0.011573	0.78676	86.410	1.2710	17.050	64.285	81.335	0.037011	0.16592	39

*Inches of mercury below one standard atmosphere.

TABLE 3 . . . PROPERTIES OF LIQUID AND SATURATED VAPOR (continued)

TEMP F	PRESSURE lb per sq in		VOLUME cu ft per lb		DENSITY lb per cu ft		ENTHALPY ** Btu per lb			ENTROPY ** Btu per (lb) (°R)		TEMP F
t	Absolute P	Gage p	Liquid v_f	Vapor v_g	Liquid $1/v_f$	Vapor $1/v_g$	Liquid h_f	Latent h_{fg}	Vapor h_g	Liquid s_f	Vapor s_g	t
40	51.667	36.971	0.011588	0.77357	86.296	1.2927	17.273	64.163	81.436	0.037453	0.16586	40
41	52.584	37.888	.011603	.76064	86.181	1.3147	17.495	64.042	81.537	.037893	.16580	41
42	53.513	38.817	.011619	.74798	86.066	1.3369	17.718	63.919	81.637	.038334	.16574	42
43	54.454	39.758	.011635	.73557	85.951	1.3595	17.941	63.796	81.737	.038774	.16568	43
44	55.407	40.711	.011650	.72341	85.836	1.3823	18.164	63.673	81.837	.039213	.16562	44
45	56.373	41.677	0.011666	0.71149	85.720	1.4055	18.387	63.550	81.937	0.039652	0.16557	45
46	57.352	42.656	.011682	.69982	85.604	1.4289	18.611	63.426	82.037	.040091	.16551	46
47	58.343	43.647	.011698	.68837	85.487	1.4527	18.835	63.301	82.136	.040529	.16546	47
48	59.347	44.651	.011714	.67715	85.371	1.4768	19.059	63.177	82.236	.040966	.16540	48
49	60.364	45.668	.011730	.66616	85.254	1.5012	19.283	63.051	82.334	.041403	.16535	49
50	61.394	46.698	0.011746	0.65537	85.136	1.5258	19.507	62.926	82.433	0.041839	0.16530	50
51	62.437	47.741	.011762	.64480	85.018	1.5509	19.732	62.800	82.532	.042276	.16524	51
52	63.494	48.798	.011779	.63444	84.900	1.5762	19.957	62.673	82.630	.042711	.16519	52
53	64.563	49.867	.011795	.62428	84.782	1.6019	20.182	62.546	82.728	.043146	.16514	53
54	65.646	50.950	.011811	.61431	84.663	1.6278	20.408	62.418	82.826	.043581	.16509	54
55	66.743	52.047	0.011828	0.60453	84.544	1.6542	20.634	62.290	82.924	0.044015	0.16504	55
56	67.853	53.157	.011845	.59495	84.425	1.6808	20.859	62.162	83.021	.044449	.16499	56
57	68.977	54.281	.011862	.58554	84.305	1.7078	21.086	62.033	83.119	.044883	.16494	57
58	70.115	55.419	.011879	.57632	84.185	1.7352	21.312	61.903	83.215	.045316	.16489	58
59	71.267	56.571	.011896	.56727	84.065	1.7628	21.539	61.773	83.312	.045748	.16484	59
60	72.433	57.737	0.011913	0.55839	83.944	1.7909	21.766	61.643	83.409	0.046180	0.16479	60
61	73.613	58.917	.011930	.54967	83.823	1.8193	21.993	61.512	83.505	.046612	.16474	61
62	74.807	60.111	.011947	.54112	83.701	1.8480	22.221	61.380	83.601	.047044	.16470	62
63	76.016	61.320	.011965	.53273	83.580	1.8771	22.448	61.248	83.696	.047475	.16465	63
64	77.239	62.543	.011982	.52450	83.457	1.9066	22.676	61.116	83.792	.047905	.16460	64
65	78.477	63.781	0.012000	0.51642	83.335	1.9364	22.905	60.982	83.887	0.048336	0.16456	65
66	79.729	65.033	.012017	.50848	83.212	1.9666	23.133	60.849	83.982	.048765	.16451	66
67	80.996	66.300	.012035	.50070	83.089	1.9972	23.362	60.715	84.077	.049195	.16447	67
68	82.279	67.583	.012053	.49305	82.965	2.0282	23.591	60.580	84.171	.049624	.16442	68
69	83.576	68.880	.012071	.48555	82.841	2.0595	23.821	60.445	84.266	.050053	.16438	69
70	84.888	70.192	0.012089	0.47818	82.717	2.0913	24.050	60.309	84.359	0.050482	0.16434	70
71	86.216	71.520	.012108	.47094	82.592	2.1234	24.281	60.172	84.453	.050910	.16429	71
72	87.559	72.863	.012126	.46383	82.467	2.1559	24.511	60.035	84.546	.051338	.16425	72
73	88.918	74.222	.012145	.45686	82.341	2.1889	24.741	59.898	84.639	.051766	.16421	73
74	90.292	75.596	.012163	.45000	82.215	2.2222	24.973	59.759	84.732	.052193	.16417	74
75	91.682	76.986	0.012182	0.44327	82.089	2.2560	25.204	59.621	84.825	0.052620	0.16412	75
76	93.087	78.391	.012201	.43666	81.962	2.2901	25.435	59.481	84.916	.053047	.16408	76
77	94.509	79.813	.012220	.43016	81.835	2.3247	25.667	59.341	85.008	.053473	.16404	77
78	95.946	81.250	.012239	.42378	81.707	2.3597	25.899	59.201	85.100	.053900	.16400	78
79	97.400	82.704	.012258	.41751	81.579	2.3951	26.132	59.059	85.191	.054326	.16396	79
80	98.870	84.174	0.012277	0.41135	81.450	2.4310	26.365	58.917	85.282	0.054751	0.16392	80
81	100.36	85.66	.012297	.40530	81.322	2.4673	26.598	58.775	85.373	.055177	.16388	81
82	101.86	87.16	.012316	.39935	81.192	2.5041	26.832	58.631	85.463	.055602	.16384	82
83	103.38	88.68	.012336	.39351	81.063	2.5413	27.065	58.488	85.553	.056027	.16380	83
84	104.92	90.22	.012356	.38776	80.932	2.5789	27.300	58.343	85.643	.056452	.16376	84
85	106.47	91.77	0.012376	0.38212	80.802	2.6170	27.534	58.198	85.732	0.056877	0.16372	85
86	108.04	93.34	.012396	.37657	80.671	2.6556	27.769	58.052	85.821	.057301	.16368	86
87	109.63	94.93	.012416	.37111	80.539	2.6946	28.005	57.905	85.910	.057725	.16364	87
88	111.23	96.53	.012437	.36575	80.407	2.7341	28.241	57.757	85.998	.058149	.16360	88
89	112.85	98.15	.012457	.36047	80.275	2.7741	28.477	57.609	86.086	.058573	.16357	89
90	114.49	99.79	0.012478	0.35529	80.142	2.8146	28.713	57.461	86.174	0.058997	0.16353	90
91	116.15	101.45	.012499	.35019	80.008	2.8556	28.950	57.311	86.261	.059420	.16349	91
92	117.82	103.12	.012520	.34518	79.874	2.8970	29.187	57.161	86.348	.059844	.16345	92
93	119.51	104.81	.012541	.34025	79.740	2.9390	29.425	57.009	86.434	.060267	.16341	93
94	121.22	106.52	.012562	.33540	79.605	2.9815	29.663	56.858	86.521	.060690	.16338	94
95	122.95	108.25	0.012583	0.33063	79.470	3.0245	29.901	56.705	86.606	0.061113	0.16334	95
96	124.70	110.00	.012605	.32594	79.334	3.0680	30.140	56.551	86.691	.061536	.16330	96
97	126.46	111.76	.012627	.32133	79.198	3.1120	30.380	56.397	86.777	.061959	.16326	97
98	128.24	113.54	.012649	.31679	79.061	3.1566	30.619	56.242	86.861	.062381	.16323	98
99	130.04	115.34	.012671	.31233	78.923	3.2017	30.859	56.086	86.945	.062804	.16319	99
100	131.86	117.16	0.012693	0.30794	78.785	3.2474	31.100	55.929	87.029	0.063227	0.16315	100
101	133.70	119.00	.012715	.30362	78.647	3.2936	31.341	55.772	87.113	.063649	.16312	101
102	135.56	120.86	.012738	.29937	78.508	3.3404	31.583	55.613	87.196	.064072	.16308	102
103	137.44	122.74	.012760	.29518	78.368	3.3877	31.824	55.454	87.278	.064494	.16304	103
104	139.33	124.63	0.012783	0.29106	78.228	3.4357	32.067	55.293	87.360	0.064916	0.16301	104

TABLE 3 . . . PROPERTIES OF LIQUID AND SATURATED VAPOR (continued)

TEMP F	PRESSURE lb per sq in		VOLUME cu ft per lb		DENSITY lb per cu ft		ENTHALPY ** Btu per lb			ENTROPY ** Btu per (lb) (^{o}R)		TEMP F
t	Absolute P	Gage p	Liquid v_f	Vapor v_g	Liquid $1/v_f$	Vapor $1/v_g$	Liquid h_f	Latent h_{fg}	Vapor h_g	Liquid s_f	Vapor s_g	t
105	141.25	126.55	0.012806	0.28701	78.088	3.4842	32.310	55.132	87.442	0.065339	0.16297	105
106	143.18	128.48	.012829	.28303	77.946	3.5333	32.553	54.970	87.523	.065761	.16293	106
107	145.13	130.43	.012853	.27910	77.804	3.5829	32.797	54.807	87.604	.066184	.16290	107
108	147.11	132.41	.012876	.27524	77.662	3.6332	33.041	54.643	87.684	.066606	.16286	108
109	149.10	134.40	.012900	.27143	77.519	3.6841	33.286	54.478	87.764	.067028	.16282	109
110	151.11	136.41	0.012924	0.26769	77.376	3.7357	33.531	54.313	87.844	0.067451	0.16279	110
111	153.14	138.44	.012948	.26400	77.231	3.7878	33.777	54.146	87.923	.067873	.16275	111
112	155.19	140.49	.012972	.26037	77.087	3.8406	34.023	53.978	88.001	.068296	.16271	112
113	157.27	142.57	.012997	.25680	76.941	3.8941	34.270	53.809	88.079	.068719	.16268	113
114	159.36	144.66	.013022	.25328	76.795	3.9482	34.517	53.639	88.156	.069141	.16264	114
115	161.47	146.77	0.013047	0.24982	76.649	4.0029	34.765	53.468	88.233	0.069564	0.16260	115
116	163.61	148.91	.013072	.24641	76.501	4.0584	35.014	53.296	88.310	.069987	.16256	116
117	165.76	151.06	.013097	.24304	76.353	4.1145	35.263	53.123	88.386	.070410	.16253	117
118	167.94	153.24	.013123	.23974	76.205	4.1713	35.512	52.949	88.461	.070833	.16249	118
119	170.13	155.43	.013148	.23647	76.056	4.2288	35.762	52.774	88.536	.071257	.16245	119
120	172.35	157.65	0.013174	0.23326	75.906	4.2870	36.013	52.597	88.610	0.071680	0.16241	120
121	174.59	159.89	.013200	.23010	75.755	4.3459	36.264	52.420	88.684	.072104	.16237	121
122	176.85	162.15	.013227	.22698	75.604	4.4056	36.516	52.241	88.757	.072528	.16234	122
123	179.13	164.43	.013254	.22391	75.452	4.4660	36.768	52.062	88.830	.072952	.16230	123
124	181.43	166.73	.013280	.22089	75.299	4.5272	37.021	51.881	88.902	.073376	.16226	124
125	183.76	169.06	0.013308	0.21791	75.145	4.5891	37.275	51.698	88.973	0.073800	0.16222	125
126	186.10	171.40	.013335	.21497	74.991	4.6518	37.529	51.515	89.044	.074225	.16218	126
127	188.47	173.77	.013363	.21207	74.836	4.7153	37.785	51.330	89.115	.074650	.16214	127
128	190.86	176.16	.013390	.20922	74.680	4.7796	38.040	51.144	89.184	.075075	.16210	128
129	193.27	178.57	.013419	.20641	74.524	4.8448	38.296	50.957	89.253	.075501	.16206	129
130	195.71	181.01	0.013447	0.20364	74.367	4.9107	38.553	50.768	89.321	0.075927	0.16202	130
131	198.16	183.46	.013476	.20091	74.209	4.9775	38.811	50.578	89.389	.076353	.16198	131
132	200.64	185.94	.013504	.19821	74.050	5.0451	39.069	50.387	89.456	.076779	.16194	132
133	203.15	188.45	.013534	.19556	73.890	5.1136	39.328	50.194	89.522	.077206	.16189	133
134	205.67	190.97	.013563	.19294	73.729	5.1829	39.588	50.000	89.588	.077633	.16185	134
135	208.22	193.52	0.013593	0.19036	73.568	5.2532	39.848	49.805	89.653	0.078061	0.16181	135
136	210.79	196.09	.013623	.18782	73.406	5.3244	40.110	49.608	89.718	.078489	.16177	136
137	213.39	198.69	.013653	.18531	73.243	5.3965	40.372	49.409	89.781	.078917	.16172	137
138	216.01	201.31	.013684	.18283	73.079	5.4695	40.634	49.210	89.844	.079346	.16168	138
139	218.65	203.95	.013715	.18039	72.914	5.5435	40.898	49.008	89.906	.079775	.16163	139
140	221.32	206.62	0.013746	0.17799	72.748	5.6184	41.162	48.805	89.967	0.080205	0.16159	140
141	224.00	209.30	.013778	.17561	72.581	5.6944	41.427	48.601	90.028	.080635	.16154	141
142	226.72	212.02	.013810	.17327	72.413	5.7713	41.693	48.394	90.087	.081065	.16150	142
143	229.46	214.76	.013842	.17096	72.244	5.8493	41.959	48.187	90.146	.081497	.16145	143
144	232.22	217.52	.013874	.16868	72.075	5.9283	42.227	47.977	90.204	.081928	.16140	144
145	235.00	220.30	0.013907	0.16644	71.904	6.0083	42.495	47.766	90.261	0.082361	0.16135	145
146	237.82	223.12	.013941	.16422	71.732	6.0895	42.765	47.553	90.318	.082794	.16130	146
147	240.65	225.95	.013974	.16203	71.559	6.1717	43.035	47.338	90.373	.083227	.16125	147
148	243.51	228.81	.014008	.15987	71.386	6.2551	43.306	47.122	90.428	.083661	.16120	148
149	246.40	231.70	.014043	.15774	71.211	6.3395	43.578	46.904	90.482	.084096	.16115	149
150	249.31	234.61	0.014078	0.15564	71.035	6.4252	43.850	46.684	90.534	0.084531	0.16110	150
151	252.24	237.54	.014113	.15356	70.857	6.5120	44.124	46.462	90.586	.084967	.16105	151
152	255.20	240.50	.014148	.15151	70.679	6.6001	44.399	46.238	90.637	.085404	.16099	152
153	258.19	243.49	.014184	.14949	70.500	6.6893	44.675	46.012	90.687	.085842	.16094	153
154	261.20	246.50	.014221	.14750	70.319	6.7799	44.951	45.784	90.735	.086280	.16088	154
155	264.24	249.54	0.014258	0.14552	70.137	6.8717	45.229	45.554	90.783	0.086719	0.16083	155
156	267.30	252.60	.014295	.14358	69.954	6.9648	45.508	45.322	90.830	.087159	.16077	156
157	270.39	255.69	.014333	.14166	69.770	7.0592	45.787	45.088	90.875	.087600	.16071	157
158	273.51	258.81	.014371	.13976	69.584	7.1551	46.068	44.852	90.920	.088041	.16065	158
159	276.65	261.95	.014410	.13789	69.397	7.2523	46.350	44.614	90.964	.088484	.16059	159
160	279.82	265.12	0.014449	0.13604	69.209	7.3509	46.633	44.373	91.006	0.088927	0.16053	160
161	283.02	268.32	.014489	.13421	69.019	7.4510	46.917	44.130	91.047	.089371	.16047	161
162	286.24	271.54	.014529	.13241	68.828	7.5525	47.202	43.885	91.087	.089817	.16040	162
163	289.49	274.79	.014570	.13062	68.635	7.6556	47.489	43.637	91.126	.090263	.16034	163
164	292.77	278.07	.014611	.12886	68.441	7.7602	47.777	43.386	91.163	.090710	.16027	164
165	296.07	281.37	0.014653	0.12712	68.245	7.8665	48.065	43.134	91.199	0.091159	0.16021	165
166	299.40	284.70	.014695	.12540	68.048	7.9743	48.355	42.879	91.234	.091608	.16014	166
167	302.76	288.06	.014738	.12370	67.850	8.0838	48.647	42.620	91.267	.092059	.16007	167
168	306.15	291.45	.014782	.12202	67.649	8.1950	48.939	42.360	91.299	.092511	.16000	168
169	309.56	294.86	0.014826	0.12037	67.447	8.3080	49.233	42.097	91.330	0.092964	0.15992	169

TABLE 7 . . . PROPERTIES OF LIQUID AND SATURATED VAPOR (continued)

TEMP F	PRESSURE lb per sq in		VOLUME cu ft per lb		DENSITY lb per cu ft		ENTHALPY ** Btu per lb			ENTROPY ** Btu per (lb) (°R)		Temp F
t	Absolute P	Gage p	Liquid v_f	Vapor v_g	Liquid $1/v_f$	Vapor $1/v_g$	Liquid h_f	Latent h_{fg}	Vapor h_g	Liquid s_f	Vapor s_g	t
-50	11.674	6.154*	0.011235	4.2224	89.004	0.23683	- 2.511	101.656	99.144	-0.00604	0.24209	-50
-49	11.996	5.498*	.011248	4.1166	88.905	.24292	- 2.262	101.519	99.257	- .00543	.24176	-49
-48	12.324	4.829*	.011261	4.0140	88.806	.24913	- 2.012	101.381	99.369	- .00483	.24143	-48
-47	12.660	4.144*	.011273	3.9145	88.707	.25546	- 1.762	101.242	99.480	- .00422	.24110	-47
-46	13.004	3.445*	.011286	3.8179	88.607	.26192	- 1.511	101.103	99.592	- .00361	.24078	-46
-45	13.354	2.732*	0.011298	3.7243	88.507	0.26851	- 1.260	100.963	99.703	-0.00301	0.24046	-45
-44	13.712	2.002*	.011311	3.6334	88.407	.27523	- 1.009	100.823	99.814	- .00241	.24014	-44
-43	14.078	1.258*	.011324	3.5452	88.307	.28207	- 0.757	100.683	99.925	- .00181	.23982	-43
-42	14.451	0.498*	.011337	3.4596	88.207	.28905	- 0.505	100.541	100.036	- .00120	.23951	-42
-41	14.833	0.137	.011350	3.3764	88.107	.29617	- 0.253	100.399	100.147	- .00060	.23919	-41
-40	15.222	0.526	0.011363	3.2957	88.006	0.30342	0.000	100.257	100.257	0.00000	0.23888	-40
-39	15.619	0.923	.011376	3.2173	87.905	.31082	0.253	100.114	100.367	.00060	.23858	-39
-38	16.024	1.328	.011389	3.1412	87.805	.31835	0.506	99.971	100.477	.00120	.23827	-38
-37	16.437	1.741	.011402	3.0673	87.703	.32602	0.760	99.826	100.587	.00180	.23797	-37
-36	16.859	2.163	.011415	2.9954	87.602	.33384	1.014	99.682	100.696	.00240	.23767	-36
-35	17.290	2.594	0.011428	2.9256	87.501	0.34181	1.269	99.536	100.805	0.00300	0.23737	-35
-34	17.728	3.032	.011442	2.8578	87.399	.34992	1.524	99.391	100.914	.00359	.23707	-34
-33	18.176	3.480	.011455	2.7919	87.297	.35818	1.779	99.244	101.023	.00419	.23678	-33
-32	18.633	3.937	.011469	2.7278	87.195	.36660	2.035	99.097	101.132	.00479	.23649	-32
-31	19.098	4.402	.011482	2.6655	87.093	.37517	2.291	98.949	101.240	.00538	.23620	-31
-30	19.573	4.877	0.011495	2.6049	86.991	0.38389	2.547	98.801	101.348	0.00598	0.23591	-30
-29	20.056	5.360	.011509	2.5460	86.888	.39278	2.804	98.652	101.456	.00657	.23563	-29
-28	20.549	5.853	.011523	2.4887	86.785	.40182	3.061	98.503	101.564	.00716	.23534	-28
-27	21.052	6.536	.011536	2.4329	86.682	.41103	3.318	98.353	101.671	.00776	.23506	-27
-26	21.564	6.868	.011550	2.3787	86.579	.42040	3.576	98.202	101.778	.00835	.23478	-26
-25	22.086	7.390	0.011564	2.3260	86.476	0.42993	3.834	98.051	101.885	0.00894	0.23451	-25
-24	22.617	7.921	.011578	2.2746	86.372	.43964	4.093	97.899	101.992	.00953	.23423	-24
-23	23.159	8.463	.011592	2.2246	86.269	.44951	4.352	97.746	102.098	.01013	.23396	-23
-22	23.711	9.015	.011606	2.1760	86.165	.45956	4.611	97.593	102.204	.01072	.23369	-22
-21	24.272	9.576	.011620	2.1287	86.061	.46978	4.871	97.439	102.310	.01131	.23342	-21
-20	24.845	10.149	0.011634	2.0826	85.956	0.48018	5.131	97.285	102.415	0.01189	0.23315	-20
-19	25.427	10.731	.011648	2.0377	85.852	.49075	5.391	97.129	102.521	.01248	.23289	-19
-18	26.020	11.324	.011662	1.9940	85.747	.50151	5.652	96.974	102.626	.01307	.23262	-18
-17	26.624	11.928	.011677	1.9514	85.642	.51245	5.913	96.817	102.730	.01366	.23236	-17
-16	27.239	12.543	.011691	1.9099	85.537	.52358	6.175	96.660	102.835	.01425	.23210	-16
-15	27.865	13.169	0.011705	1.8695	85.431	0.53489	6.436	96.502	102.939	0.01483	0.23184	-15
-14	28.501	13.805	.011720	1.8302	85.326	.54640	6.699	96.344	103.043	.01542	.23159	-14
-13	29.149	14.453	.011734	1.7918	85.220	.55810	6.961	96.185	103.146	.01600	.23133	-13
-12	29.809	15.113	.011749	1.7544	85.114	.56999	7.224	96.025	103.250	.01659	.23108	-12
-11	30.480	15.784	.011764	1.7180	85.008	.58207	7.488	95.865	103.353	.01717	.23083	-11
-10	31.162	16.466	0.011778	1.6825	84.901	0.59436	7.751	95.704	103.455	0.01776	0.23058	-10
- 9	31.856	17.160	.011793	1.6479	84.795	.60685	8.015	95.542	103.558	.01834	.23033	- 9
- 8	32.563	17.867	.011808	1.6141	84.688	.61954	8.280	95.380	103.660	.01892	.23008	- 8
- 7	33.281	18.585	.011823	1.5812	84.581	.63244	8.545	95.217	103.762	.01950	.22984	- 7
- 6	34.011	19.315	.011838	1.5491	84.473	.64555	8.810	95.053	103.863	.02009	.22960	- 6
- 5	34.754	20.058	0.011853	1.5177	84.366	0.65887	9.075	94.889	103.964	0.02067	0.22936	- 5
- 4	35.509	20.813	.011868	1.4872	84.258	.67240	9.341	94.724	104.065	.02125	.22912	- 4
- 3	36.277	21.581	.011884	1.4574	84.150	.68615	9.608	94.558	104.166	.02183	.22888	- 3
- 2	37.057	22.361	.011899	1.4283	84.042	.70012	9.874	94.391	104.266	.02241	.22864	- 2
- 1	37.850	23.154	.011914	1.4000	83.933	.71431	10.142	94.224	104.366	.02299	.22841	- 1
0	38.657	23.961	0.011930	1.3723	83.825	0.72872	10.409	94.056	104.465	0.02357	0.22817	0
1	39.476	24.780	.011945	1.3453	83.716	.74336	10.677	93.888	104.565	.02414	.22794	1
2	40.309	25.613	.011961	1.3189	83.606	.75822	10.945	93.718	104.663	.02472	.22771	2
3	41.155	26.459	.011976	1.2931	83.497	.77332	11.214	93.548	104.762	.02530	.22748	3
4	42.014	27.318	.011992	1.2680	83.387	.78865	11.483	93.378	104.860	.02587	.22725	4
5	42.888	28.192	0.012008	1.2434	83.277	0.80422	11.752	93.206	104.958	0.02645	0.22703	5
6	43.775	29.079	.012024	1.2195	83.167	.82003	12.022	93.034	105.056	.02703	.22680	6
7	44.676	29.980	.012040	1.1961	83.057	.83608	12.292	92.861	105.153	.02760	.22658	7
8	45.591	30.895	.012056	1.1732	82.946	.85237	12.562	92.688	105.250	.02818	.22636	8
9	46.521	31.825	.012072	1.1509	82.835	.86892	12.833	92.513	105.346	.02875	.22614	9
10	47.464	32.768	0.012088	1.1290	82.724	0.88571	13.104	92.338	105.442	0.02932	0.22592	10
11	48.423	33.727	.012105	1.1077	82.612	.90275	13.376	92.162	105.538	.02990	.22570	11
12	49.396	34.700	.012121	1.0869	82.501	.92005	13.648	91.986	105.633	.03047	.22548	12
13	50.384	35.688	.012138	1.0665	82.389	.93761	13.920	91.808	105.728	.03104	.22527	13
14	51.387	36.691	.012154	1.0466	82.276	.95544	14.193	91.630	105.823	.03161	.22505	14

*Inches of mercury below one standard atmosphere.

TABLE 7 . . . PROPERTIES OF LIQUID AND SATURATED VAPOR (continued)

TEMP F	PRESSURE lb per sq in		VOLUME cu ft per lb		DENSITY lb per cu ft		ENTHALPY ** Btu per lb			ENTROPY ** Btu per (lb) (°R)		TEMP F
t	Absolute P	Gage p	Liquid v_f	Vapor v_g	Liquid $1/v_f$	Vapor $1/v_g$	Liquid h_f	Latent h_{fg}	Vapor h_g	Liquid s_f	Vapor s_g	t
15	52.405	37.709	0.012171	1.0272	82.164	0.97352	14.466	91.451	105.917	0.03218	0.22484	15
16	53.438	38.742	.012188	1.0082	82.051	0.99188	14.739	91.272	106.011	.03275	.22463	16
17	54.487	39.791	.012204	0.98961	81.938	1.0105	15.013	91.091	106.105	.03332	.22442	17
18	55.551	40.855	.012221	0.97144	81.825	1.0294	15.288	90.910	106.198	.03389	.22421	18
19	56.631	41.935	.012238	0.95368	81.711	1.0486	15.562	90.728	106.290	.03446	.22400	19
20	57.727	43.031	0.012255	0.93631	81.597	1.0680	15.837	90.545	106.383	0.03503	0.22379	20
21	58.839	44.143	.012273	.91932	81.483	1.0878	16.113	90.362	106.475	.03560	.22358	21
22	59.967	45.271	.012290	.90270	81.368	1.1078	16.389	90.178	106.566	.03617	.22338	22
23	61.111	46.415	.012307	.88645	81.253	1.1281	16.665	89.993	106.657	.03674	.22318	23
24	62.272	47.576	.012325	.87055	81.138	1.1487	16.942	89.807	106.748	.03730	.22297	24
25	63.450	48.754	0.012342	0.85500	81.023	1.1696	17.219	89.620	106.839	0.03787	0.22277	25
26	64.644	49.948	.012360	.83978	80.907	1.1908	17.496	89.433	106.928	.03844	.22257	26
27	65.855	51.159	.012378	.82488	80.791	1.2123	17.774	89.244	107.018	.03900	.22237	27
28	67.083	52.387	.012395	.81031	80.675	1.2341	18.052	89.055	107.107	.03958	.22217	28
29	68.328	53.632	.012413	.79604	80.558	1.2562	18.330	88.865	107.196	.04013	.22198	29
30	69.591	54.895	0.012431	0.78208	80.441	1.2786	18.609	88.674	107.284	0.04070	0.22178	30
31	70.871	56.175	.012450	.76842	80.324	1.3014	18.889	88.483	107.372	.04126	.22158	31
32	72.169	57.473	.012468	.75503	80.207	1.3244	19.169	88.290	107.459	.04182	.22139	32
33	73.485	58.789	.012486	.74194	80.089	1.3478	19.449	88.097	107.546	.04239	.22119	33
34	74.818	60.122	.012505	.72911	79.971	1.3715	19.729	87.903	107.632	.04295	.22100	34
35	76.170	61.474	0.012523	0.71655	79.852	1.3956	20.010	87.708	107.719	0.04351	0.22081	35
36	77.540	62.844	.012542	.70425	79.733	1.4199	20.292	87.512	107.804	.04407	.22062	36
37	78.929	64.233	.012561	.69221	79.614	1.4447	20.574	87.316	107.889	.04464	.22043	37
38	80.336	65.640	.012579	.68041	79.495	1.4697	20.856	87.118	107.974	.04520	.22024	38
39	81.761	67.065	.012598	.66885	79.375	1.4951	21.138	86.920	108.058	.04576	.22005	39
40	83.206	68.510	0.012618	0.65753	79.255	1.5208	21.422	86.720	108.142	0.04632	0.21986	40
41	84.670	69.974	.012637	.64643	79.134	1.5469	21.705	86.520	108.225	.04688	.21968	41
42	86.153	71.457	.012656	.63557	79.013	1.5734	21.989	86.319	108.308	.04744	.21949	42
43	87.655	72.959	.012676	.62492	78.892	1.6002	22.273	86.117	108.390	.04800	.21931	43
44	89.177	74.481	.012695	.61448	78.770	1.6274	22.558	85.914	108.472	.04855	.21912	44
45	90.719	76.023	0.012715	0.60425	78.648	1.6549	22.843	85.710	108.553	0.04911	0.21894	45
46	92.280	77.584	.012735	.59422	78.526	1.6829	23.129	85.506	108.634	.04967	.21876	46
47	93.861	79.165	.012755	.58440	78.403	1.7112	23.415	85.300	108.715	.05023	.21858	47
48	95.463	80.767	.012775	.57476	78.280	1.7398	23.701	85.094	108.795	.05079	.21839	48
49	97.085	82.389	.012795	.56532	78.157	1.7689	23.988	84.886	108.874	.05134	.21821	49
50	98.727	84.031	0.012815	0.55606	78.033	1.7984	24.275	84.678	108.953	0.05190	0.21803	50
51	100.39	85.69	.012836	.54698	77.909	1.8282	24.563	84.468	109.031	.05245	.21785	51
52	102.07	87.38	.012856	.53808	77.784	1.8585	24.851	84.258	109.109	.05301	.21768	52
53	103.78	89.08	.012877	.52934	77.659	1.8891	25.139	84.047	109.186	.05357	.21750	53
54	105.50	90.81	.012898	.52078	77.534	1.9202	25.429	83.834	109.263	.05412	.21732	54
55	107.25	92.56	0.012919	0.51238	77.408	1.9517	25.718	83.621	109.339	0.05468	0.21714	55
56	109.02	94.32	.012940	.50414	77.282	1.9836	26.008	83.407	109.415	.05523	.21697	56
57	110.81	96.11	.012961	.49606	77.155	2.0159	26.298	83.191	109.490	.05579	.21679	57
58	112.62	97.93	.012982	.48813	77.028	2.0486	26.589	82.975	109.564	.05634	.21662	58
59	114.46	99.76	.013004	.48035	76.900	2.0818	26.880	82.758	109.638	.05689	.21644	59
60	116.31	101.62	0.013025	0.47272	76.773	2.1154	27.172	82.540	109.712	0.05745	0.21627	60
61	118.19	103.49	.013047	.46523	76.644	2.1495	27.464	82.320	109.785	.05800	.21610	61
62	120.09	105.39	.013069	.45788	76.515	2.1840	27.757	82.100	109.857	.05855	.21592	62
63	122.01	107.32	.013091	.45066	76.386	2.2190	28.050	81.878	109.929	.05910	.21575	63
64	123.96	109.26	.013114	.44358	76.257	2.2544	28.344	81.656	110.000	.05966	.21558	64
65	125.93	111.23	0.013136	0.43663	76.126	2.2903	28.638	81.432	110.070	0.06021	0.21541	65
66	127.92	113.22	.013159	.42981	75.996	2.3266	28.932	81.208	110.140	.06076	.21524	66
67	129.94	115.24	.013181	.42311	75.865	2.3635	29.228	80.982	110.209	.06131	.21507	67
68	131.97	117.28	.013204	.41653	75.733	2.4008	29.523	80.755	110.278	.06186	.21490	68
69	134.04	119.34	.013227	.41007	75.601	2.4386	29.819	80.527	110.346	.06241	.21473	69
70	136.12	121.43	0.013251	0.40373	75.469	2.4769	30.116	80.298	110.414	0.06296	0.21456	70
71	138.23	123.54	.013274	.39751	75.336	2.5157	30.413	80.068	110.480	.06351	.21439	71
72	140.37	125.67	.013297	.39139	75.202	2.5550	30.710	79.836	110.547	.06406	.21422	72
73	142.52	127.83	.013321	.38539	75.068	2.5948	31.008	79.604	110.612	.06461	.21405	73
74	144.71	130.01	.013345	.37949	74.934	2.6351	31.307	79.370	110.677	.06516	.21388	74
75	146.91	132.22	0.013369	0.37369	74.799	2.6760	31.606	79.135	110.741	0.06571	0.21372	75
76	149.15	134.45	.013393	.36800	74.664	2.7174	31.906	78.899	110.805	.06626	.21355	76
77	151.40	136.71	.013418	.36241	74.528	2.7593	32.206	78.662	110.868	.06681	.21338	77
78	153.69	138.99	.013442	.35691	74.391	2.8018	32.506	78.423	110.930	.06736	.21321	78
79	155.99	141.30	.013467	.35151	74.254	2.8449	32.808	78.184	110.991	.06791	.21305	79

TABLE 7 . . . PROPERTIES OF LIQUID AND SATURATED VAPOR (continued)

TEMP F	PRESSURE lb per sq in		VOLUME cu ft per lb		DENSITY lb per cu ft		ENTHALPY ** Btu per lb			ENTROPY ** Btu per (lb) (^{o}R)		TEMP F
t	Absolute P	Gage p	Liquid v_f	Vapor v_g	Liquid $1/v_f$	Vapor $1/v_g$	Liquid h_f	Latent h_{fg}	Vapor h_g	Liquid s_f	Vapor s_g	t
80	158.33	143.63	0.013492	0.34621	74.116	2.8885	33.109	77.943	111.052	0.06846	0.21288	80
81	160.68	145.99	.013518	.34099	73.978	2.9326	33.412	77.701	111.112	.06901	.21271	81
82	163.07	148.37	.013543	.33587	73.839	2.9774	33.714	77.457	111.171	.06956	.21255	82
83	165.48	150.78	.013569	.33083	73.700	3.0227	34.018	77.212	111.230	.07011	.21238	83
84	167.92	153.22	.013594	.32588	73.560	3.0686	34.322	76.966	111.288	.07065	.21222	84
85	170.38	155.68	0.013620	0.32101	73.420	3.1151	34.626	76.719	111.345	0.07120	0.21205	85
86	172.87	158.17	.013647	.31623	73.278	3.1622	34.931	76.470	111.401	.07175	.21188	86
87	175.38	160.69	.013673	.31153	73.137	3.2100	35.237	76.220	111.457	.07230	.21172	87
88	177.93	163.23	.013700	.30690	72.994	3.2583	35.543	75.968	111.512	.07285	.21155	88
89	180.50	165.80	.013727	.30236	72.851	3.3073	35.850	75.716	111.566	.07339	.21139	89
90	183.09	168.40	0.013754	0.29789	72.708	3.3570	36.158	75.461	111.619	0.07394	0.21122	90
91	185.72	171.02	.013781	.29349	72.564	3.4073	36.466	75.206	111.671	.07449	.21106	91
92	188.37	173.67	.013809	.28917	72.419	3.4582	36.774	74.949	111.723	.07504	.21089	92
93	191.05	176.35	.013836	.28491	72.273	3.5098	37.084	74.690	111.774	.07559	.21072	93
94	193.76	179.06	.013864	.28073	72.127	3.5621	37.394	74.430	111.824	.07613	.21056	94
95	196.50	181.80	0.013893	0.27662	71.980	3.6151	37.704	74.168	111.873	0.07668	0.21039	95
96	199.26	184.56	.013921	.27257	71.833	3.6688	38.016	73.905	111.921	.07723	.21023	96
97	202.05	187.36	.013950	.26859	71.685	3.7232	38.328	73.641	111.968	.07778	.21006	97
98	204.87	190.18	.013979	.26467	71.536	3.7783	38.640	73.375	112.015	.07832	.20989	98
99	207.72	193.03	.014008	.26081	71.386	3.8341	38.953	73.107	112.060	.07887	.20973	99
100	210.60	195.91	0.014038	0.25702	71.236	3.8907	39.267	72.838	112.105	0.07942	0.20956	100
101	213.51	198.82	.014068	.25329	71.084	3.9481	39.582	72.567	112.149	.07997	.20939	101
102	216.45	201.76	.014098	.24962	70.933	4.0062	39.897	72.294	112.192	.08052	.20923	102
103	219.42	204.72	.014128	.24600	70.780	4.0651	40.213	72.020	112.233	.08107	.20906	103
104	222.42	207.72	.014159	.24244	70.626	4.1247	40.530	71.744	112.274	.08161	.20889	104
105	225.45	210.75	0.014190	0.23894	70.472	4.1852	40.847	71.467	112.314	0.08216	0.20872	105
106	228.50	213.81	.014221	.23549	70.317	4.2465	41.166	71.187	112.353	.08271	.20855	106
107	231.59	216.90	.014253	.23209	70.161	4.3086	41.485	70.906	112.391	.08326	.20838	107
108	234.71	220.02	.014285	.22875	70.005	4.3715	41.804	70.623	112.427	.08381	.20821	108
109	237.86	223.17	.014317	.22546	69.847	4.4354	42.125	70.338	112.463	.08436	.20804	109
110	241.04	226.35	0.014350	0.22222	69.689	4.5000	42.446	70.052	112.498	0.08491	0.20787	110
111	244.25	229.56	.014382	.21903	69.529	4.5656	42.768	69.763	112.531	.08546	.20770	111
112	247.50	232.80	.014416	.21589	69.369	4.6321	43.091	69.473	112.564	.08601	.20753	112
113	250.77	236.08	.014449	.21279	69.208	4.6994	43.415	69.180	112.595	.08656	.20736	113
114	254.08	239.38	.014483	.20974	69.046	4.7677	43.739	68.886	112.626	.08711	.20718	114
115	257.42	242.72	0.014517	0.20674	68.883	4.8370	44.065	68.590	112.655	0.08766	0.20701	115
116	260.79	246.10	.014552	.20378	68.719	4.9072	44.391	68.291	112.682	.08821	.20684	116
117	264.20	249.50	.014587	.20087	68.554	4.9784	44.718	67.991	112.709	.08876	.20666	117
118	267.63	252.94	.014622	.19800	68.388	5.0506	45.046	67.688	112.735	.08932	.20649	118
119	271.10	256.41	.014658	.19517	68.221	5.1238	45.375	67.384	112.759	.08987	.20631	119
120	274.60	259.91	0.014694	0.19238	68.054	5.1981	45.705	67.077	112.782	0.09042	0.20613	120
121	278.14	263.44	.014731	.18963	67.885	5.2734	46.036	66.767	112.803	.09098	.20595	121
122	281.71	267.01	.014768	.18692	67.714	5.3498	46.368	66.456	112.824	.09153	.20578	122
123	285.31	270.62	.014805	.18426	67.543	5.4272	46.701	66.142	112.843	.09208	.20560	123
124	288.95	274.25	.014843	.18163	67.371	5.5058	47.034	65.826	112.860	.09264	.20542	124
125	292.62	277.92	0.014882	0.17903	67.197	5.5856	47.369	65.507	112.877	0.09320	0.20523	125
126	296.33	281.63	.014920	.17648	67.023	5.6665	47.705	65.186	112.891	.09375	.20505	126
127	300.07	285.37	.014960	.17396	66.847	5.7486	48.042	64.863	112.905	.09431	.20487	127
128	303.84	289.14	.014999	.17147	66.670	5.8319	48.380	64.537	112.917	.09487	.20468	128
129	307.65	292.95	.015039	.16902	66.492	5.9164	48.719	64.208	112.927	.09543	.20449	129
130	311.50	296.80	0.015080	0.16661	66.312	6.0022	49.059	63.877	112.936	0.09598	0.20431	130
131	315.38	300.68	.015121	.16422	66.131	6.0893	49.400	63.543	112.943	.09654	.20412	131
132	319.29	304.60	.015163	.16187	65.949	6.1777	49.743	63.206	112.949	.09711	.20393	132
133	323.25	308.55	.015206	.15956	65.766	6.2674	50.087	62.866	112.953	.09767	.20374	133
134	327.23	312.54	.015248	.15727	65.581	6.3585	50.432	62.523	112.955	.09823	.20354	134
135	331.26	316.56	0.015292	0.15501	65.394	6.4510	50.778	62.178	112.956	0.09879	0.20335	135
136	335.32	320.63	.015336	.15279	65.207	6.5450	51.125	61.829	112.954	.09936	.20315	136
137	339.42	324.73	.015381	.15059	65.017	6.6405	51.474	61.477	112.951	.09992	.20295	137
138	343.56	328.86	.015426	.14843	64.826	6.7374	51.824	61.123	112.947	.10049	.20275	138
139	347.73	333.04	.015472	.14629	64.634	6.8359	52.175	60.764	112.940	.10106	.20255	139
140	351.94	337.25	0.015518	0.14418	64.440	6.9360	52.528	60.403	112.931	0.10163	0.20235	140
141	356.19	341.50	.015566	.14209	64.244	7.0377	52.883	60.038	112.921	.10220	.20214	141
142	360.48	345.79	.015613	.14004	64.047	7.1410	53.238	59.670	112.908	.10277	.20194	142
143	364.81	350.11	.015662	.13801	63.848	7.2461	53.596	59.298	112.893	.10334	.20173	143
144	369.17	354.48	.015712	.13600	63.647	7.3529	53.955	58.922	112.877	.10391	.20152	144

TABLE 15 . . . PROPERTIES OF LIQUID AND SATURATED VAPOR (continued)

TEMP F	PRESSURE lb per sq in		VOLUME cu ft per lb		DENSITY lb per cu ft		ENTHALPY ** Btu per lb			ENTROPY ** Btu per (lb) (°R)		TEMP F
t	Absolute P	Gage p	Liquid v_f	Vapor v_g	Liquid $1/v_f$	Vapor $1/v_g$	Liquid h_f	Latent h_{fg}	Vapor h_g	Liquid s_f	Vapor s_g	t
- 35	21.42	6.72	0.01089	1.8637	91.85	0.5366	1.30	74.50	75.80	0.0031	0.1785	- 35
- 34	21.93	7.24	.01090	1.8226	91.74	.5487	1.56	74.37	75.93	.0037	.1784	- 34
- 33	22.46	7.76	.01091	1.7825	91.63	.5610	1.82	74.24	76.06	.0043	.1783	- 33
- 32	23.00	8.30	.01093	1.7436	91.52	.5735	2.07	74.11	76.18	.0049	.1781	- 32
- 31	23.54	8.84	.01094	1.7056	91.41	.5863	2.34	73.97	76.31	.0055	.1780	- 31
- 30	24.10	9.40	0.01095	1.6687	91.30	0.5993	2.60	73.84	76.44	0.0061	0.1779	- 30
- 29	24.66	9.97	.01097	1.6328	91.19	.6124	2.85	73.71	76.56	.0067	.1778	- 29
- 28	25.24	10.54	.01098	1.5978	91.08	.6259	3.12	73.57	76.69	.0073	.1777	- 28
- 27	25.82	11.13	.01099	1.5637	90.97	.6395	3.38	73.44	76.82	.0079	.1776	- 27
- 26	26.42	11.72	.01101	1.5305	90.85	.6534	3.64	73.30	76.94	.0085	.1775	- 26
- 25	27.02	12.33	0.01102	1.4982	90.74	0.6675	3.90	73.17	77.07	0.0091	0.1774	- 25
- 24	27.64	12.95	.01103	1.4667	90.63	.6818	4.16	73.03	77.19	.0097	.1773	- 24
- 23	28.27	13.57	.01105	1.4360	90.52	.6964	4.43	72.89	77.32	.0103	.1772	- 23
- 22	28.91	14.21	.01106	1.4061	90.40	.7112	4.69	72.75	77.44	.0109	.1771	- 22
- 21	29.56	14.86	.01108	1.3770	90.29	.7262	4.95	72.62	77.57	.0115	.1770	- 21
- 20	30.22	15.52	0.01109	1.3486	90.18	0.7415	5.21	72.48	77.69	0.0121	0.1769	- 20
- 19	30.89	16.19	.01110	1.3209	90.06	.7571	5.48	72.34	77.82	.0127	.1768	- 19
- 18	31.57	16.88	.01112	1.2939	89.95	.7729	5.74	72.20	77.94	.0133	.1767	- 18
- 17	32.27	17.57	.01113	1.2676	89.83	.7889	6.01	72.06	78.07	.0139	.1766	- 17
- 16	32.97	18.28	.01115	1.2419	89.72	.8052	6.27	71.92	78.19	.0145	.1766	- 16
- 15	33.69	18.99	0.01116	1.2169	89.60	0.8218	6.54	71.78	78.32	0.0151	0.1765	- 15
- 14	34.42	19.72	.01117	1.1925	89.49	.8386	6.80	71.64	78.44	.0156	.1764	- 14
- 13	35.16	20.46	.01119	1.1686	89.37	.8557	7.06	71.50	78.56	.0162	.1763	- 13
- 12	35.91	21.22	.01120	1.1454	89.26	.8731	7.33	71.36	78.69	.0168	.1762	- 12
- 11	36.68	21.98	.01122	1.1227	89.14	.8907	7.60	71.21	78.81	.0174	.1761	- 11
- 10	37.46	22.76	0.01123	1.1006	89.02	0.9086	7.86	71.07	78.93	0.0180	0.1760	- 10
- 9	38.25	23.55	.01125	1.0790	88.91	.9268	8.13	70.93	79.06	.0186	.1760	- 9
- 8	39.05	24.35	.01126	1.0579	88.79	.9453	8.40	70.78	79.18	.0192	.1759	- 8
- 7	39.86	25.17	.01128	1.0373	88.67	.9640	8.66	70.64	79.30	.0198	.1758	- 7
- 6	40.69	26.00	.01129	1.0172	88.55	.9831	8.93	70.49	79.42	.0204	.1757	- 6
- 5	41.53	26.84	0.01131	0.9976	88.44	1.0024	9.20	70.34	79.54	0.0209	0.1756	- 5
- 4	42.39	27.69	.01132	.9784	88.32	1.0220	9.47	70.20	79.67	.0215	.1756	- 4
- 3	43.26	28.56	.01134	.9597	88.20	1.0420	9.74	70.05	79.79	.0221	.1755	- 3
- 2	44.14	29.44	.01135	.9414	88.08	1.0622	10.00	69.91	79.91	.0227	.1754	- 2
- 1	45.03	30.33	.01137	.9236	87.96	1.0828	10.27	69.76	80.03	.0233	.1753	- 1
0	45.94	31.24	0.01138	0.9061	87.84	1.1036	10.54	69.61	80.15	0.0239	0.1753	0
1	46.86	32.16	.01140	.8891	87.72	1.1248	10.81	69.46	80.27	.0244	.1752	1
2	47.79	33.10	.01142	.8724	87.60	1.1463	11.08	69.31	80.39	.0250	.1751	2
3	48.74	34.05	.01143	.8561	87.48	1.1681	11.35	69.16	80.51	.0256	.1751	3
4	49.71	35.01	.01145	.8402	87.36	1.1902	11.62	69.01	80.63	.0262	.1750	4
5	50.68	35.99	0.01146	0.8247	87.24	1.2126	11.89	68.86	80.75	0.0268	0.1749	5
6	51.68	36.98	.01148	.8094	87.12	1.2354	12.16	68.70	80.86	.0273	.1749	6
7	52.68	37.99	.01149	.7946	87.00	1.2585	12.43	68.55	80.98	.0279	.1748	7
8	53.70	39.01	.01151	.7800	86.88	1.2820	12.70	68.40	81.10	.0285	.1747	8
9	54.74	40.04	.01153	.7658	86.76	1.3058	12.98	68.24	81.22	.0291	.1747	9
10	55.79	41.09	0.01154	0.7519	86.63	1.3300	13.25	68.08	81.33	0.0296	0.1746	10
11	56.86	42.16	.01156	.7383	86.51	1.3545	13.52	67.93	81.45	.0302	.1745	11
12	57.94	43.24	.01158	.7250	86.39	1.3793	13.80	67.77	81.57	.0308	.1745	12
13	59.03	44.34	.01159	.7120	86.26	1.4045	14.06	67.62	81.68	.0314	.1744	13
14	60.14	45.45	.01161	.6992	86.14	1.4301	14.34	67.46	81.80	.0319	.1743	14
15	61.27	46.57	0.01163	0.6868	86.02	1.4561	14.62	67.30	81.92	0.0325	0.1743	15
16	62.41	47.72	.01164	.6746	85.89	1.4824	14.89	67.14	82.03	.0331	.1742	16
17	63.57	48.88	.01166	.6626	85.77	1.5091	15.16	66.98	82.14	.0336	.1742	17
18	64.75	50.05	.01168	.6510	85.64	1.5362	15.44	66.82	82.26	.0342	.1741	18
19	65.94	51.24	.01169	.6395	85.52	1.5637	15.71	66.66	82.37	.0348	.1740	19
20	67.14	52.45	0.01171	0.6283	85.39	1.5915	15.99	66.50	82.49	0.0354	0.1740	20
21	68.37	53.67	.01173	.6174	85.26	1.6198	16.26	66.34	82.60	.0359	.1739	21
22	69.61	54.91	.01175	.6066	85.14	1.6485	16.54	66.17	82.71	.0365	.1739	22
23	70.86	56.17	.01176	.5961	85.01	1.6775	16.81	66.01	82.82	.0371	.1738	23
24	72.13	57.44	.01178	.5858	84.88	1.7070	17.10	65.84	82.94	.0376	.1738	24
25	73.42	58.73	0.01180	0.5757	84.76	1.7369	17.37	65.68	83.05	0.0382	0.1737	25
26	74.73	60.04	.01182	.5659	84.63	1.7672	17.65	65.51	83.16	.0388	.1736	26
27	76.06	61.36	.01183	.5562	84.50	1.7980	17.93	65.34	83.27	.0393	.1736	27
28	77.40	62.70	.01185	.5467	84.37	1.8292	18.21	65.17	83.38	.0399	.1735	28
29	78.76	64.06	.01187	.5374	84.24	1.8608	18.48	65.01	83.49	.0405	.1735	29

TABLE 15 . . . PROPERTIES OF LIQUID AND SATURATED VAPOR (continued)

TEMP F	PRESSURE lb per sq in		VOLUME cu ft per lb		DENSITY lb per cu ft		ENTHALPY ** Btu per lb			ENTROPY ** Btu per (lb) (°R)		TEMP F
t	Absolute P	Gage p	Liquid v_f	Vapor v_g	Liquid $1/v_f$	Vapor $1/v_g$	Liquid h_f	Latent h_{fg}	Vapor h_g	Liquid s_f	Vapor s_g	t
30	80.13	65.44	0.01189	0.5283	84.11	1.8928	18.76	64.84	83.60	0.0410	0.1734	30
31	81.53	66.83	.01191	.5194	83.98	1.9253	19.04	64.67	83.71	.0416	.1734	31
32	82.94	68.24	.01193	.5106	83.85	1.9583	19.32	64.49	83.81	.0422	.1733	32
33	84.37	69.67	.01194	.5021	83.72	1.9917	19.60	64.32	83.92	.0427	.1733	33
34	85.82	71.12	.01196	.4937	83.59	2.0256	19.88	64.15	84.03	.0433	.1732	34
35	87.28	72.59	0.01198	0.4854	83.46	2.0600	20.17	63.97	84.14	0.0438	0.1732	35
36	88.77	74.07	.01200	.4774	83.33	2.0948	20.44	63.80	84.24	.0444	.1731	36
37	90.27	75.58	.01202	.4695	83.20	2.1301	20.73	63.62	84.35	.0450	.1730	37
38	91.80	77.10	.01204	.4617	83.07	2.1659	21.01	63.44	84.45	.0455	.1730	38
39	93.34	78.64	.01206	.4541	82.93	2.2022	21.29	63.27	84.56	.0461	.1729	39
40	94.90	80.20	0.01208	0.4466	82.80	2.2390	21.57	63.09	84.66	0.0466	0.1729	40
41	96.48	81.78	.01210	.4393	82.67	2.2763	21.86	62.91	84.77	.0472	.1728	41
42	98.08	83.38	.01212	.4321	82.53	2.3142	22.14	62.73	84.87	.0478	.1728	42
43	99.70	85.00	.01214	.4251	82.40	2.3525	22.42	62.55	84.97	.0483	.1727	43
44	101.3	86.64	.01216	.4182	82.26	2.3914	22.71	62.36	85.07	.0489	.1727	44
45	103.0	88.30	0.01218	0.4114	82.13	2.4308	22.99	62.18	85.17	0.0494	0.1726	45
46	104.7	89.97	.01220	.4047	81.99	2.4708	23.28	61.99	85.27	.0500	.1726	46
47	106.4	91.67	.01222	.3982	81.86	2.5113	23.57	61.81	85.38	.0505	.1725	47
48	108.1	93.39	.01224	.3918	81.72	2.5524	23.85	61.62	85.47	.0511	.1725	48
49	109.8	95.13	.01226	.3855	81.58	2.5940	24.14	61.43	85.57	.0517	.1724	49
50	111.6	96.89	0.01228	0.3793	81.44	2.6362	24.42	61.25	85.67	0.0522	0.1724	50
51	113.4	98.66	.01230	.3733	81.31	2.6790	24.71	61.06	85.77	.0528	.1723	51
52	115.2	100.5	.01232	.3673	81.17	2.7224	25.00	60.87	85.87	.0533	.1723	52
53	117.0	102.3	.01234	.3615	81.03	2.7664	25.29	60.67	85.96	.0539	.1722	53
54	118.8	104.1	.01236	.3557	80.89	2.8110	25.58	60.48	86.06	.0544	.1722	54
55	120.7	106.0	0.01238	0.3501	80.75	2.8562	25.87	60.28	86.15	0.0550	0.1721	55
56	122.6	107.9	.01241	.3446	80.61	2.9020	26.16	60.09	86.25	.0555	.1721	56
57	124.5	109.8	.01243	.3392	80.47	2.9485	26.44	59.90	86.34	.0561	.1720	57
58	126.4	111.7	.01245	.3338	80.33	2.9956	26.73	59.70	86.43	.0566	.1720	58
59	128.4	113.7	.01247	.3286	80.18	3.0434	27.02	59.50	86.52	.0572	.1719	59
60	130.3	115.6	0.01249	0.3234	80.04	3.0918	27.32	59.30	86.62	0.0578	0.1719	60
61	132.3	117.6	.01252	.3184	79.90	3.1409	27.61	59.10	86.71	.0583	.1718	61
62	134.3	119.6	.01254	.3134	79.76	3.1907	27.91	58.89	86.80	.0589	.1717	62
63	136.4	121.7	.01256	.3085	79.61	3.2411	28.19	58.69	86.88	.0594	.1717	63
64	138.4	123.7	.01258	.3037	79.47	3.2923	28.48	58.49	86.97	.0600	.1716	64
65	140.5	125.8	0.01261	0.2990	79.32	3.3442	28.78	58.28	87.06	0.0605	0.1716	65
66	142.6	127.9	.01263	.2944	79.18	3.3968	29.08	58.07	87.15	.0611	.1715	66
67	144.8	130.1	.01265	.2898	79.03	3.4502	29.37	57.86	87.23	.0616	.1715	67
68	146.9	132.2	.01268	.2854	78.88	3.5043	29.67	57.65	87.32	.0622	.1714	68
69	149.1	134.4	.01270	.2810	78.74	3.5591	29.96	57.44	87.40	.0627	.1714	69
70	151.3	136.6	0.01272	0.2766	78.59	3.6147	30.25	57.23	87.48	0.0633	0.1713	70
71	153.5	138.8	.01275	.2724	78.44	3.6712	30.55	57.01	87.56	.0638	.1712	71
72	155.8	141.1	.01277	.2682	78.29	3.7284	30.85	56.80	87.65	.0644	.1712	72
73	158.0	143.3	.01280	.2641	78.14	3.7864	31.15	56.58	87.73	.0649	.1711	73
74	160.3	145.6	.01282	.2601	77.99	3.8452	31.45	56.36	87.81	.0655	.1711	74
75	162.7	148.0	0.01285	0.2561	77.84	3.9049	31.74	56.14	87.88	0.0660	0.1710	75
76	165.0	150.3	.01287	.2522	77.68	3.9654	32.04	55.92	87.96	.0665	.1709	76
77	167.4	152.7	.01290	.2483	77.53	4.0268	32.34	55.70	88.04	.0671	.1709	77
78	169.8	155.1	.01292	.2446	77.38	4.0890	32.64	55.47	88.11	.0676	.1708	78
79	172.2	157.5	.01295	.2408	77.22	4.1522	32.94	55.25	88.19	.0682	.1707	79
80	174.6	159.9	0.01298	0.2372	77.07	4.2162	33.24	55.02	88.26	0.0687	0.1707	80
81	177.1	162.4	.01300	.2336	76.91	4.2812	33.54	54.79	88.33	.0693	.1706	81
82	179.6	164.9	.01303	.2300	76.76	4.3471	33.84	54.56	88.40	.0698	.1706	82
83	182.1	167.4	.01305	.2266	76.60	4.4140	34.14	54.33	88.47	.0704	.1705	83
84	184.7	170.0	.01308	.2231	76.44	4.4819	34.45	54.09	88.54	.0709	.1704	84
85	187.2	172.5	0.01311	0.2197	76.29	4.5507	34.75	53.86	88.61	0.0715	0.1703	85
86	189.8	175.1	.01314	.2164	76.13	4.6206	35.06	53.62	88.68	.0720	.1703	86
87	192.5	177.8	.01316	.2132	75.97	4.6915	35.36	53.38	88.74	.0726	.1702	87
88	195.1	180.4	.01319	.2099	75.80	4.7634	35.67	53.14	88.81	.0731	.1701	88
89	197.8	183.1	.01322	.2068	75.64	4.8364	35.97	52.90	88.87	.0737	.1701	89
90	200.5	185.8	0.01325	0.2036	75.48	4.9105	36.28	52.65	88.93	0.0742	0.1700	90
91	203.2	188.5	.01328	.2006	75.32	4.9856	36.59	52.40	88.99	.0747	.1699	91
92	206.0	191.3	.01331	.1976	75.15	5.0619	36.89	52.16	89.05	.0753	.1698	92
93	208.8	194.1	.01334	.1946	74.99	5.1394	37.20	51.91	89.11	.0758	.1697	93
94	211.6	196.9	.01337	.1916	74.82	5.2180	37.51	51.65	89.16	.0764	.1697	94

TABLE 15 . . . PROPERTIES OF LIQUID AND SATURATED VAPOR (continued)

TEMP F	PRESSURE lb per sq in		VOLUME cu ft per lb		DENSITY lb per cu ft		ENTHALPY ** Btu per lb			ENTROPY ** Btu per (lb) (°R)		TEMP F
t	Absolute P	Gage p	Liquid v_f	Vapor v_g	Liquid $1/v_f$	Vapor $1/v_g$	Liquid h_f	Latent h_{fg}	Vapor h_g	Liquid s_f	Vapor s_g	t
95	214.4	199.7	0.01340	0.1888	74.65	5.2979	37.82	51.40	89.22	0.0769	0.1696	95
96	217.3	202.6	.01343	.1859	74.48	5.3789	38.13	51.14	89.27	.0775	.1695	96
97	220.2	205.5	.01346	.1831	74.32	5.4612	38.44	50.88	89.32	.0780	.1694	97
98	223.1	208.4	.01349	.1804	74.15	5.5447	38.75	50.62	89.37	.0786	.1693	98
99	226.1	211.4	.01352	.1776	73.97	5.6296	39.06	50.36	89.42	.0791	.1692	99
100	229.1	214.4	0.01355	0.1750	73.80	5.7157	39.37	50.10	89.47	0.0796	0.1692	100
101	232.1	217.4	.01358	.1723	73.63	5.8033	39.68	49.83	89.51	.0802	.1691	101
102	235.1	220.4	.01361	.1697	73.45	5.8921	40.00	49.56	89.56	.0807	.1690	102
103	238.2	223.5	.01365	.1672	73.28	5.9824	40.31	49.29	89.60	.0813	.1689	103
104	241.3	226.6	.01368	.1646	73.10	6.0741	40.62	49.02	89.64	.0818	.1688	104
105	244.4	229.7	0.01371	0.1621	72.92	6.1673	40.94	48.74	89.68	0.0824	0.1687	105
106	247.6	232.9	.01375	.1597	72.74	6.2620	41.25	48.47	89.72	.0829	.1686	106
107	250.7	236.0	.01378	.1573	72.56	6.3582	41.57	48.18	89.75	.0834	.1685	107
108	254.0	239.3	.01382	.1549	72.38	6.4560	41.88	47.90	89.78	.0840	.1684	108
109	257.2	242.5	.01385	.1525	72.20	6.5554	42.20	47.62	89.82	.0845	.1683	109
110	260.5	245.8	0.01389	0.1502	72.01	6.6564	42.52	47.33	89.85	0.0851	0.1682	110
111	263.8	249.1	.01392	.1480	71.83	6.7590	42.83	47.04	89.87	.0856	.1680	111
112	267.1	252.4	.01396	.1457	71.64	6.8634	43.15	46.75	89.90	.0862	.1679	112
113	270.5	255.8	.01400	.1435	71.45	6.9695	43.47	46.45	89.92	.0867	.1678	113
114	273.9	259.2	.01403	.1413	71.26	7.0775	43.79	46.15	89.94	.0872	.1677	114
115	277.3	262.6	0.01407	0.1391	71.07	7.1872	44.11	45.85	89.96	0.0878	0.1676	115
116	280.8	266.1	.01411	.1370	70.87	7.2988	44.43	45.55	89.98	.0883	.1674	116
117	284.3	269.6	.01415	.1349	70.68	7.4124	45.75	45.24	89.99	.0889	.1673	117
118	287.8	273.1	.01419	.1328	70.48	7.5279	45.07	44.93	90.00	.0894	.1672	118
119	291.4	276.7	.01423	.1308	70.28	7.6454	45.39	44.62	90.01	.0899	.1671	119
120	295.0	280.3	0.01427	0.1288	70.08	7.7649	45.71	44.31	90.02	0.0905	0.1669	120
121	298.6	283.9	.01431	.1268	69.88	7.8866	46.04	43.99	90.03	.0910	.1668	121
122	302.2	287.5	.01435	.1248	69.68	8.0105	46.36	43.67	90.03	.0916	.1666	122
123	305.9	291.2	.01439	.1229	69.47	8.1365	46.68	43.35	90.03	.0921	.1665	123
124	309.7	295.0	.01444	.1210	69.26	8.2648	47.00	43.02	90.02	.0926	.1663	124
125	313.4	298.7	0.01448	0.1191	69.05	8.3955	47.33	42.69	90.02	0.0932	0.1662	125
126	317.2	302.5	.01453	.1173	68.84	8.5285	47.65	42.36	90.01	.0937	.1660	126
127	321.0	306.3	.01457	.1154	68.62	8.6639	47.97	42.03	90.00	.0942	.1659	127
128	324.9	310.2	.01462	.1136	68.41	8.8019	48.29	41.69	89.98	.0948	.1657	128
129	328.8	314.1	.01467	.1118	68.19	8.9424	48.62	41.35	89.97	.0953	.1655	129
130	332.7	318.0	0.01471	0.1101	67.96	9.0855	48.95	41.00	89.95	0.0958	0.1654	130
131	336.6	321.9	.01476	.1083	67.74	9.2313	49.27	40.65	89.92	.0964	.1652	131
132	340.6	325.9	.01481	.1066	67.51	9.3798	49.59	40.30	89.89	.0969	.1650	132
133	344.7	330.0	.01486	.1049	67.28	9.5312	49.91	39.95	89.86	.0974	.1648	133
134	348.7	334.0	.01491	.1032	67.05	9.6854	50.24	39.59	89.83	.0979	.1646	134
135	352.8	338.1	0.01497	0.1016	66.81	9.8425	50.56	39.23	89.79	0.0985	0.1644	135
136	357.0	342.3	.01502	.09997	66.58	10.003	50.88	38.87	89.75	.0990	.1642	136
137	361.1	346.4	.01508	.09837	66.33	10.166	51.21	38.50	89.71	.0995	.1640	137
138	365.3	350.6	.01513	.09679	66.09	10.332	51.53	38.13	89.66	.1000	.1638	138
139	369.6	354.9	.01519	.09522	65.84	10.502	51.86	37.75	89.61	.1006	.1636	139
140	373.8	359.1	0.01525	0.09368	65.59	10.674	52.17	37.38	89.55	0.1011	0.1634	140
141	378.2	363.5	.01531	.09216	65.33	10.850	52.49	37.00	89.49	.1016	.1632	141
142	382.5	367.8	.01537	.09067	65.07	11.030	52.81	36.62	89.43	.1021	.1630	142
143	386.9	372.2	.01543	.08919	64.81	11.212	53.13	36.23	89.36	.1026	.1627	143
144	391.3	376.6	.01549	.08773	64.54	11.398	53.45	35.84	89.29	.1031	.1625	144
145	395.8	381.1	0.01556	0.08630	64.27	11.588	53.77	35.45	89.22	0.1036	0.1622	145
146	400.3	385.6	.01563	.08489	63.99	11.780	54.08	35.06	89.14	.1041	.1620	146
147	404.8	390.1	.01570	.08350	63.71	11.977	54.39	34.67	89.06	.1046	.1617	147
148	409.4	394.7	.01577	.08213	63.42	12.176	54.70	34.27	88.97	.1051	.1615	148
149	414.0	399.3	.01584	.08078	63.13	12.379	55.01	33.87	88.88	.1056	.1612	149
150	418.6	403.9	0.01591	0.07946	62.84	12.585	55.32	33.47	88.79	0.1061	0.1610	150
151	423.3	408.6	.01599	.07816	62.53	12.794	55.62	33.08	88.70	.1065	.1607	151
152	428.1	413.4	.01607	.07688	62.22	13.007	55.92	32.68	88.60	.1070	.1604	152
153	432.8	418.1	.01615	.07563	61.91	13.222	56.22	32.28	88.50	.1075	.1601	153
154	437.6	422.9	.01624	.07441	61.59	13.439	56.51	31.88	88.39	.1079	.1599	154
155	442.5	427.8	0.01632	0.07321	61.26	13.659	56.80	31.48	88.28	0.1084	0.1596	155
156	447.4	432.7	.01641	.07204	60.92	13.882	57.10	31.08	88.18	.1088	.1593	156
157	452.3	437.6	.01651	.07089	60.58	14.106	57.38	30.69	88.07	.1093	.1590	157
158	457.2	442.5	.01661	.06978	60.22	14.331	57.66	30.30	87.96	.1097	.1587	158
159	462.3	447.6	.01671	.06869	59.86	14.557	57.94	29.91	87.85	.1101	.1585	159
160	467.3	452.6	0.01681	0.06764	59.49	14.784	58.21	29.53	87.74	0.1105	0.1582	160

MISCELLANEOUS DATA

Metric Conversions

The factors for converting a number of refrigeration properties from English to Metric units and vice versa, are given in Table A.

Temperature Conversion

Table B gives the corresponding Fahrenheit and Centigrade temperatures for a wide range of temperature.

Table A. Metric Conversions

To Convert From	To	Multiply By
Atmospheres of pressure	Bars	1.013
	Inches of mercury	29.92
	Kilograms/sq centimeter	1.033
	Pounds/sq inch	14.696
Bars	Atmospheres	0.987
	Inches of mercury	29.53
	Kilograms/sq centimeter	1.020
	Pounds/sq inch	14.504
Btu	Calories	252
	Horsepower-minutes	0.01757
	Watt-hours	0.293
	Joules	1054
Btu/cubic foot	Calories/cubic centimeter	0.00890
	Calories/liter	8.90
	Kilocalories/cubic meter	8.90
Btu/minute	Calories/minute	252
	Horsepower	0.02357
	Kilowatts	0.01757
Btu/lb	Calories/gram	0.556
	Watt-hours/gram	0.000646
	Joules/gram	2.324
Btu/(lb) (°F)	Calories/(gram) (°C)	1
Calories	Btu	0.00397
	Watt-hours	0.00116
	Joules	4.184
Calories/cubic centimeter	Btu/cubic foot	112.4
	Calories/liter	0.001
	Kilocalories/cubic meter	0.001
Calories/gram	Btu/lb	1.8
	Watt-hours/gram	0.00116
	Joules/gram	4.184

Table A. Metric Conversions (cont'd)

To Convert From	To	Multiply By
Calories/(gram) (°C)	Btu/(lb) (°F)	1
Calories/liter	Btu/cubic foot	0.112
	Calories/cubic centimeter	1000
	Kilocalories/cubic meter	1
Calories/minute	Btu/minute	0.00397
	Horsepower	9.35×10^{-5}
	Kilowatts	69.7
Cubic centimeters	Cubic feet	3.53×10^{-5}
	Cubic meters	1×10^{-6}
	Gallons (U.S. Liq.)	0.000264
	Liters	0.001
Cubic centimeters/gram	Cubic feet/pound	0.0160
	Gallons (U.S. Liq.)/pound	0.120
	Liters/gram	0.001
Cubic feet	Cubic centimeters	28,320
	Cubic meters	0.0283
	Gallons (U.S. Liq.)	7.48
	Liters	28.3
Cubic feet/minute	Cubic meters/minute	0.0283
	Liters/second	0.472
Cubic feet/pound	Cubic centimeters/gram	62.43
	Cubic meters/kilogram	0.0624
	Gallons (US. Liq.)/pound	7.481
	Liters/kilogram	62.43
Cubic meters	Cubic centimeters	1×10^{6}
	Cubic feet	35.3
	Gallons (U.S. Liq.)	264
	Liters	1000
Cubic meters/kilogram	Cubic centimeters/gram	1000
	Cubic feet/pound	16.0
	Gallons (U.S. Liq.)/pound	120
	Liters/kilogram	1000
Cubic meters/minute	Cubic feet/minute	35.3
	Liters/second	16.7
Degrees K	Degrees R	1.8
Gallons	Cubic centimeters	3790
	Cubic feet	0.134
	Cubic meters	0.00379
	Liters	3.79
Grams	Pounds	0.00220
Grams/cubic centimeter	Kilograms/liter	1
	Pounds/cubic foot	62.43
	Pounds/gallon	8.345

Table A. Metric Conversions (cont'd)

To Convert From	To	Multiply By
Horsepower	Btu/minute	42.44
	Calories/minute	10,690
	Kilocalories/minute	10.7
	Kilowatts	0.746
Inches of Mercury	Atmospheres	0.0334
	Bars	0.0339
	Kilograms/sq centimeter	0.0345
	Pounds/sq inch	0.491
Joules	Btu	0.000948
	Calories	0.239
	Watt-hours	0.000278
Joules/gram	Btu/lb	0.430
	Calories/gram	0.239
Kilocalories	Calories	1000
Kilocalories/cubic meter	Btu/cubic foot	0.112
	Calories/cubic centimeter	1000
	Calories/liter	1
Kilograms	Grams	1000
Kilograms/cubic meter	Grams/cubic centimeter	0.001
	Kilograms/liter	0.001
	Pounds/cubic foot	0.0624
	Pounds/gallon (U.S. Liq.)	0.00835
Kilograms/sq centimeter	Atmospheres	0.968
	Bars	0.981
	Inches of mercury	28.96
	Pounds/sq inch	14.22
Kilowatts	Btu/minute	56.9
	Calories/minute	14,340
	Horsepower	1.34
Liters	Cubic centimeters	1000
	Cubic feet	0.0353
	Cubic meters	0.001
	Gallons (U.S. Liq.)	0.264
Liters/kilogram	Cubic centimeters/gram	1
	Cubic feet/pound	0.0160
	Cubic meters/kilogram	0.001
	Gallons (U.S. Liq.)/pound	0.120
Liters/second	Cubic feet/minute	2.12
	Cubic meters/minute	0.060
Pounds	Grams	453.6

Table A. Metric Converstions (cont'd)

To Convert From	To	Multiply By
Pounds/cubic foot	Grams/cubic centimeter	0.0160
	Kilograms/cubic meter	16.0
	Kilograms/liter	0.0160
	Pounds/gallon (U.S. Liq.)	0.134
Pounds/gallon (U.S. Liq.)	Grams/cubic centimeter	0.120
	Kilograms/cubic meter	120
	Kilograms/liter	0.120
	Pounds/cubic foot	7.48
Pounds/square inch	Atmospheres	0.0680
	Bars	0.0689
	Inches of mercury	2.036
	Kilograms/sq centimeter	0.0703
Tons of Refrigeration	Btu/minute	200
	Horsepower	4.716
	Kilocalories/minute	50.4

Temperature Conversions

$$°K = °C + 273.15 = \frac{°F + 459.67}{1.8}$$

$$°R = °F + 459.67 = (°C + 273.15)\ 1.8$$

$$°C = (°F - 32)\frac{5}{9}$$

$$°F = (°C \times \frac{9}{5}) + 32$$

Vacuum Conversions

Units of vacuum are frequently used in refrigeration and are expressed as:

Inches of mercury vacuum
Inches of mercury below one atmosphere

These units can be converted to pressure units as follows:

Inches of mercury vacuum = 29.921 − inches of mercury pressure.

Inches of mercury vacuum = 29.921 − (29.921) (atmospheres)

Inches of mercury vacuum = 29.921 − (2.036) (pounds/sq inch, abs.)

Inches of mercury vacuum = 29.921 − (28.96) (kilograms/sq centimeter)

$$\text{Atmospheres} = \frac{29.921 - \text{inches of mercury vacuum}}{29.921}$$

Pounds/sq inch, abs = 0.491 (29.921 − inches of mercury vacuum)

Kilograms/sq centimeter = 0.0345 (29.921 − inches of mercury vacuum)

$$\text{Inches of mercury} = \frac{\text{centimeters of mercury}}{2.54}$$

Centimeters of mercury vacuum = 76 − centimeters of mercury pressure

$$\text{Atmospheres} = \frac{76 - \text{centimeters of mercury}}{76}$$

Pounds/sq inch, abs = 0.193 (76 − centimeters of mercury vacuum)

Kilograms/sq centimeter = 0.0136 (76 − centimeters of mercury vacuum)

Table B. Temperature Conversion

°F	Temp. to be Converted	°C
−112.0	−80	−62.2
−110.2	−79	−61.7
−108.4	−78	−61.1
−106.6	−77	−60.6
−104.8	−76	−60.0
−103.0	−75	−59.4
−101.2	−74	−58.9
− 99.4	−73	−58.3
− 97.6	−72	−57.8
− 95.8	−71	−57.2
− 94.0	−70	−56.7
− 92.2	−69	−56.1
− 90.4	−68	−55.6
− 88.6	−67	−55.0
− 86.8	−66	−54.4
− 85.0	−65	−53.9
− 83.2	−64	−53.3
− 81.4	−63	−52.8
− 79.6	−62	−52.2
− 77.8	−61	−51.7
− 76.0	−60	−51.1
− 74.2	−59	−50.6
− 72.4	−58	−50.0
− 70.6	−57	−49.4
− 68.8	−56	−48.9
− 67.0	−55	−48.3
− 65.2	−54	−47.8
− 63.4	−53	−47.2
− 61.6	−52	−46.7
− 59.8	−51	−46.1
− 58.0	−50	−45.6
− 56.2	−49	−45.0
− 54.4	−48	−44.4
− 52.6	−47	−43.9
− 50.8	−46	−43.3
− 49.0	−45	−42.8
− 47.2	−44	−42.2
− 45.4	−43	−41.7
− 43.6	−42	−41.1
− 41.8	−41	−40.6
− 40.0	−40	−40.0
− 38.2	−39	−39.4
− 36.4	−38	−38.9
− 34.6	−37	−38.3
− 32.8	−36	−37.8

°F	Temp. to be Converted	°C
− 31.0	−35	−37.2
− 29.2	−34	−36.7
− 27.4	−33	−36.1
− 25.6	−32	−35.6
− 23.8	−31	−35.0
− 22.0	−30	−34.4
− 20.2	−29	−33.9
− 18.4	−28	−33.3
− 16.6	−27	−32.8
− 14.8	−26	−32.2
− 13.0	−25	−31.7
− 11.2	−24	−31.1
− 9.4	−23	−30.6
− 7.6	−22	−30.0
− 5.8	−21	−29.4
− 4.0	−20	−28.9
− 2.2	−19	−28.3
− 0.4	−18	−27.8
1.4	−17	−27.2
3.2	−16	−26.7
5.0	−15	−26.1
6.8	−14	−25.6
8.6	−13	−25.0
10.4	−12	−24.4
12.2	−11	−23.9
14.0	−10	−23.3
15.8	− 9	−22.8
17.6	− 8	−22.2
19.4	− 7	−21.7
21.2	− 6	−21.1
23.0	− 5	−20.6
24.8	− 4	−20.0
26.6	− 3	−19.4
28.4	− 2	−18.9
30.2	− 1	−18.3
32.0	0	−17.8
33.8	1	−17.2
35.6	2	−16.7
37.4	3	−16.1
39.2	4	−15.6
41.0	5	−15.0
42.8	6	−14.4
44.6	7	−13.9
46.4	8	−13.3
48.2	9	−12.8

°F	Temp. to be Converted	°C
50.0	10	−12.2
51.8	11	−11.7
53.6	12	−11.1
55.4	13	−10.6
57.2	14	−10.0
59.0	15	− 9.4
60.8	16	− 8.9
62.6	17	− 8.3
64.4	18	− 7.8
66.2	19	− 7.2
68.0	20	− 6.7
69.8	21	− 6.1
71.6	22	− 5.6
73.4	23	− 5.0
75.2	24	− 4.4
77.0	25	− 3.9
78.8	26	− 3.3
80.6	27	− 2.8
82.4	28	− 2.2
84.2	29	− 1.7
86.0	30	− 1.1
87.8	31	− 0.6
89.6	32	0.0
91.4	33	0.6
93.2	34	1.1
95.0	35	1.7
96.8	36	2.2
98.6	37	2.8
100.4	38	3.3
102.2	39	3.9
104.0	40	4.4
105.8	41	5.0
107.6	42	5.6
109.4	43	6.1
111.2	44	6.7
113.0	45	7.2
114.8	46	7.8
116.6	47	8.3
118.4	48	8.9
120.2	49	9.4
122.0	50	10.0
123.8	51	10.6
125.6	52	11.1
127.4	53	11.7
129.2	54	12.2

Table B. Temperature Conversion (cont'd)

°F	Temp. to be Converted	°C
131.0	55	12.8
132.8	56	13.3
134.6	57	13.9
136.4	58	14.4
138.2	59	15.0
140.0	60	15.6
141.8	61	16.1
143.6	62	16.7
145.4	63	17.2
147.2	64	17.8
149.0	65	18.3
150.8	66	18.9
152.6	67	19.4
154.4	68	20.0
156.2	69	20.6
158.0	70	21.1
159.8	71	21.7
161.8	72	22.2
163.4	73	22.8
165.2	74	23.3
167.0	75	23.9
168.8	76	24.4
170.6	77	25.0
172.4	78	25.6
174.2	79	26.1
176.0	80	26.7
177.8	81	27.2
179.6	82	27.8
181.4	83	28.3
183.2	84	28.9
185.0	85	29.4
186.8	86	30.0
188.6	87	30.6
190.4	88	31.1
192.2	89	31.7
194.0	90	32.2
195.8	91	32.8
197.6	92	33.3
199.4	93	33.9
201.2	94	34.4
203.0	95	35.0
204.8	96	35.6
206.6	97	36.1
208.4	98	36.7
210.2	99	37.2
212.0	100	37.8
213.8	101	38.3
215.6	102	38.9
217.4	103	39.4
219.2	104	40.0
221.0	105	40.6
222.8	106	41.1
224.6	107	41.7
226.4	108	42.2
228.2	109	42.8
230.0	110	43.3
231.8	111	43.9
233.6	112	44.4
235.4	113	45.0
237.2	114	45.6

Pressure-temperature Relationships

TEMP F	"Freon" Refrigerants 11	12	13	22	113	114	500	502
−50	28.9	15.4	57.0	6.2	—	27.1	13.1	0.0
−48	28.8	14.6	60.0	4.8	—	26.9	12.1	0.7
−46	28.7	13.8	63.1	3.4	—	26.7	11.1	1.5
−44	28.6	12.9	66.2	2.0	—	26.5	10.1	2.3
−42	28.5	11.9	69.4	0.5	—	26.3	9.0	3.2
−40	28.4	11.0	72.7	0.5	—	26.0	7.9	4.1
−38	28.3	10.0	76.1	1.3	—	25.8	6.7	5.1
−36	28.2	8.9	79.7	2.2	—	25.5	5.4	6.0
−34	28.1	7.8	83.3	3.0	—	25.2	4.2	7.0
−32	27.9	6.7	87.0	3.9	—	25.0	2.8	8.1
−30	27.8	5.5	90.9	4.9	29.3	24.6	1.4	9.2
−28	27.7	4.3	94.9	5.9	29.3	24.3	0.0	10.3
−26	27.5	3.0	98.9	6.9	29.2	24.0	0.8	11.5
−24	27.4	1.6	103.0	7.9	29.2	23.6	1.5	12.7
−22	27.2	0.3	107.3	9.0	29.1	23.2	2.3	14.0
−20	27.0	0.6	111.7	10.1	29.1	22.9	3.1	15.3
−18	26.8	1.3	116.2	11.3	29.0	22.4	4.0	16.7
−16	26.6	2.1	120.8	12.5	28.9	22.0	4.9	18.1
−14	26.4	2.8	125.5	13.8	28.9	21.6	5.8	19.5
−12	26.2	3.7	130.4	15.1	28.8	21.1	6.8	21.0
−10	26.0	4.5	135.4	16.5	28.7	20.6	7.8	22.6
− 8	25.8	5.4	140.5	17.9	28.6	20.1	8.8	24.2
− 6	25.5	6.3	145.8	19.3	28.5	19.6	9.9	25.8
− 4	25.3	7.2	151.1	20.8	28.4	19.0	11.0	27.5
− 2	25.0	8.2	156.5	22.4	28.3	18.4	12.1	29.3
0	24.7	9.2	162.1	24.0	28.2	17.8	13.3	31.1
2	24.4	10.2	167.8	25.6	28.1	17.2	14.5	32.9
4	24.1	11.2	173.7	27.3	28.0	16.5	15.7	34.8
6	23.8	12.3	179.7	29.1	27.9	15.8	17.0	36.9
8	23.4	13.5	185.8	30.9	27.7	15.1	18.4	38.9
10	23.1	14.6	192.1	32.8	27.6	14.4	19.8	41.0
12	22.7	15.8	198.5	34.7	27.5	13.6	21.2	43.2
14	22.3	17.1	205.7	36.7	27.3	12.8	22.7	45.4
16	21.9	18.4	211.9	38.7	27.1	12.0	24.2	47.7
18	21.5	19.7	218.7	40.9	27.0	11.1	25.7	50.0
20	21.1	21.0	225.7	43.0	26.8	10.2	27.3	52.5
22	20.6	22.4	232.9	45.3	26.6	9.3	29.0	54.9
24	20.1	23.9	240.2	47.6	26.4	8.3	30.7	57.5
26	19.7	25.4	247.7	49.9	26.2	7.3	32.5	60.1
28	19.1	26.9	255.4	52.4	26.0	6.3	34.3	62.8
30	18.6	28.5	263.2	54.9	25.8	5.2	36.1	65.6
32	18.1	30.1	271.2	57.5	25.6	4.1	38.0	68.4
34	17.5	31.7	279.4	60.1	25.3	2.9	40.0	71.3
36	16.9	33.4	287.7	62.8	25.1	1.7	42.0	74.3
38	16.3	35.2	296.2	65.6	24.8	0.6	44.1	77.4

Pressure-temperature Relationships

TEMP F	"Freon" Refrigerants							
	11	12	13	22	113	114	500	502
40	15.6	37.0	304.9	68.5	24.5	0.4	46.2	80.5
42	15.0	38.8	313.9	71.5	24.2	1.0	46.4	83.8
44	14.3	40.7	322.9	74.5	23.9	1.7	50.7	87.0
46	13.6	42.7	332.2	77.6	23.6	2.4	53.0	90.4
48	12.8	44.7	341.5	80.8	23.3	3.1	55.4	93.9
50	12.0	46.7	351.2	84.0	22.9	3.8	57.8	97.4
52	11.2	48.8	360.9	87.4	22.6	4.6	60.3	101.1
54	10.4	51.0	371.0	90.8	22.2	5.4	62.9	104.8
56	9.6	53.2	381.2	94.3	21.8	6.2	65.5	108.6
58	8.7	55.4	391.6	97.9	21.4	7.0	68.2	112.4
60	7.8	57.7	402.3	101.6	21.0	7.9	71.0	116.4
62	6.8	60.1	413.3	105.4	20.6	8.8	73.8	120.5
64	5.9	62.5	424.3	109.3	20.1	9.7	76.7	124.6
66	4.9	65.0	435.4	113.2	19.7	10.6	79.7	128.9
68	3.8	67.6	446.9	117.3	19.2	11.6	82.7	133.2
70	2.8	70.2	458.7	121.4	18.7	12.6	85.8	137.6
72	1.6	72.9	470.6	125.7	18.2	13.6	89.0	142.2
74	0.5	75.6	482.9	130.0	17.6	14.6	92.3	146.8
76	0.3	78.4	495.3	134.5	17.1	15.7	95.6	151.5
78	0.9	81.3	508.0	139.0	16.5	16.8	99.0	156.3
80	1.5	84.2	520.8	143.6	15.9	18.0	102.5	161.2
82	2.2	87.2	534.0	148.4	15.3	19.1	106.1	166.2
84	2.8	90.2	—	153.2	14.6	20.3	109.7	171.4
86	3.5	93.3	—	158.2	13.9	21.6	113.4	176.6
88	4.2	96.5	—	163.2	13.2	22.8	117.3	181.9
90	4.9	99.8	—	168.4	12.5	24.1	121.2	187.4
92	5.6	103.1	—	173.7	11.8	25.5	125.1	192.9
94	6.4	106.5	—	179.1	11.0	26.8	129.2	198.6
96	7.1	110.0	—	184.6	10.2	28.2	133.3	204.3
98	7.9	113.5	—	190.2	9.4	29.7	137.6	210.2
100	8.8	117.2	—	195.9	8.6	31.2	141.9	216.2
102	9.6	120.9	—	201.8	7.7	32.7	146.3	222.3
104	10.5	124.6	—	207.7	6.8	34.2	150.9	228.5
106	11.3	128.5	—	213.8	5.9	35.8	155.4	234.9
108	12.3	132.4	—	220.0	4.9	37.4	160.1	241.3
110	13.1	136.4	—	226.4	4.0	39.1	164.9	247.9
112	14.2	140.5	—	232.8	3.0	40.8	169.8	254.6
114	15.1	144.7	—	239.4	1.9	42.5	174.8	261.5
116	16.1	148.9	—	246.1	0.8	44.3	179.9	268.4
118	17.2	153.2	—	252.9	0.1	46.1	185.0	275.5
120	18.2	157.7	—	259.9	0.7	48.0	190.3	282.7
122	19.3	162.2	—	267.0	1.3	49.9	195.7	290.1
124	20.5	166.7	—	274.3	1.9	51.9	201.2	297.6
126	21.6	171.4	—	281.6	2.5	53.8	206.7	305.2
128	22.8	176.2	—	289.1	3.1	55.9	212.4	312.9
130	24.0	181.0	—	296.8	3.7	58.0	218.2	320.8
132	25.2	185.9	—	304.6	4.4	60.1	224.1	328.9
134	26.5	191.0	—	312.5	5.1	62.3	230.1	337.1
136	27.8	196.1	—	320.6	5.8	64.5	236.3	345.4
138	29.1	201.3	—	328.9	6.5	66.7	242.5	353.9

GLOSSARY

GLOSSARY OF TECHNICAL TERMS AS APPLIED TO REFRIGERATION OR AIR-CONDITIONING

A

Absolute Humidity: Amount of moisture in the air, indicated in grains per cubic foot.

Absolute Pressure: Gauge pressure plus atmospheric pressure (14.7 lb per in^2).

Absolute Temperature: Temperature measured from absolute zero.

Absolute Zero Temperature: temperature at which all molecular motion ceases (—460° F and —275° C).

Absorbent: Substance which has the ability to take up or absorb another substance.

Absorber: A device containing liquid for absorbing refrigerant vapor or other vapors. In an absorption system, that part of the low side used for absorbing refrigerant vapor.

Absorption Refrigerator: Refrigerator which creates low temperatures by using the cooling effect formed when a refrigerant is absorbed by chemical substance.

Accelerate: To add to speed; hasten progress of development.

Accumulator: Storage tank which receives liquid refrigerant from evaporator and prevents it from flowing into suction line.

Acid Condition In System: Condition in which refrigerant or oil in system is mixed with fluids which are acid in nature.

ACR Tubing: Tubing used in refrigeration which has ends sealed to keep tubing clean and dry.

Activated Alumina: Chemical used as a drier or desiccant.

Activated Carbon: Specially processed carbon used as a filter-drier; commonly used to clean air.

Acoustical: Pertaining to sound.

Acoustical Duct Lining: Duct with a lining designed to control or absorb sound and prevent transmission of sound from one room to another.

Adiabatic Compression: Compressing refrigerant gas without removing or adding heat.

Adsorbent: Substance which has property to hold molecules of fluids without causing a chemical or physical change.

Aerosol: An assemblage of small particles, solid or liquid, suspended in air. The diameters of the particles may vary from 100 microns down to 0.01 micron or less, e.g., dust, fog, smoke.

Agitator: Device used to cause motion in confined fluid.

Air (saturated): A mixture of dry air and saturated water vapor all at the same dry-bulb temperature.

Air (specific heat of): The quantity of heat absorbed by a unit weight of air per unit temperature rise.

Air (standard): Air with a density of 0.075 lb per ft^3 and an absolute viscosity of 0.0379×10^{-5} lb mass per (ft) (sec). This is substantially equivalent to dry air at 70°F and 29.92 in. Hg barometric pressure.

Air Blast: Forced air circulation.

Air Changes: A method of expressing the amount of air leakage into or out of a building or room in terms of the number of building volumes or room volumes exchanged.

Air Cleaner: Device used for removal of airborne impurities.

Air Coil: Coil used with some types of heat pumps which may be used either as an evaporator or as a condenser.

Air Conditioner: Device used to control temperature, humidity, cleanliness, and movement of air in conditioned space.

Air-Conditioning: The simultaneous control of all or at least the first three, of the following factors affecting the physical and chemical conditions of the atmosphere within a structure: Temperature, humidity, motion, distribution, dust, bacteria, odors, toxic gases and ionization—most of which affect in greater or lesser degree human health or comfort.

Air-Cooled Condenser: Heat of compression is transferred from condensing coils to surrounding air. This may be done either by convection or by a fan or blower.

Air Cooler: Mechanism designed to lower temperature of air passing through it.

Air Cycle, Air-Conditioning: System which removes heat from air and transfers this heat to air.

Air Diffuser: Air distribution outlet designed to direct airflow into desired patterns.

Air Return: Air returned from conditioned or refrigerated space.

Air Sensing Thermostat: Thermostat unit in which sensing element is located in refrigerated space.

Air Spill-Over: Refrigerating effect formed by cold air from freezing compartment in refrigerator spilling over, or flowing into normal storage area of refrigerator.

Air Washer: Device used to clean air, which may increase or decrease humidity.

Alcohol Brine: Water and alcohol solution which remains a liquid at below 32°F.

Allen-Type Screw: Screw with recessed head designed to be turned with hex shaped wrench.

Altitude Adjustment: Adjusting refrigerator controls so unit will operate efficiently at altitude in which it is to be used.

Ambient Temperature: Temperature of fluid (usually air) which surrounds object on all sides.

Ammeter: An electric meter used to measure current, calibrated in amperes.

Ammonia: Chemical combination of nitrogen and hydrogen (NH_3). Ammonia refrigerant is identified by R-117.

Amperage: Electron or current flow of one coulomb per second past given point in circuit.

Ampere: Unit of electric current equivalent to flow of one coulomb per second.

Ampere Turns: Term used to measure magnetic force. Represents product of amperes times number of turns in coil of electromagnet.

Amplifier: Electrical device which increases electron flow in a circuit.

Anemometer: Instrument for measuring the rate of flow of air.

Anhydrous Calcium sulphate: Dry chemical made of calcium, sulphur and oxygen ($CaSO_4$).

Annealing: Process of heat treating metal to obtain desired properties of softness and ductility (easy to form into new shape).

Anode: Positive terminal of electrolytic cell.

Anticipating Control: One which is artificially forced to cut in or cut out before it otherwise would, thus starting the cooling before needed or stopping the heating before control point is reached, to reduce the temperature fluctuation or override.

A.S.A.: Formerly, abbreviation for American Standards Association. Now known as United States of America Standards Institute.

Aspect Ratio: Ratio of length to width of rectangular air grille or duct.

Aspirating Psychrometer: A device which draws sample of air through it for humidity measurement purposes.

Aspiration: Movement produced in a fluid by suction.

Atmospheric Pressure: Pressure that gases in air exert upon the earth; measured in pounds per square inch.

Atom: Smallest particle of element that can exist alone or in combination.

Atomize: Process of changing a liquid to minute particles, or a fine spray.

Attenuate: Decrease or lessen in intensity.

Attic Fan: An exhaust fan to discharge air near the top of a building while cooler air is forced (drawn) in at a lower level.

Automatic Defrost: System of removing ice and frost from evaporators automatically.

Automatic Expansion Valve (AEV): Pressure controlled valve which reduces high pressure liquid refrigerant to low pressure liquid refrigerant.

Automatic Ice Cube Maker: Refrigerating mechanism designed to produce ice cubes in quantity automatically.

Autotransformer: A transformer in which both primary and secondary coils have turns in common. Step-up or stepdown of voltage is accomplished by taps on common winding.

Automatic Refrigerating System: One which regulates itself to maintain a definite set of conditions by means of automatic controls and valves usually responsive to temperature or pressure.

Azeotropic Mixture: Example of azeotropic mixture—refrigerant R-502 is mixture consisting of 48.8% refrigerant R-22, and 51.2% R-115. The refrigerants do not combine chemically, yet azeotropic mixture provides refrigerant characteristics desired.

B

Back Pressure: Pressure in low side of refrigerating system; also called suction pressure or low side pressure.

Baffle: Plate or vane used to direct or control movement of fluid or air within confined area.

Balance Point: The outdoor temperature at which the output of the heat pump in a specific application is equal to the heat requirement of the structure.

Ball Check Valve: Valve assembly call which permits flow of fluid in one section only.

Balloon Type Gasket: Flexible refrigerator door gasket having a large cross section.

Barometer: Instrument for measuring atmospheric pressure. It may be calibrated in pounds per square inch or in inches of mercury in column.

Bath: A liquid solution used for cleaning, plating, or maintaining a specified temperature.

Baudelot Cooler: Heat exchanger in which water flows by gravity over the outside of the tubes or plates.

Bearing: Low friction device for supporting and aligning a moving part.

Bellows: Corrugated cylindrical container which moves as pressures change, or provides a seal during movement of parts.

Bending Spring: Coil spring which is mounted on inside or outside to keep tube from collapsing while bending it.

Bernoulli's Theorem: In stream of liquid, sum of elevation head, pressure head and velocity remains constant along any line of flow provided no work is done by or upon liquid in course of its flow, and decreases in proportion to energy last in flow.

Bimetal Strip: Temperature regulating or indicating device which works on principle that two dissimilar metals with unequal expansion rates, welded together, will bend as temperatures change.

Blast Heater: A set of heat-transfer coils or sections used to heat air which is drawn or forced through it by a fan.

Bleed-Valve: Valve with small opening inside which permits a minimum fluid flow when valve is closed.

Blow (Throw): The distance an air stream travels from an outlet to a position at which air motion along the axis is reduced to a velocity of 50 ft per minute.

Blower: A fan used to force air under pressure.

Boiling Temperature: Temperature at which a fluid changes from a liquid to a gas.

Bore: Inside diameter of a cylindrical hole.

Bourdon Tube: As used in pressure gauges. Thin walled tube of elastic metal flattened and bent into circular shape, which tends to straighten as pressure inside is increased.

Bowden Cable: Tube containing a wire used to regulate a valve or control from a remote point.

Boyle's Law: Law of physics—volume of a gas varies as pressure varies, if temperature remains the same. Examples: If pressure is doubled on quantity of gas, volume becomes one half. If volume becomes doubled, gas has its pressure reduced by one half.

Brazing: Method of joining metals with nonferrous filler (without iron) using heat between 800°F. and melting point of base metals.

Breaker Strip: Strip of wood or plastic used to cover joint between outside case and inside liner of refrigerator.

Brine: Water saturated with chemical such as salt.

British Thermal Unit (Btu): Quantity of heat required to raise temperature of one pound of water one degree F.

Bulb, Sensitive: Part of sealed fluid device which reacts to temperature to be measured, or which will control a mechanism.

Bunker: In commercial installations, space in which ice or cooling element is installed.

Bypass: Passage at one side of, or around regular passage.

C

Cadmium Plated: Parts coated with thin corrosion-resistant covering of cadmium metal.

Calcium Sulfate: Chemical compound ($CaSO_4$) which is used as a drying agent or desiccant in liquid line driers.

Calibrate: To determine position indicators as required to obtain accurate measurements.

Calorie: Heat required to raise temperature of one gram of water one degree centigrade.

Calorimeter: Device used to measure quantities of heat or determine specific heats.

Capacitance (C): Property of nonconductor (condenser or capacitor) that permits storage of electrical energy in an electrostatic field.

Capacitor: Type of electrical storage device used in starting and/or running circuits on many electric motors.

Capacitor-Start Motor: Motor which has a capacitor in the starting circuit.

Capacitor-Start Motor: Motor which has a capacitor in the starting circuit.

Capacity, Refrigerating: The ability of a refrigerating system, or part thereof, to remove heat expressed as a rate of heat removal, usually measured in Btu/hr or tons/24 hr.

Capacity Reducer: In a compressor, a device such as a clearance pocket, movable cylinder head.

Capillary Tube: A type of refrigerant control. Usually consists of several feet of tubing having small inside diameter. Friction of liquid refrigerant and bubbles of vaporized refrigerant within tube serve to restrict flow so that correct high side and low side pressures are maintained while the compressor is operating. A capillary tube refrigerant control allows high side and low side pressures to balance during off cycle. Also; a small diameter tubing used to connect temperature control bulbs to control mechanisms.

Carbon Dioxide (CO_2): Compound of carbon and oxygen which is sometimes used as a refrigerant. Refrigerant number is R-744.

Carbon Filter: Air filter using activated carbon as air cleansing agent.

Carbon Tetrachloride: A colorless nonflammable liquid used as solvent and in fire extinguishers. Very toxic. Should never be allowed to touch skin, or fumes inhaled.

Carnot Cycle: A sequence of operations forming the reversible working cycle of an ideal heat engine of maximum thermal efficiency. It consists of isothermal expansion, adiabatic expansion, isothermal compression and adiabatic compression to the initial state.

Carrene: A refrigerant in group 1 (R-11). Chemical combination of carbon, chlorine and fluorine.

Cascade System: One having two or more refrigerant circuits, each with a pressure imposing element, condenser, and evaporator, where the evaporator of one circuit cools the condenser of the other (lower-temperature) circuit.

Cascade Systems: Arrangement in which two or more refrigerating systems are used in series; uses cooling coil of one machine to cool condenser of other machine. Produces ultra-low temperatures.

Casehardened: Heat treating ferrous metals (iron) so surface layer is harder than interior.

Cathode: Negative terminal of an electrical device. Electrons leave the device at this terminal.

Celsius: German language word for centigrade, the metric system temperature scale

Centigrade Scale: Temperature scale used in metric system. Freezing point of water is 0; boiling point 100.

Centimeter: Metric unit of linear measurement which equals .3937 in.

Centrifugal Compressor: Compressor which compresses gaseous refrigerants by centrifugal force.

Central Fan System: A mechanical indirect system of heating, ventilating, or air-conditioning, in which the air is treated or handled by equipment located outside the rooms served, usually at a central location, and is conveyed to and from the rooms by means of a fan and a system of distributing ducts.

Charge: the amount of refrigerant in a system.

Charging: Putting in a charge.

Charging Board: Specially designed panel or cabinet fitted with gauges, valves and refrigerant cylinders used for charging refrigerant and oil into refrigerating mechanisms.

Charles' Law: The volume of a given mass of gas at a constant pressure varies according to its temperature.

Check Valve: A device which permits fluid flow only in one direction.

Chemical Refrigeration: A system of cooling using a disposable refrigerant.

Chimney Effect: The tendency of air or gas in a duct or other vertical passage to rise when heated due to its lower density compared with that of the surrounding air or gas. In buildings, the tendency toward displacement (caused by the difference in temperature) of internal heated air by unheated outside air due to the difference in density of outside and inside air.

Choke Tube: Throttling device used to maintain correct pressure difference between high side and low side in refrigerating mechanism. Capillary tubes are sometimes called choke tubes.

Circuit: A tubing, piping or electrical wire installation which permits flow from the energy source back to energy source.

Clearance: Space in cylinder not occupied by piston at end of compression stroke, or volume of gas remaining in cylinder at same point. Measured in percentage of piston displacement.

Clearance Pocket Compressor: A small space in cylinder from which compressed gas is not completely expelled. This space is called the compressor clearance space or pocket. For effective operation, compressors are designed to have as small clearance space as possible.

Code Installation: A refrigeration or air conditioning

installation which conforms to the local code and/or the national code for safe and efficient installations.

Closed Cycle: Any cycle in which the primary medium is always enclosed and repeats the same sequence of events.

Coefficient of Conductivity: The measure of the relative rate at which different materials conduct heat. Copper is a good conductor of heat and therefore has a high coefficient of conductivity.

Coefficient of Expansion: The change in length per unit length or the change in volume per unit volume per degree change in temperature.

Coefficient of Performance (COP): The ratio of work or energy applied as compared to the energy used.

Coil: Any heating or cooling element made of pipe or tubing connected in series.

Cold: Cold is the absence of heat; a temperature considerably below normal.

Cold Junction: That part of a thermoelectric system which absorbs heat as the system operates.

Cold Wall: Refrigerator construction which has the inner lining of refrigerator serving as the cooling surface.

Colloids: Miniature cells in meat, fish and poultry.

Comfort Air-Conditioning: The simultaneous control of all, or at least the first three, of the following factors affecting the physical and chemical conditions of the atmosphere within a structure for the purpose of human comfort: temperature, humidity, motion, distribution, dust, bacteria, odors, toxic gases and ionization, most of which affect in greater or lesser degree human health or comfort.

Comfort Chart: Chart used in air-conditioning to show the dry bulb temperature and humidity for human comfort conditions.

Comfort Cooler: A system used to reduce the temperature in the living space in homes. These systems are not complete air conditioners as they do not provide complete control of heating, humidifying, dehumidification, and air circulation.

Comfort Zone: Area on psychrometric chart which shows conditions of temperature, humidity, and sometimes air movement, in which most people are comfortable.

Commutator: Part of electric motor rotor which conveys electric current to rotor windings.

Compound Gauge: Instrument for measuring pressures both above and below atmospheric pressure.

Compound Refrigerating Systems: System which has several compressors or compressor cylinders in series. The system is used to pump low pressure vapors to condensing pressures.

Compression: Term used to denote increase of pressure on a fluid by using mechanical energy.

Compression Gauge: Instrument used to measure positive pressures (pressures above atmospheric pressures) only. These gauges are usually calibrated from 0 to 300 pounds per square inch of pressure, gauge, (psig).

Compressor: The pump of a refrigerating mechanism which draws a vacuum or low pressure on cooling side of refrigerant cycle and squeezes or compresses the gas into the high pressure or condensing side of the cycle.

Compressor, Hermetic: Compressor in which driving motor is sealed in the same dome or housing that contains the compressor.

Compressor, Open-Type: Compressor in which the crankshaft extends through the crankcase and is driven by an outside motor.

Compressor, Reciprocating: Compressor which uses a piston and cylinder mechanism to provide pumping action.

Compressor, Rotary: A compressor which uses vanes, eccentric mechanisms, or other rotating devices to provide pumping action.

Compressor Seal: Leakproof seal between crankshaft and compressor body.

Condensate: Fluid which forms on an evaporator.

Condensate Pump: Device used to remove fluid condensate that collects beneath an evaporator.

Condensation: Liquid or droplets which form when a gas or vapor is cooled below its dew point.

Condense: Action of changing a gas or vapor to a liquid.

Condenser: The part of refrigeration mechanism which receives hot, high pressure refrigerant gas from compressor and cools gaseous refrigerant until it returns to liquid state.

Condenser, Air-Cooled: A heat exchanger which transfers heat to surrounding air.

Condenser Comb: Comb-like device, metal or plastic, which is used to straighten the metal fins on condensers or evaporators.

Condenser Fan: Forced air device used to move air through air-cooled condenser.

Condenser, Water-Cooled: Heat exchanger which is designed to transfer heat from hot gaseous refrigerant to water.

Condenser Water Pump: Forced water moving device used to move water through condenser.

Condensing Unit: That part of refrigerating mechanism which pumps vaporized refrigerant from evaporator, compresses it, liquefies it in the condenser and returns the liquid refrigerant to refrigerant control.

Condensing Unit Service Valves: Shutoff hand valves mounted on condensing unit to enable serviceman to install and/or service unit.

Conductance (surface film): The time rate of heat flow per unit area under steady conditions between a surface and the ambient fluid for a unit temperature difference between the surface and the fluid. In English units its value is usually expressed in Btu per (hour) (square foot) (Fahrenheit degree tomperature difference between surface and fluid).

Conductance (thermal): "C" factor—The time rate of heat flow per unit area under steady conditions through a body from one of its bounding surfaces to the other for a unit temperature difference between the two surfaces. In English units its value is usually expressed in Btu per (hour) (square foot) (Fahrenheit degree). The term is applied to specific bodies or constructions as used, either homogeneous or heterogeneous.

Conduction (thermal): The process of heat transfer through a material medium in which kinetic energy is transmitted by the particles of the material from particle to particle without gross displacement of the particles.

Conductivity (thermal): "k" factor—The time rate of heat flow through unit area of a homogeneous material under steady conditions when a unit temperature gradient is maintained in the direction perpendicular to the area. In English units its value is usually expressed in Btu per (hour) (square foot) (Fahrenheit degree per inch of thickness) Materials are considered homogeneous when the value of "k" is not affected by variation in thickness or in size of sample within the range normally used in construction.

Conductor (thermal): A material which readily transmits heat by means of conduction.

Connecting Rod: That part of compressor mechanism which connects piston to crankshaft.

Constrictor: Tube or orifice used to restrict flow of a gas or a liquid.

Contaminant: A substance (dirt, moisture, etc.) foreign to refrigerant or refrigerant oil in system.

Continuous Cycle Absorption System: System which has a continuous flow of energy input.

Control: Automatic or manual device used to stop, start and/or regulate flow of gas, liquid, and/or electricity.

Control, Compressor: (See Motor Control)

Control, Defrosting: Device to automatically defrost evaporator. It may operate by means of a clock, door cycling mechanism, or during "off" portion of refrigerating cycle.

Control, Low Pressure: Cycling device connected to low pressure side of system.

Control, Motor: A temperature or pressure operated device used to control running of motor.

Control, Pressure Motor: A high or low pressure control which is connected into the electrical circuit and used to start and stop motor when there is need for refrigeration or for safety purposes.

Control, Refrigerant: Device used to regulate flow of liquid refrigerant into evaporator, such as capillary tube, expansion valves, high and low side float valves.

Control, Temperature: A thermostatic device which automatically stops and starts motor, operation of which is based on temperature changes.

Controlled Evaporator Pressure: Controlled system which maintains definite pressure or range of pressures in evaporator.

Convection: Transfer of heat by means of movement or flow of a fluid or gas.

Convection, Forced: Transfer of heat resulting from forced movement of liquid or gas by means of fan or pump.

Convection, Natural: Circulation of a gas or liquid due to difference in density resulting from temperature differences.

Conversion Factors: Force and power may be expressed in more than one way. A horse power is equivalent to 33,000 ft lbs of work per minute, 746 W, or 2,546 Btu per hour. These values can be used for changing horsepower into foot pounds, Btu or watts.

Cooling Tower: Device which cools water by water evaporation in air. Water is cooled to wet bulb temperature of air.

Copper Plating: Condition developing in some units in which copper is electrolytically deposited on compressor part surfaces.

Counterflow: Flow in opposite direction.

"Cracking" a Valve: Opening valve a small amount.

Crankshaft Seal: Leakproof joint between crankshaft and compressor body.

Crank Throw: Distance between center line of main bearing journal and center line of the crankpin or eccentric.

Crisper: Drawer or compartment in refrigerator designed to provide high humidity along with low temtemperature to keep vegetables, especially leafy vegetables, cold and crisp.

Critical Pressure: Condition of refrigerant at which liquid and gas have same properties.

Critical Temperature: Temperature at which vapor and liquid have same properties.

Critical Vibration: Vibration which is noticeable and harmful to structure.

Cross Charged: Sealed container containing two fluids which together create a desired pressure-temperature curve.

Cryogenic Fluid: Substance which exists as a liquid or gas at ultra-low temperatures (—250°F or lower).

Cryogenics: Refrigeration which deals with producing temperatures of 250°F below zero and lower.

Cut-In: Temperature or pressure valve which closes control circuit.

Cut-Out: Temperature or pressure valve which opens control circuit.

Cycle: Series of events which have tendency to repeat same events in same order.

Cylinder Head: Part which encloses compression end of compressor cylinder.

Cylinder, Refrigerant: Cylinder in which refrigerant is purchased and dispensed. Color code painted on cylinder indicates kind of refrigerant cylinder contains.

Cylindrical Commutator: Commutator with contact surfaces parallel to the rotor shaft.

D

Dalton's Law: Vapor pressure exerted on container by a mixture of gases is equal to sum of individual vapor pressures of gases contained in mixture.

Damper: Valve for controlling airflow.

Decibel: Unit used for measuring relative loudness of sounds. One decibel is equal to approximate difference of loudness ordinarily detectable by human ear, the range of which is about 130 decibels on scale beginning with one for faintest audible sound.

Defrost Cycle: Refrigerating cycle in which evaporator frost and ice accumulation is melted.

Defrost Timer: Device connected into electrical circuit which shuts unit off long enough to permit ice and frost accumulation on evaporator to melt.

Defrosting: Process of removing frost accumulation from evaporators.

Defrosting Type Evaporator: An evaporator operating at such temperatures that ice and frost on surface melts during off part of operating cycle.

Degreasing: Solution or solvent used to remove oil or grease from refrigerator parts.

Degree-Day: Unit that represents one degree of difference from given point in average outdoor temperature of one day and is often used in estimating fuel requirements for a building. Degree-days are based on average temperature over a 24 hour period. As an example, if an average temperature for a day is 50°F, the number of degree-day for that day would be equal to 65°F minus 50°F or 15 degree-days (65 — 50 = 15). Degree-days are useful when calculating requirements for heating purposes.

Dehumidify: To remove water vapor from the atmosphere. To remove water or liquid from stored goods.

Dehumidifier: Device used to remove moisture from air in enclosed space.

Dehumidifier (surface): An air-conditioning unit, designed primarily for cooling and dehumidifying air through the action of passing the air over wet cooling coils.

Dehumidifying Effect: The difference between the moisture contents, in pounds per hour, of the entering and leaving air, multiplied by 1.060.

Dehydrate: To remove water in all forms from matter. Liquid water, hygroscopic water, and water of crystallization or water of hydration are included.

Dehydrated Oil: Lubricant which has had most of water content removed (a dry oil).

Dehydration: The removal of water vapor from air by the use of absorbing or adsorbing materials; the removal of water from stored goods.

Dehydrator: (See Drier)

Dehydrator-Receiver: A small tank which serves as liquid refrigerant reservoir and which also contains a desiccant to remove moisture. Used on most automobile air conditioning installations.

Deice Control: Device used to operate refrigerating system in such a way as to provide melting of the accumulated ice and frost.

Delta Transformer: A three-phase electrical transformer which has ends of each of three windings electrically connected.

Demand Meter: An instrument used to measure kilowatt-hour consumption of a particular circuit or group of circuits.

Density: Closeness of texture or consistency.

Deodorizer: Device which absorbs various odors, usually by principle of absorption. Activated charcoal is a common substance used.

Desiccant: Substance used to collect and hold moisture in refrigerating system. A drying agent. Common desiccants are activated alumina, silica gel.

Detector, Leak: Device used to detect and locate refrigerant leaks.

Dew Point: Temperature at which vapor (at 100 percent humidity) begins to condense and deposit as liquid.

Dialectric Fluid: Fluid with high electrical resistance.

Diaphragm: Flexible membrane usually made of thin metal, rubber, or plastic.

Dichlorodifluoromethane: Refrigerant commonly known as R-12. Chemical formula is CCl_2F_2. Cylinder color code is white. Boiling point at atmospheric pressure is −21.62°F.

Die Cast: A process of moulding low melting temperature metals in accurately shaped metal moulds.

Die Stock: Tool used to hold dies with external threads.

Dies (Thread): Tool used to cut external threads.

Differential: As applied to refrigeration and heating: difference between "cut-in"-and "cut-out" temperature or pressure of a control.

Direct Expansion Evaporator: An evaporator coil using either an automatic expansion valve (AEV) or a thermostatic expansion valve (TEV) refrigerant control.

Displacement, Piston: Volume obtained by multiplying area of cylinder bore by length of piston stroke.

Distilling Apparatus: Fluid reclaiming device used to reclaim used refrigerants. Reclaiming is usually done by vaporizing and then recondensing refrigerant.

Dome-Hat: Sealed metal container for the motor-compressor of a refrigerating unit.

Double Duty Case: Commercial refrigerator which has part of it for refrigerated storage and part equipped with glass windows for display purposes.

Double Thickness Flare: Copper, aluminum or steel tubing end which has been formed into two-wall thickness, 37 to 45 deg. bell mouth or flare.

Dowel Pin: Accurately dimensioned pin pressed into one assembly part and slipped into another assembly part to insure accurate alignment.

Draft: A current of air. Usually refers to the pressure difference which causes a current of air or gases to flow through a flue, chimney, heater, or space.

Draft Gauge: Instrument used to measure air movement.

Draft Indicator: An instrument used to indicate or measure chimney draft or combustion gas movement. Draft is measured in units of .1 in. of water column.

Drier: A substance or device used to remove moisture from a refrigeration system.

Drift: Entrained water carried from a cooling tower by air movement.

Drip Pan: Pan-shaped panel or trough used to collect condensate from evaporator coil.

Dry Bulb: An instrument with sensitive element which measures ambient (moving) air temperature.

Dry Bulb Temperature: Air temperature as indicated by ordinary thermometer.

Dry Ice: A refrigerating substance made of solid carbon dioxide which changes directly from a solid to a gas (sublimates). Its subliming temperature is 109°F below zero.

Dry System: A refrigeration system which has the evaporator liquid refrigerant mainly in the atomized or droplet condition.

Duct: Heating and air-conditioning. A tube or channel through which air is conveyed or moved.

Dust: An air suspension (aerosol) of particles of any solid material, usually with particle size less than 100 microns.

Dynamometer: Device for measuring power output or power input of a mechanism.

E

Ebulator: A pointed or sharp edged solid substance inserted in flooded type evaporators to improve evaporation (boiling) of refrigerant in coil.

Eccentric: A circle or disk mounted off center. Eccentrics are used to adjust controls and connect compressor driveshafts to pistons.

Effective Area: Actual flow area of an air inlet or outlet. Gross area minus area of vanes or grille bars.

Effective Temperature: Overall effect on a human of air temperature, humidity and air movement.

Effective Temperature Difference: The difference between the room air temperature and the supply air temperature at the outlet to the room.

Ejector: Device which uses high fluid velocity such as a venturi, to create low pressure or vacuum at its throat to draw in fluid from another source.

Electric Defrosting: Use of electric resistance heating coils to melt ice and frost off evaporators during defrosting.

Electric Heating: House heating system in which heat from electrical resistance units is used to heat rooms.

Electric Water Valve: Solenoid type (electrically operated) valve used to turn water flow on and off.

Electronics: Field of science dealing with electron devices and their uses.

Electronic Leak Detector: Electronic instrument which measures electronic flow across gas gap. Electronic flow changes indicates presence of refrigerant gas molecules.

Electronic Sound Tracer: Instrument used to detect leaks by locating source of high frequency sound caused to leak.

Electrostatic Filter: Type of filter which gives particles of dust electric charge. This causes particles to be attracted to plate so they can be removed from air stream or atmosphere.

End Bell: End structure of electric motor which usually holds motor bearings.

End Play: Slight movement of shaft along center line.

Enthalpy: Total amount of heat in one pound of a substance calculated from accepted temperature base.

Temperature of 32°F is accepted base for water vapor calculation. For refrigerator calculations, accepted base is −40°F.

Entropy: Mathematical factor used in engineering calculations. Energy in a system.

Enzyme: A complex organic substance originating from living cells that speeds up chemical changes in foods. Enzyme action is slowed by cooling.

Epoxy (Resins): A synthetic plastic adhesive.

Equalizer Tube: Device used to maintain equal pressure or equal liquid levels between two containers.

Eutectic Mixture or Solution: A mixture which melts or freezes completely at constant temperature and with constant composition. Its melting point is the lowest possible for mixtures of the given substances.

Evaporation: A term applied to the changing of a liquid to a gas. Heat is absorbed in this process.

Evaporative Condenser: A device which uses open spray or spill water to cool a condenser. Evaporation of some of the water cools the condenser water and reduces water consumption.

Evaporator: Part of a refrigerating mechanism in which the refrigerant evaporizes and absorbs heat.

Evaporator Coil: Device made of a coil of tubing which functions as a refrigerant evaporator.

Evaporator, Dry Type: An evaporator into which refrigerant is fed from a pressure reducing device. Little or no liquid refrigerant collects in the evaporator.

Evaporator Fan: Fan which cools extended heat exchange surface of evaporator.

Evaporator, Flooded: An evaporator containing liquid refrigerant at all times.

Exfiltration: Air flow outward through a wall, leak, membrane, etc.

Exhaust Opening: Any opening through which air is removed from a space which is being heated or cooled, or humidified or dehumidified, or ventilated.

Expansion Valve: A device in refrigerating system which maintains a pressure difference between the high side and low side and is operated by pressure.

Expendable Refrigerant System: System which discards the refrigerant after it has evaporated.

Extended Surface: Heat transfer surface, one side of which is increased in area by the use of fins, ribs, pins, etc.

External Equalizer: Tube connected to low pressure side of an expansion valve diaphragm and to exit of evaporator.

F

Fahrenheit Scale: On a Fahrenheit thermometer, under standard atmospheric pressure, boiling point of water is 212° and freezing point is 32° above zero on its scale.

Fail Safe Control: Device which opens circuit when sensing element fails to operate.

Fan: A radial or axial flow device used for moving or producing artificial currents of air.

Fan (Centrifugal): A fan rotor or wheel within a scroll type of housing and including driving mechanism supports for either belt drive or direct connection.

Fan (Propeller): A propeller or disc-type wheel within a mounting ring or plate and including driving mechanism supports for either belt drive or direct connection.

Fan (Tubeaxial): A propeller or disc-type wheel within a cylinder and including driving mechanism supports for either belt drive or direct connection.

Fan (Vaneaxial): A disc-type wheel within a cylinder, a set of air guide vanes located either before or after the wheel and including driving mechanism supports for either belt drive or direct connection.

Faraday Experiment: Silver chloride absorbs ammonia when cool and releases ammonia when heated. This is basis on which some absorption refrigerators operate.

Field Pole: Part of stator of motor which concentrates magnetic field of field winding.

File Card: Tool used to clean metal files.

Filter: Device for removing small particles from a fluid.

Fin: An extended surface to increase the heat transfer area, as metal sheets attached to tubes.

Finned Tubes: Heat transfer tube or pipe with extended surface in the form of fins, discs, or ribs.

Flame Test for Leaks: Tool which is principally a torch and when an air-refrigerant mixture is fed to flame, this flame will change color in presence of heated copper.

Flapper Valve: The type of valve used in refrigeration compressors which allows gaseous refrigerants to flow in only one direction.

Flare: Copper tubing is often connected to parts of refrigerating system by use of flared fittings. These fittings require that the end of tube be expanded at about 45° angle. This flare is firmly gripped by fittings to make a strong leakproof seal.

Flare Nut: Fitting used to clamp tubing flare against another fitting.

Flared Single Thickness Connection: Tube ending formed into 37 1/2° or 45° bell mouth or flare.

Flash Gas: This is the instantaneous evaporation of some liquid refrigerant in evaporator which cools remaining liquid refrigerant to desired evaporation temperature.

Flash Point: Temperature at which an oil will give off sufficient vapor to support a flash flame but will not support continuous combustion.

Flash Weld: A resistance type weld in which mating parts are brought together under considerable pressure and a heavy electrical current is passed through the joint to be welded.

Float Valve: Type of valve which is operated by sphere or pan which floats on liquid surface and controls level of liquid.

Flooded System: Type of refrigerating system in which liquid refrigerant fills evaporator.

Flooded System, Low-Side Float: Refrigerating system which has a low side float refrigerant control.

Flooding: Act of filling a space with a liquid.

Flow Meter: Instrument used to measure velocity or volume of fluid movement.

Fluid: Substance in a liquid or gaseous state; substance containing particles which move and change position without separation of the mass.

Fluid Coupling: Device which transmits drive energy to energy absorber through a fluid.

Flush: An operation to remove any material or fluids from refrigeration system parts by purging them to the atmosphere using refrigerant or other fluids.

Flux-Brazing, Soldering: Substance applied to surfaces to be joined by brazing or soldering to free them from oxides and facilitate good joint.

Flux, Magnetic: Lines of force of a magnet.

Foam Leak Detector: A system of soap bubbles or special foaming liquids brushed over joints and connections to locate leaks.

Foaming: Formation of a foam in an oil-refrigerant mixture due to rapid evaporation of refrigerant dissolved in the oil. This is most likely to occur when the compressor starts and the pressure is suddenly reduced.

Foot Pound: A unit of work. A foot pound is the amount of work done in lifting one pound one foot.

Force: Force is accumulated pressure and is expressed in pounds. If the pressure is 10 psi on a plate of 10 sq. in. area, the force is 100 lbs.

Forced Convection: Movement of fluid by mechanical force such as fans or pumps.

Force-Feed Oiling: A lubrication system which uses a pump to force oil to surfaces of moving parts.

Free Area: The total minimum area of the openings in a grille, face, or register through which air can pass.

Freezer Alarm: Device used in many freezers which sounds an alarm (bell or buzzer) when freezer temperature rises above safe limit.

Freezer Burn: A condition applied to food which has not been properly wrapped and that has become hard, dry, and discolored.

Freeze-Up: 1-The formation of ice in the refrigerant control device which may stop the flow of refrigerant into the evaporator. 2-Frost formation on a coil may stop the airflow through the coil.

Freezing: Change of state from liquid to solid.

Freezing Point: The temperature at which a liquid will solidify upon removal of heat. The freezing temperature for water is 32°F at atmospheric pressure.

Freon: Trade name for a family of synthetic chemical refrigerants manufactured by DuPont De Nemours Inc.

Frost Back: Condition in which liquid refrigerant flows from evaporator into suction line; indicated by frost formation on suction line.

Frost Control, Automatic: A control which automatically cycles refrigerating system based on frost formation on evaporator.

Frost Control, Manual: A manual control used to change refrigerating system to produce defrosting conditions.

Frost Control, Semiautomatic: A control which starts defrost part of a cycle manually and then returns system to normal operation automatically.

Frost Free Refrigerator: A refrigerated cabinet which operates with an automatic defrost during each cycle.

Frosting Type Evaporator: A refrigerating system which maintains the evaporator at frosting temperatures during phases of cycle.

Full Floating: A mechanism construction in which a shaft is free to turn in all the parts in which it is inserted.

Fumes: Smoke; aromatic smoke; odor emitted, as of flowers; a smoky or vaporous exhalation, usually odorous, as that from concentrated nitric acid. The word fumes is so broad and inclusive that its usefulness as a technical term is very limited. Its principal definitive characteristic is that it implies an odor. The terms vapor, smoke, fog, etc., which can be more strictly defined, should be used whenever possible. Also defined as solid particles generated by condensation from the gaseous state, generally, after volatilization from molten metals, etc., and often accompanied by a chemical reaction such as oxidation. Fumes flocculate and sometimes coalesce.

Fuse: Electrical safety device consisting of strip of fusible metal in circuit which melts when current is overloaded.

Fusible Plug: A plug or fitting mode with a metal of a known low melting temperature, used as safety device to release pressures in case of fire.

G

Galvanic Action: Corrosion action between two metals of different electronic activity. The action is increased in the presence of moisture.

Gas: Vapor phase or state of a substance.

Gasket: A resilient or flexible material used between mating surfaces of refrigerating unit parts or of refrigerator doors to provide a leakproof seal.

Gasket, Foam: A joint sealing device made of rubber or plastic foam strips.

Gas—Noncondensible: A gas which will not form into a liquid under pressure-temperature conditions.

Gas Valve: Device for controlling flow of gas.

Gauge, Compound: Instrument for measuring pressures both below and above atmospheric pressure.

Gauge, High Pressure: Instrument for measuring pressures in range of 0 psig to 500 psig.

Gauge, Low Pressure: Instrument for measuring pressures in range of 0 psig and 50 psig.

Gauge Manifold: A device constructed to hold compound and high pressure gauges and valved to control flow of fluids through it.

Gauge, Vacuum: Instrument used to measure pressures below atmospheric pressure.

Grain: A unit of weight and equal to one 7000th of a pound. It is used to indicate the amount of moisture in the air.

Gravity (Specific): The specific gravity of a solid or liquid is the ratio of the mass of the body to the mass of an equal volume of water at some standard temperature. At the present time a temperature of 4 C (39°F) is commonly used by physicists, but the engineer uses 60°F. The specific gravity of a gas is usually expressed in terms of dry air at the same temperature and pressure as the gas.

Grille: An ornamental or louvered opening placed at the end of an air passageway.

Grommet: A plastic metal or rubber doughnut-shaped protector for wires or tubing as they pass through hole in object.

Ground Coil: A heat exchanger buried in the ground which may be used either as an evaporator or as a condenser.

Ground, Short Circuit: A fault in an electrical circuit allowing electricity to flow into the metal parts of the structure.

Ground Wire: An electrical wire which will safely conduct electricity from a structure into the ground.

H

Halide Refrigerants: Family of refrigerants containing halogen chemicals.

Halide Torch: Type of torch used to detect halogen refrigerant leaks.

Hastelloy: Trade name for a hard, noncorroding metal alloy.

Head (Total): In flowing fluid, the sum of the static and velocity pressures at the point of measurement.

Head Pressure: Pressure which exists in condensing side of refrigerating system.

Head-Pressure Control: Pressure operated control which opens electrical circuit if high side pressure becomes excessive.

Head, Static: Pressure of fluid expressed in terms of height of column of the fluid, such as water or mercury.

Head, Velocity: In flowing fluid, height of fluid equivalent to its velocity pressure.

Heat: Form of energy the addition of which causes substances to rise in temperature; energy associated with random motion of molecules.

Heat (Latent): Heat characterized by a change of state of the substance concerned, for a given pressure and always at a constant temperature for a pure substance, i.e., heat of vaporization or of fusion.

Heat (Sensible): A term used in heating and cooling to indicate any portion of heat which changes only the temperature of the substances involved.

Heat (Specific): The heat absorbed (or given up) by a unit mass of a substance when its temperature is increased (or decreased) by 1-degree Common Units: Btu per (pound) (Fahrenheit degree), calories per (gram) (Centigrade degree). For gases, both specific heat at constant pressure (c_p) and specific heat at constant volume (c_v) are frequently used. In air-conditioning, c_p is usually used.

Heat Exchanger: Device used to transfer heat from a warm or hot surface to a cold or cooler surface. Evaporators and condensers are heat exchangers.

Heat Lag: When a substance is heated on one side, it takes time for the heat to travel through the substance. This time is called heat lag.

Heat Leakage: Flow of heat through a substance is called heat leakage.

Heat Load: Amount of heat, measured in Btu, which is removed during a period of 24 hrs.

Heat of Compression: Mechanical energy of pressure transformed into energy of heat.

Heat of Fusion: The heat released in changing a substance from a liquid state to a solid state. The heat of fusion of ice is 144 Btu per pound.

Heat of Respiration: The process by which oxygen and carbohydrates are assimilated by a substance; also when carbon dioxide and water are given off by a substance.

Heat Pump: A compression cycle system used to supply heat to a temperature controlled space, which can also remove heat from the same space.

Heat Transfer: Movement of heat from one body or substance to another. Heat may be transferred by radiation, conduction, convection or a combination of these three methods.

Heating Coil: A heat transfer device which releases heat.

Heating Control: Device which controls temperature of heat transfer unit which releases heat.

Heating Value: Amount of heat which may be obtained by burning a fuel. It is usually expressed in Btu per pound or Btu per gallon.

Heavy Ends, Hydrocarbon Oils: The heavy molecules or larger molecules of hydrocarbon oils.

Hermetic Motor: Compressor drive motor sealed within same casing which contains compressor.

Hermetic System: Refrigeration system which has a compressor driven by a motor contained in compressor dome or housing.

Hermetically Sealed Unit: A sealed hermetic-type condensing unit is a mechanical condensing unit in which the compressor and compressor motor are enclosed in the same housing with no external shaft or shaft seal, the compressor motor operating in the refrigerant atmosphere. The compressor and compressor motor housing may be of either the fully welded or brazed type, or of the service-sealed type. In the fully welded or brazed type, the housing is permanently sealed and is not provided with means of access for servicing internal parts in the field. In the service-sealed type, the housing is provided with some means of access for servicing internal parts in the field.

Hg (Mercury): Heavy silver-white metallic element; only metal that is liquid at ordinary room temperature. Symbol, Hg.

High Pressure Cut-Out: Electrical control switch operated by the high side pressure which automatically opens electrical circuit if too high head pressure or condensing pressure is reached.

High Side: Parts of a refrigerating system which are under condensing or high side pressure.

High Side Float: Refrigerant control mechanism which controls the level of the liquid refrigerant in the high pressure side of mechanism.

High Vacuum Pump: Mechanism which can create vacuum in 1000 to 1 micron range.

Hollow-Tube Gasket: Sealing device made of rubber or plastic with tubular cross section.

Hone: Fine-grit stone used for precision sharpening.

Horsepower: A unit of power equal to 33,000 foot pounds of work per minute. One electrical horsepower equals 746 watts.

Hot Gas Bypass: Piping system in refrigerating unit which moves hot refrigerant gas from condenser into low pressure side.

Hot Gas Defrost: A defrosting system in which hot refrigerant gas from the high side is directed through evaporator for short period of time and at predetermined intervals in order to remove frost from evaporator.

Hot Junction: That part of thermoelectric circuit which releases heat.

Hot Wire: A resistance wire in an electrical relay which expands when heated and contracts when cooled.

Humidifiers: Device used to add to and control the humidity in a confined space.

Humidistat: An electrical control which is operated by changing humidity.

Humidity: Moisture; dampness. Relative humidity is ratio of quantity of vapor present in air to greatest amount possible at given temperature.

Hydrolen—Tar: A hydrocarbon by-product of oil industry. Used as a low melting temperature, waterproof sealing compound.

Hydrometer: Floating instrument used to measure specific gravity of a liquid. Specific gravity is ratio of weight of any volume of a substance to weight of equal volume of substance used as a standard.

Hygrometer: An instrument used to measure degree of moisture in the atmosphere.

Hygroscopic: Ability of a substance to absorb and retain moisture and change physical dimensions as its moisture content changes.

I

ICC—Interstate Commerce Commission: A government body which controls the design and construction of pressure containers.

Ice Cream Cabinet: Commercial refrigerator which operates at approximately 0°F and is used for storage of ice cream.

Ice Melting Equivalent (I.M.E.) (Ice Melting Effect): Amount of heat absorbed by melting ice at 32°F is 144 Btu per pound of ice or 288,000 Btu per ton.

Idler: A pulley used on some belt drives to provide the proper belt tension and to eliminate belt vibration.

Ignition Transformer: A transformer designed to provide a high voltage current. Used in many heating systems to ignite fuel.

Impeller: Rotating part of a centrifugal pump.

Induction Motor: An AC motor which operates on principle of rotating magnetic field. Rotor has no electrical connection, but receives electrical energy by transformer action from field windings.

Induction Units (Low-Pressure Type): Essentially induction-type convectors. They use a jet of conditioned air (or primary air) to induce into the unit a flow of room or secondary air which mixes with the primary air. The mixture is discharged into the room through a grille at the top of the unit. Heating coils are located in the secondary air stream for use in heating.

Industrial Air-Conditioning: Air-conditioning for other uses than comfort.

Infrared Lamp: An electrical device which emits infrared rays; invisible rays just beyond red in the visible spectrum.

Insulation, Thermal: Substance used to retard or slow flow of heat through wall or partition.

Isothermal: Changes of volume or pressure under conditions of constant temperature.

Isothermal Expansion and Contraction: An action which takes place without a temperature change.

J

Joint (Brazed, High-Temperature): A gas tight joint obtained by the joining of metal parts with metallic mixtures or alloys which melt at temperatures below 1500°F but above 1000°F.

Joint (Soldered): A gas-tight joint obtained by the joining of metal parts with metallic mixtures or alloys which melt at temperatures below 1000°F.

Joint (Welded): A gas-tight joint obtained by the joining of metal parts in the plastic or molten state.

Joule-Thomson Effect: Change in temperature of a gas on expansion through a porous plug from a high pressure to a lower pressure.

Journal, Crankshaft: Part of shaft which contacts the bearing.

Junction Box: Group of electrical terminals housed in protective box or container.

K

Kata Thermometer: Large bulb alcohol thermometer used to measure air velocities or atmospheric conditions by means of cooling effect.

Kelvin Scale (K): Thermometer scale on which unit of measurement equals the centigrade degree and according to which absolute zero is 0°, the equivalent of −273.16° C. Water freezes at 273.16° and boils at 373.16°.

Kilometer: A metric unit of linear measurement = 1000 meters.

Kilowatt: Unit of electrical power, equal to 1000 watts.

L

Lacquer: A protective coating or finish which dries to form a film by evaporation of a volatile constituent.

Lamps, Steri: A lamp which gives forth a high intensity ultraviolet ray and is used to kill bacteria. It is often used in food storage cabinets.

Lapping: Smoothing a metal surface to high degree of refinement or accuracy using a fine abrasive.

Latent Heat: Heat energy absorbed in process of changing form of substance (melting, vaporization, fusion) without change in temperature or pressure.

Leak Detector: Device or instrument such as a halide torch, an electronic sniffer; or soap solution used to detect leaks.

Limit Control: Control used to open or close electrical circuits as temperature or pressure limits are reached.

Liquid Absorbent: A chemical in liquid form which has the property to "take on" or absorb moisture.

Liquid Indicator: Device located in liquid line which provides a glass window through which liquid flow may be observed.

Liquid Line: The tube which carries liquid refrigerant from the condenser or liquid receiver to the refrigerant control mechanism.

Liquid Nitrogen: Nitrogen in liquid form which is used as a low temperature refrigerant in chemical (or expendable) refrigerating systems.

Liquid Receiver: Cylinder connected to condenser outlet for storage of liquid refrigerant in a system.

Liquid-Vapor Valve Refrigerant Cylinder: A dual hand valve on refrigerant cylinders which is used to release either gas or liquid refrigerant from the cylinder.

Litharge: Lead powder mixed with glycerine to seal pipe thread joints.

Liquor: Solution used in absorption refrigeration.

Liter: Metric unit of volume which equals 61.03 in.^{3}.

Load: The amount of heat per unit time imposed on a refrigerating system, or the required rate of heat removal.

Louvers: Sloping, overlapping boards or metal plates intended to permit ventilation and shed falling water.

Low Side: That portion of a refrigerating system which is under the lowest evaporating pressure.

Low Side Float Valve: Refrigerant control valve operated by level of liquid refrigerant in low pressure side of system.

Low Side Pressure: Pressure in cooling side of refrigerating cycle.

Low Side Pressure Control: Device used to keep low side evaporating pressure from dropping below certain pressure.

M

Manifold, Service: A device equipped with gauges and manual valves, used by serviceman to service refrigerating systems.

Manometer: Instrument for measuring pressure of gases and vapors. Gas pressure is balanced against column of liquid such as mercury, in U-shaped tube.

Mass: A quantity of matter cohering together to make one body which is usually of indefinite shape.

Mean Effective Pressure (M.E.P.): Average pressure on a surface when a changing pressure condition exists.

Mechanical Cycle: Cycle which is a repetitive series of mechanical events.

Melting Point: Temperature at atmospheric pressure, at which a substance will melt.

Mercoid Bulb: An electrical circuit switch which uses a small quantity of mercury in a sealed glass tube to make or break electrical contact with terminals within the tube.

Meter: Metric unit of linear measurement equal to 39.37 in.

Methanol Drier: Alcohol type chemical used to change water in refrigerating system into a nonfreezing solution.

Methyl Chloride (R-40): A chemical once commonly used as a refrigerant. The chemical formula is CH_3Cl. Cylinder color code is orange. The boiling point at atmospheric pressure is −10.4°F.

Metric System: A decimal system of measures and weights, based on the meter and gram. Length of one meter, 39.37 in.

Micrometer: A precision measuring instrument used for making measurements accurate to .001 to .0001 in.

Micron: Unit of length in metric system; a thousandth part of one millimeter.

Micron Gauge: Instrument for measuring vacuums very close to a perfect vacuum.

Milli: A combining form denoting one thousandth; example, millivolt, one thousandth of a volt.

Modulating: A type of device or control which tends to adjust by increments (minute changes) rather than by either full on or full off operation.

Modulating Refrigeration Cycle: Refrigerating system of variable capacity.

Moisture Determination: An action using instruments and calculations to measure the relative or absolute moisture in an air conditioned space.

Moisture Indicator: Instrument used to measure moisture content of a refrigerant.

Molecule: Smallest partion of an element or compound that retains chemical identity with the substance in mass.

Molliers Diagram: Graph of refrigerant pressure, heat and temperature properties.

Monel: A trademark name for metal alloy consisting chiefly of copper and nickel.

Monitor Top: Unit built by General Electric which had a cylindrical condenser surrounding the motor-compressor, mounted on top of the cabinet.

Monochlorodifluoromethane: A refrigerant better known as Freon 12 or R-22. Chemical formula is $CHClF_2$. Cylinder color code is green.

Motor—2-Pole: A 3600 rpm electric motor (synchronous speed).

Motor—4-Pole: A 1800 rpm electric motor (synchronous speed).

Motor, Capacitor: A single-phase induction motor with an auxiliary starting winding connected in series with a condenser (capacitor) for better starting characteristics.

Motor Burnout: Condition in which the insulation of electric motor has deteriorated by overheating.

Motor Control: Device to start and/or stop a motor at certain temperature or pressure conditions.

Motor Starter: High capacity electric switches usually operated by electromagnets.

Muffler, Compressor: Sound absorber chamber in refrigeration system used to reduce sound of gas pulsations.

Mullion: Stationary part of a structure between two doors.

Multiple Evaporator System: Refrigerating system with two or more evaporators connected in parallel.

Multiple System: Refrigerating mechanism in which several evaporators are connected to one condensing unit.

N

Natural Convection: Movement of a fluid caused by temperature differences (density changes).

Neoprene: A synthetic rubber which is resistant to hydrocarbon oil and gas.

Neutralizer: Substance used to counteract acids, in refrigeration system.

No-Frost Freezer: A low temperature refrigerator-cabinet in which no frost or ice collects on produce stored in cabinet.

Nominal Size Tubing: Tubing measurement which has an inside diameter the same as iron pipe of the same stated size.

Non-Code Installation: A functional refrigerating system installed where there are no local, state, or national refrigeration codes in force.

Noncondensable Gas: Gas which does not change into a liquid at operating temperatures and pressures.

Nonferrous: Group of metals and metal alloys which contain no iron.

Nonfrosting Evaporator: An evaporator which never collects frost or ice on its surface.

Normal Charge: The thermal element charge which is part liquid and part gas under all operating conditions.

North Pole, Magnetic: End of magnet from which magnetic lines of force flow.

O

Off Cycle: That part of a refrigeration cycle when the system is not operating.

Oil Binding: Physical condition when an oil layer on top of refrigerant liquid hinders it from evaporating at its normal pressure-temperature condition.

Oil, Refrigeration: Specially prepared oil used in refrigerator mechanism circulates to same extent with refrigerant. The oil must be dry (entirely free of moisture), otherwise, moisture will condense out and freeze in the refrigerant control and may cause refrigerant mechanism to fail. An oil classified as a refrigerant oil must be free of moisture and other contaminants.

Oil Rings: Expanding rings mounted in grooves and piston; designed to prevent oil from moving into compression chamber.

Oil Separator: Device used to remove oil from gaseous refrigerant.

Open Circuit: An interrupted electrical circuit which stops flow of electricity.

Open Display Case: Commercial refrigerator designed to maintain its contents at refrigerating temperatures even though the contents are in an open case.

Open Type System: A refrigerating system which uses a belt-driven compressor or a coupling-driven compressor.

Orifice: Accurate size opening for controlling fluid flow.

Oscilloscope: A flourescent coated tube which visually shows an electrical wave.

Outside Air: External air; atmosphere exterior to refrigerated or conditioned space; ambient (surrounding) air.

Overload: Load greater than load for which system or mechanism was intended.

Overload Protector: A device, either temperature, pressure, or current operated, which will stop operation of unit if dangerous conditions arise.

Ozone: A gaseous form of oxygen usually obtained by silent discharge of electricity in oxygen or air.

P

Partial Pressures: Condition where two or more gases occupy a space and each one creates part of the total pressure.

Pascal's Law: A pressure imposed upon a fluid is transmitted equally in all directions.

Perm: The unit of permeance. A perm is equal to 1 grain per (sq ft) (hr) (inch of mercury vapor pressure difference).

Permeance: The water vapor permeance of a sheet of any thickness (or assembly between parallel surfaces) is the ratio of water vapor flow to the vapor pressure difference between the surfaces. Permeance is measured in perms.

Permanent Magnet: A material which has its molecules aligned and has its own magnetic field; bar of metal which has been permanently magnetized.

Photoelectricity: A physical action wherein an electrical flow is generated by light waves.

Pinch-Off Tool: Device used to press walls of a tubing together until fluid flow ceases.

Piston: Close fitting part which moves up and down in a cylinder.

Piston Displacement: Volume displaced by piston as it travels length of stroke.

Pitot Tube: Tube used to measure air velocities.

Plenum Chamber: Chamber or container for moving air or other gas under a slight positive pressure.

Polyphase Motor: Electrical motor designed to be used with three-phase electrical circuit.

Polystyrene: Plastic used as an insulation in some refrigerator cabinet structures.

Ponded Roof: Flat roof designed to hold quantity of water which acts as a cooling device.

Porcelain: Ceramic china-like coating applied to steel surfaces.

Pour Point (Oil): Lowest temperature at which oil will pour or flow.

Power: Time rate at which work is done or energy emitted; source or means of supplying energy.

Power Element: Sensitive element of a temperature operated control.

Pressure: An energy impact on a unit area; force or thrust exerted on a surface.

Pressure Drop: The pressure difference at two ends of a circuit, or part of a circuit, the two sides of a filter, or the pressure difference between the high side and low side in a refrigerator mechanism.

Pressure Limiter: Device which remains closed until a certain pressure is reached and then opens and releases fluid to another part of system.

Pressure-Heat Diagram: Graph of refrigerant pressure, heat and temperature properties. (Mollier's diagram.)

Pressure Motor Control: A device which opens and closes on electrical circuit as pressures change to desired pressures.

Pressure Operated Altitude (POA) Valve: Device which maintains a constant low side pressure independent of altitude of operation.

Pressure Regulator, Evaporator: An automatic pressure regulating valve. Mounted in suction line between evaporator outlet and compressor inlet. Its purpose is to maintain a predetermined pressure and temperature in the evaporator.

Pressure Suction: Pressure in low pressure side of a refrigerating system.

Pressure Water Valve: Device used to control water flow which is responsive to head pressure of refrigerating system.

Primary Control: Device which directly controls operation of heating system.

Process Tube: Length of tubing fastened to hermetic unit dome, used for servicing unit.

Protector, Circuit: An electrical device which will open an electrical circuit if excessive electrical conditions occur.

psi: A symbol or initials used to indicate pressure measured in pounds per square inch.

psia: A symbol or initials used to indicate pressure measured in pounds per square inch absolute. Absolute pressure equals gauge pressure plus atmospheric pressure.

psig: A symbol or initials used to indicate pressure in pounds per square inch gauge. The "g" indicates that it is gauge pressure and not absolute pressure.

Psychrometer or Wet Bulb Hygrometer: An instrument for measuring the relative humidity of atmospheric air.

Psychrometric Chart: A chart that shows relationship between the temperature, pressure and moisture content of the air.

Psychrometric Measurement: Measurement of temperature pressure, and humidity using a psychrometric chart.

Pull Down: An expression indicating action of removing refrigerant from all or a part of refrigerating system.

Pump Down: The act of using a compressor or a pump to reduce the pressure in a container or a system.

Purging: Releasing compressed gas to atmosphere through some part or parts for the purpose of removing contaminants from that part or parts.

Pyrometer: Instrument for measuring high temperatures.

Q

Quenching: Submerging hot solid object in cooling fluid.

Quick Connect Coupling: A device which permits easy, fast, connecting of two fluid lines.

R

R-11, Trichloromonofluoromethane: Low pressure, synthetic chemical refrigerant which is also used as a cleaning fluid.

R-12, Dichlorodifluoromethane: A popular refrigerant known as Freon 12.

R-22, Monochlorodifluoromethane: Synthetic chemical refrigerant.

R-40, Methyl Chloride: Refrigerant which was used extensively in the 1920's and 1930's.

R-113, Trichlorotrifluoroethane: Synthetic chemical refrigerant.

R-160, Ethyl Chloride: Refrigerant which is seldom used at present time.

R-170, Ethane: Low temperature application refrigerant.

R-290, Propane: Low temperature application refrigerant.

R-500: Refrigerant which is azeotropic mixture of R-12 and R-152a.

R-502: Refrigerant which is azeotropic mixture of R-22 and R-115.

R-503: Refrigerant which is azeotropic mixture of R-23 and R-13.

R-504: Refrigerant which is azeotropic mixture of R-32 and R-115.

R-600, Butane: Low temperature application refrigerant; also used as a fuel.

R-611, Methyl Formate: Low pressure refrigerant.

R-717, Ammonia: Popular refrigerant for industrial refrigerating systems; also a popular absorption system refrigerant.

R-764, Sulphur Dioxide: Low pressure refrigerant used extensively in 1920's and 1930's. Not in use at present; chemical is often used as an industrial bleaching agent.

Radial Commutator: Electrical contact surface on a rotor which is perpendicular or at right angles to the shaft center line.

Radiation: Transfer of heat by heat rays.

Range: Pressure or temperature settings of a control; change within limits.

Rankin Scale: Name given the absolute (Fahrenheit) scale. Zero on this scale is −460°F.

Receiver-Drier: A cylinder in a refrigerating system for storing liquid refrigerant and which also holds a quantity of desiccant.

Receiver Heating Element: Electrical resistance mounted in or around liquid receiver, used to maintain head pressures when ambient temperature is at freezing or below freezing.

Reciprocating: Action in which the motion is back and forth in a straight line.

Recording Ammeter: Electrical instrument which uses a pen to record amount of current flow on a moving paper chart.

Recording Thermometer: Temperature measuring instrument which has a pen marking a moving chart.

Rectifier, Electric: An electrical device for converting AC into DC.

Reed Valve: Thin flat tempered steel plate fastened at one end.

Refrigerant: Substance used in refrigerating mechanism to absorb heat in evaporator coil by change of state from a liquid to a gas, and to release its heat in a condenser as the substance returns from the gaseous state back to a liquid state.

Refrigerant Charge: Quantity of refrigerant in a system.

Refrigerant Control: Device which meters refrigerant and maintains pressure difference between high pressure and low pressure side of mechanical refrigerating system while unit is running.

Register: Combination grille and damper assembly covering on an air opening or end of an air duct.

Relative Humidity: Ratio of amount of water vapor present in air to greatest amount possible at same temperature.

Relay: Electrical mechanism which uses small current in control circuit to operate a valve switch in operating circuit.

Relief Valve: Safety device designed to open before dangerous pressure is reached.

Remote Power Element Control: Device with sensing element located apart from operating mechanism.

Remote System: Refrigerating system which has condensing unit located outside and separate from refrigerator cabinet.

Repulsion-Start Induction Motor: Type of motor which has an electrical winding on the rotor for starting purposes.

Reverse Cycle Defrost: Method of heating evaporator for defrosting purposes by using valves to move hot gas from compressor into evaporator.

Reversing Valve: Device used to reverse direction of the refrigerant flow depending upon whether heating or cooling is desired.

Ringelmann Scale: Measuring device for determining smoke density.

Riser Valve: Device used to manually control flow of refrigerant in vertical piping.

Rotary Blade Compressor: Mechanism for pumping fluid by revolving blades inside cylindrical housing.

Rotary Compressor: Mechanism which pumps fluid by using rotating motion.

Rotor: Rotating part of a mechanism.

Running Winding: Electrical winding of motor which has current flowing through it during normal operation of motor.

S

Saddle Valve (Tap-A-Line): Valve body shaped so it may be silver brazed to refrigerant tubing surface.

Safety Control: Device which will stop the refrigerating unit if unsafe pressures and/or temperatures are reached.

Safety Motor Control: Electrical device used to open circuit if the temperature, pressure, and/or the current flow exceed safe conditions.

Safety Plug: Device which will release the contents of a container above normal pressure conditions and before rupture pressures are reached.

Saturation: Condition existing when substance contains maximum of another substance for that temperature and pressure.

Scavenger Pump: Mechanism used to remove fluid from sump or container.

Schrader Valve: Spring loaded device which permits fluid flow in one direction when a center pin is depressed; in other direction when a pressure difference exists.

Scotch Yoke: Mechanism used to change reciprocating motion into rotary motion or vice-versa. Used to connect crankshaft to piston in refrigeration compressor.

Sealed Unit: (See Hermetic System.) A motor-compressor assembly in which motor and compressor operate inside sealed dome or housing.

Seal Leak: Escape of oil and/or refrigerant at the junction where shaft enters housing.

Seal, Shaft: A device used to prevent leakage between shaft and housing.

Secondary Refrigerating System: Refrigerating system in which condenser is cooled by evaporator of another or primary refrigerating system.

Second Law of Thermodynamics: Heat will flow only from material at certain temperature to material at lower temperature.

Sensible Heat: Heat which causes a change in temperature of a substance.

Sensor: A material or device which goes through a physical change or an electronic characteristic change as the conditions change.

Separator, Oil: A device used to separate refrigerant oil from refrigerant gas and return the oil to crankcase of compressor.

Sequence Controls: Group of devices which act in series or in time order.

Servel System: One type of continuous operation absorption refrigerating system.

Serviceable Hermetic: Hermetic unit housing containing motor and compressor assembled by use of bolts or threads.

Service Valve: A device to be attached to system which provides opening for gauges and/or charging lines. Also provides means of shutting off or opening gauge and charging ports, and controlling refrigerant flow in system.

Shaded Pole Motor: A small AC motor used for light start loads. Has no brushes or commutator.

Sharp Freezing: Refrigeration at temperature slightly below freezing, with moderate air circulation.

Shell-and-Tube Flooded Evaporator: Device which flows water through tubes built into cylindrical evaporator or vice-versa.

Shell Type Condenser: Cylinder or receiver which contains condensing water coils or tubes.

Short Cycling: Refrigerating system that starts and stops more frequently than it should.

Shroud: Housing over condenser or evaporator.

Sight Glass: Glass tube or glass window in refrigerating mechanism which shows amount of refrigerant, or oil in system; or, pressure of gas bubbles in liquid line.

Silica Gel: Chemical compound used as a drier, which has ability to absorb moisture when heated. Moisture is released and compound may be reused.

Silver Brazing: Brazing process in which brazing alloy contains some silver as part of joining alloy.

Sintered Oil Bearing: Porous bearing metal, usually bronze, and which has oil in pores of bearing metal.

Sling Psychrometer: Humidity measuring device with wet and dry bulb thermometers, which is moved rapidly through air when measuring humidity.

Slug: A unit of mass equal to the weight (English units) of object divided by 32.2 (acceleration due to the force of gravity).

Smoke: An air suspension (aerosol) of particles, usually but not necessarily solid, often originating in a solid nucleus, formed from combustion or sublimation. Also defined as carbon or soot particles less than 0.1 micron in size which result from the incomplete combustion of carbonaceous materials such as coal, oil, tar, and tobacco.

Smoke Test: Test made to determine completeness of combustion.

Solar Heat: Heat from visible and invisible energy waves from the sun.

Soldering: Joining two metals by adhesion of a low melting temperature metal (less than 800°F).

Solenoid Valve: Electromagnet with a moving core which serves as a valve, or operates a valve.

Solid Absorbent Refrigeration: Refrigerating system which uses solid substance as absorber of the refrigerant during cooling part of cycle and releases refrigerant when heated during generating part of cycle.

South Pole, Magnetic: That part of magnet into which magnetic flux lines flow.

Specific Gravity: Weight of a liquid compared to water which is assigned value of 1.0.

Specific Heat: Ratio of quantity of heat required to raise temperature of a body one degree to that required to raise temperature of equal mass of water one degree.

Specific Volume: Volume per unit mass of a substance.

Splash System, Oiling: Method of lubricating moving parts by agitating or splashing oil.

Split-Phase Motor: Motor with two stator windings. Winding in use while starting is disconnected by centrifugal switch after motor attains speed, then motor operates on other winding.

Split System: Refrigeration or air-conditioning installation which places condensing unit outside or remote from evaporator. Also applicable to heat pump installations.

Spray Cooling: Method of refrigerating by spraying refrigerant inside of evaporator or by spraying refrigerated water.

Squirrel Cage: Fan which has blades parallel to fan axis and moves air at right angles or perpendicular to fan axis.

Standard Atmosphere: Condition when air is at 14.7 psia pressure, at 68°F temperature.

Standard Conditions: Used as a basis for air-conditioning calculations. Temperature of 68°F, pressure of 29.92 in. of Hg and relative humidity of 30 percent.

Starting Relay: An electrical device which connects and/or disconnects starting winding of electric motor.

Starting Winding: Winding in electric motor used only during brief period when motor is starting.

Stationary Blade Compressor: A rotary pump which uses blade inside pump to separate intake chamber from exhaust chamber.

Stator, Motor: Stationary part of electric motor.

Steam: Water in vapor state.

Steam-Heating: Heating system in which steam from a boiler is conducted to radiators in space to be heated.

Steam Jet Refrigeration: Refrigerating system which uses a steam venturi to create high vacuum (low pressure) on a water container causing water to evaporate at low temperature.

Stethoscope: Instrument used to detect sounds.

Stoker: Machine used to supply a furnace with coal.

Strainer: Device such as a screen or filter used to retain solid particles while liquid passes through.

Stratification of Air: Condition in which there is little or no air movement in room; air lies in temperature layers.

Strike: Door part of a door latch.

Subcooling: Cooling of liquid refrigerant below its condensing temperature.

Sublimation: Condition where a substance changes from a solid to a gas without becoming a liquid.

Suction Line: Tube or pipe used to carry refrigerant gas from evaporator to compressor.

Suction Pressure Control Valve: Device located in the suction line which maintains constant pressure in evaporator during running portion of cycle.

Suction Service Valve: A two-way manual-operated valve located at the inlet to compressor, which controls suction gas flow and is used to service unit.

Sulfur Dioxide: Gas once commonly used as a refrigerant. Refrigerant number is R-764; chemical formula is SO_2. Cylinder color code, black; boiling point at atmospheric pressure 14°F.

Superheat: Temperature of vapor above boiling temperature of its liquid at that pressure.

Superheater: Heat exchanger arranged to cool liquid going to evaporator using this heat to superheat vapor leaving evaporator.

Surface Plate: Tool with a very accurate flat surface, used for measuring purposes, and for lapping flat surfaces.

Surge: Modulating action of temperature or pressure before it reaches its final value or setting.

Surge Tank: Container connected to a refrigerating system which increases gas volume and reduces rate of pressure change.

Swaging: Enlarging one tube end so end of other tube of same size will fit within.

Swash Plate-Wobble Plate: Device used to change rotary motion to reciprocating motion, used in some refrigeration compressors.

Sweating: This term is used two different ways in refrigeration work: 1—Condensation of moisture from air on cold surface. 2—Method of soldering in which the parts to be joined are first coated with a thin layer of solder.

Sweet Water: Term sometimes used to describe tap water.

Sylphon Seal: Corrugated metal tubing used to hold seal ring and provide leak-proof connection between seal ring and compressor body or shaft.

Synthetic Rubber, Neoprene: Soft resilient material made of a synthetic chemical compound.

T

Tap-A-Line: Device used to puncture or tap a line where there are no service valves available; sometimes called a saddle valve.

Tap Drill: Drill used to form hole prior to placing threads in hole. The drill is the size of the root diameter of tap threads.

Tap (Screw Thread): Tool used to cut internal threads.

Teflon: Synthetic rubber material often used for O rings.

Temperature: Degree of hotness or coldness as measured by a thermometer; measurement of speed of motion of molecules.

Temperature Humidity Index: Actual temperature and humidity of sample of air, compared to air at standard conditions.

Test Light: Light provided with test leads, used to test or probe electrical circuits to determine if they are alive.

Thermal Relay (Hot Wire Relay): Electrical control used to actuate a refrigeration system. This system uses a wire to convert electrical energy into heat energy.

Thermistor: Material called a semiconductor, which is between a conductor and an insulator, which has electrical resistance that varies with temperature.

Thermocouple: Device which generates electricity, using principle that if two dissimilar metals are welded together and junction is heated, a voltage will develop across open ends.

Thermocouple Thermometer: Electrical instrument using thermocouple as source of electrical flow, connected to milliammeter calibrated in temperature degrees.

Thermodisk Defrost Control: Electrical switch with bimetal disk which is controlled by electrical energy.

Thermodynamics: Science which deals with mechanical action or relations of heat.

Thermoelectric Refrigeration: A refrigerator mechanism which depends on Peletier effect. Direct current flowing through electrical junction between dissimilar metals provides heating or cooling effect depending on direction of flow of current.

Thermometer: Device for measuring temperatures.

Thermomodule: Number of thermocouples used in parallel to achieve low temperatures.

Thermostat: Device responsive to ambient temperature conditions.

Thermostatic Control: Device which operates system or part of system based on temperature changes.

Thermostatic Expansion Valve: A control valve operated by temperature and pressure within evaporator coil, which controls flow of refrigerant. Control bulb is attached to outlet of coil.

Thermostatic Motor Control: Device used to control cycling of unit through use of control bulb attached to evaporator.

Thermostatic Valve: Valve controlled by thermostatic elements.

Thermostatic Water Valve: Valve used to control flow of water through system, actuated by temperature difference. Used in units such as water-cooled compressor or condenser.

Throttling: Expansion of gas through orifice or controlled opening without gas performing any work in expansion process.

Timers: Mechanism used to control on and off times of an electrical circuit.

Timer-Thermostat: Thermostat control which includes a clock mechanism. Unit automatically controls room temperature and changes it according to time of day.

Ton of Refrigeration: Refrigerating effect equal to the melting of one ton of ice in 24 hours. This may be expressed as follows:

288,000 Btu/24 hr
12,000 Btu/1 hr
200 Btu/min

Ton Refrigeration Unit: Unit which removes same amount of heat in 24 hours as melting of one ton of ice.

Torque: Turning or twisting force.

Torque Wrenches: Wrench which may be used to measure torque or pressure applied to a nut or bolt.

Transducer: Device actuated by power from one system and supplies power in another form to second system.

Trichlorotrifluoroethane: Complete name of refrigerant R-113. Group 1 refrigerant in rather common use. Chemical compounds which make up this refrigerant are chlorine, fluorine, and ethane.

Tube, Constricted: Tubing that is reduced in diameter.

Tube-Within-A-Tube: A water-cooled condensing unit in which a small tube is placed inside large unit. Refrigerant passes through one tube; water through the other.

Tubing: Fluid carrying pipe which has a thin wall.

Triple Point: Pressure temperature condition in which a substance is in equilibrium in solid, liquid and vapor states.

Truck, Refrigerated: Commercial vehicle equipped to maintain below atmospheric temperatures.

Two-Temperature Valve: Pressure opened valve used in suction line on multiple refrigerator installations which maintains evaporators in system at different temperatures.

U

Ultraviolet: Invisible radiation waves with frequencies shorter than wave lengths of visible light and longer than X-Ray.

Universal Motor: Electric motor which will operate on both AC and DC.

Urethane Foam: Type of insulation which is foamed in between inner and outer walls of display case.

V

Vacuum: Reduction in pressure below atmospheric pressure.

Vacuum Pump: Special high efficiency compressor used for creating high vacuums for testing or drying purposes.

Valve: Device used for controlling fluid flow.

Valve, Expansion: Type of refrigerant control which maintains pressure difference between high side and low side pressure in refrigerating mechanism. Valve is caused to operate by pressure in low or suction side. Often referred to as an automatic expansion valve or AEV.

Valve Plate: Part of compressor located between top of compressor body and head which contains compressor valves.

Valve, Service: Device used by service technicians to check pressures and charge refrigerating units.

Valve, Solenoid: Valve actuated by magnetic action by means of an electrically energized coil.

Valve, Suction: Valve in refrigeration compressor which allows vaporized refrigerant to enter cylinder from suction line and prevents its return.

Valves, Water: Most water cooling units are supplied with water valves. These valves provide a flow of water to cool the system while it is running. Most water valves are controlled by solenoids.

Vapor: Word usually used to denote vaporized refrigerant rather than the word gas.

Vapor Barrier: Thin plastic or metal foil sheet used in air conditioned structures to prevent water vapor from penetrating insulating material.

Vapor Charged: Lines and component parts of system which are charged at the factory.

Vapor Lock: Condition where liquid is trapped in line because of bend or improper installation which prevents the vapor from flowing.

Vapor Pressure: Pressure imposed by either a vapor or gas.

Vapor Pressure Curve: Graphic presentation of various pressures produced by refrigerant under various temperatures.

Vapor, Saturated: A vapor condition which will result in condensation into droplets of liquid as vapor temperature is reduced.

Variable Pitch Pulley: Pulley which can be adjusted to provide different pulley ratios.

V-Belt: Type of belt that is commonly used in refrigeration work. It has a contact surface which is in the shape of letter V.

V-Block: V-shaped groove in metal block used to hold shaft.

Velocimeter: Instrument used to measure air velocities using a direct reading air speed indicating dial.

Velocity: A vector quantity which denotes at once the time rate and the direction of a motion. $V = ds/dt$. For uniform linear motion $V = s/t$. Common units are feet per second or feet per minute.

Viscosity: Term used to describe resistance of flow of fluids.

Volatile Liquid: Liquid which evaporates at low temperature and pressure.

Voltage Control: It is necessary to provide some electrical circuits with uniform or constant voltage. Electronic devices used for this purpose are called voltage controls.

Voltmeter: Instrument for measuring voltage action in electrical circuit.

Volume (Specific): The volume of a substance per unit mass; the reciprocal of density. Units: cubic feet per pound, cubic centimeters per gram, etc.

Volumetric Efficiency: Term used to express the relationship between the actual performance of a compressor or of a vacuum pump and calculated performance of the pump based on its displacement versus its actual pumping ability.

Vortex Tube: Mechanism for cooling or refrigerating which accomplishes cooling effect by releasing compressed air through specially designed opening. Air expands in rapidly spiraling column of air which separates slow moving molecules (cool) from fast moving molecules (hot).

Vortex Tube Refrigeration: Refrigerating or cooling devices using principle of vortex tube, as in mining suits.

W

Walk-In Cooler: Large commercial refrigerated space kept below room temperature. Often found in large supermarkets or wholesale meat distribution centers.

Water-Cooled Condenser: Condensing unit which is cooled through use of water.

Water Defrosting: Use of water to melt ice and frost from evaporator during off-cycle.

Wax: Ingredient in many lubricating oils which may separate out if cooled sufficiently.

Wet Bulb: Device used in measurement of relative humidity. Evaporation of moisture lowers temperature of wet bulb compared to dry bulb temperature in same area.

Wet Cell Battery: Cell or connected group of cells that converts chemical energy into electrical energy by reversible chemical reactions.

Window Unit: Commonly used when referring to air conditioners which are placed in a window. Normally a domestic application.

GLOSSARY OF TECHNICAL TERMS AS APPLIED TO HEATING

A

Air Binding or Air Bound: A condition in which a bubble or other pocket of air is present in a pipeline or item of equipment and, by its presence, prevents or reduces the desired flow or movement of the liquid or gas in the pipeline or equipment.

Air Cushion Tank: A closed tank, generally located above the boiler and connected to a hydronic system in such a manner that when the system is initially filled with water, air is trapped within the tank. When the water in the system is heated it expands and compresses the air trapped within the air cushion tank, thus providing space for the extra volume of water without creating excessive pressure. Also called expansion tank.

Air-Gas Ratio: The ratio of combustion air supply flow rate to the fuel gas supply flow rate.

Air Shutter: An adjustable shutter on the primary air openings of a burner, which is used to control the amount of combustion air introduced into the burner body.

Air Vent: A valve installed at the high points in a hot water system to permit the elimination of air from the system.

Aldehyde: A class of compounds, which can be produced during incomplete combustion of a fuel gas. They have a pungent distinct odor.

Ambient Temperature: The temperature of the air in the area of study or consideration.

Available Head: The difference in pressure which can be used to circulate water in the system. The difference in pressure which may be used to overcome friction within the system. (See Pump Head, Head)

Atmospheric Burner: (See Burner)

Atmospheric Pressure: The pressure exerted upon the earth's surface by the weight of atmosphere above it.

Atom: The smallest unit of an element which retains the particular properties of that element.

Automatic Gas Pilot Device: A gas pilot incorporating a device, which acts to automatically shut off the gas supply to the appliance burner if the pilot flame is extinguished.

B

Backfire Protection: (See Flashback Arrestor)

Baffle: A surface used for deflecting fluids, usually in the form of a plate or wall.

Balancing Fit: (See Balance Fitting)

Balance Fitting: A pipe fitting or valve designed so that its resistance to flow may be varied. These are used to balance the pressure drop in parallel circuits.

Balancing Valve: (See Balance Fitting)

Baseboard: A terminal unit resembling the base trim of a house. These units are the most popular terminal unit for residential systems.

Boiler, Heating: That part of a hydronic heating system in which heat is transferred from the fuel to the water. If steam is generated it is a steam boiler. If the temperature of the water is raised without boiling, it is classed as a hot-water boiler.

Boiler Horsepower: The equivalent evaporation of 34.5 lb of water per hr from and at 212°F. This is equal to a heat output of $970.3 \times 34.5 = 33{,}475$ Btu/h.

Bonnet: The part of the furnace casing which forms a plenum chamber from where supply ducts receive warmed air. Also called supply plenum.

Branch: That portion of the piping system which connects a terminal unit to the circuit.

Btu or British Thermal Unit: The quantity of heat required to raise the temperature of one pound of water one degree Fahrenheit.

Bull Head: The installation of a pipe tee in such a way that water enters (or leaves) the tee at both ends of the run (the straight through section of the tee) and leaves (or enters) through the side connection only.

Bunsen-Type Burner: A gas burner in which combustion air is injected into the burner by the gas jet emerging from the gas orifice, and this air is premixed with the gas supply within the burner body before the gas burns on the burner port.

Burner: A device for the final conveyance of gas, or a mixture of gas and air, to the combustion zone. (See also specific type of burner).

(1) Injection Burner. A burner employing the energy of a jet of gas to inject air for combustion into the burner and mix it with gas.

(a) Atmospheric Injection Burner. A burner in which the air injected into the burner by a jet of gas is supplied to the burner at atmospheric pressure.

(2) Power Burner. (See also Forced Draft Burner, Induced Draft Burner, Premixing Burner, and Pressure Burner). A burner in which either gas or air or both are supplied at pressure exceeding, for gas, the line pressure, and for air, atmospheric pressure.

(3) Yellow-Flame Burner. A burner in which secondary air only is depended on for the combustion of the gas.

Burner Flexibility: The degree at which a burner can operate with reasonable characteristics with a variety of fuel gases and/or variations in input rate (gas pressure).

Burner Head: That portion of a burner beyond the outlet of the mixer tube which contains the burner ports.

Burner Port: (See Port)

Burning Speed: (See Flame Velocity)

Butane: A hydrocarbon fuel gas heavier than methane and propane and a major constituent of liquefied petroleum gases.

C

Calorimeter: Device for measuring heat quantities, such as machine capacity, heat of combustion, specific heat, vital heat, heat leakage, etc.; also device for measuring quality (or moisture content) of steam or other vapor.

Cfm: Cubic feet per minute.

Chap.: Chapter

Chimney Effect: The upward movement of warm air or gas, compared with the ambient air or gas, due to the lesser density of the warmed air or gas.

Circuit: The piping extending from the boiler supply tapping to the boiler return tapping.

Circuit Main: The portion of the main in a multiple circuit system that carries only a part of the total capacity of the system.

Circulator: A motor driven device used to mechanically circulate water in the system. Also called Pump.

Colorimetric Detection Device: A device for detecting the presence of a particular substance, such as carbon monoxide, in which the presence of that substance will cause a color change in a material in the detector.

Combustion: The rapid oxidation of fuel gases accompanied by the production of heat or heat and light.

Combustion Air: Air supplied in an appliance specifically for the combustion of a fuel gas.

Combustion Chamber: The portion of an appliance within which combustion normally occurs.

Combustion Products: Constituents resulting from the combustion of a fuel gas with the oxygen in air, including the inerts, but excluding excess air.

Commercial Buildings: Such buildings as stores, shops, restaurants, motels, and large apartment buildings.

Compression Tank: (See Air Cushion Tank)

Compound: A distinct substance formed by the chemical combination of two or more elements in definite proportions.

Condensable: A gas which can be easily converted to

liquid form, usually by lowering the temperature and/or increasing pressure.

Connected Load: The total load in Btu/h attached to the boiler. It is the sum of the outputs of all terminal units and all heat to be supplied by the boiler for process applications.

Controls: Devices designed to regulate the gas, air, water or electricity supplied to a gas appliance. They may be manual, semi-automatic or automatic.

Control Valves: Any valve used to control the flow of water in a hydronic system.

Convection: The movement of a fluid set up by a combination of differences in density and the force of gravity. For example, warm water at the bottom of a vertical tank will rise and displace cooler water at the top. The cooler water will sink to the bottom as the result of its greater density.

Convector: A terminal unit surrounded on all sides by an enclosure having an air outlet at the top or upper front. Convectors operate by gravity recirculated room air.

Converter: A heat exchange unit designed to transfer heat from one distributing system to another. These may be either steam to water or water to water units. They are usually of shell and tube design.

Counterflow: In heat exchange between two fluids, opposite direction of flow, coldest portion of one meeting coldest portion of the other.

Cubic Foot of Gas: (Standard Conditions). The amount of gas which will occupy 1 cubic foot when at a temperature of 60°F, and under a pressure equivalent to that of 30 in. of mercury.

D

Damper: A valve or plate which is installed in the cold and warm air ductwork and used to regulate the amount of air flowing through the duct. A damper may also be used in the flue of a furnace.

Dead Space: The short distance between a burner port and the base of a flame.

Degree Day: A unit used to estimate fuel consumption and to specify the heating load in winter, based on temperature difference and time. There are as many degree days for any one day as there are degrees F difference in temperature between the mean temperature for the day and 65°F.

Density: The weight of a substance per unit volume. As applied to gases, the weight in pounds of a cubic foot of gas at standard pressure and temperature.

Design Heat Loss: The heat loss of a building or room at design indoor-outdoor temperature difference.

Design Load: The design heat loss plus all other heating requirements to be provided by the boiler.

Design Temperature Difference: The difference between the design indoor and outdoor temperatures.

Design Water Temperature: The average of the temperature of the water entering and leaving the boiler (or sub-circuit) when the system is operating at design conditions.

Design Water Temperature Drop: The difference between the temperature of the water leaving the boiler and returning to the boiler when the system is operating at design conditions. In large systems employing sub-circuits the design temperature drop is usually taken as the difference in the temperature of the water entering and leaving each sub-circuit.

Dilution Air: Air which enters a draft hood and mixes with the flue gases.

Direct-Indirect Heating Unit: A heating unit located in the room or space to be heated and partially enclosed, the enclosed portion being used to heat air which enters from outside the room.

Direct Return: A two-pipe system in which the first terminal unit taken off the supply main is the first unit connected to the return main.

Discharge Coefficient: The ratio of the actual flow rate of a gas from an orifice or port to the theoretical, calculated flow rate. Always less than 1.0.

Distillation: Removal of gaseous substances from solids or liquids by applying heat.

D.M.S.: Drill Manufacturer's Standard—equivalent to Standard Twist Drill or Steel Wire Gauge Numbers.

Domestic Hot Water: The heated water used for domestic or household purposes such as laundry, dishes, bathing, etc.

Double Heat Transfer: The transfer of heat from the plant to the heated medium (usually liquid) and from the liquid to the air in the conditioned space.

Down-feed One-pipe Riser (Steam): A pipe which carries steam downward to the heating units and into which condensate from the heating units drains.

Down Feed System: A Hydronic system in which the main is located above the level of the terminal units.

Down-feed System (Steam): A steam heating system in which the supply mains are above the level of the heating units which they serve.

Drain Cock: A valve installed in the lowest point of a boiler or at low points of a heating system to provide for complete drainage of water from the system.

Draft: A current of air, usually referring to the difference in pressure which causes air or gases to flow through a chimney flue, heating unit or space.

Draft Hood: (Draft Diverter) A device built into an appliance, or made part of a vent connector from an appliance, which is designed to: (1) assure the ready escape of the products of combustion in the event of no draft, backdraft, or stoppage beyond the draft hood; (2) prevent a backdraft from entering the appliance; and (3) neutralize the effect of stack action of a chimney or gas vent upon the operation of the appliance.

Downdraft: Excessive high air pressure existing at the outlet of chimney or stack which tends to make gases flow downward in the stack.

Drilled Port Burner: A burner in which the ports have been formed by drilled holes in a thick section in the burner head or by a manufacturing method which results in holes similar in size, shape and depth.

Duct: Round or rectangular sheet metal pipes through which heat is carried from the furnace to the various rooms in the building.

E

Eccentric Reducer: A pipe fitting designed to change from one pipe size to another and to keep one edge of both pipes in line. These fittings should be installed so that the "in line" section of pipe is at the top.

Effective Heat Allowance: An allowance added to the test output of certain designs of radiation to compensate for a better distribution of heat within the heated space. Some agencies do not permit the use of effective heat allowance.

Electric Heating Element: A unit assembly consisting of a resistor, insulated supports, and terminals for connecting the resistor to electric power.

Element: One of the 96 or more basic substances of which all matter is composed.

Excess Air: Air which passes through an appliance and the appliance flues in excess of that which is required for complete combustion of the gas. Usually expressed as a percentage of the air required for complete combustion of the gas.

Expansion Tank: (See Air Cushion Tank)

Exposed Area: The area of any wall, window, ceiling, floor, or partition separating a heated room from the out-of-doors or from an unheated space.

Extinction Pop: (See Flashback)

F

Fahrenheit: The common scale of temperature measurement in the English system of units. It is based on the freezing point of water being 32°F and the boiling point of water being 212°F at standard pressure conditions.

Fan-Coil: A terminal unit consisting of a finned-tube coil and a fan in a single enclosure. These units may be designed for heating, cooling, or a combination of the two. Some fan-coil units are designed to receive duct work so that the unit may serve more than one room.

Ferrous: As used in this course, ferrous relates to objects made of iron or steel.

Filter: A porous material (fiberglass or foam plastic) which is installed in the air circulation system of a furnace to remove dust particles and pollen. Some are disposable, whereas some may be cleaned and re-used.

Fig: Figure.

Finned-Tube: A heat exchange device consisting of a metal tube through which water or steam may be circulated. Metal plates or fins are attached to the outside of the tube to increase the heat transfer surface. Finned tube or fin tube, may consist of one, two, or three tiers and are designed for installation bare, or with open type grilles, covers, or enclosures having top, front, or inclined outlets. Usually finned-tube units are for use in other than residential buildings.

Fire Tube Boiler: A steel boiler in which the hot gasses from combustion are circulated through tubes which are surrounded by boiler water which fills the space between the boiler shell and the tubes.

Firing Device: The burner, either oil, gas, or coal.

Fixed Orifice: (See Orifice Spud)

Flame Arrestor: (See Flashback Arrestor)

Flame Retention Device: A device added to a burner which aids in holding the flame base close to the burner ports.

Flame Rollout: A condition where flame rolls out of a combustion chamber when the burner is turned on.

Flame Velocity: The speed at which a flame moves through a fuel-air mixture.

Flammability Limits: The maximum percentages of a fuel in an air-fuel mixture which will burn.

Flashback: An undesirable flame characteristic in which burner flames strike back into a burner to burn there or to create a pop after the gas supply has been turned off.

Flashback Arrestor: A gauze, grid or any other portion of a burner assembly used to avert flashback.

Flashtube: An ignition device, commonly used for igniting gas on range top burners. An air-gas mixture from the burner body is injected into the end of a short tube. The mixture moves along the tube, is ignited by a standing pilot flame at the other open end of the tube and the flame travels back through the mixture in the flashtube to ignite the gas at the burner ports.

Flash Boiler: A boiler with very limited water capacity. Usually about one gal. of water per 1000 Btu/h net rating.

Floating Flames: An undesirable burner operating condition, usually indicating incomplete combustion in which flames leave the burner ports to "reach" for combustion air.

Flow Control Valve: A specially designed check valve, usually installed in the supply pipe, to prevent gravity circulation of hot water within the heating system when the pump is not in operation.

Flue: An enclosed passage in the chimney to carry exhaust smoke and fumes of the heating plant to escape to the outer air.

Flue Gases, Flue Products: Products of combustion and excess air in appliance flues or heat exchangers before the draft hood.

Flue Loss: The heat lost in flue products exiting from the flue outlet of an appliance.

Flue Outlet: The opening provided in an appliance for the escape of flue gases.

Fluid: A gas or liquid, as opposed to a solid.

Foot of Water: A measure of pressure. One foot of water is the pressure created by a column of water one foot in height. It is equivalent to 0.433 $lb/in.^2$.

Forced Hot Water: Or forced circulation hot water. Hot water heating systems in which a pump is used to create the necessary flow of water.

Forced Draft Burner: A burner in which combustion air is supplied by a fan or blower.

Friction Head: In a hydronic system the friction head is the loss in pressure resulting from the flow of water in the piping system.

ft: Foot or feet.

Fuel: Any substance used for combustion.

Fuel Gas: Any substance in a gaseous form when used for combustion.

Fuel-oil Burner, Pressure Atomizing or Gun Type: A burner designed to atomize the oil for combustion under an oil supply pressure of 100 psig.

Fuel-oil Burner, Rotary Type: A burner employing a thrower ring that mixes the oil and the air.

Fuel-oil Burner, Vaporizing or Pot Type: These burners use the heat of combustion to vaporize the oil in a pool beneath the vaporizing ring, and this vapor rising through the ring ignites and maintains combustion in the burner.

Furnace: That part of a warm air heating system in which combustion takes place.

G

Gal: Gallon or gallons.

Gate Valve: A valve designed in such a way that the opening for flow, when the valve is fully open, is essentially the same as the pipe and the direction of flow through the valve is in a straight line.

gpm: The abbreviation for "gallons per minute" which is a measure of rate of flow.

Grate Area: Grate surface area measured in square feet, used in estimating the fuel burning rate.

Gravity Warm Air Heating System: See "Warm air heating system."

Gravity Hot Water: Hot water heating systems in which the circulation of water through the system is due to the difference in the density of the water in the supply and return sides of the system.

Gross Output: A rating applied to boilers. It is the total quantity of heat which the boiler will deliver and at the same time meet all limitations of applicable testing and rating codes.

H

Hard Flame: A flame with a hot, tight, well-defined inner cone.

Head: As used in this course, head refers to a pressure difference. See pressure head, pump head, available head.

Header: A piping arrangement for inter-connecting two or more supply or return tappings of a boiler. Also a section of pipe, usually short in length, to which a number of branch circuits are attached.

Heat: A form of energy.

Heat Distributing Units: (See Terminal Units)

Heating Element: (See Terminal Unit)

Heat Flow: (See Heat Loss)

Heating Effect Factor: An arbitrary allowance added to the test output of some types of terminal units when establishing the catalog ratings. This allowance is intended to give credit for improved heat distribution obtained from the terminal unit.

Heat, Latent: The heat which changes the form of a substance without changing its temperature.

Heat, Sensible: Heat which changes the temperature of a substance without changing its form.

Heat Exchanger: Any device for transferring heat from one fluid to another.

Heat Loss: As used in this course, the term applies to the rate of heat transfer from a heated building to the outdoors.

Heat Loss Factor: A number assigned to a material or

construction indicating the rate of heat transmission through that material or construction for a one degree temperature difference.

Heating Element (Electric): A unit assembly consisting of a resistor, insulated supports, and terminals for connecting the resistor to electric power.

Heat Transmission: Any time-rate of heat flow; usually refers to conduction, convection, and radiation combined.

Heat Transmission Coefficient: Any one of a number of coefficients used in calculating heat transmission through different materials and structures, by conduction, convection, and radiation.

Heating Surface: All surfaces which transmit heat from flames or flue gases to the medium being heated.

Heating Unit (Electric): A structure containing one or more heating elements, electrical terminals or leads, electric insulation, and a frame or casing, all assembled together in one unit.

Heating Value: The number of British thermal units produced by the complete combustion at constant pressure of one cubic foot of gas. Total heating value includes heat obtained from cooling the products to the initial temperature of the gas and air and condensing the water vapor formed during combustion.

High Limit Control: A switch controlled by the temperature of the water in the boiler and used to limit burner operation whenever the boiler water temperature reaches the maximum to be permitted. A safety control.

High-Temperature Water System (HTW): A hot water system operating at temperatures over 350°F and usual pressures of about 300 psi.

High Voltage Controls: Also called "line voltage controls." Controls designed to operate at normal line voltage, usually 115 V.

Hot Water Heating Systems: Hydronic systems in which heated water is circulated through the terminal units.

Humidistat: An instrument that is used to regulate the operation of a humidifier to control the amount of humidity in the conditioned air.

Humidity, Absolute: The amount of moisture actually in a given unit volume of air.

Humidity, Relative: A ratio of the weight of moisture that air actually contains at a certain temperature as compared to the amount that it could contain if it were saturated.

Hydrocarbon: Any of a number of compounds composed of carbon and hydrogen.

Hydronics: Pertaining to heating or cooling with water or vapor.

I

Ignition: The act of starting combustion.

Ignition Temperature: The minimum temperature at which combustion can be started.

Ignition Velocity: (See Flame Velocity)

Impingement Target Burner: A burner consisting simply of a gas orifice and a target, with the gas jet from the orifice entraining combustion air in the open and the mixture striking and burning on the target surface. No usual burner body is used.

Inches of Mercury Column: A unit used in measuring pressures. One inch of mercury column equals a pressure of 0.491 lb/in.^{2}.

Inchers of Water Column: A unit used in measuring pressures. One inch of water column equals a pressure of 0.578 oz/in.^{2}. One inch mercury column equals about 13.6 in. water column.

Incomplete Combustion: Combustion in which the fuel is only partially burned.

Indirect Water Heater: A coil or bundle of tubes, usually copper, surrounded by hot boiler water. The domestic water is within the tube and is heated by transfer of heat from the hot boiler water surrounding the tube.

Indoor Design Temperature: The indoor air temperature used when calculating the design heat loss. The indoor design temperature is usually assumed to be 70°F.

Indoor-Outdoor Temperature Difference: The temperature of the indoor air minus the temperature of the outdoor air.

Industrial Buildings: Such buildings as small manufacturing plants, garage, and storehouses.

Induced Draft Burner: A burner which depends on draft induced by a fan or blower at the flue outlet to draw in combustion air and vent flue gases.

Inerts: Non-combustible substances in a fuel, or in flue gases, such as nitrogen or carbon dioxide.

Infiltration: Air leakage into a building from the out-of-doors as a result of wind and indoor-outdoor temperature difference.

Infrared Burner: (Radiant Burner). A burner which is designed to operate with a hot, glowing surface. A substantial amount of its energy output is in the form of infrared radiant energy.

Injection: Drawing primary air into a gas burner by means of a flow of fuel gas.

Input Rate: The quantity of heat or fuel supplied to an appliance, expressed in volume or heat units per unit time, such as cubic feet per hour or Btu per hour.

Input Rating: The gas-burning capacity of an appliance in Btu per hour as specified by the manufacturer. Appliance input ratings are based on sea level operation up to 2,000 feet elevation. For operation at elevations above 2,000 ft, input ratings should be reduced at the rate of 4 percent for each 1,000 ft above sea level.

Instantaneous Water Heater: See tankless water heater.

J

Joint, Expansion, Bellows: An item of equipment used to compensate for the expansion and contraction of a run of pipe. The device is built with a flexible bellows that stretches or is compressed as necessary to accept the movement of the piping.

Joint, Expansion, Slip: A joint in which the provision for expansion and contraction consists of a cylinder that moves in and out of the main body of the device.

Jet Burner: A burner in which streams of gas or air-gas mixtures collide in air at some point above the burner ports and burn there.

L

Lean Mixture: An air-gas mixture which contains more air than the amount needed for complete combustion of the gas.

Lifting Flames: An unstable burner flame condition in which flames lift or blow off the burner port(s).

Liquefied Petroleum Gases: The terms "Liquefied Petroleum Gases," "LPG" and "LP Gas" mean and include any fuel gas which is composed predominantly of any of the following hydrocarbons, or mixtures of them: propane, propylene, normal butane or isobutane and butylenes.

LNG: Liquefied natural gas. Natural gas which has been cooled until it becomes a liquid.

Low Link Control: A switch operated by the temperature of the water in the boiler and used to start the burner at any time the water temperature drops to some prescribed minimum. This control is used if the boiler is supplying domestic hot water as well as heat for the building.

Low-Temperature Water System (LTW): A hot water heating system operating at design water temperatures of 250°F or less and a maximum working pressure of 160 psi.

Low Voltage Control: Controls designed to operate at voltages of 20 to 30 V.

LP Gas-Air Mixtures: Liquefied petroleum gases distributed at relatively low pressures and normal atmospheric temperatures which have been diluted with air to produce desired heating value and utilization characteristics.

Luminous Flame Burner: (See Burner, Yellow Flame)

M

Main: The pipe used to carry water between the boiler and the branches of the terminal units.

Make-up Air: The air which is supplied to a building to replace air that has been removed by an exhaust system.

Make Up Water Line: The water connection to the boiler or system for filling or adding water when necessary.

Manifold: The conduit of an appliance which supplies gas to the individual burners.

Manifold Pressure: The gas pressure in an appliance manifold, upstream of burner orifices.

Manufactured Gas: A fuel gas which is artificially produced by some process, as opposed to natural gas, which is found in the earth. Sometimes called town gas.

Medium, Heating: A substance used to convey heat from the heat source to the point of use. It is usually air, water, or steam.

Medium-Temperature Water System: A hot water system operating at temperatures of 350°F or less, with pressures not exceeding 150 psi.

Methane: A hydrocarbon gas with the formula CH_4, the principal component of natural gases.

Mixed Gas: A gas in which the heating value of manufactured gas is raised by co-mingling with natural or LPG (except where natural gas or LPG is used only for "enriching" or "reforming").

Mixer: That portion of a burner where air and gas are mixed before delivery to the burner ports.

1. Mixer Face. The air inlet end of the mixer head.
2. Mixer Head. That portion of an injection type burner, usually enlarged, into which primary air flows to mix with the gas stream.
3. Mixer Throat (Venturi throat). That portion of the mixer which has the smallest cross-sectional area, and which lies between the mixer head and the mixer tube.

4. Mixer Tube. That portion of the mixer which lies between the throat and the burner head.

Molecule: The smallest portion of an element or compound which retains the identity and characteristics of the element or compound.

Multiple Circuit: A system in which the main, or mains, form two or more parallel loops between the boiler supply and the boiler return.

Multiple Zone: A system controlled by two or more thermostats.

N

Natural Draft: The motion of flue products through an appliance generated by hot flue gases rising in a vent connected to the furnace flue outlet.

Natural Gas: Any gas found in the earth, as opposed to gases which are manufactured.

Needle, Adjustable: A tapered projection, coaxial with and movable with respect to a fixed orifice used to regulate the flow of gas.

Needle, Fixed: A tapered projection, the position of which is fixed, coaxial with an orifice which can be moved with respect to the needle to regulate flow of gas.

Net Rating: A rating applied to boilers. It is the quantity of heat available in Btu/h for the connected load.

Non-Ferrous: Metals other than iron or steel. In heating systems the principal non-ferrous metals are copper and aluminium.

O

Odorant: A substance added to an otherwise odorless, colorless and tasteless gas to give warning of gas leakage and to aid in leak detection.

Oil Burner Relay: A special, multi-purpose control used with oil burners. The device controls the operation of the oil burner and also acts as a safety to prevent operation in the event of malfunction.

One-Pipe Fitting: A specially designed tee for use in a one-pipe system to connect the supply or return branch into a circuit. These fittings cause a portion of the water flowing through the circuit to pass through the terminal unit.

One-Pipe System: A forced hot-water system using one continuous pipe or main from the boiler supply to the boiler return. The terminal units are connected to this pipe by two smaller pipes known as supply and return branches.

Orifice: An opening in an orifice cap (hood), orifice spud or other device through which gas is discharged, and whereby the flow of gas is limited and/or controlled. (See also Universal Orifice)

Orifice Cap (Hood): A movable fitting having an orifice which permits adjustment of the flow of gas by changing its position with respect to a fixed needle or other device extending into the orifice.

Orifice Discharge Coefficient: (See Discharge Coefficient)

Orifice Spud: A removable plug or cap containing an orifice which permits adjustment of the gas flow either by substitution with a spud having different sized orifices (fixed orifice) or by motion of an adjustable needle into or out of the orifice (adjustable orifice).

Outdoor Design Temperature: The outdoor temperature on which design heat losses are based.

Overrating: Operation of a gas burner at a greater rate than it was designed for.

Oxygen: An elemental gas that comprises approximately 21 percent of the atmosphere by volume. Oxygen is one of the elements required for combustion.

P

Packaged Boiler: A boiler having all components, including burner, boiler, controls, and auxiliary equipment, assembled as a unit.

Panel Heating: A heating system in which heat is transmitted by both radiation and convection from panel surfaces to both air and surrounding surfaces.

Panel Radiator: A heating unit placed on or flush with a flat wall surface, and intended to function essentially as a radiator.

Panel Systems: Or radiant system. A heating system in which the ceiling or floor serves as the terminal unit.

Peak Load: The maximum load carried by a system or a unit of equipment over a designated period of time.

pH or pH Value: A term based on the hydrogen ion concentration in water, which denotes whether the water is acid, alkaline, or neutral. A pH value of 8 or more indicates a condition of alkalinity: of 6 or less, acidity. A pH of 7 means the water is neutral.

Pilot: A small flame which is used to ignite the gas at the main burner.

Pilot Switch: A control used in conjunction with gas

burners. Its function is to prevent operation of the burner in the event of pilot failure.

Piping and Pick-up Allowance: That portion of the gross boiler output that is allowed for warming up the heating system and for taking care of the heat emission from a normal amount of piping.

Pitch: The amount of slope given to a horizontal pipe when it is installed in a heating system.

Plenum Chamber: An air compartment maintained under pressure, and connected to one or more distributing ducts.

Pressure Head: The force available to cause circulation of water or vapor in a hydronic system. See head, pump head, available head.

Port: Any opening in a burner head through which gas or an air-gas mixture is discharged for ignition.

Port Loading: The input rate of a gas burner per unit of port area, obtained by dividing input rate by total port area. Usually expressed in terms of Btu per hour per square inch of port area.

Power Burner: (See Burner)

Premixing Burner: A burner in which all, or nearly all, combustion air is mixed with the gas as primary air.

Pressure Burner: A burner in which an air and gas mixture under pressure is supplied, usually at 0.5 to 14 in. water column.

Pressure Regulator: A device for controlling and maintaining a uniform outlet gas pressure.

Pressure Reducing Valve: A diaphragm operated valve installed in the make-up water line of a hot water heating system to introduce water into the system and to prevent the system from possible exposure to city water pressures higher than the working pressure of the boiler.

Pressure Relief Valve: A device for protecting a hot water boiler (or a hot water storage tank) from excessive pressure by opening at a pre-determined pressure and discharging water, or steam, at a rate sufficient to prevent further build-up of pressure.

Primary Air: The combustion air introduced into a burner which mixes with the gas before it reaches the port. Usually expressed as a percentage of air required for complete combustion of the gas.

Primary Air Inlet: The opening or openings through which primary air is admitted into a burner.

Propane: A hydrocarbon gas heavier than methane but lighter than butane. It is used as a fuel gas alone, mixed with air or as a major constituent of liquefied petroleum gases.

psig: Pounds per square inch gauge pressure.

Pulsation: A panting of the flames in a boiler or furnace, indicating cyclic and rapid changes in the pressure in the combustion space.

Pump: A motor driven device used to mechanically circulate water in the system. Also called a circulator.

Pump Head: The difference in pressure on the supply and intake sides of the pump created by the operation of the pump.

Q

Quenching: A reduction in temperature whereby a combustion process is retarded or stopped.

R

Radiant Burner: (See Infrared Burner)

Radiant Heating: A heating system in which only the heat radiated from panels is effective in providing the heating requirements. The term radiant heating is frequently used to include both panel and radiant heating.

Radiation: The transmission of energy by means of electromagnetic waves.

Radiator: A heating unit exposed to view within the room or space to be heated. A radiator transfers heat by radiation to objects within visible range, and by conduction to the surrounding air which in turn is circulated by natural convection; a so-called radiator is also a convector, but the term radiator has been established by long usage.

Radiator (Concealed): A heating device located within, adjacent to, or exterior to the room being heated, but so covered or enclosed or concealed that the heat transfer surface of the device, which may be either a radiator or a convector, is not visible from the room. Such a device transfers its heat to the room largely by convection air currents.

Radiator Valve: A valve installed on a terminal unit to manually control the flow of water through the unit.

Rate: (See Input)

Recirculated Air: Return air passed through the conditioner before being again supplied to the conditioned space.

Reducing Fitting: A pipe fitting designed to change from one pipe size to another.

Regulator: (See Pressure Regulator)

Relay: An electrically operated switch. Usually the control circuit of the switch uses low voltage while the switch makes and breaks a line voltage circuit. However, both the control and load circuits are of the same voltage in some instances.

Relief Opening: The opening in a draft hood to permit ready escape to the atmosphere of flue products from the draft hood in event of no draft, back draft or stoppage beyond the draft hood, and to permit inspiration of air into the draft hood in the event of a strong chimney updraft.

Residential Buildings: Single family homes, duplexes, apartment buildings.

Return Branch: The piping used to return water from a terminal unit to the main, circuit main, or trunk.

Return Main: The pipe used to carry water from the return branches of the terminal units to the boiler.

Return Mains: Pipes or conduits which return the heating or cooling medium from the heat transfer unit to the source of heat or refrigeration.

Return Piping: That portion of the piping system that carries water from the terminal units back to the boiler.

Return Tapping: The opening in a boiler into which the pipe used for returning condensate or water to the boiler is connected.

Reverse Acting Control: A switch controlled by temperature and designed to open on temperature drop and close on temperature rise.

Reverse Return: A two-pipe system in which the return connections from the terminal units into the return main are made in the reverse order from that in which the supply connections are made in the supply main.

Rich Mixture: A mixture of gas and air containing too much fuel or too little air for complete combustion of the gas.

Riser: This generally refers to the vertical portion of the supply or return branches. However, any vertical piping in the heating system might be termed a riser.

Run-Out: This term generally applies to the horizontal portion of branch circuits.

S

Safety Valve: A device for protecting a steam boiler from excessive pressure by opening at a predetermined pressure setting and allowing steam to escape at a rate equal to or greater than the steam generating capacity of the boiler.

Secondary Air: Combustion air externally supplied to a burner flame at the point of combustion.

Series Loop: A forced hot water heating system with the terminal units connected so that all the water flowing through the circuit passes through each series-connected unit in the circuit.

Single Circuit System: A hydronic system composed of only one circuit.

Single Port Burner: A burner in which the entire air-gas mixture issues from a single port.

Soft Flame: A flame partially deprived of primary air such that the combustion zone is extended and inner cone is ill-defined.

Soot: A black substance, mostly consisting of small particles of carbon, which can result from incomplete combustion and appear as smoke.

SNG: Supplementary natural gas. Gases which are manufactured to duplicate natural gas.

Spud: (See Orifice)

Square Foot (Steam): A term used to express the output of boilers and radiation. When applied to boilers, it is 240 Btu/h; when applied to terminal units, it represents the amount of radiation which will emit 240 Btu/h when supplied with steam at 215°F and air at 65°F.

Square Head Cock: A type of valve often used as a balancing valve. In place of the valve handle, the stem is made square. A wrench is used to adjust the valve setting.

Specific Gravity: Specific gravity is the ratio of the weight of a given volume of gas to that of the same volume of air, both measured at the same temperature and pressure.

Spoiler Screw: (Breaker Bolt) A screw or bolt moved in or out of the gas jet in a burner to control primary air injection.

Standard Conditions: Pressure and temperature conditions selected for expressing properties of gases on a common basis. In gas appliance work, these are normally 30 in. of mercury and 60°F.

Static Pressure: The normal force per unit area at a small hole in a wall of the pipe through which the fluid (water) flows.

Steam Heating System: A hydronic system in which steam is circulated through the terminal units.

Sub-circuits: A term applied to circuits taken off of the primary distribution loop of a complex hydronic system.

Supply Branch: The piping used to supply heated water from a main, circuit main, or trunk to the terminal unit.

Supply Main: The pipe used to distribute water from the boiler to the supply branches of the terminal units.

Supply Piping: That portion of the piping system that carries water from the boiler to the terminal units or to the point of use.

Supply Tapping: The opening in a boiler into which the supply main is connected.

System Temperature: The average of the temperatures of the water leaving the boiler and returning to the boiler.

T

Tankless Water Heater: An indirect water heater designed to operate without a hot water storage tank in the system. Also called an instantaneous heater.

Tee: A pipe fitting designed to connect three sections of pipe together. Two of the connections are in line, the third is at right angles to the other two.

Terminal Units: That part of a hydronic system in which heat is transferred from the water to the air in the air conditioned space. Common terminal units include radiators, convectors, baseboard, unit heaters, finned tube, etc.

Thermal Conductivity: A term indicating the ability of a material to transmit heat. Thermal conductivity is the reciprocal of thermal resistance.

Thermal Head: The head produced by the difference in weight of the heated water in the supply side of the system and the cooler water in the return side. This is the only head available to cause circulation of water in a gravity system.

Thermal Radiation: The transmission of heat from a hot surface to a cooler one in the form of invisible electromagnetic waves, which on being absorbed by the cooler surface, raise the temperature of that surface.

Thermal Resistance: The resistance a material offers to the transmission of heat. Insulating materials have high thermal resistance. Materials such as metals have low thermal resistance.

Therm: A unit of heat having a value of 100,000 Btu.

Thermostat: A control (switch) wihch is operated by the temperature of the air.

Throat: (See Venturi)

Tie Rod: The sections of cast-iron sectional boilers are held in tight contact by means of tie rods that pass entirely through the sections.

Transformer: A device designed to change voltage. In heating controls the transformer usually converts line voltage (115 V) to low voltage (24 V).

Trunk: Or trunk main. The section of the main in a multiple circuit system that carries the combined capacity of two or more of the circuits.

Two-Pipe System: A hot-water heating system using one pipe from the boiler to supply heated water to the terminal units, and a second pipe to return the water from the terminal units back to the boiler.

Total Air: The total amount of air supplied to a burner. It is the sum of primary, secondary, and excess air.

Total Pressure: Also called impact pressure. The pressure measured in a moving fluid by an impact tube. It is the sum of the velocity pressure and the static pressure.

Town Gas: (See Manufactured Gas)

Turndown: The ratio of maximum to minimum input rates.

U

Ultimate CO_2: The percentage of carbon dioxide in dry combustion products when a fuel (gas) is completely burned with exactly the amount of air needed for complete combustion. This is the theoretical maximum CO_2 which can be obtained for a given gas in burning the gas in air.

Unit Heater: Also see fan coil. The term applies to a terminal unit designed to heat a given space. It consists of a fan and motor, a heating element, and an enclosure.

Universal Orifice: A combination fixed and adjustable orifice designed for the use of two different gases, such as LPG and natural gas.

Updraft: Excessively low air pressure existing at the outlet of a chimney or stack which tends to increase the velocity and volume of gases passing up the stack.

Up-Feed System: A hydronic system in which the supply main is located below the level of the terminal units.

Unit Ventilator: A terminal unit in which a fan is used to mechanically circulate air over the heating coil. These units are so constructed that both outdoor and room air may be circulated so as to provide ventilation as well as heat. These units may contain a cooling coil for summer operation.

Utility Gases: Natural gas, manufactured gas, liquefied petroleum gas-air mixtures or mixtures of any of these gases.

V

Vapor: The gaseous form of a substance that, under other conditions of pressure, temperature, or both, is a solid or a liquid.

Vapor Barrier: A material that is impervious to the passage of water vapor through it.

Velocity Pressure: Pressure exerted by a flowing gas by virtue of its movement in the direction of its motion. It is the difference between total pressure and static pressure.

Vent: A device, such as a pipe, to transmit flue products from an appliance to the outdoors. This term also is used to designate a small hole or opening for the escape of a fluid (such as in a gas control).

Vents: See air vents.

Ventilation: The introduction of outdoor air into a building by mechanical means.

Vent Gases: Products of combustion from gas appliances plus excess air, plus dilution air in the venting system above a draft hood.

Venturi: A section in a pipe or a burner body that narrows down and then flares out again.

Viscosity: The property of a fluid to resist flow.

W

Water Column: Abbreviated as W.C. A unit used for expressing pressure. One inch water column equals a pressure of 0.578 oz/in.^{2}.

Water Tube Boiler: A steel, hot-water boiler in which the water is circulated through the tubes and the hot gases from combustion of the fuel are circulated around the tubes inside the shell.

Y

Yellow Flame Burner: (See Burner)

Yellow Tips: (Yellow Tipping) The appearance of yellow tips in an otherwise blue flame, indicating the need for additional primary air.

Z

Zone: That portion of a hydronic system, the operation of which is controlled by a single thermostat.

Zoned System: A hydronic system in which more than one thermostat is used. This permits independent control of room air temperature at more than one location.

Zone Valve: A valve, the operation of which is controlled by a thermostat. They are used in hydronic systems to control the flow of water in localized parts of the system, thus making it possible to independently control the temperature in different zones, or areas, of the building.

GLOSSARY OF TECHNICAL TERMS AS APPLIED TO ELECTRICITY AND ELECTRICAL DATA

A

Alternating Current: Abbreviated ac. Current that reverses polarity or direction periodically. It rises from zero to maximum strength and returns to zero in one direction then goes to similar variation in the opposite direction. This is a cycle which is repeated at a fixed frequency. It can be single phase, two phase, three phase, and poly phase. Its advantage over direct or undirectional current is that its voltage can be stepped up by transformers to the high values which reduce transmission costs.

Alternator: A machine which converts mechanical energy into alternating current.

American Wire Gauge: Abbreviated AWG. A system of numbers which designate cross-sectional area of wire. As the diameter gets smaller, the number gets larger, e.g., AWG #14 = 0.0641 in. AWG #12 = 0.0808.

Ammeter: An instrument for measuring the quality of electron flow in amperes.

Ampere: Abbreviated amp. A basic unit which designates the amount of electricity passing a certain point at a specific time.

Ampere-Turn: Abbreviated AT or NI. Unit of magnetizing force pıoduced by a current flow of one ampere through one turn of wire in a coil.

Amplitude: The maximum instantaneous value of alternating current or voltage. It can be in either a positive or negative direction.

Angle of Lag or Lead: Phase angle difference between two sinusoidal wave forms having the same frequency.

Armature: The moving or rotating component of a motor, generator, relay or other electromagnetic device.

Atom: The smallest particle of matter which exhibits the properties of an element.

B

Battery: Two or more primary or secondary electrically interconnected cells.

Break: Electrical discontinuity in the circuit generally resulting from the operation of a switch or circuit breaker.

Breaker Points: Metal contacts that open and close a circuit at timed intervals.

Bridge Circuit: A circuit for determining an unknown value or resistance. A power source and three known values or resistors are interconnected in a series-parallel network along with the unknown value or resistor. If all resistors are equal, a galvanometer bridged across the parallel legs and connected between each set of resistors will give no reading. If the unknown resistor is lesser or greater than the three known resistors, the galvanometer will indicate the degree and direction of the imbalance.

Brush: A conducting material, usually of carbon or graphite, which makes continuous contact with a moving commutator through which current can flow for the remainder of the circuit.

Bus Bar: A primary power distribution source which is connected to a main power source—usually a heavy conductor consisting of rigid bar or strap.

Burnout: Accidental passage of high voltage through an electrical circuit or device which causes damage.

C

Cable: A stranded single-conductor cable or a combination multiple-conductor cable. Cable refers to larger sizes. Small cable is called *stranded wire* or *cord.* Cable can be bare or insulated. Insulated cable may be armored with lead or with steel wire or bands.

Capacitor: Two conductors or electrodes in the form of plates separated from each other by a dielectric (insulator). The capacitor blocks the flow of direct current and partially impedes the flow of alternating current. This impedance decreases as the current frequency increases.

Cation: A positively charged ion, which is attracted to the cathode during electrolysis.

Charge: The electrostatic charge is the quantity of electricity held by a capacitor or insulated object. The charge is negative if it has excess electrons, and positive if it has less electrons than normal.

Choke Coil: A coil which has low ohmic resistance and high impedance to alternating current flow. It allows direct current to pass, but limits the flow of alternating current.

Circuit: A closed circuit provides a complete path for current flow. It is an open circuit if current flow is interrupted. It may contain a number of electrical components designed to perform one or more specific functions.

Circuit Breaker: A device that is actuated by electromagnetic or thermal energy to open a circuit when current exceeds a predetermined setting. A reset method is usually provided.

Coil: A helically or spirally wound conductor, which creates a strong magnetic field with current passage.

Commutator: A ring of copper segments insulated from each other and connecting the armature and brushes of a motor or generator. It passes power into or from the brushes.

Contact: The part of a switch or relay that carries current. Current is controlled by touching or separating contacts.

Conductance: The ability of material to carry electrical current. It is the reciprocal of resistance—therefore, it is expressed in mhos (ohms spelled backward).

Conductivity: The ease of electrical transmission through a substance.

Conductor: Material or substance which readily passes electricity.

Connected Load: The sum of the capacities or continuous ratings of the load-consuming apparatus connected to a supplying system.

Contactor: A device for making or breaking load-carrying contacts by a pilot circuit through a magnetic coil.

Controller: Measures the difference between sensed output and desired output and initiates a response to correct the difference.

Core: A magnetizable portion of a device, which affords an easy path for the magnetic flux lines of a coil.

Counter EMF: Counter electromotive force; the EMF induced in an armature or coil, which opposes applied voltage.

Coulomb: An electrical unit of charge, one coulomb per second equals one ampere or 6.25×10^{18} electrons past a given point in one second.

Current: The transfer of electrical charge through a conductor between points of different voltage potential.

Current Limiter: A protective device in a high amperage circuit.

Cut-In: Switch action in a conducting mode.

Cut-Out: Switch action to the OFF position.

Cycle: A complete positive and a complete negative alternation of voltage or current. Cycles per seconds (CPS) or hertz (Hz) denotes frequency.

D

Delta Connection: The connection in a three-phase system in which terminal connections are triangular similar to the Greek letter delta. With the inductors of one phase opposite the middle of the poles, and assuming maximum current to be induced at this moment, only one half the same current value will be induced at the same moment in the other two phases.

Delta Transformer Connection: Connection with both primaries and secondaries connected in delta grouping.

Delta-Y Connection: Primaries connected in delta grouping and secondaries in star grouping.

Demand: The size of any load generally averaged over a specified interval of time but occasionally instantaneously. Demand is expressed in kilowatts, kilovolt-amperes, or other suitable units. Occasionally used interchangeably with load.

Demand (billing): The demand upon which billing to a customer is based, as specified in a rate schedule or contract. The billing demand need not coincide with the actual measured demand of the billing period.

Demand Charge: The specified charge to be billed on the basis of the billing demand, under an applicable rate schedule or contract.

Demand Meter: A device which indicates or records the demand or maximum demand. (NOTE: Since demand involves both an electrical factor and a time factor, mechanisms responsive to each of these factors are required as well as an indicating or recording mechanism. These mechanisms may be either separate from or structurally combined with one another. Demand meters may be classified as follows: Class 1—Curve-drawing meters, Class 2—Integrated-demand meters, Class 3—Lagged-demand meters.)

Device: A component that primarily carries, but does not utilize, electrical energy, does not perform useful work; e.g., controller or switch.

Dielectric: An insulator-nonconductor. The material between plates of a capacitor, which may be used to store electrostatic energy.

Dielectric Strength: Maximum voltage a dielectric can take without a breakdown.

Direct Current: Abbreviated dc. Electric current that flows only in one direction—undirectional. The varying current in one direction is a pulsating direct current, which may be derived from rectified alternating current.

Disconnecting Switch: A knife switch that opens a circuit after the load has been disconnected by some other means.

Double-Pole Switch: Simultaneously opens and closes two wires of a circuit. Completes a live circuit on each side of the OFF position.

E

E: Symbol for volts.

Eddy Currents: The induced circulating currents in a conducting material that are erected by a varying magnetic field.

Efficiency: A percentage value denoting the ratio of power output to power input.

Electrical Angle: The method specifying the exact instant in an alternating current cycle. Each cycle is 360°; therefore, 180° would indicate one-half cycle, 45° one-eighth cycle.

Electric Circuit: The complete path for electric current.

Electric Field: A magnetic region in space which has force and direction.

Electric Filament Lamp: A light source consisting of a glass bulb containing a filament electrically maintained at incandescence. (NOTE: A lighting unit, consisting of an electric filament lamp with shade, reflector, enclosing globe, housing, or other accessories, is also commonly called a lamp. In such cases, in order to distinguish between the assembled lighting unit and the light source within it, the latter is often called a bulb.)

Electrical Degree: One-360th of an alternating current or voltage cycle.

Electricity: The effect created by interaction of positive and negative electrical charges. Electrostatic attraction and repulsion cause motion and movement of current carriers, which, when given force and direction, become electrical current flow.

Electrodes: The plates in a bath to accomplish electrolysis. Current enters the solution through the anode and leaves through the cathode. Also, each plate in a storage battery.

Electrolysis: Changing the chemical characteristics of a compound through application of an electrical current.

Electrolyte: A solution of a substance (liquid or paste) that is capable of conducting electricity.

Electromagnet: A temporary magnet that creates a magnetic field only during current flow.

Electromotive Force: Abbreviated EMF. The force or electrical pressure (voltage) that produces current flow. The difference in potential electrical energy between two points.

Electron: The elemental negative charge of electricity.

Electrostatic: Electricity at rest—static electricity.

Electrostatic Charge: The electrical energy located on the surface of insulating material or in a capacitor because of an excess or deficiency of electrons.

Energy Levels: The energy states of electrons in an atom.

F

Farad: The unit of capacitance. The storage in a capacitor of one coulomb when one volt is applied.

Feedback: The transfer of energy output back to input.

Field: The space involving the magnetic lines of force.

Fluorescent Lamp: An electric discharge lamp in which the radiant energy from the electric discharge is transferred by suitable materials (phosphors) into wave lengths giving higher luminous efficiency.

Flux: The electric or magnetic lines of force in a region.

Frequency: The number of complete cycles or vibrations in a unit of time.

Four-Wire, Three-Phase Transformer: The secondaries of three transformers are star connected, the fourth wire is connected from a neutral point. The voltage between any main wire and the neutral will be 57% of the voltage between any two main wires.

Fuse: An element designed to melt or dissipate at a predetermined current value, and intended to protect against abnormal conditions of current.

G

Galvanic: Current generated chemically, which results when two dissimilar conductors are immersed in an electrolyte. In the closed electrical circuit, current will flow, which will oxidize the most easily oxidized conductor.

Gang: Mechanical connection of two or more circuit devices so they may be varied simultaneously.

Galvanometer: An instrument for measuring small dc currents.

Gauss: Electromagnetic unit of flux density: one maxwell per square centimeter.

Generator: A machine that converts mechanical energy into electrical energy. See *alternator*.

Ground: Intentional or accidental connection between an electrical circuit and earth onto a common point of zero potential.

H

Henry: Abbreviated H. The basic unit of inductance, one henry results from the current variation of one ampere per second and produces one volt.

Horsepower: Abbreviated hp. A unit of power equivalent to raising 33,000 lbs one foot in one minute, or 550 ft. lbs per second, or 776 W. of electrical power.

Hysteresis: A lag—subjecting ferromagnetic material to a varying magnetic field causes a lagging of the magnetic flux behind the magnetizing force.

I

Illumination: The density of the luminous flux on a surface; it is the quotient of the flux by the area of the surface when the latter is uniformly illuminated. (NOTE: The term illumination is also commonly used in a qualitative or general sense to designate the act of illuminating or the state of being illuminated. Usually the context will indicate which meaning is intended, but occasionally it is desirable to use the expression, level of illumination, to indicate that the quantitative meaning is intended.)

Impedance: Total opposition to alternating current flow. It can consist of any combination of inductive reactance, capacitive reactance, and resistance. Z is the symbol for impedance and the ohm its unit.

Induced: Current or voltage in a conductor which results when the conductor is moved perpendicularly to a magnetic field or subjected to a varying magnetic field.

Inductance: The characteristic of an alternating current circuit to oppose a change in current flow. The change in current through the circuit causes a change in the magnetic field. The change in magnetic field induces a counter-voltage, which tends to oppose the change in current. It is a fly-wheel effect in which the counter-voltage causes current retardation with current build-up and prolongs current flow when current decreases. It is symbolized by the letter L.

Induction: The act that produces electrification, magnetization, or induced voltage in an object by exposure to a magnetic field.

Inductive Reactance: Opposition, measured in ohms, to an alternating or pulsating current.

In Phase: The condition existing when two waves of the same frequency have their maximum and minimum values of like polarity at the same instant.

Interlock: A safety device that allows power to a circuit only after a predetermined function has taken place.

Ion: An atom or molecule that has more or less electrons than normal. The negative ion has more electrons than normal and the positive ion less.

IR Drop: Voltage drop resulting from current flow through a resistor.

J

Joule: Abbreviated J. The unit of energy or work resulting from one ampere of current flowing through one ohm of resistance in one second.

Jumper: A conductor used to bypass a switch or a break in a circuit or to make electrical connection between terminals.

Junction: A place in the circuit where two or more wires are joined.

Junction Box: A metal box into which cables or wires are inserted and joined.

K

Kilowatt: A unit for measuring electrical power. One kilowatt is equal to 1,000 W. One kilowatt is equal to 3,413 Btu per hour.

Kilowatt-Hour: A unit for measuring electrical energy. One kilowatt-hour is equal to 3,413 Btu.

Kilovolt-Ampere: Abbreviated kva. The unit of apparent power as differentiated from kilowatts or true power.

Kirchhoff's Current Law: A basic electrical law stating that the sum of all the currents flowing into a point in the circuit must be equal to the sum of the currents flowing away from the point.

Kirchhoff's Voltage Law: States that the sum of all the voltage rises in a complete circuit must equal the sum of the circuit's voltage drops.

L

Lag: The time one wave is behind another of the same frequency—usually expressed in electrical degrees.

Laws of Electric Charge: Like charges repel, unlike charges attract.

Laws of Magnetism: Like poles repel, unlike poles attract.

Lead: The phase of one alternating quantity that is ahead of the other measured in angular degrees. Also a wire connection.

Left-Hand Rule: Used to determine the direction of magnetic lines of force around a current-carrying wire. If the fingers of the left hand are placed so the thumb points in the direction of electron flow, the fingers will point in the direction of the magnetic field.

Linear: A proportional ratio of change in two related quantities. Output that varies in direct proportion to input.

Line Drop: The voltage drop between two points on a power or transmission line due to leakage, resistance, or reactance.

Line of Force: A line in an electric or magnetic field that shows direction of force.

Line Voltage: Voltage existing at wall outlets or terminals of a power line system above 30 V.

Load: Power that is being delivered by any power-producing source.

Load Factor: The ratio of the average load in kilowatts supplied during a designated period to the peak or maximum load occurring in that period. Typical formula:

$$\text{Load factor} = \frac{\text{Kwhr supplied in period}}{\text{Peak kw in period} \times \text{hours in period}}$$

The kilowatt-hours and peak generally are on a net output basis. The peak generally is for a 60-minute demand interval.

M

Magnetic Circuit: The closed path of magnetic lines of force.

Magnetic Field: The space in which a magnetic force exists.

Magnetic Flux: The sum of the lines of force issuing from the magnetic pole.

Magnetic Hysteresis: Internal friction resulting from subjecting a ferromagnetic substance to a varying magnetic field.

Magnetic Poles: Field concentrations in a magnet consisting of a north and south pole of equal strength.

Magnetize: Molecular rearrangement in a material, which converts it into a magnet.

Mega: Prefix for one million; e.g., megohm—one million ohms.

Micro: Prefix for one-millionth; e.g., microampere—one-millionth of an amp; microfarad—one-millionth of a farad.

Micron: A unit of length. One-thousandth of a millimeter or one-millionth of a meter.

Milli: A prefix for one-thousandth; e.g., milliampere—one-thousandth of an ampere.

Modulate: Output response, which varies proportionally with the input signal.

Module: A combination of components in a circuit that are in a unitized package and capable of performing a complete function.

Molecule: The smallest particle of a substance that retains all the characteristics of the substance.

Mutual Inductance: A circuit condition that exists when the relative positions of two inductors cause magnetic lines of force from one to link with the turns of the other.

N

National Electrical Code: Abbreviated NEC. A code of electrical rules based on fire underwriters' requirements for interior electric wiring. Wiring must be in conformance to NEC rules to meet insurance and municipality requirements.

Negative: The terminal or electrode with excess electrons.

Network: Two or more interconnected electrical circuits; an electrical distribution system.

Neutron: An atomic particle in the nucleus having the weight of a proton but having no electrical charge.

Normally Closed: Switch contacts closed with the circuit deenergized.

Normally Open: Switch contacts open with the circuit deenergized.

Null: Zero.

O

Ohm: A unit of resistance; the resistance that allows one ampere to flow with the potential of one volt.

Ohm's Law: Current is directly proportional to voltage and inversely proportional to resistance.

Open Circuit: Noncontinuous circuit.

Oscillator: An electrical component that generates an alternating voltage.

Overload Protection: A component to interrupt current when it reaches a predetermined high.

P

Panelboard: A single panel, or a group of panel units, designed for assembly in the form of a single panel; including buses, and with or without switches and/or automatic overcurrent protective devices for the control of light, heat, or power circuits of small individual as well as aggregate capacity; designed to be placed in a cabinet or cutout box placed in or against a wall or partition and accessible only from the front.

Parallel: Circuit connected so current has two or more paths to follow.

Permeability: The ease with which magnetic lines of force can flow through a material as compared with air, which is considered unity.

Phase: Angular relationship between waves and expressed in degrees.

Polarity: The condition denoting direction of current flow; the condition of being positive or negative, or having a magnetic north and south pole.

Pole: The part of a magnet where flux lines are concentrated or where they enter and leave the magnet; also, the electrode of a battery.

Polyphase: Two or more alternating currents that differ in phase by a predetermined number of degrees.

Positive: A point of attraction for electrons.

Potential: The amount of voltage or charge between points of a circuit.

Potentiometer: Abbreviated Pot. A resistor with three contacts. One contact is on a movable arm, which can move across the resistor. This adds resistance to one leg of a Wheatstone bridge circuit while simultaneously decreasing resistance in the other, thereby varying current to each leg of the bridge.

Power: The rate of doing work.

Power Factor: The rate of actual power as measured by a wattmeter in an alternating circuit to the apparent power determined by multiplying amperes by volts.

Primary Voltage: The voltage of the circuit supplying power to a transformer is called the primary voltage, as opposed to the output voltage of load-supply voltage which is called secondary voltage. In power supply practice the primary is almost always the high-voltage side and the secondary, the low-voltage side of a transformer.

Pull-In Voltage: The voltage value that causes the relay armature to seat on the pole face.

Q

Quick Disconnect: Quick attaching and releasing connecter halves.

R

Reactance: Opposition to alternating current by either inductance or capacitance or both. Symbolized by X and measured in ohms.

Reed Switch: Two or more highly conductive reeds encapsulated in a glass enclosure. Energy from a magnetic field will cause the reed contacts to open or close simultaneously.

Relay (Contactor): A two-circuit device. A pilot duty circuit controls a load-carrying circuit.

Reluctance: The opposition a material offers to magnetic lines of force. Equals the magnetomotive force divided by the magnetic flux.

Resistance: The opposition to current flow by a physical conductor when voltage is applied.

Resistor: A circuit element that offers resistance to current flow.

Retentivity: A material's ability to hold magnetism.

Rheostat: An adjustable or variable resistor.

Rotor: A rotating armature or member of an electrical motor or generator.

S

Secondary Voltage: The output, or load-supply voltage, of a transformer or substation is called the secondary voltage.

Self-Induction: The process by which an EMF is induced in the circuit by its own magnetic field.

Series: A circuit with one continuous path for current flow.

Series Wound: A motor or generator with armature and field windings wired in series.

Service: The conductors and equipment for delivering energy from the electricity supply system to the wiring system of the premises served.

Service Conductors: That portion of the supply conductors which extends from the street main or duct or from transformers to the service equipment of the premises supplied. For overhead conductors this includes the conductors between the last pole or other aerial support and the service equipment.

Service Drop: That portion of overhead service conductors between the last pole or other aerial support and the first point of attachment to the building.

Service Equipment: The necessary equipment, usually consisting of circuit breaker or switch and fuses, and their accessories, located near point of entrance of supply conductors to a building and intended to constitute the main control and means of cutoff for the supply to that building.

Service Raceway: The rigid metal conduit, electrical metallic tubing, or other raceway, that encloses service entrance conductors.

Servo: A control mechanism that converts a small force into a greater force.

Servo Mechanism: A closed-loop system which initiates an input signal with deviation from a desired condition; the signal is fed back into the control system until a continued response eliminates the signal.

Short Circuit: A low-resistance connection (usually accidental and undesirable) between two parts of an electrical circuit.

Shunt: A part connected in parallel with another part.

Sine: In a right-angle triangle the ratio of the opposite of any angle to the hypotenuse.

Single Phase: A single alternating voltage.

Single-Pole Switch: A switch with movement from an OFF position to a live contact.

Sinusoidal: Current that varies in proportion to the sine of an angle or time function; e.g., ordinary alternating current.

Snap Switch: One with contacts that make and break quickly through mechanical linkage.

Solenoid: A movable plunge activated by an electromagnetic coil.

Space Charge: The electrical space charge resulting from ions and electrons; the cloud of electrons around a hot vacuum-tube cathode.

Stator: The stationary part of the magnetic circuit of a motor or generator.

Switch: A mechanical device for making or breaking an electrical circuit.

Switchboard: A large single panel, frame, or assembly of panels, on which are mounted, on the face or back or both, switches, overcurrent and other protective devices, buses, and usually instruments. Switchboards are generally accessible from the rear as well as from the front and are not intended to be installed in cabinets.

T

Terminal: A point of connection for electrical conductors.

Thermal Cutout: An overcurrent protective device which contains a heater element in addition to and effecting a renewable fusible member which opens the circuit. It is not designed to interrupt short circuits.

Thermocouple: A junction of two dissimilar metals that develops a voltage when heated.

Thermoelectric: Conversion of heat to electrical energy or vice versa.

Three-Phase Delta-Connected System: In wiring, this arrangement employs three wires. Assuming 100 amps. and 1,000 Vs in each phase winding, the pressure between any two conductors is the same as the pressure in the winding, and the current in any conductor is equal to the current in the winding multiplied by the square root of 3, that is, $100 \times 1.723 = 173.2$ amps., or, disregarding the fraction, 173 amp.

Three-Phase Delta Grouping: A method of grouping a three-phase winding, in which the three circuits are connected together in the form of a triangle, and the three corners are connected to the three terminals.

Three-Phase Four-Wire System: A three-phase, three-wire system having a fourth wire connected to the neutral point of the source, which may be grounded.

Three-Phase Y (Star)-Connected Systems: In wiring, there are systems with star connections employing three wires or four wires. Assuming winding, the pressure between any two conductors is equal to the pressure in one winding multiplied by $\sqrt{3}$, that is, 1,000 $\times$ 1.732 $=$ 1,732 Vs. The current in each conductor is equal to the current in the winding, or 100 amps.

Three-Phase Y (Star) Grouping: A method of grouping a three-phase winding in which one end of each of the three circuits is brought to a common junction, usually insulated, and the three other ends are connected to three terminals. It is commonly called a Y connection or grouping, owing to the resemblance of its diagrammatic representation to the letter Y.

Three-Phase System Wiring: There are various ways of arranging the circuit for three-phase current giving numerous three-phase systems, which may be classed:

1. With respect to the number of wires used as: (a) three wire; (b) four wire.
2. With respect to the connections as: (a) star; (b) delta; (c) star delta; (d) delta star.

Transformer: An electromagnetic device, with two or more coils linked by magnetic lines of force, used to increase or decrease ac voltage. With voltage increase, current is decreased, and vice versa.

Transmission Lines: A conductor or system of conductors for the purpose of carrying electrical energy from source to load.

U

Underwriters Laboratories: Abbreviated UL. Underwriters Laboratories, Inc. maintain and operate laboratories for the examination and testing of devices, systems, and materials.

Universal Motor: A motor that can operate on ac or dc current.

Utility Rate Structure: A utility's approved schedules of charges to be made in billing for utility service rendered to various classes of customers.

V

Vector: A line that represents the direction and magnitude of a quantity.

Volt: The unit of electrical potential or pressure. The electromotive force (EMF) that will send one ampere through one ohm of resistance.

Voltage Relay: One that functions at a predetermined voltage value.

Volt-Ampere: Abbreviated VA. Volts times amperes.

W

Watt: The electrical unit of power or rate of doing work. In its simplest terms, it is the rate of energy transfer equivalent to one ampere flowing under the pressure of one volt at unity power factor. It is analogous to horsepower or foot-pounds per minute of mechanical power. One horsepower equals 746 watts.

Watt-hour Meter: An electricity meter which measures and registers the integral, with respect to time, of the active power of the circuit in which it is connected. This power integral is the energy delivered to the circuit during the interval over which the integration extends, and the unit in which it is measured is usually the kilowatt-hour.

Waveform: The graphical representation of an electromagnetic wave, which shows variations in amplitude plotted against time.

Wheatstone Bridge: A bridge with four main arms containing a resistor in each arm. With known values for three resistors, the value for the unknown resistor can be computed.

Winding: One or more wire turns, which form a continuous coil or conductive path.

ANSWERS TO PROBLEMS

ANSWERS TO PROBLEMS IN REFRIGERATION

CHAPTER 1

1. China

2. 1900

3. True

4. 13

5. False

6. False

7. Compressor, condenser, evaporator, and metering device

8. absorbs

9. Compressor and condenser

10. Compressor and evaporator

CHAPTER 2

1. Solid, liquid or gas

2. Molecule

3. Intensity

4. It is the point where, theoretically, there is no molecular action and there is a complete absence of heat. This is believed to occur at about −460°F or −273°C.

5. It is the amount of heat necessary to change the temperature of one pound of water one degree Fahrenheit at sea level.
6. $°C = \frac{5}{9}(F - 32) = \frac{5}{9} \times 36 = 20°C$
7. $Btu = W \times TD = 100 \times 50 = 5{,}000$ Btu
8. $Btu = W \times TD$ or $TD = Btu/W = \frac{750}{15} + 72 = 122°F$
9. 1.0
10. False
11. Conduction, radiation and convection
12. Poor

CHAPTER 3

1. 561.6 lbs. and 62.4 psf or 0.433 psi
2. 3.464 psi and 1.732 psi
3. 131.67 psf or 0.914 psi
4. 75 lbs. and 37.5 psi
5. Approximately 125 psi
6. (a) 2 ft. (b) 0.866 psi (c) 0.433 psi
7. 183.8 psi
8. 382.3 psi
9. 5.46 cu. ft.
10. 98.57 psig

CHAPTER 4

1. Copper
2. False
3. K or L
4. soft and hard drawn
5. Mechanical couplings and heat bonding
6. 45 degree angle
7. Single and double
8. Soft and hard soldering (brazing)
9. metal alloy, 95% tin and 5% antimony
10. 1100 to 1200°F
11. False
12. Oxygen and acetylene gas mixture

CHAPTER 5

1. Flare nut wrench
2. True—6 and 12 points are the most common.
3. False
4. Allen wrenches
5. Flat blade tip and Phillips tip
6. Single and Double Cut
7. False
8. 1/thousandth (0.001)
9. False
10. Occupational Safety and Health Act

CHAPTER 6

1. Increase the surface area of the coil by adding fins and use of a fan or blower to increase movement of air across coil.
2. When the evaporator is to operate in a space where high humidity must be maintained.
3. Submerged coils, shell-and-tube and tube-in-tube configurations, and Baudelot coolers.
4. Reciprocating, rotary, centrifugal.
5. The driving motor is sealed in the same housing as the compressor in a hermetic unit; the driving motor is outside the housing of an open-type compressor, connected to it by a shaft projecting through the housing.
6. To dissipate the heat picked up by the refrigerant in the evaporator.
7. Air-cooled, water-cooled, and evaporatively cooled.
8. Automatic expansion valve, thermostatic expansion valve, capillary tube, low-side float, and high-side float.
9. The capillary tube, because it has no moving parts to wear out and require replacement.
10. So that the refrigerant oil will move along with the vapor on the return trip to the compressor.

CHAPTER 7

1. Chemical and physical.
2. Flammability, toxicity, and explosiveness.
3. A soap solution, a halide torch or an electronic leak detector.
4. Green.
5. It reduces possible damage resulting from friction.
6. A means of telling how thick an oil is.
7. The splash system and the pressure or forced feed system.
8. The return of the oil to the compressor crankcase positively.
9. Paraffin base, Naphthene base, or Mixed base.
10. When hermetic compressors are used, the suction vapor and oil vapor traveling with it usually passes across the insulated windings of the motor.

CHAPTER 8

1. It is the amount of heat that each lb. of refrigant absorbs in the evaporator.
2. The heat content of the liquid refrigerant entering the metering device and the total heat content of the vapor leaving the evaporator.
3. Weight expressed in lb./min./ton is found by dividing the N.R.E. into 200 btu/min the equivalent of one ton of refrigeration.
4. Saturated vapor is that which is in contact with the liquid it was generated from in the evaporator and it is at the same temperature as the liquid; whereas superheated vapor is not in contact with the generat-

ing liquid, and has a temperature that is higher than the saturated vapor at the same pressure.

5. Its total displacement and its volumetric efficiency.

6. It is a graphic representation on a Mollier diagram of the various processes within a refrigeration circuit or cycle.

7. The central area is that which shows changes in the state of the refrigerant, to the left is the subcooled liquid area, while the superheated vapor is at the right of the chart.

8. They are *enthalpy* or *heat content*, *entropy*, *temperature*, *pressure*, and *volume*.

CHAPTER 9

1. Glass and Dial type

2. Thermistor

3. above and below

4. Saturated temperature

5. testing manifold

6. purging

7. False—*never*

8. Deep vacuum and triple evacuation

9. 1/25.400 of an inch

10. 500 microns

11. False

12. Charging cylinder

CHAPTER 10

1. Pressure

2. above

3. Bottom

4. fixed

5. High

6. low

7. falls

8. False

CHAPTER 11

1. heat

2. lower

3. Water

4. Lithium bromide

5. Generator, Absorber, Evaporator and Condenser

6. evaporator

7. generator

8. non-condensible

9. manual

10. True

CHAPTER 12

1. protons and neutrons
2. negative
3. True
4. chemical
5. Electromotive force
6. volt
7. True
8. Ohm
9. ampere
10. series and parallel
11. True
12. True
13. False
14. inductive reactance
15. Voltage
16. Capicator

CHAPTER 13

1. True
2. Sine curve
3. False—60 cycles in one second.
4. True
5. 120 and 240 volts—Single Phase
6. Commercial and industrial customers
7. Conductor
8. National Electrical Code
9. Mils
10. Raceway
11. electrical conductors
12. True

CHAPTER 14

1. Step-up and step-down transformers.
2. In a step-up transformer the secondary coil has a greater number of turns of wire than the primary coil. The step-down type has fewer turns on the secondary coil.
3. If the transformer is a step-down type, the voltage of the secondary coil will be lower and the amperage higher than that of the primary coil. The opposite is true of a step-up transformer: The voltage of the secondary coil will be higher (but the amperage lower) in the same ratio as the number of turns on the secondary coil to the number of turns on the primary coil.
4. (a) to start motors, and (b) to protect motors and equipment from thermal overload.
5. Single phase and polyphase (3 phase).
6. A force or combination of forces that produces, or tends to produce, rotation.
7. A second set of windings on the stator with more turns than the primary or running winding. The purpose of the second set is to provide a second phase so that the motor will start always rotating.
8. (a) Voltmeter; (b) ammeter; (c) ohmeter.
9. These are used when checking out an electrical circuit having low voltage and current measurements.
10. To prevent damage to the motor caused by overheating.

CHAPTER 15

1. Ammonia (NH_3).
2. By maintaining the required percentage of relative humidity.
3. About 0° F.
4. (a) Those which cause food to deteriorate and (b) those used in preservation of foods.
5. Any delay after harvesting in placing fruits and vegetables in refrigeration permits them to continue their ripening process, and may reduce the time they may be preserved in storage.
6. "Refrigerator cars" of an earlier era relied on ice, which was packed in bunkers in the cars. Melting of the ice removed the heat from the interior of the cars and their contents.
7. Devices which may be cooled to a low temperature and which later may be used, as in a refrigerated truck, as evaporators for cooling during a subsequent period.
8. The surface of a large number of flakes will cool the liquid more quickly.
9. Most products have "optimum" recommended storage conditions with respect to temperature and humidity. If products with different requirements are stored under common conditions, some of them will be affected differently from others.
10. "Mixed storage," which usually calls for compromise temperature and humidity conditions which may not be optimum for all of the products stored in the area.

CHAPTER 16

1. (a) Bare tube, (b) finned tube, and (c) plate.
2. Fins provide additional surface area, insuring a higher degree of heat transfer from air surrounding the coil, thus increasing its capacity.
3. Conduction.
4. Baffles help direct the air currents set up by convection so that they move in paths designed to give the optimum heat transfer.
5. Whenever it appears that the natural convection currents are not providing, or will not provide, adequate exposure of air in the refrigerated space to the evaporator.
6. Refrigerating effect (R.E.) is a measure of the amount of heat each pound of refrigerant picks up as it travels through the evaporator.
7. Frost acts as an insulator, reducing the heat transfer between the air and the refrigerant in the coil, thus reducing the efficiency of the system.
8. A hot-gas defrost removes frost from the evaporator while the compressor is running, but when the gas gives up its heat in the process it condenses into a liquid, a large "slug" of which can flow down the suction line and damage the compressor.
9. Unless the system is constructed so that oil will return with the refrigerant to the compressor, this vital part of the system may be damaged because of improper lubrication.
10. In addition to the fact that it doesn't help the compressor when it is in the evaporator, oil in the evaporator takes up space that should be used for vaporization of the refrigerant, decreasing the efficiency of the system.

CHAPTER 17

1. reciprocating, rotary, helical and centrifugal
2. False
3. open drive and hermetic
4. False

5. Field Serviceable and Full Hermetic
6. Heat
7. 3%
8. Energy Efficiency Ratio
9. Capacity
10. high ratio
11. life expectancy
12. True

CHAPTER 18

1. air cooled, water cooled and evaporative
2. True
3. natural draft and forced draft
4. Latent
5. wet bulb
6. Natural draft and forced draft
7. True
8. Drift
9. 5
10. 2″
11. 6.5 feet of pipe
12. feet of head

CHAPTER 19

1. Poor
2. False
3. 1500 feet per minute
4. 3 psi
5. False
6. 3 psi
7. True
8. 1 psi
9. away
10. False
11. True
12. color

CHAPTER 20

1. operating, regulators, safety and application
2. automatic expansion valve, thermostat expansion valve, capillary tube, low side float, high side float
3. superheat
4. False
5. bi-metal strips
6. Bourdon
7. safety
8. False
9. Evaporator
10. manual or automatic

CHAPTER 21

1. Heat transmission, air infiltration, Product Loads, Supplementary Heat.
2. sensible and latent
3. high
4. 875 Btu/h
5. loose fill, flexible, rigid, reflective or foamed-in-place
6. False
7. air infiltration load
8. True
9. False
10. water

CHAPTER 22

1. National Fire Protection Association
2. False
3. False
4. Scale on copper tubing
5. sweating
6. canvas
7. pinch-off tube
8. False
9. False—must be in the range of the operating chart.
10. undercharge
11. False
12. False

CHAPTER 23

1. High head pressure
2. False
3. False
4. False
5. low suction pressure
6. True
7. overheating
8. overfeeding, underfeeding, erratic operation
9. more
10. True
11. a leakage
12. False—it could be empty.

ANSWERS TO PROBLEMS IN AIR-CONDITIONING

CHAPTER 1

1. Fire, clothing, caves.
2. Dr. Willis Carrier
3. Freon (F-12 now R-12)
4. 1935
5. Unitary and Central Station
6. Air Conditioning & Refrigeration Institute
7. Air Conditioning & Refrigeration Equipment Manufacturers
8. Eight to nine billion dollars
9. 75%
10. 11

CHAPTER 2

1. Temperature, moisture, air movement, cleanliness, and ventilation.
2. 98.6°F dry bulb
3. Humidity
4. The combination of temperature and humidity in which most people would feel comfortable.
5. 68° to 70°F
6. 78° to 80°F and 50% relative humidity

7. FPM (feet per minute)
8. 36 lbs.
9. Smoke, dust, dirt, chemicals, and plant pollen
10. 95%

CHAPTER 3

1. The science and understanding of the properties of air.
2. 77% nitrogen, 23% oxygen with traces of other rare gases.
3. Wetted sock on a thermometer usually called a "sling psychrometer".
4. The temperature at which condensation will occur.
5. The actual weight of water vapor in the air measured in pounds or grains per pound of dry air.
6. 100%
7. Sensible and latent heat
8. Psychrometric Chart
9. 0.113 lbs. per pound dry air
10. DB = 81°F, WB = 68.7°F, Humidity Ratio = 0.0122 lbs. per pound dry air, DP = 62.5°F.
11. Total Heat Change = 13.0 BTU per pound of dry air; Sensible heat reduction = 4.0 BTU per pound of dry air; Latent heat reduction = 9.0 BTU per pound of dry air; Humidity ratio reduction = 0.082 lbs. per pound of dry air

CHAPTER 4

1. Ranch, split level, and two story
2. Basement, crawl space, slab, and post and pier.
3. Floor joists
4. Studs
5. $1\frac{1}{2}'' \times 3\frac{1}{2}''$
6. False
7. Trusses
8. 4/12
9. Double or insulated
10. Built-up roof, bonded
11. Drop or suspended
12. Architect's scale

CHAPTER 5

1. Sensible and latent.
2. Transmission load
3. "Resistance (R) value"
4. Degrees Fabrenheit temperature difference per Btu per hour per square foot.
5. Overall heat transfer coefficient, area, and temperature difference
6. High
7. Flexible blankets (rolls), batts, and loose fill
8. 50%
9. Ventilation
10. External and internal shading devices
11. People, lights, cooking, and bathing
12. 3.4 Btu/h

CHAPTER 8

1. Coal, oil, gas, and electricity
2. #1 and #2, approximately 140,000 Btu/Gallon
3. Methane
4. Liquid Petroleum, Propane, and Butane
5. Heavier
6. 10 parts air to 1 part gas
7. 3400 Btu's
8. Baseboard
9. Upflow high boy, low boy, counterflow, and horizontal
10. False
11. Combustible surfaces
12. False—very important
13. Difference between temperature of air in and air out.
14. No, unless there is a choice of two equal control points
15. High pressure atomizing gun type
16. More
17. Hot water
18. Sequencer control

CHAPTER 9

1. Throw-away
2. 75%
3. False! It can cause most service complaints.
4. Ionizing type
5. 12,000 volts d.c.
6. Air flow (CFM)
7. A very small unit of measure—1/25,400 of an inch
8. 50 watts
9. Generally no, unless it is self-washing variety
10. Soaking in a tub or sink with detergent cleaner.

CHAPTER 10

1. 80°F DB and 67°F WB indoor, 95°F DB outdoors
2. Approximately 0.75
3. 115°F DB
4. Mild weather operation
5. Low ambient operation
6. Cycling outdoor condenser fans
7. Parallel flow
8. Glycol or antifreeze solution
9. Up-flow, horizontal, and down-flow
10. Quick-connect, compression, and flared
11. Vertical lift
12. Change from single phase to three phase current
13. Cylinder unloading and hot gas bypass
14. To insure proper oil flow and to protect compressor against slugging
15. Air lock

CHAPTER 11

1. Incremental systems
2. Single-zone and multi-zone
3. Economizer cycle
4. Electronic (spark or glow plug)
5. All external electrical power to a single connection on the unit.
6. Supply and return air through a combination diffuser.
7. 2 to 2.5
8. Reversing
9. The defrost cycle
10. Coefficient of Performance (as expressed in Item 7)
11. The outdoor temperature at which the heat pump (compressor only) heating capacity balances the structure heat loss . . . beyond which supplemental heating will be needed.
12. Almost always
13. Cooling

CHAPTER 12

1. Applied systems, applied equipment, applied machinery, and engineered systems.
2. Field
3. Refrigerant to water to air
4. 40 to 45°F
5. To permit air to go over or by-pass the heating, cooling coils, and humidifiers.
6. Reciprocating
7. The number of times the water will travel the length of shell before being discharged.
8. Fouling factor is an assumed safety margin anticipating scale build-up on water tubing.
9. Direct driven or gear driven
10. Sections

CHAPTER 13

1. Mechanical pressure or temperature change
2. Turbulence, changes in shape or direction, and air control devices
3. Static and velocity pressure
4. Manometer
5. Anemometer
6. 16″ diameter round duct
7. Approximately 975 feet per minute
8. 12″ × 18″
9. Velocity in FPM

	Main	*Branch*
Supply	1000	600
Return	800	600

10. Forward curved multi-blade wheel

CHAPTER 14

1. Individual radial runs, extended plenum, loop perimeter
2. Degree days
3. Equal friction, velocity reduction, static regain
4. Not acceptable
5. The starting duct chamber on top of a furnace or air conditioner.
6. In feet of equivalent length of straight duct.
7. No
8. Supply = 0.3; return = 0.1
9. 0.15″ W.C.
10. Wet coil

CHAPTER 15

1. Wrong. It is most important.
2. 25—35 feet per minute
3. High side wall, low side wall, baseboard, floor, and ceiling
4. CFM (cubic feet per minute)
5. Spread is horizontal distance of the air pattern measured in feet to a terminal velocity of 50 FPM.
6. Throw is the vertical distance of the air pattern measured in feet to a terminal velocity of 50 FPM.
7. Terminal velocity—the average air stream velocity at the end of the throw (usually 50 FPM).
8. Drop is the vertical distance below the air outlet of a projected air stream at the end of its throw.
9. Wrong—very poor
10. 12″ × 6″ size

CHAPTER 16

1. Low, medium, high pressure
2. Free-blow
3. Single-zone, multi-zone, constant volume, variable volume, and double duct.
4. No
5. 2000 to 5000 FPM
6. 38° to 40°F
7. Variable volume or constant volume
8. Single or double induction
9. True
10. True

CHAPTER 17

1. Controller and controlled device
2. Thermostat, humidistat, and pressure stat
3. Relay, valve, and damper
4. Step down transformer
5. 0.58 amps
6. Two dissimilar metals welded together that have

different coefficients of expansion causing a change in shape with change in temperature

7. True

8. Electrical resistors

9. Fixed or variable

10. Relative humidity

11. Human hair or nylon strips

12. Electrical resistance

CHAPTER 18

1. 230 volts, single phase, 60 hertz
2. 100 to 200 amps, 200 amps would be assumed for an all electric home.
3. 3%
4. Legends and symbols
5. L_1 (black) and L_2 (white)
6. L_1 (hot) and L_2 (neutral)
7. Panel and schematic diagrams
8. To stop the heating process if air gets too hot.
9. Cycles the fan on and off in response to air temperature.
10. Ladder diagrams have numerals in the margin to identify line numbers and component locations in the system.
11. Indoor fan switch contacts in lines 9 and 10
12. Line 37
13. Broken or dash lines

CHAPTER 19

1. Single-phase to three-phase power.
2. Equipment manufacturers
3. Consulting Engineer or Control System Contractor
4. Electric, Electronic, and Pneumatic
5. 75 VA
6. Line 11/12, wrong, it is adjustable
7. Wheatstone Bridge concept
8. Bleed and non-bleed
9. 20 to 35 psi
10. Control automation

CHAPTER 20

1. U tube manometer
2. Input rating, output rating, model and serial number, air capacity rating in CFM and maximum external static pressure, air temperature rise based on CFM rating.
3. Wrong—service clearance *and* necessary distance from combustible surfaces
4. $\frac{3}{4}$″ to 1″ play for each 12″ between pulley shafts.
5. $\frac{3}{8}$″ to $\frac{1}{2}$″ of thermocouple tip
6. 95,000 Btu/h
7. 240 cubic feet per hour
8. 3 to 3$\frac{1}{2}$″ W.C. manifold pressure
9. 11″ W.C. manifold pressure
10. CAC means constant or near constant air circulation

CHAPTER 21

1. Dirty filters
2. Flashback, delayed ignition, low gas pressure, clogged orifice, defective thermocouple
3. Pilot runners
4. A small electrical current
5. Burner tube venturi
6. Insufficient primary air, low manifold pressure, clogged or dirty burner ports
7. Dirty filters, overfiring, low fan speed, or a combination of all three
8. Heat anticipator
9. False—low air flow or fan failure is primary cause.
10. Soot and carbon build-up inside heat exchanger

CHAPTER 22

1. Concrete slab foundation
2. Wrong—coil in a horizontal plane
3. Arms' reach by service technician when working on the unit
4. In approved conduit
5. Service and unrestricted air flow
6. +10% on high side and −5% on low side
7. 18° to 20°F
8. A gauge manifold
9. Low indoor air flow and low coil temperature, dirty filters, or low fan speed could be the cause.
10. Never. Always to an open sink or floor drain.

CHAPTER 23

1. Part failure, improper adjustment, poor installation, and poor design
2. 1%
3. high, capacity
4. Wrong—they are difficult and usually expensive
5. Never
6. False
7. True
8. Increases condenser pressure and operating cost
9. d
10. d

CHAPTER 24

1. Very important. Knowledge without tools is not enough.
2. False
3. True
4. False

5. (a)
6. (a)
7. Starved
8. Wrong—just the opposite
9. Overcharge
10. Yes—for the basic refrigerant cycle

CHAPTER 25

1. Wrong
2. Yes, and good communicators
3. Wrong—both personal and property (tools, trucks, etc.)
4. Wrong—he has responsibility to protect same
5. Wrong—they should remain with the homeowner
6. Keeping the filters clean
7. Wrong—they are important and should be filed
8. Wrong—the homeowner should be advised of his coverage
9. Never—it creates a lasting distrust
10. Never
11. Yes
12. Yes

INDEX

INDEX

Index

B

C

Index

D

E

F

M

N

O

P

Q

R

S

T

W

Z